DIFCO MANUAL

DEHYDRATED CULTURE MEDIA and REAGENTS

for

MICROBIOLOGY

TENTH EDITION

DIFCO LABORATORIES
DETROIT MICHIGAN 48232 USA

First Edition	1927
Second Edition	1929
Third Edition	1931
Fourth Edition	1933
Fifth Edition	1935
Sixth Edition	1939
Seventh Edition	1943
Eighth Edition	1948
Ninth Edition	1953
Reprinted	1953
Reprinted	1956
Reprinted	1958
Reprinted	1960
Reprinted	1962
Reprinted	1963
Reprinted	1964
Reprinted	1965
Reprinted	1966
Reprinted	1967
Reprinted	1969
Reprinted	1971
Reprinted	1972
Reprinted	1974
Reprinted	1977
Tenth Edition	1984
Reprinted	1985

ISBN 9-9613169-9-3

Library of Congress Catalog Card Number: 84-070512

TABLE OF CONTENTS

FOREWORD

This edition of the DIFCO MANUAL, the tenth published since 1927, has been completely revised and rewritten. The purpose of the Manual is to provide information about products used in microbiology. The Manual has never been intended to replace any official compendium or the many excellent standard text books of scientific organizations or individual authors.

Difco is perhaps best known as the pioneer in bacteriological culture media. Numerous times one will find the trademarks Difco® or Bacto® preceding the names of materials used by scientists in their published papers. Because Difco products have been readily available worldwide longer than any others, Difco products have become the common-language reagents of the microbiological community. Standardized products readily available worldwide are essential for corroborative studies demanded by rigorous science.

Recommendation and approval have been extended to our products by the authors of many standard text books and by the committees on methods and procedures of scientific societies throughout the world. Difco products continue to be prepared according to applicable standards or accepted formulae. It is expected that they will be used only by or under the supervision of microbiologists or other professionals qualified by training and experience to handle pathogenic microorganisms. Further, it is expected that the user will be thoroughly familiar with the intended uses of the formulations and will follow the test procedures outlined in the applicable official compendia and the standard text book or procedures manual of the using laboratory.

Grateful acknowledgment is made of the support we have received from microbiologists throughout the world. It is the desire of our organization to continue and extend our services to the advancement of microbiology and related sciences.

DIFCO LABORATORIES

INTRODUCTION

Microbiology, through the study of bacteria, emerged as a defined branch of modern science as the result of the monumental and immortal research of Pasteur and Koch. When in 1876, Robert Koch, for the first time in history, propagated a pathogenic bacterium in pure culture outside the host's body, he not only established *Bacillus anthracis* as the etiological agent for anthrax in cattle, but he inaugurated a method of investigating disease which ushered in the golden age of medical bacteriology.

Early mycologists, A. de Bary and O. Brefeld, and bacteriologists, R. Koch and J. Schroeter, pioneered investigations of pure culture techniques for the colonial isolation of fungi and bacteria on solid media. Koch, utilizing state-of-the-art clear liquid media which he solidified with gelatin, developed both streak and pour plate methods for isolating bacteria. Gelatin was soon replaced with agar, a solidifying agent from red algae. It was far superior to gelatin in that it was resistant to microbial digestion and liquefaction.

The capability of Koch to isolate disease-producing bacteria on solidified culture media was further advanced by manipulating the cultural environment using meat extracts and infusions so as to reproduce, as closely as possible, the infected host's tissue. The decade immediately following Koch's epoch-making introduction of solid culture media for the isolation and growth of bacteria ranks as one of the brightest in the history of medicine because of the number, variety, and brilliance of the discoveries made in that period. These discoveries, which, as Koch himself expressed it, came "as easily as ripe apples fall from a tree," were all dependent upon and resulted from the evolution of correct methods for the *in vitro* cultivation of bacteria.

The fundamental principles of pure culture isolation and propagation still constitute the foundation of microbiological practice and research. Nevertheless, it has become more and more apparent that a successful attack upon problems unsolved is closely related to, if not dependent upon, a thorough understanding of the subtle factors influencing bacterial metabolism. With a suitable culture medium, properly used, advances in microbiology are more readily made than when either the medium or method of use is inadequate. The microbiologist of today is, therefore, largely concerned with the evolution of methods for the development and maintenance of microbial growth upon which an understanding of their unique and diversified biological and biochemical characteristics can be investigated. To this end, microbiologists have developed innumerable enrichment culture techniques for the isolation and cloning of microorganisms with specific nutritional requirements. These organisms and their unique characteristics have been essential to progress in basic biological research and modern applied microbiology.

The study of microorganisms is not easy using microscopic single cells. It is general practice to study pure cultures of a single cell type. In the laboratory, microbiological culture media are utilized which contain various nutrients that favor the growth of particular microorganisms in pure cultures. These media may be of simple and defined chemical composition or may contain complex ingredients such as digests of plant and animal tissue. In particular, the cultivation of bacteria is dependent upon nutritional requirements which are known to vary widely. Autotrophic bacteria are cultivated on chemically defined or synthetic media while heterotrophic bacteria, for optimal growth, may require more complex nutrients such as peptones, meat or yeast extracts. These complex mixtures of nutrients readily supply fastidious heterotrophic bacteria with vitamins and other growth-promoting substances necessary for desired cultivation. The scientific literature abounds with descriptions of enriched, selective and differential culture media necessary for the proper isolation, recognition and enumeration of various bacterial types.

Almost without exception whenever bacteria occur in nature, and this is particularly true of the pathogenic forms, nitrogenous compounds and carbohydrates are present. These are utilized in the maintenance of growth and for the furtherance of bacterial activities. So complex is the structure of many of these substances, however, that before they can be utilized by bacteria they must be dissimilated into simpler compounds then assimilated into cellular material. Such metabolic alterations are affected by enzymatic processes of hydrolysis, oxidation, reduction, deamination, etc., and are the result of bacterial activities of primary and essential importance. These changes are ascribed to the activity of bacterial enzymes which are both numerous and varied. The processes involved, as well as their end-products, are exceedingly complex; those of fermentation, for example, result in the production of such end-products as acids, alcohols, ketones, and gases including hydrogen, carbon dioxide, methane, etc. The study of bacterial metabolism, which defines the organized chemical activities of a cell, has led to the understanding of both catabolic or degradative activities and anabolic or synthetic activities. From these studies has come a better understanding of the nutritional requirements of bacteria, and in turn, the development of culture media capable of producing rapid and luxuriant growth, both essential requisites for the isolation and study of specific organisms.

Studies to determine the forms of carbon, hydrogen, and nitrogen which could most easily be utilized by bacteria for their development were originally carried on by Naegeli[1] between 1868 and 1880, and were published by him in the latter year. Naegeli's report covered the use of a large variety of substances including carbohydrates, alcohols, amino acids, organic nitrogen compounds, and inorganic nitrogen salts.

The first reference to the use of peptone for the cultivation of microorganisms is that made by Naegeli in the report referred to above, when in 1879, he compared peptone and ammonium tartrate. Because of its content amino acids and other nitrogenous compounds which are readily utilized by bacteria, peptone soon became one of the most important constituents of culture media, as it still remains. In the light of our present knowledge, proteins are known to be complex compounds composed of amino acids joined together by means of the covalent peptide bond linkage. When subjected to hydrolysis, proteins yield polypeptides of various molecular sizes, metapeptones, proteoses, peptones and peptides, down to the level of simple amino acids. The intermediate products should be considered as classes of compounds, rather than individual substances, for there exists no sharp lines of demarcation between the various classes. One group shades by imperceptible degrees into the next. All bacteriological peptones, thus, are mixtures of various products of protein hydrolysis. Not all the products of protein decomposition are equally utilizable by all bacteria. In their relation to proteins, bacteria may be divided into two classes; those which decompose naturally occurring proteins, and those which require simpler nitrogenous compounds such as peptones and amino acids.

The relation of amino acids to bacterial metabolism, and the ability of bacteria to use these compounds, have been studied by many workers. Duval,[2,3] for example, reports that cysteine and leucine are essential in the cultivation of *Mycobacterium leprae*. Kendall, Walker, and Day[4] and Long[5] report that the growth of *M. tuberculosis* is dependent upon the presence of amino acids. Many other workers have studied the relation of amino acids to the growth of other organisms, as for example, Hall, Campbell, and Hiles[6] to the meningococcus and *Streptococcus;* Cole and Lloyd[7] and Cole and Onslow[8] to the gonococcus; and Jacoby and Frankenthal[9] to the influenza bacillus. More recently, Feeley, et al.[34] demonstrated that the nonsporeforming aerobe, *Legionella pneumophila* requires L-cysteine · HCl for growth on laboratory media. Indispensable as amino acids are to the growth of many organisms, certain of them in sufficient concentration may exert an inhibitory effect upon bacterial development.

From the data thus far summarized, it is apparent that the problem of bacterial metabolism is indeed complicated, and that the phase concerned with bacterial growth and nutrition is of the utmost practical importance. It is not improbable that bacteriological discoveries such as those with *Legionella pneumophila* await merely the evolution of suitable culture media and methods of utilizing them, just as in the past important discoveries were long delayed because of a lack of similar requirements. Bacteriologists are therefore continuing to expend much energy on the elucidation of the variations in bacterial metabolism, and are continuing to seek methods of applying, in a practical way, the results of their studies.

While the importance of nitrogenous substances for bacterial growth was recognized early in the development of bacteriological technique, it was also realized, as has been indicated, that bacteria could not always obtain their nitrogen requirements directly from protein. It is highly desirable, in fact essential, to supply nitrogen in readily assimilable form, or in other words to incorporate in media proteins which have already been partially broken down into their simpler and more readily utilizable components. Many laboratory methods, such as hydrolysis with alkali,[10] acid,[11,12,13] enzymatic digestion,[8,14,15,16,17,18] and partial digestion of plasma[10] have been described for the preparation of protein hydrolysates.

The use of protein hydrolysates, particularly gelatin and casein, has led to especially important studies related to bacterial toxins by Mueller, et al.[20-25] on the production of diphtheria toxin; that of Tamura, et al.[26] of toxin of *Clostridium welchii;* that of Bunney and Loerber[27,28] on scarlet fever toxin, and of Favorite and Hammon[29] on *Staphylococcus* enterotoxin. In addition, the work of Snell and Wright[30] on the microbiological assay of vitamins and amino acids was shown to be dependent upon the type of protein hydrolysate utilized. Closely associated with research on this nature are such studies as those of Mueller[31,32] on pimelic acid as a growth factor for *Corynebacterium diphtheriae,* and those of O'Kane[33] on synthesis of riboflavin by staphylococci. More recently, the standardization of antibiotic susceptibility testing has been shown to be influenced by peptones of culture media. Bushby and Hitchings[35] have shown that the antimicrobial activities of trimethoprim and sulfamethoxazole are influenced considerably by the thymine and thymidine found in peptones of culture media.

In this brief discussion of certain phases of bacterial nutrition, we have attempted to indicate the complexity of the subject and to emphasize the importance of continued study of bacterial nutrition. Difco Laboratories has been engaged in research closely allied to this problem in its broader aspects since 1914 when Bacto Peptone was first introduced. Difco dehydrated culture media, and ingredients of such media, have won universal acceptance as useful and dependable laboratory adjuncts in all fields of microbiology.

REFERENCES

1. Sitz'ber, math-physik. Klasse Akad. Wiss. Muenchen, 10:277, 1880.
2. J. Exp. Med., 12:46, 1910.
3. J. Exp. Med., 13:365, 1911.
4. J. Infectious Diseases, 15:455, 1914.
5. Am. Rev. Tuberculosis, 3:86, 1919.
6. Brit. Med. J., 2:398, 1918.
7. J. Path. Bact., 21:267, 1917.
8. Lancet, II:9, 1916.
9. Biochem, Zeit, 122:100, 1921.
10. Centr. Bakt., 1:29:617, 1901.
11. Indian J. Med. Research, 5:408, 1917–18.
12. Compt. rend. soc. biol., 78:261, 1915.
13. J. Bact., 25:209, 1933.
14. Ann. de L'Inst., Pasteur, 12:26, 1898.
15. Indian J. Med. Research, 7:536, 1920.
16. Sperimentale, 72:291, 1918.
17. J. Med. Research, 43:61, 1922.
18. Can. J. Pub. Health, 32:468, 1941.
19. Centr. Bakt., 1:77:108, 1916.
20. J. Bact., 29:515, 1935.
21. Brit. J. Exp. Path., 27:335, 1936.
22. Brit. J. Exp. Path., 27:342, 1936.
23. J. Bact., 36:499, 1938.
24. J. Immunol., 37:103, 1939.
25. J. Immunol., 40:21, 1941.
26. Proc. Soc. Expl. Biol. Med., 47:284, 1941.
27. J. Immunol., 40:449, 1941.
28. J. Immunol., 40:459, 1941.
29. J. Bact., 41:305, 1941.
30. J. Biol. Chem., 139:675, 1941.
31. J. Biol. Chem., 119:121, 1937.
32. J. Bact., 34:163, 1940.
33. J. Bact., 41:441, 1941.
34. J. Clin. Microbiol., 8:320, 1978.
35. Brit. J. Pharmacol., 33:72, 1968.

HISTORY OF DIFCO LABORATORIES

In 1895, the company now known worldwide as Difco Laboratories was founded to produce high quality enzymes, dehydrated tissues and glandular products. Pepsin and pancreatin were two of the earliest products.

In 1899, the first national meeting of the Society of American Bacteriologists (SAB) was attended by 30 persons at Yale University to promote "the science of bacteriology, the bringing together of American bacteriologists, and the demonstration of bacteriological methods."

Although the above events did not at first appear to be related, it was soon apparent that each group would support the rapid progress of the other. SAB (now known as the American Society for Microbiology) has become the largest single group of microbiologists in the world, and Difco Laboratories has become the largest, most widely known producer of microbiological products.

An early name of the company was Digestive Ferments Company, because it produced digestive enzymes. "Difco" is an acronym of that name. The knowledge acquired during the early years of processing animal tissues, purifying enzymes and performing dehydration procedures was to prove invaluable in the transition to preparing dehydrated bacteriological culture media.

Difco's current extensive product line began with the development of pepsin. After 1895, meat and other protein digests were produced for the cultivation of bacteria, yeasts and molds. The many experiments which were performed to standardize the activity of pepsin, pancreatin and trypsin and the adaptation of some of the peptic and tryptic digests led to the introduction of Bacto Peptone to the bacteriological community in 1912. Bacto Peptone was made regularly available in 1913.

From 1912 through 1917, the Difco staff collaborated with experts in water bacteriology to develop new formulations of dehydrated media for determining the potability of drinking water. During these years, Difco was an active member of subcommittees involved with the standardization of methods for water analysis.

Continued research of peptic and tryptic digests of animal tissues resulted in a number of superior growth promoting peptones. Proteose Peptone was developed in 1919 for preparing diphtheria toxin. Bacto Tryptose was the first peptone prepared which did not require the addition of infusions or other enrichments for the isolation and cultivation of fastidious bacteria, especially *Brucella* and streptococci. Proteose Peptone No. 3 was later developed and has been especially valuable in media for isolating and cultivating *Neisseria gonorrhoeae*. In 1939, Difco first prepared Bacto Casamino Acids and other acid hydrolyzed casein products.

Over the years, many new and improved culture media formulations were developed. The development of superior culture media for isolating and culturing pathogenic and nonpathogenic bacteria from clinical, industrial and other natural specimens continues to be a major objective.

Although Difco's major emphasis has been in the development of bacteriological culture media, extensive development and preparation of diagnostic reagents began in 1923. Desiccated beef hearts for preparing antigen for use in the Kahn and Kolmer test for

5

syphilis were prepared with the direct involvement of Dr. Kahn and Dr. Kolmer. An essential part of our philosophy, then and now, in the development of diagnostic reagents, includes the constant search for improving the sensitivity, specificity and ease of use of the product. The isolation and purification of cardiolipin and lecithin by Dr. Pangborn for preparing VDRL antigen for syphilis was promptly followed by Difco's preparation of these materials and the antigen under Dr. Pangborn's supervision.

In the early 1930's, Dr. Quick, a world renowned expert in coagulation, guided the development of Bacto Thromboplastin, the first commercial reagent for diagnosing blood coagulation anomalies and controlling anticoagulant therapy.

Following World War II, Difco's emphasis turned toward the growing needs of microbiologists as rapid advancements were made in biomedical and industrial applications. Difco's pharmaceutical preparations were gradually phased out of the product line and significant attention was given to the development and preparation of microbiological and immunological products.

The continued expansion of Difco products in other areas of clinical laboratory diagnosis resulted in additional "firsts" in the 1950's including C-Reactive Protein Antiserum, Treponemal Antigen reagents, Streptolysin O Reagent and many others.

The laboratory for tissue culture was established in 1948. This included the study of nutritional requirements of mammalian cells and the preparation of reagents and media required for tissue culture and virus propagation procedures.

In the late 1940's, Difco embarked on a program to make available the most complete line of bacterial antisera and antigens in the world.

The introduction of antibiotics was an important landmark in bacteriology. In 1945, only three antibiotics were available for use in treating a limited number of bacterial infections. The need for a simple test to determine if a bacterial isolate was susceptible to one or more of these antibiotics became obvious. A disk diffusion procedure had been described in the literature about three years earlier, but such disks were not available for general laboratory use. Difco initiated developmental programs to prepare disks for use in determining in vitro susceptibility to antimicrobics. This program resulted in the introduction of Bacto Sensitivity Disks in 1946 and Dispens-O-Discs in 1965.

Throughout the 1950's and 1960's, new products were continually added to fulfill the needs of microbiologists and immunologists. Bacto Blood Culture Bottles provided improved broth culture techniques for both aerobic and anaerobic cultivation of microorganisms. TC Chromosome Culture Kits and related reagents were provided for studying chromosomes. The PKU Test Kit supplied reagents for estimating phenylalanine in blood as an aid in detecting phenylketonuria.

In 1972, Bactrol quality control disks containing pure bacterial cultures were introduced to stimulate and facilitate quality control in the laboratory. These disks were the beginning of a new series of products which emphasized the importance of quality control in the laboratory and reinforced Difco's dedication to quality.

Throughout its history, Difco has realized that no one person nor one organization has the knowledge and experience in all our areas of interest to develop the best products available. For this reason, this Company has sought the guidance of experts to help us develop and test many of our products. The rapid advances now being made in all areas of Difco's involvement demand that such guidance remain part of our philosophy. With the help of experts, we have provided diagnostic reagents of the highest quality, and with the continued help of such individuals we will continue to make available reagents which reflect the "state of the art."

A. L. Lane

THE ORIGIN
OF DEHYDRATED CULTURE MEDIA

It is a pleasure to include, as a part of this book, the abstract given below. This is believed to be the earliest reference to the preparation and use of dehydrated culture media, at least in this country. It is to be noted that Dr. Frost's arguments in favor of these preparations are just as forceful today as they were in 1909.

DESICCATED CULTURE MEDIA

W. D. Frost
University of Wisconsin

Abstract of paper at Boston (1909) Meeting of the Society of American Bacteriologists. Science, 31:555: (Apr. 8) 1910.

In order to overcome the generally recognized faults of bacterial culture media, such as variation in composition of small batches, time consumed in preparation, rapidity with which it deteriorates, its unavailability in small institutions or private practice, the preparation of culture media in large batches in establishments especially equipped for it and then desiccated is suggested.

The author's work on this problem, covering nearly a decade of time, is considered and samples are submitted.

There is apparently no reason why the different culture media cannot be put on the market in the form which requires merely the addition of water and sterilization to make it ready for use. Not only the ordinary, but probably most of the special media, can be prepared in this way and could be put up where desired, in the form of tablets, these to be of such size that they could be put directly in test tubes and when the proper amount of water is added they would be ready for sterilization and use.

It is interesting to note that Doerr, in Kraus and Uhlenhut: *Handbuch der Mikcobiologischen Technik*, states he also prepared powdered culture media by drying on glass in 1909.

The practical application of the dehydration of culture media was initiated and pioneered by Difco in 1915, under the direction of Dr. J. W. M. Bunker. Since these early days, the application of dehydrated culture media has been so universally accepted in microbiology laboratories throughout the world that its use and test results are referred to as the "classical method." It is the standard by which new tests and procedures can be judged.

GENERAL CONDITIONS
FOR CULTURING MICROORGANISMS

The development of microorganisms upon culture media is dependent upon a number of very important factors:

    (a) The proper nutrients must be available.
    (b) Oxygen or other gases must be available as required.
    (c) A certain degree of moisture is necessary.
    (d) The medium must be of the proper reaction.
    (e) Proper temperature relations must prevail.
    (f) The medium must be sterile.
    (g) Contamination must be prevented.

A satisfactory microbiological culture medium must contain available sources of hydrogen donors and acceptors, carbon, nitrogen, sulfur, phosphorus, inorganic salts and, in certain cases, vitamins or other growth-promoting substances. These were originally supplied in the form of the meat infusions which were, and still are in certain cases, widely used in culture media. Beef or yeast extracts frequently replace meat infusions, but the preparation of these substances subject them to the loss of their heat labile nutritive factors in much the same way as infusions are affected. The addition of peptone provides a readily available source of nitrogen and carbon.

Peptone is used in culture media to supply an available form of nitrogen since native proteins are not generally attacked by bacteria. Most organisms are capable of utilizing the amino acids and other simpler nitrogenous compounds present in peptone. Continued investigations in our laboratories indicate that for the isolation and propagation of many organisms, the complicated infusion media can be replaced by simpler media prepared by using the proper peptones in place of the meat infusions previously employed.

Certain bacteria required additions of other nutrients such as serum, blood, or ascites to the culture medium upon which they are to be propagated. Carbohydrates may also be desirable at times, and certain salts such as those of calcium, manganese, magnesium, sodium, and potassium seem to be required. Dyes may be added to media as indicators of metabolic activity or because of their selective inhibitory powers. Growth promoting substances of a vitamin-like nature are essential or assist greatly in the development of certain types of bacteria. Many of these substances are given for individual bacteria in *Bergey's Manual of Determinative Bacteriology*, 8th Edition, Williams and Wilkins Co.

The consistency of a liquid medium may be modified by the addition of agar, gelatin or albumin in order to change it into a solid or semisolid state. Solid media, which were originally devised for the isolation of organisms in pure culture, are now universally used for almost all general cultural work. The semi-solid media are used chiefly for carrying stock cultures or for propagating the anaerobes.

One of the principal landmarks in bacteriology was the preparation of a satisfactory medium by Hesse's introduction of agar as a solidifying agent for bacteriological culture media, Hueppe: Die Methoden der Bakterienforschung, 250: 1891. Previous to that time infusions of plant and animal tissues, solutions of organic compounds, and gelatin media only were employed. Until the introduction of gelatin media by Koch in 1881, the only method for obtaining pure cultures was the very unsatisfactory dilution procedure devised by Lister. The solid media used after the introduction of gelatin were only partially effective since many of the organisms under investigation would not develop satisfactorily at temperatures below the melting point of the gelatin, while others liquefied the gelatin. Bacteriology as a science began with the development of methods for the cultivation of bacteria, and the use of agar was a step of greatest importance.

Most bacteria are capable of growth under ordinary conditions of oxygen tension. Certain types, however, are capable of deriving their oxygen from various substrates. The aerobic organisms require the free admission of air, while the anaerobes grow only in the exclusion of atmospheric oxygen. Between these two groups are the microaerophiles which develop best under partial anaerobic conditions and the facultative anaerobes which develop over a wide Eh range. Anaerobic conditions for growth of microorganisms are obtained in a number of ways:

(a) Addition of small amounts of agar to liquid media.
(b) Addition of fresh tissue to the medium.
(c) Culturing in the presence of aerobic organisms.
(d) Addition of a reducing substance to the medium.
(e) Displacement of the air by carbon dioxide.
(f) Absorption of the oxygen by chemicals.
(g) Removal of oxygen by direct oxidation of readily oxidizable substances such as burning a candle, heating of palladiumized asbestos, copper, hydrogen, phosphorus or other readily oxidizable metals.
(h) Incubation in the presence of germinating grain or pieces of potato.
(i) Inoculation into the deeper layers of solid media, or under a layer of oil in liquid media.
(j) Combinations of these methods.

A complete description of anaerobic methods can be found in *Laboratory Methods in Anaerobic Bacteriology CDC Laboratory Manual*, V. R. Dowell, Jr. and T. M. Hawkins, HEW Publication No. (CDC) 77-8272, 1977.

Proper moisture conditions must prevail in the culture media employed for the propagation of microorganisms. A moist medium and a moist atmosphere are necessary for the continued luxuriant growth of the vegetative cells. For example, media in plates inoculated with the gonococcus, may fail to show growth of the organism if placed in an ordinary incubator at $35 - 37°C$. A duplicate inoculation, in contrast, in a sealed container in which is placed moist cotton or a wet towel to provide moisture, will show profuse growth. Incubators, if not humidity controlled, should have open containers filled with water at all times to provide sufficient moisture for growth and prevent drying of media. Growth of most microorganisms is obtained in the absence of light. Sunlight is to be avoided unless it is essential to growth as in photosynthetic bacteria.

The pH or reaction of the culture medium, expressing its hydrogen ion concentration, is extremely important for the growth of microorganisms. The majority of the microorganisms prefer culture media which are approximately neutral, while others may require a medium which is distinctly acid. The pH or reaction of the culture medium is determined by colorimetric or electrometric measurement of its hydrogen ion concentration. It should be noted that additions of acid or alkali which are insufficient to prevent the growth of bacteria in a medium may inhibit or prevent them from proceeding with the normal functions of their metabolic processes.

The usual range of temperature suitable for the growth of mesophilic microorganisms lies between $15 - 43°C$. Psychrophilic microorganisms have, however, been known to develop at $0°C$ and others, such as the thermophilic soil organisms, may grow at $80°C$.

The pathogenic organisms in general are limited by a comparatively narrow range of temperature around 37°C while the saprophytes usually have a much broader latitude.

All organisms exhibit three cardinal points in their thermic relations:
(a) A minimum below which development ceases.
(b) An optimum at which growth is luxuriant.
(c) A maximum above which growth ceases and death occurs.

The media upon which microorganisms are grown must be sterile or free from all other forms whose development might influence or prevent the normal growth of the inoculated type. The usual method for immediate sterilization of culture media is by means of the autoclave in which steam under pressure is the sterilizing agent. The proper operation of the autoclave to insure sterilization of media requires careful manipulation.

Autoclave sterilization for 15 minutes at 15 pounds pressure (121°C) is recommended for quantities of liquid media up to one liter. If larger volumes are to be sterilized in one container, and especially if the medium is not hot when placed in the autoclave, a longer period should be employed. The medium is prepared according to formula, distributed in tubes or flasks which are then plugged with nonabsorbent cotton or loosely capped and placed in the autoclave. Tubes should be placed in racks or packed loosely in baskets. Flasks should never be more than two-thirds full.

In the operation of the autoclave, all the enclosed air must be allowed to escape and must be completely replaced by steam. Pressure-temperature relations of a properly operated autoclave are shown in the table below.

Pressure-Temperature Relations in Autoclave
(Figures based upon complete replacement of air by steam.)

Pressure in Pounds	Temperature °C	°F
5	108	226
10	116	240
15	121	250
20	127	260
25	131	267
30	134	274

If all the air is not removed from the sterilizing chamber, which condition is best shown by use of a thermometer in the exhaust line of the autoclave, an entirely different pressure-temperature relationship exists. Through the courtesy of Dr. F. W. Tanner[2] we are able to reproduce the following chart which plainly shows the actual temperature in the autoclave when the air is not completely exhausted.

When the operator is assured that all the air is replaced by steam, which is best indicated by a thermometer placed in the exhaust line, the outlet valve of the autoclave is closed and the steam pressure is raised to 15 pounds. When the thermometer indicates a temperature of 121°C heating is continued for 15 minutes. A maximum of 15 minutes is recommended for the sterilization of carbohydrate media in tubes to be used for fermentation studies. After the sterilization period has been completed, the source of steam is cut off and the autoclave is allowed to return to atmospheric pressure. Pressure should not drop too rapidly or the media will boil over, blowing the plugs from the tubes or flasks. Pressure should, however, drop rapidly enough to prevent excessive exposure of the media to heat after the sterilization period. Ordinarily about 8, and not more than 12, minutes should be required for the usual bacteriological laboratory autoclave to reach atmospheric pressure without danger of prolonged heating. The media should be removed from the autoclave shortly after sterilization and should not be permitted to remain in the autoclave for any appreciable length of time after the sterilization period.

EFFECT OF ENTRAPPED AIR ON TEMPERATURE OF AUTOCLAVE

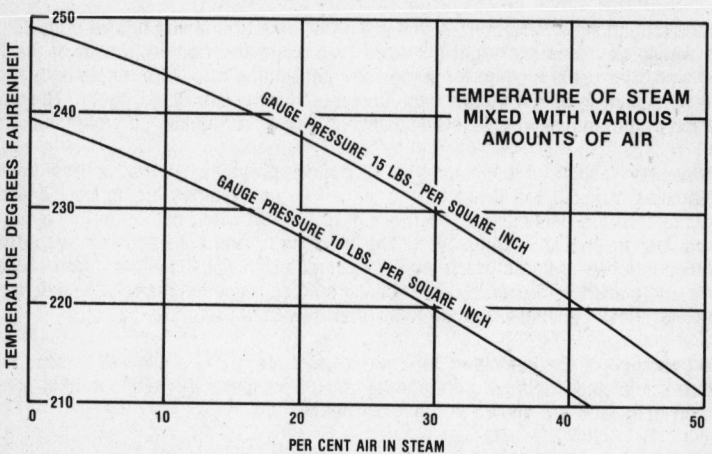

For the sterilization of coagulable material such as serum, see the method given for Bacto Loeffler Blood Serum.

Oversterilization or prolonged heating will change the composition of the medium. For example, in our laboratories we have shown that phenol red lactose broth which has been sterilized for 15 minutes at 15 lbs pressure (121°C), or sterilized by filtration, produces no demonstrable amount of acid when inoculated with *Salmonella typhi*. The same lot of medium sterilized for 30 or 45 minutes at 121°C showed appreciable acid production under the same conditions of testing. This demonstrates that oversterilization resulted in a breakdown of the lactose.

Agar media on prolonged sterilization or heating are apt to show a precipitate. Repeated melting of solidified agar, or long holding of melted agar at high temperature, may likewise cause a precipitate to form in the media. Media containing agar may also form a flocculent precipitate if the liquid medium is held in the water bath at 43 – 45°C for longer than 30 minutes. This flocculent agar precipitate, however, may be dispersed by reheating the medium.

Excessive heating of media also results in an increase in acidity. The reaction of the media will become more acid as heating is prolonged. Some media which are acid, such as wort agar (pH 4.8) will, upon prolonged heating, cause destruction of the agar. It is possible to destroy completely the gelling properties of agar by prolonged heating, and this destruction is hastened as the acidity increases.

Culture media which may be injured by autoclaving are sometimes sterilized by the discontinuous or intermittent method. This procedure consists of heating the medium in a chamber of flowing steam for a period of 20 or 30 minutes, or longer, on several successive days. Body fluids and sera are sometimes sterilized in the inspissator at 53° to 70°C for one hour on six successive days. Liquid media may be sterilized by filtration through membrane or molecular filters or unglazed porcelain or earthenware candles.

External contamination of culture media is prevented by plugging the tubes or flasks with nonabsorbent cotton before sterilization. Plugs should fit neither too loosely nor too tightly and should protect the lip of the container against the accumulation of dust. Screw cap tops or metal covers may also be used to close the tubes or flasks. Marcus and Greaves[3] have called attention to the fact that atypical cultural reactions may be obtained in sealed tubes of media used to test biochemical activities due to anaerobic conditions. Tubes of Kligler iron agar and Russell double sugar agar, for example, gave aberrant reactions in tubes sealed with screw caps or rubber stoppers. The same medium with tubes loosely capped, or caps replaced with cotton plugs, showed typical reactions.

Media should always be stored in a cool moist atmosphere to prevent evaporation preferably in screw-capped tubes or bottles. Prolonged storage of sterile media cannot, however, be recommended unless stability is established. If tubes of media have been kept for any length of time, they should be reheated just before use. Liquid media should be heated in a boiling water bath or in flowing steam for a few minutes, to drive off dissolved gases, and then cooled quickly in cold water without agitation just prior to inoculation. Agar tubes should be melted and allowed to solidify in order to secure a moist surface which is desired by most microorganisms. These precautions for both liquid and solid media are extremely important for the initiation of growth of highly parasitic organisms such as those encountered in blood culture work.

REFERENCES

1. Am. J. Med. Tech., 14:214, 1948.
2. Am. J. Pub. Health, 25:301, 1935.
3. J. Lab. Clin. Med., 36:134, 1950.

PREPARATION OF DEHYDRATED CULTURE MEDIA

Preparation of culture media for laboratory use from dehydrated media is, essentially, a simple manipulation governed by a few common-sense principles. The procedure requires precision, accuracy and attention to purity and sterility which are routine considerations for the microbiology staff.

Those who prepare media in their laboratories have the distinct advantage of using fresh media of known quality and predictable performance, available on short notice and limited only by the variety of dehydrated media retained in stock.

The following comments are included here for those who wish to prepare limited quantities of culture media but are unfamiliar with their preparation from the dehydrated form. A check list is provided to prevent or correct the problems that may result from inexperience.

Preparation of large-scale commercial batches of culture media is beyond the scope of this discussion, although the principles remain the same. Customers requiring specific advice on large-scale production may obtain information from Difco Technical Service.

The following should be a summary of current Standard Operating Procedures established within each laboratory.

DEHYDRATED MEDIA
Record on the container of dehydrated medium the date of receipt and of opening. Between uses, store as directed on the label in a low light, low humidity environment with the container tightly closed.

Ascertain that the physical characteristics of the powder are typical and appropriate before use.

EQUIPMENT
Use undamaged glassware that is thoroughly clean and has been rinsed with distilled or deionized water.

Use measuring devices, scales, pH meters, autoclaves and other equipment that are frequently and accurately calibrated.

WATER
Use distilled or deionized water that is free of chlorine, copper, lead and detergents and has been determined to have suitable conductivity and ion content, according to the specifications for Purified Water, USP.

DISSOLVING THE MEDIUM
Carefully weigh the appropriate amount of dehydrated medium and place it in a clean, dry flask two to three times larger than the volume of the final amount of medium.

Add the appropriate amount of water (and other ingredients). Swirl the flask to mix thoroughly. Most broth media are clear at this point and do not require boiling.

Using an open flame, hot plate, boiling water or steam bath, or flowing steam in an unpressurized autoclave, heat the medium to boiling with frequent agitation (stirring or swirling the flask) to prevent overheating the medium or scorching the flask.

CAUTION: Agar media, particularly those with low agar content, may boil unexpectedly soon and flow out of the flask. Agitate frequently but gently as the medium approaches boiling and the agar begins to dissolve.

If the medium requires no further manipulation and is to be used in tubes or flasks, dispense and cap it at this point. Otherwise, place a cotton-type plug in the neck of the flask.

STERILIZING THE MEDIUM
Sterilize the medium in the autoclave as directed on the package label, carefully timing the procedure from the point at which 121°C is reached. Quantities of media in excess of two liters may require a somewhat extended autoclave time to achieve sterilization. In such cases, take care to prevent overheating which will adversely affect the pH and other characteristics of the medium.

Upon completing the sterilization cycle, bring the autoclave slowly to room pressure and promptly remove the sterile medium.

ADDING ENRICHMENTS AND SUPPLEMENTS
Heat-labile enrichments and supplements are added to media after sterilization to avoid heat related deterioration of the substance.

Generally, cool sterile agar-base media to approximately 45 – 50°C in a water bath. Aseptically add the enrichment to the medium as directed on the product label or by standard procedure. Swirl the flask to mix thoroughly, avoiding the formation of bubbles.

Sterile broths may be cooled to room temperature before adding enrichment.

DISPENSING THE MEDIUM
Cool sterile media below 50 – 55°C before dispensing to avoid the formation of condensation.

Aseptically dispense the medium into sterile tubes, flasks or Petri dishes, as desired. Immediately recover or recap to reduce the chance of contamination. Leave Petri dish covers slightly open for 1 – 2 hours if a dry surface is required for a particular medium.

STORING PREPARED MEDIA
Store each prepared medium at the temperature indicated in its product description. Tighten caps or tubes and place prepared plates, inverted, in moisture-proof containers to prevent excessive moisture evaporation from the medium.

EXPIRATION DATE
Generally, properly stored media are suitable for use as long as they retain their physical and performance characteristics. As a guideline, prepared plates retain high quality for approximately one week and prepared tubes for about six months. Prepare sufficient media to fill only the needs of those time frames.

QUALITY CONTROL
Incubate a portion (4%) of prepared tubes/plates at an appropriate temperature (30 – 35°C or 20 – 25°C) for 5 – 7 days to check for sterility, then discard.

Determine that the color, clarity, pH and cultural characteristics of the prepared medium are typical and appropriate, as listed in the individual product description, before using the medium.

The following chart summarizes the importance of the above points of discussion to the quality of a prepared medium and the problems that can be avoided by using proper technique.

Problem	Deteriorated Dehydrated Medium	Improperly Washed Glassware	Impure Water	Incorrect Weighing	Incomplete Mixing	Overheating	Repeated Remelting	Dilution by a Too Large Inoculum	Other Causes
Abnormal color of medium	•	•	•						
Incorrect pH	•	•	•	•	•	•	•		Storage at high temperature Hydrolysis of ingredients pH determined at wrong temperature
Nontypical precipitate	•	•	•	•	•	•			
Incomplete solubility					•				Inadequate heating Inadequate convection in a too small flask
Darkening or carmelization	•			•	•	•			
Toxicity		•	•						Burning or scorching
Trace substances (vitamins)		•							Airborne or environmental sources of vitamins
Loss of gelation property				•	•	•		•	Hydrolysis of agar due to pH shift
Loss of nutritive value or selective or differential properties	•			•	•	•	•	•	Burning or scorching Presence of strong electrolytes, sugar solutions, detergents, antiseptics, metalic poisons, protein materials or other substances that may inhibit the inoculum
Contamination									Improper sterilization Poor technique in adding enrichments and pouring plates

DIFCO QUALITY CONTROL

The organic nature of most of the components of dehydrated culture media is variable and subject to change. Difco Laboratories has been aware of the importance of quality control since its earliest days and introduced quality control into all manufacturing processes so that end products are defined and predictable.

During the past several years, needs of microbiologists have more precisely defined product specifications and standards have been updated to include recommendations published in official and non-official compendia. The need for materials to cultivate and identify microorganisms seems to be ever increasing, ranging from biomedical needs through sanitation, food technology, research and the elaboration of substances produced by microorganisms. A growing collection of official and non-official regulations and compendia exists to further define product specifications.

Within this broad context, quality control of Difco products has three all-encompassing aspects:
1. testing of product components prior to use in manufacturing to determine their suitability for the intended use and their compatibility with other components to be used;
2. intermediate in-line testing during production to assure that the components and procedures being employed actually produce the desired product; and,
3. final testing to ascertain that the product fulfills its intended use, meeting defined standards and customer needs.

An example of the comprehensive nature of Difco quality control is seen in the production of dehydrated culture media. The typical differential medium contains, among other ingredients:

a nitrogen source
a carbon source
inorganic salts
selective agents
pH indicators
buffers
a solidifying agent

Prior to manufacturing, each component is tested in parallel with approved reference components for toxicity, purity and compatibility with the other components to be used. A sample batch is then prepared from approved components and assayed in parallel with a previously approved reference lot of the complete medium. If the color, clarity, pH, solubility, solidity and cultural response of the preliminary medium meet all internal and external specifications and demonstrate the compatibility of component ingredients, then manufacturing proceeds.

17

Upon completion of manufacturing, the potential product is examined in the dehydrated form for proper color, odor, appearance, moisture content and solubility and is use-tested in parallel with the ready-to-use standard reference medium to reconfirm color, clarity, pH, solidity and cultural response. Only after approval of this final assay is the product made available for sale.

Quality control does not stop here. Final product samples and records of all data generated during product evaluation are retained for future reference so that product integrity can be maintained from lot to lot over the ensuing years. Thus, when a product is supplied by Difco, it is defined and predictable.

At this point, responsibility for controlling product quality transfers to the customer who must determine through at least minimal testing that product integrity has been maintained through the delivery procedure and that, in the hands of the user, the product performs as desired and intended.

Similar extensive quality control testing assures that product use will meet established specifications for all Difco products, including bacterial antisera and antigens, Dispens-O-Discs, Bacto Blood Culture Bottles, Bactrol quality control organisms, etc.

MICROBIOLOGY FOR CLINICAL LABORATORY TESTING

The clinical microbiology laboratory is charged with the responsibility of investigating cases of infectious disease as directed by appropriate medical personnel. The scope of such testing is extensive and involves the many aspects of handling patient specimens. Among these are:
1. Collection of specimens
2. Transport of specimens to the laboratory
3. Detection of disease-producing microorganisms
4. Primary isolation of causative agents
5. Identification
6. Determination of antimicrobic susceptibility

Difco products intended for use in the collection, preparation, and examination of specimens taken from the human body are in vitro diagnostic products. The technical description for each product contains sufficient information for a qualified professional to use it for its intended purpose. For specific methodologies, consult any of the recognized texts devoted to this purpose, such as *Manual of Clinical Microbiology* published by the American Society for Microbiology (ASM), *Diagnostic Procedures for Bacterial, Mycotic and Parasitic Infections,* published by the American Public Health Association (APHA), and *Performance Standards for Antimicrobic Disc Susceptibility Tests* published by The National Committee For Clinical Laboratory Standards (NCCLS).

THE COLLECTION OF SPECIMENS
The collection of specimens may or may not be a clinical microbiology laboratory responsibility, depending on the nature of the specimen. Usually specimens are collected by the hospital medical staff. However, it is important for laboratory personnel to be knowledgable about collection techniques and other aspects, such as; microbial etiology, type, location and duration of disease, prior antibiotic therapy and possible contamination of the specimen by indigenous flora.[1] In addition, site preparation to minimize contamination with commensal organisms is a very important consideration, especially when obtaining samples for blood culture. Refer to Blood Culture section for specific information. Proper collection of specimens is the important first step in accurate diagnosis of disease.

TRANSPORT OF SPECIMENS TO THE LABORATORY
The condition of the specimen received by the laboratory for culture is a significant variable in recovery and final identification of the suspect pathogen. An unsatisfactory specimen can lead to erroneous or inconclusive results.

Optimal conditions dictate direct inoculation of the specimen to a suitable medium or enrichment broth. If this is not possible, the specimen should be forwarded immediately to the laboratory for evaluation. This may not be possible in all instances; therefore, it is necessary to provide a method which will maintain viability of significant pathogens and yet control proliferation of contaminants during extended periods of transit. One means of preservation is the use of transport media which are designed to provide environments suitable for maintaining the integrity of specimens.

Transport media are essentially nonnutrient, semi-solid, highly reductive media which inhibit self destructive enzymatic reactions within the cells. They also prevent the lethal effects of oxidation, thus maintaining the organisms present at a *status quo* condition.

19

Among the more widely used transport media are Bacto Transport Medium Stuart and Bacto Transport Medium Amies.

Swabs, often used to collect specimens, are suitable for collection of material from skin, mucous membranes, cervix, vagina, urethra and rectum.[2] They are unsuitable for use with pus or exudates, surgical specimens or for culturing mycobacteria.

Swabs should be carefully checked to insure they are not toxic to cultures suspected of being contained in the specimen. Once the specimen is obtained, the swab is placed into the transport medium. Convenient combinations of swab and transport medium in a mailable package are supplied by Difco as Swab Transport Packs, Amies or Stuart. For additional information refer to Transport Media for Microbiological Specimens.

DETECTION OF DISEASE PRODUCING MICROORGANISMS

Once the specimen has arrived in the laboratory it is promptly examined for evidence of infecting agents. This may be accomplished in several ways depending on the nature of the specimen. Staining and serological testing are methods commonly used.

Stain	Purpose	Difco Product
Gram Stain	General use	Gram Stain Set
Acid Fast Stain	Mycobacteria in sputum	TB Stain Set ZN TB Stain Set K
TB Fluorescent Stains	Mycobacteria in sputum	TB Fluorescent Stain Set T TB Fluorescent Stain Set M
Acridine Orange Stain	Blood and Cerebrospinal Fluid	Acridine Orange Stain

Serological Detection	Source	Difco Product
Febrile Antibodies	Blood Serum	Febrile Antigen Set
Streptococcus Exoenzyme Antibodies	Blood Serum	Antistreptolysin O Titration Reagents Antihyluronidase Titration Reagents
Syphilis Antibodies	Blood Serum	USR Reagents VDRL Reagents FTA Reagents
Bacterial Antigens	Blood and Cerebrospinal Fluid	Pyrotest (LAL)

PRIMARY ISOLATION OF CAUSATIVE AGENTS

Once it has been determined that an infectious process is present through patient signs and symptoms or detection of the presence of a causative agent, isolation of that organism is essential to the final diagnosis. The first step taken involves choosing the proper isolation culture media. This choice must take into consideration the following: source of the specimen, transport medium used, status of the specimen, Gram reaction of the culture, relative presence of commensal organisms and nature of the suspected agent.[3]

There are several types of primary isolation media.

General Purpose Media

These are nonselective primary isolation media used for culturing a wide variety of fastidious microorganisms including *Actinomyces, Brucella, Clostridium, Enterobacteriaceae, Neisseria, Pseudomonas, Staphylococcus* and *Streptococcus.* Often these formulations are enriched with such materials as blood, hemoglobin, sera or other growth accessary factors including glutamine, coenzyme (V factor) and hematin (X factor).

Selective Media
Selective media have been modified in some manner to suppress or prevent the growth of one group of organisms while permitting the growth of another group from a mixed flora. The modification of the medium may be made by altering the pH to take advantage of the ability of some organisms to grow at one pH and not at another. Another principle invoked in the selection of a particular type of organism from a mixed population is to utilize the inhibitory (bacteriostatic) property of a specific chemical without which the medium would be suitable for many species in the sample. For example, addition of crystal violet in a final concentration of 1:100,000 will inhibit most gram-positive types without affecting the gram-negative group. Similarly, antibiotics may be used for selective inhibition. Sodium azide (0.03%) has bacteriostatic action for gram-negative bacteria, the aerobic gram-positive spore forming bacilli, and certain other aerobes. The degree of selectivity may also be chosen. Moderately selective media inhibit unwanted organisms to the point where they do not interfere with growth of the wanted organism. Highly selective media, on the other hand, strongly block the growth to completely eliminate undesirable cultures.

Enrichment Media
Enrichment media suppress the growth of competitive normal flora while enhancing the growth of the desired culture. GN broth Hajna, selenite cystine broth and tetrathionate broth are a few of this type. Specimens known to be highly contaminated with normal flora often require exposure to enrichment media to be able to isolate the primary pathogen present.

Specialized Isolation Media
This group includes those formulations that satisfy nutritional needs of specific groups of organisms thereby providing colonial morphology, differentiation and identification. Examples of this type would be:

Staphylococci	— Mannitol Salt Agar
Mycobacteria	— Lowenstein Medium
Corynebacteria	— Tinsdale Agar Base/Enrichment
Candida	— BiGGY agar.

IDENTIFICATION OF CAUSATIVE AGENTS
Detection and isolation of pathogens from clinical specimens are important parts of developing a diagnosis. Accurate identification of the isolate, however, determines specific therapy. The bacteriologist is afforded a simple and effective cultural means of identifying pure cultures by the use of ordinary differential media.

Differential Media
Bacteria display all possible variations in their nutritional requirements and advantage can be made of this fact by using carefully selected differential media for identification. Some media contain one or more fermentable carbohydrates and a suitable indicator for the detection of acid or alkali production; some contain substances rich in sulfur and an indicator to detect hydrogen sulfide formation; still others contain a combination of both. Semi-solid media are used to demonstrate motility.

Some of the more common physiological properties determined on microorganisms and the media used for such determinations are as follows:

Carbohydrate Fermentation

A carbohydrate, polyhydric alcohol or glucoside is added to a basal medium (either liquid, semi-solid or solid) to which an indicator has been added to detect changes in pH which develop during growth.

Production of gas is detected in liquid media by placing Durham tubes (small inverted vials which will fill with liquid during sterilization) in the tubes at the time the medium is dispensed. The tubes are unnecessary if a solid or semi-solid medium is used. Semi-solid agar is prepared by adding 0.3 – 0.5% of agar to a satisfactory liquid medium and making stab inoculations in the column of medium with a straight inoculating needle. In such a medium (or in full-strength agar) gas production will be denoted by the appearance of gas bubbles and cracks in the medium; the same semi-solid medium may also be used to determine motility. Full strength (1.5%) agar should be cooled in slanting position with a generous butt and inoculated on the surface on the slant.

Carbohydrate Broths	Semi-solids
OF Basal Medium	Cystine Tryptic Agar/w Carbohydrates
Phenol Red Broth w/Carbohydrates	Tryptic Agar Base w/Carbohydrates
Purple Broth w/Carbohydrates	

Agars

EMB Agar Base
MacConkey Agar Base
Phenol Red Agar w/Carbohydrates
Purple Agar w/Carbohydrates

Special forms of carbohydrate fermentation media are the double and triple sugar agars. These media are of particular value in the rapid identification of the gram-negative enteric bacteria. These media are tubed in columns sufficiently deep to allow a 3 cm butt with a straight inoculating needle. The following formulations are examples of these special forms: Russell double sugar agar, Kligler iron agar and triple sugar iron agar.

Biochemical Substrate Utilization

Another means of identifying organisms is through determination of their ability to produce specific enzymes or metabolites which in turn react with specific substrates. Culture media containing these substrates along with indicators are available. In addition certain of these determinations have been simplified by the use of a disk type test or a specific reagent. The following is a list of these single determinant tests.

Determinant	Culture Media/Reagent
Acetate Utilization	Acetate Differential Agar
Acid Fast Stain	TB Stain Set K
	TB Stain Set ZN
Bacitracin Susceptibility	Differentiation Disks
	Bacitracin
Bile Solubility	Oxgall
	Sodium Desoxycholate
Catalase (*M. tuberculosis*)	Lowenstein Medium,
	Jensen Deeps
Chlamydospore Production	Chlamydospore Agar
	Corn Meal Agar
Citrate	Simmons Citrate Agar
	Koser Citrate Medium
Coagulase	Coagulase Plasma or
	Coagulase Plasma EDTA

SEROLOGY

...re valuable tools in identifying causative agents of disease. They
...isms, demonstrate antibodies in sera, and detect and quantitate the
...ies present. These serological diagnostic methods require carefully
...dardized materials. Before distribution, each lot of antigen and anti-
...d approved for its sensitivity and specificity. The table below dem-
...e of the Difco serology line.

...anism	Conventional Serology	Fluorescent Antibody Technique
	X	
...pertussis	X	X
...ussis	X	X
...s	X	
...nsis	X	
	X	
...ns	X	X
...m diphtheriae		X
	X	X
...rensis	X	
...fluenzae	X	X
...moniae	X	
...cola	X	X
...ohemorrhagiae	X	X
...ona	X	
...additional serotypes)	X	
...togenes	X	X
...rhoeae		X
...gitidis	X	X
...Lancefield groups)	X	X
...neumoniae	X	X
	X	X (polyvalent panvalent)
	X	
	X	

...serological tests and products available from Difco are discussed

...ATION

...tion technique is used in rapid screening tests and in the diagnosis
...l mostly by the gram-negative intestinal bacteria. In this reaction,
...lutinate and settle out when it reacts with its specific antibody.

...:

...contains preserved horse cells as a presumptive test and enzyme
...as a confirmatory test.

...est

Determinant	Culture Media/Reagent
Decarboxylation	Decarboxylase Medium w/ Lysine or Ornithine
	Decarboxylase Broth Moeller w/ Lysine or Ornithine
Dihydrolation	Decarboxylase Broth Moeller w/Arginine
	Decarboxylase Medium w/Arginine
DNase	DNase Test Agar w/Methyl Green
	DNase Test Agar
Esculin Hydrolysis	Bile Esculin Agar
	Bile Esculin Azide Agar
Flagella Stain	Flagella Stain
Gelatin Liquefaction	Nutrient Gelatin
Hemolysis	TSA Blood Agar Base
	Tryptose Blood Agar Base w/5% Blood (See other Blood Agar Bases also)
	Tryptic Soy Agar w/5% Blood
Hippurate	Heart Infusion Broth and Sodium Hippurate
Hydrogen Sulfide	Triple Sugar Iron Agar
	Kligler Iron Agar
	H_2S Test Strips
Indole	Indole Test Strips
KCN	KCN Broth Base
Lecithinase	Egg Yolk Enrichment 50% w/Baird Parker Agar Base
	TPEY Agar Base (*Staphylococcus*)
	SFP Agar Base
	McClung Toabe Agar Base (*Clostridium*)
Lipase	Spirit Blue Agar and Lipase Reagent
Litmus Reduction	Litmus Milk
Lysozyme	Lysozyme
	Lysozyme Substrate
	Lysozyme Buffer, Dried
Malonate Utilization	Malonate Broth
Methyl Red	MR-VP Medium
Motility	Motility Test Medium
Niacin Production	TB Niacin Test Strips
Nitrate Reduction	Nitrate Broth w/gas tube
	Nitrate Agar
	Nitrate Test Strips
ONPG	Differentiation Disks ONPG
Optochin Susceptibility	Differentiation Disks Optochin
Oxidase	p-Aminodimethylaniline Oxalate
	Differentiation Disks Oxidase
Phenylalanine Deamination	Phenylalanine Agar

Determinant	Culture Media/Reagent
Pigment Production	Loeffler Blood Serum
Proteolysis	
Egg	Egg Albumin, Soluble
Milk	Litmus Milk
Serum	Loeffler Blood Serum
Salt Tolerance	6.5%NaCl Broth
	(Heart Infusion Broth +
	6.5% NaCl)
Toxin Production	KL Virulence Agar
	KL Antitoxin Strips
Urease	Urea Agar or Urea Broth
	Differentiation Disks Urea
Voges-Proskauer	MR-VP Medium
X and V Requirements	Differentiation Disks
	BX, BV and BVX

Multiple determinant tests which combine reactions are illustrated on the following media.

Determinant	Culture Media/reagent
Motility-Indole-Ornithine Decarboxylation	MIO Medium
Motility-Indole-Lysine Decarboxylation	MIL Medium
Motility-Indole-H_2S	SIM Medium
Lactose-Sucrose-Dextrose-H_2S	Triple Sugar Iron Agar
Lactose-Dextrose-H_2S	Kligler Iron Agar
Lysine Decarboxylation-H_2S	Lysine Iron Agar
Tellurite Reduction-Lecithinase	Baird Parker Agar Base and
	TPEY Agar Base
Non-fermentation of Xylose-Lysine-Decarboxylation-H_2S	XLD Agar

Causative agent identification can be made once a profile of the isolate has been completed. It includes cell morphology, cultural and biochemical characteristics followed by serological identification.

DETERMINATION OF ANTIMICROBIC SUSCEPTIBILITY

The clinical laboratory has two major roles in aiding the physician dealing with infectious disease. The first is to isolate and identify pathogens from clinical specimens and the second, to provide antimicrobic susceptibility patterns so that appropriate therapy can be initiated.[4] Several methods for in vitro antimicrobic susceptibility testing have evolved. These include the agar disk diffusion test, the agar dilution test and the broth microdilution method. The latter two are used to determine the minimal inhibitory concentration (MIC) of the agent being tested.

Agar Disk Diffusion Test

Variations of the disc diffusion technique have been in use since the late 1940's. The need for standardization of this procedure was long recognized when Bauer, Kirby, Sherris and Turck[1] suggested a standardized single-disc method in 1966. In 1971, Ericsson and Sherris[2] published the results of an extensive series of international collaborative studies of testing techniques using broth dilution, agar dilution and agar diffusion. The National Committee for Clinical Laboratory Standards (NCCLS)[5] issued approved performance standards in 1976. They were followed by World Health Organization (WHO)[6] requirements in 1977. The test as outlined by NCCLS utilizes the

principle of standardized methodology and z
minimal inhibitory concentration.

MIC Determinations

Minimal inhibitory concentration testing is a
crobial susceptibility. Agents to be tested a
priate concentrations, then are inoculated
noted and the MIC is determined.

The original tube tests were seldom used
Today, convenient packages of prediluted a
able and ready for use.

Blood Cultures

The blood culture is among the highest p
biology laboratory. The reason for this pri
know quickly and accurately the presence
is one of the most serious forms of infectic
mortality ranging from 2 – 80% depending
must be initiated as early as possible.

Blood cultures are not only important from
a severe or life-threatening infection, but t
and extent of an infection. Therefore, it is
to use good technique and a well selecte
nation to insure adequate patient care.

Refer to Blood Culture Products for a co

REFERENCES

1. Washington, 1981, Laboratory Procedures in Clinica
2. Washington, 1981, Laboratory Procedures in Clinica
3. Lennette, 1980, Manual of Clinical Microbiology, 3r
4. Balows and Hausler, 1981, Diagnostic Procedures f
   APHA.
5. Performance Standards for Antimicrobic Disc Susc
6. World Health Organization Technical Report Series
7. Sonnenwirth, A. C., Bacteremia, 1973, Chas. C. Th

Serological tes
identify microo
amount of anti
prepared and s
serum is teste
onstrates the s

Micr

Arizona Grou
Bordetella p
Bordetella p
Brucella ab
Brucella me
Brucella sui
Candida alb
Corynebact
Escherichia
Francisella
Haemophilu
Klebsiella p
Leptospira c
Leptospira i
Leptospira p
Leptospira (
Listeria mor
Neisseria gc
Neisseria m
Streptococc
Streptococc
Salmonella

Shigella
Vibrio choler

The major types
briefly.

SLIDE AGGLUT

The slide agglu
of diseases cau
the antigen will

Available Prod
Hetrol Slide Tes
treated sheep ce

Rheumatoid Slid
Widal Antigen S

Febrile Antigen Set — contains antigens to

B. abortus	Salmonella H d
Proteus OX19	Salmonella O Group D
Salmonella H a	Febrile Positive and
Salmonella H b	Negative Controls

Febrile Antigens

B. abortus	Proteus OX2, OXK, OX19
F. tularensis	Salmonella H and O Groups
Leptospira	

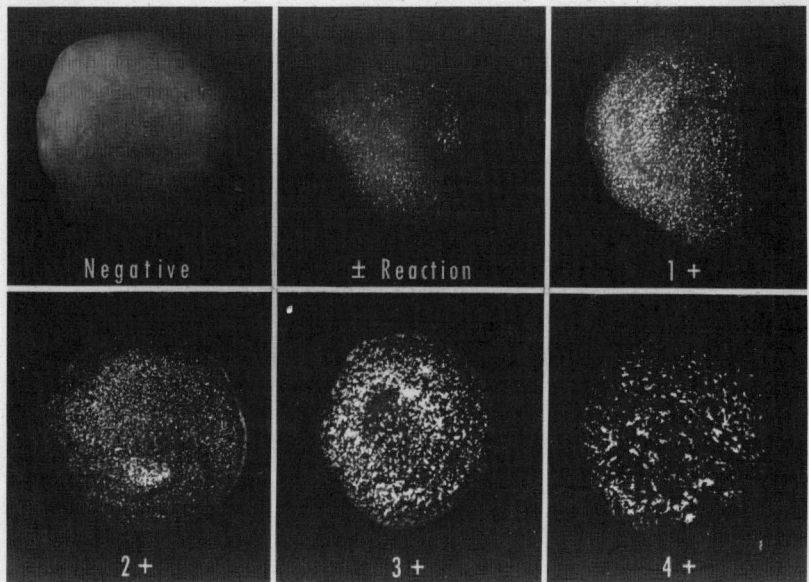

TUBE AGGLUTINATION

The tube agglutination test focuses on the quantity of antibodies present in a serum. The actual quantitation of the sera is accomplished by the addition of increasing dilutions of the antiserum with a constant amount of antigen. After incubation, the tubes are examined for the amount of agglutination (clumping) that has occurred.

Available Products:
Widal Antigen Set — contains antigens to

Salmonella H a	Salmonella O Group D
Salmonella H b	Febrile Positive and Negative Controls
Salmonella H d	

Rheumatoid Titration Test

Davidsohn's Differential Test
Beef Cell Antigen, Desiccated
Guinea Pig Kidney Antigen, Desiccated
Heterophile Forssman Antiserum
Infectious Mononucleosis Positive Serum

Febrile Antigen Set

PRECIPITATION TEST

One of the most important applications of the precipitin test is its use in differentiating among types of blood stains in forensic medicine. It is also used to detect inflammatory changes associated with various diseases in their early stages. To assure accurate results and reproducibility, all components used in the test system must be sparkling crystal clear. The slightest trace of turbidity will give ambiguous results; therefore, it is not uncommon for the reagents to be filtered or centrifuged. Precipitin tests require the antigen be serially diluted followed by the addition of a high concentration of antibodies, which may or may not be diluted. Precautions should be taken not to dilute the antiserum more than 1:5, if at all, since over-dilution will cause the amount of precipitate to decrease rapidly.

Available Products:
Animal Species Globulin Antisera
Prepared by injecting species globulins into the host animal
Example: Human Globulin Antiserum (Rabbit)

Animal Species Sera Antisera
Prepared by injecting species sera into the host animal
Example: Bovine Serum Antiserum (Rabbit)

C Protein Determination
C Protein Antiserum
C Protein Standard

Animal Species Normal Sera Control
Normal sera obtained from the various animal species
Example: Deer Serum Normal

QUELLUNG TEST

Quellung (capsular swelling) test is used for serological identification of *Haemophilus influenzae*, meningitis in spinal fluids, serotyping of *Klebsiella pneumoniae*, grouping *Neisseria meningitidis*, and typing of pneumococci. At one time, the Quellung reaction was one of the most important procedures in the medical diagnostic laboratory. However, with the onset of antibiotic therapy, it is not used as frequently. Nevertheless, the Quellung test is still the most rapid and satisfactory method for the identification of pneumococcal organisms in clinical specimens. The procedure is brief and easily performed by the placement of a loopful of a cultured suspension onto a slide, followed by the addition of an equal portion of antiserum. The antigen and antiserum are mixed, covered with a glass coverslip, then examined microscopically for the presence or absence of capsule formation. For the enhancement of capsular observation, methylene blue may be added. The size of capsule formation will vary among the different types of organisms tested. Therefore, it is necessary that all capsular swelling be checked against a negative control.

NEUTRALIZATION

Neutralization tests are valuable aids in differential diagnosis of streptococcal infections. In this type of a reaction the antigens are neutralized by their specific antibodies. The reaction is a macroscopic one and may be observed in the form of a clot, agglutination, flocculation, or lysis.

Available Products:
Streptolysin O
    ASO Standard
    Streptolysin O Reagent
    Streptolysin O Buffer

Antihyaluronidase Titration
    AHT Enzyme
    AHT Standard
    AHT Substrate

THE BASIC TECHNIQUE

1. Add a constant volume of antigen to decreasing amounts of serum.
2. Incubate.
3. Add a constant amount of a substrate to the tubes.
4. Reincubate.
5. a) In the case of Streptolysin O Reagent, centrifuge, then observe the tubes for hemolysis.
   b) With AHT, add an acidifying reagent (2N acetic acid), immediately observe for clot formation.

COMPLEMENT FIXATION

Complement Fixation has been used frequently in the Wasserman test for syphilis, as modified by Kolmer. It is also used in the diagnosis of viral and rickettsial diseases. This technique requires the addition of complement to an initial antigen-antibody reaction, allowing the complement to "fix", followed by the addition of a hemolytic indicator. Lysis occurs in a negative reaction where free complement is present, but does not occur in a positive reaction, where binding of the complement has taken place.

Available Products:
Complement (Guinea Pig)
Antisheep Hemolysin

FLOCCULATION

Flocculation is a variation of the precipitation test where the reaction distinguishes itself through the formation of clumps instead of a precipitate. The methodology may be macroscopic or microscopic. General procedure would include the following:

1. Serially dilute inactivated serum in a buffer or saline solution.
2. Prepare the antigen.
3. Add a drop of each serum dilution to wells on a test slide.
4. Add a drop of the prepared antigen to each serum dilution.
5. Rotate on a shaker for a specified time.
6. Examine for clumps macroscopically or microscopically.

Available Products:

Trichinella Antibody Detection
  Trichinella Antigen
  Trichinella Antiserum
  Latex 0.81

Syphilis Serology
  USR Antigen   VDRL Antigen
  USR Test Control Serum Set   VDRL Test Control Serum Set

PASSIVE HEMAGGLUTINATION

When bacterial antigens in nature are attached to red blood cells, these erythrocytes acquire new serologic specificities. The sensitized erythrocytes agglutinate with the addition of homologous bacterial antibodies. There may be several antigens fixed simultaneously to the surface of the erythrocytes. This type of passive or indirect hemagglutination test is used in the titration of bacterial antibodies or antigens.

FLUORESCENT ANTIBODY PROCEDURE

The fluorescent antibody technique (FA) is a specialized serological procedure which consists of an antigen-antibody reaction made visible by a fluorescent dye incorporated into the system. See the section on Fluorescent Antibody Procedure for a complete discussion.

Available Products:

FA Bordetella Pertussis/Parapertussis
FA C. Albicans
FA C. Diphtheriae
FA E. Coli Polys
FA H. Influenzae types
FA Leptospira types
FA Listeria Poly
FA Meningococcus Poly

FA N. Gonorrhoeae
FA Pneumococcus Poly
FA Salmonella Panvalent
FA Salmonella Poly
FA Staphylococcus Aureus
FA Streptococcus Groups
Fluorescent Treponemal Antibody (FTA-ABS) Test Reagents

SEROLOGICAL SUPPLEMENTARY REAGENTS

Various reagents are used to facilitate serological reactions. The products themselves are not responsible for the reactions but enhance the reading or interpretation of the tests. They may also be used to enhance an antigen-antibody response.

Available Products:

Adjuvants
  Adjuvant, Complete (Freund) — contains *M. butyricum*
  Adjuvant, Complete H-37 Ra — contains *M. tuberculosis* H-37 Ra
  Adjuvant, Incomplete (Freund)

M. Butyricum ⎫
M. Tuberculosis H-37 Ra ⎬

Add to Adjuvant, Incomplete
(Freund) as required

FA Reagents Supplementary
FA Animal Globulin Antiglobulins
FA Buffer, Dried
FA Mounting Fluid pH 7.2
FA Mounting Fluid pH 9
FA Papain

FA Rabbit Globulin
FA Rabbit Serum Blocking
FA Rhodamine Counterstain
FA Slides Stained/Unstained

Inert Carriers
Formocells Sheep
Latex 0.81

MICROBIOLOGY FOR THE EXAMINATION OF WATER AND WASTEWATER

The microbiological examination of water and wastewater is of such importance that even as early as the 1880's, bacteriologists saw a need to standardize procedures. As a result, there are today, agencies throughout the world, national and local, that regulate and standardize methods for the bacteriological examination of water. Many of these organizations follow the recommendations and specifications given in *Standard Methods for the Examination of Water and Wastewater,* 1981[1] and *Microbiological Methods for Monitoring the Environment, Water and Wastes,* 1978[2].

The dehydrated culture media listed in the outline conform to the formulae or perform similarly as specified by the above texts. Carefully selected standardized ingredients are used in the exact proportions specified in the *Standard Methods* formulae. The pH of each medium is carefully adjusted so that the final reaction will fall within the range of pH recommended in *Standard Methods.*

REFERENCES

1. Standard Methods for the Examination of Water and Wastewater, 15th ed., American Public Health Association, Washington, D.C., 1981.
2. Microbiological Methods for Monitoring the Environment, Water and Wastes, U.S. Environmental Protection Agency, EPA-600/8-78-017, December 1978.

TOTAL PLATE COUNTS

Tryptone Glucose Extract Agar
Plate Count Agar
Nutrient Agar

COLIFORM ORGANISMS

| Multiple Tube Fermentation Tests | | Membrane Filter Procedure | Differentiation |
Total Coliforms	Fecal Coliforms		
Presumptive: Lauryl Tryptose Broth Confirmed: Brilliant Green Bile 2% Completed: Lauryl Tryptose Broth Levine EMB Agar Endo Agar Nutrient Agar	EC Medium A-1 Medium	m Endo Agar LES m Endo Broth MF® m FC Broth Base Rosolic Acid	EMB Agar Tryptone MR-VP Medium Koser Citrate Medium Simmons Citrate Agar Motility Test Medium

FECAL STREPTOCOCCI

Multiple Tube Technique	Membrane Filter Technique	Plate Count	Presumptive-Confirmed Tests (EPA)	
Bile Esculin Azide Agar Azide Dextrose Broth SF Broth	KF Streptococcus Agar/Broth m Azide Agar	KF Streptococcus Agar Bile Esculin Azide Agar	Azide Dextrose Broth EVA Broth	Nutrient Gelatin TTC 1%

SALMONELLA-QUALITATIVE

Enrichment	Differential	Selective	Biochemical Reactions	Serological	Immunofluorescent
Tetrathionate Broth Base Selenite Broth GN Broth Hajna	EMB Agar Base + Sucrose	Bismuth Sulfite Agar Brilliant Green Agar Desoxycholate Agar Desoxycholate Citrate Agar MacConkey Agar SS Agar	Decarboxylase Medium Base w/Amino Acids Lysine Iron Agar MR-VP Medium Malonate Broth Motility Sulfide Medium Phenylalanine Agar	Brain Heart Infusion Agar H Broth MinESS Antiserum Set II Salmonella O Antiserum Poly A-I and Vi	FA Kirkpatrick Fixative FA Mounting Fluid FA Salmonella Panvalent FA Salmonella Polyvalent

SALMONELLA-QUALITATIVE (continued)

Enrichment	Differential	Selective	Biochemical Reactions	Serological	Immunofluorescent
		XLD Agar	Purple Broth Base w/ Carbohydrates SIM Medium Simmons Citrate Agar Triple Sugar Iron Agar Urea Agar	Salmonella Vi Anti-serum	

SALMONELLA-QUANTITATIVE

MPN Technique	Membrane Filter Technique
Tetrathionate Broth Base Brilliant Green Agar XLD Agar	m Bismuth Sulfite Broth m Brilliant Green Broth

SHIGELLA	VIBRIO CHOLERAE	STAPHYLOCOCCI	FUNGI	YEASTS
XLD Agar MacConkey Agar Shigella Antisera GN Broth Hajna	TCBS Agar Vibrio Cholera Antisera	Azide Blood Agar Base Azide Dextrose Broth Mannitol Salt Agar with Egg Yolk Phenylethanol Agar m Staphylococcus Broth	BiGGY Agar Brain Heart Infusion Agar w/Antimicrobics Cooke Rose Bengal Agar Corn Meal Agar Czapek Dox Agar Littman Oxgall Agar Malt Extract Agar/Broth Mycobiotic Agar Mycological Agar/Broth Potato Infusion Agar Sabouraud Maltose/Dextrose Agar	Carbohydrate Differentiation Disks Chlamydospore Agar Corn Meal Agars Orange Serum Agar Sabouraud Agars YM Agar/Broth Yeast Carbon Base w/Nitrogen Yeast Nitrogen Base w/Dextrose

GENERAL USE MEDIA

Blood Agar Base
Brain Heart Infusion/Agar
Dextrose Agar
Heart Infusion Agar/Broth

Nutrient Agar/Broth
Plate Count Agar
Tryptic Soy Agar/Broth
Tryptone Glucose Extract Agar

Tryptose Agar/Broth
Tryptose Blood Agar Base
Tryptose Phosphate Broth
Veal Infusion Agar/Broth

MICROBIOLOGY FOR FOODS
AND ALCOHOLIC BEVERAGES

Difco prepares a wide variety of media and reagents used in the microbiological examination of foods and alcoholic beverages. In many cases, these products are recommended in standard texts used by regulatory agencies to ensure the quality and safety of food supplies.

In the United States, a number of agencies on the local, state and national levels have jurisdiction over the quality of food released for public consumption. Many of these groups have published standard methodologies for the determination of food quality. These involve the use of standardized media and specific tests for the organisms expected to be encountered in foods under test.

Similar regulatory agencies exist throughout the world and efforts have been made in many cases to establish standardized methods. Therefore, in many cases, the same testing protocols, media and reagents are used throughout the world. Several of these organizations are: The International Commission of Microbiological Specifications for Foods, International Association of Microbiological Societies; Codex Alimentarius Commission, Expert Committee on Food Hygiene and Joint FAO/WHO Expert Committee on Food Hygiene, and Sub Committee 6 (Meat and Meat Products), International Organization for Standardization (ISO).[1]

Since it is beyond the scope of this manual to discuss these methods in detail, the user should consult appropriate manuals and compendia for test procedures as well as interpretation. Several of these include: the FDA *Bacteriological Analytical Manual;*[2] the *U.S. Department of Agriculture Microbiology Laboratory Guidebook;*[3] the *Compendium of Methods for the Microbiological Examination of Foods;*[4] *Official Methods of Analysis of the Association of Official Analytical Chemists;*[5] *Salmonella in Foods and Feeds, Review of Isolation Methods and Recommended Procedures.*[6]

The table following the discussion lists media and reagents prepared by Difco which meet the specifications or perform similarly as outlined in these compendia.

Apart from the microbiological procedures that are identical to those used for water, dairy and food products (refer to those sections), the following products are unique to the brewing industry:

Orange Serum Agar
UBA (Universal Beer Agar)
Wort Agar
WL Nutrient Medium (Wallerstein Laboratories)
WL Nutrient Broth
WL Differential Medium
YM Agar (Yeast Malt Extract)
YM Broth

REFERENCES

1. *Compendium of Methods for the Microbiological Examination of Foods,* Washington, D.C.: American Public Health Association, p 309, 1976.
2. *FDA Bacteriological Analytical Manual,* Washington, D.C.: AOAC, 1978.

3. *Microbiology Laboratory Guidebook*, Washington D.C.: Scientific Services, Animal and Plant Health Inspection Service, USDA, Memorandum No. 5, 1974.
4. *op.cit.*, Speck, M.L.
5. *Official Methods of Analysis of the Association of Official Analytical Chemists*, 13th Ed., Washington, D.C.: AOAC, 1980.
6. *Salmonella in Foods and Feeds, Review of Isolation Methods and Recommended Procedures*, Atlanta: U.S. Centers for Disease Control.

COLIFORM ORGANISMS

Presumptive Tests	Confirmed Tests	Fecal Coliforms	
Lauryl Tryptose Broth	Brilliant Green Bile 2%	Violet Red Bile Agar Brilliant Green Bile 2%	EC Broth m FC Broth

Confirmed Tests	Rapid Recovery Tests
EC Medium Levine EMB Agar Plate Count Agar Tryptone MR-VP Medium Lauryl Tryptose Broth Koser Citrate Medium	Lauryl Tryptose Broth Levine EMB Agar A-1 Medium Tryptone MR-VP Medium Koser Citrate Medium

ESCHERICHIA COLI

		Enteropathogenic	
Enrichment	**Isolation**	**Identification**	**Serology**
MacConkey Broth Lauryl Tryptose Broth Nutrient Broth EE Broth Mossel	Levine EMB Agar MacConkey Agar E. Coli OK Antisera Poly Groups VRB Agar	Triple Sugar Iron Agar Urea Broth Differentiation Disks ONPG Lysine Decarboxylase Decarboxylase Medium Base KCN Broth Base MR-VP Medium Indole Nitrate Medium Heart Infusion Agar Purple Broth Base w/Carbohydrates	Heart Infusion Agar Veal Infusion Agar Veal Infusion Broth E. Coli OK Antisera Pools A,B,C E. Coli OK Antisera Monovalent

SALMONELLA

Preenrichment	Selective Enrichment	Selective Differential	Biochemical Confirmation	Serology	Immunofluorescence
Lactose Broth Tryptic Soy Broth Nutrient Broth Selenite Broth GN Broth Hajna Buffered Peptone Water	Selenite Cystine Broth Tetrathionate Broth EE Broth Mossel Lauryl Tryptose Broth	Bismuth Sulfite Agar XLD Agar Hektoen Enteric Agar MacConkey Agar Brilliant Green Agar SS Agar Desoxycholate Agar Desoxycholate Citrate Agar	Triple Sugar Iron Agar Lysine Iron Agar Urea Broth/Agar Urea R Broth Lysine Decarboxylase Medium Phenol Red Broth Base w/Carbohydrates	Brain Heart Infusion Tryptic Soy Broth H Broth Salmonella H Antiserum Poly A-Z	FA Salmonella Panvalent FA Salmonella Polyvalent

SALMONELLA (continued)

Preenrichment	Selective Enrichment	Selective Differential	Biochemical Confirmation	Serology	Immunofluorescence
			Purple Broth Base w/Carbohydrates KCN Broth Base MR-VP Medium Simmons Citrate Agar Tryptone Malonate Broth Motility Test Medium LICNR Broth	Salmonella H Antiserum Spicer-Edwards Set Salmonella O Antiserum Poly A-I and Vi Salmonella O Grouping Antisera Salmonella Vi Antiserum Arizona Antiserum Arizona Antiserum Diphasic Arizona Antiserum Monophasic MinESS Antisera Set II	

SHIGELLA

Enrichment	Isolation	Biochemical Indentification	Serology
GN Broth Hajna Selenite Cystine Broth Brain Heart Infusion Buffered Peptone Water GN Broth Hajna	XLD Agar MacConkey Agar Desoxycholate Citrate Agar Levine EMB Agar Tergitol 7 Agar SS Agar Hektoen Enteric Agar	Triple Sugar Iron Agar Brain Heart Infusion Agar KCN Broth Base MR-VP Medium Purple Broth Base w/Carbohydrates Decarboxylase Medium Base w/ Amino Acids Phenol Red Broth Base w/Carbohydrates Urea Broth Urea R Broth Simmons Citrate Agar Tryptone Malonate Broth Motility Test Medium SIM Medium	Nutrient Agar Shigella Antisera Sets Alkalescens-Dispar Antiserum Poly

YERSINIA ENTEROCOLITICA

Enrichment	Isolation	Identification
Selenite Broth	MacConkey Agar	Triple Sugar Iron Agar
Tetrathionate Broth Base	SS Agar	Decarboxylase Medium Base w/Amino Acids
	EMB Agar	Urea Agar Base
	Tergitol 7 Agar	Purple Agar Base w/Carbohydrates
	Desoxycholate Citrate Agar	Purple Broth Base w/Carbohydrates
	Yersinia Selective Agar	Veal Infusion Broth
		Veal Infusion Agar
		MR-VP Medium
		KCN Broth Base
		Simmons Citrate Agar
		SIM Medium
		Bovine Albumin Fraction V
		TC Fetal Calf Serum

STAPHYLOCOCCUS AUREUS

Enrichment	Isolation/Enumeration	Confirmation	MPN Procedures	Enterotoxin Detection
Tryptic Soy Broth	Baird Parker Agar Base	Coagulase Plasma	Tryptic Soy Broth	McFarland Barium Sulfate Standard Set
Giolitti-Cantoni Broth	EY Tellurite Enrichment	Coagulase Plasma EDTA	Baird Parker Agar Base	Staphylococcus Medium 110
	VJ Agar		EY Tellurite Enrichment	Baird Parker Agar Base
	Staphylococcus Medium 110		Brain Heart Infusion	EY Tellurite Enrichment
	Mannitol Salt Agar		TPEY Agar Base w/TPEY Enrichment	Nutrient Agar
	Azide Blood Agar Base			Agar Purified, Difco
	SPS Agar			Agar Noble, Difco
	DNase Test Agar			Agar
	Phenylethanol Agar			Brain Heart Infusion
	TPEY Agar Base w/TPEY Enrichment			Brain Heart Infusion Agar

VIBRIO

Isolation/Differentiation	Biochemical Identification	Serology
TCBS Agar	Triple Sugar Iron Agar	Vibrio Cholerae Antisera
Transport Medium w/o Charcoal	Tryptic Soy Broth (+3% NaCl)	
Peptone Water	Tryptic Soy Agar (+3% NaCl)	
	Decarboxylase Medium Base	
	Nutrient Gelatin	
	MR-VP Medium	
	Motility Test Medium	
	Phenol Red Broth Base	
	Purple Broth Base w/Carbohydrates	

ENTEROCOCCI

KF Streptococcus Agar
Bile Esculin Agar
Brain Heart Infusion
Azide Dextrose Broth

BACILLUS	Growth	CLOSTRIDIUM			SULFATE REDUCERS	YEASTS AND MOLDS
		Isolation	Confirmation	Toxin Detection		
Nutrient Agar	Cooked Meat Medium	SFP Agar Base	Fluid Thioglycollate Medium	Trypsin 1:250, Difco	Sulfite Agar	Potato Dextrose Agar
Egg Yolk Enrichment 50%		Egg Yolk Enrichment 50%				Plate Count Agar
Phenol Red Broth Base		SPS Agar				Malt Agar
Phenol Red Dextrose Broth		Reinforced Clostridial Medium/Agar				Mycological Agar
Nitrate Broth		Anaerobic Agar				
Tryptic Soy Agar		Liver Veal Agar				
Antimicrobic Vial P		Thioglycollate Gelatin Medium				
MR-VP Medium						
Basic Fuchsin						
Dextrose Tryptone Agar						
Nutrient Broth						
Phenol Red Agar Base						
Litmus Milk						
Nutrient Gelatin						

CITRUS FRUIT	LACTOBACILLI	
Orange Serum Agar	APT Agar	Tomato Juice Media
Orange Serum Broth Concentrate	APT Broth	Lactobacilli Media AOAC
Mycobiotic Agar/Broth	Elliker Broth	Micro Inoculum Broth
	Lactobacilli MRS Broth	Litmus Milk
	Rogosa SL Agar	
	Rogosa SL Broth	

MICROBIOLOGY FOR DAIRY PRODUCTS

Microbiological test procedures for the examination of dairy products have been standardized and regulated by a number of agencies throughout the world. Among these are the United States Food and Drug Administration (USFDA) and the United States Department of Agriculture (USDA). Details of recommended test procedures, media selection and interpretation of results can be found in such appropriate reference texts as: *Standard Methods for the Examination of Dairy Products*,[1] *Bacteriological Analytical Manual*,[2] *Compendium of Methods for the Microbiological Examination of Foods*[3] and *Official Methods of Analysis of the Association of Official Chemists*.[4]

Carefully selected standardized ingredients are used in these exact proportions specified in the formulae described in these compendia. The reaction of each medium is adjusted so that the final rection of each preparation made from the dehydrated product will fall within the recommended pH range.

A guide for the selection of media follows:

STANDARD PLATE COUNT
Plate Count Agar

COLIFORM ORGANISMS
Violet Red Bile Agar
Brilliant Green Bile 2%
Levine EMB Agar
Lactose Broth
Nutrient Agar

YEASTS AND MOLDS
Potato Dextrose Agar

ANTIBIOTIC RESIDUES
Bacillus Subtilis Disk Assay for Penicillin
Antibiotic Medium 1
Bacillus Subtilis Spore Suspension
Concentration Disks Sterile Blanks
Concentration Disks Penicillin 0.1 Unit
Concentration Disks Penicillin 0.05 Unit
Concentration Disks Penase

Sarcina Lutea Cylinder Plate Assay for Penicillin
Antibiotic Medium 1
Antibiotic Medium 4
Penase Concentrate

Bacillus Stearothermophilus Disk Assay for Penicillin
PM Indicator Agar
Antibiotic Medium 4
Penase Concentrate
Concentration Disks Penase
Concentration Disks Sterile Blanks
Thermospore Suspension PM
PM Positive Control
PM Negative Control

International Standard Fil-IDF 57:1970 Disk Assay Technique for Penicillin
Plate Count Agar
Penase Concentrate
Thermospore Suspension

Modified IDF Disk Test for Penicillin
Thermospore Suspension
Plate Count Agar
Concentration Disks Sterile Blanks
Concentration Disks Penase
Penase Concentrate

Bacillus Megaterium Disk Assay for Sulfa Drugs and Antibiotics
Mueller Hinton Medium
Concentration Disks Sterile Blanks

TESTS OF EQUIPMENT, CONTAINERS, WATER AND AIR
Plate Count Agar
Violet Red Bile Agar
m Endo Broth MF®
Neutralizing Buffer

REFERENCES

1. *Standard Methods for the Examination of Dairy Products*, 14th Ed., Washington, D.C.: American Public Health Association, 1978.
2. *Bacteriological Analytical Manual*, 5th Ed. FDA, Washington, D.C.: Association of Official Analytical Chemists, 1978.
3. *Compendium of Methods for the Microbiological Examination of Foods*, Washington, D.C., American Public Health Association, 1976.
4. *Official Methods of Analysis of the Association of Official Chemists*, 13th Ed., Washington, D. C., AOAC, 1980.

PHARMACEUTICAL MICROBIOLOGY

To ensure the safety of pharmaceutical and biological products, standards for their quality have been developed by numerous government or government-allied agencies throughout the world. Among the tests performed on such products are those used to determine the effectiveness of antimicrobial preservatives, microbial limits, sterility of the products, and presence of pyrogens.

The microbiological assay of vitamins, amino acids and antibiotics constitutes an entirely separate testing protocol. Reference texts that offer complete, detailed test procedures and interpretation include: the *U.S. Pharmacopeia XX/National Formulary XV*,[1] the *U.S. Code of Federal Regulations, Food & Drugs*,[2] the *Federal Register*,[3] the *Manual of Methods of the AOAC*,[4] the *British Pharmacopoeia*,[5] the *European Pharmacopoeia*,[6] *Standards for Sterile Therapeutic Goods*,[7] the *Pharmacopoeia of Japan*[8] and *Minimum Requirements of Biological Products*[9] of the Ministry of Health and Welfare, Japanese Government. The World Health Organization has also established requirements for biological products and offers periodic bulletins.

Difco prepares a wide variety of media and reagents which meet the specifications outlined in these references. They are as follows:

Microbial Limits Testing

Tryptic Soy Agar
  (Soybean-Casein Digest Agar Medium)
Tryptic Soy Broth
  (Soybean-Casein Digest Medium)
Mannitol Salt Agar
Baird Parker Agar Base
VJ Agar
  (Vogel Johnson Agar Medium)
Pseudomonas Agar F
  (Pseudomonas Agar Medium for detection of fluorescein)
Pseudomonas Agar P
  (Pseudomonas Agar Medium for detection of pyocyanin)
Lactose Broth
  (Fluid Lactose Medium)
Selenite Cystine Broth
  (Fluid Selenite Cystine Medium)
Tetrathionate Broth Base
  (Fluid Tetrathionate Medium)
Brilliant Green Agar
XLD Agar
  (Xylose-Lysine Desoxycholate Agar Medium)
Bismuth Sulfite Agar
Triple Sugar Iron Agar
MacConkey Agar
EMB Agar, Levine
  (Eosin-Methylene Blue Agar Medium, Levine)
TAT Broth Base
  (Fluid Casein Digest-Soy Lecithin Polysorbate 20 Medium)
EY Tellurite Enrichment
Cetrimide Agar Base

Microbial Limits Testing
Nutrient Broth
Selenite Broth
Desoxycholate Citrate Agar
Urea Broth

Antimicrobial Preservatives — Effectiveness Test
Tryptic Soy Agar
(Soybean-Casein Digest Agar Medium)
Sabouraud Dextrose Agar

Sterility Testing
Fluid Thioglycollate Medium
Tryptic Soy Broth
(Soybean-Casein Digest Medium)
NIH Thioglycollate Broth
(Alternative Thioglycollate Medium)

Pyrogen Testing
Pyrotest — Limulus Amebocyte Lysate Test for In-Process Testing
(Fed. Reg. Vol. 38, No. 8, 1973)

Microbiological Assays
Difco has an extensive line of media and reagents for the microbiological assay of antibiotics, amino acids and vitamins, including many test cultures, maintenance media for test cultures, media for preparing inocula and media for the assay. Refer to appropriate section for a complete discussion on Vitamin Assay Media, Amino Acid Assay Media, or Antibiotic Assay Media.

REFERENCES
1. The United States Pharmacopeia XX/The National Formulary XV, Rockville, Md, The US Pharmacopeia Convention, Inc., 1980.
2. US Code of Federal Regulations, Food and Drugs, Part 141 to Part 599, 1981.
3. US Federal Register, Vol. 38, No. 8, 1973.
4. Official Methods of Analysis of the AOAC, 13th Ed., Washington D.C., AOAC, 1980.
5. British Pharmacopoeia 1980, Vol. II, Cambridge, England: University Press, Appendix XIVA, XVI, 1980.
6. European Pharmacopoeia, 2nd Ed., Sainte-Ruffine, France: Maisonneuve S. A., VIII.3 + VIII 4, 1980.
7. Standard for Sterile Therapeutic Goods, National Biological Standards Laboratory, Maulson, A.C.T. Australia.
8. The Pharmacopoeia of Japan, Tokyo: Society of Japanese Pharmacopoeia.
9. Minimum Requirements of Biological Products, Tokyo: Ministry of Health and Welfare, p. 341 – 413, 1982.

GUIDE FOR THE SELECTION OF CULTURE MEDIA

Microorganism	Isolation	Differentiation	Propagation
Actinomyces	Actinomyces Broth Actinomycete Isolation Agar Brain Heart Infusion Brain Heart Infusion Agar Czapek Dox Agar Sabouraud Dextrose/Maltose Agar Tryptic Soy Agar/Broth	Differentiation Disks Carbohydrates Nutrient Gelatin Tryptic Nitrate Medium	Actinomyces Broth Actinomycete Isolation Agar Brain Heart Infusion Agar Sabouraud Dextrose/Maltose Agar Tryptic Soy Agar
Alga	Euglena Broth		Euglena Broth
Anaerobes	AC Medium Anaerobic Agar Anaerobic Agar, Brewer Brewer Thioglycollate Medium Cooked Meat Medium Egg Meat Medium Fluid Thioglycollate Medium McClung Toabe Agar Base Reinforced Clostridial Medium/Agar SFP Agar Base SPS Agar Veal Infusion Agar/Broth Wilkins-Chalgrin Agar	Differentiation Disks Carbohydrates Egg Yolk Enrichment Litmus Milk Nitrate Agar/Broth Phenol Red Broth Base w/Carbohydrates SIM Medium	AC Medium Anaerobic Agar Brewer Thioglycollate Medium Cooked Meat Medium Egg Meat Medium Fluid Thioglycollate Medium Veal Infusion Agar/Broth
Bordetella	Bordet Gengou Agar Base w/Blood Charcoal Agar	Heart Infusion Agar Litmus Milk Nitrate Agar/Broth Urea Agar Base w/Urea	Bordet Gengou Agar Base w/Blood Tryptic Soy Agar w/Blood and Supplement B
Brucella	Brucella Agar/Broth Eugon Agar/Broth Tryptic Soy Agar/Broth Tryptose Agar	Brucella Agar w/Fuchsin & Thionin Differentiation Disks Carbohydrates H₂S Test Strips	Brucella Agar/Broth Tryptic Soy Agar/Broth Tryptose Agar
Campylobacter	Campylobacter Agar Base w/Antimicrobic Supplements B or S	Campylobacter Agar Base w/Antimicrobic Supplements B or S	Blood Agar Base No. 2 Cooked Meat Medium

Microorganism	Isolation	Differentiation	Propagation
Candida	BiGGY Agar Candida BCG Agar Base Chlamydospore Agar Corn Meal Agar Levine EMB Agar Pagano-Levin Agar/Base Rice Extract Agar	BiGGY Agar Candida BCG Agar Base Chlamydospore Agar Rice Extract Agar	Corn Meal Agar Malt Agar Sabouraud Dextrose/Maltose Agar
Clostridium difficile	Brain Heart Infusion Agar/Clostridium Difficile Antimicrobic Supplement CC	Brain Heart Infusion Agar/Clostridium Difficile Antimicrobic Supplement CC	Brain Heart Infusion/Agar
Coliforms	A-1 Medium Brilliant Green Agar Brilliant Green Bile 2% Desoxycholate Agar Desoxycholate Lactose Agar EC Medium EE Broth, Mossel EMB Agar Endo Agar GN Broth Hajna Hektoen Enteric Agar Lactose Broth Levine EMB Agar Lauryl Tryptose Broth m Endo Broth MF® m FC Broth Base MacConkey Agar Tergitol 7 Agar Violet Red Bile Agar	Acetate Differential Agar Brilliant Green Agar Desoxycholate Lactose Agar Differentiation Disks Carbohydrates Kligler Iron Agar Koser Citrate Medium Lauryl Tryptose Broth Lysine Iron Agar MR-VP Medium Malonate Broth Motility Test Medium Purple Agar Base/Broth Base w/Carbohydrates Simmons Citrate Agar Triple Sugar Iron Agar	Brain Heart Infusion/Agar Heart Infusion Agar/Broth Lactose Broth Nutrient Agar Tryptic Soy Agar/Broth

Microorganism	Isolation	Differentiation	Propagation
Corynebacterium	Columbia Blood Agar Base w/Tinsdale Enrichment Dextrose Proteose No. 3 w/Tellurite Blood Solution Loeffler Blood Serum Mueller Tellurite Base w/Mueller Tellurite Serum Tinsdale Base w/Tinsdale Enrichment	Brain Heart Infusion Chapman Tellurite Solution 1% Dextrose Proteose No. 3 w/Tellurite Blood Solution KL Virulence Agar/Enrichment/Strips Loeffler Blood Serum Mueller Tellurite Base w/Mueller Tellurite Serum Nitrate Broth Tinsdale Base w/Tinsdale Enrichment	Brain Heart Infusion Agar Loeffler Blood Serum
Haemophilus	Blood Agar Base w/Hemoglobin and Supplements A, B, C, or VX Casman Medium Base GC Medium Base w/Hemoglobin and Supplements A, B, C, or VX GC Medium Base w/Fildes Enrichment	Differentiation Disks BX, BV and BVX	Blood Agar Base w/Hemoglobin and Supplements A, B, C or VX GC Medium Base w/Hemoglobin and Supplements A, B, C or VX GC Medium Base w/Fildes Enrichment
Klebsiella	EMB Agar MacConkey Agar MacConkey Agar CS Worfel Ferguson Agar	Decarboxylase Base Differentiation Disks Lysine, ONPG, Ornithine H_2S Test Strips Indole Test Strips Kligler Iron Agar Lysine Decarboxylase Broth Lysine Iron Agar Phenol Red Agar Base/Broth Base w/Carbohydrates Purple Agar Base/Broth Base w/Carbohydrates SIM Medium Triple Sugar Iron Agar Worfel Ferguson Agar	Brain Heart Infusion Agar Nutrient Agar Tryptic Soy Agar/Broth

Microorganism	Isolation	Differentiation	Propagation
Lactobacilli	APT Agar/Broth Eugon Broth/Agar Lactobacilli Broth (Elliker Broth) Lactobacilli MRS Broth Micro Assay Culture Agar Micro Inoculum Broth Orange Serum Agar Orange Serum Broth Concentration 10X Rogosa SL Broth/Agar Tomato Juice Agar/Broth	Differentiation Disks Carbohydrates Lactobacilli MRS Broth Litmus Milk Snyder Test Agar Tomato Juice Agar	APT Agar/Broth Orange Serum Agar Tomato Juice Agar/Broth
Legionella	Legionella Agar Base w/Legionella Agar Enrichment	Legionella Agar Base w/Legionella Agar Enrichment	Legionella Agar w/Legionella Agar Enrichment
Leptospira	Fletcher Medium Base w/Leptospira Enrichment Leptospira Medium Base EMJH w/Leptospira Enrichment EMJH Stuart Medium Base w/Leptospira Enrichment		Leptospira Medium Base EMJH w/Leptospira Enrichment EMJH Stuart Medium Base w/Leptospira Enrichment
Listeria	Brain Heart Infusion Tryptic Soy Agar Tryptose Blood Agar Base w/Blood	Litmus Milk MR-VP Medium Motility Test Medium Nitrate Agar/Broth Phenol Red Broth Base w/Carbohydrates	Tryptic Soy Agar

Microorganism	Isolation	Differentiation	Propagation

Molds	Actinomycete Isolation Agar
	Brain Heart CC Agar
	Cooke Rose Bengal Agar
	Eugon Agar
	Littman Oxgall Agar
	Malt Agar
	Mycobiotic Agar
	Mycological Agar/Broth
	OGYE Agar Base/Enrichment
	Potato Dextrose Agar
	Rose Bengal Agar Base w/Rose
	  Bengal Antimicrobic Supplement C
	SABHI Agar Base
	Sabouraud Dextrose/Maltose Agar
	Trichophyton Agars 1,2,3,4,5,6,7
		Brain Heart CC Agar
		Corn Meal Agar
		Rose Bengal Agar Base w/Rose
		  Bengal Antimicrobic Supplement C
		SABHI Agar Base
		Sabouraud Agars
		Trichophyton Agars 1,2,3,4,5,6,7
		WL Nutrient Medium
			Corn Meal Agar
			Malt Extract Agar
			Potato Dextrose Agar
			Sabouraud Dextrose/Maltose Agar

Mycobacterium	ATS Medium Dubos Oleic Agar Lowenstein Medium Gruft Lowenstein Medium, Jensen Middlebrook 7H9 Broth and 7H10 Agar Mycobacteria 7H11 Agar Petragnani Medium	Dubos Medium Albumin Lowenstein Medium Gruft Lowenstein Medium, Jensen Mycobacteria 7H11 Agar Nitrite Test Strips TB Niacin Test Strips TB Stains	ATS Medium Dubos Media Lowenstein Medium Gruft Lowenstein Medium, Jensen Middlebrook 7H9 Broth/7H10 Agar Mycobacteria 7H11 Agar Petragnani Medium
Mycoplasma	PPLO Agar w/Mycoplasma Supplement and Supplement S PPLO Broth w/CV and PPLO Serum Fraction	Dienes Stain	PPLO Agar/Broth w/PPLO Serum Fraction PPLO Agar/Broth w/Mycoplasma Supplement Tryptic Digest Broth w/PPLO Serum Fraction

Microorganism	Isolation	Differentiation	Propagation
Neisseria	Brain Heart Infusion Agar Casman Medium Base w/Blood GC Medium Base w/Hemoglobin and Supplements A, B, C or VX GC Medium Base w/Hemoglobin Supplement A, B, C or VX and Antimicrobic Vial CNVT or Antimicrobic Vial CNV	p-Aminodimethylaniline Oxalate (Oxidase Reagent) Cystine Tryptic Agar w/Carbohydrates Differentiation Disks Carbohydrates Differentiation Disks Oxidase Phenol Red Broth Base w/Carbohydrates	Casman Medium Base w/Blood Dextrose Starch Agar GC Medium Base or Proteose No. 3 Agar w/Hemoglobin and Supplements A, B, C or VX
Pasteurella	Blood Agar Base w/Blood Cystine Heart Agar w/Hemoglobin and Supplement B Eugon Agar/Broth Tryptose Agar	Differentiation Disks Oxidase OF Basal Medium w/Carbohydrates Triple Sugar Iron Agar Tryptic Nitrate Medium Urea Agar/Base	Blood Agar Base w/Blood Cystine Heart Agar w/Hemoglobin and Supplement B Tryptic Soy Agar Tryptose Agar
Proteus	Blood Agar Base CLED Agar EMB Agar Hektoen Enteric Agar MacConkey Agar MacConkey Agar CS Tryptic Soy Agar	Decarboxylase Base w/Amino Acids Differentiation Disks Lysine, ONPG, and Urea H_2S Test Strips Indole Test Strips Kligler Iron Agar Motility Test Medium Purple Agar Base/Broth Base w/Carbohydrates SIM Medium Triple Sugar Iron Agar Urea Agar/Broth	Brain Heart Infusion Agar Nutrient Agar Tryptic Soy Agar/Broth
Protozoa	Endamoeba Medium Kupferberg Trichomonas Broth		Endamoeba Medium Kupferberg Trichomonas Broth
Pseudomonas	Cetrimide Agar Base EMB Agar MacConkey Agar Pseudomonas Isolation Agar Tryptic Soy Agar	Decarboxylase Medium Base Differentiation Disks Lysine, ONPG, Ornithine, Oxidase, Urea OF Basal Medium w/Carbohydrates Pseudomonas Agars F & P	Brain Heart Infusion Agar Tryptic Soy Agar/Broth

Microorganism	Isolation	Differentiation	Propagation

Salmonella	*Enrichment Media*
	GN Broth Hajna
	SBG Enrichment
	SBG Sulfa Enrichment
	Selenite Broth
	Selenite Cystine Broth
	Tetrathionate Broth Base
	
	Plating Media
	BG Sulfa Agar
	Bismuth Sulfite Agar
	Brilliant Green Agar
	DCLS Agar
	Desoxycholate Citrate Agar
	EMB Agar
	Hektoen Enteric Agar
	MacConkey Media
	SS Agar
	XL Agar Base
	XLD Agar
		Decarboxylase Medium Base
		Differentiation Disks Lysine, ONPG,
		  Ornithine, Urea
		H Broth
		H$_2$S Test Strips
		Lysine Iron Agar
		Lysine Decarboxylase Broth
		Motility Test Medium
		Phenol Red Tartrate Agar
		Purple Agar Base/Broth Base
		  w/Carbohydrates
		Russell Double Sugar Agar
		SIM Medium
		Triple Sugar Iron Agar
		Urea Agar/Broth
			Blood Agar Bases
			Brain Heart Infusion Agar
			Tryptic Soy Agar/Broth

Shigella	Desoxycholate Agar Desoxycholate Citrate Agar EMB Agar Hektoen Enteric Agar MacConkey Agar CS MacConkey Agar w/o CV SS Agar XL Agar Base XLD Agar	Acetate Differential Agar Decarboxylase Medium Base Differentiation Disks Lysine, ONPG, Urea H Broth H$_2$S and Indole Test Strips KCN Broth Base Kligler Iron Agar MIL Medium Motility Test Medium Purple Broth Base w/Carbohydrates Triple Sugar Iron Agar	Brain Heart Infusion Agar Nutrient Agar/Broth Tryptic Soy Agar/Broth

Microorganism	Isolation	Differentiation	Propagation
Staphylococci	Azide Blood Agar Base Baird Parker Agar Base w/EY Tellurite Enrichment Blood Agar Base Media Chapman Stone Medium Columbia CNA Agar w/Blood Dnase Test Agar Mannitol Salt Agar Phenylethanol Agar Staphylococcus Medium 110 TPEY Agar Base w/TPEY Enrichment and Antimicrobial Vial P Tryptic Soy Agar Tryptose Phosphate Broth VJ Agar	Baird Parker Agar Base w/EY Tellurite Enrichment Blood Agar Base Media w/Blood Chapman Stone Medium DNase Test Agar Mannitol Salt Agar Phenol Red Agar/Broth w/Carbohydrates Staphylococcus Medium 110 TPEY Agar Base/TPEY Enrichment Tryptic Soy Agar VJ Agar	Blood Agar Base Brain Heart Infusion Agar Dextrose Starch Agar Tryptic Soy Agar/Broth Tryptose Agar Tryptose Phosphate Broth
Streptococci (Including Enterococci)	Azide Blood Agar Base Azide Dextrose Broth BAGG Broth Blood Agar Base w/Blood Columbia CNA Agar w/Blood m Enterococcus Agar EVA Broth KF Streptococcus Agar/Broth Mitis Salivarius Agar Phenyl Ethanol Agar SF Medium Tryptic Soy Agar, Blood Agar Base w/Blood Tryptic Soy Agar w/Blood Tryptose Blood Agar Base w/Blood	Bile Esculin Agar Bile Esculin Azide Agar Blood Agar Base w/Blood Differentiation Disks Bacitracin/Optochin EVA Broth Enterococci Confirmatory Agar/Broth Enterococci Presumptive Broth Phenol Red Broth Base w/Carbohydrates SF Medium Tryptic Soy Agar, Blood Agar Base w/Blood	Blood Agar Base Tryptic Soy Agar/Broth
Streptomyces	ISP Medium 1,2,3,4		ISP Medium 1,2,3,4

Microorganism	Isolation	Differentiation	Propagation
Vibrio	TCBS Agar	TCBS Agar Tryptic Nitrate Medium Triple Sugar Iron Agar	Tryptic Soy Agar
Veillonella	Veillonella Agar		Blood Agar Base Tryptic Soy Agar
Yeasts	BiGGY Agar Corn Meal Agar Czapek Dox Broth Littman Oxgall Agar Malt Extract Agar Mycobiotic Agar Potato Dextrose Agar Sabouraud Dextrose Agar WL Nutrient Medium Wort Agar YM Agar/Broth	Cystine Heart Agar Rice Extract Agar Yeast Carbon Base Yeast Morphology Agar Yeast Nitrogen Base	Sabouraud Dextrose/Maltose Agar Tryptic Soy Agar
Yersinia enterocolitica	Yersinia Selective Agar Base w/Yersinia Antimicrobic Supplement CN	Decarboxylase Base Differentiation Disks Carbohydrates, Lysine, Ornithine Kligler Iron Agar Motility Test Medium Triple Sugar Iron Agar Yersinia Selective Agar Base w/Yersinia Antimicrobic Supplement CN	Tryptic Soy Agar

BACTO A - 1 MEDIUM

INTENDED USE
Bacto A - 1 Medium is used for determining the presence of fecal coliforms in water samples according to *Standard Methods for the Examination of Water and Wastewater*, 15th Edition.

HISTORY
The rapid and reliable determination of the bacterial quality of water is of great importance from a public health standpoint. The Standard Methods procedure recommended by APHA[1] utilizes a 48 hour enrichment with lactose broth or lauryl tryptose broth to obtain optimum recovery of fecal coliforms. Positive tubes are confirmed for *E. coli* by the use of E C medium. The total time required for the complete procedure is 72 hours. Andrews and Presnell in 1972 reported on the development of a new medium, A - 1 medium, which was capable of recovery of *E. coli* from estuarine water in 24 hours instead of 72, in greater numbers and without the preenrichment step.[2] Andrews, Diggs and Wilson demonstrated the specificity of A - 1 medium for *E. coli* in two types of shellfish and confirmed its productivity compared to the APHA method. They also found recovery to be faster and the occurrence of false positives to be substantially reduced.[3]

PRINCIPLES
Since the early 1900's the method used to determine water purity has been the enumeration of coliform organisms present in a sample or, more specifically, *E. coli*. One of the limiting factors in using *E. coli* is the length of time required for complete identification.[3] A - 1 medium was formulated to hasten the recovery of *E. coli* and to reduce the incidence of false positive cultures.

FORMULA
BACTO A - 1 MEDIUM
DEHYDRATED
Ingredients per liter

Bacto Tryptone	20 g
Bacto Lactose	5 g
Sodium Chloride	5 g
Bacto Salicin	0.5 g
Triton X-100	1 ml

Final pH 6.9 ± 0.1 at 25°C

Five hundred grams will make 15.8 liters of medium.

METHOD OF PREPARATION
1. Suspend 31.5 g Bacto A - 1 Medium in 1 liter distilled or deionized water and heat to boiling to dissolve completely.
2. Dispense 10 ml amounts into tubes containing inverted fermentation vials.
3. Sterilize in the autoclave for 10 minutes at 15 lbs pressure (121°C)

STORAGE
Bacto A - 1 Medium   Below 30°C
Prepared medium   15 – 30°C

streptococci. Bailey et al.[5] employed AC Medium in assaying for potency of strepto-mycin products using *Clostridium perfringens* as a test organism and reported excellent and rapid results. Schneiter and Kolb[6] used Bacto AC Medium for the growth of *Bacillus anthracis* and related mesophilic aerobic bacilli in their studies of the heat resistance of these organisms from hair and bristles. They reported that the medium permitted the distinctive cottonball appearance of the anthrax colonies. Kolb and Schneiter[7] used Bacto AC Medium to test the viability of *B. anthracis* following exposure to methyl bromide to test the efficiency of this compound as a germicidal and sporicidal agent.

Bacto AC Broth has the same formula as Bacto AC Medium except that the small amount of agar has been omitted.

For the sterility testing of biologicals and solutions containing mercurials as a preservative, Bacto Fluid Thioglycollate Medium should be employed.

FORMULAE

BACTO AC MEDIUM
DEHYDRATED

Ingredients per liter

Proteose Peptone No. 3, Difco	20 g
Bacto Beef Extract	3 g
Bacto Yeast Extract	3 g
Malt Extract, Difco	3 g
Bacto Dextrose	5 g
Ascorbic Acid	0.2 g
Bacto Agar	1 g

Final pH 7.2 ± 0.2 at 25°C.

Rehydrate with 35 g/liters.
Five hundred grams will make
14.2 liters of medium.

BACTO AC BROTH
DEHYDRATED

Ingredients per liter

Proteose Peptone No. 3, Difco	20 g
Bacto Beef Extract	3 g
Bacto Yeast Extract	3 g
Malt Extract, Difco	3 g
Bacto Dextrose	5 g
Ascorbic Acid	0.2 g

Final pH 7.2 ± 0.2 at 25°C.

Rehydrate with 34 g/liters.
Five hundred grams will make
14.7 liters of medium.

METHOD OF PREPARATION
1. Suspend the appropriate amount of medium in 1 liter distilled or deionized water.
2. Heat to boiling to dissolve completely.
3. Sterilize in the autoclave for 15 minutes at 15 lbs pressure (121°C).

If the medium is not used the same day it is sterilized, place in flowing steam or a boiling water bath for a few minutes to drive off dissolved gases, and allow to cool without agitation.

STORAGE
Bacto AC Media   Below 30°C
Prepared media   15 – 30°C

QUALITY CONTROL
Identity Specifications

	Bacto AC Medium	Bacto AC Broth
Dehydrated powder:	tan, homogeneous, free-flowing	light tan, homogeneous, free-flowing
% solution:	3.5%	3.4%
pH at 25°C:	7.2 ± 0.2	7.2 ± 0.2
Prepared medium:	medium amber, very slightly opalescent	medium to dark amber, clear to very slightly opalescent

Typical Cultural Response in Bacto AC Medium
After 18 – 48 Hours at 35°C

Organism	Growth
Clostridium perfringens ATCC® 12919	good to excellent
Neisseria meningitidis ATCC® 13090	good to excellent
Streptococcus mitis ATCC® 9895	good to excellent
Streptococcus pneumoniae ATCC® 6303	good to excellent

*Incubated anaerobically

Typical Cultural Response in Bacto AC Broth
After 18 – 48 Hours at 35°C

Organism	Growth
Corynebacterium diphtheriae Type *mitis*	good to excellent
Streptococcus mitis ATCC® 9895	good to excellent
Streptococcus pneumoniae ATCC® 6303	good to excellent
Streptococcus pyogenes ATCC® 19615	good to excellent

REFERENCES
1. Paper read at New York Meeting Am. Pub. Health Assoc., 1944.
2. J. Bact., 62:349, 1951.
3. J. Bact., 45:309, 1943.
4. Pub. Health Reports, 60:789, 1945.
5. Personal Communication, 1947.
6. Pub. Health Reports, Sup. No. 207, June 1948.
7. J. Bact., 59:401, 1950.

PACKAGING

Bacto AC Broth	500 g	0317-17-9
Bacto AC Medium	500 g	0316-17-0

BACTO ATS MEDIUM

INTENDED USE
Bacto ATS Medium is a prepared, coagulated egg yolk medium used for isolating and cultivating mycobacteria.

HISTORY
Bacto ATS Medium is prepared according to the formula described by the American Trudeau Society.[1]

PRINCIPLES
Bacto ATS Medium contains a minimal amount of the inhibitory agent, malachite green, thus permitting relatively early detection of mycobacteria. However, the medium is quite susceptible to proteolytic contaminating microorganisms and requires a carefully de-contaminated specimen as an inoculum.

FORMULA
BACTO ATS MEDIUM
Ingredients per liter

Egg Yolk Suspension . . . . 500 ml
Glycerol Extract of Potatoes . . . . . 500 ml
Malachite Green . . . . 0.2 ml

This product is available as a prepared tube medium.

STORAGE

Bacto ATS Medium Tubes $2 - 8°C$

QUALITY CONTROL

Identity Specifications

Reaction of medium: pH 6.5 – 7.5
Prepared medium: light green, opaque slants

Typical Cultural Response on Bacto ATS Medium
After 2 Weeks at 35°C Under 10% CO_2

Organism	Growth
Bacillus subtilis ATCC® 6633	inhibited
Mycobacterium kansasii ATCC® 12478	good to excellent
Mycobacterium tuberculosis H37Rv	good to excellent
Staphylococcus aureus ATCC® 25923	inhibited

REFERENCE

1. Am. Rev. Tuberc., 54:428, 1946.

PACKAGING

Bacto ATS Medium	12 tubes	1019-34-7
	144 tubes	1019-37-4

BACTO ACETATE DIFFERENTIAL AGAR

INTENDED USE

Bacto Acetate Differential Agar is a tube medium for differentiating members of the *Shigella* genus from members of the *Escherichia* genus.

HISTORY/PRINCIPLES

The formulation is that described by Trabulsi and Ewing[1] and is essentially the same as Bacto Simmons Citrate Agar[2] with sodium acetate replacing sodium citrate. The differentiation of groups on this medium is based on ability or failure of the test cultures to utilize acetate. Acetate utilization is indicated by a blue reaction.

Trabulsi and Ewing[1] demonstrated that none of the *Shigella* tested grew on the acetate differential agar whereas a high percentage of *E. coli* strains utilized acetate as the sole source of carbon. Of 186 *E. coli* cultures that belonged to various O antigen groups, 84% were positive within 1 or 2 days incubation and an additional 9.6% produced growth in 3 – 7 days. Similarly, 93.7% of the anaerogenic, nonmotile biotypes of *E. coli* that belonged to the specified O antigen groups (the *Alkalescens-Dispar* or A - D group) grew on the acetate agar. Members of the *Alkalescens-Dispar* O antigen groups 1 and 2 showed growth in 99% and 89% of the cultures respectively while 80% of the *Alkalenscens-Dispar* O groups 3 – 8 responded within 7 days incubation of the acetate differential agar.

The subjudice serotypes mentioned by Ewing, Reavis and Davis[3] failed to grow on the acetate medium. Other anaerogenic serotypes utilized acetate, thus differentiating them from the *Shigella*.

The majority of members of the *Salmonella, Citrobacter, Klebsiella, Enterobacter* and *Serratia* groups utilized acetate and grew on Bacto Acetate Differential Agar within 1 – 7 days whereas the *Proteus* and *Providence* groups failed to do so.

FORMULA
BACTO ACETATE DIFFERENTIAL AGAR
DEHYDRATED
Ingredients per liter

Sodium Acetate	2 g
Magnesium Sulfate	0.1 g
Sodium Chloride	5 g
Mono Ammonium Phosphate	1 g
Dipotassium Phosphate	1 g
Bacto Brom Thymol Blue	0.08 g
Bacto Agar	20 g

Final pH 6.7 ± 0.1 at 25°C.

One pound will make 15.5 liters of medium.

METHOD OF PREPARATION
1. Suspend 29.2 g in 1 liter cold distilled or deionized water and heat to boiling to dissolve completely.
2. Dispense into 13 × 100 mm tubes in sufficient amounts to give a 1 cm butt and 30 cm slant.
3. Sterilize in the autoclave for 15 minutes at 15 lbs pressure (121°C).
4. Allow to cool in the slanted position to give recommended butt and slant size.

PROCEDURE
Inoculate the surface of the slants using a 3 mm loopful of a small amount of a 16 – 18 hour culture emulsified in 1 ml of 0.85% sodium chloride solution. Incubate at 37°C for at least 7 days with intermittent observations. A positive culture is evidenced by growth and a change in the indicator from green to blue.

STORAGE
Bacto Acetate Differential Agar	Below 30°C
Prepared tubes	2 – 8°C

QUALITY CONTROL
Identity Specifications

Dehydrated powder:	beige, homogeneous, free-flowing
Reaction of 2.92% solution:	pH 6.7 ± 0.1 at 25°C
Prepared medium:	emerald green, slightly opalescent

Typical Culture Response on Bacto Acetate Differential Agar
After Up to 7 Days at 25 ± 2°C

Organism	Growth	Acetate Utilization*
Citrobacter freundii ATCC® 8090	good	+
Enterobacter cloacae ATCC® 23355	good	+
Klebsiella pneumoniae ATCC® 13883	good	+
Salmonella arizonae ATCC® 13314	good	+
Salmonella typhi ATCC® 19430	good	−
Shigella sonnei ATCC® 25931	good	−

*+ = color change of medium to blue
− = no change, medium remains green

REFERENCES

1. Publ. Hlth. Lab., 20:137, 1962.
2. Difco Manual, 1953.
3. Can. J. Microbiol., 4:89, 1958.

PACKAGING

Bacto Acetate Differential Agar   1 lb (454 g)   0742-01-2
                                         1/4 lb (114 g)   0742-02-1

BACTO ACRIDINE ORANGE STAIN

INTENDED USE

Bacto Acridine Orange Stain is recommended for use in the fluorescent microscopic detection of microorganisms in direct smears prepared from clinical and nonclinical materials. It is particularly useful in the rapid screening of normally sterile specimens such as cerebrospinal fluid, where few organisms may be present, and in the rapid examination of blood smears or smears containing proteinaceous material where differentiation of organisms from background material may be more difficult.

HISTORY OF THE TEST

Fluorochromatic staining of microorganisms using acridine orange was first described by Strugger and Hilbrich in 1942[1] and has been used in the examination of soil and water for microbial content.[2,3] In 1977, Kronvall and Myhre showed that acridine orange possessed differential staining properties with regard to clinical materials when prepared at a low pH.[4] They reported that under these conditions bacteria stained bright orange and could be easily differentiated from human cells and tissue debris which stained pale green to yellow.

In 1980, McCarthy and Senne compared acridine orange staining with blind subcultures for the detection of positive blood cultures.[5] Their results showed acridine orange staining to be a rapid, inexpensive alternative to blind subcultures. They also reported that the acridine orange stain appeared to be more sensitive than the Gram stain for detecting microorganisms and was able to detect bacteria in concentrations of approximately 1×10^4 colony-forming units per ml. Recently, Lauer, Reller and Mirrett compared acridine orange with the Gram stain for detecting microorganisms in cerebrospinal fluid and other clinical materials.[6] Their results were in agreement with those reported by McCarthy and Senne and showed acridine orange to be a simple, rapid staining procedure which was more sensitive than the Gram stain in detecting microorganisms in clinical materials.

Acridine orange at a low pH has also been used for the detection of *Trichomonas vaginalis*[7] and *Neisseria gonorrhoeae*[8] in clinical materials and for the enumeration of mycoplasmas.[9]

PRINCIPLES OF THE PROCEDURE

Acridine orange is a fluorochromatic dye which binds to nucleic acids of bacteria and other cells.[10] Under UV light, acridine orange stains RNA and single-stranded DNA orange; double-stranded DNA appears green.

When buffered at pH 3.5 to 4.0, acridine orange differentially stains microorganisms from cellular materials. Bacteria and fungi uniformly stain bright orange, whereas human epithelial and inflammatory cells and background debris stain pale green to yellow. Nuclei of activated leukocytes stain yellow, orange or red due to increased RNA production resulting from activation. Erythrocytes either do not stain or appear pale green.

Due to this differential staining property, acridine orange-stained smears prepared from clinical materials may be rapidly screened using fluorescent microscopy at 100X to 400X magnification for the presence of microorganisms fluorescing bright orange against a black or pale green to yellow background.

REAGENTS

BACTO ACRIDINE ORANGE STAIN

Formula per liter

Acridine Orange . . . . 0.1 g
Acetate Buffer, 0.5 M . . . . 1 liter

Storage Instructions

Store at 15 – 30°C. The expiration date applies to the product in its intact container when stored as directed.

Product Deterioration

Do not use if there is evidence of a precipitate or the solution shows other signs of deterioration.

SPECIMEN COLLECTION

Specimens should be collected in sterile containers or with sterile swabs and transported immediately to the laboratory in accordance with recommended guidelines.[11]

PROCEDURE

Materials Provided:
Bacto Acridine Orange Stain

Materials Required but not Provided:
Fluorescent microscope suitable for use with acridine orange
Glass microscope slides
Methanol

Preparation, Staining and Examination of Smears

1. Prepare a smear of the specimen to be stained on a clean glass slide.
2. Allow to air dry.
3. Fix smear with 50% or 100% methanol for 1 to 2 minutes.
4. Drain excess methanol and allow smear to dry.
5. Flood slide with acridine orange stain for 2 minutes.
6. Rinse thoroughly with tap water and allow to dry.
7. Smears may be initially examined at 100X to 400X magnification using a fluorescent microscope. Findings should be confirmed by examination at 1000X with an oil immersion objective.

USER QUALITY CONTROL

1. Examine the acridine orange staining solution for color and clarity. The solution should be clear, orange and without evidence of a precipitate.
2. Determine the pH of the solution. The pH should be 3.5 to 4.0.
3. Check the performance of the stain using 4 to 6 hour Tryptic Soy Broth with 5% sheep blood cultures of the organisms indicated below. Prepare smears, one culture per slide, and proceed as described under PREPARATION, STAINING AND EXAMINATION OF SMEARS.

Expected Results

Organisms	Bacteria	Background
Escherichia coli ATCC® 25922	Bright orange	Pale green erythrocytes and
Streptococcus faecalis ATCC® 33186	Bright orange	yellow, yellow-green or orange leukocytes against a black field. Green, yellow, orange or red staining debris may be observed.

RESULTS

Bacteria and fungi stain bright orange. The background appears black to yellow green. Human epithelial and inflammatory cells and tissue debris stain pale green to yellow. Activated leukocytes will stain yellow, orange or red depending upon the level of activation and amount of RNA produced, whereas erythrocytes either do not stain or stain pale green.

LIMITATIONS OF THE PROCEDURE

1. Acridine orange staining provides presumptive information on the presence and identification of microorganisms which may be present in the specimen. Since microorganisms seen in smears, including nonviables, may arise from external sources, i.e., specimen collection devices, slides, or water used for rinsing, all positive smears should be confirmed by culture.
2. Approximately 10^4 colony-forming units per ml are required for detection by this method.
3. Acridine orange staining does not distinguish between gram-positive and gram-negative organisms. The gram reaction may be determined by Gram staining directly over the acridine orange after removal of the immersion oil.
4. Nuclei or granules from disintegrated activated leukocytes may resemble cocci at lower magnifications, i.e., 100X – 400X. They may be distinguished on the basis of morphology at higher magnifications, i.e., 1000X.
5. Certain types of debris may fluoresce in acridine orange stained smears. This debris may be distinguished from microorganisms on the basis of morphology when viewed at higher magnification.

REFERENCES

1. Strugger, S., and P. Hilbrich, 1942. Die fluoreszenzmikroskopische unterscheidung lebender und toten bakterienzeillen mit hilfe des akridinorangefärbung. Deut. Teirarztl. Wochscher. 50:121 – 130.
2. Strugger, S. 1948. Fluorescence microscope examination of bacteria in soil. Can. J. Research, 26:188 – 193.
3. Jones, J. G., and B. M. Simon. 1975. An investigation of errors in direct counts of aquatic bacteria by epifluorescence microscopy, with reference to a new method for dyeing membrane filters. J. Appl. Bacteriol. 39:317 – 329.
4. Kronvall, G., and E. Myhre. 1977. Differential staining of bacteria in clinical specimens using acridine orange buffered at low pH. Acta. Path. Microbiol. Scand. Sect. B 85:249 – 254.
5. McCarthy, L. R., and J. E. Senne. 1980. Evaluation of acridine orange stain for detection of microorganisms in blood cultures. J. Clin. Microbiol. 11:281 – 285.
6. Lauer, B. A., L. B. Reller, and S. Mirrett. 1981. Comparison of acridine orange and Gram stains for detection of microorganisms in cerebrospinal fluid and other clinical specimens. J. Clin. Microbiol. 14:201 – 205.
7. Greenwood, J. R., and K. Kirk-Hillaire. 1981. Evaluation of acridine orange stain for detection of *Trichomonas vaginalis* in vaginal specimens. J. Clin. Microbiol. 14:699.
8. Forsum, U., and A. Hallen. 1979. Acridine orange staining of urethral and cervical smears for the diagnosis of gonorrhoea. Acta. Derm. Venereol. 59:281 – 282.
9. Rosendal, S., and A. Valdivieso-Garcia. 1981. Enumeration of mycoplasmas after acridine orange staining. Appl. Environ. Microbiol. 41:1000 – 1002.
10. Kasten, F. H. 1967. Cytochemical studies with acridine orange and the influence of dye contaminants in the staining of nucleic acids. Internat. Rev. Cytol. 21:141 – 202.
11. Isenberg, H. D., J. A. Washington II, A. Balows, and A. C. Sonnenwirth. 1980. Collection, handling, and processing of specimens, p. 52 – 80. *In* E. H. Lennette, A. Balows, W. J. Hausler, Jr., and J. P. Truant (ed.). Manual of clinical microbiology, 3rd Ed. American Society for Microbiology, Washington, D.C.

PACKAGING
Bacto Acridine Orange Stain 6 × 250 ml 3336-76-8

BACTO ACTINOMYCES BROTH

INTENDED USE
Bacto Actinomyces Broth is recommended for the cultivation and maintenance of the anaerobic *Actinomyces* species as described by Pine and Watson[1] and modified by Ajello, Georg, Kaplan and Kaufman.[2]

FORMULA

BACTO ACTINOMYCES BROTH
DEHYDRATED

Ingredients per liter

Bacto Heart Infusion Broth	25 g
Bacto Yeast Extract	5 g
Bacto Casitone	4 g
Cysteine Hydrochloride	1 g
Bacto Dextrose	5 g
Soluble Starch	1 g
Monopotassium Phosphate	15 g
Ammonium Sulfate	1 g
Magnesium Sulfate	0.2 g
Calcium Chloride	0.02 g

Final pH 7.2 ± 0.2 at 25°C.

One pound will make 7.9 liters of medium.

METHOD OF PREPARATION
1. To rehydrate, suspend 57 g in 1 liter distilled or deionized water and dissolve completely.
2. Dispense into tubes and autoclave for 15 minutes at 15 lbs pressure (121°C).

STORAGE

Bacto Actinomyces Broth	Below 30°C
Prepared medium	15 – 30°C

QUALITY CONTROL

Identity Specifications

Dehydrated powder:	light tan, homogeneous, free-flowing
Reaction of 5.7% solution:	pH 7.2 ± 0.2 at 25°C
Prepared medium:	medium to dark amber, slightly opalescent, may have a slight precipitate

Typical Cultural Response in Bacto Actinomyces Broth
After 18 – 48 Hours at 30 ± 2°C

Organism	Growth
Streptomyces achromogenes ATCC® 12767	good
Streptomyces albus ATCC® 3004	good
Streptomyces lavendulae ATCC® 8664	good

REFERENCES
1. J. Lab. & Clin. Med., 54:107, 1959.
2. Lab. Manual Med. Mycology, USDHEW, PHS, CDC, Atlanta, Ga.

PACKAGING
Bacto Actinomyces Broth   1 lb (454 g)   0840-01-3

BACTO ACTINOMYCETE ISOLATION AGAR

INTENDED USE
Bacto Actinomycete Isolation Agar, prepared according to the formulation of Olson,[1] is recommended for isolating and propagating *Actinomycetes* from soil and water.

FORMULA

BACTO ACTINOMYCETE ISOLATION AGAR
DEHYDRATED
Ingredients per liter

Sodium Caseinate . . . . 2 g
Asparagine . . . . 0.1 g
Sodium Propionate . . . . 4 g
Dipotassium Phosphate . . . . 0.5 g
Magnesium Sulfate . . . . 0.1 g
Ferrous Sulfate . . . . 0.001 g
Bacto Agar . . . . 15 g

Final pH 8.1 ± 0.2 at 25°C.

One pound will make 20.6 liters of medium.

PROCEDURE
1. Suspend 22 g in 1 liter distilled or deionized water and heat to boiling to dissolve completely.
2. Add 5 g of Bacto Glycerol, distribute into tubes or flasks and sterilize in the autoclave for 15 minutes at 15 lbs pressure (121°C).
3. Allow medium to cool to 55 – 60°C and pour into Petri dishes as desired.
4. Inoculate with 1 drop diluted culture or specimen and spread over the surface using a sterile bent glass rod.
5. Incubate at 30°C for 40 – 72 hours.

STORAGE
Bacto Actinomycete Isolation Agar   Below 30°C
Prepared medium   2 – 8°C

QUALITY CONTROL
Identity Specifications

Dehydrated powder: light beige, homogeneous, free-flowing
Reaction of 2.2% solution with
  0.5% glycerol: pH 8.1 ± 0.2 at 25°C
Prepared medium: medium amber, opalescent

**Typical Cultural Response on Bacto Actinomycete Isolation Agar
After 40 – 72 Hours at 30°C**

Organism	Growth
Escherichia coli ATCC® 25922	inhibited
Streptomyces albus ATCC® 3004	good to excellent
Streptomyces lavendulae ATCC® 8664	good to excellent

REFERENCE

1. Olson, E. H., Personal Communication, 1960.

PACKAGING

Bacto Actinomycete Isolation Agar	1 lb (454 g)	0957-01-2
Bacto Glycerol	100 g	0282-15-2
	500 g	0282-17-0

ADJUVANTS

BACTO ADJUVANT, COMPLETE FREUND
BACTO ADJUVANT, COMPLETE H37 Ra
BACTO ADJUVANT, INCOMPLETE FREUND

INTENDED USE

Bacto Adjuvants are intended for use in the preparation of antigen-adjuvant emulsions used in immunological studies with laboratory animals. They are not intended for human use or for therapeutic use. The immune response to antigens, especially soluble antigens, is enhanced and prolonged by using antigens emulsified in the adjuvants.

The incomplete adjuvant is used when it is desired to avoid the use of acid-fast bacilli. However, the antibody response is not as marked as with the complete adjuvants.

PRINCIPLES

The adjuvants provide a pool of the emulsified antigen at the injection site, allowing the antigen to be released slowly over an extended period of time. The addition of mycobacteria to these adjuvants further enhances the immune response to antigens, particularly soluble antigens.[1]

FORMULAE

BACTO ADJUVANT, COMPLETE FREUND

Arlacel A (Mannide Monooleate)	. . 1.5 ml
Bayol F (paraffin oil)	. . . . 8.5 ml
Mycobacterium butyricum (killed and dried)	. . . . 5 mg

BACTO ADJUVANT, COMPLETE H37 Ra

Arlacel A (Mannide Monooleate)	. . 1.5 ml
Bayol F (paraffin oil)	. . . . 8.5 ml
Mycobacterium tuberculosis H37 Ra (killed and dried)	. . . . 10 mg

BACTO ADJUVANT, INCOMPLETE FREUND

Arlacel A (Mannide Monooleate) . . 1.5 ml
Bayol F (paraffin oil) . . . . 8.5 ml

Bacto Adjuvant, Complete Freund is a suspension of *Mycobacterium butyricum* in a mixture of paraffin oil (Bayol F) and an emulsifying agent Arlacel A (Mannide Monooleate).

Bacto Adjuvant, Complete H37 Ra is a complete adjuvant prepared in accordance with Freund's technique using *M. tuberculosis* H37 Ra instead of *M. butyricum.*

Bacto Adjuvant, Incomplete Freund is similar to Bacto Adjuvant, Complete Freund except that *M. butyricum* has been omitted.

Bacto M. Butyricum, Desiccated, Code 0640 or Bacto M. Tuberculosis H37 Ra, Desiccated, Code 3114, are intended for use with Bacto Adjuvant, Incomplete Freund to give the complete adjuvant.

PREPARATION OF ANTIGEN EMULSION
The antigen-adjuvant emulsion is prepared by mixing equal volumes of the aqueous antigen solution or suspension and adjuvant. Thorough emulsification is effected by mixing in a blender or by drawing the mixture into a syringe through a needle and forceably ejecting the mixture a number of times into a beaker to obtain a smooth emulsion. The emulsion is satisfactory if a drop placed on the surface of water will not spread.

STORAGE
Bacto Adjuvants	15 – 30°C
Bacto M. Butyricum, Desiccated	2 – 8°C
Bacto M. Tuberculosis H37 Ra, Desiccated	2 – 8°C

Antigen-adjuvant emulsions remain stable several weeks if stored in the refrigerator at 2 – 8°C.

QUALITY CONTROL
Identity Specifications
Bacto Adjuvant, Complete Freund:	oily, yellow liquid with easily dispersed suspension
Bacto Adjuvant, Complete H37 Ra:	oily, yellow liquid with easily dispersed suspension
Bacto Adjuvant, Incomplete Freund:	oily, yellow liquid

REFERENCE
1. Immunological Adjuvants, World Health Organization, Technical Report Series, No. 595, Geneva, pp. 9 – 12, 16 – 19, 1976.

PACKAGING
Bacto Adjuvant, Complete Freund	6 × 10 ml	0638-60-7
Bacto Adjuvant, Complete H37 Ra	6 × 10 ml	3113-60-5
Bacto Adjuvant, Incomplete Freund	6 × 10 ml	0639-60-6
Bacto M. Butyricum, Desiccated	6 × 100 mg	0640-33-7
Bacto M. Tuberculosis H37 Ra, Desiccated	6 × 100 mg	3114-33-8

BACTO AGAR

Bacto Agar is a purified agar from which the extraneous matter, pigmented portions, and salts have been removed or reduced to a minimum. Bacto Agar is available in the form of fine, light colored granules, which are convenient for weighing and handling. It dissolves rapidly in purified water, eg., distilled, deionized or water purified by reverse osmosis, yielding clear solutions. It is usually employed in solid culture media in concentrations of 1 – 2%. The use of small quantities of agar (0.05 – 0.3%) in media is used frequently for determining motility and growth of anaerobes and microaerophiles. The addition of such amounts of agar to liquid media permits all degrees of oxygen tension to exist and, thus, aids in the development of many fastidious aerobic and anaerobic organisms.

The value of the use of small quantities of agar in media for sterility testing was pointed out by Falk, Bucca and Simmons[1] and has been incorporated in the thioglycollate medium for sterility testing of biologics and antibiotics by official procedures.[2,3,4]

REFERENCES
1. J. Bact., 37:121:1939.
2. National Institute of Health Circular: Culture Media for the Sterility Test, and Revision, Feb. 25, 1946.
3. Compilation of Regulations for Tests and Methods of Assay and Certification of Antibiotic Drugs, Federal Security Agency, Food and Drug Administration.
4. United States Pharmocopeia XX/National Formulary XV, 1206, 1980.

PACKAGING
Bacto Agar

	1 lb (454 g)	0140-01-0
	1/4 lb (114 g)	0140-02-9
	5 kg	0140-03-8
	5 lb (2.27 kg)	0140-05-6
	10 kg	0140-08-3

AGAR BACTERIOLOGICAL, TECHNICAL

Agar Bacteriological, Technical may be used as a solidifying agent in bacteriological culture media and in other applications. Although it has not been as carefully standardized as is Bacto Agar or other bacteriological agars, certain parameters such as solubility, gelation temperatures, solidity and other parameters are carefully monitored to permit its use.

PACKAGING
Agar Bacteriological, Technical

	500 g	0812-17-9
	5 lb (2.27 kg)	0812-05-3
	10 kg	0812-08-0

AGAR FLAKE

Agar Flake is agar in a flake form, and is used in bacteriological culture media.

PACKAGING
Agar Flake

	1 lb (454 g)	0970-01-5

AGAR NOBLE

Agar Noble is a carefully washed agar purified according to the method of Noble and Tonney[1] and is essentially free from impurities. It is used extensively in electrophoretic procedures, nutrutional studies and wherever an agar of increased purity is needed. The above authors specify the use of Agar Noble as an ingredient of their brilliant green lactose bile agar for the direct plate count of the coliform group of bacteria in water. Bacto Brilliant Green Bile Agar, using Agar Noble, is based on the formulation recommended by these authors.

REFERENCE
1. J. Am. Water Works Assoc., 27:108, 1935.

PACKAGING
Agar Noble

1 lb (454 g)	0142-01-8
1/4 lb (114 g)	0142-02-7
5 lb (2.27 g)	0142-05-4
10 kg	0142-08-1

AGAR PURIFIED

Agar Purified is highly purified agar which as been exhaustively extracted with purified water and an organic solvent resulting in a solidifying agent which is essentially free from materials such as nitrogenous compounds, inorganic salts and vitamins.

It is recommended for use in nutritional studies, tissue culture procedures, immunological diffusion and electrophoretic techniques and in other procedures requiring a highly purified gel.

PACKAGING
Agar Purified

1 lb (454 g)	0560-01-1
1/4 lb (114 g)	0560-02-0

p-AMINOBENZOIC ACID, DIFCO

p-Aminobenzoic Acid, Difco is recommended for use in liquid culture media for the prevention of bacterial stasis due to any sulfonamide drug. It should be added to all liquid media used for blood cultures of patients under sulfonamide therapy, and in the culture of exudates or other materials containing sulfonamide compounds. As pointed out by Lockwood[1] and McLeod[2] most culture media normally contain some sulfonamide inhibitors which generally are not sufficient to neutralize completely the sulfonamides likely to be encountered. The addition of 5 mg of p-aminobenzoic acid to 100 ml of medium will more than suffice to neutralize the bacteriostatic effect of 1.5 mg % sulfonamide drug. This quantity of p-aminobenzoic acid is not toxic to fastidious pathogens even though the inocula contain only a few organisms. The report of Janeway[3] shows the value of p-aminobenzoic acid in culture media wherever sulfonamides are encountered.

REFERENCES
1. J. Immunol., 35:155, 1938.
2. J. Exp. Med., 72:217, 1940.
3. J. Am. Med. Assoc., 116:941, 1941.

PACKAGING

p-Aminobenzoic Acid, Difco	100 g	0240-15-3

p-AMINODIMETHYLANILINE OXALATE, DIFCO

p-Aminodimethylaniline Oxalate, Difco is recommended for detecting oxidase production by microorganisms. It has the advantage over the monohydrochloride salt in that it is more stable in the powdered form and also in solution. The oxidase reagent is prepared by adding 1 g of p-Aminodimethylaniline Oxalate, Difco to 100 ml distilled or deionized water and heating gently. *Do not overheat.* This solution is used to flood plates containing isolated colonies to determine the oxidase reaction. Positive colonies assume a pink color, turning to maroon and finally to black. Carpenter[1] reported that the dry crystalline oxalate salt is more stable than the monohydrochloride salt. No change was observed after six months storage. Aqueous solutions were also more stable. Both salts showed about the same toxicity for microorganisms. The oxalate salt is slightly less soluble in cold distilled water, but solution is hastened by warming gently. Carpenter[2] further stated the oxalate salt possesses the additional advantages over the monohydrochloride in that it does not form the marked black precipitate on chocolate agar sometimes observed with the use of the monohydrochloride, especially when freshly prepared solutions are not employed.

REFERENCES
1. Science, 105:649, 1948.
2. Diagnostic Procedures and Reagents, 3rd Ed.:107, 1950.

PACKAGING

p-Aminodimethylaniline Oxalate, Difco	25 g	0329-13-9
	100 g	0329-15-7

AMINO ACIDS

Difco Amino Acids are pure and are used as standards for chemical and microbiological procedures. These products are used for preparing chemically defined media and in media for nutritional studies for determining amino acid requirements. See section on Vitamin and Amino Acid Assay Media.

Product	Code No.	Pkg. Size	Intended Use
Asparagine	0144-13-2	25 g	• Ingredient of synthetic culture media.
	0144-15-0	100 g	• Employed in studies of bacterial nutrition.

Product	Code No.	Pkg. Size	Intended Use
Asparagine	0144-17-8	500 g	• Source of organic nitrogen of known chemical compound and is readily available for bacterial energy and growth when used in culture media. • Widely employed in media for cultivation of *Mycobacterium tuberculosis*.
DL-Alanine	0182-13-5	25 g	• Source of nitrogen in studies of the metabolism of various organisms.
L-Cystine	0184-08-0	10 kg	• Used extensively in synthetic culture media, particularly those employed in studies of bacterial metabolism.
	0184-13-3	25 g	• NIH[1] sterility testing of biologicals.
	0184-15-1	100 g	• USP XX/NF XV[2] sterility testing.
L-Tryptophane	0188-12-0	10 g	• An amino acid essential for the growth of many microorganisms. • Utilized by bacteria in the elaboration of indole, therefore has been employed in media devised for testing indole production. NOTE: Destroyed by acid hydrolysis, therefore synthetic media using acid hydrolysates or chemically pure amino acid mixtures as sources of nitrogen require the addition of tryptophane for the growth of most bacteria.
L-Tyrosine	0189-13-8	25 g	• Preparation of culture media.
L-Ornithine HCl	0293-11-3	5 g	• Used in culture media for the differentiation of *Enterobacteriaceae* on basis of ornithine decarboxylase reaction.
	0293-12-2	10 g	
	0293-13-1	25 g	
L-Arginine HCl	● 0583-12-1	10 g	• Used in culture media for the differentiation of *Enterobacteriaceae* on basis of arginine dihydrolase reaction.
L-Lysine HCl	0705-11-5	5 g	• Used in culture media for the differentiation of *Enterobacteriaceae* on basis of lysine decarboxylase reaction.
Histidine HCl	0707-11-3	5 g	• Preparation of culture media.

REFERENCES

1. National Institute of Health Circular: Culture Media for the Sterility Test, and Revision: Feb. 5, 1946.
2. USP XX/NF XV: 879, 1979.

HISTORY
Refer to *Difco Manual*, 9th Ed., p. 268 – 269, 1953.

BACTO ANAEROBIC AGAR

INTENDED USE
Bacto Anaerobic Agar is a modified Brewer formula formerly recommended for use with Brewer anaerobic Petri dish covers for the surface cultivation of anaerobic organisms. It is now recommended as a general purpose medium for anaerobic cultures. For a discussion of Brewer's formulations refer to Bacto Brewer Anaerobic Agar.

FORMULA
BACTO ANAEROBIC AGAR
DEHYDRATED

Ingredients per liter

Bacto Casitone	20 g	Sodium Thioglycollate, Difco	2 g
Sodium Chloride	5 g	Sodium Formaldehyde	
Bacto Dextrose	10 g	Sulfoxylate	1 g
Bacto Agar	20 g	Bacto Methylene Blue	0.002 g

Final pH 7.2 ± 0.2 at 25°C.

One pound will make 7.8 liters of medium.

METHOD OF PREPARATION
1. To rehydrate, suspend 58 g in 1 liter distilled or deionized water.
2. Heat to boiling to dissolve completely.
3. Sterilize in the autoclave for 15 minutes at 15 lbs pressure (121°C).

PROCEDURE
Brewer Anaerobic Agar Plates
1. Dispense the medium into Petri dishes using 50 – 60 ml of medium for the 95 × 20 mm dish.
   For best results use porous tops for the dishes during solidification in order to obtain a dry surface.
2. Inoculate the surface of the medium by streaking or smearing.
3. Cover inoculated dish with a sterile Brewer anaerobic Petri dish cover.

   It is essential that the sealing ring inside the cover makes perfect contact with the medium. This seal must not be broken before the end of the incubation period.

   Poured plates may be made in a similar manner except that the inoculum is placed in the plate before adding the sterile medium.
4. Incubate aerobically as desired.

Standard Plates
1. If pour plates are desired, dispense 0.1 to 1 ml inoculum into plate and cover with 20 – 25 ml of media. Swirl to mix and allow to solidify.
2. If streak plates are desired, pour 20 – 25 ml of medium into plates. Allow to solidify on a flat surface. Inoculate with 0.1 ml of diluted organism. Cover layers (5 ml) generally give better results.
3. Incubate anaerobically for amount of time and temperature as required by organism(s) being tested.

STORAGE

Bacto Anaerobic Agar	Below 30°C
Prepared media	2 – 8°C

QUALITY CONTROL

Identity Specifications

Dehydrated powder:	light beige, homogeneous, free-flowing
Reaction of 5.8% solution:	pH 7.2 ± 0.2 at 25°C
Prepared medium:	light amber, slightly opalescent
Prepared plates:	light green, slightly opalescent

Typical Cultural Response on Bacto Anaerobic Agar
After 18 – 48 Hours at 35°C Anaerobically

Organism	Growth
Clostridium butyricum ATCC® 9690	good to excellent
Clostridium perfringens ATCC® 12919	good to excellent
Clostridium sporogenes ATCC® 11437	good to excellent

PACKAGING

Bacto Anaerobic Agar	1 lb (454 g)	0536-01-2

ANIMAL SPECIES ANTISERA

PROTEIN ANTISERA
NORMAL CONTROLS

Bacto Animal Species Antisera are prepared for use in the serological detection of animal blood globulins employing the precipitin tube technique.

Bacto Protein Antisera are prepared against some of the more commonly encountered proteins. They, too, are employed in the precipitin tube technique.

Bacto Normal Sera are prepared for use as controls for the above antisera.

The group of antisera commonly known as "medical-legal" or "forensic" sera have been used for many years to detect blood globulins in such cases as rape, murder and food adulteration employing a "precipitin" or "ring test." Kraus[1] probably was the first to describe this reaction when he found that serum from animals immunized with organisms such as *Vibrio cholerae* formed precipitates with culture filtrates of the homologous organism.

It was subsequently found that the precipitin reaction could be applied to nonbacterial systems such as the identification of animal globulins[2,3,4] as well as to detect the adulteration of meats and meat products.[5]

REAGENTS

Bacto Animal Species Antisera and Protein Antisera are stable, desiccated, absorbed (when necessary) antisera for use in the precipitin tube test. They are prepared in the designated species as indicated in parenthesis following the product name.

Bacto Normal Sera are clarified, stable, desiccated, normal sera for use in homologous and heterologous control systems. To rehydrate, add 1 ml sterile distilled or deionized water and rotate the vial gently to dissolve the contents completely. The resultant solution contains sufficient Merthiolate® to render the sera bacteriostatic when stored at 2 – 8°C. In most cases the final concentration is approximately 1:10,000.

Discard any serum which is cloudy or has a precipitate upon rehydration, unless the serum has been clarified by centrifugation and has been shown to be of proper reactivity using validated positive and negative controls. Normal controls should be diluted 1:1000 prior to use.

These reagents are stable to the expiry date on the label when stored at 2 – 8°C.

Do not expose rehydrated serum to room temperature for prolonged periods of time. Discard any antiserum which becomes cloudy during storage. They should not be subjected to repeated freezing and thawing. Such treatment reduces the antibody content.

PRECAUTION
These reagents have been prepared for use in the tube test only. They have been tested with homologous and heterologous only globulins listed under PACKAGING below. The antiglobulins have not been tested for use in meat adulteration procedures. If these sera are used in techniques other than those recommended herein, the user MUST test each lot of antiserum under his (her) test conditions employing positive and negative homologous and heterologous antigens for complete control of the system.

SPECIMEN PREPARATION
Blood Stains
1. Place the dried blood specimen, powder or a section of blood stained material in 0.85% NaCl (saline) solution overnight at refrigerator temperatures (2 – 8°C). The ratio of blood to NaCl solution is recommended at 4 – 5 mg of dried blood or 1 square centimeter of stained material per ml of saline.
2. Prepare a 1:50 dilution of the extract by adding 1 part of the blood specimen extract to 49 parts of NaCl solution. The resultant dilution is approximately 1:1,000. The extract should be clear and free from precipitate. If it is not clear, filtration or high speed centrifugation is necessary. The clarified extract dilution is the antigen to be tested with specific globulin antisera.

Serum
1. Collect 5 – 10 ml whole blood.
2. Allow the blood to clot and obtain the syneresed serum with a Pasteur pipette. If the serum is not free of erythrocytes, clarify by centrifugation. The serum must be sparklingly clear and free of fat or lipids, which may be misinterpreted as a precipitate.
3. Prepare a 1:1,000 dilution of the serum to be tested by adding 1 part of serum to 999 parts of 0.85% NaCl solution.

PRECAUTIONS
1. Check purity (bacterial) and pH of the distilled or deionized water used in rehydration if the antiserum rehydrates cloudy. Discard any serum which is cloudy and/or has a precipitate unless it has been clarified and shown to react properly with known control antigens.
2. Make certain the proper dilution is prepared for a given specimen.
3. Both the antigen and antiglobulin must be sparklingly clear to avoid false positive reactions. It may be necessary to clarify them by centrifugation prior to their use in the test.

4. It is extremely important to have a clear line of separation of the antigen and antiserum at the interface. If mixing of the 2 occurs during the overlaying process, a positive reaction may be obscured.
5. It is necessary to examine the tubes under a proper light source.
6. It is important to use clean, detergent-free glassware since detergents may cloud the mixture resulting in a false-positive reaction.
7. In the test performance, the antigen should be added to the tube first so that the antiserum might be layered below the extract.
8. Discard the test tubes at the end of 30 minutes since false positive or negative reactions might occur after this time.
9. The antigen should not be frozen nor subject to prolonged exposure to room or elevated temperatures.
10. The antiserum should not be subjected to repeated freezing and thawing. Such treatment is detrimental to the antibody content.
11. If at any time an antigen or antiserum becomes turbid and it is proven by a Gram stain to be bacterially contaminated, the reagent must be discarded.
12. Adhere strictly to the time and temperature limitations in the test.
13. The antigens and antisera must be stored at $2 - 8°C$ when not in actual use.
14. The rehydrated serum has a 1 year expiry date when stored at $2 - 8°C$ and has not been contaminated from an external source. The expiry date after rehydration does not exceed the expiry date on the label.
15. The control antigens have an expiry date equivalent to that which appears on the label, when stored under proper conditions in an uncontaminated state.
16. Discard the 1:1,000 antigen dilution at the end of the working day. Highly diluted protein solutions are not stable.
17. As is in any serological test, known positive and negative controls should be employed to ascertain validity of test results.

PROCEDURE
Materials Provided:
Bacto Animal Species Antisera
Bacto Protein Antisera
Bacto Normal Control Sera

Materials Required but not Provided:
0.85% NaCl	Tuberculin syringes
Centrifuge	25 gauge cannulas
Refrigerator	7 × 75 mm tubes
Pasteur pipettes	Test tube racks
Sterile distilled or deionized water	37°C water bath
1 ml and 0.2 ml pipettes	Fluorescent desk lamp

It is necessary to layer the species globulin antiserum under the antigen to be tested, in a test tube in such a manner as to form a distinct line at the interface of the 2 liquids. This may be done by several methods. Some laboratories prefer first to introduce the antiserum into the tube and layer the antigen over the antiserum. Others prefer to place the antigen in the tube and carefully layer the antiserum under the antigen. The latter method may be done with ease by using a 1 ml tuberculin syringe and 25 gauge cannula. The antigen is added by means of a pipette, the sides of the tube allowed to drain free of the antigen and the tube tilted to a near horizontal position whereupon the cannula is inserted into the tube and allowed to rest on the side of the tube approximately 1 cm from the antigen. The antiserum is released very slowly from the syringe and allowed to run down the side of the tube and layer itself under the antigen. The tube is then carefully brought to the upright position and should show a clear line of demarcation between the antigen and antiserum.

1. Place 0.1 ml of the antigen extract dilution to be tested in a 7 × 75 mm tube using 0.2 ml Kahn or 1 ml serological pipette. Allow the antigen to drain from the side of the tube.
2. Place 0.1 ml of the desired antiserum in the tube using a tuberculin syringe as described above.
3. Incubate the tube at 37°C for 30 minutes.
4. Observe the tube for a precipitate at the interface by examining the tube using a desk lamp. This may be done by insuring the light passage through the tube but observing from a slight angle, thus not permitting the light to shine directly into the observer's eyes.

INTERPRETATION

Preliminary Test

Tube	Antiserum (0.1 ml)	Antigen (1:1,000)	Result
1	Bovine Globulin	0.1 ml	No Reaction
2	Human Globulin	0.1 ml	No Reaction
3	Rabbit Globulin	0.1 ml	Precipitate

Interpretation: Tentatively rabbit protein

Final Test

Tube	Antiserum (0.1 ml)	Antigen (0.1 ml of 1:1,000)	Result
1	Rabbit Globulin	Rabbit Serum Normal	Precipitate
2	Rabbit Globulin	Extract of Suspected Material	Precipitate
3	Rabbit Globulin	0.85% NaCl	No Reaction
4	Rabbit Serum Normal	Extract of Suspected Material	No Reaction

Number 3 above is a control of the species globulin antiserum and number 4 is an antigen control. If either are not absolutely clear, misleading interpretations could result.

LIMITATIONS OF THE PROCEDURE

Serum proteins are often antigenically related according to species phylogenesis. Thus, closely related species antiglobulins may and do often cross-react strongly.

Absorption of a given antiglobulin with a serum of a closely related species will remove the homologous precipitins to the point as to render the serum useless. Therefore, it is necessary to remember that there is a reciprocal antigenic relationship between human and monkey antiglobulins and sera, between sheep and goats, and between rat and mouse reagents. These are the most obvious major relationships. It must be realized that other weaker relationships may exist.

REFERENCES

1. Kraus, R., Wien. Klin. Wochschr., 10:736, 1897.
2. Bordet, J., Ann. d Inst. Pasteur, 13:225, 1899.
3. Heidelberger, M., J. Am. Chem. Soc., 60:242, 1938.
4. Gradwohl's Legal Medicine, Year Book Medical Publ., Chicago, IL, 1976.
5. Oswald, E. J., J.A.O.A.C., 36:107, 1953.

PACKAGING

Animal Species Globulin Antisera	Size	Code
Bacto Bovine Globulin Antiserum (Rabbit)	1 ml	2362-50-7
Bacto Chicken Globulin Antiserum (Rabbit)	1 ml	2363-50-6
Bacto Deer Globulin Antiserum (Rabbit)	1 ml	2433-50-2
Bacto Human Globulin Antiserum (Rabbit)	1 ml	2457-50-3
Bacto Porcine Globulin Antiserum (Rabbit)	1 ml	2410-50-9
Bacto Rabbit Globulin Antiserum (Goat)	1 ml	2366-50-3
Bacto Rat Globulin Antiserum (Rabbit)	1 ml	2413-50-6
Bacto Sheep Globulin Antiserum (Rabbit)	1 ml	2364-50-5
Animal Species Sera Antisera		
Bacto Bovine Serum Antiserum (Rabbit)	1 ml	2456-50-4
Bacto Human Serum Antiserum (Rabbit)	1 ml	2452-50-8
Protein Antisera		
Bacto Egg Albumin Antiserum (Rabbit)	1 ml	2451-50-9
Animal Species Normal Sera Controls		
Bacto Bovine Serum Normal	1 ml	2417-50-2
Bacto Chicken Serum Normal	1 ml	2419-50-0
Bacto Deer Serum Normal	1 ml	2438-50-7
Bacto Goat Serum Normal	1 ml	2429-50-8
Bacto Guinea Pig Serum Normal	1 ml	2427-50-0
Bacto Horse Serum Normal	1 ml	2421-50-6
Bacto Human Serum Normal	1 ml	2426-50-1
Bacto Mouse Serum Normal	1 ml	2422-50-5
Bacto Porcine Serum Normal	1 ml	2428-50-9
Bacto Rabbit Serum Normal	1 ml	2423-50-4
Bacto Rat Serum Normal	1 ml	2424-50-3
Bacto Sheep Serum Normal	1 ml	2425-50-2

Note: All the above listed Animal Species Normal Sera are supplied in an undiluted state and are to be diluted 1:1,000 prior to use as controls in the precipitin test with the exception of rat sera, which are supplied in dilutions of 1:10 and need a further dilution of only 1:100 prior to use.

ANTIBIOTIC ASSAY MEDIA

INTENDED USE

Bacto Antibiotic Assay Media are prepared for use in the microbiological assay of antibiotics in pharmaceutical products, body fluids, feeds and other sample materials.

BACKGROUND

Bacto Antibiotic Assay Media are prepared according to the specifications prescribed in the *U.S. Pharmacopeia XX*[1] and by the FDA.[2] To simplify and unify the nomenclature of these media, Bacto Antibiotic Media 1 – 12 are identified by the names employed by Grove, and Randall in *Assay Methods of Antibiotics*.[3] Each medium carries the former identification code number, and if previously prepared, the name formerly used is also indicated. Bacto Antibiotic Medium 19 corresponds to that used in *Outline of Details for Official Microbiological Assays of Antibiotics*.[4] Bacto Mycin Assay Agar duplicates the formula of the medium specified by the US/FDA[5] for assay of streptomycin in the cylinder plate method. The use of this medium assures well defined zones of inhibition of the test organism.

The assay of antibiotics can be performed by either the cylinder plate method or turbidimetric procedure. The methodologies for the microbiological assay of antibiotics have been clearly delineated. The use of standardized culture media in the test is one of the fundamental requisites for satisfactory results. All conditions must be carefully controlled in the microbiological assay of antibiotics, as is true of all assays.

It is not within the scope of this discussion to give the detailed use of each medium in the assay of the various antibiotics. The details are available in official publications such as *Tests and Methods of Assay of Antibiotic and Antibiotic Containing Drugs*, FDA, CFR[2] and USP XX.[1] However, there are certain general methods used in the assay procedures that are employed for most antibiotics and these are described briefly.

CYLINDER PLATE ASSAY

The cylinder plate method was first described by Abraham, et al[6] for the assay of penicillin. It was later modified by Foster and Woodruff[7] and by Schmidt and Moyer[8] and others. This method depends upon the diffusion of the antibiotic from steel cylinders placed on the surface of inoculated agar medium. Inhibition of growth of the test organism occurs in the proximity of the cylinder and the diameter of the zone of inhibition depends, within limits, upon the concentration of the antibiotic under test. This method is commonly employed in the assay of commercial preparations of penicillin and other antibiotics. It has also been adapted for the quantitative determination of antibiotics in body fluids and other materials.

Petri dishes 20 × 100 mm are used in this test. The depth of the dish is important since the cylinder used in the assay must not be pushed into the medium by the cover. Porcelain covers, glazed on the outside only, are recommended. Stainless steel 10 mm cylinders (8 mm ± 0.1 mm o.d., 6 mm ± 0.1 mm i.d.) are recommended for use as they have sufficient weight to effect a good seal on the agar surface.

A level surface is required to obtain uniform layer thickness of base and seed agar in the Petri dishes for accurate assays. For assays requiring a base and seed layer, the base layer is allowed to solidify and then is overlaid with the seed layer containing the proper concentration of the test organism. The amount of base layer varies for the different antibiotics, most requiring 21 ml of medium with 4 ml of inoculated seed layer. In these cases dishes with flat bottoms must be used to assure complete coverage of the bottom of the dish when small amounts of base medium are employed. The plate should be tilted to obtain an even coverage of the base layer with the seed layer, following inoculation, and allowed to solidify in a level position. Plates should be used the same day as prepared. Four, or generally 6, cylinders are used per plate. The cylinders are placed on each inoculated plate so that they are evenly spaced on a 2.8 cm radius.

TURBIDIMETRIC ASSAY

Turbidimetric determinations where applicable have the advantage of requiring a short incubation period, making reading possible after 3 or 4 hours. The presence of solvents or other inhibitory materials may influence turbidimetric readings more markedly than cylinder plate assays. Samples should be clear for satisfactory turbidimetric readings.

For the turbidimetric method, working dilutions of the antibiotic reference standards are prepared in specific concentrations. To 1 ml quantities of these solutions in suitable tubes are added 9 ml of inoculated broth as required. Similar solutions of the assay materials containing approximately the same amount of antibiotic activity are prepared and placed in tubes. The tubes are then incubated for 3 – 4 hours at the required temperature, generally in a water bath. At the end of the incubation period, growth is halted by the addition of 0.5 ml of formalin 1:3. The amount of growth is determined by measuring light transmittance with a suitable spectrophotometer. The concentration of the antibiotic is determined by comparing amounts of growth obtained with that given by the reference standard solutions.

CULTURE

Generally, organisms used for assay are carried in stock on agar slants and transfers made at 1 or 2 week intervals. The inoculum used for assay is generally prepared by washing the growth from a fresh 24 – 48 hour agar slant culture of the test organism using either sterile distilled water, saline or antibiotic medium 3, and further diluting the culture to obtain the desired concentration of organisms. In some turbidimetric assays an 18 – 24 hour culture of the test organism in antibiotic medium 3, diluted to obtain the optimal number of organisms, is used as the inoculum.

When *Bacillus subtilis* is used as the test organism, the organism may be grown on antibiotic medium 1 for 1 week at 37°C, the spores washed from the agar surface and heated at 56°C for 30 minutes. The spores are washed 3 times in distilled water and again heated at 65°C for 30 minutes prior to dilution to the optimal concentration to give a sharp zone in the assay.

Spore suspension of *B. subtilis* may also be prepared in antibiotic medium 6 with incubation at 26°C for 4 – 6 days until 80% or more cells are in the spore state as shown by microscopic examination of the culture. During incubation the culture is placed on a mechanical (reciprocal) shaker. Upon satisfactory sporulation of the culture, it is centrifuged and cells reconstituted to 30% of the original volume with sterile saline and immersed in a water bath at 70°C for 30 minutes.

The addition of 300 mg manganese sulfate ($MnSO_4 \cdot H_2O$) per liter of Bacto Antibiotic Medium 1 often aids the sporulation of *B. subtilis*. The modified antibiotic medium 1 may be used to advantage in preparing the spore suspension. Standardized spore suspension prepared from *B. subtilis* ATCC® 6633, which is labeled Bacto Subtilis Spore Suspension, is available from Difco.

Spore suspension of *B. cereus var. mycoides* are prepared, using antibiotic medium 1 with an incubation temperature of 30°C for 1 week. Washing and preparation of spores is completed as for *B. subtilis*. Bacto Cereus Spore Suspension is a standardized spore suspension of *B. cereus var. mycoides* available for immediate use.

STORAGE

Bacto Antibiotic Media 1 – 19	Below 30°C
Bacto Mycin Assay Agar	Below 30°C
Bacto Antibiotic Medium 10	2 – 8°C
Prepared media	2 – 8°C

Guide to the Selection of Media for the Microbiological Assay of Antibiotics[2]

Antibiotic	Method of Assay	Organism	Maintenance	Inoculum	Bacto Antibiotic Media		Turbidimetric Assay Medium
					Cylinder Plate		
					Base Layer	Seed Layer	
Amikacin	Turbidimetric	Staphylococcus aureus ATCC® 6538P					3
Amoxicillin	Cylinder Plate	Micrococcus luteus ATCC® 9341	1	1	11	11	
Amphotericin B	Cylinder Plate	Saccharomyces cerevisiae ATCC® 9763	19	13	11	19	
Ampicillin	Cylinder Plate	Micrococcus luteus ATCC® 9341	1	1	11	11	
Bacitracin	Cylinder Plate	Micrococcus luteus ATCC® 7468 or ATCC® 10240	1	1	2	1	
Capreomycin	Turbidimetric	Klebsiella pneumoniae ATCC® 10031	1	1			3
Carbenicillin	Cylinder Plate	Pseudomonas aeruginosa ATCC® 25619	1	1	9	10	
Cefaclor	Cylinder Plate	Staphylococcus aureus ATCC® 6538P	1	1	2	1	
Cefadroxil	Cylinder Plate	Staphylococcus aureus ATCC® 6538P	1	1	2	1	
Cefamandole	Cylinder Plate	Staphylococcus aureus ATCC® 6538P	1	1	2	1	
Cefazolin	Cylinder Plate	Staphylococcus aureus ATCC® 6538P	1	1	2	1	
Cefotaxime	Cylinder Plate	Staphylococcus aureus ATCC® 6538P	1	1	2	1	
Cefoxitin	Cylinder Plate	Staphylococcus aureus ATCC® 6538P	1	1	2	1	
Cephacetrile	Cylinder Plate	Staphylococcus aureus ATCC® 6538P	1	1	2	1	
Cephalexin	Cylinder Plate	Staphylococcus aureus ATCC® 6538P	1	1	2	1	
Cephaloglycin	Cylinder Plate	Staphylococcus aureus ATCC® 6538P	1	1	2	1	
Cephaloridine	Cylinder Plate	Staphylococcus aureus ATCC® 6538P	1	1	2	1	
Cephalothin	Cylinder Plate	Staphylococcus aureus ATCC® 6538P	1	1	2	1	
Cephapirin	Cylinder Plate	Staphylococcus aureus ATCC® 6538P	1	1	2	1	
Cephradine	Cylinder Plate	Staphylococcus aureus ATCC® 6538P	1	1	2	1	
Chloramphenicol	Turbidimetric	Escherichia coli ATCC® 10536	1	1			3
Chlortetracycline	Turbidimetric	Staphylococcus aureus ATCC® 6538P	1	1			3
Clindamycin	Cylinder Plate	Micrococcus luteus ATCC® 9341	1	1	11	11	
Cloxacillin	Cylinder Plate	Staphylococcus aureus ATCC® 6538P	1	1	2	1	
Colistimethate, sodium	Cylinder plate	Bordetella bronchiseptica ATCC® 4617	1	1	9	10	
Cyclacillin	Cylinder Plate	Micrococcus luteus ATCC® 9341	1	1	11	11	
Cycloserine	Cylinder Plate	Staphylococcus aureus ATCC® 6538P	1	1	2	1	
Cycloserine	Turbidimetric	Staphylococcus aureus ATCC® 6538P	1	1			3
Dactinomycin	Cylinder Plate	Bacillus subtilis ATCC® 6633	1	1	5	5	
Demeclocycline	Turbidimetric	Staphylococcus aureus ATCC® 6538P	1	1			3
Dicloxacillin	Cylinder Plate	Staphylococcus aureus ATCC® 6538P	1	1	2	1	

Guide to the Selection of Media for the Microbiological Assay of Antibiotics[2]
(Continued)

Antibiotic	Method of Assay	Organism	Maintenance	Inoculum	Bacto Antibiotic Media Cylinder Plate Base Layer	Seed Layer	Turbidimetric Assay Medium
Dihydrostreptomycin	Cylinder Plate	Bacillus subtilis ATCC® 6633	1	1	5	5	
Dihydrostreptomycin	Turbidimetric	Klebsiella pneumoniae ATCC® 10031	1	1			3
Doxycycline	Turbidimetric	Staphylococcus aureus ATCC® 6538P	1	1			3
Erythromycin	Cylinder Plate	Micrococcus luteus ATCC® 9341	1	1	11	11	
Gentamicin	Cylinder Plate	Staphylococcus epidermidis ATCC® 12228	1	1	11	11	
Gramicidin	Turbidimetric	Streptococcus faecium ATCC® 10541	3	3			3
Kanamycin	Turbidimetric	Staphylococcus aureus ATCC® 6538P	1	1			3
Kanamycin B	Cylinder Plate	Bacillus subtilis ATCC® 6633	1	1	5	5	
Lincomycin	Turbidimetric	Staphylococcus aureus ATCC® 6538P	1	1			3
Meclocycline	Turbidimetric	Staphylococcus aureus ATCC® 6538P	1	1			3
Methacycline	Turbidimetric	Staphylococcus aureus ATCC® 6538P	1	1			3
Methicillin	Cylinder Plate	Staphylococcus aureus ATCC® 6538P	1	1	2	1	
Minocycline	Turbidimetric	Staphylococcus aureus ATCC® 6538P	1	1			3
Mithramycin	Cylinder Plate	Staphylococcus aureus ATCC® 6538P	1	1	8	8	
Mitomycin	Cylinder Plate	Bacillus subtilis ATCC® 6633	1	1	8	8	
Nafcillin	Cylinder Plate	Staphylococcus aureus ATCC® 6538P	1	1	2	1	
Natamycin	Cylinder Plate	Saccharomyces cerevisiae ATCC® 9763	19	13	2	19	
Neomycin	Cylinder Plate	Staphylococcus aureus ATCC® 6538P	1	1	11	11	
		or					
Neomycin	Cylinder Plate	Staphylococcus epidermidis ATCC® 12228	1	1	11	11	
Novobiocin	Cylinder Plate	Staphylococcus epidermidis ATCC® 12228	1	1	2	1	
Nystatin	Cylinder Plate	Saccharomyces cerevisiae ATCC® 2601	19	19	19	19	
Oleandomycin	Cylinder Plate	Staphylococcus epidermidis ATCC® 12228	1	1	11	11	
Oxacillin	Cylinder Plate	Staphylococcus aureus ATCC® 6538P	1	1	2	1	
Oxytetracycline	Turbidimetric	Staphylococcus aureus ATCC® 6538P	1	1			3
Paromomycin	Cylinder Plate	Staphylococcus epidermidis ATCC® 12228	1	1	11	11	
Penicillin G & V	Cylinder Plate	Staphylococcus aureus ATCC® 6538P	1	1	2	1	
Polymyxin B	Cylinder Plate	Bordetella bronchiseptica ATCC® 4617	1	1	9	10	
Rifampin	Cylinder Plate	Bacillus subtilis ATCC® 6633	1	1	2	2	
Roliteracycline	Turbidimetric	Staphylococcus aureus ATCC® 6538	1	1			3
Sisomycin	Cylinder Plate	Staphylococcus epidermidis ATCC® 12228	1	1	11	11	

Antibiotic	Method	Test Organism				
Spectinomycin	Turbidimetric	Escherichia coli ATCC® 10536	1			3
Streptomycin	Cylinder Plate	Bacillus subtilis ATCC® 6633	1	5	5	3
Streptomycin	Turbidimetric	Klebsiella pneumoniae ATCC® 10031	1			3
Tetracycline	Turbidimetric	Staphylococcus aureus ATCC® 6538P	1			3
Tobramycin	Turbidimetric	Staphylococcus aureus ATCC® 6538P	1			3
Troleandomycin	Turbidimetric	Klebsiella pneumoniae ATCC® 10031	1			
Tyrothricin	Turbidimetric	Streptococcus faecium ATCC® 10541	3			
Vancomycin	Cylinder Plate	Bacillus subtilis ATCC® 6633	1	8	8	

FORMULAE

BACTO ANTIBIOTIC ASSAY MEDIA

Ingredients per liter

	MEDIUM 1 (Penassay Seed Agar)	MEDIUM 2 (Penassay Base Agar)	MEDIUM 3 (Penassay Broth)	MEDIUM 4 (Yeast Beef Agar)	MEDIUM 5 (Streptomycin Assay Agar)	MEDIUM 6	MEDIUM 8
Bacto Beef Extract	1.5 g	1.5 g	1.5 g	1.5 g	1.5 g		1.5 g
Bacto Yeast Extract	3 g	3 g	1.5 g	3 g	3 g		3 g
Bacto Casitone	4 g					17 g	
Bacto Peptone	6 g	6 g	5 g	6 g	6 g		6 g
Bacto Dextrose	1 g		1 g	1 g		2.5 g	
Bacto Agar	15 g	15 g		15 g	15 g		15 g
Sodium Chloride			3.5 g			5 g	
Dipotassium Phosphate			3.68 g			2.5 g	
Monopotassium Phosphate			1.32 g				
Soytone						3 g	
Manganous Sulfate						0.03 g	
Grams/liter	30.5	25.5	17.5	26.5	25.5	30	25.5
Final pH at 25°C.	6.6 ± 0.1	6.6 ± 0.1	7.0 ± 0.05	6.55 ± 0.05	7.9 ± 0.1	7.0 ± 0.1	5.9 ± 0.1
Yield (liters/pound)	14.8	17.8	25.9	17.1	17.8	15.1	17.8

FORMULAE

BACTO ANTIBIOTIC ASSAY MEDIA (continued)

Ingredients per liter

	MEDIUM 9 (Polymyxin Base Agar)	MEDIUM 10 (Polymyxin Seed Agar)	MEDIUM 11 (Neomycin Assay Agar)	MEDIUM 12	MEDIUM 19	BACTO MYCIN ASSAY AGAR
Bacto Beef Extract			1.5 g	2.5 g	2.4 g	3 g
Bacto Yeast Extract			3 g	5 g	4.7 g	
Bacto Casitone	17 g	17 g	4 g			
Bacto Peptone			6 g	10 g	9.4 g	5 g
Bacto Dextrose	2.5 g	2.5 g	1 g	10 g	10 g	
Bacto Agar	20 g	12 g	15 g	25 g	23.5 g	15 g
Sodium Chloride	5 g	5 g		10 g	10 g	
Dipotassium Phosphate	2.5 g	2.5 g				
Soytone	3 g	3 g				
Sorbitan Monooleate Complex . . .		10 g				
Grams/liter	50	52	30.5	62.5	60	23
Final pH at 25°C	7.2 ± 0.1	7.3 ± 0.2	8.0 ± 0.1	6.0 ± 0.2	6.1 ± 0.1	7.9 ± 0.1
Yield (liters/pound)	9.0	8.7	14.8	7.2	7.5	21.7 liters / 500 grams

QUALITY CONTROL

Identity Specifications

Antibiotic Medium

Dehydrated powder: all homogeneous, free-flowing except as noted
Prepared medium:

	1	2	3	4	5	6
Dehydrated powder	beige	light tan	tan	light tan	light tan	light beige
Prepared medium	light-medium amber, slightly opalescent	light-medium amber, very slightly opalescent	light to medium amber, clear	light amber, very slightly opalescent	light amber, slightly opalescent	light amber, clear may have a slight precipitate

Antibiotic Medium

Dehydrated powder: all homogeneous, free-flowing except as noted
Prepared medium:

	8	9	10	11	12	19	Mycin Assay Agar
Dehydrated powder	light tan	off-white	beige, may not be free-flowing	beige	tan	light tan	tan
Prepared medium	light amber, slightly opalescent	light amber, slightly opalescent, may have a slight flocculant precipitate	light-medium amber, slightly opalescent	medium amber, very slightly to slightly opalescent	light-medium amber, very slightly opalescent	medium amber, slightly opalescent	medium amber, opalescent

REFERENCES

1. Pharmacopeia of the United States/National Formulary, USP/NF XX, Washington, D.C.: US Pharmacopeal Convention, p. 882 – 888, 1980.
2. Tests and Methods of Assay of Antibiotics and Antibiotic-Containing Drugs, FDA, CFR, Title 21, Part 436, Subpart D, Washington, D.C.: U.S. Government Printing Office, paragraphs 436.100 – 436.106, p. 242 – 259, (April 1) 1983.
3. Grove, D. C., and Randall, W. A., Assay Methods of Antibiotics, Medical Encyclopedia Inc., New York, 1955.
4. Kirshbaum, A., and Arret, B., Outline of Details for Official Microbiological Assays of Antibiotics, J. Pharm. Sci., 56:512, (April), 1967.
5. Regulations for Tests & Methods of Assay of Antibiotic Drugs, 1971.
6. Lancet, 2:177, 1941.
7. J. Bact., 46:187, 1943.
8. ibid., 47:199, 1944.

PACKAGING

Product	Size	Code
Bacto Antibiotic Medium 1	1/4 lb (114 g)	0263-02-0
	1 lb (454 g)	0263-01-1
	5 lb (2.27 kg)	0263-05-7
	10 kg	0263-08-4
Bacto Antibiotic Medium 2	500 g	0270-17-4
	10 kg	0270-08-5
Bacto Antibiotic Medium 3	1/4 lb (114 g)	0243-02-5
	1 lb (454 g)	0243-01-6
	5 lb (2.27 kg)	0243-05-2
Bacto Antibiotic Medium 4	1 lb (454 g)	0244-01-5
Bacto Antibiotic Medium 5	1/4 lb (114 g)	0277-02-4
	1 lb (454 g)	0277-01-5
	5 lb (2.27 kg)	0277-05-1
Bacto Antibiotic Medium 6	1 lb (454 g)	0660-01-0
Bacto Antibiotic Medium 8	1 lb (454 g)	0667-01-3
Bacto Antibiotic Medium 9	500 g	0462-17-2
Bacto Antibiotic Medium 10	500 g	0463-17-1
Bacto Antibiotic Medium 11	1/4 lb (114 g)	0593-02-1
	1 lb (454 g)	0593-01-2
	10 kg	0593-08-5
Bacto Antibiotic Medium 12	1 lb (454 g)	0669-01-1
Bacto Antibiotic Medium 19	500 g	0043-17-0
Bacto Mycin Assay Agar	500 g	0281-17-1
	10 kg	0281-08-2

Other products used in the microbiological assay of antibiotics:

Product	Size	Code
Bacto Fluid Sabouraud Medium	1 lb (454 g)	0642-01-3
	5 lb (2.27 kg)	0642-05-9
Bacto Fluid Thioglycollate Medium	1/4 lb (114 g)	0256-02-9
	1 lb (454 g)	0256-01-0
	12 tubes	0256-34-1
	144 tubes	0256-37-8
	6 × 100 ml	0256-73-3
Bacto Heart Infusion Agar	1/4 lb (114 g)	0044-02-6
	1 lb (454 g)	0044-01-7
	5 lb (2.27 kg)	0044-05-3
	10 kg	0044-08-0
Bacto Penase Concentrate	6 × 20 ml	0346-63-7
	100 ml	0346-72-6
	6 × 100 ml	0346-73-5

ANTIHYALURONIDASE TITRATION TEST (AHT)

BACTO AHT KIT
BACTO AHT ENZYME
BACTO AHT STANDARD
BACTO AHT SUBSTRATE

INTENDED USE
Bacto AHT Kit and Bacto AHT Reagents are used in the determination of the antihyaluronidase titer of sera as an adjunctive test to the antistreptolysin O titer (ASO) determination for the detection of Group A *Streptococcus* infections.

HISTORY
Quinn and Liao[1] demonstrated that the antihyaluronidase titration (AHT) test is highly specific for determining Group A *Streptococcus* infections and that normal individuals may exhibit titers up to 1:250. The occurence of an antihyaluronidase titer of 1:1024 and higher is significant in rheumatic fever as was pointed out by Harris, Harris and Nagle.[2] Similarly to the antistreptolysin O determinations, the significance of an elevated antihyaluronidase titer is increased when more than one specimen is examined from each patient. If a base line titer is first determined and a second specimen obtained 3 – 5 weeks later shows an increase in titer, it is obvious that the patient has or has recently had a *Streptococcus* infection.

Wannamaker and Ayoub[3] recently pointed out that the antistreptolysin O test was the most efficient procedure available for detecting Group A *Streptococcus* infections. McCarty[4] and Stetson[5] have demonstrated that about 80% of Group A *Streptococcus* infections can be detected by the antistreptolysin O titer alone. The detection of such *Streptococcus* infection can be increased from 80% with ASO to 90% by using the antihyaluronidase titration as an adjunct to the antistreptolysin O determination as shown by Stetson[5] and Taranta.[6]

PRINCIPLES AND SUMMARY OF THE TEST
The test is based on the inhibition of *Streptococcus* hyaluronidase, by antihyaluronidase in the patient's serum, and the measurement of the excess hyaluronidase by its ability to hydrolyze potassium hyaluronate.

The test consists essentially of:
1. Combining a constant volume of standardized hyaluronidase (AHT enzyme) with various dilutions of the patient's serum.
2. Incubating at 37°C for 15 minutes and then refrigerating for 10 minutes.
3. Adding potassium hyaluronate solution (AHT substrate).
4. Incubating at 37°C for 20 minutes and refrigerating for 30 minutes.
5. Adding 0.1 ml 2N acetic acid and shaking vigorously.
6. Observing the tubes for presence or absence of clot formation and clarity.

The antihyaluronidase titer is the reciprocal of the highest serum dilution showing a clot (not a thread). The technique of the test is a modification of the mucin clot prevention procedure described by Robertson, Ropes and Bauer[7] and essentially similar to that used by Harris and Harris,[8] and by Quinn.[9]

REAGENTS
Bacto AHT Enzyme is a standardized desiccated streptococcal hyaluronidase.

To rehydrate add 4 ml distilled water and rotate the vial in an end-over-end motion to effect complete solution. Do not rehydrate until ready to use. Unused portion should be discarded at the completion of the test procedure.

Bacto AHT Substrate is a standardized desiccated potassium hyaluronate.

To rehydrate add 8 ml distilled water, stopper and shake vigorously to dissolve completely. The substrate may be used up to 1 week after rehydration if it is stored at 2 – 8°C and is not allowed to become contaminated.

Bacto AHT Standard is a desiccated serum control which when rehydrated and used in the described procedure, yields an antihyaluronidase titer of 128 units. To rehydrate add 1 ml distilled water and dissolve the contents by gentle end-over-end rotation.

Bacto AHT Reagents and Kit are stable to the expiry date when unreconstituted and stored at 2 – 8°C.

SPECIMEN COLLECTION
1.5 ml whole blood is collected aseptically from a patient. The blood is allowed to clot and the syneresed serum is transferred to a clean, dry test tube. Store serum in the refrigerator until ready to use. Fresh or inactivated sera are equally satisfactory for the test.

Contaminated, chylous or hemoglobin containing serum should not be used.

PROCEDURE
Materials Provided:
Bacto AHT Enzyme
Bacto AHT Substrate
Bacto AHT Standard

Materials Required but not Provided:
Acetic acid 2N (prepared by adding 10 ml glacial acetic acid to 70 ml distilled water)
13 × 75 mm test tubes (7 for each serum to be tested, 5 for the standard and 2 for reagent controls)
Distilled water
5 ml serological pipettes
1 ml serological pipettes graduated in 0.01 ml increments
Water bath at 37°C
Refrigerator 2 – 8°C
Interval timer

1. Set up a row of chemically clean 13 × 75 mm tubes numbered from 1 – 7 for each serum to be tested. Behind this row, set up another row numbered 8 – 14. The front row is used for 7 dilutions of a test serum; the first 5 tubes of the second row are used for 5 dilutions of the AHT Standard; the 13th and 14th tubes are the hyaluronidase and substrate controls respectively.
2. Serum dilutions: Add 0.25 ml distilled water to tubes numbered 2 – 7 and 9 – 13. Add 0.6 ml distilled water to tube number 14. Prepare a 1:32 dilution of the patient's serum by adding 0.1 ml of serum to 3.1 ml distilled water and mixing thoroughly. Add 0.25 ml of the 1:32 dilution to tubes 1 and 2. Mix the contents of tube 2 by drawing the solution into a 1 ml serological pipette, graduated in 0.01 ml and then forcing it back out. Repeat 2 more times. Transfer 0.25 ml from tube 2 to tube 3 and mix as in tube 2. Continue the dilutions through tube 7 and discard 0.25 ml from tube 7. The 7 tubes will now contain 0.25 ml of dilutions 1:32 through 1:2048.

3. Add 0.25 ml of the rehydrated standard to tubes 8 and 9. The contents of tube 9 are mixed by drawing it into a 1 ml serological pipette graduated in 0.01 ml and delivering it back into the tube. This is repeated 2 more times and then 0.25 ml is transferred to tube 10. Continue the dilutions through tube 12 and discard 0.25 ml from tube 12.
4. Add exactly 0.25 ml Bacto AHT Enzyme solution to tubes numbered 1 – 13.
5. Shake the tubes to obtain an even mixture and incubate in a water bath at 37°C for 15 minutes.
6. Cool in the refrigerator at 2 – 8°C for 10 minutes.
7. Add 0.5 ml Bacto AHT Substrate to all tubes and shake the tubes thoroughly.
8. Incubate at 37°C for 20 minutes.
9. Cool in the refrigerator at 2 – 8°C for 30 minutes.
10. Add 0.1 ml of 2N acetic acid with a 1 ml serological pipette to all tubes and shake the rack vigorously to obtain thorough mixing of the contents of the tubes.
11. Observation of the presence of a clot is made by holding the tube slightly above eye level and shaking gently. Occasionally, the clot will adhere to the side of the tube and will be loosened by gentle shaking of the tube. The end-point is read visually by observing the highest dilution tube having a clot. The end-point is also confirmed by visually comparing the turbidity in each tube containing the assay serum with the 2 control tubes; i.e., the AHT enzyme, tube 13, which should show no clot, and the AHT substrate, tube 14, which should have a definite clot. The liquid in the tubes in which a clot is present will be clear; those tubes in which no clot is present will be opalescent.

RESULTS
The antihyaluronidase titer is the reciprocal of the highest dilution having a clot. An example of a titration result follows:

Patient's Serum		AHT Standard		AHT Enzyme Control	AHT Substrate Control
1:32	+	1:32	+	–	+
1:64	+	1:64	+		
1:128	+	1:128	+		
1:256	+	1:256	–		
1:512	–	1:512	–		
1:1024	–				
1:2048	–				

A positive (+) recording indicates the presence of a clot. Similarly, a negative (–) recording shows absence of a clot. The patient's serum in this titration has a titer of 256 units; the standard has a titer of 128 units; the AHT enzyme control tube contains enzyme and substrate only, therefore, no clot should be present and the AHT substrate control tube contains substrate only and should show a clot.

LIMITATIONS OF THE PROCEDURE
Some patients with a Group A *Streptococcus* infection do not produce a significant antihyaluronidase titer. Repeatedly negative results with this test cannot rule out a diagnosis of Group A streptococcal infection.

EXPECTED VALUES
Quinn and Liao[1] demonstrated that normal individuals may exhibit titers up to 1:250. The occurence of an antihyaluronidase titer of 1:1024 and higher is significant in rheumatic fever.[2]

Tabulated Summary of the Antihyaluronidase Test

	Tube No.	Serum Dilution	ml Diluted Serum	ml AHT Enzyme	ml Distilled Water	Step 1	ml AHT Substrate	Step 2	ml 2N Acetic Acid	Step 3
Patient's Serum	1	1:32	.25	.25	—		.5		.1	
	2	1:64	.25	.25	—		.5		.1	
	3	1:128	.25	.25	—		.5		.1	
	4	1:256	.25	.25	—		.5		.1	
	5	1:512	.25	.25	—	Shake thoroughly and incubate at 37°C for 15 minutes. Cool in the refrigerator at 2–8°C for 10 minutes.	.5	Shake thoroughly and incubate at 37°C for 20 minutes. Cool in the refrigerator at 2–8°C for 30 minutes.	.1	Shake vigorously and record presence or absence of clot.
	6	1:1024	.25	.25	—		.5		.1	
	7	1:2048	.25	.25	—		.5		.1	
AHT Standard	8	1:32	.25	.25	—		.5		.1	
	9	1:64	.25	.25	—		.5		.1	
	10	1:128	.25	.25	—		.5		.1	
	11	1:256	.25	.25	—		.5		.1	
	12	1:512	.25	.25	—		.5		.1	
AHT Enzyme Control	13	—	—	.25	.25		.5		.1	
AHT Substrate Control	14	—	—	—	.5		.5		.1	

SPECIFIC PERFORMANCE CHARACTERISTICS

When used alone, the AHT test is less sensitive than the ASO test for detecting Group A streptococcal infection. The reason the AHT test improves the detection rate for Group A streptococcal infection from 80 – 90% when used in conjunction with the ASO test versus the ASO test used alone, is because approximately 20% of individuals with a Group A streptococcal infection do not produce a significant ASO titer, but some of this same group do produce a significant antihyaluronidase titer.[5,6]

REFERENCES

1. Quinn, R. W. and Liao, S. J., J. Clin. Invest., 29:1156, 1950.
2. Harris, T. N., Harris, S., and Nagle, R. L., Pediatrics, 3:482, 1949.
3. Wannamaker, L. W. and Ayoub, E. M., Circulation, 21:598, 1960.
4. McCarty, M., Rheumatic Fever, edited by Thomas, L., Univ. of Minn. Press, p. 136, 1952.
5. Stetson, C. A., Jr., Streptococcal Infections, Edited by McCarty, M., Columbia Univ. Press, p. 208, 1954.
6. Taranta, A., paper read at the Symposium on "The Streptococcus Rheumatic Fever and Glomerulonephritis," New York Univ., November 1962.
7. Robertson, W., Ropes, M. W., and Bauer, W., J. Biol. Chem., 133:261, 1940.
8. Harris, T. N., and Harris, S., Am. J. Med. Sci., 217:127, 1949.
9. Quinn, R. W., Personal Communication, 1963.

PACKAGING

Bacto AHT Kit	2 tests	0934-32-3
Contains:		
Bacto AHT Enzyme	2 × 4 ml	
Bacto AHT Standard	2 × 1 ml	
Bacto AHT Substrate	2 × 8 ml	
Bacto AHT Enzyme	6 × 4 ml	0935-33-1
Bacto AHT Standard	6 × 1 ml	0946-51-5
Bacto AHT Substrate	6 × 8 ml	0945-33-9

BACTO ANTIMICROBIC VIAL A

INTENDED USE

Bacto Antimicrobic Vial A contains 25 mg sterile desiccated chlortetracycline (Aureomycin®) for use in preparing DTM agar described by Taplin, Azias, Rebell and Blank[1] for the detection and isolation of dermatophytes. It is also applicable for use in other media requiring this antibiotic.

METHOD OF PREPARATION

1. Aseptically add 10 ml sterile distilled or deionized water to the Bacto Antimicrobic Vial A.
2. Agitate gently to effect complete solution. The resulting concentration is 2.5 mg per ml of rehydrated solution.
3. The rehydrated contents of the vial may be added to sterile culture media cooled to 50°C in desired concentration. When used to prepare DTM agar, 1 vial is added to 250 ml of sterile basal medium.
4. Swirl to obtain a homogeneous solution.
5. Dispense as desired.

STORAGE

Bacto Antimicrobic Vial A   2 – 8°C

QUALITY CONTROL

Identity Specifications

Appearance: yellow button or powder
Rehydrated appearance: yellow, clear

For complete discussion on typical cultural response, refer to Bacto DTM Agar.

REFERENCE

1. Arch. Dermatology, 99:203, 1969.

PACKAGING

Bacto Antimicrobic Vial A   6 × 10 ml   3333-60-9

BACTO ANTIMICROBIC VIAL CNV
BACTO ANTIMICROBIC VIAL CNVT

INTENDED USE

Bacto Antimicrobic Vial CNV and Bacto Antimicrobic Vial CNVT are sterile desiccated preparations employed as inhibitors in selective media to culture *Neisseria gonorrhoeae* and *Neisseria meningitidis*.

HISTORY

Attempts to formulate a selective medium for isolating *N. gonorrhoeae* resulted in the development of Thayer-Martin medium. Thayer and Martin[1,2] explored the use of antibiotics as inhibitors of bacterial and fungal contaminants usually found in specimens obtained from the urogenital tract. In 1966, these authors reported on a medium composed of chocolate agar, enriched, as described by Christensen and Schoenlein,[3] to which vancomycin, Colistimethate® and nystatin were added. The Thayer-Martin medium proved to be selective for *N. gonorrhoeae* and *N. meningitidis* while maintaining the cultural value of the chocolate agar, enriched. Thayer-Martin medium has been recommended as the primary isolation medium of *N. gonorrhoeae*[1] from specimens suspected of containing this organism.

Seth[4] reported that the addition of trimethoprim lactate to a gonococcal culture medium containing the above inhibitors suppressed the growth of *Proteus* species without interferring with the growth of gonococci.

Martin and Lester[5] described transgrow, a selective transport and growth medium containing all four inhibitors, to reduce or eliminate the growth of commensal organisms usually encountered in clinical specimens containing *N. gonorrhoeae* and *N. meningitidis*. The Centers for Disease Control[6] recommended that transgrow, thereafter to be called modified Thayer-Martin medium, be the standard for the transport and culture of *N. gonorrhoeae*.

The introduction of Thayer-Martin medium does not negate, necessarily, the use of chocolate agar, enriched. It has been found that some strains of *N. gonorrhoeae* may be inhibited by the antibiotics incorporated in this medium. Therefore, to insure optimal recovery of gonococci, the use of both the selective medium and chocolate agar, enriched, are recommended when cultivating gonococci from urogenital specimens.[7]

PRINCIPLES

The addition of vancomycin suppresses growth of most gram-positive organisms, colistin inhibits gram-negatives, nystatin is effective against yeasts and the trimethoprim lactate suppresses the *Proteus* species.

FORMULAE

BACTO ANTIMICROBIC VIAL CNV, DESICCATED

	10 ml Vial	100 ml Vial
Colistin Sulfate	7,500 µg	75,000 µg
Nystatin	12,500 units	125,000 units
Vancomycin	3,000 µg	30,000 µg

BACTO ANTIMICROBIC VIAL CNVT, DESICCATED

	10 ml Vial	100 ml Vial
Colistin Sulfate	7,500 µg	75,000 µg
Nystatin	12,500 units	125,000 units
Vancomycin	3,000 µg	30,000 µg
Trimethoprim Lactate	5,000 µg	50,000 µg

METHOD OF PREPARATION

1. Rehydrate Bacto Antimicrobic Vial CNV or Bacto Antimicrobic Vial CNVT with sterile distilled or deionized water to the amount stated on the label.
2. Rotate the vial in an end-over-end motion to dissolve the contents completely.
3. Add 1 ml of rehydrated solution per 100 ml of sterile medium cooled to 50°C.

Final Concentration Per ML of Medium

	Bacto Antimicrobic Vial	
	CNV	CNVT
Colistin Sulfate	7.5 µg	7.5 µg
Nystatin	12.5 units	12.5 units
Vancomycin	3 µg	3 µg
Trimethoprim Lactate	—	5 µg

STORAGE

Bacto Antimicrobic Vial CNV	2 – 8°C
Bacto Antimicrobic Vial CNVT	2 – 8°C

QUALITY CONTROL

Identity Specifications

Appearance:	pale yellow button or powder
Rehydrated appearance:	off-white to pale yellow, opalescent to opaque even suspension.

For complete discussion on typical cultural response refer to Bacto Thayer Martin Medium.

REFERENCES

1. Publ. Hlth. Reports, 79:49, 1964.
2. Publ. Hlth. Reports, 81:599, 1966.
3. Ann. Meeting CA. Publ. Assn., 1947.
4. Brit. J. Vener. Dis., 46:201, 1970.
5. Public Health Rep., 86:30, 1971.
6. Public Health Service CDC Memo, 1975.
7. Proficiency Test Rep., CDC, Atlanta, GA.

PACKAGING

Bacto Antimicrobic Vial CNV	6 × 10 ml	3260-60-6
	100 ml	3260-72-2
Bacto Antimicrobic Vial CNVT	6 × 10 ml	3198-60-3
	100 ml	3198-72-9

BACTO ANTIMICROBIC VIAL K

INTENDED USE

Bacto Antimicrobic Vial K contains 25,000 µg sterile kanamycin for use as a bacterial inhibitor in media prepared for the detection and enumeration of specific bacterial species and/or strains. It is used as an additive to Bacto SFP Agar Base for the isolation of *Clostridium perfringens* from foods. The addition of polymyxin B and kanamycin will inhibit the growth of most undesirable bacteria and permit unrestricted growth of clostridia.

METHOD OF PREPARATION

1. Aseptically add 10 ml sterile distilled or deionized water to the Bacto Antimicrobic Vial K.
2. Shake to dissolve contents. Each ml of this solution contains 2.5 mg kanamycin.
3. Add the rehydrated contents of the vial to sterile culture media cooled to 50°C in desired concentration. When used to prepare Bacto SFP Agar Base, add 4.8 ml to each liter of base and cover layer as described in the directions for preparation of this medium.
4. Swirl to mix uniformly.

STORAGE

Bacto Antimicrobic Vial K   2 – 8°C

QUALITY CONTROL

Appearance: white button or powder
Rehydrated appearance: colorless, clear solution

For complete discussion on typical cultural response, refer to Bacto SFP Agar Base.

PACKAGING

Bacto Antimicrobic Vial K	6 × 10 ml	3339-60-3

BACTO ANTIMICROBIC VIAL P

INTENDED USE

Bacto Antimicrobic Vials P are sterile vials containing 30,000 units Polymyxin B per vial. They are used as inhibitors in selective media such as Bacto TPEY Agar described by Crisley[1] for the detection and enumeration of coagulase positive staphylococci. Bacto Antimicrobic Vials P are also used in the preparation of Bacto SFP Agar Base for the selective growth of clostridia.

METHOD OF PREPARATION
1. Aseptically add 10 ml sterile distilled or deionized water to the Bacto Antimicrobic Vial P.
2. Rotate in an end-over-end motion to dissolve the contents completely.
3. Add the rehydrated contents of the vial to sterile culture media that has been cooled to 50 – 55°C in desired concentration.

   When used to prepare Bacto TPEY Agar Base or Bacto SFP Agar Base, add 10 ml Bacto Antimicrobic Vial P to 900 ml of the basal medium.
4. Swirl to mix uniformly.

STORAGE
Bacto Antimicrobic Vial P   2 – 8°C

QUALITY CONTROL

Appearance: white button or powder
Rehydrated appearance: colorless, clear solution

For complete discussion on typical cultural response, refer to Bacto TPEY Agar Base and Bacto SFP Agar Base.

REFERENCE
1. Publ. Hlth. Serv. Publ., No. 1142, 1964.

PACKAGING
Bacto Antimicrobic Vial P   6 × 10 ml   3268-60-8

BACTO ANTISHEEP HEMOLYSIN, GLYCERINATED

Bacto Antisheep Hemolysin is a highly potent antisheep rabbit serum preserved with 50% glycerin. It is recommended for use in complement fixation procedures and other laboratory techniques requiring a stable, highly potent antisheep hemolysin.

STORAGE
Bacto Antisheep Hemolysin, Glycerinated   2 – 8°C

PACKAGING
Bacto Antisheep Hemolysin, Glycerinated   5 ml   0217-56-2
                                            10 ml   0217-59-9

ARIZONA ANTISERUM

BACTO ARIZONA ANTISERUM POLY MONOPHASIC
BACTO ARIZONA ANTISERUM POLY DIPHASIC

INTENDED USE
Bacto Arizona Antiserum Poly Monophasic and Bacto Arizona Antiserum Poly Diphasic are stable high titered, desiccated polyvalent antisera prepared in rabbits for the ser-

ological detection and identification of the *Arizona* (*Salmonella arizonae*) group of microorganisms.

SUMMARY AND EXPLANATION OF THE TEST

Biochemically the *Arizona* group of organisms ferments lactose but otherwise is closely related to the genus *Salmonella*.

Serologically, some members of the *Arizona* group have antigens in common with certain *Salmonella* serotypes. This in turn may give rise to some cross reactivity in polyvalent antisera. These antisera however, are useful where cross reactions do not exist. They are also useful in identifying the *Arizona* group of organisms on the basis of intensity of reaction even in the presence of some cross reactivity with *Salmonella* organisms. These antisera are used in the flagellar tube agglutination procedure.

REAGENTS

Bacto Arizona Antiserum Poly Monophasic is prepared from organisms possessing the following H antigens[1]: 1, 2, 3, 5, 6, 7, 8, 9, 10, 11, 13, 14, 15, 16, 17, 18, and 20. Bacto Arizona Antiserum Poly Diphasic is prepared from organisms possessing the following H antigens[2]: 21, 22, 23, 24, 25, 26, 27, 28ab, 28ac, 28ad, 29, 30, 31, 32ab, 32ac, 33, 34, 35, 37, 38, 39, 40ab, 40ac, 41, 42 and 43.

When properly rehydrated, each antiserum contains approximately 1:5000 Merthiolate®, sufficient to maintain a bacteriostatic condition at a storage of 2 – 8°C.

When used according to the suggested procedure (0.5 ml diluted antiserum per test) each 3 ml vial of Bacto Arizona Antiserum Poly Monophasic will perform approximately 200 tests. Bacto Arizona Antiserum Poly Diphasic will perform approximately 54 tests.

REHYDRATION

To rehydrate Bacto Arizona Antisera Monophasic or Diphasic add 3 ml sterile 0.85% NaCl solution and rotate in an end-over-end motion until dissolved completely.

STORAGE

Store both desiccated and rehydrated Bacto Arizona Monophasic or Diphasic Antisera at 2 – 8°C.

Prolonged exposure to room temperature or repeated freezing and thawing are detrimental to antisera.

EXPIRATION DATE

Both the desiccated and rehydrated Bacto Arizona Monophasic or Diphasic Antisera are stable to the expiry date on the label when stored at 2 – 8°C.

SPECIMEN PREPARATION

It is emphasized that biochemical testing must be employed to characterize and confirm the suspected pathogen as a member of the genus *Arizona* prior to applying serological techniques.

Suspect fecal specimens should be immediately plated to appropriate media or enrichments as soon as possible. When specimens cannot be cultured soon after collection they should be placed in a transport or preservative medium until proper cultural methods can be undertaken. Table I shows a schema for the examination of suspect specimens.

It should be remembered that some strains of *Arizona* may not grow well on highly selective media. Therefore, both highly selective and slightly selective differential plating media should be employed in cultural procedures for optimal recovery of *Arizona*.

Table I Isolation and Identification Schema

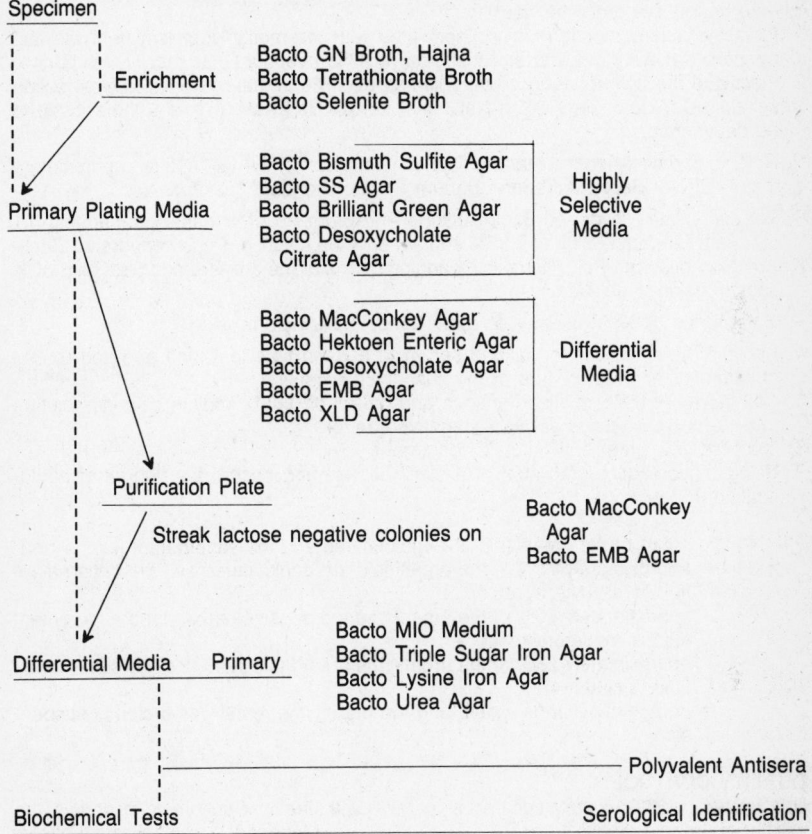

PROCEDURE
Materials Provided:
Bacto Arizona Antiserum Poly Monophasic
Bacto Arizona Antiserum Poly Diphasic

Materials Required but not Provided:

Sodium chloride 0.85%   Fluorescent light source
Formalin   Bacto Motility GI Medium
Kahn type test tubes   Bacto Veal Infusion Broth
Serological pipettes   37°C Incubator
Water bath (50°C)

Occasionally it is necessary to increase the motility of the test organisms. This is accomplished by making several consecutive transfers in Bacto Motility GI Medium. Inoculate the tube slightly below the surface of the medium by the stab method. Incubate the tubes at 37°C for 18 – 20 hours. Transfer only those organisms that have migrated to the bottom of the tube when making successive cultures. After several transfers if the motility of the culture is such that they travel 50 – 60 mm through the medium in 18 – 20 hours, it is ready for use.

1. Inoculate a Bacto Veal Infusion Broth tube with the motile organism from the last transfer to motility medium and incubate at 37°C overnight.
2. Inactivate the culture using equal volumes of 0.6% formalized physiological saline solution (6 ml formalin + 8.5 g NaCl in 1 liter distilled water). This is the antigen to be determined.

   Note: It is recommended that the final concentration of the test antigen approximate the density of Bacto McFarland Barium Sulfate Standard No. 2 or No. 3.

3. For use, dilute rehydrated Bacto Arizona Antiserum Poly Monophasic by adding 0.1 ml of antiserum to 25 ml of 0.85% sodium chloride solution. Dilute rehydrated Bacto Arizona Antiserum Poly Diphasic by adding 0.1 ml of the antiserum to 6.25 ml 0.85% sodium chloride solution.

   Note: Bring all test materials to room temperature before using.

4. Add 0.5 ml of the desired diluted serum to a Kahn type test tube and add 0.5 ml of a formalized broth culture of the suspected organism.
5. Add 0.5 ml of the formalized antigen and 0.5 ml of 0.85% sodium chloride solution to a Kahn type tube to act as a negative control.
6. Incubate in a water bath at 50°C for 1 hour.
7. Using a fluorescent lamp against a black background, record the degree of agglutination as follows:

   4+   100% of the organisms are agglutinated and the supernatant fluid is clear
   3+   approximately 75% of the organisms are agglutinated and the supernatant fluid is slightly cloudy
   2+   approximately 50% of the organisms are agglutinated and the supernatant fluid is moderately cloudy
   1+   approximately 25% of the organisms are agglutinated and the supernatant fluid is cloudy
   –   no agglutination is visible and the organisms remain as a cloudy suspension

QUALITY CONTROL
Use known positive control cultures in parallel with the test culture to ascertain the validity of the test results.

It is also recommended that a control tube be run in parallel using 0.5 ml of 0.85% sodium chloride solution and 0.5 ml of the organism suspension to determine whether the culture spontaneously agglutinates, indicating a rough culture. If agglutination occurs, subculture the organism successively on Bacto Veal Infusion Agar until the culture is smooth (no longer agglutinates in this procedure). Then pass the culture through successive transfers in Bacto Motility GI Medium.

RESULTS
A 3+ or greater reaction is considered positive.

REFERENCES
1. Identification of *Enterobacteriaceae*, Burgess Publishing Co., 1962.
2. Personal Communication, Jan. 1968.

PACKAGING

| Bacto Arizona Antiserum Poly Monophasic | 3 ml | 2750-47-3 |
| Bacto Arizona Antiserum Poly Diphasic | 3 ml | 2749-47-7 |

BACTO ASCITIC FLUID

INTENDED USE
Bacto Ascitic fluid, prepared from serous fluids removed aseptically from peritoneal cavities, is used as an enrichment for culture media. The product is sterility tested.

PRINCIPLES
Bacto Ascitic Fluid is used an an enrichment for culture media. The usual proportion for use is one 10 ml ampule of Bacto Ascitic Fluid for each 20 or 30 ml of sterile medium. To prevent coagulation of the ascitic fluid, the medium should be cooled to 45 – 50°C before the enrichment is added. Agar media should be prepared with concentrations of agar sufficiently high to allow for dilution of the media with the fluid. In order to prevent contamination, great care should always be taken to observe aseptic precautions when Bacto Ascitic Fluid is added to culture media. The enriched media should always be incubated to insure their sterility before they are used.

METHOD OF PREPARATION
1. Prepare Bacto Brain Heart Infusion per label instructions and dispense in 6 ml amounts in tubes.
2. Apply caps and sterilize for 15 minutes at 15 lbs presure (121°C). Allow to cool to below 50°C.
3. Aseptically add 2 ml Bacto Ascitic Fluid to each tube.
4. Incubate tubes overnight to assure sterility then inoculate as desired.

STORAGE
Bacto Ascitic Fluid   2 – 8°C

QUALITY CONTROL
Identity Specifications
Appearance: light amber, clear

Typical Cultural Response in Bacto Brain Heart Infusion w/25% Bacto Ascitic Fluid After 18 – 24 Hours at 35°C

Organism	Growth
Neisseria meningitidis ATCC® 13090	excellent
Streptococcus pneumoniae ATCC® 6303	excellent
Streptococcus pyogenes ATCC® 19615	excellent

PACKAGING

| Bacto Ascitic Fluid | 6 × 10 ml | 0135-60-5 |
| | 12 × 10 ml | 0135-61-4 |

BACTO AUTOLYZED YEAST

Bacto Autolyzed Yeast is a dessicated product containing both the soluble and insoluble portions of autolyzed baker's yeast. It is recommended for preparation of yeast supplements used in microbiological assay for riboflavin and pantothenic acid.[1,2]

REFERENCES
1. J. Ind. Eng. Chem., 13:567, 1941. (Anal. Ed.)
2. J. Ind. Eng. Chem., 14:909, 1942. (Anal. Ed.)

PACKAGING
Bacto Autolyzed Yeast	1 lb (454 g)	0229-01-4
	10 kg	0229-08-7

BACTO AZIDE BLOOD AGAR BASE

INTENDED USE
Bacto Azide Blood Agar Base is used for isolating streptococci and staphylococci from a variety of specimens. When blood is added, the medium is suitable for determining typical hemolytic reactions.

HISTORY
Edwards,[1] Hartman,[2] and Bryan, Devereux, Hirschey and Corbett[3] reported on the use of sodium azide as a selective agent for isolating streptococci causing mastitis. Snyder and Lichstein[4] and Lichstein and Snyder[5] reported that 0.01% sodium azide in blood agar prevented the swarming of *Proteus* and permitted the isolation of streptococci from known mixtures of bacteria; gram-negative organisms being inhibited.

PRINCIPLES
Mallmann, Botwright and Churchill,[6] studying the selective bacteriostatic effect of slow oxidizing agents, reported that sodium azide exerted a bacteriostatic effect on gram-negative bacteria, permitting growth of gram-positive organisms.

FORMULA
BACTO AZIDE BLOOD AGAR BASE
DEHYDRATED
Ingredients per liter

Bacto Tryptose	10 g
Bacto Beef Extract	3 g
Sodium Chloride	5 g
Sodium Azide	0.2 g
Bacto Agar	15 g

Final pH 7.2 ± 0.2 at 25°C.

One pound will make 13.7 liters of single-strength medium.

METHOD OF PREPARATION
1. Suspend 33 g in 1 liter distilled or deionized water and heat to boiling to dissolve completely.
2. Sterilize in the autoclave for 15 minutes at 15 lbs pressure (121°C). Cool to 45 – 50°C.

3. If preparing blood agar, aseptically add 5% sterile, defibrinated, room temperature blood volume/volume. Mix well.
4. Dispense into Petri dishes or tubes, as desired.

STORAGE

Bacto Azide Blood Agar Base   Below 30°C
Prepared plates or tubes   2 – 8°C

QUALITY CONTROL
Identity Specifications

Dehydrated powder: tan, homogeneous, free-flowing
Reaction of 3.3% solution: pH 7.2 ± 0.2 at 25°C
Plates prepared with 5% sheep blood: cherry red, may darken on standing

Typical Cultural Response on Bacto Azide Blood Agar Base Prepared with 5% Sheep Blood After 40 – 48 Hours at 35°C

Organism	Growth	Hemolysis
Escherichia coli ATCC® 25922	none	—
Neisseria meningitidis ATCC® 13090	none	—
Staphylococcus epidermidis ATCC® 12228	excellent	none
Streptococcus faecalis ATCC® 19433	excellent	alpha/gamma
Streptococcus pneumoniae ATCC® 6603	excellent	alpha
Streptococcus pyogenes ATCC® 19615	excellent	beta

REFERENCES
1. J. Comp. Path. Therap., 46:211, 1933.
2. Milchw. Forsch., 18:116, 1936.
3. North Am. Vet., 20:424, 1939.
4. J. Infectious Diseases, 67:113, 1940.
5. J. Bact., 42:653, 1941.
6. J. Infectious Diseases, 69:215, 1941.

PACKAGING

Bacto Azide Blood Agar Base   1 lb (454 g)   0409-01-6

BACTO AZIDE DEXTROSE BROTH

INTENDED USE

Bacto Azide Dextrose Broth is a selective Standard Methods[1] medium for the detection of streptococci in water, sewage and milk.

HISTORY/PRINCIPLES

Bacto Azide Dextrose Broth is prepared according to the formula of Rothe[2] of the Illinois State Health Department.

The use of sodium azide as an inhibitor of gram-negative organisms in an attempt to detect streptococci has been pointed out by a number of investigators. Edwards[3] in 1933 used a liquid medium containing crystal violet and sodium azide as a selective broth in the isolation of mastitis streptococci. Hartman[4] reported the value of sodium azide as a selective agent for the isolation of streptococci causing mastitis. Bryan, Devereux, Hirschey and Corbett[5] reported that sodium azide in a concentration of 1:5000 was a better selective preservative for milk cultures and gave more accurate results for the microscopic and Hotis tests for *Streptococcus mastitis* than 1:50,000 brilliant green.

Mallmann, Botwright and Churchill,[6] in studying the selective bacteriostatic effect of slow oxidizing agents, reported that sodium azide exerted a bacteriostatic effect on gram-negative bacteria and permitted the growth of gram-positive organisms.

In a comparative study of media for the detection of streptococci in water and sewage, Mallmann and Seligmann[7] used Bacto Azide Dextrose Broth. They reported the medium offered a new means of measuring the presence of streptococci in water, sewage, shellfish and other materials in which sewage pollution is suspected.

FORMULA

BACTO AZIDE DEXTROSE BROTH
DEHYDRATED

Ingredients per liter

Bacto Beef Extract . . . . 4.5 g
Bacto Tryptose . . . . 15 g
Bacto Dextrose . . . . 7.5 g
Sodium Chloride . . . . 7.5 g
Sodium Azide . . . . 0.2 g

Final pH 7.2 ± 0.2 at 25°C.

One pound will make 13.1 liters of medium.

METHOD OF PREPARATION
1. Dissolve 34.7 g in 1 liter distilled or deionized water.
2. Dispense into test tubes.
3. Sterilize in the autoclave for 15 minutes at 15 lbs pressure (121°C).

If inoculum is larger than 1 ml per 10 ml of medium, the medium should be prepared in double or multiple strength.

Consult *Standard Methods for the Examination of Water and Wastewater* for presumptive test procedure.[1]

STORAGE
Bacto Azide Dextrose Broth   Below 30°C
Prepared medium   2 – 8°C

QUALITY CONTROL
Identity Specifications

Dehydrated powder: light beige, homogeneous, free-flowing
Reaction of 3.47% solution: pH 7.2 ± 0.2 at 25°C
Prepared medium: light to medium amber, clear

Typical Cultural Response in Bacto Azide Dextrose Broth
After 18 – 24 Hours at 35°C

Organism	Growth
Escherichia coli ATCC® 25922	inhibited
Staphylococcus aureus ATCC® 25923	inhibited
Streptococcus faecalis ATCC® 19433	good

REFERENCES

1. Standard methods for the examination of water and wastewater, 15th Ed., American Public Health Association, Inc., Washington, D.C., 1980.
2. Rothe, 1948, Personal Communication.
3. Edwards, 1933, Journal of Comp. Path. Therap., 46:211.
4. Hartman, 1936, Milchw. Forsch., 18:166.
5. Bryan, Devereux, Hirschey and Corbett, 1939, North Am. Vet., 20:424.
6. Mallman, Botwright and Churchill, 1941, Journal of Infectious Disease, 69:215.
7. Mallman and Seligmann, 1950, American Journal of Public Health, 40:286.

PACKAGING

Bacto Azide Dextrose Broth	1 lb (454 g)	0387-01-2
	10 kg	0387-08-5

BACTO BAGG BROTH

INTENDED USE

Bacto BAGG Broth is recommended for presumptive and confirmation tests for fecal streptococci.

HISTORY/PRINCIPLES

Bacto BAGG Broth (Buffered Azide Glucose Glycerol Broth) is a selective liquid medium for the detection of fecal streptococci from all types of specimens. It is prepared according to the formula given by Hajna[1] and is a modification of SF (*Streptococcus faecalis*) medium as described by Hajna and Perry.[2] Coliform and other gram-negative organisms and buccal streptococci are inhibited on this medium as they are on SF medium while fecal streptococci develop unrestricted.

In using SF medium for the detection and isolation of fecal streptococci, particularly *Streptococcus faecalis*, Hajna noted that acid production in the medium was much more rapid when the inoculum consisted of a stool received in buffered glycerol saline. Investigation of this observation showed that glycerol facilitated the fermentation of dextrose by members of *Streptococcus* Group D and Hajna[1] described the addition of glycerol to SF medium to give the same rapid acid production with pure cultures of *S. faecalis* as with stool specimens. The concentration of brom cresol purple was decreased to detect more readily the color change from purple to yellow within 24 hours incubation.

Hajna[1] reported that growth with acid production is almost definite evidence of the presence of fecal streptococci. *Streptococcus faecalis* subsp. *zymogenes* and *liquefaciens* develop at 37°C or 45°C. *Streptococcus lactis* does not grow at 45°C, making possible a differentiation of these organisms using this incubation temperature. An incubation temperature of 37°C is suggested for the detection of fecal streptococci from swimming pools, water samples, food products such as oysters and crab meat and from pathological material such as catheterized urine and exudates.

FORMULA

BACTO BAGG BROTH
DEHYDRATED

Ingredients per liter

Bacto Tryptose . . . . 20 g
Bacto Dextrose . . . . 5 g
Dipotassium Phosphate . . . . 4 g
Monopotassium Phosphate . . . . 1.5 g
Sodium Chloride . . . . 5 g
Sodium Azide . . . . 0.5 g
Bacto Brom Cresol Purple . . . . . 0.015 g

Final pH 6.9 ± 0.2 at 25°C.

Five hundred grams will make 13.8 liters of medium.

METHOD OF PREPARATION

1. To rehydrate the medium, dissolve 36 g in 1 liter of distilled or deionized water containing 5 ml Bacto Glycerol.
2. Distribute in tubes in 10 ml amounts and sterilize in the autoclave for 15 minutes at 10 lbs pressure (116°C). Sterilization at 121°C is not recommended.

NOTE: When the inoculum is larger than 1 ml, particular care must be taken to preserve the correct concentration of the ingredients after dilution with the sample. For example, if 10 ml of water are to be added to 10 ml of medium, the medium should be prepared in double strength.

STORAGE

Bacto BAGG Broth   Below 30°C
Prepared media   2 – 8°C

QUALITY CONTROL

Identity Specifications

Dehydrated powder: light beige, homogeneous, free-flowing
Reaction of 3.6% solution
   w/5% Glycerol: pH 6.9 ± 0.2 at 25°C
Prepared medium: purple, clear

Typical Cultural Response in Bacto BAGG Broth
After 18 – 24 Hours at 45 ± 1°C

Organism	Growth	Acid
Enterobacter aerogenes ATCC® 13048	inhibited	–
Escherichia coli ATCC® 25922	inhibited	–
Streptococcus bovis ATCC® 27960	good to excellent	+
Streptococcus faecalis ATCC® 29212	good to excellent	+
Streptococcus pyogenes ATCC® 19615	inhibited	–

+ = positive, yellow
– = negative, no change, purple

REFERENCES

1. Am. J. Pub. Health, 33:550, 1943.
2. J. Inf. Dis., 67:113, 1940.
3. J. Bact., 42:653, 1941.
4. Sewage Works J., 12:875, 1940.
5. J. Bact., 42:137, 1941.
6. J. Bact., 46:343, 1943.

PACKAGING
Bacto BAGG Broth   500 g   0442-17-7

BACTO BCP-D AGAR

INTENDED USE

Bacto BCP-D Agar (Brom Cresol Purple Desoxycholate Agar) is a differential plating medium containing lactose and sucrose for the isolation of gram-negative enteric bacilli. It is prepared according to the formula of Hajna and Damon,[1] using sodium desoxycholate as the selective agent with brom cresol purple as the indicator of acid production.

HISTORY/PRINCIPLES

Bacto BCP-D Agar is recommended for the isolation of members of the *Salmonella-Shigella* group and other organisms not capable of fermenting lactose and/or sucrose. These organisms form transparent, colorless or bluish colonies while coliform organisms, capable of fermenting these carbohydrates, produce yellow, opaque or white colonies surrounded by a zone of precipitated desoxycholate. Coliform organisms develop unrestricted and, accordingly, the plates should be inoculated so as to avoid overgrowth. The authors reported *Shigella sonnei* (types I, II and R) grew well on the medium, while spreading by *Proteus vulgaris* and *P. mirabilis* was rarely observed. The medium is especially recommended for use in proctoscopic work, using rectal swabs for transporting the specimen and for inoculating the plate. Colonies suspected of being enteric pathogens are picked and transferred to Bacto Triple Sugar Iron Agar, Bacto Kligler Iron Agar or other differential tube media to determine further identifying characteristic biochemical reactions.

FORMULA

BACTO BCP-D AGAR
DEHYDRATED

Ingredients per liter

Bacto Yeast Extract	2 g	Sodium Citrate	2 g
Bacto Casitone	7.5 g	Sodium Desoxycholate	1 g
Bacto Thiopeptone	7.5 g	Sodium Chloride	5 g
Bacto Lactose	10 g	Bacto Agar	25 g
Saccharose, Difco	10 g	Bacto Brom Cresol Purple	0.02 g

Final pH 7.2 ± 0.2 at 25°C.

One pound will make 6.5 liters of medium.

METHOD OF PREPARATION

1. Suspend 70 g in 1 liter distilled or deionized water. Heat to boiling to dissolve completely.
2. If to be used the same day as rehydrated, plates may be poured without autoclave sterilization. If not to be used the same day, sterilize in the autoclave for 15 minutes at 15 lbs pressure (121°C).
3. Allow the media to cool to 55 – 60°C then pour plates as desired. To minimize swarming, allow surface of medium to dry prior to streaking.

STORAGE

Bacto BCP-D Agar   Below 30°C
Prepared plates   2 – 8°C

QUALITY CONTROL
Identity Specifications

Dehydrated powder: tan, homogeneous, free-flowing
Reaction of 7.0% solution: pH 7.2 ± 0.2 at 25°C
Prepared medium: purple, slightly opalescent, may have slight precipitate

Typical Cultural Response on Bacto BCP-D Agar
After 18 – 24 Hours at 35°C

Organism	Growth	Color of Colony
Escherichia coli ATCC® 25922	good to excellent	yellow w/bile ppt
Proteus mirabilis ATCC® 25933	good to excellent	colorless
Proteus vulgaris ATCC® 6380	good to excellent	yellow, no swarming
Salmonella typhimurium ATCC® 14028	good to excellent	colorless
Shigella dysenteriae ATCC® 13313	good to excellent	colorless
Staphylococcus aureus ATCC® 25923	inhibited	—

REFERENCE
1. Appl. Microbiol., 4:341:1956.

PACKAGING
Bacto BCP-D Agar   1 lb (454 g)   0487-01-1

BACTO BG SULFA AGAR

INTENDED USE
Bacto BG Sulfa Agar (Brilliant Green) is a highly selective medium for the isolation of Salmonella. It has the same composition as Bacto Brilliant Green Agar and in addition contains 0.1% sodium sulfapyridine, as suggested by Osborne and Stokes.[1] A complete discussion on the use of this product in the isolation of *Salmonella* can be found under Bacto Brilliant Green Agar.

HISTORY/PRINCIPLES
Osborne and Stokes[1] reported that the addition of 0.1% sodium sulfapyridine to brilliant green agar (Bacto BG Sulfa Agar) enhanced the selective properties of this medium for *Salmonella*. Bacto BG Sulfa Agar duplicates the formulation of these authors and is recommended as a selective isolation medium for *Salmonella* following enrichment in Bacto SBG Enrichment and Bacto SBG Sulfa Enrichment. It is also recommended for direct inoculation with primary specimens for *Salmonella* isolation.

FORMULA
BACTO BG SULFA AGAR
DEHYDRATED

Ingredients per liter

Bacto Yeast Extract	3 g
Proteose Peptone No. 3, Difco	10 g
Bacto Lactose	10 g
Saccharose, Difco	10 g
Sodium Sulfapyridine	1 g
Sodium Chloride	5 g
Bacto Agar	20 g
Bacto Brilliant Green	0.0125 g

Final pH 6.9 ± 0.2 at 25°C.

One pound will make 7.7 liters of final medium.

METHOD OF PREPARATION
1. Suspend 59 g in 1 liter distilled or deionized water. Heat to boiling to dissolve.
2. Sterilize in the autoclave for 15 minutes at 15 lbs pressure (121°C). Selectivity will be decreased if the medium is sterilized for a longer period.
3. Pour into Petri dishes as desired. Allow surface to dry thoroughly before using.

STORAGE
Bacto BG Sulfa Agar — Below 30°C
Prepared plates — 2 – 8°C

QUALITY CONTROL
Identity Specifications
Dehydrated powder: pink/light tan, homogeneous, free-flowing
Reaction of 5.9% solution: pH 6.9 ± 0.2 at 25°C
Prepared medium: dark, reddish amber, slightly opalescent

Typical Cultural Response on Bacto BG Sulfa Agar
After 18 – 48 Hours at 35°C

Organism	Growth	Color of Colony
Escherichia coli ATCC® 25922	poor to fair	yellow-green
Proteus vulgaris ATCC® 13315	none	—
Salmonella enteritidis ATCC® 13076	good	pink-white
Salmonella typhimurium ATCC® 14028	good	pink-white

REFERENCE
1. Appl. Microbiol., 3:295, 1955.

PACKAGING
Bacto BG Sulfa Agar — 1 lb (454 g) — 0717-01-3

BACTO BTB LACTOSE AGAR

INTENDED USE
Bacto BTB Lactose Agar (Brom Thymol Blue) is recommended for isolation of pathogenic staphylococci according to the procedure of Chapman et al.[1]

FORMULA
BACTO BTB LACTOSE AGAR
DEHYDRATED
Ingredients per liter

Bacto Beef Extract .... 3 g
Proteose Peptone, Difco .... 5 g
Bacto Lactose .... 10 g
Bacto Brom Thymol Blue .... 0.17 g
Bacto Agar .... 15 g

Final pH 8.6 ± 0.2 at 25°C.

One pound will make 13.7 liters of medium.

METHOD OF PREPARATION
1. Suspend 33 g in 1 liter distilled or deionized water. Heat to boiling to suspend medium completely.
2. Sterilize in the autoclave for 15 minutes at 15 lbs pressure (121°C).

STORAGE

Bacto BTB Lactose Agar	Below 30°C
Prepared medium	2 – 8°C

QUALITY CONTROL

Identity Specifications

Dehydrated powder:	greenish tan, homogeneous, free-flowing
Reaction of 3.3% solution:	pH 8.6 ± 0.2 at 25°C
Prepared medium:	blue, clear to very slightly opalescent

Typical Cultural Response on Bacto BTB Lactose Agar
After 24 – 48 Hours at 35 – 37°C

Organism	Growth	Color of Colony
Escherichia coli ATCC® 25922	good to excellent	yellow
Salmonella typhi ATCC® 6539	good to excellent	blue/colorless
Staphylococcus aureus ATCC® 25923	good to excellent	yellow
Staphylococcus aureus ATCC® 6538	good to excellent	yellow
Staphylococcus epidermidis ATCC® 12228	good to excellent	blue/colorless

REFERENCE

1. Chapman, Lieb, Berens and Curcio. J. Bact., 33:533, 1937.

PACKAGING

Bacto BTB Lactose Agar	500 g	0078-17-8

BACTROL™ DISKS

INTENDED USE

Bactrol™ Disks provide a variety of known microbial cultures for the following uses:
1. controlling the quality of prepared culture media, bacterial stains and biochemical reagents used for isolating, identifying and differentiating microorganisms;
2. monitoring the precision and accuracy of procedures and equipment employed in the microbiological laboratory; and,
3. controlling the materials and procedures used in antimicrobial disc diffusion and dilution (MIC) susceptibility testing.

HISTORY

Bactrol Disks were initially introduced by Difco Laboratories in 1972 to make cultures of expected cultural, biochemical and serological characteristics readily available without the cost and variability of maintaining stock cultures or frozen or lyophilized microorganisms.

PRINCIPLES

When quality control cultures are used in parallel with a test culture or unknown, variables that could cause erroneous test results are detected and defined.

REAGENTS

Bactrol™ Disks are stabilized, water-soluble disks containing derivatives of American Type Culture Collection® (ATCC®) cultures.

The following microorganisms are available as Bactrol Disks:

Acinetobacter calcoaceticus ATCC® 19606	1650-35-0
Branhamella catarrhalis ATCC® 8176	1668-35-0
Citrobacter freundii ATCC® 8090	1651-35-9
Edwardsiella tarda ATCC® 15947	1658-35-2
Enterobacter aerogenes ATCC® 13048	1652-35-8
Enterobacter cloacae ATCC® 23355	1641-35-2
Escherichia coli ATCC® 25922 (Seattle strain)	1639-35-6
Escherichia coli ATCC® 35218	1659-35-9
Haemophilus influenzae ATCC® 35056	1665-35-3
Klebsiella pneumoniae ATCC® 13883	1647-35-6
Neisseria gonorrhoeae ATCC® 19424	1661-35-7
Neisseria lactamicus ATCC® 23970	1663-35-5
Neisseria meningitidis ATCC® 13090	1667-35-1
Neisseria sicca ATCC® 9913	1662-35-6
Proteus vulgaris ATCC® 13315	1648-35-5
Pseudomonas aeruginosa ATCC® 27853 (Boston strain)	1649-35-4
Salmonella typhimurium ATCC® 14028	1644-35-9
Serratia marcescens ATCC® 8100	1643-35-0
Shigella flexneri ATCC® 12022	1653-35-7
Shigella sonnei ATCC® 25931	1654-35-6
Staphylococcus aureus ATCC® 25923 (Seattle strain)	1637-35-8
Staphylococcus aureus ATCC® 29213	1642-35-1
Staphylococcus epidermidis ATCC® 12228	1638-35-7
Streptococcus agalactiae ATCC® 13813	1666-35-2
Streptococcus faecalis ATCC® 19433	1655-35-5
Streptococcus faecalis ATCC® 29212	1646-35-7
Streptococcus faecalis ATCC® 33186	1657-35-3
Streptococcus pneumoniae ATCC® 6303	1664-35-4
Streptococcus pyogenes ATCC® 19615	1645-35-8
Yersinia enterocolitica ATCC® 27729	1669-35-9

"ATCC" and "American Type Culture Collection" are registered trademarks of The American Type Culture Collection. Bactrol Disks are derivatives of ATCC® cultures.

PRECAUTIONS
Bactrol Disks contain viable microorganisms and should be used only in laboratories by microbiologists or persons under the supervision of microbiologists qualified by training or experience to work with bacteria.

PROCEDURE IN CASE OF BREAKAGE OR SPILLAGE
1. Obtain a suitable bactericidal solution such as 70 – 90% isopropyl alcohol, 1 – 2% phenol, quaternary compound or a liquid chlorine laundry bleach.
2. Remove unbroken vials with forceps and wash their surfaces with the above bactericidal solution. Unbroken vials may be retained for future use.
3. Collect exposed disks and contaminated packaging in a suitable container and autoclave for 30 minutes at 15 pounds pressure (121°C) or thoroughly incinerate.
4. Wash all exposed surface areas with the above bactericidal solution.

STORAGE
1. Store Bactrol Disks at 2 – 8°C. Unnecessary exposure during use to temperatures above this range may result in reduction or loss of viability of the bacteria in the disks.

2. Protect Bactrol Disks from moisture uptake, which will cause reduction or loss of viability, by promptly and tightly replacing the stopper and cap on each vial after removing a disk.

PROCEDURE
Reconstitution of Bactrol Disks for Use on Solid Media
1. Using flamed and cooled sterile forceps or needle, aseptically remove a disk from a vial and place into 1 – 2 ml sterile brain heart infusion, veal infusion broth or tryptic soy broth.
2. Shake or vortex occasionally to completely dissolve the disk (up to approximately five minutes). Discoloration and/or slight opalescence of the broth will be noted when dissolving those disks containing blood. Do not use this discoloration or opalescence as an indication of growth of the culture.
3. Inoculate the dissolved culture on the desired solid media, being sure to include blood or factors V and X (Bacto Supplement B or Bacto Supplement VX) in the media for those cultures with such requirements for growth. See **TABLE 1.**

Reconstitution of Bactrol Disks for Use in Liquid Media
1. Using flamed and cooled sterile forceps or needle, aseptically remove a disk from a vial and place directly into the desired broth, being sure the medium includes blood or factors V and X (Bacto Supplement B or Bacto Supplement VX) for those cultures with such requirements for growth. See **TABLE 1.**
2. Shake or vortex to dissolve the disk and uniformly distribute the organisms. A subtle discoloration and/or opalescence of the medium may be noted when dissolving those disks containing blood. Do not use this discoloration or opalescence as an indication of growth of the culture.

Reconstitution of Bactrol Disks for Use on Slides
1. Dissolve the disk and inoculate an appropriate growth medium as described above in **Reconstitution of Bactrol Disks for Use on Solid Media.**
2. Incubate the culture overnight at the optimum growth temperature. See **TABLE 1.**
3. Transfer 1 – 2 loopfuls of the growth to a well-cleaned slide, spread with the loop to about 0.5 – 1 cm diameter circle and fix as required.
4. Proceed with staining as required.

Reconstitution of Bactrol Disks *Pseudomonas aeruginosa* ATCC® 27853
Bactrol Disks *Pseudomonas aeruginosa* ATCC® 27853, because it grows slowly, requires alternate reconstitution to achieve usable growth.
1. Using flamed and cooled sterile forceps or needle, aseptically remove a disk from a vial and place in 1 – 2 ml sterile brain heart infusion, veal infusion broth or tryptic soy broth. Use 1 ml sterile tryptic soy broth when reconstituting for use in diffusion or dilution susceptibility testing.
2. Incubate at 35 – 37°C for 2 – 3 hours with occasional shaking. Longer incubation may be required before increased turbidity is evident.
3. Streak a loopful of the incubated suspension on a tryptic soy agar plate in a manner to obtain isolated colonies.
4. Incubate the inoculated plate overnight at 35 – 37°C.
5. Proceed as desired to transfer colonies.

Reconstitution of Bactrol Disks for Use in Quality Control of Diffusion and Dilution Susceptibility Testing
1. Using flamed and cooled sterile forceps or needle, aseptically remove a disk from a vial and place into 1 ml sterile tryptic soy broth.

2. Incubate at 35 – 37°C for 2 – 3 hours with occasional shaking. *Pseudomonas*, because it grows slowly, may require longer incubation before the broth is slightly turbid.
3. Streak a loopful of the incubated suspension on a tryptic soy agar plate in a manner to obtain isolated colonies.
4. Incubate the inoculated plate overnight at 35 – 37°C.
5. Transfer colonies as described in the standard tests in ASM-2: 4.1.1 or M7-T: 3.4.2.[1,2]

INTERPRETATION OF RESULTS

1. Compare quality control results with expected values on the following tables:
   TABLE 1. Morphology and Growth Characteristics
   TABLE 2. Biochemical Reactions
   TABLE 3. Antimicrobial Disc Diffusion
              Susceptibility Test Control Zones
   TABLE 4. Dilution Antimicrobial Susceptibility Tests (MIC)
   Consult the QUALITY CONTROL section of each culture medium to determine the cultural appearance of particular organisms on that medium.
2. If control results are not as expected, determine the source of error and take corrective action to eliminate it. Incorrect control culture results give tentative evidence of systematic error or product failure that could cause misinterpretation of test culture results.
3. Expected results using control cultures indicate that test culture values will be accurate.

LIMITATIONS

1. Reconstitute new Bactrol Disks daily or as often as cultures are needed. Repeated subculturing may lead to the development of mutant strains which may, in turn, cause aberrant test results.
2. *Pseudomonas aeruginosa* ATCC® 27853 develops resistance to carbenicillin after repeated transfers on laboratory media. To prevent development and use of a resistant microorganism, reconstitute a new disk daily.

   Occasionally, different colonial types, particularly of *Pseudomonas*, are seen. This has not been found to interfere with the susceptibility test.

REFERENCES

1. NCCLS Approved Standard: ASM-2, Performance standards for antimicrobic disc susceptibility tests, 2nd Ed. Villanova, Pa., National Committee for Clinical Laboratory Standards, 1979.
2. Tentative Standard, Methods for dilution antimicrobial susceptibility tests for bacteria that grow aerobically. Villanova, Pa., National Committee for Clinical Laboratory Standards, 1983.

PACKAGING

Bactrol Disks are packaged 25 disks per vial. Please consult Difco PRODUCT LIST III for current availability and product codes.

Bactrol™ Disks Set A†	1628-32-2
Bactrol™ Disks Set B†	1629-32-1
Bactrol™ Disks Set C†	1656-32-7
Bactrol™ Disks Set D†	1660-32-1
Bactrol™ MIC Set†	1640-32-6

†"ATCC" and "American Type Culture Collection" are registered trademarks of the American Type Culture Collection. Bactrol Disks are derivatives of ATCC® cultures.

Table 1. Morphology and Growth Characteristics

	ATCC®	Cell Morphology	Motility	Gram Reaction	O₂/CO₂ Requirement	Hemolysis	Optimum Growth Temperature	Growth on:	Remarks
Acinetobacter calcoaceticus	19606	Coccus or coccobacillary rod, 1.0 – 1.5 by 1.5 – 2.5 μ, occurs in pairs and short chains predominantly	Nonmotile	–	Strictly aerobic		35 – 37°C		var. Anitratus
Branhamella catarrhalis	8176	Coccus, occurs in pairs	Nonmotile	–	Aerobic		35 – 37°C		Grows at 22°C
Citrobacter freundii	8090	Rod, 1.0 by 1.5 – 2.0 μ	Motile, peritrichous flagella	–	Aerobic, facultatively anaerobic		35 – 37°C		
Edwardsiella tarda	15947	Short rods	Motile, peritrichous flagella	–	Aerobic, facultatively anaerobic		35 – 37°C		
Enterobacter aerogenes	13048	Rod, 0.5 – 0.8 by 1.2 μ	Motile, peritrichous flagella	–	Aerobic, facultatively anaerobic		35 – 37°C		
Enterobacter cloacae	23355	Rod, 0.5 – 1 by 1 – 2 μ	Motile, peritrichous flagella	–	Aerobic, facultatively anaerobic		35 – 37°C		
Escherichia coli	25922	Rod, 0.5 by 1.2 μ, of varying lengths, occurs singly, in pairs and short chains.	Usually motile, peritrichous flagella	–	Aerobic, facultatively anaerobic		35 – 37°C		
Escherichia coli	35218	Short rods	Motile, peritrichous flagella	–	Aerobic, facultatively anaerobic		35 – 37°C		
Haemophilus influenzae	35056	Small rods, filamentous	Nonmotile	–	Aerobic, facultatively anaerobic		35 – 37°C		Requires factors V and X for growth
Klebsiella pneumoniae	13883	Rod, 0.3 – 0.5 by 5 μ, occurs singly and in pairs, produces large capsules	Nonmotile	–	Aerobic, facultatively anaerobic		35 – 37°C		
Neisseria gonorrhoeae	19424	Cocci, occur in pairs, adjacent sides usually flattened	Nonmotile	–	Aerobic, 5 – 10% CO₂ required		35 – 37°C	Nutrient Agar –	FA N. gonorrhoeae + Requires blood for growth

Organism	ATCC #	Morphology	Motility		Oxygen	Hemolysis	Temp	Growth	Notes
Neisseria lactamicus	23970	Cocci, occur in pairs, adjacent sides usually flattened	Nonmotile	–	Aerobic		35 – 37°C	Nutrient Agar +	
Neisseria meningitidis	13090	Cocci, occur in pairs, adjacent sides usually flattened	Nonmotile	–	Aerobic		35 – 37°C	Nutrient Agar –	Serogroup B FA N. gonorrhoeae – FA N. meningitidis + FA N. meningitidis –
Neisseria sicca	9913	Cocci, occur in pairs, adjacent sides usually flattened	Nonmotile	–	Aerobic		35 – 37°C	Nutrient Agar +	
Proteus vulgaris	13315	Rod, 0.5 – 1 by 1 – 3 μ, occurs singly, in pairs and long chains	Actively motile, peritrichous flagella	–	Aerobic, facultatively anaerobic		35 – 37°C		
Pseudomonas aeruginosa	27853	Rod, 0.5 – 0.6 by 1.5 μ, occurs singly, in pairs and short chains	Motile, 1 – 3 polar flagella	–	Aerobic, facultatively anaerobic		35 – 37°C		Produces pyocyanin and fluorescein Gives off a grape-like odor
Salmonella typhimurium	14028	Short plump rod, 0.5 by 1 – 1.5 μ, occurs singly	Motile, peritrichous flagella	–	Aerobic, facultatively anaerobic		35 – 37°C		Biochemical designation: *Salmonella enteritidis* Serological designation: *Salmonella typhimurium*
Serratia marcescens	8100	Short rod, 0.5 by 0.5 – 1 μ, occurs singly or occasionally in 5 or 6 element chains	Motile, peritrichous flagella	–	Aerobic, facultatively anaerobic		35 – 37°C		
Shigella flexneri	12022	Rod, 0.5 by 1.0 – 1.5 μ, occurs singly, often filamentous in old cultures	Nonmotile	–	Aerobic, facultatively anaerobic		35 – 37°C		
Shigella sonnei	25931	Rod, 0.5 by 1.0 – 2.0 μ, occurs singly	Nonmotile	–	Aerobic, facultatively anaerobic		35 – 37°C		
Staphylococcus aureus	25923	Coccus, 0.8 – 1 μ, occurs singly, in pairs, short chains or grape-like clusters	Nonmotile	+	Aerobic, facultatively anaerobic	Beta (48 hrs)	35 – 37°C	6.5% NaCl +	
Staphylococcus aureus	29213	Coccus, occurs singly, in pairs or irregular clusters	Nonmotile	+	Aerobic, facultatively anaerobic	Beta (48 hrs)	35 – 37°C	6.5% NaCl +	

+ Reaction is positive or present; or, organism is susceptible.
– Reaction is negative or absent; or, organism is resistant.
Omission of data describing reaction indicates that such information is not relevant or appropriate for the particular strain.

Table 1. continued

	ATCC®	Cell Morphology	Motility	Gram Reaction	O₂/CO₂ Requirement	Hemolysis	Optimum Growth Temperature	Growth on:	Remarks
Staphylococcus epidermidis	12228	Coccus, 0.5 – 0.6 μ, occurs singly, in pairs, short chains or irregular grape-like clusters	Nonmotile	+	Aerobic, facultatively anaerobic	Gamma	35 – 37°C	6.5% NaCl +	
Streptococcus agalactiae	13813	Coccus, occurs in pairs or chains	Nonmotile	+	Aerobic, facultatively anaerobic	Gamma	35 – 37°C		Serogroup B
Streptococcus faecalis	19433	Coccus, 0.5 – 1.0 μ, occurs mostly in pairs or short chains	Nonmotile	+	Aerobic, facultatively anaerobic	Gamma	35 – 37°C	Bile Esculin Agar + 6.5% NaCl +	Serogroup D Grows at 45°C
Streptococcus faecalis	29212	Coccus, 0.5 – 1.0 μ, occurs mostly in pairs or short chains	Nonmotile	+	Aerobic, facultatively anaerobic	Gamma	35 – 37°C	Bile Esculin Agar + 6.5% NaCl +	Serogroup D Grows at 45°C
Streptococcus faecalis	33186	Coccus, 0.5 – 1.0 μ, occurs mostly in pairs or short chains	Nonmotile	+	Aerobic, facultatively anaerobic	Gamma	35 – 37°C	Bile Esculin Agar + 6.5% NaCl +	Serogroup D Control organism for Mueller Hinton Medium (Bushby strain)
Streptococcus pneumoniae	6303	Coccus, occurs in pairs or chains, lancet shaped	Nonmotile	+	Aerobic, facultatively anaerobic	Alpha	35 – 37°C	Bile Esculin Agar – Nutrient Agar –	Capsular reaction (Quellung) with type 3 antiserum
Streptococcus pyogenes	19615	Coccus, 0.6 – 1 μ, occurs in pairs or short to long chains	Nonmotile	+	Aerobic, facultatively anaerobic	Beta	35 – 37°C	6.5% NaCl –	Serogroup A
Yersinia enterocolitica	27729	Rod, coccoid in young cultures	Motile at 25°C, peritrichous flagella; nonmotile at 36°C	–	Aerobic, facultatively anaerobic		25 – 30°C		

+ Reaction is positive or present; or, organism is susceptible.
− Reaction is negative or absent; or, organism is resistant.
Omission of data describing reaction indicates that such information is not relevant or appropriate for the particular strain.

Table 2. Biochemical Reactions

	ATCC®	Glucose Gas	Glucose Acid	Lactose	Maltose	Sucrose	Fructose	Arabinose
Acinetobacter calcoaceticus	19606			−				
Branhamella catarrhalis	8176		−	−	−	−	−	
Citrobacter freundii	8090	+	+	+		+		+
Edwardsiella tarda	15947	+	+	−				−
Enterobacter aerogenes	13048	+	+	+		+		+
Enterobacter cloacae	23355	+	+	+		+		+
Escherichia coli	25922	+	+	+		−		+
Escherichia coli	35218	+	+	+		−		+
Haemophilus influenzae	35056		+					
Klebsiella pneumoniae	13883	+	+	+		+		+
Neisseria gonorrhoeae	19424		+	−	−	−	−	
Neisseria lactamicus	23970		+	+	+	−	−	
Neisseria meningitidis	13090		+	−	+	−	−	
Neisseria sicca	9913		+	−	+	+	+	
Proteus vulgaris	13315	+	+	−		+		−
Pseudomonas aeruginosa	27853		−	−	−	−		
Salmonella typhimurium	14028	+	+	−		−		+
Serratia marcescens	8100	+	+	−		+		−
Shigella flexneri	12022		−	−		−		+
Shigella sonnei	25931		−	(+)		(+)		+
Staphylococcus aureus	25923		+					
Staphylococcus aureus	29213		+					
Staphylococcus epidermidis	12228		+					
Streptococcus agalactiae	13813							
Streptococcus faecalis	19433		+					−
Streptococcus faecalis	29212		+					−
Streptococcus faecalis	33186		+					−
Streptococcus pneumoniae	6303							
Streptococcus pyogenes	19615		+					
Yersinia enterocolitica	27729				(+)	+		

+ Reaction is positive or present; or, organism is susceptible.
− Reaction is negative or absent; or, organism is resistant.
( ) Delayed reaction.
Omission of data describing reaction indicates that such information is not relevant or appropriate for the particular strain

Table 2. continued

	ATCC®	Raffinose	Rhamnose	Xylose	Mannitol	Dulcitol	Salicin	Adonitol
Acinetobacter calcoaceticus	19606			+	−			
Branhamella catarrhalis	8176							
Citrobacter freundii	8090	−	+		+	−	−	−
Edwardsiella tarda	15947				−			
Enterobacter aerogenes	13048	+	+		+	−	+	+
Enterobacter cloacae	23355	+	+		+	−	(+)	−
Escherichia coli	25922	−	+		+	+	−	−
Escherichia coli	35218							
Haemophilus influenzae	35056							
Klebsiella pneumoniae	13883	+	+		+	+	+	+
Neisseria gonorrhoeae	19424							
Neisseria lactamicus	23970							
Neisseria meningitidis	13090							
Neisseria sicca	9913							
Proteus vulgaris	13315	−	−		−	−	−	−
Pseudomonas aeruginosa	27853							
Salmonella typhimurium	14028	−	+		+	−	−	−
Serratia marcescens	8100	−	−		+	−	+	(+)
Shigella flexneri	12022	−	−	−	+	−	−	−
Shigella sonnei	25931	−	+	−	+	−	−	−
Staphylococcus aureus	25923				+			
Staphylococcus aureus	29213				+			
Staphylococcus epidermidis	12228				−			
Streptococcus agalactiae	13813						+	
Streptococcus faecalis	19433	−			+			
Streptococcus faecalis	29212	−			+			
Streptococcus faecalis	33186	−			+			
Streptococcus pneumoniae	6303						−	
Streptococcus pyogenes	19615							
Yersinia enterocolitica	27729				+		−	

+ Reaction is positive or present; or, organism is susceptible.
− Reaction is negative or absent; or, organism is resistant.
( ) Delayed reaction.
Omission of data describing reaction indicates that such information is not relevant or appropriate for the particular strain

Table 2. continued

	ATCC®	Inositol	Sorbitol	Inulin Fermentation	Polysaccharide from Sucrose	Catalase	Coagulase	DNase
Acinetobacter calcoaceticus	19606					+		
Branhamella catarrhalis	8176					+		
Citrobacter freundii	8090	−	+					
Edwardsiella tarda	15947					+		
Enterobacter aerogenes	13048	+	+					
Enterobacter cloacae	23355	−	+					
Escherichia coli	25922	−	+					
Escherichia coli	35218					+		
Haemophilus influenzae	35056					+		
Klebsiella pneumoniae	13883	+	+					
Neisseria gonorrhoeae	19424				−	+		
Neisseria lactamicus	23970				−	+		
Neisseria meningitidis	13090				−	+		
Neisseria sicca	9913				+	+		
Proteus vulgaris	13315	−	−					
Pseudomonas aeruginosa	27853					+		
Salmonella typhimurium	14028	+	+					
Serratia marcescens	8100	+	+					
Shigella flexneri	12022	−	−					
Shigella sonnei	25931	−	−					
Staphylococcus aureus	25923					+	+	+
Staphylococcus aureus	29213						+	+
Staphylococcus epidermidis	12228					+	−	−
Streptococcus agalactiae	13813			−		−		
Streptococcus faecalis	19433		+			−	−	
Streptococcus faecalis	29212		+			−	−	
Streptococcus faecalis	33186		+			−	−	
Streptococcus pneumoniae	6303			+		−		
Streptococcus pyogenes	19615		−			−	−	+
Yersinia enterocolitica	27729					+		

+ Reaction is positive or present; or, organism is susceptible.
− Reaction is negative or absent; or, organism is resistant.
( ) Delayed reaction.
Omission of data describing reaction indicates that such information is not relevant or appropriate for the particular strain

Table 2. continued

	ATCC®	Gelatinase (22°C)	β-Lactamase	Oxidase	Urease	Arginine	Lysine	Ornithine
Acinetobacter calcoaceticus	19606	–		–	–			–
Branhamella catarrhalis	8176			+				
Citrobacter freundii	8090	–			(+)	–	–	–
Edwardsiella tarda	15947	–			–		+	+
Enterobacter aerogenes	13048	–			–	–	+	+
Enterobacter cloacae	23355	(+)			–	+	–	+
Escherichia coli	25922	–			–	+	+	+
Escherichia coli	35218		+		–		+	
Haemophilus influenzae	35056		+	–				
Klebsiella pneumoniae	13883	–			+	–	+	–
Neisseria gonorrhoeae	19424		–	+				
Neisseria lactamicus	23970			+				
Neisseria meningitidis	13090		–	+				
Neisseria sicca	9913			+				
Proteus vulgaris	13315	(+)			+	–	–	–
Pseudomonas aeruginosa	27853	(+)		+	(+)			
Salmonella typhimurium	14028	–			–	+	+	+
Serratia marcescens	8100	(+)			–	–	+	+
Shigella flexneri	12022	–			–	–	–	–
Shigella sonnei	25931	–			–	–	–	+
Staphylococcus aureus	25923							
Staphylococcus aureus	29213							
Staphylococcus epidermidis	12228				–			
Streptococcus agalactiae	13813			–		+		
Streptococcus faecalis	19433	–				+		
Streptococcus faecalis	29212					+		
Streptococcus faecalis	33186					+		
Streptococcus pneumoniae	6303			–	–			
Streptococcus pyogenes	19615							
Yersinia enterocolitica	27729				+			+

+ Reaction is positive or present; or, organism is susceptible.
– Reaction is negative or absent; or, organism is resistant.
( ) Delayed reaction.
Omission of data describing reaction indicates that such information is not relevant or appropriate for the particular strain

Table 2. continued

	ATCC®	Phenyl-alanine	Indole	Methyl Red	Voges-Proskauer	ONPG	CAMP Test	Citrate
Acinetobacter calcoaceticus	19606	−						+
Branhamella catarrhalis	8176							
Citrobacter freundii	8090	−	−	+	−			+
Edwardsiella tarda	15947		+	+	−			−
Enterobacter aerogenes	13048	−	−	−	+			+
Enterobacter cloacae	23355	−	−	−	+			+
Escherichia coli	25922	−	+	+	−			−
Escherichia coli	35218	−	+	+	−			−
Haemophilus influenzae	35056							
Klebsiella pneumoniae	13883	−	−	+	−			+
Neisseria gonorrhoeae	19424							
Neisseria lactamicus	23970					+		
Neisseria meningitidis	13090					−		
Neisseria sicca	9913							
Proteus vulgaris	13315	+	+	+	−			−
Pseudomonas aeruginosa	27853	−	−	−				+
Salmonella typhimurium	14028	−	−	+	−			+
Serratia marcescens	8100	−	−	−	+			+
Shigella flexneri	12022	−	−	+	−			−
Shigella sonnei	25931	−	−	+	−			−
Staphylococcus aureus	25923							
Staphylococcus aureus	29213							
Staphylococcus epidermidis	12228							
Streptococcus agalactiae	13813						+	
Streptococcus faecalis	19433							
Streptococcus faecalis	29212							
Streptococcus faecalis	33186							
Streptococcus pneumoniae	6303						−	
Streptococcus pyogenes	19615							
Yersinia enterocolitica	27729							

+ Reaction is positive or present; or, organism is susceptible.
− Reaction is negative or absent; or, organism is resistant.
( ) Delayed reaction.
Omission of data describing reaction indicates that such information is not relevant or appropriate for the particular strain

Table 2. continued

	ATCC®	Nitrate	Sodium Hippurate	Bile Solubility	H₂S	KCN	Bacitracin Susceptibility	Optochin Susceptibility
Acinetobacter calcoaceticus	19606	–			–^{a}			
Branhamella catarrhalis	8176	+			–^{b}			
Citrobacter freundii	8090				+^{a}	+		
Edwardsiella tarda	15947				+^{a}			
Enterobacter aerogenes	13048				–^{a}	+		
Enterobacter cloacae	23355				–^{a}	+		
Escherichia coli	25922				–^{a}	–		
Escherichia coli	35218				–^{a}			
Haemophilus influenzae	35056	+						
Klebsiella pneumoniae	13883				–^{a}	–		
Neisseria gonorrhoeae	19424				–^{b}			
Neisseria lactamicus	23970				–^{b}			
Neisseria meningitidis	13090				–^{b}			
Neisseria sicca	9913				+^{b}			
Proteus vulgaris	13315				+^{a}	+		
Pseudomonas aeruginosa	27853	+						
Salmonella typhimurium	14028				+^{a}	–		
Serratia marcescens	8100				–^{a}	+		
Shigella flexneri	12022				–^{a}	–		
Shigella sonnei	25931				–^{a}	–		
Staphylococcus aureus	25923		+					
Staphylococcus aureus	29213		+					
Staphylococcus epidermidis	12228		+					
Streptococcus agalactiae	13813	+	–				–	–
Streptococcus faecalis	19433	–	–				–	–
Streptococcus faecalis	29212	–	–				–	–
Streptococcus faecalis	33186	–	–				–	–
Streptococcus pneumoniae	6303	–	+				–	+
Streptococcus pyogenes	19615	–					+	
Yersinia enterocolitica	27729							

+ Reaction is positive or present; or, organism is susceptible. a Using TSI.
– Reaction is negative or absent; or, organism is resistant. b Using lead acetate strip.
( ) Delayed reaction.
Omission of data describing reaction indicates that such information is not relevant or appropriate for the particular strain

Table 3. Antimicrobial Disc Diffusion Susceptibility Test Control Zones

Antimicrobial Agent	Disc Content	S. aureus ATCC® 25923	E. coli ATCC® 25922	P. aeruginosa ATCC® 27853	E. coli ATCC® 35218	S. faecalis ATCC® 33186
Amikacin	30 mcg	20 – 26	19 – 26	18 – 26		
Ampicillin	10 mcg	27 – 35	16 – 22		no zones	
Augmentin	20/10 mcg				18 – 22	
Carbenicillin	100 mcg		23 – 29	18 – 24		
Cefamandole	30 mcg	26 – 34	24 – 30			
Cefoperazone	75 mcg	24 – 33	28 – 34	23 – 29		
Cefotaxime	30 mcg	25 – 31	29 – 35	18 – 22		
Cefoxitin	30 mcg	23 – 29	23 – 29			
Cephalothin	30 mcg	29 – 37	17 – 21			
Chloramphenicol	30 mcg	19 – 26	21 – 27			
Clindamycin	2 mcg	24 – 30				
Colistin	10 mcg		11 – 15			
Doxycycline	30 mcg	23 – 29	18 – 24			
Erythromycin	15 mcg	22 – 30				
Gentamicin	10 mcg	19 – 27	19 – 26	16 – 21		
Kanamycin	30 mcg	19 – 26	17 – 25			
Methicillin	5 mcg	17 – 22				
Mezlocillin	75 mcg		23 – 29	19 – 25		
Minocycline	30 mcg	25 – 30	19 – 25			
Moxalactam	30 mcg	18 – 24	28 – 35	17 – 25		
Nafcillin	1 mcg	16 – 22				
Nalidixic Acid	30 mcg		22 – 28			
Neomycin	30 mcg	18 – 26	17 – 23			
Netilmicin	30 mcg	22 – 31	22 – 30	17 – 23		
Nitrofurantoin	300 mcg		21 – 26			
Oxacillin	1 mcg	18 – 24				
Penicillin G	10 units	26 – 37				
Piperacillin	100 mcg		24 – 30	25 – 33		
Polymyxin B	300 units	7 – 13	12 – 16			
Streptomycin	10 mcg	14 – 22	12 – 20			
Sulfisoxazole	250 or 300 mcg	24 – 34	18 – 26			
Tetracycline	30 mcg	19 – 28	18 – 25			
Ticarcillin	75 mcg		24 – 30		21 – 27	
Tobramycin	10 mcg	19 – 29	18 – 26	19 – 25		
Trimethoprim	5 mcg	21 – 28	21 – 28			25 – 29*
Trimethoprim/ Sulfamethoxazole	1.25/23.75 mcg	24 – 32	24 – 32			24 – 30*
Vancomycin	30 mcg	15 – 19				

*If the zone is out of range or nondistinct or has innerzonal colonies, the Mueller Hinton medium in use is unsuitable for testing with trimethoprim because of unacceptably high levels of thymidine.

Table 4.   Dilution Antimicrobial Susceptibility Tests (MIC ± 1 log$_2$ dilution)

Antimicrobic	Staphylococcus aureus ATCC® 29213	Streptococcus faecalis ATCC® 29212	Escherichia coli ATCC® 25922	Pseudomonas aeruginosa ATCC® 27853
Amoxicillin	0.5	1	8	—
Ampicillin	0.5	1	4	—
Carbenicillin	4	32	8	32
Methicillin	1 – 2	>16	—	—
Nafcillin	0.25	4 – 8	—	—
Oxacillin	0.25	8	—	—
Penicillin	0.25	2	—	—
Ticarcillin	4	32	2 – 4	16
Cefamandole	1	32	0.25 – 0.5	—
Cefoxitin	4	>128	2	—
Cephalothin	0.12 – 0.25	16	8	—
Amikacin	2	128	1 – 2	4
Gentamicin	0.5 – 1	8	0.5	2 – 4
Kanamycin	≤1	32 – 64	2 – 4	>128
Tobramycin	1	16	0.5	1
Chloramphenicol	4	8	4	>32
Clindamycin	0.06 – 0.12	8 – 16	—	—
Colistin	>4	>128	0.5 – 1	2 – 4
Erythromycin	0.12 – 0.25	1	32	—
Tetracycline	0.5	32	2	16 – 32
Vancomycin	1	2	—	—
Nalidixic Acid	128	—	2	—
Nitrofurantoin	16	8	8	—
Sulfisoxazole	64	64	16	>256
Trimethoprim/ Sulfamethoxazole (1/19)‡	<0.5/9.5	<0.5/9.5	<0.5/9.5	16/304

†These MICs obtained in several reference laboratories by agar dilution or by broth microdilution with cation supplemented broth.
‡Very medium dependent especially with enterococci.

BACTO BAIRD-PARKER AGAR BASE

INTENDED USE
Bacto Baird-Parker Agar Base when supplemented with Bacto EY Tellurite Enrichment is recommended for the detection and enumeration of coagulase positive staphylococci in foods.

HISTORY/PRINCIPLES
Baird-Parker agar originally formulated by Baird-Parker[1] has been recommended for microbial limits tests by The United States Pharmacopeial Convention[2] and the FDA.[3]

The complete medium is prepared by adding, aseptically, Bacto EY Tellurite Enrichment to the sterile Bacto Baird-Parker Agar Base. The medium permits the detection, enumeration and isolation of coagulase positive staphylococci after 24 hours incubation, from a variety of specimens such as food products, air, dust and soil.

The medium contains lithium and potassium tellurite to suppress growth of undesired organisms while allowing *Staphylococcus aureus* to grow. In addition, pyruvate and glycine are included to enhance growth of staphylococci. The tellurite and egg yolk components are responsible for the differentiation of coagulase positive staphylococci by the formation of black, shiny, convex colonies surrounded by a clear zone from the coagulase negative staphylococci. Growth of the latter is seen only occasionally and the colonies, with clear wide opaque zones, which appear after 24 hours incubation at 37°C, are easily distinguished by their irregular appearance. *Proteus* or *Bacillus* species may also grow but as brown colonies.

FORMULA

BACTO BAIRD-PARKER AGAR BASE
DEHYDRATED

Ingredients per liter

Bacto Tryptone	10 g
Bacto Beef Extract	5 g
Bacto Yeast Extract	1 g
Glycine	12 g
Sodium Pyruvate	10 g
Lithium Chloride	5 g
Bacto Agar	20 g

Final pH 7.0 ± 0.2 at 25°C.

One pound will make 7.2 liters of medium.

METHOD OF PREPARATION

1. Suspend 63 g in 950 ml distilled or deionized water. Heat to boiling to dissolve completely.
2. Sterilize in the autoclave for 15 minutes at 15 lbs pressure (121°C).
3. Allow to cool to 45 – 50°C. Meanwhile, warm Bacto EY Tellurite Enrichment to 45 – 50°C.
4. Shake the enrichment thoroughly to resuspend the precipitate.
5. Aseptically add 50 ml enrichment to prepared base. Mix thoroughly and dispense as desired. Approximately 15 ml is recommended for standard size Petri dishes to be sufficiently thin to observe clear zones.
6. Allow surfaces to dry, then spread 0.1 ml inoculum over the plates.
7. Incubate at 37°C and read after 24 – 26 hours. If negative for staphylococci, reincubate for an additional 24 hours and read again.

STORAGE

Bacto Baird-Parker Agar Base	Below 30°C
Bacto EY Tellurite Enrichment	2 – 8°C
Prepared medium	2 – 8°C

QUALITY CONTROL
Identity Specifications

Dehydrated powder:	light tan, homogeneous, free-flowing
Enrichment:	canary yellow, opaque with a resuspendable precipitate
Reaction of 6.3% solution:	pH 7.0 ± 0.2 at 25°C
Prepared base:	light to medium amber, very slightly opalescent
Prepared plates:	yellow, opaque

Typical Cultural Response on Bacto Baird-Parker Agar
After 24 – 48 Hours at 37°C

Organism	Growth	Color of Colony	Lecithinase/ Halos
Bacillus subtilis ATCC® 6633	none to poor	brown	–
Escherichia coli ATCC® 25922	none	–	–
Proteus mirabilis ATCC® 25933	good to excellent	brown	–
Staphylococcus aureus ATCC® 6538	good to excellent	black	+
Staphylococcus aureus ATCC® 25923	good to excellent	black	+
Staphylococcus epidermidis ATCC® 12228	poor to good	black	–

REFERENCES
1. J. Appl. Bact., 25:12, 1962.
2. U.S. Pharmacopeia National Formulary XX, 1980, Mark Publ. Co. Easton, Pa.
3. FDA Bacteriological Analytical Manual, 5th Ed., 1978.

PACKAGING

Bacto Baird-Parker Agar Base	1 lb (454 g)	0768-01-1
	1/4 lb (114 g)	0768-02-0
Bacto EY Tellurite Enrichment	6 × 100 ml	0779-73-1

BACTO BALAMUTH MEDIUM

INTENDED USE
Bacto Balamuth Medium is prepared according to the method given by Balamuth[1] for culturing *Entamoeba histolytica*. It consists of an aqueous extract of egg yolk and liver. The medium is supplied in screw cap test tubes, ready for inoculation.

PROCEDURE
A loopful of sterile Bacto Rice Powder is added along with the inoculum. The rice powder is sterilized by dry heat at 150°C for 1 hour. Following inoculation, tubes are incubated at 37°C for up to 72 hours before being considered negative. The culture is examined periodically microscopically to determine the growth of amoeba.

STORAGE
Bacto Balamuth Medium   2 – 8°C

REFERENCE
1. Am. J. Clin. Path., 16:380, 1946.

PACKAGING

Bacto Balamuth Medium	12 tubes	1021-34-3
Bacto Rice Powder	100 g	0146-15-8

BACTO BEEF

Bacto Beef is a desiccated powder of fresh lean beef. It is prepared especially for use in making beef infusion media. Large quantities of beef are processed at one time in order to secure a uniform and homogeneous product. The nutritive qualities of the fresh beef are retained in Bacto Beef and may be preserved in the infusions prepared from it. Media prepared from Bacto Beef are superior to beef extract media and are equal to infusion media made from market beef.

One hundred g of Bacto Beef is equivalent to 500 g of fresh lean beef. An excellent infusion, however, can be prepared by using 50 g of Bacto Beef per 1 liter of distilled or deionized water. This mixture is infused at 50°C for one hour. It is then heated to boiling for a few minutes to coagulate some of the proteins and is filtered. Peptone and the other ingredients of the medium are then added to the filtrate. After adjustment of the reaction to pH 7.5 – 7.8 and subsequent boiling for a few minutes, additional coagulation will result which is removed by filtration before the medium is sterilized.

Infusion media prepared from Bacto Beef have been recommended for use in the microbial examination of butter.[1] Herrold[2] has also used an infusion from Bacto Beef in preparing blood agar for primary cultivation of the gonococcus.

REFERENCES

1. J. Dairy Science, 16:289, 1933.
2. J. Infectious Diseases, 42:79, 1928.

PACKAGING

Bacto Beef	1 lb (454 g)	0131-01-1

BACTO BEEF EXTRACT

Bacto Beef Extract is prepared and standardized for use in microbiological culture media, where it is generally used to replace infusions of meat. It is standard in its composition and reaction and does not require adjustment of reaction or filtration. For many years, beef extract media have been recommended as standard for use in the routine bacteriological examination of water, milk and other materials where it is important to have media of uniform composition.

Bacto Beef Extract may be relied upon for biochemical studies, particularly fermentation reactions, because of its freedom from fermentable substances which would interfere with the accuracy of such determinations.

In culture media, Bacto Beef Extract is usually employed in concentrations of 0.3% as in the standard media recommended for water and for milk analysis.[1,2] Concentrations may vary slightly according to the requirements of individual formulae but do not often exceed 0.5%. In 0.3% concentration, Bacto Beef Extract forms clear solutions, rich in the nutriments required for bacterial metabolism. After autoclave sterilization, this solution has a reaction of pH 6.9.

REFERENCES

1. Standard Methods for Examination of Water and Sewage, 15th Ed., 1980.
2. Standard Methods for the Examination of Dairy Products, 9th Ed., 1948.

PACKAGING

Bacto Beef Extract	1 lb (454 g)	0126-01-8
	1/4 lb (114 g)	0126-02-7

BEEF EXTRACT, TECHNICAL

Beef Extract, Technical is used in bacteriological culture media when a standardized beef extract is not essential.

PACKAGING
Beef Extract, Technical   500 g   0884-17-2

BACTO BEEF HEART FOR INFUSION

Bacto Beef Heart for Infusion is prepared from fresh beef heart tissue and is particularly recommended for preparing heart infusion media. Bacto Beef Heart for Infusion is processed from large quantities of raw material, retaining all the nutritive and growth-stimulating properties of the fresh tissues.

One hundred grams of Bacto Beef Heart for Infusion are the equivalent of 500 g of fresh heart tissue. Generally, excellent infusions can be prepared using 50 g of Bacto Beef Heart for Infusion per 1 liter of distilled water or deionized water. For best results, infuse at 50°C for 1 hour, heat to boiling for a few minutes to coagulate some of the proteins and filter. Peptone and the other ingredients of the medium should then be added to the filtrate, the reaction adjusted to pH 7.5 – 7.8 and the medium boiled and filtered before sterilizing.

PACKAGING
Bacto Beef Heart for Infusion   1 lb (454 g)   0132-01-0

BACTO BEEF SERUM

INTENDED USE
Bacto Beef Serum is filter sterilized fresh normal beef serum. It is recommended as an enrichment for use in bacteriological culture media and in other procedures requiring beef serum.

PRINCIPLES
Bacto Beef Serum may be added aseptically to a prepared basal medium such as Bacto Tryptose Blood Agar Base in 10% concentration.

METHOD OF PREPARATION
1. Suspend 3.5 g of Bacto Tryptose Blood Agar Base in 90 ml distilled or deionized water.
2. Sterilize in the autoclave for 15 minutes at 15 lbs pressure (121°C).
3. Allow to cool to 50 – 55°C.
4. Aseptically add 1 ampule (10 ml) to prepared base.
5. Swirl to mix well then pour plates as desired.

STORAGE
Bacto Beef Serum   2 – 8°C

QUALITY CONTROL
Appearance: light amber, clear to very slightly opalescent.

Typical Cultural Response on Bacto Tryptose Blood Agar Base
w/10% Bacto Beef Serum After 18 – 48 Hours at 35°C

Organism	Growth
Staphylococcus aureus ATCC® 25923	excellent
Streptococcus pneumoniae ATCC® 6303	excellent
Streptococcus pyogenes ATCC® 19615	excellent

PACKAGING
Bacto Beef Serum   12 × 10 ml   0260-61-1

BACTO BiGGY AGAR

INTENDED USE
Bacto BiGGY (Bismuth sulfite-Glucose-Glycine-Yeast) Agar, also referred to as Nickerson Medium, is a selective medium recommended for the detection, isolation and differentiation of *Candida* species.

HISTORY/PRINCIPLES
Bacto BiGGY Agar is a slight modification of the Nickerson[1] formulation, employing a selective and differential mechanism similar to that of Bacto Bismuth Sulfite Agar. Bismuth sulfite in the formula inhibits bacterial growth while most of the *Candida* grow rapidly. *Candida* reduce the bismuth sulfite and become brown to black in color. Brown to black colonies with a pasty appearance are usually yeasts.

Species differentiation according to Nickerson after 48 hours would give the following results:

C. albicans: intensely brown-black colonies with slight mycelial fringe, medium sized, no diffusion.

C. tropicalis: discrete dark brown colonies with black centers and sheen, medium sized, diffuse blackening of the surrounding medium after 72 hours of incubation.

C. pseudotropicalis: large, dark reddish-brown colonies, flat with slight mycelial fringe.

C. krusei: large flat wrinkled colonies with silvery black top, brown edge and yellow halo.

C. parakrusei: medium sized flat wrinkled colonies with reddish-brown color and yellow mycelial fringe.

FORMULA

BACTO BiGGY AGAR
DEHYDRATED
Ingredients per liter

Bacto Yeast Extract . . . . 1 g
Glycine . . . . 10 g
Bacto Dextrose . . . . 10 g
Bismuth Sulfite Indicator . . . . 8 g
Bacto Agar . . . . 20 g

Final pH 6.8 ± 0.2 at 25°C.

One pound will make 9.26 liters of medium.

METHOD OF PREPARATION
1. Suspend 49 g in 1 liter distilled or deionized water. Heat to boiling to dissolve completely. Do not boil for longer than a few minutes as over heating will destroy the selective properties. DO NOT AUTOCLAVE.
2. The medium contains a flocculent precipitate that should be evenly dispersed by swirling medium in flasks just prior to dispensing in tubes or plates.

STORAGE
Bacto BiGGY Agar   Below 30°C
Prepared medium   2 – 8°C

QUALITY CONTROL
Identity Specifications

Dehydrated powder: light beige, homogeneous, free-flowing
Reaction of 4.9% solution: pH 6.8 ± 0.2 at 25°C
Prepared medium: very light amber, opalescent with flocculant precipitate

Typical Cultural Response on Bacto BiGGY Agar
After 18 – 48 Hours at 30°C

Organism	Growth	Colony Description
Candida albicans ATCC® 10231	good to excellent	brown to black, no diffusion
Candida krusei	good to excellent	dark reddish brown, yellow halo
Candida pseudotropicalis	good to excellent	reddish brown, flat
Candida tropicalis ATCC® 750	good to excellent	brown to black, sheen, diffusion
Escherichia coli ATCC® 25922	inhibited	—
Staphylococcus aureus ATCC® 25923	inhibited	—

REFERENCE
1. J. Inf. Dis., 93:43, 1953.

PACKAGING
Bacto BiGGY Agar

1 lb (454 g)	0635-01-2
1/4 lb (114 g)	0635-02-1
12 tubes	0635-34-3
144 tubes	0635-37-0

BACTO BILE ESCULIN AGAR
BACTO BILE ESCULIN AZIDE AGAR
BACTO BILE ESCULIN AGAR BASE

INTENDED USE
Bacto Bile Esculin Agar Base and Bacto Bile Esculin Agar are differential media whereas Bacto Bile Esculin Azide Agar is a selective medium. All are recommended for the isolation and presumptive identification of group D streptococci.

HISTORY
Bacto Bile Esculin Agar and Bacto Bile Esculin Agar Base are prepared according to the formulation described by Swan[1] and further evaluated by Facklam and Moody.[2]

Rochaix[3] first drew attention to the value of esculin hydrolysis in the identification of enterococci. Meyer and Schonfeld[4] demonstrated that esculin hydrolysis in a bile medium gave the best single differential test for the enterococci. They found that 61 to 62 strains of enterococci hydrolyzed esculin in a bile containing medium. Williams and Hirsch[5] demonstrated that 83.6% of 140 strains of bile tolerant streptococci hydrolyzed esculin.

Swan[1] compared esculin hydrolysis by streptococci in his bile medium with results obtained by the Lancefield serological method of grouping. Of 121 presumptive enterococci, 117 were of Lancefield's group D. All the group D strains were heat resistant, bile tolerant and all hydrolized esculin. Of 21 strains of presumptive Viridans group streptococci, none reacted with *Streptococcus* group D antiserum. Of the 21 strains, 19 were neither heat resistant nor bile tolerant. All *Streptococcus bovis* and *Streptococcus durans* strains were bile tolerant and hydrolyzed esculin. None of 8 strains of micrococci were heat resistant or hydrolyzed esculin.

Facklam and Moody[2] tested 700 strains of streptococci representing all known serological groups on the bile esculin medium described by Swan. Positive reactions were observed only with *Streptococcus* group D and group Q. One hundred percent (100%) of the Group D streptococci were bile esculin positive. The bile esculin test of Swan appeared to be the most reliable presumptive test for identification of the group D streptococci isolated from clinical sources.

Bacto Bile Esculin Azide Agar is a modification of the medium reported by Isenberg.[6] The formula modifies Bacto Bile Esculin Agar making the medium more selective and still providing for the rapid growth and efficient recovery of group D streptococci. Esculin hydrolysis and bile tolerance, as shown by Swan[1] and by Facklam and Moody,[2] permit the isolation and identification of group D streptococci in 24 hours.

PRINCIPLES
Group D streptococci grow readily on the bile esculin agar and hydrolyze the esculin which imparts a dark brown color to the medium. This reaction denotes their bile tolerance, ability to hydrolyze esculin and constitutes a positive reaction. A few streptococci tolerate the bile, but do not hydrolyze esculin and thus impart no color to the medium.

Gram-negative bacteria are inhibited by sodium azide and gram-positive bacteria other than group D streptococci are inhibited by bile salts. Group D streptococci grow readily on the medium and hydrolyze esculin resulting in a dark brown color around the colonies after 18 – 24 hours incubation at 35 – 37°C.

FORMULAE

BACTO BILE ESCULIN AGAR
DEHYDRATED

Ingredients per liter

Bacto Beef Extract	3 g
Bacto Peptone	5 g
Bacto Oxgall	40 g
Bacto Esculin	1 g
Ferric Citrate	0.5 g
Bacto Agar	15 g

Final pH 6.6 ± 0.2 at 25°C.

One pound will make 7.2 liters of final medium.
Rehydrate with 64 g/liter.

BACTO BILE ESCULIN AZIDE AGAR
DEHYDRATED

Ingredients per liter

Bacto Yeast Extract	5 g	Ferric Ammonium Citrate	0.5 g
Bacto Proteose Peptone No. 3	3 g	Sodium Chloride	5 g
Bacto Tryptone	17 g	Sodium Azide	0.15 g
Bacto Oxgall	10 g	Bacto Agar	15 g
Bacto Esculin	1 g		

Final pH 7.1 ± 0.2 at 25°C.

One pound will make 7.9 liters of final medium.
Rehydrate with 57 g/liter.

BACTO BILE ESCULIN AGAR BASE

Bacto Bile Esculin Agar Base has the same formulation as Bacto Bile Esculin Agar except the Esculin has been omitted. Rehydrate with 63 g/liter.

METHOD OF PREPARATION

1. Suspend the appropriate amount in 1 liter distilled or deionized water and heat to boiling to dissolve completely. Add desired amount of esculin to Bacto Bile Esculin Agar Base.
2. Dispense into tubes or flasks, as desired.
3. Sterilize in the autoclave for 15 minutes at 15 lbs pressure (121°C). Overheating may cause darkening of the media.
4. Cool to 50 – 55°C and dispense into sterile Petri dishes. If tubed, allow the medium to solidify in the slanted position.

The addition of 50 ml of filter sterilized horse serum to each liter of medium is optional since in some laboratories the growth of group D streptococci and hydrolysis of esculin were not improved by the addition of the serum. Swirl the flask to obtain a uniform solution and dispense into sterile plates or tubes as desired.

STORAGE

Bacto Bile Esculin Media	Below 30°C
prepared media	2 – 8°C

QUALITY CONTROL
Identity Specifications

Dehydrated powder: tan, homogeneous, free-flowing
Reaction of appropriate solution: see above for appropriate pH
Prepared medium: medium to dark amber, slightly opalescent; media with esculin have bluish tinge

Typical Cultural Response on Bacto Bile Esculin Media
After 18 – 24 Hours at 35°C

Organism	Growth	*Esculin hydrolysis
Proteus mirabilis ATCC® 25933	good	−
Streptococcus Group A	none to poor	−
Streptococcus bovis	good	+
Streptococcus faecalis	good	+
Streptococcus faecium	good	+

*+ = blackening of media
− = no change

REFERENCES
1. J. Clin. Path. 7:160, 1954.
2. Bact. Proceedings M33, 1969.
3. Cr. Soc. Biol. Paris 90:771, 1924.
4. Centralb. Bact. Abt. I:99:402, 1926.
5. J. Hyg. Lond. 48:504, 1950.
6. Clin. Lab. Forum/July, 1970.

PACKAGING

Bacto Bile Esculin Agar	1 lb (454 g)	0879-01-7
	1/4 lb (114 g)	0879-02-6
	5 lb (2.27 kg)	0879-05-3
	12 tubes	0879-34-8
Bacto Bile Esculin Agar Base	1 lb (454 g)	0878-01-8
Bacto Bile Esculin Azide Agar	1/4 lb (114 g)	0525-02-4
	1 lb (454 g)	0525-01-5
	5 lb (2.27 kg)	0525-05-1
	12 tubes	0525-34-6

BACTO BILE SALTS

Bacto Bile Salts was originally developed for use in the lactose bile salt agar of MacConkey,[1] one of the best known of the plating media for the isolation of organisms of the colon-typhoid group. As originally described by MacConkey, the medium was a peptone lactose or glucose agar containing 0.5% sodium glycocholate. Bacto Bile Salts fulfilled the requirements of the original sodium glycocholate. However, in recent studies, we have developed Bacto Bile Salts No. 3 which has been found to be more suitable than the Bacto Bile Salts previously employed in the MacConkey agar.

Bacto Bile Salts has proven particularly valuable in the preparation of Bacto Tetrathionate Broth.

Bacto Bile Salts is readily soluble in distilled water and is neutral in reaction. It may, therefore, be used in preparing media without adjustment of reaction or filtration.

REFERENCE
1. J. Hyg., 5:333, 1905.

PACKAGING
Bacto Bile Salts

1 lb (454 g)	0129-01-5
1/4 lb (114 g)	0129-02-4
10 kg	0129-08-8

BACTO BILE SALTS NO. 3

Bacto Bile Salts No. 3 is a modification of Bacto Bile Salts and is prepared especially for use in Bacto MacConkey Agar, Bacto SS Agar, and Bacto Violet Red Bile Agar.

In early investigations on these media, it was found that Bacto Bile Salts gave a somewhat heavy precipitate around lactose fermenting colonies, which, under certain conditions, made it difficult to detect colonies of the pathogenic lactose nonfermenters. Bacto Bile Salts No. 3 obviated this fault, giving a clearer medium and also reducing the precipitation around the coliform colonies, thus facilitating the detection of lactose nonfermenting colonies.

Bacto Bile Salts No. 3 inhibits the growth of gram-positive organisms and spore forming bacilli without affecting the development of gram-negative enteric bacilli. It is readily soluble in distilled or deionized water and is neutral in reaction. It may, therefore, be used in preparing media without filtration or adjustment of reaction.

PACKAGING
Bacto Bile Salts No. 3

1 lb (454 g)	0130-01-2
1/4 lb (114 g)	0130-02-1

BACTO BISMUTH SULFITE AGAR

INTENDED USE
Bacto Bismuth Sulfite Agar is a highly selective medium for the isolation of *Salmonella typhi* and other *Salmonella* from feces, urine, sewage and other materials harboring this organism. It is particularly recommended after preliminary enrichment of the specimen in Bacto Tetrathionate Broth or other enrichment broths.

HISTORY
Bacto Bismuth Sulfite Agar is a modification of the Wilson and Blair[1,2,3] formula. Wilson[4,5] and Wilson and Blair[6,7,8] clearly demonstrated the superiority of this type of medium over others in the isolation of *S. typhi*.

Cope and Kasper[9] increased their positive findings of typhoid from 1.2 to 16.8% among food handlers and from 8.4 to 17.5% among contacts by the use of Bacto Bismuth Sulfite Agar. Employing this medium in the routine laboratory examination of fecal and urine specimens these same authors[10] obtained 40% more positive isolations of *S. typhi* than were obtained on endo medium. Gunther and Tuft,[11] employing various media in a comparative way for the isolation of typhoid from stool and urine specimens, found Bacto Bismuth Sulfite Agar most efficient. Upon this medium they obtained 38.4% more positives than on endo, 33% more positives than on eosin methylene blue agar, and 80% more positives on bismuth sulfite agar than on the desoxycholate media. These workers found Bacto Bismuth Sulfite Agar to be superior to Wilson's original medium, being easier to prepare, relatively more stable and more sensitive. Green and Beard[12] using Bacto Bismuth Sulfite Agar in their studies on the "Survival of *E. typhi* in Sewage Treatment Plant Processes," claimed that this medium so successfully inhibited sewage organisms that their interference was negligible. Beard[13] stated that such a highly selective medium as bismuth sulfite agar made possible the study of the survival of typhoid in nature. The value of bismuth sulfite agar as a plating medium subsequent to enrichment has been demonstrated by Hajna and Perry.[14]

Since these earlier references to the use of bismuth sulfite agar, this medium has been generally accepted as routine for the detection of incitants of enteric disease, especially by most *Salmonella*. The value of the medium is demonstrated by the many references to the use of bismuth sulfite agar in scientific publications, laboratory manuals and texts. The use of bismuth sulfite agar is specified in *Standard Methods for the Examination of Dairy Products*[15] for the isolation of pathogenic bacteria from cheese, and in *Diagnostic Procedures and Reagents*[16] of the American Public Health Association for the examination of specimens for evidence of infection with *Salmonella*.

The *Manual of Clinical Microbiology* of the American Society for Microbiology recommends the use of bismuth sulfite agar in any case where salmonellae are suspected as it is "the most efficient medium yet devised for the isolation of this microorganism".[17]

The use of bismuth sulfite agar is also recommended in *Laboratory Procedures in Clinical Microbiology*,[18] *Standard Methods for the Examination of Water and Wastewater*,[19] the FDA *Bacteriological Analytical Manual*, 5th Edition,[20] and *Compendium of Methods for the Microbiological Examination of Foods*.[21]

PRINCIPLES

Bacto Bismuth Sulfite Agar closely approaches the ideal medium for the isolation of *S. typhi* and other *Salmonella* from feces, urine, sewage and other infectious materials. The typhoid organism grows luxuriantly on the medium forming characteristic black colonies, while the gram-positive bacteria and members of the coliform group are inhibited. This unique inhibitory action of Bacto Bismuth Sulfite Agar toward gram-positive and coliform organisms permits the use of a much larger inoculum than has been possible with other media employed for similar purposes in the past. The use of larger inocula greatly increases the possibility of recovering the pathogens, especially when they are present in relatively small numbers, such as may be encountered in the early course of the disease or in the checking of carriers and releases.

In Bacto Bismuth Sulfite Agar, bismuth sulfite and brilliant green are complementary in inhibiting gram-positive bacteria and members of the coliform group, while allowing *Salmonella* to grow luxuriantly. H_2S production occurs due to the presence of sulfur compounds in the formula. When H_2S is present, the iron in the formula is precipitated, giving positive cultures the characteristic brown to black color with metallic sheen.

FORMULA

BACTO BISMUTH SULFITE AGAR
DEHYDRATED
Ingredients per liter

Bacto Beef Extract	5 g	Ferrous Sulfate	0.3 g
Bacto Peptone	10 g	Bismuth Sulfite Indicator	8 g
Bacto Dextrose	5 g	Bacto Agar	20 g
Disodium Phosphate	4 g	Bacto Brilliant Green	0.025 g

Final pH 7.7 ± 0.2 at 25°C.

One pound will make 8.7 liters of medium.

METHOD OF PREPARATION

1. Suspend 52 g Bacto Bismuth Sulfite Agar in 1 liter distilled or deionized water and heat to boiling no longer than 1 – 2 minutes. **Avoid overheating** and **do not auto-clave.**
2. Cool to 50 – 55°C.
3. Gently swirl flask to evenly disperse the heavy precipitate just prior to pouring and dispense into sterile Petri dishes.
   NOTE: Best results are obtained when the medium is dissolved and used immediately. If it is necessary to prepare the medium several days before using, it should be poured into plates and stored in a cold moist atmosphere to prevent drying. The melted medium should not be allowed to solidify in flasks and be remelted.

METHOD OF INOCULATION
Streak or Smear Plate
Streak or smear the surface of a plate with a heavy inoculum of the fecal material in such a way that on some portion of the plate the inoculum will be light, permitting the development of discrete colonies.

Poured Plate

1. Transfer about 2 or more g of the fecal material to a test tube, add 12 – 15 ml water, and mix well, being careful to break up all the larger particles of the material. Specimens preserved in glycerol must be diluted with water to reduce the glycerol content since *S. typhi* in poured plates is inhibited in the presence of 2% glycerol.
2. Insert a loosely packed cotton plug, about 1 inch long, into the tube and slowly force it down through the fecal mixture by means of a glass rod or pipette so that all the gross particles are carried to the bottom of the tube on the cotton plug and an opaque fluid rises through the cotton. A second cotton filtration may be necessary, since it is essential that the supernatant fluid be free from gross particles. Such solid particles in the medium may support growth of the extraneous organisms giving pseudo-blackening which may be mistaken for typhoid colonies.
   Some workers may prefer to allow the gross solid particles of fecal suspension to settle by gravity instead of removing them by filtration with cotton. In such cases it is not advisable to allow the suspension to stand longer than 30 minutes in order to obtain a supernatant fluid free from gross particles. Other methods of preparing fecal suspensions that will give a liquid free from gross solid suspended material without removing typhoid may also be employed.
3. Transfer about 5 ml of the prepared fecal suspension to one Petri dish and 1 drop to a second dish. Add 20 ml of bismuth sulfite agar, cooled to 45°C, to each dish and mix thoroughly. It is necessary to use at least 20 ml of the medium to each 5 ml of inoculum, for dilution of the medium beyond this point will allow the development of extraneous fecal forms.

4. Incubate at 37°C and observe after 24 hours for typical colonies as described below. Frequently, typical colonies develop within 24 hours incubation. However, in all cases the plates should be incubated for at least 48 hours to allow the development of most *Salmonella* strains, before considering the specimen negative. Specimens containing only a small number of *Salmonella* should show isolated colonies from the 5 ml inoculum, while those specimens containing increasingly large numbers of *Salmonella* should show isolated colonies from the 1 drop inoculum in the poured plate or on the smear plate.

In the examination of samples of urine, blood, sewage or other material, either the poured plate or smear method with bismuth sulfite agar may be used. It is suggested that in examining blood specimens, the specimen first be inoculated into a tube of broth and after preliminary incubation of 8 – 12 hours, smeared onto the bismuth sulfite agar plate. For additional details about specimen preparation and inoculation, consult relevant reference texts.[15-20]

DESCRIPTION OF COLONIES
Streak or Smear Plate
The typical discrete surface *S. typhi* colony is black and is surrounded by a black or brownish-black zone which may be several times the size of the colony. By reflected light, preferably daylight, this zone exhibits a distinctly characteristic metallic sheen. Plates heavily seeded with *S. typhi* may not show this reaction except possibly near the margin of the mass inoculation. In these congested areas, this organism frequently appears as small light green colonies. This fact emphasizes the importance of inoculating plates in such a manner as to have some sparsely populated areas with discrete *S. typhi* colonies.

Poured Plate
Well isolated subsurface *S. typhi* colonies are circular, jet black and well defined. The size of the black colony may vary from 1 to 4 mm in diameter depending upon the particular strain, length of incubation and position of the colony in the agar. Only those colonies growing very close to the surface or on the surface will show a decided black metallic sheen. Plates containing *S. typhi* too numerous to permit the development of individual colonies give a black plate or a plate dotted with black areas. Plates with about 300 to 1000 typhoid colonies will exhibit this appearance. When this organism develops in a plate in still larger numbers, typical blackening does not occur and the appearance is that of a negative plate.

Ordinarily *S. typhi* will develop well isolated colonies showing typical round jet black colonies with or without sheen, from either the 5 ml or 1 drop inoculation of cotton-filtered fecal suspension using the poured plate method. However, the typhoid organisms developing from the specimens containing large numbers of *S. typhi* may be so numerous that the blackening cannot occur typically and the plate may appear dotted black or greenish gray. From such heavily seeded specimens the direct smear on bismuth sulfite agar from feces should demonstrate *S. typhi*, while the poured plate should give positive results from specimens containing lesser numbers of this organism.

DESCRIPTION OF COLONIES OTHER THAN *S. TYPHI*
S. schottmuelleri (Paratyphoid B) and *S. enteritidis* grow luxuriantly upon Bacto Bismuth Sulfite Agar forming black surface and subsurface colonies slightly more moist, but otherwise similar to those produced by *S. typhi*.

S. paratyphi (Paratyphoid A), S. typhimurium, S. choleraesuis and *Proteus morganii* develop upon Bacto Bismuth Sulfite Agar, yielding flat or only slightly raised green colonies.

Generally, the members of the dysentery group other than *S. flexneri* and *S. sonnei* are inhibited. The Flexner and Sonne strains that do develop upon this medium produce brownish raised colonies with depressed centers and exhibit a crater-like appearance.

Coli are usually completely inhibited. Occasionally a strain will be encountered that will develop small black, brown or greenish glistening surface colonies. This color is confined entirely to the colony itself and shows no metallic sheen. Likewise a few strains of aerogenes may develop on this medium forming raised, mucoid colonies. These may exhibit a silvery sheen, appreciably lighter in color than that produced by typhoid. Subsurface colonies of the coliform group, when they develop, are green or brown in color, generally lenticular in shape, and not at all to be confused with the typical round black typhoid subsurface colony. There are some members of the coliform group capable of producing hydrogen sulfide that may develop on the medium, giving colonies similar in appearance to *S. typhi.* These may be readily differentiated in that they produce gas from lactose in differential media — Bacto Russell Double Sugar Agar, Bacto Kligler Iron Agar or Bacto Triple Sugar Iron Agar, for example. The hydrolysis of urea as demonstrated on Bacto Urea Broth or Bacto Urea Agar may be used to identify *Proteus* or paracolon organisms.

The isolation and purification of *S. typhi* for agglutination or fermentation studies may be accomplished readily by picking characteristic black colonies from smeared or poured plates of bismuth sulfite agar, and subculturing them upon Bacto MacConkey Agar. The purified colonies thus obtained may then be picked to differential tube media such as Bacto Russell Double Sugar Agar, Bacto Kligler Iron Agar, Bacto Triple Sugar Iron Agar or other satisfactory differential media for partial identification. Agglutination tests may be made from the fresh growth on the differential tube media or from the growth on nutrient agar slants inoculated from the differential media. The growth on the differential tube media may also be used for inoculating carbohydrate media for fermentation studies. It is a common practice among many bacteriologists to pick colonies typical of *S. typhi* directly from bismuth sulfite agar onto the differential tube media. This may be permissible if the colonies are discrete and well isolated, but it must be remembered that although coliform bacteria are inhibited they are not destroyed by the medium.

It is recommended that Bacto MacConkey Agar, Bacto Hektoen Enteric Agar or Bacto XLD Agar (nonselective media) and Bacto SS Agar (a selective medium supporting luxuriant growth of all *Shigella* and *Salmonella* strains) be used in conjunction with Bacto Bismuth Sulfite Agar for the routine examination of stool and urine specimens. Bacto Tetrathionate Broth Base, a fluid enrichment medium for *Salmonella*, is also recommended for use in conjunction with Bacto Bismuth Sulfite Agar. Results in our laboratory show that the number of positive isolations obtained from tetrathionate broth enrichment is decidedly greater when the enriched specimen is plated on bismuth sulfite agar rather than on MacConkey agar or SS agar.

A procedure designed to show the largest number of pathogens from a specimen would be:
1. Streak or smear a large inoculum on one plate of SS agar and one plate of bismuth sulfite agar.
2. Streak or smear a light inoculum on one plate of any of the above listed nonselective media.

3. Prepare bismuth sulfite agar poured plates with a 5 ml and a one drop inoculum.
4. Enrich for 12 – 18 hours in tetrathionate broth, follow by streaking on one plate of bismuth sulfite agar and one plate of SS agar.

STORAGE

Bacto Bismuth Sulfite Agar	Below 30°C
Prepared plates[22]	2 – 8°C for up to 4 days

QUALITY CONTROL
Identity Specifications

Dehydrated powder:	light green, homogeneous, free-flowing
Reaction of 5.2% solution:	pH 7.7 ± 0.2 at 25°C
Prepared medium:	light green, opaque with a flocculent precipitate that must be dispersed on swirling contents of flask

Typical Cultural Response on Bacto Bismuth Sulfite Agar
After 40 – 48 Hours at 35°C

Organism	Growth	Color of Colony
Enterobacter aerogenes ATCC® 13048	none to poor	brown to green*
Escherichia coli ATCC® 25922	none to poor	brown to green*
Salmonella enteritidis ATCC® 13076	excellent	black w/metallic sheen
Salmonella typhi ATCC® 19430	excellent	black w/metallic sheen
Shigella flexneri ATCC® 12022	none to poor	brown*
Streptococcus faecalis ATCC® 29212	none	—

*depends on the density of inoculum

REFERENCES

1. J. Path. Bact., 29:310, 1926.
2. J. Hyg., 26:374, 1929.
3. J. Hyg., 31:139, 1931.
4. J. Hyg., 21:392, 1923.
5. Brit. Med. J., 1:1061, 1928.
6. op. cit., J. Path. Bact., 1926.
7. op. cit., J. Hyg., 1929.
8. op. cit., J. Hyg., 1931.
9. J. Bact., 34:565, 1937.
10. Am. J. Publ. Hlth., 28:1065, 1938.
11. J. Lab. Clin. Med., 24:461, 1939.
12. Am. J. Publ. Hlth., 28:762, 1938.
13. J. Am. Water Works Assoc., 30:124, 1938.
14. J. Lab. Clin. Med., 23:1185, 1938.
15. Standard Methods for the Examination of Dairy Products, 10th Ed., APHA, 1953.
16. Diagnostic Procedures and Reagents, 3rd Ed., APHA, 1950.
17. Lennette, E. H., Balows, A., Hausler, W. J. Jr., and Truant, J. P., Manual of Clinical Microbiology, 3rd Ed., American Society for Microbiology; Washington, D.C., pp. 26 and 202, 1980.
18. Washington, J. A., Laboratory Procedures in Clinical Microbiology, Springer-Verlag: New York, pp. 112 and 114, 1981.
19. Greenberg, A. E., Connors, J. J., and Jenkins, D., Standard Methods for the Examination of Water and Wastewater, 15th Ed., APHA, Washington, D.C., pp. 832 – 833, 1980.
20. FDA, Bacteriological Analytical Manual, US Food and Drug Administration: Washington, D.C., pp. VI - 15 – VI - 17, (Aug.), 1980.
21. Speck, Marvin L., Compendium of Methods for the Microbiological Examination of Foods, APHA, Washington, D.C., pp. 305 – 306, 316 – 318, 1980.
22. D'Aoust, J. Y., 1977, Effect of storage conditions on the performance of bismuth sulfite agar. J. Clin. Micro., 5:122 – 124.

PACKAGING

Bacto Bismuth Sulfite Agar	1 lb (454 g)	0073-01-1
	1/4 lb (114 g)	0073-02-0
	10 kg	0073-08-4

BACTO m BISMUTH SULFITE BROTH

INTENDED USE

Bacto m Bismuth Sulfite Broth is a selective medium for use in the detection of *Salmonella* by the membrane filter technique.

HISTORY/PRINCIPLES

Bacto m Bismuth Sulfite Broth is prepared in accordance with the formula published by Clark, Geldreich, Jeter and Kabler[1] consisting of Bacto Bismuth Sulfite Agar with all the constituents in double concentration and with the omission of Bacto Agar. This medium is particularly recommended for the detection of *S. typhi* from water supplies and other specimens.

The value of bismuth sulfite broth in the membrane filter technique for the detection of *S. typhi* has been repeatedly mentioned.[1,2,3,4,5,6] Preliminary enrichment on a nonselective medium is not required; in fact, is undesirable.[6] Excellent growth of *S. typhi* is obtained generally after an incubation of 30 hours at 35°C. Collet, Johnston, Ey and Croft,[7] using Bacto m Bismuth Sulfite Broth, were able to isolate *S. typhi* from well water during the investigation of a case of suspected typhoid fever. They used an incubation temperature of 37°C in an atmosphere saturated with humidity for 40 – 48 hours. After 24 hours incubation, *S. typhi* colonies began to darken, but sheen and black-brown halos were not well developed before 40 hours.

FORMULA

BACTO m BISMUTH SULFITE BROTH
DEHYDRATED

Ingredients per liter

Ferrous Sulfate	0.6 g
Bacto Beef Extract	10 g
Bacto Peptone	20 g
Bacto Dextrose	10 g
Disodium Phosphate	8 g
Bismuth Sulfite Indicator	16 g
Bacto Brilliant Green	0.05 g

Final pH 7.7 ± 0.2 at 25°C.

One pound will make 7.1 liters of medium.

METHOD OF PREPARATION

1. Suspend 64 g in 1 liter distilled or deionized water.
2. Heat to boiling.

Bacto m Bismuth Sulfite Broth, like Bacto Bismuth Sulfite Agar, must not be sterilized in the autoclave since excessive heating destroys the selective properties of the medium. As in other membrane filter media 2.0 – 2.2 ml are added to each pad. The medium normally contains a flocculent precipitate which may be dispersed evenly by twirling the tube or flask of medium just prior to applying to the pad.

STORAGE

Bacto m Bismuth Sulfite Broth   Below 30°C
Prepared medium   2 – 8°C

QUALITY CONTROL

Identity Specifications

Dehydrated powder: greenish beige, homogeneous, free-flowing
Reaction of 6.4% solution: pH 7.7 ± 0.2 at 25°C
Prepared medium: greenish medium amber, opaque with a flocculent precipitate

Typical Cultural Response on Bacto m Bismuth Sulfite Broth
After 40 – 48 Hours at 35°C in Humidified Atmosphere

Organism	Growth	Color of Colony
Escherichia coli ATCC® 25922	inhibited	—
Salmonella typhi ATCC® 6539	good to excellent	black w/sheen
Salmonella typhimurium ATCC® 14028	good to excellent	black w/sheen
Staphylococcus aureus ATCC® 25923	inhibited	—

REFERENCES

1. Pub. Hlth. Repts., 66:951, 1951.
2. J. Am. Water Works Assoc., 43:943, 1951.
3. *ibid.*, 44:471, 1952.
4. Am. J. Pub. Hlth., 42:390, 1952.
5. Bact. Proc. SAB, Boston, p. 33, 1952.
6. J. Am. Water Works Assoc., 45:1196, 1953.
7. Am. J. Pub. Hlth., 44:55, 1954.

PACKAGING

Bacto m Bismuth Sulfite Broth   1 lb (454 g)   0416-01-7

BACTO BLOOD AGAR BASE

INTENDED USE

Bacto Blood Agar Base is an infusion medium to which blood may be added for isolating and cultivating fastidious microorganisms. The slightly acid pH of the base favors distinct hemolytic reactions.

HISTORY

Bacto Blood Agar Base is essentially, Huntoon's[1] medium with an acid reaction. Norton[2] found the pH of 6.8 to be advantageous in culturing streptococci and pneumococci.

FORMULA

BACTO BLOOD AGAR BASE
DEHYDRATED

Ingredients per liter

Beef Heart, Infusion from . . . . .500 g
Bacto Tryptose . . . . 10 g
Sodium Chloride . . . . 5 g
Bacto Agar . . . . 15 g

Final pH 6.8 ± 0.2 at 25°C.

One pound will make 11.35 liters of single-strength medium.

METHOD OF PREPARATION
1. Suspend 40 g in 1 liter distilled or deionized water and heat to boiling to dissolve completely.
2. Sterilize in the autoclave for 15 minutes at 15 lbs pressure (121°C). Cool to 45 – 50°C.
3. Aseptically add 5% sterile, defibrinated, room-temperature blood volume/volume. Mix well.
4. Dispense into Petri dishes or tubes, as desired.

STORAGE
Bacto Blood Agar Base   Below 30°C
Prepared plates or tubes   2 – 8°C

QUALITY CONTROL
Identity Specifications

Dehydrated powder: tan, free-flowing, homogeneous
Reaction of 4% solution: pH 6.8 ± 0.2 at 25°C
Plates prepared with 5% sheep blood: cherry red, opaque, firmly solid

Typical Cultural Response on Bacto Blood Agar Prepared with 5% Sheep Blood After 18 – 24 Hours at 35 ± 2°C

Organism	Growth w/o Blood	Growth w/Blood	Hemolysis
Neisseria meningitidis			
ATCC® 13090	good	excellent	none
Staphylococcus aureus			
ATCC® 25923	good to excellent	excellent	beta
Staphylococcus epidermidis			
ATCC® 12228	good to excellent	excellent	none
Streptococcus pneumoniae			
ATCC® 6303	fair to good	good to excellent	alpha
Streptococcus pyogenes			
ATCC® 19615	fair to good	good to excellent	beta

REFERENCES
1. J. Inf. Dis., 23:170, 1918.
2. J. Lab. Clin. Med., 17:558, 1932.

PACKAGING
Bacto Blood Agar Base   1 lb (454 g)   0045-01-6
                                        1/4 lb (114 g)   0045-02-5
                                        5 lb (2.27 kg)   0045-05-2

BACTO BLOOD AGAR BASE NO. 2

INTENDED USE
Bacto Blood Agar Base No. 2 is a nutritionally rich medium to which blood may be added for maximum recovery of fastidious microorganisms. Growth and hemolytic reactions of streptococci and pneumococci are enhanced on this medium.

Supplemented with 7% lysed horse blood and Bacto Campylobacter Antimicrobic Supplement S, Bacto Blood Agar Base No. 2 may be used to recover *Campylobacter fetus.*

Supplemented with 10% hemolyzed sheep blood, *Haemophilus* may be recovered.

FORMULA

BACTO BLOOD AGAR BASE NO. 2
DEHYDRATED

Ingredients per liter

Proteose Peptone, Difco	15 g
Liver Digest	2.5 g
Bacto Yeast Extract	5 g
Sodium Chloride	5 g
Bacto Agar	12 g

Final pH 7.4 ± 0.1 at 25°C.

Five hundred grams will make 12.7 liters of single-strength medium.

METHOD OF PREPARATION

1. Suspend 39.5 g in 1 liter distilled or deionized water and heat to boiling to dissolve completely.
2. Sterilize in the autoclave for 15 minutes at 15 lbs pressure (121°C). Cool to 45 – 50°C.
3. Aseptically add 5% sterile, defibrinated, room-temperature blood volume/volume. Mix well.
4. Dispense into Petri dishes or tubes, as desired.

STORAGE

Bacto Blood Agar Base No. 2   Below 30°C
Prepared plates or tubes   2 – 8°C

QUALITY CONTROL
Identity Specifications

Dehydrated powder: beige, homogeneous, free-flowing
Reaction of 3.95% solution: pH 7.4 ± 0.1 at 25°C
Plates prepared with 5% sheep blood: cherry red, opaque, no hemolysis

Typical Cultural Response on Bacto Blood Agar Base No. 2 Prepared with 5% Sheep Blood After 40 – 48 Hours at 35 ± 2°C

Organism	Growth	Hemolysis
Neisseria meningitidis ATCC® 13090	good to excellent	none
Streptococcus pneumoniae ATCC® 6303	good to excellent	alpha
Streptococcus pyogenes ATCC® 19615	good to excellent	beta

PACKAGING

Bacto Blood Agar Base No. 2	500 g	0696-17-0
	5 lb (2.27 kg)	0696-05-4

BACTO BLOOD CULTURE PRODUCTS

BACTO BLOOD CULTURE BOTTLES
BLOOD COLLECTOR
BCB VENT/SUB UNIT
BCB RACK

INTENDED USE

Bacto Blood Culture Bottles are used for culturing both aerobic and anaerobic microorganisms from blood by the broth culture technique.

The Blood Collector is used to obtain blood specimens for culture.

BCB Vent/Sub Units are used for venting and subculturing blood culture bottles.

The BCB Rack is used to hold and to transport blood culture bottles while being processed and to allow proper distribution of heat during incubation.

SUMMARY AND EXPLANATION OF THE TEST

Septicemia is one of the most serious forms of infections in man and is associated with a significant mortality depending on associated conditions.[1] Blood is normally sterile in the healthy individual. Microorganisms enter the blood stream from a variety of sources such as intravascular sites via the lymphatic system, directly into the blood from intravascular infections or infected intravenous instrumentation and devices.[2] A sudden influx of bacteria is ordinarily cleared from the blood stream within minutes to hours by the normal body defenses. However, if the patient is debilitated or the rate of organism entry exceeds the ability to clear them from the blood, septicemia results. The introduction of bacteria into the blood stream may be continuous as in the case of endocarditis but in the majority of bacteremias it is intermittent.

A bacteremia does not necessarily mean that the patient is ill. A transient bacteremia can result from such incidents as dental prophylaxis[3] or a barium enema.[4] However, because a blood culture is done only on a patient suspected of having an infection or on a patient highly susceptible to an infection, i.e., after open heart surgery, the finding of a bacteremia (excluding contamination) must be considered significant.

The number, timing and volume of samples taken for blood cultures are extremely important aspects in achieving optimal results. Washington[5] reported that 97 – 99% of the bacteremias he studied were detected within the first three separate blood cultures collected within a 24 hour period. Where intermittent bacteremia is involved, samples should be spaced at least an hour apart, with the first one being taken as soon as possible after a chill episode.[6] Blood culture studies have demonstrated that as the volume of blood cultured is increased from 2 to 20 ml, the yield of positive cultures increases by 30 – 50%.[2]

The blood culture method, where blood is withdrawn aseptically from a suitable vein and is added to a blood culture bottle is widely used to determine if a septicemia is present. Bacto Blood Culture Bottles constitute a culture media system designed to provide optimal nutritional and environmental requirements for organisms commonly encountered in these infections.

Because of the diverse nature of the organisms found in septicemias and because it is possible to have more than one type of organism involved at a time, two bottles should be used for each sample.[2] To encourage the growth of aerobic organisms, one bottle should be vented to replace the vacuum in the bottle with sterile air. The other bottle should remain unvented, thus providing an environment favorable for anaerobic growth.

A ratio of blood to medium of 1:10 is optimal and influential in neutralizing the bactericidal properties of the blood itself as well as the inhibitory effects of antimicrobial agents.

A 1:10 ratio or greater may be used in pediatric cases because the order of magnitude of organisms in a child's blood is usually greater than in an adult's. Kennaugh, Gregory and Hendley[7] reported 63 – 80% better recovery of common organisms found in infant specimens when blood to broth ratios were increased from 1:10 to 1:100.

Inoculated blood culture bottles should be incubated at 35 – 37°C and examined as early as 6 hours for visual evidence of growth (turbidity, hemolysis, gas formation and the formation of discrete colonies). Visibly negative bottles should be returned to the incubator and held for at least 7 days with a visual inspection daily. Incubation beyond 7 days may be necessary with specimens from seriously ill patients who are receiving antimicrobial agents and who have not responded to initial therapy or for patients with endocarditis due to fastidious organisms.[8]

Detection time in positive aerobic bottles has been decreased by the use of a blind subculture between 6 – 18 hours of incubation. Subculture should be made onto both chocolate and blood agar and incubated for 2 days at 35 – 37°C in 5 – 10% CO_2. The use of blind subcultures beyond 2 days of incubation is not routinely recommended.[2] The value of a blind subculture in the anaerobic bottle is uncertain.

Microscopic examination of smears from 24 hour macroscopically negative bottles with acridine orange stain has been reported to improve detection time and was more sensitive than the Gram stain or methylene blue stain.[9,10]

It is necessary to include an anticoagulant as a part of the Blood Culture Bottle system. Anticoagulants avoid the possibility of the blood sample clotting, thus tying up any bacteria present. Sodium polyanetholsulfonate (SPS) has been found not only to provide anticoagulant activity but also to improve overall recovery and speed of isolation due to its ability to inhibit complement and lysozyme activity, interfere with phagocytosis and inactivate aminoglycosides.[2]

Blood culture samples should be obtained, whenever possible, before therapy with antimicrobial agents has been initiated. This is not always possible or practical and the blood sample must be cultured with the knowledge that a potentially inhibitive agent is present. An inoculum ratio of 1:10 has been found to help this situation by diluting out the antimicrobic agent to noninhibitory concentrations. In addition, SPS has an inactivating effect on aminoglycosides. Bacto Blood Culture Bottles containing thiol broth have the unique ability to neutralize a wide range of antimicrobial agents including penicillin G, carbenicillin, nafcillin, oxacillin, gentamicin, and sulfonamides.[11,12]

A culture medium of high solute content which is obtained by the addition of sucrose to a final concentration of up to 15% has been shown to prevent the rupture and lysis of bacteria in various stages of cell wall deterioration. The protected cells are thus able to metabolize and grow.[13]

FORMULA

Bacto Blood Culture Bottles are available containing a culture medium in 50 ml or 100 ml amounts. The formulation of each culture medium is essentially as outlined on the label of the corresponding Difco dehydrated culture medium. Every Bacto Blood Culture Bottle contains a partial vacuum and added CO_2. When sodium polyanetholsulfonate (SPS) and/or sucrose are present in the medium, their final concentrations are 0.025% and 15%, respectively.

Bacto Blood Culture Bottles

Brain Heart Infusion w/PAB and CO_2, Under Vacuum
Brain Heart Infusion w/PAB, SPS and CO_2, Under Vacuum
Columbia Broth w/CO_2, Under Vacuum
Columbia Broth w/SPS and CO_2, Under Vacuum
Columbia Broth w/SPS, Sucrose and CO_2, Under Vacuum
Fluid Thioglycollate Medium w/CO_2, Under Vacuum
Fluid Thioglycollate Medium w/SPS and CO_2, Under Vacuum
Thiol Broth w/CO_2, Under Vacuum
Thiol Broth w/SPS and CO_2, Under Vacuum
Thiol Broth w/SPS, Sucrose and CO_2, Under Vacuum
Tryptic Soy Broth w/CO_2, Under Vacuum
Tryptic Soy Broth w/SPS and CO_2, Under Vacuum
Tryptic Soy Broth w/SPS, Sucrose and CO_2, Under Vacuum

Bacto Blood Culture Bottles Banded Sets

Banded Set 1
   10 sets containing 2 bottles each:
      Tryptic Soy Broth w/SPS and CO_2, Under Vacuum

Banded Set 2
   10 sets containing 2 bottles each:
      Tryptic Soy Broth w/SPS and CO_2, Under Vacuum

Banded Set 3
   10 sets containing 1 bottle each:
      Tryptic Soy Broth w/SPS and CO_2, Under Vacuum
      Thiol Broth w/SPS and CO_2, Under Vacuum

Banded Set 4
   10 sets containing 1 bottle each:
      Tryptic Soy Broth w/SPS and CO_2, Under Vacuum
      Thiol Broth w/SPS and CO_2, Under Vacuum

Blood Collector

Used to obtain blood specimens. Each Blood Collector unit consists of 15″ sterile tubing with a 21 gauge winged venipuncture needle on one end and a stopper perforating assembly on the opposite end.

BCB Vent/Sub Unit

For venting and subculturing blood culture bottles.

BCB Rack

A vinyl coated wire rack that conveniently holds and protects 10 × 100 ml or 14 × 50 ml Bacto Blood Culture Bottles.

PRECAUTIONS

1. The media in the bottles should be clear except FTM which may be slightly opalescent due to the agar in the medium. Some media to which CO_2 has been added, and particularly those containing SPS, may be slightly opalescent or contain a trace of precipitate. Do not confuse slight opalescence with turbidity. Do not use a turbid medium.

2. When examining a blood culture bottle by visual means and prior to subculturing, it is recommended that all bottles showing signs of growth, particularly during the initial 18 – 24 hours incubation period, be treated as potentially having excess gas formation.

   Do not shake such bottles. A trial venting may be advisable to relieve excess pressure before a subculture is taken. The BCB Vent/Sub Unit is recommended for this purpose. The operation is best performed under a biological safety hood.

3. Bacto Blood Culture Bottles, Blood Collector, BCB Vent/Sub Unit should be used by personnel qualified by training or experience to work with blood samples and microorganisms.

STORAGE

Bacto Blood Culture Bottles are stable to the expiry date on the label when stored at 15 – 30°C and protected from light. Refrigeration is not required.

SPECIMEN COLLECTION

Proper skin disinfection is an essential requirement to reduce the incidence of contamination with blood cultures. The venipuncture site should be cleansed with 70% isopropyl or ethyl alcohol, then swabbed with 1 – 2% tincture of iodine in concentric circles starting at the center of the site. An iodophore such as povidone-iodine solution may be used where iodine irritation or hypersensitivity is a problem. The iodine solution should be allowed to act for one minute then be removed with alcohol soaked swabs. Simultaneously the protective cap on the Bacto Blood Culture Bottle should be removed and the rubber stopper swabbed in the same manner. If palpation of the vein is necessary, the venipuncturist's finger should also be decontaminated.

Blood is added to the bottle(s) using either an open or closed system. In the open system, blood is collected with a sterile needle and syringe and either injected into the blood culture bottle at the bedside or the specimen is mixed with an anticoagulant in an evacuated tube and transported to the laboratory where it is then injected into the bottle. In the closed system, the blood collector is used, permitting blood to enter the bottle directly from the patient. The closed system is less prone to introduce contaminants into the culture system.

PROCEDURE
Materials Provided:
Bacto Blood Culture Bottles
Blood Collector
BCB Vent/Sub Unit
BCB Rack

Materials Required but not Provided:
Tourniquet
Tincture of Iodine, 2% solution or povidone-iodine
Sterile gauze sponges saturated with 70% isopropyl alcohol or ethyl alcohol
Sterile swabs

Forceps
Sterile syringes and needles (for open system)
Wire loop
Containers for used blood collectors or syringes and needles
Sterile air filter
This filter may be prepared simply in the laboratory by asceptically filling the hub of a sterile disposable 20 – 22 gauge needle with a generous amount of sterile absorbent cotton.
Incubator 35°C
Sterile screw cap centrifuge tubes
Centrifuge
Autoclave
Glass slide 1″ × 3″
Usual culture media for a wide variety of fastidious microorganisms
Bandages
Fluorescent microscope (for use with Bacto Acridine Orange Stain)
Methanol

Needle and Syringe Method (Open System)

1. Identify the patient.
2. Label the aerobic blood culture bottle and anaerobic blood culture bottle with the patient's name and identification number, date and time.
3. Break the seals and remove the caps from the two Bacto Blood Culture Bottles.
4. Wipe the stoppers with the 70% isopropyl or ethyl alcohol swabs or an iodine preparation. Allow to dry.
5. Inspect both arms of the patient to find a suitable vein. The venipuncture site should be free of infection or other skin disorder. Apply the tourniquet to the patient's arm and select an optimal venipuncture site, then release the tourniquet.
6. Vigorously cleanse the venipuncture site with 70% isopropyl or ethyl alcohol swabs. Allow to dry.
7. Apply 2% tincture of iodine or 10% povidone-iodine to the venipuncture site starting at the center and moving in concentric circles to the periphery. This covers about a two-inch radius around the site. Allow to dry 60 seconds.
8. Carefully open the sterile packages containing the needle and syringe.
9. Assemble the needle and syringe making sure the needle is tight in the syringe and the plunger is completely pushed in.
10. Reapply the tourniquet and have the patient open and close his hand until the vein previously selected is prominent again.
11. Carefully withdraw the needle protector and set it aside.
12. Using your hand not holding the syringe, immobilize the patient's vein by placing your index finger above the prepared site and your thumb below and press. Do not touch the prepared venipuncture site.
13. With the needle at about a 15° angle with the skin and the needle bevel up, carefully insert the needle into the vein. As the needle enters the vein, a little give will be felt. Release the tourniquet.
14. Carefully draw the required amount of blood by slowly pulling back on the plunger.
15. Once the desired blood is obtained, carefully remove the needle and syringe assembly while placing a sterile gauze over the venipuncture site and applying pressure. Have the patient secure the gauze by bending the arm upward.
16. Inject the required amount of blood into each bottle. NOTE: As a further step to eliminate contamination from the normal flora of the skin, it is good practice to change the needle before injecting blood into the bottle.
17. Replace the needle protector on the needle or cut the needle with needle cutters.

18. Inspect the patient's arm to assure that all blood flow has ceased and apply a bandage.
19. Dispose of needle and syringe according to hospital and/or laboratory procedures.

Blood Collector Method (Closed System)

1. Identify the patient.
2. Label the aerobic blood culture bottle and anaerobic blood culture bottle with the patient's name and identification number, date and time.
3. Break the seals and remove the caps from the two Bacto Blood Culture Bottles.
4. Wipe the bottle stoppers with the 70% isopropyl or ethyl alcohol swabs or an iodine preparation. Allow to dry.
5. Inspect both arms of the patient to find a suitable vein. The venipuncture site should be free of infection or other skin disorder. Apply the tourniquet to the patient's arm and select an optimal venipuncture site, then release the tourniquet.
6. Vigorously cleanse the venipuncture site with 70% isopropyl or ethyl alcohol swabs. Allow to dry.
7. Apply 2% tincture of iodine or 10% povidone-iodine to the venipuncture site starting at the center and moving in concentric circles to the periphery. This covers about a two inch radius around the site. Allow to dry 60 seconds.
8. Prepare a 2-1/2″ piece of surgical tape and place in a convenient place.
9. Carefully open the sterile package containing the blood collector.
10. Reapply the tourniquet and have the patient open and close his hand until the vein previously selected is prominent again.
11. Carefully withdraw the winged needle from its protector using a twisting motion and discard the protector. Squeeze the wings together so that the ribbed sides are in contact with the fingers and the smooth sides face inward. This will result in the needle bevel facing upward.
12. Immobilize the vein with the opposite hand by placing your index finger above the prepared site and your thumb below and press. Do not touch the prepared venipuncture site.
13. Carefully, but firmly, insert the winged needle with the bevel up. Successful entry into the vein results in a slight blood flow into the tubing.
14. Carefully place the surgical tape across a portion of the wings, perpendicular to the tubing, below the venipuncture site and then loosen the tourniquet.
15. Remove the clear cotton-plugged protector shield from the stopper-perforating assembly but do not remove the blue slide valve. Plunge the needle through the blue slide valve into the target area on the stopper of one Bacto Blood Culture Bottle keeping the bottle in an upright position. Draw approximately 5 ml of blood into the 50 ml Bacto Blood Culture Bottle or 10 ml of blood into the 100 ml bottle, filling the bottles to the 55 or 110 ml positions, respectively, on the volume guide.
16. Remove the stopper-perforating assembly from the first bottle. The blood flow ceases when the needle is removed from the bottle due to the blue slide valve. Fill subsequent bottles as required. Gently mix each bottle.
17. Carefully remove the surgical tape and winged needle while placing a sterile gauze over the venipuncture site and applying pressure. Have the patient secure the gauze by bending the arm upward.
18. Either cut the winged and stopper-perforating assembly needles with cutters or insert the winged needle into the cotton end of the stopper-perforating assembly needle protector and the stopper-perforating assembly needle into the opposite end of the protector. The latter will create a closed system.
19. Inspect the patient's arm to assure that all blood flow has ceased and apply a bandage, as required.
20. Dispose of blood collecting units according to hospital and/or laboratory procedures.

Venting
BCB Vent/Sub Unit Method
1. If present, remove screw cap from inoculated Bacto Blood Culture Bottle.
2. Disinfect stopper on bottle.
3. Aseptically remove blue sheath from BCB Vent/Sub Unit, exposing short beveled needle, and insert needle through center of stopper.
4. Do not disturb white plastic sheath.
5. Incubate bottle according to instructions in Insert 1415 packaged with Bacto Blood Culture Bottles. Alternately, a sterile cotton-plugged syringe needle may be used.

INCUBATION
Incubate bottles at 35 – 37°C until growth appears or for 7 days, longer if necessary. (See SUMMARY section.) The BCB Rack may be used to facilitate the handling of Bacto Blood Culture Bottles while they are being processed. In addition, the BCB Rack allows proper heat distribution during incubation.

EXAMINATION
Examine daily during the first week for appearance of turbidity throughout the medium, hemolysis, signs of gas production or growth directly above the blood. Examine twice during the second week. Visually positive bottles should be Gram stained and then subcultured onto blood agar, chocolate agar, and other plating media (MacConkey, EMB, phenylethanol, or Columbia CNA, etc.). Plates should be incubated under both aerobic and anaerobic conditions.

Once isolated colonies have been attained, their antimicrobic susceptibility and identification should be determined.

If an aerobic bottle is still visually negative after 24 hours of incubation, a sample should be aseptically taken, followed by:
1. A smear made and stained with Bacto Acridine Orange Stain (See Bacto Acridine Orange Stain for complete information).
2. A blind subculture.

If a BCB Vent/Sub Unit has been used to vent the bottle, proceed as follows:
1. Gently agitate vented Bacto Blood Culture Bottle to evenly disperse microorganisms.
2. Aseptically remove the white plastic sheath, exposing the blunt-ended subculture needle.
3. Tip bottle and allow 1 – 2 drops of broth culture to flow out of the unit onto the surface of the plating medium.
4. Securely replace the sterile white plastic sheath.
5. Return bottle to incubator if further incubation is desired.

Results, either positive or negative, should be promptly reported to the requesting physician.

QUALITY CONTROL
Bacto Blood Culture Bottles should be examined before use for evidence of deterioration. (See Precautions Section.)

There is no practical or inexpensive way to provide the desired confidence level that negative results obtained in a given blood culture bottle are due to its proper or improper performance. There are just too many variables involved, including the probability that a given sample from a bacteremic patient will not contain viable organisms.

Probably the best controls are result oriented. For example, repeatedly negative results from multiple samples (at least 3 – 4) taken over a period of several days and timed as close to a chill as possible is usually good evidence that an infecting agent is not involved. Confidence in this result is bolstered if, using similar bottles, the overall positive blood culture rate for the hospital, excluding contamination, is 6% or more.[1]

LIMITATIONS

1. Although every effort is made to eliminate nonviable but stainable bacteria from Bacto Blood Culture Bottles, it is impossible to guarantee that every bottle will be completely free of an occasional nonviable bacterium. When such bacteria are found, they will generally (or most likely) be of the genus *Bacillus* and may stain irregularly — gram-positive or gram-negative. However, since bacteria have also been known to occur on clear but nonsterile auxiliary reagents and devices used in performing Gram stains or used in collection of specimens, the finding of a rare bacterium on Gram stained smears of blood culture media cannot always be interpreted as originating from the medium. If an occasional bacterium is found, we recommend that the culture be reexamined after an additional 2 hours incubation for increase in number of cells. This is usually indicative of growth, but the significance of this increase must be determined by laboratorians, using their experience and expertise in interpreting these results and taking into account the characteristics and morphology (i.e., gram-positive or gram-negative rod or cocci) of the suspect agent.

2. A preliminary Gram stain may provide useful presumptive evidence of the identification of the organism causing the bacteremia. However, because of the great variety of organisms that may be found in a blood culture and their sometimes similar appearance in a Gram stained smear but often dissimilar nutritional and growth condition requirements, a wide variety of culture media and incubation conditions should be utilized when subculturing to expedite the isolation and identification of any organism present.

3. As mentioned, it is difficult to avoid an occasional contaminant in a blood culture. This situation is further complicated by the fact that some of the common contaminants such as *Staphylococcus epidermidis* or *Proprionibacterium acnes* have been reported as etiological agents of endocarditis and of septicemia. Finding the organism consistently in multiple blood samples from a patient is the best evidence that the organism is not a contaminant.

4. It is possible for a patient to have a bacteremia caused by an organism that will not grow in the culture media system being used routinely. The *Leptospira* and *F. tularensis* would be good examples of this situation. For *Leptospira* a special medium such as Bacto Leptospira Medium EMJH is recommended. In the case of *F. tularensis*, because of the organism's great virulence, it is best to use the "febrile" agglutination test or other indirect evidence for detection.

REFERENCES

1. Sonnenwirth, A. C., 1973, Bacteremia, Chas. E. Thomas, Springfield IL.
2. Reller, Murray and MacLowry, 1982, Cumitech 1A, ASM, Washington, D.C.
3. Frock et al, 1974, ASM Meeting, Chicago.
4. Leach et al, 1974, ASM Meeting, Chicago.
5. Washington, J. A., 1975, Blood Cultures: Principles and Techniques, Mayo Clinic Proceedings, 50:91 – 98.
6. Tilton, R. C., 1982, Annual Review of Microbiology, 36:467 – 493.
7. Kennaugh, Gregory and Hendley, 1983, Abstracts of ASM Annual Meeting, #C84.
8. Lennette, 1980, Manual of Clinical Microbiology, 3rd Edition, Page 55, ASM, Washington, D.C.
9. McCarthy and Senne, 1980, Journal of Clinical Microbiology, 11:281 – 285.
10. Mirrett, Lauer, Miller, and Reller, 1982, Journal of Clinical Microbiology, 15:562 – 566.
11. Murray and Niles, 1982, Journal of Clinical Microbiology, 16:982 – 984.
12. Szawatkowski, 1976, Medical Laboratory Science, 33:5 – 12.
13. Rosner, 1970, Applied Microbiology, 19:281, ASM.

PACKAGING

Brain Heart Infusion w/PAB and CO_2, Under Vacuum	10 × 50 ml	0652-37-8
Brain Heart Infusion w/PAB, SPS and CO_2, Under Vacuum	10 × 50 ml	1700-37-8
	10 × 100 ml	1700-74-2
Columbia Broth w/CO_2, Under Vacuum	10 × 50 ml	1727-37-7
	10 × 100 ml	1727-74-1
Columbia Broth w/SPS and CO_2, Under Vacuum	10 × 50 ml	1729-37-5
	10 × 100 ml	1729-74-9
Columbia Broth w/SPS, Sucrose and CO_2, Under Vacuum	10 × 100 ml	1739-74-7
Fluid Thioglycollate Medium w/CO_2, Under Vacuum	10 × 50 ml	0726-37-0
	10 × 100 ml	0726-74-4
Fluid Thioglycollate Medium w/SPS and CO_2, Under Vacuum	10 × 50 ml	1705-37-3
Thiol Broth w/CO_2, Under Vacuum	10 × 50 ml	0355-37-8
	10 × 100 ml	0355-74-2
Thiol Broth w/SPS and CO_2, Under Vacuum	10 × 50 ml	1723-37-1
	10 × 100 ml	1723-74-5
Thiol Broth w/SPS, Sucrose and CO_2, Under Vacuum	10 × 100 ml	1738-74-8
Tryptic Soy Broth w/CO_2, Under Vacuum	10 × 50 ml	0936-37-6
	10 × 100 ml	0936-74-0
Tryptic Soy Broth w/SPS and CO_2, Under Vacuum	10 × 50 ml	1712-37-4
	10 × 100 ml	1712-74-8
Tryptic Soy Broth w/SPS, Sucrose and CO_2, Under Vacuum	10 × 50 ml	1737-37-5
	10 × 100 ml	1737-74-9

Bacto Blood Culture Bottles Banded Sets

Banded Set 1	10 sets (100 ml)	1750-32-2
10 sets containing 2 bottles each: Tryptic Soy Broth w/SPS and CO_2, Under Vacuum		
Banded Set 2	10 sets (50 ml)	1764-32-6
10 sets containing 2 bottles each: Tryptic Soy Broth w/SPS and CO_2, Under Vacuum		
Banded Set 3	10 sets (50 ml)	1778-32-0
10 sets containing 1 bottle each: Tryptic Soy Broth w/SPS and CO_2, Under Vacuum Thiol Broth w/SPS and CO_2, Under Vacuum		
Banded Set 4	10 sets (100 ml)	1795-32-9
10 sets containing 1 bottle each: Tryptic Soy Broth w/SPS and CO_2, Under Vacuum Thiol Broth w/SPS and CO_2, Under Vacuum		
Blood Collector	10 collectors	1832-31-5
BCB Vent/Sub Unit	100 units	0976-31-3
BCB Rack	2 racks	1822-32-6
Bacto Acridine Orange Stain	6 × 250 ml	3336-76-8
Bacto Gram Stain Set	4 × 250 ml	3328-32-1
Bacto Gram Crystal Violet	6 × 250 ml	3329-76-7
Bacto Gram Decolorizer	6 × 250 ml	3330-76-4
Bacto Gram Iodine	6 × 250 ml	3331-76-3
Bacto Gram Safranin	6 × 250 ml	3332-76-2

BACTO BORDET GENGOU AGAR BASE

INTENDED USE

Bacto Bordet Gengou Agar Base, enriched with sterile defibrinated blood, is used for the isolation of *Bordetella pertussis* in the diagnosis of whooping cough.

HISTORY/PRINCIPLES

Bacto Bordet Gengou Agar Base is a modification of the medium originally described by Bordet and Gengou[1] in 1906 for the cultivation of *Haemophilus pertussis* now *Bordetella pertussis*. It is prepared according to the formula recommended in Diagnostic Procedures and Reagents[2] of the American Public Health Association for the isolation of this organism. The addition of 1% proteose peptone to the medium is suggested if employed for mass culture of *B. pertussis* as in vaccine production.

The "cough plate" method for the diagnosis of whooping cough was originally reported by Chievitz and Meyer.[3] Lawson and Mueller,[4] and Sauer and Hambrecht[5] used modifications of the Bordet Gengou medium to demonstrate the value of the cultural diagnosis of this disease. This medium has been applied routinely as a diagnostic procedure for public health laboratories as a result of the thorough and painstaking investigations of Kendric, and her associates. Kendrick and Eldering[6] first used a modified Bordet Gengou medium for the isolation and propagation of *B. pertussis*. Eldering and Kendrick[7] reported that the addition of 1% proteose peptone or neopeptone increased the growth of *B. pertussis*, thereby increasing the yield of vaccine.

With this modification of the Bordet Gengou medium, enriched with 15 to 20% blood, the appearance of colonies of *B. pertussis* is typical, being smooth, raised, glistening and not over 1 mm in diameter. They are of a pearly, almost transparent appearance, and are surrounded by a characteristic zone of hemolysis which is not sharply defined, but which merges diffusely into the medium. The zone of hemolysis usually is absent if 30% or more blood is added to the medium. Sterile sheep, rabbit or human blood may be used in preparing the medium. Horse blood should not be used in preparing vaccine.

Kendrick, Miller and Lawson[7] and Kendrick, Lawson and Miller[8] recommended that, after exposure, cough plates prepared from the modified Bordet Gengou blood agar should be incubated at 37°C. During the first 48 hours incubation they are examined for contamination by molds and spreaders, which are cut aseptically from the medium. The plates are examined twice daily, using moderate magnification until typical colonies of *B. pertussis* are found or until discarded after 6 days of incubation.

Maclean[9] used Bacto Bordet Gengou Agar Base and reported it to be efficient in the isolation of *B. pertussis*. He further reported that this medium was a valuable standard for the comparison of various lots of media prepared from ingredients.

Tarshis and Frisch[10] investigated the addition of bank blood to various media for the cultivation of tubercle bacilli in pure culture and directly from sputa under routine diagnostic conditions. Three standard tuberculosis media were used in the comparative study. They recommended the addition of 25% bank blood to Bacto Bordet Gengou Agar Base or Bacto Blood Agar Base with 1% glycerol added, since media of this type grew tubercle bacilli from small inocula producing colonies that were readily recognized. These media were easily prepared and in addition were economical. They were also satisfactorily employed in streptomycin sensitivity tests.

FORMULA

BACTO BORDET GENGOU AGAR BASE
DEHYDRATED
Ingredients per liter

Potato, Infusion from	125 g
Sodium Chloride	5.5 g
Bacto Agar	20 g

Final pH 6.7 ± 0.2 at 25°C.

One pound will make 15 liters of medium.

METHOD OF PREPARATION
1. To rehydrate, suspend 30 g in 1 liter distilled or deionized water containing 10 g of glycerol.
2. Heat to boiling to dissolve the medium completely.
3. Sterilize in the autoclave for 15 minutes at 15 lbs pressure (121°C).
4. Cool to 45 – 50°C and add sterile blood to a final concentration of 15% or more if desired. Rabbit blood is preferred for routine testing.
5. Mix and pour into sterile Petri dishes.

STORAGE
Bacto Bordet Gengou Agar Base	Below 30°C
Prepared medium	2 – 8°C

QUALITY CONTROL
Identity Specifications

Dehydrated powder:	beige, homogeneous, free-flowing
Reaction of 3.0% solution:	pH 6.7 ± 0.2 at 25°C
Prepared medium:	light to medium amber, opalescent, may have a slight precipitate
Prepared plates:	cherry red, opaque

Typical Cultural Response on Bacto Bordet Gengou Agar Base
After 48 – 72 Hours at 35°C

Organism	Growth*
Bordetella bronchiseptica ATCC® 4617	good to excellent
Bordetella parapertussis	good to excellent
Bordetella pertussis ATCC® 8467	good to excellent

*With or without 15% rabbit blood

REFERENCES
1. Am. Inst. Pasteur, 20:731, 1906.
2. Diagnostic Procedures and Reagents, 3rd Ed.:141, 1950.
3. Ann. Inst. Pasteur, 30:503, 1916.
4. J. Am. Med. Assoc., 89:275, 1927.
5. J. Am. Med. Assoc., 95:263, 1930.
6. Am. J. Pub. Health, 24:309, 1934.
7. Am. J. Pub. Health, 26:506, 1936.
8. Sixth Annual Year Book (1935 – 36), p. 200, Suppl., Am. J. Pub. Health, 26:No. 3, 1936.
9. J. Path. Bact., 45:472, 1937.
10. Am. J. Clin. Path., 21:101, 1951.

PACKAGING
Bacto Bordet Gengou Agar Base	1 lb (454 g)	0048-01-3
	1/4 lb (114 g)	0048-02-2

BACTO BORDETELLA PERTUSSIS ANTIGEN

PRODUCT DESCRIPTION AND INTENDED USE

Bacto Bordetella Pertussis Antigen is a suspension of a selected strain of *B. pertussis*, etiological agent of whooping cough, recommended for use in a rapid tube agglutination test for detecting agglutinins in serum. It may also be used either in the rapid tube test or slide test as a positive control antigen for checking isolated colonies suspected of being *B. pertussis* in conjunction with Bacto Bordetella Pertussis Antiserum. The antigen has been prepared according to Kendrick.[1]

REAGENT

When used according to the suggested procedure, each 5 ml vial is sufficient to perform approximately 6 tests.

The antigen is liquid and ready for use.

STORAGE AND EXPIRATION DATE

Store Bacto Bordetella Pertussis Antigen at 2 – 10°C. This reagent is stable through the expiration date on the label when properly stored, provided they have not been exposed to temperature extremes (excessive heat or freezing) and have not become contaminated chemically or bacterially during routine usage. Exposure to repeated freezing and thawing is particularly detrimental to antigen stability.

PROCEDURE

Materials Provided:
Bacto Bordetella Pertussis Antigen

Materials Required but not Provided:
Kahn serological tubes
0.85% NaCl solution
0.2 ml serological pipettes
Test tube rack
Suitable light source

PRECAUTIONS

1. Shake antigen vial well before use to insure a smooth, uniform suspension. Occasionally, upon storage, *Bordetella* suspensions may tend to settle out. If an antigen demonstrates a roughness in the negative control, (sterile saline), the organism may be resuspended by vigorous shaking. If this is necessary, allow the antigen to stand until the entrapped air particles are released.
2. Discard any antigen demonstrating positive reactions in the negative control which cannot be resuspended (No. 1 above). *Bordetella* antigens will demonstrate irreversible autoagglutination if at any time during shipment or storage they are subjected to freezing temperatures. Do not allow to freeze.
3. Discard any antigen which does not react properly in the positive control serum.

METHODOLOGY

1. Place eight Kahn serological tubes in a Kahn rack for each serum to be tested.
2. Prepare serial dilutions of the serum in 0.85% sodium chloride solution from 1:2 to 1:128 in 0.1 ml quantities in the first 7 tubes using a 0.2 ml Kahn serological pipette.
3. Add 0.1 ml saline solution to the 8th tube which serves as a negative control for the antigen.

4. Add 0.1 ml of Bacto Bordetella Pertussis Antigen to each of the eight tubes. Shake the antigen bottle thoroughly before use.
5. Rock the rack for 3 minutes tilting the racks side to side in such a manner that the antigen-serum mixture flows approximately three quarters of the distance up the tube. The rocking should approximate 60 side to side motions per minute.
6. After rocking, add 0.5 ml of 0.85% sodium chloride solution to each tube and read for agglutination. Record results as follows:

    4+   Complete agglutination
    3+   75% of the cells agglutinated
    2+   50% of the cells agglutinated
    1+   25% of the cells agglutinated
    −   No agglutination

QUALITY CONTROL

Bacto Bordetella Pertussis Antiserum may be used as a positive control when following the methodology above.

RESULTS

The end point is the tube containing the highest dilutions of serum in which a 2+ reaction is observed.

INTERPRETATION

The tube of antigen control as prepared in step 3 must be negative to validate the remainder of the test.

Persons possessing a high concentration of antibody are considered immune but the converse is not necessarily true.[2,3]

Occasionally infected subjects may fail to demonstrate antibodies by this technique.

REFERENCES

1. Diagnostic Procedures & Reagents, 4th Ed., APHA, 1963.
2. J. Pediat., 22:644, 1943.
3. Ibid., 30:29, 1947.

PACKAGING

Bacto Bordetella Pertussis Antigen	5 ml	2585-56-2
Bacto Bordetella Pertussis Antiserum	1 ml	2309-50-3

BORDETELLA ANTISERA
BACTO BORDETELLA PERTUSSIS ANTISERUM
BACTO BORDETELLA PARAPERTUSSIS ANTISERUM

INTENDED USE

Bacto Bordetella Pertussis and Bacto Bordetella Parapertussis Antisera are used in the serological identification of *Bordetella pertussis* and *Bordetella parapertussis* by the slide agglutination technique.

SUMMARY AND EXPLANATION OF THE TEST

Bordetella pertussis is the primary etiological agent of whooping cough while *Bordetella parapertussis* may cause pertussis-like respiratory infections.

Bordet and Gengou[1] isolated *B. pertussis* in 1906, on glycerine-potato-blood-agar. Bacto Bordet Gengou Agar has remained the medium of choice for the cultivation of *B. pertussis* and other *Bordetella*.

Other media may be used for the cultivation of *Bordetella*; however, they should not be employed if serological procedures are to be performed. Alternative media may cause the organisms to dissociate from smooth cultures to rough variants, resulting in antigenic inconsistencies.

Serological procedures may be employed to provide corroborative evidence and is a valuable technique in differentiating *B. pertussis* from *B. parapertussis*.

REAGENTS

Bacto Bordetella Pertussis and Bacto Bordetella Parapertussis Antisera are high titered, desiccated antisera prepared in rabbits for use in the slide agglutination technique.

When properly rehydrated each antiserum contains approximately 1:5000 Merthiolate®, sufficient to maintain a bacteriostatic condition at storage of 2 – 8°C.

When used according to the suggested PROCEDURE each 1 ml vial diluted 1:10 is sufficient to perform approximately 200 tests.

REHYDRATION

To rehydrate the antisera, add 1 ml sterile, distilled or deionized water to each vial and rotate gently to dissolve contents completely.

STORAGE

Store both unrehydrated and rehydrated Bordetella antisera at 2 – 8°C.

Prolonged exposure to room temperature or to repeated freezing and thawing is detrimental to antisera.

EXPIRATION DATE

Both unrehydrated and rehydrated but not diluted antisera are stable through the expiration date on the label.

Antisera which has been diluted 1:10 should be discarded after 1 – 2 days.

Discard any antiserum that is cloudy or has a precipitate after rehydration and storage unless it can be clarified by centrifugation and demonstrates proper reactivity with validated positive and negative control cultures.

SPECIMEN PREPARATION

For optimal recovery, specimens should be obtained from the nasopharynx. Nasopharyngeal swabs should be emulsified in 0.2 – 0.5 ml sterile 1% Bacto Casamino Acids solution. The specimen may be held in this solution for a maximum of 2 hours. Use the casamino acids emulsion to inoculate plates of Bacto Bordet Gengou Agar Base enriched with 15 – 20% sterile blood. To assure recovery of *Bordetella*, two isolation plates should be used, one containing only Bacto Bordet Gengou Agar Base and another with Bacto Bordet Gengou Agar Base with 0.5 unit penicillin per ml.[2] Incubate plates at 37°C for at least 24 – 48 hours. Colonies may be picked for serological procedures as soon as growth is apparent.

The appearance of typical colonies of *B. pertussis* is smooth, raised, glistening and not over 1 mm in diameter. They are of a pearly, almost transparent appearance, and are surrounded by a characteristic zone of hemolysis which is not sharply defined, but which merges diffusely into the medium. *B. parapertussis* colonies are usually larger than *B. pertussis* and the surface is not as glistening and may have a slightly brown color. Both organisms are small gram-negative bacilli and occur single, in pairs, or in clumps.

The remaining emulsion may be used for the direct fluorescent antibody technique. For complete information regarding this procedure refer to Bacto FA Bordetella Pertussis and Bacto FA Bordetella Parapertussis.

PROCEDURE
Materials Provided:
Bacto Bordetella Pertussis Antiserum
Bacto Bordetella Parapertussis Antiserum

Materials Required but not Provided:
Wax pencil
Microscope slides or glass plate marked into squares
Bacteriological loop

1. Dilute a small portion of the rehydrated antiserum 1:10 with 0.85% NaCl solution. Dilute only that amount of antiserum to be used in 1 or 2 days test.
2. Place a drop of the diluted antiserum on a clean dry microscope slide or a glass plate which has been marked into squares with a wax pencil.
3. Obtain a small portion of the suspected colony with a bacteriological loop and rub the culture into the antiserum until complete emulsification is accomplished.
4. Repeat steps 2 and 3 using 0.85% NaCl solution instead of the antiserum. This represents the antigen control and should not agglutinate.
5. Rock the slide gently for 1 – 2 minutes and observe for agglutination.
6. Record the reaction as follows:
   - 4+ agglutination = all of the cells agglutinate
   - 3+ agglutination = 75% of the cells agglutinate
   - 2+ agglutination = 50% of the cells agglutinate
   - 1+ agglutination = 25% of the cells agglutinate
   - ± agglutination = less than 25% of the cells agglutinate
   - − agglutination = none of the cells agglutinate

QUALITY CONTROL
Use known positive and negative control cultures in parallel with the test culture to ascertain the validity of test results.

RESULTS
A 4+ reaction is considered a positive result. Agglutination of the test culture should be rapid and complete. No agglutination should be visible in the saline control. If agglutination does occur the culture is rough and not suitable for serological techniques.

LIMITATIONS OF THE PROCEDURE
Final identification of *B. pertussis* and *B. parapertussis* cannot be based solely on serological results. The results obtained from serological techniques can serve only to corroborate cultural, biochemical and clinical findings.

REFERENCES
1. Bordet, J., and Gengou, O., Ann. Inst. Pasteur, 20:731, 1906.
2. Lennette, Spaulding, Truant, Manual of Clinical Microbiology, 3rd Ed., Washington, D.C.: ASM, Chapter 26, 1981.

PACKAGING
Bacto Bordetella Parapertussis Antiserum	1 ml	2310-50-0
Bacto Bordetella Pertussis Antiserum	1 ml	2309-50-3

BACTO BOVINE ALBUMIN 5%

INTENDED USE
Bacto Bovine Albumin 5% is a filter sterilized solution of bovine albumin (fraction V) and is suggested for use as an enrichment in media for culturing a large variety of microorganisms and tissue cells.

Davis and Dubos[1] recommended the use of bovine albumin in a final concentration of 0.5% in liquid media for culturing *M. tuberculosis*. They demonstrated that bovine albumin neutralized the toxicity of fatty acids and permitted more luxuriant growth of *M. tuberculosis* than was obtained in the medium without albumin. Bacto Dubos Medium Albumin is a modification of Bacto Bovine Albumin 5% in that sodium chloride and dextrose are included, thus eliminating the need to add these ingredients individually as originally suggested by Dubos et al. They also recommended the use of albumin in a final concentration of 0.5% in Long's synthetic medium for culturing mycobacteria.

Ellinghausen and McCullough[2] used a filter sterilized solution of bovine albumin fraction V in a final concentration of 1% in liquid, semisolid and solid media for culturing leptospires. Enhanced growth is obtained with bovine albumin as compared with normal rabbit serum when added to the Ellinghausen-McCullough medium.

The stabilization of the metabolic activity of rickettsiae in media containing 0.1 – 1% bovine albumin was reported by Bovarnick et al.[3] and by Rees and Weiss.[4]

Morton et al.[5] demonstrated that 1% bovine albumin stimulated the growth of *Mycoplasma* (PPLO). The addition of either bovine albumin or Bacto PPLO Serum Fraction in a final concentration of 1% yielded superior growth of mycoplasma from clinical specimens.

Improved growth of *Trichonomas vaginalis* in the presence of trace amounts of serum albumin and linoleic acid was obtained by Sprince and Kupferberg et al.[6]

The vast literature on the growth of tissue cells contains numerous references to the stimulating properties of albumin to the growth of a large variety of mammalian cells.

STORAGE
Bacto Bovine Albumin 5%   2 – 8°C

QUALITY CONTROL
Identity Specifications
Appearance: light amber, clear solution
Reaction of solution: pH 7.0 ± 0.2 at 25°C

REFERENCES
1. J. Bact., 55:11, 1945.
2. Bact. Proc., 62:54, 1962.
3. J. Bact., 59:9, 1950.
4. J. Bact., 95:389, 1968.
5. J. Dental Research, 30:415, 1951.
6. Proc. Soc. Exper. Biol. Med., 67:304, 1948.

PACKAGING
Bacto Bovine Albumin 5%   12 × 20 ml   0668-64-6

BACTO BOVINE PLASMA FRACTION V

INTENDED USE
Bacto Bovine Plasma Fraction V is albumin powder (Fraction V) from bovine plasma for use as an enrichment for culture media.

METHOD OF PREPARATION
1. Rehydrate 2 vials with 5 ml 0.85% saline in each. Adjust pH to 6.8 – 7.0 and filter sterilize.
2. Prepare and sterilize 90 ml Bacto Dubos Broth Base per label instructions.
3. Allow medium to cool to 50°C and aseptically add 10 ml rehydrated Bacto Bovine Plasma Fraction V. Mix well and dispense into sterile tubes in 5 ml amounts.
4. Inoculate tubes with *Mycobacterium* cultures and incubate as required.

STORAGE
Bacto Bovine Plasma Fraction V   2 – 8°C

QUALITY CONTROL
Identity Specifications
Dehydrated powder: off white, fine flakes
Reaction of 5.0% solution
 in 0.85% saline: adjust pH as desired
Prepared medium: very light amber, clear to very slightly opalescent

Typical Cultural Response in Bacto Dubos Broth Base
After 2 Weeks at 35°C Under 10% CO_2

Organism	Growth
Mycobacterium fortuitum ATCC® 6841	good to excellent
Mycobacterium tuberculosis H37Rv (ATCC® 26518)	good to excellent

PACKAGING
Bacto Bovine Plasma Fraction V   10 g   0497-12-6
Bacto Dubos Broth Base   500 g   0385-17-6

BACTO BRAIN HEART CC AGAR

INTENDED USE
Bacto Brain Heart CC Agar is a selective medium used for isolating and cultivating fastidious pathogenic fungi such as *Histoplasma capsulatum* and *Blastomyces dermatiditis*.

PRINCIPLES
Bacto Brain Heart CC Agar is prepared according to the recommendation of George.[1] Chloramphenicol and cycloheximide (Actidione®) are included in the formulation as selective agents to restrict growth of bacteria and saprophytic yeasts and molds.

FORMULA
BACTO BRAIN HEART CC AGAR
DEHYDRATED
Ingredients per liter

Calf Brains, Infusion from . . . . . . 200 g	Disodium Phosphate . . . . 2.5 g
Beef Hearts, Infusion from . . . . . 250 g	Bacto Agar . . . . 15 g
Proteose Peptone, Difco . . . . 10 g	Chloramphenicol . . . . 50 mg
Bacto Dextrose . . . . 2 g	Actidione® . . . . 500 mg
Sodium Chloride . . . . 5 g	

Final pH 7.4 at 25°C.

One pound will make 8.7 liters of final medium.

METHOD OF PREPARATION
1. Suspend 52 grams in 1 liter distilled or deionized water and heat to boiling to dissolve completely.
2. Dispense as desired.
3. Sterilize in the autoclave for 15 minutes at 15 lbs pressure (121°C). Avoid excessive exposure to heat which may reduce the selectivity of the medium.

STORAGE
Bacto Brain Heart CC Agar   Below 30°C
Prepared medium   2 – 8°C

QUALITY CONTROL
Identity Specifications

Dehydrated powder:	beige, homogeneous, free-flowing
Reaction of 5.2% solution:	pH 7.4 ± 0.2 at 25°C
Prepared medium:	light to medium amber, clear to slightly opalescent with no precipitate

Typical Cultural Response on Bacto Brain Heart CC Agar
After 96 Hours at 25°C

Organism	Recovery
Aspergillus niger ATCC® 16404	inhibited
Candida albicans ATCC® 26790	fair to good
Candida tropicalis	inhibited
Escherichia coli ATCC® 25922	inhibited
Trichophyton megninii ATCC® 12106	good
Trichophyton tonsurans ATCC® 10220	good
Trichophyton verrucosum	good

REFERENCE
1. Personal communications, 1964.

PACKAGING
Bacto Brain Heart CC Agar

1 lb (454 g)   0483-01-5
12 tubes   0483-34-6

BACTO BRAIN HEART INFUSION

INTENDED USE
Bacto Brain Heart Infusion is a highly nutritive broth medium used for cultivating a variety of fastidious microorganisms including streptococci, pneumococci and meningococci. Because of its nutritive qualities, Bacto Brain Heart Infusion is appropriate for the culture of blood.

HISTORY
In 1919, Rosenow[1] devised an excellent medium for culturing streptococci by adding pieces of brain tissue to dextrose broth. In this medium, he cultured organisms from focal infections in the teeth and other tissues. Hayden,[2] using Rosenow's procedure but adding crushed marble to the medium, reported that the medium favored growth of organisms from tooth infections, especially those showing a close relationship to eye infections.

PRINCIPLES
Bacto Brain Heart Infusion duplicates the media of Rosenow and Hayden in principle. It contains essential nutrients in a clear broth formulation. An infusion of brains has replaced the brain tissue and disodium phosphate has replaced the calcium carbonate buffer. If desired, 0.1% agar can be added to the broth to yield conditions of oxygen tension, similar to those produced by the brain tissue.

FORMULA
BACTO BRAIN HEART INFUSION
DEHYDRATED

Ingredients per liter

Calf Brains, Infusion from	200 g
Beef Heart, Infusion from	250 g
Proteose Peptone, Difco	10 g
Bacto Dextrose	2 g
Sodium Chloride	5 g
Disodium Phosphate	2.5 g

Final pH 7.4 ± 0.2 at 25°C

One pound will make 12.2 liters of final medium.

METHOD OF PREPARATION
1. Dissolve 37 grams in 1 liter distilled or deionized water.
2. Dispense as desired.
3. Sterilize in the autoclave for 15 minutes at 15 pounds pressure (121°C).

STORAGE
Bacto Brain Heart Infusion   Below 30°C
Prepared medium   15 – 30°C

QUALITY CONTROL

Identity Specifications

Dehydrated powder: light tan, homogeneous, free flowing
Reaction of 3.7% solution: pH 7.4 ± 0.2 at 25°C
Prepared medium: light to medium amber, clear without significant precipitate

Typical Cultural Response in Bacto Brain Heart Infusion
After 18 – 24 Hours at 35 ± 2°C

Organism	Recovery
Neisseria meningitidis ATCC® 13090	good to excellent
Streptococcus pneumoniae ATCC® 6303	good to excellent
Streptococcus pyogenes ATCC® 19615	good to excellent

REFERENCES

1. J. Dental Research, 1:205, 1919.
2. Arch. Internal Med., 32:828, 1923.

PACKAGING
Bacto Brain Heart Infusion

	1 lb (454 g)	0037-01-6
	1/4 lb (114 g)	0037-02-5
	5 lb (2.27 kg)	0037-05-2
	10 kg	0037-08-9
	12 tubes	0037-34-7
	144 tubes	0037-37-4

BRAIN HEART INFUSION MEDIA

BACTO BRAIN HEART INFUSION w/PAB
BACTO BRAIN HEART INFUSION w/PAB AND AGAR

INTENDED USE
Bacto Brain Heart Infusion w/PAB and Bacto Brain Heart Infusion w/PAB and Agar are modifications of brain heart infusion used for culturing blood, particularly from patients under sulfonamide therapy.

HISTORY
In 1921, Hitchens[1] described the advantages of a medium with a low agar concentration and its influence on the growth of bacteria, particularly anaerobes. In a broth containing uniformly distributed 0.1% agar, good growth of aerobes is obtained in the upper portion of the medium. Most anaerobes will grow in the lowest part of the medium.

Falk, Bucca and Simmons,[2] using media containing small amounts of agar (0.06 – 0.25%) to test the sterility of biologicals, demonstrated that even common bacteria exhibit growth in a much shorter incubation time in such media.

PRINCIPLES
Bacto Brain Heart Infusion w/PAB and Agar contains 5 mg percent p-aminobenzoic acid (0.05 g/liter) to neutralize the maximum amount of sulfonamide that could be carried over to the medium in the blood inoculum and inhibit growth of the pathogenic

organisms inoculated. In addition, the entire formulation will inactivate streptomycin in the ratio of 10 ml medium to 100 units of streptomycin.

The addition of 0.1% agar to the medium provides optimum conditions for aerobic organisms, microaerophiles and obligate anaerobes.

FORMULAE

BACTO BRAIN HEART INFUSION w/PAB
DEHYDRATED

Ingredients per liter

Calf Brains, Infusion from	200 g
Beef Heart, Infusion from	250 g
Proteose Peptone, Difco	10 g
Bacto Dextrose	2 g
Sodium Chloride	5 g
Sodium Phosphate, Dibasic	2.5 g
p-Aminobenzoic Acid	0.05 g

Final pH 7.4 ± 0.2 at 25°C.

One pound will make 12.2 liters.
Rehydrate with 37 grams/liter.

BACTO BRAIN HEART INFUSION w/PAB AND AGAR
DEHYDRATED

Ingredients per liter

Calf Brains, Infusion from	200 g
Beef Heart, Infusion from	250 g
Proteose Peptone, Difco	10 g
Bacto Dextrose	2 g
Sodium Chloride	5 g
Sodium Phosphate, Dibasic	2.5 g
p-Aminobenzoic Acid	0.05 g
Bacto Agar	1 g

Final pH 7.4 ± 0.2 at 25°C.

One pound will make 11.9 liters.
Rehydrate with 38 grams/liter.

METHOD OF PREPARATION

1. Suspend appropriate amount in 1 liter distilled or deionized water and heat to boiling to dissolve completely.
2. Dispense as desired.
3. Sterilize in the autoclave for 15 minutes at 15 lbs pressure (121°C).

STORAGE

Bacto Brain Heart Infusion w/PAB	Below 30°C
Bacto Brain Heart Infusion w/PAB and Agar	Below 30°C
Prepared medium	15 – 30°C

QUALITY CONTROL
Identity Specifications

Dehydrated powder:	light tan, homogeneous, free-flowing
Reaction of solution:	pH 7.4 ± 0.2 at 25°C
Prepared agar medium:	light to medium amber, slightly opalescent. Agar may settle upon standing.
Prepared broth medium:	light to medium amber, clear without significant precipitate

Typical Cultural Response After 18 – 24 Hours at 35°C

Organisms	Recovery
Neisseria meningitidis ATCC® 13090	fair to good
Streptococcus pneumoniae ATCC® 6303	good to excellent
Streptococcus pyogenes ATCC® 19615	good to excellent

REFERENCES

1. J. Infectious Diseases, 29:390, 1921.
2. J. Bact., 37:121, 1939.

PACKAGING

Bacto Brain Heart Infusion w/PAB	1 lb (454 g)	0498-01-8
Bacto Brain Heart Infusion	1 lb (454 g)	0499-01-7
w/PAB and Agar	12 tubes	0499-34-8

BACTO BRAIN HEART INFUSION AGAR

INTENDED USE
Bacto Brain Heart Infusion Agar is a complete medium used for cultivating a variety of fastidious microorganisms, fungi and yeasts. The medium is used in combination with penicillin and streptomycin for isolating fungi.

HISTORY
Roseburg, Epps and Clark[1] reported that Bacto Brain Heart Infusion with 2% agar was more satisfactory than 1% dextrose infusion agar for the isolation and cultivation of *Actinomyces israeli*. Incubation in an atmosphere of 5% carbon dioxide was required for best results. The addition of sheep blood to the medium offered no growth advantage. Howell[2] used Bacto Brain Heart Infusion to which was added 2% Bacto Agar and 10% sterile defibrinated horse blood for the cultivation of *Histoplasma capsulatum*. A selective medium for the isolation of this organism was prepared by adding 40 μg streptomycin and 20 units penicillin per ml of medium. In comparison with a blood agar similarly prepared from potato dextrose agar, the brain heart infusion agar gave a greater number of positive isolations. Incubation at room temperature was more efficient than at 37°C. Colonies of *H. capsulatum* isolated on brain heart infusion agar must be transferred to a medium such as potato dextrose agar to obtain the characteristic tuberculate chlamydospores typical of this fungus. Conant[3] recommended that a plate of Bacto Brain Heart Infusion Agar be streaked and incubated at 37°C under anaerobic conditions with the addition of 5% carbon dioxide to obtain growth of the microaerophilic *A. bovis* in culturing this organism from infected mucous membranes and subcutaneous tissues.

Kotcher, Robinson and Miller[4] compared various media for the isolation of *H. capsulatum* from tissues of experimentally infected mice. Their results showed that brain heart infusion blood agar gave the highest percentage recovery of *H. capsulatum* from the tissues of the infected mice.

FORMULA
BACTO BRAIN HEART INFUSION AGAR
DEHYDRATED

Ingredients per liter

Calf Brains, Infusion from	200 g
Beef Hearts, Infusion from	250 g
Proteose Peptone, Difco	10 g
Bacto Dextrose	2 g
Sodium Chloride	5 g
Disodium Phosphate	2.5 g
Bacto Agar	15 g

Final pH 7.4 ± 0.2 at 25°C.

One pound will make 8.7 liters of final medium.

METHOD OF PREPARATION

1. Suspend 52 g in 1 liter distilled or deionized water and heat to boiling to dissolve completely.
2. Sterilize in the autoclave for 15 minutes at 15 lbs pressure (121°C).
3. If the selective medium for fungi is desired, cool medium to 50 – 55°C and aseptically add 20 units penicillin and 40 μg streptomycin per ml of sterile medium.
4. Dispense as desired.

STORAGE

Bacto Brain Heart Infusion Agar	Below 30°C
Prepared medium	15 – 30°C

QUALITY CONTROL

Identity Specifications

Dehydrated powder:	medium tan, homogeneous, free-flowing.
Reaction of 5.2% solution:	pH 7.4 ± 0.2 at 25°C
Prepared medium:	light to medium amber, opalescent with flocculent precipitate

Typical Cultural Response on Bacto Brain Heart Infusion Agar
After 24 – 96 Hours at 35°C or *30°C

Organism	Recovery w/o Blood	Recovery w/5% Sheep Blood**
*Aspergillus niger ATCC® 16404	excellent	excellent
Clostridium difficile ATCC® 17858	not tested	good
Neisseria meningitidis ATCC® 13090	fair to good	good
*Saccharomyces cerevisiae ATCC® 9763	excellent	excellent
Streptococcus pneumoniae ATCC® 6303	fair to good	good
Streptococcus pyogenes ATCC® 19615	fair to good	good

**This medium is not recommended for determining hemolytic reactions due to the high dextrose content.

REFERENCES

1. J. Infectious Diseases, 74:131, 1944.
2. Public Health Reports, 63:173, 1948.
3. Diagnostic Procedures and Reagents, 3rd Ed., 452, 1950.
4. J. Bact., 62:613, 1951.

PACKAGING

Bacto Brain Heart Infusion Agar	1 lb (454 g)	0418-01-5
	1/4 lb (114 g)	0418-02-4
	5 lb (2.27 kg)	0418-05-1
	12 tubes	0418-34-6
	144 tubes	0418-37-3

BACTO BREWER ANAEROBIC AGAR

INTENDED USE

Bacto Brewer Anaerobic Agar is used for cultivating anaerobic and microaerophilic organisms.

HISTORY
In 1942, Brewer[1] described a special Petri dish cover to permit surface growth of anaerobes and microaerophiles on agar with a low oxidation-reduction potential without the use of anaerobic jars or other special apparatus.

PRINCIPLES
While originally formulated and modified for use in the procedure described by Brewer, Bacto Brewer Anaerobic Agar is a suitable plating medium for cultivating anaerobes by techniques currently in use.

FORMULA
BACTO BREWER ANAEROBIC AGAR
DEHYDRATED
Ingredients per liter

Bacto Tryptone	5 g	Bacto Agar .... 20 g
Proteose Peptone No. 3, Difco	10 g	Sodium Thioglycollate, Difco ..... 2 g
Bacto Yeast Extract	5 g	Sodium Formaldehyde
Bacto Dextrose	10 g	Sulfoxylate .... 1 g
Sodium Chloride	5 g	Resazurin, Certified .... 0.002 g

Final pH 7.2 ± 0.2 at 25°C.

One pound will make 7.8 liters of medium.

METHOD OF PREPARATION
1. Suspend 58 g in 1 liter distilled or deionized water and heat to boiling to dissolve completely.
2. Sterilize in the autoclave for 15 minutes at 15 lbs pressure (121°C).
3. Dispense as desired.

STORAGE
Bacto Brewer Anaerobic Agar   Below 30°C
Prepared medium   2 – 8°C

QUALITY CONTROL
Identity Specifications

Dehydrated powder: light beige, homogeneous, free-flowing
Reaction of 5.8% solution: pH 7.2 ± 0.2 at 25°C
Prepared medium: light amber, slightly opalescent, becoming red due to aeration on standing.

Typical Cultural Response on Bacto Brewer Anaerobic Agar
After 18 – 48 Hours at 35°C, Anaerobically

Organism	Growth
Clostridium beijerincki ATCC® 17795 (formerly *C. multifermentans*)	good to excellent
Clostridium botulinum ATCC® 19397	good to excellent
Clostridium perfringens ATCC® 12924	good to excellent
Clostridium sporogenes ATCC® 11437	good to excellent

REFERENCE
1. Science, 95:587, 1942.

PACKAGING
Bacto Brewer Anaerobic Agar   1 lb (454 g)   0279-01-3

BACTO BREWER THIOGLYCOLLATE MEDIUM
BACTO BREWER THIOGLYCOLLATE MODIFIED

INTENDED USE

Bacto Brewer Thioglycollate Medium and Bacto Brewer Thioglycollate Modified are prepared for use in testing the sterility of biological products and other materials. They are particularly suited for the cultivation of anaerobic and microaerophilic microorganisms.

HISTORY

Bacto Brewer Thioglycollate Medium is the original Brewer formula[1,2] as formerly specified in National Institute of Health Bulletin, "Fluid Thioglycollate Medium for Sterility Tests."[3]

Linden thioglycollate medium was also formerly specified in the National Institute of Health[3] for sterility testing. Bacto Brewer Thioglycollate Modified is a modification of Linden thioglycollate medium.

PRINCIPLES

The thioglycollate in these formulations along with the small amount of agar present helps to ensure reduced aerobic conditions in the media. The thioglycollate present in these products helps to neutralize the toxicity of mercurial preservatives which might be present in the test materials.

These formulae contain Bacto Methylene Blue as an oxidation-reduction indicator. An increase in oxidation of the medium changes this indicator's color from colorless to blue.

FORMULAE

	BACTO BREWER THIOGLYCOLLATE MEDIUM DEHYDRATED Ingredients per liter	BACTO BREWER THIOGLYCOLLATE MODIFIED DEHYDRATED Ingredients per liter
Proteose Peptone, Difco	10 g	—
Beef, Infusion from	500 g	—
Tryptic Digest of Casein	—	17 g
Bacto Soytone	—	3 g
Dextrose	5 g	10 g
Sodium Chloride	5 g	5 g
Dipotassium Phosphate	2 g	2 g
Sodium Thioglycollate, Difco	0.5 g	1 g
Bacto Agar	0.5 g	0.5 g
Bacto Methylene Blue	0.002 g	0.002 g
Final pH at 25°C	7.2 ± 0.2	7.2 ± 0.2
Grams per liter	40.5	38.5
One pound will make liters of medium	11.2	11.79

METHOD OF PREPARATION

1. Suspend the appropriate amount of medium in 1 liter distilled or deionized water and heat to boiling to dissolve the medium completely.
2. Dispense into tubes or other desired culture vessels.
3. Sterilize in the autoclave for 15 minutes at 15 lbs pressure (121°C).

NOTE: If medium is not used immediately after preparation and more than 20% of the uppermost portion of the medium has changed to a green color, it is not suitable for use. Under such circumstances, one reheating in a boiling water bath is permissible to drive off the absorbed oxygen.

STORAGE
Bacto Brewer Thioglycollate Media   Below 30°C
Prepared media   15 – 30°C in the dark

QUALITY CONTROL
Identity Specifications

Dehydrated powder: beige, homogeneous, free-flowing
Reaction of appropriate solution: pH 7.2 ± 0.2 at 25°C
Prepared medium: light to medium amber, clear to very slightly opalescent, with upper 10% or less medium green

Typical Cultural Response on Bacto Brewer Thioglycollate Media
After 18 – 48 Hours at 35°C

Organism	Growth
Bacteroides melaninogenicus ATCC® 25845	good to excellent
Clostridium sporogenes ATCC® 11437	good to excellent
Streptococcus mitis ATCC® 9895	good to excellent

REFERENCES
1. J. Bact., 39:10, 1940.
2. JAMA, 115:598, 1940.
3. "Fluid Thioglycollate Medium for the Sterility Test," National Institute of Health Bulletin, Dec. 30, 1941.

PACKAGING
Bacto Brewer Thioglycollate Medium	1 lb (454 g)	0236-01-5
Bacto Brewer Thioglycollate Modified	1 lb (454 g)	0237-01-4

BACTO BRILLIANT GREEN AGAR

INTENDED USE
Bacto Brilliant Green Agar is a highly selective medium recommended for the isolation of *Salmonella*, other than *Salmonella typhi*, directly from stools or other materials suspected of containing these organisms or after preliminary enrichment in Bacto Tetrathionate Broth.

HISTORY
The use of a brilliant green agar as a primary plating medium for the isolation of *Salmonella* was first described by Kristensen, Lester and Jurgens[1] who reported it useful for the differentiation of "paratyphoid B" from other intestinal gram-negative bacilli. Later, Kauffmann[2] modified their formula and used the brilliant green agar in conjunction with a tetrathionate broth for the isolation of *Salmonella* from stools. Galton and Quan[3] increased their positive *Salmonella* findings 164% by the use of tetrathionate broth and plating on brilliant green agar. Broh-Kahn[4] and Edwards[5] similarly employed the Kauffmann modification of brilliant green agar with superior results.

PRINCIPLES

Bacto Brilliant Green Agar is a slight modification of the medium as described by Kauffmann. The outstanding selectivity of this medium permits the use of moderately heavy inocula, which should be evenly distributed over the surface. Inoculation with heavy suspensions of stools or other materials suspected of containing *Salmonella* usually results in an almost pure culture of these organisms. Growth of other bacteria is almost completely inhibited by the presence of the brilliant green dye. The typical *Salmonella* colonies appear as slightly pink-white opaque colonies surrounded by a brilliant red medium. The few lactose or sucrose fermenting organisms are readily differentiated due to the formation of a yellow-green colony surrounded by an intense yellow-green zone.

Bacto Brilliant Green Agar is not suitable for the isolation of *S. typhi* or *Shigella* organisms, however some strains of *S. typhi* and *Proteus* may grow forming red colonies. In the routine examination of stools, rectal swabs, or other materials for the gram-negative intestinal pathogens, other primary plating media such as Bacto SS Agar, Bacto Bismuth Sulfite Agar and Bacto MacConkey Agar, as well as fluid enrichments such as Bacto Tetrathionate Broth and Bacto Selenite Broth, should be used with Bacto Brilliant Green Agar.

FORMULA

BACTO BRILLIANT GREEN AGAR
DEHYDRATED

Ingredients per liter

Proteose Peptone No. 3, Difco	10 g
Bacto Yeast Extract	3 g
Bacto Lactose	10 g
Bacto Saccharose	10 g
Sodium Chloride	5 g
Bacto Agar	20 g
Bacto Brilliant Green	0.0125 g
Bacto Phenol Red	0.08 g

Final pH 6.9 ± 0.2 at 25°C.

One pound will make 7.8 liters of medium.

METHOD OF PREPARATION

1. To rehydrate, suspend 58 g in 1 liter distilled or deionized water and heat to boiling to dissolve completely.
2. Sterilize in the autoclave for 15 minutes at 15 lbs pressure (121°C). AVOID OVERHEATING.
3. Dispense as desired.
4. Inoculate the sterilized medium by evenly distributing over surface.
5. Incubate at 35°C for 18 – 24 hours and examine plates for growth.

STORAGE

Bacto Brilliant Green Agar	Below 30°C
Prepared plates	2 – 8°C

QUALITY CONTROL
Identity Specifications

Dehydrated powder:	pink, homogeneous, free-flowing
Reaction of 5.8% solution:	pH 6.9 ± 0.2 at 25°C
Prepared medium:	brownish green, very slightly opalescent

Typical Cultural Response on Bacto Brilliant Green Agar
After 18 – 24 Hours at 35 ± 2°C

Organism	Recovery	Color of Colony
Escherichia coli ATCC® 25922	none to poor	yellow-green
Salmonella enteritidis ATCC® 13076	good	pink-white
Salmonella typhi ATCC® 19430	none to poor	red
Salmonella typhimurium ATCC® 14028	good	pink-white
Staphylococcus aureus ATCC® 25923	none	—

REFERENCES

1. Brit. J. Exp. Path., 6:291, 1925.
2. Zeit. Hyg., 117:26, 1935.
3. Am. J. Pub. Health, 34:1071, 1944.
4. Military Surgeon, 99:770, 1946.
5. Personal Communication, 1947.

PACKAGING

Bacto Brilliant Green Agar	1 lb (454 g)	0285-01-5
	1/4 lb (114 g)	0285-02-4

BACTO BRILLIANT GREEN BILE AGAR

INTENDED USE

Bacto Brilliant Green Bile Agar is a selective and differential medium for determining the relative density of coliform bacteria in water,[1] sewage and foods.[2]

HISTORY

Bacto Brilliant Green Bile Agar duplicates the medium described by Noble and Tonney[1] for determining the relative density of coliform bacteria in water and sewage. This medium is not recommended for the determination of the absolute density of coliform organisms in water samples, but rather as an indication of the degree of contamination of the sample.

PRINCIPLES

Brilliant green dye and bile in the medium are the selective agents for the gram-negative bacilli. Those gram-negatives that ferment lactose form colonies that are deep red at the center with a pink halo against a blue green background. The colonies vary from 0.4 to 0.8 mm in diameter.

FORMULA

BACTO BRILLIANT GREEN BILE AGAR
DEHYDRATED
Ingredients per liter

Bacto Peptone	8.25 g	Monopotassium Phosphate	0.0153 g
Bacto Lactose	1.9 g	Agar Noble	10.15 g
Bacto Oxgall	0.00295 g	Erioglaucine	0.0649 g
Sodium Sulfite	0.205 g	Bacto Basic Fuchsin	0.0776 g
Ferric Chloride	0.0295 g	Bacto Brilliant Green	0.0000295 g

Final pH 6.9 ± 0.2 at 25°C.

Five hundred grams will make 24.2 liters of medium.

METHOD OF PREPARATION
1. Suspend 20.6 g in 1 liter distilled or deionized water and heat to boiling to dissolve completely.
2. Sterilize in the autoclave for 15 minutes at 15 lbs pressure (121°C).

STORAGE
Bacto Brilliant Green Bile Agar   Below 30°C
Prepared medium   2 – 8°C

The medium is rather sensitive to light, particularly direct sunlight, which produces a decrease in the productivity of the medium and change in color from deep blue to purple or red. It is recommended that the medium be prepared just prior to use and when necessary to store the medium, it should be kept in the dark.

QUALITY CONTROL
Identity Specifications
Dehydrated powder: light purple, homogeneous, free-flowing
Reaction of 2.06% solution: pH 6.9 ± 0.2 at 25°C
Prepared medium: bluish purple, slightly opalescent

Typical Cultural Response on Bacto Brilliant Green Bile Agar
After 18 – 24 Hours at 35°C

Organism	Growth	Color of Colony
Enterobacter aerogenes ATCC® 13048	good to excellent	pink
Escherichia coli ATCC® 25922	good to excellent	deep red w/bile ppt.
Salmonella enteritidis ATCC® 13076	good to excellent	colorless to light pink
Staphylococcus aureus ATCC® 25923	inhibited	—

REFERENCES
1. Noble and Tonney, 1935, Journal of American Water Works Association, 27:108.
2. Compendium of Methods for the Microbiological Examination of Foods, American Public Health Association Inc., Washington D.C., 1976.

PACKAGING
Bacto Brilliant Green Bile Agar   500 g   0014-17-5

BACTO BRILLIANT GREEN BILE 2%

INTENDED USE
Bacto Brilliant Green Bile 2% is a Standard Methods medium for the confirmed and completed tests for coliform bacteria in water,[1] dairy[2] and other food products.[3]

HISTORY/PRINCIPLES
The development of a selective medium which would inhibit organisms other than members of the coliform group has long been of interest to sanitary bacteriologists. The principal interest has been in media containing bile and brilliant green. The American Water Works Associaton made extensive studies of this problem through its Committee No. 1 on Standard Methods of Water Analysis. Dunham and Schoenlein[4] have recorded their investigations of the proportions of bile and brilliant green giving optimum results. They reported that, under the conditions of their investigations, a reduced bile

content and a dilution of dye higher than that originally suggested by Muer and Harris,[5] improved conditions for the development of *Escherichia coli*. The necessity of maintaining the proper concentration of ingredients after the water sample is added was emphasized. Jordan[6] indicated that this medium is slightly superior to lactose broth and to the more concentrated bile medium in the detection of the coliform group in water. McCrady and Langevin[7] reported that Bacto Brilliant Green Bile 2% is satisfactory for the detection of the coliform group in controlling the pasteurization of milk. McCrady[8] in studies on media for the detection of the presence of coliform organisms in water, found that while selective media were not as satisfactory as lactose broth for the presumptive phase, they could be recommended for confirmation of the presumptive phase and that for this purpose Bacto Brilliant Green Bile 2% was most satisfactory.

Bacto Brilliant Green Bile 2% is prepared according to the formula specified in *Standard Methods for the Examination of Water and Wastewater*,[1] *Standard Methods for the Examination of Dairy Products*,[2] and *Compendium of Methods for the Microbiological Examination of Foods*.[3]

When the medium is to be used in water purification plant control where the inoculum is greater than 1 ml, particular care must be taken to preserve the correct concentration of dye and bile in the medium after dilution with the sample. The table given below indicates the quantity of dehydrated medium to use per 1 liter distilled water to maintain the correct concentrations of dye and bile.

Concentrations of Dehydrated Medium Required to Maintain the Proper Concentration of Ingredients

Inoculum	Volume Medium in Tube	Volume Medium and Inoculum	Bacto Brilliant Green Bile 2% used per liter
1 ml or less	10 ml	10 ml	40 g
10 ml	20 ml	30 ml	60 g
10 ml	30 ml	40 ml	53 g

In the presumptive phase for members of the coliform group in the examination of dairy products, a series of tubes of Bacto Brilliant Green Bile 2% is inoculated with appropriate dilutions of the sample. Use 5 tubes of each solution. Select dilutions to provide at least one positive and one negative tube in the series inoculated. Incubate tubes for 48 hours at 35 – 37°C. Gas formation constitutes a positive presumptive phase.

FORMULA
BACTO BRILLIANT GREEN BILE 2%
DEHYDRATED

Ingredients per liter

Bacto Peptone	10 g
Bacto Oxgall	20 g
Bacto Lactose	10 g
Brilliant Green	0.0133 g

Final pH 7.2 ± 0.2 at 25°C.

One pound will make 11.3 liters of medium.

METHOD OF PREPARATION
1. Suspend 40 g in 1 liter distilled or deionized water and warm slightly to dissolve completely.
2. Dispense required amount in test tubes.

3. Place an inverted fermentation vial in each tube.
4. Place closure on tubes and sterilize in the autoclave for 15 minutes at 15 lbs pressure (121°C).
5. Before opening the autoclave, allow the temperature to drop below 75°C to avoid entrapment of air bubbles in the inverted vials.
6. Follow the recommended procedure in the Standard Methods compendium for desired use.

STORAGE
Bacto Brilliant Green Bile 2%   Below 30°C
Prepared tubes   2 – 8°C

QUALITY CONTROL
Identity Specifications
Dehydrated medium: greenish beige, homogeneous, free-flowing
Reaction of 4.0% solution: 7.2 ± 0.2 at 25°C
Prepared medium: emerald green, clear without significant precipitate

Typical Cultural Response in Bacto Brilliant Green Bile 2%
After 18 – 48 Hours at 35°C

Organisms	Recovery	Gas
Enterobacter aerogenes ATCC® 13048	good to excellent	+
Escherichia coli ATCC® 25922	good to excellent	+
Staphylococcus aureus ATCC® 25923	inhibited	–
Streptococcus faecalis ATCC® 19433	inhibited	–

REFERENCES
1. Standard Methods for the Examination of Water and Wastewater, 15th Ed., American Public Health Association, Inc., Washington, D.C., 1980.
2. Standard Methods for the Examination of Dairy Products, 14th Ed., American Public Health Association Inc., Washington, D.C., 1978.
3. Compendium of Methods for the Microbiological Examination of Foods, American Public Health Association Inc., Washington, D.C., 1976.
4. Dunham and Schoenlein, 1926, Stain Technology, 1:129.
5. Muer and Harris, 1920, American Journal of Public Health, 10:874.
6. Jordan, 1927, J. American Water Works Association, 18:337.
7. McCrady and Langevin, 1932, J. Dairy Science, 15:321.
8. McCrady, 1937, American Journal of Public Health, 27:1243.

PACKAGING
Bacto Brilliant Green Bile 2%		
1 lb (454 g)	0007-01-2
1/4 lb (114 g)	0007-02-1
5 lb (2.27 kg)	0007-05-8

BACTO m BRILLIANT GREEN BROTH

INTENDED USE
Bacto m Brilliant Green Broth is a selective differential medium for primary screening for *Salmonella* in polluted water by the membrane filter technique.

HISTORY/PRINCIPLES
Bacto m Brilliant Green Broth is a modification of Kauffmann's[1] brilliant green agar, in that it contains no agar and all other constituents of the medium are in double strength.

Kabler and Clark[2] mentioned the application of this medium in a membrane filter screening procedure originally developed by Geldreich and Jeter.[3] In this technique, an appropriate volume of the water sample is filtered through the membrane filter and the filter placed on an absorbent pad saturated with Bacto m Tetrathionate Broth Base. After incubation at 35°C for 3 hours in a moisture saturated atmosphere, the membrane is transferred to another absorbent pad saturated with Bacto m Brilliant Green Broth and incubation at 35°C is continued for 15 more hours. Following this incubation period on the tetrathionate and brilliant green media, the membrane is transferred to a fresh pad saturated with urease test reagent.[*] Within 15 – 20 minutes the reactions on the urease test reagent are observed and recorded.

Purple colonies are urease positive and lactose and saccharose negative — probably *Proteus*. Yellow colonies are urease negative and lactose or sucrose positive — coliforms. Red colonies are urease negative and lactose and sucrose negative — probably *Salmonella*.

*Urease test reagent — (urea 20 g, brom thymol blue 0.16 g, phenol red 0.2 g) per liter of water.

FORMULA

BACTO m BRILLIANT GREEN BROTH
DEHYDRATED
Ingredients per liter

Proteose Peptone No. 3, Difco	20 g
Bacto Yeast Extract	6 g
Bacto Lactose	20 g
Bacto Saccharose	20 g
Sodium Chloride	10 g
Bacto Phenol Red	0.16 g
Bacto Brilliant Green	0.025 g

Final pH 6.9 ± 0.2 at 25°C.

One pound will make 16 liters of medium.

METHOD OF PREPARATION
1. Suspend 76 g in 1 liter distilled or deionized water and heat to boiling to dissolve completely.
2. Cool to room temperature.
3. Add 2 ml to each sterile absorbent pad being used.

NOTE: The medium should be used within 24 hours of rehydration.

STORAGE
Bacto m Brilliant Green Broth   Below 30°C
Prepared medium   Storage not recommended

QUALITY CONTROL
Identity Specifications

Dehydrated powder: pink, homogeneous, free-flowing
Reaction of 7.6% solution: pH 6.9 ± 0.2 at 25°C
Prepared medium: greenish red, slightly opalescent

Typical Cultural Response on Bacto m Brilliant Green Broth
After 18 – 24 Hours at 35°C in Humid Atmosphere

Organism	Growth	Color of Colony
Escherichia coli ATCC® 25922	good to excellent	yellow
Salmonella enteritidis ATCC® 13076	good to excellent	pink to red
Salmonella typhimurium ATCC® 14028	good to excellent	pink to red

REFERENCES

1. Kauffmann, 1935, Zeit. Hyg., 117:26.
2. Kabler and Clark, 1952, American Journal of Public Health, 42:390.
3. Geldreich and Jeter, 1952, Bact. Proc. SAB, Boston, p. 33.

PACKAGING

Bacto m Brilliant Green Broth	1 lb (454 g)	0494-01-2
Bacto m Tetrathionate Broth Base	1 lb (454 g)	0580-01-7

BACTO BRUCELLA ANTIGENS
BACTO BRUCELLA ANTISERA

PRODUCT DESCRIPTION AND INTENDED USE

Bacto Brucella Antigens are chemically inactivated suspensions of widely recognized strains of commonly encountered *Brucella* for use in the slide and tube agglutination procedures to detect antibodies (agglutinins) in serum.

The rapid slide procedure is a screening test designed to detect agglutinins, whereas the tube test is a confirmatory procedure designed to quantitate agglutinin compositions. It is therefore necessary that any positive results obtained in the screening (slide test) of specimens be confirmed by a tube test.

Bacto Brucella Antisera are hyperimmune antisera prepared in rabbits designed to demonstrate agglutination for use as a positive control employing these test procedures.

These reagents are for use only in the test procedure described herein. They are not recommended for the direct diagnosis of a disease but, rather, only for the detection of the presence of agglutinins, particularly in cases of seroconversion.

SUMMARY AND BACKGROUND

Brucella is representative of a number of species of pathogenic microorganisms which, upon invasion, produce a fever in its host. Consequently, it is often referred to as a "febrile antigen."

For a complete discussion refer to Bacto Febrile Antigens.

There are 6 recognized species of *Brucella* as listed in Table I. Of these only the first 3 are generally associated with diseases in man.[1]

Table I *Brucella* Species

Species		Human Suscept.	A & M* Antigens	Basic Fuchsin A	Basic Fuchsin B	Thionin C	Thionin D	CO₂ Req'd.	H₂S
1. *B. abortus*	biotype 1	Yes	Both	+	+	−	−	d	+
	2			−	−	−	−	d	+
	3			+	+	+	+	d	+
	4			+	+	−	−	d	+
	5			+	+	−	+	−	−
	6			+	+	−	+	−	−
	7			+	+	−	+	−	d
	8			+	+	−	+	+	−
	9			+	+	−	+	−	+

Table I *Brucella* Species (continued)

Species	Human Suscept.	A & M* Antigens	Growth on Basic Fuchsin		Thionin		CO₂ Req'd.	H₂S
			A	B	C	D		
2. *B. melitensis* biotype 1	Yes	Both	+	+	−	+	−	−
2			+	+	−	+	−	−
3			+	+	−	+	−	−
3. *B. suis* biotype 1	Yes	Both	−	−	+	+	−	+
2			−	−	−	+	−	−
3			+	+	+	+	−	−
4			+	+	+	+	−	−
4. *B. neotomae*	No	A	−	−	−	−	−	+
5. *B. ovis*	No	Neither	+	+	+	+	+	−
6. *B. canis*	Rare	Neither	−	±	+	+	−	−

*A = abortus, M = melitensis A = 1:50,000 B = 1:100,000 C = 1:25,000 D = 1:50,000 d = delayed

The Rapid Slide Test is the most widely used procedure employing bacterial antigens because of the simplicity with which the results may be reported. Negative reactions with this test can usually be reported as such if all 5 serum dilutions have been used. **Note:** Even though the slide test is not quantitative, it is necessary to run the series of dilutions in order to detect agglutinin content of a serum which might be overlooked in the case of a "prozone phenomenon." This often occurs in the sera containing a high titer of *Brucella* agglutinins where higher concentrations of the serum may yield negative results but a dilution of the serum is positive.

In this test, varying amounts of the sera to be tested are distributed to the corresponding square of a previously marked glass slide. The antigen is added and mixed with the sera. Positive or negative results are read 1 minute after mixing. Any positive reaction in the slide test must be confirmed using the tube test.

The Macroscopic Tube Test[2] should be used to confirm the presence of antibodies demonstrated by the slide technique and to quantitate their titer in suspected sera. In this test, the patient's serum is serially diluted in test tubes, a constant amount of the appropriate dilution of the antigen is added to each tube, the resultant mixture is incubated according to directions and the agglutination pattern is read and recorded.

REAGENTS
Antigens
Bacto Brucella Abortus Antigen (slide and tube) are prepared from *B. abortus* 1119 - 3.[2,3,4]

Bacto Brucella Melitensis Antigen (slide) is prepared from *Brucella melitensis* 2500 obtained from Huddleson.[5]

Bacto Brucella Suis Antigen (slide) is prepared from *Brucella suis* 1820 obtained from Huddleson.[5]

When used according to the suggested procedure, each 5 ml vial of Bacto Brucella Antigen (slide) is sufficient to perform approximately 20 slide tests.

Each 25 ml vial of Bacto Brucella Abortus Antigen (tube) is sufficient to perform approximately 6 tube tests.

Concentration of Antigens

Bacto Brucella Abortus Antigen (tube) is adjusted to a density approximating a Mc-Farland Barium Sulfate Standard No. 3 (9×10^8 organisms/ml).

Bacto Brucella Antigens (slide) are adjusted to a density of approximately 10% suspension v/v.

Note: The density of the antigens may vary from the above values. They are adjusted to perform at their optimum when standardized with hyperimmune sera obtained from laboratory animals. In some antigens glycerin to a final volume of 20% is added when necessary to adjust the suspension sensitivity.

Preservatives

Bacto Brucella Abortus Antigen (tube) contains a final concentration of 0.5% phenol (5 g/liter). Bacto Brucella Antigens (slide) contain a final concentration of 0.5% phenol (5 g/liter), crystal violet in approximately 0.002% final concentration, and brilliant green in approximately 0.005% concentration.

Due to various factors, including the genus of the organism involved, the density of each lot of suspension etc., the color intensity of each lot of antigen may vary and is normal. Such differences in color will not affect the outcome of the test.

Control Sera

Bacto Brucella Antisera are unabsorbed antisera prepared in rabbits. All 3 species of the *Brucella* listed are antigenically related since they contain common A (*abortus*) and M (*melitensis*) substances. It is obvious from this that serological cross reactions exist in unabsorbed sera from these species. Monospecific sera prepared by absorption result in sera that are weak, making interpretation of agglutination results difficult and producing a fairly unstable reagent.

Merthiolate® is added as a preservative in the amount of 1:5000 (w/v) final concentration.

Bacto Febrile Negative Control is a standardized protein solution, containing Merthiolate® in the amount of 1:5000 (w/v) final concentration.

When used according to the recommended procedure, each 3 ml vial of Bacto Brucella Antisera is sufficient to perform approximately 19 slide tests. Each 3 ml vial of Bacto Brucella Abortus Antiserum is sufficient to perform approximately 30 tube tests.

Each 5 ml vial of Bacto Febrile Negative control is sufficient to perform approximately 32 slide or 48 tubes tests.

INSTRUCTIONS FOR REHYDRATION AND STORAGE
1. The antigens for both tests are liquid and ready for use.
2. The control serum is rehydrated by adding 3 ml sterile 0.85% NaCl solution at room temperature.
3. Bacto Febrile Negative Control is rehydrated by adding 5 ml sterile distilled or deionized water.

4. Store all reagents at 2 – 8°C. These reagents are stable to the expiry date on the label when stored at 2 – 8°C provided they have not been exposed to temperature extremes (excessive heat or freezing) and have not become contaminated chemically or bacterially during routine usage.

Bacto Brucella Antigens are not to be used for immunization of man or animals.

SPECIMEN COLLECTION
1. Collect 5 – 10 ml whole blood aseptically from the patient.
2. Allow the blood to clot and obtain the syneresed serum with a Pasteur pipette. If the serum is not free of erythrocytes, clarify by centrifugation. Note: Do not heat the serum since significant antibodies may be thermolabile.

PROCEDURE
Materials Provided:
Bacto Brucella Abortus Antigen (slide)
Bacto Brucella Abortus Antigen (tube)
Bacto Brucella Abortus Antiserum
Bacto Brucella AMS Antiserum Poly
Bacto Brucella Melitensis Antigen (for slide procedure)
Bacto Brucella Melitensis Antiserum
Bacto Brucella Suis Antigen (for slide procedure)
Bacto Brucella Suis Antiserum

Droppers for the above antigens (supplied droppers deliver approximately 0.05 ml per drop.)

Materials Required but not Provided:
Rapid Slide Test
Wax marking pencils
Ruler
Glass plate 8″ × 8″
0.2 ml serological pipettes
Applicator sticks
Suitable light source
Isotonic saline (0.85 g NaCl/100 ml distilled or deionized water)

Purchase Separately:
Bacto Febrile Negative Control

Tube Test:
Isotonic saline (0.85 g NaCl/100 ml distilled or deionized water)
Kahn tubes 12 mm × 75 mm
Kahn type tube supports
Serological pipettes, 1 ml and 5 ml capacity
Water bath at 37°C
Suitable light source

PRECAUTIONS
All reagents and equipment used in the slide test must be at room temperature prior to use.

Specimen
1. The specimen must be clear and free of visible fat. It must be free of excessive hemolysis and not bacterially contaminated.
2. The specimen must be used in the unheated state. If inactivated by heat, some thermolabile agglutinins may be destroyed.

3. In mixing the tubes in the the tube test, the specimen should not be mixed too vigorously since foaming may result in agglutinin denaturization.

Glassware
1. All glassware that is employed in the preparation, testing and storage of these reagents must be free of detergents or other harmful residues.
2. Pipettes employed must be clean (acid washed if necessary) to deliver proper volumes.

Antigens
1. Shake antigen vial well before use to insure a smooth, uniform suspension.
2. Discard any antigen demonstrating positive reactions in the negative control or demonstrating spontaneous agglutination in the vial. Febrile antigens will demonstrate autoagglutination if at any time during shipment or storage they are subject to freezing temperatures. Do not allow to freeze.
3. Discard any antigen which does not react properly in the positive control serum.
4. Variations in the color intensity of the antigen is normal and will not affect the outcome of the test.

Control Sera
Discard any control serum in which a precipitate forms. If bacterial contamination is suspected, autoclave the reagent or use some suitable chemical procedure to inactivate the contaminant before discarding the reagent.

METHODOLOGY
SPECIAL NOTE: Because a change or no change in titer over a period of time are the best indicators of active infection or noninfection, respectively, and because the accuracy and precision of the tests can be affected not only by test conditions but also by the subjectivity of the person in reading the end-point, the following protocol is recommended.

A preliminary test using either rapid slide test and/or the macroscopic tube test may be performed on the initial serum specimen and reported to the physician at that time. An aliquot of the serum should be transferred to a sterile test tube, sealed tightly, and kept in the freezer. When the second serum is obtained, it should be run in parallel with the original specimen. In this manner the original serum will serve as a control and any difference in titer will be more credible since the bias associated with the performance of the test, as well as determining the end-point, will be minimized. Obviously, it would be best if the person who performed the original test also performed the second test.

Exposure to heat (from external sources, light source, burner flame, etc.) may cause evaporation of the test mixture and may result in false positive interpretations. Avoid such evaporation of the test mixture.

Rapid Slide Test
Note: The rapid slide test should be used for screening only. Any positive results in the slide procedure should be confirmed by the tube test.

In some cases of brucellosis, sera may demonstrate a prozone (inability of an antigen to react in higher serum concentrations). It is advised that all 5 serum dilutions be run in the rapid slide test rather than just 1 dilution to eliminate the possibility of missing positive reactions due to the prozone.

1. Prepare the glass plate 8″ × 8″ by making a series of 1 - 1/2″ ruled squares with the ruler and wax marking pencil. Each row of 5 squares is sufficient to test 1 antigen against the sera diluted to 1:320. Note: The glass plate should be thoroughly cleaned and dried after each use.
2. Pipette 0.08, 0.04, 0.02, 0.01, and 0.005 ml serum onto a row of squares on the ruled glass plate using a 0.2 ml serological pipette.
3. Place a drop of the appropriate antigen for the slide test on each drop of serum. Note: Shake the antigen well before using.
4. Mix the serum-antigen composite with an applicator stick, starting with the 0.005 ml serum dilution and working towards the 0.08 ml dilution. The final dilutions are correlated approximately to the macroscopic tube dilutions and are 1:20, 1:40, 1:80, 1:160 and 1:320, respectively.
5. Hold the glass plate in both hands and gently "rotate" it 15 – 20 times. Observe the agglutination within 1 minute over a suitable light source. Reactions occurring later may be due to the reactants drying on the slide.

Macroscopic Tube Test
Prepare serial dilutions of serum to be tested and the control sera in 0.5 ml amounts in Kahn tubes in the following manner:
1. Place 8 Kahn tubes in a rack for each serum to be tested.
2. Pipette 0.9 ml of 0.85% NaCl into the first tube of each row and 0.5 ml into each of the remaining tubes.
3. Add 0.1 ml of the serum to the first tube containing 0.9 ml of NaCl solution.
4. Mix well with a pipette and transfer 0.5 ml to the second tube. Mix thoroughly.
5. Continue carrying the 0.5 ml of the serum dilution through tube 7. Discard 0.5 ml from tube 7 after mixing thoroughly. Tube 8 is the antigen control tube.
6. Add 0.5 ml of Bacto Brucella Abortus Antigen (Tube) to each of the 8 tubes. Shake the racks to mix the antigen and antiserum. The resultant dilutions are 1:20 through 1:1280, respectively.
7. Incubate in a water bath at 37°C for 48 ± 3 hours (changed to conform to NADL recommended technique). It is important in this test to use the recommended time and temperature of incubation and to make certain the water bath is in a location free of mechanical vibration.

CALIBRATION AND QUALITY CONTROL
For greater proficiency in test interpretation always include a positive and negative serum control in each test protocol.

Both positive and negative controls are diluted in the same proportion as the patient's serum and processed in exactly the same manner following procedure above for the rapid slide test or for the macroscopic tube test. Bacto Febrile Negative Control may be used as the negative control in both tests.

An antigen is considered to be satisfactory if it does not clump with the negative control and it reacts to a titer of 1:80 or more with the positive control.

Discard
1. Any control serum in which a precipitate forms or which does not dissolve completely upon rehydration.
2. Any antigen when clumps appear on the cap or dropper or which agglutinates spontaneously or which agglutinates in the negative control.
3. All antigens and control sera when past the end of the month of the expiry date.

RESULTS
Rapid Slide Test
Record results as follows:

4+	complete agglutination
3+	approximately 75% of the cells are clumped
2+	approximately 50% of the cells are clumped
1+	approximately 25% of the cells are clumped
+	trace agglutination
−	no agglutination

The titer of the serum is recorded as that dilution of the specimen in which at least 2+ (50%) agglutination occurs. See Tables II and III.

Macroscopic Tube Test
Using a fluorescent lamp against a black background, record the degree of agglutination as follows:

4+	100% of the organisms are agglutinated and the supernatant fluid is clear
3+	approximately 75% of the organisms are agglutinated and the supernatant fluid is slightly cloudy
2+	approximately 50% of the organisms are agglutinated and the supernatant fluid is moderately cloudy
1+	approximately 25% of the organisms are agglutinated and the supernatant fluid is cloudy
−	no agglutination is visible and the organisms remain as a cloudy suspension. (This reaction should be observed in tube 8.)

Sample Calculations
Table II Rapid Slide Test

ml Serum	Correlated Dilution	Reactions Specimen 1	Specimen 2	Specimen 3
0.08	1:20	3+	4+	4+
0.04	1:40	2+	4+	3+
0.02	1:80	1+	3+	2+
0.01	1:160	−	3+	+
0.005	1:320	−	1+	−
Serum Titer		1:40	1:160	1:80

Table III Macroscopic Tube Test

Serum Dilution	Specimen 1	Reaction Specimen 2	Specimen 3
1:20	4+	3+	4+
1:40	4+	2+	4+
1:80	3+	1+	4+
1:160	2+	−	4+
1:320	1+	−	3+
1:640	−	−	2+
1:1280	−	−	1+
Serum Titer	1:160	1:40	1:640

INTERPRETATION
Report the serum titer as the reciprocal of the highest dilution showing a 2+ reaction. See Tables II and III.

A difference in titer of 1 dilution, plus or minus, between replicate samples or between several samples drawn 1 – 2 weeks apart and run in parallel cannot be considered significant. A 1 dilution, plus or minus, deviation is within the limits of laboratory error.[6,7]

Past history in the use of *Brucella* suspensions has resulted in a pattern of titers which has been considered "significant." A 1:80 titer is considered a weakly positive serum while most patients with acute undulant fever demonstrate a titer of 1:320 or greater.

LIMITATIONS[8,9,10]

1. The major limitation of the febrile tests is that of interpretation. Again it must be stressed that the antigens are to be used for the detection and quantitation of agglutinin content in human sera. They are not to be used for making a diagnosis of a specific disease. A definitive diagnosis must be made by the physician, taking into consideration the history and physical state of the patient as well as data obtained from other laboratory tests.

   Although the *Brucella* antigens are useful for screening purposes, this technique should not be a complete substitute for the conventional isolation and serological identification of the etiological agent. Isolation of the organism is necessary for a definitive diagnosis.

2. It is necessary to run several serum specimens taken at different times, from the same patient, in parallel, to detect quantitative differences in agglutinin content.

3. The slide test is for screening only. Any quantitation must be done using the tube test.

4. There are many known antigenic similarities and cross-reactions. A definite serological relationship exists between *Brucella* and *Francisella tularensis*. Cross reactions may also occur between *Brucella* positive sera and *Proteus* OX19 antigen as well as with *Vibrio cholerae* and *Yersinia enterocolitica* serotype 9.

5. Serum specimens from patients suffering with acute brucellosis demonstrate little or no antibody titer during the first 10 days of the disease.

6. Serologic interpretation of an agglutinin titer in vaccinated individuals should be avoided since antibody levels may persist for years.

7. An antibody titer in excess of 1:160 may occur in healthy individuals with a past history of the disease.

8. Individuals recovered from brucellosis may demonstrate a nonspecific anamnestic agglutinin response upon infection with an etiological agent of a heterologous febrile species.

9. The severity of the disease is not correlated with agglutinin titers.

REFERENCES

1. Bergey's Manual of Determinative Bacteriology, 8th Ed., Williams and Wilkins Co., 1974.
2. Spink, W. W., McCullough, N. B., Hutchings, L. H., Amer. J. Clin. Path., 24:486, 1954.
3. White, P. B., Brit. J. Exper. Path., 14:145, 1933.
4. USDA, ANS Publications, 1959.
5. Personal Communication.
6. Syphilis, A synopsis, HEW, PHS, Publ. No. 1660, Jan. 1968.
7. Quality Control in Clinical Micro., Revised, ADCP Comm. on Continuing Education, Coun. on Micro., 1968.
8. Alton, G. G., Jones, L. M., and Pietz, D. E., Lab. Tech. in Brucellosis, WHO, 1975.
9. Bodily, H. L., et al, Diagnostic Procedures for Bacterial, Mycotic and Parasitic Infections, APHA, Inc. 5th Ed., 1970.
10. McCullough, N. B., Manual of Clinical Immunology, ASM, 1976.

PACKAGING

Bacto Brucella Abortus Antigen (slide)	5 ml	2909-56-1
Bacto Brucella Abortus Antigen (tube)	25 ml	2466-65-5
Bacto Brucella Abortus Antiserum	3 ml	2871-47-7
Bacto Brucella AMS Antiserum Poly	3 ml	2890-47-4
Bacto Brucella Melitensis Antigen	5 ml	2916-56-2
Bacto Brucella Melitensis Antiserum	3 ml	2889-47-7
Bacto Brucella Suis Antigen	5 ml	2915-56-3
Bacto Brucella Suis Antiserum	3 ml	2888-47-8

BACTO BRUCELLA BROTH
BACTO BRUCELLA AGAR

INTENDED USE

Bacto Brucella Broth and Bacto Brucella Agar are recommended for the cultivation of *Brucella* and other fastidious organisms.

FORMULAE

BACTO BRUCELLA BROTH
DEHYDRATED

Ingredients per liter

Bacto Tryptone	10 g
Bacto Peptamin	10 g
Bacto Dextrose	1 g
Bacto Yeast Extract	2 g
Sodium Chloride	5 g
Sodium Bisulfite	0.1 g

Final pH 7.0 ± 0.2 at 25°C.

One pound will make 10.5 liters of medium. Rehydrate with 28 g/liter.

BACTO BRUCELLA AGAR
DEHYDRATED

Ingredients per liter

Bacto Tryptone	10 g
Bacto Peptamin	10 g
Bacto Dextrose	1 g
Bacto Yeast Extract	2 g
Sodium Chloride	5 g
Sodium Bisulfite	0.1 g
Bacto Agar	15 g

Final pH 7.0 ± 0.2 at 25°C.

One pound will make 16.2 liters of medium. Rehydrate with 43 g/liter.

METHOD OF PREPARATION

1. Suspend appropriate amount of medium in 1 liter distilled or deionized water. Heat to boiling to dissolve completely.
2. Distribute into tubes or flasks as desired and sterilize in the autoclave for 15 minutes at 15 lbs pressure (121°C).
3. Allow agar medium to cool to 50 – 55°C and dispense into Petri dishes.
4. Inoculate as desired and incubate at 35°C for 24 – 72 hours under 10% CO_2.

STORAGE

Bacto Brucella media	Below 30°C
Prepared tubes	15 – 30°C
Prepared plates	2 – 8°C

QUALITY CONTROL

Identity Specifications

Dehydrated powder:	light beige, homogeneous, free-flowing
Reaction of appropriate solution:	pH 7.0 ± 0.2 at 25°C
Prepared medium:	light amber, clear to slightly opalescent

Typical Cultural Response in Bacto Brucella Media
After 24 – 72 Hours at 35°C Under 10% CO_2

Organism	Growth
Brucella abortus ATCC® 4315	good to excellent
Brucella melitensis ATCC® 4309	good to excellent
Brucella suis ATCC® 4314	good to excellent

PACKAGING

Bacto Brucella Broth	1 lb (454 g)	0495-01-1
	1/4 lb (114 g)	0495-02-0
	10 kg	0495-08-4
Bacto Brucella Agar	1 lb (454 g)	0964-01-3
	5 lb (2.27 kg)	0964-05-9
	10 kg	0964-08-6

BACTO BUFFERED PEPTONE WATER

INTENDED USE

Bacto Buffered Peptone Water is a preenrichment medium used for increasing recovery of injured *Salmonella* species from foods prior to selective enrichment and isolation.

PRINCIPLES

Sublethal injury to salmonellae may result from food preservation techniques involving heat, desiccation, preservatives, high osmotic pressure or pH changes.[1] Enriching injured cells in lactose broth (pH 6.9 ± 0.2) may be further detrimental to their recovery.[2] Preenrichment with buffered peptone water (pH 7.2 ± 0.2) insures maintaining a high pH over the 24-hour incubation period, resulting in repair of cells that may have an increased sensitivity to low pH.[3] This is particularly important for vegetable specimens which have a low buffering capacity.

FORMULA

BACTO BUFFERED PEPTONE WATER
DEHYDRATED

Ingredients per liter

```
Peptone . . . . . . . . . . . . . . . . . . . 10 g
Sodium Chloride . . . . . . . . . . . . . . 5 g
Sodium Phosphate, Dibasic . . . . . . 3.5 g
Potassium Phosphate, Monobasic . . 1.5 g
```

Final pH 7.2 ± 0.2 at 25°C.

Five hundred grams will make 25 liters of final medium.

PROCEDURE
For Preparation of the Medium

1. Rehydrate Bacto Buffered Peptone Water by dissolving 20 grams in 1 liter distilled or deionized water.
2. Dispense in 50 ml amounts.
3. Sterilize in the autoclave for 15 minutes at 15 lbs pressure (121°C).

For Isolation of *Salmonella*[4]

1. Aseptically inoculate 10 grams of the specimen into 50 ml sterile Bacto Buffered Peptone Water.
2. Incubate at 35°C for 18 hours.
3. Transfer 10 ml of the above medium to 100 ml Bacto Tetrathionate Broth Base.
4. Incubate at 43°C.
5. After 24 and 48 hours incubation, subculture to Bacto Brilliant Green Agar plates.
6. Incubate plates at 35°C for 18 hours.
7. Examine plates for colonies of *Salmonella*.

STORAGE

Bacto Buffered Peptone Water   Below 30°C
Prepared medium   15 – 30°C

QUALITY CONTROL
Identity Specifications

Dehydrated medium: cream white to light tan, homogeneous free flowing
Reaction of 2% solution: 7.2 ± 0.2 at 25°C
Prepared medium: light amber, clear with no significant precipitate

Typical Cultural Response in Bacto Buffered Peptone Water
After 18 – 24 Hours at 35°C

Organism	Recovery
Salmonella enteritidis ATCC® 13076	good to excellent
Salmonella typhi ATCC® 19430	good to excellent
Salmonella typhimurium ATCC® 14028	good to excellent

REFERENCES

1. Edel, W., and E. H. Kampelmacher. 1973. Bull. Wld. Hlth. Org. 48:167 – 174.
2. Angelotti, R. 1963. Microbiological quality of foods. Academic Press, New York.
3. Sadovski, A. Y. 1977. J. Fd. Technol. 12:85 – 91.
4. Poelma, P. L., and J. H. Silliker. 1976. *Salmonella*, p. 301 – 304. *In* M. L. Speck (ed.), Compendium of methods for the microbiological examination of foods. American Public Health Association, Washington, D.C.

PACKAGING

Bacto Buffered Peptone Water   500 g   1810-17-9

BACTO BUSHNELL-HAAS BROTH

INTENDED USE

Bacto Bushnell-Haas Broth is prepared for examining fuels for microbial contamination and for studying hydrocarbon utilization by microorganisms.

HISTORY

Bacto Bushnell-Haas Broth is prepared according to the formula for Bushnell and Haas[1] and is recommended for the microbiological examination of fuels by the SIM Committee on Microbiological Deterioration of Fuels.[2]

PRINCIPLES

Bacto Bushnell-Haas Broth contains all the nutrients necessary for the growth of these bacteria with the exception of the hydrocarbon. These bacteria are capable of decomposing a variety of hydrocarbons among which are kerosene, light and heavy mineral oils, paraffin wax and gasoline.

FORMULA

BACTO BUSHNELL-HAAS BROTH
DEHYDRATED

Ingredients per liter

Magnesium Sulfate . . . . 0.2 g
Calcium Chloride . . . . 0.02 g
Monopotassium Phosphate . . . . 1 g
Dipotassium Phosphate . . . . 1 g
Ammonium Nitrate . . . . 1 g
Ferric Chloride . . . . 0.05 g

Final pH 7.0 ± 0.2 at 25°C.

One pound will make 140 liters of medium.

METHOD OF PREPARATION

1. To rehydrate, dissolve 3.25 g in 1 liter distilled or deionized water.
2. Dispense into flasks and sterilize in the autoclave for 15 minutes at 15 lbs pressure (121°C). A precipitate, white prior to sterilization becoming yellow to orange after sterilization, is normal.

3. For liquid cultures the hydrocarbon is added to the broth or layered on the broth surface.

The specific procedure for the microbiological examination of fuels is described in SIM Special Publication No. 1, July, 1963.

STORAGE

Bacto Bushnell-Haas Broth	Below 30°C
Prepared medium	15 – 30°C

QUALITY CONTROL

Identity Specifications

Dehydrated powder:	pinkish beige, homogeneous, free-flowing
Reaction of 0.325% solution:	pH 7.0 ± 0.2 at 25°C
Prepared medium:	colorless to very light amber, clear supernatant over yellow-orange precipitate.

Typical Cultural Response in Bacto Bushnell-Haas Broth
After up to 1 Week at 25 – 30°C

Organism	Growth (plain)	Growth (w/Mineral Oil)
Pseudomonas aeruginosa ATCC® 9027	poor	good to excellent
Pseudomonas aeruginosa ATCC® 10145	poor	good to excellent
Pseudomonas aeruginosa ATCC® 14207	poor	good to excellent
Pseudomonas aeruginosa ATCC® 27853	poor	good to excellent

REFERENCES
1. J. Bact., 41:653, 1941.
2. SIM Special Publication, No. 1, July, 1963.

PACKAGING

Bacto Bushnell-Haas Broth	1 lb	0578-01-1

BACTO C PROTEIN ANTISERUM
BACTO C PROTEIN STANDARD

INTENDED USE
Bacto C Protein Antiserum is used in the detection of C-reactive protein (C protein) by the semiquantitative precipitin test.

Bacto C Protein Standard is used as a control on the methodology for the determination of C protein using Bacto C Protein Antiserum.

HISTORY
C-reactive protein was first reported by Tillet and Francis[1] as appearing in the blood during the acute phase of pneumococcal pneumonia. Lofstrom[2] found C protein in serum from patients in the acute diseases associated with tissue degeneration. Kroop and Shackman[3] demonstrated that serial C protein determinations were of value in the subsidence of cardiac inflammatory processes and could be used to differentiate coronary

insufficiency from myocardial infarction. The Journal of the American Medical Association[4,5] points to the extreme sensitivity and significance of the test as a measure of the extent of the disease process in rheumatic fever and its value in the detection of inflammatory change. Bunin, Kuttner, Baldwin and McEwen[6] studying the efficacy of therapy in rheumatic fever found that the detection of C protein was a more sensitive index of rheumatic activity than the sedimentation rate.

The crystallization and characterization of C protein by McCarty[7] and its further purification by Wood, McCarty and Slater[8] permitted the preparation of the highly specific antiserum. Bacto C Protein Antiserum is prepared in accordance with the procedures recommended by these authors.

The capillary procedure recommended by Anderson and McCarty[9] for the detection of the protein is readily performed and the results are semiquantitative when conditions of the test are followed. Bacto C Protein Antiserum is prepared with a distinctive dye which accentuates any antigen-antibody precipitate thereby permitting a reading to be made more readily and accurately.

PRINCIPLES

C-reactive protein is present in minimal nonsignificant concentrations in sera from "normal" individuals, which are not detected by the capillary test. However, concentrations of C-reactive protein which are of value as indicating acute phase infection or tissue damage can be detected by the capillary test. C-reactive protein is serologically distinct from normal protein and can be detected readily by the use of Bacto C Protein Antiserum.

Essentially, the semiquantitative precipitin test for the determination of C protein consist of taking up the patient's serum into a capillary tube which has just been charged with C protein antiserum. The 2 sera are mixed, placed upright in a plasticine block and incubated for a specified time and temperature. If C protein is present in the patient's serum it will react with the C protein antiserum causing a precipitate to occur. The amount of precipitate formed is then measured and recorded.

REAGENTS

Bacto C Protein Antiserum is a desiccated specific immune serum prepared from an animal which has been immunized with C protein. The antiserum will provide 2 – 3 mm of precipitate with Bacto C Protein Standard. The antiserum contains 1:10,000 Merthiolate® as a preservative and a blue dye for greater ease in observing the precipitate.

Bacto C Protein Standard is a standardized solution of C protein for use as a positive control. When used in the test procedure, it shows at least 2 – 3 mm of precipitate (2 – 3 plus).

REHYDRATION

Bacto C Protein Antiserum is rehydrated by adding 1 ml of sterile distilled water and rotating the vial gently to dissolve completely.

STORAGE AND EXPIRATION

Bacto C Protein Antiserum and Bacto C Protein Standard are stable to the expiry date on the label when stored at 2 – 8°C.

SPECIMEN COLLECTION

1. Collect 5 ml whole blood asceptically from the patient.
2. Allow the blood to clot and obtain the syneresed serum with a Pasteur pipette.

NOTE: Both the patient's serum and the Bacto C Protein Antiserum should be centrifuged if particles are evident. Plasma is not satisfactory for the determination of C-reactive protein since free calcium ion is essential for this reaction.

C-reactive protein may also be determined with blood collected from finger tip, ear lobe or heel as was described by Goldin and Kaplan.[10] The blood is collected in a capillary tube having a diameter sufficiently large so that the capillary tube used for the test will fit into the tube used for collection of blood. Allow the blood to clot in the capillary tube and seal one end. Centrifuge lightly to obtain a serum layer. Insert the test capillary tube containing the Bacto C Protein Antiserum into the patient's serum and proceed as below.

PROCEDURE
Materials Provided:
Bacto C Protein Antiserum
Bacto C Protein Standard
Bacto Capillary Pipettes

Materials Required but not Provided:
Centrifuge
Plasticine or modeling clay block
Incubator at 37°C
Refrigerator at 2 – 8°C

PROCEDURE FOR DETERMINING C-REACTIVE PROTEIN
1. A capillary tube, 0.7 – 1 mm inside diameter and 75 – 90 mm long, is dipped into rehydrated Bacto C Protein Antiserum. The fluid is allowed to rise about 25 – 30 mm in the tube. It is essential that the capillary be dipped into the antiserum first to prevent contamination of the antiserum with patient's serum.
2. Place the finger over the top of the tube and remove from the antiserum.
3. Wipe off the excess antiserum.
4. Without removing the finger from the end of the tube, dip the capillary into the patient's serum and permit an equal volume of the patient's serum (or Bacto C Protein Standard) to enter the tube. It is important that the antiserum and patient's serum be in contact. Air bubbles at the serum-antiserum interface are to be avoided.
5. Invert the capillary tube several times to obtain a mixing of the 2 sera, and allow the column of serum to move toward one end of the capillary. Place the finger at the end, leaving an air space of about 1 cm at the opposite end.
6. Wipe off the excess serum and finger marks from the tube, since these will interfere with the accurate reading of the test.
7. Insert the tube into a plasticine or modeling clay block, making sure that the bottom of the serum column is about 1 cm above the block.
8. Place in the incubator at 37°C for 2 hours. It is also permissable to leave the tube at room temperature if an incubator is not available.
9. If C-reactive protein is present in the patient's serum, a precipitate will be noticeable at the end of 30 minutes and a qualitative report may be made at this time. Continue the incubation for a total of 2 hours and then place the block into the refrigerator at 2 – 8°C overnight, at which time the final reading is made. Readings should be made by examining capillaries in front of a light source without the use of a background. The dye present in the antiserum accentuates any antigen-antibody precipitate permitting a reading to be made readily and accurately.

CONTROL
Bacto C Protein Standard is recommended as a control of the methodology of the determination of C protein using Bacto C Protein Antiserum. The same procedure for the performance of the test is employed as described for patient's serum. The capillary tube should be placed first into rehydrated Bacto C Protein Antiserum and the solution permitted to rise 25 – 30 mm in the tube. It is most important that the outside of the

capillary be wiped free from antiserum prior to placing in Bacto C Protein Standard to avoid any addition of the antiserum to the Bacto C Protein Standard. A similar amount of Bacto C Protein Standard is permitted to rise in the tube by capillary attraction, and the capillary tube inverted several times. The test is read following incubation as described above.

RESULTS
As a rule, the amount of precipitate present at the end of 30 minutes to 2 hours incubation at 37°C is similar to that obtained after overnight refrigeration. However, all negative reactions should be confirmed by overnight refrigeration. This preliminary reading, however, yields only qualitative results since it is necessary that the precipitate be allowed to settle in the tube so that the height of the column of precipitate can be measured to obtain a semiquantitative reading.

If the precipitate has not completely settled after overnight refrigeration, an approximation can be made of the height of the precipitate column by summing up the total of the dispersed precipitates. Readings may be recorded as follows:

Precipitate	Reading
None	Negative
1 mm	1+
2 mm	2+
3 mm	3+
4 mm	4+
5 mm	5+
6 mm or more	6+ or more

LIMITATIONS OF THE PROCEDURE
The serological detection of C protein, while indicative of active inflammation, is not diagnostically specific for any disease entity. For example, although C protein is found in the serum from acute rheumatics, it is not diagnostic of this disease nor does it indicate the presence of a Group A *Streptococcus* infection.

EXPECTED VALUES
C-reactive protein is not found in normal bloods.

REFERENCES
1. J. Exp. Med., 52:561, 1930.
2. Acta. Med. Scan., Suppl., 141:1, 1943.
3. Proc. Soc. Exp. Biol. Med., 86:95, 1954.
4. J. Am. Med. Assoc., 156:667, 1954.
5. ibid., 157:625, 1955.
6. ibid., 150:1273, 1952.
7. J. Exp. Med., 85:491, 1947.
8. ibid., 100:71, 1954.
9. Am. J. Med., 8:445, 1950.
10. Am. J. Clin. Path., 25:1432, 1955.

PACKAGING

Bacto C Protein Antiserum	1 ml	0671-50-7
	6 × 1 ml	0671-51-6
Bacto C Protein Standard	0.5 ml	0763-48-1

BACTO CHO MEDIUM BASE

INTENDED USE
Bacto CHO Medium Base is a basal medium to which carbohydrates may be added for use in fermentation studies of anaerobic bacteria.

HISTORY/PRINCIPLES
Bacto CHO Medium Base is prepared according to the formula described in "Laboratory Methods in Anaerobic Bacteriology," CDC, Atlanta, Georgia.[1]

Clinical recognition of the potential pathogenicity of anaerobes and the severity of their infections has stimulated the necessity to isolate, identify and establish susceptibility patterns for these organisms. These organisms can invade any organ or tissue of the body causing a wide variety of infections.

Identification of anaerobes is based on cellular morphology, colonial morphology on blood agar and biochemical characteristics.[1] Although procedures for isolation and identification of anaerobes are similar to those for aerobic organisms, there are some significant differences. Proper collection and transport of suspect specimens is of primary importance. Specimens should be collected in such a manner as to minimize exposure to oxygen. Specimens should be transported in appropriate containers of oxygen-free gas or media with a reducing agent. Once in the laboratory, specimens should be cultured promptly under proper atmospheric conditions.

As with aerobic organisms, carbohydrate utilization patterns play an important role in identification of anaerobes. Yet, the less efficient metabolism of anaerobes requires carbohydrate utilization media to be richer in auxillary growth factors and to contain a higher concentration of peptone as well as carbohydrate. Bacto CHO Medium Base meets the requirements of even fastidious anaerobes.

FORMULA

BACTO CHO MEDIUM BASE
DEHYDRATED

Ingredients per liter

Bacto Tryptone	15 g
Bacto Yeast Extract	7 g
Bacto L-Cystine	0.25 g
Sodium Chloride	2.5 g
Ascorbic Acid	0.1 g
Sodium Thioglycollate	0.5 g
Bacto Brom Thymol Blue	0.01 g
Bacto Agar	0.75 g

Final pH 7.0 ± 0.1 at 25°C

One pound will make 17.4 liters of medium.

METHOD OF PREPARATION
1. Suspend 26 g base in 1 liter cold distilled or deionized water. Heat to boiling to dissolve completely.
2. Distribute in 100 ml amounts in flasks.
3. Sterilize in the autoclave for 15 minutes at 15 lbs pressure (121°C).
4. Allow to cool to 45 – 50°C. Aseptically add 6.25 ml of 10% sterile solution of the desired carbohydrate to the sterile basal medium.

5. Mix thoroughly and dispense 8.5 ml aseptically into sterile tubes (15 × 125 mm).
6. Bacto CHO Medium Base w/dextrose should contain a fermentation vial. The medium is prepared by adding 0.6 g dextrose to 100 ml Bacto CHO Medium Base. Dispense 8.5 ml into 15 × 125 mm tubes containing a fermentation vial and autoclave as above. Allow autoclave temperature to drop slowly to room temperature before removing tubes.

PROCEDURE
A 24 hour culture of the isolate taken from blood agar should be Gram-stained and checked for purity before attempting to establish fermentation patterns.
1. Expel air from capillary pipette before aspirating pure culture.
2. Inoculate tubes of Bacto CHO Medium w/carbohydrates near the bottom, with a drop of culture.
3. While withdrawing pipette, expel a small amount of inoculum along the line of the stab. Expel air from pipette before inoculating each tube.
4. Incubate tubes at 35 – 37°C in an aerobic or anaerobic atmosphere depending on requirements of the organism. Refer to reference 1 for specific information.
5. Incubate for 7 days. Read tubes daily and record results.

Observe control tube for gas or indicator reduction (some clostridia will reduce the indicator). Observe carbohydrate tubes for acid, gas, indicator reduction or no change. A change in color of the medium from blue-green to yellow indicates an acid reaction.

STORAGE
Bacto CHO Medium Base   Below 30°C
Prepared tubes   15 – 30°C*

*Media used for anaerobes should be freshly prepared, if not, the tubes must be heated prior to inoculation in a boiling water bath for 10 minutes and cooled in tap water.

QUALITY CONTROL
Identity Specifications

Dehydrated powder: light beige, homogeneous, free-flowing
Reaction of 2.6% solution: pH 7.0 ± 0.1 at 25°C
Prepared medium: green, slightly opalescent

Typical Cultural Response in Bacto CHO Medium
After up to 7 days at 35°C Anaerobically

Organism	Growth	Fermentation with	
		Dextrose	Lactose
Bacteroides fragilis ATCC® 25285	good	+	+
Bacteriodes melaninogenicus ATCC® 25611	good	–	+
Bacteriodes vulgatus ATCC® 8482	good	–	–
Clostridium botulinum ATCC® 25763	good	+	–
Clostridium perfringens ATCC® 12919	good	+	+
Escherichia coli ATCC® 25922	good	+	+

REFERENCE
1. Laboratory Methods in Anaerobic Bacteriology, CDC Laboratory Manual, U.S. Dept. HEW, Pub. No. (CDC) 74-8262, January 1974.

PACKAGING
Bacto CHO Medium Base   1 lb (454 g)   0841-01-2

BACTO CLED AGAR

INTENDED USE
Bacto CLED Agar (Cystine Lactose-Electrolyte-Deficient Agar) is designated for the cultivation and estimation of bacteria from urine. It is particularly suited for use as a dip inoculum transport medium.

HISTORY/PRINCIPLES
Bacto CLED Agar is prepared according to the formula of Mackey and Sandys[1] employing its electrolyte deficiency to inhibit the swarming of *Proteus* which otherwise would obscure the observation of colonies. Lactose is included in the medium to detect lactose fermenting coliform contaminants which are easily recognized by a color change of the medium from green to yellow.

When used as a dip inoculum transport medium, the method indicated by Dixon and Clarke,[2] is recommended. The sterile medium is dispensed into suitable sterile spoons and solidified. The prepared medium in spoons can be dipped into samples of urine and subsequently transported to the laboratory for incubation and examination. The spoons after preparation are kept in sterile one ounce screw-capped glass containers containing a small piece of damp cotton. Inoculated spoons are incubated overnight at 35 – 37°C and the total number of colonies are recorded. If more than 200 colonies grow on a spoon, the growth may become semiconfluent and the presence of more than 100,000 bacteria per ml is indicated. If fewer than 20 colonies grow on the spoon, less than 10,000 bacteria is indicated.

FORMULA

BACTO CLED AGAR
DEHYDRATED

Ingredients per liter

Bacto Beef Extract	3 g
Bacto Peptone	4 g
Bacto Tryptone	4 g
Bacto L-Cystine	0.128 g
Bacto Lactose	10 g
Bacto Agar	15 g
Bacto Brom Thymol Blue	0.02 g

Final pH 7.3 ± 0.2 at 25°C.

One pound will make 12.6 liters of medium.

METHOD OF PREPARATION
1. Suspend 36 g in 1 liter distilled or deionized water. Heat to boiling to dissolve completely.
2. Dispense into tubes or flasks and sterilize in the autoclave for 15 minutes at 15 lbs pressure (121°C).
3. The sterile medium may be aseptically dispensed into suitable sterile spoons or Petri dishes as desired.

STORAGE
Bacto CLED Agar	Below 30°C
Prepared plates or spoons	2 – 8°C

QUALITY CONTROL

Identity Specifications

Dehydrated powder: beige with slight green tinge, homogeneous, free-flowing
Reaction of 3.6% solution: 7.3 ± 0.2 at 25°C
Prepared medium: yellowish green, very slightly opalescent

Typical Cultural Response on Bacto CLED Agar
After 18 – 24 Hours at 35 ± 2°C

Organism	Growth	Color of Medium
Enterobacter aerogenes ATCC® 13048	good to excellent	slight yellow to blue
Escherichia coli ATCC® 25922	good to excellent	yellow
Proteus vulgaris ATCC® 13315	good to excellent, swarming inhibited	blue to blue-green
Staphylococcus aureus ATCC® 25923	good to excellent	slight yellow or no change
Streptococcus faecalis ATCC® 19433	good to excellent	slight yellow or no change

REFERENCES

1. Brit. Med. J. 2:1286,1965.
2. Can. Med. Assoc. J. 99:741,1968.

PACKAGING

Bacto CLED Agar	1 lb (454 g)	0971-01-4
	1/4 lb (114 g)	0971-02-3
	10 kg	0971-08-7

CAMPYLOBACTER MEDIA

BACTO CAMPYLOBACTER AGAR KIT SKIRROW
BACTO CAMPYLOBACTER AGAR KIT BLASER

INTENDED USE

Bacto Campylobacter Agar Kit Skirrow and Bacto Campylobacter Agar Kit Blaser are used for preparing Campylobacter Agar according to the formulations of Skirrow[1] and Blaser et al.[2,3]

The respective kits contain Bacto Campylobacter Agar Base and either Bacto Campylobacter Antimicrobic Supplement S or Bacto Campylobacter Antimicrobic Supplement B in lyophilized premeasured quantities to prepare Campylobacter Agar.

HISTORY OF THE TEST

Gastroenteritis in humans due to *Campylobacter fetus* subsp. *jejuni* is well recognized.[1-5] Selective isolation of this organism from fecal specimens was first reported by Dekeyser et al. in 1972, who used membrane filtration followed by inoculation of the filtered specimen onto selective media.[6] In 1977, Skirrow described a selective medium for *Campylobacter* species which consisted of Bacto Blood Agar Base No. 2 supplemented with 7% lysed horse blood and vancomycin, polymyxin B and trimethoprim.[1] Skirrow found that this combination of selective agents obviated the need for filtration as previously described by Dekeyser. Blaser et al. in 1978 reported on the use of brucella agar supplemented with 10% defibrinated sheep blood and vancomycin, polymyxin B, trimethoprim and amphotericin B for the selective isolation of *C. fetus* subsp. *jejuni*.[2] Cephalothin was later incorporated into the Blaser et al. formula for improved inhibition of normal enteric flora.[3]

Bacto Blood Agar Base No. 2 was selected by Difco Laboratories as the base medium for use with Bacto Campylobacter Antimicrobic Supplements in preparing Campylobacter Agar since test results in our laboratories showed somewhat better growth of *Campylobacter* on this medium than on brucella agar. Further, trimethoprim was observed to be more active in Bacto Blood Agar Base No. 2 than in brucella agar. In order to provide for optimal growth and recovery of *Campylobacter,* the Bacto Blood Agar Base No. 2 provided in our Campylobacter Agar kits has been interstandardized with our Campylobacter Antimicrobic Supplements through careful piloting of the raw materials used in the base. This interstandardized Bacto Blood Agar Base No. 2 has been designated as Bacto Campylobacter Agar Base.

PRINCIPLES OF THE PROCEDURE
Bacto Campylobacter Agar Base is a nutritionally rich medium based upon Bacto Blood Agar Base No. 2 and supports good growth of *Campylobacter* species. Supplementation of this medium with antimicrobial agents as described by Skirrow[1] and Blaser et al.[2,3] provides for markedly reduced growth of normal enteric bacteria and improved growth and recovery of *C. fetus* subsp. *jejuni* from fecal specimens. Growth of fungi is markedly to completely inhibited on Campylobacter Agar prepared with Bacto Campylobacter Antimicrobic Supplement B due to the presence of amphotericin B.

FORMULAE

BACTO CAMPYLOBACTER AGAR BASE
Ingredients per liter

Proteose Peptone, Difco	15 g
Liver Digest	2.5 g
Bacto Yeast Extract	5 g
Sodium Chloride	5 g
Bacto Agar	12 g

Final pH 7.4 ± 0.2 at 25°C.

BACTO CAMPYLOBACTER ANTIMICROBIC SUPPLEMENT S

	10 ml vial	5 ml vial
Vancomycin	10 mg	5 mg
Polymyxin B	2,500 units	1,250 units
Trimethoprim	5 mg	2.5 mg

BACTO CAMPYLOBACTER ANTIMICROBIC SUPPLEMENT B

	10 ml vial	5 ml vial
Vancomycin	10 mg	5 mg
Polymyxin B	2,500 units	1,250 units
Trimethoprim	5 mg	2.5 mg
Cephalothin	15 mg	7.5 mg
Amphotericin B	2 mg	1 mg

PRECAUTIONS
Use aseptic technique in rehydrating the supplements and in adding the supplements to the basal medium.

Follow proper, established laboratory procedures in handling and disposing of infectious materials.

STORAGE AND REHYDRATION INSTRUCTIONS
Store Bacto Campylobacter Agar Kits at 2 – 8°C.

To rehydrate Bacto Campylobacter Agar Base, suspend the contents of the Unipack in 1 liter or 500 ml distilled or deionized water as directed on the label. Heat to boiling to dissolve completely. Sterilize in the autoclave for 15 minutes at 15 lbs pressure (121°C).

To rehydrate Bacto Campylobacter Antimicrobic Supplements, aseptically add 10 ml or 5 ml sterile distilled or deionized water as directed on the label. Invert the vial gently several times to dissolve the powder. Store rehydrated vials at 2 – 8°C. Use within 24 hours after rehydration.

Do not open or rehydrate reagents until ready to use.

The expiration date applies to the products in their intact containers when stored as directed.

PRODUCT DETERIORATION
Do not use the rehydrated supplement if it is contaminated, partially or totally evaporated or shows other signs of deterioration. Do not use the agar base if it is caked, discolored or shows other signs of deterioration.

SPECIMEN COLLECTION
Fecal specimens should be collected in sterile containers or with a sterile rectal swab and transported immediately to the laboratory. If the specimen cannot be inoculated onto appropriate media within 4 hours after collection, the specimen should be maintained or transported in Cary Blair Transport Medium.[7] Food and environmental specimens should be collected in sterile containers and transported to the laboratory in accordance with recommended guidelines.[8]

PROCEDURE
Materials Provided:
Bacto Campylobacter Agar Base
Bacto Campylobacter Antimicrobic Supplement S or
Bacto Campylobacter Antimicrobic Supplement B

Materials Not Provided:

Specimen containers or sterile rectal swabs	Inoculating loop
Suitable system for providing a microaerophilic environment	Incubator (42°C)
Bunsen burner or incinerator	User quality control cultures
Sterile defibrinated sheep blood or sterile lysed horse blood	Sterile Petri dishes

Preparation of Campylobacter Agar
1. To Bacto Campylobacter Agar Base, rehydrated and sterilized according to label directions and cooled to 45 – 50°C, aseptically add 5 – 7% sterile lysed horse blood or 10% sterile defibrinated sheep blood.

2. Aseptically add 1% Bacto Campylobacter Antimicrobic Supplement S or Bacto Campylobacter Antimicrobic Supplement B (10 ml supplement per 1 liter base or 5 ml supplement per 500 ml base). Mix thoroughly, avoiding the formation of air bubbles, and dispense into sterile Petri dishes.

Inoculation and incubation
1. Inoculate the specimen directly onto the surface of a Campylobacter Agar plate and streak for isolation.
2. Incubate inoculated plates at 42°C under a microaerophilic atmosphere containing 5 – 6% oxygen and 3 – 10% carbon dioxide. A number of methods are available for establishing this microaerophilic environment. Consult appropriate references for specific information.[7,10,11]
3. Examine plates for growth after 24 and 48 hours incubation.

RESULTS
Campylobacter fetus subsp. *jejuni* colonies on Campylobacter Agar appear nonhemolytic, flat and gray with an irregular edge or raised and round with a mucoid appearance; some strains may appear tan or slightly pink. Swarming or spreading may be observed on moist surfaces. Plates examined after 24 hours incubation should be examined quickly and reincubated under microaerophilic conditions to maintain the viability of the more oxygen sensitive strains. Growth of normal enteric bacteria is markedly to completely inhibited. Growth of fungi is markedly to completely inhibited on Campylobacter Agar prepared with Campylobacter Antimicrobic Supplement B.

USER QUALITY CONTROL
1. Examine the agar base for color and texture. The powder should be beige, free-flowing and homogeneous.
2. Determine the pH of the medium after preparation and cooling to 25°C. The pH should be 7.4 ± 0.2
3. Examine the lyophilized and rehydrated antimicrobic supplement for evidence of deterioration as described under PRODUCT DETERIORATION.
4. Check the performance of the base and antimicrobic supplement by testing in the complete medium. Plates should be inoculated with approximately 100 colony forming units of the test cultures and incubated at 42°C in a reduced oxygen atmosphere. Examine plates for growth after 24 and 48 hours incubation.

| | Expected Resuts on Campylobacter Agar Prepared with: | |
| | Bacto Campylobacter Antimicrobic Supplement S (Skirrow) | Bacto Campylobacter Antimicrobic Supplement B (Blaser et al.) |
Organism		
Campylobacter fetus subsp. *jejuni* ATCC® 29428	good growth	good growth
Escherichia coli ATCC® 25922	marked to complete inhibition	marked to complete inhibition
Streptococcus faecalis ATCC® 33186	marked to complete inhibition	marked to complete inhibition
Candida albicans ATCC® 10231	growth	marked to complete inhibition

LIMITATIONS OF THE PROCEDURE
1. Bacto Campylobacter Agar Base and Bacto Campylobacter Antimicrobic Supplements S and B are intended for use in the preparation of Campylobacter Agar. Although these Campylobacter Agars are selective primarily for *Campylobacter*, biochemical testing using pure cultures is necessary for complete identification. Consult appropriate references for further information.[7,10,11]

2. Growth of *Campylobacter fetus* subsp. *intestinalis* may be dramatically inhibited on Campylobacter Agar prepared with Bacto Campylobacter Antimicrobic Supplement B due to the presence of cephalothin. The use of Bacto Campylobacter Antimicrobic Supplement S with incubation at 35°C is suggested when isolation of this organism from mixed populations is desired.
3. Due to the selective properties of the complete medium and the organisms themselves, some strains of *C. fetus* subsp. *jejuni* may be encountered that fail to grow or grow poorly on this medium; similarly, some strains of normal enteric organisms may be encountered that are not inhibited or only partially inhibited on this medium.

REFERENCES

1. Skirrow, M. D. 1977. Campylobacter enteritis: A "new" disease. Br. Med. J. 2:9–11.
2. Blaser, M., J. Cravens, B. W. Powers, and W. L. Wang. 1978. Campylobacter enteritis associated with canine infection. Lancet (ii):979–980.
3. Blaser, M. J., V. Berkowitz, F. M. LaForce, J. Cravens, L. B. Reller, and W. L. L. Wang. 1979. Campylobacter enteritis: clinical and epidemiologic features. Ann. Intern. Med. 91:179–185.
4. Drake, A. A., M. J. R. Gilchrist, J. A. Washington, K. A. Huizenga, and R. E. Van Scoy. 1981. Diarrhea due to *Campylobacter fetus* subspecies *jejuni*: A clinical review of 63 cases. Mayo Clin. Proc. 56:414–423.
5. Blaser, M. J., R. B. Parsons, and W. L. L. Wang. 1980. Acute colitis caused by *Campylobacter fetus* ssp. *jejuni*. Gastroenterology. 78:448–453.
6. Dekeyser, P., M. Gossvin-Detrain, J. P. Butzler, and J. Sternon. 1972. Acute enteritis due to related vibrio: first positive stool cultures. J. Infect. Dis. 125:390–392.
7. *Campylobacter*. Laboratory methods for isolation and identification. September, 1980. U.S. Department of Health and Human Services. Center for Disease Control, Atlanta, GA.
8. Bryan, F. L. 1980. Procedures to use during outbreaks of food-borne disease, p. 40–51. *In* E. H. Lennette, A. Balows, W. J. Hausler, Jr., and J. P. Truant (ed.), Manual of clinical microbiology, 3rd Ed. American Society for Microbiology, Washington, D.C.
9. Kaplan, R. L. 1980. *Campylobacter*, p. 235–251. *In* E. H. Lennette, A. Balows, W. J. Hausler, Jr., and J. P. Truant (ed.), Manual of clinical microbiology, 3rd Ed. American Society for Microbiology, Washington, D.C.
10. Karmali, M. A., and P. C. Fleming. 1979. Application of the Fortner principle to isolation of *Campylobacter* from stools. J. Clin. Microbiol. 10:245–247.
11. Sack, R. B., R. C. Tilton, and A. S. Weissfeld, 1980. Cumitech 12, Laboratory diagnosis of bacterial diarrhea. Coordinating ed., S. J. Rubin. American Society for Microbiology, Washington, D.C.

PACKAGING

Bacto Campylobacter Agar Kit Skirrow Contains:	6 × 1 liter	3280-32-7
Bacto Campylobacter Agar Base	6 × 39.5 g	
Bacto Campylobacter Antimicrobic Supplement S	6 × 10 ml	
Bacto Campylobacter Agar Kit Skirrow Contains:	6 × 500 ml	3280-40-7
Bacto Campylobacter Agar Base	6 × 19.75 g	
Bacto Campylobacter Antimicrobic Supplement S	6 × 5 ml	
Bacto Campylobacter Agar Kit Blaser Contains:	6 × 1 liter	3279-32-0
Bacto Campylobacter Agar Base	6 × 39.5 g	
Bacto Campylobacter Antimicrobic Supplement B	6 × 10 ml	
Bacto Campylobacter Agar Kit Blaser Contains:	6 × 500 ml	3279-40-0
Bacto Campylobacter Agar Base	6 × 19.75 g	
Bacto Campylobacter Antimicrobic Supplement B	6 × 5 ml	

BACTO CANDIDA ALBICANS ANTISERUM

INTENDED USE

Bacto Candida Albicans Antiserum is prepared for use in the slide agglutination test for the presumptive identification of *C. albicans*.

REAGENTS
Bacto Candida Albicans Antiserum is an adsorbed desiccated rabbit antiserum prepared in accordance with the procedure described by Rosenthal and Gurnari.[1] It is prepared from *C. albicans*, is adsorbed with other *Candida* species, and is specific for *C. albicans* with the exception that it reacts slightly with *C. tropicalis* and *C. stellatoidea*.

When properly rehydrated, the antiserum contains approximately 1:10,000 Merthiolate®, sufficient to maintain a bacteriostatic condition when stored at 2 – 8°C.

When used according to the suggest PROCEDURE, each 3 ml vial is sufficient to perform 60 tests.

REHYDRATION
To rehydrate Bacto Candida Albicans Antiserum, add 3 ml 0.85% sodium chloride per vial and agitate gently to effect complete solution.

STORAGE
Store both desiccated and rehydrated Bacto Candida Albicans Antiserum at 2 – 8°C. Prolonged exposure to room temperature or repeated freezing and thawing is detrimental to antisera.

EXPIRATION DATE
Bacto Candida Albicans Antiserum, desiccated and rehydrated, are stable to the expiration date on the label when stored as directed.

Discard any serum that is cloudy or has a precipitate after rehydration or storage unless it can be clarified by centrifugation and demonstrates proper reactivity with validated positive and negative control cultures.

SPECIMEN PREPARATION
1. Grow the culture to be tested on Bacto Sabouraud Dextrose Agar at room temperature.
2. Prepare a heavy suspension of the suspected culture in 0.85% sodium chloride solution containing 0.5% phenol.
3. Filter through nylon cloth or gauze.

PROCEDURE
Materials Provided:
Bacto Candida Albicans Antiserum

Materials Required but not Provided:
Bacto Sabouraud Dextrose Agar   Phenol
Glass microscope slide   Fluorescent light source
Bacteriological loop   Serological pipette
Bunsen Burner   Bacto Rabbit Serum Normal
0.85% NaCl Solution

Macroscopic Slide Agglutination Test
1. Add one drop of the prepared culture to each end of a microscope slide.
2. To one drop of the suspension add one drop of Bacto Candida Albicans Antiserum.
3. To the second drop add one drop 0.85% sodium chloride or Bacto Rabbit Serum Normal as a control for culture roughness.

4. Mix each droplet with a fresh end of an applicator stick.
5. Rotate the slide by hand for 20 – 30 seconds.
6. Observe for agglutination.

QUALITY CONTROL

Use known positive and negative control cultures in parallel with the test culture to ascertain the validity of the test results.

RESULTS

The control droplet should show no spontaneous agglutination, otherwise, the antigen is unsatisfactory for the agglutination technique.

Agglutination of test suspension should be rapid and complete if positive for *C. albicans*.

LIMITATIONS OF THE PROCEDURE

Serological techniques for the identification of *C. albicans* can only serve to corroborate cultural and biochemical findings.

REFERENCE

1. J. Invest. Derm., 31:251, 1958.

PACKAGING

Bacto Candida Albicans Antiserum   3 ml   2281-47-1

BACTO CANDIDA BCG AGAR BASE

INTENDED USE

Bacto Candida BCG (Brom Cresol Green) Agar Base is used for the primary isolation and identification of *Candida* species. It is especially applicable for the diagnostic laboratory dependent primarily on cultural methods for demonstrating morphological and biochemical reactions characterizing the different *Candida* species of medical importance and for which speed and accuracy are essential.

HISTORY/PRINCIPLES

Bacto Candida BCG Agar Base is prepared according to the formulation of Harold and Snyder[1] and employed as the authors suggest. They demonstrated that the triphenyl-tetrazolium chloride (TTC) employed in the Pagano Levin medium retarded the growth of some species of *Candida* and completely restricted the growth of others. The authors overcame this growth impedance by using brom cresol green as the indicator. The brom cresol green had the advantage of being nontoxic. The color change is a function of pH and quite specific color patterns appear both in the base and surface of colonies for differentiation of *Candida* species. Neomycin is added to the medium in a concentration of 500 μg/ml as a selective agent.

The procedure recommended by Harold and Snyder[1] is best illustrated in their "Scheme for cultural identification of *Candida* species of medical importance" which follows.

Scheme for cultural identification of *Candida* species of medical importance

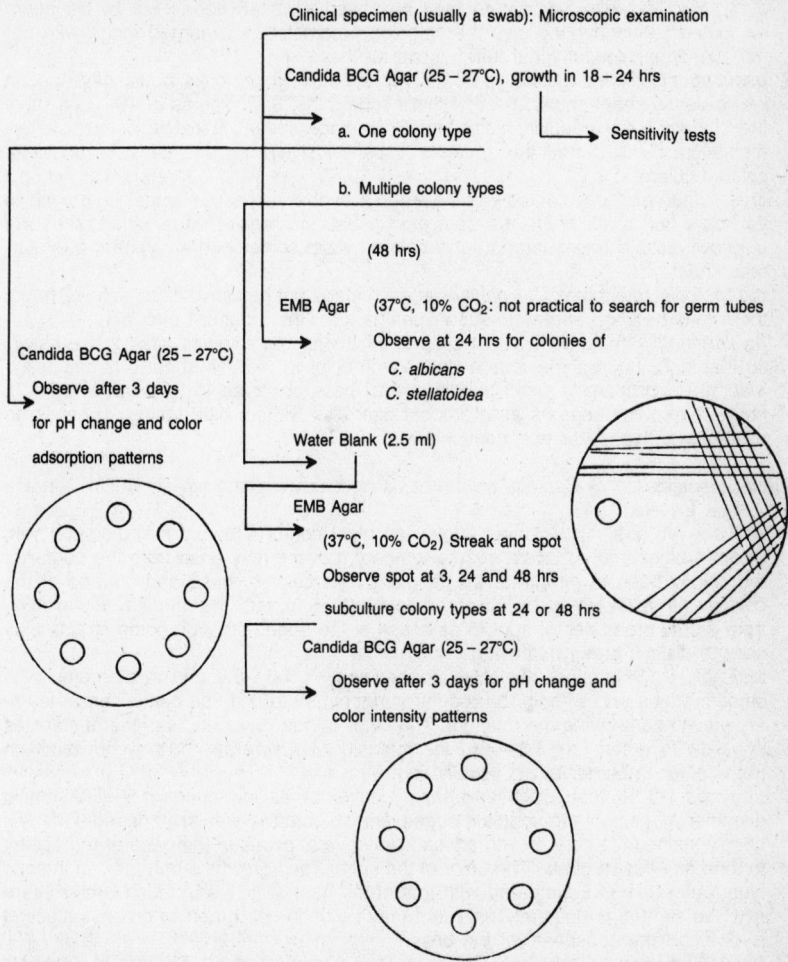

Clinical specimen (usually a swab): Microscopic examination

Candida BCG Agar (25 – 27°C), growth in 18 – 24 hrs

a. One colony type → Sensitivity tests

b. Multiple colony types

(48 hrs)

EMB Agar (37°C, 10% CO_2: not practical to search for germ tubes

Observe at 24 hrs for colonies of
C. albicans
C. stellatoidea

Candida BCG Agar (25 – 27°C)
Observe after 3 days
for pH change and color
adsorption patterns

Water Blank (2.5 ml)

EMB Agar
(37°C, 10% CO_2) Streak and spot
Observe spot at 3, 24 and 48 hrs
subculture colony types at 24 or 48 hrs

Candida BCG Agar (25 – 27°C)
Observe after 3 days for pH change and
color intensity patterns

The various *Candida* species on Bacto EMB Agar incubated for 2 – 3 hours at 37°C in an atmosphere of 10% carbon dioxide can be differentiated as follows:

C. albicans: a high percentage of the yeast cells show germ tubes in two hours, pseudohyphal forms are present in 3 hours, and in 24 hours there are enormous number of blastospores massed in clusters along the hyphae.

C. stellatoidea: a smaller percentage of germ tubes appear in 3 hours than in 2 hours for *C. albicans* but there are not pseudohyphae; in 24 hours there is characteristic extensive filamentous growth with little sporulation.

C. *tropicalis:* large round to ovoid refractile cells without germ tubes in 3 hours but showing well developed pseudohyphal buds many of which are arranged in chains of 3 cells. No definitive feature may be observed in 48 hours, but by 72 hours elongated pseudohyphae give the colonies a complete submerged fringe with numerous large pseudohyphal tufts across the base.

C. *pseudotropicalis:* elongating yeast cells in 2 hours which in 24 hours have grown into a dense sheet showing a few fine hyphal tufts; the colonies in 48 hours have the characteristic greenish black metallic surface sheen indicative of lactose fermentation. By 72 hours the sheen is usually lost leaving an intensely red violet colored colony.

C. *krusei:* the cells are markedly ovoid without hyphal or pseudohyphal outgrowth in 24 hours but by 72 hours the spot growth has in irregular edge with a sheetlike outgrowth and shows a large, highly colored violet center with a lavender grey surface sheet.

C. *parapsilosis:* resembles C. *tropicalis* at each stage but has smaller cells; in 72 hours there may be short sparse pseudohyphal outgrowths around a pink disc.

C. *guilliermondii:* small round yeast cells at 24 hours; the colonies adsorb dye slowly so that in 72 hours there is a lavender pink color to the thin sheet of small cells.

T. *glabrata:* exceptionally small budding yeast cells observed in 3 hours; in 24 – 72 hours the colony appears as a smooth pink disk without blue toning. There is no evidence of hyphae or pseudohyphae.

Further identification of *Candida* on the basis of colony morphology on Bacto Candida BCG Agar follows:

C. *albicans* (45 strains): Colonies appear as blunt cones 4.5 – 5.5 mm diameter with smooth edges and surfaces; coarse feathery growths may arise from the center of the colony base to penetrate the medium. The color of base and surface of the colonies is yellowish to bluish green (pH 4.2 – 5.0) with the intensity diminishing from a grey green center spot to paleness at the edge, although some strains may show a distinct green outer ring.

C. *stellatoidea* (36 strains): Colonies appear convex 4.0 – 5.0 mm in diameter, with smooth edges and smooth to irregular surfaces; there is a fine central basal feathery growth penetrating the medium. The color of both base and surface of colonies is yellow to green (pH 3.8 – 4.6) the intensity of which may or may not diminish from center to border but is usually light.

C. *tropicalis* (45 strains): Colonies appear convex or as low cones 4.5 – 5.0 mm in diameter with smooth to undulate edges, and smooth to granular or ridged surfaces; deeply stained feathery growth arises from several points in the base of the colony to form an effusive cloud. The color of the submerged growth is normally an intense blue green (pH 5.0) compared with that of the base which is of much lighter intensity; the surface is uniformly pale and may be yellowish green to green, reflecting a lower pH than observed of the base.

C. *pseudotropicalis* (27 strains): Colonies appear convex, 4.5 – 5.5 mm in diameter with undulate to smooth edges, and smooth surfaces; occasionally the surface is membranous but all colonies are shiny in appearance, and there is feathering growth emerging from several points in the base of the colony. The color of a large central area in the base of the colony is a medium green (pH 4.4 – 4.8) which diminishes in intensity toward the edge; a similar distribution of color occurs on the surface, but this green is bright in hue and is never grayed as it is with C. *tropicalis.*

C. *krusei* (27 strains): Colonies appear as low cones 4.5 – 5.0 mm in diameter with pseudohyphal edges, which may be weakly contractile or spreading, and have dull floccose surfaces. There is abundant lightly colored growth penetrating the medium from the base of the colony. The base of colony is a medium blue green (pH 5.0) in the center diminishing in intensity to paleness at the edge; the surface is usually a light green to yellow green without much concentration in any part.

C. parapsilosis (54 strains): Colonies appear as convex to low cones 3.5 – 4.5 mm in diameter with smooth or slightly spreading edges, but vary from smooth to granular or rough surfaces; there is no submerged growth. The color for both base and surface of the colony is blue green (pH 4.6 – 5.4) over much of the colony, being more intense in the base than the surface which is modified by a thin grayish film of cells; the intensity in color fades abruptly leaving a broad pale edge.

C. guilliermondii (36 strains): Colonies appear as low cones 4.0 – 5.0 mm in diameter with very smooth edges and highly glossy surfaces; there maybe a weak, fine feathered submerged growth. Both base and surface of the colony tend to have blue centers (pH 5.0 – 5.4) of medium intensity fading into a pale edge; however, the surface may be blue green with the central third lightened with gray.

T. glabrata* (8 strains): Colonies are smooth and convex, 4.6 – 5.0 mm diameter; the surface color pattern is pale green in the center which becomes medium green at the edge (pH 4.6 – 5.0) and the base has the same color pattern but of less intensity.

FORMULA

BACTO CANDIDA BCG AGAR BASE
DEHYDRATED
Ingredients per liter

Bacto Peptone	10 g
Bacto Yeast Extract	1 g
Bacto Dextrose	40 g
Bacto Agar	15 g
Brom Cresol Green	0.02 g

Final pH 6.1 ± 0.1 at 25°C.

One pound will make 6.9 liters of medium.

METHOD OF PREPARATION
1. Suspend 66 g in 1 liter distilled or deionized water. Heat to boiling to dissolve completely.
2. Sterilize in the autoclave for 15 minutes at 15 lbs pressure (121°C).
3. Allow to cool to 50 – 55°C and add sterile neomycin to a concentration of 500 μg/ml. Mix thoroughly to obtain a uniform solution.
4. Dispense into sterile Petri dishes or tubes.

STORAGE
Bacto Candida BCG Agar Base	Below 30°C
Prepared medium	2 – 8°C

QUALITY CONTROL
Identity Specifications

Dehydrated powder:	bluish green, homogeneous, free-flowing
Reaction of 6.6% solution:	pH 6.1 ± 0.1 at 25°C
Prepared medium:	bluish green, opalescent

---

*T. glabrata is being encountered with increasing frequency in clinical specimens and probably for the same reason as the *Candida;* widespread use of antibiotics and steroids.[2] Because *T. glabrata* can be confused with *Candida* species when grown on Bacto EMB Agar but not on Bacto Candida BCG Agar, its description is included to illustrate the problem of early identification of this and other yeasts.

Typical Cultural Response on Bacto Candida BCG Agar
After 24 – 48 Hours at 25 – 30°C

Organism	Growth	Color of Medium
Candida albicans ATCC® 26790	good to excellent	yellow
Candida krusei	good to excellent	yellow
Candida tropicalis	good to excellent	yellow
Eschericha coli ATCC® 25922	inhibited	green
Staphylococcus aureus ATCC® 25923	inhibited	green

REFERENCES

1. Personal Communication, 1968.
2. Manual of Clinical Microbiology, ASM, Lennette et al, 1982.

PACKAGING

Bacto Candida BCG Agar Base	1 lb (454 kg)	0835-01-0

BACTO CAPSULE INK

TEST SUMMARY

Bacto Capsule Ink is a suspension of carbon particles of optimal size for use in the Quellung reaction in determining capsule size of members of the genus *Klebsiella* as described by Edwards and Ewing.[1]

PROCEDURE

1. Prepare a suspension of the test organism in 0.85% NaCl solution containing form-aldehyde in a final concentration of 0.5% by volume. Dilute to density equivalent to a Bacto McFarland Barium Sulfate Standard No. 3.
2. Place a drop of the *Klebsiella* suspension on a clean microscope slide.
3. Place a very small drop of Bacto Capsule Ink next to the bacterial droplet. Do not mix.
4. Place a coverglass over the 2 droplets avoiding entrapment of air bubbles. The two suspensions will run together forming a varying degree of carbon concentrations. Scan the slide to find the area of the slide having an optimal density of ink and examine for capsule size.

STORAGE

Bacto Capsule Ink	15 – 30°C

REFERENCE

1. Identification of *Enterobacteriaceae*, 2nd Ed., Burgess Pub. Co.

PACKAGING

Bacto Capsule Ink	6 × 1 ml	3335-51-8

CARBOHYDRATES

Carbohydrates are extensively employed in culture media as a source of energy for bacteria and, more particularly, for differentiating genera and identifying species. The ability of an organism to attack a particular carbohydrate is a definite characteristic of bacterial species and under controlled conditions remains constant for the organism throughout generations of cultivation on media.

For testing the fermentative reactions of bacteria, carbohydrates in the dehydrated form are generally added to a sugar-free medium in a concentration of 0.5% or 1%. The medium is then inoculated with test organisms and incubated at 35 ± 2°C. Readings are generally made at 18 – 24 hours and at 40 – 48 hours. The earlier reading is performed to detect possible reversion of the reactions and the later reading is performed for the slower-reacting bacteria.

Carbohydrate solutions, sterilized by filtration, are prepared for the use of laboratories desiring fermentation media in which the carbohydrate has not been heat sterilized. These carbohydrates are ampuled in 10 ml amounts, each ampul containing 1 gram of the carbohydrate in sterile solution. Carbohydrate solutions in ampuls are convenient and economical when small quantities of fermentation media are required.

For the preparation of carbohydrate media, the basic sugar-free medium is prepared and sterilized as usual. The desired carbohydrate solution is then added aseptically and the medium is dispensed into sterile containers with aseptic precautions. To prepare a medium containing 1% of the carbohydrate, the contents of one ampul are added to 90 ml of medium; to prepare a medium with 0.5% of the carbohydrate, the contents of one ampul are added to 190 ml of the sterile sugar-free base. The final medium should always be incubated before use to insure its sterility. It is then inoculated with test organisms and incubated at 35 ± 2°C. Again, readings are made at 18 – 24 hours and at 40 – 48 hours.

Bacto Differentiation Disks Carbohydrates are sterile paper disks containing 17 – 23 mg of carbohydrate for use in the differentiation and presumptive identification of bacteria based on their fermentation reactions. Final confirmation of identity, especially the pathogens, should be made using further biochemical tests and serological procedures.

Disks are transferred to either the liquid or semisolid media in tubes and inoculated heavily with test organisms. Readings are made after 18 – 24 hours and again after 40 – 48 hours incubation at 35 ± 2°C. Solid media are inoculated by smearing, streaking, pouring, or superimposing upon the surface a thin inoculated seed layer of the test culture before applying the disks to the surface. Readings on solid media are also made after 18 – 24 and 40 – 48 hours incubation at 35 ± 2°C.

For a complete discussion on the use of carbohydrates refer to the following product descriptions:

Liquid Media	Semisolid Media	Solid Media
Bacto Phenol Red Broth Base	Bacto Cystine Tryptic Agar	Bacto Phenol Red Agar Base
Bacto Purple Broth Base	Bacto Tryptic Agar Base	Bacto Purple Agar Base

Carbohydrates

Product	Package Size	Code	Test Organisms	
			Positive	Negative
Adonitol Differentiation Disks Adonitol (Ad)	10 g 1 × 25 disks 6 × 25 disks	0157-12-7 1600-35-1 1600-33-3	*Klebsiella pneumoniae* ATCC® 13883	*Escherichia coli* ATCC® 25922
L-Arabinose Differentiation Disks Arabinose (Ar)	10 g 25 g 100 g 1 × 25 disks 6 × 25 disks	0159-12-5 0159-13-4 0159-15-2 1601-35-0 1601-33-2	*Escherichia coli* ATCC® 25922	*Proteus vulgaris* ATCC® 6380
Cellobiose	10 g 25 g	0160-12-2 0160-13-1	*Klebsiella pneumoniae* ATCC® 13883	*Escherichia coli* ATCC® 25922
Cellulose	25 g	0778-13-5	—	—
Dextrose[1] Dextrose Solution 10% Dextrose Solution 50%[2] Differentiation Disks Dextrose (D)	5 lb (2.27 kg) 10 kg 100 g 500 g 12 × 10 ml 6 × 10 ml 12 × 10 ml 1 × 25 disks 6 × 25 disks	0155-05-8 0155-08-5 0155-15-6 0155-17-4 0155-61-9 0973-60-0 0973-61-9 1602-35-9 1602-33-1	*Salmonella typhimurium* ATCC® 14028 *Escherichia coli* ATCC® 25922	*Alcaligenes faecalis* ATCC® 8750
Dulcitol Differentiation Disks Dulcitol (Du)	25 g 100 g 1 × 25 disks 6 × 25 disks	0162-13-9 0162-15-7 1603-35-8 1603-33-0	*Klebsiella pneumoniae* ATCC® 13883	*Proteus vulgaris* ATCC® 6380
D-Galactose Differentiation Disks Galactose (Ga)	100 g 500 g 1 × 25 disks 6 × 25 disks	0163-15-6 0163-17-4 1604-35-7 1604-33-9	*Salmonella arizonae* *(Arizona arizonae)* ATCC® 13314	*Neisseria meningitidis* ATCC® 13090

Product	Size	Catalog No.	Control Organism 1	Control Organism 2
Inositol Differentiation Disks Inositol (I)	25 g 100 g 1 × 25 disks 6 × 25 disks	0164-13-7 0164-15-5 1605-35-6 1605-33-8	*Klebsiella pneumoniae* ATCC® 13883	*Escherichia coli* ATCC® 25922
Inulin Differentiation Disks Inulin (In)	10 g 25 g 100 g 1 × 25 disks 6 × 25 disks	0165-12-7 0165-13-6 0165-15-4 1606-35-5 1606-33-7	*Streptococcus pneumoniae* ATCC® 6303	*Streptococcus pyogenes* ATCC® 19615
Lactose[1] Lactose Solution 10% Differentiation Disks Lactose (L)	5 lb (2.27 kg) 10 kg 100 g 500 g 12 × 10 ml 1 × 25 disks 6 × 25 disks	0156-05-7 0156-08-4 0156-15-5 0156-17-3 0156-61-8 1607-35-4 1607-33-6	*Escherichia coli* ATCC® 25922	*Salmonella typhimurium* ATCC® 14028
Levulose (D-Fructose) Differentiation Disks Levulose (Le)	25 g 100 g 1 × 25 disks 6 × 25 disks	0167-13-4 0167-15-2 1608-35-3 1608-33-5	*Streptococcus pneumoniae* ATCC® 6303	*Neisseria meningitidis* ATCC® 13090
Maltose[1] Maltose Solution 10% Differentiation Disks Maltose (M) Maltose Technical[3]	10 kg 100 g 500 g 12 × 10 ml 1 × 25 disks 6 × 25 disks 10 kg 100 g 500 g	0168-08-0 0168-15-1 0168-17-9 0168-61-4 1609-35-2 1609-33-4 0169-08-9 0169-15-0 0169-17-8	*Escherichia coli* ATCC® 25922	*Shigella dysenteriae* ATCC® 13313
D-Mannitol[1] Mannitol Solution 10% Differentiation Disks Mannitol (Mn)	10 kg 100 g 500 g 12 × 10 ml 1 × 25 disks 6 × 25 disks	0170-08-6 0170-15-7 0170-17-5 0170-61-0 1610-35-9 1610-33-1	*Staphylococcus aureus* ATCC® 25923	*Pseudomonas aeruginosa* ATCC® 27853

Carbohydrates (continued)

Product	Package Size	Code	Test Organisms	
			Positive	**Negative**
D-Mannose Differentiation Disks Mannose (Ma)	10 g 25 g 1 × 25 disks 6 × 25 disks	0171-12-9 0171-13-8 1611-35-8 1611-33-0	*Salmonella arizonae* *(Arizona arizonae)* ATCC® 13314	*Proteus vulgaris* ATCC® 6380
Melezitose	10 g 25 g	0172-12-8 0172-13-7	—	*Saccharomyces cerevisiae* ATCC® 9763
Melibiose Differentiation Disks Melibiose (Me)	5 g 10 g 1 × 25 disks 6 × 25 disks	0173-11-8 0173-12-7 1612-35-7 1612-33-9	*Salmonella typhimurium* ATCC® 14028	*Providencia alcalifaciens* ATCC® 9886
Raffinose Differentiation Disks Raffinose (Ra)	25 g 100 g 1 × 25 disks 6 × 25 disks	0174-13-5 0174-15-3 1613-35-6 1613-33-8	*Klebsiella pneumoniae* ATCC® 13883	*Proteus vulgaris* ATCC® 6380
Rhamnose Differentiation Disks Rhamnose (Rh)	10 g 25 g 100 g 1 × 25 disks 6 × 25 disks	0175-12-5 0175-13-4 0175-15-2 1614-35-5 1614-33-7	*Salmonella arizonae* *(Arizona arizonae)* ATCC® 13314	*Edwardsiella tarda* ATCC® 15947
Saccharose[1] (Sucrose) Saccharose Solution 10% Differentiation Disks Sucrose (Su)	10 kg 100 g 500 g 12 × 10 ml 1 × 25 disks 6 × 25 disks	0176-08-0 0176-15-1 0176-17-9 0176-61-4 1617-35-2 1617-33-4	*Klebsiella pneumoniae* ATCC® 13883	*Salmonella typhimurium* ATCC® 14028

Product	Quantity	Catalog No.		
Salicin	10 g	0177-12-3	Klebsiella pneumoniae ATCC® 13883	Shigella sonnei ATCC® 25931
	25 g	0177-13-2		
	100 g	0177-15-0		
Differentiation Disks Salicin (Sa)	1 × 25 disks	1615-35-4		
	6 × 25 disks	1615-33-6		
D-Sorbitol	100 g	0179-15-8	Salmonella arizonae (Arizona arizonae) ATCC® 13314	Edwardsiella tarda ATCC® 15947
	500 g	0179-17-6		
Differentiation Disks Sorbitol (So)	1 × 25 disks	1616-35-3		
	6 × 25 disks	1616-33-5		
Trehalose	5 g	0180-11-9	Escherichia coli ATCC® 25922	Providencia alcalifaciens ATCC® 9886
	10 g	0180-12-8		
Differentiation Disks Trehalose (Tr)	1 × 25 disks	1618-35-1		
	6 × 25 disks	1618-33-3		
D-Xylose	25 g	0181-13-6	Salmonella typhimurium ATCC® 14028	Providencia alcalifaciens ATCC® 9886
Differentiation Disks Xylose (X)	1 × 25 disks	1619-35-0		
	6 × 25 disks	1619-33-2		

[1] A purified carbohydrate prepared especially for use in bacteriological culture media.
[2] Dextrose solution 50% is also used with Sellers differential agar for the differentiation and identification of nonfermentative gram-negative bacilli.
[3] A carbohydrate for general use.

BACTO CASAMINO ACIDS

Bacto Casamino Acids is acid hydrolyzed casein recommended for use in microbiological culture media which require a completely hydrolyzed protein as a nitrogen source. Hydrolysis is carried on until all the nitrogen in the casein is converted to amino acids or other compounds of relative chemical simplicity. The sodium chloride content is satisfactory for diphtheria toxin production,[1,2] and has had all but the last traces of iron removed. Bacto Casamino Acids is also particularly well suited for nutritional studies, microbiological assays, the preparation of "synthetic" or chemically defined media, preparation of tetanus toxins and pertussis vaccines, and media used for sulfonamide inhibitor studies.

REFERENCES
1. J. Immunol., 40:21, 1941.
2. J. Immunol., 34:103, 1939.

HISTORY
Refer to *Difco Manual*, 9th Edition, p. 265 – 266, 1953.

PACKAGING
Bacto Casamino Acids		
	1 lb (454 g)	0230-01-1
	1/4 lb (114 g)	0230-02-0
	5 lb (2.27 kg)	0230-05-7
	10 kg	0230-08-4

CASAMINO ACIDS, TECHNICAL

Casamino Acids, Technical is acid hydrolyzed casein. The hydrolysis is carried out as in the preparation of Bacto Casamino Acids, but the sodium chloride and iron content of this product have not been decreased to the same extent. Casamino Acids, Technical is recommended for use in culture media where amino acid mixtures are required for a nitrogen source and the sodium chloride content need not be low. It is particularly valuable in studying the growth requirements of bacteria.

HISTORY
Refer to *Difco Manual*, 9th Edition, p. 267, 1953.

PACKAGING
Casamino Acids, Technical		
	1 lb (454 g)	0231-01-0
	5 lb (2.27 kg)	0231-05-6
	10 kg	0231-08-3

CASEIN, PURIFIED

Casein, Purified is recommended for all procedures requiring a casein of high purity. Repeated precipitation and washing of casein at the isoelectric point produces a purified casein of uniform quality. Casein, Purified is soluble in dilute solutions of sodium hydroxide forming the sodium salt.

PACKAGING

Casein, Purified	100 g	0336-15-8
	500 g	0336-17-6

CASEIN, TECHNICAL

Casein, Technical is a technical grade of this protein recommended for use in the preparation of culture media, and in other techniques not requiring a purified product.

PACKAGING

Casein, Technical	500 g	0337-17-5

BACTO CASITONE

Bacto Casitone is a specially prepared pancreatic digest of casein, conforming to the specifications given by the National Institute of Health[1] and the U.S. Pharmacopeia/National Formulary.[2] It is recommended for preparing media for sterility testing according to the National Institute of Health,[1] the Food and Drug Administration,[3] and the U.S. Pharmacopeia/National Formulary.[2] It is also recommended for preparing media where an enzymatically hydrolyzed casein is desired, as, for example, the liquid medium for the cultivation of *Mycobacterium tuberculosis* as described by Dubos and Middlebrook.[4]

Bacto Casitone is soluble in distilled or deionized water, yielding a clear solution that is neutral in reaction. It has a high tryptophane content, making it satisfactory for the detection of indole production.

REFERENCES

1. National Institute of Health Circular: Culture Media for the Sterility Test, and Revision: February 5, 1946.
2. USP XX/NF XV: 879:1980.
3. The Compilation of Tests and Methods of Assay for Antibiotic Drugs, Federal Security Agency, Food and Drug Administration.
4. Am. Rev. Tuber., 56:335:1947.

PACKAGING

Bacto Casitone	1 lb (454 g)	0259-01-7
	1/4 lb (114 g)	0259-02-6
	5 lb (2.27 kg)	0259-05-3
	10 kg	0259-08-0

BACTO CASMAN MEDIUM BASE

INTENDED USE

Bacto Casman Medium Base is used with blood for isolating fastidious microorganisms under reduced oxygen tension.

HISTORY

Casman[1-3] described an infusion-free medium enriched with 5% blood and incubated anaerobically, that supported growth of pure cultures and clinical specimens equally as well as Wright's[4] infusion base medium.

Adjustment of the formulation produced a medium that supported satisfactory growth of *Haemophilus influenzae* and increased growth of pathogenic cocci, favored typical hemolysis (particularly of beta-hemolytic streptocci) and slightly stimulated growth of some strains of *Neisseria meningitidis*.

PRINCIPLES

Improved growth of pathogenic cocci results from the addition of a small amount of dextrose to the FORMULA. The addition of 0.005% nicotinamide retards the removal of coenzyme (V factor) by nucleotidase of the added erythrocytes and, thus, permits growth of *H. influenzae*. The addition of 0.1% corn starch and use of washed agar provides for increased growth of *Neisseria* strains without interfering with typical hemolysis. The addition of lysed blood stimulates growth of some strains of *N. gonorrhoeae*.

FORMULA

BACTO CASMAN MEDIUM BASE
DEHYDRATED

Ingredients per liter

Proteose Peptone No. 3, Difco	10 g
Bacto Tryptose	10 g
Bacto Beef Extract	3 g
Nicotinamide	0.05 g
p-Aminobenzoic Acid	0.05 g
Bacto Dextrose	0.5 g
Corn Starch	1 g
Sodium Chloride	5 g
Agar Noble	14 g

Final pH 7.3 ± 0.2 at 25°C.

One pound will make 10.5 liters of single-strength medium.

METHOD OF PREPARATION

1. Suspend 43 g in 1 liter distilled or deionized water and heat to boiling to dissolve completely.
2. Sterilize in the autoclave for 15 minutes at 15 lbs pressure (121°C). Cool to 50°C.
3. Aseptically add 5% sterile blood and 0.15% sterile water-lysed blood solution (one part blood to three parts water). Omit water-lysed blood if sterile blood is partially lysed due to storage. Mix well.
4. Dispense into Petri dishes or tubes, as desired.

STORAGE

Bacto Casman Medium Base	Below 30°C
Prepared plates or tubes	2 – 8°C

QUALITY CONTROL
Identity Specifications

Dehydrated powder:

light tan, homogeneous, free-flowing

Reaction of 4.3% solution:

pH 7.3 ± 0.2 at 25°C

Plates prepared with 5% sheep blood plus 0.15% water-lysed sheep blood solution:

cherry red, opaque, no hemolysis

Typical Cultural Response on Bacto Casman Medium Base Prepared with Sheep Blood After 40 – 48 Hours at 35 ± 2°C

Organism	Growth	Hemolysis
Haemophilus influenzae ATCC® 19418	good to excellent	none
Neisseria meningitidis ATCC® 13090	excellent	none
Streptococcus mitis ATCC® 9895	excellent	beta
Streptococcus pneumoniae ATCC® 6303	excellent	alpha
Streptococcus pyogenes ATCC® 19615	excellent	beta

REFERENCES
1. Am. J. Clin. Path., 17:28, 1947.
2. J. Bact., 43:33, 1942.
3. J. Bact., 53:561, 1947.
4. J. Path. Bact., 37:257, 1933.

PACKAGING
Bacto Casman Medium Base   1 lb (454 g)   0290-01-8

BACTO CEREUS SPORE SUSPENSION

Bacto Cereus Spore Suspension is an aqueous suspension of spores of *Bacillus cereus var. mycoides* ATCC® 11778 for use in the microbiological assay of several antibiotics in accordance with FDA or USP guidelines.

STORAGE
Bacto Cereus Spore Suspension   2 – 8°C

QUALITY CONTROL
Identity Specifications

Appearance: milky white suspension
Concentration of spores: $2.5 \times 10^7 \pm 20\%$ spores per 1 ml ampule

PACKAGING
Bacto Cereus Spore Suspension	10 × 1 ml	0959-52-8
	25 × 1 ml	0959-36-9

BACTO CETRIMIDE AGAR BASE

INTENDED USE
Bacto Cetrimide Agar Base is a selective medium recommended for the isolation and identification of *Pseudomonas aeruginosa*.

HISTORY/PRINCIPLES

Bacto Cetrimide Agar Base is prepared according to a slight modification of the medium A formulation of King, Ward and Raney[1] and is the same formula as the Cetrimide Agar Medium formula given in the United States Pharmacopeia XX (1980).[2]

Cetrimide (cetyltrimethylammonium bromide) is added to inhibit bacteria other than *Pseudomonas aeruginosa*. Its action as a quaternary ammonium, cationic detergent causes nitrogen and phospherous to be released from bacterial cells other than *P. aeruginosa*.

Brown and Lowbury[3] utilized Cetrimide in the medium B formulation of King, Ward and Raney[1] to obtain a selective medium for *P. aeruginosa* which would show the production of fluorescein. Bacto Cetrimide Agar Base utilized the selectivity of Cetrimide in the medium A formulation of King, Ward and Raney.[1] This provided a selective medium which showed the production of pyocyanin by *P. aeruginosa*. When prepared according to the formulation as given, the medium is complete except for glycerol, which must be added separately.

FORMULA

BACTO CETRIMIDE AGAR BASE
DEHYDRATED

Ingredients per liter

Bacto Peptone	20 g
Magnesium Chloride	1.4 g
Potassium Sulfate	10 g
Cetrimide	0.3 g
(Cetyltrimethylammonium Bromide)	
Bacto Agar	13.6 g

Final pH 7.2 ± 0.2 at 25°C.

One pound will make 10 liters of final medium.

METHOD OF PREPARATION

1. Suspend 45.3 g in 1 liter distilled or deionized water. Add 10 ml Bacto Glycerol and heat to boiling to dissolve.
2. Dispense into tubes or flasks.
3. Sterilize in the autoclave for 15 minutes at 15 lbs pressure (121°C).
4. For the isolation of *P. aeruginosa,* plates of Bacto Cetrimide Agar Base may be streak inoculated from nonselective enrichment medium such as Bacto Brain Heart Infusion, Bacto Tryptic Soy Broth or Bacto Lactose Broth or it may be streak inoculated directly from test material if the count is sufficiently high.

STORAGE

Bacto Cetrimide Agar Base — Below 30°C
Prepared plates — 2 – 8°C

QUALITY CONTROL

Identity Specifications

Dehydrated powder: beige
Reaction of 4.53% solution with 1% Bacto Glycerol: pH 7.2 ± 0.2 at 25°C
Prepared medium: light amber, opalescent with a precipitate

Typical Cultural Response on Bacto Cetrimide Agar Base
After 24 – 48 Hours at 35 ± 2°C

Organism	Growth
Escherichia coli ATCC® 25922	inhibited
Pseudomonas aeruginosa ATCC® 27853	good to excellent*
Staphylococcus aureus ATCC® 25923	inhibited

*The colonies which appear may be pigmented blue, blue-green, yellow-green or nonpigmented.

REFERENCES
1. J. Lab. & Clin. Med., 44:301, 1954.
2. USP XX, 1980, page 875.
3. J. Clin. Path., 18:752, 1965.

PACKAGING
Bacto Cetrimide Agar Base	1 lb (454 g)	0854-01-6
	1/4 lb (114 g)	0854-02-5

BACTO CHAPMAN STONE MEDIUM

INTENDED USE
Bacto Chapman Stone Medium is a selective medium used for isolating staphylococci.

HISTORY
Bacto Chapman Stone Medium is prepared according to the formula described by Chapman.[1]

PRINCIPLES
This medium is similar to Bacto Staphylococcus Medium 110 except that the amount of sodium chloride is reduced to 5.5% and ammonium sulfate is included in the formulation, negating the need to flood the plate with ammonium sulfate solution for determining liquefaction of gelatin by Stone's method.

Sodium chloride acts as a selective agent in this medium.

After 48 hours incubation at 30°C, any mannitol-positive, yellow or orange colonies surrounded by a clear zone are likely to be *S. aureus*. White or nonpigmented colonies, with or without a clear zone, are probably *S. epidermidis*.

Mannitol fermentation can be determined by placing a drop of brom cresol purple on several areas of an agar plate from which typical pigmented colonies were removed for use in the coagulase test. Any change in color of the indicator compared with that of the uninoculated medium indicates fermentation of mannitol.

FORMULA
BACTO CHAPMAN STONE MEDIUM
DEHYDRATED
Ingredients per liter

Bacto Yeast Extract	2.5 g	Sodium Chloride	55 g
Bacto Tryptone	10 g	Ammonium Sulfate	75 g
Bacto Gelatin	30 g	Dipotassium Phosphate	5 g
D-Mannitol, Difco	10 g	Bacto Agar	15 g

Final pH 7.0 ± 0.2 at 25°C.

One pound will make 2.2 liters of medium.

METHOD OF PREPARATION
1. Suspend 20.2 g in 100 ml distilled or deionized water and heat to boiling to dissolve completely.
2. Sterilize in the autoclave for 10 minutes at 15 lbs pressure (121°C). Sterilization may be omitted if medium is to be inoculated the same day as prepared.
3. Dispense as desired.

STORAGE
Bacto Chapman Stone Medium   Below 30°C
Prepared medium   2 – 8°C

QUALITY CONTROL
Identity Specifications

Dehydrated powder: light beige, homogeneous, free-flowing
Reaction of 20.2% solution: pH 7.0 ± 0.2 at 25°C
Prepared medium: light to medium amber, opalescent
                             with a precipitate

Typical Cultural Response on Bacto Chapman Stone Medium
After 18 – 48 Hours at 30°C

Organism	Growth	Clearing or Halo
Escherichia coli ATCC® 25922	inhibited	–
Staphylococcus aureus ATCC® 25923	good to excellent	+
Staphylococcus epidermidis ATCC® 12228	good to excellent	+

REFERENCE
1. Food Research, 13:100, 1948.

PACKAGING
Bacto Chapman Stone Medium   1 lb (454 g)   0313-01-1
                                         10 kg   0313-08-4

BACTO CHAPMAN TELLURITE SOLUTION 1%

INTENDED USE
Bacto Chapman Tellurite Solution 1% is a sterile 1% solution of potassium tellurite prepared and standardized especially for use with Bacto Mitis Salivarius Agar and Bacto Tellurite Glycine Agar.

HISTORY
Chapman[1,2] described resulting selectivity when this solution was used in combination with Mitis Salivarius Agar.

METHOD OF PREPARATION
1. Add exactly 1 ml to 1 liter of sterile melted Bacto Mitis Salivarius Agar at 50 – 55°C.
2. Mix thoroughly and pour into plates.

STORAGE
Bacto Chapman Tellurite Solution 1%   15 – 30°C

The solution must be stored in a glass stoppered container free from organic matter.

QUALITY CONTROL
Identity Specifications
Prepared enrichment: clear, colorless solution

Typical Cultural Response on Bacto Mitis Salivarius Agar
After 18 – 24 Hours at 35 ± 2°C

Organism	Results
Escherichia coli ATCC® 25922	markedly inhibited, may be brown
Staphylococcus aureus ATCC® 25923	markedly inhibited
Streptococcus mitis ATCC® 6249	small blue colonies
Streptococcus salivarius ATCC® 9759	blue "gum drop" colonies
Streptococcus faecalis ATCC® 19433	dark blue/black shiny colonies

Typical Cultural Response on Bacto Tellurite Glycine Agar
After 18 – 48 Hours at 35 ± 2°C

Staphylococcus aureus ATCC® 25923	black colonies
Staphylococcus epidermidis ATCC® 14990	inhibited, grey colonies if present
Escherichia coli ATCC® 25922	markedly to completely inhibited

REFERENCES
1. Am. J. Digestive Diseases, 13:105, 1946.
2. Trans. N.Y. Acad. Sciences, 10:45, 1947.

PACKAGING
Bacto Chapman Tellurite Solution 1%	6 × 1 ml	0299-51-8
	6 × 25 ml	0299-66-1

BACTO CHARCOAL AGAR

INTENDED USE
Bacto Charcoal Agar is recommended for use in the cultivation of fastidious organisms, especially *Bordetella pertussis,* for vaccine production and stock culture maintenance.

HISTORY/PRINCIPLES
Bacto Charcoal Agar is prepared according to the method of Mishulow, Sharpe and Cohen.[1] They found this medium to be an efficient substitute for Bordet-Gengou agar in the production of *B. pertussis* vaccines.

FORMULA
BACTO CHARCOAL AGAR
DEHYDRATED
Ingredients per liter

```
Beef Heart, Infusion from  . . . . . . . .500 g
Bacto Peptone  . . . . . . . . . . . . . . . 10 g
Sodium Chloride  . . . . . . . . . . . . . . 5 g
Bacto Soluble Starch  . . . . . . . . . . . 10 g
Bacto Yeast Extract . . . . . . . . . . . . 3.5 g
Norit® SG  . . . . . . . . . . . . . . . . . . 4 g
Bacto Agar . . . . . . . . . . . . . . . . . 18 g
```

Final pH 7.3 ± 0.2 at 25°C.

One pound will make 7.26 liters of medium.

METHOD OF PREPARATION
1. Suspend 62.5 g in 1 liter cold distilled or deionized water. Heat to boiling with frequent agitation.
2. Sterilize in the autoclave for 15 minutes at 15 lbs pressure (121°C).
3. Allow to cool to 45 – 50°C. Swirl the medium to obtain a uniform suspension of charcoal and dispense into sterile tubes for slants or Petri dishes.
4. The medium is inoculated by surface streaks with specimens or cultures.

STORAGE
Bacto Charcoal Agar   Below 30°C
Prepared medium   2 – 8°C

QUALITY CONTROL
Identity Specifications
Dehydrated powder: grey, homogeneous, free-flowing
Reaction of 6.25% solution: pH 7.3 ± 0.2 at 25°C
Prepared medium: black, opaque with a precipitate
                            (undissolved charcoal)

Typical Cultural Response on Bacto Charcoal Agar
After 24 – 48 Hours at 35 ± 2°C

Organism	Growth
Bordetella bronchiseptica ATCC® 4617	good to excellent
Bordetella parapertussis	good to excellent
Bordetella pertussis ATCC® 9340	good to excellent

REFERENCE
1. Am. J. Publ. Hlth., 43:1466, 1953.

PACKAGING
Bacto Charcoal Agar   1 lb (454 g)   0894-01-8

BACTO CHICKEN PANCREAS

Bacto Chicken Pancreas is standardized desiccated chicken pancreas used in the enzymatic liberation of folic acid from its conjugated state. It is prepared according to the method described by the Official Analytical Chemists.[1] Bacto Chicken Pancreas is recommended for use in techniques requiring this enzyme, among which the following may be cited: *Official Methods of Analysis of the Association of Official Analytical Chemists,*[2] *Methods of Vitamin Assay,*[3] Sreenivasan, Harper and Elvehjem,[4] Burkholder, McVeigh and Wilson,[5] Mimms and Laskowski,[6] Bird, Bressler, Brown, Campbell and Emmett[7] and Laskowski, Mimms and Day.[8]

REFERENCES
1. A.O.A.C., Seventh Ed.:785, 1950.
2. A.O.A.C., Seventh Ed.:786, 1950.
3. The Association of Vitamin Chemists, Inc., Second Ed.:231, 1951.
4. J. Biol. Chem., 177:117, 1949.
5. Arch. of Biochem., 7:287, 1945.
6. J. Biol. Chem., 160:493, 1945.
7. J. Biol. Chem., 159:631, 1945.
8. J. Biol. Chem., 157:731, 1945.

PACKAGING
Bacto Chicken Pancreas   10 g   0459-12-2

BACTO CHLAMYDOSPORE AGAR

INTENDED USE
Bacto Chlamydospore Agar is used in differentiating *C. albicans* from other species of *Candida* on the basis of chlamydospore formation.

HISTORY/PRINCIPLES
Bacto Chlamydospore Agar, prepared according to the formula of Nickerson and Mankowski,[1] encourages chlamydospore formation by *C. albicans* and affords a rapid means of identifying this organism.

One of the most important differential characteristics of *C. albicans* is its ability to form chlamydospores on certain media. This property is perhaps the best single criterion for identification; however, since *Candida tropicalis* and *Candida stellatoidea* have also been demonstrated to produce chlamydospores under some conditions, additional differential tests must be employed where these latter species might be encountered.

Bacto Tween® 80 added in 0.01 to 0.05% concentration to Bacto Chlamydospore Agar enhances chlamydospore formation not only in *C. albicans* but also in *C. tropicalis* and *C. stellatoidea* and necessitates extreme caution in its use. Other differential criteria include the budding elongated cells, fine spreading of mycelia from line of streak, fermentation of sugars and reduction of indicators.

A selective medium may be prepared from Bacto Chlamydospore Agar by the addition of antibiotics to the sterile cooled medium just prior to dispensing into tubes or plates. This permits the direct isolation of *Candida* from clinic specimens and simultaneous identification of *C. albicans*. Broom[2] reported excellent results by the addition of 40 units of penicillin, 40 μg of streptomycin and 20 μg of Aureomycin® (Chlortetracycline) per ml to chlamydospore agar for the isolation and detection of *C. albicans* from sputum and other specimens. The antibiotics were added aseptically to the sterile medium after cooling to 45–50°C.

FORMULA
BACTO CHLAMYDOSPORE AGAR
DEHYDRATED
Ingredients per liter

Ammonium Sulfate	1 g
Monopotassium Phosphate	1 g
Biotin	5 μg
Trypan Blue	0.1 g
Purified Polysaccharide	20 g
Bacto Agar	15 g

Final pH is 5.1 ± 0.2 at 25°C.

One pound will make 12.2 liters of medium.

METHOD OF PREPARATION
1. Suspend 37 g in 1 liter cold distilled or deionized water and heat to boiling to dissolve completely.

2. Dispense into tubes or flasks as desired.
3. Autoclave for 15 minutes at 15 lbs pressure (121°C).
4. If plates are desired, cool to 50 – 55°C and dispense into sterile Petri dishes.

PROCEDURE
1. Using a sterile needle, scratch an "X" onto the agar surface with inoculum.
2. Aseptically place an alcohol-flamed and cooled cover slip onto the agar surface, placing it directly over the intersecting lines of the "X." Avoid entrapping air under the cover slip. Replace cover of Petri dish.
3. Incubate plates inverted at 20 – 25°C for 2 – 6 days. Temperatures higher than 25°C will not permit chlamydospore production.
4. Examine plates under low power of microscope after 3 or more days.
5. Alternatively, an inoculating needle may be made to cut through the medium to the bottom surface, permitting direct microscopic examination through the glass by inverting the plate.

Chlamydospore formation proceeds best under the slightly reduced oxygen tension provided by the above mentioned methods of inoculation. Chlamydospores selectively absorb the trypan blue and are blue in color, whereas the filaments are colorless.

STORAGE
Bacto Chlamydospore Agar   Below 30°C
Prepared plates or tubes   2 – 8°C

QUALITY CONTROL
Identity Specifications
Dehydrated powder: bluish tan, homogeneous, free-flowing
Reaction of 3.7% solution: pH 5.1 ± 0.2 at 25°C
Prepared medium: dark blue, slightly opalescent

Typical Cultural Response in Bacto Chlamydospore Agar
After 2 – 6 Days at 20 – 25°C

Organism	Growth	Chlamydospores
Candida albicans ATCC® 26790	excellent	+
Candida albicans isolate	excellent	+
Candida krusei isolate	excellent	−
Candida minosa isolate	excellent	−
Candida tropicalis isolate	excellent	−
Candida zeylanoides isolate	excellent	−

REFERENCES
1. J. Inf. Dis., 92:20, 1953.
2. Personal Communication, 1954.

PACKAGING
Bacto Chlamydospore Agar   1 lb (454 g)   0513-01-9

BACTO CHOLESTEROL (ASH FREE)

INTENDED USE
Bacto Cholesterol (Ash Free) is prepared especially for cholesterolizing antigens. It is of the highest purity and is ash free.

The addition of pure cholesterol to reinforce antigens was first recommended by Sachs.[1] By this means, antigens are rendered more sensitive without increasing their anticomplementary or hemolytic powers.

REFERENCE
1. Klin. Wochschr., Berlin, 48:2066, 1911.

PACKAGING

Bacto Cholesterol (Ash Free)	10 g	0224-12-6

BACTO CHRISTENSEN AGAR

INTENDED USE
Bacto Christensen Agar is used for the differentiation of enteric pathogens and coliforms based on citrate utilization.

HISTORY
Bacto Christensen Agar is a modification of the Christensen iron agar (citrate-sulfide medium),[1] having the same composition except that ferric ammonium citrate and sodium thiosulfate have been omitted. This modification was described by Edwards and Ewing.[2] These authors prefer to determine hydrogen sulfide production on triple sugar iron agar and for that reason recommend the use of the Christensen Agar without the addition of the hydrogen sulfide indicator, for the differentiation of *Shigella* from *Alkalescens-Dispar*.

PRINCIPLES
Citrate utilization in Christensen iron agar is denoted by the production of a deep red color in the slanted portion of the tube. Strong typical reactions are obtained after 4 to 8 hours incubation and occur aerobically. Tubes should not be stoppered or tightly capped. Organisms attacking citrate less vigorously may require as long as 3 days to produce this color change.

Hydrogen sulfide production in Christensen iron agar is shown by the development of a black color along the line of the stab. Strong positive cultures, upon prolonged incubation (24 hours) turn the entire butt a jet black. Some cultures, considered hydrogen sulfide positive, may require from 2 to 5 days to show a positive reaction. Christensen[2] reported that all members of the genera *Escherichia, Enterobacter, Citrobacter* (*Bethesda-Ballerup*) and *Salmonella*, as well as *Alkalescens* and *Dispar* organisms were capable of utilizing citrate as a source of energy in his medium. *Shigella dysenteriae, S. flexneri* and *S. sonnei* failed to utilize citrate. Thus this medium, on the basis of citrate utilization, differentiates *Shigella* from most members of the *Alkalescens-Dispar*. On the medium containing iron, strong hydrogen sulfide production is shown only by most members of the genus *Salmonella*. Some, including *S. typhosa*, are weakly positive following 2 to 5 days incubation. Organisms usually considered hydrogen sulfide positive, such as *Citrobacter* (*Bethesda-Ballerup*), are also weakly positive in 2 to 5 days.

Edwards and Ewing[1] recommended the use of triple sugar iron agar for the determination of hydrogen sulfide production and the Christensen agar for citrate utilization in the differentiation of *Alkalescens* and *Dispar* organisms from *Shigella*. The former organisms change the slant of the medium to a deep red, while *Shigella* strains grow without changing the color of the medium.

FORMULA

BACTO CHRISTENSEN AGAR
DEHYDRATED

Ingredients per liter

```
Bacto Yeast Extract . . . . . . . . . . . 0.5 g
Cysteine Hydrochloride . . . . . . . . . 0.1 g
Sodium Citrate . . . . . . . . . . . . . . . 3 g
Bacto Dextrose . . . . . . . . . . . . . . 0.2 g
Monopotassium Phosphate . . . . . . . . 1 g
Sodium Chloride . . . . . . . . . . . . . . 5 g
Bacto Agar . . . . . . . . . . . . . . . . . 15 g
Bacto Phenol Red . . . . . . . . . . . 0.012 g
```

Final pH 6.9 ± 0.2 at 25°C.

One pound will make 18.3 liters of medium.

METHOD OF PREPARATION

1. Suspend 24.8 g in 1 liter cold distilled water and heat to boiling to dissolve completely.
2. Distribute into tubes and sterilize in the autoclave for 15 minutes at 15 lbs pressure (121°C).
3. Allow to solidify in a slanted position to give a deep butt.

STORAGE

Bacto Christensen Agar　　　　　　　　Below 30°C
Prepared tubes　　　　　　　　　　　　2 – 8°C

QUALITY CONTROL

Identity Specifications

Dehydrated powder:　　　　　light pink, homogeneous, free-flowing
Reaction of 2.48% solution:　pH 6.9 ± 0.2 at 25°C
Prepared medium:　　　　　　orange-red, very slightly opalescent

Typical Cultural Response on Bacto Christensen Agar
After 18 – 72 Hours at 35 ± 2°C

Organism	Growth	Color Reaction
Enterobacter aerogenes ATCC® 13048	good	cerise
Escherichia coli ATCC® 25922	good	cerise
Salmonella enteritidis ATCC® 13076	good	cerise
Shigella dysenteriae ATCC® 13313	good	no change
Shigella flexneri ATCC® 12022	good	no change
Shigella sonnei ATCC® 25931	good	no change

REFERENCES

1. Christensen, W. B., "Hydrogen Sulfide Production and Citrate Utilization in the Differentiation of Enteric Pathogens and Coliform Bacteria," Research Bull. No. 1, Weld County Health Dept. Greeley, Co., 1:3 – 16, 1949.
2. Edwards & Ewing, Identification of Enterobacteriaceae, p. 179, 1955 and p. 242, 1962.

PACKAGING

Bacto Christensen Agar　　　　　　1 lb (454 g)　　　　　0695-01-9

TC CHROMOSOME CULTURE KIT

INTENDED USE
The TC Chromosome Culture Kit provides a qualitative method for culturing leukocytes and examination of chromosomes from human lymphocytes.

HISTORY
The TC Chromosome Culture Kit was designed for culturing the lymphocytes present in the plasma from 10 ml of heparinized human peripheral blood for the purpose of examining their propagated chromosomes. The procedure is a development of the discovery by Nowell,[1] and the investigations by Moorhead, Nowell, Mellman, Batipps, and Hungerford,[2] that phytohemagglutinin stimulates lymphocytes of human peripheral blood to undergo mitoses *in vitro*.

The sample of blood is initially placed in the TC Chromosome Blood Separation Vial for the rapid sedimentation by gravity of the erythrocytes. The supernatant suspension of lymphocytes in the plasma is used to inoculate the TC Chromosome Medium bottle which has been freshly rehydrated with the TC Chromosome Reconstituting Fluid. The addition of colchicine after an incubation of 3 days terminates the mitoses at the metaphase, and the resultant chromosomes are then swollen by treatment with a hypotonic solution as reported by Hughes[3] and Hsu and Pomerat.[4] After fixation with 3:1 methanol: acetic acid and attachment to microscope slides, in accordance with the procedures of Moorhead and Nowell[5] and Scherz,[6] the chromosomes may be stained with Giemsa, orcein, or other stains as developed by Rothfels and Siminovitch.[7] For more detailed identification, the chromosomes may also be stained by a variety of procedures to demonstrate banding.[8,9,10,11,12,13,14]

The assembly of the components of the TC Chromosome Culture Kit as a single unit was suggested by Jacobson,[15] and the announcement of the results obtained from its use was made by Jacobson and Telford[16] at the American Institute of Biological Sciences in Corvallis, Oregon, August 1962. Shortly thereafter Difco Laboratories first made available the TC Chromosome Culture Kit.

All the reagents of the kit are pretested on at least 4 different normal human bloods and they must demonstrate mitotic indices of at least 30 on each sample before they are approved for release. The kit contains sufficient reagents to perform cultures from 2 patients or duplicate cultures from 1 patient.

REAGENTS
The active ingredient, phytohemagglutinin, is an extract from beans. An optimal amount of purified phytohemagglutinin is present in the TC Chromosome Medium so that a mitotic index of at least 30 will be obtained from each of 4 normal human bloods when the test procedure is followed. TC Chromosome Reconstituting Fluid is used to rehydrate the TC Chromosome Medium. The reconstituted culture medium contains approximately 160 units of penicillin G and 160 µg of streptomycin per ml before inoculation and sufficient tissue culture medium to sustain viability of the cells.

The colchicine at a concentration of 1×10^{-5} M in the TC Chromosome Arresting Solution has been checked to confirm its ability to terminate mitoses at the metaphase with the concomitant accumulation of chromosomes.

Bacto Giemsa Stain, which has been certified previously by the Biological Staining Commission, has been tested to demonstrate its ability to stain chromosomes after the 20-fold dilution of the 1 ml stock solution.

TC Chromosome Blood Separation Vial facilitates the separation of the plasma-lymphocytes suspension from the erythrocytes.

TC Hanks Solution is a balanced salt solution in a concentration necessary to maintain the osmotic integrity of the cells during the harvesting procedure.

STORAGE AND EXPIRATION
TC Chromosome Culture Kit and its components are stable to the expiry date on the label when stored at 2 – 8°C, preferably at low humidity. Evidence of deterioration is indicated if the TC Chromosome Medium shrinks perceptibly from its typical appearance as a dried cake, or if the TC Chromosome Reconstituting Fluid, TC Chromosome Arresting Solution, or TC Hanks Solution change in color from reddish pink to yellow or magenta.

SPECIMEN COLLECTION AND PRESERVATION
The test is performed on the lymphocyte suspension present in the plasma isolated from peripheral blood. Aseptic techniques should be observed from the collection of the blood sample until the addition of the TC Chromosome Arresting Solution. The plasma suspension is conveniently obtained by aseptically withdrawing 10 ml of blood by syringe and transferring the sample immediately to one of the TC Chromosome Blood Separation Vials, which contain heparin. Within an hour the erythrocytes will settle by gravity to provide sufficient plasma suspension (4 – 5 ml) to inoculate 2 parallel lymphocyte cultures. (Eventually, the lymphocytes themselves will settle onto the erythrocyte layer, making it necessary to resuspend them by gentle inversion. A storage period of a day or more in the TC Chromosome Blood Separation Vial at 2 – 8°C is permissible, but not recommended for optimal results.) Further instructions regarding the handling of the blood specimen are given in the steps of the procedure.

Interfering Substances
Inasmuch as a number of substances have been reported deleterious to the initiation of mitoses *in vitro*, it is recommended that the patient should not be under medication. Also, since irradiation has been reported to cause "breaks" in chromosomes, it is best that the patient should not have been X-rayed recently.

PROCEDURE
Materials Provided:
The TC Chromosome Culture Kit contains the following reagents which permit cultures from 2 patients or duplicate cultures from 1 patient:

TC Chromosome Blood Separation Vials TC Chromosome Arresting Solution
TC Chromosome Medium   TC Hanks Solution
TC Chromosome Reconstituting Fluid   Bacto Giemsa Stain (Stock Solution)

Materials Required but not Provided:
Syringe, 10 ml (sterile)   Microscope (12.5X eyepiece with 10X low
Pasteur pipette with bulb   power, 40X high dry, and 100X oil
Incubator at 37°C   immersion objectives)
Centrifuge, small   Camera, if pictures desired
Water aspirator   Methanol, reagent grade, 15 ml
Microscope slides   Acetic acid, reagent grade, 5 ml

For laboratories performing chromosomal analyses on a large scale, the TC Chromosome Medium packaged with Reconstituting Fluid (code 5102) in a set of six is also available.

For cultures using a few drops of whole blood, as from infants, small animals, or the screening of adults, the TC Chromosome Microtest Kit (code 5060) is recommended.

LYMPHOCYTE SEPARATION AND INOCULATION

1. Withdraw 10 ml of blood from the patient who has abstained from eating for at least 3 hours and aseptically transfer it to a TC Chromosome Blood Separation Vial. Mix by inversion and allow it to stand at room temperature or 37°C for 1 – 2 hours until at least 4 ml of plasma-lymphocyte suspension has separated.
2. Rehydrate each required bottle of TC Chromosome Medium with a vial of TC Chromosome Reconstituting Fluid. This can be accomplished conveniently by holding the lip of the TC Chromosome Reconstituting Fluid vial inside the mouth of the TC Chromosome Medium bottle while pouring slowly. The TC Chromosome Medium should not be rehydrated with the TC Chromosome Reconstituting Fluid more than an hour before its inoculation with blood.
3. Inoculate each bottle of rehydrated TC Chromosome Medium with 1.5 – 2.0 ml of plasma-lymphocyte suspension from the TC Chromosome Blood Separation Vial. This transfer is conveniently performed with a sterile Pasteur pipette and bulb.

INCUBATION OF CULTURES

4. Incubate the inoculated TC Chromosome Medium bottle, preferably in a vertical position, for 3 – 4 days at 37°C. The color of the reconstituted medium should be kept pink before and after inoculation, and during the incubation period, by loosening the cap temporarily in order to allow escape of CO_2 if the color becomes amber.

   A significant increase in the mitotic index is often obtained by incubating the culture 4 days instead of 3. Disperse the cells by swirling the contents twice daily. It is very important to maintain the proper pH range of the culture at all times. The indicator should not become more yellow than a light amber and not redder than a light pink. The color is observed more readily when the culture bottle is kept in a vertical, rather than horizontal, position. If the indicator becomes amber, loosen the cap on the bottle by a quarter of a turn for an hour or so to allow the excess CO_2 to escape.

5. Add 1 vial of TC Chromosome Arresting Solution to the incubated culture and swirl it to insure thorough mixing; this will terminate the mitoses at the metaphase.
6. Incubate an additional 4 – 6 hours at 37°C.

   The exposure of the chromosomes to the TC Chromosome Arresting Solution should not be less than 4 hours nor more than 6.

HARVESTING AND FIXATION OF CELLS

7. Transfer the entire culture to a 15 ml graduated conical centrifuge tube and centrifuge for 6 – 8 minutes at 600 – 800 rpm.
8. Carefully aspirate off all but 0.1 ml of the supernatant fluid.
9. Add 5 ml of warm (37°C) TC Hanks Solution and resuspend the cells in the centrifuge tube with a Pasteur pipette.
10. Centrifuge at 600 – 800 rpm for 6 – 8 minutes.
11. Carefully aspirate off all but 0.1 ml of supernatant with the pipette and add 1 ml of TC Hanks Solution.
12. Resuspend the packed cells with the Pasteur pipette.

13. Add 3 ml of warm (37°C) distilled water in 1 ml portions with momentary agitation after each addition to produce a hypotonic solution.
14. Incubate the suspension at 37°C for 10 minutes only. The exposure of the cells to this hypotonic, diluted TC Hanks Solution should not exceed 10 minutes.
15. Centrifuge the lymphocyte suspension at 600 – 800 rpm for 6 – 8 minutes.
16. Carefully aspirate off the supernatant.
17. Add slowly, without disturbing the button of cells, 4 ml of freshly prepared fixative consisting of 1 part glacial acetic acid and 3 parts methanol (reagent grade only).
18. Let the cells soak in the fixative for 15 – 30 minutes. Cells should be treated gently during this stage of fixation.
19. Resuspend with the Pasteur pipette.
    NOTE: At this point, cells may be stored in the refrigerator overnight at 2 – 8°C, if desired.
20. Centrifuge at 600 – 800 rpm for 6 – 8 minutes and carefully remove the supernatant by aspiration.
21. Resuspend in 4 ml fresh fixative with the Pasteur pipette, let stand for 5 minutes, and centrifuge at 600 – 800 rpm for 6 – 8 minutes. Repeat this step again if it is necessary to disperse clumps of cells.
22. Carefully aspirate the supernatant.
23. Add 0.05 – 1.0 ml of fresh fixative to the button of cells and resuspend with the Pasteur pipette to get a hazy suspension.

PREPARATION OF SLIDES
24. Label clean microscope slides and place them in clean, chilled, distilled water.
25. In rapid succession, shake the excess water off a chilled slide, wipe the water off its underside, add 3 – 4 drops of the cell suspension by means of the Pasteur pipette, tip the slide several times to spread the suspension, and ignite the fixative by bringing it momentarily in contact with a flame. As soon as the fixative is burned off, wave the slide vigorously to hasten drying. The slide should not get hot, but drying should be accomplished as rapidly as possible.

STAINING OF SLIDES
Slides may be stained with Giemsa, orcein, or other stains. The procedure using Giemsa is given below:

The 1 ml of stock Bacto Giemsa Stain should be used the same day it is diluted 20-fold with water.
26. Dilute the 1 ml of stock Bacto Giemsa Stain with 19 ml of distilled water.
27. Place the slides in a small staining dish or Petri dish and cover them with 20 ml of the staining solution for 10 – 20 minutes.
28. Rinse the slides gently in distilled water and air dry.
29. Examine the slides under the microscope. The mitotic spreads may be scanned at a total magnification of 125 X, or examined more closely at 500 X, or photographed under oil immersion at 1000 X. Slides may be protected by cover slips and made permanent by conventional procedures.
    Alternatively, the chromosomes may be treated by staining procedures to demonstrate G-banding as reported by others.[8,9,10,11,12,13,14]

RESULTS
A mitotic index of at least 30 may be expected from the lymphocytes from the heparinized peripheral blood of a healthy individual. The mitotic index is usually considerably higher. Ample mitotic spreads are produced for karyotyping the chromosomes.

LIMITATIONS OF THE PROCEDURE

The presence of a number of drugs in the culture, either inadvertently introduced by way of the individual's blood sample, or intentionally added to the culture for the study of their effects on cellular physiology, may be deleterious to the development of normal mitoses and chromosomes. The *in vitro* effects of some drugs are due to their inhibition of one of the sequential steps occurring in cell division, such as DNA-replication, RNA-synthesis, protein synthesis, deposition of messenger RNA on the ribosomes, and for other reasons. The *in vitro* determination of the step in lymphocyte cell division affected by a drug producing abnormal mitoses and chromosomes suggests a new method for the study of its mode of action. Cytotoxic effects in culture are also shown by some plastic materials. Irradiation of patients often produce "breaks" in the chromosomes from their cultured lymphocytes.

The use of the TC Chromosome Culture Kit only provides chromosomes in metaphase for subsequent studies either by banding techniques or by karyotyping. For details on these procedures refer to the referenced literature or a standard text on Cytogenetics.[17]

REFERENCES

1. Nowell, Cancer Research, 20:462, 1960.
2. Moorhead, Nowell, Mellman, Batipps, and Hungerford, Exp. Cell Res., 20:613, 1960.
3. Hughes, Quarterly J. Microscopic Sci., 93:207, 1952.
4. Hsu and Pomerat, J. Heredity, 44:23, 1953.
5. Moorhead and Nowell, Personal Communication, 1961.
6. Scherz Stain Tech., 37:386, 1962.
7. *ibid.*, Rothfels and Siminovitch, 33:73, 1958.
8. Sumner, Evans, and Buckland, Nature (New Biol.), 232:31, 1971.
9. *ibid.*, Schnedl, 233:93, 1971.
10. Patil, Merrick and Lubs, Science, 173:821, 1971.
11. Seabright, Lancet, ii:971, 1971.
12. Wang and Federoff, Nature (New Biol.), 235:53, 1972.
13. Dev., Warburton, and Miller, Lancet, i:1285, 1972.
14. Arrighi, Hsu, Pathak, Shirley, and Stock, Mammalian Chromosome Newsletter, 15(2):88, 1974.
15. Jacobson, Personal Communication, 1962.
16. Jacobson and Telford, Paper Delivered at AIBS Meeting, Corvallis, Oregon, August 1962.
17. Roberts, Fraser, J. A., An Introduction to Medical Genetics, Oxford University Press, 1963.

PACKAGING

TC Chromosome Culture Kit	2 tests	5842-32-3
Contains:		
TC Chromosome Blood Separation Vial	2	
TC Chromosome Medium	2	
TC Chromosome Reconstituting Fluid	2	
TC Chromosome Arresting Solution	2	
TC Hanks Solution	2	
Bacto Giemsa Stain	1	

The contents of the Kit provide sufficient reagents for blood samples from 2 patients, a single patient in duplicate, or a patient and a normal control.

Components Available Individually:

TC Chromosome Arresting Solution	6 × 1 ml	5845-51-6
TC Chromosome Blood Separation Vials	6 vials	5855-33-6
TC Chromosome Medium w/		
Reconstituting Fluid	6 tests	5102-32-8
TC Hanks Solution	6 × 12 ml	5508-33-7
	100 ml	5508-72-9
	400 ml	5508-78-3
Bacto Giemsa Stain	6 × 1 ml	0846-51-6

TC CHROMOSOME MICROTEST KIT

INTENDED USE

The TC Chromosome Microtest Kit provides a qualitative method for the propagation and examination of chromosomes from lymphocytes of humans and animals.

HISTORY

The TC Chromosome Microtest Kit was especially designed for culturing the lymphocytes from 4 – 6 drops (0.15 – 0.20 ml) of whole, heparinized blood for the examination of their propagated chromosomes. This small sample is particularly advantageous when the amount of blood is limited, as in the case of infants, small animals, or in the screening of adults. The sample is readily obtained from a finger, heel, or ear using one of the sterile lancets and capillary tubes provided. Since no preliminary separation from erythrocytes is required, a culture can be set up in 5 minutes.

A variety of techniques using small amounts of whole blood for chromosomal studies have been reported by Arakaki and Sparkes,[1] Kwei-Chi Ho and Smith,[2] and Bishun and Robinson.[3] The TC Chromosome Microtest Kit incorporates the best features of the various techniques into a single unit. All of the reagents of the kit are pretested on at least 4 normal human bloods and must demonstrate mitotic indices of at least 30 on each sample before they are approved for release. The kit contains sufficient reagents to perform cultures from 2 patients or duplicate cultures from one. The kit has also been used successfully for obtaining the chromosomes of a variety of animals, including crocodile, beaver, hedgehog, snake, guinea pig, and rabbit.

This procedure is a development of the discovery by Nowell,[4] and the investigations by Moorhead, Nowell, Mellman, Batipps and Hungerford,[5] that phytohemagglutinin stimulates lymphocytes of human peripheral blood to undergo mitoses *in vitro*. The addition of colchicine after an incubation of 3 days terminates the mitoses at the metaphase with its accompanying accumulation of chromosomes. The earlier contributions of Hughes[6] and of Hsu and Pomerat[7] in the hypotonic swelling of chromosomes, of Moorhead and Nowell[8] and Scherz[9] in the preparation of slides, and of Rothfels and Siminovitch[10] in staining, were important developments in this field. After the chromosomes are treated with fixative and attached to the slides they may be stained with Giemsa, orcein, or other stains. For more detailed identification, the chromosomes may be stained by a variety of procedures to demonstrate banding.[11,12,13,14,15,16,17]

REAGENTS

The active ingredient, phytohemagglutinin, is an extract from beans. An optimal amount of purified phytohemagglutinin is present in the TC Chromosome Microtest Medium so that a mitotic index of at least 30 will be obtained from each of 4 different normal human bloods when the test procedure is followed. TC Chromosome Microtest Medium is rehydrated with its accompanying TC Chromosome Microtest Reconstituting Fluid. The reconstituted medium contains approximately 80 units of penicillin G and 80 μg of streptomycin per ml and sufficient tissue culture medium to sustain viability of the cells.

The colchicine at a concentration of 5×10^{-6} M in the TC Chromosome Microtest Arresting Solution has been checked to confirm its ability to terminate mitoses at the metaphase with the concomitant accumulation of chromosomes.

Bacto Giemsa Stain, which has been previously certified by the Biological Staining Commission, has been tested to demonstrate its ability to stain chromosomes after the 20-fold dilution of the 1 ml stock solution.

TC Hanks Microtest Solution is a balanced salt solution in a concentration necessary to maintain the osmotic integrity of the cells during the harvesting procedure.

Sterile blood lancets are supplied for obtaining finger tip blood. Sterile heparinized capillary pipettes are provided for collection of the blood sample.

STORAGE AND EXPIRATION
TC Chromosome Microtest Kit and its components are stable to the expiry date on the label when stored at 2 – 8°C, preferably at low humidity. Evidence of deterioration is indicated if the TC Chromosome Microtest Medium shrinks from a cake to a button, or if the TC Chromosome Microtest Reconstituting Fluid, TC Chromosome Microtest Arresting Solution, or TC Hanks Microtest Solution change color from reddish-pink to yellow or magenta.

SPECIMEN COLLECTION AND PRESERVATION
The test is performed on approximately 3 – 5 drops (0.15 – 0.20 ml) of whole blood. Aseptic techniques should be observed from the collection of the blood sample until the addition of the TC Chromosome Microtest Arresting Solution. Successful results have been obtained on smaller quantities. Sterile lancets for pricking the alcohol-sterilized ear, finger, or heel are provided, as well as sterile TC Capillary Pipettes for transferring the drops of blood aseptically from the skin into the TC Chromosome Microtest Medium, which should be freshly rehydrated with TC Chromosome Microtest Reconstituting Fluid. If there is a delay in this transfer, the sterile sample of blood may be heparinized to preclude clotting and kept at 0 – 20°C for as long as 24 hours before adding it to the culture. The heparin, which is used at a level of 30 units for 1 ml of blood, must be free of phenol.

Interfering Substances
Inasmuch as a number of substances have been reported deleterious to the initiation of mitoses *in vitro*, it is recommended that the patient should not be under medication. Also, since irradiation has been reported to cause "breaks" in chromosomes, it is best that the patient should not have been X-rayed recently.

The components of the TC Chromosome Microtest Kit have been used successfully in propagating the lymphocytes and examining the chromosomes of a number of mammals and reptiles. However, certain animal species have peripheral blood lymphocytes which are refractory and consequently specific variations in conditions are required to initiate their mitoses.

PROCEDURE
Materials Provided:
TC Chromosome Microtest Medium
TC Chromosome Microtest Reconstituting Fluid
TC Chromosome Microtest Arresting Solution
TC Hanks Microtest Solution
Bacto Giemsa Stain (Stock Solution)
Sterile blood lancets
Sterile capillary tubes with bulbs
Pasteur pipettes (2) with bulb (1)

Materials Required but not Provided:
Incubator at 37°C
Centrifuge (small)
Water aspirator

Microscope slides
Microscope (12.5X eyepiece with 10X low power, 40X high dry, and 100X oil immersion objectives)
Camera, if pictures desired
Methanol, reagent grade, 25 ml
Acetic Acid, reagent grade, 10 ml

For laboratories performing chromosomal analyses on a large scale, the TC Chromosome Microtest Medium packaged only with TC Chromosome Microtest Reconstituting Fluid, in a set of 6 (code 5398) is also available.

PREPARATION OF CULTURE

1. Rehydrate 1 vial of TC Chromosome Microtest Medium by adding the contents of 1 vial of TC Chromosome Microtest Reconstituting Fluid. The TC Chromosome Microtest Medium should not be rehydrated with the TC Chromosome Microtest Reconstituting Fluid more than an hour before its inoculation with blood.
2. Prick the clean dry finger, ear, or heel with the enclosed blood lancet and, by means of the capillary tube, transfer 4 – 6 drops of blood (2 – 3 full capillary tubes) into the TC Chromosome Microtest Medium. Mix the contents by gently inverting the closed vial 3 times.
3. Incubate the culture at 37°C for 3 – 4 days.

   The color of the reconstituted medium should be kept pink before and after inoculation and during the incubation period by loosening the cap temporarily to allow escape of CO_2 if the color becomes amber. A significant increase in mitotic index is often obtained by incubating the culture 4 days instead of 3. Disperse the cells by shaking the contents twice a day. If the color of the medium becomes light amber (not yellow), loosen the cap for an hour or so to allow it to become faintly pink by the loss of CO_2.
4. Add the contents of the TC Chromosome Microtest Arresting Solution to the culture and swirl to ensure even mixing. Continue the incubation for another 4 – 6 hours. Exposure to the TC Chromosome Microtest Arresting Solution should not be less than 4 hours nor more than 6.

HARVESTING AND FIXATION OF CELLS

5. Pour the entire culture into a 15 ml conical centrifuge tube and collect the cells at 600 – 800 rpm for 6 – 8 minutes.
6. Carefully aspirate off all but 0.1 ml of the supernatant fluid.
7. Add 4 ml of TC Hanks Microtest Solution and resuspend the cells using the Pasteur pipette.
8. Collect the button of cells by centrifugation at 600 – 800 rpm for 6 – 8 minutes.
9. Carefully aspirate off all but 0.1 ml of the supernatant fluid and add 1 ml of TC Hanks Microtest Solution.
10. Resuspend the cells with the Pasteur pipette.
11. Add a total of 3 ml of warm (37°C) distilled water in 3 equal portions, with immediate agitation after each addition. Resuspend the cells gently with the pipette and allow them to stand in the hypotonic solution at 37°C for 10 minutes. Exposure of the cells to the hypotonic solution should not exceed 10 minutes.
12. Collect the cells by centrifugation at 600 – 800 rpm for 6 – 8 minutes.
13. Aspirate off, very carefully, almost all the supernatant hypotonic solution.
14. Add 4 ml of a fresh mixture of 3 parts of methanol and 1 part of glacial acetic acid, both of reagent grade, carefully without disturbing the button of cells. This operation is best done with a 5 ml pipette. Allow the cells to soak in the fixative for 15 – 30 minutes.

    The cells should be treated gently during this period in the fixative.

15. By means of the Pasteur pipette gently resuspend the small button of cells.
NOTE: Cells may be stored overnight in the refrigerator at $2-8°C$ at this point, if desired.
16. Collect the cells by centrifugation at $600-800$ rpm for $6-8$ minutes.
17. Aspirate off, very carefully, most of the fixative. Care must be exercised not to draw off the cells. Add 4 ml fresh fixative and repeat steps 15, 16, and 17, once.
18. Add 0.5 ml of fresh fixative and resuspend the cells by means of the Pasteur pipette.

PREPARATION OF SLIDES
19. Label clean microscope slides and place them in clean, chilled distilled water.
20. In rapid succession, shake the excess water off a chilled slide, wipe the water off its underside, add $3-4$ drops of the cell suspension by means of the Pasteur pipette, tip the slide several times to spread the suspension, and ignite the fixative by bringing it momentarily in contact with a flame. As soon as the fixative is burned off, wave the slide vigorously to hasten drying. The slide should not get hot, but drying should be accomplished as rapidly as possible.

STAINING OF SLIDES
Slides may be stained with Giemsa, orcein, or other stains. The procedure using Giemsa is given below:
21. Dilute the 1 ml of Bacto Giemsa Stain stock solution with 19 ml of distilled water. Diluted Giemsa Stain must be used the same day.
22. Place the slides in a small staining dish or Petri dish and cover them with the staining solution for $10-20$ minutes.
23. Rinse the slides gently in distilled water and air dry.
24. Examine the slides under the microscope. The mitotic spreads may be scanned at a total magnification of 125X, or examined more closely at 500X, or photographed under oil immersion at 1000X. Slides may be protected by cover slips and made permanent by conventional procedures.

Alternatively, the chromosomes may be treated by staining procedures to demonstrate G-banding as reported by others.[11,12,13,14,15,16,17]

RESULTS
A mitotic index of at least 30 may be expected from the lymphocytes from the heparinized peripheral blood of a healthy individual. The mitotic index is usually considerably higher than this minimum value. Ample mitotic spreads are produced for karyotyping the chromosomes.

LIMITATIONS OF THE PROCEDURE
The presence of a number of drugs in the culture, either inadvertantly introduced by way of the individual's blood sample, or intentionally added to the culture for the study of their effects on cellular physiology, may be deleterious to the development of normal mitoses and chromosomes. The *in vitro* effects of some drugs are due to their inhibition of one of the sequential steps occurring in cell division, such as DNA replication, RNA synthesis, protein synthesis, deposition of messenger RNA on the ribosomes, and for other reasons. The *in vitro* determination of the step in lymphocyte cell division affected by a drug producing abnormal mitoses and chromosomes suggests a new method for the study of its mode of action. Cytotoxic effects in culture are also shown by some plastic materials. Irradiation of patients often produce "breaks" in the chromosomes from their cultured lymphocytes.

The use of the TC Chromosome Microtest Kit only provides chromosomes in metaphase for subsequent studies either by banding techniques or by karyotyping. For details on these procedures refer to the referenced literature or a standard text on Cytogenetics.[18]

REFERENCES

1. Arakaki and Sparkes, Cytogenetics, 2:57, 1963.
2. Kwei-Chi Ho and Smith, Human Chromosome Newsletter, 11:20, December, 1963.
3. Bishun and Robinson, Human Chromosome Newsletter, 12:16, April, 1964.
4. Nowell, P. C., Cancer Research, 20:462, 1960.
5. Moorhead, Nowell, Mellman, Batipps, and Hungerford, Exp. Cell. Res., 20:613, 1960.
6. Hughes, Quart. J. Microscopic Sci., 93:207, 1952.
7. Hsu and Pomerat, J. Heredity, 44:23, 1953.
8. Moorhead and Nowell, Personal Communication.
9. Scherz, Stain Tech., 37:386, 1962.
10. ibid., Rothfels and Siminovitch, 33:73, 1958.
11. Summer, Evans, and Buchland, Nature (New Biol.), 232:31, 1971.
12. ibid., Schnedl, W., 233:93, 1971.
13. Patil, Merrick, and Lubs, Science, 173:821, 1971.
14. Seabright, M., Lancet, ii:971, 1971.
15. Wang and Federoff, Nature (New Biol.), 235:52, 1972.
16. Dev. Warburton, and Miller, Lancet, i:1285, 1972.
17. Arrighi, Hsu, Pathak, Shirley, and Stock, Mammalian Chromosome Newsletter, 15:(2):88, 1974.
18. Roberts, Fraser, J. A., An Introduction to Medical Genetics, Oxford University Press, 1963.

PACKAGING

TC Chromosome Microtest Kit	2 tests	5060-32-8
Contains:		
TC Chromosome Microtest Medium	2	
TC Chromosome Microtest		
Reconstituting Fluid	2	
TC Chromosome Microtest		
Arresting Solution	2	
TC Hanks Microtest Solution	2	
Bacto Giemsa Stain	1	
Sterile Blood Lancets	2	
Pasteur pipettes with bulb	2	
Sterile capillary tubes w/bulbs	2	

The contents of the kit provide sufficient reagents for blood samples from 2 patients, a single patient in duplicate, or a patient and a normal control.

Components Available Individually:

TC Chromosome Microtest		
Arresting Solution	6 × 1 ml	5063-51-1
TC Chromosome Microtest Medium	6 tests	5398-32-1
w/Reconstituting Fluid		
TC Hanks Microtest Solution	6 × 5 ml	5064-57-4
Bacto Giemsa Stain	6 × 1 ml	0846-51-6

BACTO COAGULASE AGAR BASE

INTENDED USE

Bacto Coagulase Agar Base is recommended as a basal medium to which Bacto Coagulase Plasma or sterile pretested human plasma from the blood bank may be added for isolating and differentiating staphylococci from clinical specimens or for classifying pure cultures, according to the method described by Esher and Faulkner.[1]

PRINCIPLES
Differentiation is based on mannitol fermentation and coagulase production by the pathogenic, but not by the nonpathogenic staphylococci. The medium may be inoculated directly by streaking with the specimen or with 3 or 4 pure cultures on a divided plate. Mannitol fermentation and coagulase production is indicated by a yellow opaque zone around the colony. *Escherichia coli* will produce a yellow slightly opaque zone, but the colony is easily differentiated from the *Staphylococcus* colonies.

FORMULA

BACTO COAGULASE AGAR BASE
DEHYDRATED

Ingedients per liter

Bacto Brain Heart Infusion . . . . 5 g
Bacto Tryptic Soy Agar . . . . 28 g
Mannitol . . . . 10 g
Bacto Agar . . . . 4 g
Brom Cresol Purple . . . . 0.016 g

Final pH 7.4 ± 0.2 at 25°C.

One pound will make 10.3 liters of medium.

METHOD OF PREPARATION
1. Suspend 47 g in 1 liter distilled or deionized water. Heat to boiling to dissolve completely.
2. Sterilize for 15 minutes at 15 lbs pressure (121°C).
3. Allow to cool to 45 – 50°C and add 7 – 15% Bacto Coagulase Plasma. Swirl to mix well.
4. Pour into plates and inoculate as desired using clinical specimens or pure cultures.

STORAGE
Bacto Coagulase Agar Base   Below 30°C
Prepared medium   2 – 8°C

QUALITY CONTROL
Identity Specifications

Dehydrated powder: tan, homogeneous, free-flowing
Reaction of 4.7% solution: pH 7.4 ± 0.2 at 25°C
Prepared medium: purple, slightly opalescent

Typical Cultural Response on Bacto Coagulase Agar
After 18 – 48 Hours at 35°C

Organism	Growth	Mannitol Fermentation	Coagulase
Staphylococcus aureus ATCC® 25923	good	+ (yellow)	halo
Staphylococcus epidermidis ATCC® 12228	good	− (purple)	no halo

REFERENCE
1. Amer. J. Clin. Path., 32:192, 1959.

PACKAGING
Bacto Coagulase Agar Base	1 lb (454 g)	0501-01-3
Bacto Coagulase Plasma (Rabbit)	6 × 3 ml	0286-46-1
	6 × 15 ml	0286-86-2
	6 × 25 ml	0286-66-6
Bacto Coagulase Plasma EDTA (Rabbit)	6 × 3 ml	0803-46-5
	6 × 15 ml	0803-86-6
	6 × 25 ml	0803-66-0

COAGULASE PLASMA
BACTO COAGULASE PLASMA
BACTO COAGULASE PLASMA EDTA

INTENDED USE
Bacto Coagulase Plasma and Bacto Coagulase Plasma EDTA are standardized, desiccated rabbit plasmas used for detecting coagulase enzyme produced by *Staphylococcus aureus*.[1]

Bacto Coagulase Plasma and Bacto Coagulase Plasma EDTA are also used as a test medium for the production of germ tubes by *Candida albicans*.[2]

SUMMARY AND EXPLANATION OF THE TESTS
Coagulase Detection
Of the clinically important species of *Staphylococcus*, three are well documented: *S. aureus*, *S. epidermidis* and *S. saprophyticus*.[3] Differentiation of the coagulase-positive *S. aureus* from the coagulase-negative species, including *S. epidermidis* and *S. saprophyticus*, is crucial not only because *S. aureus* is a health risk of prime importance but also because the latter species are increasingly associated with septicemia, bacterial endocarditis, colonization of prostheses and urinary tract infections.[4] Identification of staphylococci is based on microscopic examination, colonial morphology and cultural and biochemical characteristics. The detection of coagulase, however, is the most widely used criterion for differentiation between species.[5 – 9]

The ability of *Staphylococcus* to produce coagulase, an enzyme capable of clotting plasma, was first reported by Loeb[10] in 1903. Dorang[11] indicated the practical significance of this test. Since that time, many investigators have tried to correlate the production of coagulase with the pathogenicity of staphylococci. Chapman, Berens, Peters and Curcio,[12] in a study of coagulase and hemolysin production by *Staphylococcus*, showed that strains producing coagulase were usually pathogenic, regardless of their hemolytic or chromogenic properties. However, more recent experience has demonstrated that the ability or inability of a *Staphylococcus* to produce coagulase cannot be relied upon to always indicate its pathogenicity.[13]

Germ Tube Development
There are several methods used for identifying *Candida albicans*. Among the more popular techniques is the culture of the organism in a medium that promotes the development of germ tubes.[14] Ogletree identified various media in which germ tubes will grow.[15] Human serum is widely used but inherent variations in its composition and handling can alter its effectiveness.[16] Smith and Elliott,[2] in searching for a suitable substitute, have recommended rabbit coagulase plasma. They found coagulase plasma EDTA to be entirely suitable, readily available and free from variation.

PRINCIPLES OF THE PROCEDURE
Coagulase Detection
Staphylococcus aureus produces two types of coagulase, free and bound. Free coagulase is an extracellular enzyme produced when the organism is cultured in broth. Bound coagulase, also known as clumping factor, remains attached to the cell wall of the organism.

The tube test consists of adding *S. aureus* from an overnight broth culture or from a noninhibitory agar plate to a tube of rehydrated coagulase plasma and then incubating at 37°C. The formation of a clot in the plasma indicates coagulase production. The tube test is the most frequently used method because of its greater accuracy and its ability to detect both bound and free coagulase.[17] The slide test is performed by adding a heavy suspension of cells to a drop of plasma on a slide, then observing for the presence or absence of a clot. This test is used infrequently because it is less accurate, is subject to time related false-negative results and requires that negative results be confirmed by the tube test. The slide test is only capable of detecting bound coagulase.[3]

Germ Tube Development
The germ tube test involves suspending suspect colonies of yeast in coagulase plasma, incubating the tubes at 37°C for two to four hours and then observing the cells microscopically for the development of germ tubes. *Candida albicans* and *C. stellatoidea* produce germ tubes under these test conditions. Other species do not. *C. albicans* and *C. stellatoidea* are easily differentiated from each other by a sucrose assimilation test and by sensitivity to cycloheximide.

REAGENTS
Bacto Coagulase Plasma is desiccated rabbit plasma to which sodium citrate has been added as the anticoagulant.

Bacto Coagulase Plasma EDTA is desiccated rabbit plasma to which EDTA (ethylenediaminetetraacetic acid) has been added as the anticoagulant. EDTA is not utilized by bacteria and therefore will not give rise to false-positive coagulase reactions by bacteria which utilize citrate.

REHYDRATION
Rehydrate Bacto Coagulase Plasma and Bacto Coagulase Plasma EDTA by adding sterile distilled or deionized water to the vial, as indicated below. Mix by gentle end-over-end rotation of the vial.

Product Size	Sterile Water	Approximate Number of Tests
3 ml	3 ml	6
15 ml	15 ml	30
25 ml	25 ml	50

If, upon rehydration, the plasma is not in complete solution or if fibrin clots or strands are evident, discard the plasma and test the pH of the distilled water. An acid pH of the water could result in an unsatisfactory reagent.

STORAGE
Store unopened Bacto Coagulase Plasmas at 2 – 8°C.

Store reconstituted plasmas at 2 – 8°C or aliquot in 0.5 ml quantities, freeze promptly and store at −20°C. Do not thaw and refreeze.

Expiration Date
Unopened Bacto Coagulase Plasmas are stable through the expiration date on the label when stored as directed.

Reconstituted plasmas, if kept uncontaminated, retain their activity for five days when stored at 2 – 8°C or for up to 30 days when aliquoted and stored at −20°C, not exceeding the expiration date on the label.

SPECIMEN PREPARATION
Coagulase Detection
1. Determine that the test culture is pure and has the following characteristics of *Staphylococcus aureus.*
   Morphology (medium dependent):

Bacto Blood Agar Base w/5% blood	Opaque, yellow to orange, with hemolysis
Bacto Coagulase Mannitol Agar	Opaque, with yellow to orange zones
Bacto DNase Test Agar w/Methyl Green	Clearing of green dye
Bacto Mannitol Salt Agar	Yellow to orange, surrounded by yellow zones
Bacto Staphyloccoccus Medium 110	Yellow to orange
Bacto Tellurite Glycine Agar	Black
Bacto VJ Agar	Black, surrounded by yellow zones
Bacto Baird Parker Agar	Grey to black shiny colonies surrounded by zones of clearing
Gram stain:	Gram-positive cocci occurring in grape-like clusters or, occasionally, in chains
Catalase test:	+
Mannitol fermentation:	+

2. Using a bacteriological loop, transfer a well isolated colony from a pure culture into a tube of sterile Bacto Brain Heart Infusion. Incubate at 37°C for 18 – 24 hours or until dense growth is observed. Alternatively, take 2 – 4 colonies (one loopful) directly from a noninhibitory agar plate as an inoculum instead of the broth culture.

Germ Tube Development
Select a well-isolated colony of a suspect yeast grown on Bacto Sabouraud Dextrose Agar for 48 – 72 hours.

PROCEDURE
Materials Provided
Bacto Coagulase Plasma
Bacto Coagulase Plasma EDTA

Materials Required but not Provided
Bacteriological inoculating loop
Sterile 1 ml pipettes
Sterile Pasteur pipette
Sterile serological pipettes 1, 5 and 10 ml
Bacteriological incubator at 37°C
Freshly distilled water
Erlenmeyer flask, 50 ml
Culture tubes, 12 × 75 mm
Test tube support for culture tubes
Interval timer
Water bath at 37°C.

COAGULASE TEST
1. Using a sterile 1 ml pipette, add 0.5 ml of the rehydrated plasma to a 12 × 75 mm test tube supported in a rack.
2. Using a sterile 1 ml serological pipette, add two drops of the overnight broth culture of the test organism to the tube of plasma or, using a sterile bacteriological loop, thoroughly emulsify 2 – 4 colonies (1 loopful) from a **noninhibitory** agar plate in the tube of plasma.

3. Mix gently.
4. Incubate in a water bath at 37°C for 4 – 24 hours. If it is necessary to use an incubator, it must be one without a CO_2 atmosphere since the presence of CO_2 may cause false-positive results.
5. Examine periodically for coagulation by gently tipping the tube after the first hour and once every hour thereafter until four hours have elapsed. If desired, reincubate and examine after 24 hours. Avoid shaking or agitating the tube during reading. Doubtful or false-negative results may occur due to breakdown of the clot.
6. Record results.

GERM TUBE TEST
1. Using a sterile 1 ml pipette, add 0.5 ml of the rehydrated plasma to a 12 × 75 mm test tube, supported in a rack.
2. Touch the tip of a sterile Pasteur pipette to a yeast colony growing on the Bacto Sabouraud Dextrose Agar plate.
3. Gently emulsify the cells which adhere to the pipette in the tube of rehydrated plasma.
4. Incubate the mixture at 37°C for 2 – 4 hours.
5. Examine one drop of the incubated mixture microscopically for germ tubes.
6. Record results.

RESULTS
Coagulase Production
Any degree of clotting of the Bacto Coagulase Plasma or Bacto Coagulase Plasma EDTA in up to 24 hours constitutes a positive test.

Germ Tube Development
The development of short lateral hyphal filaments (germ tubes) on the individual yeast cells is considered a positive test.

QUALITY CONTROL
Use known positive and negative control cultures in parallel with the test culture to ascertain the validity of test results.

Organism	Expected Results
Staphylococcus aureus ATCC® 25923	clot in COAGULASE TEST
Staphylococcus epidermidis ATCC® 12228	no clot in COAGULASE TEST
Candida albicans ATCC® 14053	germ tube development
Candida tropicalis ATCC® 750	no germ tube development

LIMITATIONS
1. The slide agglutination technique for determining the coagulase activity of staphylococci is not recommended since false-positive reactions may occur with some strains when animal plasmas are used. In addition, spontaneous agglutination may occur when rough cultures are used. When this test is employed, all negative slide reactions must be confirmed by the tube test.
2. Some species of organisms utilize citrate in their metabolism and will yield false-positive reactions for coagulase activity. Normally, this would not cause problems since the coagulase test is performed on staphylococci almost exclusively. However, it is possible for bacteria which utilize citrate to contaminate Staphylococcus cultures on which the coagulase test is being performed and they may, upon prolonged incubation, give false-positive results due to utilization of the citrate.
3. When checking results of the COAGULASE TEST, tubes should be observed hourly during the first four hours of incubation. Some strains of S. aureus produce fibrinolysin which may lyse clots formed earlier. If the tubes are not read until 24 hours of incubation, reversion to a false negative might result.

REFERENCES

1. Subcommittee on taxonomy of staphylococci and micrococci. Int. Bu. Bact. Nomen. and Taxon, 15:109 – 110, 1965.
2. Diag. Med., May – June:91 – 93, 1983.
3. Bergey's manual of determ. bact., 8th Ed., Williams and Wilkins, Baltimore:484, 1974.
4. Manual of clin. micro., 3rd Ed., ASM:84, 1980.
5. J. Path. Bact., 45:295 – 303, 1937.
6. J. Bact., 35:311 – 333, 1938.
7. J. Path. Bact., 50:83 – 88, 1940.
8. Can. J. Micro., 2:703 – 714, 1956.
9. Diagnostic Proc., APHA, 6th Ed., 596 – 597, 1981.
10. J. Med. Res., 10:407 – 419, 1903.
11. Znetraibl. f. Baict. Labt orig., 99:74, 1926.
12. Applied Micro., 23:725 – 733, 1972.
13. Manual of clin. micro., 3rd Ed., ASM:568, 1980.
14. Antonie Van Leeuwenhook, 44:15, 1978.
15. Mycopathologia, 61:183, 1977.
16. Biochem. test for ID of med. import. bact., 2nd Ed., Williams and Wilkins, Baltimore, 1980.
17. J. Bact., 47:211, 1944.

PACKAGING

Bacto Coagulase Plasma	6 × 3 ml	0286-46-1
	6 × 15 ml	0286-86-2
	6 × 25 ml	0286-66-6
Bacto Coagulase Plasma EDTA	6 × 3 ml	0803-46-4
	6 × 15 ml	0803-86-6
	6 × 25 ml	0803-66-0

BACTO COLUMBIA BLOOD AGAR BASE

INTENDED USE

Bacto Columbia Blood Agar Base is an infusion-free basal medium to which blood may be added for isolating and cultivating fastidious microorganisms or to which other enrichments may be added for special purposes.

HISTORY

The basal medium and specialized variations were described by Ellner et al. in 1966.[1]

PRINCIPLES

The developers of the medium combined two types of peptones, a casein hydrolysate and an infusion peptone, to obtain both rapid and luxuriant growth and sharply defined hemolytic reactions, typical colonial morphology and improved pigment production.

FORMULA

BACTO COLUMBIA BLOOD AGAR BASE
DEHYDRATED

Ingredients per liter

Bacto Pantone	10 g
Bacto Bitone	10 g
Tryptic Digest of Beef Heart	3 g
Corn Starch	1 g
Sodium Chloride	5 g
Bacto Agar	15 g

Final pH 7.3 ± 0.2 at 25°C.

One pound will make 10.2 liters of single-strength medium.

PROCEDURE

1. Suspend 44 g in 1 liter distilled or deionized water and heat to boiling to dissolve completely.
2. Sterilize in the autoclave for 15 minutes at 15 lbs pressure (121°C).
3. Aseptically prepare one of the following variations after cooling to 45°C.
   BLOOD AGAR: Add 5% sterile defibrinated blood and mix well.
   CHOCOLATE AGAR: Add 10% sterile defibrinated sheep blood. Heat to 85°C with constant agitation. Cool to 45°C, then add 1% Bacto Supplement B.
   THAYER-MARTIN MEDIUM: Prepare chocolate agar (above), then add 10 μg ristocetin and 25 units polymyxin per ml of medium.
   SELECTIVE MEDIUM FOR GRAM-POSITIVE COCCI: Prepare blood agar, then add 10 μg colistin and 15 μg nalidixic acid per ml of medium. (The complete formulation is available as Bacto Columbia CNA Agar.)
   CORYNEBACTERIUM DIPHTHERIAE VIRULENCE TEST MEDIUM[1]
   TINSDALE MEDIUM[1]
   LACTOSE-EGG-YOLK-MILK AGAR[1]
   NEISSERIA MEDIUM[1]
   SEMISOLID MEDIUM FOR MOTILITY[1]
4. Dispense into Petri dishes or tubes, as desired.

STORAGE

Bacto Columbia Blood Agar Base   Below 30°C
Prepared plates or tubes   2 – 8°C

QUALITY CONTROL

Identity Specifications

Dehydrated powder: beige, free-flowing, homogeneous
Reaction of 4.4% solution: 7.3 ± 0.2 at 25°C
Plates prepared with 5% sheep blood: cherry red, opaque, firmly solid

Typical Response on Bacto Columbia Blood Agar Base Prepared with 5% Sheep Blood After 40 – 48 Hours at 35 ± 2°C

Organism	Growth	Hemolysis
Neisseria meningitidis ATCC® 13090	excellent	none
Staphylococcus aureus ATCC® 25923	excellent	beta/gamma
Staphylococcus epidermidis ATCC® 12228	excellent	gamma
Streptococcus pneumoniae ATCC® 6303	good to excellent	alpha
Streptococcus pyogenes ATCC® 19615	good to excellent	beta

REFERENCE

1. Am. J. Clin. Path., 45:502, 1966.

PACKAGING

Bacto Columbia Blood Agar Base		
	1 lb (454 g)	0792-01-1
	1/4 lb (114 g)	0792-02-0
	5 lb (2.27 kg)	0792-05-7
	10 kg	0792-08-4

BACTO COLUMBIA BROTH

INTENDED USE

Bacto Columbia Broth is used for culturing clinical specimens suspected of containing bacteria which are fastidious in their growth requirements and as a general purpose broth.

HISTORY/PRINCIPLES

Bacto Columbia Broth is prepared according to the formulation described by Morello and Ellner.[1]

The authors demonstrated columbia broth to be superior to a commonly used general purpose broth in that the growth rates of *Staphylococcus aureus*, *Escherichia coli* and *Streptococcus* (viridans and enterococcus Groups) were faster in this medium.

The authors emphasized that the addition of sodium polyanetholsulfonate (SPS) was an essential step for enhancing bacterial growth by virtue of its anticomplimentary and antiphagocytic properties. They recommended that it be added to the medium in a final concentration of 0.01%. Greater concentration of SPS than this while not causing bacterial inhibition did interfere with the usual appearances of growth by causing the inoculated medium to become slightly opalescent after several days incubation. Also the settling of the blood cells was delayed. These observations on the effects of SPS concentration on aspects of blood culturing are a confirmation of similar findings by other authors.

Lowrance and Traub[2] demonstrated that 0.025% SPS inhibited the coagulation and bactericidal activity of 20% blood, twice the percent recommended in blood culturing, and that this concentration was more effective for isolating small numbers of pathogens. The same authors[3] also observed that SPS in the above concentration will precipitate beta lipoproteins, IgG and other serum fractions resulting in slight opalescence of the supernatant fluid in a blood culture. Although, on further standing and without disturbing the bottle, the precipitated proteins will usually settle on the buffy coat layer there will be sufficient cases where some precipitate will become attached to the sides of the bottle resulting in the appearance of an opalescent medium.

For the reasons cited above, opalescence in a medium, then, cannot always be relied upon as evidence of bacterial growth in the bottle. Conversely, it is possible for significant numbers of viable bacteria to be present in an inoculated and incubated blood culture bottle and yet give none of the usual signs of bacterial growth. For obvious reasons, the performance of the Gram stain or subculture of a blood culture should not be contingent upon the usual visual criteria of bacterial growth. A "blind" subculture is always recommended after the first 18 – 24 hours of incubation.

FORMULA

BACTO COLUMBIA BROTH
DEHYDRATED

Ingredients per liter

Bacto Pantone	10 g	Magnesium Sulfate Anhydrous	0.1 g
Bacto Bitone	10 g	Ferrous Sulfate	0.02 g
Tryptic Digest of Beef Heart	3 g	Sodium Carbonate	0.6 g
L-Cysteine Hydrochloride	0.1 g	Tris (Hydroxymethyl)	
Bacto Dextrose	2.5 g	Aminomethane	0.83 g
Sodium Chloride	5 g	Tris (Hydroxymethyl)	
		Aminomethane HCl	2.86 g

Final pH 7.5 ± 0.2 at 25°C.

One pound will make 13 liters of medium.

METHOD OF PREPARATION

1. Suspend 35 g in 1 liter distilled or deionized water and agitate to dissolve completely. The desired concentration of SPS may be added at this time with agitation to insure uniform solution.

2. Distribute in suitable containers and sterilize in the autoclave for 15 minutes at 15 lbs pressure (121°C).
3. Allow to cool to room temperature before inoculating.

STORAGE

Bacto Columbia Broth   Below 30°C
Prepared medium   15 – 30°C

QUALITY CONTROL

Identity Specifications

Dehydrated powder: light beige, homogeneous, free-flowing
Reaction of 3.5% solution: pH 7.5 ± 0.2 at 25°C
Prepared medium: light amber, clear to very slightly opalescent, may have
                                a fine precipitate

Typical Cultural Response in Bacto Columbia Broth
After 18 – 48 Hours at 35°C

Clostridium perfringens ATCC® 12924   good to excellent
Neisseria meningitidis ATCC® 13090   good to excellent
Staphylococcus aureus ATCC® 25923   good to excellent
Streptococcus mitis ATCC® 9895   good to excellent
Streptococcus pyogenes ATCC® 19615   good to excellent

*Incubated anaerobically

REFERENCES

1. Applied Microbiology, 17:68, 1969.
2. *ibid.*, 17:839, 1969.
3. *ibid.*, 20:465, 1970.

PACKAGING

Bacto Columbia Broth   1 lb (454 g)   0944-01-8
                                        5 lb (2.27 kg)   0944-05-4
                                        10 kg   0944-08-1

BACTO COLUMBIA CNA AGAR

INTENDED USE

Bacto Columbia CNA Agar is a selective basal medium to which blood is added for selectively isolating gram-positive cocci.

HISTORY

Ellner et al.[1] described a variation of columbia blood agar base for selecting gram-positive cocci. Bacto Columbia CNA Agar is the complete basal formulation of that variation.

PRINCIPLES

The authors found that the basal columbia blood agar base, enriched with 5% sheep blood and 10 μg colistin and 15 μg nalidixic acid per ml of medium, suppressed growth of *Proteus*, *Klebsiella* and *Pseudomonas* species while permitting unrestricted growth of staphylococci, hemolytic streptococci and enterococci.

FORMULA

BACTO COLUMBIA CNA AGAR
DEHYDRATED

Ingredients per liter

Bacto Pantone .... 10 g
Bacto Bitone .... 10 g
Tryptic Digest of Beef Heart ...... 3 g
Corn Starch .... 1 g
Sodium Chloride .... 5 g
Colistin Sulfate .... 10 mg
Nalidixic Acid .... 15 mg
Bacto Agar .... 15 g

Final pH 7.3 ± 0.2 at 25°C.

One pound will make 10.2 liters of final medium.

METHOD OF PREPARATION

1. Suspend 44 g in 1 liter distilled or deionized water and heat to boiling to dissolve completely.
2. Sterilize in the autoclave for 15 minutes at 15 lbs pressure (121°C). Avoid overheating. Cool to 45 – 50°C.
3. Aseptically add 5% sterile, defibrinated, room temperature blood volume/volume.
4. Dispense into Petri dishes or tubes, as desired.

STORAGE

Bacto Columbia CNA Agar   Below 30°C
Prepared plates or tubes   2 – 8°C

QUALITY CONTROL

Identity Specifications

Dehydrated powder: beige, free-flowing, homogeneous
Reaction of 4.4% solution: 7.3 ± 0.2 at 25°C
Plates prepared with 5% sheep blood: cherry red, opalescent, firmly solid

Typical Cultural Response on Bacto Columbia CNA Agar Prepared with 5% Sheep Blood After 40 – 48 Hours at 35 ± 2°C

Organism	Growth	Hemolysis
Escherichia coli ATCC® 25922	inhibited	none
Neisseria meningitidis ATCC® 13090	inhibited	none
Staphylococcus aureus ATCC® 25923	excellent	beta/gamma
Staphylococcus epidermidis ATCC® 12228	excellent	gamma
Streptococcus pneumoniae ATCC® 6303	excellent	alpha
Streptococcus pyogenes ATCC® 19615	excellent	beta

REFERENCE
1. Am. J. Clin. Path., 45:502, 1966.

PACKAGING

Bacto Columbia CNA Agar

1 lb (454 g)	0867-01-1
1/4 lb (114 g)	0867-02-0
5 lb (2.27 kg)	0867-05-7

BACTO COMPLEMENT WITH
BACTO COMPLEMENT RECONSTITUTING FLUID

INTENDED USE
Bacto Complement is used in the One-Fifth Volume Kolmer Complement Fixation Test for syphilis[1,2] in conjunction with Bacto Antisheep Hemolysin, Glycerinated. It is also used in other complement fixation procedures and techniques requiring a stable, highly potent complement.

REAGENTS
Bacto Complement is pooled fresh guinea pig serum dried from the frozen state. The pooling of the large numbers of sera as used in the preparation of Bacto Complement eliminates the possibility of nonspecific reactions with antigens used in the complement-fixation test.

Rehydration
Bacto Complement is rehydrated by adding the specified amount of Bacto Complement Reconstituting Fluid and gently rotating the vial to dissolve the contents completely.

Storage and Expiration Date
Store unrehydrated Bacto Complement at 2 – 8°C. The unrehydrated Bacto Complement is stable until the expiration date printed on its label. If not all of the rehydrated complement is to be used the day on which it is rehydrated, the quantity required for the day's tests should be removed to a clean container and the remainder refrozen and stored in the freezing compartment of the refrigerator at −20°C. The rehydrated complement if kept properly stoppered and frozen will retain its potency for 2 – 3 weeks. After thawing the frozen solution, avoid unnecessary exposure to room temperature.

PROCEDURE
Variables encountered in performing complement fixation tests necessitate that the complement be titered just prior to use with the same reagents employed in the test. Dilutions of the rehydrated complement for titration of the complement and hemolysin and in performing complement fixation tests are made with Kolmer Saline,[3] prepared by dissolving 8.5 g NaCl and 0.1 g $MgCl_2$ reagent grade or 1 unit (8.6 g) Kolmer Saline in 1 liter distilled water. It has been shown that the use of a satisfactory saline is of utmost importance in the performance of complement fixation tests. In many localities the use of tap water for the preparation of saline is preferred over distilled water. The titer of complement is often improved by incorporating 0.1 g of magnesium chloride or magnesium sulfate per liter of tap or distilled water saline, resulting in more clear-cut complement fixation reactions.

The hemolytic activity of the complement may be enhanced, if desired, by adding 0.04 g calcium chloride ($CaCl_2 \cdot 2H_2O$) reagent grade per liter Kolmer Saline as described by Browne, Michelbacher and Coffey[4] and by Kolmer and Lynch.[5]

REFERENCES
1. J. Clin. Path., 12:109 – 115, 1942.
2. Pub. Hlth. Serv. Publ. No. 411, 1964.
3. J. Lab. Clin. Med., 12:153, 1926.
4. Am. J. Clin. Path., 24:934, 1954.
5. Am. J. Clin. Path., 24:946, 1954.

PACKAGING

Bacto Complement	2 ml	0383-53-3
	5 ml	0383-56-0
	10 ml	0383-59-7

BACTO CONCANAVALIN A

TEST SUMMARY AND BACKGROUND

Bacto Concanavalin A is a purified lectin isolated from jack beans.[1] By virtue of its ability to form precipitates with certain polysaccharides and glycoproteins it has become a very useful reagent to isolate, characterize, and study these substances.

Because of its reaction with many glycoproteins, concanavalin A has been utilized in studies of the Arthus reaction,[2] serum globulins,[3] and the delayed hypersensitivity reaction.[4,5] It has produced a marked inhibition of the delayed hypersensitivity reaction in tuberculin-sensitive guinea pigs.

This property of selective precipitation has also been applied to the characterization of a number of polysaccharides elaborated in cultures of mycobacteria.[6]

Concanavalin A has been employed in the sensitization of erythrocytes.[7] For example, sheep erythrocytes coated with concanavalin A were reported to produce higher titers in the hemagglutination procedure for determining antisera titers than erythrocytes treated with bis-diazotized benzidine (BDB). Also, the erythrocytes sensitized with concanavalin A were more stable than those treated with BDB.

Concanavalin A has been used to separate blood group substances in hog gastric juice,[8] and to isolate the antihemophilic factor.[9]

Its ability to stimulate lymphocytes to the transformed state has allowed it to join Bacto Phytohemagglutinin P and M as another useful tool for further investigations of cellular reactions and interactions. Concanavalin A binds to human lymphocytes in culture media and stimulates their incorporation of thymidine and uridine.[10]

Trypsin-digested fragments of concanavalin A were found to restore virus-transformed fibroblast tumor cells in culture to cells exhibiting normal behavior.[11]

Most of the selective precipitations of concanavalin A stem from its unique ability to react with branched chain polysaccharides having an alpha-D-glucopyranosyl, alpha-D-mannopyranosyl, or beta-D-fructofuranosyl residue at their nonreducing terminal ends. Thus, it reacts with certain polysaccharides, glyco-proteins, glycogens, amylopectins, alpha-D-mannans, and some levans and dextrans.[4] Many of these reactions may be reversed by displacement with methyl alpha-D-mannoside which has a great affinity for concanavalin A.

A study of the migration inhibitory factor (MIF) from guinea pig lymphocytes stimulated by concanavalin A has shown the MIF to have a number of physicochemical characteristics similar to those demonstrated by MIF from lymphocytes stimulated by antigen.[12]

REHYDRATION AND STORAGE

Bacto Concanavalin A should be stored at 2 – 8°C. Once rehydrated it may be stored at 2 – 8°C for a limited period or aliquoted into useable amounts and frozen at −20°C.

Bacto Concanavalin A is available in 50 mg amounts as a powder which has been lyophilized from 1M sodium chloride. It is to be rehydrated by the addition of 5 ml of sterile distilled water. Because concanavalin A is easily denatured, a small amount of precipitate sometimes remains after rehydration. This may be removed readily by centrifugation.

PROCEDURE

SENSITIZATION OF ERYTHROCYTES

In the procedure recommended by Leon and Young[7] for the sensitization of erythrocytes by concanavalin A, a 10% suspension of sheep erythrocytes previously washed in phosphate buffered saline (PBS) is mixed with an equal volume of 0.005% concanavalin A. After incubation and centrifugation, the cells are resuspended in their original volume of PBS. One ml of the cell suspension is mixed with 5 ml of the desired glycoprotein or polysaccharide. After incubation the mixture is centrifuged, washed, and resuspended in 10 ml of PBS. The presence of 2% fetal calf serum is beneficial in the performance of the subsequent reactions of the sensitized erythrocytes with agglutinating sera.

TRANSFORMATION OF LYMPHOCYTES

For the transformation of lymphocytes from peripheral blood and the demonstration of their mitogenic activity the following procedure is very convenient: Collect 10 ml of blood in Bacto TC Chromosome Blood Separation Vials. After the erythrocytes have settled by gravity (1 – 2 hours) transfer aseptically one-third to one-half of the supernatant plasma-lymphocyte suspension to 7 ml of Bacto TC Medium RPMI #1640 containing 700 units of penicillin and 700 μg of streptomycin sulfate. (A concentration of $1 – 1.2 \times 10^6$ lymphocytes per ml is desirable.) Add 0.1 ml of a rehydrated vial of Bacto Concanavalin A and incubate the culture for three days at 37°C. The caps should be loosened when necessary in order to maintain the color of the phenol red between amber and a light pink. This precaution is especially important during the first six hours of the incubation. The incubation is terminated by the addition of Bacto TC Chromosome Arresting Solution, followed by six hours more incubation at 37°C. The cells are collected and rinsed by centrifugation, swollen in hypotonic solution, and fixed in 1:4 acetic acid:methanol. Drops of the cell suspension in the fixative are placed on clean, wet microscope slides and ignited by momentary contact with a flame, as described by Scherz.[13] The cells may be stained with Bacto Giemsa Stain or orcein according to the procedures of Rothfels and Siminovitch[14] and examined under the microscope for mitotic spreads.

DEMONSTRATION OF THE INCORPORATION OF THYMIDINE BY LYMPHOCYTES STIMULATED BY CONCANAVALIN A

Powell and Leon[10] used a suspension of saline-washed tonsillar cells in TC medium 199 containing penicillin and streptomycin. After incubating one hour at 37°C a volume containing 1.5×10^6 cells was dispensed in tubes. Ten micrograms of concanavalin A were added to each tube, and after incubating for 30 minutes, 0.4 ml of autologous plasma was added, followed by more TC medium 199 to a total volume of 2 ml. (Because of the aforementioned reaction of concanavalin A with certain serum factors, it is sometimes desirable to preincubate the lymphocytes with concanavalin A before adding the plasma.) The cultures are incubated for 3 days, and the incorporation of thymidine is observed by a 4-hour pulse with tritiated thymidine (^{3}HTdR).

By a similar procedure the incorporation of uridine was observed after a pulse with tritiated uridine (^{3}HU).

REFERENCES

1. Agrawal and Goldstein, Biochim., Biophys. Acta., 133:376, 1967; *ibid.*, 147:262, 1967.
2. Kind and Peterson, Science, 160:312, 1968.
3. Goldstain, So, Yang, and Callies, J. Immunol., 103:694, 1969.
4. Leon and Schwartz, Soc. Exp. Biol. Med., 131:735, 1969.
5. Schwartz, Leon, and Pelley, J. Immunol., 104:265, 1970.
6. Daniel and Wisnieski, Amer. Rev. Resp. Dis., 101:762, 1970.
7. Leon and Young, J. Immunol., 104:1556, 1970.
8. Lloyd, Kabat, and Beychok, *ibid.*, 102:1354, 1969.
9. Kass, Ratnoff, and Leon, J. Clin. Invest., 48:351, 1969.
10. Powell and Leon, Exp. Cell. Res., 62:315, 1970.
11. Burger and Noonan, Nature, 228:512, 1970.
12. Remold and David, Sixth Leucocyte Culture Conf., Eastsound, Wash., 1971.
13. Scherz, Stain Tech., 37:386, 1962.
14. Rothfels and Siminovitch, *ibid.*, 33:73, 1958.

PACKAGING

Bacto Concanavalin A	5 ml	3351-56-2
	6 × 5 ml	3351-57-1
TC Medium RPMI #1640	100 ml	5087-72-8
TC Medium RPMI #1640, Dried	10 × 1 liter	5085-37-4
TC Medium 199	100 ml	5477-72-6
	400 ml	5477-78-0
	1 liter	5477-81-5
TC Medium 199, Dried	10 × 1 liter	5701-37-8
	10 liters	5701-24-3
TC Chromosome Blood Separation Vials	6 vials	5855-33-6
TC Chromosome Arresting Solution	6 × 1 ml	5845-51-6
TC Hanks Solution	6 × 12 ml	5508-33-7
	100 ml	5508-72-9
	400 ml	5508-78-3
Bacto Giemsa Stain	6 × 1 ml	0846-51-6

CONCENTRATION DISKS 1/4″ AND 1/2″
PENICILLIN 0.05 UNITS
PENICILLIN 0.1 UNITS
PENASE
STERILE BLANKS

INTENDED USE

Bacto Concentration Disks Penicillin, Penase and Sterile Blanks are used in disk procedures for detecting antimicrobial agents in dairy products as described by *Standard Methods for the Examination of Dairy Products*[1] and *Official Methods of Analysis of the Association of Official Analytical Chemists.*[2]

REAGENTS

Bacto Concentration Disks are standardized 1/4″ and 1/2″ filter paper disks available as follows:

Penicillin 0.05 units ⎫
Penicillin 0.1 units ⎭   Used as positive controls to compare penicillin activity;

Penase (penicillinase)   Used to inactivate up to 25 units penicillin/ml milk to differentiate penicillin from other inhibitors potentially present in a milk sample;

Sterile Blanks   Used for preparing the test sample and, in the AOAC procedure, the positive control.

STORAGE
Bacto Concentration Disks 1/4″ and 1/2″

Penicillin 0.05 units	2 – 8°C
Penicillin 0.1 units	2 – 8°C
Penase	2 – 8°C
Sterile Blanks	15 – 30°C

PROCEDURE
Refer to PENICILLIN IN MILK: AOAC RAPID DETECTION PROCEDURE and PENICILLIN IN MILK: STANDARD METHODS PROCEDURE for entire descriptions of test procedures.

QUALITY CONTROL
Identity Specifications
Concentration Disks Penicillin
Appearance: white paper disks, size as indicated on label, without ragged edges, printed "P" with the concentration under the letter.

Concentration Disks Penase
Appearance: white paper disks, size as indicated on label, without ragged edges, printed "PSE"

Concentration Disks Sterile Blanks
Appearance: white paper disks, size as indicated on label, without ragged edges or printing.

REFERENCES
1. Martin, E. H. (ed.): Standard Methods for the Examination of Dairy Products, 14th Ed., Washington, D.C., American Public Health Association, p. 141, 1978.
2. Horwitz, W. (ed.): Official Methods of Analysis of the Association of Official Analytical Chemists, 3rd supp. to 13th Ed., Washington, D.C., Association of Official Analytical Chemists, p. 466, 1980.

PACKAGING

	Bacto Concentration Disks			
	1/4″		1/2″	
Penicillin 0.05	6 × 25 disks	1563-33-8	6 × 25 disks	1588-33-9
	25 disks	1563-35-6	25 disks	1588-35-7
Penicillin 0.1			6 × 25 disks	1589-33-8
			25 disks	1589-35-6
Penase	6 × 100 disks	1598-33-7	6 × 50 disks	1595-33-0
Sterile Blanks	6 × 100 disks	1599-33-6	6 × 50 disks	1571-33-8
			50 disks	1571-35-6

BACTO CONRADI DRIGALSKI AGAR

INTENDED USE
Bacto Conradi Drigalski Agar is used for the isolation of gram-negative enteric bacilli.

HISTORY/PRINCIPLES
Bacto Conradi Drigalski Agar is a noninhibitive medium, prepared to duplicate the original formula of Conradi and Drigalski.[1]

The differential restraining action exerted by certain dyes on the growth of bacteria is a well-known property and is utilized frequently in culture media. Crystal violet, for example, when present in a dilution of 1:250,000 in agar media, inhibits quite generally the development of gram-positive organisms but has no appreciable effect on the growth of gram-negative bacteria. This property is employed in Bacto Conradi Drigalski Agar for the isolation of the gram-negative intestinal bacteria from contaminated material such as water, feces, etc.

Crystal violet is retained because of its selective action but the litmus, which was originally used as the indicator has been replaced with Bacto Brom Cresol Purple. This indicator is much more satisfactory than litmus. Colonies fermenting the lactose in the medium are surrounded by a yellow zone. Bacto Isoelectric Casein, prepared in our own laboratories, has been substituted for the nutrose specified in the original formula.

FORMULA
BACTO CONRADI DRIGALSKI AGAR
DEHYDRATED
Ingredients per liter

Bacto Peptone	10 g
Bacto Isoelectric Casein	10 g
Bacto Lactose	10 g
Sodium Chloride	5 g
Bacto Agar	15 g
Bacto Brom Cresol Purple	0.03 g
Bacto Crystal Violet	0.004 g

Final pH 6.8 ± 0.2 at 25°C.

One pound will make 9 liters of medium.

METHOD OF PREPARATION
1. To rehydrate the medium, suspend 50 g in 1 liter distilled or deionized water and heat to boiling to dissolve completely.
2. Distribute in tubes or flasks and sterilize in the autoclave for 15 minutes at 15 lbs pressure (121°C).

STORAGE
Bacto Conradi Drigalski Agar	Below 30°C
Prepared medium	2 – 8°C

QUALITY CONTROL
Identity Specifications

Dehydrated powder:	blue, homogeneous, free-flowing
Reaction of 5% solution:	pH 6.8 ± 0.2 at 25°C
Prepared medium:	dark blue, opalescent

Typical Cultural Response on Bacto Conradi Drigalski Agar
After 18 – 24 Hours at 35 ± 2°C

Organism	Growth	Color of Colony	Lactose* Fermentation
Enterobacter aerogenes ATCC® 13048	good to excellent	colorless	+
Escherichia coli ATCC® 25922	good to excellent	white	+
Salmonella typhimurium ATCC® 14028	good to excellent	colorless	–
Streptococcus faecalis ATCC® 19433	inhibited	—	—

*Lactose fermentation + = positive, yellow zone around colony
           – = negative, blue or no change

REFERENCE
1. Zeit, Hyg., 39:283, 1902.

PACKAGING
Bacto Conradi Drigalski Agar   500 g   0083-17-1

BACTO COOKE ROSE BENGAL AGAR

INTENDED USE
Bacto Cooke Rose Bengal Agar, a selective medium for the isolation of fungi, is pre-pared according to the formula given by Cooke.[1] The selectivity of the medium may be increased by the addition of antibiotics.

HISTORY/PRINCIPLES
A variety of materials and methods have been used to inhibit bacteria in an attempt to isolate fungi from mixed flora. The selective action of acid reactions obtained with various acids has been widely employed. Waksman[2] described an acid medium con-sisting of peptone, dextrose, inorganic salts and agar adjusted to pH 4.2 with sulfuric acid for the isolation of fungi from soil. Smith and Dawson[3] used rose bengal in a dilution of 1:15,000 for the inhibition of bacteria in media with nearly neutral reactions without retarding the development of fungi. Martin[4] used 1:30,000 rose bengal and 30 μg streptomycin per ml and demonstrated inhibition of a wide variety of bacteria at reactions between pH 5.5 to 6.5 without inhibition of fungi. Cooke,[1] using the Waksman[2] medium without adjustment, investigated the media for the isolation of fungi from sew-age and reported that Bacto Soytone was particularly suitable for use in the medium, and that the combination of chlortetracycline, or oxytetracycline, with rose bengal in-creased the selective action over the use of rose bengal alone, or the dye with strep-tomycin or chloramphenicol. He recommended the use of nonacidified Waksman soil mold plate count agar prepared with Bacto Soytone, with 0.035 g rose bengal per liter to which is added 35 μg chlortetracycline or oxytetracycline per ml for mold counts of polluted water and sewage. The use of chlortetracycline was preferred due to the greater stability of the aqueous solution of the antibiotic.

FORMULA
BACTO COOKE ROSE BENGAL AGAR
DEHYDRATED
Ingredients per liter

Bacto Soytone	5 g	Magnesium Sulfate	0.5 g
Bacto Dextrose	10 g	Bacto Agar	20 g
Monopotassium Phosphate	1 g	Rose Bengal	0.035 g

Final pH 6.0 ± 0.2 at 25°C.

One pound will make 12.6 liters of final medium.

METHOD OF PREPARATION

1. To rehydrate, suspend 36 g Bacto Cooke Rose Bengal Agar in 1 liter distilled or deionized water. Heat to boiling to dissolve the medium completely.
2. Sterilize in the autoclave for 15 minutes at 15 lbs pressure (121°C).
3. To obtain a concentration of 35 µg chlortetracycline per ml, dissolve 250 mg of chlortetracycline in 100 ml sterile distilled water. Add 14 ml of this solution to 1 liter sterile melted Bacto Cooke Rose Bengal Agar. Cooke[1] reported that the solution of chlortetracycline retained its activity for a period of three months.

STORAGE

Bacto Cooke Rose Bengal Agar   Below 30°C
Prepared plates   2 – 8°C, in the dark

QUALITY CONTROL

Identity Specifications

Dehydrated powder: pinkish tan, homogeneous, free-flowing
Reaction of 3.6% solution: pH 6.0 ± 0.2 at 25°C
Prepared medium: pinkish red, slightly opalescent

Typical Cultural Reactions on Bacto Cooke Rose Bengal Agar
After 1 – 4 Days at 25 – 30°C in The Dark

Organism	Plain	w/chlortetracycline
Aspergillus niger ATCC® 16404	good	good
Candida albicans ATCC® 26790	good to excellent	good to excellent
Escherichia coli ATCC® 25922	good to excellent	inhibited
Saccharomyces cerevisiae ATCC® 9763	good to excellent	good to excellent
Streptococcus faecalis ATCC® 19433	inhibited	inhibited

REFERENCES

1. Antibiotics and Chemotherapy, 4:657:1954.
2. J. Bact., 7:339:1922.
3. Soil Sci., 58:467:1944.
4. Ibid., 69:215:1950.

PACKAGING

Bacto Cooke Rose Bengal Agar   1 lb (454 g)   0703-01-9

BACTO COOKED MEAT MEDIUM

INTENDED USE

Bacto Cooked Meat Medium is used for cultivating anaerobic microorganisms and maintaining stock cultures.

HISTORY

Theobald Smith[1] first made use of fresh unheated animal tissue for cultivating anaerobic organisms. Tarozzi[2] confirmed Smith's findings on the value of unheated tissue in broths for anaerobic culture, and discovered further that he could heat the meat-broth to 104 – 105°C. for 15 minutes without destroying its capacity to support anaerobic growth. A steam sterilized emulsion of brain tissue in water was employed by von Hibler[3,4] for cultivating anaerobic bacilli, and found to be particularly valuable in culturing and classifying these organisms. It was further noted by von Hibler that organisms growing in the cooked brain mash were less susceptible to the harmful effects of toxic metabolic products than were those cultured in milk or carbohydrate serum media. Robertson[5] carefully analyzed von Hibler's results and substituted beef heart for the brain tissue. She found the cooked meat medium to be equally as satisfactory as the cooked brain medium. Henry[6] employed the cooked meat medium successfully for culturing the anaerobes and recommended it for differentiating between various putrefactive and saccharolytic species. Holman[7] used cooked meat medium for general culture purposes and commented: "Perhaps the most favorable characteristic of the medium, after its general growth stimulating influence, is that the products of growth do not rapidly destroy the various forms, and of all media in common use it is to the meat medium that one can constantly return to reisolate bacteria which have died out or have become hopelessly overgrown in other media. The meat is the best single medium we have for studying the anaerobes from war wounds, the reactions are useful for rapid differentiation of groups, as well as for individual identification, and no other medium we have can so readily indicate the presence of anaerobes in mixed cultures where they are often not expected. The cooked meat medium is the most useful medium we have at present for obtaining growth of both anaerobic and aerobic bacteria, for storing mixed cultures for later isolation, as well as pure cultures for further investigation."

Official Methods of Analysis of the Association of Official Analytical Chemists specifies cooked meat medium for use in detecting *Clostridium botulinum* and its toxins.[8]

PRINCIPLES
Cooked meat medium, containing solid meat particles, has the capacity to initiate growth of bacteria from minute inocula and maintain viability of cultures over long periods of time. In mixed cultures, slower-growing organisms proliferate without great danger of overgrowth by faster-growing organisms.

FORMULA
BACTO COOKED MEAT MEDIUM
DEHYDRATED

Ingredients per liter

```
Beef Heart . . . . . . . . . . . . . . . . . . 454 g
Proteose Peptone, Difco . . . . . . . . .  20 g
Bacto Dextrose . . . . . . . . . . . . . . .   2 g
Sodium Chloride . . . . . . . . . . . . . .   5 g
```

Final pH 7.2 ± 0.2 at 25°C.

One pound will make 3.6 liters of medium.

METHOD OF PREPARATION
1. Suspend 12.5 grams in 100 ml of distilled or deionized water (in tubes, 1.25 grams/10 ml).
2. Allow to stand for 15 minutes until all particles are thoroughly wetted and form an even suspension.
3. Sterilize in the autoclave for 15 minutes at 15 lbs pressure (121°C). Avoid rapid release of pressure after sterilization to prevent expelling the medium from the container.
4. If medium is not used the same day it is sterilized, place the tubes in flowing steam or a boiling water bath for a few minutes to drive off dissolved gases. Allow to cool without agitation.

STORAGE
Bacto Cooked Meat Medium	Below 30°C
Prepared medium	15 – 30°C

QUALITY CONTROL

Identity Specifications

Dehydrated powder: brown pellets
Reaction of 12.5% solution: pH 7.2 ± 0.2 at 25°C
Prepared medium: medium amber, clear supernatant over insoluble granules.

Typical Cultural Response in Bacto Cooked Meat Medium
After 40 – 48 Hours at 35°C

Organism	Growth
Clostridium botulinum ATCC® 25763	good to excellent
Clostridium perfringens ATCC® 12924	good to excellent
Clostridium sporogenes ATCC® 11437	good to excellent
Streptococcus pneumoniae ATCC® 6303	good to excellent
Streptococcus faecalis ATCC® 19433	good to excellent

REFERENCES

1. Centr. Bakt., 7:509, 1890.
2. Centr. Bakt., 38:619, 1905.
3. Centr. Bakt., 25:513, 1899.
4. Von Hibler: Untersuchungen ueber die Pathogen Anaeroben, 1908.
5. J. Path. Bact., 20:327, 1916.
6. J. Path. Bact., 21:344, 1917.
7. J. Bact., 4:149, 1919.
8. Horwitz, W. 1980. Official methods of analysis of the Association of Official Analytical Chemists, 13th Ed. Association of Official Analytical Chemists, Washington, D.C.

PACKAGING

Bacto Cooked Meat Medium	1 lb (454 g)	0267-01-7
	1/4 lb (114 g)	0267-02-6
	12 tubes	0267-34-8
	144 tubes	0267-37-5

BACTO CORN MEAL AGAR

INTENDED USE

Bacto Corn Meal Agar is recommended for the production of chlamydospores by *Candida albicans* and for the cultivation of phytopathological fungi. The medium is prepared from an infusion of ground yellow corn and solidified by the addition of 1.5% Bacto Agar.

FORMULA

BACTO CORN MEAL AGAR
DEHYDRATED

Ingredients per liter

Corn Meal, Infusion from . . . . 50 g
Bacto Agar . . . . 15 g

Final pH 6.0 ± 0.2 at 25°C.

One pound will make 26.7 liters of final medium.

PROCEDURE

1. To rehydrate, suspend 17 g in 1 liter distilled or deionized water and heat to boiling to dissolve completely. Add 1% Tween 80®, if desired, to encourage chlamydospore production.

2. Sterilize in the autoclave for 15 minutes at 15 lbs pressure (121°C).
3. Dispense as desired.

STORAGE
Bacto Corn Meal Agar   Below 30°C
Prepared plates   2 – 8°C

QUALITY CONTROL
Identity Specifications
Dehydrated powder: light yellow, homogeneous, free-flowing
Reaction of 1.7% solution: pH 6.0 ± 0.2 at 25°C
Prepared medium: light amber, opalescent with no significant precipitate

Typical Cultural Response on Bacto Corn Meal Agar
After Up to 4 Days at 23 – 25°C

Organism	Growth	Chlamydospores
Aspergillus niger ATCC® 16404	excellent	−
Candida albicans ATCC® 26790	excellent	+
Candida albicans ATCC® 10231	excellent	±
Saccharomyces cerevisiae ATCC® 9763	excellent	−
Saccharomyces uvarum ATCC® 9080	excellent	−

PACKAGING
Bacto Corn Meal Agar		
	1 lb (454 g)	0386-01-3
	1/4 lb (114 g)	0386-02-2
	12 tubes	0386-34-4

BACTO CORN MEAL AGAR w/DEXTROSE

INTENDED USE
Bacto Corn Meal Agar w/Dextrose is recommended for the cultivation of phytopath-ological and other fungi.

PRINCIPLES
The medium produces a more luxuriant growth of some fungi when compared to Bacto Corn Meal Agar without the added dextrose. It is prepared from an infusion of ground yellow corn to which is added 0.2% Bacto Dextrose and 1.5% Bacto Agar.

FORMULA
BACTO CORN MEAL AGAR w/DEXTROSE
DEHYDRATED

Ingredients per liter

Corn Meal Infusion from . . . . 50 g
Bacto Dextrose . . . .   2 g
Bacto Agar . . . . 15 g

Final pH 6.0 ± 0.2 at 25°C.

One pound will make 23.8 liters of medium.

PROCEDURE
1. To rehydrate, suspend 19 g in 1 liter distilled or deionized water and heat to boiling to dissolve completely.
2. Sterilize in the autoclave for 15 minutes at 15 lbs pressure (121°C).

STORAGE

Bacto Corn Meal Agar w/Dextrose   Below 30°C
Prepared plates   2 – 8°C

QUALITY CONTROL

Identity Specifications

Dehydrated powder: light yellow, homogeneous, free-flowing
Reaction of 1.9% solution: pH 6.0 ± 0.2 at 25°C
Prepared medium: very light amber, opalescent with no significant
                            precipitate

Typical Cultural Response on Bacto Corn Meal Agar w/Dextrose
After 40 – 48 Hours at 25 – 30°C

Organism	Growth
Aspergillus niger ATCC® 16404	excellent
Saccharomyces cerevisiae ATCC® 9763	excellent
Saccharomyces uvarum ATCC® 9080	excellent

PACKAGING

Bacto Corn Meal Agar w/Dextrose   500 g   0114-17-4

BACTO CRYSTAL VIOLET AGAR

INTENDED USE

Bacto Crystal Violet Agar is recommended for the differentiation of staphylococci.

HISTORY

Bacto Crystal Violet Agar was recommended by Chapman[1] for use in the differentiation of pure cultures of pathogenic from nonpathogenic strains of staphylococci.

Chapman, Berens, Peters, and Curcio[2] studied 500 strains of staphylococci attempting to correlate pigment production, hemolytic activity, and coagulase activity of the organisms. They concluded from a further study of 100 strains that it was possible to estimate the toxicity of staphylococci on the basis of their pigment production, hemolytic, and coagulating characteristics; hemolytic activity alone being insufficient as a guide to toxicity.

In a study of the correlation between hemolytic and coagulase activities, animal inoculation, and other tests, Chapman and Berens[3] reported that staphylococci produced different colored growths when cultured on crystal violet agar. There was a correlation between the color of the growths and the hemolytic and the coagulating properties of the strains. After incubation for 36 hours on crystal violet agar, hemolytic and coagulating strains produced purple or violet growths, or golden growths often tinged with violet, while nonhemolytic, noncoagulating strains produced white colonies, some of which were mottled with violet while others possessed edges which were sometimes violet. From over 1000 strains tested, this correlation between hemolytic and coagulase tests and color of the growth on crystal violet agar was 93%. Further, animal inoculation tests showed a correlation of 96.4% between the crystal violet agar reaction and the lethal effect of killed cultures on rabbits.

Chapman[4] also studied the color of growths of members of the *Micrococcus catarrhalis* group on crystal violet agar. The results showed a relationship between the color produced by a strain and its dissociative properties.

PRINCIPLES

Crystal violet is markedly inhibitory to staphylococci. A fair growth, however, can be obtained in a 1:300,000 concentration of the dye if the medium is inoculated heavily. Because of its inhibitory properties, crystal violet agar is not suitable for primary isolation but is well adapted to the study of pure cultures where a mass inoculation can be used.

The sterile medium, in tubes or plates, is heavily inoculated and incubated 36 – 48 hours at 35°C. At the end of this time the colored colonies are easily recognized; longer incubation or storage in the refrigerator tend to cause all the colonies to become colored. Strains producing violet or golden growths are likely to be pathogenic for rabbits while about 95% of those producing white or pale violet growths are not pathogenic.

FORMULA

BACTO CRYSTAL VIOLET AGAR
DEHYDRATED

Ingredients per liter

Bacto Beef Extract .... 3 g
Proteose Peptone, Difco .... 5 g
Bacto Lactose .... 10 g
Bacto Agar .... 15 g
Bacto Crystal Violet .... 0.0033 g

Final pH 6.8 ± 0.1 at 25°C.

One pound will make 13.7 liters of medium.

METHOD OF PREPARATION

1. Suspend 33 g in 1 liter distilled or deionized water and heat to boiling to dissolve completely.
2. Sterilize in the autoclave for 15 minutes at 15 lbs pressure (121°C).

STORAGE

Bacto Crystal Violet Agar   Below 30°C
Prepared medium   2 – 8°C

QUALITY CONTROL

Identity Specifications

Dehydrated powder: light tan, homogeneous, free-flowing
Reaction of 3.3% solution: 6.8 ± 0.1 at 25°C
Prepared medium: purple, clear to slightly opalescent

Typical Cultural Response on Bacto Crystal Violet Agar
After 40 – 48 Hours at 35°C

Organism	Recovery	Color of Colony
Escherichia coli ATCC® 25922	good to excellent	purple
Staphylococcus aureus ATCC® 25923	fair to good	yellow
Staphylococcus epidermidis ATCC® 14990	fair to good	purple/very slight yellow
Streptococcus pyogenes ATCC® 19615	inhibited	—

REFERENCES
1. J. Bact., 32:199, 1936.
2. J. Bact., 28:343, 1934.
3. J. Bact., 29:437, 1935.
4. Strain Tech., 11:25, 1936.

PACKAGING
Bacto Crystal Violet Agar   1 lb (454 g)   0077-01-7

BACTO CYSTINE HEART AGAR

INTENDED USE
Bacto Cystine Heart Agar, enriched with Bacto Hemoglobin, is recommended for the cultivation of *Francisella tularensis*. This medium is suggested for this purpose by *Diagnostic Procedures and Reagents*[1] of the American Public Health Association. Bacto Cystine Heart Agar without enrichment supports excellent growth of gram-negative cocci and other pathogenic microorganisms.

HISTORY/PRINCIPLES
Since *Francisella tularensis* was first isolated by McCoy and Chapin,[2] many media have been described for its cultivation. A large number of the media first employed were difficult to prepare and contained egg or serum. Francis[3] reported blood dextrose cystine agar in his later investigation as being a satisfactory medium for cultivating this fastidious organism. Shaw[4] added 0.05% cystine and 1% dextrose to Bacto Heart Infusion Agar for the cultivation of *F. tularensis*. Shaw[5] also showed that the amount of destruction of cystine during autoclaving was minimal and did not adversely affect growth.

Rhamy[6] found Francis' blood dextrose cystine agar to be excellent but often it became contaminated due to the difficulties attendant to its preparation. In his experience an autoclaved solution of Bacto Hemoglobin added to Bacto Cystine Heart Agar proved to be entirely satisfactory for the cultivation of *F. tularensis*. In three or four days the growth is sufficient for the preparation of bacterial antigens. Because of its nutritional value, this medium may also be used for cultivating many other organisms ordinarily difficult to grow.

Bacto Cystine Heart Agar was originally developed in collaboration with Rhamy. As mentioned in the paper by Rhamy, referred to above, W. M. Simpson found this formula, with a reaction of pH 6.8, a most satisfactory medium for the cultivation of this organism. Also cooperating in these preliminary trial studies of the medium, Francis found a culture medium made with Bacto Cystine Heart Agar and Bacto Hemoglobin entirely satisfactory for growing *F. tularensis*.

FORMULA

BACTO CYSTINE HEART AGAR
DEHYDRATED
Ingredients per liter

Beef Heart, Infusion from	500 g
Proteose Peptone, Difco	10 g
Bacto Dextrose	10 g
Sodium Chloride	5 g
L-Cystine, Difco	1 g
Bacto Agar	15 g

Final pH 6.8 ± 0.2 at 25°C.

Five hundred grams will make 9.8 liters of medium.

METHOD OF PREPARATION

When used with Bacto Hemoglobin, the medium is prepared for use as follows:

1. Suspend 10.2 g Bacto Cystine Heart Agar in 100 ml cold distilled or deionized water and heat to boiling to dissolve completely. Sterilize in the autoclave for 15 minutes at 15 lbs pressure (121°C).
2. Place 2 g of Bacto Hemoglobin in a dry flask and add 100 ml cold distilled or deionized water, while the flask is being agitated vigorously. The hemoglobin suspension is shaken intermittently for 10 – 15 minutes to break up all aggregates and effect complete solution and sterilized in the autoclave for 15 minutes at 15 lbs pressure (121°C).
3. Both solutions are cooled to 50 – 60°C, mixed and poured into sterile Petri dishes or tubes.

When a plain cystine dextrose agar, without hemoglobin, is desired:

1. Suspend 51 g of Bacto Cystine Heart Agar in 1 liter distilled or deionized water and heat to boiling to dissolve the medium completely.
2. Distribute in tubes or flasks and sterilize in the autoclave for 15 minutes at 15 lbs pressure (121°C).
3. Dispense into Petri dishes as desired.

STORAGE

Bacto Cystine Heart Agar	Below 30°C
Prepared media	2 – 8°C

QUALITY CONTROL

Identity Specifications

Dehydrated powder:	beige, homogeneous, free-flowing
Reaction of 5.1% solution:	pH 6.8 ± 0.2 at 25°C
Prepared medium:	medium amber, slightly opalescent

Typical Cultural Response on Bacto Cystine Heart Agar with 2% Bacto Hemoglobin After 40 – 48 Hours at 35°C

Organism	Growth
Francisella tularensis ATCC® 29684	good to excellent
Neisseria meningitidis ATCC® 13090	good to excellent
Streptococcus pneumoniae ATCC® 6303	good to excellent
Streptococcus pyogenes ATCC® 19615	good to excellent

REFERENCES

1. Diagnostic Procedures and Reagents, 3rd Ed., 259, 1950.
2. J. Infectious Diseases, 10:61, 1912.
3. J. Am. Med. Assoc., 91:1155, 1928.
4. Zentr. Bakt. I. Abt. Orig., 118:216, 1930.
5. J. Lab. Clin. Med., 16:294, 1930.
6. Am. J. Clin. Path., 3:121, 1933.

PACKAGING

Bacto Cystine Heart Agar	500 g	0047-17-6

BACTO CYSTINE TRYPTIC AGAR

INTENDED USE

Bacto Cystine Tryptic Agar is a semisolid basal medium recommended for fermentation studies with the addition of various carbohydrates. Also, its low agar content makes it well suited for motility studies.

FORMULA

BACTO CYSTINE TRYPTIC AGAR
DEHYDRATED
Ingredients per liter

Bacto Tryptose	20 g
L-Cystine	0.5 g
Sodium Chloride	5 g
Sodium Sulfite	0.5 g
Bacto Agar	2.5 g
Bacto Phenol Red	0.017 g

Final pH 7.3 ± 0.2 at 25°C.

One pound will make 15.3 liters of medium.

METHOD OF PREPARATION
1. Suspend 29.5 g in 1 liter distilled or deionized water. Heat to boiling to dissolve completely.
2. Dispense in tubes in 8 – 10 ml amounts and sterilize in the autoclave for 15 minutes at 15 lbs pressure (121°C).
3. Cool to 50 – 55°C and aseptically add appropriate Bacto Differentiation Disk Carbohydrate. Allow to cool to below 37°C in an unslanted, upright position.
4. Inoculate with wire loop or needle using a heavy inoculum, stabbing into the medium.
5. Incubate for 4 – 18 hours at 35°C. If increased CO_2 is necessary, do not use CO_2 incubator. This could lead to erroneous results. Use a candle jar or can. Longer incubation may be necessary for slower growing organisms.

Carbohydrates may be added in other forms as well. Refer to Purple Agar Base for discussion on use of carbohydrates with culture media.

STORAGE
Bacto Cystine Tryptic Agar   Below 30°C
Prepared medium   2 – 8°C

QUALITY CONTROL
Identity Specifications
Dehydrated powder: pink, homogeneous free-flowing
Reaction of 2.85% solution: pH 7.3 ± 0.2 at 25°C
Prepared medium: red, very slightly opalescent

Typical Cultural Response in Bacto Cystine Tryptic Agar
After 4 – 18 Hours or Longer if Necessary at 35°C

Organism	Growth	Motility	Acid Production w/Dextrose
Escherichia coli ATCC® 25922	good	+	+
Neisseria gonorrhoeae ATCC® 19424	good	–	+
Neisseria meningitidis ATCC® 13090	good	–	+
Streptococcus pneumoniae ATCC® 6303	good	–	+

+ = positive, yellow for acid or diffuse growth for motility
– = negative, no change

PACKAGING
Bacto Cystine Tryptic Agar		
	1 lb (454 g)	0523-01-7
	1/4 lb (114 g)	0523-02-6
	12 tubes	0523-34-8

Bacto Cystine Tryptic Agar w/Adonitol	12 tubes	1177-34-5
Bacto Cystine Tryptic Agar w/Dextrose	12 tubes	1056-34-1
Bacto Cystine Tryptic Agar w/Dulcitol	12 tubes	1252-34-3
Bacto Cystine Tryptic Agar w/Lactose	12 tubes	1057-34-0
Bacto Cystine Tryptic Agar w/Levulose	12 tubes	1249-34-9
Bacto Cystine Tryptic Agar w/Malonate	12 tubes	1253-34-2
Bacto Cystine Tryptic Agar w/Maltose	12 tubes	1058-34-9
Bacto Cystine Tryptic Agar w/Mannitol	12 tubes	1248-34-0
Bacto Cystine Tryptic Agar w/Salicin	12 tubes	1059-34-8
Bacto Cystine Tryptic Agar w/Sorbitol	12 tubes	1250-34-5
Bacto Cystine Tryptic Agar w/Sucrose	12 tubes	1060-34-5
Bacto Cystine Tryptic Agar w/Xylose	12 tubes	1061-34-4

BACTO CZAPEK DOX BROTH

INTENDED USE
Bacto Czapek Dox Broth is a defined liquid medium, nearly neutral in reaction, designed for the cultivation of fungi and bacteria capable of utilizing inorganic nitrogen.

HISTORY/PRINCIPLES
Bacto Czapek Dox Broth is a modification of the Czapek formula of Dox[1] prepared according to the directions given by Thom and Raper.[2]

Media prepared with only inorganic sources of nitrogen and chemically defined compounds as sources of carbon are useful for a variety of microbiological procedures. They are of principal value in soil microbiology, for the enrichment, cultivation and identification of soil bacteria and fungi or for mildew resistance tests, as well as for other tests wherein a simple chemically defined medium is desired. Sodium nitrate is the sole source of nitrogen and saccharose serves as a source of carbon. Bacto Czapek Dox Broth will support a moderately vigorous growth of nearly all saprophytic aspergilli.[2]

FORMULA
BACTO CZAPEK DOX BROTH
DEHYDRATED

Ingredients per liter

Bacto Saccharose	30 g
Sodium Nitrate	3 g
Dipotassium Phosphate	1 g
Magnesium Sulfate	0.5 g
Potassium Chloride	0.5 g
Ferrous Sulfate	0.01 g

Final pH 7.3 ± 0.2 at 25°C.

One pound will make 13 liters of medium.

METHOD OF PREPARATION
1. To rehydrate, dissolve 35 g in 1 liter distilled or deionized water.
2. Distribute into tubes or flasks and sterilize in the autoclave for 15 minutes at 15 lbs pressure (121°C).

STORAGE
Bacto Czapek Dox Broth	Below 30°C
Prepared tubes	15 – 30°C

QUALITY CONTROL

Identity Specifications

Dehydrated powder: white, homogeneous, free-flowing
Reaction of 3.5% solution: pH 7.3 ± 0.2 at 25°C
Prepared medium: colorless, clear, may have a slight precipitate

Typical Cultural Response in Bacto Czapek Dox Broth
After 40 – 72 Hours at 30 ± 2°C

Organism	Growth
Aspergillus niger ATCC® 16404	good to excellent
Candida albicans ATCC® 10231	good to excellent
Saccharomyces cerevisiae ATCC® 9763	good to excellent

REFERENCES

1. U.S. Dept. Ag. Bur. Anim. Ind. Bull., 120:70, 1910.
2. Thom and Raper, Manual of the Aspergilli, 39, 1945.

PACKAGING

Bacto Czapek Dox Broth	1 lb (454 g)	0338-01-2
	1/4 lb (114 g)	0338-02-1

BACTO CZAPEK SOLUTION AGAR

INTENDED USE

Bacto Czapek Solution Agar is a solid neutral medium of known chemical composition with nitrate as the sole source of nitrogen. It is used in the cultivation of saprophytic fungi, soil bacteria and other microorganisms.

PRINCIPLES

Bacto Czapek Solution Agar is prepared according to the formula given by Thom and Church.[1] The medium is of value in many microbiological procedures, such as the cultivation and identification of fungi, growth of soil bacteria, mildew tests and others. Thom and Raper[2] state that this medium will produce a moderately vigorous growth of nearly all saprophytic aspergilli and will yield characteristic mycelia and conidia useful in comparative studies. Standard Methods for Examination of Water and Wastewater, 15th Edition (1980) recommends this medium for isolation of *Aspergillus, Penicillium* and related fungi.

FORMULA

BACTO CZAPEK SOLUTION AGAR
DEHYDRATED

Ingredients per liter

Bacto Saccharose	30 g
Sodium Nitrate	2 g
Dipotassium Phosphate	1 g
Magnesium Sulfate	0.5 g
Potassium Chloride	0.5 g
Ferrous Sulfate	0.01 g
Bacto Agar	15 g

Final pH 7.3 ± 0.2 at 25°C.

One pound will make 9.2 liters of medium.

METHOD OF PREPARATION

1. To rehydrate, suspend 49 g in 1 liter distilled or deionized water. Heat to boiling to dissolve the medium completely.
2. Distribute into tubes or flasks and sterilize in the autoclave for 15 minutes at 15 lbs pressure (121°C).

STORAGE

Bacto Czapek Solution Agar   Below 30°C
Prepared medium   2 – 8°C

QUALITY CONTROL

Identity Specifications

Dehydrated powder: very light beige, homogeneous, free-flowing
Reaction of 4.9% solution: pH 7.3 ± 0.2 at 25°C
Prepared medium: light amber, opalescent with uniform flocculent precipitate

Typical Cultural Response in Bacto Czapek Solution Agar
After 40 – 48 Hours at 30 ± 2°C

Organism	Growth
Aspergillus niger ATCC® 16404	good to excellent
Saccharomyces cerevisiae ATCC® 9763	none to fair

REFERENCES

1. Thom and Church, The Aspergilli, 39:1926.
2. Thom and Raper, Manual of the Aspergilli, 39:1945.

PACKAGING

Bacto Czapek Solution Agar	1 lb (454 g)	0339-01-1
	1/4 lb (114 g)	0339-02-0

BACTO DCLS AGAR

INTENDED USE

Bacto DCLS Agar (Desoxycholate Citrate Lactose Saccharose Agar) is a selective medium, containing lactose and saccharose (sucrose), for the isolation of gram-negative enteric bacilli from stools, urines and other specimens.

HISTORY/PRINCIPLES

Bacto DCLS Agar is a modification of the desoxycholate citrate agar described by Leifson[1] and Bacto SS Agar, using neutral red as an indicator. Coliform organisms, capable of fermenting lactose or sucrose are generally inhibited as are all gram-positive microorganisms. It is suggested that Bacto MacConkey Agar, or other nonselective medium, be used in conjunction with Bacto DCLS Agar.

Holt, Harris and Teague[2] used both lactose and sucrose in a plating medium for the isolation of gram-negative enteric pathogens, since some coliform organisms fermented sucrose more readily than lactose thus eliminating many false positive reactions. Laboratories interested in enteric bacilli capable of fermenting sucrose and not lactose should employ plating media prepared without sucrose.

Generally, coliform organisms are markedly to completely inhibited on Bacto DCLS Agar permitting the use of a heavy inoculum without danger of overgrowth. Organisms capable of fermenting lactose or sucrose, when they grow, form red opaque colonies. *Salmonella* and *Shigella* organisms form colorless to slightly pink transparent colonies following 18 to 24 hours incubation at 35 – 37°C. Colonies suspected of being enteric pathogens are picked into differential media such as Bacto Kligler Iron Agar, Bacto Triple Sugar Iron Agar or other desired medium for the determination of further distinguishing biochemical reactions.

FORMULA

BACTO DCLS AGAR
DEHYDRATED

Ingredients per liter

Bacto Beef Extract	3 g	Sodium Thiosulfate	5 g
Proteose Peptone No. 3, Difco	7 g	Sodium Desoxycholate	2.5 g
Bacto Lactose	5 g	Bacto Agar	12 g
Saccharose, Difco	5 g	Neutral Red	0.03 g
Sodium Citrate	10 g		

Final pH 7.2 ± 0.1 at 25°C.

One pound will make 9.1 liters of medium.

METHOD OF PREPARATION

1. Suspend 49.5 g in 1 liter distilled or deionized water. Heat to boiling to dissolve completely.
2. DO NOT AUTOCLAVE.
3. Allow to cool to 50 – 55°C and pour about 20 ml of medium into standard Petri dishes.
4. Allow to dry for about 2 hours with the covers partially removed.

STORAGE

Bacto DCLS Agar	Below 30°C
Prepared medium	2 – 8°C

QUALITY CONTROL

Identity Specifications

Dehydrated powder:	light pink, homogeneous, free-flowing
Reaction of 4.95% solution:	pH 7.2 ± 0.1 at 25°C
Prepared medium:	orange-red, slightly opalescent

Typical Cultural Response on Bacto DCLS Agar
After 18 – 48 Hours at 35°C

Organism	Growth	Color of Colony
Escherichia coli ATCC® 25922	inhibited	red, opaque, if any
Proteus vulgaris ATCC® 13315	good to excellent	red
Salmonella typhimurium ATCC® 14028	good to excellent	colorless to slightly pink
Shigella flexneri ATCC® 12022	poor to good	colorless to slightly pink

REFERENCES

1. J. Path. Bact., 40:581:1935.
2. J. Infectious Diseases, 18:596:1916.

PACKAGING
Bacto DCLS Agar   1 lb (454 g)   0759-01-2

D/E NEUTRALIZING MEDIA
BACTO D/E NEUTRALIZING AGAR
BACTO D/E NEUTRALIZING BROTH
BACTO D/E NEUTRALIZING BROTH BASE

INTENDED USE
Bacto D/E Neutralizing Agar, Bacto D/E Neutralizing Broth and Bacto D/E Neutralizing Broth Base are prepared for the neutralizing and testing of antiseptics and disinfectants according to the procedure of Engley and Dey.[1]

PRINCIPLES
Bacto D/E Neutralizing Agar and Bacto D/E Neutralizing Broth are especially suited for environmental sampling where neutralization of the chemical is important to determine its bactericidal activity. A strongly bacteriostatic substance may contains bacteria held in bacteriostasis but which may still be able to cause infection. The media will neutralize a broad spectrum of antiseptic and disinfectant chemicals including quaternary ammonium compounds, phenolics, iodine and chlorine preparations, mercurials (Merthiolate®), formaldehyde and gluteraldehyde. Bacto D/E Neutralizing Broth Base has the same formula as Bacto D/E Neutralizing Broth but does not contain the neutralizing components.

FORMULAE

	BACTO D/E NEUTRALIZING BROTH BASE DEHYDRATED	BACTO D/E NEUTRALIZING BROTH DEHYDRATED	BACTO D/E NEUTRALIZING AGAR DEHYDRATED
		Ingredients per liter	
Bacto Tryptone	5 g	5 g	5 g
Bacto Yeast Extract	2.5 g	2.5 g	2.5 g
Bacto Dextrose	10 g	10 g	10 g
Sodium Thioglycollate	—	1 g	1 g
Sodium Thiosulfate	—	6 g	6 g
Sodium Bisulfite	—	2.5 g	2.5 g
Lecithin (Soybean)	—	7 g	7 g
Polysorbate 80	—	5 g	5 g
Bacto Brom Cresol Purple	0.02 g	0.02 g	0.02 g
Bacto Agar	—	—	15 g
Solution (g/liter)	17.5 g/liter	39 g/liter	54 g/liter
500 g will make:	28.2 liters	12.8 liters	9.2 liters

Final pH of each medium is 7.6 ± 0.2 at 25°C.

METHOD OF PREPARATION
1. Suspend the appropriate amount of medium in 1 liter distilled or deionized water.
2. Heat to boiling to dissolve completely to obtain an even suspension.
3. Dispense 9 ml amounts of the broth media into tubes and dispense the agar medium into flasks.
4. Sterilize in the autoclave for 15 minutes at 15 lbs pressure (121°C).

PROCEDURE

For testing disinfectants, prepare two sets of test tubes, one containing 9 ml of sterile Bacto D/E Neutralizing Broth and another set containing 9 ml amounts of Bacto D/E Neutralizing Broth Base. Add 1 ml of disinfectant solution to each tube, mix thoroughly and let stand for 15 minutes. The disinfectant may require diluting to obtain the neutralized concentration. Inoculate the tubes with 0.1 ml of a 1:100,000 dilution of overnight broth cultures.

Incubate both sets of tubes for 48 hours at 37°C and observe for growth, which is indicated by a color change from purple to yellow or by the formation of a pellicle. Growth in Bacto D/E Neutralizing Broth with no growth in Bacto D/E Neutralizing Broth Base indicates neutralization of the disinfectant and a possible bacteriostatic substance. To determine if viable organisms are present and to indicate bactericidal activity, inoculate samples from the broth tubes onto plates of Bacto D/E Neutralizing Agar. Incubate for 48 hours at 37°C. Positive growth from negative tubes of Bacto D/E Neutralizing Broth Base indicates a bacteriostatic substance whereas negative growth indicates a bactericidal disinfectant. All positive broth tubes should show growth on the Bacto D/E Neutralizing Agar plates.

A disk plate method may also be used for testing disinfectants using Bacto D/E Neutralizing Agar. In this procedure, pour plates are prepared with inocula from the test cultures as needed, depending on the disinfectant to be tested. Prepare plates (100 × 15 mm) by aseptically dispensing 20 ml of inoculated medium into each plate. Place one 12.55 mm white filter paper disk for each control and test substance on the plates and dispense 0.1 ml control and test substance solution on the disks. Also prepare plates of standard methods agar in the same manner. Incubate the plates for 24 – 48 hours at 37°C. Zones of inhibition occurring around disks on the standard methods agar plates with no zones occurring on the Bacto D/E Neutralizing Agar plates indicates neutralization of the disinfectant by Bacto D/E Neutralizing Agar.

The following control disinfectants may be used in the test procedure: chlorine 2% (chlorine bleach), formaldehyde 2%, gluteraldehyde 1%, iodine 2%, Merthiolate® 1/1000, phenol 2% and quaternary ammonium compounds 1/750. One or more of these controls may be used depending on the types of disinfectant to be tested.

STORAGE

Bacto D/E Neutralizing Agar and Broth	2 – 8°C
Bacto D/E Neutralizing Broth Base	Below 30°C
Prepared media	2 – 8°C

QUALITY CONTROL

Identity Specifications

	D/E Neutralizing Broth Base	D/E Neutralizing Broth	D/E Neutralizing Agar
Dehydrated powder:	purple, homogeneous, free-flowing	bluish grey, homogeneous, moist, lumpy	bluish grey, homogeneous, moist, lumpy
pH at 25°C:	7.6 ± 0.2	7.6 ± 0.2	7.6 ± 0.2
Prepared medium:	purple, clear to slightly opalescent	purple, opaque with an even suspension of particles	lavender, opaque

**Typical Cultural Response in Bacto D/E Neutralizing Media
After 40 – 48 Hours at 35 ± 2°C**

Organism	Growth
Bacillus subtilis ATCC® 6633	good to excellent
Escherichia coli ATCC® 25922	good to excellent
Pseudomonas aeruginosa ATCC® 27853	good to excellent
Salmonella typhimurium ATCC® 14028	good to excellent
Staphylococcus aureus ATCC® 25923	good to excellent

REFERENCE

1. CSMA Proceedings, 1970.

PACKAGING

Bacto D/E Neutralizing Broth	500 g	0819-17-2
Bacto D/E Neutralizing Agar	500 g	0686-17-2
Bacto D/E Neutralizing Broth Base	500 g	0823-17-6

BACTO DNase TEST AGAR
BACTO DNase TEST AGAR w/METHYL GREEN

INTENDED USE

Bacto DNase Test Agar is recommended for determining deoxyribonuclease activity of microorganisms, particularly staphylococci, and for the isolation and differentiation of *Serratia marcescens* from non DNase producing organisms.

Bacto DNase Test Agar w/Methyl Green as described by Smith, Hancock, and Rhoden[1] is a modification of Bacto DNase Test Agar. It is also used for detecting DNase activity in microorganisms and for the isolation and differentiation of *Staphylococcus aureus* and *Serratia marcescens* from other organisms having similar characteristics except DNase production.

HISTORY/PRINCIPLES

Weckman and Catlin[2] demonstrated the increased DNase activity of *Staphylococcus aureus* cultures isolated from clinical specimens and the close correlation with coagulase production. They suggested that DNase activity could be used to identify potentially pathogenic staphylococci. Jeffries, Holtman and Guse[3] incorporated DNA in an agar medium for studying DNase production by bacteria and fungi. Microorganisms elaborating deoxyribonuclease, when streaked on the Bacto DNase Test Agar surface and incubated, depolymerized the DNA. DNase activity was demonstrated by flooding the plates with 0.1 N HCl. Organisms which degraded the DNA gave clear zones around the streaks whereas those which produced no DNase showed no clearing. These authors also showed a close correlation between coagulase production and DNase activity. DiSalvo[4] obtained excellent correlation between the coagulase and DNase activities of staphylococci isolated from clinical specimens. Fusillo and Weiss[5] studied the calcium requirements of staphylococci for DNase production and concluded that additional calcium is unnecessary when a complete nutritive medium is used.

The increased importance of *S. marcescens* as the etiologic agent in a variety of infections was reported by Ewing, Johnson, and Davis.[6] Davis[7] added toluidine blue and crystal violet to Bacto DNase Test Agar to isolate and differentiate *S. marcescens* from other bacteria usually encountered in clinical specimens. He concluded that this modified medium proved extremely valuable for detecting this species. Botticher[8] studied the classification of the *Serratia* genus and pointed out that DNase active gram-negative rods could be accepted as *Serratia* species.

Kurnick[9] showed that methyl green combines only with highly polymerized DNA and at pH 7.5. When combination does not take place, the color fades. This observation was applied to a test for determining DNase in body fluids and blood serum. Smith, Hancock and Rhoden[1] applied this principle to modify DNase test agar for detecting staphylococci, streptococci, and *Serratia*. The method has the advantage that DNase activity can be detected by a clearing of the dye around DNase producing colonies. Acid does not have to be added and colonies can be picked directly from the plate for additional studies.

Mannitol fermentation can be determined simultaneously with DNase production by adding 10 g mannitol and 0.025 g phenol red to the Bacto DNase Test Agar before sterilization.

FORMULAE

BACTO DNase TEST AGAR
DEHYDRATED

Ingredients per liter

Bacto Tryptose	20 g
Deoxyribonucleic Acid	2 g
Sodium Chloride	5 g
Bacto Agar	15 g

Final pH 7.3 ± 0.2 at 25°C.

Five hundred grams will make 11.9 liters of medium.

BACTO DNase TEST AGAR w/METHYL GREEN
DEHYDRATED

Ingredients per liter

Bacto Tryptose	20 g
Deoxyribonucleic Acid	2 g
Sodium Chloride	5 g
Bacto Agar	15 g
Methyl Green	0.05 g

Final pH 7.3 ± 0.2 at 25°C.

One pound will make 10.8 liters of medium.

PROCEDURE

1. Suspend 42 g in 1 liter distilled or deionized water. Heat to boiling to dissolve completely.
2. Sterilize in the autoclave for 15 minutes at 15 lbs pressure (121°C).
3. Dispense into sterile Petri dishes or as desired. Allow to solidify.
4. Plates are inoculated by streaking or spotting with the material or culture being tested. Using a 1 – 2 cm streak or spot inocula of approximately 5 mm diameter, several cultures may be tested on one plate.
5. Incubate at 35°C for 18 – 24 hours.
6. Flood Bacto DNase Test Agar plates with 0.1N hydrochloric acid and observe for clearing around the streak indicating DNase activity. Observe Bacto DNase Test Agar w/Methyl Green plates for clearing around streak.

STORAGE

Bacto DNase Test Media	Below 30°C
Prepared plates	2 – 8°C

QUALITY CONTROL

Identity Specifications

	Bacto DNase Test Agar	DNase Test Agar w/Methyl Green
Dehydrated powder:	light beige, free-flowing	light green w/white particles, free-flowing
Reaction of 4.2% solution:	pH 7.3 ± 0.2	pH 7.3 ± 0.2
Prepared medium:	light to medium amber, slightly opalescent	green, slightly opalescent, may have a fine precipitate

Typical Cultural Response on Bacto DNase Test Agar
After 18 – 24 Hours at 35°C

Organism	Growth	Zone of Clearing
Serratia marcescens ATCC® 8100	good to excellent	+
Staphylococcus aureus ATCC® 25923	good to excellent	+
Staphylococcus epidermidis ATCC® 12228	good to excellent	−
Streptococcus pyogenes ATCC® 19615	good to excellent	+

REFERENCES

1. Applied Microbiology, 18:991, 1969.
2. J. Bact., 73:747, 1957.
3. J. Bact., 73:590, 1957.
4. Medical Tech. Bulletin, 9:191, 1958.
5. J. Bact., 78:520, 1959.
6. CDC, 1962.
7. Bact. Proceedings, 1967.
8. The Filter, March, p. 9, 1970.
9. Archives of Biochemistry, 9:41, 1950.

PACKAGING

Bacto DNase Test Agar	5 lb (2.27 kg)	0632-05-1
	500 g	0632-17-7
	100 g	0632-15-9
Bacto DNase Test Agar w/Methyl Green	1 lb (454 g)	0220-01-3
	1/4 lb (114 g)	0220-02-2

BACTO DNA (DEOXYRIBONUCLEIC ACID)

Bacto DNA, a polymerized preparation of deoxyribonucleic acid, is used in bacteriological culture media for determining the ability of certain bacteria to depolymerize this compound by means of their deoxyribonucleic (DNAse) activity.

PACKAGING

Bacto DNA (Deoxyribonucleic Acid)	1 g	3231-10-3
	10 g	3231-12-1

BACTO DTM AGAR

INTENDED USE

Bacto DTM (Dermatophyte Test Medium) Agar is used in the isolation and identification of dermatophytes.

HISTORY AND TEST SUMMARY

Dermatomycoses are perhaps the most common fungal infections of man. These infections are most often caused by members of the genera *Microsporum, Epidermophyton,* and *Trichophyton.* Occasionally, *Candida albicans* is implicated. The dermatophytes are distributed world wide but some species are found in only limited geographic locations.

The dermatophytes are mycelial fungi possessing keratinolytic properties which enable them to invade skin, nails and hair. Infections caused by these organisms are commonly referred to as ringworm and usually classified by the Latin word tinea (worm) followed by the area of the body infected, i.e., tinea pedis, tinea cruris.

Identification of these fungi is based on direct examination of clinical material and by cultural methods. Because the dermatophytes are not sensitive to the antibiotics cycloheximide or chloramphenicol, media containing these agents (i.e., Bacto Mycobiotic Agar) can be used successfully for their isolation and cultivation. However, differentiation and identification of these fungi, based on their gross morphological characteristics, requires specialized training.

The frequency of occurrence and wide distribution of these pathogens dictate the need for a simplified, rapid method of identification. Taplin, Zaias, Rebell and Blank[1] formulated a medium, now known as dermatophyte test medium, which meets these requirements, enabling those individuals and labs not specifically trained in mycology to detect these fungi on a routine basis.

The selective medium contains the antibiotics cycloheximide, chlortetracycline and gentamicin, a combination which the authors found to be more effective than antibiotics previously used to suppress the growth of contaminant bacteria, saprophytic yeasts and molds.

Further differentiation and identification of dermatophytes on this medium is based upon the use of phenol red as a pH indicator. Nickerson[2] determined that dermatophytes produce alkaline metabolites which change the color of the indicator from yellow to red. The reliability of this characteristic as a differential mechanism has been confirmed in studies of Stockdale,[3] Baxter,[4] Goldfarb and Hermann.[5] Baxter[4] and Wiegand, Ulrich and Winkelmann[6] explored the use of an ink blue agar for identification of dermatophytes, however, this indicator was found to be too sensitive and resultant decolorization of the agar difficult to distinguish. Taplin et al.[1] favored the use of phenol red which proved nontoxic and its reaction more distinctive and discernable for detection of dermatophytes.

These authors report that successful identification of dermatophytes can be made with 97% accuracy by unskilled personnel. DTM is suitable for field surveys as well as routine laboratory work. The combination of antibiotics in the medium obviates the need for cleansing infected areas to remove contaminants. The authors do not, however, recommend the use of the medium for grossly contaminated specimens, such as nails, which may negate the effects of antibiotics in the medium or produce erroneous pH reactions.

REAGENTS
Bacto DTM Agar is prepared according to the formulation of Taplin, Zaias, Rebell and Blank.[1] The ready to use medium is packaged in 1 oz screw capped bottles, which provide sufficient surface for direct inoculation and immediate incubation of the culture.

FORMULA
BACTO DTM AGAR
Ingredients per liter

Bacto Soytone	10 g	Bacto Agar	20 g
Bacto Dextrose	10 g	Bacto Phenol Red	0.2 g
Actidione® (Cycloheximide)	0.5 g	Chlortetracycline	0.1 g
Gentamicin	0.1 g	Distilled Water	1 liter

PRECAUTIONS
Discard bottles which show visible contamination or possible contamination as indicated by a change in the color of the medium.

STORAGE
Bacto DTM Agar   2 – 8°C

PROCEDURE
Materials provided: Materials required but not provided:
  Bacto DTM Agar   Forceps with fine points
                           25 – 30°C incubator

1. Hair and skin scrapings should be placed firmly on the surface of the medium with flamed forceps.
2. Cap bottles loosely to avoid accumulation of moisture, which could lead to erroneous results.
3. Incubate at 25 – 30°C for 14 days. Examine each day for change in color of medium.
4. Examine culture after 14 days for colonial growth and pH change.

RESULTS
Cultures should be observed for flat or fluffy, slow-growing, heaped and folded, glabourous colonies typical of dermatophytes. These fungi will change the color of the medium from yellow to red. Occasionally a bacterial contaminant will produce this color change but these contaminants can be distinguished by gross morphology.

LIMITATIONS of the PROCEDURE
Bacto DTM Agar is suitable for the isolation and identification of dermatophytes from hair or skin. Generic or species differentiation and identification cannot be accomplished on this medium. The intense red color produced by dermatophytes prevents its use in determining color of the colony and pigmentation of the reverse side of colony necessary to complete the study of gross morphology. For complete morphological determinations, colonies should be transferred to Bacto Sabouraud Dextrose Agar.

Species identification of *Trichophyton* is accomplished by nutritional studies using Bacto Trichophyton Agars 1 – 7.

Bacto DTM Agar should not be employed for identification of dermatophytes from nail specimens or other grossly contaminated clinical material.

REFERENCES

1. Archiv. Derm., 99:203 – 209, 1969.
2. Nickerson, W. J., Biology of Pathogenic Fungi, Chronical Botanica Co., Int. Plant Science Pub., Mass 1947.
3. Biol. Rev., 28:84 – 104, 1953.
4. J. Invest. Derm., 44:23 – 25, 1965.
5. J. Invest. Derm., 27:193 – 201, 1956.
6. Mayo Clin. Proc., 43:795 – 802, 1968.

PACKAGING

Bacto DTM Agar   12 × 1 oz   0942-44-9

DECARBOXYLASE DIFFERENTIAL MEDIA

BACTO DECARBOXYLASE BASE MOELLER
BACTO DECARBOXYLASE MEDIUM BASE
BACTO LYSINE DECARBOXYLASE BROTH

INTENDED USE

Bacto Decarboxylase Differential Media with the addition of 0.5 – 1% appropriate L-amino acids are used to differentiate bacteria on the basis of their ability to decarboxylate the amino acids.

HISTORY

Moeller[1,2,3] can be credited with the first practical application of the amino acid decarboxylase test for distinguishing between various microorganisms. Impressed with the work of Gale[4,5], and Gale and Epps[6,7,8] on bacterial amino acid decarboxylases, Moeller studied these enzyme systems to determine their usefulness for differentiating the *Enterobacteriaceae*. He observed that the production of lysine, arginine, ornithine and glutamic acid decarboxylase by various members of this family afforded a valuable adjunct to other biochemical tests in differentiation of bacteria with closely related physiological characteristics.

Calquist[9] subsequently developed a medium utilizing the lysine decarboxylase reaction to differentiate *Salmonella arizonae* (*Arizona*) from *Citrobacter* (Bethesda-Ballerup). Falkow[10] studied the value of the lysine decarboxylase test as an aid to differentiation and identification of *Salmonella* and *Shigella* and developed a lysine decarboxylase medium which gave valid and reliable results. Although the Falkow medium was originally formulated for use in the lysine decarboxylase test only, further study by the author and by Edwards, Davis and Ewing[11] substantiated the use of the medium in ornithine and arginine decarboxylase reactions.

Ewing, Davis and Edwards[11] thoroughly investigated the decarboxylase activity of 2937 cultures of *Enterobacteriaceae* on lysine, arginine and ornithine using different basal-media. Particular interest was given to results obtained from the Moeller basal medium and the Falkow medium. The authors obtained comparable results with the Moeller and Falkow media. However, the results obtained with the Moeller formulation proved more reliable for cultures of *Klebsiella* and *Enterobacter*. They concluded that the Moeller method should be considered the standard or reference method but acknowledged the suitability of the Falkow medium in determining decarboxylase reactions for members of the *Enterobacteriaceae* other than *Klebsiella* or *Enterobacter*.

Bacto Decarboxylase Base Moeller is the Moeller formulation while Bacto Decarboxylase Medium Base is the formula described by Falkow. Bacto Lysine Decarboxylase Broth is the Falkow medium with L-lysine added in 0.5% concentration.

PRINCIPLES

After inoculation, enteric bacteria will begin fermenting dextrose resulting in an acid pH. Bacteria which produce the decarboxylase under test, i.e., lysine decarboxylase, ornithine decarboxylase or arginine decarboxylase (or dihydrolase) will produce alkaline products and increase the pH. Cadaverine is produced from lysine, putrescine is produced from arginine. The resulting reaction after 25 – 96 hours will show an alkaline reaction (purple or violet) for the decarboxylase producing bacteria and an acid pH (yellow tube) for bacteria not producing decarboxylase. The color of positive tubes may vary in shade from a deep purple or violet to a light purple due to partial destruction of the indicator by some bacteria. Negative tubes will be a definite yellow. Control tubes of basal media should be inoculated to verify the reaction.

To obtain proper reactions, the inoculated tubes must be protected from air to avoid false alkalinization at the surface of the medium. If not protected, decarboxylase negative bacteria may appear to be positive. Protection from air can be accomplished by overlaying the medium with sterile mineral oil as suggested by Ewing et al.[11]

FORMULAE

BACTO DECARBOXYLASE BASE MOELLER
DEHYDRATED

Ingredients per liter

Bacto Peptone	5 g
Bacto Beef Extract	5 g
Bacto Dextrose	0.5 g
Bacto Brom Cresol Purple	0.01 g
Cresol Red	0.005 g
Pyridoxal	0.005 g

Final pH 6.0 ± 0.2 at 25°C.

One pound will make 43.2 liters of medium.

BACTO DECARBOXYLASE BROTH MOELLER w/ARGININE
BACTO DECARBOXYLASE BROTH MOELLER w/LYSINE
BACTO DECARBOXYLASE BROTH MOELLER w/ORNITHINE

The above broths are prepared according to the formulation described by Moeller[1,2,3] using Bacto Decarboxylase Base Moeller and 1% of the appropriate L-amino acid.

BACTO DECARBOXYLASE MEDIUM BASE (FALKOW)
DEHYDRATED

Ingredients per liter

Bacto Peptone	5 g
Bacto Yeast Extract	3 g
Bacto Dextrose	1 g
Bacto Brom Cresol Purple	0.02 g

Final pH 6.8 ± 0.2 at 25°C.

One pound will make 50.4 liters of medium.

BACTO DECARBOXYLASE MEDIUM w/ARGININE
BACTO DECARBOXYLASE MEDIUM w/LYSINE
BACTO DECARBOXYLASE MEDIUM w/ORNITHINE

The above broths are prepared according to the formulation described by Falkow[10] using Bacto Decarboxylase Medium and 0.5% of the appropriate L-amino acids.

BACTO LYSINE DECARBOXYLASE BROTH
DEHYDRATED

Ingredients per liter

```
Bacto Peptone ................. 5 g
Bacto Yeast Extract ............. 3 g
Bacto Dextrose ................. 1 g
L-Lysine .................... 5 g
Bacto Brom Cresol Purple ...... 0.02 g
```

Final pH 6.8 ± 0.2 at 25°C.

One pound will make 32.4 liters of medium.

PREPARATION
Decarboxylase Broth Moeller

Rehydrate Bacto Decarboxylase Base Moeller by suspending 10.5 g in 1 liter distilled or deionized water and heat to dissolve completely. Add 10 g of L-lysine, L-arginine, L-ornithine or other L-amino acids to be studied per liter of basal medium and agitate to effect complete solution. Where DL-amino acids are employed use 2% concentration rather than 1%. No further adjustment of reaction will be required when lysine or arginine are used. Ornithine being highly acidic requires the readjustment of pH with NaOH before sterilization. Usually 1 liter of the medium in which 10 g L-ornithine has been dissolved requires the addition of 4.6 ml 1N NaOH. Dispense the medium in 5 ml amounts into screw capped tubes and sterilize in the autoclave for 10 minutes at 15 lbs pressure (121°C).

Decarboxylase Broth (Falkow)

Rehydrate Bacto Decarboxylase Medium Base by suspending 9 g in 1 liter distilled or deionized water and warm to dissolve completely. Add 5 g L-lysine, L-ornithine, L-arginine or other L-amino acids as desired per liter of medium and warm to dissolve completely. Where DL-amino acids are employed use 1% instead of 0.5%. No further adjustment of reaction is required when lysine or arginine are used. Ornithine HCl being highly acidic requires the adjustment of pH with NaOH before sterilization. Usually 1 liter of medium in which 5 g of ornithine HCl is dissolved requires the addition of 2.1 ml 1N NaOH. Dispense in 5 ml amounts into screw capped culture tubes and sterilize in the autoclave for 15 minutes at 15 lbs pressure (121°C).

The final reaction of the autoclaved medium will be pH 6.8 ± 0.2 as indicated by the purplish color of the indicator in the medium.

Bacto Lysine Decarboxylase Broth

Rehydrate the medium by suspending 14 g in 1 liter distilled or deionized water and heat to boiling to dissolve completely. Dispense in 5 ml amounts into screw capped test tubes. Sterilize in the autoclave at 15 lbs pressure (121°C) for 15 minutes. Cool, inoculate and screw caps on tightly. Incubate at 37°C for 24 hours and read reactions.

NOTE: Tubes of Bacto Decarboxylase Broth Moeller w/amino acids should be light purple exhibiting a slightly green tinge. Tubes of Bacto Decarboxylase Medium w/amino acids should be purple. Any distinctive change in color of these media is indicative of a change in the pH, of medium. In suspect cases, the pH of the medium should be checked. If pH has changed, the tubes should be discarded.

SPECIMEN PREPARATION
Only pure cultures of enteric bacteria taken from purification plates or agar slants are to be used for biochemical tests. A presumptive identification of the bacteria under investigation should be made on the basis of morphological and cultural characteristics prior to biochemical testing.

PROCEDURE
Materials Provided:

Bacto Decarboxylase Broth Moeller
  w/Arginine
Bacto Decarboxylase Broth Moeller
  w/Lysine
Bacto Decarboxylase Broth Moeller
  w/Ornithine

or:

Bacto Decarboxylase Medium
  w/Arginine
Bacto Decarboxylase Medium
  w/Lysine
Bacto Decarboxylase Medium
  w/Ornithine

Materials Required but not Provided:
Wire loops (bacteriological)
37°C incubator
Mineral oil
1. Inoculate decarboxylase broths Moeller or decarboxylase media with a light inoculum of a 24 hour agar slant culture using a bacteriological loop. A control tube should also be inoculated.
2. Aseptically overly the inoculated broth tubes and controls with 4 – 5 mm sterile mineral oil.
3. Incubate tubes at 37°C for up to 4 days. Tubes must be read daily after 24 hours incubation and observed for a change in color from purple to yellow to purple.

RESULTS
The decarboxylase reactions of the *Enterobacteriaceae* and other organisms are given in Table 1.

Table 1 Decarboxylase Reactions of Various *Enterobacteriaceae* and Closely Related Organisms on Lysine, Arginine and Ornithine

	Lysine	Arginine	Ornithine
Salmonella arizonae	+	(+) or +	+
Citrobacter freundii			
(Bethesda-Ballerup)	–	d	d
Citrobacter diversus	–	+ or (+)	+
Providencia	–	–	–
P. rettgeri	–	–	–
Shigella			
S. dysenteriae	–	– or (+)	–
S. flexneri	–	– or (+)	–
S. boydii	–	– or (+)	–
S. sonnei	–	d	+

Table 1 (continued)

	Lysine	Arginine	Ornithine
Salmonella			
S. typhi	+	(+) or −	−
S. paratyphi A	−	(+) or +	+
S. gallinarum	+	−	−
Other types	+	(+)	+
Klebsiella species*†	+	−	−
Edwardsiella tarda	+	−	+
Enterobacter (Aerobacter)			
E. cloacae	−	+	+
E. aerogenes	+	−	+
Hafnia alvei	+	d	+
Serratia marcescens	+	−	+
Serratia liquifaciens	+ or (+)	−	+
Serratia rubidaea	+ or (+)	−	−
Proteus			
P. vulgaris	−	−	−
P. mirabilis	−	−	+
Morganella morganii	−	−	+
Escherichia coli*	d	d	d
Alkalescens Dispar*	d	d	d
Organisms other than *Enterobacteriaceae*			
Pseudomonas	−	+	−
Alcaligenes	−	−	−
Flavobacterium	−	−	−

+ = Purple color, positive reaction
− = Yellow color or no change in color, negative reaction
(+) = Delayed positive reaction
d = Different biochemical types (some strains positive, some negative)

*Ewing, Davis and Edwards studied the decarboxylase reactions of 830 cultures of *E. coli* and obtained 226 negative reactions on lysine, 455 negative on arginine and 405 negative on ornithine. Of 412 *Klebsiella* cultures studied, 91.4% were positive on lysine, 74.2% were positive on arginine and 37.7% were positive on ornithine.
†Falkow medium should not be used for determining reactions of these organisms.

LIMITATIONS OF THE PROCEDURE
Biochemical characteristics of the *Enterobacteriaceae* serve to confirm presumptive identification based on cultural, morphological and/or serological findings. Therefore, biochemical testing should be attempted on pure cultural isolates only and subsequent to differential determinations.

The decarboxylase reactions are part of a total biochemical profile for members of the *Enterobacteriaceae* and related organisms. Results obtained from these reactions, therefore, can be considered indicative of a given genus or species but conclusive and final identification of these organisms cannot be made solely on the basis of the decarboxylase reactions.

STORAGE
Bacto Decarboxylase Differential Media	Below 30°C
Prepared tubes	15 − 30°C
Bacto Decarboxylase Base w/amino acids	
(prepared tubes)	2 − 8°C

QUALITY CONTROL

Identity Specifications

	Decarboxylase Base Moeller	Decarboxylase Medium Base	Lysine Decarboxylase Base
Dehydrated medium:	light to medium tan, homogeneous, free-flowing	light beige, homogeneous, free-flowing	light beige, homogeneous, free-flowing
% solution:	1.05%	0.9%	1.4%
pH at 25°C:	6.0 ± 0.2	6.8 ± 0.2	6.8 ± 0.2
Prepared medium:	yellowish red, slightly opalescent	purple, clear	purple, clear

REFERENCES

1. Acta Path. et Micro. Scand., 34:102, 1954.
2. ibid., 35:259, 1954.
3. ibid., 36:158, 1955.
4. Biochem. J., 34:392,846,583, 1940.
5. ibid., 35:66, 1941.
6. ibid., 36:600, 1942.
7. Nature, 152:327, 1943.
8. Biochem. J., 38:250, 1944.
9. J. Bact., 71:339, 1956.
10. Tech. Bull, Reg. Med. Tech., 28:106, 1958 or Amer. J. Clin. Path., 29:598, 1958.
11. Pub. Health Lab., 18:77, 1960.

PACKAGING

Ready to use Media:

Bacto Decarboxylase Broth Moeller w/Arginine	12 tubes	1271-34-0
Bacto Decarboxylase Broth Moeller w/Lysine	12 tubes	1273-34-8
Bacto Decarboxylase Broth Moeller w/Ornithine	12 tubes	1272-34-9

Dehydrated Media & Reagents:

Bacto Decarboxylase Base Moeller	1/4 lb (114 g)	0890-02-1
	1 lb (454 g)	0890-01-2
Bacto Decarboxylase Medium Base	1/4 lb (114 g)	0872-02-3
	1 lb (454 g)	0872-01-4
	10 kg	0872-08-7
Bacto Lysine Decarboxylase Broth	1 lb (454 g)	0215-01-0
	1/4 lb (114 g)	0215-02-9
Bacto L-Arginine HCl	10 g	0583-12-1
Bacto L-Lysine HCl	5 g	0705-11-5
Bacto L-Ornithine HCl	5 g	0293-11-3
	10 g	0293-12-2
	25 g	0293-13-1

BACTO DESOXYCHOLATE AGAR

INTENDED USE
Bacto Desoxycholate Agar is a differential and slightly selective plating medium for the isolation of gram-negative enteric bacilli. The medium is also used for enumeration of coliform organisms in milk and dairy products.[1]

HISTORY/PRINCIPLES
The formula of Bacto Desoxycholate Agar is essentially that described by Leifson[2] with differentiation of enteric bacilli based on lactose fermentation. Coliform colonies are red in contrast to light, colorless colonies produced by enteric organisms not capable of attacking lactose. Organisms other than those of the enteric group are inhibited by the citrates and sodium desoxycholate.

Saccharose in 1% concentration may be added to isolation media such as Bacto Desoxycholate Agar to permit the detection of certain members of the coliform group which ferment saccharose more readily than lactose. This principle was described by Holt-Harris and Teague[3] and has been employed by many other bacteriologists. In some laboratories pathogenic significance is assigned to these organisms and, under such conditions, saccharose should not be added to the medium.

FORMULA

BACTO DESOXYCHOLATE AGAR
DEHYDRATED

Ingredients per liter

Bacto Peptone	10 g
Bacto Lactose	10 g
Sodium Desoxycholate	1 g
Sodium Chloride	5 g
Dipotassium Phosphate	2 g
Ferric Citrate	1 g
Sodium Citrate	1 g
Bacto Agar	15 g
Neutral Red	0.03 g

Final pH 7.3 ± 0.2 at 25°C.

One pound will make 10 liters of medium.

METHOD OF PREPARATION
1. Suspend 45 g in 1 liter distilled or deionized water.
2. Heat to boiling for approximately 1 minute with frequent and careful swirling to dissolve completely. Avoid over heating as this medium is heat sensitive. **DO NOT AUTOCLAVE OR REMELT.**
3. Cool to 45 – 50°C and dispense into sterile Petri dishes.

For presumptive determination of coliforms in dairy products, a pour plate of Desoxycholate Agar is prepared. After the medium has solidified, many investigators prefer to add a thin cover layer of uninoculated medium.

For the isolation of enteric pathogens, streak or smear plates are prepared. Plates should be dry before inoculation for best results.

STORAGE
Bacto Desoxycholate Agar   Below 30°C
Prepared medium   2 – 8°C

QUALITY CONTROL
Identity Specifications

Dehydrated powder: pinkish beige, homogeneous, free-flowing
Reaction of 4.5% solution: 7.3 ± 0.2 at 25°C
Prepared medium: reddish orange, very slightly opalescent

Typical Cultural Response on Bacto Desoxycholate Agar (streak plates)
After 18 – 24 Hours at 35°C

Organism	Growth	Color of Colony
Escherichia coli ATCC® 25922	good	pink w/bile precipitate
Salmonella typhimurium ATCC® 14028	good	colorless
Staphylococcus aureus ATCC® 25923	inhibited	colorless, if any

REFERENCES

1. Standard Methods for The Examination of Dairy Products, 9th Ed. American Public Health Association, Inc., Washington, D.C., 1948.
2. Leifson, 1935, Journal of Pathogenic Bacteriology 40:581.
3. Holt-Harris and Teague, 1916, Journal of Infectious Diseases, 18:596.

PACKAGING

Bacto Desoxycholate Agar	1 lb (454 g)	0273-01-9
	1/4 lb (114 g)	0273-02-8

BACTO DESOXYCHOLATE CITRATE AGAR

INTENDED USE
Bacto Desoxycholate Citrate Agar is a highly selective medium for the isolation of enteric pathogens, particularly *Salmonella* and many *Shigella*.

HISTORY
Bacto Desoxycholate Citrate Agar, a modification of the original Leifson[1] formula, is recommended for use in all cases where a selective desoxycholate citrate agar is specified or desired. The use of desoxycholate citrate agar is recommended as a plating procedure for examination of specimens for evidence of infection with *Salmonella* and *Shigella* as given in *Diagnostic Procedures and Reagents*[2] of the American Public Health Association.

For the routine examination of stool and urine specimens it is recommended that Bacto MacConkey Agar, a nonselective medium, and Bacto Bismuth Sulfite Agar or Bacto SS Agar be run in conjunction with Bacto Desoxycholate Citrate Agar.

Saccharose (sucrose) in 1% concentration may be added to isolation media, such as Bacto Desoxycholate Citrate Agar to permit the detection of certain members of the coliform group which ferment saccharose more readily than lactose. This principle was described by Holt-Harris and Teague[3] and has been employed by many other bacteriologists. In some laboratories, pathogenic significance is assigned to these organisms, and under such conditions, saccharose should not be added to the medium.

PRINCIPLES
On desoxycholate citrate agar the growth of coliform bacteria and gram-positive bacteria are inhibited or greatly suppressed due to the sodium desoxycholate and sodium citrate in the formula. Gram-positive bacteria are generally inhibited. *Salmonella* and

Shigella organisms grow quite unrestricted. The selectivity of this medium permits the use of fairly heavy inocula without danger of overgrowth of the *Shigella* and *Salmonella* by extraneous organisms. Occasionally, however, coliform strains are encountered that persist on desoxycholate citrate agar. Such strains, if present in large numbers, produce acid from the lactose, precipitate the bile salt, and give an opaque red medium which makes it difficult to detect the pathogens. Distribution of the inoculum over the surface of the medium to give a sparsely populated section helps to insure against complete masking of the pathogens by such coliform organisms.

If the lactose in the formula is fermented by the test organism it causes an acidification of the medium around these colonies. This pH shift causes the indicator Bacto Neutral Red to change to red and for bile to be precipitated. The reduction of ferric ammonium citrate to iron sulfide by H_2S producing organisms is indicated by blackening of the central portion of the colony.

FORMULA
BACTO DESOXYCHOLATE CITRATE AGAR
DEHYDRATED
Ingredients per liter

Pork, Infusion from	330 g	Ferric Ammonium Citrate	2 g
Proteose Peptone No. 3, Difco	10 g	Sodium Desoxycholate	5 g
Bacto Lactose	10 g	Bacto Agar	13.5 g
Sodium Citrate	20 g	Neutral red	0.02 g

Final pH 7.5 ± 0.2 at 25°C.

One pound will make 6.5 liters of medium.

METHOD OF PREPARATION
1. Suspend 70 g in 1 liter distilled or deionized water.
2. Heat just to boiling with frequent and careful swirling to dissolve completely. Avoid overheating. Do not autoclave.
3. Cool to 45 – 50°C and dispense into sterile Petri dishes.
4. Allow the surface to dry for two hours with the covers partially removed before inoculation.

PROCEDURE
Use a heavy inoculum to streak plates. Incubate streaked plates at 35 – 37°C for 18 – 24 hours. If no lactose fermentors are observed, incubate for an additional 24 hours.

Lactose positive colonies are pink surrounded by a zone of bile precipitation. Lactose negative colonies are colorless. Organisms which produce H_2S will reduce ferric ammonium citrate to iron sulfide resulting in black centered colonies.

STORAGE
Desoxycholate Citrate Agar   Below 30°C
Prepared medium   2 – 8°C

QUALITY CONTROL
Identity Specifications

Dehydrated powder: pinkish beige, homogeneous, free-flowing
Reaction of 7.0% solution: pH 7.5 ± 0.2 at 25°C
Prepared medium: orange-red, very slightly opalescent

**Typical Cultural Response on Bacto Desoxycholate Citrate Agar
After 18 – 24 Hours at 35°C**

Organism	Growth	Color of Colony	H₂S
Escherichia coli ATCC® 25922	fair to good	pink w/bile precipitate	−
Salmonella enteritidis ATCC® 13076	fair to good	colorless	+
Salmonella typhimurium ATCC® 14028	fair to good	colorless	+
Shigella flexneri ATCC® 12022	fair	colorless	−
Streptococcus faecalis ATCC® 19433	inhibited	—	−

H_2S − = negative, no blackening
H_2S + = positive, black centers or colonies

REFERENCES

1. J. Path. Bact., 40:581, 1935.
2. Diagnostic Procedures and Reagents, 3rd Ed., 212, 1950.
3. J. Infectious Diseases, 18:596, 1916.

PACKAGING

Bacto Desoxycholate Citrate Agar	1 lb (454 g)	0274-01-8
	1/4 lb (114 g)	0274-02-7

BACTO DESOXYCHOLATE LACTOSE AGAR

INTENDED USE

Bacto Desoxycholate Lactose Agar is a differential and slightly selective medium for isolating gram-negative enteric bacilli. The medium may also be used for enumerating coliform organisms from water, wastewater, milk and dairy products.

HISTORY

Bacto Desoxycholate Lactose Agar is a modification of the medium described by Leifson.[1] It is prepared according to the formula specified in *Standard Methods for the Examination of Dairy Products, 11th Edition,*[2] and *Standard Methods for the Examination of Water and Wastewater, 11th Edition.*[3]

PRINCIPLES

Gram-positive organisms are generally inhibited by the sodium desoxycholate and sodium citrate. Essentially Bacto Desoxycholate Lactose Agar differs from Bacto Desoxycholate Agar in that it contains less sodium desoxycholate and is accordingly slightly less selective against gram-positive organisms.

As the lactose in the formula is utilized, acid is produced resulting in a change of the pH indicator, Bacto Neutral Red, from faint pink to red and causing the precipitation of bile around coliform colonies. Lactose fermenting coliform organisms grow unrestricted and form typical red subsurface colonies, surrounded by a zone of precipitated bile. Lactose nonfermenters form colorless colonies. Incubation periods should be 24 hours or less since other organisms may develop on longer incubation causing confusion.

FORMULA

BACTO DESOXYCHOLATE LACTOSE AGAR
DEHYDRATED

Ingredients per liter

Bacto Peptone	10 g
Bacto Lactose	10 g
Sodium Desoxycholate	0.5 g
Sodium Chloride	5 g
Sodium Citrate	2 g
Bacto Agar	15 g
Bacto Neutral Red	0.03 g

Final pH 7.1 ± 0.2 at 25°C.

One pound will make 10.7 liters of medium.

PROCEDURE

1. Suspend 42.5 g in 1 liter distilled or deionized water and heat to boiling to dissolve completely. The medium requires no further heating if used at once.
2. If the medium is to be stored, distribute into tubes or flasks and sterilize in the autoclave for 15 minutes at 15 lbs pressure (121°C). Avoid overheating of the medium.
3. The medium may be inoculated by the spread-plate or pour-plate techniques. A thin layer of sterile medium may be used as a cover, thus having only subsurface colonies. Following solidification of the inoculated plate a small amount of sterile medium (about 4 – 5 ml per 95 mm plate) is added and allowed to solidify, forming a thin cover layer.

STORAGE

Bacto Desoxycholate Lactose Agar   Below 30°C
Prepared medium   2 – 8°C

QUALITY CONTROL

Identity Specifications

Dehydrated powder: pinkish beige, homogeneous, free-flowing
Reaction of 4.25% solution: pH 7.1 ± 0.2 at 25°C
Prepared medium: pinkish red, very slightly opalescent

Typical Cultural Response on Bacto Desoxycholate Lactose Agar
After 18 – 24 Hours at 35°C

Organism	Growth	Color of Colony
Bacillus subtilis ATCC® 6633	inhibited	—
Enterobacter aerogenes ATCC® 13048	good to excellent	pink w/bile precipitate
Escherichia coli ATCC® 25922	good to excellent	pink w/bile precipitate
Salmonella typhimurium ATCC® 14028	good to excellent	colorless
Streptococcus faecalis ATCC® 19433	inhibited	—

REFERENCES

1. J. Path. Bact., 40:581, 1935.
2. Standard Methods for the Examination of Dairy Products, 11th Ed., APHA Inc., New York, 1960.
3. Standard Methods for the Examination of Water and Wastewater, 11th Ed., APHA, Inc., New York, 1960.

PACKAGING

Bacto Desoxycholate Lactose Agar   1 lb (454 g)   0420-01-1

BACTO DESOXYCHOLIC ACID

Bacto Desoxycholic Acid is a bile acid used in bacteriological culture media. It inhibits gram-positive cocci and spore-forming organisms but does not inhibit gram-negative enteric bacilli.

PACKAGING
Bacto Desoxycholic Acid 25 g 0531-13-3

DEXTRIN

Dextrin is a polysaccharide derived from high quality starch and is soluble in 1% or higher concentrations in distilled or deionized water with gentle warming. As a 1% concentration in fermentation base culture media, such as Bacto Phenol Red Broth Base and Bacto Purple Broth Base, it can be used to differentiate bacteria with otherwise similar biochemical characteristics.

PACKAGING
Dextrin 100 g 0161-15-8
500 g 0161-17-6

BACTO DEXTROSE AGAR

INTENDED USE
Bacto Dextrose Agar is recommended for cultivation of a wide variety of microorganisms. It is especially adapted for preparing dextrose blood agar.

HISTORY
In 1932, Norton[1] recommended an agar medium containing 0.5 to 1.0% dextrose and about 5% defibrinated blood for isolating microorganisms from pus, as well as for cultivating *Neisseria meningitidis, N. gonorrhoeae, Haemophilus influenzae* and *Bordetella pertussis.*

PRINCIPLES
Dextrose is a readily available source of energy utilized by a large number of organisms. This ingredient makes Bacto Dextrose Agar suitable for the production of early and abundant growth, the shortening of lag periods of old cultures and the initiation of growth of bacteria capable of utilizing dextrose.

FORMULA
BACTO DEXTROSE AGAR
DEHYDRATED
Ingredients per liter

Bacto Beef Extract . . . . 3 g
Bacto Tryptose . . . . 10 g
Bacto Dextrose . . . . 10 g
Sodium Chloride . . . . 5 g
Bacto Agar . . . . 15 g

Final pH 7.3 ± 0.2 at 35°C.

One pound will make 10.5 liters of final medium.

METHOD OF PREPARATION
1. Suspend 43 g in 1 liter distilled or deionized water and heat to boiling to dissolve completely.
2. Sterilize in the autoclave for 15 minutes at 15 lbs pressure (121°C).
3. Dispense as desired.

STORAGE
Bacto Dextrose Agar   Below 30°C
Prepared medium   2 – 8°C

QUALITY CONTROL
Identity Specifications

Dehydrated powder: medium beige, homogeneous, free-flowing
Reaction of 4.3% solution: pH 7.3 ± 0.2 at 25°C
Prepared medium: medium amber, slightly opalescent

Typical Cultural Response on Bacto Dextrose Agar
After 18 – 48 Hours at 35°C

Organism	Recovery Plain	With 5% Sheep Blood
Bordetella pertussis ATCC® 9797	good to excellent	good to excellent
Clostridium perfringens ATCC® 12919	fair to good	good to excellent
Neisseria gonorrhoeae ATCC® 19424	good to excellent	good to excellent
Neisseria meningitidis ATCC® 13090	good to excellent	good to excellent
Streptococcus pneumoniae ATCC® 6303	good to excellent	good to excellent
Streptococcus pyogenes ATCC® 19615	good to excellent	good to excellent

REFERENCE
1. J. Lab. Clin. Med., 17:585, 1932.

PACKAGING
Bacto Dextrose Agar   1 lb (454 g)   0067-01-9

BACTO DEXTROSE BROTH

INTENDED USE
Bacto Dextrose Broth is used for cultivating fastidious microorganisms.

HISTORY
Waisbren, Carr and Dunnett[1] used dextrose broth for testing the sensitivity of micro-organisms to antibiotics by the tube dilution method. They demonstrated that a soy bean peptone medium inhibited the action of neomycin, Aureomycin® (chlortetracycline) Terramycin® (oxytetracycline) and polymyxin against the test organism while tryptose phosphate broth, dextrose broth and nutrient broth were suitable for the test.

PRINCIPLES
Bacto Dextrose Broth contains Bacto Tryptose and 0.5% dextrose (d-glucose), a readily available source of energy utilized by a large number of organisms. The broth gives rapid growth and hastens the early development of injured forms.

The addition of 0.1 to 0.2% agar to the broth medium improves the productivity of the medium for most purposes. The low agar concentration provides suitable conditions for aerobic growth in the clear upper zone and for microaerophilic and anaerobic growth in the lower agar zones.

FORMULA

BACTO DEXTROSE BROTH
DEHYDRATED

Ingredients per liter

Bacto Beef Extract .... 3 g
Bacto Dextrose .... 5 g
Sodium Chloride .... 5 g
Bacto Tryptose .... 10 g

Final pH 7.2 ± 0.2 at 25°C.

One pound will make 19.7 liters of final medium.

METHOD OF PREPARATION
1. Dissolve 23 g in 1 liter distilled or deionized water.
2. If a medium containing 0.1% agar is desired, suspend 1 g agar with above ingredients and heat to boiling to dissolve completely.
3. Dispense as desired.
4. Sterilize in the autoclave for 15 minutes at 15 lbs pressure (121°C).

STORAGE
Bacto Dextrose Broth	Below 30°C
Prepared medium	15 – 30°C

QUALITY CONTROL
Identity Specifications

Dehydrated powder:	light tan, homogeneous, free-flowing
Reaction of 2.3% solution:	pH 7.2 ± 0.2 at 25°C
Prepared medium:	light to medium amber, clear

Typical Cultural Response in Bacto Dextrose Broth
After 18 – 24 Hours at 35°C

Organism	Recovery	Gas
Escherichia coli ATCC® 25922	excellent	+
Salmonella typhi ATCC® 6539	excellent	−

Typical Cultural Response in Bacto Dextrose Broth with 0.1% Bacto Agar
After 18 – 48 Hours at 35°C

Organism	Recovery
Neisseria meningitidis ATCC® 13090	good to excellent
Streptococcus pneumoniae ATCC® 6303	good to excellent
Streptococcus pyogenes ATCC® 19615	good to excellent

REFERENCE
1. Am. J. Clin. Path., 21:884, 1951.

PACKAGING
Bacto Dextrose Broth	1 lb (454 g)	0063-01-3

BACTO DEXTROSE PROTEOSE NO. 3 AGAR

INTENDED USE

Bacto Dextrose Proteose No. 3 Agar is used in combination with Bacto Tellurite Blood Solution for isolating *Corynebacterium diphtheriae*.

HISTORY/PRINCIPLES

Since 1912, a variety of serum media containing tellurite have been described for isolating *C. diphtheriae* and differentiating between its various strains.[1-8] In each of the media, tellurite inhibited growth of staphylococci and streptococci, permitting *C. diphtheriae* to grow. The occasional strain of *Staphylococcus* able to grow could be differentiated on the basis of colonial morphology.

FORMULA

BACTO DEXTROSE PROTEOSE NO. 3 AGAR
DEHYDRATED

Ingredients per liter

Proteose Peptone No. 3, Difco . . . . 20 g
Bacto Dextrose . . . . 2 g
Sodium Chloride . . . . 5 g
Bacto Agar . . . . 13 g

Final pH 7.4 ± 0.2 at 25°C.

One pound will make 11.3 liters of medium.

METHOD OF PREPARATION

1. Suspend 4 g in 100 ml distilled or deionized water and heat to boiling to dissolve completely. Sterilize in the autoclave for 15 minutes at 15 lbs pressure (121°C). Cool to 75 – 80°C.
2. Aseptically add 5 ml Bacto Tellurite Blood Solution to the sterile agar and swirl gently.
3. Maintain the mixture at 70 – 80°C until it takes on the appearance of chocolate agar (10 – 15 minutes).
4. Cool slowly to 50°C.
5. Dispense into sterile Petri dishes and allow to solidify.
6. Occasional strains and small numbers of *C. diphtheriae* may be inhibited by potassium tellurite. Maximum recovery of the pathogen is insured by inoculating a nonselective medium (blood agar or Loeffler medium) in parallel with the selective tellurite medium.
7. *C. diphtheriae* cultured on selective tellurite media may not exhibit typical morphology or staining characteristics. Suspect black colonies should be subcultured on a nonselective medium for these determinations. Contaminating organisms are ordinarily inhibited or killed during preparation (heating) and incubation of the final medium.

STORAGE

Bacto Dextrose Proteose No. 3
  Agar   Below 30°C

Storage of prepared medium is not recommended.

QUALITY CONTROL
Identity Specifications

Dehydrated powder: light beige, homogeneous, free-flowing
Reaction of 4% solution: pH 7.4 ± 0.2 at 25°C
Prepared tellurite medium: dark reddish brown, opaque

Typical Cultural Response on Tellurite Medium
After 24 – 48 Hours at 35°C

Organism	Recovery	Colony Color
Corynebacterium diphtheriae Type *gravis*	good to excellent	black
Corynebacterium diphtheriae Type *intermedius*	good to excellent	black
Corynebacterium diptheriae Type *mitis*	good to excellent	black
Staphylococcus aureus ATCC® 25923	inhibited	—
Streptococcus pyogenes ATCC® 19615	inhibited	—

REFERENCES

1. Muench. Wochschr., 59:1652, 1912.
2. Zentr. Bakt., 114:539, 1929.
3. J. Path. Bact., 34:667, 1931.
4. J. Hyg., 32:544, 1932.
5. J. Path. Bact., 38:114, 1934.
6. J. Infectious Diseases, 59:22, 1936.
7. J. Infectious Diseases, 60:99, 1937.
8. Am. J. Pub. Health., 29:664, 1939.

PACKAGING

Bacto Dextrose Proteose No. 3 Agar   500 g   0068-17-0

BACTO DEXTROSE STARCH AGAR

INTENDED USE

Bacto Dextrose Starch Agar is used for propagating pure cultures of *Neisseria gonorrhoeae* and other fastidious microorganisms.

PRINCIPLES

Bacto Dextrose Starch Agar is a nutritionally rich medium which without enrichment, yields luxuriant growth of many fastidious microorganisms.

FORMULA

BACTO DEXTROSE STARCH AGAR
DEHYDRATED

Ingredients per liter

Proteose Peptone No. 3, Difco	15 g
Bacto Dextrose	2 g
Soluble Starch, Difco	10 g
Sodium Chloride	5 g
Disodium Phosphate	3 g
Bacto Gelatin	20 g
Bacto Agar	10 g

Final pH 7.3 ± 0.2 at 25°C.

One pound will make 6.9 liters of final medium.

METHOD OF PREPARATION

1. Suspend 65 g in 1 liter distilled or deionized water and heat to boiling to dissolve completely.
2. Dispense into tubes or flasks, agitating the medium to evenly distribute the normal flocculent precipitate.
3. Sterilize in the autoclave for 15 minutes at 15 lbs pressure (121°C).

STORAGE

Bacto Dextrose Starch Agar	Below 30°C
Prepared medium	2 – 8°C

QUALITY CONTROL

Identity Specifications

Dehydrated powder:	beige, homogeneous, free-flowing
Reaction of 6.5% solution:	pH 7.3 ± 0.2 at 25°C
Prepared medium:	light amber, opalescent with flocculent precipitate

Typical Cultural Response on Bacto Dextrose Starch Agar
After 18 – 48 Hours at 35°C

Organism	Growth
Neisseria gonorrhoeae ATCC® 19424	good to excellent
Neisseria meningitidis ATCC® 13090	good to excellent
Streptococcus pneumoniae ATCC® 6303	good to excellent
Streptococcus pyogenes ATCC® 19615	good to excellent

PACKAGING

Bacto Dextrose Starch Agar	500 g	0066-17-2
	10 kg	0066-08-3

BACTO DEXTROSE TRYPTONE AGAR

INTENDED USE

Bacto Dextrose Tryptone Agar is used for the determination of thermophilic "flat sour" organisms associated with the spoilage of food products.[1]

HISTORY/PRINCIPLES

Bacto Dextrose Tryptone Agar was developed in collaboration with the research laboratories of the National Canners Association and is recommended for the above use in their *Bacterial Standards for Sugar*.[2] For plate count of mesophilic or thermophilic aerobes in sweetening agents used in frozen desserts, *Standard Methods for the Examination of Dairy Products*[3] recommends the use of Bacto Dextrose Tryptone Agar.

Bacto Dextrose Tryptone Agar is primarily used as a plating medium and, when employed for determining thermophiles, it should be incubated at 55°C for 36 – 48 hours in an incubator sufficiently humid to prevent drying of the medium.

FORMULA

BACTO DEXTROSE TRYPTONE AGAR
DEHYDRATED

Ingredients per liter

Bacto Tryptone	10 g
Bacto Dextrose	5 g
Bacto Agar	15 g
Bacto Brom Cresol Purple	0.04 g

Final pH 6.7 ± 0.2 at 25°C.

Five hundred grams will make 16.6 liters of medium.

METHOD OF PREPARATION
1. Suspend 30 g in 1 liter of distilled or deionized water and heat to boiling to dissolve completely.
2. Dispense as desired.
3. Sterilize in the autoclave for 15 minutes at 15 lbs pressure (121°C).
4. Inoculate sterile Petri plates by the pour plate technique.

STORAGE
Bacto Dextrose Tryptone Agar   Below 30°C
Prepared medium   15 – 30°C

QUALITY CONTROL
Identity Specifications

Dehydrated powder:	light greenish beige, homogeneous, free-flowing
Reaction of 3% solution:	pH 6.7 ± 0.2 at 25°C
Prepared medium:	purple, slightly opalescent

Typical Cultural Response on Bacto Dextrose Tryptone Agar
After 36 – 48 Hours at 55°C

Organism	Growth	Color of Colony
Bacillus coagulans ATCC® 8038	good	yellow

REFERENCES
1. National Canners Assoc., 1968, Laboratory Manual for Food Canners and Processors, Vol. 1:13.
2. National Canners Assoc., 1933, Bacterial Standards for Sugar.
3. Standard Methods for the Examination of Dairy Products, 9th Ed., American Public Health Assoc., Inc., Washington, D.C., 1948.

PACKAGING
Bacto Dextrose Tryptone Agar   500 g   0080-17-4

BACTO m DEXTROSE TRYPTONE BROTH

INTENDED USE
Bacto m Dextrose Tryptone Broth is used in the membrane filter technique for the cultivation of thermophilic "flat sour" organisms associated with the spoilage of food products.

HISTORY/PRINCIPLES
This nonselective medium, containing brom cresol purple as an indicator, is well suited for the cultivation of a variety of microorganisms. The inoculated membranes on this medium are incubated at 55°C in sealed dishes for the detection and enumeration of thermophilic "flat sour" sporulating microorganisms.

FORMULA
BACTO m DEXTROSE TRYPTONE BROTH
DEHYDRATED
Ingredients per liter

Bacto Tryptone .... 20 g
Bacto Dextrose .... 10 g
Bacto Brom Cresol Purple ...... 0.04 g

Final pH 6.7 ± 0.2 at 25°C.

One pound will make 15.1 liters of medium.

METHOD OF PREPARATION

1. Suspend 30 g in 1 liter distilled or deionized water and heat to boiling to dissolve completely.
2. Sterilize in the autoclave for 15 minutes at 15 lbs pressure (121°C).

STORAGE

Bacto m Dextrose Tryptone Broth	Below 30°C
Prepared medium	15 – 30°C

QUALITY CONTROL

Identity Specifications

Dehydrated powder:	greenish beige, homogeneous, free-flowing
Reaction of 3% solution:	pH 6.7 ± 0.2 at 25°C
Prepared medium:	purple, clear

Typical Cultural Response on Bacto m Dextrose Tryptone Broth
After 36 – 48 Hours at 55°C in Humid Atmosphere

Organism	Growth
Bacillus stearothermophilus	good to excellent

PACKAGING

Bacto m Dextrose Tryptone Broth	1 lb (454 g)	0706-01-6

BACTO DIENES STAIN

INTENDED USE

Bacto Dienes Stain is recommended for use in staining colonies of *Mycoplasma*.

HISTORY

Two methods of staining have been described in the literature. Method 1 stains colonies directly on the agar surface in the plate.[1,2] Method 2 stains the colonies on the microscope slide.[3]

METHOD OF STAINING
Method 1

1. Moisten a cotton swab with Bacto Dienes Stain and apply the stain to the agar adjacent to suspected *Mycoplasma* colonies.
2. Invert the Petri dish on the stage of a microscope and examine using approximately 100X magnification.

Method 2

1. Add a drop of Bacto Dienes Stain to a clean dry coverglass and spread it over the entire surface.
2. Allow the stain on the coverglass to air dry.
3. Cut a small block of agar of the suspected culture with a sterile scalpel and place, colony side up, on a microscope slide.
4. Place the stained side of the coverglass on the surface of the agar block.
5. Ring the coverglass with paraffin or vaspar.
6. Examine microscopically using approximately 100X magnification.

RESULTS

The *Mycoplasma* colonies will be stained yielding blue centers with a lighter blue periphery. Colonies of bacteria will stain initially but will decolorize in approximately 1/2 hour while colonies of *Mycoplasma* retain the stain indefinitely.

STORAGE

Bacto Dienes Stain   15 – 30°C

REFERENCES

1. J. Inf. Dis., 65:24, 1939.
2. J. Bact., 44:37, 1942.
3. *ibid.*, 50:441, 1945.

PACKAGING

Bacto Dienes Stain   100 ml   0780-72-9

BACTO DIFFERENTIATION DISKS BV, BX, AND BVX

INTENDED USE

Bacto Differentiation Disks BV, BX, and BVX are used for the presumptive identification of *Haemophilus* species on the basis of their requirements for X or V factors or both.

HISTORY/PRINCIPLES

Members of the genus *Haemophilus* are fastidious organisms requiring an enriched medium especially chocolate agar containing Bacto Supplement B, C, Bacto Supplement VX, or Bacto Fildes Enrichment for optimal growth. It is these exacting growth requirements which aid in differentiation and species identification. Members of this genus are characterized by their dependence on one or both growth factors X and V. The requirement for X, V, or VX factors, which are present in the above mentioned enrichments is most significant in isolation and differentiation of *Haemophilus* species, particularly *H. influenzae* and *H. parainfluenzae*.

Davis[1] and Thjotta and Avery[2] recognized that the type species *H. influenzae* required two growth factors, one from blood which they called X factor and another which they called V factor. The X factor has been identified as iron protoporphyrin and hemin has been used as a source of this factor. Lwoof and Lwoof[3] prepared V factor from baker's yeast. It has been determined that V factor is diphosphopyridine neucleotide.

Isolation of *Haemophilus* by conventional cultural methods is best accomplished on an enriched medium which supplies both these accessory growth factors. Bacto Hemoglobin may be employed for X factor while Bacto Supplement A, B, or C will provide V factor. Bacto Fildes Enrichment, usually employed in a basal medium (i.e., Bacto Heart Infusion Agar or Bacto Tryptic Soy Agar) in a 5% concentration, is a source of both X and V factors.

The *Staphylococcus* satellite method is also employed as a means of isolating *Haemophilus*. *S. aureus* provides V factor and when streaked on blood or chocolate blood agar (w/o enrichment) will enhance growth of *Haemophilus* around the *Staphylococcus* streak. This method is also used as a means of species identification and differentiation.

When attempting to isolate *Haemophilus,* particularly *H. influenzae,* from throat or nasopharyngeal specimens, it is desirable to inhibit the growth of contaminating microorganisms.

Parket and Hoeprich[4] added 0.2 units penicillin, 5 μg ristocetin, 5 μg vancomycin and 25 units nystatin to each ml of medium for isolating *Haemophilus* from clinical material. More recently, Crawford, Borden and Kukman,[5] Klein and Blazevic,[6] and Young[7] proposed the use of an enriched medium to which bacitracin is added to suppress normal flora from throat and nasopharyngeal specimens and facilitate growth of *Haemophilus.* These authors found the selectivity of bacitracin to be consistent and to yield normal growth patterns or gross morphology of *Haemophilus.*

Haemophilus influenzae and *H. parainfluenzae* are both commonly found in the nasopharynx and throat. *H. influenzae* has been implicated in several types of respiratory infections usually in the role of opportunist. *H. parainfluenzae* is rarely implicated in pathological processes. In culture, these two species are morphologically and colonially similar, making differentiation difficult. Their need for accessory growth factors have therefore been used as a method of differentiating these species.

Parket and Hoeprich[4] prepared disks containing the X and V factors and used them to differentiate *H. influenzae,* which requires X and V factor, from *H. parainfluenzae,* which requires only V factor for growth. Cultures were inoculated on heart infusion agar, tryptic soy agar or nutrient agar. Kohn[8] suggested the addition of bacitracin to X and V disks to inhibit most of the non-*Haemophilus* organisms usually found in the throat and nasopharynx. Russel[9] used disks impregnated with X and V factors to identify organisms belonging to the genus *Haemophilus* based on their requirement for these factors and reported the suitability of this procedure.

Bacto Differentiation Disks BVX, which contains bacitracin, V and X factors, may be used for preliminary isolation of *Haemophilus* organisms. Isolation of *Haemophilus* using Bacto Differentiation Disks BVX obviates the need for preparation of complex enriched media and the use of live cultures of *Staphylococcus aureus.*

Differentiation and species identification should be done on pure cultures which have been isolated by one of the conventional cultural methods previously described or by the differentiation disk method.

REAGENTS
Bacto Differentiation Disks BVX are paper disks 3/8 inch diameter containing bacitracin and sufficient iron protoporphyrin and diphosphopyridin neucleotide to support the growth of *Haemophilus.*

Bacto Differentiation Disks BV are paper disks 3/8 inch diameter containing bacitracin and sufficient diphosphopyridin neucleotide to support the growth of *Haemophilus* species which require V factor.

Bacto Differentiation Disks BX are paper disks 3/8 inch diameter containing bacitracin and sufficient iron protoporphyrin to support the growth of *Haemophilus* species which require X factor.

SPECIMEN COLLECTION
Specimens for the isolation of *Haemophilus* may be obtained from cerebrospinal fluid, blood, the throat or nasopharynx and conjunctivae.

SPECIMEN PREPARATION
As mentioned previously, when employing Bacto Differentiation Disks BVX, it is not necessary to use media containing blood or supplements. Bacto Differentiation Disks BV may be used when blood agar is employed. Conversely, differentiation disks BVX and BV are not required for isolation if an enriched medium such as, enriched chocolate agar or tryptic soy agar with 5% Fildes enrichment is employed. However, because bacitracin has been found to be effective in the isolation of *Haemophilus* from specimens containing mixed flora, it is recommended that a BV disk be used as an additional aid in the isolation of *Haemophilus* even when an enriched medium is used.

Wire or swab specimens should be streaked for isolation on 1) Bacto Heart Infusion Agar or Bacto Tryptic Soy Agar or 2) Bacto Blood Agar Base. *Haemophilus* does not grow on sheep or human blood agar. Rabbit or horse blood agar should be used when attempting to isolate and identify these microorganisms.

PROCEDURE
Materials Provided:
Bacto Differentiation Disks BVX
Bacto Differentiation Disks BV
Bacto Differentiation Disks BX

Purchase Separately:
Bacto Heart Infusion Agar
Bacto Tryptic Soy Agar
Bacto Tryptic Soy Broth

Materials Required but not Provided:
Forceps
37°C incubator

Isolation
1. Using aseptic technique, remove a Bacto Differentiation Disk BVX or BV from vial with flamed forceps.
2. a) Place BVX disk onto inoculated heart infusion agar or tryptic soy agar and press gently.
   b) Where blood agar or enriched chocolate agar is employed, place BV disk on surface of inoculated medium and press gently.
3. Incubate at 37°C for 18 – 24 hours.
4. Examine plates for growth around BVX or BV disks.

Results
Observe growth typical of *Haemophilus* species. On chocolate agar, colonies of *H. influenzae* and *H. parainfluenzae* appear as colorless, transparent, moist colonies while *H. parahaemolyticus* is soft, pearly and transluscent. On blood agar, *H. parahaemolyticus* will exhibit beta hemolysis. By contrast, no hemolysis is displayed by *H. influenzae* or *H. parainfluenzae*.

Differentiation
Only pure cultures should be used for differentiation of *Haemophilus* species.

Isolated and presumptively identified colonies may be picked either from 1) the area around BVX disk or BV disk, in the case of blood agar, or 2) an enriched medium such as enriched chocolate agar or tryptic soy agar with 1% Fildes enrichment.

If conventional cultural procedures, requiring an enriched medium, have been employed for isolating *Haemophilus*, it will be necessary to dilute the inoculum 1:100 in Bacto Tryptic Soy Broth to prevent carryover of growth factors to differential plate.

Streak cultures onto Bacto Heart Infusion Agar or Bacto Tryptic Soy Agar in a manner that will insure uniform distribution of the inoculum over the entire surface of the medium.

Method I

1. Using aseptic technique, remove a Bacto differentiation Disk BVX, BV and BX from vials with flamed forceps.
2. Place the disks on surface of the medium in the form of an equilateral triangle with approximately 30 – 35 mm between each disk.
3. Incubate at 37°C for 18 – 24 hours.
4. Examine plate for growth around the disks.

Method II

1. Using aseptic technique, remove a Bacto Differentiation Disk BV and BX from vials with flamed forceps.
2. Place on surface of the medium about 3 – 5 mm apart.
3. Incubate at 37°C for 18 – 24 hours.
4. Examine plates for growth between or around the disks.

Results

The pattern of reaction of *Haemophilus* species is as follows:

Haemophilus Species	Growth Factor Requirement	
	X	V
H. aegyptius	+	+
H. haemoglobinophilus	+	−
H. influenzae	+	+
H. parahaemolyticus	−	+
H. parainfluenzae	−	+

INTERPRETATION OF RESULTS

If Method I has been employed in differential test, growth patterns will be around individual disks in accordance with the requirements listed in the table. For example, *H. influenzae,* which requires both X and V factors, will grow around the BVX disk only. No growth should be apparent around the BV or BX disk. For species requiring only one factor, growth will be apparent around the BVX disks and around a BV or BX disk, i.e., *H. parainfluenzae* will grow around BVX and BV disk only. No growth should be apparent around the BX disk.

If Method II has been employed in the differential test, growth patterns will appear between both disks and/or around an individual disk. For example, *H. influenzae* will grow between the BX and BV disks only. No growth should be apparent around either disk. On the other hand, *H. parainfluenzae* will grow between the BV and BX and around the BV disks.

Results which are incongruent with the growth requirements stated in the Table may be indicative of one or both of following:

1. Improper dilution of cultures taken from enriched media. An inoculum that is too dilute may yield no growth. One which has been under-diluted may contain sufficient growth factors to affect growth over entire plate or in areas around one or more inappropriate disks on the plate.
2. Mixed cultures may result in growth outside the zone of inhibition of bacitracin of BV, BVX, or BX disk.

It should be remembered that only presumptive identification of the *Haemophilus* species is possible using Bacto Differentiation Disks BV, BVX and BX.

Determination of hemolysis and characteristics other than hemolysis may be required for confirmation of a presumptive identification particularly if *H. aegyptius* is involved as it has growth factor requirements and hemolytic characteristics identical to *H. influenzae*. Its slower growth rate, nonproduction of indole and usual source (conjunctivae) helps to distinguish *H. aegyptius* from *H. influenzae*. Biochemical and serological testing may also be used for species identification.

LIMITATIONS OF THE PROCEDURE
The use of Bacto Differentiation Disks BVX only serves as a means of primary isolation of *Haemophilus*. This isolation procedure cannot yield species identification.

Differential procedures employing Bacto Differentiation Disks BV, BX and BVX serve to identify the growth requirements of the *Haemophilus* species under study. Its contribution to overall cultural evaluation yields presumptive species identification. Due to the similarities in requirements for X and V factors among *Haemophilus* species, this procedure cannot be used as the sole criterion of presumptive identification.

Final species identification requires further biochemical testing and, in the case of *H. influenzae*, serological typing using Bacto H. Influenzae Antisera types.

STORAGE
Bacto Differentiation Disks BV, BX or BVX   2 – 8°C

QUALITY CONTROL
Identity Specifications

Bacto Differentiation Disks BV	Bacto Differentiation Disks BX	Bacto Differentiation Disks BVX
Appearance: 3/8″ white paper disks printed "BV" on both sides in black	3/8″ white paper disks, printed "BX" on both sides in black	3/8″ white paper disks, printed "BVX" on both sides in black

Typical Cultural Response using Method II and Bacto Heart Infusion Agar After 18 – 24 Hours at 35°C

Organism	BV	BX	Growth around disks BVX or BV-BX (3 – 5 mm apart)
Haemophilus influenzae ATCC® 35056	–	–	+
Haemophilus haemoglobinophilus ATCC® 19416	–	+	+
Haemophilus parainfluenzae ATCC® 7901	+	–	+
Staphylococcus aureus ATCC® 25923			zone of inhibition around all disks

REFERENCES
1. J. Infect. Dis., 21:392, 1917.
2. J. Exp. Med., 34:97, 1921.
3. Proc. Royal Soc. London, 122:352, 1937.
4. Am. J. Clin. Path., 37:319, 1962.

5. Applied Micro., 18:646 – 649, 1969.
6. Am. J. Med. Tech., 36:97 – 106, 1970.
7. Manual of Clinical Microbiology, 3rd Ed., Ch. 25, ASM, Wash., D.C., 1980.
8. Personal Communication, June, 1963.
9. Am. J. Clin. Path., 44:86, 1965.

PACKAGING

Bacto Differentiation Disks BV	6 vials	1622-33-7
	1 vial	1622-35-5
Bacto Differentiation Disks BVX	6 vials	1623-33-6
	1 vial	1623-35-4
Bacto Differentiation Disks BX	6 vials	1621-33-8
	1 vial	1621-35-6
Bacto Differentiation Disks BV-BX		
(3 vials each BV and BX)	6 vials	1624-32-6
Bacto Heart Infusion Agar	1 lb (454 g)	0044-01-7
	1/4 lb (114 g)	0044-02-6
	5 lb (2.27 kg)	0044-05-3

BACTO DIFFERENTIATION DISKS BACITRACIN

INTENDED USE

Bacto Differentiation Disks Bacitracin, offered as 1/4″ or 3/8″ paper disks which have 0.04 units bacitracin per disk, are used for the differentiation of group A streptococci from other beta-hemolytic streptococci.

HISTORY/PRINCIPLES

Because of the frequency of occurrence and potential pathogenicity of group A streptococci, it is essential that such isolates be identified as accurately and rapidly as possible. This may be accomplished by a combination of methods including the determination of their biochemical, hemolytic, and serologic characteristics. Hemolytic reactions alone, although valuable, are inadequate in differentiating beta-hemolytic streptococci (i.e., group A, B, C and G are beta-hemolytic).

It is generally accepted that a final identification of group A streptococci is considered incomplete unless specific serological grouping has been accomplished.[1] Serological grouping can be accomplished by the Lancefield precipitin test or fluorescent antibody technique.

However, because serological methods may be inconvenient for some laboratories, alternative methods for differentiating group A streptococci have been sought. One such method is based on the susceptibility of this group to minimal concentrations of bacitracin.

Maxted[2] reported a method of differentiating group A streptococci from other beta-hemolytic streptococci based on their susceptibility to bacitracin. Filter paper squares were dipped into a solution of the antibiotic and the bacitracin-impregnated squares placed on the surface of media inoculated with the test organisms. Following 24 hours incubation, group A streptococci, being susceptible to the amount of bacitracin in the square, showed a zone of inhibition of growth. Other hemolytic streptococci, resistant to the bacitracin, lacked or showed minimal zones of inhibition. He tested over 3,000 strains and reported 99.0% of group A streptococci were inhibited and 95.3% of the beta-hemolytic streptococci other than group A streptococci were not inhibited.

Levinson and Frank[3] continued Maxted's studies and confirmed his results. These authors used 6.4 mm paper disks impregnated with bacitracin. They recommended the use of this technique in laboratories where serological methods cannot be employed or when laboratories must differentiate large numbers of isolates.

It should be noted that some strains of groups B, C and G streptococci are also sensitive to bacitracin. Additional biochemical and/or serological tests must be used to confirm a false-positive result. Therefore, a positive bacitracin differentiation test should be considered as presumptive identification only of group A streptococci.

SPECIMEN PREPARATION
Mixed cultures or clinical specimens should not be used to determine susceptibility to bacitracin.

Blood agar prepared with Bacto Tryptic Soy Agar, Bacto Heart Infusion Agar or Bacto Tryptose Blood Agar Base is recommended for the presumptive identification of group A streptococci using Bacto Differentiation Disks Bacitracin.

The surface of the medium is inoculated with an overnight culture of the test organism from Bacto Brain Heart Infusion. A sterile swab is dipped into the culture and excess fluid squeezed out against the size of the tube. The medium is inoculated to obtain confluent growth by moving the swab back and forth across the surface while rotating the plate. The hemolytic reactions of the test organism should be determined prior to the performance of the bacitracin test. It must be remembered that even a small concentration of bacitracin present in differentiation disks is sufficient to inhibit many alpha-hemolytic streptococci including *S. pneumoniae.*

PROCEDURE
1. Streak plates with loop or swab so as to obtain confluent growth.
2. Using aseptic technique, place disk onto inoculated surface by dispensing 1/4" disk from magazine or removing 3/8" disk from vial with sterile forceps. Press disk down gently with forceps.
3. Invert and incubate plates at 35 – 37°C in CO_2 for 18 – 24 hours or until growth appears.
4. Observe for a zone of inhibition around the disks.

RESULTS
Group A streptococci susceptible to bacitracin are inhibited and show a zone of inhibition of growth around the disk while most other beta-hemolytic streptococci grow to the edge of the disk. Occasionally a strain belonging to groups B, C, F or G may show a minimal zone of inhibition by bacitracin. Any zone of inhibition, regardless of size, is positive. Biochemical and serological tests will reduce the possibility of those few erroneous results.

Facklam[1,4] has outlined 5 tests as criteria for identifying pathogenic streptococci: 1) hemolytic activity, 2) sensitivity to bacitracin in differentiation disks, 3) hydrolysis of hippurate, 4) hydrolysis of esculin and 5) tolerance to 6.5% NaCl. Using these criteria, bacitracin-susceptible group B streptococci can be differentiated from group A streptococci by their ability to hydrolyze hippurate.

The occasional minimal inhibition of strains belonging to groups B, C, F and G by bacitracin can still lead to reporting errors. Facklam has noted, however, that 99.4% of group A streptococci can be presumptively identified using the 5 tests described above.

Final identification of group A streptococci should be based on the serological test using Bacto streptococcal grouping antisera by the Lancefield precipitin test or by the direct fluorescent antibody technique using Bacto FA Streptococcus Groups.

STORAGE
Bacto Differentiation Disks Bacitracin 1/4" or 3/8"   Below 8°C

QUALITY CONTROL
Identity Specifications

	Bacto Differentiation Disks Bacitracin 1/4"	Bacto Differentiation Disks Bacitracin 3/8"
Appearance:	1/4" white paper disks, printed with "A" on both sides	3/8" white paper disks, printed with "B" on both sides

Typical Cultural Response On Bacto Tryptic Soy Agar w/5% Sheep Blood using Bacto Differentiation Disks Bacitracin
After 18 – 24 Hours at 35 – 37°C under 10% CO_2

Organism	Zone Size (1/4")	Zone Size (3/8")
Streptococcus pyogenes ATCC® 19615	≥14 mm	≥15 mm
Streptococcus agalactiae ATCC® 13813	≤14 mm	≤15 mm

REFERENCES
1. Manual of Clinical Microbiology, Second Ed., Ch. 8, ASM, Washington, D.C., 1974.
2. J. Clin. Path., 6:224, 1953.
3. J. Bact., 69:284, 1955.
4. CDC pamphlet "Presumptive Identification of Streptococcus Groups."

PACKAGING

Bacto Differentiation Disks Bacitracin 3/8"	6 vials	1631-33-6
	1 vial	1631-35-4
Bacto Differentiation Disks Bacitracin 1/4"	6 magazines	1631-91-5
Bacto Differentiation Disks Set O-B 3/8"	6 vials	1634-32-4
Bacto Brain Heart Infusion	1 lb (454 g)	0037-01-6
	1/4 lb (114 g)	0037-02-5
	5 lb (2.27 kg)	0037-05-2
	10 kg	0037-08-9
Bacto Tryptic Soy Agar	1 lb (454 g)	0369-01-4
	1/4 lb (114 g)	0369-02-3
	5 lb (2.27 kg)	0369-05-0
	10 kg	0369-08-7

BACTO DIFFERENTIATION DISKS ONPG

INTENDED USE
Bacto Differentiation Disks ONPG are recommended for detecting the presence of galactosidase in cultures of enteric bacilli. They are particularly useful in the rapid differentiation of late lactose fermenting Salmonella arizonae and Citrobacter from Salmo-

nella sp. For complete differentiation, additional biochemical and serological tests are required.

HISTORY/PRINCIPLES

Lactose fermentation as reported by Lubin and Ewing[1] depends on the action of 2 enzymes, a beta-galactosidase permease and a beta-galactosidase. In slow lactose fermenting microorganisms, lactose enters the cells slowly due to a permease deficiency. Bulow[2] reported that beta-galactosidase could be present in bacteria even if they were nonlactose fermentors, and that these organisms are often more or less impermeable to lactose. A rapid test for galactosidase is possible using ONPG because it enters the cell rapidly. The reaction is based on the hydrolysis of O-nitrophenyl beta-D-galactopyranoside by galactosidase to form O-nitrophenol and is characterized by the formation of a yellow color.

LeMinor and Ben Hamida[3] studied the ONPG reactions of the *Enterobacteriaceae* and found slow lactose fermenting *E. coli, Hafnia, Citrobacter, Klebsiella, E. cloacae, Serratia, S. arizonae, Shigella dysenteriae* and *S. sonnei* all gave positive ONPG tests. *Salmonella, Proteus, Providencia* and *Alcalescens* were found to always give negative ONPG tests. They determined that the ONPG test is essential for rapid differentiation of the *Enterobacteriaceae.*

Lubin and Ewing[1] performed ONPG tests on 452 cultures of *Salmonella* which included 166 stock cultures representing 151 serotypes. Out of these cultures, only 9 (1.95%) gave a positive ONPG test and 4 of these were lactose positive biotypes. They also tested 446 cultures of the *S. arizonae* group and found that 7.2% gave a negative ONPG test. The majority of the cultures gave a positive test within 20 minutes.

Pickett and Goodman[4] studied the ONPG reaction of the *Citrobacter* and stated that 84% may be expected to give a positive test within 60 minutes and 94% should give a positive test within 4 hours.

Wilson, Padron and Dockstader[5,6] devised an effective screening procedure for *Salmonella* which employs only the ONPG test and an agglutination test using Salmonella polyvalent flagellar antiserum. By this procedure most of the non-*Salmonella* are eliminated within 24 hours and the *Salmonella* are rarely discarded. They tested ONPG disks in this procedure and found them to give results in complete agreement with the conventional test procedure.

METHOD OF PREPARATION

1. To test for galactosidase using Bacto Differentiation Disks ONPG, dispense 0.2 ml of 0.85% NaCl in a Kahn tube.
2. Aseptically suspend a loopful of culture into the saline.
3. Place a Bacto Differentiation Disk ONPG in the tube and incubate for 20 minutes at 37°C.
4. Make observations and hold for 4 hours before making a negative determination. A positive reaction is indicated by the formation of a yellow color.

The ONPG Test Reactions of Enteric Bacteria

Organisms	ONPG Reaction	Organisms	ONPG Reaction
Salmonella arizonae	+	Enterobacter (Aerobacter)	+
Citrobacter	+	Klebsiella	+
Salmonella sp.	−	Serratia	+
Escherichia	+	Proteus	−
Shigella	±	Providencia	−

STORAGE
Bacto Differentiation Disks ONPG   2 – 8°C

QUALITY CONTROL
Identity Specifications
Appearance: white paper disks 1/4" in diameter, no printing on disk

Typical Cultural Response of Bacto Differentiation Disks, ONPG in 0.85% Saline After up to 4 Hours at 37°C

Organism	Reaction
Salmonella arizonae ATCC® 13314	+
Salmonella typhimurium ATCC® 14028	−

+ = positive, yellow color
− = negative, no color change

REFERENCES
1. Publ. Hlth. Lab., 22:83, 1964.
2. Acta. Pathol. Microbiol. Scand., 60:387, 1964.
3. Annal. Inst. Pasteur, 102:267, 1962.
4. Appl. Microbiol., 14:178, 1966.
5. Bact. Proc., A7:p. 2, 1969.
6. Paper presented at 69th Ann. Meeting of the ASM, 1969.

PACKAGING
Bacto Differentiation Disks ONPG	6 × 50 disks	1635-33-2
	1 × 50 disks	1635-35-0

BACTO DIFFERENTIATION DISKS OPTOCHIN

INTENDED USE
Bacto Differentiation Disks Optochin, offered as 1/4" or 3/8" paper disks which have been impregnated with a solution of ethylhydro-cupreine hydrochloride are used for differentiating *Streptococcus pneumoniae* (pneumococci) from alpha-hemolytic streptococci.

HISTORY/PRINCIPLES
Streptococcus pneumoniae, like other streptococci, is a common inhabitant of the upper respiratory tract. Pneumococci have been implicated in respiratory infections, subacute bacterial endocarditis and pneumonia. On blood agar, pneumococci yield hemolytic reactions which are indistinguishable from those of other alpha-hemolytic streptococci. Differentiation and identification of these microorganisms, therefore, require serological or biochemical methods. Serological tests for the identification of pneumococci, particularly the Neufeld Quellung reaction, are generally accepted as the method of choice.

Biochemical tests also prove valuable in differentiating and identifying pneumococci. Specifically, the three biochemical tests used are fermentation of inulin, bile solubility and optochin sensitivity.

The susceptibility of pneumococci to optochin has long been established[1,2] but its application to diagnostic procedures was not explored until 1955. Bowers and Jeffries[3] reported on the use of dried paper disks saturated with a solution of optochin for differentiating pneumococci from other streptococci, basing the differentiation on the susceptibility of these organisms to the chemical. The disks were placed on the inoculated surface of a suitable medium containing blood. Pneumococci were inhibited as indicated by a zone of inhibition of growth around the disks. No inhibition was observed with other alpha-hemolytic streptococci. The authors reported on 695 strains of alpha-hemolytic streptococci, 243 being bile soluble. Only 2 strains showed lack of correlation between bile solubility and optochin susceptibility. They pointed out that the method is less time consuming than the bile solubility test for identifying pneumococci and that the test can be performed using overnight plate cultures of sputum or pus containing mixed bacterial flora as well as from pure cultures. However, the use of pure cultures is preferred.

Bowen, Thiele, Stearman and Schaub[4] also examined the usefulness of optochin disks and corroborated the findings of Bowers and Jeffries. These authors also found that although some strains of alpha-hemolytic streptococci gave small zones of inhibition, they were clearly distinguishable from the sharp, well-defined, wider zones of the pneumococci.

SPECIMEN PREPARATION
Mixed cultures or clinical specimens should not be used to determine susceptibility to optochin.

Blood agar prepared with Bacto Tryptic Soy Agar, Bacto Heart Infusion Agar or Bacto Tryptose Blood Agar Base is recommended for the presumptive identification of pneumococci using Bacto Differentiation Disks Optochin.

The surface of the medium is inoculated with an overnight culture of the test organism from Bacto Brain Heart Infusion. A sterile swab is dipped into the culture and excess fluid squeezed out against the side of the tube. The medium is inoculated to obtain confluent growth by moving the swab back and forth across the surface while rotating the plate. The hemolytic reactions of the test organism should be determined prior to the performance of the optochin test.

PROCEDURE
1. Steak plates with loop or swab so as to obtain confluent growth.
2. Using aseptic technique, place disk onto inoculated surface by dispensing 1/4" disk from magazine or removing 3/8" disk from vial with sterile forceps. Press disk down gently with forceps.
3. Invert and incubate plates at 35 – 37°C in CO_2 for 18 – 24 hours or until growth appears.
4. Observe for a zone of inhibition around the disks.

The susceptibility of pneumococci to optochin can be evaluated as providing only presumptive identification of this species. It has been shown that some strains of other alpha-hemolytic streptococci are susceptible (but only slightly) to the concentration of optochin contained in the differentiation disks.

Final identification of *S. pneumoniae* is made by serogrouping the culture by the Neufeld Quellung reaction with Bacto pneumococcal grouping sera.

STORAGE
Bacto Differentiation Disks Optochin 1/4" or 3/8"   Below 8°C

QUALITY CONTROL
Identity Specifications

Bacto Differentiation Disks Optochin 1/4"	Bacto Differentiation Disks Optochin 3/8"
Appearance: 1/4" white, paper disks printed with "P" on both sides	3/8" white, paper disks, printed with "O" on both sides.

Typical Cultural Response on Bacto Tryptic Soy Agar w/5% Sheep Blood using Bacto Differentiation Disks Optochin
After 18 – 24 hours at 35 – 37°C

Organism	Zone Size (1/4")	Zone Size (3/8")
Streptococcus pneumoniae ATCC® 6303	≥14 mm	≥15 mm
Streptococcus pyogenes ATCC® 19615	≤14 mm	≤15 mm

REFERENCES
1. J. Exp. Med. 22:269, 1915.
2. Serological Studies on the Pneumococci, Munksgaard, Copenhagen, Oxford University Press, London, 1943.
3. J. Clin. Path., 8:58, 1955.
4. J. Lab. Clin. Med., 49:641, 1957.

PACKAGING

Bacto Differentiation Disks Optochin 3/8"	6 vials	1632-33-5
	1 vial	1632-35-3
Bacto Differentiation Disks Optochin 1/4"	6 magazines	1632-91-4
Bacto Differentiation Disk Set O-B 3/8"	6 vials	1634-32-4
Bacto Heart Infusion	1 lb (454 g)	0037-01-6
	1/4 lb (114 g)	0037-02-5
	5 lb (2.27 kg)	0037-05-2
	10 kg	0037-08-9
Bacto Tryptic Soy Agar	1 lb (454 g)	0369-01-4
	1/4 lb (114 g)	0369-02-3
	5 lb (2.27 kg)	0369-05-0
	10 kg	0369-08-7

BACTO DIFFERENTIATION DISKS OXIDASE

INTENDED USE
Bacto Differentiation Disks Oxidase are paper disks containing para-aminodimethylaniline to detect oxidase production by microorganisms.

HISTORY/PRINCIPLES
See discussion under Bacto Oxidase Reagent.

PROCEDURE
1. Obtain an 18 – 48 hour culture of organism(s) to be tested. Colonies should be well isolated but do not have to be in pure culture. Suspected colonies of *Neisseria gonorrhoeae* will appear small, raised, convex, translucent, greyish-white, with undulate margins.
2. Apply a Bacto Differentiation Disk Oxidase using sterile forceps touching the colony in question.

3. Apply one drop distilled or deionized water to disk to supply moisture.
4. Observe for formation of a pink to maroon to almost black color of the colony within 20 minutes.

STORAGE
Bacto Differentiation Disks Oxidase $2-8°C$

QUALITY CONTROL
Identity Specifications
Appearance: tan to grey paper disks, 3/8″ in diameter, no printing

Typical Cultural Response with 18 – 48 Hour Culture
After up to 20 Minutes at 25 – 30°C

Organism	Reaction
Escherichia coli ATCC® 25922	−
Neisseria gonorrhoeae ATCC® 19424	+
Pseudomonas aeruginosa ATCC® 27853	+
Staphylococcus aureus ATCC® 25923	−

+ = positive, pink to maroon to almost black
− = negative, no color change

PACKAGING
Bacto Differentiation Disks Oxidase	6 × 50 disks	1633-33-4
	1 × 50 disks	1633-35-2

BACTO DISINFECTANT TEST BROTH AOAC
BACTO SYNTHETIC BROTH AOAC

INTENDED USE
Bacto Disinfectant Test Broth AOAC and Bacto Synthetic Broth AOAC are used for determining phenol coefficients of disinfectants.

HISTORY/PRINCIPLE
Bacto Disinfectant Test Broth AOAC is prepared according to the formula of the A.O.A.C.[1,2] Bacto Synthetic Broth AOAC is a chemically defined broth as recommended by A.O.A.C.[3] Both media contain the nutrients necessary for the growth of the test organisms, *Salmonella typhi*, *Staphylococcus aureus* and *Pseudomonas aeruginosa*.

FORMULAE
BACTO DISINFECTANT TEST BROTH AOAC
DEHYDRATED

Ingredients per liter

Bacto Peptamin . . . . 10 g
    Peptic Digest Animal Tissue USP
Bacto Beef Extract . . . . 5 g
Sodium Chloride . . . . 5 g

Final pH 6.8 ± 0.2 at 25°C.

One pound will make 22.7 liters of medium.

BACTO SYNTHETIC BROTH AOAC
DEHYDRATED

Ingredients per liter

L-Cystine	0.05 g	L-Glutamic Acid	1.3 g
DL-Methionine	0.37 g	L-Aspartic Acid	0.45 g
L-Arginine	0.4 g	DL-Phenylalanine	0.26 g
DL-Histidine	0.3 g	DL-Tryptophane	0.05 g
L-Lysine	0.85 g	L-Proline	0.05 g
L-Tyrosine	0.21 g	Sodium Chloride	3 g
DL-Threonine	0.5 g	Potassium Chloride	0.2 g
DL-Valine	1 g	Magnesium Sulfate	
L-Leucine	0.8 g	Anhydrous Reagent	0.05 g
DL-Isoleucine	0.44 g	Potassium Phosphate	1.5 g
Amino Acetic Acid	0.06 g	Disodium Phosphate	4 g
DL-Serine	0.61 g	Thiamine Hydrochloride	0.01 g
DL-Alanine	0.43 g	Nicotinamide	0.01 g

Final pH 7.1 ± 0.1 at 25°C.

One pound will make 26.7 liters of medium.

METHOD OF PREPARATION
Bacto Disinfectant Broth AOAC

1. Suspend 20 g in 1 liter distilled or deionized water and dissolve completely.
2. Distribute into tubes.
3. Sterilize in the autoclave for 15 minutes at 15 lbs pressure (121°C).
4. Use for transferring daily cultures for test per A.O.A.C.[3]

Bacto Synthetic Broth AOAC

1. Suspend 17 g in 1 liter distilled or deionized water and heat to boiling for 1 – 2 minutes.
2. Dispense in 10 ml amounts into 20 × 150 mm culture tubes.
3. Sterilize in the autoclave for 20 minutes at 15 lbs pressure (121°C).
4. Add aseptically 0.1 ml of Bacto Dextrose Solution 10% before inoculating the tubes.

STORAGE

Bacto Disinfectant Broth AOAC	Below 30°C
Bacto Synthetic Broth AOAC	Below 30°C
Prepared media	2 – 8°C

QUALITY CONTROL

Identity Specifications

	Disinfectant Test Broth AOAC	Synthetic Broth AOAC
Dehydrated powder:	light tan, homogeneous, free-flowing	white, homogeneous, free-flowing
% solution:	2.0%	1.7%
pH at 25°C:	6.8 ± 0.2	7.1 ± 0.1
Prepared medium:	light to medium amber, clear to very slightly opalescent	colorless, clear

Typical Cultural Response After 40 – 48 Hours at 35°C

Organism	Growth
Salmonella typhi ATCC® 6539	good to excellent, good neutralization of disinfectant when added

REFERENCES
1. AOAC, Ninth Ed., 1960.
2. Journal of AOAC 47:176:1964.
3. Methods of Analysis AOAC, 13th Ed., 1980.

PACKAGING

Bacto Disinfectant Test Broth AOAC	1 lb (454 g)	0351-01-4
Bacto Synthetic Broth AOAC	1 lb (454 g)	0352-01-3
Bacto Dextrose Solution 10%	12 × 10 ml	0155-61-9

BACTO DORSET EGG MEDIUM

INTENDED USE
Bacto Dorset Egg Medium is a prepared, coagulated egg medium used for culturing mycobacteria.

HISTORY/PRINCIPLES
Bacto Dorset Egg Medium, a modification of the whole egg medium described by Dorset,[1] is a nonselective medium well suited for maintenance of pure cultures of *Mycobacterium*.

FORMULA

BACTO DORSET EGG MEDIUM
Ingredients per liter

Whole Egg Suspension .... 950 ml
Glycerol .... 50 ml

This product is available as a prepared tube medium.

STORAGE
Prepared medium   2 – 8°C

QUALITY CONTROL

Identity Specifications

Reaction of medium: pH 6.8 – 7.4 at 25°C
Prepared medium: yellow, opaque slant

Typical Cultural Response on Bacto Dorset Egg Medium
After 2 Weeks at 35°C under 10% CO_2

Organism	Growth
Mycobacterium bovis ATCC® 19210	good to excellent
Mycobacterium fortuitum ATCC® 6841	good to excellent
Mycobacterium tuberculosis H37Rv (ATCC® 25618)	good to excellent

REFERENCE
1. Am. Med., 3:555, 1902.

PACKAGING

Bacto Dorset Egg Medium	12 tubes	1006-34-2

BACTO DROSOPHILA MEDIUM M

INTENDED USE

Bacto Drosophila Medium M is a complete dehydrated medium for the cultivation of *Drosophila*. It is prepared according to a slight modification of the formula of Mickey[1] and is similar to the formulations of Demeric[2] and Lewis.[3]

FORMULA

BACTO DROSOPHILA MEDIUM M
DEHYDRATED

Ingredients per liter

Dried Brewers Yeast . . . . 13.3 g
Corn Meal . . . . 133 g
Karo® Syrup . . . . 180 g
Methyl Parahydroxybenzoate . . . . 0.093 g
Bacto Agar . . . . 13.3 g

Final pH 6.0 ± 0.2 at 25°C.

One pound will make 1.3 liters of medium.

PROCEDURE

1. To rehydrate, suspend 34 g in 100 ml distilled water.
2. Heat to boiling with constant stirring for several minutes. Dispense the medium into containers as desired.
3. For the culture of *Drosophila,* pour molten medium into bottles or flasks to a depth of about 2 inches for quart and liter size containers, and to about 0.5 inches for ounce size containers.
4. Cut strips of filter paper to the length of the bottle. Insert strips to provide a place for flies to pupate.
5. Inoculate the surface of medium with a heavy suspension of live yeast to induce adult flies to lay eggs.
6. Add a male and female fly possessing desired genetic characters to the bottles for the purpose of mating.
7. Incubate cultures for about 12 days at which time the flies should begin to emerge.

STORAGE

Bacto Drosophila Medium M   15 – 30°C

QUALITY CONTROL

Identity Specifications

Dehydrated powder: light beige, homogeneous
Reaction of 34% solution: pH 6.0 ± 0.2 at 25°C
Prepared medium: light to medium amber, opaque, insoluble

Typical Cultural Response on Bacto Drosophila Medium M
After 40 – 48 Hours at 25 – 30°C

Organism	Growth
Saccharomyces uvarum ATCC® 9080	good to excellent

REFERENCES

1. Personal Communication.
2. Biology of Drosophila by Demeric.
3. Personal Communication.

PACKAGING
Bacto Drosophila Medium M   1 lb (454 g)   0683-01-3

DUBOS MEDIA

BACTO DUBOS BROTH BASE
BACTO DUBOS MEDIUM ALBUMIN
BACTO DUBOS MEDIUM SERUM

INTENDED USE
Bacto Dubos Broth Base is used with Bacto Dubos Medium Albumin or Bacto Dubos Medium Serum for rapidly cultivating pure cultures of *Mycobacterium tuberculosis.*

HISTORY
Dubos broth is prepared according to the Dubos, Fenner and Pierce[1] modification of the medium originally described by Dubos and Davis[2] and Dubos and Middlebrook.[3]

Dubos and Davis, in a study of factors influencing the growth of *M. tuberculosis,* described a liquid medium in which growth of the tubercle bacillus was obtained from dilute inocula in 3 – 5 days. While suitable for cultivating pure cultures, the authors cautioned against using the medium for diagnostic purposes.[6–8]

PRINCIPLES
The addition of albumin fraction V to the medium was found to facilitate growth of tubercle bacilli but not to markedly increase the total amount of growth produced. Less pure preparations of the protein increased the amount of growth as well as initiated growth from minute inocula.

Previous media had demonstrated growth of *M. tuberculosis* as a surface pellicle. Dubos broth provided readily dispersible, subsurface growth, with pellicle formation occurring only in old cultures. The authors found this type of growth valuable in obtaining a uniform suspension of organisms for use in other tests.

Bacto Dubos Broth Base prepared with Bacto Dubos Medium Serum will generally initiate growth from a smaller inoculum and yield more luxuriant growth than the basal medium enriched with Bacto Dubos Medium Albumin. Growth is generally more granular with the serum enrichment, while it is more diffuse with the albumin enrichment. With either enrichment, growth is evident in 3 – 5 days from a 10^{-3} inoculum of *M. tuberculosis* and in 10 – 15 days from a 10^{-7} inoculum.

FORMULA
BACTO DUBOS BROTH BASE
DEHYDRATED
Ingredients per liter

Bacto Casitone	0.5 g	Ferric Ammonium Citrate	50 mg
Bacto Asparagine	2 g	Magnesium Sulfate	10 mg
Tween 80	0.2 g	Calcium Chloride	0.5 mg
Monopotassium Phosphate	1 g	Zinc Sulfate	0.1 mg
Disodium Phosphate (Anhyd.)	2.5 g	Copper Sulfate	0.1 mg

Final pH 6.6 ± 0.2 at 25°C.

Five hundred grams will make 76.9 liters of medium.

Bacto Dubos Medium Albumin is a filter-sterilized, 5% solution of albumin fraction V from bovine plasma and 7.5% dextrose in normal saline.

Bacto Dubos Medium Serum is a filter-sterilized preparation of beef serum with 7.5% dextrose added.

METHOD OF PREPARATION
1. Dissolve 1.3 g in 180 ml distilled or deionized water (or 170 ml water with 10 ml Bacto Glycerol added).
2. Sterilize in the autoclave for 15 minutes at 15 lbs pressure (121°C). Cool below 50°C.
3. Aseptically add 20 ml Bacto Dubos Medium Albumin or Bacto Dubos Medium Serum. Mix thoroughly.
4. Aseptically dispense 5 – 7 ml amounts into 16 – 20 mm diameter test tubes.
5. Incubate for 24 hours to test sterility.

STORAGE
Bacto Dubos Broth Base	Below 30°C
Bacto Dubos Medium Enrichments	2 – 8°C
Prepared media	2 – 8°C

QUALITY CONTROL
Identity Specifications
Dehydrated powder:	light beige, homogeneous, free-flowing
Enrichments:	very light amber, clear
Reaction of .65% solution:	pH 6.6 ± 0.2 at 25°C
Prepared medium:	very light amber, clear, may show a slight precipitate

Typical Cultural Response After 2 weeks at 35°C
Organism	Growth
Mycobacterium tuberculosis H37Rv	excellent
Mycobacterium kansasii ATCC® 12478	excellent
Mycobacterium gordonae ATCC® 14470	excellent
Mycobacterium avium ATCC® 25291	excellent
Mycobacterium smegmatis ATCC® 14468	excellent

REFERENCES
1. Am. Rev. Tuberculosis, 61:66, 1950.
2. J. Expl. Med., 83:409, 1946.
3. Am. Rev. Tuberculosis, 56:334, 1947.
4. Proc. Soc. Exp. Biol. Med., 62:298, 1946.
5. J. Lab. Clin. Med., 32:842, 1947.
6. Proc. Soc. Exp. Biol. Med., 65:210, 1947.

PACKAGING
Bacto Dubos Broth Base	500 g	0385-17-6

Enrichments
Bacto Dubos Medium Albumin	12 × 20 ml	0309-64-1
Bacto Dubos Medium Serum	12 × 20 ml	0292-64-0

Prepared Media
Bacto Dubos Albumin Broth	12 tubes	1022-34-2
Bacto Dubos Serum Broth	12 tubes	0385-34-5

DUBOS OLEIC MEDIA
BACTO DUBOS OLEIC AGAR BASE
BACTO DUBOS OLEIC ALBUMIN COMPLEX

INTENDED USE
Bacto Dubos Oleic Agar Base is used with Bacto Dubos Oleic Albumin Complex and penicillin for isolating *Mycobacterium tuberculosis* and determining its sensitivity to chemotherapeutic agents.

HISTORY
Dubos and Middlebrook[1] described an agar medium suitable for primary isolation and cultivation of the tubercle bacillus and for studying colony morphology. In comparative studies, Wallace and Erlich[2] and Byham[3] reported that Dubos oleic albumin agar medium was superior to other media studied for primary isolation work.

PRINCIPLES
The basal medium is prepared without glycerol or dextrose to discourage growth of commensal organisms that might have survived the process of concentrating the specimen. Asparagine is added in minimal quantities to provide maximum selectivity.

FORMULA
BACTO DUBOS OLEIC AGAR BASE
DEHYDRATED
Ingredients per liter

Bacto Casitone	0.5 g	Magnesium Sulfate	10 mg
Bacto Asparagine	1 g	Calcium Chloride	0.5 mg
Monopotassium Phosphate	1 g	Zinc Sulfate	0.1 mg
Disodium Phosphate (Anhyd.)	2.5 g	Copper Sulfate	0.1 mg
Ferric Ammonium Citrate	50 mg	Bacto Agar	15 g

Final pH 6.6 ± 0.2 at 25°C.

Five hundred grams will make 25 liters of medium.

Bacto Dubos Oleic Albumin Complex is a sterile enrichment consisting of a 0.05% solution of alkalinized oleic acid in a 5% solution of albumin fraction V in normal saline (0.85% NaCl).

METHOD OF PREPARATION
1. Suspend 4 g in 180 ml distilled or deionized water and heat to boiling to dissolve completely.
2. Sterilize in the autoclave for 15 minutes at 15 lbs pressure (121°C). Cool to 50 – 55°C.
3. Aseptically add 20 ml Bacto Dubos Oleic Albumin Complex and 5,000 – 10,000 units of penicillin (25 – 50 units/ml). Mix thoroughly.
4. Aseptically dispense into sterile tubes and allow to solidify in a slanted position.

STORAGE
Bacto Dubos Oleic Agar Base	Below 30°C
Bacto Dubos Oleic Albumin Complex	2 – 8°C
Prepared medium	2 – 8°C

QUALITY CONTROL

Identity Specifications

Dehydrated powder:	beige, homogeneous, free-flowing
Enrichment:	light amber, clear
Reaction of 2% solution:	pH 6.6 ± 0.2 at 25°C
Prepared medium:	light amber, opalescent, with a fine precipitate

Typical Cultural Response on Bacto Dubos Oleic Agar
After 2 – 3 Weeks at 35°C

Organism	Growth	Morphology
Mycobacterium tuberculosis H37Rv	Excellent growth	flat, rough, corded, dry and usually non-pigmented colonies
Mycobacterium kansasii ATCC® 12478	Excellent growth	photochromogenic with flat, smooth or somewhat granular surfaces and regular or slightly undulating margins.
Mycobacterium gordonae ATCC® 14470	Excellent growth	Smooth yellow to orange colonies; occasionally rough.
Mycobacterium avium ATCC® 25291	Excellent growth	Smooth, thin, non-pigmented colonies
Mycobacterium smegmatis ATCC® 14468	Excellent growth	Rough or smooth white, domed colonies

REFERENCES

1. Am. Rev. Tuberculosis, 56:334, 1942.
2. Am. Rev. Tuberculosis, 61:563, 1950.
3. Am. J. Clin. Path., 20:678, 1950.

PACKAGING

Bacto Dubos Oleic Agar Base	500 g	0373-17-0
Bacto Dubos Oleic Albumin Complex	12 × 20 ml	0375-64-0
Bacto Dubos Oleic Agar	12 tubes	0373-34-9

DYES AND INDICATORS

Dyes and indicators are essential to the preparation of most differential culture media. In such media, the dyes may act as bacteriostatic agents, as inhibitors of growth, or as indicators of changes in the degree of acidity or alkalinity of the substrate. Therefore, only dyes of known purity and known dye content should be used in the preparation of culture media.

Great care is necessary in the preparation of differential or selective media, even with standardized dyes. All ingredients of the medium must be used in the amounts specified in the formula and the reaction of the medium must be adjusted with extreme care in order to obtain proper results. These dyes and indicators, subject to both bacteriological and chemical assays, are nontoxic in the concentration usually employed in media.

The actual dye content of each dye is shown on each label.

REAGENTS

Bacto Basic Fuchsin (For Endo Agar and Other Basic Fuchsin Media)

Bacto Basic Fuchsin is recommended for the preparation of Endo agar and other basic fuchsin media. In Endo agar, this dye is partially decolorized by sodium sulfite. On this medium lactose-fermenting organisms form red colonies and change the color of the surrounding medium from faint pink to red; typical *Escherichia coli* colonies, in addition to being red in color, generally exhibit a brilliant metallic sheen. Organisms not fermenting lactose form uncolored colonies which do not alter the appearance of the medium. Bacto Basic Fuchsin is used in the preparation of many staining formulations for mycobacteria.

Bacto Brilliant Green (For Brilliant Green Media)

Bacto Brilliant Green is recommended for use in preparing the brilliant green bile media and media not containing bile acids. This dye, in the proper concentration in media, exhibits the desired selectivity, being inhibitive to gram-positive, nonfermenting organisms, permitting the unrestricted development of the colon-aerogenes group.

Bacto Crystal Violet (Gentian Violet)

Crystal violet and its impure form, gentian violet, have long been used in culture media because of their selective inhibitory action toward the gram-positive bacteria. Although Bacto Crystal Violet has a wide range over which it is not significantly toxic to the gram-negative bacteria and is still definitely bacteriostatic toward the gram-positive organisms, it should be used with extreme care to preserve the proper ratio of dye to medium and to maintain the correct reaction. Bacto Crystal Violet is used in the preparation of the Gram Crystal Violet stains.

In tryptose agar, Bacto Crystal Violet may be used in a concentration of 1:700,000 for isolation of *Brucella* strains from contaminated milk. Bacto Crystal Violet is used in Bacto Supplement A, a selective enrichment for gonococci and meningococci.

Bacto Eosin Y

Bacto Eosin Y is recommended for use in conjunction with Bacto Methylene Blue in the preparation of eosin methylene blue agar according to the Levine formula.[1]

1. Bull. 62, Iowa Eng. Exp. Sta., 1921.

Bacto Methylene Blue

Bacto Methylene Blue is recommended for use in conjunction with Bacto Eosin Y in the preparation of eosin methylene blue agar according to the Levine formula.[1] Bacto Methylene Blue is used for preparing a general bacteriological staining solution.

1. Bull. 62, Iowa Eng. Exp. Sta., 1921.

Product Name	Synonym	Color Index Number	Packaging	
Bacto Basic Fuchsin	Pararosanilin	C.I. 42500	25 g	0191-13-4
			100 g	0191-15-2
Bacto Brilliant Green	C.I. Basic Green 1	C.I. 42040	25 g	0192-13-3
			100 g	0192-15-1
Bacto Crystal Violet (Gentian Violet)	C.I. Basic Violet 3	C.I. 42555	25 g	0193-13-2
			100 g	0193-15-0
Bacto Eosin Y	C.I. Acid Red 87	C.I. 45380	25 g	0194-13-1
			100 g	0194-15-9
Bacto Methylene Blue	C.I. Basic Blue 9	C.I. 52015	25 g	0195-13-0
			100 g	0195-15-8

Product	Chemical Name	A*	pH Range	pK	Acid Color	Alkali Color	Packaging
Bacto Brom Phenol Blue	3,3',5,5'-Tetrabromophenol-sulfonphthalein	14.9	3.0 – 4.9	3.98	Yellow	Blue	5 g 0199-11-8
Bacto Brom Cresol Green	3,3',5,5'-Tetrabromo-m-cresolsulfonphthalein	14.3	3.8 – 5.4	4.67	Yellow	Blue	5 g 0200-11-5
Bacto Brom Cresol Purple	5,5'-Dibromo-o-cresol-sulfonphthalein	18.5	5.2 – 6.8	6.3	Yellow	Purple	1 g 0201-10-5 / 5 g 0201-11-4
Bacto Brom Thymol Blue	3,3'-Dibromothymolsulfonphthalein	16.0	6.0 – 7.6	7.0	Yellow	Blue	5 g 0202-11-3
Bacto Cresol Red	o-Cresolsulfonphthalein	26.2	0.2 – 1.8 / 7.2 – 8.8	— / 8.3	Red / Yellow	Yellow / Red	5 g 0204-11-1
Bacto Phenol Red	Phenolsulfonphthalein	28.2	6.8 – 8.4	7.9	Yellow	Red	5 g 0203-11-2

* ____ ml 0.01N NaOH required per 0.1 g indicator. Dilute to 250 ml with distilled or deionized water for 0.04% solution for use as indicator for colorimetric pH determinations.

Product	pH Range	Acid Color	Alkali Color	Preparation	Packaging
Litmus C.I. (ED. 1) 1242	5.6 – 8.3	Red	Blue	1 g in 100 ml water	25 g 0209-13-4
Methyl Red C.I. 13020	4.4 – 6.2	Red	Yellow	0.1 g in 300 ml ethanol and dilute to 500 ml with distilled water	25 g 0207-13-6
Neutral Red C.I. 50040	6.8 – 8.0	Red	Yellow	0.1 g in 60 ml ethanol and dilute to 100 ml with distilled water	25 g 0208-13-5
Malachite Green C.I. 42000	Bacteriostatic Agent			As required	25 g 0629-13-6 / 100 g 0629-15-4
Resazurin	Redox Indicator —	Oxidized blue	Reduced pink	As required	25 g 0704-13-4
Rosolic Acid C.I. 43800	6.8 – 8.2	Yellow	Red	As required with m FC Broth Base and m FC Agar	6 × 1 g 3228-09-1
TTC Triphenyl-tetrazolium chloride	Redox Indicator —	Oxidized colorless	Reduced red	As required	10 g 0643-12-9 / 25 g 0643-01-8

E COLI O/OK ANTISERA
BACTO E COLI O ANTISERA
BACTO E COLI OK ANTISERA

INTENDED USE

Bacto E. Coli Antisera are prepared for use in serological screening, presumptive identification and confirmation of members of the genus and species *Escherichia coli.* Bacto E. Coli OK Antisera are employed in the identification of the K or envelope antigens in the slide agglutination technique, while Bacto E. Coli O Antisera are employed in the identification of the somatic antigens in the tube agglutination test.

SUMMARY AND EXPLANATION OF THE TESTS

E. coli are isolated repeatedly from both human and animal sources and are found widely distributed in nature (foods, water, soil, etc.).[1,2] Although one of the more common groups of organisms in our environment, and usually nonpathogenic, they can cause clinical infections in man. *E. coli* have been incriminated in or have been found to be the etiological agents in cases of cystitis, pyelitis, pyelonephritis, gall bladder infections, septicemia, meningitis, endocarditis, summer diarrhea of children and adults and epidemic infant diarrhea.[3] Of these infections, epidemic infant diarrhea is of foremost clinical significance.

Prior to the development of serological methods for the identification of pathogenic *E. coli,* biochemical tests were used in an attempt to differentiate from *E. coli* isolated from normal adults and children. Biochemical tests alone were inadequate in making such a distinction; nonpathogenic and pathogenic *E. coli* may possess different antigenic identities but they may have identical biochemical characteristics.

The delineation of "enteropathogenic" cultures from the "nonpathogens" was facilitated by the work of Kauffmann et al[4] which resulted in a serological typing system for *E. coli.* Later work of Orskov[5] and studies of Ewing and Edwards[1] and Ewing[6] contributed to the development of an antigenic schema for this group of organisms which made possible the identification and characterization of nonpathogenic *E. coli* and those serotypes isolated from cases of diarrheal disease.

Note: *E. coli* has recently been the subject of attention in acute diarrhea in older children and adults. Involved are the enteroinvasive *E. coli,* (*Shigella*-like syndrome) and the enterotoxigenic *E. coli* (toxin producers).

The following are some of the serological O groups which have been found by investigators to be incitants of acute infant, child and adult diarrhea:

Enteropathogenic:	026	0114	0127
	055	0119	0128
	086	0125	0142
	0111	0126	

Enteroinvasive:	028ac	0136	0152
	0112ac	0143	0164
	0124	0144	

Enterotoxigenic:

06	025	0109
08	027	0148
015	075	0159
023	078	

Note: It has been suggested, from several sources, that the serotyping of *E. coli*, employing commercially available antisera, be discontinued. However, since antitoxins are not available and since techniques such as the rabbit ilial loop technique, the YI adrenal cell test, the hemagglutination system, the Chinese hamster ovary cell test and ganglioside immunosorbent assay are not readily available in most laboratories, and further, since serotyping is currently the only reasonable and available tool, the philosophy of discontinuance of serotyping is open to question.

Although the above serogroups of enteropathogenic *E. coli* have been established through epidemiological investigation, they should not be considered as the only serogroups responsible for epidemic infant diarrhea. The possibility that other serogroups may be implicated still remains.

Due to the nature of the disease it is imperative that the etiological agent be properly and accurately identified. Epidemic or sporadic infant diarrhea may be caused by a variety of microorganisms. *Salmonella* and *Shigella* have been incriminated in this type of infection and occasionally bacteria from other groups or viral agents have been suspect. However, the majority of such outbreaks have been attributed to enteropathogenic *E. coli* (EEC). The use of serological methods provide evidence of the presence of EEC thus permitting prompt and specific therapeutic action. These methods may also assist in the establishment of possible additional serotypes of *E. coli* involved in gastroenteritis and are an invaluable aid in epidemiological studies of nosocomial outbreaks.

PRINCIPLES OF THE PROCEDURE

The serological technique is based upon the reaction of specific antiserum with its homologous antigen. Although the validity of this technique has been proven through immunological studies, serological techniques, when applied to identification of microorganisms, are not absolute. While the specificity of serological methods is not absolute, serogrouping or serotyping of *E. coli*, taken with biochemical characteristics can provide an accurate identification of the etiological agent of the infection.

E. coli serology is based on the presence or absence of 3 kinds of antigens: "K" or envelope, "O" or somatic and "H" or flagellar.

K antigens are envelope or sheath antigens which are heat-labile and inhibit agglutination of a living organism in somatic antisera. There are approximately 93 different K antigens whose detection serves as a basis for the preliminary identification of the organisms. Sera prepared from K antigens are used in the slide technique employing living cultures (unheated). Several K sera have been pooled to prepare polyvalents to minimize the number of sera necessary to screen cultures.

Note: One must be aware of the cross-reactions which will occur using the polyvalent antisera, due to the presence of common agglutinins:

Common Antigen	Serogroups	Polyvalent Antisera
K61	086a:K61, 020a020b:K61	B and C
018	018a018c:K77, 018a018b:K76	C and D
0112	0112ac:K66, 0112a0112b:K68	C and E
0127	0127a:K63, 0127a0127b:K65	A and E

There are 3 varieties of K antigens designated as L, A and B which may be differentiated by differences in their heat stability as shown in Table I.

Table I

K Antigen	Characteristics
L	1. Agglutinable in L serum inactivated at 100°C for 1 hour.
	2. Suspension rendered agglutinable in O serum at 100°C for 1 hour.
	3. Antibody binding power is inactivated at 100°C for 1 hour.
	4. Antigenicity inactivated at 100°C for 1 hour.
	5. Occur as envelopes or sheaths. Occasionally as capsules.
A	1. Agglutinable in A serum inactivated at 121°C for 2-1/2 hours.
	2. Suspension rendered agglutinable in O serum at 121°C for 2-1/2 hours.
	3. Antibody binding power is not inactivated at 100°C for 2-1/2 hours or at 121°C for 2 hours.
	4. Antigenicity inactivated at 121°C for 2-1/2 hours.
	5. Occur as capsules.
B	1. Agglutinable in B serum inactivated at 100°C for 1 hour.
	2. Suspensions rendered agglutinable in O serum at 100°C for 1 hour.
	3. Antibody binding power is not inactivated at 100°C for 2-1/2 hours or at 121°C for 2 hours.
	4. Antigenicity inactivated at 100°C for 1 hour.
	5. Occur as envelopes or sheaths.

Courtesy of Dr. William H. Ewing

In the past, nomenclature describing the serogroups of *E. coli* reflects the differences in the K antigen, i.e., 055:B5. Current nomenclature uses the K antigen rather than the variety, L, A, or B, therefore, E. coli 055:B5 is now designated 055:K59.
Following is a list of the B antigens together with their corresponding K antigen designations:

B Antigen	K Antigen	B Antigen	K Antigen
B1	K56	B12	K67
B2	K26	B13	K68
B3	K57	B14	K69
B4	K58	B15	K70
B5	K59	B16	K71
B6	K60	B17	K72
B7	K61	B18	K73
B8	K63	B19	K75
B9	K64	B20	K76
B10	K65	B21	K77
B11	K66	B22	K78

Somatic O antigens are heat-stable and are not inactivated by heating as high as 121°C. There are approximately 157 entities designated by numbers from 01 to 150; 31, 47, 67, 72, 94 and 122 have been deleted from the schema. Detection of the O antigen by tube titration serves as presumptive identification of the *E. coli* strain.

It must be emphasized that *E. coli* antisera are used as screening reagents and reagents for final identification. Since numerous serological relationships and serological identities exist in the family *Enterobacteriaceae*, one must characterize the isolate bio-chemically to at least the generic level before serological final identification is reported.

If Bacto E. Coli Antisera are to be used in techniques other than those recommended, the user must test each lot of antiserum under conditions using known positive and negative cultures.

REAGENTS

Bacto E. Coli OK Polyvalent and individual OK Antisera are stable, desiccated rabbit antisera prepared whenever possible from nonmotile strains of *E. coli*. No attempt has been made to remove H agglutinins from the polyvalent sera since they are used in a screening procedure. For the individual OK sera, however, H agglutinins have been removed by absorption procedure when necessary. The OK sera have not been tested with all ±93 K antigens, nor have they been tested and/or absorbed with members of the O groups. The individual and polyvalent antisera are only tested and absorbed with the following 11 OK serogroups:

026:K60	0112:K66	0126:K71
055:K59	0119:K69	0127:K63
086:K61	0124:K72	0128:K67
0111:K58	0125:K70	

Bacto E. Coli O Antisera are stable, desiccated rabbit antisera prepared from heated cultures. They have been cross-absorbed with the 11 O groups listed above under E. Coli OK Antisera. No attempt has been made to render them specific for the other ± 157 O serogroups of *E. coli*.

Bacto E. Coli Antisera are available for most levels of serological methods, screening, serogrouping and serotyping. Bacto E. Coli OK Polyvalent Antisera are employed in the screening of suspect specimens for evidence of the presence of *E. coli*. Five poly-valent antisera are available encompassing the 20 most commonly encountered *E. coli* serogroups. Strong, positive agglutination with one of these polyvalent antisera in con-junction with typical biochemical reactions may be considered presumptive evidence of *E. coli* in the specimens. These antisera are employed in the slide agglutination test.

Serogrouping of *E. coli* can be accomplished with Bacto E. Coli OK and Bacto E. Coli O Antisera. Individual O antisera in the tube titration test using bacterial suspensions in serial dilutions of the O antisera to determine the O antigen group provide preliminary identification of *E. coli*. Serogrouping of suspect *E. coli* is recommended in cases of a possible outbreak of epidemic infant diarrhea.[7,8]

Each vial of antiserum contains approximately 1:10,000 Merthiolate,® sufficient to main-tain a bacteriostatic condition at a storage of 2 – 8°C.

When used according to the suggested PROCEDURE, each 3 ml vial of Bacto E. Coli OK Antiserum will be sufficient to perform approximately 60 tests and each vial of Bacto E. Coli O Antiserum will be sufficient to perform approximately 30 tests.

Rehydration

To rehydrate Bacto E. Coli OK and O Antisera add 3 ml 0.85% NaCl solution and rotate gently to dissolve completely. Upon rehydration the resultant antiserum will be a 1:2 working dilution (RTD).

Bacto E. Coli OK Antisera, upon rehydration should be used without further dilution. Bacto E. Coli O Antisera should be prepared in serial dilutions (1:20 – 1:1280) for use. The methodology for this test is given below under "Tube Technique for O Antigen Titration."

Any antiserum in which a precipitate forms or which does not dissolve completely upon rehydration should be discarded.

Check purity (bacterial) and pH of the saline used in rehydration if the antiserum rehydrates cloudy. Discard any serum which is cloudy and/or has a precipitate unless it has been clarified and shown to react properly with known control cultures.

SPECIMEN COLLECTION
Specimens should be taken from feshly soiled diapers, rectal swabs or stool specimens prior to antibiotic therapy. If swabs are employed they should be treated in a buffer solution (Sorenson's pH 7.4) which may be prepared as follows:

Stock Solution A: Dissolve 9.464 g of anhydrous Na_2PO_4 which has been previously dried at 130°C in 1 liter distilled water.

Stock Solution B: Dissolve 9.073 g of anhydrous KH_2PO_4 previously dried at 110°C in 1 liter distilled water.

Mix solutions A and B in the following manner:

80 ml stock solution A
+20 ml stock solution B
Final pH 7.38 ± 0.02.

Preparation of the Buffered Swabs:
1. Place 500 swabs in a 1000 ml beaker containing 100 ml Sorenson's Phosphate Buffer.
2. Boil for 5 minutes.
3. Drain off excess solution.
4. Place swabs in cotton stoppered tubes, dry in an oven.
5. Sterilize.

Preservation of Specimens:
Specimens which cannot be immediately plated should be preserved in a buffered glycerol solution prepared as follows:

Sodium Chloride	4.2 g
Dipotassium Phosphate Anhydrous (K_2HPO_4)	3.1 g
Monopotassium Phosphate Anhydrous (KH_2PO_4)	1.0 g
Glycerol	300.0 ml
Distilled Water	1000.0 ml

Dispense 10 ml buffered glycerol solution in screw cap test tubes. Autoclave for 15 minutes at 116°C. Add enough phenol red to give a distinct red color. Deterioration of the buffer is detectable when phenol red indicator changes from red to yellow.

Specimens should be emulsified in tubes containing 0.5 – 1 ml buffered glycerol broth and stored at 2 – 8°C until further examination can be performed.

Specimens which must be mailed should be placed in buffered glycerol solution, Bacto Transport Medium Amies or Bacto Transport Medium Stuart.

It is essential that appropriate media be used for detecting, isolating and differentiating microorganisms on morphological and physiological characteristics before applying serological techniques. Specimens suspected of containing *E. coli* should be plated on Bacto EMB Agar and/or Bacto MacConkey Agar. In addition to these differential media, plates of Bacto Blood Agar Base or Bacto Veal Infusion Agar should be inoculated. The differential media, Bacto EMB Agar and Bacto MacConkey Agar will inhibit the growth of gram-positive microorganisms for at least 24 hours. Bacto Blood Agar Base and Bacto Veal Infusion Agar are nonselective media and are used to assure growth of those strains of *E. coli* which do not grow well on differential media. The use of these media is also important in obtaining accurate slide agglutination results. Bile salts in MacConkey agar or acid formed from lactose fermentation on EMB agar may result in erroneous slide agglutination patterns. Therefore, a minimum of 2 plates should be used for specimens suspect of *E. coli*, Bacto Blood Agar Base or Bacto Veal Infusion Agar and Bacto EMB Agar or Bacto MacConkey Agar. Incubate for 18 – 24 hours at 35 – 37°C. Colonies picked for serological procedures should not be taken directly from differential media. The following schema outlines the cultivation, biochemical and serological methods employed in the isolation and identification of *E. coli*.

Table II

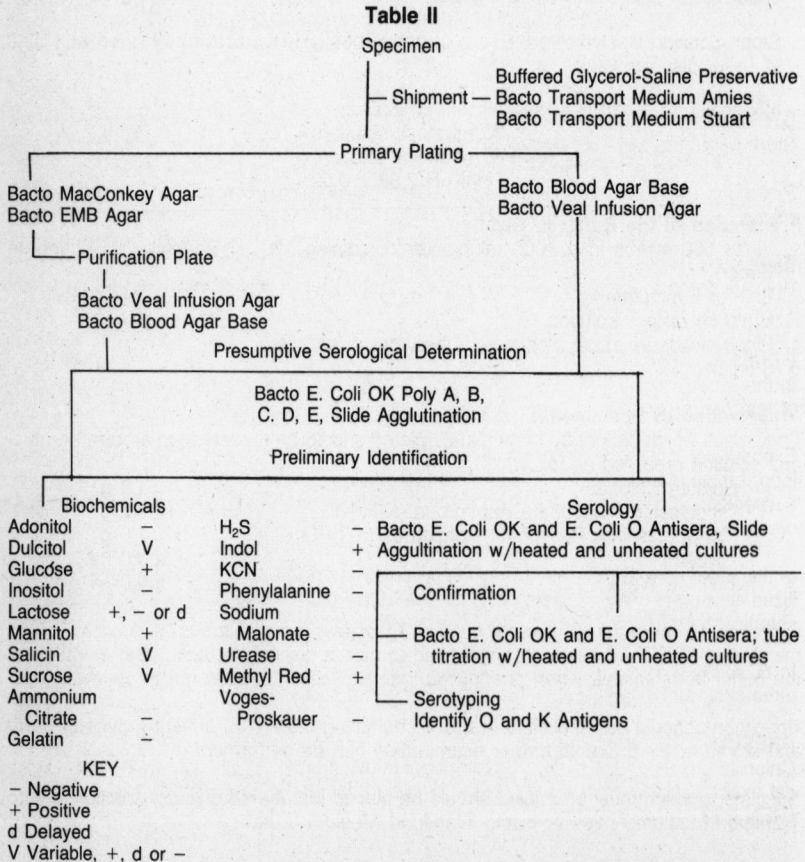

Biochemicals				Serology
Adonitol	–	H₂S	–	Bacto E. Coli OK and E. Coli O Antisera, Slide
Dulcitol	V	Indol	+	Agglutination w/heated and unheated cultures
Glucose	+	KCN	–	
Inositol	–	Phenylalanine	–	Confirmation
Lactose	+, – or d	Sodium		
Mannitol	+	Malonate	–	Bacto E. Coli OK and E. Coli O Antisera; tube
Salicin	V	Urease	–	titration w/heated and unheated cultures
Sucrose	V	Methyl Red	+	
Ammonium		Voges-		Serotyping
Citrate	–	Proskauer	–	Identify O and K Antigens
Gelatin	±			

KEY
– Negative
+ Positive
d Delayed
V Variable, +, d or –

The test organism must be identified to at least the generic and in some cases species level biochemically prior to serological serotyping of *E. coli*.

The test culture must be checked in a saline control (both in the tube and slide technique) for smoothness, or preferably in normal rabbit serum in the slide test. Often stock cultures and sometimes isolated cultures are rough and will agglutinate spontaneously in a normal serum. Therefore, it is necessary to select smooth colonies for serological testing.

PROCEDURE
Materials Provided:
Bacto E. Coli OK Antisera and droppers
Bacto E. Coli O Antisera

Purchase Separately:
Bacto McFarland Barium Sulfate Standard No. 3.

Materials Required but not Provided:
Slide Test
0.85% NaCl
Bacteriological loop
Glass plates marked into 1" squares
Burner
Fluorescent desk lamp

Tube Test
Kahn tubes (12 × 75 mm)
    and rack
0.85% NaCl
Water bath at 50°C
Formalin
1 ml serological pipettes

STORAGE
Store both unrehydrated and rehydrated Bacto E. Coli O and OK Antisera at 2 – 8°C.

Prolonged exposure to room temperature or repeated freezing and thawing are detrimental to antisera.

Expiration Date
Both desiccated and rehydrated Bacto E. Coli O and OK Antisera are stable to the expiration date on the label when stored as directed.

GENERAL INSTRUCTIONS
In the slide test, all materials and equipment must be at room temperature at the time of test performance.

Exposure of the organism or plate to heat from external sources (a hot bacteriological loop, burner flame, light source, etc.) may result in either a culture which cannot be suspended readily or evaporation and/or precipitation of the test mixture which may result in false-positive reactions.

In the slide test, all equipment and materials (glass plates, loop, culture) must be sterilized after use by flame (loop) or autoclave (glass plate and culture) or by an adequate chemical method, since living organisms are used as the antigen.

In *E. coli* serology, as in any serological test, known positive and negative control cultures should be employed to ascertain validity of test results.

PRESUMPTIVE SCREENING
Colonies grown on blood agar base or veal infusion agar may be subjected to serological examination immediately following overnight incubation as described under SPECIMEN PREPARATION.

This initial serological screening procedure is employed to detect those strains of *E. coli* which do not grow on differential media[1] and to give a rapid preliminary indication of the presence of *E. coli*. (Since serological procedures cannot be performed directly on colonies grown on differential media the loss of valuable time for subculture to a nonselective medium can be avoided.)

For screening specimens for presumptive evidence of *E. coli*, portions of 10 or more individual colonies from Bacto Blood Agar Base or Bacto Veal Infusion Agar should be tested directly on Bacto E. Coli OK Antiserum Poly by the following technique:
1. Place a drop of the desired OK polyvalent antiserum on a square of the glass plate for each colony to be tested.
2. Pick a portion of an individual colony from an agar plate and place into the drop of antiserum using a bacteriological loop.
3. Mix to obtain a uniform suspension.
4. Repeat same procedure (1 – 3 above) for each colony under test.
5. Care must be taken not to cross-mix the ingredients of the droplets.
6. Rock the glass plate to insure a thorough mixture; continue to rotate by hand for 1 minute.
7. Record agglutination as follows:

$$++++ \quad \text{agglutination} = \text{all of the cells agglutinate}$$
$$+++ \quad \text{agglutination} = 75\% \text{ of the cells agglutinate}$$
$$++ \quad \text{agglutination} = 50\% \text{ of the cells agglutinate}$$
$$+ \quad \text{agglutination} = 25\% \text{ of the cells agglutinate}$$
$$\pm \quad \text{agglutination} = <25\% \text{ of the cells agglutinate}$$
$$- \quad \text{agglutination} = \text{none of the cells agglutinate}$$

RESULTS
A +++ reaction or greater should be considered as the endpoint at the RTD.

If strong slide agglutination reactions are obtained, it may be considered as presumptive evidence that one of the serogroups of *E. coli* represented in the polyvalent antiserum is present in the specimen.

LIMITATIONS OF THE PROCEDURE
Because *Salmonella, Shigella* and *E. coli* organisms are not readily differentiated on these nonselective media and cross reactions among these organisms may occur, the user is advised to test colonies taken from MacConkey agar or EMB agar according to procedure outlined below.

PRESUMPTIVE SEROGROUPING
Transfer 3 or more isolated, smooth, opaque, *Escherichia*-like colonies from selective media to blood agar base slants. The slants are inoculated in such a manner that growth occurs over the entire surface of the agar. Incubate at 35 – 37°C overnight. Additional transfers should be made and reserved for further serological testing and biochemical characterization.
1. Place a drop of the desired OK polyvalent antiserum on a square of the glass plate for each slant culture to be tested.
2. Place a drop of 0.85% NaCl solution on a square next to the droplet of antiserum.
3. Transfer a portion of a loopful of the slant cultures to the droplets (serum and saline) and mix to obtain a uniform suspension. Care must be taken not to cross-mix the ingredients of the droplets.
4. Rock the plate to insure thorough mixing of the suspension and continue to rotate by hand for 1 minute.

5. Record agglutination as follows:

++++ agglutination = all of the cells agglutinate
+++ agglutination = 75% of the cells agglutinate
++ agglutination = 50% of the cells agglutinate
+ agglutination = 25% of the cells agglutinate
± agglutination = <25% of the cells agglutinate
− agglutination = none of the cells agglutinate

RESULTS

Agglutination in the antiserum should be rapid and complete. There should be no agglutination of the suspension in the 0.85% NaCl droplet. A +++ reaction or greater should be considered as the endpoint at the RTD.

If agglutination occurs with the polyvalent antiserum, test the culture in the slide test from an agar plate or slant as above employing the individual OK antisera which is contained in the polyvalent antiserum.

NOTE: An adequate K antigen should be used for K serology. It is recommended that an O inagglutinable colony be used for this purpose.

At this point, biochemical tests should be run with appropriate media. Results of these tests, typical of *E. coli* (see Table I) in conjunction with positive slide agglutination, can be considered presumptive identification of the particular *E. coli* serogroup.

Confirmation of the O antigen content is then done by use of Bacto E. Coli O Antiserum in a tube agglutination technique. Final identification of the culture is based upon biochemical reactions as well as the presence of the O and K antigens.

TUBE TECHNIQUE for O Antigen Titration

Pure cultures of the test organism are prepared on blood agar base by incubating at 37°C for 16 – 18 hours.

1. Suspend some of the growth from the solid medium in 0.85% NaCl solution to give a homogenous suspension.
2. Heat the bacterial suspension in a boiling water bath for 30 – 60 minutes. Cultures should be homogenous; a precipitate indicates a rough culture and should be discarded.
3. Allow suspension to cool and dilute to a density approximately that of a Bacto McFarland Barium Sulfate Standard No. 3 using 0.85% NaCl solution.
4. Add formalin to a final concentration of 0.5% by volume.
5. Prepare serial dilutions of the specific Bacto E. Coli O Antiserum in Kahn serological tubes (12 × 75 mm) to give dilutions of 1:20 – 1:1280. Set up 8 tubes in a rack. Add 0.9 ml of 0.85% NaCl to the first tube and 0.5 ml to the remaining 7 tubes. Add 0.1 ml of the specific antiserum to the first tube and mix thoroughly. Transfer 0.5 ml from tube 1 to tube 2 and mix. Transfer 0.5 ml from tube 2 to tube 3, etc., on through tube 7, discard 0.5 ml of the mixture from tube 7; tube 8 serves as an antigen control.
6. Place 0.5 ml of heated bacterial suspension into each of the 8 tubes. Final dilutions 1:40 – 1:2560.
7. Incubate in a water bath at 50°C for 18 – 20 hours.
8. Examine for agglutination. Cultures showing 2+ or greater agglutination in dilutions of 1:320 or greater are considered to contain the same O antigens as the antiserum and thus constitutes final identification of the somatic antigen of the organism.

LIMITATIONS OF THE PROCEDURE

While serological procedures, as applied to microorganisms, provide corroborative evidence and can be used to identify particular antigenic sites of the genus or species under study, they cannot be used alone to identify the etiological agent of disease. Cultural methods should preceed any serological examination, and final identification should not be attempted without biochemical characterization. This is particularly true of members of *Enterobacteriaceae*. Members of this family are antigenically related, and cross-reactions can occur; therefore, identification of genera within the family cannot be based on serological methods alone.

Specifically, antigens of *E. coli* are antigenically related to or identical with *Shigella* and to Alkalescens-Dispar. It has been shown that *E. coli* 0111 is identical to *Salmonella* 035. Cross-reactions with these genera are documented in the literature. There are also intrageneric antigenic similarities and therefore particular cultures of an *E. coli* serogroup may result in a heterologous reaction if only weakly, with an antiserum of a different group (O antisera) or serogroup (OK antisera). The pattern of these heterologous intrageneric reactions is stated in *Identification of Enterobacteriaceae*.[1]

It must be borne in mind that an antiserum which has not been tested for and absorbed for all known antigens of that particular serological schema, may contain unexpected heterologous agglutinins, in some cases equal to or greater than its homologous titers even though the major cross-reactions in a given serum are known.

Rough strains detected by spontaneous agglutination in saline reactions in numerous antisera, or those which precipitate upon heating, cannot be serotyped.

REFERENCES

1. Identification of *Enterobacteriaceae*, Burgess Pub. Co., 3rd Ed., 1972.
2. The O Antigen Groups of *Escherichia coli* Cultures from Various Sources, USPHS-CDC, July 1971.
3. Bailey, W. R., and Scott, E. G., Diagnostic Microbiology, 3rd Ed., C. V. Mosby Co., St. Louis, 1970.
4. Kauffmann, F., J. Immunology, 57:71 – 100, 1947.
5. Orskov, F., Acta Pathol. Microbiol. Scand., 29:373, 1956.
6. Ewing, W. H., et al., Studies on the Serology of *Escherichia coli* Group, CDC Monograph, Communicable Dis. Cen., Atlanta, GA, 1956.
7. Ewing, W. H., Public Hlth. Lab., 27:19 – 30, 1969.
8. Isolation and Preliminary Identification of *Escherichia coli* Serotypes Associated with Diarrheal Dis., CDC Monograph, Atlanta, GA, 1956.

PACKAGING

Bacto E. Coli Antisera are prepared in 3 ml vials and supplied singly and in sets.

Bacto E. Coli OK Antiserum   Serotype and Code listed below

Serotype	Antiserum Code 3 ml vial
Antiserum 02:K56(B1)	2967-47-2
Antiserum 08:K25(B2)	2968-47-1
Antiserum 09:K57(B3)	2969-47-0
Antiserum 018a018b:K76(B20)	2630-47-9
Antiserum 018a018c:K77(B21)	2631-47-8
Antiserum 020a020c:K61(B7)	2632-47-7
Antiserum 020a020b:K84(B)	2633-47-6
Antiserum 026:K60(B6)	2676-47-4
Antiserum 028:K73(B18)	2634-47-5

Serotype	Antiserum Code 3 ml vial
Antiserum 044:K74	2635-47-4
Antiserum 055:K59(B5)	2674-47-6
Antiserum 086a:K61(B7)	2730-47-8
Antiserum 0111:K58(B4)	2673-47-7
Antiserum 0112a0112b:K68(B13)	2586-47-3
Antiserum 0112a0112c:K66(B11)	2341-47-9
Antiserum 0113:K75(B19)	2638-47-1
Antiserum 0114:K90	2925-47-3
Antiserum 0119:K69(B14)	2725-47-5
Antiserum 0124:K72(B17)	2728-47-2
Antiserum 0125:K70(B15)	2726-47-4
Antiserum 0126:K71(B16)	2727-47-3
Antiserum 0127a:K63(B8)	2675-47-5
Antiserum 0127a0127b:K65(B10)	2972-47-5
Antiserum 0128:K67(B12)	2729-47-1
Antiserum 0136:K78(B22)	2639-47-0
Antiserum 0138:K81	3040-47-1
Antiserum 0139:K82	3041-47-0
Antiserum 0141:K85	3042-47-9
Antiserum 0141:K87	3043-47-8
Antiserum 0141:K88	3044-47-7

Bacto E. Coli OK Poly A 2677-47-3
Contains 026:K60(B6), 055:K59(B5), 0111:K58(B4), 0127a:K63(B8)

Bacto E. Coli OK Poly B 2731-47-7
Contains 086a:K61(B7), 0119:K69(B14), 0124:K72(B17), 0125:K70(B15),
0126:K71(B16), 0128:K67(B12)

Bacto E. Coli OK Poly C 2629-47-2
Contains 018a018c:K77(B21), 020a020c:K61(B7), 020a020b:K84(B),
028:K73(B18), 044:K74, 0112a0112c:K66(B11)

Bacto E. Coli OK Poly D 2964-47-5
Contains 02:K56(B1), 08:K25(B2), 09:K57(B3), 018a018b:K76(B20)

Bacto E. Coli OK Poly E 2965-47-4
Contains 0112a0112b:K68(B13), 0113:K75(B19), 0127a0127b:K65(B10),
0136:K78(B22)

Bacto E. Coli OK Set A .... 5 × 3 ml . . . 2678-32-9
Contains 1 vial each of Poly A, 026:K60(B6), 055:K59(B5), 0111:K58(B4),
0127a:K63(B8)

Bacto E. Coli OK Set B .... 7 × 3 ml . . . 2747-32-6
Contains 1 vial each of Poly B, 086a:K61(B7), 0119:K69(B14), 0124:K72(B17),
0125:K70(B15), 0126:K71(B16), 0128:K67(B12)

Bacto E. Coli OK Set C .... 7 × 3 ml . . . 2640-32-4
Contains 1 vial each of Poly C, 018a018c:K77(B21), 020a020c:K61(B7),
020a020b:K84(B), 028:K73(B18), 044:K74, 0112a0112c:K66(B11)

Bacto E. Coli O Antiserum

Serotype	Antiserum Code 3 ml vial
Antiserum 01a01b	2498-47-0
Antiserum 02a02b	2499-47-9
Antiserum 04	2501-47-5
Antiserum 05	2597-47-0
Antiserum 06	2502-47-4
Antiserum 07	2503-47-3
Antiserum 08	2598-47-9
Antiserum 09	2504-47-2
Antiserum 011	2505-47-1
Antiserum 012	2600-47-5
Antiserum 013	2601-47-4
Antiserum 014	2602-47-3
Antiserum 015	2506-47-0
Antiserum 017	2604-47-1
Antiserum 018ab018ac	2507-47-9
Antiserum 019a019b	2605-47-0
Antiserum 020a020b	2606-47-9
Antiserum 021	2607-47-8
Antiserum 022	2608-47-7
Antiserum 024	2610-47-3
Antiserum 025	2611-47-2
Antiserum 026a026b	2682-47-6
Antiserum 028ab028ac	2612-47-1
Antiserum 034	2927-47-1
Antiserum 036	2928-47-0
Antiserum 040	2929-47-9
Antiserum 044	2508-47-8
Antiserum 045	2613-47-0
Antiserum 055	2680-47-8
Antiserum 060	2946-47-8
Antiserum 062	2509-47-7
Antiserum 068	2510-47-4
Antiserum 069	2931-47-5
Antiserum 071	2932-47-4
Antiserum 075	2511-47-3
Antiserum 078	2934-47-2
Antiserum 083	2936-47-0
Antiserum 086a	2737-47-1
Antiserum 0102	2614-47-9
Antiserum 0111ab0111ac	2679-47-1
Antiserum 0112ab0112ac	2382-47-9
Antiserum 0113	2786-47-1
Antiserum 0119	2732-47-6
Antiserum 0120	2939-47-7
Antiserum 0124	2735-47-3
Antiserum 0125ab0125ac	2733-47-5
Antiserum 0126	2734-47-4
Antiserum 0127a	2681-47-7
Antiserum 0128ab0128ac	2736-47-2
Antiserum 0140	2788-47-9

Serotype	Antiserum Code 3 ml vial
Antiserum 0145	2940-47-4
Antiserum 0150	2966-47-3

Bacto E. Coli O Antisera Set A .... 4 × 3 ml . . . 2683-32-2
   Contains 1 vial each of serogroups 026a026b, 055, 0111ab0111ac, 0127a

Bacto E. Coli O Antisera Set B .... 6 × 3 ml . . . 2748-32-5
   Contains 1 vial each serogroups 086a, 0119, 0214, 0125ab0125ac, 0126,
   0128ab0128ac

Bacto E. Coli O Antisera Set C .... 5 × 3 ml . . . 2643-32-1
   Contains 1 vial each of serogroups 018ab018ac, 020a020b, 028ab028ac,
   0112ab0112ac, 044

BACTO EC MEDIUM

INTENDED USE
Bacto EC Medium is a Standard Methods[1,2] medium for differentiating and enumerating fecal and nonfecal coliform organisms in water, wastewater and shellfish.

HISTORY/PRINCIPLES
In an effort to improve further the methods for the detection of members of the coliform group and *E. coli* Hajna and Perry[3] developed EC medium. This medium consisted of a buffered lactose broth to which was added 0.15% of Bacto Bile Salts No. 3. Growth of spore formers and fecal streptococci was inhibited by the bile salts while growth of coli was enhanced by its presence. The medium can be used at 37°C for the detection of coliform organisms or at 45.5°C for the isolation of *E. coli.* In a further evaluation study of EC medium and lauryl tryptose broth, Perry and Hajna[4] reported the results obtained from eleven different laboratories examining a variety of waters, milk, and shellfish. The results indicated that these media were highly specific for coliform bacteria. A presumptive test reading with the EC medium or lauryl tryptose broth seemed more dependable than the usual "confirmed" or "completed" test for coliform bacteria.

Bacto EC Medium (Escherichia Coli Medium) is usually prepared with fermentation tubes. Lactose fermentation with gas production serves as presumptive evidence for the presence of coliform organisms.

FORMULA
BACTO EC MEDIUM
DEHYDRATED
Ingredients per liter

Bacto Tryptose	20 g
Bacto Lactose	5 g
Bacto Bile Salts No. 3	1.5 g
Dipotassium Phosphate	4 g
Monopotassium Phosphate	1.5 g
Sodium Chloride	5 g

Final pH 6.9 ± 0.2 at 25°C.

One pound will make 12.2 liters of medium.

METHOD OF PREPARATION
1. Suspend 37 g in 1 liter distilled or deionized water and warm slightly to dissolve completely.
2. Dispense into test tubes each containing an inverted fermentation vial.
3. Sterilize in the autoclave for 15 minutes at 15 lbs pressure (121°C). Before opening the autoclave, allow temperature to drop below 75°C to avoid entrapping air bubbles in the inverted vials.

Follow the methods and procedures as stated in *Standard Methods for the Examination of Water and Wastewater*[1] and *Compendium of Methods for the Microbiological Examination of Foods.*[2]

STORAGE
Bacto EC Medium   Below 30°C
Prepared medium   15 – 30°C

QUALITY CONTROL
Identity Specifications

Dehydrated powder: light beige, homogeneous, free-flowing
Reaction of 3.7% solution: pH 6.9 ± 0.2 at 25°C
Prepared medium: light amber, clear

Typical Cultural Response in Bacto EC Medium
After 24 ± 2 Hours at 44.5 ± 0.2°C

Organism	Growth	Gas
Bacillus subtilis ATCC® 6633	inhibited	–
Enterobacter aerogenes ATCC® 13048	inhibited	–
Escherichia coli ATCC® 25922	good	+
Streptococcus faecalis ATCC® 19433	inhibited	–

REFERENCES
1. American Public Health Association, 1980, Standard Methods for the Examination of Water and Wastewater, 15th Ed., American Public Health Association, Inc., Washington, D.C.
2. American Public Health Association, 1976, Compendium of Methods for the Microbiological Examination of Foods, American Public Health Association, Inc., Washington, D.C.
3. Perry and Hajna, 1943, American Journal of Public Health, 33:550.
4. Perry and Hajna, 1944, American Journal of Public Health, 34:735.

PACKAGING
Bacto EC Medium		
1 lb (454 g)	0314-01-0	
1/4 lb (114 g)	0314-02-9	
10 kg	0314-08-3	

BACTO EE BROTH MOSSEL

INTENDED USE
Bacto EE Broth Mossel is a selective enrichment broth for *Enterobacteriaceae* in bacteriological examinations of foods.

HISTORY/PRINCIPLES
Bacto EE Broth Mossel is prepared according to the formula of Mossel, Visser and Cornelissen.[1] It contains dextrose to allow the development of most *Enterobacteri-*

aceae, thus insuring the detection of *Salmonella* and other lactose negative organisms along with the lactose positive types. Also, the use of growth rather than gas formation, to indicate the presence of *Enterobacteriaceae,* avoids missing anaerogenic organisms. Bacto Brilliant Green and Bacto Oxgall are employed as the selective agents.

FORMULA

BACTO EE BROTH MOSSEL
DEHYDRATED

Ingredients per liter

Bacto Tryptose	10 g
Bacto Dextrose	5 g
Disodium Phosphate	8 g
Monopotassium Phosphate	2 g
Bacto Brilliant Green	0.0135 g
Bacto Oxgall, Dehydrated	20 g

Final pH 7.2 ± 0.2 at 25°C.

One pound will make 10.1 liters of final medium.

METHOD OF PREPARATION

1. Suspend 45 g in 1 liter distilled or deionized water. Heat to boiling for 1 – 2 minutes.
2. Dispense in 120 ml amounts in 250 ml flasks and stopper with cotton plugs or loose fitting caps. Nine ml dispensed into tubes may be used for small inocula.
3. Heat in flowing steam (100°C) or boiling water bath for 30 minutes to sterilize.

PROCEDURE

1. Inoculate flasks with approximately 10 g of homogenized food or other material to be tested.
2. Shake the inoculated medium thoroughly for a few seconds to mix well.
3. Incubate for 20 – 24 hours at 35 – 37°C. Shake again at the end of the first 3 hours of incubation.
4. Prepare plates such as Bacto Violet Red Bile Agar for streaking. To insure recovery of dextrose-only fermenters, add 1% dextrose before boiling.
5. Streak a loopful of the enrichment culture onto prepared plates.
6. Incubate plates for 18 – 24 hours at 35 – 37°C.
7. Coliforms appear pink to purplish red on violet red bile agar. If dextrose has been added salmonellae should also appear purplish.

STORAGE

Bacto EE Broth Mossel	Below 30°C
Prepared medium	15 – 30°C

QUALITY CONTROL

Identity Specifications

Dehydrated powder:	light green, homogeneous, free-flowing
Reaction of 4.5% solution:	pH 7.2 ± 0.2 at 25°C
Prepared medium:	emerald green, clear

Typical Cultural Response in Bacto EE Broth Mossel
After 20 – 24 Hours at 35°C

Organism	Growth	Acid*
Enterobacter aerogenes ATCC® 13048	good to excellent	+
Escherichia coli ATCC® 25922	good to excellent	+

Organism	Growth	Acid*
Proteus mirabilis ATCC® 25933	good to excellent	+
Salmonella enteritidis ATCC® 13076	good to excellent	± (may be delayed)
Salmonella sp. serotype panama ATCC® 7378	good to excellent	± (may be delayed)
Shigella boydii ATCC® 12030	good to excellent	−
Staphylococcus aureus ATCC® 25923	inhibited	−

*+ = acid production, yellow
− = no change, green

REFERENCE
1. J. Appl. Bact., 26:444, 1963.

PACKAGING
Bacto EE Broth Mossel   1 lb (454 g)   0566-01-5

BACTO EMB AGAR

INTENDED USE
Bacto EMB Agar, a differential plating medium, is recommended for the detection and isolation of the gram-negative enteric bacteria.

HISTORY AND PRINCIPLES
The original eosin methylene blue agar was devised by Holt-Harris and Teague.[1] They used a combination of eosin and methylene blue as an indicator which gave sharp and distinct differentiation between colonies of lactose fermenting organisms and those which did not ferment lactose. Sucrose was included in the medium to detect those members of the coliform group which ferment this carbohydrate more readily than lactose. Lactose positive colonies are either black or possess dark centers with transparent colorless peripheries, while those that are lactose or sucrose negative are colorless. The eosin methylene blue agar of Holt-Harris and Teague possessed definite advantages over the fuchsin sulfite agar of Endo, in that it was more sensitive, more accurate, more stable and gave an earlier differentiation between the lactose fermenters and lactose and sucrose nonfermenters.

Two years after Holt-Harris and Teague had introduced their new medium, Levine[2] described an eosin methylene blue agar for the purpose of differentiating fecal and nonfecal types of the coli aerogenes group. Levine's medium, as discussed above, also differentiates *salmonellae* and other nonlactose fermenters from the coliform organisms.

Bacto EMB Agar is a combination of the Levine and the Holt-Harris and Teague formulae. It contains Bacto Peptone and phosphate as recommended by Levine and retains the 2 carbohydrates, lactose and sucrose, as suggested by Holt-Harris and Teague. The ratio of the 2 indicator dyes has been worked out to give the best differentiation and minimum toxicity. This medium has all the advantages of the original media referred to above.

FORMULA

BACTO EMB AGAR
DEHYDRATED

Ingredients per liter

Bacto Peptone .... 10 g
Bacto Lactose .... 5 g
Bacto Sucrose .... 5 g
Dipotassium Phosphate .... 2 g
Bacto Agar .... 13.5 g
Bacto Eosin Y .... 0.4 g
Bacto Methylene Blue .... 0.065 g

Final pH 7.2 ± 0.2 at 25°C.

One pound will make 12.6 liters of medium.

METHOD OF PREPARATION

1. Suspend 36 g in 1 liter distilled or deionized water and heat to boiling to dissolve completely. Bacto EMB Agar inoculated the same day as rehydrated may be used without autoclave sterilization. Under these conditions, the medium need be heated only to boiling to dissolve it completely before pouring into Petri dishes.
2. Distribute into tubes or flasks.
3. Sterilize in the autoclave for 15 minutes at 15 lbs pressure (121°C). Avoid overheating.
4. Sterilization reduces the methylene blue, leaving the medium orange color. The normal purple color of the medium may be restored by gentle shaking. If the reduced medium is not shaken to oxidize the methylene blue, a dark zone beginning at the top and extending downward through the medium will gradually appear. The sterilized medium normally contains a flocculant precipitate which should not be removed. By cooling to 50°C and gently agitating the medium before pouring it into plates, this flocculation will be finely dispersed.

PROCEDURE

It is recommended that for the detection of the more fastidious *Shigella* and *Salmonella* organisms, a nonselective but differential medium such as Bacto MacConkey Agar, Bacto Hektoen Enteric Agar and a selective medium such as Bacto SS Agar, Bacto Bismuth Sulfite Agar (especially for *Salmonella*) and Bacto Desoxycholate Citrate Agar be run in parallel with Bacto EMB Agar.

Use a medium density inoculum to streak the plates to obtain isolated colonies. Incubate for 18 – 24 hours at 35 – 37°C.

Salmonella and *Shigella* organisms are translucent and amber colored or colorless. Coliforms that utilize the lactose and/or sucrose are blue-black with a greenish metallic sheen. Other coliforms such as *Enterobacter* form mucoid, pink colonies, which are often confluent.

STORAGE

Bacto EMB Agar   Below 30°C
Prepared medium   2 – 8°C

QUALITY CONTROL

Identity Specifications

Dehydrated powder: pinkish purple, homogeneous, free-flowing
Reaction of 3.6% solution: pH 7.2 ± 0.2 at 25°C
Prepared medium: green with an orange cast, opalescent with a flocculant precipitate
Prepared plates: purple with greenish orange cast, opalescent, may have a fine precipitate

Typical Cultural Response on Bacto EMB Agar
After 18 – 24 Hours at 35°C

Organism	Growth	Color of Colony
Enterobacter aerogenes ATCC® 13048	good	pink, no sheen
Escherichia coli ATCC® 25922	good to excellent	purple w/black centers and green metallic sheen
Klebsiella pneumoniae ATCC® 13883	good	green metallic sheen, dark centers
Proteus mirabilis ATCC® 25933	good to excellent	colorless
Salmonella typhimurium ATCC® 14028	good to excellent	colorless
Staphylococcus aureus ATCC® 25923	inhibited	—

REFERENCES

1. J. Infectious Diseases, 18:596, 1916.
2. *ibid.*, 23:43, 1918.

PACKAGING

Bacto EMB Agar	1 lb (454 g)	0076-01-8
	1/4 lb (114 g)	0076-02-7
	5 lb (2.27 kg)	0076-05-4
	10 kg	0076-08-1

BACTO EMB AGAR BASE

INTENDED USE
Bacto EMB Agar Base is an eosin methylene blue agar basal medium to which carbohydrates and other test substances may be added for culturing, differentiating and studying bacterial variation of the gram-negative bacterial rods.

HISTORY
Bacto EMB Agar Base has the same formulation as Bacto Levine EMB Agar[1] with the lactose omitted. For a complete discussion see Bacto Levine EMB Agar.

FORMULA

BACTO EMB AGAR BASE
DEHYDRATED

Ingredients per liter

```
Bacto Peptone  . . . . . . . . . . . . . .  10 g
Dipotassium Phosphate  . . . . . . . . . .  2 g
Bacto Agar . . . . . . . . . . . . . . . . .  15 g
Bacto Eosin Y  . . . . . . . . . . . . . .  0.4 g
Bacto Methylene Blue  . . . . . . . .  0.065 g
```

Final pH 7.3 ± 0.2 at 25°C.

One pound will make 16.5 liters of medium.

METHOD OF PREPARATION
1. Suspend 27.5 g Bacto EMB Agar Base in 1 liter distilled or deionized water and heat to boiling to dissolve completely.

2. Add carbohydrate or other test substance in desired concentration and agitate to dissolve.
3. Sterilize in the autoclave for 15 minutes at 15 lbs pressure (121°C).

The carbohydrate or other test substances may be sterilized separately by autoclaving or filtering and added to the sterile basal medium after cooling it to 45 – 50°C.

STORAGE
Bacto EMB Agar Base   Below 30°C

REFERENCE
1. Bull. 62, Iowa Eng. Exp. Sta., 1921.

PACKAGING
Bacto EMB Agar Base   1 lb (454 g)   0511-01-1

BACTO EVA BROTH

INTENDED USE
Bacto EVA Broth is a selective confirmative medium used for detecting and confirming enterococci and as an indicator of fecal pollution in water and other specimens.

HISTORY/PRINCIPLES
Bacto EVA (Ethyl Violet Azide) Broth is prepared according to the formulation of Litsky, Mallmann and Fifield[1] with the reduced amount of dextrose and increased dye as suggested by the authors,[2] and as used by Litsky and Mallman.[3] The presence of enterococci is particularly valuable in detecting fecal or sewage pollution of well water, untreated water and river water, and for the examination of beach waters. Bacto EVA Broth is used in conjunction with Bacto Azide Dextrose Broth.

Mallmann and Seligmann[4] compared a variety of media for the detection of fecal streptococci and reported azide dextrose broth to be highly productive as well as a satisfactory enrichment medium for a presumptive test for streptococci. Using this medium, confirmation was necessary since it permitted the development of a few gram-positive rods and gram-positive microorganisms other than enterococci. Litsky, Mallmann and Fifield[1] studied a variety of dyes and selective agents for streptococci and developed a confirmatory medium using ethyl violet and sodium azide as selective agents. Their results indicated that the medium as described was specific for enterococci, and that gram-positive bacilli and gram-positive cocci other than enterococci that gave "false positive" presumptive tests with azide dextrose broth, were inhibited in the ethyl violet medium, making it an ideal confirmative medium for enterococci. Their report showed that their method detected and confirmed 100 to 1000 times as many enterococci from river water and sewage samples as did the other methods used in the survey. Enterococci in water are detected by adding appropriate dilutions of the specimens to azide dextrose broth and incubating at 37°C for 48 hours. Three loopfuls of all cultures showing growth, a positive presumptive test, are then transferred to the ethyl violet azide broth. A positive confirmed test is indicated by growth following incubation at 37°C within 48 hours. A large inoculum, as indicated, is required for this selective medium since EVA Broth was not designed to support growth from dilute inocula.

Further investigation of the medium by the authors[2] indicated that 0.5% dextrose was fully as productive as the medium originally described containing 1.5% dextrose, and that the medium with the smaller amount of carbohydrate was less adversely affected by heat during sterilization. As pointed out by Litsky, Mallmann and Fifield,[1] the actual dye concentration of different batches of ethyl violet have varied and standardization of the dye content of the medium is necessary. Bacto EVA Broth contains the decreased amount of dextrose and the amount of dye as used by Litsky and Mallmann[3] producing the correct selective action for the medium as reported in their comparison of the MPN of *Escherichia coli* and the presumptive and confirmed test for enterococci for the examination of river waters for sewage pollution. Their results showed that a positive correlation of +0.84 existed between the numbers of *E. coli* and enterococci in river water samples taken over a 2 year period. The density of enterococci was approximately 7.6 times that of *E. coli*. These authors stressed the value of enterococci in sanitary bacteriology.

In their study of the conditions for the destruction of fecal streptococci and coliform organisms in the preparation of frozen foods, Larkin, Litsky and Fuller[5] used azide dextrose broth as a presumptive medium and ethyl violet azide broth for confirmation for the streptococci. Their results indicated that generally fecal streptococci were recovered more consistently and in greater numbers than were coliforms and could be used in preference to coliforms as indicator bacteria in frozen foods.

FORMULA

BACTO EVA BROTH
DEHYDRATED

Ingredients per liter

Bacto Tryptose	20 g
Bacto Dextrose	5 g
Dipotassium Phosphate	2.7 g
Monopotassium Phosphate	2.7 g
Sodium Chloride	5 g
Sodium Azide	0.4 g
Ethyl Violet	0.00083 g

Final pH 7.0 ± 0.2 at 25°C.

One pound will make 12.8 liters of medium.

METHOD OF PREPARATION
1. To rehydrate, dissolve 35.8 g in 1 liter distilled or deionized water.
2. Dispense in tubes in 10 ml amounts.
3. Sterilize in the autoclave for 15 minutes at 15 lbs pressure (121°C).

STORAGE
Bacto EVA Broth   Below 30°C
Prepared medium   15 – 30°C

QUALITY CONTROL

Identity Specifications

Dehydrated powder: light beige, homogeneous, free-flowing
Reaction of 3.58% solution: pH 7.0 ± 0.2 at 25°C
Prepared medium: light amber, clear

Typical Cultural Response in Bacto EVA Broth
After 18 – 48 Hours at 35°C

Organism	Growth
Escherichia coli ATCC® 25922	inhibited
Staphylococcus aureus ATCC® 25923	inhibited
Streptococcus faecalis ATCC® 29212	good to excellent
Streptococcus faecalis ATCC® 19433	good to excellent
Streptococcus pyogenes ATCC® 19615	inhibited

REFERENCES

1. Am. J. Pub. Health, 43:873, 1953.
2. Personal Communication, 1954.
3. Paper read at APHA Meeting, Buffalo, 1954.
4. Am. J. Pub. Health, 40:286, 1950.
5. Appl. Microbiol., 3:98, 102, 104, 107, 1955.

PACKAGING
Bacto EVA Broth   1 lb (454 g)   0606-01-7

BACTO EY TELLURITE ENRICHMENT

Bacto EY Tellurite Enrichment is a tellurite, egg yolk emulsion prepared for use with Bacto Baird Parker Agar Base in the detection and enumeration of coagulase positive staphylococci.

For a complete discussion refer to Bacto Baird Parker Agar Base.

PACKAGING
Bacto EY Tellurite Enrichment   6 × 100 ml   0779-73-1

BACTO EGG ALBUMEN SOLUBLE

Bacto Egg Albumen Soluble is desiccated egg albumen and can be used in the preparation of culture media or for studies of bacterial metabolism. When rehydrated in 13.5% concentration, Bacto Egg Albumen Soluble will give the equivalent of the original form of the product.

PACKAGING
Bacto Egg Albumen Soluble   100 g   0255-15-5
                                                500 g   0255-17-3

BACTO EGG YOLK COAGULATED

Bacto Egg Yolk Coagulated is prepared for use in media for culturing and differentiating anaerobic bacteria as recommended by Spray.[1]

REFERENCE
1. Lab. Clin. Med., 18:512, 1932 – 3.

PACKAGING

Bacto Egg Yolk Coagulated	100 g	0148-15-6

BACTO EGG YOLK ENRICHMENT 50%

INTENDED USE

Bacto Egg Yolk Enrichment 50% is a sterile concentrated egg yolk emulsion recommended for use in a variety of media such as Bacto SFP Agar Base, Bacto Baird Parker Agar Base, Bacto TPEY Agar Base and Bacto McClung Toabe Agar Base for the isolation and identification of *Clostridium, Bacillus,* and *Staphylococcus* species on the basis of their lecithinase activity.

HISTORY/PRINCIPLES

MacFarlane, Oakley, and Anderson[1] reported that crude lecithovitellin from egg yolk gave a stronger reaction with *C. perfringens* toxin than did human serum. Hayward[2] reported that *C. perfringens* on an agar medium containing human serum produced well defined opacity extending from the edge of the colony and that this reaction was inhibited by the addition of homologous antitoxin. The opacity produced by the serum was not as satisfactory as that produced by egg yolk. McClung and Toabe[3] compared the reactions obtained with serum with those obtained with 50% egg yolk emulsion and found that serum gave considerably less dense precipitation and zones that were not as easily recognized. They showed that the use of McClung Toabe agar with 50% egg yolk emulsion made possible the presumptive identification of *Clostridium* species. Shahidi and Ferguson[4] added 50% egg yolk emulsion to SFP agar and utilized both the lecithinase reaction and sulfite reduction to identify *Clostridium perfringens*.

Gillespie and Alder[5] observed that most strains of coagulase positive staphylococci produced opacity when grown on media containing egg yolk. This reaction was strongly inhibited by *Staphylococcus* antitoxin. The egg yolk reaction, tellurite reduction and mannitol fermentation was utilized by Innes[6] to detect coagulase positive staphylococci. The colonies appeared after 24 hours and were easily recognized by their dark gray color and opaque zones around the colonies. Crisley[7] obtained excellent recovery of coagulase positive staphylococci on TPEY agar enriched with egg yolk. Baird Parker[8] included the egg yolk reaction, tellurite reduction and pyruvate utilization in Baird Parker agar and was able to confirm all positive isolates on Baird Parker agar as coagulase positive staphylococci.

Bacto Egg Yolk Enrichment 50% is added aseptically to prepared and sterilized culture media cooled to 50°C. In Bacto SFP Agar Base, Bacto McClung Toabe Agar Base, and Bacto TPEY Agar Base, 100 ml Bacto Egg Yolk Enrichment 50% is added to 900 ml or 10 ml to 90 ml of prepared base. Bacto Baird Parker Agar is prepared by adding either 50 ml Bacto Egg Yolk Enrichment 50% to 950 ml or 5 ml to 95 ml base cooled to 50°C.

STORAGE

Bacto Egg Yolk Enrichment 50%	2 – 8°C
Prepared plates	2 – 8°C

QUALITY CONTROL

Identity Specifications

Appearance: canary yellow, opaque solution with a resuspendable precipitate

Typical Cultural Response on Bacto McClung Toabe Agar with Bacto Egg Yolk Enrichment 50% After 18 – 48 Hours at 35°C

Organism	Growth	Lecithinase/ Halos
Bacillus subtilis ATCC® 6633	good to excellent	+
Clostridium perfringens ATCC® 12919	good to excellent	+
Staphylococcus aureus ATCC® 25923	good to excellent	+
Staphylococcus epidermidis ATCC® 12228	good to excellent	−

REFERENCES

1. J. Path. Bact., 52:99, 1941.
2. Brit. Med. J., 1:811 – 814, 916, 1941.
3. J. Bact., 53:139, 1947.
4. Applied Microbiology, 21:500, 1971.
5. J. Pathol. Bacteriol., 64:187, 1952.
6. J. Appl. Bact., 23:108, 1960.
7. Public Hlth. Serv., Publ. No. 1142, 1964.
8. J. Appl. Bacteriol., 5112, 1962.

PACKAGING

Bacto Egg Yolk Enrichment 50%	6 × 100 ml	3347-73-8
	12 × 10 ml	3347-61-2

BACTO EIJKMAN LACTOSE MEDIUM

INTENDED USE

Bacto Eijkman Lactose Medium is used to differentiate *Escherichia coli* from other coliform organisms based on their ability to grow and liberate gas from lactose.

HISTORY/PRINCIPLES

In 1904, Eijkman[1] described a method of separating the strains of coli originating from the feces of warm-blooded animals from the strains characteristic of cold-blooded animals. His method consisted essentially of placing the water under investigation in fermentation tubes or flasks and adding one-eighth its volume of a sterile solution containing 10% dextrose, 10% peptone, and 5% sodium chloride. This mixture was then incubated at 46°C. The presence of a uniform turbidity and gas production was considered indicative of the presence of fecal coli strains. Many investigators have studied this method with water samples and pure cultures with varying results. One of the factors limiting the value of this method was the inability to obtain growth of subcultures from positive tubes incubated at 46°C. Undoubtedly, the acidity at the relatively high temperature of incubation was responsible for the death of the culture within 24 to 48 hours.

Perry and Hajna[2] modified Eijkman's original method by decreasing the carbohydrate content and adding a phosphate buffer. Their study demonstrated that with 0.3% dextrose and a potassium phosphate buffer the reaction of the medium, after inoculation with *E. coli* and incubation, was pH 5.6, while under the conditions described by Eijkman the reaction of the medium was pH 4.5. As a result, they were able to culture *E. coli* in every instance after incubation at 46°C for 96 hours and for longer periods.

Perry[3] reported on the use of a modified Eijkman medium using 0.3% lactose. This medium had been used successfully and routinely for the isolation of *E. coli* for a number of years. In a personal communication, Perry[4] recommended that Bacto Tryptose replace Bacto Peptone in this formula. The specificity of this medium for the detection of *E. coli* of fecal origin requires that the composition of the medium be uniform and exact, and that the incubation temperature be properly controlled at 45.5 to 46°C at all times. The formula for Bacto Eijkman Lactose Medium conforms to the formula for Eijkman broth as specified in *Standard Methods for the Examination of Water and Sewage.*[5]

FORMULA

BACTO EIJKMAN LACTOSE MEDIUM
DEHYDRATED

Ingredients per liter

Bacto Tryptose . . . . 15 g
Bacto Lactose . . . . 3 g
Dipotassium Phosphate . . . . 4 g
Potassium Dihydrogen Phosphate . . 1.5 g
Sodium Chloride . . . . 5 g

Final pH 6.8 ± 0.1 at 25°C.

Five hundred grams will make 17.5 liters of medium.

METHOD OF PREPARATION
1. Dissolve 28.5 g in 1 liter of distilled or deionized water.
2. Dispense into test tubes each containing an inverted fermentation vial.
3. Sterilize in the autoclave for 15 minutes at 15 lbs pressure (121°C).

If inoculum is larger than 1 ml per 10 ml of medium, the medium should be prepared in double or multiple strength to maintain the proper ratio of ingredients to medium.

STORAGE
Bacto Eijkman Lactose Medium   Below 30°C
Prepared medium   15 – 30°C

QUALITY CONTROL
Identity Specifications

Dehydrated powder: very light beige, homogeneous, free-flowing
Reaction of 2.85% solution: pH 6.8 ± 0.1 at 25°C
Prepared medium: very light amber, clear to very slightly opalescent

Typical Cultural Response in Bacto Eijkman Lactose Medium
After 18 – 48 Hours at 45.5 – 46°C

Organism	Growth
Enterobacter aerogenes ATCC® 13048	fair
Escherichia coli ATCC® 13762	good to excellent
Escherichia coli ATCC® 25922	good to excellent

REFERENCES

1. Eijkman, 1904, Centr. Bakt. II Abt., 37:742.
2. Perry and Hajna, 1933, Journal of Bacteriology, 26:419.
3. Perry, 1939, Food Research, 4:381.
4. Perry, 1939, Personal Communication.
5. American Public Health Association, 1946, Standard Methods for the Examination of Water and Sewage, 9th Ed., American Public Health Association, Inc., Washington, D.C.

PACKAGING

Bacto Eijkman Lactose Medium   500 g   0017-17-2

BACTO ELLIKER BROTH

INTENDED USE

Bacto Elliker Broth is used for culturing streptococci and lactobacilli of importance in the dairy industry.

HISTORY/PRINCIPLES

Bacto Elliker Broth, prepared according to the formulation of Elliker, Anderson and Hannesson,[1] is a slightly acidic medium containing nutrients that support the growth of recommended organisms. The medium is a modification of McLaughlin's[2] formulation.

FORMULA

BACTO ELLIKER BROTH
DEHYDRATED

Ingredients per liter

Bacto Tryptone	20 g	Bacto Saccharose	5 g
Bacto Yeast Extract	5 g	Sodium Chloride	4 g
Bacto Gelatin	2.5 g	Sodium Acetate	1.5 g
Bacto Dextrose	5 g	Ascorbic Acid	0.5 g
Bacto Lactose	5 g		

Final pH 6.8 ± 0.2 at 25°C.

One pound will make 9.36 liters of medium.

METHOD OF PREPARATION

1. To rehydrate the medium, suspend 48.5 g in 1 liter distilled or deionized water.
2. Heat to boiling to dissolve completely.
3. Dispense into tubes or flasks.
4. Sterilize in the autoclave for 15 minutes at 15 lbs pressure (121°C).

STORAGE

Bacto Elliker Broth   Below 30°C
Prepared medium   15 – 30°C

QUALITY CONTROL

Identity Specifications

Dehydrated powder: light beige, homogeneous, free-flowing
Reaction of 4.85% solution: pH 6.8 ± 0.2 at 25°C
Prepared medium: light amber, clear

Typical Cultural Response on Bacto Elliker Broth
After 18 – 48 Hours at 35°C

Organism	Growth
Lactobacillus casei ATCC® 7469	good to excellent
Lactobacillus lactis ATCC® 8000	good to excellent
Lactobacillus sp ATCC® 11506	good to excellent
Streptococcus cremoris	good to excellent

REFERENCES

1. J. Dairy Science, 39:1611, 1956.
2. J. Bact., 51:560, 1946.

PACKAGING

Bacto Elliker Broth	1 lb (454 g)	0974-01-1

BACTO EMERSON AGAR

INTENDED USE

Bacto Emerson Agar is a nutritive dextrose agar well suited for the cultivation of *Actinomyces*, *Streptomyces* and other fungi.

FORMULA

BACTO EMERSON AGAR
DEHYDRATED

Ingredients per liter

Bacto Beef Extract	4 g
Bacto Yeast Extract	1 g
Bacto Peptone	4 g
Bacto Dextrose	10 g
Sodium Chloride	2.5 g
Bacto Agar	20 g

Final pH 7.0 ± 0.1 at 25°C.

One pound will make 10.9 liters of medium.

METHOD OF PREPARATION

1. To rehydrate, suspend 41.5 g in 1 liter distilled or deionized water. Heat to boiling to dissolve completely.
2. Distribute in tubes or flasks and sterilize in the autoclave for 15 minutes at 15 lbs pressure (121°C).

STORAGE

Bacto Emerson Agar	Below 30°C
Prepared plates	2 – 8°C

QUALITY CONTROL

Identity Specifications

Dehydrated powder:	light tan, homogeneous, free-flowing
Reaction of 4.15% solution:	pH 7.0 ± 0.1 at 25°C
Prepared medium:	medium amber, slightly opalescent

Typical Cultural Response on Bacto Emerson Agar
After 40 – 72 Hours at 30 ± 2°C

Organism	Recovery
Aspergillus niger ATCC® 16404	good to excellent
Saccharomyces cerevisiae ATCC® 9763	good to excellent
Streptomyces achromogenes ATCC® 12767	good to excellent
Streptomyces albus ATCC® 3004	good to excellent
Streptomyces lavendulae ATCC® 8664	good to excellent

PACKAGING

Bacto Emerson Agar   1 lb (454 g)   0587-01-0

BACTO EMERSON Yp Ss AGAR

INTENDED USE

Bacto Emerson Yp Ss Agar is recommended for the cultivation of *Allomyces* and other fungi.

HISTORY/PRINCIPLES

Bacto Emerson Yp Ss Agar is prepared according to the formula given by Emerson.[1] Emerson and Wilson[2] used the medium in half strength for streaking zygotes or zoospores to obtain single germlings. Following solidification of the diluted medium in Petri dishes, three or four drops of sterile water were placed in the center of the plate. The swarmers were pipetted into this water and then spread evenly over the surface with a glass rod.

FORMULA

BACTO EMERSON Yp Ss AGAR
DEHYDRATED

Ingredients per liter

```
Bacto Yeast Extract . . . . . . . . . . . . .  4 g
Soluble Starch, Difco . . . . . . . . . . .  15 g
Dipotassium Phosphate  . . . . . . . . . .  1 g
Magnesium Sulfate  . . . . . . . . . . . .  0.5 g
Bacto Agar . . . . . . . . . . . . . . . . .  20 g
```

Final pH 7.0 ± 0.2 at 25°C.

One pound will make 10.9 liters of final medium.

METHOD OF PREPARATION

1. To rehydrate, suspend 40.5 g in 1 liter distilled or deionized water (20.2 g per liter for diluted medium).
2. Heat to boiling to dissolve completely.
3. Sterilize in the autoclave for 15 minutes at 15 lbs pressure (121°C).

STORAGE

Bacto Emerson Yp Ss Agar   Below 30°C
Prepared medium   2 – 8°C

QUALITY CONTROL
Identity Specifications

Dehydrated powder: light beige, homogeneous, free-flowing
Reactions of 4.05% solution: pH 7.0 ± 0.2 at 25°C
Prepared medium: light to medium amber, slightly opalescent, may
                              have a slight flocculent precipitate

Typical Cultural Response on Bacto Emerson Yp Ss Agar
After 40 – 72 Hours at 30°C

Organism	Growth
Aspergillus niger ATCC® 16404	good to excellent
Saccharomyces cerevisiae ATCC® 9763	good to excellent
Saccharomyces uvarum ATCC® 9080	good to excellent

REFERENCES
1. Lloydia, 4:77, 1941.
2. Mycologia, 46:393, 1954.

PACKAGING
Bacto Emerson Yp Ss Agar   1 lb (454 g)   0739-01-7

BACTO ENDAMOEBA MEDIUM

INTENDED USE
Bacto Endamoeba Medium is a liver infusion agar used for the cultivation of *Endamoeba histolytica* from specimens.

HISTORY/PRINCIPLES
The formula for Bacto Endamoeba Medium corresponds to that recommended by Cleveland and Sanders[1] and Cleveland and Collier.[2]

Cleveland and his associates made a comprehensive study of the cultivation of *E. histolytica*. They used egg, serum and various other materials for cultivating the amoeba and state[1]: "We feel however, that we really did not cultivate *Entamoeba histolytica* until we grew it on slants of liver infusion agar covered with fresh horse serum-saline 1:6 and containing rice flour." They used about half strength Bacto Liver Infusion Agar in this study. Bacto Endamoeba Medium has been prepared to duplicate the modified liver infusion agar described by Cleveland and his co workers.

This medium, furthermore, is reported by them to be almost specific for *E. histolytica*, as far as the intestinal amoebae of man are concerned. They attempted to cultivate *Endolimax nana, Dientamoeba fragilis,* and *E. coli* and failed, while *E. histolytica* grew abundantly. In a committee report Spector,[3] referee, states that the method of Cleveland and Collier using Bacto Liver Infusion Agar overlaid with sterile serum-saline 1:6 was one of the best for practical diagnostic purposes and that Wassermann-negative human inactivated serum, horse, or rabbit serum may be used. This report also mentions that other intestinal amoebae do not grow as readily in this culture medium as does *E. histolytica*.

FORMULA

BACTO ENDAMOEBA MEDIUM
DEHYDRATED
Ingredients per liter

Bacto Liver, Infusion from . . . . .272 g
Proteose Peptone, Difco . . . . 5.5 g
Sodium Glycerophosphate . . . . 3 g
Sodium Chloride . . . . 2.7 g
Bacto Agar . . . . 11 g

Final pH 7.0 ± 0.2 at 25°C.

Five hundred grams will make 15.1 liters of medium.

METHOD OF PREPARATION
1. To rehydrate, suspend 33 g in 1 liter distilled or deionized water and heat to boiling to dissolve completely.
2. Dispense in tubes and sterilize in the autoclave for 15 minutes at 15 lbs pressure (121°C).
3. Allow tubes to solidify in a slanted position.
4. Cover about 1/2 of the slant with fresh sterile horse serum-saline (1:6) and add a 5 mm loopful of Bacto Rice Powder which has been sterilized in a dry heat oven at 160°C for one hour. Scorching must be prevented.

STORAGE
Bacto Endamoeba Medium   Below 30°C
Prepared medium   2 – 8°C

QUALTIY CONTROL
Identity Specifications

Dehydrated powder: tan, free-flowing
Reaction of 3.3% solution: pH 7.0 ± 0.2 at 25°C
Prepared medium: dark amber, slightly opalescent

Typical Cultural Response on Bacto Endamoeba Medium
After 68 – 72 Hours at 25 – 30°C

Organism	Growth
Endamoeba histolytica	good
Endamoeba invadens	good
Endamoeba moskkovskii	good

REFERENCES
1. Arch. Protiskenkunde, 70:223, 1930.
2. Am. J. Hyg., 12:606, 1930.
3. Sixth Annual Year Book (1935 – 36), p. 130, Suppl., Am. J. Pub. Health, 26:No. 3, 1936.

PACKAGING
Bacto Endamoeba Medium   500 g   0053-17-7

BACTO ENDO AGAR

INTENDED USE
Bacto Endo Agar is used for the confirmation of the presumptive test for members of the coliform group of organisms.

HISTORY/PRINCIPLES

Endo[1] originally described a medium using a fuchsin sulfite indicator to differentiate lactose fermenting and lactose nonfermenting organisms from the intestinal tract. Upon plates of this medium, in which the fuchsin has been decolorized by sodium sulfite, salmonellae and other lactose nonfermenting organisms appear as clear, colorless, glistening drops against the faint pink background of the medium. Coliform organisms fermenting lactose yield red colonies and color the surrounding medium. The typical reactions of this medium are not caused by acid production but by the intermediate product acetaldehyde, which is fixed by the sodium sulfite as was shown by Margolena and Hansen[2] and Neuberg and Nord.[3]

Endo's original formula has been subjected to many modifications, due largely to variations in the available dyes and sulfites and to new uses of the medium advocated by individual investigators. The result has been a multiplicity of variations of the formula. Harris[4] investigated the problem of Endo agar, studying various ingredients, reaction, and available dyes. He reported that by using Levine's modification[5] several sources of error were eliminated and that Bacto Peptone as recommended by Levine gave satisfactory results. In this modification Harris found that a basic fuchsin composed of almost equal parts of rosanilin and pararosanilin gave color reactions which were exceedingly sensitive and consistent. Bacto Endo Agar was developed in cooperation with Harris, conforming to the Levine modification with the dye combination proposed by Harris.

Endo agar was originally developed for the isolation of typhoid bacilli. Since that time more satisfactory media have been developed for this problem and Endo agar has proved of most value in the bacteriological examination of water.

It has been recommended by the American Public Health Association as a "Standard Methods" medium for use in both water[6] and dairy products.[7]

FORMULA

BACTO ENDO AGAR
DEHYDRATED

Ingredients per liter

Bacto Peptone .... 10 g
Bacto Lactose .... 10 g
Dipotassium Phosphate .... 3.5 g
Bacto Agar .... 15 g
Bacto Basic Fuchsin .... 0.5 g
Sodium Sulfite .... 2.5 g

Final pH 7.5 ± 0.2 at 25°C.

One pound will make 10.9 liters of medium.

METHOD OF PREPARATION

1. Suspend 41.5 g in 1 liter distilled or deionized water and heat to boiling to dissolve completely.
2. Sterilize in the autoclave for 15 minutes at 15 lbs pressure (121°C).
3. Following autoclaving, twirl or gently shake the flask to evenly disperse the characteristic flocculant precipitate that forms during heating.
4. Pour into sterile Petri dishes.

Bacto Endo Agar inoculated the same day as rehydrated may be used without autoclave sterilization. Under these conditions the medium need be heated only to boiling to dissolve it completely before pouring into plates.

STORAGE

Bacto Endo Agar   Below 30°C
Prepared medium   2 – 8°C

QUALITY CONTROL

Identity Specifications

Dehydrated powder: medium purple, homogeneous, free-flowing
Reaction of 4.15% solution: pH 7.5 ± 0.2 at 25°C
Prepared medium: pinkish orange, opalescent with a precipitate

Typical Cultural Response on Bacto Endo Agar
After 24 ± 2 Hours at 35°C

Organism	Recovery	Color of Colony
Enterobacter aerogenes ATCC® 13048	good to excellent	red
Escherichia coli ATCC® 25922	good to excellent	red w/metallic sheen
Salmonella typhi ATCC® 6539	good to excellent	colorless
Shigella sonnei ATCC® 25931	good to excellent	colorless

REFERENCES

1. Endo, 1914, Centr. Bakt., Abt I, Orig., 35:109.
2. Margolena and Hansen, 1933, Stain Tech., 8:131.
3. Neuberg and Nord, 1919, Biochem. Zeit., 96:133.
4. Harris, 1925, Military Surgeon, 57:280.
5. Levine, 1918, Abst. Bact., 2:13.
6. American Public Health Association, 1975, Standard Methods for the Examination of Water and Wastewater, 14th Ed., American Public Health Association, Inc., Washington, D.C.
7. American Public Health Association, 1972, Standard Methods for the Examination of Dairy Products, 13th Ed., American Public Health Association, Inc., Washington, D.C.

PACKAGING

Bacto Endo Agar

1 lb (454 g)	0006-01-3
1/4 lb (114 g)	0006-02-2
5 lb (2.27 kg)	0006-05-9

BACTO m ENDO AGAR LES

INTENDED USE

Bacto m Endo Agar LES is a Standard Methods Medium[1] used for the enumeration of coliform organisms in water by the membrane filter technique.

HISTORY/PRINCIPLES

Bacto m Endo Agar LES is prepared according to the formulation of McCarthy, Delaney and Grasso[2] for determining the coliform contents of water by a two step membrane filter procedure using Bacto Lauryl Tryptose Broth as a preliminary enrichment. The authors claim higher coliform counts and better sheen than by the one step technique using Bacto m Endo Broth.

FORMULA

BACTO m ENDO AGAR LES
DEHYDRATED
Ingredients per liter

Bacto Yeast Extract	1.2 g	Sodium Chloride	3.7 g
Casitone	3.7 g	Sodium Desoxycholate	0.1 g
Thiopeptone	3.7 g	Sodium Lauryl Sulphate	0.05 g
Tryptose	7.5 g	Sodium Sulfite	1.6 g
Lactose	9.4 g	Bacto Basic Fuchsin	0.8 g
Potassium Phosphate, Dibasic	3.3 g	Bacto Agar	15 g
Potassium Phosphate, Monobasic	1 g		

Final pH 7.2 ± 0.2 at 25°C.

One pound will make 9 liters of medium.

METHOD OF PREPARATION

1. Suspend 51 g in 1 liter distilled or deionized water to which has been added 20 ml ethyl alcohol.
2. Heat to boiling to dissolve completely.
3. Cool to 45 – 50°C.
4. Dispense 4 ml amounts into the lower half of 50 – 60 mm Petri dishes and allow to solidify.

NOTE: If larger dishes are used, dispense sufficient medium to give a depth equivalent to that in the 60 mm dish.

TECHNIQUE

1. Prepare Bacto Lauryl Tryptose Broth according to label directions, dispensing the medium in flasks or tubes in quantities best suited for one's particular needs.
2. Rehydrate Bacto *m* Endo Agar LES, dispense into 60 mm Petri dishes in 4 ml amounts and allow to solidify.
3. Invert the plates containing the solidified medium and place a membrane filter absorbent pad inside the cover.
4. Add 1.8 – 2.0 ml Bacto Lauryl Tryptose Broth to each pad.
5. Carefully place a filter membrane through which the water sample has been filtered top side up on the pad in a rolling motion to avoid entrapping air bubbles.
6. Incubate at 35°C for 1-1/2 – 2-1/2 hours. Transfer the membrane from the pad to the surface of the medium in the Petri dish bottom keeping the side on which the bacteria have been collected facing upward.
7. Leave the filter pad in the lid and incubate the plates in the inverted positions at 35°C for 22 ± 2 hours.
8. The coliform colonies will be red and have the characteristic metallic sheen. Observe and count the colonies.

STORAGE

Bacto m Endo Agar LES	Below 30°C
Prepared medium	2 – 8°C

QUALITY CONTROL
Identity Specifications

Dehydrated powder:	purple, homogeneous, free-flowing
Reaction of 5.1% solution w/2% ethanol:	pH 7.2 ± 0.2 at 25°C
Prepared medium:	rose colored, slightly opalescent, may have a precipitate

Typical Cultural Response on Bacto m Endo Agar LES (as described above)
After 20 – 24 Hours at 35°C

Organism	Growth	Color of Colony
Enterobacter aerogenes ATCC® 13048	good	red to black w/sheen
Escherichia coli ATCC® 25922	good	red to black w/sheen
Salmonella typhi ATCC® 6539	good	colorless
Staphylococcus aureus ATCC® 25923	inhibited	—

REFERENCES

1. Standard Methods for the Examination of Water and Wastewater, 15th Ed., American Public Health Association, Inc., Washington, D.C., 1980.
2. McCarthy, DeLaney and Grasso, Water & Sewage Works, 108:238, 1961.

PACKAGING

Bacto m Endo Agar LES	1 lb (454 g)	0736-01-0
	1/4 lb (114 g)	0736-02-9
Bacto Lauryl Tryptose Broth	1 lb (454 g)	0241-01-8
	1/4 lb (114 g)	0241-02-7
	5 lb (2.27 kg)	0241-05-4
	10 kg	0241-08-1

BACTO m ENDO BROTH MF®

INTENDED USE

Bacto m Endo Broth MF® is a Standard Methods Medium[1] used in the one step membrane filter technique for the enumeration of coliform organisms in water.

HISTORY/PRINCIPLES

Bacto m Endo Broth MF® is a selective differential medium prepared according to the formulation of the Millipore Filter Corporation.[2] This medium, a combination of the former Bacto m HD Endo Medium and Bacto Lauryl Tryptose Broth, has proven to be especially well suited for the detection of coliforms by the membrane filter technique. Preliminary enrichment on a nonselective medium is not required in Bacto m Endo Broth MF®. The productivity, together with its ability to produce typical coliform colonies with sheen, make this the medium of choice for the coliform determination from water and other specimens.

FORMULA

BACTO m ENDO BROTH MF®
DEHYDRATED

Ingredients per liter

Bacto Yeast Extract . . . . 1.5 g	Dipotassium Phosphate . . . . 4.375 g
Bacto Casitone . . . . 5 g	Monopotassium Phosphate . . . . . 1.375 g
Bacto Thiopeptone . . . . 5 g	Sodium Chloride . . . . 5 g
Bacto Tryptose . . . . 10 g	Sodium Lauryl Sulfate . . . . 0.05 g
Bacto Lactose . . . . 12.5 g	Sodium Sulfite . . . . 2.1 g
Sodium Desoxycholate . . . . 0.1 g	Bacto Basic Fuchsin . . . . 1.05 g

Final pH 7.2 ± 0.2 at 25°C.

One pound will make 9.4 liters of medium.

METHOD OF PREPARATION
1. Suspend 48 g in 1 liter distilled or deionized water containing 20 ml of ethanol.
2. Heat to boiling. Do not prolong heating or heat in autoclave.
3. Cool to room temperature.
4. Dispense onto sterile absorbant pads.

NOTE: The medium should be used the same day as rehydrated and it should be protected from bright light.

STORAGE
Bacto m Endo Broth MF®   Below 30°C
Prepared medium   2 – 8° in the dark

QUALITY CONTROL
Identity Specifications

Dehydrated powder: Pink to purple, homogeneous, free-flowing
Reaction of 4.8% solution
  with 2% ethanol: 7.2 ± 0.2 at 25°C
Prepared medium: pinkish red, opalescent, may have a precipitate

Typical Cultural Response on Bacto m Endo Broth MF®
After 24 ± 2 Hours at 35°C

Organism	Growth	Color of Colony
Escherichia coli ATCC® 25922	good to excellent	pink w/metallic sheen
Salmonella typhimurium ATCC® 14028	good to excellent	colorless to slightly pink
Staphylococcus aureus ATCC® 25923	inhibited	—

REFERENCES
1. Standard Methods for the Examination of Water and Wastewater, 15th Ed., American Public Health Assoc., Inc., Washington, D.C., 1980.
2. Millipore Filter Corporation, Personal Communication, 1956.

PACKAGING
Bacto m Endo Broth MF®   1 lb (454 g)   0749-01-5
                                        1/4 lb (114 g)   0749-02-4
                                        25 × 2 ml   0749-36-4

BACTO ENTERIC FERMENTATION BASE

INTENDED USE
Bacto Enteric Fermentation Base with the addition of carbohydrates and Andrade's Indicator is used for determining fermentation reactions.

HISTORY/PRINCIPLES
Bacto Enteric Fermentation Base is prepared according to the formula described by Edwards and Ewing[1] and in *Diagnostic Procedures for Bacterial, Mycotic and Parasitic Infections*, 6th Edition published by the American Public Health Association.[2]

FORMULA

BACTO ENTERIC FERMENTATION BASE
DEHYDRATED

Ingredients per liter

Beef Extract . . . . 3 g
Peptone . . . . 10 g
Sodium Chloride . . . . 5 g

Final pH 7.2 ± 0.1 at 25°C.

Five hundred grams will make 27.78 liters of medium.

METHOD OF PREPARATION

1. Suspend 18 g in 1 liter distilled or deionized water.
2. Add 10 ml Andrade's Indicator.
3. Heat to boiling to dissolve completely. Add appropriate amounts of sterile carbo-hydrates as indicated below.
4. Dispense into 3 ml amounts into test tubes with inverted fermentation vials (Durham tubes).
5. Sterilize in the autoclave for 15 minutes at 15 lbs pressure (121°C).

Carbohydrate	Final Concentration	Add Before Autoclaving	Add After Autoclaving
Adonitol	0.5%	X	
Arabinose	0.5%		X
Cellobiose	0.5%		X
Dextrose (Glucose)	1%	X	
Dulcitol	0.5%	X	
Glycerol*	0.5%	X	
Inositol	0.5%	X	
Lactose	1%		X
Mannitol	1%	X	
Salicin	0.5%	X	
Sucrose	1%		X
Xylose	0.5%		X

However, it is recommended that all carbohydrates be added as filter sterilized solutions in appropriate volumes to the previously autoclaved and cooled media. Mixing is accomplished by rotating the tubes between the palms of the hands.

*Medium containing glycerol should be autoclaved for 10 minutes at 15 lbs pressure (121°C).

PROCEDURE

Inoculate from a culture which has been incubated overnight (18 – 20 hours). Incubate the tubes at 35 – 37°C for 4 – 5 days, examining daily. Note acid production indicated by formation of a dark pink to red color, and gas production in at least 3% of the volume of the fermentation vial. Negative tubes remain colorless. They should be observed regularly for a total of 30 days.

STORAGE

Bacto Enteric Fermentation Base	Below 30°C
Prepared medium	15 – 30°C

QUALITY CONTROL

Identity Specifications

Dehydrated powder:	light tan, homogeneous, free-flowing
Reaction of 1.8% solution:	pH 7.2 ± 0.1 at 25°C
Prepared medium:	light amber, clear

Typical Cultural Response on Bacto Enteric Fermentation Base w/Dextrose and Andrade's Indicator After 18 – 24 Hours at 35°C

Organism	Growth	Acid	Gas
Escherichia coli ATCC® 25922	good to excellent	+	+
Klebsiella pneumoniae ATCC® 13883	good to excellent	+	+
Salmonella typhimurium ATCC® 14028	good to excellent	+	+
Serratia marcescens ATCC® 8100	good to excellent	+	–
Shigella flexneri ATCC® 12022	good to excellent	+	–

REFERENCES

1. Edwards, P. R. and Ewing, W. H., Identification of Enterobacteriaceae, 3rd Ed., Minneapolis, Burgess Publishing Co., p. 342 – 343, 1972.
2. Balows, A. and Hausler, W. J., Diagnostic Procedures for Bacteria, Mycotic and Parasitic Infections, 6th Ed., Washington, D.C., American Public Health Association, p. 793, 1981.

PACKAGING

Bacto Enteric Fermentation Base 500 g 1828-17-9

ENTEROCOCCI MEDIA

BACTO ENTEROCOCCI PRESUMPTIVE BROTH
BACTO ENTEROCOCCI CONFIRMATORY AGAR
BACTO ENTEROCOCCI CONFIRMATORY BROTH

INTENDED USE

Bacto Enterococci Presumptive Broth, Bacto Enterococci Confirmatory Agar and Broth are used for both the presumptive detection and confirmation of enterococci in water supplies.

HISTORY/PRINCIPLES

These enterococci media are prepared according to the formulae published by Sandholzer and Winter[1] for the detection of enterococci in water supplies, swimming pools, sewage or other specimens suspected of containing these organisms. The procedure as described, consists of a presumptive phase in which the production of acid and turbidity (growth) in an azide presumptive broth after incubation at 45°C is considered positive presumptive evidence for the presence of enterococci. The positive presumptive tests are then confirmed by inoculating a slant-broth combination prepared with an azide agar medium overlaid with a salt-azide-penicillin broth. Pinpoint colonies on the slant, growth sediment in the broth, the presence of gram-positive ovoid streptococci in the broth, and a negative catalase test is considered confirmed positive evidence of the presence of enterococci.

FORMULAE

BACTO ENTEROCOCCI PRESUMPTIVE BROTH
DEHYDRATED

Ingredients per liter

Bacto Yeast Extract . . . . 5 g
Bacto Tryptone . . . . 5 g
Bacto Dextrose . . . . 5 g
Sodium Azide . . . . 0.4 g
Bacto Brom Thymol Blue . . . . . . 0.032 g

Final pH 8.4 ± 0.2 at 25°C.

Five hundred grams will make 32.4 liters of medium.

BACTO ENTEROCOCCI CONFIRMATORY AGAR
DEHYDRATED

Ingredients per liter

Bacto Yeast Extract	5 g
Bacto Tryptone	5 g
Bacto Dextrose	5 g
Sodium Azide	0.4 g
Bacto Agar	15 g
Bacto Methylene Blue	0.01 g

Final pH 8.0 ± 0.2 at 25°C.

Five hundred grams will make
16.4 liters of medium.

BACTO ENTEROCOCCI CONFIRMATORY BROTH
DEHYDRATED

Ingredients per liter

Bacto Yeast Extract	5 g
Bacto Tryptone	5 g
Bacto Dextrose	5 g
Sodium Azide	0.4 g
Sodium Chloride	65 g
Bacto Methylene Blue	0.01 g

Final pH 8.0 ± 0.2 at 25°C.

Five hundred grams will make
6.2 liters of medium.

METHOD OF PREPARATION
Bacto Enterococci Presumptive Broth
1. Dissolve the quantity of Bacto Enterococci Presumptive Broth specified in the table below in 1 liter of distilled or deionized water.
2. Dispense into tubes as directed.
3. Sterilize in the autoclave for 15 minutes at 15 lbs pressure (121°C).

	Medium g/liter	Dispense volume/tube	Inoculum
Single strength	15.4 g	8 ml	1 loop – 2 ml
5X strength	77 g	2 ml	10 ml

4. Allow to cool to room temperature and inoculate as desired.
5. Incubation of inoculated medium is at 45°C.

Bacto Enterococci Confirmatory Agar
1. Suspend 30.4 g in 1 liter distilled or deionized water and heat to boiling to dissolve completely.
2. Distribute into tubes.
3. Sterilize in the autoclave for 15 minutes at 15 lbs pressure (121°C).
4. Allow tubes to cool in a slanted position.

Bacto Enterococci Confirmatory Broth
1. Dissolve 80.4 g in 1 liter distilled or deionized water.
2. Distribute into flasks in 100 ml quantities.
3. Sterilize in the autoclave for 15 minutes at 15 lbs pressure (121°C).
4. Allow to cool to room temperature.
5. Add 65 units of penicillin to each 100 ml of medium.
6. Add enough Enterococci Confirmatory Broth containing penicillin to each Enterococci Confirmatory Agar slant to cover approximately one-half the surface of the slant.
7. Incubation of inoculated confirmatory media is at 35 – 37°C or at 45°C.

STORAGE

Bacto Enterococci Presumptive Broth	Below 30°C
Bacto Enterococci Confirmatory Agar	Below 30°C
Bacto Enterococci Confirmatory Broth	Below 30°C
Prepared Media	2 – 8°C

QUALITY CONTROL

Identity Specifications

	Presumptive Broth	Confirmatory Agar	Confirmatory Broth
Dehydrated powder:	light beige to green, homogeneous, free-flowing	greyish tan, homogeneous, free-flowing	very light beige, homogeneous, free-flowing
% solution:	1.54%	3.04%	8.04%
pH at 25°C:	8.4 ± 0.2	8.0 ± 0.2	8.0 ± 0.2
Prepared medium:	blue, clear	blue slant with amber butt, slightly opalescent (liquid medium is emerald green)	light blue, clear

Typical Cultural Response On Bacto Enterococci Media
After 18 – 24 Hours at 35°C

Organism	Growth	Acid	Catalase
Escherichia coli ATCC® 25922	inhibited	–	–
Streptococcus faecalis ATCC® 19433	good to excellent	+	–
Streptococcus faecalis ATCC® 29212	good to excellent	+	–

REFERENCE

1. Sandholzer and Winter, 1946, Commercial Fisheries Leaflet T1a.

PACKAGING

Bacto Enterococci Presumptive Broth	500 g	0300-17-8
Bacto Enterococci Confirmatory Agar	500 g	0301-17-7
Bacto Enterococci Confirmatory Broth	500 g	0302-17-6

BACTO m ENTEROCOCCUS AGAR

INTENDED USE

Bacto m Enterococcus Agar is a selective medium for use in membrane filtration procedures or as a direct plating medium for isolation and enumeration of enterococci in water, food or other materials.

HISTORY/PRINCIPLES

Bacto m Enterococcus Agar is a complete medium prepared according to the improved formulation of Slanetz and Bartley[1] with TTC. This medium was demonstrated by the authors to be superior to any membrane filter medium thus far devised for enumeration of enterococci. Higher counts and larger colonies were obtained by incubating the inoculated membranes on the agar surface than on pads saturated with liquid medium. Furthermore, the medium proved to be 100% selective for fecal streptococci even when heavily polluted water samples were tested.

Using Bacto *m* Enterococcus Agar, the authors obtained higher counts than by the MPN procedure, the mean ratio being 1.9:1. Many of the samples tested demonstrated the medium to give results comparable with those obtained by the techniques of Litsky, Mallmann and Fifield[2] for the detection of enterococci. The membrane filter method had the advantage in being simpler to perform, did not require confirmation and permitted a direct count of enterococci in 48 hours.

Burkwall and Hartman[3] compared the productivity of 15 agar media frequently used for isolation and enumeration of enterococci and concluded that the addition of 0.2% sodium carbonate and 0.05% Tween® 80 to Bacto *m* Enterococcus Agar resulted in a medium superior to the other media in the direct plating method. Briefly, the procedure used by Burkwall and Hartman consists of homogenizing, at high speed for 2 – 3 minutes, 50 g of the sample in 450 ml sterile 0.1% Bacto Peptone and plating the homogenate or dilutions in duplicate.

A modification of the above procedure is described by the Nordic Committee on Food Analysis.[4] In this method, the sample is homogenized and diluted with sterile 0.85% saline and the surface of the medium inoculated with either the homogenate or the dilutions.

FORMULA

BACTO m ENTEROCOCCUS AGAR
DEHYDRATED

Ingredients per liter

Bacto Tryptose .... 20 g	Sodium Azide .... 0.4 g
Bacto Yeast Extract .... 5 g	Bacto Agar .... 10 g
Bacto Dextrose .... 2 g	2,3,5 – Triphenyl Tetrazolium
Dipotassium Phosphate .... 4 g	Chloride .... 0.1 g

Final pH 7.2 ± 0.2 at 25°C.

One pound will make 7.15 liters of medium.

METHOD OF PREPARATION
1. Suspend 42 g in 1 liter distilled or deionized water and heat to boiling to dissolve completely. Do not overheat.
2. Dispense into Petri dishes as desired.

TECHNIQUE
Place the membranes through which the fluid specimens have been filtered, inoculum side up, directly on the medium surface. If the Burkwall and Hartman or the Nordic Committee on Food Analysis modifications are used, inoculate the surface of the medium with homogenate or dilutions. Incubate for 48 hours at 35 – 37°C and make observations. The enterococci appear on the membranes or on the surface of the medium as pink to dark maroon colonies from 0.5 – 3 mm in diameter.

STORAGE
Bacto m Enterococcus Agar	Below 30°C
Prepared medium	2 – 8°C

QUALITY CONTROL
Identity Specifications

Dehydrated powder:	light beige, homogeneous, free-flowing
Reaction of 4.2% solution:	7.2 ± 0.2 at 25°C
Prepared medium:	light amber, slightly opalescent

**Typical Cultural Response on Bacto m Enterococcus Agar After
40 – 48 Hours at 35°C**

Organism	Growth	Color of Colony
Escherichia coli ATCC® 25922	inhibited	—
Streptococcus faecalis ATCC® 19433	good to excellent	pink to red
Streptococcus faecalis ATCC® 29212	good to excellent	pink to red

REFERENCES

1. Slanetz and Bartley, 1957, Journal of Bacteriology, 74:591.
2. Litsky, Mallman and Fifield, 1953, American Journal of Public Health, 43:873.
3. Burkwall and Hartman, 1964, Applied Microbiology, 12:18.
4. Nordic Committee on Food Analysis, 1968, Leaflet #68.

PACKAGING

Bacto m Enterococcus Agar	1 lb (454 g)	0746-01-8
	1/4 lb (114 g)	0746-02-7

BACTO ERIOCHROME BLACK

Bacto Eriochrome Black is recommended for use as a counterstain in fluorescent antibody techniques, especially when used in the serological identification of *Listeria*.

Hall and Hansen[1] recommended that it be used as a counterstain when tissue impression smears are being stained or when excessive extraneous material is present.

For a complete discussion on procedure, refer to Listeria Serology.

STORAGE

Bacto Eriochrome Black	15 – 30°C

REFERENCE

1. Zbl. Bakt. Para. Inf. and Hyg., 184:548, 1962.

PACKAGING

Bacto Eriochrome Black	5 ml	3249-56-8

ESCULIN

Esculin, a water soluble glycoside is added as a 0.1% concentration to culture media containing a ferric salt. These media are used to differentiate certain streptococci on the basis of their ability to hydrolyze esculin. The hydrolytic product reacts with the ferric compound yielding a grey to black color. The test is used for differentiating certain streptococci such as *S. faecalis*, which hydrolyzes esculin, from *S. agalactiae*, which does not hydrolyze this glycoside.

PACKAGING

Esculin	10 g	0158-12-6

BACTO EUGON AGAR
BACTO EUGON BROTH

INTENDED USE
Bacto Eugon media are recommended for culturing a large variety of microorganisms especially for procedures which require mass cultivation as in vaccine production.

HISTORY/PRINCIPLES
Bacto Eugon Broth and Bacto Eugon Agar contain dextrose as a ready source of energy thereby supporting rapid and luxuriant growth of most microorganisms. They may be used with or without enrichments as required. When enriched with blood they support luxuriant growth of pathogenic fungi including *Nocardia, Histoplasma* and *Blastomyces*. When enriched with Bacto Supplement B or C, they support excellent growth of *Neisseria, Francisella* and *Brucella*. The unenriched media support rapid growth of lactobacilli associated with cured meat products, dairy products and other foods.

Niven[1] reported the use of eugon agar for the detection of lactic acid in cured meats and recommended it as an all purpose medium for investigating spoilage problems in meats and other food products. Harrison and Hansen[2] employed the medium successfully for making plate counts of the intestinal flora of turkeys. Frank[3] demonstrated its usefulness in germinating anaerobic spores that had been pasteurized at 104°C.

Bacto Eugon Agar is not recommended as a blood agar base for hemolytic reactions because of its high sugar content.

FORMULAE

BACTO EUGON AGAR
DEHYDRATED

Ingredients per liter

Bacto Tryptose	15 g
Bacto Soytone	5 g
Bacto Dextrose	5 g
L-Cystine, Difco	0.2 g
Sodium Chloride	4 g
Sodium Sulfite	0.2 g
Bacto Agar	15 g

Final pH 7.0 ± 0.2 at 25°C.

One pound will make 10 liters of medium.
Rehydrate with 45.4 g/liter.

BACTO EUGON BROTH
DEHYDRATED

Ingredients per liter

Bacto Tryptose	15 g
Bacto Soytone	5 g
Bacto Dextrose	5 g
L-Cystine, Difco	0.2 g
Sodium Chloride	4 g
Sodium Sulfite	0.2 g

Final pH 7.0 ± 0.2 at 25°C.

One pound will make 14.9 liters of medium.
Rehydrate with 30.4 g/liter.

METHOD OF PREPARATION
1. Suspend appropriate amount in 1 liter distilled or deionized water. Heat to boiling to dissolve completely.
2. Sterilize in the autoclave for 15 minutes at 15 lbs pressure (121°C).
3. Allow to cool to 50 – 55°C and add any desired enrichments. Dispense aseptically into Petri dishes or tubes as required.

NOTE: Since most microorganisms prefer a freshly prepared medium with a moist surface, it is suggested that Bacto Eugon Agar be prepared as required or melted and resolidified just prior to use. Do not melt and resolidify media containing enrichments.

STORAGE

Bacto Eugon media	Below 30°C
Prepared media	2 – 8°C

QUALITY CONTROL

Identity Specifications

Dehydrated powder:	beige, homogeneous, free-flowing
Reaction of appropriate solution:	pH 7.0 ± 0.2 at 25°C
Prepared medium:	light amber, clear to slightly opalescent, may have a very slight precipitate

Typical Cultural Response on/in Bacto Eugon Media
After 40 – 48 Hours 35°C*

Organism	Growth (Plain)
*Aspergillus niger ATCC® 16404	good
Brucella abortus ATCC® 4315	good
*Candida albicans ATCC® 26790	good
Lactobacillus fermentum ATCC® 9338	good
Lactobacillus sp ATCC® 11506	good
Neisseria meningitidis ATCC® 13090	good
Shigella flexneri ATCC® 12022	good to excellent
Streptococcus pneumoniae ATCC® 6303	good to excellent
Streptococcus pyogenes ATCC® 19615	good to excellent

*Incubate these cultures at 30°C

REFERENCES

1. J. Bact., 58:633, 1949.
2. ibid., 59:197, 1950.
3. ibid., 70:269, 1955.

PACKAGING

Bacto Eugon Agar	1 lb (454 g)	0589-01-8
Bacto Eugon Broth	1 lb (454 g)	0590-01-5

BACTO F35M HAJNA

INTENDED USE

Bacto F35M Hajna (Urease Indole Test Broth) is a differential medium for determining urease and indole production by *Enterobacteriaceae*.

HISTORY

It is prepared according to the formulation suggested by Hajna.[1] The broth is used in combination with Bacto Motility Sulfide Medium after the motility and sulfide production have been recorded.

METHOD OF PREPARATION

1. Suspend 30 g in 1 liter distilled or deionized water and agitate to dissolve completely.
2. Distribute in 100 ml amounts into sterile screw capped bottles (or 9 ml amounts into tubes) and heat at 100°C in flowing steam for 5 minutes.
3. Final pH 6.9 ± 0.2 at 25°C. One pound will make 15.1 liters of medium.

PROCEDURE

Cultures to be tested are inoculated by stabbing into Bacto Motility Sulfide Medium and after overnight incubation at 30°C, motility and sulfide production are recorded. The medium is overlayed with 1 ml of the sterile Bacto F35M Hajna (Urease Indole Test Broth) and incubated at 35°C for no longer than 4 hours. Readings for urease production are read at the end of 30 minutes and 4 hours. A change of color from green to yellow indicates a negative urease test and deep blue to purple is a positive urease test.

After the urease test is read, add 0.3 ml Kovacs' Reagent, shake and allow the reagent to come to the top of the broth. Read immediately. A red color of the Kovacs' Reagent indicates a positive test for indole.

STORAGE

Bacto F35M Hajna   Below 30°C
Prepared medium   15 – 30°C

QUALITY CONTROL
Identity Specifications

Dehydrated powder: light pinkish beige, homogeneous, free-flowing
Reaction of 3.0% solution: pH 6.9 ± 0.2 at 25°C
Prepared medium: reddish green, clear

Typical Cultural Response in Bacto F35M Hajna

Organism	Growth	Bacto Motility Sulfite Medium Motility	H$_2$S	Indole	Urease
Enterobacter aerogenes ATCC® 13048	good	+	−	−	−
Escherichia coli ATCC® 25922	good	+	−	+	−
Proteus mirabilis ATCC® 25933	good	+	+	−	+
Proteus vulgaris ATCC® 13315	good	+	+	+	+

REFERENCE

1. Personal Communication, 1967.

PACKAGING

Bacto F35M Hajna   1 lb (454 g)   0675-01-3

BACTO m FC MEDIA

BACTO m FC BROTH BASE
BACTO m FC AGAR

INTENDED USE

Bacto m FC Agar and Bacto m FC Broth Base are used for the detection and enumeration of fecal coliform organisms by the membrane filter technique at elevated temperatures.

HISTORY/PRINCIPLES

Bacto m FC Broth Base is prepared according to the formula of Geldreich, Clark and Kabler[1] as published by Geldreich, Clark, Huff and Bert.[2] Bacto m FC Agar is the same formula with agar added as a solidifying agent. On these media, supplemented with Bacto Rosolic Acid, colonies of fecal coliforms appear blue and all other colonies grey.

FORMULAE

BACTO m FC AGAR
DEHYDRATED
Ingredients per liter

Bacto Tryptose . . . . 10 g
Proteose Peptone No. 3, Difco . . . . . 5 g
Bacto Yeast Extract . . . . 3 g
Bacto Lactose . . . . 12.5 g
Bacto Bile Salts No. 3 . . . . 1.5 g
Sodium Chloride . . . . 5 g
Bacto Agar . . . . 15 g
Aniline Blue . . . . 0.1 g

Final pH 7.4 ± 0.2 at 25°C.

One pound will make 8.7 liters
of medium.

BACTO m FC BROTH BASE
DEHYDRATED
Ingredients per liter

Bacto Tryptose . . . . 10 g
Proteose Peptone No. 3, Difco . . . . . 5 g
Bacto Yeast Extract . . . . 3 g
Sodium Chloride . . . . 5 g
Bacto Lactose . . . . 12.5 g
Bacto Bile Salts No. 3 . . . . 1.5 g
Aniline Blue . . . . 0.1 g

Final pH 7.4 ± 0.2 at 25°C.

One pound will make 12.3 liters
of medium.

METHOD OF PREPARATION
Bacto m FC Agar
1. Suspend 52 g in 1 liter distilled or deionized water and heat to boiling to dissolve completely.
2. Add 10 ml 1% Bacto Rosolic Acid in 0.2N NaOH solution and continue heating for one minute.
3. Cool to 50°C, pour plates and allow to solidify.

Bacto m FC Broth Base
1. Suspend 37 g in 1 liter distilled or deionized water.
2. Add 10 ml 1% Bacto Rosolic Acid in 0.2N NaOH solution and heat to boiling.
3. Cool to room temperature and add 2 ml of broth to each sterile absorbant pad placed in a Petri dish.

TECHNIQUE
1. Place membrane filter through which the sample has been filtered on the surface of the agar or onto the top of the saturated absorbant pad.
2. Place cover of Petri dish on tightly.
3. Submerge the closed dishes in a 44.5°C water bath for 24 hours.
4. Remove from water bath and observe for coliforms and count colonies.

STORAGE
Bacto m FC media   Below 30°C
Prepared media   2 – 8°C

QUALITY CONTROL
Identity Specifications

Dehydrated powder: beige, homogeneous, free-flowing
Reaction of appropriate solution
  without rosolic acid: pH 7.4 ± 0.2 at 25°C
Prepared medium (plain):   blue, slightly opalescent
Prepared medium (w/rosolic acid):   cranberry red, slightly opalescent

Typical Cultural Response on Bacto m FC Media After 22 – 24 Hours

Organism	Growth at 44.5°C	Growth at 35°C	Color of colony
Escherichia coli ATCC® 25922	good	good	blue
Salmonella typhimurium ATCC® 14028	inhibited	good	grey
Shigella flexneri ATCC® 12022	inhibited	good	grey
Streptococcus faecalis ATCC® 19433	inhibited	inhibited	—

REFERENCES

1. Geldreich, Clark and Kabler, 1963, USPHS, HEW. Personal Communication.
2. Geldreich, Clark, Huff and Bert, 1965, Journal of American Water Works Association, 57:208.

PACKAGING

Bacto m FC Agar	1 lb (454 g)	0677-01-1
	1/4 lb (114 g)	0677-02-0
Bacto m FC Broth Base	1 lb (454 g)	0883-01-1
	1/4 lb (114 g)	0883-02-0
Bacto Rosolic Acid	6 × 1 g	3228-09-1

BACTO FEBRILE ANTIGEN SET

PRODUCT DESCRIPTION AND INTENDED USE

Bacto Febrile Antigens are chemically inactivated, dyed suspensions of internationally recognized strains of bacteria, for use in the slide and tube agglutination procedures to detect antibodies (agglutinins) implicated in certain febrile diseases.

The rapid slide procedure is a screening test for detecting agglutinins whereas the tube test is a confirmatory procedure for quantitating agglutinin composition. It is therefore necessary that any positive results obtained in the screening (slide test) of human sera that the tube test be employed to establish the titer.

Bacto Febrile Positive Control Polyvalent is a hyperimmune antiserum prepared in goats, designed to demonstrate agglutination employing these test procedures, to a "significant titer" or greater with all the bacterial antigens provided in this set.

Bacto Febrile Negative Serum is a serum preparation of animal origin, designed to demonstrate a negative reaction with the antigens included in this set.

These reagents are for use only in the test procedure described herein. They are not recommended for the direct diagnosis of a disease but rather only for the detection of the presence of agglutinins particularly in cases of seroconversion.

Past history in the use of bacterial suspensions in the detection of febrile agglutinins, has resulted in a pattern of titers which have been considered "significant." These titers may be found in Table I. It must be emphasized that, although the information obtained is not absolute, a four-fold increase in titer generally between 2 consecutive specimens is regarded clinically significant. The value of febrile antigens is in the detection of an increase in the titer of agglutinins in sera from the same patient in successive bleedings when the test is performed at the same time, using the same reagents, by the same technician.

HISTORY/PRINCIPLES

"Febrile Antigen" is a term which has been accepted generally as referring to bacterial suspensions representative of a number of species of microorganisms pathogenic to man and which are accompanied by a fever in the host. There are a number of such organisms, but by classical definition, the most prominent include the genera *Salmonella, Brucella, Francisella (Pasteurella), Leptospira* and some members of the *Rickettsia*.[1,2,3,4]

Grunbaum[5] and Widal[6] introduced techniques which were to be the first practical application of the new science of immunology. Their test has become known universally as the Widal Test. Their tests quantitively measured the agglutinins in the sera of patients with Typhoid Fever. It has been extended to include *Salmonella* other than *S. typhi* by the use of a variety of Salmonella O and H Antigens.

Later, Weil and Felix[7] discovered a fortuitous cross-reaction between the agglutinins produced by the causative agent of Typhus Fever, *Rickettsia prowazekia,* and certain strains of *Proteus vulgaris* isolated from patients suffering from Typhus Fever. These *Proteus* strains, designated OX19 and OX2 seem to share an alkali stable polysaccharide with *R. prowazekia* and therefore, produce agglutinins identical to it.[8,9]

The well-known Weil-Felix reaction utilizing the Proteus Antigen was extended when in 1923, Kingsbury isolated another Proteus strain, OXK, which gave specific reaction with serum from patients convalescing from scrub typhus.

Although the febrile antigen agglutination technique is useful for purposes described above, it should not be a substitute for conventional methods for the isolation and identification of the etiological agent when such isolation is possible.

When an organism invades its host, the body responds to an antigenic stimulus with the production of antibodies. The titer or concentration of the antibody produced forms very slowly in the early stages of the disease, rises to a maximum level and then falls to a low or even undetectable level.

The level and rate of antibody formation and the rise and fall of titer depends upon the antigenicity of the antigen, the amount and route of introduction of the antigen into the host, the age and past history of the host as well as upon its physiological state at the time of stimulation. If the host has had previous exposure to the same antigen, the secondary response is usually more rapid and the titer rises to a higher level. The latter phenomenon is known as a specific anamnestic reaction. The titer of antibody so produced may be measured qualitatively or quantitatively using a homologous antigen.

The immune response of an individual to the etiological agent has an important influence on the usefulness of the test. A negative reaction does not preclude an active infection. Such a reaction may mean that the specimen was taken before the host produced antibodies to the etiological agent. Conversely, a positive reaction with a given antigen may not be diagnostic, since a patient's serum may exhibit a rise in heterologous agglutinins during the course of a disease, i.e., an infection by a *Brucella* can result in a rise of *Francisella (Pasteurella) tularensis* or *Salmonella* agglutinins to significant titers. In some cases, titers to heterologous strains of *Leptospira* may be even greater than that of the etiological agent. Such reactions are known as nonspecific anamnestic reactions, i.e., a host response to antigenic stimulus resulting in production of agglutinins reacting to antigens to which the host might not have been exposed previously.

The febrile antigen rapid slide test, including the Widal and Weil-Felix reactions usually employs 6 or more antigens from bacteria commonly causing fever in the host. A Proteus OX19 suspension is used to detect rickettsial antibodies. Varying amounts of the sera to be tested are distributed to the corresponding square of a previously marked glass slide. A febrile antigen is added and mixed with the sera. Positive or negative results are read 1 minute after mixing.

The rapid slide test is the most widely used procedure employing febrile antigens because of the simplicity with which the results may be reported. Negative reactions with this test can usually be reported as such if all 5 serum dilutions have been used. (Note: Even though the slide test is not quantitative it is necessary to run the series of dilutions in order to detect agglutinin content of a serum which might be overlooked in the case of a "prozone phenomenon.") This often occurs in the sera containing *Brucella* agglutinins and to a lesser extent typhoid agglutinins where higher concentrations of the serum may yield negative results but a dilution of the serum is positive.

The macroscopic tube test[10] should be used to confirm the presence of antibodies demonstrated by the slide technique and to quantitate their titer in suspected sera. In this test, the patient's serum is serially diluted in test tubes, a constant amount of the appropriate dilution of a febrile antigen is added to each tube, the resultant mixture is incubated according to directions and the agglutination pattern is read and recorded.

When quantitative determination of *Rickettsia* or *Brucella* agglutinins are necessary, tube antigens are to be used. These antigens are not supplied in the set but are available from Difco.

REAGENTS
Antigens
Bacto Brucella Abortus Antigen is prepared from strain 1119-3 obtained from the USDA[11,12,13]

Bacto Proteus OX19 Antigen — *Proteus vulgaris* OX19
Bacto Salmonella H Antigen a — *Salmonella enteritidis bioser Paratyphi A*
Bacto Salmonella H Antigen b — *Salmonella enteritidis ser Paratyphi B*
Bacto Salmonella H Antigen d — *Salmonella typhi* H901
Bacto Salmonella O Antigen Group D — *Salmonella typhi* O901

Concentration of Antigen[14,15]
Bacto Brucella Abortus Antigen is adjusted to a cell concentration of approximately 10% by volume.

Bacto Salmonella Antigens (both O and H) have a density of approximately 20X a McFarland Barium Sulfate Standard No. 3 (1.8×10^{10} organisms per ml) for use in the undiluted state employing the rapid slide screening and diluted 1:20 for use in the tube test.

Note: The density of the antigens may vary from the above values. They are adjusted to perform at their optimum when standardized with hyperimmune sera obtained from laboratory animals. In some antigens, glycerin to a final volume of 20% is added when necessary to adjust the suspension's sensitivity.

Bacto Proteus Antigens are adjusted to a concentration of approximately 7.5% by volume.

Due to various factors, including the genus of the organism involved, the density of each lot of suspension etc., the color intensity of the dye in each lot of antigen may vary and is normal. Such differences in color will not affect the outcome of the test.

CONTROL SERA
Bacto Febrile Positive Control Polyvalent contains agglutinins for those antigens listed above to a titer of 1:80 or greater as determined in our laboratories.

Merthiolate® is added as a preservative in the amount of 1:5,000 (w/v) final concentration.

Bacto Febrile Negative Control is a standardized protein solution containing Merthiolate® as a preservative in the amount of 1:5,000 (w/v) final concentration.

REHYDRATION AND STORAGE
1. The antigens are liquid and ready for use in the slide test. The Salmonella Antigens may be used in the tube test also, provided they are first diluted 1:20 with formalized saline. Note: Slight variations in the color intensity of the antigen is normal and will not affect the outcome of the test.
2. The control sera are rehydrated by adding 5 ml sterile distilled or deionized water at room temperature.
3. Store all reagents at 2 – 8°C.

These reagents are stable to the expiry date on the label when stored at 2 – 8°C provided they have not been exposed to temperature extremes (excessive heat or freezing) and have not become contaminated chemically or bacterially during routine usage.

Bacto Febrile Antigens are not to be used for immunization of man or animals. When used according to the suggested procedures, each 5 ml vial of antigen is sufficient to perform approximately 20 slide tests. Each 5 ml vial of Salmonella Antigens used in the tube test, is sufficient to perform approximately 25 tests.

Each 5 ml vial of Bacto Febrile Positive Control Polyvalent and Bacto Febrile Negative Control are sufficient to perform approximately 32 slide tests or 50 tube tests using a single antigen or approximately 5 slide tests using all the antigens in the kit.

PRESERVATIVES
Bacto Proteus OX19 Antigen contains:
A final concentration of approximately 0.25% formaldehyde
Crystal Violet in approximately 0.002% final concentration
Brilliant Green in approximately 0.004% final concentration.

Bacto Salmonella H Antigens contain:
A final concentration of approximately 0.3% formaldehyde
Crystal Violet in approximately 0.002% final concentration
Brilliant Green in approximately 0.005% final concentration.

Bacto Salmonella O Antigen Group D contains:
A final concentration of approximately 0.5% phenol (5 g/liter)
Crystal Violet in approximately 0.002% final concentration
Brilliant Green in approximately 0.005% final concentration.

Bacto Brucella Abortus Antigen contains:
A final concentration of 0.5% phenol (5 g/liter)
Crystal Violet in approximately 0.002% final concentration
Brilliant Green in approximately 0.004% final concentration.

PROCEDURE, RESULTS AND INTERPRETATION

Materials Provided:

Bacto Brucella Abortus Antigen
Bacto Proteus OX19 Antigen (slide)
Bacto Salmonella O Antigen Group D
Bacto Salmonella H Antigen a
Bacto Salmonella H Antigen b
Bacto Salmonella H Antigen d
Bacto Febrile Positive Control Polyvalent
Bacto Febrile Negative Control
Droppers for the above antigens (the supplied droppers deliver approximately 0.05 ml/drop)

Purchase Separately:

Bacto Brucella Abortus Antigen (tube)
Bacto Proteus OX19 Antigen (tube)

Materials Required but not Provided:

Rapid Slide Test

Wax marking pencils
Ruler
Glass plate 8″ × 8″
0.2 ml serological pipettes
Distilled or deionized water
Erlenmeyer flask, 50 ml
Applicator sticks
Suitable light source

Tube Test

Formalized saline (0.5 ml formaldehyde and 0.85 g NaCl per 100 ml distilled or deionized water)
Isotonic saline (0.85 g NaCl per 100 ml distilled or deionized water)
Distilled or deionized water
Erlenmeyer flasks
Kahn type tubes and supports
Serological pipettes — 1 ml and 5 ml capacity
Water baths at 37°C and 50°C
Refrigerator 2 – 8°C
Suitable light source

For complete description of the test procedures, results, limitations and interpretation refer to Bacto Salmonella H and O Antigens, Bacto Brucella Antigens, and Bacto Proteus Antigens.

An illustration and explanation for interpretation of macroscopic tube reactions can be found in the Serology introductory section in the front of this manual.

EXPECTED VALUES

Table I presents data that will be helpful in making serological tests with the febrile antigens. The values tabulated will vary somewhat in certain cases.

Table I

Bacto Antigen	Pathology	Time of Maximum Titer	Significant Titer
B. Abortus	Brucellosis	3 – 5 weeks	1:80 over 1:160 indicative
Proteus OX19	Rocky Mt. Spotted Fever	2 – 3 weeks	1:160 – 1:320
Proteus OX19	Typhus	2 – 3 weeks	1:160 and higher
Salmonella H Antigen d (Typhoid H)	Typhoid Fever	4 – 5 weeks	1:180
Salmonella O Antigen Group D (Typhoid O)	Typhoid Fever	3 – 5 weeks	1:80* over 1:160 indicative
Salmonella H Antigen a (Para A)	Paratyphoid Fever	3 – 5 weeks	1:80*
Salmonella H Antigen b (Para B)	Paratyphoid Fever	3 – 5 weeks	1:80*
Leptospira	Leptospirosis	1 – 2 weeks	1:100
F. Tularensis	Tularemia	4 – 8 weeks	1:160

*Significant in nonvaccinated individuals.

NOTE: A difference in titer of 1 dilution, plus or minus, between replicate samples or between several samples drawn 1 – 2 weeks apart and run in parallel cannot be considered significant. A 1 dilution ± deviation is within the limits of laboratory error.[16,17]

REFERENCES

1. Schubert, J. H., Holdeman, L. and Martin, D. S., J. Lab. Clin. Med., 44:194, 1964.
2. Welch, H. and Stuart, C. A., J. Lab. and Clin. Med., 21:411, 1936.
3. Welch, H. and Mickle, F. L., Amer. J. Pub. Hlth., 26:248, 1936.
4. Diamond, B. E., Pub. Hlth, Lab., 6:74, 1948.
5. Grunbaum, A. S., Lancet, 2:806, 1896.
6. Widal, F., Bull. Soc. Med. Hop. de Patis, 13, 1896.
7. Weil, E. and Felix, A., Wein Klin, 29:974, 1916.
8. Castaneda, M. R., J. Exper. Med., 62:289, 1935.
9. White, P. B., Brit. J. Exper. Path., 14:145, 1933.
10. Spink, W. W., McCullough, N. B., Hutchings, L. H., Amer. J. Clin. Path., 24:486, 1954.
11. USDA, ARS Publication, 1959.
12. Spink, W. W., Amer. J. Clin. Path., 22:201, 1952.
13. Feinberg, R. J., and Wright, G. G., J. Immunol., 67:115 (Aug), 1951.
14. Identification of Entero., Edwards, P. R., and Ewing, W. H., 3rd Ed., Burgess Publ. Co.
15. Huddleson, J. F., and Abell, E., J. Infect. Dis., 42:242, 1928.
16. Syphilis, A Synopsis, US Dept. of Hlth., Ed. and Welf., Publ. Hlth. Serv. Publ. No. 1660, Jan 1968.
17. Quality Control in Clinical Micro., Revised, ASCP Comm. on Continuing Education, Coun. on Micro., 1968.

PACKAGING

Bacto Febrile Antigen Set	8 × 5 ml	2407-32-7
Contains: (Also available individually)		
Bacto Brucella Abortus Antigen (slide)	5 ml	2909-56-1
Bacto Proteus OX19 Antigen (slide)	5 ml	2234-56-7
Bacto Salmonella O Antigen Group D (Typhoid O)	5 ml	2842-56-1
Bacto Salmonella H Antigen a (Paratyphoid A)	5 ml	2844-56-9
Bacto Salmonella H Antigen b (Paratyphoid B)	5 ml	2845-56-8
Bacto Salmonella H Antigen d (Typhoid H)	5 ml	2847-56-6
Bacto Febrile Positive Control Polyvalent	5 ml	3238-56-1
Bacto Febrile Negative Control	5 ml	3239-56-0

Other Febrile Antigens not supplied in the set but useful under conditions described in this literature:

Bacto Brucella Abortus Antigen (tube)	25 ml	2466-65-5
Bacto Proteus OX19 Antigen (tube)	25 ml	2247-65-1
Bacto Proteus OX2 Antigen (slide)	5 ml	2243-56-6
Bacto Proteus OX2 Antigen (tube)	25 ml	2248-65-0
Bacto Proteus OXK Antigen (slide)	5 ml	2244-56-5
Bacto Proteus OXK Antigen (tube)	25 ml	2249-65-9
Bacto Francisella Tularensis Antigen (slide)	5 ml	2240-56-9
Bacto Francisella Tularensis Antigen (tube)	25 ml	2251-65-4
Bacto Salmonella O Antigen Group C	5 ml	2841-56-2
Bacto Salmonella O Antigen Group E	5 ml	2843-56-0
Bacto Salmonella H Antigen c	5 ml	2846-56-7
Bacto Leptospira Antigens	See complete listing in Product List	

BACTO FILDES ENRICHMENT

INTENDED USE

Bacto Fildes Enrichment is a sterile digest of sheep blood recommended for use in both liquid and solid media for culturing *Haemophilus influenzae* and other microorganisms requiring blood derivatives for optimal growth.

Bacto Fildes Enrichment is recommended for use in Bacto Brain Heart Infusion, Bacto Heart Infusion Broth, Bacto Brain Heart Infusion Agar or Bacto Heart Infusion Agar.

HISTORY/PRINCIPLES

Bacto Fildes Enrichment, prepared as described by Fildes[1,2], is rich in hematin and coenzyme required by *H. influenzae* and other members of the *Haemophilus* group and contains additional factors from the blood stimulatory to many fastidious microorganisms.

Bacto Fildes Enrichment is recommended for use in Bacto Brain Heart Infusion, Bacto Heart Infusion Broth and Bacto Brain Heart Infusion Agar or Bacto Heart Infusion Agar. Optimal results obtained when the enrichment is employed in 5% concentration in the above media. Since many factors in the enrichment are heat labile, it cannot be heated but must be added aseptically in the proper amounts to media which have been sterilized in the autoclave and cooled to 50 – 55°C.

STORAGE

Bacto Fildes Enrichment   2 – 8°C

QUALITY CONTROL

Appearance: very dark brown solution

Typical Cultural Response on Bacto Heart Infusion Agar w/5% Bacto Fildes Enrichment After 18 – 24 Hours at 35°C

Organism	Growth
Haemophilus influenzae ATCC® 19418	excellent
Haemophilus parainfluenzae ATCC® 7901	excellent

REFERENCES

1. Brit. J. Exp. Path., 1:129 – 130, 1920.
2. *ibid.*, 2:16 – 25, 1921.
3. Stokes, Clinical Bacteriology, 2nd Ed., London, Edward Arnold, 1960.
4. Willis, Anaerobic Bact. in Clin. Med., London, Butterworth & Co., Ltd., 1960.

PACKAGING

Bacto Fildes Enrichment	6 × 5 ml	0349-57-2
	100 ml	0349-72-3
	6 × 100 ml	0349-73-2

BACTO FLAGELLA STAIN

Bacto Flagella Stain, Dehydrated is prepared according to the formulation described by Leifson[1] and is recommended for staining flagella of bacteria.

REAGENTS
Preparation of Flagella Stain

Bacto Flagella Stain	1.9 g
Alcohol, Ethyl, 95%	33 ml
Distilled or Deionized Water	67 ml

1. Prepare 100 ml solution, or larger quantities as required by suspending the dry ingredients in a mixture of the alcohol and water.
2. Allow to stand for 20 minutes with frequent shaking to obtain complete solution.
3. Determine the pH of the solution which should be 5.0 ± 0.2.
4. If the pH is below 4.8, adjust carefully by adding drop by drop of a 1% aqueous solution of sodium carbonate with thorough shaking. If the pH is above 5.2 adjust carefully with a 1% aqueous solution of tannic acid.

Performance of Flagella Stain
Precaution: The use of scrupulously clean slides is essential for best results.

PROCEDURE
1. Soak slides for 1 week at room temperature in acid dichromate solution or a 3% (concentrated) hydrochloric acid in 95% ethanol solution.
2. Wash slides 10 times with tap water followed by 2 rinsings with distilled or deionized water.
3. Pass the slide through the blue tip of the bunsen burner flame and lay heated side up on a clean paper towel.
4. Using a wax pencil make a heavy wax outline around 2/3 of the flamed side of slide. Handle slide by unmarked end only.
5. Add a large loopful of the suspension of the organism to be examined for flagella to one end of the rectangular area. Best results are obtained using a suspension of distilled or deionized water and the growth from an agar medium.
6. Tilt the slide to permit the suspension to run down to the other end of the slide. Discard the slide if the drop does not run down evenly.
7. Air dry the film for several hours on a level surface. **Do not heat.**
8. Place the slide on a horizontal staining rack.
9. Add 1 ml of the flagella stain to the slide and stain for 5 – 15 minutes. The correct staining time must be determined for each lot of solution.
10. Carefully flood off the stain by adding tap water to the slide while it remains on the rack. **Do not tilt slide.**

COUNTERSTAIN AS FOLLOWS
1. Add 1 ml Bacto Methylene Blue, Loeffler solution to 2 ml distilled or deionized water and apply to the marked area. Allow to remain on the slide for 1 – 2 minutes.
2. Flood off the stain with tap water and air dry.
3. Examine the slide under oil immersion lens of microscope.

RESULTS
Flagella will be red to maroon and the body of the organism blue.

REFERENCE
1. J. Bact., 36:656, 1938.

PACKAGING

Bacto Flagella Stain	10 g	3117-12-0
	6 × 10 g	3117-33-5

BACTO FLETCHER MEDIUM BASE

INTENDED USE
Bacto Fletcher Medium Base prepared according to the formulation of Fletcher[1] is recommended for the isolation, cultivation and maintenance of cultures of *Leptospira*.

HISTORY
Fletcher Medium Base was the medium of choice for use in the isolation of *Leptospira* from blood, urine and kidney specimens by Galton, Acree, Lewis and Prather.[2]

FORMULA
BACTO FLETCHER MEDIUM BASE
DEHYDRATED
Ingredients per liter

Bacto Peptone . . . . 0.3 g
Bacto Beef Extract . . . . 0.2 g
Sodium Chloride . . . . 0.5 g
Bacto Agar . . . . 1.5 g

Final pH 7.9 ± 0.1 at 25°C.

One pound will make 181.6 liters of final medium.

METHOD OF PREPARATION
1. To rehydrate, suspend 2.5 g in 920 ml distilled or deionized water.
2. Heat to boiling to dissolve completely.
3. Autoclave for 15 minutes at 15 lbs pressure (121°C).
4. Allow the medium to cool to 50 – 55°C and add 80 ml rehydrated Bacto Leptospira Enrichment or sterile rabbit serum.
5. Aseptically dispense into sterile screw capped tubes in 5 – 7 ml amounts. Store at room temperature overnight.
6. Inactivate the whole medium the day following its preparation by placing the tubes in a water bath at 56°C for one hour.
7. Allow to cool before inoculating.

STORAGE
Bacto Fletcher Medium Base   Below 30°C
Prepared medium   15 – 30°C

QUIALITY CONTROL
Identity Specifications
Dehydrated powder: beige, homogeneous, free-flowing
Reaction of 0.25% solution: pH 7.9 ± 0.1 at 25°C
Prepared medium: very light amber, very slightly opalescent

Typical Cultural Response on Bacto Fletcher Medium Base
After up to 5 Days at 30°C

Organism	Growth
Leptospira interrogans serotype *australis*	good to excellent
Leptospira interrogans serotype *canicola*	good to excellent
Leptospira interrogans serotype *grippotyphosa*	good to excellent

REFERENCES
1. Trans. Roy. Soc. Trop. Med. & Hyg., 21:265, 1927 – 28.
2. J. Amer. Vet. Med. Assoc., 128:87, 1956.

PACKAGING

Bacto Fletcher Medium Base	1 lb (454 g)	0987-01-6
Bacto Leptospira Enrichment	6 × 10 ml	0452-60-0

FLUORESCENT ANTIBODY PROCEDURES

The classical method for determining the infective agent in an infectious disease is that set forth by Koch in his postulates:

1. The organism should be found in all cases of the disease in question, and its distribution in the body should be in accordance with the lesions observed.
2. The organism should be cultivated outside the body of the host, in pure culture, for several generations.
3. The organisms so isolated should reproduce the same disease in other susceptible animals.

Fulfilling these postulates will establish the identity of the organism involved, but over the course of years other investigators have attempted to simplify the detection and identification of the etiological agent of a disease for the benefit of the patient. Behring's discovery of antitoxin, followed by Landsteiner's work on the specificity of serological reactions, paved the way to development of practical serological methods for identifying microorganisms. Subsequently, great strides were made in simplifying these techniques as investigators unraveled the web of serological relationships among pathogenic microorganisms.

Even though many groups of organisms have been well defined serologically, the final identification still involves culturing, isolating, examining and applying biochemical and serological techniques — all of which are time consuming. Clinicians prefer methods which allow rapid identification of infective agents or antibodies elicited by them directly from the specimen. The fluorescent antibody technique is only one of many such rapid methods available.

What is the fluorescent antibody technique? To answer this question and to place this method in proper relationship in serology, it will help to recall that there are six major antibody groups recognized by serologists. These are agglutinins, precipitins, opsonins, lysins, bacteriocidins and complement-fixing antibodies. The fluorescent antibody is a seventh type. Thus, the fluorescent antibody (FA) technique is a specialized serological procedure which consists of an antigen-antibody reaction made visible by a fluorescent dye incorporated into the system.

A substance is said to be fluorescent if it absorbs light energy of one wave length and emits light of another wave length. The use of such a color label on proteins is not new. In 1933 Heidelberger[1] prepared such a modified protein molecule by conjugating egg albumin with the salt of benzidine for its quantitative measurement by spectrophotometric methods. Not much more was done with this technique until 1941, when Coons[2] described the immunological properties of an antibody protein containing a fluorescent group. The second World War caused a lag in the use of this technique. In post-war years many factors increased interest in rapid identification procedures and brought great advances in the application of fluorescent antibody methods. Factors which stimulated interest in FA methods are:

1. The wide use of rapid transportation that has introduced communicable diseases into areas never before experiencing them, thus, increasing the spread of epidemics.
2. The introduction of antibiotic therapy requiring the establishment of the true identity of an organism and its susceptibility prior to the administration of the antibiotic of choice.
3. The evolution of specific antisera to numerous pathogens and the knowledge of their cross reactivity.
4. The description of a stable labeling agent — fluorescein isothiocyanate — by Riggs.[3]

It must be emphasized that the fluorescent antibody test is still developing and it is not yet the complete answer. We cannot yet eliminate conventional cultural and serological methods. Its use has proven highly successful in some instances but there are a number of problems yet to be solved. Despite this, the fluorescent antibody technique has become established as a useful adjunct procedure in the serology laboratory.

ADVANTAGES OF THE FA METHOD

1. The main advantage of the FA method over other procedures is the rapidity with which results are obtained. In some cases, a specimen may be examined and reported the same day as received.
2. In most cases, cultural methods may be eliminated or shortened considerably.
3. Isolation of the pure culture is not necessary.
4. Small numbers of organisms in a clinical specimen are easily found even though the specimen may contain a large number of contaminating organisms.
5. Specimens unsatisfactory for other procedures which contain nonviable pathogens, or pathogens which are difficult to cultivate, are satisfactory for use in the FA technique.
6. Microorganisms in a positive specimen fluoresce even though they may be found situated intracellularly.
7. In addition to its use in identifying microorganisms, the FA technique is applicable to the detection of soluble antigens, the identification of native proteins and autoantibodies and to other studies such as the site of antibody formation and hypersensitivity.
8. The sensitivity of the FA method is comparable to that of other serological tests in most cases; in others, the sensitivity and specificity may actually surpass that of conventional methods.
9. Simplicity of performance for routine work after the technique has been successfully mastered.
10. The relatively low cost per test after basic equipment has been obtained.

DISADVANTAGES OF THE FA METHOD

Over and above the general problems encountered in any immunoserologic test the disadvantages of the FA method include:

1. The initial investment in fluorescent microscopic equipment.
2. The necessity to obtain adequately trained personnel.

STAINING TECHNIQUES

Two main staining techniques employed in fluorescent microscopy are the direct and indirect methods. Two others which have not been employed as often, but nevertheless are important, are the inhibition and the complement staining techniques.

Direct FA Technique

The direct FA technique is easily performed and most commonly employed. It consists of a direct antigen-antibody combination on a microscope slide to which the antigen has been fixed and the homologous conjugated antiglobulin added. This is generally the method of choice since fewer steps and precautions are necessary. A schematic representation of the direct method is shown in Figure 1.

Figure 1* Schematic Representation of a Direct Staining Test with Fluorescent Antibody

LABELED ANTIBODY
BACTO FA E. COLI

ANTIGEN
E. COLI

LABELED PRODUCT
FLUORESCENT E. COLI

The control or test of specificity of staining in the direct test is illustrated in Figure 2, Steps 1 and 2.

Figure 2*

Step 1

UNLABELED ANTIBODY
BACTO E. COLI OK ANTISERUM

ANTIGEN
E. COLI

UNLABELED PRODUCT
NONFLUORESCENT E. COLI

Step 2

UNLABELED PRODUCT
NONFLUORESCENT
E. COLI

LABELED ANTIBODY
BACTO FA E. COLI

UNLABELED PRODUCT
NONFLUORESCENT
E. COLI

Indirect FA Technique

The indirect technique may be used both for the identification of unknown antigens and unknown antibodies in sera or other body fluids. This is schematically illustrated in Figure 3*, Steps 1 and 2.

**Figure 3* Schematic Representation of the Indirect Fluorescent
Antibody Staining Reaction**

Step 1

 + =

UNLABELED ANTIBODY UNLABELED ANTIGEN UNLABELED PRODUCT
PATIENT'S SERUM   LEPTOSPIRA   NONFLUORESCENT
                   LEPTOSPIRA

Step 2

 + =

LABELED ANTIBODY UNLABELED PRODUCT LABELED PRODUCT
BACTO FA HUMAN NONFLUORESCENT   FLUORESCENT
GLOBULIN ANTIGLOBULIN LEPTOSPIRA   LEPTOSPIRA

The control or test of specificity of a positive indirect test is illustrated in Figure 4, Steps 1 and 2.

Figure 4*

Step 1

 + =

ANTIBODY-FREE SERUM UNLABELED ANTIGEN UNLABELED ANTIGEN
NORMAL HUMAN SERUM LEPTOSPIRA   NONFLUORESCENT
                   LEPTOSPIRA

Step 2

LABELED ANTIBODY UNLABELED ANTIGEN UNLABELED ANTIGEN
BACTO FA HUMAN NONFLUORESCENT NONFLUORESCENT
GLOBULIN ANTIGLOBULIN LEPTOSPIRA   LEPTOSPIRA

A flow sheet illustrating the FA indirect test for identifying unknown antiserum is shown in Figure 5.

Figure 5* Indirect Test for Antibody in Unknown Serum

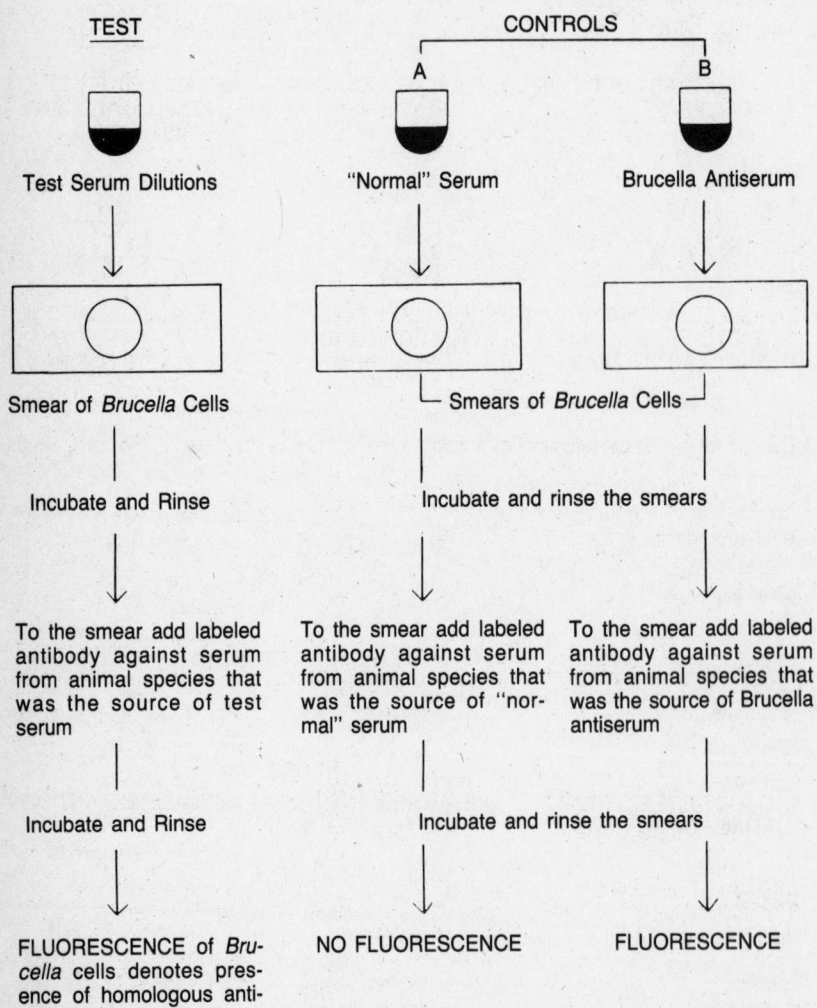

A flow sheet of the procedure identifying an antigen by the indirect FA test is illustrated in Figure 6.

Figure 6* Indirect Test for Unknown Antigen

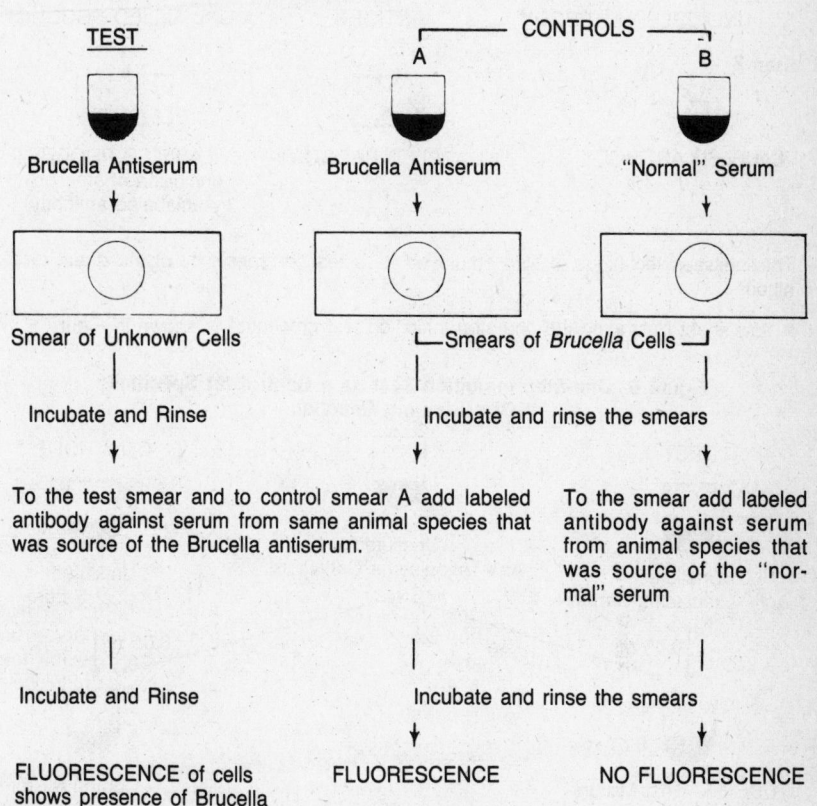

TEST | CONTROLS
A | B

Brucella Antiserum | Brucella Antiserum | "Normal" Serum

Smear of Unknown Cells | Smears of *Brucella* Cells

Incubate and Rinse | Incubate and rinse the smears

To the test smear and to control smear A add labeled antibody against serum from same animal species that was source of the Brucella antiserum. | To the smear add labeled antibody against serum from animal species that was source of the "normal" serum

Incubate and Rinse | Incubate and rinse the smears

FLUORESCENCE of cells shows presence of Brucella | FLUORESCENCE | NO FLUORESCENCE

Inhibition Technique

The fluorescent antibody inhibition staining procedure is based on the immunological phenomenon of blocking specific antigen-antibody reactions by first exposing the antigen to a different aliquot of homologous unlabeled antibody solution. For example, if a smear of *Streptococcus* group A is exposed to specific unlabeled *Streptococcus* group A antibody, the bacteria will become saturated with antibody. If the same smear is then exposed to fluorescent labeled *Streptococcus* group A antibody, no reaction will occur and the organisms will remain nonfluorescent. In actual practice, one encounters bright staining if no blocking has occurred and less bright staining if blocking has occurred. A schematic representation of the fluorescent antibody inhibition reaction is illustrated in Figure 7.

Figure 7* Schematic Representation of Fluorescent Antibody Inhibition Staining Reaction

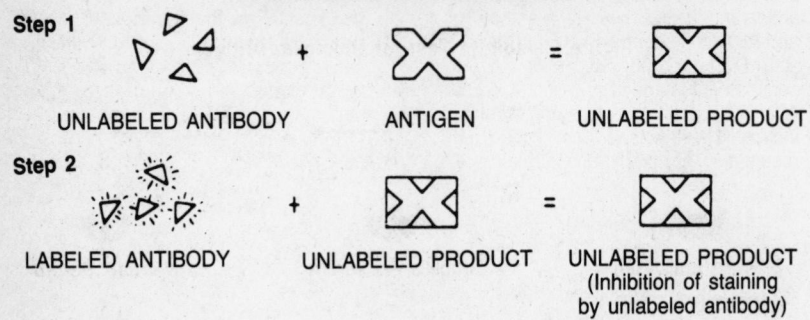

Step 1

UNLABELED ANTIBODY   +   ANTIGEN   =   UNLABELED PRODUCT

Step 2

LABELED ANTIBODY   +   UNLABELED PRODUCT   =   UNLABELED PRODUCT
(Inhibition of staining by unlabeled antibody)

The inhibition technique is also employed as a test for specificity of the direct technique.

A flow sheet illustrating the one-step inhibition test procedure is shown in Figure 8.

Figure 8* One-Step Inhibition Test as a Control for Specificity Of a Staining Reaction

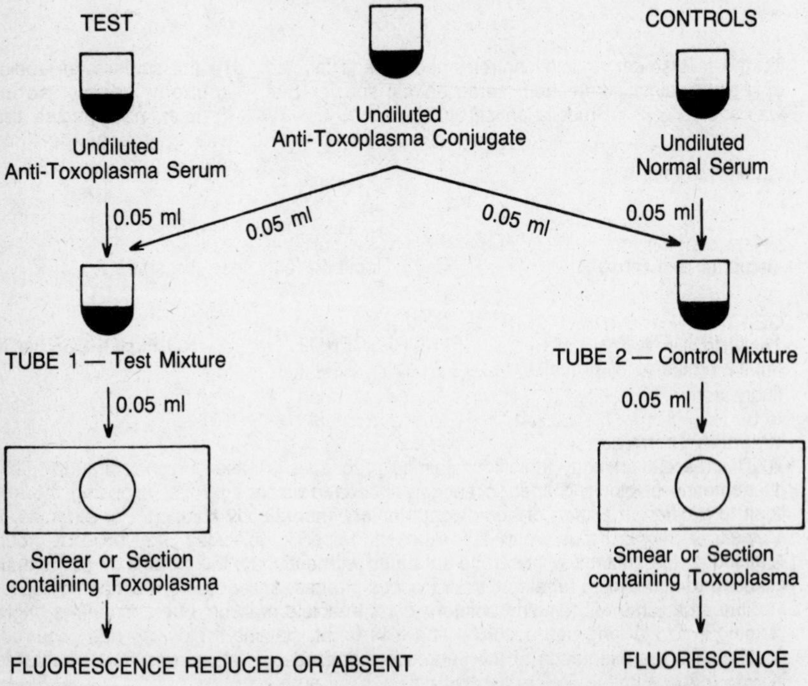

TEST   CONTROLS

Undiluted
Anti-Toxoplasma Serum

Undiluted
Anti-Toxoplasma Conjugate

Undiluted
Normal Serum

0.05 ml   0.05 ml   0.05 ml   0.05 ml

TUBE 1 — Test Mixture   TUBE 2 — Control Mixture

0.05 ml   0.05 ml

Smear or Section
containing Toxoplasma

Smear or Section
containing Toxoplasma

FLUORESCENCE REDUCED OR ABSENT   FLUORESCENCE

INTERPRETATION: Staining of toxoplasma by labeled antibody is inhibited by presence of anti-toxoplasma serum in mixture. Normal serum does not inhibit since it contains no specific antibody.

Complement Staining Technique

Complement staining permits identification of either an unknown antigen or an unknown antiserum. It is similar to the indirect procedure except that the fluorescent labeled antiglobulin (secondary reagent) is not directed against the species supplying the antiserum (primary reagent) but rather against the species supplying the complement. The complement is added to the antiserum and antigen during the first stage of the reaction. After the usual incubation and washing, labeled anti-guinea pig complement is applied, incubated and washed free from excess reagent. Figure 9 illustrates a schematic representation of complement staining with fluorescent antibody.

Figure 9* Schematic Representation of Complement Staining With Fluorescent Antibody

Step 1

O UNLABELED COMPLEMENT   ANTIGEN   UNLABELED PRODUCT
A UNLABELED ANTIBODY

Step 2

LABELED   UNLABELED PRODUCT   LABELED PRODUCT
ANTI-COMPLEMENT

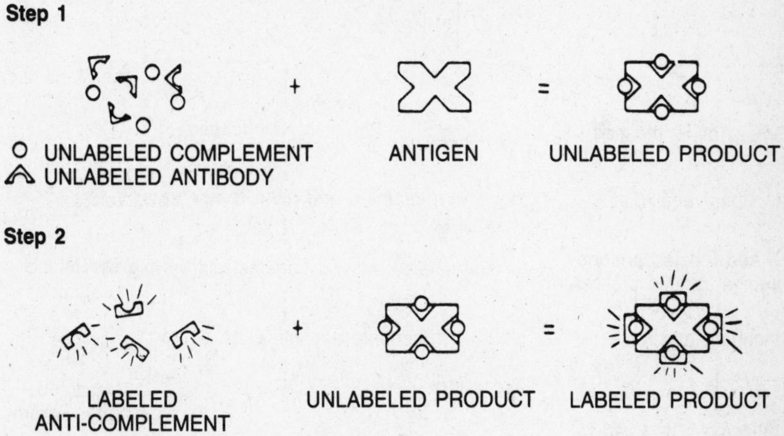

A flow sheet illustrating the FA complement test for identifying an antibody in unknown serum is shown in Figure 10 (as shown on page 370).

GENERAL PRECAUTIONS AND CONSIDERATIONS

The following list of precautions was compiled from actual problems encountered in numerous FA workshops and from experiences related by investigators in the field of fluorescent microscopy. Many of the comments listed below may seem, at first glance, to be too obvious. Many times the "too obvious" is overlooked.

A. The Specimen

1. If the specimen has too few organisms, it should be concentrated by centrifugation.
2. If too many organisms are present, the specimen should be diluted, particularly if a pure isolated culture is employed.
3. Care must be exercised to prevent the loss of the antigen from the microscope slide during processing. To detect this possibility, observe the slide under the tungsten light source of the microscope to determine the relative numbers of organisms present prior to examination with the ultraviolet light.
4. If excessive amounts of tissue or mucous are present, counterstain with Bacto Rhodamine or Bacto Eriochrome Black to reduce auto and background fluorescence. The methodology may be found under the description of these reagents.

Figure 10* Complement Test for Antibody in Unknown Serum

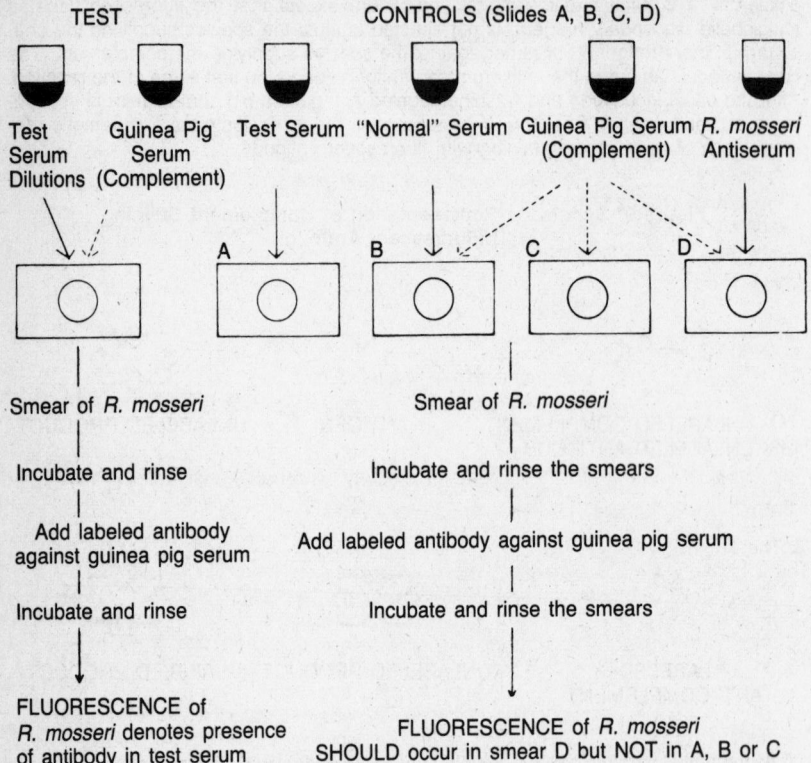

TEST   CONTROLS (Slides A, B, C, D)

Test   Guinea Pig   Test Serum   "Normal" Serum   Guinea Pig Serum   *R. mosseri*
Serum   Serum   (Complement)   Antiserum
Dilutions   (Complement)

A   B   C   D

Smear of *R. mosseri*   Smear of *R. mosseri*

Incubate and rinse   Incubate and rinse the smears

Add labeled antibody   Add labeled antibody against guinea pig serum
against guinea pig serum

Incubate and rinse   Incubate and rinse the smears

FLUORESCENCE of
R. mosseri denotes presence   FLUORESCENCE of *R. mosseri*
of antibody in test serum   SHOULD occur in smear D but NOT in A, B or C

*Figures 1 through 10 in this section are courtesy of The Center for Disease Control, U.S.P.H.S., Atlanta, Georgia.

5. In some cases, cross reactions might occur if mixed cultures are encountered. This is particularly true with the *Enterobacteriaceae*.
6. The state of the antigen is important. When possible it is advisable to use fresh specimens. Dead or autolyzed cultures could result in minimal fluorescence.
7. In some clinical specimens abundant extraneous material might be present. These artifacts may adsorb the stain nonspecifically making examination of the slide difficult or impossible.

B. The Slide
1. The microscope slide must be grease-free. Wash in Bon Ami®, or a similar cleanser, rinse, wash in acetone or alcohol and dry with a lint-free cloth.
2. Some microscopic set-ups are more demanding than others and may require a slide of 1 mm thickness. In most cases, this thickness is not critical but should be remembered if difficulties in bringing the specimen into focus are encountered.

3. The specimen must be adequately fixed to the slide. Under-fixation may result in the mechanical loss of the antigen. Over-fixation with heat or chemicals could result in the destruction or alteration of the protein. For proper fixation follow the directions given with each conjugate since one system may vary from another.

C. The Conjugate

1. Various conjugates may cross react with antigenically related or similar groups of organisms. The reaction spectrum of the conjugate must be known.
2. The titer of the conjugate should be known so that it might be diluted prior to use. Optimal dilution of the conjugate reduces the possible degree of cross reaction fluorescence and nonspecific background fluorescence. This dilution may vary from laboratory to laboratory due to fluorescent microscope equipment, bulb age, etc., and may be determined as follows:

Dilution of Conjugate	Fluorescence
1:10	4+
1:20	4+
1:40	4+
1:80	4+
1:160	2+

In this example, the last 4+ fluorescence is found in a 1:80 dilution of the conjugate. One less dilution is chosen for a margin of safety. The optimal dilution, therefore, in this case is 1:40

D. The Technique

The following are points which should be considered regarding technique:

1. Over-fixation of the antigen, particularly if heat is used.
2. Fixation of too many slides per volume of alcohol or acetone. It is recommended that not more than 50 slides be fixed per 200 ml of alcohol or acetone.
3. In spreading the conjugate over the smear, a clean piece of applicator stick must be used. The stick should not touch and thereby disturb the smear.
4. Drying of the conjugate on the slide could result in what might be interpreted as false positive reactions. The slide should be incubated for the prescribed time and temperature in a moist atmosphere. A piece of moist filter paper affixed to a lid of a Petri dish is satisfactory.
5. If the same slide is used for staining a smear using two different conjugates, care must be taken not to run the two conjugates together. It is apparent that false positive or negative results could occur.
6. The slides should be rinsed in Bacto FA Buffer for a sufficient length of time to remove all the free excess conjugate.
7. The reaction of the buffered saline should be checked periodically to insure a proper pH of around 7.2. This can be done easily by transferring a few drops of the buffer to a clean spot plate and adding a drop of 0.04% phenol red indicator. A red color, not cerise, denotes a satisfactory pH. This can also be done by the use of a pH meter.
8. In blotting the slides, lint-free paper must be used to avoid addition of extraneous material which might increase artifacts. One is also cautioned to use a clean blotting surface for each slide to avoid carrying over fluorescing organisms from slide to slide.
9. The pH of the mounting fluid should also be checked on occasion to insure proper pH. An acid pH could result in a "quenching" of fluorescence. The pH may be checked as in step 7 above. Avoid adding an excessive amount of mounting fluid to the microscope slide. A very small drop is most satisfactory.

10. When adding the mounting fluid, care is taken to avoid bubble formation which will interfere with microscopic examination.
11. Avoid formation of bubbles also when the coverglass is added.
12. To prepare a semipermanent reference mount, ring the coverglass with clear nail polish and store the slide at 2 – 8°C. Avoid excessive exposure to light.

E. Controls
1. For the preparation of control slides used in the various techniques, follow the methodology for each specific test. Positive and negative controls must be prepared for each day's run to assure proper technique, microscope alignment, bulb efficiency and proper filter selection. Controls also serve as a guide for comparison of the degree of fluorescence to that of the unknown specimen.
2. *Nonspecific Staining Control*
In the direct test this is accomplished by observing fluorescence or lack of fluorescence using a heterologous culture but one which is closely related to that from which the conjugate was prepared. In the indirect test, the nonspecific staining control is a slide consisting of the fixed antigen to which the test dilution of the conjugated antianimal species globulin, FA animal globulin antiglobulin has been added without the use of the unconjugated antibody. Staining of the antigen indicates the conjugate to be unsatisfactory at the dilution being used, that free dye is present or the fluorescein isothiocyanate used for conjugation was unsatisfactory because of its impurity.
3. The description of other controls for a specific test is found in the discussion of each reagent.

F. Cross Reactions, Nonspecific Staining and Autofluorescence
1. Cross reaction is a term applied to a reaction of a conjugate with heterologous antigens having some antigenic similarities. Thus, a conjugate for *Streptococcus* group A may cross react with certain strains of streptococci groups C and G. Similarly, conjugates of *Neisseria gonorrhoeae* may cross react with some strains of *Neisseria meningitidis*. Adsorption of the antiserum or conjugate with the offending organism may not remove all the cross reactions without reducing the homologous titer. If such a residual cross reaction of a given conjugate is noted, several methods are recommended for its removal.
   a. Add varying amounts of Bacto FA Rabbit Serum, Blocking to the conjugate.
   b. If a more potent blocking reagent is needed, unconjugated immune sera prepared for use in conventional tests may be employed.
   c. Bacto FA Papain is successfully employed for enzymatic alternation of the antigen to remove certain cross reactions.
2. *Nonspecific Staining*
Nonspecific staining results in a positive staining of an antigen in the absence of specific antibodies in the conjugate. This may be due to unreacted fluorescent material[4] or by conjugated serum proteins other than antibody globulins. The unreacted fluorescent material may be removed by further dialysis of the conjugate or by passing through a Sephadex column or by extraction with activated charcoal.
3. *Autofluorescence*
Various body tissues will produce a natural fluorescence when examined microscopically. The color of fluorescence varies from tissue to tissue. Some emit blue, blue-green and green colors. Others emit yellow and red colors. Selection of proper filter combination will reduce most autofluorescence to a minimum to provide sufficient color contrast between autofluorescence and immunofluorescence.

Some tissues in clinical specimens may absorb varying amounts of the fluorescein and result in nonspecific fluorescence.

G. Microscopic Equipment
1. *Microscope*. It is beyond the scope of this treatise to recommend a fluorescent microscopic setup. Several excellent combinations are available. Contact one or more of the major microscope makers for more details. For more information on fluorescent-antibody microscopy consult the chapter on Immunofluorescence Techniques in the *Manual of Clinical Microbiology,* 3rd Edition.[5]
2. *Optics*. The darkfield condenser is more satisfactory than brightfield for FA methods since it filters out less ultraviolet light, concentrates illumination in the focal plane of the specimen and affords a dark background for a more satisfactory examination of the specimen. It is recommended, therefore, that a darkfield cardiod-type condenser together with 10X or 20X eyepieces be used. For scanning, an objective lens of 10 – 16X is very useful. 45X and 54X oil immersion lenses have been demonstrated to be very useful. A third objective of 100X with an iris diaphragm has also been proven to be most valuable. For complete information, it is recommended to consult representatives from the various microscope manufacturers.
3. *Light Source*. Both the arc burner and tungsten filament lamp light sources are needed for the complete examination of the specimen. For fluorescent microscopy, a very intense light is necessary, emitting the wave length of energy within the range of that absorbed by the dye used in the conjugation process. There are several light sources available today. One is the Osram HB0 200 mercury arc lamp. Its light transmission spectrum is seen in Figure 11. Halogen lamps and incident lighting are recent innovations in fluorescent microscopy and are widely used. The *Manual of Clinical Microbiology* previously refered to should be consulted for the details.

Figure 11 Osram HB0 200 Emission Spectrum Chart for the Mercury Vapor ARC Lamp versus a Low Voltage Lamp 30 W

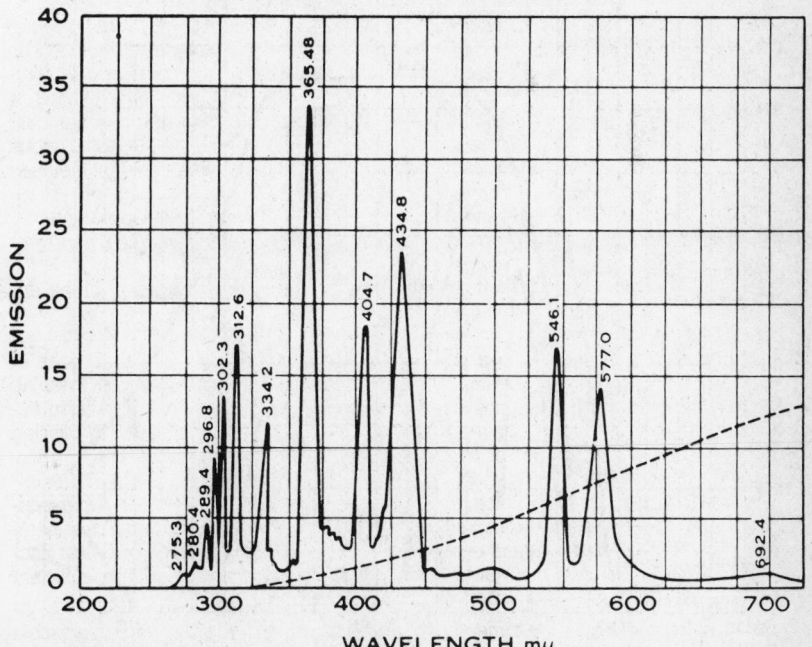

The length of the life of a bulb varies considerably depending upon line current fluctuations and usage. Bulb life may be prolonged if the burner is kept on over the period of time the system is to be used rather than turning the bulb on and off when needed. The frequency of the initial ignition of the bulb appears to be the most important factor in bulb life prolongation. One word of caution. If it is necessary to replace the mercury arc burner, make sure the bulb to be replaced is allowed to return to room temperature prior to removal. Do not open the lamp housing when the bulb is hot! Many laboratories keep a log on the length of time a bulb is in operation. Others routinely check the bulb output with a light meter. In any case, it is essential to run a positive control slide in each series.

H. Filter Systems

There are three filter systems employed in the fluorescent antibody technique: namely, a heat filtering system, the excitation light filter system and the barrier light filter system. The heat filters are located between the light source and the excitor filters and perform the function of removing heat energy from the light source. The light energy transmission spectrum of a common heat filter, BG22, is shown in Figure 12.

The excitor filters are located between the heat filters and the specimen. Their function is to transmit the wave length of light which the fluorochrome is capable of absorbing and removing other unwanted wave lengths. Two main excitor filters used in bacterial fluorescence are the UG1 and BG12 filters. Their transmission spectra are shown in Figure 12.

Figure 12 Transmission Spectrum of Excitor Filters

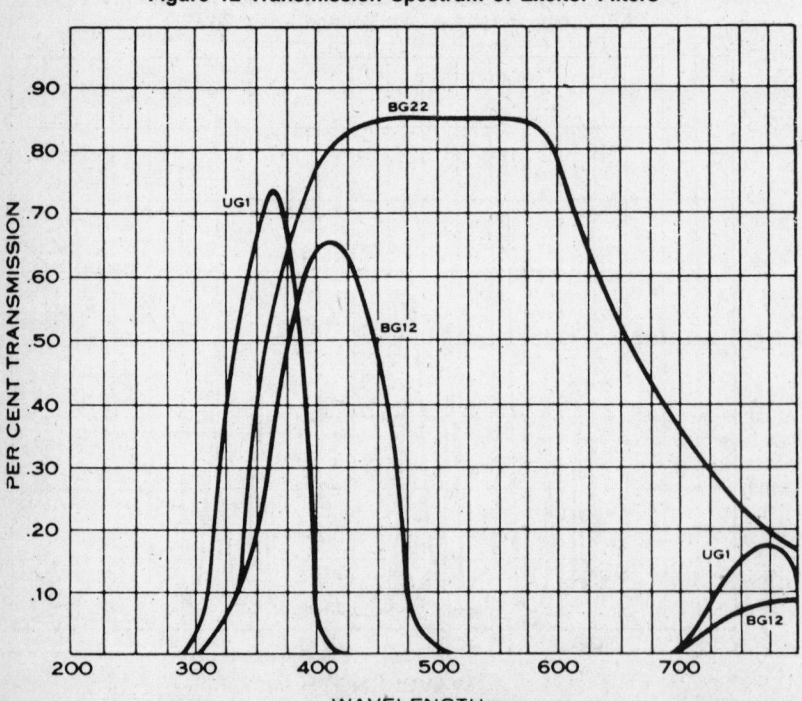

It may be seen that they permit wave lengths of light in the ultraviolet range to pass (350 – 400 mμ) but remove its visible wave lengths (500 – 700 mμ).

The third filter system consists of the barrier filters. They are located between the specimen and the observer. The function of this system is to transmit the visible wave lengths of light emitted by the specimen and hold back wave lengths below the 500 mμ range. The two filters often employed are the GG – 9 and OG1 filters whose transmission spectra are shown in Figure 13.

Figure 13 Transmission Spectrum of Barrier Filters

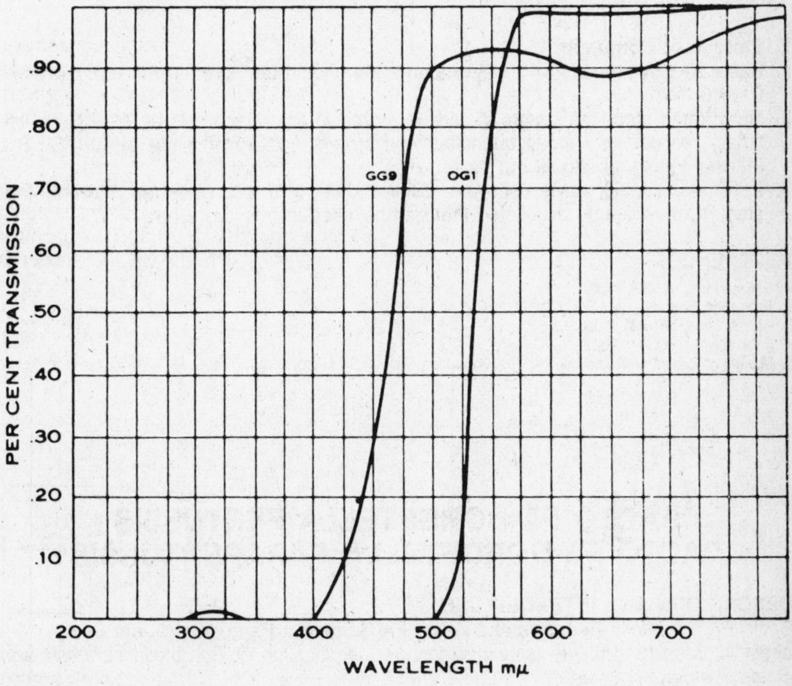

I. Alignment
At times it is necessary to realign the fluorescent light source, particularly if the bulb has been changed or if the microscopic equipment is used intermittently for other purposes. In some cases, nonalignment is one of the most troublesome features of this technique. Recently introduced fluorescent equipment permits rapid and easy alignment of the light source. Known positive slides such as Bacto FA Slide Stained Streptococcus Group A are used as controls for microscopic alignment and filter selection.

J. Examination of Slides
1. Examine the specimen first by using the tungsten light source for numbers and kinds of organisms present.
2. Using the ultraviolet light source examine the specimen slides for number of fluorescing organisms and brilliancy as well as their morphology and compare with that obtained on the control slides.

3. Use the recommended magnification. The total magnification may vary from one system to another. Refer to the individual reagents for recommended magnification.
4. A microscopic field, when left exposed to the ultraviolet light, will rapidly fade. If this occurs, it is necessary to change fields to continue examination of the specimen.

K. Storage of Slides
1. Optimally, prepared slides should be examined immediately after preparation. Those not examined immediately may be stored 1 – 2 hours in the dark. Exposure to either direct light or ultraviolet light will result in partial or complete fading of fluorescence.
2. For prolonged storage, unstained slides and stained control slides should be be stored in the freezer below 0°C.

L. Storage of Conjugate
1. Bacto FA conjugates in the desiccated state should be stored in the refrigerator at 2 – 8°C.
2. Rehydrated stock conjugate should be stored in the refrigerator at 2 – 8°C or, optimally, divided into small quantities and stored in a frozen state at −20°C. The thawed conjugate should not be refrozen.
3. If the stock conjugate is used in a diluted state, it is recommended that only that amount of conjugate be diluted that can be used in one day's run.

REFERENCES
1. J. Exp. Med. 58:137, 1933.
2. Proc. Soc. Exp. Biol. & Med., 47:200, 1941.
3. Am. J. Path., 34:1081, 1958.
4. J. Immunol., 89:124, 1962.
5. Manual of Clinical Microbiology, 3rd Ed., American Society for Microbiology, Wash., D.C., 1980.

BACTO FA BORDETELLA PERTUSSIS
BACTO FA BORDETELLA PARAPERTUSSIS

DESCRIPTION AND INTENDED USE
Bacto FA Bordetella Pertussis and Bacto FA Bordetella Parapertussis are used in the direct fluorescent antibody technique for the identification of *Bordetella pertussis* and *Bordetella parapertussis*.

BACKGROUND
Bordetella pertussis is the primary agent in whooping cough. It is rarely found in the nasopharynx of normal individuals although it has an 85% communicability rate.[1] This high transmission rate makes careful monitoring of any reported case, of great public health concern, even though vaccination against *B. pertussis* is recommended and usually occurs within the first 3 months of infant life.

Bordetella parapertussis has a much lower infection rate but may cause pertussis-like syndromes.

B. pertussis was isolated in 1906 by Bordet and Gengou on glycerol-potato-blood agar.[2] Bacto Bordet-Gengou Agar, with slight modification, is still the medium of choice for culturing and isolating *B. pertussis*.

Bordetella parapertussis also grows well on this medium and resembles *B. pertussis,* however, colonies of this organism appear larger than *B. pertussis* and exhibit a brown pigment.

B. pertussis and *B. parapertussis* are slow growing organisms, developing in 3 – 4 days. By employing the fluorescent antibody technique, the time required to detect these organisms can be significantly reduced. The FA procedure may be applied to direct nasopharynx smears or may be used to identify young cultures of *B. pertussis* or *B. parapertussis.*

Eldering, Eveland and Kendrick,[3,4] and Holwerda and Eldering,[5] indicated the usefulness of the FA procedure, although complete correlation between the agglutination method and FA technique was not obtained. Nonetheless, the FA procedure was able to detect both smooth and rough cultures of *B. pertussis* and *B. parapertussis* and could also be applied to direct specimens. Further data obtained indicated little or no cross reactions between conjugates prepared from *B. pertussis* and *B. parapertussis* cultures.

REAGENTS
Bacto FA Bordetella Pertussis and Bacto FA Bordetella Parapertussis are desiccated fluorescein conjugated anti-Bordetella chicken globulins. They have been prepared according to the method of Eldering, Eveland and Kendrick[3,4] and Holwerda and Eldering.[5]

Bacto FA Bordetella Pertussis and Bacto FA Bordetella Parapertussis are rehydrated by adding 5 ml distilled or deionized water to each vial and rotated gently to dissolve contents completely.

The working dilution of the conjugate should be determined shortly after its rehydration. The titer of a conjugate varies with the technique used, the fluorescent microscope used, the filter used, and the age of the bulb.

The conjugate should be titrated using a known culture of *B. pertussis* or *B. parapertussis* homologous to the conjugate. Dilutions of the conjugate are made in rehydrated Bacto FA Buffer. The titer is determined as follows:

Dilution of Conjugate	Fluorescence
1:5	4+
1:10	4+
1:20	4+
1:40	4+
1:80	2+

In this example, the last 4+ fluorescence is found in a 1:40 dilution of the conjugate. One less dilution is chosen for a margin of safety. The working dilution, therefore, in this case is 1:20.

STORAGE AND EXPIRATION DATE
Bacto FA Bordetella Pertussis and Bacto FA Bordetella Parapertussis in the desiccated state are stable to the expiry date on the label when stored at 2 – 8°C. Aliquots of the titered conjugate should be put into small vials, frozen in the undiluted state and stored below −20°C for optimal stability. The conjugate should not be exposed to repeated freezing and thawing.

PRECAUTIONS

1. It is assumed that the user of these reagents is familiar with fluorescent antibody techniques. Those desiring information concerning the advantages and disadvantages of the FA technique, the various staining techniques, and general precautions and considerations concerning FA procedures should refer to FA General Procedures.
2. All glassware that is employed in the preparation testing and storage of these reagents must be free of detergents or other harmful residues.

SPECIMEN PREPARATION

Direct nasopharyngeal specimens or cultures from Bacto Bordet-Gengou Agar may be employed in this direct FA procedure:

Direct Nasopharyngeal Smears

1. Obtain nasopharyngeal swabs and emulsify them in 0.2 – 0.5 ml of sterile 1% Bacto Casamino Acids.
2. Hold specimen in casamino acids solution for no more than 2 hours.
3. Smear emulsified specimen on a clean microscope slide.
4. Allow smear to air dry and fix by gentle heating or by 1 minute immersion in 95% ethanol.

Cultural Isolates

1. Use same emulsion in step 1 above to inoculate plates of Bacto Bordet-Gengou Agar Base Enriched with 15 – 20% blood, and Bacto Bordet-Gengou Agar Base with 15 – 20% blood and 0.5 unit penicillin per ml.
2. Incubate plates at 35°C for at least 24 hours.
3. Examine plates microscopically for growth of *B. pertussis* or *B. parapertussis* colonies.
   The appearance of typical colonies of a *B. pertussis* is smooth, raised, glistening and not over 1 mm in diameter. They are of a pearly, almost transparent appearance, and are surrounded by a characteristic zone of hemolysis which is not sharply defined, but which merges diffusely into the medium. *B. parapertussis* and the surface is not as glistening and may have a slightly brown color. Both organisms are small gram-negative bacilli and occur singly, in pairs, or in clumps.
4. Pick appropriate colonies and emulsify in sterile casamino acids solution described in step 2 above.
5. Prepare smear as stated under **Direct Nasopharyngeal Smears.**

PROCEDURE

Materials Provided:
Bacto FA Bordetella Pertussis
Bacto FA Bordetella Parapertussis
Bacto FA Buffer, Dried
Bacto FA Mounting Fluid pH 7.2

Materials Required but not Provided:
Moisture chamber or a Petri dish which contains a moistened piece of filter paper
Bibulous paper
Fluorescent microscope assembly

1. Add several drops of the appropriate Bacto FA Bordetella conjugate to the fixed smear.
2. Spread the conjugate over the surface of the smear.
3. Place the slides in a moisture chamber.

4. Incubate at room temperature for 30 minutes.
5. Remove the excess conjugate and place the slide in a staining jar containing Bacto FA Buffer for 10 minutes with 2 changes of the buffer followed by 1 rinse in distilled water.
6. Remove the slide and blot gently with bibulous paper. Allow to air dry.
7. Add a small drop of Bacto FA Mounting Fluid to the center of the stained area and cover with a cover glass.
8. Examine each smear using a magnification of 900 – 1000X and record presence or absence and degree of fluorescence.

QUALITY CONTROL
Positive — Bacto FA Bordetella Pertussis or Bacto FA Bordetella Parapertussis with known homologous strain.
Negative — Bacto FA Bordetella Pertussis or Bacto FA Bordetella Parapertussis with a heterologous strain of Bordetella.

Brilliant, 4+ fluorescence should result from the positive control. No fluorescence should result from the negative.

Microscope Control — It is recommended that an additional control be incorporated. A known positive fluorescing slide or Bacto FA Slide Stained Streptococcus Group A as a check on microscope alignment, etc.

RESULTS
Read and record results based on intensity of fluorescence – to 4+ as outlined below. Cultures exhibiting a fluorescence of 2+ or greater should be considered positive.
- 4+ Maximum fluorescence; brilliant yellow-green; clear-cut cell outline; sharply defined cell center
- 3+ Less brilliant yellow-green fluorescence; clear-cut cell outline; sharply defined cell center
- 2+ Definite but dim fluorescence; cell outline less well defined
- 1+ Very subdued fluorescence; cell outline indistinguishable from cell center in most instances
- – Negligible or complete lack of fluorescence (negative)

INTERPRETATION
A 2+ or greater fluorescence in the unknown smear and no fluorescence in the other conjugate is evidence that the unknown organism is homologous to the fluorescing conjugate.

LIMITATIONS OF THE PROCEDURE
The fluorescent antibody technique can provide only presumptive identification of Bordetella pertussis or Bordetella parapertussis. A negative result should not be considered conclusive as this type of reaction may occur when only a few organisms are present in the specimen. Final identification can only be made after consideration of cultural, morphological and serological characteristics.

REFERENCES
1. Smith, Conant, and Overman, Zinsser Microbiology, 13th Ed., pg. 482, Appleton-Century-Crofts, New York, 1964.
2. Bordet, J., and Gengou, O., Ann. Inst. Pasteur, 20:731, 1906.
3. Am. J. Dis. Child., 101:149, 1961.
4. J. Bact., 83:745, 1962.
5. ibid., 86:449, 1963.

PACKAGING

Bacto FA Bordetella Parapertussis	5 ml	2378-56-3
Bacto FA Bordetella Pertussis	5 ml	2359-56-6
Bacto FA Buffer, Dried	6 × 10 g	2314-33-8
	100 g	2314-15-0
Bacto FA Mounting Fluid pH 7.2	6 × 5 ml	2329-57-2

BACTO FA BUFFER, DRIED

Bacto FA Buffer, Dried is a phosphate buffer-NaCl mixture which, upon rehydration, yields a 0.85% saline solution buffered to pH 7.2.

It is used for making dilutions of rehydrated FA Antimicrobial Globulins for test purposes as well as washing slides in FA staining processes. It is also recommended as a general purpose phosphate buffered saline.

To rehydrate, add the contents of a 10 g vial to 1 liter, or the contents of a 100 g bottle to 10 liters freshly distilled water and stir until completely dissolved. Rehydrated Bacto FA Buffer, Dried should be stored at 2 – 8°C. Solution showing turbidity or mold growth should be discarded.

Detailed directions for the use of Bacto FA Buffer, Dried in conjunction with other Bacto FA Reagents are in the package inserts accompanying the major FA Reagents for a given procedure.

PACKAGING

Bacto FA Buffer, Dried	6 × 10 g	2314-33-8
	100 g	2314-15-0
	10 kg	2314-08-9

BACTO FA C. ALBICANS

INTENDED USE

Bacto FA C. Albicans is used in the direct fluorescent antibody technique for the identification of *Candida albicans*.

BACKGROUND

C. albicans has been implicated in a variety of infections, ranging from thrush and cutaneous infections to septicemia, endocarditis, meningitis and vaginitis. Because this organism is endogenous and widely distributed through normal populations, it is necessary to recover it repeatedly and in significant numbers before it can be established as the etiological agent. These criteria make conventional techniques time consuming. Further, demonstration of chlamydospores by *C. albicans* (a primary criterion for its identification) may be difficult to achieve.

Application of the FA technique in the detection of *C. albicans* was explored by Gordon[1] as a means of simplifying the identification process and reducing the time necessary to identify the organism. This author obtained satisfactory results in distinguishing *Candida* from other yeasts and specifically identifying *C. albicans*. The fluorescent antibody technique was shown to be capable of differentiating *C. albicans* from all other species except *C. tropicalis* and *C. stellatoidea* in some cases.

REAGENTS
Bacto FA C. Albicans is a desiccated, fluorescein conjugated rabbit globulin prepared against *Candida albicans* serogroups A and B. Methods of preparing and using the reagents are described by Gordon[1].

To rehydrate the conjugate, add 5 ml sterile, distilled or deionized water to the vial and rotate gently to dissolve contents completely.

The working dilution of the conjugate should be determined shortly after its rehydration. The titer of a conjugate varies with the technique used, the fluorescent microscope used, the filter used and the age of the bulb.

The conjugate should be titrated using a known culture of *C. albicans*. Dilutions of the conjugate are made in rehydrated Bacto FA Buffer. The titer is determined as follows:

Dilution of Conjugate	Fluorescence
1:5	4+
1:10	4+
1:20	4+
1:40	4+
1:80	2+

In this sample, the last 4+ fluorescence is found in a 1:40 dilution of the conjugate. One less dilution is chosen for a margin of safety. The working dilution, therefore, in this case is 1:20.

Bacto FA Rabbit Globulin is a desiccated fluorescein conjugate used as a control for possible nonspecific staining of some organisms particularly *Staphylococcus* in the direct fluorescent antibody technique.

To rehydrate, add 5 ml distilled or deionized water to a vial and rotate gently to dissolve the contents completely. Before use, a dilution equivalent to that used for the particular FA C. Albicans conjugate being used is made with Bacto FA Buffer. Prepare only a sufficient amount of diluted globulin for use in each days run. Store the desiccated or rehydrated reagent at 2 – 8°C.

STORAGE AND EXPIRATION
Bacto FA C. Albicans in the desiccated state is stable to the expiry date on the label when stored at 2 – 8°C.

Aliquots of the titered conjugate should be put into small vials and frozen in the undiluted state and stored below –20°C for optimal stability.

The conjugate should not be subjected to repeated freezing and thawing.

PRECAUTIONS
1. It is assumed that the user of these reagents is familiar with fluorescent antibody techniques. Those desiring information concerning the advantages and disadvantages of the FA technique, the various staining techniques, and general precautions

and considerations concerning FA procedures should refer to FA GENERAL PRO-
CEDURES.
2. Discard the conjugate if bacterially contaminated.
3. All glassware that is employed in the preparation, testing and storage of these reagents must be free of detergents and other harmful residues.

SPECIMEN PREPARATION
Specimens should be cultured on Bacto Sabouraud Dextrose Agar and incubated at
room temperature for at least 24 hours. Colonies may be picked for the FA procedure
as soon as growth is apparent. Smears are prepared as follows:
1. Prepare 2 smears of the organism by placing a drop of the emulsified culture from
Bacto Sabouraud Dextrose Agar on microscope slides.
2. Allow to air dry.
3. Fix gently with heat.

PROCEDURE
Materials Provided:
Bacto FA C. Albicans
Bacto FA Buffer, Dried
Bacto FA Mounting Fluid pH 7.2
Bacto FA Rabbit Globulin

Materials Required but not Provided:
Microscope slides
Applicator sticks
Bibulous paper
Moisture chamber
Fluorescent microscope assembly
FA Slide Stained Streptococcus Group A

FA TECHNIQUE
1. Using 2 smears of the specimen, place several drops of Bacto FA C. Albicans on
one slide and several drops of Bacto FA Rabbit Globulin on the second as a control
for nonspecific fluorescence.
2. Spread the conjugate over the surface of the smear with a clean applicator stick
without disturbing the cells.
3. Incubate the slides under a Petri dish which contains a moistened piece of filter
paper for 30 minutes at room temperature.
4. Drain off the excess conjugate and place in Bacto FA Buffer for 10 minutes with 2
changes of buffer, followed by 1 rinse in distilled water.
5. Remove the slides and blot gently. Allow to air dry.
6. Add a small drop of Bacto FA Mounting Fluid pH 7.2 in the center of the stained
area and mount with a cover glass.
7. Examine each smear and record the intensity or lack of fluorescence.

QUALITY CONTROL
The following controls must be performed in each test series:
Positive Control — Use a known culture homologous to the conjugate.
Negative Control — Use appropriate dilution of FA Rabbit Globulin equivalent to the
dilution of the Bacto FA C. Albicans
Microscope Control — It is recommended that an additional control be incorporated.
A known positive fluorescing slide or Bacto FA Slide Stained Streptococcus Group
A as a check on microscope alignment, etc.

RESULTS
Brilliant fluorescence should result with the positive control and homologous test culture. No fluorescence of the organism should result on the smear to which Bacto FA
Rabbit Globulin was added. If a positive reaction did occur in the negative control, the
organism stained nonspecifically and the test results are not valid.

LIMITATIONS OF THE PROCEDURE

Results obtained by the direct FA technique can only be considered as presumptive identification of *Candida albicans*, as cross reactions can occur with *C. tropicalis* (antigenically similar to serogroup A, *C. albicans*) and with *C. stellatoidea* (antigenically similar to serogroup B, *C. albicans*). These cross reactions are minimized when the conjugate is properly titrated. Two useful additional tests to corroborate the presumptive fluorescent antibody identification of *C. albicans* are:

1. Formation of "germ tubes" by *C. albicans* when exposed to blood serum for 4 hours at 37°C.
2. Organisms action on sucrose in a fermentation medium such as Bacto Phenol Red Sucrose Broth containing a Durham tube. *C. albicans* forms acid only, *C. stellatoidea* does not utilize the sugar, and *C. tropicalis* forms acid and gas.

REFERENCE

1. Gordon, M. A., Proc. of the Soc. for Exp. Biol. and Med., 97:694 – 698, 1958.

PACKAGING

Bacto FA C. Albicans	5 ml	3234-56-5
Bacto FA Buffer, Dried	6 × 10 g	2314-33-8
	100 g	2314-15-0
Bacto FA Mounting Fluid pH 7.2	6 × 5 ml	2329-57-2
Bacto FA Rabbit Globulin	5 ml	2379-56-2

BACTO FA C. DIPHTHERIAE

INTENDED USE

Bacto FA C. Diphtheriae is recommended for use in the direct fluorescent antibody technique for the identification of *Corynebacterium diphtheriae*.

BACKGROUND

C. diphtheriae was first described by Klebs in 1883, and isolated in pure culture by Loeffler in 1884. With the discovery of the diphtheriae exotoxin by Roux and Yersin in 1888, the organism was firmly established as the etiological agent of diphtheria. Because virulent *C. diphtheriae* is difficult to purge from the nasopharynx of healthy carriers, diphtheria remains a major disease of man.

Primary isolation of *C. diphtheriae* is effected on a differential tellurite medium such as Bacto Tinsdale Medium. Morphological studies are best accomplished on subcultures from Bacto Loeffler Blood Serum. Morphological, biochemical, and colonial characteristics are necessary to identify *C. diphtheriae* yet the ultimate diagnostic criterion is the toxigenicity of the strain under study. This may be shown by the in vitro technique of Elek[1] using Bacto KL Virulence Agar, Bacto KL Virulence Enrichment and Bacto KL Antitoxin Strips.

The eventual diagnosis of *C. diphtheriae* as the etiological agent may require 2 – 3 days of bacteriological work, but utilization of the fluorescent antibody technique can provide rapid presumptive identification of the organism and significantly reduce the time required to firmly establish it as the etiological agent.

Proper use of Bacto FA C. Diphtheriae can provide positive identification of *C. diphtheriae* within 4 hours. Success of the method is dependent on proper technique and interpretation of results. Present information indicates that the FA test for detecting and identifying *C. diphtheriae* is at least as specific and sensitive as conventional methods.[1] However, it is recommended that conventional isolation be used in parallel with the FA procedure. These isolation methods presently cannot be superseded by the FA technique as toxigenic strains of *C. ulcerans* will stain with *C. diphtheriae* conjugate, though atoxic strains will not.

REAGENTS
FA C. Diphtheriae is desiccated fluorescein conjugated rabbit anti-*C. diphtheriae* globulin for use in the direct fluorescent technique as described by Moody and Jones.[2,3,4]

To rehydrate the conjugate add 5 ml sterile distilled or deionized water to the vial and rotate gently to dissolve contents completely. The working dilution of the conjugate should be determined shortly after its titration.

The titer of a conjugate varies with the technique used, the fluorescent microscope used, the filter and the age of the bulb.

The conjugate should be titrated using a known culture of *C. diphtheriae*. Dilutions of the conjugate are made in rehydrated Bacto FA Buffer. The titer is determined as follows:

Dilution of Conjugate	Fluorescence
1:5	4+
1:10	4+
1:20	4+
1:40	4+
1:80	2+

In this example, the last 4+ fluorescence is found in a 1:40 dilution of the conjugate. One less dilution is chosen for a margin of safety. The working dilution, therefore, in this case is 1:20.

Bacto FA Rabbit Globulin is a desiccated, fluorescein conjugate used as control for possible nonspecific staining of some organisms particularly *Staphylococcus,* in the direct fluorescent antibody technique.

To rehydrate add 5 ml sterile distilled or deionized water to a vial and rotate gently to dissolve the contents completely. Before use, a dilution equivalent to that used for the *C. diphtheriae* conjugate being used is made with Bacto FA Buffer. Prepare only a sufficient amount of diluted globulin for use in each day's run.

STORAGE AND EXPIRATION DATE
Bacto FA C. Diphtheriae in the desiccated state is stable to the expiry date on the label when stored at 2 – 8°C.

Aliquots of the titered conjugate should be put into small vials and frozen in the undiluted state and stored below −20°C for optimal stability.

The conjugate should not be subjected to repeated freezing and thawing.

PRECAUTIONS
1. It is assumed that the user of these reagents is familiar with fluorescent antibody techniques. Those desiring information concerning the advantages and disadvan-

tages of the FA technique, the various staining techniques, and general precautions and considerations concerning FA procedures should refer to FA GENERAL PROCEDURES.
2. Discard the conjugate if bacterially contaminated.
3. All glassware that is employed in the preparation, testing and storage of these reagents must be free of detergents and other harmful residues.

SPECIMEN PREPARATION

Specimens may be taken from the nasopharynx, conjunctivae, throat and wounds.
1. Specimen may be handled in one of 3 ways:
   a. When tests are performed within 4 hours, swab must be placed in 1 ml Bacto Heart Infusion Broth containing 0.05% dextrose and incubated for 4 hours at 37°C.
   b. When tests are performed between 4 and 24 hours, use 5 ml of the broth. Incubate 18 – 20 hours at 37°C.
   c. Where Bacto Loeffler Medium is used, transfer growth to the broth upon receipt in the laboratory and incubate as above. Incubate 18 – 20 hours at 37°C.
2. Prepare 2 slides from each clinical specimen:
   a. Centrifuge and discard supernatant from broth and resuspend sediment in a few drops of buffered saline.
   b. Suspend growth from Bacto Loeffler Medium tubes in buffered saline.
3. Prepare 1 slide from known *C. diphtheriae* cultures.
4. Allow the slides to air dry and fix immediately in 95% ethanol for 2 minutes.

PROCEDURE

Materials Provided:
Bacto FA C. Diphtheriae
Bacto FA Buffer, Dried
Bacto FA Rabbit Globulin
Bacto FA Mounting Fluid pH 7.2

Materials Required but not Provided:
Microscope slides
Moisture chamber
Bibulous paper
Applicator sticks
Cover glasses
Fluorescent microscope assembly

FA TECHNIQUE

1. To one specimen slide and the known culture slide, add 1 drop of appropriately diluted Bacto FA C. Diphtheriae. To the other specimen slide, add 1 drop of diluted Bacto FA Rabbit Globulin. Use a fresh end of an applicator stick to spread the drops over the smears without disturbing the cells.
2. Incubate the slides for 30 minutes at room temperature in a moisture chamber. Such a chamber may be conveniently prepared for a few slides by affixing a piece of moist filter paper to the inside surface of a Petri dish and inverting the dish over the slides.
3. Remove the excess conjugate and place the slides in Bacto FA Buffer for 8 – 10 minutes. Change the buffer twice during this period. Rinse the slides in distilled water.
4. Remove the slides and blot gently. Allow to air dry.
5. Add a drop of Bacto FA Mounting Fluid in the center of the stained area and mount with a cover glass.
6. Examine each smear using a suitable fluorescent microscope with an oil immersion objective giving a total magnification of 900 – 1000X. Slides not examined immediately may be kept in the dark for 1 to 2 hours before reading.

QUALITY CONTROL
As in most serological tests, certain cross reactions occur with *C. diphtheriae* antiglobulins. The following controls must be performed on duplicate smears in each test series:

Positive — Use a known culture of *C. diphtheriae* in place of the specimen.

Negative — Use Bacto FA Rabbit Globulin in place of the Bacto FA C. Diphtheriae globulin.

Microscope Control — It is recommended that an additional control be incorporated. A known positive fluorescing slide or Bacto FA Slide Stained Streptococcus Group A as a check on microscope alignment, etc.

RESULTS
Read and record results based on intensity of fluorescence 1+ through 4+ as compared with positive and negative control slides. Cultures exhibiting 2+ fluorescence or greater should be considered positive.

4+ Maximum fluorescence; brilliant yellow-green; clear-cut cell outline; sharply defined cell center
3+ Less brilliant yellow-green fluorescence; clear-cut cell outline; sharply defined cell center
2+ Definite but dim fluorescence; cell outline less well defined
1+ Very subdued fluorescence; cell outline indistinguishable from cell center in most instances
— Negligible or complete lack of fluorescence (negative)

If organisms appear fluorescent on test, positive and negative control slides, observe test and positive slides using tungsten light and examine the cellular morphology.

LIMITATIONS OF THE PROCEDURE
The fluorescent antibody technique can only provide rapid presumptive identification of *C. diphtheriae.* It must be remembered that the FA technique cannot distinguish atoxigenic and toxigenic strains of this organism and that cross reactions can occur with *C. ulcerans;* therefore, its use as a diagnostic determinant is limited to preliminary findings.

Final identification of *C. diphtheriae* and evidence of its toxigenicity requires further morphological, biochemical and toxinogenic testing.

REFERENCES
1. Am. J. Clin. Path., 29:181, 1958.
2. J. Bacto., 86:285, 1963.
3. Bact. Proc. 145, 1960.
4. *ibid.,* 141, 1960.

PACKAGING
Bacto FA C. Diphtheriae	5 ml	3207-56-8
Bacto FA Buffer, Dried	6 × 10 g	2314-33-8
	100 g	2314-15-0
Bacto FA Mounting Fluid pH 7.2	6 × 5 g	2329-57-2
Bacto FA Rabbit Globulin	5 ml	2379-56-2

BACTO FA E. COLI POLY A, B, AND C

DESCRIPTION AND INTENDED USE

Bacto FA E. Coli Poly A, B, and C are desiccated, fluorescein conjugated, polyvalent rabbit anti-E. coli globulins for use in direct fluorescent technique for the rapid preliminary identification of enteropathogenic E. coli (EEC) associated with infantile diarrhea.

BACKGROUND

The use of the fluorescent method for detecting E. coli serogroups in stool specimens was first described by Whitaker, Page, Stulberg and Zuelzer.[1] Further evidence substantiating the usefulness of the direct FA method for detecting pathogens in fecal specimens was reported by Cohen, Page and Stulberg,[2] Nelson, Whitaker, Hempstead and Harris,[3] Page and Stulberg[4] and Cherry and Moody.[5]

Investigations of Thomason, Cherry and Ewing[6] indicated that the fluorescent antibody technique results are similar to those obtained with type specific antibodies.

Later, Cherry and Moody,[5] noted that the sensitivity of the test established it as an excellent tool for screening throat and fecal specimens for EEC associated with epidemic or institutional infant diarrhea. The studies, comparing cultural and FA methods, cited in these authors' evaluation indicated that "no specimen was found to be positive by culture when negative by immunofluoresence." Further the authors noted that the greater number of positives by FA over cultural methods are due to the increased sensitivity of the FA technique rather than a lack of specificity. The immunofluorescent method can detect organisms which are culturally negative due to drug therapy, lytic bacteriophage or bacterial antagonism.

Although few heterologous staining reactions were observed with other members of the Enterobacteriaceae in studies surveyed by Cherry and Moody, it is important to remember that intergeneric and intrageneric relationships do exist in this family. Therefore, as with conventional serological methods the fluorescent antibody technique should be corroborated by conventional culture and biochemical identification techniques for enteropathogenic E. coli.

Methods of preparing and using FA reagents are described by Thomason, Cherry, Davis, and Pomales-Lebron,[7] Thomason, Cherry and Pomales-Lebron[8] and Cherry, Thomason, Pomales-Lebron and Ewing.[9]

REAGENTS

Bacto FA E. Coli Poly A, Poly B and Poly C are fluorescein conjugated rabbit antiglobulins used in the preliminary screening of fecal specimens for the presence of E. coli.

To rehydrate the conjugate add 5 ml sterile distilled or deionized water to each vial and rotate gently to dissolve the contents completely. No further dilution is required for the polyvalent conjugates.

STORAGE AND EXPIRATION

Bacto FA E. Coli Poly A, B, C, in the desiccated state are stable to the expiry date on the label when stored at 2 – 8°C.

Aliquots of a rehydrated conjugate should be put into small vials and frozen in the undiluted state and stored below −20°C for optimal stability. These frozen aliquots are stable to the expiry date on the original vial. The conjugate should not be subjected to repeated freezing and thawing.

PRECAUTIONS

1. It is assumed that the user of these reagents is familiar with fluorescent antibody techniques. Those desiring information concerning the advantages and disadvantages of the FA technique, the various staining techniques, and general precautions and considerations concerning FA procedures should refer to FA GENERAL PROCEDURES.
2. Discard the conjugate if bacterially contaminated.
3. All glassware that is employed in the preparation, testing and storage of these reagents must be free of detergents or other harmful residues.

SPECIMEN PREPARATION

1. Obtain rectal swabs or a stool specimen and smear directly on prepared microscope slides.
2. Air dry the smear, then fix for 1 – 2 minutes in 95% ethanol.
3. Drain the slide free from fixative and rinse in Bacto FA Buffer.

PROCEDURE

Materials Provided:
Bacto FA E. Coli Poly A, B and C
Bacto FA Mounting Fluid pH 7.2
Bacto FA Buffer, Dried

Materials Required but not Provided:
Filter Paper
Petri dish
Applicator sticks
Microscope slides
Cover glass
Bibulous paper
Fluorescent microscope assembly

FA TECHNIQUE

1. Apply several drops of the appropriate conjugate to the smeared slides.
2. Spread the conjugate over the surface of the smear with a clean applicator stick without disturbing the cells.
3. Place the slides under a Petri dish containing moistened filter paper to retard evaporation of the conjugate, for 30 minutes.
4. Remove the excess conjugate and place the slides in Bacto FA Buffer for 8 – 10 minutes. Change the buffer once during this period. Rinse the slides in distilled water.
5. Remove the slides and blot gently with bibulous paper. Allow to air dry.
6. Add a small drop of Bacto FA Mounting Fluid pH 7.2 in the center of the stained area and mount with a cover glass.
7. Examine each smear using a suitable fluorescent microscope with an oil immersion objective giving a total magnification of 900 – 1000X. Slides not examined immediately should be stored in the dark for not more than 1 – 2 hours.

QUALITY CONTROL

The following controls must be performed in each test:

Positive Control — Use a known culture homologous to the conjugate being used.
Negative Control — Use a known *E. coli* culture heterologous to the conjugate being used.
Microscope Control — It is recommended that an additional control be incorporated. A known positive fluorescing slide or Bacto FA Slide Stained Streptococcus Group A as a check on microscope alignment, etc.

RESULTS
Read and record results based on intensity of fluorescence 1+ to 4+ as outlined below. Cross reactions with other antigenicially related enterics may be encountered. Cultures exhibiting a fluorescence of 2+ or greater should be considered positive.

INTERPRETATION OF RESULTS
A 2+ or greater fluorescence in the unknown smear with the homologous conjugate and no fluorescence in the unknown smear in other FA E. Coli Poly conjugates is evidence that the unknown organism is homologous to one of the strains incorporated into that conjugate.

LIMITATIONS OF THE PROCEDURE
Bacto FA E. Coli Poly A, B and C are used for rapid preliminary screening of fecal specimens for the presence of EEC. However, due to common antigens within the *Enterobacteriaceae*, cross reactions may occur. Following are a list of antigenic relationships of *E. coli* and other members of the *Enterobacteriaceae*. These relationships should be kept in mind when interpreting results of the FA procedure:

$$E.\ coli\ 0111\ =\ Salmonella\ 035\ \ \ =\ Arizona\ 020$$
$$E.\ coli\ 025\ \neq\ E.\ coli\ 026$$
$$E.\ coli\ 090\ \neq\ E.\ coli\ 086a$$
$$E.\ coli\ 01\ \ =\ Shigella\ dysenteriae\ 1$$
$$E.\ coli\ 0112\ =\ Shigella\ dysenteriae\ 1$$
$$E.\ coli\ 0124\ =\ Shigella\ dysenteriae\ 3$$

Key:
  = identical
  ≠ similar

For further information regarding the antigenic relationhips of *E. coli* refer to *Identification of Enterobacteriaceae*.[8]

Because of the intergeneric and intrageneric relationships of *E. coli* final identification cannot be made without additional cultural and biochemical studies. Pure cultures under study may be more fully characterized by conventional serological methods described in "E. Coli O and OK Antisera."

REFERENCES
1. Whitaker, J. A., Page, R. H., Stulberg, C. S., and Zuelzer, W. W., Am. J. Dis. Child., 95:1 – 8, 1958.
2. Cohen, J. O., Page, R. H., and Stulberg, C. S., Am. J. Dis. Child., 23:159 – 164, 1962.
3. Nelson, J. D., Whitaker, J. A., Hempstead, B., Harris, M., J. Am. Med. Assoc., 176:26 – 30, 1961.
4. Page, R. H., and Stulberg, C. S., Am. J. Dis. Child., 4:149 – 156, 1962.
5. Cherry, W. B. and Moody, M. D., Bact. Rev., 29:222 – 250, 1965.
6. Thomason, B. M., Cherry, W. B., and Ewing, W. H., Bact. Proc., 90, 1959.
7. Thomason, B. M., Cherry, W. B., Davis, B. R., and Pomales-Lebron Bull., WHO, 25:137 – 152, 1961.
8. Edwards, P. R., and Ewing, W. H., Identification of *Enterobacteriaceae*, 3rd Ed., Burgess Pub. Co., 1971.

PACKAGING

Bacto FA E. Coli Poly A	5 ml	2334-56-6
Contains 026:K60(B6), 055:K59(B5), 0111:K58(B4), 0127a:K63(B8)		
Bacto FA E. Coli Poly B	5 ml	2339-56-1
Contains 086a:K61(B7), 0119:K69(B14), 0124:K72(B17), 0125:K70(B15), 0126:K71(B16), 0128:K67(B12)		
Bacto FA E. Coli Poly C	5 ml	3299-56-7
Contains 018a018c:K77(B21), 020a020c:K61(B7), 020a020b:K84(B), 028:K73(B18), 044:K74		
Bacto FA Buffer, Dried	6 × 10 g	2314-33-8
	100 g	2314-15-0
Bacto FA Mounting Fluid pH 7.2	6 × 5 ml	2329-57-2

FA H. INFLUENZAE

BACTO FA H. INFLUENZAE TYPES A AND B
BACTO FA MENINGOCOCCUS POLY
BACTO FA PNEUMOCOCCUS POLY

INTENDED USE

The fluorescein conjugated antiglobulins listed above are used in the direct FA technique for rapid screening of cerebrospinal fluid for the presence of *Streptococcus (Diplococcus) pneumoniae, Haemophilus influenzae,* and *Neissera meningitidis* in suspect cases of bacterial meningitis.

BACKGROUND

Although bacterial meningitis may be caused by a variety of microorganisms, *H. influenzae, N. meningitidis* and *S. pneumoniae* are the 3 most common etiologic agents of this type of infection. *H. influenzae* is the most frequent cause of meningitis in children, with *H. influenzae* type B being the predominant incitant. The majority of the cases occur in children between the ages of 3 months and 3 years.

H. influenzae is rarely the cause of bacterial meningitis in individuals younger than 3 months (neonates) or beyond 3 years. On the other hand, while *S. pneumoniae* and *N. meningitidis* rarely occur in the neonate, they can occur at any age thereafter.

The use of the FA technique for the demonstration of pneumococcal antigen in tissues was described by Coons, Creech, Jones and Berliner.[1] It was later recommended for the detection of *S. pneumoniae* in spinal fluids and pure culture techniques.

Page, Caldroney and Stulberg[2] in their studies also used the FA technique for detection of *H. influenzae* in cerebrospinal fluid smears and declared this technique to be of value as an adjunct to conventional techniques in the rapid identification of *H. influenzae.*

The application of the FA technique in the identification of *N. meningitidis* was first described by Metzger and Smith[3] who discussed its feasibility in clinical laboratories despite the knowledge of the existence of cross reactions between the serological groups in conjugated globulins. This latter fact prompted preparation of an "over all polyvalent conjugate" prepared from an antiserum consisting of all four serological groups A, B, C and D.

Bielgeleisen, Mitchell, Marcus, Rhoden and Blumberg[4] evaluated the fluorescent antibody technique for its clinical application in identification of pathogens associated with cerebrospinal meningitis. Particular emphasis was placed upon detection and identification of *N. meningitidis, H. influenzae* and *S. pneumoniae.*

These authors found the immunofluorescent technique to be as sensitive as conventional cultural techniques and superior to gram stains in detecting these pathogens. Further, the FA technique was particularly valuable in detecting partially treated cases of meningitis. Such prior antibiotic therapy, in many cases resulted in culturally negative specimens, but did not adversely affect immunofluorescent staining. Indeed, the FA technique may be the only means of detecting these pathogens when intense antibiotic therapy precedes microbiological procedures.

Rapid screening of cerebrospinal fluid by the direct FA technique using Bacto FA H. Influenzae Types A and B, Bacto FA Meningococcus Poly and Bacto Pneumococcus Poly can provide presumptive identification of the 3 most common etiologic agents of meningitis.

REAGENTS

Bacto FA H. Influenzae Types are desiccated, high titered, specific fluorescein conjugated rabbit antiglobulins prepared against *H. influenzae* types A and B for use in the direct FA technique.

Bacto FA Pneumococcus Poly is a desiccated, fluorescein conjugated, polyvalent, rabbit antiglobulin prepared against *S. pneumoniae* serotypes 1 – 33 excluding serotypes 26 and 30.

Bacto FA Meningococcus Poly is a desiccated, fluorescein conjugated, polyvalent anti-Meningococcus rabbit globulin recommended for use in the rapid screening of *N. meningitidis*.

To rehydrate the conjugates, add 5 ml distilled water to each vial and rotate gently to dissolve contents completely. No further dilution is required for Bacto FA Meningococcus Poly or Bacto FA Pneumococcus Poly.

A working dilution of Bacto FA Influenzae Types A and B should be determined shortly before rehydration. The titer of a conjugate varies with the technique used, the fluorescent microscope used, the filter used and the age of the bulb.

The conjugate should be titrated with a known culture of *H. influenzae* homologous to the conjugate. Dilutions of the conjugate are made in rehydrated Bacto FA Buffer. The titer is determined as follows:

Dilution of Conjugate	Fluorescence
1:5	4+
1:10	4+
1:20	4+
1:40	4+
1:80	2+

In this example, the last 4+ fluorescence is found in a 1:40 dilution of the conjugate. One less dilution is chosen for a margin of safety. The working dilution, therefore, in this case is 1:20. Dilute only the amount of conjugate for use in each day's run.

STORAGE AND EXPIRATION DATE

The conjugates in the desiccated state are stable to the expiry date on the label when stored at 2 – 8°C.

Aliquots of a rehydrated conjugate should be put into small vials and frozen in the undiluted state and stored below −20°C for optimal stability. The conjugates should not be subjected to repeated freezing and thawing.

PRECAUTIONS

1. It is assumed that the user of these reagents is familiar with fluorescent antibody techniques. Those desiring information concerning the advantages and disadvantages of the FA technique, the various staining techniques, and general precautions and considerations concerning FA procedures should refer to "FA General Procedures."

2. Discard any conjugate that becomes bacterially contaminated.
3. All glassware employed in the preparation, testing and storage of these reagents must be free of detergents or other harmful residues.

SPECIMEN PREPARATION
Four slides should be prepared for immunofluorescent staining with Bacto FA Meningococcus Poly, Bacto FA Pneumococcus Poly, and Bacto FA H. Influenzae Types A and B.
1. Centrifuge the cerebrospinal fluid specimen, not used for other tests, in an aseptic manner at 2500 rpm for 10 – 15 minutes.
2. Aspirate off the supernatant fluid and discard.
3. Add 1 drop of the sediment to each of 3 microscopic slides and spread over approximately 1 cm area with a capillary pipette.
4. Allow to air dry.
5. Fix by gentle heating.

PROCEDURE
Materials Provided:

Bacto FA H. Influenzae Types A and B
Bacto FA Meningococcus Poly
Bacto FA Pneumococcus Poly
Bacto FA Mounting Fluid pH 7.2
Bacto FA Buffer, Dried
Bacto FA Rabbit Globulin

Materials Required but not Provided:

Centrifuge
Capillary pipettes
Microscope slides
Cover glasses
Moisture chamber
Fluorescent microscope assembly

FA TECHNIQUE
1. Apply several drops of the conjugate over the fixed smear.
2. Place in a moisture chamber and incubate at room temperature for 15 minutes. Such a chamber may be conveniently prepared for a few slides by affixing a piece of moist filter paper to the inside surface of a Petri dish and inverting the dish over the slides.
3. Drain off excess conjugate and rinse in Bacto FA Buffer.
4. Place in a Coplin jar containing Bacto FA Buffer for 10 minutes. Make 2 changes of buffer. Give a final rinse in distilled water.
5. Remove the slide from the jar and allow to drain and air dry.
6. Add 1 small drop of Bacto FA Mounting Fluid pH 7.2 and mount with a cover glass.
7. Examine with a suitable fluorescent microscope for fluorescing organisms of typical morphology.

QUALITY CONTROL
The following controls must be performed in each test series:
Positive Control — use known cultures homologous to the conjugates being used.
Negative Control — use appropriate dilution of Bacto FA Rabbit Globulin (same dilution as working dilution of conjugate).
Microscope Control — it is recommended that an additional control be incorporated. A known positive fluorescing slide or Bacto FA Slide Stained Streptococcus Group A as a check on microscope alignment, etc.

RESULTS
Read and record results based on intensity of fluorescence 1+ to 4+ as outlined below. All 3 smears should be examined for comparison of intensity of fluorescence and possible cross reaction. Cultures exhibiting a fluorescence of 2+ or greater should be considered positive.

4+   Maximum fluorescence; brilliant yellow-green; clear-cut cell outline; sharply
      defined cell center
3+   Less brilliant yellow-green fluorescence; clear-cut cell outline; sharply defined
      cell center
2+   Definite but dim fluorescence; cell outline less well defined
1+   Very subdued fluorescence; cell outline indistinguishable from cell center in
      most instances
−   Negligible or complete lack of fluorescence (negative)

If a negative result is obtained with above conjugates the possibility of an *H. influenzae* type other than types A or B should be considered and tested before ruling out *H. influenzae* as the etiological agent.

LIMITATIONS OF THE PROCEDURE

The direct FA technique for the detection of *S. pneumoniae*, *N. meningitidis* and *H. influenzae* should only be performed on spinal fluid and pure cultures. Their use for the identification of *H. influenzae*, *N. meningitidis* and *S. pneumoniae* in mixed cultures is not recommended. This procedure only serves to provide presumptive identification of these organisms and is intended to complement but not replace conventional bacteriological techniques. Further cultural and biochemical studies are necessary to confirm results obtained in the use of this immunofluorescent technique and serogrouping of these organisms should be done by microscopic slide agglutination or by the microscopic slide Quellung (Neufeld) test.

REFERENCES

1. Coons, A. H., Creech, H. J., Jones, R. N. and Berliner, E., J. Immunol., 45:159 – 170, 1942.
2. Page, R. H., Caldroney, G. L., and Stulberg, C. S., Am. J. Dis. Child., 101:155 – 159, 1961.
3. Metzger, J. F., and Smith, C. W., U.S. Armed Forces Med. J., 11:1185 – 1189, 1960.
4. Beigeleisen, J. Z., Mitchell, M. S., Marcus, B. B., Rhoden, D. L., and Blumberg, R. W., J. Lab. and Clin. Med. 65:976 – 989, 1965.

PACKAGING

Bacto FA H. Influenzae Type A	5 ml	3303-56-1
Bacto FA H. Influenzae Type B	5 ml	3235-56-4
Bacto FA Meiningococcus Poly	5 ml	3272-56-8
Bacto FA Pneumococcus Poly	5 ml	3270-56-0
Bacto FA Mounting Fluid pH 7.2	6 × 5 ml	2329-57-2
Bacto FA Buffer, Dried	6 × 10 ml	2314-33-8
	100 g	2314-15-0
Bacto FA Rabbit Globulin	5 ml	2379-56-2

BACTO FA KIRKPATRICK FIXATIVE MODIFIED

Bacto FA Kirkpatrick Fixative Modified is an alcohol based fixative prepared by a modification of Kirkpatrick's method described by Goepfert and Hicks[1] which contains 6 parts isopropyl alcohol, 3 parts chloroform and 1 part formaldehyde 37% solution. It is stable at room temperature to the expiry date on the label if tightly closed. Do not refrigerate.

It is recommended for fixation of bacterial cells to the microscope slide in the fluorescent antibody technique particularly in the FA staining of *Salmonella*.

PREPARATION OF THE SPECIMEN
1. Smear a loopful of broth culture of the organism on a clean, dry microscope slide.
2. Air dry the smear at room temperature.
3. Fix the slide for 3 minutes in Bacto FA Kirkpatrick Fixative Modified in a Coplin jar or other suitable container.
4. Rinse the slide with 95% ethyl alcohol.
5. Air dry the slide at room temperature before adding the conjugate.

REFERENCE
1. Applied Microbiology, 18:612, 1969.

PACKAGING
Bacto FA Kirkpatrick Fixative Modified   6 × 100 ml   3188-73-0

FA LEPTOSPIRA CONJUGATES
BACTO FA LEPTOSPIRA CANICOLA
BACTO FA LEPTOSPIRA ICTEROHEMORRHAGIAE

INTENDED USE
Bacto FA Leptospira conjugates are desiccated, fluorescent labeled anti-leptospira globulins for detecting and identifying *Leptospira*. They are rehydrated by adding 5 ml sterile distilled or deionized water per vial.

SUMMARY AND BACKGROUND
Since normal cross-reactions exist between sera prepared against these species of *Leptospira* it is necessary to determine the optimal dilution of each conjugate to be used for differentiation prior to use. This may vary from laboratory to laboratory depending upon the U.V. microscopic equipment in use.

The working dilution of the conjugate should be determined upon rehydration. The titer of a conjugate varies with the technique, the fluorescent microscope, the filter and the age of the bulb and therefore may vary from laboratory to laboratory. Dilutions of the conjugate are made in rehydrated Bacto FA Buffer. The titer is determined as follows:

Dilution of Conjugate	Fluorescence
1:5	4+*
1:10	4+
1:20	4+
1:40	4+
1:80	2+

*A 4+ fluorescence is defined as brilliant yellow-green cocci with sharp cell outlines and nonstaining centers.

In this example, the last 4+ fluorescence is found in a 1:40 dilution of the conjugate. One less dilution is chosen for a margin of safety. The working dilution, therefore, in this case is 1:20.

Aliquots of the titered conjugate should be distributed into small vials and frozen in the undiluted state and stored below −20°C for optimal stability. Prepare only a sufficient amount of diluted conjugate for each day's run.

Unrehydrated Bacto FA Leptospira should be stored at 2 − 8°C.

PRECAUTIONS

1. It is assumed that the user of these reagents is familiar with fluorescent antibody techniques. Those desiring information concerning these techniques should refer to "FA General Procedures" in the introductory Serology discussion.
2. The test organism must be identified to at least the generic, and in some cases the species level, biochemically prior to serological confirmation.
3. Discard the conjugate if bacterially contaminated.
4. The conjugate should not be subjected to repeated freezing and thawing. Such treatment is detrimental to the antibody content.
5. All glassware that is employed in the preparation, testing and storage of these reagents must be free of detergents or other harmful residues.

SPECIMEN PREPARATION

1. Smear specimen or culture directly on a microscope slide.
2. Allow to air dry.
3. Fix by a one minute immersion in 95% ethyl alcohol or by passing lightly through a flame.

PROCEDURE

1. Place a drop of the desired Bacto FA Leptospira on the fixed slide and distribute it evenly over the specimen.
2. Place the slide in a Petri dish containing a moistened absorbent paper to prevent drying and allow to react at room temperature for 30 minutes.
3. Rinse the excess labeling reagent off the slide with rehydrated Bacto FA Buffer and then place it in a Bacto FA Buffer solution in a Coplin jar for 10 minutes with two changes of buffer.
4. Remove slide from jar and gently blot between layers of absorbent paper.
5. When all moisture has been removed, mount the slide by placing a drop of Bacto FA Mounting Fluid pH 7.2 on the stained area and cover it with a cover slip.
6. Examine the slide under a fluorescent microscope with U.V. light for fluorescent *Leptospira* cells.

QUALITY CONTROL

It is highly recommended that both positive and negative controls be run in parallel with the test specimen and examined at the same time. A *Leptospira* culture homologous to the Bacto FA Leptospira would constitute a positive control while a *Leptospira* culture of a different serotype would be a negative control.

It is recommended that a microscope control also be incorporated. A known positive fluorescing slide or Bacto FA Slide Stained Streptococcus Group A will provide a check on microscope alignment, etc.

RESULTS

Read and record results based on intensity of fluorescence – to 4+ as outlined below. Cultures exhibiting a fluorescence of 2+ or greater should be considered positive.

4+	Maximum fluorescence; brilliant yellow-green; clear-cut cell outline; sharply defined cell center
3+	Less brilliant yellow-green fluorescence; clear-cut cell outline; sharply defined cell center
2+	Definite but dim fluorescence; cell outline less well defined
1+	Very subdued fluorescence; cell outline indistinguishable from cell center in most instances
–	Negligible or complete lack of fluorescence (negative)

INTERPRETATION OF RESULTS
A 2+ or greater fluorescence in the unknown smear is evidence that the unknown organism is homologous to the conjugate used.

LIMITATIONS OF THE PROCEDURE
While serological procedures, as applied to microorganisms, provide corroborative evidence and can be used to identify particular antigenic sites of the genus or species under study, they cannot be used alone to identify the etiological agent of disease. Cultural isolation and at least preliminary biochemical differentiation must precede any serological examination and final identification cannot be made without biochemical and serological characterization.

Particularly important in fluorescent microscopy is the knowledge that the microscopic system is in alignment and completely functional in the use of adequate and proper control systems and in the optimal storage of the conjugates.

PACKAGING
Bacto FA Leptospira Canicola	5 ml	2396-56-1
Bacto FA Leptospira Icterohemorrhagiae	5 ml	2397-56-0

BACTO FA LISTERIA POLY

INTENDED USE
Bacto FA Listeria Poly is recommended for use in the direct fluorescent antibody technique for the identification of Listeria monocytogenes. This conjugate will detect approximately 98% of the Listeria cultures. Bacto FA Listeria Poly will detect all 4 Listeria serotypes.

SUMMARY
Listeria monocytogenes was suspected of being an animal pathogen as early as 1910 by Helphers[1] and positively identified as such in 1924 by Murray, Webb and Swann.[2] Nyfeldt recorded the isolation of Listeria monocytogenes in humans as early as 1929, yet listeriosis is considered one of the latest recognized infections of man.[3] Many listeria infections were attributed to other organisms. Specimens containing L. monocytogenes were often discounted as negative and when the organism was recovered, it was considered a laboratory contaminant.

Gray[4] reported listeriosis in man and animals to manifest itself in a number of disease states, including meningitis, septicemia, bacteremia, abortion, endocarditis, meningoencephalitis, and conjunctivitis. Winn, Cherry and King[5] emphasized that as clinicians have become more cognizant of the pathogenic potential of L. monocytogenes, there is a marked increase in the number of cases of listeriosis being reported. Hood[6] stated that it is necessary therefore, for the modern laboratory to be on the alert for this potential pathogen.

The Listeria are short gram-positive, nonspore-forming, noncapsulated motile rods. Motility is most pronounced at 20°C. They are microaerophilic and therefore thrive best under reduced oxygen tension. L. monocytogenes has been found to be present either in the infective state or in the asymptomatic carrier state in most warm blooded animals both domestic and wild. They are believed to be widely distributed in nature.[7]

Identification of Listeria is based on successful isolation of the organism, biochemical characterization and serological confirmation.

There are four recognized serological types of *L. monocytogenes* designated 1 – 4 according to their somatic antigen components. Gray[4] reported that the serological analysis of available Listeria cultures indicated that within the United States, 32.8% were of type 1, 65.3% were type 4 and less than 1% were of types 2 or 3.

Listeria are generally slow growing and delicate organisms especially on primary isolation. Attempts at direct isolation from positive specimens may, at times, be met with failure. The likelihood of recovery is improved when they are preincubated at 4°C, (cold enrichment). This process may require a few days or several months with weekly to monthly transfers before positive cultures are obtained. Yet the potential severity of listeriosis (meningitis, meningoencephalitis) necessitates rapid identification of the organism. In this respect the direct FA technique is particularly valuable. Cerebrospinal fluid or tissue impression smears may be examined directly using the FA technique. Further, this technique can detect nonviable organisms in specimens or detect organisms in specimens from which culture is impossible.

REAGENTS
Bacto FA Listeria Poly is a stable, desiccated, high titered rabbit Listeria antiglobulin conjugated with fluorescein as recommended by Eveland.[6,7]

Bacto FA Rabbit Globulin is a desiccated, fluorescein conjugate used as a control for possible nonspecific staining of some organisms, particularly *Staphylococcus* in the direct fluorescent antibody technique.

When used according to the suggested procedure, each 5 ml vial is sufficient to perform approximately 50 tests.

REHYDRATION AND STORAGE
To rehydrate Bacto FA Listeria Poly, add 5 ml sterile, distilled or deionized water to the vial and rotate gently to dissolve contents completely. The rehydrated conjugate should be stored at 2 – 8°C.

Aliquots of the conjugate should be put into small vials and frozen in the undiluted state and stored below −20°C for optimal stability.

Bacto FA Listeria Poly is ready to use and requires no further dilution.

To rehydrate Bacto FA Rabbit Globulin, add 5 ml distilled water to a vial and rotate gently to dissolve the contents completely.

EXPIRATION
Bacto Listeria O Antiserum Poly will remain stable to the expiry date on the label when stored at 2 – 8°C.

SPECIMEN PREPARATION
Cerebrospinal fluid, meconium, tissue impression smears or isolated cultures may be used in the direct FA technique. If tissue impression smears are employed in the test, it is recommended that the slides be counterstained with Bacto Eriochrome Black.
1. Prepare a smear of the specimen or a drop of the emulsified culture on a microscope slide. Use Bacto FA Buffer to prepare emulsion. Avoid the use of saline as a diluent for *Listeria* cultures.

2. Allow to air dry.
3. Fix for 1 minute by immersion in 95% ethanol.

PROCEDURE

Materials Provided:
Bacto FA Listeria Poly
Bacto FA Mounting Fluid
Bacto FA Buffer Dried
Bacto FA Rabbit Globulin
Bacto Eriochrome Black

Materials Required but not Provided:
Microscope slides
Applicator sticks
Moisture chamber
Bibulous paper
Fluorescent microscope assembly
Cover glasses

1. Using one smear of the specimen and a known control slide place several drops of the appropriate conjugate on each.
2. Add several drops of Bacto FA Rabbit Globulin on a second specimen slide.
3. Spread the conjugate over the surface of the smear with a clean applicator stick without disturbing the cells.
4. Incubate the slides for 30 minutes at room temperature in a moisture chamber. Such a chamber may be conveniently prepared for a few slides by affixing a moist filter paper to the inside surface of a Petri dish and inverting the dish over the slides.
5. Drain off excess conjugate and place in Bacto FA Buffer for 10 minutes with 2 changes of buffer. Rinse the slides in distilled water.
6. Remove slides and blot gently with bibulous paper. Allow to air dry.
7. Add a small drop of Bacto FA Mounting Fluid pH 7.2 in the center of the stained area and mount with a cover glass.
8. Examine each smear and record the intensity or lack of fluorescence.

QUALITY CONTROL

Use known positive and negative control cultures in parallel with the test culture to ascertain the validity of test results.

RESULTS

Brilliant fluorescence should result with the positive control. Very slight to no fluorescence should result on the smear to which Bacto FA Rabbit Globulin was added. If equal intensity of the positive and negative control results, the test is not valid since nonspecific staining has occurred. If the Bacto FA Rabbit Globulin stains the specimen to the point where the background fluorescence interferes with interpretation, it is recommended that the specimen be counterstained with Bacto Eriochrome Black.

Counterstaining Technique

It is recommended that Bacto Eriochrome Black be used as a counterstain when tissue impression smears are being stained or when excessive extraneous material is present as described by Hall and Hansen.[8] After the FA Listeria serotype has been added to the slide, it is flooded with a 1:10 dilution of Bacto Eriochrome Black and allowed to stain for 10 – 15 seconds. The 1:10 dilution is prepared by addition of 1 part of the dye solution to 9 parts Bacto FA Buffer. The dilution should be used the same day prepared. The excess dye is removed, the slide is blotted and mounted with a cover glass and Bacto FA Mounting Fluid. The slide is ready for examination. Tissue and extraneous materials are stained a pinkish red in contrast to the yellow-green color of specific immunofluorescence.

LIMITATIONS OF THE PROCEDURE

The direct FA procedure can provide rapid presumptive identification of *L. monocytogenes*. Results obtained from the FA procedure must be confirmed by isolation of the organism in pure culture and biochemical characterization.

REFERENCES
1. Svenk. Vet. Tidskr., 2:265, 1911.
2. J. Path. & Bact., 29:407, 1926.
3. Gray, M. L. and Killinger, H. H., Bact. Revs., 30:309–382, 1966.
4. Second Symposium on Listeria Infection, Montana State Coll., 1962.
5. Annals N.Y. Acad. Sci., 70:624, 1958.
6. Paper presented APHA, 1963.
7. Paper presented APHA, 1964.
8. Zbl. Bakt. Para. Inf. & Hyg., 184:548, 1962.

PACKAGING

Bacto FA Listeria Poly	5 ml	2469-56-3
Bacto FA Mounting Fluid pH 7.2	6 × 5 ml	2329-57-2
Bacto FA Buffer Dried	6 × 10 g	2314-33-8
	100 g	2314-15-0
Bacto FA Rabbit Globulin	5 ml	2379-56-2
Bacto Eriochrome Black	5 ml	3249-56-8

BACTO FA MOUNTING FLUID pH 7.2

INTENDED USE
Bacto FA Mounting Fluid pH 7.2 is a standardized reagent grade glycerin adjusted to pH 7.2 for use in mounting specimens on slides to be viewed under the fluorescent microscope.

STORAGE
Bacto FA Mounting Fluid pH 7.2   Room temperature

PROCEDURE
For use, place a small drop of the mounting fluid over the fluorescent labeled smear on the microscope slide. Place a cover glass over the droplet; care being taken not to entrap air bubbles.

Detailed directions for the use of Bacto FA Mounting Fluid pH 7.2 in conjunction with other Bacto FA Reagents are in the package inserts accompanying the major FA Reagents for a given procedure.

PACKAGING
Bacto FA Mounting Fluid pH 7.2   6 × 5 ml   2329-57-2

BACTO FA MOUNTING FLUID pH 9

INTENDED USE
Bacto FA Mounting Fluid pH 9 is a standardized reagent grade glycerin adjusted to pH 9 for use in the FA technique particularly with Bacto FA N. Gonorrhoeae and Bacto FA Salmonella Poly and Panvalent conjugates.

HISTORY
Numerous reports in the literature indicate the preference of investigators and clinicians for a more alkaline mounting fluid than the conventional type of pH 7.2 especially when the slide is rinsed in distilled water rather than FA Buffer.

STORAGE
Bacto FA Mounting Fluid pH 9   15 – 30°C

TEST PROCEDURE
For use, place a small drop of the mounting fluid over the fluorescent labeled smear on the microscope slide. Place a cover glass over the droplet, care being taken not to entrap air bubbles.

Detailed directions for the use of Bacto FA Mounting Fluid pH 9 in conjunction with other Bacto FA Reagents are in the package inserts accompanying the major FA Reagents for a given procedure.

PACKAGING
Bacto FA Mounting Fluid pH 9   6 × 5 ml   3340-57-5

BACTO FA PAPAIN

INTENDED USE
Bacto FA Papain is a standardized, desiccated, cysteine activated solution for use in the elimination of some commonly occurring cross-reactions in the FA technique.

HISTORY
Bacto FA Papain is prepared according to the method of Komninos and Tompkins.[1]

In their investigations it was found that by pretreating a smear with papain prior to staining with a specific conjugate nonspecific reactions to the staphylococci were removed. This removal of cross-reactions to the staphylococci by papain was successfully employed in the detection of *Neisseria meningitidis*, *Streptococcus* Group A and enteropathogenic *E. coli* employing both the direct and indirect FA technique.

PROCEDURE
1. Rehydrate Bacto FA Papain by adding 1 ml distilled water per vial. Rotate gently to effect solution. Bacto FA Papain will precipitate if stored overnight at 2 – 10°C and therefore should be rehydrated and completely used on the day prepared.
2. Prior to staining, the fixed slides are covered with several drops of the rehydrated Bacto FA Papain.
3. Place in a moist chamber at 37°C for 20 minutes.
4. Remove the slide and wash with FA buffer.
5. Complete the staining procedure (either direct or indirect) as recommended in the writeup for each specific conjugate.

STORAGE
Bacto FA Papain in the desiccated state is stable indefinitely at 2 – 8°C. It must be used the same day as rehydrated.

REFERENCE
1. Tech. Bull. Reg. Med. Tech., 33:129, 1963.

PACKAGING
Bacto FA Papain   6 × 1 ml   3232-51-2

BACTO FA RABBIT GLOBULIN

Bacto FA Rabbit Globulin is used to detect nonspecific staining in FA procedures. Rehydrate Bacto FA Rabbit Globulin by adding 5 ml distilled or deionized water and rotating to dissolve contents completely.

Store unrehydrated Bacto FA Rabbit Globulin at 2 – 8°C. Once rehydrated, Bacto FA Rabbit Globulin should be stored at 2 – 8°C or preferably aliquoted into usable quantities and stored at −20°C.

PACKAGING

Bacto FA Rabbit Globulin   5 ml   2379-56-2

BACTO FA RABBIT GLOBULIN ANTIGLOBULIN (GOAT)

INTENDED USE

Bacto FA Rabbit Globulin Antiglobulin (Goat), a desiccated anti-rabbit globulin prepared in goats and conjugated with a highly fluorescent dye is used in the indirect test for detecting specific antibodies in body fluids and for identifying microbial as well as other antigens. It is also applicable to the FA inhibition and complement staining procedures.

REHYDRATION

Rehydrate by adding 5 ml distilled or deionized water and rotating to dissolve contents completely.

STORAGE

Bacto FA Rabbit Globulin Antiglobulin (Goat)   2 – 8°C
Rehydrated material   2 – 8°C

PROCEDURE

A small drop of the conjugate is applied to the specimen on a microscope slide and allowed to react for 30 minutes in a moist atmosphere at room temperature. The excess reagent is removed from the slide by washing in Bacto FA Buffer solution in a Coplin jar for 10 minutes with 2 changes of buffer solution. The slide is then blotted dry between layers of absorbent paper and mounted by placing 1 drop of Bacto FA Mounting Fluid on the stained area and superimposing a cover glass. It is then ready for observing under the fluorescent microscope with U.V. light. Fluorescence indicates a reaction between the antigen and antibody globulin of the primary reagent.

PACKAGING

Bacto FA Rabbit Globulin Antiglobulin (Goat)   5 ml   2351-56-4

BACTO FA RABBIT SERUM (BLOCKING)

INTENDED USE

Bacto FA Rabbit Serum (Blocking) is a standardized desiccated normal rabbit serum used in blocking cross-reactions in FA conjugates.

TEST SUMMARY
Most available conjugated antiglobulins used in fluorescent antibody methods have been prepared in rabbits. It was noted that cross-reactions to some strains of staphylococci occurred with most conjugates regardless of the antigen used for immunization. It was observed that most normal fluorescent conjugated rabbit globulin would stain those strains nonspecifically. Redys, Ross and Borman[1] blocked group C cross-reactions occurring in *Streptococcus* Group A conjugates with unconjugated *Streptococcus* Group C immune serum. This reduced staphylococcal cross-reactions. Moody found most normal rabbit serum when added to the conjugate in varying amounts (depending upon the degree of cross-reaction) would block nonspecific cross-reactions. To rehydrate Bacto FA Rabbit Serum (Blocking), add 5 ml distilled or deionized water and rotate to dissolve.

STORAGE
Bacto FA Rabbit Serum (Blocking)   2 – 8°C
Rehydrated material   2 – 8°C

PROCEDURE
Bacto FA Rabbit Serum (Blocking) is selected normal rabbit serum which has the capacity to reduce cross-reactions to a minimum. Different conjugates may require different quantities of FA Rabbit Serum (Blocking). The procedure requires the preparation of various ratios of conjugate and FA Rabbit Serum (Blocking) from 1:10 to 1:1. The various mixtures are then employed in the FA technique with known homologous antigens and cultures of staphylococci known to cross-react with the conjugate. The optimal ratio for blockage is the dilution where a minimal amount of FA Rabbit Serum (Blocking) is employed, and nonspecific fluorescence is reduced to a minimum without altering the sensitivity of the conjugate.

REFERENCE
1. Bacto. Proc., 139, 1960.

PACKAGING
Bacto FA Rabbit Serum (Blocking)   5 ml   3233-56-6

BACTO FA RHODAMINE COUNTERSTAIN

INTENDED USE
Bacto FA Rhodamine Counterstain is a desiccated normal bovine serum, conjugated with Lissamine Rhodamine, used as a counterstain in the fluorescent antibody (FA) technique.

HISTORY AND TEST SUMMARY
Bacto Rhodamine Counterstain is prepared according to the procedure of Smith, Marshall and Eveland.[1]

In many FA procedures, tissues and/or cellular debris are present in the specimen to be examined. The tissues often adsorb nonspecifically the fluorescein of the conjugate, making it difficult and often impossible to differentiate between nonspecific and immunofluorescence. When such a slide is pretreated with Bacto FA Rhodamine Counterstain, or if a conjugate to be used in the presence of tissue is mixed with the counterstain, the tissues are stained a reddish-orange in contrast to a yellow-green specific staining.

STORAGE
Bacto FA Rhodamine Counterstain $2 - 8°C$

PROCEDURE
Bacto FA Rhodamine Counterstain is recommended in the dilution of 1 part counterstain to 20 parts of the conjugate.

To rehydrate the counterstain, add 5 ml distilled or deionized water and rotate to dissolve.

REFERENCE
1. Proc. Soc. Exp. Biol. & Med., 102:179, 1959.

PACKAGING
Bacto FA Rhodamine Counterstain   5 ml   2340-56-8

FA SALMONELLA
BACTO FA SALMONELLA PANVALENT
BACTO FA SALMONELLA POLY

DESCRIPTION
Bacto FA Salmonella Panvalent is a desiccated, fluorescein conjugated goat anti-*Salmonella* globulin prepared from selected strains of *Salmonella* and *Arizona* representing most of the known somatic and flagellar antigens in the genus *Salmonella*.[1,2]

Bacto FA Salmonella Poly is a desiccated, fluorescein conjugated goat anti-*Salmonella* globulin prepared from motile organisms representative of somatic groups $A - S^1$ in accordance with Thomason[3] and Insalata.[4]

Bacto FA Salmonella Panvalent and Bacto FA Salmonella Poly are recommended for use in the direct fluorescent antibody technique for detecting *Salmonella*.

BACKGROUND
These reagents are used in screening food, water and clinical specimens for the presence of *Salmonella*.[5,6,7,8,9,10,11,12] Since these polyvalent antiglobulins are not absorbed for cross-reactivity with antigens of other members of the *Enterobacteriaceae*, false positive results can be expected. All positive specimens, therefore, must be confirmed by cultural, biochemical and conventional serological methods.

PRINCIPLES OF THE PROCEDURE
The direct fluorescent antibody technique, for use in the detection of microorganisms, is the simplest of the immunofluorescence methods and may be defined as follows: The organism (antigen) is fixed to a microscopic slide by heat or chemical methods. The fixed antigen is overlayed with a specific antibody which is labeled with a fluorescent marker, fluorescein isothiocyanate (FITC). The resultant antigen-antibody reaction is then observed microscopically, utilizing a suitable wave length of light compatible to the fluorescent marker employed.

REAGENTS
Bacto FA Salmonella Panvalent contains somatic groups A – Z, 51 – 64 and all flagellar factors.

Bacto FA Salmonella Poly contains somatic groups A – S including 0 factors 1 – 25, 27, 28, 30, 34 – 41, 46 and Vi and H antigens a – i, k – p, r – z, z_4, z_6, z_{10}, z_{13}, z_{15}, z_{23}, z_{24}, z_{27}, z_{28}, z_{29}, z_{32}, z_{35}, z_{38}, z_{42}, and 1, 2, 5, 6 and 7.

To rehydrate the conjugate, add 5 ml sterile distilled or deionized water to the vial and rotate gently to dissolve contents completely. No further dilution is required. The resultant solution contains 1:5,000 Merthiolate®.

STORAGE AND EXPIRATION
These reagents are stable to the expiry date on the label when stored at 2 – 8°C.

Aliquots of a rehydrated conjugate should be put into small vials and stored below –20°C for optimal stability.

Bacto FA Kirkpatrick Fixative Modified is an alcohol based fixative prepared by a modification of Kirkpatrick's method described by Goepfert and Hicks[13] which contains 6 parts isopropyl alcohol, 3 parts chloroform and 1 part formaldehyde 37% solution. It is stable at room temperature to the expiry date on the label if tightly closed. Do not refrigerate.

It is recommended for fixation of bacterial cells to the microscope slide in the fluorescent antibody technique particularly in the FA staining of *Salmonella*.

Bacto FA Mounting Fluid pH 9 is a standardized reagent grade glycerin adjusted to approximately pH 9. It is stable to the expiry date on the label when stored at 15 – 30°C.

SPECIMEN PREPARATION
All specimens to be screened by fluorescent antibody method must be pre-enriched in a suitable broth. For food samples, the media employed and methods of preparing specimens will depend on the type of product under test. Refer to AOAC[12] for appropriate preparation of the specimen and suitable media for pre-enrichment broths. In most instances, pre-enrichment incubation is for 24 hours ±2 hours at 35°C.

Where selenite cystine broth or tetrathionate broth is not used for pre-enrichment, transfer 1 ml of the pre-enriched samples to 10 ml of selenite broth and 1 ml to 10 ml of tetrathionate broth. Incubate for 24 hours at 35°C. Subsequently, transfer 1 ml of this sample to 10 ml selenite cystine broth and incubate for 4 hours.

When screening clinical specimens for *Salmonella*, pre-enrichment in selenite cystine broth or tetrathionate broth, as outlined above, is essential. Transfer 1 ml of pre-enriched sample into 10 ml selenite cystine broth and incubate for 4 hours at 35 – 37°C. The specimen may be diluted or concentrated by centrifugation to approximate a McFarland Barium Sulfate Standard No. 1.

PRECAUTIONS
1. It is assumed that the user of these reagents is familiar with fluorescent antibody techniques. Those desiring information concerning the advantages and disadvantages of the FA technique, the various staining techniques, and in particular, the general precautions and considerations concerning FA procedures, should refer to "FA General Procedures" in the introductory Serology discussion.

2. The test organism must be identified to at least the generic, and in some cases the specific, level biochemically prior to serological confirmation.
3. The density of the antigen suspension is important. Too dilute a suspension may result in overlooking a positive reaction. Too dense a suspension may result in weak staining.
4. Discard the conjugate if bacterially contaminated.
5. The conjugate should not be subjected to repeated freezing and thawing. Such treatment is detrimental to the antibody content.
6. All glassware that is employed in the preparation, testing and storage of these reagents must be free of detergents or other harmful residues.

PROCEDURE

Materials Provided:
Bacto FA Salmonella Panvalent
Bacto FA Salmonella Poly

Materials Available Separately from Difco:
Bacto FA Mounting Fluid pH 9
Bacto FA Buffer, Dried
Bacto FA Kirkpatrick Fixative Modified
Bacto McFarland Barium Sulfate Standards
Appropriate Culture Media

Materials Required but not Provided:
Distilled or deionized water
Serological pipettes
Culture tubes and plates
Centrifuge and tubes
Timer
Refrigerator 2 – 8°C
Freezer −20°C or lower
Incubator 35°C
Autoclave

Bacteriological loop
Microscope slides
Applicator sticks
Known *Salmonella* cultures
Bibulous paper
Moisture chamber
Coplin jars (or equivalent)
Cover glasses
Fluorescent microscope assembly

FA Technique

1. Using a 2 mm bacteriological loop transfer a loopful of properly prepared sample onto a microscope slide.
2. Smear the specimen on the slide and allow to air dry.
3. Fix the specimen on the slide by immersion for 1 minute in FA Kirkpatrick Fixative Modified.
4. Allow the slide to drain dry.
5. Apply several drops of conjugate over the fixed smear.
6. Place in a moist atmosphere at room temperature for 30 minutes.
7. Drain off excess conjugate and rinse in FA Buffer.
8. Place in a Coplin jar containing FA Buffer for 10 minutes changing the buffer twice during this period.
9. Remove slide from the buffer and rinse one time in distilled or deionized water to remove residual salt.
10. Drain off excess distilled or deionized water and add a small drop of FA Mounting Fluid pH 9.
11. Mount with a cover glass.
12. Examine same day with suitable fluorescent microscope for both somatic and flagellar staining. Avoid contact with direct sunlight.

QUALITY CONTROL

A known positive sample should be included in each run. The sample should be subjected to the treatment described under "Specimen Preparation" before proceeding to the FA procedure.

A known positive fluorescing slide should be available at all times to attest to the proper functioning of the fluorescent microscope. Difco prepares Bacto FA Slide Stained Streptococcus Group A for this purpose.

RESULTS
Read and record results based on intensity of fluorescence (− to 4+) as outlined below. Cultures exhibiting a fluorescence of 3+ or greater should be considered positive.

4+   Maximum fluorescence; brilliant yellow-green; clear cut cell outline; sharply defined cell center

3+   Less brilliant yellow-green fluorescence; clear cut cell outline; sharply defined cell center

2+   Definite but dim fluorescence; cell outline less well defined

1+   Very subdued fluorescence; cell outline indistinguishable from cell center in most instances

−   Negligible or complete lack of fluorescence (negative)

INTERPRETATION
Specimens exhibiting a positive reaction of 3+ or greater with Bacto FA Salmonella Panvalent or Bacto FA Salmonella Poly should be confirmed by cultural, biochemical and serological methods. Negative FA reactive specimens may be eliminated from further examination.

LIMITATIONS OF THE PROCEDURE
Bacto FA Salmonella Panvalent and Bacto FA Salmonella Poly are used for screening suspect specimens. Immunospecific cross-reactions with other members of the *Enterobacteriaceae* will occur because of similar antigenicity and biochemical characteristics. A brief outline on the specificity of these FA conjugates can be found in "Standard Methods for the Microbiological Examination of Foods."[5] These reagents should not be used in confirmatory tests.

While serological procedures, as applied to microorganisms, provide corroborative evidence and can be used to identify particular antigenic sites of the genus or species under study, they cannot be used alone to identify the etiological agent of a disease. Cultural isolation and at least preliminary biochemical differentiation must preceed any serological examination and final identification cannot be made without biochemical and serological characterization.

REFERENCES
1. Edwards, P. R., and Ewing, W. H., Identification of *Enterobacteriaceae*, Burgess, Minneapolis, 3rd Ed., 1972.
2. Kauffmann, F., The Bacteriology of *Enterobacteriaceae*, Williams and Wilkins, Baltimore, 1966.
3. Personal Communication, 1969.
4. *ibid.*
5. Compendium of Methods for the Microbiological Examination of Foods, APHA, Washington, D.C., 1976.
6. Laramore, C. R., and Mority, C. W., Applied Microbiology, 17:352 – 354, 1969.
7. Hilker, J. S., and Solberg, M., Applied Microbiology, 26:751 – 756, 1973.
8. Hoben, D. A., Ashton, D. H., and Peterson, A. C., Applied Microbiology, 25:123 – 129, 1973.
9. Mohr, H. K., Trenk, H. L., Yeterian, M., Applied Microbiology, 27:324 – 328, 1974.
10. Thomason, B. M., and Wells, J. G., Applied Microbiology, 22:876 – 884, 1971.
11. Insalata, N. F., Schulte, S. J., and Berman, J. H., Applied Microbiology, 15:1145 – 1149, 1967.
12. Official Methods of Analysis of the Association of Official Analytical Chemists, AOAC, Washington, D.C., 12th Ed., 1975.
13. Goepfert, J. M., and Hicks, Applied Microbiology, 18:612 – 617, 1969.

PACKAGING

Bacto FA Salmonella Panvalent	5 ml	3185-56-4
Bacto FA Salmonella Poly	5 ml	3187-56-2
Bacto FA Buffer, Dried	6 × 10 g	2314-33-8
	100 g	2314-15-0
Bacto FA Kirkpatrick Fixative Modified	6 × 100 ml	3188-73-0
Bacto FA Mounting Fluid pH 9	6 × 5 ml	3340-57-5
Bacto McFarland Barium Sulfate Standards	1 set	0691-32-6

BACTO FA STAPHYLOCOCCUS AUREUS

INTENDED USE AND DESCRIPTION
Bacto FA Staphylococcus Aureus is a desiccated fluorescent labeled anti-*S. aureus* globulin for the detection of coagulase-positive staphylococci employing the fluorescent antibody technique.

REAGENT
Rehydrate Bacto FA Staphylococcus Aureus with 5 ml distilled or deionized water per vial.

Before use, a dilution equivalent to 1:200 is prepared by adding 0.1 ml of the conjugate to 19.9 ml of Bacto FA Buffer. Prepare only a sufficient amount of diluted conjugate for each day's use.

STORAGE AND EXPIRATION
Bacto FA Staphylococcus Aureus stored at 2 – 8°C is stable through the expiration date on the label. Once rehydrated, store the undiluted conjugate at −20°C in aliquots of useable amounts.

SPECIMEN COLLECTION
There are a number of media used for the isolation, selection and differentiation of staphylococci. Table I below lists media which can be used for isolation of *S. aureus* and its typical colonial appearance.

Table I

Bacto Blood Agar Base w/5% blood	Opaque, yellow to orange, with hemolysis
Bacto Coagulase Mannitol Agar	Opaque, with yellow to orange zones
Bacto DNase Test Agar w/Methyl Green	Clearing of green dye
Bacto Mannitol Salt Agar	Yellow to orange, surrounded by yellow zones
Bacto Staphylococcus Medium 110	Yellow to orange
Bacto Tellurite Glycine Agar	Black
Bacto VJ Agar	Black, surrounded by yellow zones
Bacto Baird Parker Agar	Grey to black shiny colonies surrounded by zone of clearing

Most pathogenic strains of staphylococci lyse erythrocytes when grown on blood agar. Lysis is variable depending upon the species of blood used and the hemolysin produced. This may be summarized as follows:

Species of Blood	Reaction Hemolysin		
	Alpha	Beta	Gamma
Human	–	–	+
Rabbit	+	–	+
Sheep	+	+	+

+ = hemolysis   – = no hemolysis

Using a bacteriological loop, transfer a well isolated colony from a pure culture of gram-positive cocci occurring in "grape-like clusters" which is both catalase and mannitol positive, into a tube of sterile Bacto Brain Heart Infusion. Incubate at 37°C for 18 – 24 hours or until a dense growth is observed. This is the specimen to be tested.

SPECIMEN PREPARATION
1. Smear a saline suspension of the growth from the specimen culture onto a microscope slide. The density of this suspension should approximate a Bacto McFarland Barium Sulfate Standard No. 3.
2. Allow to air dry.
3. Fix for 1 minute in 95% alcohol.

PROCEDURE
Materials Provided:
Bacto FA Staphylococcus Aureus

Materials Available from Difco:
Bacto FA Buffer, Dried
Bacto FA Mounting Fluid pH 7.2
Bacto McFarland Barium Sulfate Standards

Materials Required but not Provided:
Coplin jar
Distilled or deionized water
Serological pipettes
Culture tubes and plates
Timer
Refrigerator 2 – 8°C
Freezer –20°C or lower
Incubator 37°C

Autoclave
Bacteriological loop
Microscope slide
Applicator sticks or toothpicks
Moisture chamber or Petri dish
Cover glasses
Fluorescent microscope assembly

FA Technique
1. Cover the smear with a 1:200 dilution of the conjugate prepared in rehydrated Bacto FA Buffer (0.1 ml of the conjugate per 19.9 ml buffer).
2. Place the slide in a Petri dish containing a moistened absorbent paper to prevent drying and allow to react at room temperature for 30 minutes.
3. Remove the excess conjugate from the slide.
4. Rinse with Bacto FA Buffer and place it in a Coplin jar containing the buffer for 10 minutes changing the buffer twice during this period, followed by one rinse of distilled or deionized water to remove residual salt.
5. Remove the slide and gently blot between layers of absorbent paper.
6. Place a small drop of Bacto FA Mounting Fluid pH 7.2 on the stained area and cover with a coverglass.
7. Examine the slide under a fluorescent microscope employing a UV light source. Determine the morphology of the cells and the intensity of the staining.

QUALITY CONTROL
It is recommended that a known coagulase-positive and a coagulase-negative *Staphylococcus* be used for positive and negative controls respectively.

RESULTS
Read and record results based on intensity of fluorescence (− to 4+) as outlined below. Cultures exhibiting a fluorescence of 3+ or greater should be considered positive.

4+ Maximum fluorescence; brilliant yellow-green; clear-cut cell outline; sharply defined cell center

3+ Less brilliant yellow-green fluorescence; clear-cut cell outline; sharply defined cell center

2+ Definite but dim fluorescence; cell outline less well defined

1+ Very subdued fluorescence; cell outline indistinguishable from cell center in most instances

−   Negligible or complete lack of fluorescence (negative)

LIMITATIONS
This procedure serves only to provide presumptive identification of these organisms, and is intended to complement but not replace conventional bacteriological techniques. Further cultural and biochemical studies are necessary to confirm results obtained in the use of this immunofluorescent technique and serogrouping of these organisms.

PACKAGING
Bacto FA Staphylococcus Aureus   5 ml   3271-56-9

BACTO FA STREPTOCOCCUS GROUPS A, B, C, D, F AND G

BACTO FA SLIDE STAINED STREPTOCOCCUS GROUP A
BACTO FA SLIDE UNSTAINED STREPTOCOCCUS GROUP A

INTENDED USE AND DESCRIPTION
Bacto FA Streptococcus Groups A – D, F and G are desiccated, rabbit antistreptococcus globulins conjugated with fluorescein isothiocyanate (FITC). They are recommended for use in the direct fluorescent antibody technique for the identification of their respective serologic groups of streptococci.

Bacto FA Slide Stained and Bacto FA Slide Unstained are microscope slides affixed with known microorganisms which may be used for controlling staining and microscopic procedures.

On Bacto FA Slide Stained, *Streptococcus* Group A organisms have been fixed and stained with homologous fluorescent conjugated antiglobulins. The slides are ready to use to check microscope alignment and efficiency of the light source.

On Bacto FA Slide Unstained, *Streptococcus* Group A organisms have been fixed and are ready to use as the positive control for the homologous fluorescent antibody conjugate or as a negative control for heterologous *Streptococcus* fluorescent antibody conjugates. These slides also serve as controls for other reagents used, as well as for the technique of the procedure.

BACKGROUND
Lancefield[1,2,3] divided the *Streptococcus* into serological groups according to the group specific somatic carbohydrate they possessed. The Lancefield groups have quite dif-

ferent clinical significance and in some cases different biochemical and hemolytic differences within the same serological group. Therefore, all *Streptococcus* exhibiting beta-hemolysis on blood agar and also those which fail to exhibit beta-hemolysis but which seem to be involved in an infection should be grouped.[4] A final identification of *Streptococcus* for diagnostic purposes is considered incomplete unless specific grouping has been accomplished.[5]

The demonstration by Moody, Ellis and Updyke,[6] that group specific conjugates could be prepared from the antiserum used in the Lancefield precipitin test, led to the development of the direct fluorescent antibody technique for the identification of *Streptococcus* groups.

The group A conjugate as well as the others are prepared essentially according to the method of Moody, Ellis and Updyke;[6] Moody, Siegel, Pittman and Winter;[7] and Redys, Ross and Borman.[8]

Warfield, Page, Zuelzer and Stulberg;[9] Peeples, Spielman and Moody;[10] and Rauch and Ranty[11] demonstrated the practicability of the fluorescent procedure. Its sensitivity for the detection and identification of group A *Streptococcus* from throat swabs was declared by Moody, Siegel, Pittman and Winter[12] and Estela and Shuey[13] to be equal to, or greater than, that of the cultural and precipitin tests.

PRINCIPLES OF THE PROCEDURE

The direct fluorescent antibody technique, for use in the detection of microorganisms, is the simplest of the immunofluorescence methods and may be defined as follows:

The organism (antigen) is fixed to a microscope slide by heat or chemical methods. The fixed antigen is overlayed with a specific antibody which is labeled with a fluorescent marker (FITC). The resultant antigen-antibody reaction is then observed microscopically, utilizing a suitable wave length of light compatable to the fluorescent marker employed.

REAGENTS

Bacto FA Streptococcus Groups A – D, F and G have been cross-absorbed to remove cross reactivity especially those known to exist between serogroups A, C and G. The normally occurring *Staphylococcus* agglutinins have been blocked by the use of unconjugated normal rabbit serum or by the use of an unconjugated *Staphylococcus* immune serum. Upon rehydration with 5 ml distilled or deionized water, the conjugates contain 1:5,000 Merthiolate® as a preservative.

The working dilution of the conjugate should be determined upon rehydration. The titer of a conjugate varies with the technique, the fluorescent microscope, the filter and the age of the bulb and therefore may vary from laboratory to laboratory. Dilutions of the conjugate are made in rehydrated Bacto FA Buffer. The titer is determined as follows:

Dilution of Conjugate	Fluorescence
1:5	4+*
1:10	4+
1:20	4+
1:40	4+
1:80	2+

*A 4+ fluorescence is defined as brilliant yellow-green cocci with sharp cell outlines and nonstaining centers.

In this example, the last 4+ fluorescence is found in a 1:40 dilution of the conjugate. One less dilution is chosen for a margin of safety. The working dilution, therefore, in this case is 1:20.

Bacto FA Rabbit Globulin is a desiccated, fluorescein conjugated, normal rabbit globulin used as control for possible nonspecific staining of some organisms, particularly *Staphylococcus*, in the direct fluorescent antibody technique.

To rehydrate add 5 ml distilled or deionized water to a vial and rotate gently to dissolve the contents completely. The resultant solution contains 1:5,000 Merthiolate®. Before use, a dilution equivalent to that used for the particular *Streptococcus* conjugate being used is made in Bacto FA Buffer. Prepare only a sufficient amount of diluted globulin for use in each day's run.

Bacto FA Slide Unstained Streptococcus Group A are slides containing organisms fixed and ready for staining. They are used as a positive control for the conjugate.

Bacto FA Slide Stained Streptococcus Group A contains Group A *Streptococcus* which has been stained to a 4+ reaction with Bacto FA Streptococcus Group A and is ready for examination. They are employed for use as a positive control for the fluorescent microscopic system.

STORAGE AND EXPIRATION

Bacto FA Streptococcus Groups A – D, F and G, in the desiccated state should be stored at 2 – 8°C. Aliquots of the titered conjugate should be distributed into small vials and frozen in the undiluted state and stored below −20°C for optimal stability. Prepare only a sufficient amount of diluted conjugate for each day's run.

Store the desiccated or undiluted, rehydrated Bacto FA Rabbit Globulin at 2 – 8°C. These reagents are stable to the expiry date on the label when stored at 2 – 8°C. Any conjugate in which a precipitate forms or which does not dissolve completely upon rehydration should be discarded.

Bacto FA Slides Unstained and Stained Streptococcus Group A should be stored at 2 – 8°C.

Bacto FA Slides are stable to the expiry date on the label if stored under the recommended conditions.

PRECAUTIONS

1. It is assumed that the user of these reagents is familiar with fluorescent antibody techniques. Those desiring information concerning the advantages and disadvantages of the FA technique, the various staining techniques, and in particular the general precautions and considerations concerning FA procedures should refer to "FA General Procedures" in the introductory Serology discussion.
2. The test organism must be identified to at least the generic, and in some cases the species level, biochemically prior to serological confirmation.
3. Discard the conjugate if bacterially contaminated.
4. The conjugate should not be subjected to repeated freezing and thawing. Such treatment is detrimental to the antibody content.
5. All glassware that is employed in the preparation, testing and storage of these reagents must be free of detergents or other harmful residues.

SPECIMEN COLLECTION
Specimens submitted to the laboratory for the possibility of containing streptococci must be obtained under proper medical guidance from nasal, nasopharyngeal or throat areas, from skin, wounds, pus, blood, cerebrospinal fluid and urine. It is imperative especially when material is obtained from the throat to culture, that it be done properly to obtain an adequate specimen. Improperly obtained specimens will yield cultures that contain minimal numbers of streptococci.

The isolation medium recommended for this species of organisms is any one of several blood agar bases containing 5% sterile, defibrinated sheep blood. Hemolytic reactions should be determined from a pure culture prior to serological examination. Sheep blood plates are recommended as they exhibit clear cut reactions for streptococci. It has been found that hemolytic reactions of *Streptococcus* Groups A, B, C and G strains were identical when sheep, rabbit, horse and human blood were used. The only difference was in the Group D streptococci, which demonstrated alpha-hemolysis in sheep blood and beta-hemolysis in rabbit horse and human blood.[14] For antigen preparation Bacto Todd Hewitt Broth is recommended.

Note: Specimens containing streptococci survive well at 4°C on tightly capped blood agar slants but survive poorly in a broth medium.

SPECIMEN PREPARATION
1. Place a swab from throat or other specimen source into a tube containing 1 ml Bacto Todd Hewitt Broth and incubate for 2 – 5 hours at 37°C.
2. Drain the swab against the side of the tube and place it into another sterile tube and store in the refrigerator for use in precipitin test when desired.
3. Centrifuge tubes containing known and unknown cultures for 3 – 5 minutes at 1500 – 2000 rpm, to sediment cells.
4. Decant and discard the supernatant broth. Resuspend the cells in 1 ml Bacto FA Buffer.
5. Repeat centrifuging for 3 – 5 minutes, decant and discard the supernatant.
6. Resuspend the cells in Bacto FA Buffer to approximate a Bacto McFarland Barium Sulfate Standard No. 3.
7. Prepare duplicate smears of the same culture on prepared microscope slides.
8. Allow the smears to air dry 15 – 20 minutes.
9. Fix the smears by placing the slides in 95% ethanol for 1 minute.

Note: Smears may also be prepared from specimens grown on blood agar and suspected of being *Streptococcus* as a result of their growth characteristics.

PROCEDURE
Materials Provided:
Bacto FA Streptococcus Groups A, B, C, D, F & G

Materials Available from Difco:
Bacto FA Rabbit Globulin
Bacto FA Buffer, Dried
Bacto FA Mounting Fluid pH 9
Bacto FA Slides Unstained Streptococcus Group A
Bacto FA Slides Stained Streptococcus Group A
Bacto McFarland Barium Sulfate Standard
Appropriate culture media

Materials Required but not Provided:

Distilled or deionized water
Serological pipettes
Culture tubes and plates
Centrifuge and tubes
Timer
Refrigerator 2 – 8°C
Freezer −20°C or lower
Incubator 37°C
Bacteriological loop

Microscope slides
Applicator sticks (or tooth picks)
95% ethanol
Known *Streptococcus* cultures
Bibulous paper
Moisture chamber
Coplin jars (or equivalent)
Cover glasses
Fluorescent microscope assembly

FA TECHNIQUE

1. Add several drops of an appropriate dilution of Bacto FA Streptococcus conjugate to each smear on one end of each microscope slide. Distribute it evenly over the entire smear with an applicator stick in such a manner as not to disturb the smear.
2. Add several drops of the same dilution of Bacto FA Rabbit Globulin to each smear on the other end of each microscope slide, care being taken not to allow the 2 conjugates to run together.
3. Prevent evaporation of the conjugates by covering the slide with a Petri dish cover to which moistened bibulous paper has been affixed. Incubate in this manner at room temperature for 30 minutes.
4. Tip the slide to allow the excess conjugate to run off.
5. Rinse the slides in Bacto FA Buffer for a total of 3 changes of buffer in 10 minutes. This may be followed by a rinse in distilled or deionized water to remove salt residue.
6. Allow the slide to air dry.
7. Place a small drop of Bacto FA Mounting Fluid pH 9 in the center of each smear and cover with a cover glass.
8. Examine each smear using a suitable fluorescent microscope with an oil immersion lens for a total magnification of 900 – 1000X. Observe for cellular morphology and intensity of stain. Slides of the clinical specimen may be held overnight at 2 – 8°C in the dark and examined if they cannot be observed the day of preparation.

QUALITY CONTROL

The following controls must be performed in each test series:

Positive Control — Use a known culture homologous to the conjugate being used. When a control culture is not available, in *Streptococcus* Group A determination, a Bacto FA Slide Unstained Streptococcus Group A should be used.

Negative Control — Use appropriate dilution of Bacto FA Rabbit Globulin instead of the indicated Bacto FA Streptococcus Group conjugate.

Microscope Control — It is recommended that an additional control be incorporated. A known positive fluorescing slide or Bacto FA Slide Stained Streptococcus Group A as a check on microscope alignment, etc.

RESULTS

Read and record results based on intensity of fluorescence − to 4+ as outlined below. Cultures exhibiting a fluorescence of 2+ or greater should be considered positive.

4+ Maximum fluorescence; brilliant yellow-green; clear-cut cell outline; sharply defined cell center

3+ Less brilliant yellow-green fluorescence; clear-cut cell outline; sharply defined cell center

2+ Definite but dim fluorescence; cell outline less well defined

1+ Very subdued fluorescence; cell outline indistinguishable from cell center in most instances

− Negligible or complete lack of fluorescence (negative)

INTERPRETATION OF RESULTS

A 4+ fluorescence in the unknown smear with the homologous conjugate, and no fluorescence in the unknown smear by the FA Rabbit Globulin, is evidence that the unknown organism is homologous to the particular *Streptococcus* conjugate used.

A typical result for the identification of Group A *Streptococcus* is illustrated.

FA Streptococcus Group A		FA Rabbit Globulin		
Unknown	Known	Unknown	Known	Results
4+	4+	−	−	Group A *Streptococcus*
−	4+	−	−	No Group A *Streptococcus*
4+	4+	4+	−	Inconclusive*

*The unknown cultures showing fluorescence in both the Group A conjugate and Bacto FA Rabbit Globulin could be other groups of either *Streptococcus* or *Staphylococcus*.

Table I displays some of the *Streptococcus* groups associated with infections in man and animals together with their hemolytic and serological characteristics and for which Bacto FA Streptococcus conjugates are available.

Table I

Serogroup	Species of Streptococcus	Type of Hemolysis	Host Association
A	S. pyogenes	β	Human
B	S. agalactiae	α,β,γ	Human and Bovine
C	S. dysgalactiae	α	Animals
C	S. equi	β	Equine
C	S. equisimilis	β	Human and Animal
C	S. zooepidemicus	β	Animals
D	S. bovis	α,γ	Human and Animal
D	S. equinus	α(weak)	Equine
D	S. faecalis	γ	Human and Animal
D	S. faecium	α	Human and Animal
F*	S. anginosus	α,γ	Human
G	S. species	β	Human

* = Extracts from Streptococcus MG reacts with Lancefield's Group F. Antiserum.
KEY: α = alpha-hemolysis
β = beta-hemolysis
γ = gamma-hemolysis

LIMITATIONS OF THE PROCEDURE

While serological procedures, as applied to microorganisms, provide corroborative evidence and can be used to identify particular antigenic sites of the genus or species under study, they cannot be used alone to identify the etiological agent of disease. Cultural isolation and at least preliminary biochemical differentiation must preceed any serological examination and final identification cannot be made without biochemical and serological characterization.

Particularly important in fluorescent microscopy is the knowledge that the microscopic system is in alignment and completely functional in the use of adequate and proper control systems, and in the optimal storage of the conjugates.

REFERENCES

1. J. Exp. Med., 57:571, 1933.
2. J. Exp. Med., 47:91, 1928.
3. Proc. Soc. Exp. Biol. and Med., 38:473, 1938.
4. Clin. Bact., 3rd Ed., E. Arold LTD, London, 1970.
5. Manual of Clin. Micro., ASM, 1970.
6. J. Bact., 75:553, 1958.
7. Am. J. Publ. Hlth., 53:1083, 1963.
8. Bact. Proc., 130, 1960.
9. J. Dis. Child., 101:160, 1961.
10. Publ. Hlth. Repts., 76:651, 1961.
11. J. Lab. Clin. Med., 61:529, 1963.
12. Am. J. Clin. Path., 40:591, 1963.
13. Proc. Soc. Exp. Biol. and Med., 96:477, 1957.
14. Manual of Clin. Micro., ASM, 3rd Ed., 1980.

PACKAGING

Bacto FA Streptococcus Group A	5 ml	2318-56-6
Bacto FA Streptococcus Group B	5 ml	2319-56-5
Bacto FA Streptococcus Group C	5 ml	2320-56-2
Bacto FA Streptococcus Group D	5 ml	2321-56-1
Bacto FA Streptococcus Group F	5 ml	2323-56-9
Bacto FA Streptococcus Group G	5 ml	2324-56-8
Bacto FA Rabbit Globulin	5 ml	2379-56-2
Bacto FA Buffer, Dried	6 × 10 g	2314-33-8
	100 g	2314-15-0
Bacto FA Mounting Fluid pH 9	6 × 5 ml	3340-57-5
Bacto FA Slide Set Streptococcus Group A	6 slides	3151-32-3
Bacto FA Slide Stained Streptococcus Group A	2 slides	3193-32-3*
Bacto Todd Hewitt Broth	1/4 lb (114 g)	0492-02-3
	1 lb (454 g)	0492-01-4
Bacto FA Slide Unstained Streptococcus Group A	6 slides	3172-33-7
Bacto McFarland Barium Sulfate Standards	1 set	0691-32-6

BACTO FORMOCELLS SHEEP

INTENDED USE

Bacto Formocells Sheep are stable formalinized erythrocytes from sheep for use in indirect hemagglutination (HA) and hemagglutination inhibition (HI) procedures.

HISTORY/PRINCIPLES

Hemagglutination and hemagglutination inhibition procedures afford a very sensitive serological means of detecting antibodies and antigens. The reliability of erythrocytes in indirect hemagglutination and hemagglutination inhibition procedures has been amply demonstrated by Salk,[1] Boyden,[2] Stavitsky[3] and others. Fresh red cell suspensions have been used widely for this purpose; however, it is not always possible when using fresh erythrocytes to obtain reproducible results due to variations in the red cell suspensions. Contamination, instability of sensitized and nonsensitized cells, autoagglutination and loss of agglutination activity of sensitized cells as described by Cole and Farrel[4] have contributed to these variations and limited the use of the HA and HI techniques.

Many of the problems encountered with fresh erythrocytes in HA and HI procedures are eliminated by using standardized stable formocell suspensions. Gross clumping as observed with fresh erythrocytes when exposed to agglutinins is not evident with Formocells. The latter yields small granular, evenly distributed clumps. They retain their native morphology and dispense freely without clumping to give a uniform suspension. They have been demonstrated to be satisfactory for more than a year for sensitization with fresh antigen or to be sensitized and stored for use as required. Formocells have been successfully sensitized with a variety of antigens among which are the following: animal serous fluids, albumins and globulins, bacterial polysaccharides and lipopolysaccharides, Vi antigen, viruses, diptheria toxoid, tuberculin, hemocyanin, Vibrio fetus extract, echinococcus, ragweed extract and others.

The use of formalinized erythrocytes was first described by Rubino[5] who demonstrated their advantages over fresh erythrocytes in hemagglutination procedures. Subsequently, Flick[6] described a procedure for formalin treating human and chicken erythrocytes for use in measuring the hemagglutinating activity of influenzae A virus. He reported that for this purpose the formalinized cells yielded results comparable to those of fresh erythrocytes. Rodriques-da-Silva and Feldman[7] also demonstrated formalinized chicken erythrocytes to be suitable for measuring influenzae A virus and its antibody. They reported that the addition of 1% or less of inactivated rabbit serum eliminated all nonspecific clumping.

The successful use of formalinized erythrocytes to titrate enterobacterial antibody was demonstrated by Feeley, Sword, Manclark, and Picket.[8] Sieburth[9] and Young, Gillen, Massey and Baker[10] demonstrated that the cells could be sensitized as easily with multiple polysaccharide antigens as with a single antigen. Csizmas[11] and Gaines[12] sensitized formalinized cells with Vi antigen and obtained similar titers a year after preparation.

Rose[13] sensitized Formocells with normal horse serum and with thyroglobulins. He could detect antibodies to horse serum in some patients about to receive tetanus antitoxin and used this as a means of predetermining the occurrence of serum sickness.

Formalinized cells were used by Ingraham[14] for coupling with bovine serum globulin, bovine serum albumin, ovalbumin, echinococcus cyst fluid and thyrotropic hormones. He used 1% rabbit serum in saline in which to suspend the sensitized cells to eliminate some nonspecific clumping. Stavitsky[15] also sensitized Formocells with a variety of albumins and globulins as well as hemocyanin, diphtheria toxoid and other antigens. The results indicated good correlation with sensitized fresh red cells.

Strausser[16] used Westergren sedimentation tubes for performing the HA and HI tests and reported the height of the column of sedimented cells at the end of 1 hour. The amount of settling correlated closely with the degree of agglutination. Ingraham[17] modified Liener's[18] photometric hemagglutinating procedure for use with formalinized cells and obtained comparable results.

The procedure for coupling proteins, polysaccharides and viruses to Formocells is identical to those used with fresh erythrocytes. Coupling with polysaccharides is described by Young et al.,[10] with proteins by Boyden[2] and Stavitsky,[3] and with viral antigens by Salk[1] and in the Communicable Disease Center Influenzae Report No. 8.[19] Procedures for performing the HA and HI tests are also described in these publications.

STORAGE

Bacto Formocells Sheep   2 – 8°C
Bacto Formocells Sheep, Sensitized   2 – 8°C

PREPARATION OF BACTO FORMOCELLS SHEEP FOR USE

Bacto Formocells Sheep, upon standing for prolonged periods, settle to form a compact sediment with a clear, slightly tinged supernate. The supernate is poured off and a similar volume of 0.85% NaCl added. The cells are resuspended by vigorous shaking and washed once in normal saline after which they are ready for sensitization and coupling procedures.

SENSITIZATION OF BACTO FORMOCELLS SHEEP

1. Wash the Bacto Formocells Sheep 4 times with 4 volumes of 0.85% NaCl (normal saline).
2. Determine the packed cell volume and dilute to 2.5% cells with normal saline.
3. For sensitization with proteins, the cells are first treated with an equal volume of 1:20,000 tannic acid at 37°C for 10 minutes. They are then centrifuged and washed once in a buffered saline of a suitable pH depending on the protein being studied.

Optimal sensitization conditions for some antigens:

Antigen	Concentration μg/ml	pH of Buffered Saline
Bovine gamma globulin	100	6.4
Bovine serum albumin	100	5.6
Human serum albumin	50	5.6
Egg albumin	1000	5.6
Beta lactoglobulin	1000	7.2

4. If cells are to be sensitized with polysaccharide use Steps 1 and 2.
5. Prepare a 5% suspension of the washed cells in the buffered saline. Add an equal volume of antigen in normal saline and allow to stand at room temperature for 30 minutes with occasional swirling.
6. Centrifuge and discard the supernate.
7. Wash cells once in 2 volumes 1:200 normal rabbit serum; discard supernate.
8. Suspend cells in an equal volume 1:200 normal rabbit serum.
9. Add Merthiolate® to 1:10,000 if desired.
10. Store at 2 – 8°C. These sensitized cells may also be stored at freezing temperatures without change in their sensitivity.

TYPICAL INDIRECT HEMAGGLUTINATION (HA) PROCEDURE
Stavitsky

1. Prepare 1:200 dilution normal rabbit serum by adding 0.1 ml serum to 19.9 ml 0.85% NaCl solution (saline). Bovine serum albumin 0.1% in saline may be used in place of the diluted rabbit serum.
2. Set up a series of clean 10 × 100 mm tubes; the number of tubes used depends upon the anticipated titer. Add 0.9 ml diluent Step 1 to the first tube and 0.5 ml to the remaining tubes.
3. Prepare a 1:10 dilution of the serum to be tested by adding 0.1 ml to 0.9 ml 1:200 rabbit serum in the first tube.

4. Mix well and transfer 0.5 ml to the second tube.
5. Mix the contents of the second tube and transfer 0.5 ml to the third tube.
6. Continue the dilution procedure as far as desired and discard 0.5 ml from the last tube.
7. The final tube is to contain 0.5 ml of diluent only.
8. Add 0.05 ml of the dispersed sensitized Bacto Formocells Sheep to each tube and mix well by shaking the rack.
9. Allow to settle at room temperature for 1 1/2 – 2 hours and read sediment by pattern:

    4+ Compact granular appearing mat
    3+ Smooth mat with folded periphery
    2+ Smooth mat with ragged edges
    +   Narrow ring around edge of smooth mat
    ±   Thick dark ring around edge of mat
    −   Uniform dark button

Stavitsky considers the last 2+ tubes as the end point.

TYPICAL HEMAGGLUTINATION-INHIBITION (HI) PROCEDURE

1. Prepare sensitized cells and serum dilutions as for the HA procedure except that 2 sets of serum dilutions are required.
2. Add 0.1 ml of the antigen solution to 1 set of serum dilutions and 0.1 ml of normal saline to the other set.
3. Add 0.05 ml of the sensitized cells to all tubes. Shake well.
4. Incubate at 37°C and read the sedimentation patterns as in the HA procedures.

LIMITATIONS OF PROCEDURE

Bacto Formocells Sheep are not sensitive to complement and hemolysins and therefore cannot be used in complement fixation tests.

REFERENCES

1. J. Immun., 49:87, 1944.
2. J. Expl. Med., 93:107, 1951.
3. J. Immun., 72:360, 1954.
4. J. Expl. Med., 102:645, 1955.
5. Annals, Inst. Pasteur, 47:147, 1931.
6. Proc. Soc. Expl. Biol. & Med., 68:448, 1948.
7. Personal Communication, 1958.
8. Am. J. Clin. Path., 30:77, 1958.
9. J. Bact. Proc., 1959.
10. Am. J. Publ. Hlth., 50:1866, 1960.
11. Proc. Soc. for Expl. Biol. & Med., 103:157, 1960.
12. Personal Communication, 1960.
13. Personal Communication, 1961.
14. Proc. Soc. for Expl. Biol. & Med., 99:452, 1958.
15. Personal Communication, 1961.
16. Serological Museum, 21:2, 1959.
17. Personal Communication, 1961.
18. Arch. Biochem. & Biophys., 54:223, 1955.
19. U.S. Dept. Hlth., Educ. & Welfare, CDC Influenzae Report #8, 1957.

PACKAGING

Bacto Formocells Sheep   25 ml   3136-65-3

BACTO FRANCISELLA TULARENSIS ANTIGENS
BACTO FRANCISELLA TULARENSIS ANTISERUM

PRODUCT DESCRIPTION AND INTENDED USE

Bacto Francisella Tularensis Antigens are chemically inactivated suspensions of *Francisella tularensis* for use in the slide and tube agglutination procedures to detect antibodies (agglutinins) in serum specimens.

The rapid slide procedure is a screening test designed to detect agglutinins whereas the tube test is a confirmatory procedure designed to quantitate agglutinin compositions. It is therefore necessary that any positive results obtained in the screening (slide test) of specimens be confirmed by a tube test.

Bacto Francisella Tularensis Antiserum is a hyperimmune antiserum prepared in rabbits, designed to demonstrate agglutination for use as a positive control employing these test procedures.

These reagents are for use only in the test procedures described. They are not recommended for the direct diagnosis of a disease but rather only for the detection of the presence of agglutinins particularly in cases of seroconversion.

SUMMARY AND BACKGROUND

"Febrile Antigen" is a term which has been accepted generally as referring to bacterial suspensions, representative of a number of species of microorganisms pathogenic to man and which are accompanied by a fever in the host. There are a number of such organisms, but by classical definition, the most prominent include the genera *Salmonella*, *Brucella*, *Francisella (Pasteurella)*, *Leptospira* and some members of the *Rickettsia*.[1,2,3,4]

According to Bergey[5] two species of the genus *Francisella* exist, *F. tularensis* and *F. novicida*, the latter species occurs rarely and is not known to infect man. There are no common antigenic relationships between the two species.

F. tularensis causes sporadic acute infections in man (tularemia). The organism is transmitted to humans by animals (its natural host), biting insects, handling of infected carcasses or contaminated meat or water. Tularemia is manifested in varying degrees of severity. Six types are generally delineated; ulceroglandular, oculoglandular, glandular, typhoidal, pulmonary and ingestion.

Identification of *F. tularensis* is based on its growth requirements, morphology and serology. Biochemical characteristics are not significant in the identification of this organism.[6]

F. tularensis will not grow on ordinary media. It requires both blood and cystine or cysteine, exhibiting slow colonial growth. Gram stains of cultural isolates also aid in the identification of this organism.

Serological identification is accomplished by the detection of agglutinin titers (*F. tularensis* antibodies) by the rapid slide or macroscopic tube agglutination techniques. Direct serological identification of *F. tularensis* is determined using Francisella Tularensis Antiserum.

The Rapid Slide Test is the most widely used procedure employing bacterial antigens because of the simplicity with which the results may be reported. Negative reactions with this test can usually be reported as such if all 5 serum dilutions have been used. Note: Even though the slide test is not quantitative, it is necessary to run the series of dilutions in order to detect agglutinin content of a serum which might be overlooked in the case of a "prozone phenomenon". This often occurs in the sera containing a high titer of tularensis agglutinins where higher concentrations of the serum may yield negative results but a dilution of the serum is positive.

In this test, varying amounts of the sera to be tested are distributed to the corresponding square of a previously marked glass slide. The antigen is added and mixed with the sera. Positive or negative results are read 1 minute after mixing. Any positive reaction in the slide test must be confirmed using the tube test.

The Macroscopic Tube Test[7] should be used to confirm the presence of antibodies demonstrated by the slide technique and to quantitate their titer in suspected sera. In this test, the patient's serum is serially diluted in test tubes, a constant amount of the appropriate dilution of the antigen is added to each tube, the resultant mixture is incubated according to directions and the agglutination pattern is read and recorded.

For additional information refer to Bacto Febrile Antigens.

Antigens
Bacto Francisella Tularensis (slide) is a phenolized suspension of *F. tularensis* prepared according to the method of Schneider, Mitchell and Hardy[8] with slight modifications and is used in the slide agglutination test.

Bacto Francisella Tularensis (tube) is a formalinized suspension of *F. tularensis* prepared in accordance with the methods of the National Institutes of Health[9] and is used in the macroscopic tube test. The antigen is adjusted for immediate use with no further dilution necessary.

When used according to the suggested procedure, each 5 ml vial of Bacto Francisella Tularensis Antigen (slide) is sufficient to perform approximately 20 tests. Each 25 ml vial of Bacto Francisella Tularensis Antigen (tube) is sufficient for approximately 6 tests.

CONCENTRATION OF ANTIGENS
Bacto Francisella Tularensis Antigen (slide) is adjusted to a density of approximately 10% v/v.

Bacto Francisella Tularensis Antigen (tube) is adjusted to a density approximating a McFarland Barium Sulfate Standard No. 3 (9×10^8 organisms per ml).

Note: The density of the antigens may vary from the above values. They are adjusted to perform at their optimum when standardized with hyperimmune sera obtained from laboratory animals. In some antigens, glycerin is added when necessary to adjust the suspension sensitivity.

PRESERVATIVES
Bacto Francisella Tularensis Antigen (slide) contains phenol to a final concentration of 0.5% (5 g per liter), crystal violet in approximately 0.002%, brilliant green in approximately 0.005% final concentration, and glycerin to a final concentration of 20% (v/v).

Bacto Francisella Tularensis Antigen (tube) contains 0.5% formaldehyde as a preservative (5 ml/liter). It contains no dye.

Due to various factors, including the genera of the organism involved, the density of each lot of suspension etc., the color intensity of the dye in each lot of slide antigen may vary and is normal. Such differences in color will not affect the outcome of the test.

CONTROL ANTISERA
Bacto Francisella Tularensis Antiserum is a desiccated, unabsorbed hyperimmune antiserum prepared in rabbits. It is designed to demonstrate agglutination employing these test procedures, to a "significant" titer or greater with its homologous antigen, significant titer being defined as a 1:160 final test dilution or greater.

Merthiolate® is added as a preservative in the amount of 1:5,000 (w/v) final concentration. Bacto Febrile Negative Control may be used as a negative control in both test procedures. It is a standardized protein solution containing Merthiolate® as a preservative in the amount of approximately 1:5,000 (w/v) final concentration.

When used according to the suggested Procedure, each 3 ml vial of control antisera is sufficient to perform approximately 19 slide tests or 30 tube tests.

INSTRUCTIONS FOR REHYDRATION AND STORAGE
1. The antigens for both tests are liquid and ready for use.
2. The control antiserum is rehydrated by adding 3 ml sterile 0.85% NaCl solution at room temperature.
3. Bacto Febrile Negative Control is rehydrated by adding 5 ml distilled or deionized water.
4. Store all reagents at 2 – 8°C. These reagents are stable to the expiry date on the label when stored at 2 – 8°C provided they have not been exposed to temperature extremes (excessive heat or freezing) and have not become contaminated chemically or bacterially during routine usage.

Bacto Francisella Tularensis Antigens are not to be used for immunization of man or animals.

SPECIMEN COLLECTION
1. Collect 5 – 10 ml whole blood aseptically from the patient.
2. Allow the blood to clot and obtain the syneresed serum with a Pasteur pipette. If the serum is not free of erythrocytes, clarify by centrifugation. Note: Do not heat the serum, since significant antibodies may be thermolabile.

PROCEDURE
Materials Provided:
Bacto Francisella Tularensis Antigen (tube)
Bacto Francisella Tularensis Antigen (slide)
Bacto Francisella Tularensis Antiserum
Droppers for the above antigen (supplied droppers deliver approximately 0.05 ml per drop)

Purchase Separately:
Bacto Brucella Abortus Antigen (tube)
Bacto Brucella Abortus Antigen (slide)

Bacto Brucella Abortus Antiserum
Bacto Febrile Negative Control

Materials Required but not Provided:
Rapid Slide Test
Wax marking pencils
Ruler
Glass plate 8″ × 8″
0.2 ml serological pipettes

Isotonic saline (0.85 g NaCl per 100 ml
    distilled or deionized water)
Applicator sticks
Suitable light source

Tube Test
Isotonic saline (0.85 g NaCl per 100 ml distilled or deionized water)
Distilled or deionized water
Kahn tubes 12 mm × 75 mm
Kahn type tube supports
Serological pipettes, 1 ml and 5 ml capacity
Water bath at 37°C
Suitable light source

PRECAUTIONS
All reagents and equipment used in the slide test must be at room temperature prior to use.

Specimen
1. The specimen must be clear and free of visible fat. It must be free of excessive hemolysis and not bacterially contaminated.
2. The specimen must be used in the unheated state. If inactivated by heat, some thermolabile agglutinins may be destroyed.
3. In mixing the tubes in the tube test, the specimen should not be mixed too vigorously since foaming may result in agglutinin denaturization.

Glassware
1. All glassware that is employed in the preparation, testing and storage of these reagents must be free of detergents or other harmful residues.
2. Pipettes employed must be clean (acid washed if necessary) to deliver proper volumes.

Antigens
1. Shake antigen vial well before use to insure a smooth, uniform suspension.
2. Discard any antigen demonstrating positive reactions in the negative control or any demonstrating spontaneous agglutination in the vial. Febrile antigens will demonstrate autoagglutination if at any time during shipment or storage they are subject to freezing temperatures. Do not allow to freeze.
3. Discard any antigen which does not react properly in the positive control serum.
4. Variations in the color intensity of the antigen is normal and will not affect the outcome of the test.

Control Sera
Discard any control serum in which a precipitate forms. If bacterial contamination is suspected, autoclave the reagent or use some suitable chemical procedure to inactivate the contaminant before discarding the reagent.

METHODOLOGY
Special Note: Because a change or no change in titer over a period of time are the best indicators of active infection or noninfection respectively, and because the accuracy and precision of the tests can be affected not only by test conditions but also by

the subjectivity of the person in reading the end-point, the following protocol is recommended.

A preliminary test using either rapid slide test and/or the macroscopic tube test may be performed on the initial serum specimen and reported to the physician at that time. An aliquot of the serum should be transferred to a sterile test tube, sealed tightly, and kept in the freezer. When the second serum is obtained, it should be run in parallel with the original specimen. In this manner the original serum will serve as a control and any difference in titer will be more credible since the bias associated with the performance of the test, as well as determining the end-point, will be minimized. Obviously, it would be best if the person who performed the original test also performed the second test.

Exposure to heat (from external source, light source, burner flame, etc.) may result in evaporation of the test mixture and in false positive interpretations. Avoid such evaporation of the test mixture.

Rapid Slide Test
Note: The antigen, serum and equipment must be at room temperature prior to use in the slide test.
1. Prepare the glass plate 8″ × 8″ by making a series of 1-1/2″ ruled squares with the ruler and wax marking pencil. Each row of 5 squares is sufficient to test one antigen against the serum diluted to 1:320. Note: The glass plate should be thoroughly cleaned and dried after each use.
2. Pipette 0.08, 0.04, 0.02, 0.01 and 0.005 ml serum onto a row of squares on the ruled glass plate using a 0.2 ml serological pipette.
3. Place a drop of the appropriate antigen for the slide test on each drop of serum. Note: Shake the antigen well before using.
4. Mix each serum-antigen composite with an applicator stick starting with the 0.005 ml serum dilution and working towards the 0.08 ml dilution. The final dilutions are correlated approximately to the macroscopic tube dilutions and are 1:20, 1:40, 1:80, 1:160 and 1:320, respectively.
5. Hold the glass plate in both hands and gently "rotate" it a minimum of 15 – 20 times. Observe the agglutination within 1 minute over a suitable light source. Reactions occuring later may be due to the reactants drying on the slide.

Macroscopic Tube Test
Prepare serial dilutions of serum to be tested and the control sera in 0.5 ml amounts in Kahn tubes in the following manner:
1. Place 8 Kahn tubes in a rack for each serum to be tested.
2. Pipette 0.9 ml of 0.85% NaCl into the first tube of each row and 0.5 ml into each of the remaining tubes.
3. Add 0.1 ml of the serum to the first tube containing 0.9 ml of NaCl solution.
4. Mix well with a pipette and transfer 0.5 ml to the second tube. Mix thoroughly.
5. Continue carrying the 0.5 ml of the serum dilution through tube 7. Discard 0.5 ml from tube 7 after mixing thoroughly. Tube 8 is the antigen control tube.
6. Add 0.5 ml of the desired antigen to each of the 8 tubes. Shake the racks to mix the antigen and antiserum. The resultant dilutions are 1:20 through 1:1280, respectively.
7. Incubate in a water bath at 37°C for 20 – 24 hours. It is important in this test to use the recommended time and temperature of incubation and to make certain the water bath is in a location free of mechanical vibration.

CALIBRATION AND QUALITY CONTROL
For greater proficiency in test interpretation always include a positive and negative serum control in each test protocol.

Both positive and negative controls are diluted in the same proportion as the patient's serum and processed in exactly the same manner following above procedure for the rapid slide test or for the macroscopic tube test. Bacto Febrile Negative Control may be used as the negative control in both tests.

An antigen is considered to be satisfactory if it does not clump with the negative control and it reacts to a titer of 1:160 or more with the positive control.

DISCARD:
1. Any control serum in which a precipitate forms or which does not dissolve completely upon rehydration.
2. Any antigen when clumps appear on the cap or dropper or which agglutinates spontaneously or which agglutinates in the negative control.
3. All antigens and control sera when past the end of the month of the expiry date. Antigen and control sera are stable up to expiry date on label if not contaminated with organisms or chemical material and stored at 2 – 8°C.

RESULTS
Rapid Slide Test
Record results as follows:

4+	complete agglutination
3+	approximately 75% of the cells are clumped
2+	approximately 50% of the cells are clumped
1+	approximately 25% of the cells are clumped
±	trace agglutination
−	no agglutination

The titer of serum is recorded as that dilution of the specimen in which at least 2+ (50%) agglutination occurs. See Tables I and II.

Macroscopic Tube Test
Using a fluorescent lamp against a black background, record the degree of agglutination as follows:

4+	100% of the organisms are agglutinated and the supernatant fluid is clear
3+	approximately 75% of the organisms are agglutinated and the supernatant fluid is slight cloudy
2+	approximately 50% of the organisms are agglutinated and the supernatant fluid is moderately cloudy
1+	approximately 25% of the organisms are agglutinated and the supernatant fluid is cloudy.
−	no agglutination is visible and the organisms remain as a cloudy suspension. This reaction should be observed in tube 8.

SAMPLE CALCULATIONS

Table I
Rapid Slide Test

		Reactions		
ml Serum	Correlated Dilution	Spec. 1	Spec. 2	Spec. 3
0.08	1:20	3+	4+	4+
0.04	1:40	2+	4+	3+
0.02	1:80	1+	3+	2+
0.01	1:160	−	3+	+
0.005	1:320	−	1+	−
Serum Titer		1:40	1:160	1:80

Table II
Macroscopic Tube Test

Serum Dilution	Specimen 1	Reactions Specimen 2	Specimen 3
1:20	4+	3+	4+
1:40	4+	2+	4+
1:80	3+	1+	4+
1:160	2+	−	4+
1:320	1+	−	3+
1:640	−	−	2+
1:1280	−	−	1+
Serum Titer	1:160	1:40	1:640

INTERPRETATION[6,10,11]

Report the serum titer as the reciprocal of the highest dilution showing a 2+ reaction. See Tables I and II.

A difference in titer of 1 dilution, plus or minus, between replicate samples or between several samples drawn 1 − 2 weeks apart and run in parallel cannot be considered significant. A 1 dilution plus or minus deviation is within the limits of laboratory error.[12,13]

Febrile antigens have been found to be a very useful tool to assist in the diagnosis of tularemia. It must be emphasized, however, that in this test, as in other febrile tests, paired specimens from the same patient indicating a rise in agglutinin titer between the acute and convalescent phases is important. The two serum specimens should be tested simultaneously. If a convalescent phase serum is not available, a titer of 1:80 − 1:160 in the acute phase specimen together with symptoms compatable with tularemia, is suggestive of this disease state.

The significant titer is generally attained in the second week after onset of the disease and rises to its maximum within a month or 6 weeks. The titer levels off after 6 − 8 weeks and begins to drop after a year. A residual agglutinin titer, however, may persist for several years thereafter.

LIMITATIONS

1. The major limitation of the febrile tests is that of interpretation. Again, it must be stressed that the antigens are to be used for the detection and quantitation of agglutinin content in human sera. They are not to be used singly for a definite diagnosis. A definitive diagnosis must be made by the physician, taking into consideration the history and physical state of the patient as well as data obtained from other laboratory tests.

2. It is necessary to run several serum specimens taken at the acute and convalescence stage from the same patient, in parallel, to detect quantitative differences in agglutinin content.

3. In some tularemia sera, cross-reactions may be evident with *Brucella* antigens.

   It is recommended that, since it has been reported that cross-reactions exist between serum specimens of brucellosis and tularemia, both antigens be employed concurrently. Generally the homologous system is of significantly higher titer than that of the heterologous system.

4. In some cases of tularemia, sera may demonstrate a prozone. (Inability of an antigen to react in higher serum concentrations.) It is advised that all 5 serum dilutions be run in the rapid slide test rather than just 1 dilution to eliminate the possibility of missing positive reactions due to the prozone.

REFERENCES

1. Schubert, J. H., Holdeman, L. and Martin, D. S., J. Lab. Clin. Med., 44:194, 1964.
2. Welch, H. and Stuart, C. A., J. Lab. and Clin. Med., 21:411, 1936.
3. Welch, H. and Mickle, F. L., Amer. J. Pub. Hlth., 26:248, 1936.
4. Diamond, B. E., Pub. Hlth. Lab., 6:74, 1948.
5. Bergey's Manual of Determinative Bacteriology, 8th Ed., Williams & Wilkins Co., 1974.
6. Lennette, Balows, Hausler, Truant, Manual of Clinical Microbiology, 3rd Ed., Washington, D.C., ASM, 1980.
7. Spink, W. W., McCullough, N. B., Hutchings, L. H., Amer. J. Clin. Path., 24:486, 1954.
8. Publ. Hlth. Lab., 8:35, 1950.
9. Federal Security Agency, NIH, Memo Jan. 15, 1947.
10. Snyder, M. J., Manual of Clinical Immunology, 1976.
11. Bodily, H. L. et al, Diagnostic Procedures for Bacterial, Mycotic and Parasitic Infections, APHA, Inc., 1970.
12. Syphilis A Synopsis, US Dept. of Hlth. Ed. and Welf., Publ. Hlth. Serv. Publ. No. 1660, Jan., 1968.
13. Quality Control in Clinical Micro., Revised, ASCP Comm. on Continuing Education, Comm. on Micro., 1968.

PACKAGING

Bacto Francisella Tularensis Antigen (Slide)	5 ml	2240-56-9
Bacto Francisella Tularensis Antigen (Tube)	5 ml	2251-56-5
	25 ml	2251-65-4
Bacto Francisella Tularensis Antiserum	3 ml	2241-47-0

BACTO GC MEDIUM BASE

INTENDED USE

Bacto GC Medium Base, in combination with Bacto Hemoglobin and Bacto Supplement A, B, C or VX is used for cultivating *Neisseria gonorrhoeae* and other similar fastidious microorganisms. If a medium selective for *N. gonorrhoeae* is desired, Bacto Antimicrobic Vial CNV or CNVT may also be added.

HISTORY

Johnston[1] described a chocolate agar containing Bacto Proteose Peptone No. 3 Agar, Bacto Hemoglobin, 30% ascitic fluid and 1:50,000 tyrothricin that demonstrated accelerated early growth (24 hours) of *N. gonorrhoeae*. As a result, Difco Laboratories designed Bacto GC Medium Base in 1947. Christensen and Schoenlein[2] demonstrated that Bacto GC Medium Base, enriched with Bacto Hemoglobin and Bacto Supplement A or B, yielded cultural results equal to or better than the Johnston medium at 24 hours. Johnston[3] found the two media comparable.

PRINCIPLES

Johnston's[1] research indicated that accelerated growth of *N. gonorrhoeae* on the chocolate agars studied was due primarily to the decreased agar content (solidity) of the media. Johnston preferred the use of Supplement A over Supplement B since the crystal violet helped in suppressing extraneous contaminating microorganisms.

FORMULA

BACTO GC MEDIUM BASE
DEHYDRATED

Ingredients per liter

Proteose Peptone No. 3, Difco	15 g
Corn Starch	1 g
Potassium Phosphate, Dibasic	4 g
Potassium Phosphate, Monobasic	1 g
Sodium Chloride	5 g
Bacto Agar	10 g

Final pH 7.2 ± 0.2 at 25°C.

One pound will make 12.6 liters of single-strength medium.

METHOD OF PREPARATION
1. Suspend 36 g Bacto GC Medium Base in 500 ml distilled or deionized water and heat to boiling to dissolve completely.
2. Sterilize in the autoclave for 15 minutes at 15 lbs pressure (121°C). Cool to 45 – 50°C.
3. Add 500 ml sterile 2% solution of Bacto Hemoglobin or Bacto Hemoglobin Solution 2%. Mix well.
4. Add 10 ml Bacto Supplement A or B to the sterile medium. Mix well.
5. If a selective chocolate agar is desired, add 10 ml rehydrated Bacto Antimicrobic Vial CNV or CNVT.
6. Dispense into sterile Petri dishes. Allow surface of medium to dry before inoculation.

STORAGE
Bacto GC Medium Base   Below 30°C
Chocolate agar plates   2 – 8°C

QUALITY CONTROL
Identity Specifications

Dehydrated medium: off white, free-flowing powder
pH of basal medium: 7.2 ± 0.2 at 25°C
Prepared plates: light to medium amber, opalescent, ground glass
                      appearance

Cultural Response on Chocolate Agar Prepared From Bacto GC Medium Base After 40 – 48 Hours at 35 ± 2°C

Organism	Growth Response
Haemophilus influenzae ATCC® 19418	good to excellent
Neisseria gonorrhoeae ATCC® 19424	good to excellent
Neisseria meningitidis ATCC® 13090	good to excellent
Streptococcus pneumoniae ATCC® 6303	good to excellent
Streptococcus pyogenes ATCC® 19615	good to excellent

REFERENCES
1. J. Venereal Disease Inform., 26:239, 1945.
2. Paper read at the annual meeting of the Canadian Pub. Health Assoc., 1947.
3. Personal Communication, 1947.

PACKAGING
Bacto GC Medium Base		
	1 lb (454 g)	0289-01-1
	1/4 lb (114 g)	0289-02-0
	5 lb (2.27 kg)	0289-05-7
	10 kg	0289-08-4

BACTO GN BROTH, HAJNA

INTENDED USE
Bacto GN Broth, Hajna is a selective liquid medium used for the cultivation of gram-negative organisms from all types of specimens.

HISTORY/PRINCIPLES
Bacto GN Broth, Hajna is prepared according to the formula of Hajna.[1] Hajna[2,3] suggested the enrichment of organisms from rectal swabs in Hajna GN Broth for 1 to 6

hours prior to plating on solid media. Bacto GN Broth, Hajna was also shown to be of value in the isolation of gram-negative organisms from urines, blood clots, swabs from eating and drinking utensils, throat swabs and from sputa specimens. An incubation temperature of 30°C was recommended for the cultivation of gram-negative organisms from sputa. Croft and Miller[4] compared direct streaking of rectal swabs with swabs which were placed in a tube of Bacto GN Broth, Hajna and then streaked on the usual isolation media within 6 to 8 hours following collection. They reported more isolations of *Shigella* by the use of the Hajna GN Broth than with direct or immediate streaking on the isolation media. Only 3 *Salmonella* strains were isolated from the 2696 specimens examined in this series. This low recovery of *Salmonella* does not indicate that Hajna GN Broth fails to enrich these organisms, but rather that *Salmonella* was not present in appreciable numbers. Hajna[2] isolated 127 *Salmonella* and 105 *Shigella* from 480 rectal swabs using Hajna GN Broth. Croft and Miller[4] suggest that Hajna GN Broth enrichment would be useful for survey operations in the field.

Bacto Tryptose serves as a nutriment in the medium. Sodium citrate and sodium desoxycholate are bactericidal to gram-positive organisms and inhibit the development of coliforms, while the phosphates serve as a buffer. The increased concentration of mannitol over dextrose assists in limiting the growth of *Proteus* and in accelerating the growth of *Salmonella* and *Shigella* capable of fermenting mannitol. *Pseudomonas aeruginosa* and *Proteus* will grow in the medium but will not overgrow *Shigella* or *Salmonella* in mixed culture.

FORMULA
BACTO GN BROTH, HAJNA
DEHYDRATED
Ingredients per liter

Bacto Tryptose	20 g	Sodium Desoxycholate	0.5 g
Bacto Dextrose	1 g	Dipotassium Phosphate	4 g
D-Mannitol	2 g	Monopotassium Phosphate	1.5 g
Sodium Citrate	5 g	Sodium Chloride	5 g

Final pH 7.0 ± 0.2 at 25°C.

One pound will make 11.7 liters of medium.

METHOD OF PREPARATION
1. To rehydrate, dissolve 39 g in 1 liter distilled or deionized water.
2. Dispense in tubes and sterilize in the autoclave for 15 minutes at 10 lbs pressure (116°C) or at 15 lbs pressure (121°C).
3. Avoid excessive heating of the medium.

STORAGE
Bacto GN Broth, Hajna   Below 30°C
Prepared medium   2 – 8°C

QUALITY CONTROL
Identity Specifications

Dehydrated powder: tan, homogeneous, free-flowing
Reaction of 3.9% solution: pH 7.0 ± 0.2 at 25°C
Prepared medium: light amber, clear to very slightly opalescent

Typical Cultural Response in Bacto GN Broth, Hajna
After 18 – 24 Hours at 35°C

Organism	Growth
Escherichia coli ATCC® 25922	good to excellent
Proteus mirabilis ATCC® 25933	good to excellent
Salmonella typhimurium ATCC® 14028	good to excellent
Shigella flexneri ATCC® 12022	good to excellent
Streptococcus faecalis ATCC® 19433	none to poor

REFERENCES
1. Publ. Hlth. Lab., 13:59, 1955.
2. *ibid.*, 13:83, 1955.
3. Air Univ. Sch. Av. Med., USAF, 56:39, 1956.
4. Am. J. Clin. Path., 26:411, 1956.

PACKAGING

Bacto GN Broth, Hajna	1 lb (454 g)	0486-01-2
	5 lb (2.27 kg)	0486-05-8

BACTO GELATIN

INTENDED USE
Bacto Gelatin is a high grade gelatin in granular form for convenience in handling. It is especially processed for use in all microbiological procedures requiring a product of high purity, solubility, clarity, and specific pH. It may be used as a solidifying agent or may be incorporated into culture media for various uses.

HISTORY
Gelatin was first employed as a solidifying agent for bacteriological culture media by Koch in 1881. This innovation paved the way for the future of the science although gelatin media were soon replaced by others containing agar as the solidifying material.

The use of Bacto Gelatin in culture media for determining gelatinolysis (elaboration of gelatinases) by bacteria is recommended by the Committee on Bacteriological Technic[1] of the Society of American Bacteriologists. Levine and Carpenter[2] and Levine and Shaw[3] also employed Bacto Gelatin in the media used in their studies of gelatin liquefaction. Garner and Tillett[4] used culture media prepared with Bacto Gelatin in their study of the fibrinolytic activity of hemolytic streptococci.

PRINCIPLES
The melting point of a 12% concentration of Bacto Gelatin is between 28 and 30°C, which allows it to be used as a solidifying agent. Certain microorganisms elaborate gelatinolytic enzymes (gelatinases) which hydrolyze the gelatin, resulting in the liquefaction of a solidified medium or preventing the gelation of a medium containing gelatin.

METHOD OF PREPARATION
Bacto Gelatin is used in a 12% concentration in the formula for nutrient gelatin as given in *Standard Methods of Water Analysis* of the American Public Health Association. Bacto Gelatin should be added to the liquid slowly, while swirling the flask frequently. Heat gently to obtain complete solution, preferably by placing in a water bath at 50 – 55°C. Other formulae may require larger or smaller quantities of gelatin or combinations of gelatin and agar.

The reaction of a 12% solution of Bacto Gelatin is pH 6.8, thus adjustment of reaction is generally not required when used in culture media.

Solutions of Bacto Gelatin are clear and colorless to light straw depending on the concentration of gelatin dissolved.

STORAGE
Bacto Gelatin   15 – 30°C

REFERENCES
1. Pure Culture Study of Bacteria, 4:No. 3:1936.
2. J. Bact., 8:297:1923.
3. J. Bact., 9:225:1924.
4. J. Exp. Med., 60:255:1934.

PACKAGING
Bacto Gelatin	1 lb (454 g)	0143-01-7
	1/4 lb (114 g)	0143-02-6
	5 lb (2.27 kg)	0143-05-3
	10 kg	0143-08-0

BACTO GELATONE

Bacto Gelatone is a pancreatic digest of gelatin for use in microbiological culture media.

PACKAGING
Bacto Gelatone	1 lb (454 g)	0657-01-5

BACTO GIOLITTI-CANTONI BROTH BASE
BACTO POTASSIUM TELLURITE SOLUTION 3.5%

INTENDED USE
Bacto Giolitti-Cantoni Broth Base and Bacto Potassium Tellurite Solution 3.5% are used in combination for selecting and enriching *Staphylococcus aureus* being isolated from foodstuffs.

HISTORY
Giolitti and Cantoni[1] described the basal medium, the addition of potassium tellurite to the basal medium and the test procedure for enriching minimal numbers of staphylococci in foods. Mossel, Harrewijn and Elzebroek[2] recommended the medium for detecting *Staphylococcus aureus* in dried milk and other infant foods where the organism should be absent from 1 g of test material.

PRINCIPLES
Mannitol and sodium pyruvate present in the basal medium act as growth stimulants for *Staphylococcus aureus*, aiding in detection of small numbers of organisms.[3] Lithium chloride[4] inhibits gram-negative lactose-fermenting bacilli. Potassium tellurite, in combination with glycine, inhibits gram-positive bacilli other than staphylococci. The an-

aerobic environment which results from overlaying the inoculated medium with sterile paraffin inhibits micrococci. As a result, bacteria (other than staphylococci) which may be present in food are markedly to completely inhibited.

The medium is suitable for the examination of meat and meat products upon reduction of the concentration of the potassium tellurite to 0.35% and of the weight of the test sample to 0.1 – 0.01 g.

FORMULA

BACTO GIOLITTI-CANTONI BROTH BASE
DEHYDRATED

Ingredients per liter

Bacto Tryptone	10 g	Sodium Chloride	5 g
Bacto Beef Extract	5 g	Lithium Chloride	5 g
Bacto Yeast Extract	5 g	Glycine	1.2 g
Mannitol	20 g	Sodium Pyruvate	3 g

Final pH 6.9 ± 0.2 at 25°C.

Five hundred grams will make 9.2 liters of medium.

Bacto Potassium Tellurite Solution 3.5% is a filter-sterilized solution of potassium tellurite in distilled water.

PREPARATION

1. Rehydrate Bacto Giolitti-Cantoni Broth Base by suspending 54.2 g in 1 liter distilled or deionized water. Warm gently to dissolve.
2. Dispense 19 ml amounts into approximately 20 mm by 200 mm test tubes.
3. Sterilize in the autoclave for 15 minutes at 15 lbs pressure (121°C). Cool rapidly to room temperature.
4. If the medium is not to be used immediately, store at 2 – 8°C for no more than 10 – 15 days. Immediately prior to use, expel absorbed oxygen by placing the tubes in free-flowing steam for 15 – 20 minutes.
5. Aseptically add 0.3 ml Bacto Potassium Tellurite Solution 3.5% to each tube (0.03 ml when testing meat or meat products) and swirl tubes gently to disperse throughout the medium.

PROCEDURE

1. Inoculate 1 g or 1 ml of test sample (0.1 g or 0.1 ml when testing meat or meat products) and 1 ml aliquots of each of a suitable decimal dilution series of the test sample into duplicate tubes.
2. Overlay each tube with 5 ml (to an approximate height of 2 cm) sterile molten paraffin wax.
3. Incubate 48 hours at 37°C, examining daily.
4. If NO BLACKENING OF THE MEDIUM OCCURS, consider the test negative for *Staphylococcus aureus*. If BLACKENING OF THE MEDIUM OCCURS, generally or at the bottom of the tubes, streak the black precipitate on a *Staphylococcus* isolation medium (Bacto Baird Parker Agar Base supplemented with Bacto EY Tellurite Enrichment or Bacto Mannitol Salt Agar) and incubate at 37°C for 24 – 48 hours.
5. Report the test sample positive for *Staphylococcus aureus* if,

   On Bacto Baird Parker Agar Base supplemented with Bacto EY Tellurite Enrichment, black colonies surrounded by a zone of clearing appear, or if;

   On Bacto Mannitol Salt Agar, colonies with a yellow zone appear. On this medium most bacteria other than *Staphylococcus aureus* are inhibited.

STORAGE

Bacto Giolitti-Cantoni Broth Base   Below 30°C
Bacto Potassium Tellurite Solution 3.5%   15 – 30°C
Prepared medium   2 – 8°C

QUALITY CONTROL

Identity Specifications

Dehydrated powder: tan, homogeneous, free-flowing
Reaction of 5.42% solution: pH 6.9 ± 0.2 at 25°C
Prepared medium: medium amber, clear
Appearance of enrichment: colorless, clear liquid

Typical Cultural Response in Bacto Giolitti-Cantoni Broth
After 40 – 48 Hours at 37°C

Organism	Growth
Escherichia coli ATCC® 25922	inhibited
Micrococcus luteus ATCC® 10240	inhibited
Staphylococcus aureus ATCC® 6538	good, blackening observed
Staphylococcus aureus ATCC® 25923	good, blackening observed

REFERENCES

1. J. Appl. Bact., 29:395, 1966.
2. UNICEF, 1973.
3. J. Appl. Bact., 25:12, 1962.
4. Lambkin, S. and German, A., Precis de Microbiologic, p. 63, Paris Masson, 1963.

PACKAGING

Bacto Giolitti-Cantoni Broth Base	500 g	1809-17-2
Bacto Potassium Tellurite Solution 3.5%	25 ml	1814-65-6
Bacto Baird Parker Agar Base	1 lb	0768-01-1
	1/4 lb	0768-02-0
Bacto EY Tellurite Enrichment	6 × 100 ml	0779-73-1
Bacto Mannitol Salt Agar	1 lb	0306-01-0
	1/4 lb	0306-02-9
	5 lb	0306-05-6
	10 kg	0306-08-3

GLUCOSE/PENASE ASCITES MEDIA

BACTO GLUCOSE ASCITES MEDIUM
BACTO PENASE ASCITES MEDIUM

INTENDED USE

Bacto Glucose Ascites Medium is prepared for the bacterial examination of root canals.

Bacto Penase Ascites Medium is Bacto Glucose Ascites Medium to which 1 ml Bacto Penase has been added per 10 ml of medium for the inactivation of penicillin carried over on the absorbent point from root canals treated with this antibiotic. Each tube of Bacto Penase Ascites Medium contains sufficient Bacto Penase to inactivate over 100,000 units of penicillin and will also inactivate up to 100 units of streptomycin.

PRINCIPLES

Bacteriological procedures have become an indispensable tool in the armamentarium of the dental prcfession in research and practice. Effective cultural methods are available for detecting and enumerating microorganisms encountered in the dental practice, establishing their etiological role in dental pathologies, determining the effectiveness of therapeutic agents used in dentistry, and testing sterility of root canals and sockets after root resection. The protection of the vital pulp and the treatment of pulpless teeth against a major source of irritation, i.e., bacterial infection, are among the primary responsibilities of the dental practitioner. A negative culture obtained after debridement, obliteration and sterilization of the canal will assure him of a successful root canal therapy.

It is essential that the medium employed in determining the sterility of root canals be capable of supporting growth of all possible contaminating microorganisms. The medium must possess, in addition to high nutrient qualities, the capacity to encourage the development of both aerobic and anaerobic bacteria and, if possible, to neutralize the inhibitory effects of medications employed in root canal therapy.

HISTORY

Bacteriological examination of the root canal as a routine procedure to determine sterility before filling the canal was probably first suggested by Onderdonk.[1] The purpose of bacteriological examination has been ably stated by Appleton[2] who pointed out that if the function of root canal therapy is to render the canal and periapical tissues sterile, the only method which can determine whether that objective has been attained is a bacteriological examination. The value of bacteriological examination has been amply demonstrated by Grossman.[3,4] Cultural examinations of 150 root canals deemed ready for filling as judged by the appearance and odor of the dressings revealed that 42% were not sterile. As pointed out by the author, the entire purpose of root canal therapy would have been defeated had the teeth been filled at that time. It is evident, therefore, that the clinical appearance is an insufficient criterion for determining when a canal is ready to be filled. Abramson[5] demonstrated a definite relationship between the degree of success in treating pulpless teeth and the technique used in the bacteriological examination of the canal. The importance of such examination was further stressed by Zeldow and Ingle[6] who reported a failure rate of 16.7% in root canal therapy when positive cultures were obtained compared to only 5.5% with negative cultures.

Bacto Glucose Ascites Medium is a prepared medium containing Bacto Brain Heart Infusion w/PAB and Agar enriched with 10% ascitic fluid. Sommers and Crowley[7] employed glucose ascites medium with success in a study of the relationship between bacterial cultures and roentgenographic findings. Crowley,[8] using the same medium, succeeded in isolating an actinomyces-like organism from root canals. Grossman[9] pointed out that the glucose ascites medium because of its agar content had the advantage of confining the growth near the cotton point being cultured. This property increased the ease of detecting early growth.

TEST PROCEDURE

Grossman[4,10] suggests the technique which follows:

The dressing from the canal is discarded. A fresh, sterile absorbent point is inserted in the canal to cleanse the canal surface of any trace of medicament. This absorbent point is also discarded and another fresh sterile point is inserted to the apex of the tooth and allowed to remain there for a minute. It is then removed and, if the tip of the absorbent point is moistened with exudate, it is dropped into a tube of culture medium. If the absorbent point comes out dry, a fresh sterile absorbent point is moistened with a little culture medium, under sterile precautions. This moistened absorbent point is then inserted into the root canal. In this entire procedure, it is assumed, of course, that principles of asepsis are strictly adhered to, that the tooth has been isolated under a rubber dam, that the field of operation has been sterilized, and that the temporary filling and dressing have been removed with sterile instruments.

Do not heat Bacto Glucose Ascites Medium or Bacto Penase Ascites Medium prior to use as they contain coagulable protein.

To inoculate the medium, unscrew the cap and, holding the tube at a 45° angle to prevent contamination from the air, drop the absorbent point used to culture the root canal into the tube and replace the cap. The tube should be incubated at 37°C for 48 hours before it is examined for growth. If a thermostatically controlled incubator is not available, a thermos bottle or a homemade incubator may be substituted. Inexpensive dental incubators are available from dental supply companies. Ingle[11] also described the fabrication of a simple incubator using an insulated box, light bulb and a thermostat.

INTERPRETATION OF RESULTS
After 48 hours incubation, the tube is removed from the incubator and examined for growth. Growth is manifested by an increase in turbidity of the medium when compared with an uninoculated tube of the same medium. Any increase in turbidity of the medium, however slight, is indicative of growth of bacteria. In this connection, it is well to repeat the test. Quite often growth will be confined to a slight area around the tip or surface of the absorbent point. If the culture is sterile, there will be no evidence of increased turbidity of the medium.

It is possible to obtain a false positive test due to contamination and this must be stringently guarded against through the use of sterile equipment and rigidly aseptic technique. A false negative test may result when the canal is "dry" and has not been moistened with a little of the culture medium or if an insufficient sample has been removed from the canal on the absorbent point. As a general rule, two consecutive negative tests are obtained before the root canal is considered ready to be filled.

STORAGE
Bacto Glucose Ascites Medium	2 – 8°C
Bacto Penase Ascites Medium	2 – 8°C

REFERENCES
1. Inter. Dent., 22:20, 1901.
2. Dent. Cosmos, 74:798, 1932.
3. J. Dent. Res., 15:364, 1936.
4. Endodontic Pract. 5th Ed., 1960.
5. Paper presented at the Ann. Meeting of the Am. Association of Endodontitis, Feb. 1961.
6. J.A.D.A., 66:9, 1963.
7. J. Am. Dent. Assoc., 27:723, 1940.
8. J. Dent. Res., 20:189, 1941.
9. ibid., 19:349, 1940.
10. Root Canal Therapy, 2nd Ed., p 273, 1946.
11. Endodontics, C. V. Mosby, 1960.

PACKAGING
Bacto Glucose Ascites Medium	12 tubes	1007-34-1
	144 tubes	1007-37-8
Bacto Penase Ascites Medium	12 tubes	1015-34-1

BACTO GLYCEROL

Bacto Glycerol is a highly purified alcohol used in the preparation of a large variety of reagents frequently used in the clinical laboratory. As a fermentable alcohol, this carbon compound is used occasionally for differentiating certain bacteria. It is used in media for isolating and culturing fastidious bacteria, in transport media to maintain viability of bacteria present in a specimen, for preparing long term bacterial preservation media, for preparing fixatives or mounting fluids and many other uses.

PACKAGING
Bacto Glycerol	100 g	0282-15-2
	500 g	0282-17-0

BACTO GRAM STAIN SET AND REAGENTS

TEST SUMMARY
Bacto Gram Stain Set and Reagents are used to stain microorganisms from cultures or specimens by the differential Gram method.

Although the Gram stain is among the least complicated and least time consuming of all microbiological tests, the information which may be gleaned from a properly stained smear of a culture or a specimen is among the most valuable aids to the clinician and microbiologist. A properly performed Gram stain on a culture or a specimen can provide important preliminary information concerning the type of organisms present, the techniques that should be pursued to characterize them, and the therapy to initiate while waiting for culture and susceptibility tests results. **The Gram stain results should never be a substitute for culture studies of the specimen.**

The mechanism of the Gram stain is not clearly understood. There is general agreement that a gram-positive organism retains the primary stain after decolorization because of a variety of factors including their possession of isoelectric points at about pH 2, the presence of a magnesium ribonucleate protein complex, a phosphoric ester and the mordant effect of iodine.

The test consists essentially of applying Bacto Gram Crystal Violet to a fixed smear from a culture or specimen, removing this primary stain after a suitable staining period and then applying Bacto Gram Iodine as a mordant. The mordant in turn is removed and Bacto Gram Decolorizer is added to remove the primary stain where possible. Bacto Gram Safranin is then added as a counterstain.

Organisms are judged to be gram-positive if they retain the primary stain after decolorization. Gram-negative organisms are decolorized and appear pink to red because they take up the counterstain.

HISTORY
The first published acknowledgement of the Gram staining procedure was in 1883, when Christian Gram was developing a staining procedure that would differentiate schizomycetes from tissue cells microscopically. Gram was a co-worker of Carl Friedlander in the municipal hospital of Berlin, who first mentioned the technique in a treatise

upon "the Micrococci of pneumonia." In the following year, Gram published his technique in detail. Although his original procedure was not presented in the exact form as we know it today, the four fundamental steps are still the same and Gram's staining procedure is recognized as a major contribution to biological science.

FORMULA
Bacto Gram Crystal Violet is a 1% aqueous solution of Bacto Crystal Violet.

Bacto Gram Decolorizer

CAUTION: Flammable

Formulation per liter:
Acetone . . . . 250 ml
Isopropyl Alcohol . . . . 750 ml

Bacto Gram Iodine

CAUTION: Poison

Formulation per 300 ml final solution:
Iodine Crystals . . . . 1 g
Potassium Iodide . . . . 2 g
Distilled or Deionized water . . . . . 300 ml

Bacto Gram Safranin

Formulation per liter:
Safranin O Powder (pure dye)   . . . . 4 g
SD3A Alcohol, Anhydrous . . . . 200 ml
Distilled or Deionized Water   . . . . . 800 ml

NOTE: Bacto Gram Crystal Violet, Decolorizer and Safranin are supplied in pliable plastic bottles. The Bacto Gram Iodine is supplied in 100X concentration in a glass ampul and is to be transferred to the plastic bottle containing diluent supplied for that purpose. Dropper caps for the plastic bottles are supplied for easy use of the reagents in the laboratory staining procedure.

Once the Gram Iodine 100X concentrate is mixed with its diluent, the solution should be exposed to the air as little as possible and kept away from heat. Studies have shown that Gram iodine solution is relatively unstable and may cause variability in the Gram stain when sufficient iodine is no longer available in the solution. For this reason, as well as for others, controls should be included in all staining runs or at least once daily (see Controls) to ensure that the iodine solution is providing the proper mordant activity.

STORAGE:
Bacto Gram Stain Set and Reagents   15 – 30°C

PREPARATION OF SMEARS FROM CULTURES OR SPECIMENS
For microorganisms in liquid suspension:
1. Spread a loopful on a clean glass slide.
2. Let smear air dry.
3. Heat-fix by passing through a low gas flame 3 times.
4. Allow fixed slide to cool to room temperature before staining.

For staining microorganisms from colonies on solid media:
1. Place a small loopful of distilled water on a clean microscope slide.
2. Touch a sterile inoculating loop to a colony, rub it into the droplet to obtain an even suspension of microorganisms and spread the suspension over the surface of the slide to obtain correct density.
3. Let the smear dry at room temperature.
4. Heat-fix by passing the slide 3 times through a low gas flame.
5. Cool to room temperature before staining.

For staining microorganisms in specimens:
1. Obtain specimen by pipette, loop or swab.

2. Smear the specimen over a clean slide surface to obtain a thin film.
3. Air dry.
4. Heat-fix by passing slide 3 times through a low gas flame.
5. Allow to cool to room temperature before staining.

STAINING PROCEDURE

Materials Provided:
Bacto Gram Crystal Violet
Bacto Gram Decolorizer
Bacto Gram Iodine
Bacto Gram Safranin

Purchase Separately:
Bactrol Disks Set B
   Containing *Escherichia coli*
   and *Staphylococcus aureus.*
                  or
Bactrol™ Gram Slide

Materials Required but not Provided:
Microscope slides
Bunsen burner
Bacteriological loop

Swabs
Blotting paper
Microscope with oil immersion lens

Method of Staining Using the Dropper Bottles Provided:

1. Flood the fixed smears with Bacto Gram Crystal Violet stain and allow to stain for 1 minute.
2. Wash off excess crystal violet lightly with cold tap water.
3. Flood slide with Bacto Gram Iodine and allow the iodine to remain on slide for 1 minute.
4. Remove iodine by gentle washing with tap water.
5. Flood off excess water with Bacto Gram Decolorizer until solvent runs colorlessly from the slide (30 – 60 seconds).
6. Wash slide gently and thoroughly in cold tap water.
7. Counterstain with Bacto Gram Safranin for 30 – 60 seconds.
8. Wash off excess stain with cold tap water.
9. Blot off excess water with paper toweling or blotting paper or allow to air dry.
10. Examine under oil immersion lens.

NOTE: It is advisable to use an 18 – 24 hour culture for best results since fresh cells have a greater affinity than old cells for most dyes. This is particularly true of many spore formers, which are strongly gram-positive when examined in fresh cultures but which later become gram-variable or gram-negative.

The Gram stain reaction, like the acid-fast reaction, is altered by physical disruption of the bacterial cell wall or protoplast. The cell walls of gram-positive bacteria interpose a barrier which prevents leaching of the dye complex from the cytoplasm. Cell walls of gram-negative bacteria contain lipids soluble in organic solvents, which are then free to decolorize the cytoplasm. Therefore, a microorganism that is physically disrupted by excess heating will not react to Gram staining as expected.

CONTROLS

It is recommended that controls be run daily using known gram-positive and gram-negative microorganisms.

Bactrol Disks *Staphylococcus aureus* ATCC® 25923 and Bactrol Disks *Escherichia coli* ATCC® 25922 or Bactrol Gram Slides are recommended for these controls.

NOTE: When performing the Gram stain on a clinical specimen, particularly when the results will be used as a guide to the selection of a therapeutic agent, it is very important that controls be included in the same staining run or, preferably, on the same slide. Such a control system provides additional assurance that the iodine solution is providing the proper mordant activity and that decolorization was performed properly.

A suggested technique is as follows:
1. Divide a microscope slide into 3 equal segments with a wax marking pencil.
2. In the center segment make a smear of the specimen (see Specimen Preparation section).
3. On the left-hand segment place a Bactrol Disk *Staphylococcus aureus*. Add a drop of sterile water and rub the disk in the water for 15 – 30 seconds. Remove the undissolved portion of the disk and sterilize by placing in a beaker containing 70 – 90% isopropyl alcohol, 1 – 2% phenol, quaternary compound, or a laundry bleach.
4. Repeat step 3 on the right-hand segment using a Bactrol Disk *Escherichia coli*.
5. Fix slides and proceed with the staining procedure.

RESULTS
Retention of the purple-black mordant treated primary stain indicates a gram-positive microorganism. Microbial cells which decolorize and which stain pink to red with the counterstain are gram-negative.

LIMITATIONS OF THE PROCEDURE
Prior treatment with antibacterial drugs may cause gram-positive organisms from a specimen to appear gram-negative.

The Gram stain provides preliminary identification information only and is not a substitute for culture studies on the specimen.

REFERENCES
1. Manual of Microbiol. Methods, N.Y., McGraw-Hill, pgs. 15 – 18, 1957.
2. Stain Tech., 37:139, 1962.
3. Diagnostic Microbiology, 2nd Ed., p. 320, 1966.
4. Principles of Bacteriology and Immunology, 5th Ed., Vol. 1, p. 43, 1964.

PACKAGING

Bacto Gram Stain Set Contains:	4 × 250 ml	3328-32-1
Bacto Gram Crystal Violet	250 ml	
Bacto Gram Decolorizer	250 ml	
Bacto Gram Iodine	250 ml	
Bacto Gram Safranin	250 ml	
Components available individually:		
Bacto Gram Crystal Violet	6 × 250 ml	3329-76-7
Bacto Gram Decolorizer	6 × 250 ml	3330-76-4
Bacto Gram Iodine	6 × 250 ml	3331-76-3
Bacto Gram Safranin	6 × 250 ml	3332-76-2
Bactrol Disks Set B	6 vials	1629-32-1
Contains 2 vials each:		
Escherichia coli	25 disks per vial	
Pseudomonas aeruginosa	25 disks per vial	
Staphylococcus aureus	25 disks per vial	
Bactrol™ Gram Slide	50 slides	3140-26-5

BACTO H BROTH

INTENDED USE
Bacto H Broth is recommended for the preparation of the "H" agglutination antigen as used in the differentiation and identification of members of the *Salmonella* group.

HISTORY/PRINCIPLES
Bacto H Broth is prepared according to the formula described by Hajna & Damon.[1] This medium has been used in the rapid method of differentiating *Enterobacteriaceae* described by Hajna.[2] The elaboration of indole is also shown on this medium. Hajna[3] reported that the combination of Bacto Peptone and Bacto Tryptone in this medium has proven to be nutritionally satisfactory for the gram-negative enteric organisms.

Hajna recommended that typical colonies be picked from primary plating media such as Bacto SS Agar Bacto Bismuth Sulfite Agar or other appropriate media and inoculated into tubes of Bacto Triple Sugar Iron Agar, Bacto Motility Test Medium and Bacto H Broth. The tubes are then incubated overnight. Cultures showing reactions suggestive of *Salmonella typhi* or *Salmonella*-like organisms are subjected to the indole test and serological determinations.

FORMULA
BACTO H BROTH
DEHYDRATED
Ingredients per liter

Bacto Beef Extract	3 g
Bacto Tryptone	5 g
Bacto Peptone	5 g
Bacto Dextrose	1 g
Dipotassium Phosphate	2.5 g
Sodium Chloride	5 g

Final pH 7.2 ± 0.2 at 25°C.

One pound will make 21.1 liters of medium.

METHOD OF PREPARATION
1. To rehydrate, dissolve 21.5 g in 1 liter distilled or deionized water.
2. Distribute in 4 ml amounts in 13 × 100 mm tubes.
3. Sterilize in the autoclave for 15 minutes at 10 lbs pressure (116°C).

STORAGE
Bacto H Broth	Below 30°C
Prepared tubes	15 – 30°C

QUALITY CONTROL
Identity Specifications

Dehydrated powder:	light tan, homogeneous, free-flowing
Reaction of 2.15% solution:	pH 7.2 ± 0.2 at 25°C
Prepared medium:	medium amber, clear

Typical Cultural Response in Bacto H Broth
After 18 – 24 Hours at 35°C

Organism	Growth
Salmonella enteritidis ATCC® 13076	good to excellent
Salmonella typhi ATCC® 19430	good to excellent
Salmonella typhimurium ATCC® 14028	good to excellent

REFERENCES
1. Pub. Hlth. Rep., 65:116, 1950.
2. Pub. Hlth. Lab., 9:23, 1951.
3. Personal Communication, 1951.

PACKAGING

Bacto H Broth   500 g   0451-17-5

HATTS

HEMAGGLUTINATION — TREPONEMAL TEST FOR SYPHILIS FOR THE SEROLOGICAL DETECTION OF SYPHILIS ANTIBODY

INTENDED USE
Bacto HATTS Kit (Hemagglutination-Treponemal Test for Syphilis) is for use as a confirmation test for qualitative detection of antibodies to *Treponema pallidum*.

HISTORY
The hemagglutination test for syphilis was first described by Rathlev[1,2] and later modified by Tomizawa and Kasamatsu.[3] A number of investigators have compared hemagglutination with fluorescent antibody techniques as confirmation tests to be used as aids in the diagnosis of syphilis and have found acceptable correlation between these methods. These investigations have been reviewed by Shore[4] and Ravel.[5]

Results of Wentworth et. al. in a three-laboratory study tend to confirm the conclusions drawn by other investigators that HATTS compares favorably with the FTA - ABS test in sensitivity and specificity as a confirmatory test.[6] The most notable finding of a study by Peter et. al. is that the HATTS test gives less than half as many false-positive reactions as the FTA - ABS test.[7] Larsen has reported that hemagglutination tests offer an advantage to those laboratories not equipped for FTA - ABS testing or that have numerous specimens to run in a short period of time.[8] The hemagglutination test for syphilis combines the sensitivity and specificity of a treponemal test with the economy and simplicity of nontreponemal tests.

PRINCIPLES OF HATTS
Test sera are diluted with an absorbing diluent that blocks nonspecific treponemal antibodies to eliminate false-positive results. The absorbed sera are combined with both *T. pallidum* (Nichols strain) sensitized erythrocytes (TPSE) and control unsensitized erythrocytes (CE) utilizing separate microtitration wells for each. Syphilitic antibody, if present in the serum, will agglutinate the TPSE, forming a smooth mat of cells in the well of the microtitration tray and will not agglutinate the CE, allowing them to settle in the other well to form a button of cells. The degree of agglutination of TPSE produced by syphilitic serum will vary with the concentration of syphilitic antibody present in the serum.

REAGENTS
Bacto HATTS Treponema Pallidum Sensitized Erythrocytes (TPSE) are lyophilized, stabilized turkey erythrocytes sensitized with *T. pallidum* antigen. When rehydrated according to directions, a 2.5% cell suspension is obtained. Lyophilized TPSE are stable through the expiry date on the label when stored at 2 – 8°C. TPSE working dilutions are stable for 5 days when stored at 2 – 8°C.

Bacto HATTS Control Erythrocytes (CE) are lyophilized, stabilized, unsensitized turkey erythrocytes from the same batch of cells as TPSE. When rehydrated according to directions, a 2.5% cell suspension will be obtained. Lyophilized CE are stable through the expiry date on the label when stored at $2 - 8°C$. CE working dilutions are stable for 5 days when stored at $2 - 8°C$.

Bacto HATTS Test Diluent is a 0.24% sodium chloride solution containing 1.54% non-syphilitic human serum, 1.6% FTA Sorbent and 0.24% sodium azide. Bacto HATTS Test Diluent is stable through the expiry date on the label when stored at $2 - 8°C$.

Bacto HATTS Test Diluent Pink for serum dilutions is a 0.24% sodium chloride solution containing 1.54% nonsyphilitic human serum, 1.6% FTA Sorbent, 0.24% sodium azide and 0.00139% FD&C Red #40. Bacto HATTS Test Diluent Pink is stable through the expiry date on the label when stored at $2 - 8°C$.

Bacto HATTS Reactive Control Serum (RCS) is a lyophilized syphilitic rabbit serum containing 0.1% sodium azide. The end-point titer is stated on the label. The lyophilized RCS is stable through the expiry date on the label when stored at $2 - 8°C$. Rehydrated RCS is stable for 3 months when stored at $2 - 8°C$.

Bacto HATTS Non-Reactive Control Serum (NCS) is a lyophilized normal human serum containing 0.1% sodium azide. The lyophilized NCS is stable through the expiry date on the label when stored at $2 - 8°C$. Rehydrated NCS is stable for 3 months when stored at $2 - 8°C$.

Bacto HATTS Rehydrating Water contains 0.1% sodium azide. Bacto HATTS Rehydrating Water is stable through the expiry date on the label when stored at $2 - 8°C$.

PRECAUTIONS AND WARNINGS
1. Use Linbro/Titertek™ Microtiter plate No. 76 - 311 - 05 or other microtitration U-bottom tray that has been determined in comparison tests to yield comparable HATTS tests results.
2. All equipment, including microtitration trays, must be clean and free of dust. Do not reuse microtitration trays. If extraneous material is present, cell autoagglutination may occur.
3. Do not pipette any reagent by mouth. Human sera may contain infectious agents, particularly Hepatitis B surface antigen. Human sera used in HATTS reagents have been tested and found negative for Hepatitis B_s Antigen by the RIA test method.
4. Grossly contaminated or hemolyzed sera should not be tested.
5. Bacto HATTS Test Diluent and Bacto HATTS Test Diluent Pink may acquire a fine precipitate which does not interfere with test results. Do not use if it becomes contaminated.
6. Do not use microdilutors for assay, since they may scratch the tray wells causing the cells to settle in a rough pattern which may be difficult to read.
7. The test sera, reagents and room should be at $26 \pm 3°C$ at the time of testing. Temperatures outside this range may influence the test results.
8. Do not contaminate reagents.
9. Do not interchange reagents from different kits having different control numbers.

MATERIALS PROVIDED
Bacto HATTS Treponema Pallidum Sensitized Erythrocytes (TPSE).
Bacto HATTS Control Erythrocytes (CE)
Bacto HATTS Test Diluent
Bacto HATTS Test Diluent Pink

Bacto HATTS Reactive Control Serum (RCS)
Bacto HATTS Non-Reactive Control Serum (NCS)
Bacto HATTS Rehydrating Water

MATERIALS REQUIRED BUT NOT PROVIDED

Microtitration U-bottom trays that have been tested by user and found to yield satisfactory results
Precision pipettes, pipette droppers or similar equipment to deliver the following: 0.020 ml, 0.025 ml, 0.050 ml and 0.10 ml
Disposable Pasteur pipettes
Microtitration test reading mirror
Marking pen
Test tubes, clear plastic or glass (12 × 75 mm)
Test tube rack of appropriate size
Disposable pipette tips
(Optional) Automated dilutor for preparation of serum dilutions in test tubes

SPECIMEN COLLECTION — TEST SERA

1. Venous blood should be collected with aseptic technique.
2. If syringe is used, remove needle from syringe and slowly expel blood into labeled tube without anticoagulant.
3. Allow the blood to clot.
4. Loosen clot with applicator stick and centrifuge for 5 minutes at approximately 2000 × g to separate serum.
5. Remove serum and store at 2 – 8°C until tested.
   Note: Untested sera may be stored at 2 – 8°C for not more than 2 days or frozen at −20°C in tightly stoppered tubes.

PREPARATION OF REAGENTS AND TEST SERA

All lyophilized reagents must be rehydrated with Bacto HATTS Rehydrating Water.

After opening TPSE, CE, Test Diluent, Test Diluent Pink and Rehydrating Water, discard the metal seal and use the rubber stopper as a cap in subsequent procedures.
1. Tap TPSE and CE vials on lab bench to force contents to bottom of vial. Remove seal and rubber stopper.
2. Rehydrate TPSE and CE by adding Bacto HATTS Rehydrating Water according to the table below:

To:	Add:	100 – 130 Test Kit	300 – 350 Test Kit
TPSE	Rehydrating Water	0.6 ml	2 ml
CE	Rehydrating Water	0.6 ml	2 ml

Cap vial with rubber stopper and swirl gently to completely rehydrate cells. Allow rehydrated vials to stand undisturbed at 26 ± 3°C for 30 minutes and then prepare working dilutions within 1 hour.
Caution: Do not use TPSE and/or CE if rough visible granules appear after rehydration.
3. Bacto HATTS Test Diluent is ready to use. Allow diluent to reach room temperature and invert or swirl vial several times prior to use.
4. Prepare working dilutions of TPSE and CE by adding Bacto HATTS Test Diluent to contents of the above rehydrated vials according to the table below.
   Note: Be sure that the TPSE and CE working dilutions are prepared with the HATTS Test Diluent, not HATTS Test Diluent Pink.

To:	Add:	100 – 130 Test Kit	300 – 350 Test Kit
TPSE	Test Diluent	3.0 ml	10 ml
CE	Test Diluent	3.0 ml	10 ml

Stopper and invert each vial several times to mix.

Allow working dilutions to stand undisturbed in individual vials for 30 minutes at 26 ± 3°C prior to use. If more than 1 vial of TPSE or CE are prepared as above, the contents of TPSE vials should be pooled and the contents of CE vials should be pooled. If working dilutions are not to be used at this point, store them at 2 – 8°C.

Note: Prior to use, GENTLY mix cell working dilutions with a disposable Pasteur pipette, or by swirling or by stoppering and inverting to ensure total cell suspension.

5. Rehydrate Bacto HATTS Reactive Control Serum (RCS) and Bacto HATTS Non-Reactive Control Serum (NCS) vials by adding 1 ml Bacto HATTS Rehydrating Water per vial. Replace rubber stopper and invert several times to completely rehydrate sera. Discard stopper and cap vial with screw cap. If it is to be used on multiple test days, divide control sera into 0.3 ml aliquots in sterile tubes, stopper tightly and store at 2 – 8°C. Rehydrated sera are stable for 3 months when stored at 2 – 8°C. Mix well before use.

6. Prepare control and test sera dilutions as follows:
   A. Inactivate control and test sera at 56°C for 30 minutes prior to testing. Reheat previously heated sera for 10 minutes at 56°C on the day of testing.
   B. Label 1 test tube (12 × 75 mm) for each serum to be tested and for each of the 2 control sera. Place all tubes in a test tube rack of appropriate size.
   Note: Steps C and D may be performed with an automatic dilutor.
   C. Add 0.3 ml Bacto HATTS Test Diluent Pink to each tube.
   D. Pipette 20 µl (0.02 ml) of test serum or control serum to the respectively labeled test tube to give a 1:16 dilution of serum in Bacto HATTS Test Diluent Pink. Mix well 8 – 10 times using the same pipette tip as was used to add serum to the pink diluent.
   Note: By using the HATTS Test Diluent Pink, the location of the wells in the microtitration trays containing reagent versus wells not containing reagent can clearly be identified. This simplifies setting up HATTS and helps eliminate errors such as missing or duplicating wells. Since only 25 µl of colored 1:16 diluted serum is utilized by the HATTS test, the color nearly disappears when 100 µl of cell working dilution is added. The final cell reading and interpretation is therefore not affected. Serum dilutions should be prepared on the same day they are tested.

HATTS PROCEDURE FOR REAGENT CONTROLS AND TEST SERA
Note: Reagent controls are to be set up concurrently with test sera and reactions read at the same time interval.

Steps 2 through 4 must be included on FIRST MICROTITRATION TRAY ONLY per assay.
1. Perform HATTS at 26 ± 3°C.
2. Pipette 25 µl (0.025 ml) of diluted 1:16 RCS (Step 6D — PREPARATION OF REAGENTS AND TEST SERA) to the wells labeled Row A, No. 1, No. 2 and Row B, No. 1, and 25 µl of diluted 1:16 NCS to the wells labeled Row A, No. 9 and Row B, No. 2 as indicated by REAGENT CONTROL DIAGRAM.

REAGENT CONTROL DIAGRAM

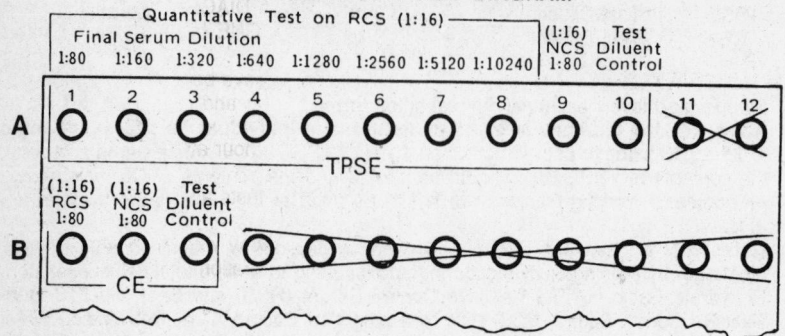

3. Using a pipette or pipette dropper, add 25 µl of Bacto HATTS Test Diluent Pink to the wells labeled Row A, No. 2 through No. 8 and No. 10 and to the well labeled Row B, No. 3 for Test Diluent Control as shown in REAGENT CONTROL DIAGRAM.
4. Mix the contents of the second well Row A containing 25 µl of RCS and 25 µl of Bacto HATTS Test Diluent Pink 8 – 10 times using a 25 µl pipette. Using the same pipette and tip, transfer 25 µl from second well into third well. Mix 8 – 10 times then transfer 25 µl into fourth well and mix. Continue transferring and mixing until eighth well contains 50 µl. Mix eighth well, then discard 25 µl.
5. Pipette 25 µl of diluted 1:16 test serum (Step 6D — PREPARATION OF REAGENTS AND TEST SERA) to each of 2 wells per serum tested as indicated in TEST SERA DIAGRAM.

TEST SERA DIAGRAM
Test Sera (Final serum dilution is 1:80)

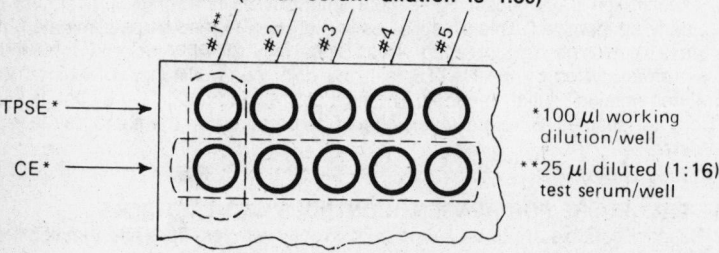

TPSE* ⟶

CE* ⟶

*100 µl working dilution/well

**25 µl diluted (1:16) test serum/well

6. GENTLY mix CE working dilution with a disposable Pasteur pipette, or by swirling or by stoppering and inverting to ensure total cell suspension. Using 100 µl pipette or 50 µl microtitration dropper, add 100 µl (0.1 ml) of the working dilution of CE (Step 3 — PREPARATION OF REAGENTS AND TEST SERA) to the designated wells containing RCS, NCS, Test Diluent Pink and/or test sera as indicated by the REAGENT CONTROL and TEST SERA DIAGRAMS.
7. Mix TPSE working dilution as above to ensure total cell suspension. Using a 100 µl pipette or 50 µl microtitration dropper, add 100 µl of the working dilution of TPSE (Step 3 — PREPARATION OF REAGENTS AND TEST SERA) to the des-

ignated wells containing RCS, NCS, Test Diluent Pink and/or test sera as indicated by the REAGENT CONTROL and TEST SERA DIAGRAMS.

Note: Add **both** CE and TPSE working dilutions to ONE microtitration tray at a time.

8. IMMEDIATELY after CE and TPSE working dilutions have been added to one tray gently agitate or tap the tray to ensure proper mixing and cover with an empty tray.

9. Allow the microtitration tray to stand undisturbed for 1 hour at $26 \pm 3°C$ in an area free from vibration and draft.

10. Place trays on microtitration reading mirror and read tests according to HATTS — READING AND RESULTS.

Note: Microtitration trays may be read the following day with comparable results if stored covered at $26 \pm 3°C$ in an area free from vibration.

HATTS — READING AND RESULTS

1. The nonreactive settling pattern of TPSE may be slightly different from the pattern produced by the CE. With CE a clean, sharply delineated, compact button of cells in center of well will be seen, whereas, with TPSE the button may not be as compact and may not be sharply delineated.

2. Read and record the reagent control results according to Table I.

3. The NCS (1:80) and the Test Diluent Controls with both TPSE and CE and the RCS with CE (1:80 final serum dilution) should be nonreactive.

4. The RCS with TPSE should give a 1+ or greater reading at the end-point dilution stated on the RCS vial, or within one doubling dilution of the stated end-point dilution.

5. If expected results are not obtained with all controls, the test run is invalid.

6. Read and record the test sera results according to Table I.

7. Test sera must always produce a negative reading (−) with CE for a valid test.

Note: Test sera producing any reaction with CE should be retested. If, on repeated analysis, the CE still shows a reaction, the HATTS should be reported as inconclusive and the sera tested with FTA-ABS.

8. All reactions of test sera interpreted as 4+ to 1+ with TPSE are reported as **Reactive.**

9. All test sera giving a reaction interpreted as ± with TPSE should be retested due to possible errors in technique. Repeat ± reactions are reported as **Non-Reactive.**

10. All reactions of test sera interpreted as − with TPSE are reported as **Non-Reactive.**

Table I

Description of Hemagglutination Pattern	Reading	Report
Mat of cells covering entire bottom of well, sometimes with folded edges	4+	Reactive
Mat of cells covering less area of well than 4+ description, surrounded by narrow red circle	3+	Reactive
Mat of cells covering less area of well than 3+ description, surrounded by slightly more defined reddish circle	2+	Reactive
Mat of cells covering less area of well than 2+ description, surrounded by more defined ring-like red circle that appears to cast a haze-like shadow around the circle	1+	Reactive
Settling of cells having a hole in the center giving the appearance of a well defined, thick, dense, red ring with a fairly clear background around this ring	±	Non-Reactive
Button of cells with or without a small hole in center	−	Non-Reactive

LIMITATIONS OF HATTS

1. HATTS is a confirmation test to detect antibodies to *T. pallidum* as an aid in the diagnosis of syphilis. It has been reported to be less sensitive than fluorescent techniques for demonstration of antibodies in early primary syphilis.[8]
2. HATTS is not to be used as a screening test or to follow efficacy of treatment.
3. Serum from patients with systemic lupus erythematosis, autoimmune diseases, leprosy, viral infections and drug addiction may on occasion react with the sensitized and unsensitized cells.[7,9]
4. Do not use plasma since studies have not been completed.

A diagnosis of syphilis cannot be made only on the results obtained with laboratory reagents but must be correlated with the patient's medical history and other clinical findings.

SPECIFIC PERFORMANCE CHARACTERISTICS

HATTS (Hemagglutination-Treponemal Test for Syphilis) is a confirmatory test using a specific antigen to test for serum antibodies to *T. pallidum.* Wentworth et. al. in a study of 1,056 syphilis cases (373 untreated and 683 treated) reported 93.4% agreement for untreated syphilis and 93.9% agreement for treated syphilis between FTA-ABS and HATTS results for combined data from 3 field study laboratories.

Additionally, 1,805 nonsyphilitic sera including 1,048 presumed normals, 502 biologic false positives (BFP) and 255 sera from cases of diseases other than syphilis were studied. HATTS was nonreactive in 99.4% of presumed normals, 86.9% of BFP sera and 84.7% of diseases other than syphilis for an overall specificity of 93.9% as compared to FTA-ABS nonreactivity in 99.2% of presumed normals, 76.9% of BFP sera and 77.3% of other diseases for an overall specificity of 90.1%.[6]

Peter et. al., in a study of 628 nonsyphilitic individuals (299 Civil Service employees, 149 prenatal patients, 100 nonsyphilitic infectious diseases and 80 BFP cases) reported overall false-positive reactions of 10.8% for RPR card and 3.3% for FTA-ABS with an additional 4.0% borderline as compared to only 1.6% for HATTS.[7]

REFERENCES

1. Rathlev, T., Hemagglutination Tests Utilizing Antigens from Pathogenic and Apathogenic *Treponema pallidum.* World Health Org. Geneva VDT/RES/77:65, 1965.
2. Rathlev, T., Hemagglutination Test Utilizing Pathogenic *Treponema pallidum* for the Serodiagnosis of Syphilis. Br. J. Ven. Dis., 43:181 – 185, 1967.
3. Tomizawa, T., Kasamatsu, S., Hemagglutination Tests for Diagnosis of Syphilis. A Preliminary Report. Jpn. J. Med. Sci. Biol., 19:305 – 308, 1966.
4. Shore, R. N., Hemagglutination Tests and Related Advances in Serodiagnosis of Syphilis. Arch. Dermatol., 109:854 – 857, 1974.
5. Ravel, R., Hemagglutination Test for Syphilis (MHA) as Alternative to the FTA-ABS. Lab. Med., 7:22 – 34, 1976.
6. Wentworth, B. B., Thompson, M. A., Peter, C. R., Bawdon, R. E., Wilson, D. L., Comparative Study of a Hemagglutination Treponemal Test for Syphilis (HATTS) with Other Serologic Methods for the Diagnosis of Syphilis. Sexually Transmitted Diseases, 103 – 111, July – Sept. 1978.
7. Peter, C. R., Thompson, M. A., Wilson, D. L., False-Positive Reactions in the Rapid Plasma Reagin-Card, Fluorescent Treponemal Antibody-Absorbed, and Hemagglutination Treponemal Syphilis Serology Tests. J. Clin. Micro., 9:369 – 372, March 1979.
8. Larsen, S. A., Syphilis Serology: 40 million Assays Annually Spark Research for Newer Diagnostic Tests. Lab World, 32 (August): 53 – 57, 1981.
9. Larsen, S. A., Hambie, E. A., Pettit, D. E., Perryman, M. W., Kraus, S. J., Specificity, Sensitivity and Reproducibility Among the Fluorescent Treponemal Antibody-Absorption Test, the Microhemagglutination Assay for *Treponema pallidum* Antibodies, and the Hemagglutination Treponemal Test for Syphilis. J. Clin. Micro., 14:441 – 445, Oct. 1981.

PACKAGING

Bacto HATTS Kit	100 – 130 tests	3425-32-3
Contains:		
Bacto HATTS Treponema Pallidum Sensitized		
Erythrocytes (TPSE)	4 × 0.6 ml	
Bacto HATTS Control Erythrocytes (CE)	4 × 0.6 ml	
Bacto HATTS Test Diluent	1 × 40 ml	
Bacto HATTS Test Diluent Pink	1 × 40 ml	
Bacto HATTS Reactive Control Serum (RCS)	1 × 1 ml	
Bacto HATTS Non-Reactive Control Serum (NCS)	1 × 1 ml	
Bacto HATTS Rehydrating Water	1 × 10 ml	
Bacto HATTS Kit	300 – 350 tests	3400-32-2
Contains:		
Bacto HATTS Treponema Pallidum Sensitized		
Erythrocytes (TPSE)	3 × 2 ml	
Bacto HATTS Control Erythrocytes (CE)	3 × 2 ml	
Bacto HATTS Test Diluent	1 × 100 ml	
Bacto HATTS Test Diluent Pink	1 × 100 ml	
Bacto HATTS Reactive Control Serum (RCS)	1 × 1 ml	
Bacto HATTS Non-Reactive Control Serum (NCS)	1 × 1 ml	
Bacto HATTS Rehydrating Water	1 × 30	

Samples of each lot are tested by the Centers for Disease Control and found to meet CDC specifications.

BACTO H$_2$S TEST STRIPS

INTENDED USE

Bacto H$_2$S Test Strips are used in culture tubes to detect hydrogen sulfide production by bacteria. They may be used in tubes of any good nutrient medium standardized for use in demonstrating hydrogen sulfide production.

PRODECURE

A Bacto H$_2$S Test Strip is aseptically removed from the vial containing the strips by using sterile forceps and inserted into a freshly inoculated tube of culture medium. Fold top of the strip over the lip of the tube and hold in place with the cotton plug or screw cap so that the end of the strip will be about 10 mm above the medium surface. Do not allow strip to come in contact with the medium as it is toxic to cultures. Incubate culture tubes at 37°C for 16 – 24 hours. A positive reaction for hydrogen sulfide will be indicated by the formation of a brownish black color, often with sheen on the H$_2$S test strip.

STORAGE

Bacto H$_2$S Test Strips   15 – 30°C

QUALITY CONTROL
Identity Specifications

Appearance: white paper strips, 1/4″ wide × 2.9″ long ± 1/32″

Typical Cultural Response on Bacto H₂S Test Strips Used as Above with
2% Proteose Peptone No. 3, Difco After 16 – 24 Hours at 35°C

Organism	Growth	H₂S
Klebsiella pneumoniae ATCC® 13883	good to excellent	–
Salmonella typhimurium ATCC® 14028	good to excellent	+

+ = positive, brownish black color on strip
– = negative, white strip

PACKAGING

Bacto H₂S Test Strips	1 × 25 strips	1626-30-6
Proteose Peptone No. 3, Difco	1 lb (454 g)	0122-01-2
	10 kg	0122-08-5

BACTO HAEMOPHILUS INFLUENZAE ANTISERA

INTENDED USE

Bacto Haemophilus Influenzae Antisera are prepared for use in the serological detection and typing of *Haemophilus influenzae* cultures employing the slide agglutination technique and/or the microscopic Quellung reaction. The polyvalent antiserum is used to screen *H. influenzae* cultures, while the monovalent antisera are used to serotype the organism.

SUMMARY AND EXPLANATION OF THE TEST

Pfeiffer[1] was first credited with the isolation of *H. influenzae* from patients during the influenza pandemic of 1890. Since this organism occurred frequently in specimens obtained from persons with the clinical disease, and since it was not, at that time, found in normal individuals, it was incriminated erroneously as the etiological agent of viral influenza.

Since that time, *H. influenzae* has been found to be in the common nasopharyngeal flora of numerous normal healthy individuals. It was also found to play a decided role as an opportunistic secondary invader — usually following viral influenza. As a primary infectious agent (especially in children 3 years of age or younger) *H. influenzae* has been incriminated in diseases such as acute epiglottis, acute laryngotracheal obstructions, otitis media, pharyngitis, pneumonia, septic arthritis, sinusitis and subacute bacterial endocarditis. Serotype b is believed to cause over 90% of all *Haemophilus* infections in children.

This species of organisms may be responsible for essentially the same disease in adults, but generally in adults suffering from debilitating diseases such as immune deficiencies, alcoholism, etc.

Positive serological results, combined with a biochemically defined culture of *H. influenzae* confirms the serotype present in a specimen. Too heavy emphasis on serology without adequate biochemicals may be misleading as demonstrated in the following examples:
1. *H. influenzae* type a cross reacts with *Streptococcus pneumoniae* type 6.
2. *H. influenzae* type b crosses with *S. pneumoniae* types 6, 15A, 29 and 35B.
3. *H. influenzae* type c crosses with *S. pneumoniae* type 11.

4. The ability of this species of organisms to occur in acapsular form (non-type specific strains).
5. They are also antigenically related to staphylococci, streptococci, *Escherichia coli* and other organisms.[2]

HAEMOPHILUS INFLUENZAE ANTIGENIC ANALYSIS

Currently there are 6 accepted serotypes, designated as types a, b, c, d, e, and f. As indicated before, the most prevalent serotype has been type b with possibly type a secondary in importance.

The principle of the serologic typing of *H. influenzae* involves the intimate mixing of a portion of a bacteriological loopful of the biochemically characterized culture (the antigen) with a droplet of the hyperimmune serum (antibody). If the serum contains agglutinins for the antigen present on the organism, a rapid clumping (agglutination) of at least 3+ reaction will occur. Such a reaction in an individual serum identifies the serotype of the organism.

Some laboratories however, prefer the capsular Quellung (swelling) reaction for serotyping *H. influenzae*. The principle of this antigen-antibody reaction is not agglutination as in the slide technique but rather an apparent increase in capsular size due to deposition of antibody on the cell surface.

If the Quellung reaction is performed, one must be aware that these organisms are often found in the acapsulated state and that capsulated strains of type e generally possess small capsules. Such strains should be defined serologically employing the slide agglutination test only.

REAGENTS

Bacto Haemophilus Influenzae Antisera are stable, desiccated, absorbed (when necessary) polyvalent or type-specific antisera prepared in rabbits and designed for use in the slide agglutination and Quellung techniques.

When rehydrated, the protein content of Bacto Haemophilus Influenzae Antisera approximates that of a normally hyperimmune animal serum.

When properly rehydrated, each antiserum contains approximately 1:10,000 Merthiolate®, sufficient to maintain a bacteriostatic condition at storage of 2 – 8°C.

When used according to the recommended procedure each 1 ml vial is sufficient to perform approximately 20 slide agglutination tests or 50 Quellung reactions.

REHYDRATION

To rehydrate Bacto Haemophilus Influenzae Antisera, add 1 ml distilled or deionized water and rotate gently to dissolve completely.

Discard any serum which is cloudy and/or has a precipitate unless it has been clarified by centrifugation and shown to react properly with known control cultures.

STORAGE

Store both desiccated and rehydrated antisera at 2 – 8°C.

Prolonged exposure to room temperature or repeated freezing and thawing are detrimental to antisera.

EXPIRATION DATE

These reagents are stable to the expiry date on the label when stored at 2 – 8°C in the desiccated state. Upon rehydration it is recommended that the rehydration date be written on the label since Bacto Haemophilus Influenzae Antisera have an expiry date of 1 year after rehydration when stored under proper conditions. The 1 year expiry date after rehydration does not exceed the expiry date stated on the label.

Discard any antiserum which becomes cloudy during storage.

SPECIMEN COLLECTION

Specimens submitted to the laboratory for *H. influenzae* must be obtained under proper medical guidance and include examination of cerebrospinal fluid, blood samples, throat swabs, sputum and skin scrapings. The proper collection together with the prompt processing of the specimen cannot be over-emphasized. Care must be exercised to keep the specimens moist since the organisms are susceptible to drying. Immediate processing of the specimens upon receipt is urged but if it is stored briefly it is recommended to allow it to stand at room temperature rather than at 2 – 8°C.

The isolation media recommended for this species of organisms are Bacto Chocolate Agar, enriched or Levinthal agar.[2] The latter is preferred for serological studies.

SPECIMEN PREPARATION
Fluid Specimens
1. Centrifuge specimens for 15 minutes at 2,500 rpm immediately upon receipt.
2. Aspirate off supernatant with a sterile capillary pipette.
3. Use remaining fluid and sediment to inoculate appropriate media.

Wire or Swab Specimens
Streak onto surface of appropriate media. Incubate at 37°C for 20 – 24 hours preferably in a candle jar (5 – 10% carbon dioxide).

Some of the species of the genus *Haemophilus* may be differentiated on their differences in hemolysis and their growth requirements for factors X (hemin) and V (diphosphopyridinenucleotide). For more detailed information on determining the X and V requirements of *Haemophilus* refer to Bacto Differentiation Disks BV, BX and BVX.

GENERAL INSTRUCTIONS
1. Serological identification of *H. influenzae* may be performed by the slide agglutination technique. However, this technique is not suitable for direct examination of spinal fluid specimens, therefore, specimens should be prepared as stated under "Specimen Preparation-Wire or Swab Specimens."
2. These antisera have been prepared for use to identify cultures which have been defined biochemically. If variations in the recommended procedures are to be used, the investigator is advised to test each lot of antiserum with known positive control cultures to insure its proper homologous and heterologous reactions under their test conditions.
3. Exposure of the organisms or plate to heat from external sources (a hot bacteriological loop, burner flame, light source, etc.) may result in a culture which cannot be suspended readily or evaporation and/or precipitation of the test mixture all of which may result in false positive reactions.
4. In both tests all materials and equipment must be at room temperature at the time of test performance.
5. In both tests, all equipment and materials (glass plates, loop, culture) must be sterilized after use by flame (loop) or autoclave (glass plate and culture) or by an adequate chemical method since living organisms are used as the antigen.

6. Encapsulated type e strains possess at best small capsules and therefore may not react in the Quellung reaction. This type also results in much finer agglutination in the slide agglutination technique when compared to the agglutination of the other types.

PROCEDURE
Materials Provided:
Bacto Haemophilus Influenzae Antiserum Poly
Bacto Haemophilus Influenzae Antisera Types a – f
Droppers

Materials Required but not Provided:
Normal Rabbit Serum	Microscope slides
0.85% NaCl solution	Cover glasses
Bacteriological loop	Bacto Loeffler Methylene Blue
Burner	Bacto Capsule Ink
Glass plate ruled into 1″ squares	Immersion oil
Fluorescent desk lamp	Microscope
Autoclave	

SLIDE AGGLUTINATION TEST
1. Mark off a glass plate into 1″ square sections with a wax pencil.
2. Place a small drop of the appropriate antiserum on the ruled section of the plate using the supplied dropper.
3. To the square next to the one containing the antiserum, place one drop of 0.85% NaCl solution or a drop of normal rabbit serum. This will serve as a negative control of the bacterial suspension. Often stock cultures and, sometimes, isolated cultures are rough and will agglutinate spontaneously in a normal serum. Therefore, it is necessary to select smooth colonies for serological testing. It is recommended that more than one colony be tested from both assay and control cultures.
4. Transfer a portion of a loopful of growth of the isolate to the section containing the NaCl solution, or normal rabbit serum and suspend thoroughly.
5. Transfer a portion of a loopful of growth to the square containing the antiserum and suspend thoroughly.
6. Rock the slide for 1 minute and avoid excessive evaporation.
7. Record agglutination as follows:

$$++++ \text{ agglutination} = \text{all the cells agglutinate}$$
$$+++ \text{ agglutination} = 75\% \text{ of the cells agglutinate}$$
$$++ \text{ agglutination} = 50\% \text{ of the cells agglutinate}$$
$$+ \text{ agglutination} = 25\% \text{ of the cells agglutinate}$$
$$\pm \text{ agglutination} = <25\% \text{ of the cells agglutinate}$$
$$- \text{ agglutination} = \text{none of the cells agglutinate}$$

QUALITY CONTROL
In *H. influenzae* serology, as in any serological test, known positive and negative control cultures should be employed to ascertain validity of test results.

RESULTS
A +++ reaction or greater should be considered as the end point at the RTD.

Positive agglutination in the polyvalent antiserum and in one of the individual typing sera combined with adequate biochemical results, confirm the serotype of *H. influenzae* present.

QUELLUNG REACTION

Note: Some strains of *H. influenzae* are acapsulated and therefore unsatisfactory for use in the Quellung reaction.

1. Prepare a suspension of the test organism by diluting growth form an overnight culture on Levinthal agar incubated at 37°C in 0.85% NaCl solution to approximate the density of 1:100 of a Bacto McFarland Barium Sulfate Standard No. 3. This density will give a suspension containing roughly 25 – 50 bacterial cells per 1000X microscopic field.

   Care should be taken not to employ too dense a bacterial suspension for the Quellung reaction. If the number of organisms is much greater than this and they appear clumped without apparent positive Quellung reaction, prepare another slide using fewer organisms before judging the slide to be negative.

2. Place a loopful of a suspension of the culture to be tested on each end of a microscope slide.

3. If a spinal fluid specimen is used, use the undiluted sediment described under Specimen Collection above.

4. To the first loopful of the culture on the slide add a loopful of the Bacto Haemophilus Influenzae Antiserum Poly and a loopful of Bacto Loeffler Methylene Blue stain or Bacto Capsule Ink. Mix thoroughly and cover with a cover glass.

5. To the second loopful of the culture add a loopful of Bacto Rabbit Serum Normal and a loopful of Bacto Loeffler Methylene Blue stain or Bacto Capsule Ink. Mix thoroughly and cover with a cover glass. This mixture serves as a control for capsule size.

6. Examine microscopically using an oil immersion lens. The reaction should occur within 3 – 5 minutes but the specimen should not be regarded as negative until it has been examined 30 minutes after its preparation.

7. The control is observed first to establish the normal size of the capsule before the mixture containing the immune serum is examined.

8. Examine several microscopic fields since, in some cases, all the cells may not exhibit capsular swelling.

QUALITY CONTROL

Known positive and negative control cultures should be employed to ascertain validity of test results.

RESULTS

A positive reaction is indicated by a definite increase in capsule size.

A positive result obtained with the Quellung reaction may be considered presumptive evidence of the presence of one of the *H. influenzae* types. The test should be repeated with Bacto Haemophilus Influenzae Antiserum Types a – f beginning with types a and b.

LIMITATIONS OF THE PROCEDURE

While serological procedures, as applied to microorganisms provide corroborative evidence and can be used to identify particular antigenic sites of genus or species under study, they cannot be used alone to identify the etiological agent of disease. Cultural isolation and at least preliminary biochemical differentiation must preceed any serological examination and final identification cannot be made without biochemical and serological characterization.

It must be remembered that *Haemophilus influenzae* antisera cross reacts with some strains of *Streptococcus pneumoniae*, *Escherichia coli,* staphylococci and streptococci.

However, the possibility of the occurence of such cross reactions will be eliminated when a Gram stain is performed on the specimen and adequate biochemical testing is done prior to or concurrently with serological identification.

REFERENCES

1. Ztschr. F. Hyg., 13:257 – 386, 1893.
2. Lennette, E. H., Balows, A., Hausler, W. J. Jr., Truant, J. P., Manual of Clinical Microbiology, 3rd Ed., Washington, D.C.: ASM, p. 330 – 336, 1980.

PACKAGING

Bacto Haemophilus Influenzae Antiserum Type a	1 ml	2250-50-2
Bacto Haemophilus Influenzae Antiserum Type b	1 ml	2236-50-1
Bacto Haemophilus Influenzae Antiserum Type c	1 ml	2789-50-2
Bacto Haemophilus Influenzae Antiserum Type d	1 ml	2790-50-9
Bacto Haemophilus Influenzae Antiserum Type e	1 ml	2791-50-8
Bacto Haemophilus Influenzae Antiserum Type f	1 ml	2792-50-7
Bacto Haemophilus Influenzae Antiserum Poly	1 ml	2237-50-0
Bacto Haemophilus Influenzae Antisera Set	7 × 1 ml	2793-32-9
Contains 1 vial each of Bacto Haemophilus		
Influenzae Antiserum Types a – f and Poly		
Bacto Loeffler Methylene Blue	5 ml	3111-56-3
Bacto Capsule Ink	6 × 1 ml	3335-51-8
Bacto Rabbit Serum Normal	1 ml	2423-50-4

BACTO HEART INFUSION BROTH
BACTO HEART INFUSION AGAR

INTENDED USE

Bacto Heart Infusion media are infusion media used for isolating and cultivating a wide variety of fastidious microorganisms. Bacto Heart Infusion Agar can be used as a base for the preparation of blood agar in determining hemolytic reactions.

HISTORY/PRINCIPLES

Bouillon, or a liquid medium containing an infusion of meat, was one of the first media used for cultivation of bacteria. Many modifications of this medium have been used from time to time for a wide variety of purposes. Huntoon[1] using fresh beef heart and Bacto Peptone, prepared a "hormone" broth in a special manner, to retain the growth-promoting substances. These "growth" accessory factors were described by Lloyd[2] and Cole and Lloyd.[3]

Huntoon was one among the many to show that highly pathogenic organisms, such as the meningococcus and pneumococcus, could be grown on an infusion medium without enrichment. Bacto Tryptose, as employed in this formula, is better suited to the nutritional requirements of pathogenic bacteria than is Bacto Peptone which was used by Huntoon in the preparation of his "hormone" agar.

Bacto Heart Infusion Broth, a liquid medium prepared from an infusion of beef heart, may be modified in many ways. The addition of dextrose, blood or other ingredients results in an unlimited number of media used for a variety of purposes. By the addition of 0.1 – 0.2% Bacto Agar, the cultural value of Bacto Heart Infusion Broth can be greatly increased.

Bacto Heart Infusion Agar is a satisfactory medium for mass culture of organisms, making it valuable in the preparation of vaccines. As early as 1939,[4] this medium was used for determining the presence and bacterial count of hemolytic streptococci in milk.

FORMULAE

BACTO HEART INFUSION AGAR
DEHYDRATED

Ingredients per liter

Beef Heart, Infusion from . . . . 500 g
Bacto Tryptose . . . . 10 g
Sodium Chloride . . . . 5 g
Bacto Agar . . . . 15 g

Final pH 7.4 ± 0.2 at 25°C.

One pound will make 11.35 liters of single strength medium. Rehydrate with 40 grams/liter.

BACTO HEART INFUSION BROTH
DEHYDRATED

Ingredients per liter

Beef Heart, Infusion from . . . . 500 g
Bacto Tryptose . . . . 10 g
Sodium Chloride . . . . 5 g

Final pH 7.4 ± 0.2 at 25°C.

One pound will make 18.1 liters of single strength medium. Rehydrate with 25 grams/liter.

METHOD OF PREPARATION

1. Suspend appropriate amount in 1 liter distilled or deionized water. Heat the agar medium to boiling to dissolve completely.
2. If broth tubes are being prepared, dispense into appropriate containers.
3. Sterilize in the autoclave for 15 minutes at 15 lbs pressure (121°C).
4. If a blood medium is desired, cool to 45 – 50°C and aseptically add 5% sterile, defibrinated, room temperature blood, volume/volume. Mix well.
5. Dispense into Petri dishes or tubes, as desired.

STORAGE

Bacto Heart Infusion media   Below 30°C
Prepared media   2 – 8°C

QUALITY CONTROL

Identity Specifications

	Bacto Heart Infusion Broth	Bacto Heart Infusion Agar
Dehydrated powder:	beige, homogeneous, free-flowing	beige, free-flowing homogeneous
Reaction of solution:	2.5%, pH 7.4 ± 0.2 at 25°C	4%, pH 7.4 ± 0.2 at 25°C
Prepared medium:	light to medium amber, clear	medium amber, slightly opalescent
Plates prepared with 5% sheep blood:		cherry red, opaque, firmly solid

Typical Cultural Response After 18 – 48 Hours at 35°C

Organism	Heart Infusion Broth	Heart Infusion Agar	Heart Infusion Agar w/5% Sheep Blood	Hemolysis
Escherichia coli ATCC® 25922	good to excellent	excellent	excellent	beta
Neisseria meningitidis ATCC® 13090	good to excellent	excellent	excellent	none
Streptococcus pneumoniae ATCC® 6303	good to excellent	good to excellent	excellent	alpha
Streptococcus pyogenes ATCC® 19615	good to excellent	good to excellent	excellent	beta

REFERENCES

1. J. Inf. Dis., 23:169, 1918.
2. J. Path. and Bact., 21 (Part 1): 113, 1916.
3. J. Path. and Bact., 21 (Part 2): 267, 1917.
4. Diagnostic Procedures and Reagents, 3rd Ed., 13: 1950.

PACKAGING

Bacto Heart Infusion Agar	1 lb (454 g)	0044-01-7
	1/4 lb (114 g)	0044-02-6
	5 lb (2.27 kg)	0044-05-3
	10 kg	0044-08-0
Bacto Heart Infusion Broth	1 lb (454 g)	0038-01-5
	1/4 lb (114 g)	0038-02-4
	5 lb (2.27 kg)	0038-05-1

BACTO HEKTOEN ENTERIC AGAR

INTENDED USE

Bacto Hektoen Enteric Agar is a differential, selective medium for the isolation and differentiation of *Salmonella* and *Shigella* from gram-negative enteric pathogens.

HISTORY/PRINCIPLES

Bacto Hektoen Enteric Agar was developed by King and Metzger.[1,2] By increasing the carbohydrate and peptone content to counteract the inhibitory effects of the bile salts and indicators, they were able to formulate a medium that would only slightly inhibit the growth of *Salmonella* and *Shigella* yet still inhibit accompanying microorganisms.

The lactose-positive colonies are differentiated from lactose-negative colonies due to the presence of the two indicators, brom thymol blue and acid fuchsin. The additional carbohydrates sucrose and salicin, which are fermented more easily than lactose, also serve to further differentiate organisms. The combination of thiosulfate with the ferric ammonium citrate causes the H_2S-producing colonies to become black.

The following table summarizes the typical appearance of the more important bacteria.

Organism	Appearance
Shigella	Green, moist, raised colonies
Salmonella, Proteus	Blue-green, with or without a black center
Coliforms (rapid lactose fermenters)	Salmon-pink to orange, surrounded by a zone of bile precipitate

FORMULA

BACTO HEKTOEN ENTERIC AGAR
DEHYDRATED

Proteose, Peptone, Difco	12 g	Sodium Chloride	5 g
Bacto Yeast Extract	3 g	Sodium Thiosulfate	5 g
Bacto Bile Salts No. 3	9 g	Ferric Ammonium Citrate	1.5 g
Bacto Lactose	12 g	Bacto Agar	14 g
Bacto Saccharose	12 g	Brom Thymol Blue	0.065 g
Bacto Salicin	2 g	Acid Fuchsin	0.1 g

Final pH 7.5 ± 0.2 at 25°C.

One pound will make 6 liters of medium.

METHOD OF PREPARATION
1. Suspend 76 g in 1 liter distilled or deionized water.
2. Heat to boiling to dissolve completely. **Avoid overheating.** DO NOT AUTOCLAVE
3. Dispense as desired.

STORAGE
Bacto Hektoen Enteric Agar   Below 30°C
Prepared plates   2 – 8°C

QUALITY CONTROL
Identity Specifications

Dehydrated powder: light purplish beige, homogeneous, free-flowing
Reaction of 7.6% solution: pH 7.5 ± 0.2 at 25°C
Prepared medium: brown with greenish cast, slightly opalescent
Prepared plates: green with yellowish cast, slightly opalescent

Typical Cultural Response on Bacto Hektoen Enteric Agar
After 18 – 24 Hours at 35°C

Organism	Growth	Color of Colony
Enterobacter aerogenes ATCC® 13048	fair to good	Salmon-orange
Escherichia coli ATCC® 25922	none to fair	Orange (may have bile precipitate)
Salmonella enteritidis ATCC® 13076	excellent	Greenish-blue*
Salmonella typhimurium ATCC® 14028	excellent	Greenish-blue*
Shigella flexneri ATCC® 12022	excellent	Greenish-blue
Streptococcus faecalis ATCC® 19433	none	—

*May have black centers (H_2S production).

REFERENCES
1. Appl. Microbiol., 16:577, 1968.
2. Ibid., 16:579, 1968.

PACKAGING
Bacto Hektoen Enteric Agar		
1/4 lb (114 g)	0853-02-6
1 lb (454 g)	0853-01-7
5 lb (2.27 kg)	0853-05-3
10 kg	0853-08-0

BACTO HEMAGGLUTINATION BUFFER

INTENDED USE
Bacto Hemagglutination Buffer, prepared according to the formula of Wheeler, Luhby, and Scholl,[1] is a desiccation consisting of sodium chloride, disodium phosphate and potassium dihydrogen phosphate. It is recommended for the preparation of isotonic buffered saline for use in hemagglutination studies, for suspending and washing erythrocytes preparatory for hemagglutination, for use in rehydrating Bacto Phytohemmagglutinin M and P as described by Nowell[2] for the differential separation of erythrocytes and leucocytes from peripheral blood and for use with Bacto FA N. Gonorrhoeae in the fluorescent technique for detecting N. gonorrhoeae in clinical specimens.[3,4,5]

FORMULA

BACTO HEMAGGLUTINATION BUFFER
DEHYDRATED

Ingredients per liter

Disodium Phosphate .... 0.765 g
Monopotassium Phosphate ..... 0.219 g
Sodium Chloride .... 7.65 g

METHOD OF PREPARATION

Bacto Hemagglutination Buffer is rehydrated by dissolving 8.6 g (1 vial) of the powder in 1 liter of distilled water. The resultant solution is isotonic for blood cells.

STORAGE

Bacto Hemagglutination Buffer	Below 30°C
Prepared solution:	2 – 8°C

QUALITY CONTROL

Identity Specifications

Appearance:	white, homogeneous, free-flowing
Reaction of 0.86% solution:	pH 7.3 ± 0.1 at 25°C
Prepared solution:	colorless, clear solution, no significant precipitate

REFERENCES

1. J. Immunol., 65:30, 1950.
2. Cancer Res., 20:462, 1960
3. Proc. Soc. Exp. Biol. & Med., 101:322, 1959.
4. Publ. Hlth. Repts., 75:125, 1960.
5. PHA Pub. No. 499, 1962.

PACKAGING

Bacto Hemagglutination Buffer	6 × 8.6 g	0512-33-2

BACTO HEMOGLOBIN

INTENDED USE

Bacto Hemoglobin is an autoclavable preparation of beef blood used to prepare chocolate agar for cultivating and isolating fastidious microorganisms.

HISTORY

Bacto Hemoglobin is prepared according to the procedure described by Spray[1].

METHOD OF PREPARATION

1. Place 10 g Bacto Hemoglobin in a dry beaker.
2. Measure 500 ml distilled or deionized water and add in approximately 100 ml amounts, stirring well after each addition. Use a spatula to break up clumps. Transfer to flasks as desired for autoclaving.
3. Sterilize in the autoclave for 15 minutes at 15 lbs pressure (121°C). Cool to 45 – 50°C.
4. Swirl flask to reestablish complete solution and then add to an equal amount of double-strength sterile agar base cooled to 45 – 50°C.

STORAGE

Bacto Hemoglobin   Below 30°C
Prepared 2% solution   15 – 30°C

QUALITY CONTROL

Identity Specifications

Dehydrated enrichment: dark brown, fine, free-flowing
Prepared 2% solution: chocolate brown, opaque with a dispersible precipitate

Determine CULTURAL RESPONSE of product from prepared chocolate agar (Bacto
GC Medium Base, etc.).

REFERENCE

1. J. Lab. Clin. Med., 16:166, 1930.

PACKAGING

Bacto Hemoglobin	1 lb (454 g)	0136-01-6
	1/4 lb (114 g)	0136-02-5
	5 lb (2.27 kg)	0136-05-2
Bacto Hemoglobin Solution 2%	100 ml	3248-72-9
	6 × 100 ml	3248-73-8

BACTO HERELLEA AGAR

INTENDED USE

Bacto Herellea Agar is a selective medium used for isolation and differentiation of gram-
negative nonfermentative and fermentative organisms. It is especially recommended
for the differentiation of *Acinetobacter* species from *Neisseria gonorrhoeae* in specimens of urethal or vaginal origin.

HISTORY/PRINCIPLES

Bacto Herellea Agar is prepared according to the formula of Mandel, Wright and
McKinnon.[1] In their study of the role of *Mima polymorpha* and *Herellea vaginicola* (now
identified as the genus *Acinetobacter*) in gonorrhea, they experienced difficulty in rec-
ognizing these organisms in the presence of large numbers of gram-positive cocci and
gram-negative rods which frequently occurred in specimens of urethral or vaginal origin.
This medium is especially designed to resolve this problem. After 24 hours incubation
at 37°C the gram-positive bacteria are inhibited and the colonies of fermentative gram-
negative bacilli are shown by their yellow color and yellow zones around the colonies.
Acinetobacter are not fermentative and their colonies are easily distinguished by their
pale lavender color.

Bile salts is the selective agent that inhibits the growth of *Neisseria* and gram-positive
organisms. Lactose and maltose, the fermentable carbohydrates, when broken down
effect a change in the brom cresol purple indicator.

FORMULA

BACTO HERELLEA AGAR
DEHYDRATED
Ingredients per liter

Bacto Tryptone . . . . 15 g
Bacto Soytone . . . . 5 g
Lactose . . . . 10 g
Maltose . . . . 10 g
Bile Salts No. 3 . . . . 1.25 g
Brom Cresol Purple . . . . 0.02 g
Sodium Chloride . . . . 5 g
Bacto Agar . . . . 16 g

Final pH 6.8 ± 0.2 at 25°C.

One pound will make 7.3 liters of medium.

METHOD OF PREPARATION
1. Suspend 62 g in 1 liter distilled or deionized water. Heat to boiling to dissolve completely.
2. Dispense into tubes or flasks as desired.
3. Sterilize in the autoclave for 15 minutes at 15 lbs pressure (121°C).

STORAGE
Bacto Herellea Agar   Below 30°C
Prepared medium   2 – 8°C

QUALITY CONTROL
Identity Specifications

Dehydrated powder: beige with green tinge, homogeneous, free-flowing
Reaction of 6.2% solution: pH 6.8 ± 0.2 at 25°C
Prepared medium: purple, slightly opalescent

Typical Cultural Response on Bacto Herellea Agar
After 18 – 24 Hours at 35°C

Organism	Growth	Color of Colony or Media
Acinetobacter calcoaceticus ATCC® 17961 (formerly H. vaginicola)	good to excellent	pale lavender
Acinetobacter lwoffii ATCC® 9957 (formerly M. polymorpha)	good to excellent	pale lavender
Escherichia coli ATCC® 25922	good to excellent	yellow
Proteus vulgaris ATCC® 6380	good to excellent	yellow, may be delayed
Salmonella enteritidis ATCC® 13076	good to excellent	yellow, may be delayed
Staphylococcus aureus ATCC® 25923	inhibited	—
Staphylococcus epidermidis ATCC® 12228	inhibited	—

REFERENCE
1. J. Bact., 88:1524, 1964.

PACKAGING
Bacto Herellea Agar   1 lb (454 g)   0909-01-1

BACTO HETROL SLIDE TEST KIT

INTENDED USE
Bacto Hetrol Slide Test Kit is recommended for the serological detection of infectious mononucleosis heterophile antibodies by the rapid macroscopic slide agglutination technique.

TEST SUMMARY AND HISTORY
The Bacto Hetrol Slide Test employs stabilized horse erythrocytes for presumptive identification of heterophile antibodies. It also incorporates the findings of Wollner[1] by the addition of specially prepared papainized sheep erythrocytes which provide confirmation of the presence of heterophile agglutinins indicative of infectious mononucleosis. Positive and negative controls are included in the kit and serve as controls on the reagents and technique employed.

Hemagglutination of sheep erythrocytes for detecting heterophile antibodies was first suggested by Paul and Bunnell.[2] Since heterophile agglutinins may be present in sera from other diseases as well as from patients with infectious mononucleosis, Davidsohn, Stern, and Kashiwagi[3] introduced the "Differential Absorption Test" to distinguish between these agglutinins. In the differential test, heterophile infectious mononucleosis antibodies are almost completely absorbed by beef erythrocytes but not by guinea pig kidney suspension. Routinely, the presumptive test is performed first and if an agglutination titer above 1:56 is obtained, the differential absorption test is done.

Wollner[4] observed that papainized sheep erythrocytes were agglutinated by infectious mononucleosis serum in much lower dilutions than were untreated cells. Conversely, these enzyme treated erythrocytes yielded higher agglutination titers with normal serum than did untreated erythrocytes. This finding was the basis for the presumptive enzyme test for infectious mononucleosis. Refinement[5] of this presumptive test resulted in a differential tube test employing papainized and native sheep erythrocytes. The Wollner Differential Enzyme Test proved to be highly specific for infectious mononucleosis and provided sharp distinction between the infectious mononucleosis heterophile antibody and other nonspecific heterophile antibodies. The sharpness of the reaction facilitated detection of infectious mononucleosis antibodies in much lower dilutions than did other serological tests, thus eliminating a source of false negative results.

Numerous attempts have been made to simplify the presumptive test to permit more rapid and less costly determination of the presence of heterophile antibodies. Reports of slide tests using fresh sheep erythrocytes for the presumptive determination of heterophile antibodies in infectious mononucleosis have appeared in the literature. Among these are the slide test procedures described by Butt and Foord,[6] Straus,[7] Rappaport and Skariton,[8] Moloney and Malzone,[9] Vaughn,[10] and Brunfitt and O'Grady.[11] More recently, Lovric,[12] reporting on a modification of the Brunfitt and O'Grady procedure, using native and enzyme treated erythrocytes in a differential slide test for detecting infectious mononucleosis antibodies. He found complete agreement between his slide test and the tube test.

Cox[13] reported the results of a slide test using stabilized erythrocytes. He concluded that there was good correlation with standard tube tests using fresh erythrocytes. However, his method did not differentiate between the heterophile antibodies of infectious mononucleosis and those found in certain other diseases.

Lane[14] reviewed the advances in laboratory aids for the detection of infectious mononucleosis and related preliminary results obtained with a modification of the Lovric test. In this modification, stabilized erythrocytes, instead of fresh erythrocytes, were used to detect the heterophile antibodies of infectious mononucleosis. The modified Lovric slide test run in parallel with the classical tube tests on more than 400 sera from patients with and without infectious mononucleosis yielded results comparable with those of the tube tests and clinical findings. Since the stabilized cells yielded results comparable with those of the presumptive tube test, Lane recommended their use in the slide test for detecting heterophile antibodies in infectious mononucleosis.

Davidsohn and Lee,[15] reporting on a differential slide test using horse erythrocytes rather than sheep erythrocytes, found the former to be equally as specific but more sensitive than sheep agglutinins. Hoff and Bauer[16] described a rapid slide test using formalized horse erythrocytes which gave 98% correlation with the Davidsohn differential test.

Bacto Hetrol Slide Test, a modification of the Lovric test, incorporates the principles of the enzyme test of Wollner and the later findings of Hoff and Bauer. The stabilized horse erythrocytees provide greater sensitivity than obtained with sheep agglutinins while the papainized sheep cells assure the specificity of the test procedure.

The preserved horse cell suspension will be agglutinated by sera containing heterophile agglutinins. This is a presumptive test as horse erythrocytes will react with both infectious mononucleosis heterophiles and nonspecific heterophile antibodies. The enzyme treated sheep cells will react with only **non-infectious mononucleosis** heterophile antibodies. Lack of agglutination of enzyme treated cells in conjunction with agglutinated horse cells is indicative of infectious mononucleosis agglutinins. Thus, the papainized sheep erythrocytes serve as a confirmatory determination.

FORMULA
Bacto Hetrol Reagent P is a stabilized horse erythrocyte suspension for routinely screening sera for heterophile antibodies as a presumptive test for infectious mononucleosis.

Bacto Hetrol Reagent C is a stabilized enzyme treated sheep erythrocyte suspension used to confirm a positive heterophile test as being that of infectious mononucleosis.

Bacto Hetrol Negative Control is desiccated, pooled, normal human sera which has been tested by the slide and classical tube tests and found free of heterophile agglutinins.

Bacto Hetrol Positive Control is a desiccated, pooled sera from clinically and serologically proven cases of infectious mononucleosis.

Bacto Agglutination Slide has 6 sections of approximately one square inch each.

Bacto Agglutination Slide should be thoroughly washed after each test to remove traces of the test serum. Thorough rinsing with tap water and distilled water and finally drying with a paper towel is sufficient. Do not use detergent.

STORAGE AND EXPIRATION DATES
The reagents are stable to the expiry date on the label when stored at $2-8°C$.

SPECIMEN COLLECTION
Serum sufficient for 2 drops (dispensed from capillary bulb pipette) is necessary for each test run. Active or inactivated serum may be used in the test. Cloudy serum or

serum containing a precipitate should be centrifuged before use. Serum may be frozen if it cannot be tested when collected. Do not use whole blood or plasma for performance of this test.

The time at which the specimen is collected has particular relevance to the results obtained in the test. If blood samples are taken too early after the onset of infectious mononucleosis, antibody formation may not be sufficient to indicate the presence of the disease. Samples taken late in the course of the illness may also prove inconclusive. It is recommended that two specimens be taken during the course of the infection; one in the acute stage of the illness and again 10 days to 2 weeks later.

PROCEDURE
Materials Provided:
Bacto Hetrol Slide Test Kit
  Bacto Hetrol Reagent P
  Bacto Hetrol Reagent C
  Bacto Hetrol Positive Control
  Bacto Hetrol Negative Control
  Bacto Agglutination Slide

Materials Required but not Provided:
Capillary bulb pipette
Applicator sticks

1. Rehydrate Bacto Hetrol Positive Control and Bacto Hetrol Negative Control by adding 0.5 ml distilled water to each vial and inverting the vials several times to effect complete solution of the contents.
2. Remove caps from Bacto Hetrol Reagent P, Bacto Hetrol Reagent C and the rehydrated control sera and replace them with the supplied dropper caps. Do not interchange droppers after they have been placed in the vials.
3. Using a capillary bulb pipette, place a drop of the serum specimen to be tested in each of the center squares on the Bacto Agglutination Slide.
4. Place a drop of Bacto Hetrol Positive Control on each of 2 squares at one end of the slide and a drop of Bacto Hetrol Negative Control on each of 2 squares at the other end of the slide. Add 1 drop of the well suspended Bacto Hetrol Reagent P to the top row of 3 drops of sera on the Bacto Agglutination Slide and a drop of Bacto Hetrol Reagent C to each drop in the bottom row.
5. Mix drops in each square with different applicator sticks or toothpicks spreading each mixture over an area of about 2 square cm.
6. Rock the slide gently for exactly 2 minutes over an indirect light and observe for macroscopic agglutination.

RESULTS
Coarse agglutination of the cells will be readily observed within 2 minutes. NOTE: Rocking the slide for longer than 2 minutes may result in very fine clumping with some sera.

Typical Results Obtained with Bacto Hetrol Slide Test:

	Hetrol Reagent P	Hetrol Reagent C
Normal Serum	No clumping	No clumping
Bacto Hetrol Negative Control	No clumping	No clumping
Infectious Mononucleosis Serum	Macroscopic clumping	No clumping
Bacto Hetrol Positive Control	Macroscopic clumping	No clumping
Serum with Heterophile Antibodies other than those of Infectious Mononucleosis	Macroscopic clumping	Macroscopic clumping

LIMITATIONS OF THE PROCEDURE
Serological tests designed to detect heterophile agglutinins cannot be considered "specific" for this disease.

Bacto Hetrol Slide Test provides a rapid macroscopic slide determination. It is a qualitative test, thus, no titer information is obtained with this procedure. Bacto Hetrol Slide Test is not designed as a substitute for clinical and hematological evaluations. Indeed, all three aspects are often necessary to make an accurate diagnosis of infectious mononucleosis. A positive heterophile test concurrent with clinical and hematological results can be considered diagnostic for infectious mononucleosis. However, absence of heterophile agglutinins in suspect cases is not sufficient evidence to rule out the possibility of infectious mononucleosis.

The Davidsohn differential test may be performed when results obtained with Bacto Hetrol Slide Test are not in agreement with the clinical and hematological results, i.e., when either a false negative or a false positive result is suspected.

SPECIFIC PERFORMANCE CHARACTERISTICS

Bacto Hetrol Reagent P, a stabilized horse erythrocyte suspension, provides for the presumptive identification of infectious mononucleosis. Preserved horse erythrocytes have shown a high degree of specificity. Hoff and Bauer[16] found that a formalinized horse erythrocyte slide test exhibited 98% correlation with Davidsohn differential test in detecting suspect cases of infectious mononucleosis. Davidsohn[17] observed that the rate of false positive results for a preserved horse erythrocyte slide test was 0.4% when testing control sera and concluded that the diagnostic accuracy closely approached that of the differential adsorbtion test.

Preserved horse erythrocytes, though highly specific, do not exhibit the same degree of sensitivity. In one study, false negative results were observed in 2% of the cases diagnosed as infectious mononucleosis.[18]

Bacto Hetrol Reagent C, a stabilized papain treated erythrocyte suspension, incorporates the principles of the enzyme sheep erythrocyte test of Wollner. This test has been shown to be highly specific and highly sensitive in various studies evaluating serological tests for infectious mononucleosis. Davidsohn and Lee[19] found that the enzyme test gave comparable results to the differential adsorbtion test. In testing 241 serums from 130 patients diagnosed as having infectious mononucleosis, these authors experienced only 1 false positive reaction with the enzyme test. In an additional 55 patient sera exhibiting elevated titers, but not diagnosed as infectious mononucleosis, 3 false negatives were obtained.

Muschel and Piper[20] also experienced favorable results with the enzyme test and concluded that the test was simple and accurate for the serological diagnosis of infectious mononucleosis.

A much more extensive study was undertaken by Springer and Callahan[21] in evaluating the enzyme test. These authors tested 599 individuals over an 8 year period for infectious mononucleosis agglutinins. They experienced not one false positive nor false negative result with the enzyme test. In several cases this was instrumental in arriving at an accurate diagnosis in cases suspected of, or misdiagnosed as, infectious mononucleosis.

Based on these findings, it is expected that Bacto Hetrol Slide Test will not exhibit false positive reactions. However, because the sensitivity of this test is necessarily based on the expected results with preserved horse erythrocytes (Bacto Hetrol Reagent P) false negative reactions can occur.

REFERENCES

1. Wollner, D., Ztschr. Imunitatsforsh., 112:290 – 308, 1955.
2. Paul, J. R., and Bunnell, W. W., Am. J. Med. Sci., 183:90 – 104, 1932.
3. Davidsohn, I., Stern, K., and Kashiwagi, C., Am. J. Clin. Path., 21:1101 – 1113, 1951.
4. Wollner, D. 1955.
5. Wollner, D., Ztschr. Immunitatsforsch., 113:301 – 318, 1956.
6. J. Lab. and Clin. Med., 20:538, 1935.
7. Am. J. Clin. Path., 6:546, 1936.
8. ibid., 19:665, 1949.
9. Blood, 4:722, 1949.
10. J. Clin. Path., 4:104, 1951.
11. ibid., 10:243, 1957.
12. Lancet, 1:142, 1961.
13. J. Lab. and Clin. Med., 48:298, 1956.
14. Paper read at Michigan SAB, September, 1962.
15. Davidsohn, I., and Lee, C. L., Am. J. of Clin. Path., 41:115 – 125, 1964.
16. Hoff, G., and Bauer, S., JAMA, 194:351 – 353, 1965.
17. Davidsohn, R. J. L., J. Clin. Path., 20:643 – 646, 1967.
18. Galloway, E., Can. J. Med. Tech., 197 – 206, 1969.
19. Davidsohn, I., and Lee, C. L., 1964.
20. Muschel, L. H., and Piper, D. R., Am. J. Clin. Path., 32:240 – 244, 1959.
21. Springer, G. F., and Callahan, H. J., J. Lab. Clin. Med., 65:617 – 627, 1965.

PACKAGING

Bacto Hetrol Slide Test Kit	50 tests	3274-32-5
Contains:		
Bacto Hetrol Reagent P	2 ml	
Bacto Hetrol Reagent C	2 ml	
Bacto Hetrol Positive Control	0.5 ml	
Bacto Hetrol Negative Control	0.5 ml	
Bacto Agglutination Slide	1 slide	

BACTO HORSE SERUM, DESICCATED

INTENDED USE

Bacto Horse Serum, Dessicated is prepared from normal horse serum. It is filter sterilized and recommended for use as an enrichment in culture media.

METHOD OF PREPARATION

Reconstitute each vial by aseptically adding 10 ml sterile distilled or deionized water and gently mixing in an end-over-end motion.

STORAGE

Bacto Horse Serum, Desiccated and reconstituted: 2 – 8°C

QUALITY CONTROL

Identity Specifications

Appearance: straw colored button or powder
Rehydrated appearance: medium amber, very slightly to slightly opalescent

Typical Cultural Response on Bacto Tryptose Blood Agar Base with 10% Rehydrated Bacto Horse Serum, Desiccated After 18 – 48 Hours at 35 – 37°C

Organism	Growth
Streptococcus mitis ATCC® 9895	good to excellent
Streptococcus pneumoniae ATCC® 6303	good to excellent

PACKAGING

Bacto Horse Serum, Desiccated	6 × 10 ml	0261-60-1
	12 × 10 ml	0261-61-0

BACTO INH TEST STRIPS
BACTO INH TEST CONTROL

INTENDED USE
Bacto INH Test Strips are used for detecting isonicotinic acid and its metabolites in urine.

Bacto INH Test Control is used as a positive control in the above procedure.

HISTORY
The need for a sensitive, specific, easily performed test for the presence of isonicotinic acid and its metabolites in urine was stressed by Short and Case,[1] Gangadharam et al.,[2] Hobby and Deuschle,[3] Kasik et al.,[4] Eidus and Hamilton,[5] Eidus and Ling,[6] and by Eidus and Harnanansingh.[7] However, prior to 1962, INH tests were either time consuming, insensitive or inaccurate. Later tests offered greater accuracy but required freshly prepared reagents and costly equipment.

Kilburn et al.[8] proposed use of the reagent-impregnated strip in 1972.

FORMULA
Bacto INH Test Strips are absorbent paper strips impregnated with chloramine T, potassium thiocyanate, citric acid and barbituric acid as described by Kilburn et al. with some modification.

Bacto INH Test Control is an INH-impregnated disk that will yield a positive result in the test procedure.

STORAGE

Bacto INH Test Strips	2 – 8°C in the dark
Bacto INH Test Control	15 – 30°C in the dark

PREPARATION OF REAGENTS
1. Obtain one sealed-tube Bacto INH Test Strip from the jar. Cut off one corner of the plastic tube at the arrow end of the strip.
2. Prepare a positive control for the test by placing one Bacto INH Test Control disk in 2 ml distilled or deionized water in a test tube. Shake 2 – 3 times over a 15 minute period to assure extraction of the INH into the water.

PROCEDURE
1. At the opposite end of the strip from the arrow, squeeze 1/2 inch of the plastic tube of a Bacto INH Test Strip between thumb and forefinger and insert open end of tube below the surface of the urine specimen. Release pressure. Sufficient specimen should rise in the tube to cover the arrow on the strip. The tube may be left to float in the urine container or may be transferred to a test tube.
2. Retain at room temperature.
3. Observe for results at 15 to 30 minutes.

QUALITY CONTROL
Perform the test using the solution prepared from Bacto INH Test Control in parallel with the urine specimen.

RESULTS
The appearance of a blue, purple or green color on the strip and in the liquid in the plastic tube is a positive test and indicates the presence of INH or its metabolites in the test specimen. Only the strip may appear colored if the reaction is weak.

REFERENCES
1. Tubercle (London) 38:288, 1957.
2. Tubercle (London) 39:191, 1958.
3. Am. Rev. Resp. Dis., 80:415, 1959.
4. Am. Rev. Resp. Dis., 85:282, 1962.
5. Am. Rev. Resp. Dis., 89:587, 1962.
6. WHO/TB76, 1969.
7. Clinical Chemistry, 17:492, 1971.
8. Am. Rev. Resp. Dis., 106:923, 1972.

PACKAGING

Bacto INH Test Strips	25 strips	3189-30-1
Bacto INH Test Control	50 disks	3190-90-5

BACTO ISP MEDIUM 3

INTENDED USE
Bacto ISP Medium 3 is recommended for characterizing *Streptomyces* species according to the International Streptomyces Project (ISP).

HISTORY
Bacto ISP Medium 3 is based on the original formula of Shirling and Gottlieb.[1]

PROCEDURE
1. Suspend 22 g in 1 liter distilled or deionized water. Heat to boiling with constant agitation to completely dissolve the agar.
2. Sterilize in the autoclave for 15 minutes at 15 lbs pressure (121°C).
3. Allow to cool slightly before pouring plates. Swirl medium intermittently while pouring to assure a uniform distribution of oatmeal. Allow to solidify on a flat surface.
4. Inoculate plates by streaking with a sterilized wire loop, 0.1 ml of the test organism diluted in sterile saline to contain appropriate desired dilution. Drop the inoculum near the edge of the plate, then streak drop from source in one direction (5 streaks) then perpendicular (4 streaks).
5. Incubate at 30°C for 48 – 96 hours. Examine for growth.

STORAGE

Bacto ISP Medium 3	Below 30°C
Prepared Medium	2 – 8°C

QUALITY CONTROL
Identity Specifications

Dehydrated powder:	beige, homogeneous, free-flowing
Reaction of 2.2% solution:	pH 7.2 ± 0.1 at 25°C
Prepared medium:	very light amber, opalescent with precipitation

Typical Cultural Response on Bacto ISP Medium 3
After 48 – 96 Hours at 30°C

Organism	Growth
Streptomyces albus ATCC® 3004	good to excellent
Streptomyces lavendulae ATCC® 8664	good to excellent

REFERENCE
1. International Journal of Systematic Bacteriology, Vol. 16:3:313 – 340, July 1966.

PACKAGING
Bacto ISP Medium 3   500 g   0771-17-8

BACTO ISP MEDIUM 4
(INORGANIC SALTS STARCH AGAR)

INTENDED USE
Bacto ISP Medium 4 (Inorganic Salts Starch Agar) is recommended for the characterization of *Streptomyces.*

HISTORY
ISP media were developed by Difco Laboratories for the International Streptomyces Project (ISP) in order to select stable properties and reproducible procedures for characterization of streptomycetes.[1]

FORMULA

BACTO ISP MEDIUM 4
(INORGANIC SALTS STARCH AGAR)
DEHYDRATED

Ingredients per liter

```
Bacto Soluble Starch . . . . . . . . . . . . . . . . . 10 g
Potassium Phosphate, Dibasic  . . . . . . . . . . . 1 g
Magnesium Sulfate USP  . . . . . . . . . . . . . . . 1 g
Sodium Chloride  . . . . . . . . . . . . . . . . . . . . . 1 g
Ammonium Sulfate  . . . . . . . . . . . . . . . . . . . 2 g
Calcium Carbonate . . . . . . . . . . . . . . . . . . . 2 g
Ferrous Sulfate (FeSO₄·7H₂O)  . . . . . . . . 0.001 g
Manganous Chloride (MnCl₂·7H₂O)  . . . . . 0.001 g
Zinc Sulfate (ZnSO₄·7H₂O)  . . . . . . . . . . 0.001 g
Bacto Agar . . . . . . . . . . . . . . . . . . . . . . . 20 g
```

Ferrous Sulfate ($FeSO_4 \cdot 7H_2O$) — 0.001 g
Manganous Chloride ($MnCl_2 \cdot 7H_2O$) — 0.001 g
Zinc Sulfate ($ZnSO_4 \cdot 7H_2O$) — 0.001 g

Final pH 7.2 ± 0.2 at 25°C.

One pound will make 12.3 liters of medium.

METHOD OF PREPARATION
1. To rehydrate, suspend 37 g in 1 liter distilled or deionized water. Heat to boiling to dissolve completely.
2. Agitate the medium to obtain a uniform suspension and distribute into flasks in desired amounts.
3. Sterilize in the autoclave for 15 minutes at 15 lbs pressure (121°C).
4. Swirl the sterile medium in flasks to obtain a uniform mixture before pouring into Petri dishes.

STORAGE

Bacto ISP Medium 4	Below 30°C
Prepared medium	2 – 8°C

QUALITY

Identity Specifications

Dehydrated powder:	white to light beige, homogeneous, free-flowing
Reaction of 3.7% solution:	pH 7.2 ± 0.2 at 25°C
Prepared medium:	very light amber, opaque, may have a precipitate

Typical Cultural Response on Bacto ISP Medium 4
After 40 – 72 Hours at 30 ± 2°C

Organism	Recovery
Streptomyces achromogenes ATCC® 12767	good to excellent
Streptomyces albus ATCC® 3004	good to excellent
Streptomyces lavendulae ATCC® 8664	good to excellent

REFERENCE

1. Shirling, E. B., and D. Gottlieb, "Methods for Characterization of *Streptomyces* Species," Internatl. J. Systematic Bact., 16:3, July 1966.

PACKAGING

Bacto ISP Medium 4	1 lb (454 g)	0772-01-5
(Inorganic Salts Starch Agar)		

BACTO INDOLE TEST STRIPS

INTENDED USE

Bacto Indole Test Strips are prepared for use with Bacto Tryptone 1% solution and other media standardized for detecting indole production by bacteria. They are particularly useful as an aid in differentiating slow lactose fermenting strains of *Escherichia coli* from *Salmonella Paratyphi A*. *E. coli* will display a similar reaction to *S. Paratyphi A* in TSI agar, but will give a positive indole test distinguishing them from *Salmonella* which do not produce indole.

PROCEDURE

The test is made by placing a Bacto Indole Test Strip in a tube of grown culture or freshly inoculated medium. The strip is removed from the vial using forceps and then folded over the lip of the tube; replace the cotton plug or screw cap so that the end of the strip will be about 10 mm above the medium surface. The culture tubes containing the strips are incubated for 5 – 30 minutes or as required for growth if inserted just after inoculation at 37°C and observed. The formation of a violet color on the strip indicates a positive test for indole.

STORAGE

Bacto Indole Test Strips   15 – 30°C

QUALITY CONTROL

Identity Specifications

Appearance: pale yellow paper strips approximately 1/4″ × 2.9″ ± 1/32″

Typical Cultural Response on Bacto Indole Test Strips When Used as
Above with 1% Bacto Tryptone After 16 – 24 Hours at 35°C

Organism	Growth	Indole
Enterobacter aerogenes ATCC® 13048	good to excellent	−
Escherichia coli ATCC® 25922	good to excellent	+

+ = positive, pink to violet color at edge of strip
− = negative, pale yellow strip

PACKAGING

Bacto Indole Test Strips	1 × 25 strips	1627-30-5
Bacto Tryptone	1 lb (454 g)	0123-01-1
	1/4 lb (114 g)	0123-02-0
	10 kg	0123-08-4

INFECTIOUS MONONUCLEOSIS AND OTHER HETEROPHILE REAGENTS

BACTO BEEF CELL ANTIGEN, DESICCATED
BACTO GUINEA PIG KIDNEY ANTIGEN, DESICCATED
BACTO HETEROPHILE FORSSMAN ANTISERUM
BACTO INFECTIOUS MONONUCLEOSIS POSITIVE SERUM

INTENDED USE

These reagents are designed for the detection of heterophile agglutinins associated with infectious mononucleosis according to the procedures of Davidsohn[1] and Paul and Bunnell.[2] Bacto Guinea Pig Kidney Antigen and Bacto Beef Cell Antigen are standardized for use in the adsorption techniques of the differential serological test for infectious mononucleosis as outlined by Davidsohn.[3] Bacto Heterophile Forssman Antiserum and Bacto Infectious Mononucleosis Positive Serum are designed for use as positive controls for the adsorption techniques and reagents employed in the Davidsohn differential test.

SUMMARY AND EXPLANATION OF THE TESTS

The Davidsohn differential test is a quantitative tube dilution technique capable of differentiating heterophile agglutinins of infectious mononucleosis from Forssman type heterophile agglutinins associated with horse serum sensitization or serum sickness, elevated titers of antisheep agglutinins caused by infections other than infectious mononucleosis, as well as normally occurring heterophilic antibodies in human sera.

Infectious mononucleosis (IM) is an acute infection caused by the Epstein-Barr virus (EB) of the Herpes family. The infection affects several organs of the body and is characterized hematologically by atypical lymphocytes in the circulating blood. Clinical manifestations are most often characterized by fever, enlargement of lymph nodes and spleen, and an elevated absorbed heterophile-antibody titer.

The clinical aspects of the disease are such that diagnosis cannot be made on history and symptomology alone. The onset of infectious mononucleosis is often difficult to discern and symptoms characteristic of the pathological processes may be confused with those of other nonrelated viral and bacterial diseases. Numerous cases have been cited[4,5] in which infectious mononucleosis has been misidentified on the basis of clinical indications alone.

Hematologic and serologic tests along with the clinical indications provide the most accurate information for diagnosing infectious mononucleosis. Classical serologic tests for detecting heterophile antibodies include the presumptive test of Paul and Bunnell and the differential test of Davidsohn. More recently, the indirect fluorescent (IF) test and the immunoperoxidase (IP) staining test have been used to detect antibodies to the EB virus.

Although heterophile antibodies in patient serum at a titer of 1:56 or higher can be indicative of infectious mononucleosis, a positive reaction with these sheep cell agglutinins may be present in cases totally unrelated to infectious mononucleosis.

Heterophile agglutinins have been known to occur in normal sera at titers of 1:28 and in some cases as high as 1:56. Other disease processes may result in titers of 1:112 to 1:224. High titers usually suggestive of infectious mononucleosis have been observed in patients with serum sickness or as a result of antigenic stimulation. For these reasons, the Paul-Bunnell test relying solely on the presence of sheep cell agglutinins should be considered a presumptive test.

In contrast, the Davidsohn differential test will distinguish the specific agglutinins of infectious mononucleosis from those outlined above. Adsorption with guinea pig kidney antigen will differentiate between sheep cell agglutinins from infectious mononucleosis and agglutinins resulting from therapeutic agents containing horse serum proteins. In cases of infectious mononucleosis this technique will reduce the antisheep agglutinins to no less than 1/8 their original titer or a drop of not more than 3 tubes in the two-fold serial dilution. However, antisheep agglutinins are completely removed by beef cell antigen. Therefore, the presence of the sheep cell agglutinins in serum after contact with guinea pig kidney antigen and a complete removal of these agglutinins after contact with beef cell antigen is indicative of infectious mononucleosis. Complete adsorption of antisheep agglutinins by guinea pig kidney antigen is evidence against the presence of infectious mononucleosis.

HISTORY

Davidsohn[6,7,8,9] established that sheep cell agglutination could be relied upon to determine the presence of heterophilic agglutinins in sera and showed that a high heterophilic titer may be present in sera from patients receiving injections of biologicals containing horse serum protein. These antibodies may persist in the blood for over a year after such therapy. Paul and Bunnell[10] showed increased titers of heterophilic agglutinins in sera from infectious mononucleosis patients, and used the Davidsohn sheep cell agglutination technique to determine these agglutinins.

This finding is the basis for their presumptive serologic test of mononucleosis. The presence in sera of heterophilic agglutinins for sheep cells at a titer of 1:56 or higher with typical hematological and clinical findings is usually suggestive of infectious mononucleosis.

Continued studies of the nature and occurence of heterophilic antibodies led to the development of a specific, differential serological test for infectious mononucleosis. Stuart, Burgess, Lawson and Wellman[11] and Stuart[12] used guinea pig kidney antigen for the differentiation of horse serum sickness and infectious mononucleosis. They showed that horse serum sickness agglutinins were adsorbed by guinea pig kidney tissue, while agglutinins due to infectious mononucleosis were adsorbed only partially, or not at all.

Stuart, Welch, Cunningham and Burgess[13] suggested the use of a beef cell antigen in addition to the guinea pig kidney antigen to detect the presence of sheep cell heterophilic antibodies due to infectious mononucleosis in sera containing normally high heterophilic agglutinins. Davidsohn,[14] continuing his investigations on heterophilic antibodies, independently carried out investigations similar to those of Stuart and associates, and reported the use of guinea pig kidney and beef cell antigen for differentiating antibodies for infectious mononucleosis from other heterophile antibodies.

REAGENTS
Bacto Guinea Pig Kidney Antigen, Desiccated is standardized for use in the adsorption techniques of the Davidsohn differential test for infectious mononucleosis. It is a sterile suspension in 0.85% NaCl dried from the frozen state. This reagent should be rehydrated by adding 1 ml distilled or deionized water to the 1 ml vial, or 5 ml distilled or deionized water to the 5 ml vial. This reagent should be stored in the refrigerator at 2 – 8°C and will retain its activity to the expiry date stated on the label of each vial. Do not allow to freeze. Do not expose to heat.

Bacto Beef Cell Antigen, Desiccated is standardized for use in the adsorption techniques of the Davidsohn differential test for infectious mononucleosis. It is a sterile, washed, and heated suspension of beef erythrocytes dried from the frozen state. This reagent should be rehydrated by adding 1 ml distilled or deionized water to the 1 ml vial, or 5 ml distilled or deionized water to the 5 ml vial. This reagent should be stored in the refrigerator at 2 – 8°C and will retain its reactivity through the expiry date stated on the label of each vial. Do not expose to heat. Do not freeze.

Bacto Infectious Mononucleosis Positive Serum is a standardized human serum which provides a presumptive and differential reaction pattern typical of infectious mononucleosis.

Bacto Heterophile Forssman Antiserum is standardized to provide a presumptive and differential reaction pattern typical of Forssman antibodies associated with serum sickness.

These control sera are desiccated products and should be rehydrated by adding 3 ml of distilled or deionized water. They are to be stored in the refrigerator at 2 – 8°C and will remain stable through the expiry date stated on the label of each vial. Do not expose these control sera to heat. Do not freeze.

PRECAUTIONS
Each human serum used in preparing Bacto Infectious Mononucleosis Positive Serum has been screened for the presence of Hepatitis B surface antigen. Only sera exhibiting a negative reaction are used in preparing these reagents.

As an additional precaution, the user should employ proper and aseptic technique in handling all human sera.

SPECIMEN COLLECTION
0.2 ml of patient serum inactivated at 56°C for 30 minutes is needed for each phase of the test; 0.4 ml total serum for differential test; or 0.6 ml total serum when presumptive and differential test is performed.

The time at which the specimen is collected has particular relevance to the results obtained in the test. If blood samples are taken too early after the onset of infectious mononucleosis, antibody formation may not be sufficient to indicate the presence of the disease. Samples taken late in the course of the illness may also prove inconclusive. It is recommended that 2 specimens be taken during the course of the infection, one in the acute stage of the illness and again 10 days to 2 weeks later.

PROCEDURE
Materials Provided:
Bacto Beef Cell Antigen, Desiccated
Bacto Guinea Pig Kidney Antigen, Desiccated
Bacto Forssman Heterophile Antiserum
Bacto Infectious Mononucleosis Positive Serum

Materials Required but not Provided:
13 × 85 mm test tubes
10 × 75 mm test tubes — set of 10 for each procedure
0.85% NaCl solution
2% sheep cell suspension (washed)

Presumptive Test (Unadsorbed Serum)
1. Washed sheep cell suspensions are not stable and should be prepared at the time test is to be performed. Sheep blood should not be less than 24 hours old or more than 1 week old. Cells are to be washed 3 times in physiological saline in the following manner:
   a. Filter an adequate quantity of preserved sheep blood through gauze into a 50 ml round-bottom centrifuge tube.
   b. Add 2 or 3 volumes of saline solution to each tube.
   c. Centrifuge tubes at a force sufficient to pellet down cells in 5 minutes.
   d. Remove supernatant fluid by suction through a capillary pipette, taking off upper white cell layer.
   e. Fill tube with saline solution and resuspend cells by inverting and gently shaking tube.
   f. Recentrifuge tube and repeat the process for a total of 3 washings. If supernatant fluid is not colorless on third washing, cells are too fragile and should not be used.
   g. After supernatant fluid is removed from third washing, cells are poured or washed into a 15 ml graduated centrifuge tube and centrifuged at previously used speed for 10 minutes in order to pack cells firmly and evenly.
   h. Read the volume of packed cells in the centrifuge tube and carefully remove supernatant fluid.
   i. Prepare a 2% suspension of sheep cells by washing them into a flask with 49 volumes of saline solution. Shake flask to insure even suspension of cells.
      Example: 2.1 ml (packed cells) × 49 = 102.9 ml (saline solution required)
2. Bacto Forssman Heterophile Antiserum (positive control) should be tested with each battery of samples. This serum has been inactivated during its preparation. No further inactivation is necessary.
3. Set up a row of 10 tubes (10 × 75 mm).
4. Add 0.4 ml 0.85% NaCl solution to tube 1 and 0.25 ml to the other tubes.
5. Inactivate the test serum at 56°C for 30 minutes.
6. Add 0.1 ml inactivated serum to tube 1, mix and transfer 0.25 ml to tube 2 through tube 10. Discard 0.25 ml from tube 10.
7. Add 0.1 ml 2% sheep red cell suspension and shake tubes. Final dilutions are 1:7, 1:14, etc.
8. Allow to stand at room temperature for 2 hours.
9. Shake the tubes to resuspend sedimented cells and record results. If no clumping is visible to the naked eye, place the tube horizontally on the stage of the microscope and examine with the low power objective.

Differential Test

Bacto Guinea Pig Kidney and Bacto Beef Cell Antigens are used similarly in the test. Control sera are to be used in the same manner as the patient serum in the test.

1. Washed sheep cell suspensions are not stable and should be prepared at the time test is to be performed. Prepare cells as stated under the Presumptive Test procedure.
2. Positive controls, Bacto Infectious Mononucleosis Positive Serum and Bacto Forssman Heterophile Antiserum should be tested with each battery of samples. These sera have been inactivated during their preparation. No further inactivation is necessary.
3. Shake vial of antigen suspension thoroughly and place 1 ml into a 13 × 85 mm test tube.
4. Add 0.2 ml of the patient's serum which has been inactivated at 56°C for 30 minutes. Shake well.
5. Allow to stand for 3 minutes.
6. Centrifuge at 1500 rpm for 10 minutes.
7. Carefully transfer the supernatant fluid with a capillary pipette to another tube.
8. To each of 10 (10 × 75 mm) tubes, except tube 1, add 0.25 ml of 0.85% NaCl solution.
9. Add 0.25 ml of the supernatant from step 7 to tube 1.
10. Add 0.25 ml of the supernatant from step 7 to tube 2 and mix well.
11. Transfer 0.25 ml from tube 2 to tube 3. Mix well and continue this procedure through tube 10. Discard 0.25 ml from tube 10 after mixing.
12. To each tube, add 0.1 ml of a 2% suspension of washed sheep cells. Shake well. The final serum dilution in each tube will be 1:7, 1:14, 1:28, 1:56, 1:112, etc.
13. Incubate for 2 hours at room temperature. After 15 minutes, shake the tubes one at a time to resuspend the sediment and observe for clumping. Final reading should be made after 2 hours. If no clumping is visible by the naked eye, observe under the low power objective of the microscope.

RESULTS AND INTERPRETATION

Clumping of the sera is indicative of a positive test. The titer is the reciprocal of the highest dilution showing clumping.

Typical Results of Presumptive and Differential Tests For Infectious Mononucleosis

Serum Source	Titer of Unadsorbed Serum	Titer After Guinea Pig Kidney Antigen	Adsorption with Beef Cell Antigen
Patients having neither infectious mononucleosis nor serum sickness	1:28 to 1:112	0	$\overset{\circ}{+}$
Patients having serum sickness	1:56 to 1:224	0	0
Patients with infectious mononucleosis	1:56 to 1:3584 and higher	+	0

+ = Titers after adsorption with kidney antigen not more than 3 tubes lower than the unadsorbed serum
0 = Negative
$\overset{\circ}{+}$ = Negative or low titer

The above chart is a summary of typical results obtained from patient serum tested by the presumptive method of Paul and Bunnell and by the differential method of Davidsohn.

Interpretation

Patients having neither infectious mononucleosis nor serum sickness: "normal" serum should exhibit no titer when tested for sheep cell agglutinins. However, "normal" serum may contain sheep cell agglutinins of 1:28 and may go as high as 1:56. The possibility of the presence of these nonspecific heterophile antibodies must be considered when analyzing the results of the presumptive test. Higher titers may be observed in viral diseases, lymphoproliferate disorders, after antitoxin administration, or after administration of blood group specific substances. These agglutinins will be completely removed upon adsorption with guinea pig kidney antigen and completely or partially removed by adsorption with beef cell antigen.

Patients having serum sickness will exhibit unadsorbed titers of 1:56 to 1:224. This titer is within the range indicative of infectious mononucleosis and therefore can be a source of false positive results. However, adsorption with guinea pig kidney antigen or beef cell antigen will completely remove these Forssman type antibodies, indicating the presence of serum sickness rather than mononucleosis.

Patients with infectious mononucleosis usually exhibit titers from 1:56 to 1:3584. However, as stated above, titers indicative of infectious mononucleosis are also found in other immune responses. Confirmation of a diagnosis based on the unadsorbed titer of 1:56 – 1:3584 or higher is achieved by: 1) adsorption with guinea pig kidney antigen, which will remove 87.5% of sheep cell agglutinins resulting in a 3 tube drop in the original unadsorbed titer; 2) removal of these sheep cell agglutinins by beef cell antigen.

NOTE: A titer of less than 1:56 can be observed in cases of infectious mononucleosis when samples are taken too soon after onset of the disease and results would therefore be inconclusive. Should this occur an additional sample should be taken 3 – 5 days later and a subsequent sample about 10 days to 2 weeks later.

LIMITATIONS OF THE PROCEDURE

Serological tests for infectious mononucleosis are not designed as a substitute for clinical and hematological investigations, but should be used in addition to these tests to obtain the most accurate diagnosis, and to detect those cases in which clinical indications are inconclusive and/or hematological studies are unavailable. A positive heterophile test concurrent with clinical and hematological results can be considered diagnostic for infectious mononucleosis. However, because of the nonspecific nature of heterophile tests, absence of these agglutinins in suspect cases does not rule out the possibility of infectious mononucleosis.

SPECIFIC PERFORMANCE CHARACTERISTICS

It is generally accepted that the Davidsohn differential test is specific for heterophile agglutinins associated with infectious mononucleosis. The high degree of specificity exhibited by this test is due to the nature of heterophilic antibodies and their unique reactions with kidney and beef cell antigens. False positives are seldom observed and are usually due to the technique employed rather than the actual reliability of the test itself.[15]

The sensitivity of this heterophilic agglutination test is somewhat less. False negatives can occur or inconclusive results may be obtained when the differential test is performed. False negative results may be obtained when attempting to detect heterophile agglutinins in children with infectious mononucleosis.[16] Titers may be low or in some cases completely undetectable.

An inconclusive result may be obtained when the unadsorbed titer is 1:28 or less. This usually arises when the test is performed too soon after the onset of the infection (usually 1 – 7 days).

REFERENCES

1. Davidsohn, I., Amer. J. of Clin. Path. Technical Suppl., 6:56 – 60, 1938.
2. Paul, J. R. and Bunnell, W. W., Amer. J. of Med. Sci., 183:90 – 104, 1932.
3. Davidsohn, I., 1938.
4. Jacobs, D. S., Amer. J. of Med. Tech., Jan – Feb, 30 – 40, 1968.
5. Davidsohn, I., Stern, K., and Kashiwagi, C., Amer. J. of Clin. Path., 21:1101 – 1113, 1951.
6. Davidsohn, I., Arch. Path. and Lab. Med., 4:776 – 806, 1927.
7. Davidsohn, I., J. of Immunology, 16:259 – 273, 1929.
8. Davidsohn, I., J. of Immunology, 18:31 – 49, 1930.
9. Davidsohn, I., J. of Infectious Diseases, 53:219 – 229, 1933.
10. Paul, J. R., and Bunnell, W. W., 1932.
11. Stuart, C. A., et al., Arch. Int. Med., 54:199, 1934.
12. Stuart, C. A., Proceedings Soc. Exp. Biol. Med., 32:861, 1935.
13. Stuart, C. A., et al., Arch. Int. Med., 58:512, 1936.
14. Davidsohn, I., JAMA, 108:289, 1937.
15. Davidsohn, I., 1951.
16. Chessin, L., Miller, J. M., and Mogabgab, W. J., Patient Care, July, 1968.

PACKAGING

Bacto Beef Cell Antigen, Desiccated	12 × 1 ml	0475-52-3
	6 × 5 ml	0475-57-8
Bacto Guinea Pig Kidney Antigen, Desiccated	12 × 1 ml	0476-52-2
	5 ml	0476-56-8
	6 × 5 ml	0476-57-7
Bacto Heterophile Forssman Antiserum	3 ml	0873-47-9
Bacto Infectious Mononucleosis Positive Serum	3 ml	0891-47-7

INVERTASE ANALYTICAL

Invertase Analytical, a stable enzymatic extract of yeast cells, is used for preparing invert sugar from sucrose. The packaged ampules contain a standardized extract, 5 ml of which will cause complete inversion of 50 ml of a 10% sucrose solution in 1 hour at room temperature. Invertase Analytical is most active at a reaction of pH 4.4 – 4.6. The K value of Invertase Analytical is K (A.O.A.C.) = 0.1.

PACKAGING

Invertase Analytical	12 × 10 ml	0154-61-0

BACTO ISOELECTRIC CASEIN

Bacto Isoelectric Casein is a protein of exceptional purity, prepared by repeated precipitation of casein at its isoelectric point. This highly purified casein is recommended for use in routine bacteriological culture media and also for the most exacting procedures. It is suitable for the Fuld Gross method for determining tryptic activity. Bacto Isoelectric Casein is soluble in a concentration of 1 – 2% in a slightly alkaline solution.

Sodium caseinate is readily prepared by suspending Bacto Isoelectric Casein in water, adjusting to a slight alkaline reaction with sodium hydroxide solution and maintaining this alkalinity until solution is complete. When prepared in this manner the solution of Bacto Isoelectric Casein is light in color and slightly opalescent.

PACKAGING

Bacto Isoelectric Casein   100 g   0145-15-9

K AGAR

K Agar is the potassium salt of kappa-carageenan. It is a gelling agent but is not a complete substitute for agar. However, it can replace agar in several bacteriological culture media. Lines[1] reported some difficulties in the preparation of blood plates, pH adjustment, and handling, but indicated its usefulness in some routine culture media. It was especially clear when prepared in semisolid motility media, thus enabling results to be more easily read.

REFERENCE

1. Lines, Anthony D. "Value of the K⁺ Salt of Carageenan as an Agar Substitute in Routine Bacteriological Media," Applied and Environmental Microbiology, p. 637 – 639, Dec. 1977.

PACKAGING

K Agar   1 lb (454 g)   0526-01-4

BACTO KCN BROTH BASE

INTENDED USE

Bacto KCN Broth Base is recommended for the differentiation of *Enterobacteriaceae* on the basis of their ability to grow in the presence of potassium cyanide.

HISTORY

Bacto KCN Broth Base is prepared according to the formula given by Moeller[1] as modified by Edwards and Ewing[2] and Edwards and Fife.[3]

PRINCIPLES

Moeller[1] showed that media containing potassium cyanide permitted differential growth of *Enterobacteriaceae*. *E. coli*, *Salmonella* and *Shigella* were inhibited in the medium while members of the *Klebsiella*, *Citrobacter* and *Proteus* groups grew unrestrictedly. *Citrobacter freundii* also grew in the medium.

Edwards and Fife[3] compared various peptones in different concentrations in the Moeller KCN medium. They reported that Proteose Peptone No. 3 in 0.3% concentration was most satisfactory in that the medium so prepared supported excellent growth of *Citrobacter* strains while it inhibited *Salmonella*. These same authors studied 1,981 cultures of the *Salmonella*, *Arizona (Salmonella arizonae)* and *Citrobacter* organisms in this medium. Of the 580 *Citrobacter* cultures, 574 or 99%, grew in 24 hours incubation. Only 10 *Salmonella* cultures out of 900 showed growth in 48 hours. Among the 501 *Arizona* strains tested, 39 grew in the medium. The authors recommended the use of this medium in the diagnostic laboratory. Their experience indicated that the complete medium could be stored for 30 days at 4°C without loss of cyanide or alteration in culture value.

FORMULA

BACTO KCN BROTH BASE
DEHYDRATED

Ingredients per liter

Proteose Peptone No. 3, Difco	3 g
Disodium Phosphate	5.64 g
Monopotassium Phosphate	0.225 g
Sodium Chloride	5 g

Final pH 7.6 ± 0.2 at 25°C.

One pound will make 32.9 liters of medium.

METHOD OF PREPARATION
1. Dissolve 13.8 g in 1 liter distilled or deionized water.
2. Distribute in 100 ml amounts and sterilize in the autoclave for 15 minutes at 15 lbs pressure (121°C).
3. Cool the sterile basal medium to room temperature.
4. Prepare a 0.5% solution of potassium cyanide in distilled or deionized water and cool to room temperature.
5. Using a syringe or bulb pipette, add 1.5 ml of the potassium cyanide solution to each 100 ml basal medium.
6. Distribute into tubes and stopper immediately with paraffined stoppers.

Caution: Extreme care should be taken in handling and disposing of potassium cyanide.

PROCEDURE
The tubes are inoculated heavily with 1 – 3 loops of a 24 hour broth culture of the test organisms in Bacto KCN Broth Base without added potassium cyanide. The tubes are quickly stoppered and incubated at 37°C for 2 days. Observations for growth are made at the end of 24 and 48 hours incubation.

STORAGE
Bacto KCN Broth Base	Below 30°C
Prepared tubes	2 – 8°C

QUALITY CONTROL
Identity Specifications

Dehydrated powder:	light beige, homogeneous, free-flowing
Reaction of 1.38% solution:	pH 7.6 ± 0.2 at 25°C
Prepared medium:	very light amber, clear

Typical Culture Response in Bacto KCN Broth
After 48 Hours at 35°C

	Growth (plain)	Growth (w/KCN)
Citrobacter freundii ATCC® 8090	good	good
Escherichia coli ATCC® 25922	good	inhibited
Klebsiella pneumoniae ATCC® 13883	good	inhibited*
Proteus vulgaris ATCC® 6380	good	good
Providencia alcalifaciens ATCC® 9886	good	good
Salmonella arizonae ATCC® 13314	good	inhibited
Salmonella enteritidis ATCC® 13076	good	inhibited
Shigella flexneri ATCC® 12022	good	inhibited

*other strains may not be inhibited

REFERENCES

1. Acta. Pathol. Microbiol. Scand., 34:115, 1954.
2. Edwards, P. R. and Ewing, W. H., Identification of *Enterobacteriaceae*. Burgess Pub. Co., 1955.
3. Appl. Microbiol., 4:46, 1956.

PACKAGING

Bacto KCN Broth Base	1 lb (454 g)	0647-01-8
	1/4 lb (114 g)	0647-02-7

BACTO KF STREPTOCOCCUS AGAR
BACTO KF STREPTOCOCCUS BROTH

INTENDED USE

Bacto KF Streptococcus Agar and Broth are selective media recommended for detecting and enumerating fecal streptococci.

HISTORY

Bacto KF Streptococcus media are prepared according to the formula of Kenner, Clark and Kabler[1,2] developed for use in detecting fecal streptococci in polluted surface waters. The authors compared KF Streptococcus media with other media and reported their formulation to be superior to all others tested in either the MPN multiple tube procedure or membrane filter technique. Currently KF Streptococcus agar is recommended by the American Public Health Association[3] for performing plate counts detecting fecal streptococci in water.

PRINCIPLES

Bacto KF Streptococcus Agar is the same formulation as the broth but with an additional 20 g of agar per liter as a solidifying agent. Maltose and lactose are the fermentable carbohydrates, sodium azide the selective agent and brom cresol purple the indicator dye. The addition of triphenyltetrazolium chloride 1% (TTC) causes the enterococci to have a deep red color as a result of tetrazolium reduction to an acid azo dye.

FORMULAE

BACTO KF STREPTOCOCCUS AGAR
DEHYDRATED

Ingredients per liter

Proteose Peptone No. 3, Difco	10 g
Bacto Yeast Extract	10 g
Sodium Chloride	5 g
Sodium Glycerophosphate	10 g
Maltose	20 g
Lactose	1 g
Sodium Azide	0.4 g
Bacto Brom Cresol Purple	0.015 g
Bacto Agar	20 g

Final pH 7.2 ± 0.2 at 25°C.

One pound will make 5.9 liters of final medium.
Use 76.4 g/liter.

BACTO KF STREPTOCOCCUS BROTH
DEHYDRATED

Ingredients per liter

Proteose Peptone No. 3, Difco	10 g
Bacto Yeast Extract	10 g
Sodium Chloride	5 g
Sodium Glycerophosphate	10 g
Maltose	20 g
Lactose	1 g
Sodium Azide	0.4 g
Bacto Brom Cresol Purple	0.015 g

Final pH 7.2 ± 0.2 at 25°C.

One pound will make 8 liters of final medium.
Use 56.4 g/liter.

METHOD OF PREPARATION
M.P.N. Procedure
To rehydrate Bacto KF Streptococcus Broth in single strength concentration, for use in the M.P.N. test on water samples of 1 ml or less, suspend 56.4 g in 1 liter cold distilled water and heat to boiling to dissolve completely. Dispense 10 ml amounts into culture tubes and sterilize for 10 minutes at 15 lbs pressure (121°C).

To rehydrate Bacto KF Streptococcus Broth for use in the M.P.N. test using 10 ml water samples, suspend 76.4 g in 1 liter cold distilled water and heat to boiling to dissolve completely. Dispense in 20 ml amounts in Public Health culture tubes and sterilize for 10 minutes at 15 lbs pressure (121°C).

Inoculated tubes are incubated at 34 – 36°C for 48 hours. M.P.N. tubes are positive when turbid growth appears imparting a yellow color to the medium without foaming. Where foaming occurs confirmation for streptococci should be made by Gram staining.

Membrane Filter Procedure
To rehydrate Bacto KF Streptococcus Broth for use in the membrane filter procedure suspend 56.4 g in 1 liter cold distilled water and heat to boiling to dissolve completely. Dispense in 100 ml amounts into flasks and sterilize for 10 minutes at 15 lbs pressure (121°C). Cool to 60°C and add one ml Bacto TTC Solution 1% (triphenyltetrazolium chloride 1%) per 100 ml sterile broth. (The authors suggested an alternate procedure of adding one ml of an unsterilized 1% TTC solution to 100 ml sterile medium at the boiling temperature and allowing it to cool.) Autoclave sterilization of the medium containing tetrazolium derivatives is not recommended. The sterile medium containing TTC is used to saturate the paper pad on which the inoculated membrane is laid. The inoculated membrane filters are incubated in an atmosphere saturated with water vapor for 48 hours at 34 – 36°C. All red or pink colonies visible with 15X magnification are counted as streptococcal colonies.

Streptococcus Plate Count
To rehydrate Bacto KF Streptococcus Agar suspend 76.4 g in 1 liter cold distilled water and heat to boiling to dissolve completely. Continue heating for an additional 5 minutes. Dispense in 100 ml quantities or as desired in flasks. Do not autoclave. Overheating will lower the pH and render the medium less productive. Cool to 50°C and add 1 ml Bacto TTC Solution 1% (triphenyltetrazolium chloride 1%) per 100 ml sterile medium. Mix to obtain uniform distribution of the TTC throughout the medium. Allow to cool to 45°C and use for making pour plates. Incubate inoculated plates for 48 hours at 34 – 36°C and with the aid of a dissecting microscope with a magnification of 15 diameters count all colonies showing a red or pink center as streptococci.

STORAGE
Bacto KF Streptococcus media	Below 30°C
Prepared media	2 – 8°C

QUALITY CONTROL
Identity Specifications
Dehydrated powder:	light greenish beige
Reaction of appropriate solution:	pH 7.2 ± 0.2 at 25°C
Prepared medium:	light purple, clear to slightly opalescent

Typical Cultural Response in/on Bacto KF Streptococcus Media
After 46 – 48 Hours at 34 – 36°C

Organism	Growth	Color of Colony
Enterobacter aerogenes ATCC® 13048	inhibited	—
Escherichia coli ATCC® 25922	inhibited	—
Streptococcus faecalis ATCC® 19433	good to excellent	red
Streptococcus faecalis ATCC® 29212	good to excellent	red

REFERENCES

1. Appl. Micro., 9:15, 1961.
2. Am. J. Publ. Hlth., 50:1553, 1960.
3. Standard Methods for the Examination of Water and Wastewater Analysis, 15th Ed., Greenberg, A. E., Connors, H. and Jenkins, G., Franson, M. A., A.P.H.A Washington, D.C., 1981.

PACKAGING

Bacto KF Streptococcus Agar	1 lb (454 g)	0496-01-0
Bacto KF Streptococcus Broth	1 lb (454 g)	0997-01-4
Bacto TTC Solution 1%	30 ml	3112-67-9

KL VIRULENCE AGAR AND ENRICHMENT

BACTO KL VIRULENCE AGAR
BACTO KL VIRULENCE ENRICHMENT
BACTO KL ANTITOXIN STRIPS

INTENDED USE

Bacto KL Virulence Agar, supplemented with Bacto KL Virulence Enrichment, and Bacto KL Antitoxin Strips, are used in testing the toxigenicity (virulence) of *Corynebacterium diphtheriae* by the agar diffusion technique.

HISTORY/PRINCIPLES

Bacto KL Virulence Agar and Bacto KL Virulence Enrichment are prepared according to the formulation of Hermann, Moore and Parsons.[1]

Elek[2] first described the agar plate diffusion technique for demonstrating the in vitro toxigenicity (virulence) of *C. diphtheriae*. King, Frobisher and Parsons[3] expanded upon Elek's technique and by using a carefully standardized medium obtained results in complete agreement with animal inoculation tests. These authors demonstrated Proteose Peptone, Difco to possess properties essential for toxin production and through its incorporation into the medium were able to obtain consistent results. They used rabbit, sheep or horse serum as enrichments but demonstrated human serum to be unsatisfactory. Broom[4] using the medium of King et al. demonstrated great differences in the results obtained with serum from different species. She obtained best results with rabbit serum, however in her experience even rabbit serum from different animals varied widely. Hermann, Moore and Parsons[1] refined the medium used in the in vitro KL Virulence Test. They simplified the basal medium and developed a nonserous enrichment which overcame the irregularities encountered in previous formulations. The medium and enrichment described by these authors have been prepared and standardized for use in KL Virulence Tests.

FORMULA

BACTO KL VIRULENCE AGAR
DEHYDRATED
Ingredients per liter

Proteose Peptone, Difco . . . . 20 g
Sodium Chloride . . . . 2.5 g
Bacto Agar . . . . 15 g

Final pH 7.8 ± 0.2 at 25°C.

One pound will make 12.1 liters of basal medium.

Bacto KL Virulence Enrichment
Use 2 ml enrichment per 10 ml agar.

METHOD OF PREPARATION
1. Suspend 37.5 g Bacto KL Virulence Agar in 1 liter distilled or deionized water and heat to boiling to dissolve completely.
2. Sterilize in the autoclave for 15 minutes at 15 lbs pressure (121°C).
3. Allow to cool to 55 – 60°C in a water bath.
4. Aseptically add 2 ml Bacto KL Virulence Enrichment and 0.5 ml Bacto Chapman Tellurite Solution 1% to as many sterile 100 mm Petri dishes as desired. Using aseptic technique quickly pipette 10 ml Bacto KL Virulence Agar to a plate and rotate 20 times to obtain a uniform mixture. An alternate procedure can be used to prepare the plates. Dispense the KL virulence agar in 10 ml amounts into test tubes, sterilize for 15 minutes at 15 lbs pressure (121°C), cool to 50°C in a water bath, add the tellurite and enrichment and agitate to obtain a uniform mixture before dispensing into sterile Petri dishes.
5. Before the medium solidifies, place one Bacto KL Antitoxin Strip or a sterile filter paper strip 1 cm × 8 cm saturated with a potent diphtheria antitoxin across the diameter of the plate and allow it to sink through the medium to the bottom of the plate. Sterile forceps may be used, if necessary.
6. Allow the medium to solidify with the cover ajar to obtain a dry surface.

PROCEDURE
Inoculate the medium by streaking a loopful of a 24 hour culture in a single line across the plate perpendicular to the paper strip. As many as 5 cultures may be tested on a single plate.

Incubate plates at 37°C for 72 hours. Examine at 24, 48 and 72 hour intervals for lines of precipitation raying out at 45° angles from line of culture streak.

Toxigenic (virulent) cultures will show fine lines of precipitation. Nontoxigenic strains will show no lines of precipitation.

STORAGE
Bacto KL Virulence Agar	Below 30°C
Bacto KL Virulence Enrichment	2 – 8°C
Bacto KL Antitoxin Strips	2 – 8°C
Prepared plates	2 – 8°C

QUALITY CONTROL
Identity Specifications

Dehydrated powder: light beige, homogeneous, free-flowing
Reaction of 3.75% solution: pH 7.8 ± 0.2 at 25°C
Prepared medium: light medium amber, slightly opalescent, may have a slight precipitate
Enrichment: colorless to very light amber, clear
Strips: white paper, 1 × 8 cm

Typical Cultural Response on Bacto KL Virulence Agar
After 24 – 72 Hours at 37°C

Organism	Growth
Bacillus subtilis ATCC® 6633	–
Corynebacterium diphtheriae Type *gravis*	+
Corynebacterium diphtheriae Type *intermedius*	+
Corynebacterium diphtheriae Type *mitis*	+
Staphylococcus aureus ATCC® 25923	–

+ = positive, black line of precipitation at a 45° angle to the strip
– = negative, no line of precipitation

REFERENCES
1. Am. J. Clin. Path., 29:181, 1958.
2. Brit. Med. J., 1:493, 1948.
3. Am. J. Pub. Health, 39:1314, 1949.
4. Personal Communications, 1958.

PACKAGING

Bacto KL Virulence Agar	1 lb (454 g)	0985-01-8
	1/4 lb (114 g)	0985-02-7
Bacto KL Virulence Enrichment	6 × 20 ml	0986-63-2
	12 × 20 ml	0986-64-1
Bacto KL Antitoxin Strips	12 strips	3101-30-6
Bacto Chapman Tellurite Solution 1%	6 × 1 ml	0299-51-8
	6 × 25 ml	0299-66-1

BACTO KINGSBURY STANDARDS

Bacto Kingsbury Standards, prepared according to the specifications of Kingsbury, Clark, Williams and Post,[1] are permanent turbidity standards prepared from colloidal glass. They are used for direct visual comparisons and standardizations of such materials as bacterial suspensions in the preparation of vaccines.

The set consists of seven standards, corresponding to 5, 10, 20, 30, 40, 50 and 75% albumin. The colloidal glass, on long standing, partially settles to the bottom of the tubes. A homogeneous suspension is readily obtained by shaking the tubes vigorously until no sediment is visible on the bottom of the tubes.

REFERENCE
1. J. Lab. Clin. Med., 11:981, 1926.

PACKAGING

Bacto Kingsbury Standards	1 set	0484-32-7

KLEBSIELLA ANTISERA

INTENDED USE
Bacto Klebsiella Antisera are used in the serological identification of *Klebsiella pneumoniae* by the Neufeld (Quellung) reaction.

SUMMARY AND EXPLANATION OF TEST
Klebsiella pneumoniae may cause a variety of infections in man and is present in the nasopharynx of approximately 1 – 5% of the normal population. Prevention and control of these infections may be difficult as the *Klebsiellae* are capable of developing resistance to antibiotics and sulfonamides.

Klebsiella grow readily on routine plating media forming mucoid colonies. Identification of this genus and differentiation from other *Enterobacteriaceae,* particularly *Enterobacter,* is accomplished biochemically.

Serologically, 72 serotypes of *Klebsiella* have been described and are found widely in nasopharyngeal infections. Since antisera from some high types tend to cross-react when employed in the slide agglutination technique, the Quellung reaction is the method of choice because of the specificity of the capsular reaction.[1]

PRINCIPLES OF THE PROCEDURE
The Neufeld (Quellung) reaction is not a true swelling of the capsule. It has been reported that the apparent increase in capsule size is due to the reaction of the capsular substance and type specific serum antibody which forms an antigen-antibody complex on the cell surface and appears as a ground glass enlarged capsule.

REAGENTS
Bacto Klebsiella Antisera are stable, desiccated antisera for use in the Neufeld (Quellung) reaction for the identification of *Klebsiella pneumoniae* serotypes.[2,3]

When properly rehydrated, each antiserum contains approximately 1:5000 Merthiolate®, sufficient to maintain a bacteriostatic condition at a storage of 2 – 8°C.

When used according to the suggested procedure (2 loopfuls/test), each 1 ml is sufficient to perform approximately 25 tests.

REHYDRATION
To rehydrate Bacto Klebsiella Antisera, add 1 ml sterile distilled or deionized water and rotate the vial gently to completely dissolve the contents.

STORAGE
Store both desiccated and rehydrated Bacto Klebsiella Antisera at 2 – 8°C.

Prolonged exposure to room temperature or repeated freezing and thawing are detrimental to antisera.

EXPIRATION DATE
Bacto Klebsiella Antisera are stable to the expiry date on the label when stored at 2 – 8°C.

SPECIMEN PREPARATION
In order to assure optimal capsule formation, cultures of *Klebsiellae* should be grown on Bacto Worfel Ferguson Agar. Incubate 18 – 24 hours at 37°C.

The culture is examined for amount and size of capsule production in Bacto Loeffler Methylene Blue wet-mounts. After the size of the capsules has been determined, the Quellung reaction is performed.

Only cultures possessing capsules of moderate size should be used in the Quellung reaction. Cultures with very small capsules render the Quellung reaction difficult to observe while extremely large capsules tend to obscure positive reactions.

PROCEDURE
Materials Provided:
Bacto Klebsiella Antisera Poly Types 1 – 6
Bacto Klebsiella Antisera Types 1 – 6

Purchase Separately:
Bacto Loeffler Methylene Blue
Bacto McFarland Barium Sulfate
   Standards
Bacto Rabbit Serum Normal

Materials Required but not Provided:
Microscope slides
Bacteriological loop

Cover slips
Microscope

1. Prepare a dense suspension of the test culture grown on Bacto Worfel Ferguson Agar. The suspension should be prepared in 0.85% NaCl solution and have a density approximating a 1:100 dilution of a suspension equal to that of a Bacto McFarland Barium Sulfate Standard No. 3. This will yield a suspension having approximately 10 to 25 bacterial cells per microscopic field.

   Caution must be exercised that the bacterial suspension is not too dense; this might result in a minimal reaction.
2. Place a large loopful of the appropriate Bacto Klebsiella Antiserum on one end of a microscope slide.
3. Place a large drop of Bacto Rabbit Serum Normal on the other end of the slide.
4. Add one loopful of the bacterial suspension to each droplet. Care must be taken not to carry over positive serum to the control slide during the mixing process or that the two preparations are not allowed to mix during examination.
5. Add one loopful Bacto Loeffler Methylene Blue to each droplet and mix well.
6. Place a cover glass over each mixture and allow to stand at room temperature for 5 minutes before examining microscopically.
7. Examine both preparations microscopically using 800 to 1000X magnification.

RESULTS
Compare the capsule size of the control mixture (organism in normal serum) to that of the test mixture (organism in Bacto Klebsiella Antiserum). An apparent increase in size of the capsule in the test mixture indicates a positive reaction.

LIMITATIONS OF THE PROCEDURE
Serological techniques for the identification of *Klebsiella* can only serve to corroborate cultural and biochemical findings. As with other members of the *Enterobacteriaceae*, antigenic relationships do exist between genera, which makes cross reactions possible. Specifically, some O antigens of *Klebsiella* are identical to some O antigens of *E. coli*. For further information on these antigenic relationships refer to *Identification of Enterobacteriaceae*.[4]

REFERENCES
1. Personal Communication, 1967.
2. Identification of *Enterobacteriaceae*, Burgess Publ. Co., 1962.
3. J. Inf. Dis., 91:92, 1952.
4. Edwards, P. R., and Ewing, W. H., Identification of *Enterobacteriaceae*, Burgess Publ. Co., 1972.

PACKAGING

Bacto Klebsiella Antiserum
Serotype

Type 1	1 ml	2828-50-5
Type 2	1 ml	2829-50-4
Type 3	1 ml	2830-50-1
Type 4	1 ml	2831-50-0
Type 5	1 ml	2832-50-9
Type 6	1 ml	2833-50-8
Bacto Klebsiella Antiserum Poly Types 1 – 6	1 ml	2254-50-8
Bacto Klebsiella Antiserum Set	6 × 1 ml	2894-32-7

BACTO KLIGLER IRON AGAR

INTENDED USE

Bacto Kligler Iron Agar, a modification of the Kligler[1] medium, is recommended for the identification of gram-negative enteric bacilli based on the fermentation of dextrose and lactose and hydrogen sulfide production. It is recommended to identify further pure cultures of colonies picked from primary plating media, such as Bacto MacConkey Agar, Bacto SS Agar, Bacto Bismuth Sulfite Agar and others.

HISTORY

Kligler's[1] original medium was a soft nutrient agar containing dextrose, Andrade indicator and lead acetate. While experimenting with this medium and other combinations, Kligler discovered that Russell's medium, containing Andrade indicator and lead acetate, gave a good differentiation and recommended it for the differentiation of the typhoid, paratyphoid and dysentery groups.[2] Bailey and Lacey,[3] attempting to simplify the formula, found that phenol red was particularly adaptable and recommended that it be used as the indicator of hydrogen ion concentration. A similar medium including saccharose and incorporating Bacto Tryptone as a nutrient with ferrous sulfate and thiosulfate as the indicator of hydrogen sulfide production was developed by Sulkin and Willett.[4]

PRINCIPLES

Bacto Kligler Iron Agar combines the principles of Russell double sugar agar and lead acetate agar into one medium which permits a differentiation of the gram-negative rods both on the basis of their ability to ferment dextrose or lactose and on their ability to produce hydrogen sulfide. It differentiates the lactose-splitting organisms from the lactose nonfermenters; distinguishes *Salmonella typhi* from other *Salmonella* and *Shigella* and differentiates *S. paratyphi* (Paratyphoid A) from *S. schottmuelleri* and *S. enteritidis*. Bacto Kligler Iron Agar is prepared with phenol red as an indicator of the production of acid and ferrous sulfate as an indicator of hydrogen sulfide production. This combination of ingredients gives sensitive, distinct, clear-cut reactions.

FORMULA

BACTO KLIGLER IRON AGAR
DEHYDRATED

Ingredients per liter

Bacto Beef Extract	3 g
Bacto Yeast Extract	3 g
Bacto Peptone	15 g
Proteose Peptone, Difco	5 g
Bacto Lactose	10 g
Bacto Dextrose	1 g
Ferrous Sulfate	0.2 g
Sodium Chloride	5 g
Sodium Thiosulfate	0.3 g
Bacto Agar	12 g
Bacto Phenol Red	0.024 g

Final pH 7.4 ± 0.2 at 25°C.

One pound will make 8.2 liters of medium.

PROCEDURE

1. To rehydrate the medium, suspend 55 g in 1 liter distilled or deionized water and heat to boiling to dissolve the medium completely.
2. Dispense as desired in tubes.
3. Sterilize in the autoclave for 15 minutes at 15 lbs pressure (121°C).
4. Allow tubes to solidify in a slanted position so that generous butt is obtained.
5. Best reactions are obtained on freshly prepared media. If the medium is not used the day it is sterilized, melt the medium and solidify before use.
6. For typical cultural reactions in 18 hours, it is recommended that tubes of Bacto Kligler Iron Agar be inoculated heavily with growth from a solid culture medium by smearing over the surface of the slant and stabbing the butt. If inoculated from a suspension of organisms, or from broth culture, typical reactions of hydrogen sulfide production and reversion may not be obtained until 36 – 40 hours at 37°C.

To obtain true differential cultural reactions of this medium, it is necessary to have a pure culture. In inoculating directly from isolation media such as Bacto MacConkey Agar, Bacto SS Agar or Bacto Bismuth Sulfite Agar plates, select well isolated colonies and pick only the very center of the colony.

If there is any question as to the ability to obtain a pure culture from a certain colony, it is recommended that the suspicious colony be purified by streaking on Bacto MacConkey Agar before inoculating into Bacto Kligler Iron Agar. This procedure is always recommended to insure culture purity when picking from poured plates of Bacto Bismuth Sulfite Agar. It is often possible to detect contaminated cultures on Bacto Kligler Iron Agar slants. When this is the case, it is necessary to isolate the organism in pure cultures before its typical cultural reaction can be determined.

Organisms capable of fermenting dextrose but not lactose, *S. typhi* for example, will show an initial acid slant in short incubation periods. As the dextrose is utilized, the reaction under aerobic conditions reverts and becomes alkaline. Under anaerobic conditions, in the butt of the tube, these same organisms are not capable of causing a reversion of the reaction and remain acid.

Fermentation reactions in Bacto Kligler Iron Agar are similar to those in Bacto Russell Double Sugar Agar, i.e., a red slant and yellow butt with or without gas indicates fermentation of the small quantity of dextrose; a yellow slant and butt with or without gas formation indicates fermentation of the lactose; and a tube showing no change indicates that neither dextrose nor lactose has been fermented.

Bacto Kligler Iron Agar also indicates whether or not hydrogen sulfide has been produced. This is shown by a blackening of the medium.

STORAGE
Bacto Kligler Iron Agar   Below 30°C
Prepared tubes   2 – 8°C

QUALITY CONTROL
Identity Specifications

Dehydrated powder: pinkish beige, homogeneous, free-flowing
Reaction of 5.5% solution: pH 7.4 ± 0.2 at 25°C.
Prepared medium: orange-red, slightly opalescent, may have a slight precipitate

Typical Cultural Response in Bacto Kligler Iron Agar
After 18 – 48 Hours at 37°C

Organism	Recovery	Slant	Butt	Gas	H₂S
Citrobacter freundii ATCC® 8090	good	A	A	+	+
Escherichia coli ATCC® 25922	good	A	A	+	−
Proteus vulgaris ATCC® 6380	good	K	A	−	+
Salmonella enteritidis ATCC® 13076	good	K	A	+	+
Shigella flexneri ATCC® 12022	good	K	A	−	−

KEY
A = acid, yellow
K = alkaline, red
+H₂S = blackening
−H₂S = no change

REFERENCES
1. Am. J. Pub. Health, 7:1042, 1917.
2. J. Exp. Med., 28:319, 1918.
3. J. Bact., 13:183, 1927.
4. J. Lab. Clin. Med., 25:649, 1940.

PACKAGING
Bacto Kligler Iron Agar	1 lb (454 g)	0086-01-6
	1/4 lb (114 g)	0086-02-5
	12 tubes	0086-34-7
	144 tubes	0086-37-4

BACTO KOSER CITRATE MEDIUM

INTENDED USE
Bacto Koser Citrate Medium is a Standard Methods[1,2] medium for the differentiation of *Escherichia coli* and *Enterobacter aerogenes* based on their utilization of citrate.

HISTORY/PRINCIPLES
A means of differentiating coli-aerogenes organisms in sanitary studies was demonstrated by Koser,[3] who showed that either citric acid or its sodium salt as the only

source of carbon, when employed in a synthetic medium, is readily utilized by *E. aerogenes*, while *E. coli* fail to grow.

In using this medium, coli-like colonies from Endo or eosin methylene blue agar plates are inoculated into tubes of Koser citrate medium and, after 24 – 48 hours incubation, tubes showing marked turbidity may be assumed to contain organisms of the aerogenes group. In as much as coli-type organisms fail to grow in this medium, all tubes inoculated from coli-like colonies on Endo or eosin methylene blue agar plates and remaining clear after 36 hours incubation may be considered as coli.

Bacto Koser Citrate Medium is prepared according to Koser's[3] original formula. Chemically pure salts are used in the preparation of the medium and it is carefully tested to be sure that no sources of carbon, other than the citrate radical, or nitrogen other than ammonium salts, are present.

FORMULA

BACTO KOSER CITRATE MEDIUM
DEHYDRATED

Ingredients per liter

Sodium Ammonium Phosphate	1.5 g
Monobasic Potassium Phosphate	1 g
Magnesium Sulfate	0.2 g
Sodium Citrate	3 g

Final pH 6.7 ± 0.2 at 25°C.

One pound will make 79.6 liters of medium.

METHOD OF PREPARATION
1. Suspend 5.7 g in 1 liter distilled or deionized water.
2. Dispense into tubes.
3. Sterilize in the autoclave for 15 minutes at 15 lbs pressure (121°C).

STORAGE
Bacto Koser Citrate Medium	Below 30°C
Prepared medium	15 – 30°C

QUALITY CONTROL

Identity Specifications

Dehydrated powder:	white, homogeneous, free-flowing
Reaction of 0.57% solution:	pH 6.7 ± 0.2 at 25°C
Prepared medium:	colorless, clear, no significant precipitate

Typical Cultural Response in Bacto Koser Citrate Medium
After 18 – 24 Hours at 35°C

Organism	Growth
Enterobacter aerogenes ATCC® 13048	good to excellent
Enterobacter cloacae ATCC® 23355	good to excellent
Escherichia coli ATCC® 25922	none

REFERENCES
1. American Public Health Association, Standard Methods for the Examination of Water and Wastewater, 15th Ed., American Public Health Association, Inc., Washington, D.C., 1980.
2. American Public Health Association, Compendium of Methods for the Microbiological Examination of Foods, American Public Health Association, Inc., Washington, D.C., 1976.
3. Koser, Journal of Bacteriology, 8:493, 1923.

PACKAGING
Bacto Koser Citrate Medium

1 lb (454 g)
1/4 lb (114 g)

0015-01-2
0015-02-1

BACTO KUPFERBERG TRICHOMONAS BASE
BACTO KUPFERBERG TRICHOMONAS BROTH

INTENDED USE

Bacto Kupferberg Trichomonas Base and Bacto Kupferberg Trichomonas Broth are recommended for the detection and cultivation of the *Trichomonas* species.

HISTORY/PRINCIPLES

Bacto Kupferberg Trichomonas media are prepared according to the formulations of Kupferberg, Johnson and Sprince.[1] The broth is the base to which 100 mg chloramphenicol per liter has been added. The superiority of the culture procedure over the microscopic method for detecting the presence of *Trichomonas* in clinical specimens was demonstrated by Williams,[2] and Kean and Day.[3] Kupferberg[4] also demonstrated the greater accuracy of the culture procedure for detecting *Trichomonas* organisms in clinical material and stressed that negative cultures are the best criteria for ascertaining the efficacy of therapy in these infections. Adler and Pulvertaft,[5] Johnson, Trussel and John,[6] Williams[2] and Kupferberg[7] further demonstrated the value of the culture procedure as an adjunct in the diagnosis of trichomonal infections. These authors demonstrated that antibiotics could be used in fairly high concentrations to suppress bacterial growth without retarding the growth of these protozoa.

FORMULAE

BACTO KUPFERBERG TRICHOMONAS BASE
DEHYDRATED

Ingredients per liter

Bacto Tryptone	20 g
Bacto Maltose	1 g
Cysteine Hydrochloride	1.5 g
Bacto Agar	1 g
Bacto Methylene Blue	0.003 g

Final pH 6.0 ± 0.2 at 25°C.

One pound will make 19.3 liters
of final medium.
Rehydrate with 23.5 g/950 ml.

BACTO KUPFERBERG TRICHOMONAS BROTH
DEHYDRATED

Ingredients per liter

Bacto Tryptose	20 g
Bacto Maltose	1 g
Cysteine Hydrochloride	1.5 g
Bacto Agar	1 g
Bacto Methylene Blue	0.003 g
Chloramphenicol	0.1 g

Final pH 6.0 ± 0.2 at 25°C.

One pound will make 19.2 liters
of final medium.
Rehydrate with 23.6 g/950 ml.

PROCEDURE

1. Suspend appropriate amount in 950 ml distilled or deionized water. Heat to boiling to dissolve completely.
2. Sterilize in the autoclave at 15 lbs pressure 121°C, the base for 15 minutes but the broth for only 10 minutes.
3. Allow to cool to 50 – 55°C in a water bath and add 50 ml sterile beef or human serum (available as Bacto TC Bovine Serum, Desiccated, or Bacto TC Human Serum, Desiccated, rehydrated per label instructions).
4. Any additional desired antibiotics should be added at this point. 250 units penicillin and 1 mg streptomycin or 1 mg chloramphenicol per ml of final medium has been used successfully for primary culture work.

5. Mix gently to obtain a uniform solution and dispense aseptically in 10 ml amounts in sterile culture tubes.
6. The prepared media are inoculated with the clinical specimen and incubated at 30 – 37°C for up to 7 days.
7. A drop of the culture is then placed on a slide and examined microscopically with the low powder magnification. If the culture is positive, a diagnosis of trichomoniasis may be made; if negative, the tubes are incubated for an additional 48 – 72 hours and reexamined. If it is again negative, it is discarded.
8. Kupferberg[4] suggests culturing a specimen 1 week after completion of therapy. If this culture is negative, a second culture is taken in 2 – 4 weeks and if negative, the patient is considered cured.

STORAGE

Bacto Kupferberg Trichomonas Base	Below 30°C
Bacto Kupferberg Trichomonas Broth	Below 30°C
Prepared medium	2 – 8°C

QUALITY CONTROL

Identity Specifications

Dehydrated powder:	light beige, homogeneous, free-flowing
Reaction of appropriate solution:	pH 6.0 ± 0.2 at 25°C
Prepared medium:	blue-green, clear to very slightly opalescent, may have a slight precipitate.
Prepared tubes:	light amber with green ring at top, clear to very slightly opalescent, may have a slight precipitate

Typical Cultural Response in Bacto Kupferberg Trichomonas Base or Broth After 72 Hours at 30°C

Organism	Growth
*Pentatrichomonas hominis ATCC® 30000	good to excellent
Trichomonas gallinae ATCC® 30002	good to excellent
Trichomonas tenax ATCC® 30207	good to excellent
*Trichomonas vaginalis ATCC® 30001	good to excellent

*Incubate anaerobically

REFERENCES

1. Kupferberg, A. B., Johnson, G., and Sprince, H., Nutritional Requirements of Trichomonas vaginalis. Proc. Soc. Exper. Biol. Med., 67:304 – 308, 1948.
2. Williams, M. H., A Study of the Trichomonal Population in Experimentally Infected Rheusus Monkeys; I. The Efficiency of Intensive Microscopic Search Compared to Culture Technique. Am. J. Obst. and Gynec., 60:224 – 225, 1950.
3. Kean, B. H. and Day, E., Trichomonas vaginalis Infection. An Evaluation of Three Diagnostic Techniques with Data on Incidence. Am. J. Obst. and Gynec., 68:1510 – 1518, 1954.
4. Kupferberg, A. B., Trichomonas vaginalis: Nutritional Requirements and Diagnostic Procedures. International Rec. Med. and Gen. Practive Clinics, 168:709 – 717, 1955.
5. Adler, S. and Pulvertaft, R. J., The Use of Penicillin for Obtaining Bacteria-Free Cultures of Trichomonas vaginalis, Am. Trop. Med., 38:188 – 189, 1944.
6. Johnson, J. G., Trussel, M., and John, F., Isolation of Trichomonas vaginalis with Penicillin, Science, 102:126 – 128, 1945.
7. Kupferberg, A. B., Personal Communication.

PACKAGING

Bacto Kupferberg Trichomonas Base	1 lb (454 g)	0910-01-8
Bacto Kupferberg Trichomonas Broth	1 lb (454 g)	0911-01-7

BACTO LICNR BROTH

(LYSINE-IRON-CYSTINE-NEUTRAL RED BROTH)

INTENDED USE

Bacto LICNR Broth is a differential medium for the rapid presumptive detection of *Salmonella* in foods, food ingredients and feed materials. It is a modification of the formula of Hoben, Ashton and Peterson[1] who established its usefulness for detecting *Salmonella* in food samples in three days, thus reducing the holding time for foods and food ingredients.

FORMULA

BACTO LICNR BROTH
DEHYDRATED

Ingredients per liter

Bacto Yeast Extract . . . . 3 g	Bacto Dextrose . . . . 1 g
Bacto Tryptone . . . . 5 g	Bacto Salicin . . . . 1 g
(Pancreatic Digest of Casein)	Ferric Ammonium Citrate . . . . 0.5 g
L-Lysine Monohydrochloride . . . . . . 10 g	Sodium Thiosulfate . . . . 0.1 g
L-Cystine . . . . 0.1 g	Bacto Neutral Red . . . . 0.025 g
Mannitol . . . . 5 g	

Final pH 6.2 + 0.2 at 25°C.

Five hundred grams will make 19.4 liters of final medium.

PREPARATION OF MEDIUM

1. Suspend 25.7 g in 1 liter distilled or deionized water and heat to boiling for 1 – 2 minutes.
2. Dispense into tubes in 10 ml amounts.
3. Sterilize in the autoclave for 15 minutes at 15 lbs pressure (121°C).
4. The pH at this point is 6.2 ± 0.2 at 25°C.
5. Aseptically add 0.1 ml filter sterilized aqueous solution containing 1500 μg/ml novobiocin to each 10 ml tube. Mix well before using.

TEST PROCEDURE

1. Add 25 g of the test sample to 225 ml Bacto Lactose Broth and blend for 2 minutes at high speed using a suitable blender. Incubate the suspension for 20 – 24 hours at 35 – 37°C.
2. Add 1 ml of the above culture suspension to a 10 ml tube of Bacto Tetrathionate Broth. Incubate for 20 – 24 hours at 35 – 37°C.
3. Add 1 ml of the above secondary culture to a 10 ml tube of LICNR Broth containing 15 μg/ml novobiocin. Incubate for 24 hours at 35 – 37°C.
4. Observe the reaction. A black precipitate resulting from H_2S formation and/or a yellow color indicating an alkaline reaction give presumptive evidence of *Salmonella* in the original sample.
5. To eliminate the possibility of non H_2S producing *Salmonella*, incubate the tubes for an additional 16 – 24 hours. Prepare a brom thymol blue solution by dissolving 0.3 g Bacto Brom Thymol Blue in 2 ml 0.1 N NaOH and diluting to 100 ml with 50% ethanol in distilled or deionized water. Add 0.1 ml of the 0.3% solution brom thymol blue to each tube. A color change of the medium from yellow to dark green or blue indicates an alkaline reaction and the presence of *Salmonella*.

STORAGE

Bacto LICNR Broth	Below 30°C
Prepared medium	2 – 8°C

QUALITY CONTROL

Identity Specifications

Dehydrated powder: light pink, homogeneous, free-flowing
Reaction of 2.57% solution: pH 6.2 ± 0.2 at 25°C
Preared medium: red, clear

Typical Cultural Response in Bacto LICNR Broth
After 24 – 48 Hours at 35°C

Organism	Growth	Color of Medium	Color of Medium After Brom Thymol Blue	H_2S
Escherichia coli ATCC® 25922	inhibited	red	red-blue	−
Salmonella enteritidis ATCC® 13076	good	yellow	dark green-blue	+
Salmonella typhi ATCC® 19430	good	yellow	dark green-blue	+
Shigella flexneri ATCC® 12022	inhibited	red	red-blue	−

REFERENCE
1. Applied microbiology, 25:123, 1973.

PACKAGING
Bacto LICNR Broth	500 g	1806-17-5
Bacto Brom Thymol Blue	5 g	0202-11-3
Bacto Lactose Broth	1 lb (454 g)	0004-01-5
	1/4 lb (114 g)	0004-02-4
	5 lb (2.27 g)	0004-05-1
Bacto Tetrathionate Broth Base	500 g	0104-17-6
	100 g	0104-15-8
	5 lb (2.27 kg)	0104-05-0

BACTO LACTALBUMIN

Bacto Lactalbumin is the nonsoluble denatured protein fraction of milk and is used as an ingredient in culture media.

PACKAGING
Bacto Lactalbumin	100 g	0693-15-5

BACTO LACTOBACILLI MRS BROTH

INTENDED USE
Bacto Lactobacilli MRS Broth is used for the cultivation of lactobacilli.

HISTORY/PRINCIPLES
Bacto Lactobacilli MRS Broth is prepared according to the formulation of deMan, Rogosa and Sharpe[1] with slight modification. This medium was demonstrated by the authors to support luxuriant growth of all lactobacilli from oral, dairy, fecal and other sources.

FORMULA

BACTO LACTOBACILLI MRS BROTH
DEHYDRATED

Ingredients per liter

Bacto Proteose Peptone No. 3	10 g	Ammonium Citrate	2 g
Bacto Beef Extract	10 g	Sodium Acetate	5 g
Bacto Yeast Extract	5 g	Magnesium Sulfate	0.1 g
Dextrose	20 g	Manganese Sulfate	0.05 g
Sorbitan Monooleate Complex	1 g	Disodium Phosphate	2 g

Final pH 6.5 ± 0.2 at 25°C.

One pound will make 8.25 liters of medium.

METHOD OF PREPARATION
1. Suspend 55 g in 1 liter of distilled or deionized water and heat to boiling to dissolve completely.
2. Distribute into tubes, bottles, or flasks as desired.
3. Sterilize in the autoclave for 15 minutes at 15 lbs pressure (121°C).

STORAGE
Bacto Lactobacilli MRS Broth   2 – 8°C
Prepared medium   2 – 8°C

QUALITY CONTROL
Identity Specifications

Dehydrated powder: tan, homogeneous, free-flowing
Reaction of 5.5% solution: pH 6.5 ± 0.2 at 25°C
Prepared medium: dark amber, clear

Typical Cultural Response in Bacto Lactobacilli MRS Broth
After 18 – 24 Hours or longer at 35°C

Organism	Growth
Lactobacillus fermentum ATCC® 9338	good to excellent
Lactobacillus leichmannii ATCC® 7830	good to excellent
Lactobacillus sp ATCC® 11506	good to excellent

REFERENCE
1. deMan, Rogosa and Sharpe, 1960, Journal of Applied Bacteriology, 23:130.

PACKAGING
Bacto Lactobacilli MRS Broth	1 lb (454 g)	0881-01-3
	1/4 lb (114 g)	0881-02-2

BACTO LACTOSE BROTH

INTENDED USE
Bacto Lactose Broth is a Standard Methods medium for detection of coliform bacteria in water,[1] foods,[2] and dairy products.[3]

HISTORY
Since 1917 Standard Methods of Water Analysis, in the interest of greater uniformity, had recommended the use of a beef extract lactose broth in place of the infusion me-

dium formerly employed. Bacto Lactose Broth is prepared according to the formulation recommended in *Standard Methods.*

PRINCIPLES
In the determination of the potability of drinking water one of the most important tests is the detection of possible fecal contamination. *Escherichia coli* is the organism most frequently used as the indicator of fecal contamination of potable water.

The demonstration of the presence of coliform bacteria in water has been reduced to a relatively simple process:[1]
1. The determination of growth and gas production in lactose broth resulting from the direct inoculation of water (presumptive phase).
2. The inoculation of differential or selective media from positive tubes of lactose broth (confirmed phase).
3. The identification of gram-negative, nonsporulating, aerobic organisms capable of producing gas when reinoculated into lactose broth (completed phase).

In testing dairy products, lactose broth is used only in the completed test.[3]

Bacto Lactose Broth yields maximum growth of the organisms of the coli-aerogenes group.[4] *Standard Methods*[1] specifies that when using inocula greater than 1 ml, multiple strength lactose broth shall be prepared to maintain the concentration of nutriments as given in the table below:

Concentrations of Dehydrated Medium Required to Maintain the Proper Concentration of Ingredients

Inoculum	Amount Medium in Tube	Vol. Medium and Inoculum	Bacto Lactose Broth used per 1 liter
1 ml	10 ml or more	11 ml or more	13 g
10 ml	30 ml	40 ml	17.3 g
10 ml	20 ml	30 ml	19.5 g
100 ml	50 ml	150 ml	39 g
100 ml	35 ml	135 ml	49.4 g
100 ml	20 ml	120 ml	78 g

FORMULA

BACTO LACTOSE BROTH
DEHYDRATED

Ingredients per liter

Bacto Beef Extract	3 g
Bacto Peptone	5 g
Bacto Lactose	5 g

Final pH 6.9 ± 0.2 at 25°C.

One pound will make 34.9 liters of final medium.

METHOD OF PREPARATION
1. Suspend 13 g in 1 liter distilled or deionized water and warm slightly to dissolve completely. A more concentrated medium may be prepared if required (see PRINCIPLES).
2. Dispense into tubes as desired.
3. Place an inverted fermentation vial in each tube.
4. Place closure on tubes and sterilize in the autoclave for 15 minutes at 15 lbs pressure (121°C).

5. Before opening autoclave, allow temperature to drop below 75°C to avoid entrapped air bubbles in inverted vials.
6. Follow the recommended procedure in the Standard Methods compendium for your use.

STORAGE
Bacto Lactose Broth   Below 30°C
Prepared medium   15 – 30°C

QUALITY CONTROL
Identity Specifications
Dehydrated powder: light beige, homogeneous, free-flowing
Reaction of 1.3% solution: pH 6.9 ± 0.2 at 25°C
Prepared medium: light to medium amber, clear, no significant precipitate

Typical Cultural Response in Bacto Lactose Broth
After 18 – 48 Hours at 35°C

Organism	Growth	Gas
Enterobacter aerogenes ATCC® 13048	good to excellent	+
Escherichia coli ATCC® 25922	good to excellent	+
Pseudomonas aeruginosa ATCC® 27853	good to excellent	−
Streptococcus faecalis ATCC® 19433	good to excellent	−

REFERENCES
1. American Public Health Association, 1975, Standard Methods for the Examination of Water and Wastewater, 14th Ed., American Public Health Association, Inc., Washington, D.C.
2. American Public Health Association, 1976, Compendium of Methods for the Microbiological Examination of Foods, American Public Health Association, Inc., Washington, D.C.
3. American Public Health Association, 1978, Standard Methods for the Examination of Dairy Products, 14th Ed., American Public Health Association, Inc., Washington, D.C.
4. Public Health Reports, 44:2863:1929.

PACKAGING
Bacto Lactose Broth	1 lb (454 g)	0004-01-5
	1/4 lb (114 g)	0004-02-4
	5 lb (2.27 kg)	0004-05-1

BACTO LASH SERUM MEDIUM

INTENDED USE
Bacto Lash Serum Medium is a complete prepared medium for the cultivation of *Trichomonas vaginalis* from clinical specimens. It is an isotonic casamino acid, serum medium prepared according to the formulation of Lash[1] with slight modifications.

STORAGE
Bacto Lash Serum Medium   2 – 8°C

PROCEDURE
Obtain specimen with a sterile swab and place in a tube of Bacto Lash Serum Medium. Incubate at 37°C for 12 to 18 hours and examine microscopically for *T. vaginalis* by wet mount slide technique using the low power and high dry objectives.

REFERENCE
1. American Journal of Tropical Medicine, 30:641, 1950.

PACKAGING

Bacto Lash Serum Medium   12 tubes   1016-34-0

BACTO LATEX 0.81

INTENDED USE

Bacto Latex 0.81, a standardized suspension of latex particles having an average diameter of 0.81 micrometers, is used in the latex fixation tests for rheumatoid arthritis (detection of Rheumatoid Factor), certain nonrheumatic diseases, and other tests for cryptococcosis, tularemia, brucellosis, leptospirosis, salmonellosis, and trichinosis.[1,2,3] Bacto Latex 0.81 has proven to be an excellent inert carrier of antibody or antigen for use in rapid detection procedures. It is particularly applicable to the RA serological procedures described by Singer and Plotz[4] and as modified by Rheins et al.,[5] Rothermich and Philips,[6] Goften et al.[7] and Singer and Plotz.[8] As high as 90% sensitivity and specificity in the RA agglutination test was obtained with the standardized latex.

Bacto Latex 0.81 with eosin solution, adapted to a latex fixation slide screening test for rheumatoid arthritis was also described by Singer and Plotz.[9] This is available in a set that includes all the reagents for 100 tests, the Bacto Rheumatoid Slide Test, code 3237. The usefulness and practicality of the test in the laboratory for the detection of rheumatoid arthritis and certain nonrheumatic diseases was described by Bianchi and Keech,[10] Lane and Decker,[11] Dresner and Trombly,[12] Caplan,[13] Howell, Malcolm and Pike,[14] Bonomo, LoSpalluto and Ziff[15] and Kunkel, Simon and Fudenberg.[16] A quantitative titration of a positive slide screening test by the multiple dilution technique is recommended, using the Bacto Rheumatoid Titration Set, code 3300.

STORAGE AND EXPIRATION DATE

The latex suspension is stable to the expiry date on the label when stored at 2 – 8°C. **Do not freeze.** Exposure to extremes in temperature may render latex unsatisfactory for use in latex fixation tests. To assure reagent is suitable for use, a spot check may be made by placing a drop of Bacto Eosin Solution on a slide and adding 2 drops of uniformly suspended Bacto Latex 0.81 over about a 1 square inch area. Tilt the slide back and forth over a light viewer for 3 minutes. Absence of agglutination indicates reagent is satisfactory. Discard any latex which autoagglutinates.

For instructions on the use of Bacto Latex 0.81, refer to Bacto Rheumatoid Slide Test and Bacto Rheumatoid Titration Test.

REFERENCES
1. Proc. Soc. Exp. Biol. & Med., 114:64, 1963.
2. Abs. Bioan. Tech., 11:133, 1963.
3. Publ. Hlth. Rpts., 78:227, 1963.
4. Am. J. Med., 21:888, 1956.
5. J. Lab. & Clin. Med., 50:113, 1957.
6. JAMA, 164:1999, 1957.
7. J. Can. Med. Assoc., 77:1098, 1957.
8. Arth. & Rheum., 1:142, 1958.
9. JAMA, 168:180, 1958.
10. ibid., 185:318, 1963.
11. ibid., 173:982, 1960.

12. New Engl. J. Med., 261:981, 1959.
13. Tufts New Engl. Med. Ctr. Bull., 4:136, 1958.
14. Am. J. Med., 29:662, 1960.
15. Arth. & Rheum., 6:104, 1963.
16. *ibid.*, 1:289, 1958.

PACKAGING

Bacto Latex 0.81	5 ml	3102-56-4
	6 × 5 ml	3102-57-3
	25 ml	3102-65-3
	1 liter	3102-81-3

BACTO LAURYL TRYPTOSE BROTH

INTENDED USE

Lauryl Tryptose Broth is a *Standard Methods* medium for detecting coliform bacteria in water, wastewater and food.[1,2]

HISTORY/PRINCIPLES

Bacto Lauryl Tryptose Broth is prepared according to the formula of Mallmann and Darby.[3]

The fermentation of lactose with the production of gas has been used as an indicator of the potability of water for many years. The formation of gas from lactose broth constitutes a presumptive test for the coliform group (this term includes all aerobic and facultative anaerobic, gram-negative nonspore-forming bacilli which are capable of producing gas from lactose).[4] Cowls[5] showed that the addition of sodium lauryl sulfate to lactose broth gave a medium selective for the coliform group.

Darby and Mallmann[6] demonstrated the value of Bacto Tryptose in the detection of coliform organisms. In a 2% concentration of Bacto Tryptose, the rate of reproduction during the early logarithmic growth phase was increased over that obtained with Bacto Peptone. The addition of phosphate buffer to the Bacto Tryptose medium caused a greater growth in the late logarithmic phase and slightly greater increase during the lag phase than did the nonbuffered medium. When sodium chloride was added to the medium, a marked increase in the rate of reproduction during the lag and early growth phases was observed. Their final medium permitted the so-called "slow lactose fermenters" to produce gas in greater quantities in a shorter period of time. The medium consisted of 2% Bacto Tryptose, 0.5% lactose, 0.4% dipotassium phosphate, 0.15% monopotassium phosphate, and 0.5% sodium chloride, and had a final reaction of pH 6.8.

In an attempt to improve the methods used to demonstrate members of the coliform group from water, Mallmann and Darby[3] investigated a large number of wetting agents, and showed that sodium lauryl sulfate gave best results as a noninhibitive selective agent for members of the coliform group. Optimum results were obtained by the addition of 1:10,000 sodium lauryl sulfate to the buffered tryptose lactose broth. In their comparative study in the checking of various types of water it was shown that the lauryl sulfate tryptose broth (Bacto Lauryl Tryptose Broth) gave a higher colon index than did the confirmatory *Standard Methods* media, and that gas production in the lauryl tryptose broth served not only as a presumptive test, but was also confirmatory of the presence of the coliform group for routine testing of water.

In a study of the coliform bacteria from chlorinated waters, Levine[7] compared lactose broth and lauryl tryptose broth. The latter medium gave fewer false positive presumptive tests than did lactose broth and suppressed the spore forming aerogenic bacteria. However, organisms showing a delayed fermentation of lactose were not eliminated by lauryl tryptose broth.

Bacto Lauryl Tryptose Broth was studied by 17 collaborating laboratories situated throughout the United States and Canada. The results of this comparative survey are reported by McCrady.[8] The study comprised the use of different types of water and embraced different methods of treatment of samples. The results showed that the substitution of lauryl tryptose broth for *Standard Methods* lactose broth would result in a reduction in the number of primary gas positives to be confirmed, and an increase in the number of positive coliforms. It was recommended that further study be made by different laboratories with particular reference to use of the medium in the examination of finished waters.

Perry and Hajna[9] in a comparative study of E C medium and lauryl tryptose broth reported both media to be highly sensitive and specific for coliform bacteria from water, shellfish and sewage. A positive presumptive test with either medium was more dependable than the usual "confirmed" or "completed" test.

The Ninth Edition of *Standard Methods for the Examination of Water and Sewage* allowed the substitution of lauryl tryptose broth for lactose broth in the standard tests for members of the coliform group in the examination of all waters except final filtered, treated and filter-treated waters: "It may be substituted for lactose broth also in the examination of final filtered, treated and filtered-treated waters provided the laboratory worker has amply demonstrated by correlation of positive completed tests (isolations of coliform organisms) secured through the use of lauryl sulfate tryptose broth with those secured through the use of lactose broth, in the examination of such waters, that the substitution results in no reduction from the density of coliform organisms indicated by the standard procedure using lactose broth."[4]

Bacto Lauryl Tryptose Broth may be prepared in single strength when examining 1 ml or less of water as an inoculum. For inocula of 10 ml consult the table given below.

Concentration of Dehydrated Medium Required to Maintain the Proper Concentration of Ingredients

Inoculum	Amt. Medium in Tube	Vol. Medium and Inoculum	Bacto Lauryl Tryptose Broth used per 1 liter
1 ml or less	10 ml	10 ml	35.6 g
10 ml	20 ml	30 ml	53.4 g
10 ml	30 ml	40 ml	47.3 g

FORMULA

BACTO LAURYL TRYPTOSE BROTH
DEHYDRATED

Ingredients per liter

Bacto Tryptose	20 g
Bacto Lactose	5 g
Potassium Phosphate, Dibasic	2.75 g
Potassium Phosphate, Monobasic	2.75 g
Sodium Chloride	5 g
Sodium Lauryl Sulfate	0.1 g

Final pH 6.8 ± 0.2 at 25°C

One pound will make 12.7 liters of single strength medium.

METHOD OF PREPARATION

1. Suspend 35.6 g or amount required in 1 liter distilled or deionized water and warm slightly to dissolve completely.
2. Dispense required amount into test tubes along with a fermentation vial in each tube.
3. Place closure on tubes and sterilize in the autoclave for 15 minutes at 15 lbs pressure (121°C). Before opening the autoclave allow the temperature to drop below 75°C to avoid entrapped air bubbles in the inverted vials.

STORAGE

Bacto Lauryl Tryptose Broth   Below 30°C
Prepared medium   15 – 30°C

QUALITY CONTROL

Identity Specifications

Dehydrated powder: light tan, homogeneous, free-flowing
Reaction of 3.56% solution: pH 6.8 ± 0.2 at 25°C
Prepared medium: light amber, clear to slightly opalescent.

Typical Cultural Response in Bacto Lauryl Tryptose Broth
After 24 ± 2 Hours at 35°C.

Organism	Growth	Gas Production
Enterobacter aerogenes ATCC® 13048	good to excellent	+
Escherichia coli ATCC® 25922	good to excellent	+
Salmonella typhimurium ATCC® 14028	good to excellent	−
Staphylococcus aureus ATCC® 25923	markedly inhibited	−

REFERENCES

1. American Public Health Association, 1980, Standard Methods for The Examination of Water and Wastewater, 15th Edition, American Public Health Association, Inc., Washington, D.C.
2. American Public Health Association, 1976, Compendium of Methods for The Microbiological Examination of Foods, American Public Health Association, Inc., Washington, D.C.
3. Mallmann and Darby, 1941, American Journal of Public Health, 31:127.
4. American Public Health Association, 1946, Standard Methods for The Examination of Water and Sewage, 9th Ed., American Public Health Association, Inc., Washington, D.C.
5. Cowls, 1938, Journal of American Water Works Association, 30:979.
6. Darby and Mallmann, 1939, Journal of American Water Works Association, 31:689.
7. Levine, 1941, American Journal of Public Health, 31:351.
8. McCrady, 1943, American Journal of Public Health, 33:1199.
9. Perry and Hajna, 1944, American Journal of Public Health, 34:735.

PACKAGING

Bacto Lauryl Tryptose Broth		
1 lb (454 g)	0241-01-8	
1/4 lb (114 g)	0241-02-7	
5 lb (2.27 kg)	0241-05-4	
10 kg	0241-08-1	

BACTO LEAD ACETATE AGAR

INTENDED USE

Bacto Lead Acetate Agar is recommended for the detection of hydrogen sulfide production for the differentiation of gram-negative intestinal bacteria.

HISTORY

Orlowski[1] noted that *Salmonella typhi* could be distinguished from the coliform organisms by culturing them in a medium containing lead acetate, an indicator of hydrogen sulfide production. Jordan and Victorson[2] showed further that *S. paratyphi* (paratyphoid A) and *S. schottmuelleri* (paratyphoid B) could be distinguished on the basis of hydrogen sulfide production by growing them in a lead acetate medium. *S. paratyphi* produced no browning, whereas *S. schottmuelleri* gave a definite browning of the medium within 18 – 24 hours after inoculation.

Bacto Lead Acetate Agar was developed based on a modification suggested by R. S. Spray. Morrison and Tanner[3] used Bacto Lead Acetate Agar in their study of hydrogen sulfide production by the thermophilic bacteria from water. They found the medium well adapted to the determination of this characteristic. Spray[4] employed it with success in his studies on semisolid media for the cultivation and identification of the sporulating anaerobes.

PRINCIPLES

Certain bacteria possess the ability of liberating hydrogen sulfide from proteins or their split products. This property has been widely used in culture media for differentiating and identifying members of the gram-negative intestinal group of bacteria, as well as for the identification of other microorganisms. Lead acetate in the medium is the indicator of the hydrogen sulfide production.

Bacto Lead Acetate Agar was developed after considerable research to obtain a medium which would give an accurate differentiation. This medium shows no inhibition due to the toxicity of lead. The non-toxicity of Bacto Lead Acetate Agar for certain bacteria is confirmed in *Pure Culture Study of Bacteria*.[5]

FORMULA

BACTO LEAD ACETATE AGAR
DEHYDRATED

Ingredients per liter

Bacto Peptone	15 g
Proteose Peptone, Difco	5 g
Bacto Dextrose	1 g
Lead Acetate	0.2 g
Sodium Thiosulfate	0.08 g
Bacto Agar	15 g

Final pH 6.6 ± 0.2 at 25°C

Five hundred grams will make 13.8 liters of final medium.

PROCEDURE

1. To rehydrate, suspend 36 grams in 1 liter distilled or deionized water and heat to boiling to dissolve completely.
2. Dispense into tubes and sterilize in the autoclave for 15 minutes at 15 pounds pressure (121°C).
3. Tubes are slanted to allow generous butt.
4. Pure cultures of gram-negative microorganisms isolated from MacConkey agar, bismuth sulfite agar, EMB agar or other plating media should be streaked upon the surface of the slant and stabbed into the butt of the lead acetate agar. With this procedure, surface browning can be observed, as well as browning along the line of puncture. Since the medium contains dextrose it will also indicate gas production from this carbohydrate by the presence of bubbles in the butt of the tube.

For the determination of hydrogen sulfide production in a medium free from dextrose, see Bacto Peptone Iron Agar.

QUALITY CONTROL

Identity Specifications

Dehydrated powder: light beige, homogeneous, free-flowing
Reaction of 3.6% solution: pH 6.6 ± 0.2 at 25°C
Prepared medium: light to medium amber, opalescent, may have a slight precipitate

Typical Cultural Response in Bacto Lead Acetate Agar
After 18 – 24 Hours at 35 ± 2°C

Organism	Growth	H_2S	Gas
Enterobacter aerogenes ATCC® 13048	good	−	+
Escherichia coli ATCC® 25922	good	−	+
Salmonella typhi ATCC® 6539	good	+	+
Salmonella typhimurium ATCC® 14028	good	+	−
Shigella dysenteriae ATCC® 13313	good	−	−
Shigella flexneri ATCC® 12022	good	−	−

STORAGE

Bacto Lead Acetate Agar   Below 30°
Prepared tubes   15 – 30°C

REFERENCES

1. Dissert. St. Petersburg, 1897.
2. J. Infectious Diseases, 21:554, 1917.
3. J. Bact., 7:343:1922.
4. J. Bact., 32:135:1936.
5. Pure Culture Study of Bacteria, 1: No. 8:1933.

PACKAGING

Bacto Lead Acetate Agar   500 g   0088-17-6

LEGIONELLA MEDIA

BACTO LEGIONELLA AGAR BASE
BACTO LEGIONELLA AGAR ENRICHMENT
(BCYEα Agar, Modified)

INTENDED USE

Bacto Legionella Agar Base and Bacto Legionella Agar Enrichment are intended for use in the preparation of Legionella Agar. The complete medium is based on the charcoal yeast extract (CYE) agar formula of Feeley et al.,[1] as modified by Edelstein,[2] and is recommended for use in the isolation and cultivation of Legionella from clinical and nonclinical materials.

HISTORY

Legionella pneumophila was first demonstrated as a causative agent of pneumonia by McDade et al., who isolated this organism from lung tissue of patients who died of

Legionnaires Disease.[3] Since that time, six additional species of Legionella have been described and have been shown to be either causative agents of pneumonia,[4-7] or to be associated with pneumonia patients by serological procedures.[8,9]

Legionellae have been frequently isolated from environmental aquatic sources including air conditioning evaporation condensers,[10] cooling towers[11] and potable water sources such as shower heads[12] or water fixtures.[13] Presumably, these and other aquatic sources serve as reservoirs for this group of organisms and represent the means by which legionellosis is acquired.

Initial isolation of L. pneumophila was performed by McDade et al., using guinea pigs and embryonated chicken eggs.[3] In 1978, Weaver reported that Mueller Hinton chocolate agar would support good growth of this organism.[14] Feeley et al. subsequently determined that ferric pyrophosphate and L-cysteine would respectively replace the hemoglobin and enrichment components of Mueller Hinton chocolate agar.[14] Feeley et al. also showed that growth of L. pneumophila was optimal at pH 6.9. These observations were incorporated into an improved legionella agar, which they called F-G agar.

In 1979, Feeley et al. described a modification of F-G agar.[1] This medium, which they called charcoal yeast extract (CYE) agar was found to provide better growth of Legionella than obtained on F-G agar. Subsequently, Pasculle et al., reported that the performance of CYE agar could be further improved by the addition of ACES buffer.[15] Recently, Edelstein described a further modification of CYE agar in which he incorporated both ACES buffer and alpha-ketoglutarate.[2] Edelstein reported that this further modification, which he referred to as BCYEα agar, provided for improved growth and recovery of Legionella. Bacto Legionella Agar from Difco Laboratories is a modification of the BCYEα agar formula of Edelstein. In the Difco formula, the concentration of ACES buffer was reduced from 10 g/liter to 6 g/liter.

PRINCIPLES OF THE PROCEDURE

Bacto Legionella Agar consists of a basal medium containing ACES buffer, charcoal, yeast extract, agar and α-ketoglutarate and an enrichment containing L-cysteine and ferric pyrophosphate. ACES buffer was added to maintain the proper pH of the medium for optimal growth.[15] Charcoal serves in this medium as a detoxifier or may be acting as a CO_2 collector or surface tension modifier.[1] Yeast extract, L-cysteine, ferric pyrophosphate and α-ketoglutarate were included in the medium as sources of nutrients or growth factors.[1,2,16] Agar was added to the medium as a solidifying agent. The concentration of ACES buffer was decreased for improved gel strength.

FORMULAE

BACTO LEGIONELLA AGAR BASE
DEHYDRATED

Ingredients per liter

Bacto Yeast Extract . . . . 10 g
Charcoal, Activated . . . . 1.5 g
ACES Buffer . . . . 6 g
α-Ketoglutarate . . . . 1 g
Potassium Hydroxide . . . . 1.5 g
Bacto Agar . . . . 17 g

BACTO LEGIONELLA AGAR ENRICHMENT

Ingredients per 5 ml vial

L-Cysteine·HCl . . . . 0.2 g
Ferric Pyrophosphate . . . . 0.125 g

PRECAUTIONS
Use aseptic technique in rehydrating Bacto Legionella Agar Enrichment and adding the supplement to the base. Follow proper, established laboratory procedures in handling and disposing of infectious materials. Since legionellosis is apparently acquired by the airborne route, care should be taken to minimize aerosols. Use of a biological safety cabinet in the handling and inoculation of specimens suspected to contain *Legionella* organisms and in the handling and examination of inoculated plates is recommended. Work surfaces should be disinfected with 5% phenol or 5% hypochlorite solutions.

STORAGE INSTRUCTIONS FOR BACTO LEGIONELLA AGAR BASE
Store below 30°C. The powder is very hygroscopic. Keep container tightly closed. The expiration date applies to the product stored in its intact container when stored as directed.

STORAGE AND REHYDRATION INSTRUCTIONS FOR BACTO LEGIONELLA AGAR ENRICHMENT
Store at 2 – 8°C. Protect from light. Use immediately after rehydration. To rehydrate, aseptically add 5 ml sterile distilled or deionized water and invert gently several times to resuspend the powder. The expiration date applies to the product in its intact container when stored as directed. Do not open or rehydrate vials until ready to use.

PRODUCT DETERIORATION
Do not use the rehydrated enrichment if it is contaminated, partially or completely evaporated or shows other signs of deterioration. Do not use the agar base if it is caked, discolored or shows other signs of deterioration.

SPECIMEN COLLECTION
Specimens should be collected in sterile containers and transported immediately to the laboratory in accordance with recommended guidelines.[13,17 – 19]

PROCEDURE
Materials Provided
Bacto Legionella Agar Base
Bacto Legionella Agar Enrichment

Materials Required but not Provided
Specimen container   User quality control cultures
Inoculating loop   Sterile Petri dishes
Bunsen burner or incinerator   Sterile distilled or deionized water
Incubator (35°C)

Preparation of Legionella Agar
1. To rehydrate the base, suspend 18.5 g of Bacto Legionella Agar Base in 500 ml distilled or deionized water. Do not heat prior to sterilization.
2. Sterilize in the autoclave for 15 minutes at 15 lbs pressure (121°C).
3. Cool to 45 – 50°C.
4. Aseptically add 5 ml of rehydrated Bacto Legionella Agar Enrichment to the molten agar.
5. Mix thoroughly.
6. Check pH. It should be 6.85 to 7.0. Adjust, if necessary, with 1N HCl or 1N KOH.
7. Dispense into sterile 90 – 100 mm Petri dishes, approximately 20 ml per dish. Maintain agitation during dispensing to prevent settling of the charcoal particles.

Inoculation and Incubation

Process specimens as appropriate for that specimen and inoculate directly onto the surface of a legionella agar plate. Streak for isolation. For information on processing of specimens intended for *Legionella* cultures, consult appropriate references.[13,17-19]

Incubate plates aerobically in a humidified atmosphere containing 2.5% CO_2 at 35°C for a minimum of 4 days. Examine daily for evidence of growth.

RESULTS

Colonies of *Legionella* should be visible after 2 to 5 days incubation and appear light blue to blue-gray in color. Upon longer incubation, colonies become larger, smoother and gray-white to white in appearance. Colonies suspected of being *Legionella* should be gram-stained and subcultured to a fresh legionella agar plate and to a blood agar plate not containing L-cysteine. Gram-negative organisms that grow on legionella agar but fail to grow on blood agar (without L-cysteine) may be presumptively identified as *Legionella* species.[17,19-21] Definitive identification is performed on the basis of growth, morphology, and biochemical and immunological reactions. Appropriate references should be consulted for further information on identification procedures.[17,19-20]

USER QUALITY CONTROL

1. Examine the agar base for color and texture. The powder should be gray-black, free-flowing and homogeneous.
2. Examine the lyophilized and rehydrated supplement for evidence of deterioration as described under PRODUCT DETERIORATION.
3. Check the performance of the base and enrichment by testing in the complete medium. Plates should be inoculated with approximately 100 – 200 colony forming units of the test cultures and incubated aerobically under humidified conditions at 35°C. Examine plates for growth after 48 and 72 hours incubation. Results should be as stated below:

Typical Cultural Response on Legionella Agar After 48 – 72 Hours at 35 ± 2°C

Organism	Growth	Colonial Morphology
Legionella pneumophila ATCC® 33153	Good	Light blue to blue-gray colonies becoming gray-white to white in appearance
Legionella pneumophila ATCC® 33155	Good	Light blue to blue-gray colonies becoming gray-white to white in appearance
Legionella dumoffii ATCC® 33343	Good	Light blue to blue-gray colonies becoming gray-white to white in appearance

LIMITATIONS OF THE PROCEDURE

1. Bacto Legionella Agar Base and Bacto Legionella Agar Enrichment are intended for use in the preparation of legionella agar. Although this medium is recommended for use in the isolation and cultivation of *Legionella* species, organisms other than *Legionella* will grow on this medium and must be differentiated from *Legionella* by appropriate testing.
2. Due to variations in the nutritional requirements of this group of organisms, some strains of *Legionella* may be encountered that fail to grow or grow poorly on this medium.

3. Growth of *Legionella* has been shown to be significantly affected by pH.[14] Care should be taken in adjusting the pH of the medium in order to obtain optimal performance.

REFERENCES

1. Feeley, J. C., R. J. Gibson, G. W. Gorman, N. L. Langford, J. K. Rasheed, D. C. Mackel, and W. B. Baine, 1979, Charcoal yeast extract agar: primary isolation medium for *Legionella pneumophila*. J. Clin. Microbiol. 10:437 – 441.
2. Edelstein, P. H., 1981, Improved semiselective medium for isolation of *Legionella pneumophila* from contaminated clinical and environmental specimens, J. Clin. Microbiol, 14:298 – 303.
3. McDade, J. E., C. C. Shepard, D. W. Fraser, T. S. Tsai, M. A. Redus, W. R. Dowdle, and the field investigation team, 1977, Legionnaires' disease: isolation of a bacterium and demonstration of its role in other respiratory disease, N. Engl. J. Med., 297:1197 – 1203.
4. Pasculle, A. W., R. L. Myerowitz, and C. R. Rinaldo, 1979, New bacterial agent of pneumonia isolated from renal transplant recipients, Lancet, 2:58 – 61.
5. Bozeman, F. M., J. W. Humphries, and J. M. Campbell, A new group of rickettsia-like agents recovered from guinea pigs, Acta. Virol., 12:87 – 93.
6. Lewallen, K. R., R. M. McKinney, D. J. Brenner, C. W. Moss, D. H. Dail, B. M. Thomason, and R. A. Bright, 1979, A newly identified bacterium phenotypically resembling but genetically distinct from *Legionella pneumophila*: an isolate in a case of pneumonia, Ann. Intern. Med., 91:831 – 834.
7. McKinney, R. M., R. K. Porschen, P. H. Edelstein, M. L. Bissett, P. P. Harris, S. P. Bondell, A. G. Steigerwalt, R. E. Weaver, M. E. Ein, D. S. Lindquist, R. S. Kops, and D. J. Brenner, 1981, *Legionella longbeachae* species nova, another etiologic agent of human pneumonia, Ann. Intern. Med., 94:739 – 743.
8. Cherry, W. B., G. W. Gorman, L. H. Orrison, C. W. Moss, A. G. Steigerwalt, H. W. Wilkinson, S. E. Johnson, R. M. McKinney, and D. J. Brenner, 1982, *Legionella jordanis*: a new species of *Legionella* isolated from water and sewage, J. Clin. Microbiol., 15:290 – 297.
9. Morris, G. K., A. Steigerwalt, J. C. Feeley, E. S. Wong, W. T. Martin, C. M. Patton, and D. J. Brenner, 1980, *Legionella gormanii* sp. nov., J. Clin. Microbiol., 12:718 – 721.
10. Cordes, L. G., D. W. Fraser, P. Skaliy, C. A. Perlino, W. R. Elsea, G. F. Mallison, and P. S. Hayes, 1980, Legionnaires disease outbreak at an Atlanta, Georgia country club: evidence for spread from an evaporative condenser, Am. J. Epidemiol., 111:425 – 431.
11. Dordero, T. J., R. C. Rendtorff, G. W. Mallison, R. M. Weeks, J. S. Levy, W. E. Wong, and W. Schaffner, 1980, An outbreak of Legionnaires' disease associated with a contaminated air conditioning cooling tower, N. Engl. J. Med., 302:365 – 370.
12. Cordes, L. G., A. M. Wiesenthal, G. W. Gorman, J. T. Phair, H. M. Sommers, A. Brown, V. L. Yu, M. H. Magrussen, R. D. Meyer, J. S. Wolf, K. N. Shands, and D. W. Fraser, 1981, Are showers a source of nosocomial Legionnaires' disease?, Ann. Intern. Med., 94:195 – 197.
13. Edelstein, P. H., 1982, Comparative study of selective media for isolation of *Legionella pneumophila* from potable water, J. Clin. Microbiol., 16:697 – 699.
14. Feeley, J. C., G. W. Gorman, R. E. Weaver, D. C. Mackel, and W. H. Smith, 1978, Primary isolation media for Legionnaires disease bacterium, J. Clin. Microbiol., 8:320 – 325.
15. Pasculle, A. W., J. C. Feeley, R. J. Gibson, L. G. Cordes, R. L. Myerowitz, C. M. Patton, G. W. Gorman, L. L. Carmack, J. W. Ezzell, and J. N. Dowling, 1980, Pittsburgh pneumonia agent: direct isolation from human lung tissue, J. Infect. Dis., 141:727 – 732.
16. Pine, L., J. R. George, M. W. Reeves, and W. K. Harrell, 1979, Development of a chemically defined liquid medium for growth of *Legionella pneumophila*, J. Clin. Microbiol., 9:615 – 626.
17. Feeley, J. C., and G. W. Gorman, 1980, Legionella, p. 318 – 324, *In* E. H. Lennette, A. Balows, W. J. Hausler, Jr., and J. P. Truant (ed), Manual of clinical microbiology, 3rd Ed., American Society for Microbiology, Washington, D.C.
18. Morris, G. K., P. Skaliy, C. M. Patton, and J. C. Feeley, 1979, Method for isolating Legionnaires' disease bacterium from soil and water samples, p. 85 – 90, *In* G. L. Jones and G. A. Hibert (ed.), "Legionnaires'", the disease, the bacterium and methodology, Center for Disease Control, Atlanta, GA.
19. Feeley, J. C., G. W. Gorman, and R. J. Gibson, 1979, Primary isolation media and methods, p. 77 – 84, *In* G. L. Jones and G. A. Hibert (ed.), "Legionnaires'": the disease, the bacterium and methodology, Center for Disease Control, Atlanta, GA.
20. Weaver, R. E., and J. C. Feeley, 1979, Cultural and biochemical characterization of Legionnaires disease bacterium, p. 19 – 25, *In* G. L. Jones and G. A. Hibert (ed.), "Legionnaires'": the disease, the bacterium and methodology, Center for Disease Control, Atlanta, GA.
21. Myerowitz, R. L., 1982, *Legionella pneumophila* and other *Legionella* species: an update, Lab. Med., 13:611 – 617.

PACKAGING

Bacto Legionella Agar Base	500 g	1830-17-5
Bacto Legionella Agar Enrichment	6 × 5 ml	3390-57-4

BACTO LEPTOSPIRA ANTIGENS
BACTO LEPTOSPIRA ANTISERA

PRODUCT DESCRIPTION AND INTENDED USE

Bacto Leptospira Antigens are chemically inactivated suspensions of internationally recognized strains of *Leptospira* for use in the macroscopic slide agglutination procedure to detect antibodies in serum specimens found in certain disease states.

Bacto Leptospira Antisera are hyperimmune rabbit sera for use as positive controls in this test.

These reagents are designed for use only in the test procedure described herein. They are not recommended for the direct diagnosis of a disease but rather only for the detection of the presence of antibodies, particularly in cases of seroconversion.

SUMMARY AND BACKGROUND

Leptospira is representative of a number of species of pathogenic microorganisms which upon invasion produce a fever in the host. Consequently, it is often referred to as a "febrile antigen."

For a complete discussion refer to Bacto Febrile Antigens.

Although the taxonomy of the genus *Leptospira* has not been completely established, it is currently divided into 2 complexes, *Leptospira interrogans* and *Leptospira biflexa*. The complex L. interrogans contains the pathogenic *Leptospira* comprising approximately 17 distinct serological groups which consist of over 150 serovars, whereas the *L. biflexa* species are saprophytic and consist of a single serological group. (See Table I for more details on *L. interrogans*.)

Table I

L. interrogans

Serogroup	Bacto Leptospira Antigens and Antisera
andamana	Leptospira Andamana
australis	Leptospira Australis
autumnalis	Leptospira Autumnalis
ballum	Leptospira Ballum
bataviae	Leptospira Bataviae
canicola	Leptospira Canicola
celledoni	Leptospira Celledoni
cynopteri	Leptospira Cynopteri
grippotyphosa	Leptospira Grippotyphosa
hebdomadis	Leptospira Georgia (LT 117)
	Leptospira Hardjo
	Leptospira Medanensis
	Leptospira Sejroe
	Leptospira Wolffii
icterohemorrhagiae	Leptospira Icterohemorrhagiae
javanica	Leptospira Javanica
panama	Leptospira Panama
pomona	Leptospira Pomona
pyrogenes	Leptospira Pyrogenes
shermani	Leptospira Shermani
tarassovi	Leptospira Hyos

Even though the agglutination-lysis technique appears to be the method of choice in the serological detection of *Leptospira* antibodies, due to its specificity, its use is limited to larger reference laboratories because of its complexity and the necessity to utilize living cultures in the test performance.

The macroscopic slide test as described by Galton et al.,[5] however, enjoys popularity since it overcomes these obstacles in that the bacterial suspension is nonviable, the test is easily performed and the results are obtained quickly.

In this test, varying amounts of the sera to be tested are distributed to the corresponding square of a previously marked glass slide. The antigen is added and mixed with the test serum. Positive or negative results are read after rotating 4 minutes on a mechanical rotator.

REAGENTS
Bacto Leptospira Antigens are stable suspensions of individual and pooled *Leptospira* serotypes prepared according to the procedure of Galton et al[5] for use in the macroscopic slide test. They contain formaldehyde in a final concentration of 0.5% (v/v), glycerol to a final concentration of 20% (v/v) and sodium chloride USP to a final concentration of 12% (w/v). They are adjusted spectrophotometrically to approximate a 25 – 26% light transmission.

Bacto Leptospira Antisera are stable desiccated, high-titered antisera prepared in rabbits. They are recommended for use as control sera in the rapid slide agglutination test.

When used according to the suggested procedure, each 5 ml vial of antigen is sufficient to perform approximately 100 tests. Each 3 ml vial of antiserum is sufficient to perform approximately 300 tests.

Merthiolate® is added as a preservative to approximately 1:10,000 (w/v) final concentration.

REHYDRATION AND STORAGE
1. The antigens are liquid and ready for use.
2. The control sera are rehydrated by adding 3 ml 0.85% sodium chloride solution at room temperature.
3. Store all reagents at 2 – 8°C. These reagents are stable through the expiry date on the label when stored at 2 – 8°C, provided they have not been exposed to temperature extremes (excessive heat or freezing) and have not become contaminated chemically or bacterially during usage. They should not be subjected to repeated freezing and thawing. Such treatment is detrimental to the antibody content of the control sera and the stability of the antigens.

Bacto Leptospira Antigens are not to be used for immunization of man or animals.

SPECIMEN COLLECTION
1. Collect 5 – 10 ml whole blood aseptically from the patient.
2. Allow the blood to clot and obtain the syneresed serum with a Pasteur pipette. If the serum is not free of erythrocytes, clarify by centrifugation. **Note:** Do not heat the serum since significant antibodies may be thermolabile.

PROCEDURE
Materials Provided:
Bacto Leptospira Antigens
Bacto Leptospira Antisera
Droppers for the above antigens (the supplied droppers deliver approximately 0.05 ml/ drop)

Materials Required but not Provided:
Macroscopic Slide Test
  Wax marking pencil
  Ruler
  Glass plate 8″ × 8″
  0.2 ml serological pipettes
  Mechanical rotator at 125 rpm
  0.85% sterile NaCl solution
  Applicator sticks
  Suitable light source

PRECAUTIONS
All reagents and equipment used in the slide test must be at room temperature prior to use.

Specimen
1. The specimen must be clear and free of visible fat. It must be free of excessive hemolysis and not bacterially contaminated.
2. The specimen must be used in the unheated state. If inactivated by heat, some thermolabile antibodies may be destroyed.

Glassware
1. All glassware that is employed in the preparation, testing and storage of these reagents must be free of detergents or other harmful residues.
2. Pipettes employed must be clean (acid washed if necessary) to delivery proper volumes.

Antigens
1. Shake antigen vial well before use to insure a smooth, uniform suspension. Occasionally, upon storage, *Leptospira* suspensions may tend to settle out. If an antigen demonstrates a roughness in the negative control (sterile saline), the organisms may be resuspended by vigorous shaking. If this is necessary, allow the antigen to stand until the entrapped air particles are released.
2. Discard any antigen demonstrating positive reactions in the negative control which cannot be resuspended (No. 1 above). *Leptospira* antigens will demonstrate irreversible autoagglutination if at any time during shipment or storage they are subjected to freezing temperatures. Do not allow to freeze.
3. Discard any antigen which does not react properly in the positive control serum.

Control Sera
1. For greater proficiency in test interpretation always include a positive control and a saline control (negative control) in each protocol.
2. Discard any positive control serum (Leptospira antisera) in which a precipitate forms.
3. If bacterial contamination is suspected, autoclave the reagent or use some suitable chemical procedure to inactivate the contaminant before discarding the reagent.

METHODOLOGY
1. Place 0.01 ml of the serum to be tested in the appropriate number of squares on a slide or ruled glass plate with 0.2 ml pipette.
2. Using supplied dropper, add 1 drop of each desired antigen pool to a separate drop of serum.
3. Using a separate piece of applicator stick, mix each drop of antigen and serum.
4. Rotate the plate by hand approximately 5 – 10 times.
5. Place on the mechanical rotator for 4 minutes at 125 rpm.
   The slide test should be read immediately after the rotation period of 4 minutes. Reactions occuring later may be due to the reactants drying on the slide.
6. Observe reaction over a light.
7. Record agglutination as follows:
   Positive   + definite clumping
   Doubtful   ± slight clumping
   Negative   − even suspension
8. When a positive or doubtful reaction is obtained in one or more of the antigen pools, test that serum with the separate antigens of the pool. Use the above procedure.
9. Record reaction of single antigens as follows:
   4+   all organisms appear clumped; supernatant is clear
   3+   75% clumped
   2+   50% clumped
   1+   25% clumped
   −   no agglutination and even suspension

An illustration of this type of reaction can be seen in the serology introduction.

10. When a positive reaction is obtained and a titer is desired, dilute the serum 1:5 with 4 parts 0.85% NaCl to 1 part serum. Place the diluted serum in 0.04, 0.02, 0.01 and 0.005 ml amounts with a 0.2 ml pipette on the slide. Add 1 drop of the appropriate antigen to each drop of serum, mix, rotate and read as described.

For convenience in reporting the reactions, the serum dilutions are numbered serially in ascending order:

Dilution Number	1	2	3	4	5
Serum	0.01	0.04	0.02	0.01	0.005
	(undiluted)	(— — — diluted 1:5 — — —)			

11. When the end-point is desired and is not reached in the above dilutions, the serum may be diluted 1:50 and titrated further as in step 10.

CALIBRATION AND QUALITY CONTROL
Control the test by using Bacto Leptospira Antiserum for each serotype in place of the serum specimen.

Both positive and negative (saline) controls are run in parallel with the serum specimen. The positive control is diluted in the same proportion as the patient's serum and processed in exactly the same manner following the procedure above for the rapid slide test.

An antigen is considered to be satisfactory if it does not clump with the negative control and it reacts:

a) Pool antigen: a 4+ reaction in 0.01 ml of each undiluted homologous control serum contained in that pool.
b) Individual antigen: must react 2+ or greater at the 5th serum dilution (step 10 above) of the 1:5 dilution of the homologous serum.

Discard
1. Any control serum in which a precipitate forms or which does not dissolve completely upon rehydration.
2. Any antigen when clumps appear on the cap or dropper or which agglutinates in the negative control.
3. All antigens and control sera when past the end of the month of the expiry date.

INTERPRETATION
1. In the past, attempts have been made to correlate titers derived from the macroscopic slide agglutination test to that obtained in the agglutination lysis tube test.[5] Further work concluded that such transposition is not recommended.[6] The true value of this test lies in the fact that results based on determination of antibodies in a single serum specimen is of limited significance; however, if a comparison of at least 2 samples are taken during the acute phase and another 2 – 5 weeks later demonstrates a 4-fold rise in titer, such sera indicates current infection. This is true in all tests designed to detect the antibody content of the host's serum.
2. Conversely, paired specimens collected during early and late convalescence stages may show a 4-fold drop in titer indicating recent infection.
3. A reaction in dilution No. 5 in the slide test (see step 10 above) on a single specimen suggests a current infection and a low titer (dilutions 2 – 4 in the slide test) may indicate only past experience with leptospirosis.[7]
4. Serum collected during the acute phase of the disease infected with one serotype may show an equal or higher titer with an antigen of a heterologous serotype.
5. Cross agglutination occurs with rapid slide test antigens and the hyperimmune control rabbit sera. The approximate degree of cross-reactivity may be found in the literature.[8]
6. A negative reaction even on paired specimens does not rule out the possibility of infection since another serotype other than the one(s) being tested may be the etiological agent.

Special Note: Because a change or no change in titer over a period of time are the best indicators of active infection or noninfection, respectively, the accuracy and precision of the tests can be affected not only by test conditions but also by the subjectivity of the person in reading the end-point, the following protocol is recommended:

A preliminary test using the rapid slide method may be performed on the initial serum specimen and reported to the physician at that time. An aliquot of the serum should be transferred to a sterile test tube, sealed tightly and kept in the freezer. When the second serum is obtained, it should be run in parallel with the original specimen. In this manner, the original serum will serve as a control and any difference in titer will be more credible since the bias associated with the performance of the test as well as determining the end-point will be minimized. Obviously, it would be best if the person who performed the original test also performed the second test.

LIMITATIONS OF THE PROCEDURE
1. The major limitation of the *Leptospira* slide test is that of interpretation. It is necessary to run several serum specimens taken at different times from the same patient in parallel to detect quantitative differences in antibody content.

2. Although Bacto Leptospira Antigens are designed for use to detect and quantitate antibody content in serum specimens, they are for screening purposes only. The test should not be used to replace conventional isolation and serological identification of the etiological agent.
3. The serotype of the etiological agent cannot be determined by serological examination of the serum specimen.
4. The antigens have been tested using hyperimmune animal sera only.
5. The antisera have been prepared for use as controls. If they are to be used in other test procedures, the investigator is advised to test each lot of antiserum with known homologous and heterologous cultures to insure proper reactivity under his/her test conditions.

REFERENCES

1. Schubert, J. H., Holdeman, L. and Martin, D. S., J. Lab. Clin. Med., 44:194, 1964.
2. Welch, H. and Stuart, C. A., J. Lab. and Clin. Med., 21:411, 1936.
3. Welch, H. and Mickle, F. L., Amer. J. Pub. Hlth., 26:248, 1936.
4. Diamond, B. E., Pub. Hlth. Lab., 6:74, 1948.
5. Galton, M. M., Powers, D. S., Hale, A. D. and Cornell, R., Amer. J. Vet. Res., 19:505 – 512, 1958.
6. Evins, G. M., West, B., and Reynolds, G. H., Health Lab Science, 3:235 – 250, 1966.
7. Personal Communication, 1959.
8. World Health Organization, WHO Tech. Rep. Serv., No. 113, 1956.

PACKAGING
Bacto

Leptospira Andamana Antiserum	3 ml	2491-47-7
Leptospira Australis Antigen	5 ml	2213-56-2
Leptospira Australis Antiserum	3 ml	2225-47-0
Leptospira Autumnalis Antigen	5 ml	2210-56-5
Leptospira Autumnalis Antiserum	3 ml	2222-47-3
Leptospira Ballum Antigen	5 ml	2204-56-3
Leptospira Ballum Antiserum	3 ml	2216-47-1
Leptospira Bataviae Antigen	5 ml	2206-56-1
Leptospira Bataviae Antiserum	3 ml	2219-47-8
Leptospira Canicola Antigen	5 ml	2202-56-5
Leptospira Canicola Antiserum	3 ml	2217-47-0
Leptospira Celledoni Antigen	5 ml	2618-56-3
Leptospira Celledoni Antiserum	3 ml	2625-47-6
Leptospira Cynopteri Antiserum	3 ml	2624-47-7
Leptospira Georgia (LT 117) Antigen	5 ml	2215-56-0
Leptospira Georgia (LT 117) Antiserum	3 ml	2227-47-8
Leptospira Grippotyphosa Antigen	5 ml	2207-56-0
Leptospira Grippotyphosa Antiserum	3 ml	2220-47-5
Leptospira Hardjo Antigen	5 ml	2942-56-0
Leptospira Hardjo Antiserum	3 ml	2943-47-1
Leptospira Hyos Antigen	5 ml	2214-56-1
Leptospira Hyos Antiserum	3 ml	2226-47-9
Leptospira Icterohemorrhagiae Antigen	5 ml	2201-56-6
Leptospira Icterohemorrhagiae Antiserum	3 ml	2218-47-9
Leptospira Javanica Antigen	5 ml	2619-56-2
Leptospira Javanica Antiserum	3 ml	2626-47-5
Leptospira Medanensis Antiserum	3 ml	2493-47-5
Leptospira Panama Antigen	5 ml	2620-56-9
Leptospira Panama Antiserum	3 ml	2627-47-4
Leptospira Pomona Antigen	5 ml	2200-56-7
Leptospira Pomona Antiserum	3 ml	2223-47-2

Leptospira Pyrogenes Antigen	5 ml	2208-56-9
Leptospira Pyrogenes Antiserum	3 ml	2221-47-4
Leptospira Sejroe Antigen	5 ml	2211-56-4
Leptospira Sejroe Antiserum	3 ml	2224-47-1
Leptospira Shermani Antigen	5 ml	2621-56-8
Leptospira Wolffii Antigen	5 ml	2494-56-2
Leptospira Wolffii Antiserum	3 ml	2495-47-3
Leptospira Antigen Pool 1	5 ml	2203-56-4
Contains L. Ballum, L. Canicola, L. Icterohemorrhagiae		
Leptospira Antigen Pool 2	5 ml	2205-56-2
Contains L. Bataviae, L. Grippotyphosa, L. Pyrogenes		
Leptospira Antigen Pool 3	5 ml	2209-56-8
Contains L. Autumnalis, L. Pomona, L. Wolffii		
Leptospira Antigen Pool 4	5 ml	2212-56-3
Contains L. Australis, L. Hyos, L. Georgia (LT 117)		
Leptospira Antigen Pool 5	5 ml	2622-56-7
Contains L. Cynopteri, L. Celledoni, L. Javanica		
Leptospira Antigen Pool 6	5 ml	2623-56-6
Contains L. Cynopteri, L. Panama, L. Shermani		

BACTO LEPTOSPIRA ENRICHMENT

INTENDED USE
Bacto Leptospira Enrichment, a lyophilized rabbit serum containing native hemoglobin, is recommended for use with either Bacto Stuart Medium Base or Bacto Fletcher Medium Base for the cultivation of *Leptospira*.

METHOD OF PREPARATION
Rehydrate with 10 ml sterile distilled or deionized water. Rotate the vial in an end over end motion to dissolve completely.

STORAGE
Bacto Leptospira Enrichment   2 – 8°C

QUALITY CONTROL
Identity Specifications

Appearance: tan button or powder
Rehydrated appearance: medium amber, clear to very slightly opalescent liquid

For a complete discussion on Typical Cultural Response, refer to Bacto Stuart Medium Base or Bacto Fletcher Medium Base.

PACKAGING
Bacto Leptospira Enrichment   6 × 10 ml   0452-60-0

BACTO LEPTOSPIRA MEDIUM
BASE/ENRICHMENT EMJH

INTENDED USE
Bacto Leptospira Medium Base EMJH supplemented with Bacto Leptospira Enrichment EMJH is recommended for the cultivation and maintenance of the *Leptospira*.

HISTORY
The basal medium and the enrichment are prepared according to the formulations described by Ellinghausen and McCullough[1] as modified by Johnson and Harris.[2,3]

FORMULA
BACTO LEPTOSPIRA MEDIUM BASE EMJH
DEHYDRATED
Ingredients per liter

Sodium Phosphate Dibasic . . . . 1 g
Potassium Phosphate Monobasic . . . 0.3 g
Sodium Chloride . . . . 1 g
Ammonium Chloride . . . . 0.25 g
Thiamine . . . . 0.005 g

Final pH 7.5 ± 0.2 at 25°C.

One pound will make 177 liters of basal medium or 197.4 liters of complete medium.

METHOD OF PREPARATION
1. Suspend 2.3 g in 900 ml distilled or deionized water. Swirl to dissolve completely.
2. Sterilize in the autoclave for 15 minutes at 15 lbs pressure (121°C).
3. Allow to cool to room temperature.
4. Aseptically add 100 ml Bacto Leptospira Enrichment EMJH. Mix uniformly.
5. Aseptically dispense into sterile tubes or bottles as desired.
6. Inoculate as desired and incubate at 29 – 30°C for up to 7 days.

STORAGE
Bacto Leptospira Medium Base	Below 30°C
Bacto Leptospira Enrichment EMJH	2 – 8°C
Prepared tubes or bottles	2 – 8°C

QUALITY CONTROL
Identity Specifications

Dehydrated powder:	white, homogeneous, free-flowing
Reaction of 0.23% solution:	pH 7.5 ± 0.2 at 25°C
Enrichment:	medium to dark amber, clear
Prepared medium:	colorless, clear

Typical Cultural Response on Bacto Leptospira Medium EMJH
After 2 – 7 Days at 29 – 30°C

Organism	Growth
Leptospira interrogans Serogroup Canicola	good to excellent
Leptospira interrogans Serogroup Grippotyphosa	good to excellent

REFERENCES
1. Am. J. Vet. Res., 26:39, 1965.
2. Personal Communication.
3. J. Bact., 94:27, 1967.

PACKAGING

Bacto Leptospira Medium Base EMJH	1 lb (454 g)	0794-01-9
Bacto Leptospira Enrichment EMJH	100 ml	0795-72-2
	6 × 100 ml	0795-73-1

BACTO LETHEEN MEDIA

INTENDED USE
Bacto Letheen media are used to determine the bactericidal activity of quaternary ammonium compounds.

HISTORY/PRINCIPLES
The value of highly nutritional media containing a neutralizing agent for quaternary ammonium compounds in sanitizers was first described by Weber and Black.[2] Bacto Letheen media are also modifications of the AOAC[1] formulae.

Bacto Letheen Agar is used to determine the bactericidal activity of quaternary ammonium compounds using *Escherichia coli* ATCC® 11229 or *Staphylococcus aureus* ATCC® 6538. It is also recommended as a solid medium containing quaternary ammonium compound neutralizer for the microbiological sampling of environmental surfaces.

Bacto Letheen Broth is used in determining the phenol coefficient of quaternary ammonium compounds.

FORMULAE

BACTO LETHEEN AGAR
DEHYDRATED
Ingredients per liter

Bacto Beef Extract . . . . 3 g
Bacto Tryptone . . . . 5 g
Bacto Dextrose . . . . 1 g
Bacto Agar . . . . 15 g
Sorbitan Monooleate . . . . 7 g
Lecithin . . . . 1 g

Final pH 7.0 ± 0.2 at 25°C.

One pound will make 14 liters of
final medium.
Rehydrate with 32 grams/liter.

BACTO LETHEEN BROTH
DEHYDRATED
Ingredients per liter

Bacto Peptamin . . . . 10 g
Bacto Beef Extract . . . . 5 g
Lecithin . . . . 0.7 g
Sorbitan Monooleate . . . . 5 g
Sodium Chloride . . . . 5 g

Final pH 7.0 ± 0.2 at 25°C.

One pound will make 17.7 liters
of final medium.
Rehydrate with 25.7 grams/liter.

METHOD OF PREPARATION
1. To rehydrate, suspend the appropriate amount in 1 liter distilled or deionized water.
2. Heat the media to boiling to dissolve completely.
3. Dispense letheen agar in flasks and letheen broth in 10 ml quantities into 20 × 150 mm test tubes.
4. Sterilize in the autoclave for 15 minutes at 15 lbs pressure (121°C).

5. Cool agar to 50 – 55°C before pouring plates.
6. See AOAC for dilution and inoculation procedure.

STORAGE
Bacto Letheen media   2 – 8°C
Prepared media   2 – 8°C

QUALITY CONTROL
Identity Specifications
Dehydrated powder: tan, homogeneous, free-flowing
Reaction of appropriate solution: pH 7.0 ± 0.2 at 25°C
Prepared medium: medium amber, clear to slightly opalescent

Typical Cultural Response on Bacto Letheen Media
After 24 – 48 Hours at 35°C

Organism	Growth
Escherichia coli ATCC® 11229	good to excellent
Staphylococcus aureus ATCC® 6538	good to excellent

REFERENCES
1. AOAC, Tenth Ed., 1965.
2. Soap and Sanitary Chemicals, Oct. 1948.

PACKAGING
Bacto Letheen Agar	1 lb (454 g)	0680-01-6
Bacto Letheen Broth	1 lb (454 g)	0681-01-5

BACTO LEVINE EMB AGAR

INTENDED USE
Bacto Levine EMB Agar is a Standard Methods plating medium for the isolation and differentiation of lactose-fermenting from lactose-nonfermenting gram-negative enteric bacilli.

HISTORY/PRINCIPLES
Holt-Harris and Teaque[1] originally devised eosin methylene blue agar. These workers employed a combination of eosin and methylene blue as an indicator which gave sharp and distinct differentiation between colonies of lactose-fermenting organisms and those which did not ferment lactose.

Levine[2,3,4] altered the original EMB agar formula by deleting the sucrose present and increasing the lactose. The resultant medium gave excellent differentiation of *Escherichia coli* from *Enterobacter aerogenes*. Coliform colonies usually show a dark center and have a greenish metallic sheen. Occasionally, variants are observed similar to the type described but having no sheen. Another variant may grow effusely in colonies somewhat larger than typical growth with a distinct metallic sheen. Colonies of *E. aerogenes* are usually much larger than typical *E. coli* and tend to run together. Their centers are usually brown in color and not as dark as *E. coli*. Metallic sheen is only occasionally observed. The following chart gives a more detailed differentiation between the two:

	E. coli[a]	*E. aerogenes*[b]
Size	Well isolated colonies are 2 – 3 mm in diameter.	Well isolated colonies are larger than *E. coli*; usually 4 – 6 mm in diameter or more.
Confluence	Neighboring colonies show little tendency to run together.	Neighboring colonies run together quickly.
Elevation	Colonies slightly raised; surface flat or slightly concave, rarely convex.	Colonies considerably raised and markedly convex.
Appearance by Transmitted Light	Dark, almost black centers which extend more than 3/4 across the diameter of the colony; internal structure of central dark portion difficult to discern.	Centers deep brown; not as dark as *E. coli* and smaller in proportion to the rest of the colony. Striated internal structure often observed in young colonies.
Appearance by Reflected Light	Colonies dark, button-like, often concentrically ringed with a greenish metallic sheen.	Much lighter than *E. coli*, metallic sheen not observed except occasionally in depressed center when such is present.

a. Two other types have been occasionally encountered. One resembles the type described, except that there is no metallic sheen, the colonies being wine colored. The other type of colony is somewhat larger (4 mm), grows effusely, and has a marked crenated or irregular edge, the central portion showing a very distinct metallic sheen. These 2 varieties constitute about 2 or 3% of the colonies observed.
b. A small type of aerogenes colony, about the size of the colon colonies, which shows no tendency to coalesce, has been occasionally encountered.

To obtain the most satisfactory reactions, especially in the differentiation of *E. coli* and *E. aerogenes*, it is very important that particular care be taken in the choice of the dyes and a meticulous determination be made of their proper proportions. The Bacto dyes in this medium are selected to satisfy the extreme delicacy of the medium. Levine recommended that the reaction should not be readjusted and the medium should not be filtered. The medium is relatively stable and may be stored for short periods of time. It is usually used in Petri dishes; however, some laboratories have found tubes to be convenient. The tubes should be prepared with long slants and no butts.

Bacto Levine EMB Agar is prepared according to the formula specified in *Standard Methods for the Examination of Water and Wastewater*,[5] *Standard Methods for the Examination of Dairy Products*,[6] and *Compendium of Methods for the Microbiological Examination of Foods*.[7] Its principle function is the confirmation of presumptive tests for members of the coliform group in the bacteriological examination of water, milk and foods.

FORMULA

BACTO LEVINE EMB AGAR
DEHYDRATED

Ingredients per liter

Bacto Peptone	10 g
Bacto Lactose	10 g
Dipotassium Phosphate	2 g
Bacto Agar	15 g
Bacto Eosin Y	0.4 g
Bacto Methylene Blue	0.065 g

Final pH 7.1 ± 0.2 at 25°C.

One pound will make 12 liters of medium.

METHOD OF PREPARATION
1. Suspend 37.5 g Bacto Levine EMB Agar in 1 liter distilled or deionized water and heat to boiling to dissolve completely.
2. Sterilize in the autoclave for 15 minutes at 15 lbs pressure (121°C). Avoid overheating.
3. Dispense into sterile Petri dishes. The characteristic flocculent precipitate present in the medium following autoclaving may be evenly dispersed by gently swirling or shaking the flask just prior to pouring into Petri dishes.

Bacto Levine EMB Agar inoculated the same day as rehydrated may be used without autoclave sterilization. Under these conditions the medium need be heated only to boiling to dissolve it completely before pouring into plates.

STORAGE
Bacto Levine EMB Agar	Below 30°C
Prepared plates or tubes	2 – 8°C

QUALITY CONTROL
Identity Specifications

Dehydrated powder:	reddish pink, homogeneous, free-flowing
Reaction of 3.75% solution:	7.1 ± 0.2 at 25°C
Prepared medium:	wine red with greenish cast, slightly opalescent with finely dispersed flocculent precipitate

Typical Cultural Response on Bacto Levine EMB Agar
After 18 – 24 Hours at 35°C

Organism	Recovery	Colony Color
Enterobacter aerogenes ATCC® 13048	good	flesh color, pink
Escherichia coli ATCC® 25922	good to excellent	green metallic sheen
Pseudomonas aeruginosa ATCC® 27853	good to excellent	colorless
Saccharomyces cerevisiae ATCC® 9763	none to poor	cream
Salmonella typhimurium ATCC® 14028	good	colorless/amber
Staphylococcus aureus ATCC® 25923	none to poor	colorless/amber

REFERENCES
1. Holt-Harris and Teague, 1916, J. Infectious Diseases, 18:596.
2. Levine, 1918, J. of Infectious Diseases, 23:43.
3. Levine, 1921, Bull. 62, Iowa Eng. Exp. Sta.
4. Levine, 1921, J. AWWA, 8:151.
5. American Pub. Hlth. Assoc., 1980, Standard Methods for the Examination of Water and Wastewater, 15th Ed., American Pub. Hlth. Assoc., Inc., Washington, D.C.
6. Am. Pub. Hlth. Assoc., 1978, Standard Methods for the Examination of Dairy Products, 14th Ed., American Pub. Hlth. Assoc., Inc., Washington, D.C.
7. Am. Pub. Hlth. Assoc., 1976, Compendium of Methods for the Microbiological Examination of Foods, Am. Pub. Hlth. Assoc., Inc., Washington, D.C.

PACKAGING
Bacto Levine EMB Agar		
	1 lb (454 g)	0005-01-4
	1/4 lb (114 g)	0005-02-3
	5 lb (2.27 kg)	0005-05-0
	10 kg	0005-08-7

BACTO LIMA BEAN AGAR

INTENDED USE
Bacto Lima Bean Agar is recommended for the cultivation of phytophathological and other fungi.

PRINCIPLES
Bacto Lima Bean Agar is prepared from an infusion of dry lima beans, and is solidified by the addition of 1.5% Bacto Agar. It possesses all the nutritive properties of an infusion of lima beans and has a final reaction of pH 5.6, making it well suited for the growth of fungi.

FORMULA
BACTO LIMA BEAN AGAR
DEHYDRATED

Ingredients per liter

Lima Beans, Infusion from ...... 62.5 g
Bacto Agar .... 15 g

Final pH 5.6 ± 0.2 at 25°C.

One pound will make 19.7 liters of medium.

METHOD OF PREPARATION
1. To rehydrate, suspend 23 g in 1 liter distilled or deionized water and heat to boiling to dissolve completely.
2. Sterilize in the autoclave for 15 minutes at 15 lbs pressure (121°C).

STORAGE
Bacto Lima Bean Agar	Below 30°C
Prepared plates	2 – 8°C

QUALITY CONTROL
Identity Specifications

Dehydrated powder:	light yellowish tan, homogeneous, free-flowing
Reaction of 2.3% solution:	pH 5.6 ± 0.2 at 25°C
Prepared medium:	light to medium amber, opalescent, may have a slight precipitate

Typical Cultural Response on Bacto Lima Bean Agar
After 40 – 48 Hours at 30°C

Organism	Growth
Aspergillus niger ATCC® 16404	good to excellent
Saccharomyces cerevisiae ATCC® 9763	good to excellent

PACKAGING
Bacto Lima Bean Agar	500 g	0117-17-1

BACTO LIPASE REAGENT

Bacto Lipase Reagent is a lipoidal emulsion for use with Bacto Spirit Blue Agar. For a complete discussion see Bacto Spirit Blue Agar.

PACKAGING

Bacto Lipase Reagent	6 × 20 ml	0431-63-3

BACTO LIPOPOLYSACCHARIDES

REAGENTS

Bacto Lipopolysaccharides are nonsterile, desiccated bacterial endotoxins of high antigenic properties obtained by extracting bacteria by one or more recognized published procedures. The Boivin[1] trichloracetic acid procedure as modified by Webster, Sagin, Landy and Johnson[2] and the phenol extraction method of Westphal[3] have yielded endotoxins of high potency. While these lipopolysaccharides are not the most highly purified preparations, they do possess the properties desired for most immunological and pathological studies.

The endotoxic potencies of lipopolysaccharides vary with the microbial species or strain employed as well as with the method of extraction. The endotoxin response has also been shown by Kessel and Braun[4] to vary greatly depending upon prior sensitization of the test animals to the parent organisms either by natural or artificial means. The lipopolysaccharides are standardized using white mice to give LD_{50}'s of 0.5 mg or less. All are highly pyrogenic. The LD_{50} values in mice and lipid A values of each lot number are available upon request.

STORAGE

Store desiccated Bacto Lipopolysaccharides at 2 – 8°C. Once reconstituted, the solution may be stored for a limited period of time at 2 – 8°C. For additional stability the reconstituted Bacto Lipopolysaccharides may be aliquoted into usable amounts and frozen at −20°C. Frozen aliquots should be thawed only once.

REFERENCES

1. Boivin, A., and Mesrobeani, L., in Comp. Rend. Soc. Biol., 113:490, 1933; 128:5, 1938.
2. Marion E. Webster, Jerome F. Sagin, Maurice Landy, and Arthur Johnson in J. Immunology, 74:455, 1955.
3. Van Otto Westphal, Otto Luderitz, and Fritz Bister in Z. Naturforsch., 7B:148 – 155, 1952.
4. Federation of Am. Soc. for Expt'l. Bio., Paper 2773, 1964.

PACKAGING

Bacto Lipopolysaccharides are lyophilized in 100 mg units. Packages of 1 g are supplied as 10 × 100 mg. Packaging includes closure with rubber stopper and aluminum seals to prevent moisture and other contamination resulting from frequent opening for withdrawal of partial quantities.

		Westphal Method	Boivin Method
Bacto Lipopolysaccharide *E. coli* 026:B6	100 mg	—	3920-25-2
	1 g	—	3920-10-9
Bacto Lipopolysaccharide *E. coli* 055:B5	100 mg	3120-25-0	3923-25-9
	1 g	3120-10-7	3923-10-6

		Westphal Method	Boivin Method
Bacto Lipopolysaccharide E. Coli 0111:B4	100 mg	3122-25-8	3922-25-0
	1 g	3122-10-5	3922-10-7
Bacto Lipopolysaccharide E. Coli 0127:B8	100 mg	3123-25-7	3880-25-0
	1 g	—	3880-10-7
Bacto Lipopolysaccharide E. Coli 0128:B12	100 mg	3131-25-7	3924-25-8
Bacto Lipopolysaccharide S. Abortus Equi	100 mg	3127-25-3	3106-25-8
Bacto Lipopolysaccharide S. Enteritidis	100 mg	3126-25-4	3105-25-9
Bacto Lipopolysaccharide S. Minnesota 9700	100 mg	3349-25-5	—
Bacto Lipopolysaccharide S. Typhosa 0901	100 mg	3124-25-6	3946-25-2
Bacto Lipopolysaccharide S. Typhimurium	100 mg	3125-25-5	3998-25-9
Bacto Lipopolysaccharide S. Marcescens	100 mg	3130-25-8	—

BACTO LISTERIA ANTISERA
BACTO LISTERIA ANTIGENS

INTENDED USE

Bacto Listeria O Antisera Types 1, 4 and Poly are recommended for use in the macroscopic tube and rapid slide test for the identification of Listeria monocytogenes.

Bacto Listeria O Antigens Types 1 and 4 (tube and slide) are used as positive controls in the macroscopic tube and rapid slide tests respectively.

SUMMARY AND EXPLANATION OF THE TEST

Listeria monocytogenes was suspected of being an animal pathogen as early as 1910 by Helphers[1] and positively identified as such in 1924 by Murray, Webb and Swann.[2] Nyfeldt recorded the isolation of Listeria monocytogenes in humans as early as 1929, yet listeriosis is considered one of the latest recognized infections of man.[3] Many Listeria infections were attributed to other organisms. Specimens containing L. monocytogenes were often discounted as negative and when the organism was recovered it was considered a laboratory contaminant.

Gray[4] reported listeriosis in man and animals to manifest itself in a number of diseases, including meningitis, septicemia, bacteremia, abortion, endocarditis, meningoencephalitis and conjunctivitis. Winn, Cherry and King[5] emphasized that, as clinicians have become more cognizant of the pathogenic potential of L. monocytogenes, there is a marked increase in the number of cases of listeriosis being reported. Hood[6] stated that it is necessary, therefore, for the modern laboratory to be on the alert for this.

The Listeria are short, gram-positive, nonspore forming, noncapsulated, motile rods. Motility is most pronounced at 20°C. They are microaerophilic and therefore thrive best under reduced oxygen tension. L. monocytogenes has been found to be present either in the infective state or in the asymptomatic carrier state in most warm blooded animals, both domestic and wild. They are believed to be widely distributed in nature.[7]

Identification of Listeria is based on successful isolation of the organism, biochemical characterization and serological confirmation.

There are four recognized serological types of L. monocytogenes designated 1 – 4 according to their somatic antigen components. Gray[4] reported that the serological analysis of available *Listeria* cultures indicated that, within the United States, 32.8% were of type 1, 65.3% were type 4 and less than 1% were of type 2 or 3.

Listeria is generally a slow growing and delicate organism especially on primary isolation. Attempts at direct isolation from positive specimens may, at times, be met with failure. The likelihood of recovery is improved when it is preincubated at 4°C (cold enrichment). This process may require a few days or several months with weekly to monthly transfers before positive cultures are obtained. Yet, the potential severity of listeriosis (meningitis, meningoencephalitis) necessitates rapid identification of the organism.

REAGENTS
Bacto Listeria O Antisera Types 1, 4 and Poly are stable, desiccated, high titered rabbit *Listeria* antiglobulins. They have been prepared in accordance with procedures recommended by Gray[8] and Eveland.[9] Bacto Listeria O Antisera Types 1 and 4 are specific for the respective serotypes of *L. monocytogenes* while Bacto Listeria O Antiserum Poly contains agglutinins for serotypes 1 and 4 of *Listeria monocytogenes*.

Bacto Listeria O Antigen Types 1 and 4 (tube) and Bacto Listeria O Antigen Types 1 and 4 (slide) are suspensions of the appropriate serotypes of *L. monocytogenes* adjusted to the recommended pH and turbidity.[10]

When used according to the suggested procedure, the reagents will yield the following:

Reagent	Vial	Number of Tests
Listeria O Antiserum	1 ml	10 tube or 400 slide
Listeria O Antigen (slide)	5 ml	100 slide
Listeria O Antigen (tube)	25 ml	5 tube

REHYDRATION
To rehydrate the antisera, add 1 ml sterile distilled or deionized water to each vial and rotate gently to dissolve contents completely.

STORAGE
Store unrehydrated and rehydrated reagents at 2 – 8°C. Prolonged exposure to room temperature or freezing and thawing are detrimental to these products.

EXPIRATION DATE
Bacto Listeria O Antisera, both desiccated and rehydrated, but not diluted and Bacto Listeria Antigens are stable through the expiration date on the label.

Discard any serum that is cloudy or has a precipitate after rehydration or storage unless it can be clarified by centrifugation and demonstrates proper reactivity with validated positive and negative control cultures.

Once Bacto Listeria Antiserum has been diluted for the slide procedure, it should be discarded after one week.

SPECIMEN PREPARATION
The following schema outlines methods for the isolation and identification of *Listeria*.

SCHEME for ISOLATION of *LISTERIA MONOCYTOGENES*
Clinical Specimens

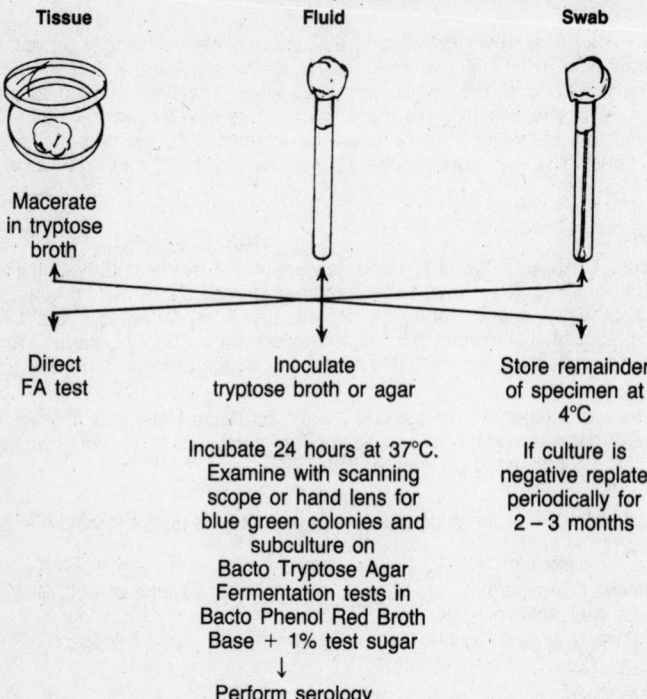

Tissue	Fluid	Swab

Macerate in tryptose broth

Direct FA test	Inoculate tryptose broth or agar	Store remainder of specimen at 4°C
	Incubate 24 hours at 37°C. Examine with scanning scope or hand lens for blue green colonies and subculture on Bacto Tryptose Agar Fermentation tests in Bacto Phenol Red Broth Base + 1% test sugar ↓ Perform serology	If culture is negative replate periodically for 2 – 3 months

After isolation of the *Listeria*, biochemical tests should be performed. The following protocol is typical for the *Listeria*:[10]

ACID ONLY	IRREGULAR	NO ACID
glucose	arabinose	dulcitol
levulose	galactose	inositol
maltose	glycerol	inulin
salicin	sorbitol	mannitol
trehalose	sucrose	raffinose

Nitrate	−	Catalase	+
MR	Generally +	Gelatin	−
VP	+	Starch	−
Indole	−	Urea	−

Macroscopic Tube Test

1. Culture the test organism in Bacto Tryptose Broth or on Bacto Tryptose Agar.
2. Make a suspension of the test organism in rehydrated Bacto FA Buffer Dried containing 0.3% formaldehyde.
3. Adjust to a density approximating that of a Bacto McFarland Barium Sulfate Standard No. 3.

Rapid Slide Test

1. Culture the test organism in Bacto Tryptose Broth or on Bacto Tryptose Agar and suspend growth in rehydrated FA Buffer Dried.
2. Heat the organism suspension at 80°C for 1 hour in a water bath.
3. Centrifuge the suspension and remove the bulk of the supernatant fluid.
4. Resuspend the organisms in the remaining portion of liquid.

PROCEDURE
Materials Provided:
Bacto Listeria O Antisera Type 1, 4 and Poly
Bacto Listeria O Antigens Type 1, 4 (slide and tube)
Bacto FA Buffer, Dried

Materials Required but not Provided:
Macroscopic Tube Test
  Kahn tubes and rack
  Serological pipette
  50°C water bath
  Refrigerator 2 – 8°C

Rapid Slide Test
  Glass plate
  Droppers
  80°C water bath

Macroscopic Tube Test Procedure
Controls:
Positive — Bacto Listeria O Antigen with homologous antiserum.
Negative — A tube consisting of rehydrated Bacto FA Buffer Dried and the test organism suspension.

1. Prepare a row of 9 Kahn serological tubes in a rack for each suspension to be tested. Add 0.9 ml of formalized FA buffer to the first tube and 0.5 ml to the remaining tubes.
2. Using a 1 ml serological pipette, add 0.1 ml of the rehydrated Bacto Listeria O Antiserum Type 1, 4 or Poly to tube 1 in each row and mix thoroughly.
3. Transfer 0.5 ml from tube 1 to tube 2 and mix thoroughly. In like manner, continue transferring 0.5 ml through tube 8 discarding 0.5 ml from tube 8 after mixing. Tube 9 is a negative control tube. Upon addition of the antigen to the tubes, the final dilutions are 1:20 through 1:2560 for tubes 1 through 8, respectively.
4. Add 0.5 ml of the antigen suspension to each of the 9 tubes and shake the rack to mix.
5. Incubate in a water bath at 50°C for 2 hours. Refrigerate overnight and read for agglutination the following morning.
6. Record as follows:
    4+   100% cells clumped
    3+   75% cells clumped
    2+   50% cells clumped
    1+   25% cells clumped
    −   No cells clumped

RESULTS OF THE MACROSCOPIC TUBE TEST
The negative control tube must be free of agglutination. If positive, the antigen is unsatisfactory. Homologous reactions using Bacto Listeria O Antisera should exceed a titer of 2+ at 1:320.

Rapid Slide Test Procedure
Controls:
Positive — Bacto Listeria O Antiserum plus homologous O Antigen (Slide).

Negative — Bacto FA Buffer Dried rehydrated containing 0.3 ml formaldehyde per 100 ml plus a drop of the organism suspension.

1. Add 1 drop of the heated organism to 1 drop of the antiserum diluted 1:20 in saline solution on a glass plate and rock back and forth for 1 – 2 minutes.
2. Similarly, add 1 drop of the organism suspension to 1 drop of negative control buffer solution and rock for 1 – 2 minutes. Read for agglutination.

RESULTS OF THE RAPID SLIDE TEST
The organism suspension should not agglutinate in the saline control but should with homologous antisera.

LIMITATIONS OF THE PROCEDURE
Serological techniques employing Bacto Listeria Antisera for the identification of *Listeria monocytogenes* serves as corroborative evidence in the determination of the organism as the etiological agent of disease. Final identification cannot be made without consideration of morphological, serological, and biochemical characterization.

REFERENCES
1. Svenk. Vet. Tidskr., 2:265, 1911.
2. J. Path. & Bact., 29:407, 1926.
3. Gray, M.L. and Killinger, H.H., Bact. Revs., 30:309 – 382, 1966.
4. Second Symposium on Listeric Infection, Montana State Coll., 1962.
5. Annals N.Y. Acad. Sci., 70:624, 1958.
6. Am. J. Clin. Path., 28:18, 1957.
7. Paper presented APHA, 1961.
8. Personal Communication, 1963.
9. Personal Communication, 1964.
10. Listeriosis, Hafner Publ. Co., N.Y., 1961.

PACKAGING
Bacto Listeria O Antigen Type 1 (Slide)	5 ml	2303-56-3
Bacto Listeria O Antigen Type 1 (Tube)	25 ml	2305-65-0
Bacto Listeria O Antigen Type 4 (Slide)	5 ml	2304-56-2
Bacto Listeria O Antigen Type 4 (Tube)	25 ml	2306-65-9
Bacto Listeria O Antiserum Type 1	1 ml	2300-50-2
Bacto Listeria O Antiserum Type 4	1 ml	2301-50-1
Bacto Listeria O Antiserum Poly	1 ml	2302-50-0
Bacto FA Buffer Dried	6 × 10 g	2314-33-8

BACTO LITTMAN OXGALL AGAR

INTENDED USE
Bacto Littman Oxgall Agar with the addition of streptomycin is a selective medium for the primary isolation and cultivation of fungi, especially dermatophytes.

HISTORY/PRINCIPLES
Littman[1] described a selective medium for the primary isolation of fungi. Crystal violet and streptomycin are used as selective bacteriostatic agents, while Bacto Oxgall is used to restrict the spreading of fungus colonies. The medium is neutral in reaction, which favors the growth of many pathogenic fungi.

Littman has shown that his medium is especially valuable for culturing the dermatophytes. Molds and yeasts form nonspreading discrete colonies, easy to isolate in pure culture. He also suggests that the medium may be used for the following purposes: estimation of the normal fungal flora of feces, sputum and other human discharges; evaluation of human disorders of the upper and lower respiratory and gastrointestinal tract caused by fungi; single cell isolation of fungi; and plate count of viable saprophytic fungi in foodstuffs and air. In a comparative study, Littman[2] compared this medium with Sabouraud dextrose agar using a large variety of pathogenic and saprophytic fungi. On the Littman oxgall agar the majority of fungi tested produced colonies at the end of the first month of incubation about half the size of the colonies on Sabouraud dextrose agar, but equal in size after 56 days of incubation. He reported the isolation of three times as many fungi from feces, sputum, skin scrapings and hair on his medium as were isolated on Sabouraud dextrose agar and four times as many pathogenic dermatophytes on the selective medium as on the Sabouraud medium. The selective oxgall agar of Littman is specified in "Diagnostic Procedures and Reagents"[3] of the American Public Health Association for the isolation of pathogenic fungi.

FORMULA

BACTO LITTMAN OXGALL AGAR
DEHYDRATED

Ingredients per liter

Bacto Peptone	10 g
Bacto Dextrose	10 g
Bacto Oxgall	15 g
Bacto Agar	20 g
Bacto Crystal Violet	0.01 g

Final pH 7.0 ± 0.2 at 25°C.

One pound will make 8.2 liters of medium.

PROCEDURE

1. To rehydrate, suspend 55 g in 1 liter distilled or deionized water and heat to boiling to dissolve completely.
2. Sterilize in the autoclave for 15 minutes at 15 lbs pressure (121°C).
3. Cool to 45 – 50°C and add 30 μg of streptomycin per ml of medium.
4. A concentration of 30 μg/ml of medium may be obtained by adding 10 ml sterile distilled water to a 1 g (1 million μg) bottle of streptomycin or dihydrostreptomycin. One ml of this solution is added to 9 ml sterile distilled water to give a solution containing 10,000 μg streptomycin per ml. To each liter of sterile melted medium at 45 – 50°C are added 3 ml of this solution to obtain 30 μg per ml (0.3 ml for 100 ml medium).
5. Dispense in sterile Petri dishes, 100 mm in diameter, or distribute in sterile tubes and slant. Let stand at room temperature until temperature equilibrates before inoculation.
6. For inoculation, skin and nail scrapings or infected hairs are placed directly on the surface of the medium. Exudates, sputa or fecal suspensions are spread over the surface with a sterile swab. The selectivity of the medium permits the use of a heavy inoculum without the danger of overgrowth by bacteria or saprophytic fungi.
7. Plates are incubated at room temperature, or preferably, at 30°C for 4 – 8 days. Do not incubate at 37°C.

STORAGE

Bacto Littman Oxgall Agar	Below 30°C
Prepared plates or tubes	2 – 8°C

QUALITY CONTROL

Identity Specifications

Dehydrated powder: greyish blue, homogeneous, free-flowing
Reaction of 5.5% solution: pH 7.0 ± 0.2 at 25°C
Prepared medium: blue, slightly opalescent

Typical Cultural Response on Bacto Littman Oxgall Agar
After 40 – 48 Hours at 25 – 30°C

Organism	Plain	w/Streptomycin
Candida albicans ATCC® 10231	good to excellent	good to excellent
Escherichia coli ATCC® 25922	good	inhibited
Saccharomyces cerevisiae ATCC® 9763	good to excellent	good to excellent
Saccharomyces uvarum ATCC® 9080	good to excellent	good to excellent
Trichophyton	fair to good	fair to good

REFERENCES

1. Science, 106:109, 1947.
2. Tech. Bull., Reg. Med. Tech., 18:409, 1948.
3. Diagnostic Procedures and Reagents, 3rd Ed., p. 452, 1950.

PACKAGING

Bacto Littman Oxgall Agar   1 lb (454 g)   0294-01-4
                                    12 tubes   0294-34-5

BACTO LIVER

Bacto Liver is prepared from large quantities of carefully trimmed fresh beef liver and is recommended for use in preparing liver infusion media. The nutritive factors of fresh liver tissue are retained in infusions prepared from Bacto Liver.

The equivalent of 500 g of fresh liver is obtained with 135 g of Bacto Liver. An excellent infusion can be prepared by using 75 g, or even 50 g, of Bacto Liver per 1 liter of distilled water. The latter concentration is recommended for media for general culture purposes. The infusion is made by warming the mixture to 50°C and holding it at this temperature with frequent agitation for 1 hour. The mixture is heated to boiling for a few minutes to coagulate a portion of the proteins and is filtered. Peptone and other ingredients of the medium are then added to the filtrate. After adjustment of the reaction to pH 7.5 – 7.8 and subsequent boiling, there will be a further coagulation, which should be removed by filtration before the medium is sterilized.

PACKAGING

Bacto Liver   1 lb (454 g)   0133-01-9

BACTO LIVER INFUSION AGAR

INTENDED USE

Bacto Liver Infusion Agar is used for the cultivation of Brucella and other pathogenic organisms.

FORMULA

BACTO LIVER INFUSION AGAR
DEHYDRATED

Ingredients per liter

Beef Liver, Infusion from . . . . .500 g
Proteose Peptone, Difco . . . . 10 g
Sodium Chloride . . . . 5 g
Bacto Agar . . . . 20 g

Final pH 6.9 ± 0.2 at 25°C.

One pound will make 8.25 liters of medium.

METHOD OF PREPARATION

1. To rehydrate, suspend 55 g in 1 liter cold distilled water and bring to a boil to dissolve completely.
2. Sterilize in the autoclave 15 minutes at 15 lbs pressure (121°C).

For cultivation of *Endamoeba histolytica* Bacto Endamoeba Medium should be used.

STORAGE

Bacto Liver Infusion Agar   Below 30°C
Prepared medium   2 – 8°C

QUALITY CONTROL
Identity Specifications

Dehydrated powder: tan, homogeneous, free-flowing
Reaction of 5.5% solution: pH 6.9 ± 0.2 at 25°C
Prepared medium: medium to dark amber, slightly opalescent

Typical Cultural Response on Bacto Liver Infusion Agar
After 18 – 48 Hours or Longer if Necessary at 35°C

Organism	Growth
Brucella abortus ATCC® 4315	good
Brucella melitensis ATCC® 4309	good
Brucella suis ATCC® 6597	good
Streptococcus pneumoniae ATCC® 6303	good

PACKAGING

Bacto Liver Infusion Agar   500 g   0052-17-8

BACTO LIVER INFUSION BROTH

INTENDED USE

Bacto Liver Infusion Broth is used for cultivating a variety of organisms, particularly *Brucella* and anaerobes.

FORMULA

BACTO LIVER INFUSION BROTH
DEHYDRATED

Ingredients per liter

Beef Liver, Infusion from . . . . .500 g
Proteose Peptone, Difco . . . . 10 g
Sodium Chloride . . . . 5 g

Final pH 6.9 ± 0.2 at 25°C.

Five hundred grams will make 14.2 liters of medium.

METHOD OF PREPARATION
1. Dissolve 35 g in 1 liter distilled or deionized water.
2. Sterilize in the autoclave for 15 minutes at 15 lbs pressure (121°C).
3. Dispense as desired.

STORAGE
Bacto Liver Infusion Broth   Below 30°C
Prepared medium   15 – 30°C

QUALITY CONTROL
Identity Specifications

Dehydrated powder: tan, homogeneous, free-flowing
Reaction of 3.5% solution: pH 6.9 ± 0.2 at 25°C
Prepared medium: dark amber, clear, may be very slightly opalescent with
                          a few particles

Typical Cultural Response in Bacto Liver Infusion Broth
After 24 – 48 Hours at 35°C

Organism	Growth
Brucella abortus ATCC® 4315	good to excellent
Brucella melitensis ATCC® 4309	good to excellent
Brucella suis ATCC® 4314	good to excellent
Clostridium sporogenes ATCC® 11439	good to excellent
Streptococcus mitis ATCC® 9895	good to excellent

PACKAGING
Bacto Liver Infusion Broth	500 g	0269-17-7
	10 kg	0269-08-8

BACTO LIVER VEAL AGAR

INTENDED USE
Bacto Liver Veal Agar is recommended for the cultivation of anaerobes.

HISTORY/PRINCIPLES
Numerous methods for the cultivation of anaerobes have been devised and many media have been proposed for this purpose. The use of the anaerobic culture dish as described by Spray[1,2] has been one of the procedures suggested for the propagation of these organisms. The medium prepared from Bacto Liver Veal Agar is identical to

the medium described by Spray[3,4] for use in his anaerobic culture dishes for cultivation of anaerobes.

Bacto Liver Veal Agar gives excellent growth of the sporulating anaerobes. In a personal communication, Spray reported the usefulness of this medium for isolation purposes since *Clostridium perfringens* colonies were recovered within 6 hours from time of inoculation and *C. tetani* within 8 hours. Gas production is inhibited when the medium is inoculated sparingly. With proper dilution giving 10 – 15 colonies per plate, primary isolations of pure cultures are readily obtained.

The medium may also be used in deep tube cultures of the sporulating anaerobes. Bacto Liver Veal Agar is also suitable for the routine isolation and cultivation of many aerobes.

FORMULA

BACTO LIVER VEAL AGAR
DEHYDRATED

Ingredients per liter

Bacto Liver, Infusion from	50 g
Veal, Infusion from	500 g
Proteose Peptone, Difco	20 g
Bacto Gelatin	20 g
Soluble Starch, Difco	10 g
Bacto Isoelectric Casein	2 g
Bacto Dextrose	5 g
Neopeptone, Difco	1.3 g
Bacto Tryptone	1.3 g
Sodium Chloride	5 g
Sodium Nitrate	2 g
Bacto Agar	15 g

Final pH 7.3 ± 0.2 at 25°C.

Five hundred grams will make 5.1 liters of medium.

METHOD OF PREPARATION
1. To rehydrate, suspend 97 g in 1 liter distilled or deionized water and heat to boiling to dissolve the medium completely.
2. Distribute in tubes or flasks and sterilize in the autoclave for 15 minutes at 15 lbs pressure (121°C).

When it is to be used in the anaerobic dish, Spray[3] recommended that unless taken directly from the sterilizer, the medium should be boiled for 10 minutes and then cooled to 50°C without agitation. Serial inoculations are then made and the medium is poured into the dishes. After solidification, 5 ml sterile Bacto Liver Veal Agar is poured over the medium as a cover layer to prevent the spreading of surface colonies.

STORAGE
Bacto Liver Veal Agar	Below 30°C
Prepared medium	2 – 8°C

QUALITY CONTROL
Identity Specifications

Dehydrated powder:	light beige, homogeneous, free-flowing
Reaction of 9.7% solution:	pH 7.3 ± 0.2 at 25°C
Prepared medium:	medium-dark amber, opalescent, may have a slight precipitate

Typical Cultural Response on Bacto Liver Veal Agar
After 18 – 24 Hours at 35°C under Atmosphere as Required for Organism

Organism	Growth
Clostridium botulinum ATCC® 25763	good to excellent
Clostridium tetani ATCC® 10779	good to excellent
Neisseria meningitidis ATCC® 13090	good to excellent
Streptococcus pneumoniae ATCC® 6303	good to excellent

REFERENCES
1. J. Lab. Clin. Med., 16:203, 1930.
2. J. Bact., 21:23, 1931.
3. Personal Communication.
4. J. Bact., 32:135, 1936.

PACKAGING
Bacto Liver Veal Agar   500 g   0059-17-1

BACTO LOEFFLER BLOOD SERUM

INTENDED USE
Bacto Loeffler Blood Serum is used for cultivating *Corynebacterium diphtheriae* from clinical specimens and in pure culture. The medium is also used for demonstrating pigment production and proteolysis.

HISTORY
Bacto Loeffler Blood Serum is a modification of the horse serum, dextrose broth medium described by Loeffler in 1887[1] for cultivating *C. diphtheriae.*

Cleveland and Sanders[2] and Spector[3] used Bacto Loeffler Blood Serum in media for cultivating *Endamoeba histolytica.* Thompson added Bacto Loeffler Blood Serum, hydrolyzed with sodium hydroxide, to a citrate agar for isolating *C. diphtheriae.* Diphtheria bacilli were stimulated and other throat organisms were inhibited.

PRINCIPLES
C. diphtheriae grows rapidly and luxuriantly on Loeffler blood serum, developing morphologically typical organisms in 18 – 24 hours.

FORMULA
BACTO LOEFFLER BLOOD SERUM
DEHYDRATED
Ingredients per liter

Beef Blood Serum . . . . 3 parts
Dextrose Broth . . . . 1 part

Final pH 7.1 ± 0.2 at 25°C.

One pound will make 5.7 liters of final medium.

METHOD OF PREPARATION
1. Dissolve 80 g in 1 liter distilled or deionized water warmed to 42 – 45°C.
2. Dispense into tubes having screw caps or other closures that provide a tight seal to keep the medium free of bubbles. Slant the tubes in the autoclave.

3. Coagulate the medium as follows:
   –Close door loosely.
   –Turn on steam and maintain flowing steam for 10 minutes. It is important to maintain a constant flow of steam during coagulation.
4. Sterilize the medium as follows:
   –Close door tightly.
   –Sterilize for 15 minutes at 15 lbs pressure (121°C).
5. Shut off autoclave and allow pressure to fall to zero before removing tubes.

STORAGE
Bacto Loeffler Blood Serum	2 – 8°C
Prepared medium	2 – 8°C

QUALITY CONTROL
Identity Specifications

Dehydrated powder:	medium beige, homogeneous, fine
Reaction of 8% solution:	pH 7.1 ± 0.2 at 25°C
Prepared medium (coagulated in tubes):	light beige, opaque, slightly transparent at apex of slant

Typical Cultural Response on Bacto Loeffler Blood Serum
After 18 – 24 Hours at 35°C

Organism	Growth
Corynebacterium diphtheriae Type mitis	good to excellent
Corynebacterium diphtheriae Type intermedius	good to excellent
Corynebacterium diphtheriae Type gravis	good to excellent

REFERENCES
1. Centr. Bakt., 2:105, 1887.
2. Arch. Protistenkunde, 70:223, 1930.
3. J. Preventive Med., 6:117, 1932.

PACKAGING
Bacto Loeffler Blood Serum	1 lb (454 g)	0070-01-4
	1/4 lb (114 g)	0070-02-3
	5 lb (2.27 kg)	0070-05-0
	12 tubes	0070-34-5
	144 tubes	0070-37-2

BACTO LOWENSTEIN MEDIUM BASE

BACTO LOWENSTEIN MEDIUM (JENSEN MODIFICATION)
BACTO LOWENSTEIN MEDIUM DEEPS
BACTO LOWENSTEIN MEDIUM w/5% NaCl
BACTO LOWENSTEIN MEDIUM GRUFT

INTENDED USE
Bacto Lowenstein Medium Base is used for preparing a variety of coagulated egg media for isolating, cultivating and differentiating mycobacteria.

HISTORY

Bacto Lowenstein Medium is prepared according to Jensen's modification of Lowenstein's medium.[1] Bacto Lowenstein Medium Gruft is prepared according to Gruft's modification of Jensen's formulation.[2,3]

PRINCIPLES

The use of an egg-base medium for primary isolation of mycobacteria has two advantages:
1. The medium supports a wide variety of mycobacteria.
2. The growth of mycobacteria on egg media is satisfactory for the niacin test.

The major disadvantage of egg-base media is that contamination by proteolytic organisms may cause liquefaction of the medium.

The Jensen modification of Lowenstein medium uses a moderate concentration of malachite green to prevent growth of the majority of contaminants surviving decontamination of the specimen while encouraging earliest possible growth of mycobacteria. The Jensen formulation is available in tube and 1 oz bottle slants for cultivation purposes and as deeps (20 × 150 mm tubes) for use in performing the semiquantitative catalase test.

The Jensen formulation may be modified by the addition of 5% NaCl for use in differentiating slowly growing mycobacteria from rapidly growing mycobacteria on the basis of sodium chloride tolerance.

The Gruft modification of the Jensen formulation incorporates 50 units/ml penicillin and 35 mcg/ml nalidixic acid to decrease contaminants while using a gentler specimen digestion procedure and includes 0.05 mcg/ml ribonucleic acid to increase the isolation rate of mycobacteria.

FORMULAE

BACTO LOWENSTEIN MEDIUM BASE
DEHYDRATED

Ingredients per 600 ml

Bacto Asparagine	3.6 g
Monopotassium Phosphate	2.4 g
Magnesium Sulfate	0.24 g
Magnesium Citrate	0.6 g
Potato Flour	30 g
Malachite Green	0.4 g

One pound will make 12.2 × 600 ml of basal medium
for preparation of 12.2 × 1600 ml of final medium containing egg.

	Bacto Lowenstein Medium	Bacto Lowenstein Medium w/5% NaCl	Bacto Lowenstein Medium Gruft
Ingredients per 1600 ml final solution			
Bacto Lowenstein Medium Base	37.2 g	37.2 g	37.2 g
Bacto Glycerol*	12 ml	12 ml	12 ml
Distilled or Deionized Water	588 ml	588 ml	588 ml
Sodium Chloride	—	80 g	—
Homogenized Egg	1 liter	1 liter	1 liter
Penicillin	—	—	80,000 units
Nalidixic Acid	—	—	56,000 mcg
Ribonucleic Acid	—	—	80 mcg

*Omit glycerol if bovis bacilli or other glycerophobic organisms are to be cultivated.

METHOD OF PREPARATION

1. Suspend 37.2 g of base in 600 ml distilled or deionized water containing 12 ml Bacto Glycerol and heat to boiling with constant agitation.
2. Sterilize in the autoclave for 15 minutes at 15 lbs pressure (121°C). Cool to 45 – 60°C.
3. Add the sterile base to 1 liter of a uniform suspension of fresh egg that was collected and suspended under aseptic conditions. Mix thoroughly, avoiding the inclusion of air bubbles.
4. Dispense the complete medium into sterile tubes or bottles, preferably screw capped. Arrange in a slant position in a suitable rack.
5. Coagulate the medium in an inspissator, water bath or autoclave at 85°C for 45 minutes.

STORAGE

Bacto Lowenstein Medium Base   Below 30°C
Prepared media   2 – 8°C

QUALITY CONTROL

Identity Specifications

Dehydrated powder: dark blue, homogeneous, free-flowing
Prepared slants: pale green, opaque, smooth slants

Typical Cultural Response After 2 – 3 Weeks at 35°C

	Lowenstein Medium Jensen	Lowenstein Medium Gruft	Lowenstein Medium w/5% NaCl
Mycobacterium tuberculosis H37Rv (ATCC® 25618)	excellent growth granular, rough, worty, dry, friable colonies	good to excellent granular, rough, worty, dry, friable colonies	inhibited, no growth
Mycobacterium kansasii ATCC® 12478	excellent growth smooth to rough, photo-chromogenic colonies	good to excellent smooth to rough, photo-chromogenic colonies	inhibited, no growth
Mycobacterium gordonae ATCC® 14470	excellent growth smooth, yellow-orange colonies	good to excellent smooth, yellow-orange colonies	inhibited, no growth
Mycobacterium avium ATCC® 25291	excellent growth, smooth, non-pigmented colonies	good to excellent smooth, non-pigmented colonies	inhibited, no growth
Mycobacterium smegmatis ATCC® 14468	excellent growth, wrinkled, creamy white colonies	good to excellent wrinkled, creamy white colonies	good to excellent wrinkled, creamy white colonies

REFERENCES

1. Enter. Bacteriol. Parasitenk. Abt. 1, 125:222, 1932.
2. J. Bact., 90:829, 1965.
3. HLS, 8:79, 1972.

PACKAGING

Bacto Lowenstein Medium Base	1 lb (454 g)	0444-01-3
	5 lb (2.27 kg)	0444-05-9
Bacto Lowenstein Medium, Jensen	12 tubes	1017-34-9
	144 tubes	1017-37-6
	12 × 1 oz.	1017-44-7
	144 × 1 oz.	1017-45-6
Bacto Lowenstein Medium, Jensen	12 tubes	1289-34-0
Deeps	144 tubes	1289-37-7
Bacto Lowenstein Medium, Jensen	12 tubes	1423-34-7
w/5% NaCl	144 tubes	1423-37-4
Bacto Lowenstein Medium Gruft	12 tubes	1417-34-5
	144 tubes	1417-37-2

BACTO LYSINE IRON AGAR

INTENDED USE

Bacto Lysine Iron Agar is a differential medium for detecting *Salmonella arizonae*, formerly *Arizona arizonae*, cultures which ferment lactose rapidly.

HISTORY/PRINCIPLES

Bacto Lysine Iron Agar is prepared according to the formulation of Edwards and Fife.[1] They studied the reactions of 1247 cultures of enteric bacteria on lysine iron agar and focused on the differentiation of *S. arizonae* and other salmonellae. Many of the organisms on triple sugar iron agar fermented the lactose so rapidly that the expected H_2S production was suppressed. There was the possibility that the organisms frequently found in food infection outbreaks could be overlooked. If the importance of such organisms in outbreaks and sporadic cases of diarrheal disease was to be assessed, it was essential to devise a method of isolation and preliminary screening in which lactose fermentation was not involved. By incorporating lysine and eliminating lactose, they found that only certain groups decarboxylate lysine rapidly and produce hydrogen sulfide abundantly.

Since *S. arizonae* strains produce *Salmonella*-like colonies on bismuth sulfite agar, and since salmonellae and *S. arizonae* cultures are the only delineated groups of *Enterobacteriaceae* which produce H_2S vigorously and produce lysine decarboxylase rapidly, it is suggested that black colonies on bismuth sulfite agar plates, inoculated directly from specimens and from enrichment media be picked to lysine iron agar. In this way the presence of *Salmonella* and *S. arizonae* strains, including cultures which ferment lactose rapidly, can thus be detected, since they produce an alkaline reaction throughout the tube and blacken the medium.

Those organisms that rapidly produce lysine decarboxylase rapidly reverse the typical acid reaction (yellow color) and cause an alkaline reaction (purple color) throughout the medium. Those without the enzyme produce an alkaline slant and an acid butt.

FORMULA

BACTO LYSINE IRON AGAR
DEHYDRATED
Ingredients per liter

Bacto Peptone .... 5 g	Ferric Ammonium Citrate .... 0.5 g
Bacto Yeast Extract .... 3 g	Sodium Thiosulfate .... 0.04 g
Bacto Dextrose .... 1 g	Bacto Brom Cresol Purple ...... 0.02 g
L-Lysine Hydrochloride .... 10 g	Bacto Agar .... 15 g

Final pH 6.7 ± 0.2 at 25°C.

One pound will make 13.1 liters of medium.

PROCEDURE
1. Suspend 34.5 g in 1 liter distilled or deionized water and heat to boiling to dissolve completely.
2. Dispense as required.
3. Sterilize in the autoclave for 12 minutes at 15 lbs pressure (121°C).
4. Allow medium to cool in a position that will provide a short slant and a deep butt.
5. Inoculate with a straight needle by stabbing to the base of the butt and streaking the slant.
6. Caps of the tubes should be replaced loosely, allowing for aerobic conditions.

STORAGE
Bacto Lysine Iron Agar — Below 30°C
Prepared medium — 2 – 8°C

QUALITY CONTROL
Identity Specifications
Dehydrated powder: beige, homogeneous, free-flowing
Reaction of 3.45% solution: ph 6.7 ± 0.2 at 25°C
Prepared medium: reddish purple, slightly opalescent

Typical Cultural Response on Bacto Lysine Iron Agar
After 18–24 Hours at 35°C

Organism	Growth	Slant	Butt	H₂S
Citrobacter freundii ATCC® 8090	good	K	A	+
Escherichia coli ATCC® 25922	good	K	K	−
Proteus mirabilis ATCC® 25933	good	R	A	−
Salmonella typhimurium ATCC® 14028	good	K	K	+
Shigella flexneri ATCC® 12022	good	K	A	−
Salmonella arizonae	good	K	K	+

K = alkaline, no change, red-purple   + = blackening
R = lysine deaminase, deep red   − = no change
A = acid, yellow

REFERENCE
1. Appl. Microbiol. 9:478 – 480, 1961.

PACKAGING
Bacto Lysine Iron Agar		
	1 lb (454 g)	0849-01-4
	1/4 lb (114 g)	0849-02-3
	5 lb (2.27 kg)	0849-05-0
	12 tubes	0849-34-5
	144 tubes	0849-37-2

LYSOZYME REAGENTS FOR USE IN ASSAY
BACTO LYSOZYME
BACTO LYSOZYME BUFFER
BACTO LYSOZYME SUBSTRATE

INTENDED USE

The assay for lysozyme consists essentially of comparing the degree of lysis of *Micrococcus lysodeikticus* (*Micrococcus luteus*) ATCC® 4698 by lysozyme present in a test sample with a standard curve prepared by adding known amounts of lysozyme to *M. lysodeikticus*. Lysis is measured on a spectrophotometer.

HISTORY

Fleming[1] first described the use of *M. lysodeikticus* as the test organism for lysozyme detection in the seed agar plate method and turbidimetrically in 1922.

Goldsworthy and Florey[2] modified Fleming's methods, using a washed, standardized suspension of cells in saline. Boasson[3] employed a phenol-killed cell suspension of *M. lysodeikticus*.

Hartsell and Smolelis[4] further modified the method by recommending use of an ultra-violet-killed suspension of *M. lysodeikticus* for assaying for lysozyme. They demonstrated that ultra-violet rays of a given frequency killed the organism without affecting the cellular membrane and that these cells yielded consistently reproducible assay results.

Shugar[5] found that lysozyme activity could be measured by adding 20 – 50 microliters lysozyme to a suspension of *M. lysodeikticus* having an optical density of 0.5 – 0.75. Spectrophotometric readings were made at 30 second intervals at 5400 Ångströms.

REAGENTS

Bacto Lysozyme Substrate is a nonviable, standardized culture of *Micrococcus lysodeikticus* prepared by a modification of the procedure described by Hartsell and Smolelis.[4] The culture is dried from the frozen state.

Bacto Lysozyme is a crystalline enzyme isolated from fresh egg white using a modification of the procedure described by Fevold and Alderton.[6]

Bacto Lysozyme Buffer contains sodium phosphate, dibasic and potassium phosphate, monobasic, yielding pH 6.2 when dissolved in distilled or deionized water.

PROCEDURE

1. Rehydrate Bacto Lysozyme Buffer by dissolving 9.6 g in 1 liter distilled or deionized water.
2. Suspend sufficient Bacto Lysozyme Substrate in rehydrated Bacto Lysozyme Buffer solution to obtain a 10% light transmission at a wave length of 5400 Ångstroms as compared with distilled water at 100% transmission (approximately 50 mg/100 ml).
3. Dissolve 10 mg Bacto Lysozyme in 100 ml rehydrated Bacto Lysozyme Buffer solution to obtain a 1:10,000 dilution. Prepare further dilutions of 1:100,000, 1:200,000, 1:400,000, 1:800,000, 1:1,600,000, 1:3,200,000 and 1:6,400,000 using Bacto Lysozyme Buffer solution as the diluent.

4. Prepare appropriate dilutions of the test samples containing unknown amounts of lysozyme using Bacto Lysozyme Buffer solution.
5. Mix in separate tubes 5 ml of each Bacto Lysozyme dilution with 5 ml of Bacto Lysozyme Substrate suspension. Mix in separate tubes 5 ml of each test sample dilution with 5 ml of the Bacto Lysozyme Substrate suspension.
6. Let stand at room temperature for **exactly** 20 minutes.
7. Read turbidity on a spectrophotometer at 540 nm.
8. Determine lysozyme content of each test sample by interpolation from the standard curve.

STORAGE

Bacto Lysozyme	2 – 8°C
Bacto Lysozyme Buffer (desiccated)	Below 30°C
Bacto Lysozyme Buffer (rehydrated)	2 – 8°C
Bacto Lysozyme Substrate	2 – 8°C

QUALITY CONTROL
Identity Specifications

Appearance of Bacto Lysozyme:	white, homogeneous, free-flowing
Rehydrated appearance:	colorless, clear
Appearance of Bacto Lysozyme Buffer:	white, homogeneous, free-flowing
Reaction of 0.96% solution:	pH 6.2 ± 0.1 at 25°C
Rehydrated appearance:	colorless, clear
Appearance of Bacto Lysozyme Substrate:	yellow, fluffy powder
Rehydrated appearance:	yellow, even suspension

REFERENCES

1. Proc. Roy. Soc. (London), 93:306, 1922.
2. Brit. J. Exp. Path., 11:192, 1930.
3. J. Immunol., 34:381, 1938.
4. J. Bact., 58:731, 1949.
5. Biochem. Biophys. Acta., 8:302, 1952.
6. Biochem. Prep., 1:67, 1949.

PACKAGING

Bacto Lysozyme Substrate	1 g	0461-10-0
Bacto Lysozyme	1 g	0465-10-6
Bacto Lysozyme Buffer	6 × 9.6 g	0464-33-0

BACTO M BROTH

INTENDED USE
Bacto M Broth is recommended for use in detecting *Salmonella* in foods and feeds by the accelerated enrichment serology (ES) procedure.

HISTORY
Bacto M Broth (Mannose) is prepared according to the formula of Deibel and Sperber[1] and contains all the nutrients necessary for good growth and flagella development of *Salmonella*.

Fantasia, Sperber and Deibel[2] compared the ES procedure with the traditional procedure outlined in the *Bacteriological Analytical Manual*[3] (BAM) and reported excellent agreement between the 2 procedures. They found the ES procedure to be more rapid and less complicated to perform, while maintaining the accuracy and sensitivity of the BAM procedure.

FORMULA

BACTO M BROTH
DEHYDRATED
Ingredients per liter

Bacto Yeast Extract	5 g	Dipotassium Phosphate	5 g
Bacto Tryptone	12.5 g	Manganese Chloride	0.14 g
Bacto D-Mannose	2 g	Magnesium Sulfate	0.8 g
Sodium Citrate	5 g	Ferrous Sulfate	0.04 g
Sodium Chloride	5 g	Tween® 80	0.75 g

Final pH 7.0 ± 0.2 at 25°C.

One pound will make 12.5 liters of medium.

METHOD OF PREPARATION
1. To rehydrate, suspend 36.2 g in 1 liter distilled or deionized water and heat to boiling for 1 – 2 minutes.
2. Sterilize in the autoclave for 15 minutes at 15 lbs pressure (121°C).
3. Bacto M Broth may be used as follows or as a culture medium.

TEST PROCEDURE
1. Prepare a 10% suspension of the test sample in Bacto Lactose Broth. Incubate for 18 – 24 hours at 35°C.
2. Aseptically transfer 1 ml of the above culture to both 9 ml of Bacto Selenite Cystine Broth and 9 ml of Bacto Tetrathionate Broth. Incubate for 24 hours at 35°C.
3. Inoculate one 10 ml tube of Bacto M Broth with one drop from each of the above cultures. Incubate for 6 – 8 hours at 35°C.
4. Prepare a formalin-salt solution by adding 4.2 g of NaCl and 3 ml of formalin to 100 ml of distilled water. Place one drop in each of two Kahn tubes.
5. Carefully insert a pipette about 1 inch below the surface of the M broth culture and transfer 0.85 ml of culture to each of the above Kahn tubes containing formalin-salt solution.
6. Rehydrate Bacto Salmonella H Antisera Spicer-Edwards Set, Salmonella H Antiserum Poly D and Salmonella H Antiserum z6. Prepare a pooled antiserum by combining together 0.5 ml of each rehydrated antiserum in 11.5 ml of 0.85% NaCl.
7. Add 0.1 ml pooled Salmonella H Antiserum to one Kahn tube (above). Add 0.1 ml 0.85% NaCl solution to the other tube. Shake the tubes gently. Incubate for 1^1/$_2$ hours in a 50°C water bath.
8. Agglutination in the Kahn tube containing antiserum indicates the presence of *Salmonella*. Agglutination in the Kahn tube containing 0.85% NaCl solution (control tube) indicates a rough culture which should be streaked out until smooth, passed through Bacto Motility GI Medium to enhance flagella and then retested with pooled antiserum.

STORAGE
Bacto M Broth	2 – 8°C
Prepared medium	2 – 8°C

QUALITY CONTROL

Identity Specifications

Dehydrated powder:	beige, homogeneous, free-flowing
Reaction of 3.62% solution:	pH 7.0 ± 0.2 at 25°C
Prepared medium:	light amber, clear

Typical Cultural Response in Bacto M Broth
After 18 – 24 Hours at 35°C

Organism	Growth
Salmonella choleraesuis ATCC® 12011	excellent
Salmonella typhimurium ATCC® 14028	excellent

REFERENCES

1. Appl. Micro., 17:533, 1969.
2. *ibid.*, 17:540, 1969.
3. Bacteriological Analytical Manual US FDA, Wash., D.C., 1969.

PACKAGING

Bacto M Broth	1 lb (454 g)	0940-01-2
Bacto Motility GI Medium	1 lb (454 g)	0869-01-9
	1/4 lb (114 g)	0869-02-8
Bacto Lactose Broth	1 lb (454 g)	0004-01-5
	1/4 lb (114 g)	0004-02-4
	5 lb (2.27 kg)	0004-05-1
Bacto Salmonella H Antisera Spicer-Edwards Set	7 × 3 ml	2328-32-3
Bacto Salmonella H Antiserum z6	3 ml	2473-47-2
Bacto Salmonella H Antiserum Poly D	3 ml	2542-47-6
Bacto Selenite Cystine Broth	1 lb (454 g)	0275-01-7
	1/4 lb (114 g)	0275-02-6
	10 kg	0275-08-0
	12 tubes	0275-34-8
	144 tubes	0275-37-5
Bacto Tetrathionate Broth Base	100 g	0104-15-8
	500 g	0104-17-6

BACTO MIL MEDIUM

INTENDED USE

Bacto MIL (Motility-Indole-Lysine) Medium is used as an aid for the identification of the *Enterobacteriaceae* on the basis of motility, lysine decarboxylase, lysine deaminase and indole production.

HISTORY/PRINCIPLES

Bacto MIL Medium, prepared according to the formula of Reller and Merrett,[1] provides a single culture medium combining 4 differentiating reactions into 1 tube. When used with Bacto Triple Sugar Iron Agar (TSI) and Bacto Urea Agar as many as 9 reactions are combined into only 3 tubes. This combination enables reliable initial recognition of enteric pathogens of the *Enterobacteriaceae*. Extensive testing of 890 enteric cultures gave essentially the same results with MIL medium as with the standard motility, indole, and lysine decarboxylase (Moeller) test media.

FORMULA

BACTO MIL MEDIUM

Ingredients per liter

Bacto Peptone	10 g	Bacto Dextrose	1 g
Bacto Tryptone	10 g	Ferric Ammonium Citrate	0.5 g
Bacto Yeast Extract	3 g	Bacto Brom Cresol Purple	0.02 g
L-Lysine Hydrochloride	10 g	Bacto Agar	2 g

Final pH 6.6 ± 0.2 at 25°C.

Five hundred grams will make 13.7 liters of medium.

METHOD OF PREPARATION

1. Suspend 36.5 g in 1 liter distilled or deionized water and heat to boiling to dissolve completely.
2. Dispense in 5 ml amounts in 13 × 100 mm screw capped test tubes.
3. Sterilize in the autoclave for 15 minutes at 15 lbs pressure (121°C).

PROCEDURE

Test cultures are inoculated into prepared tubes of MIL medium by stabbing with a straight wire to the bottom of the medium. The tubes are then incubated for 18 – 24 hours at 35°C. Lysine deaminase, motility, and lysine decarboxylase reactions are read before adding the indole reagent for the indole test. Lysine deaminase is indicated by a red or red-brown color reaction in the top centimeter of the medium. Motility is indicated by a clouding of the medium or by growth extending from the inoculation line. Lysine decarboxylase is indicated by a purple color throughout the medium. This color may vary in intensity and may be bleached out to a pale light color due to reduction of the indicator. Lysine negative cultures produce a yellow tube which may be purple or red on the top. Tubes which show a purple reaction with a red color on top should be incubated for a longer period of time. For the indole test 3 or 4 drops of Kovac's reagent are added to the top of the tube. The appearance of a pink to red color in the reagent is interpreted as a positive indole test.

The following table gives the typical reactions of the *Enterobacteriaceae* in MIL medium:

	Lysine Decarboxylase	Motility	Lysine Deaminase	Indole Production
Citrobacter diversus	−	+	−	+
Citrobacter freundii	−	+	−	−
Edwardsiella	+	+	−	+
Enterobacter aerogenes	+	+	−	−
Enterobacter agglomerans	−	+ or −	− or +	− or +
Enterobacter cloacae	−	+	−	−
Escherichia	+ or −	+ or −	−	+
Hafnia alvei	+	+	−	−
Klebsiella pneumoniae	+	−	−	− (+)
Morganella morganii	−	+	+ or −	+
Proteus mirabilis	−	+	+	−
Proteus vulgaris	−	+	+	+
Providencia	−	+	+	+ or −
Salmonella	+	+	−	−
Serratia liquefaciens	+	+	−	−
Serratia marcescens	+	+	−	−
Serratia rubidaea	+	+	−	−
Shigella	−	−	−	− or +
Yersinia entercolitica	−	−	−	− or +
Yersinia enterocolitica @25°C	−	+	−	− or +

KEY: + = Positive   − = Negative   ( ) = Occasional reaction

On occasion a reaction may occur which differs from that indicated in the table or published information. Such results may occur if an inadequate inoculum is used or if the inoculum is not a pure culture. It is also to be understood that positive and negative reactions are based on 90% or more occurrences. When aberrant reaction occurs, subcultures should be replated on differential media to insure purity of the culture.

STORAGE

Bacto MIL Medium — Below 30°C
Prepared tubes — 15 – 30°C

QUALITY CONTROL

Identity Specifications

Dehydrated powder: light beige, homogeneous, free-flowing
Reaction of 3.65% solution: pH 6.6 ± 0.2 at 25°C
Prepared medium: reddish purple, clear

Typical Cultural Response in Bacto MIL Medium
After 18 – 24 Hours at 35°C

Organism	Growth	Lysine Decarboxylase	Motility	Lysine Deaminase	Indole Production
Escherichia coli ATCC® 25922	good to excellent	+	+	−	+
Klebsiella pneumoniae ATCC® 13883	good to excellent	+	−	−	−
Providencia alcalifaciens ATCC® 9886	good to excellent	−	+	+	−
Salmonella enteritidis ATCC® 13076	good to excellent	+	+	−	−
Shigella flexneri ATCC® 12022	good to excellent	−	−	−	−

REFERENCE

1. J. Clin. Microbiology, 2:247, 1975.

PACKAGING

Bacto MIL Medium — 500 g — 1804-17-7
144 tubes — 1804-37-3

BACTO MIO MEDIUM

INTENDED USE

Bacto MIO Medium is used for the identification of the *Enterobacteriaceae* on the basis of motility, ornithine decarboxylase activity and indole production.

HISTORY

Bacto MIO Medium, prepared according to the formula of Ederer and Clark,[1] provides a single culture medium combining 3 differentiating reactions into 1 tube. When used with Bacto Triple Sugar Iron Agar (TSI) as many as 7 reactions are combined into only 2 tubes. The results of an extensive study comparing the cultural reactions of *Enterobacteriaceae* on the MIO Medium and on the classical media were reported by Ederer and Clark, who stressed the advantages of the MIO Medium.

PRINCIPLES

Motility is demonstrated by a clouding of the medium or growth diffusing out from the stab inoculation line. Nonmotile cultures do not demonstrate this diffuse growth but rather grow only along the inoculation line.

The ornithine reaction is indicated by a purple color throughout the medium. This color may vary in intensity and may be bleached out to a pale light color due to reduction of the indicator. The reaction is due to the test organism fermenting the dextrose in the formula, reducing its pH. This acid condition causes the pH indicator, brom cresol purple, to turn to yellow and provides optimal conditions for the decarboxylation of the ornithine in the formula. If this reaction occurs, the pH rises as a result and the indicator becomes purple. Ornithine-negative cultures produce a yellow tube which may be purple on the top.

The indole test is based upon the reaction of indole, if present, with the aldehyde present in Kovacs reagent which is added to the tube after the motility and ornithine decarboxylase reactions are determined. Indole-positive cultures produce a pink to red color in the reagent. Indole is produced by an organism from the tryptophane present in the peptones in the formula.[2]

FORMULA

BACTO MIO MEDIUM
DEHYDRATED

Ingredients per liter

```
Bacto Yeast Extract . . . . . . . . . . . . . 3 g
Bacto Peptone  . . . . . . . . . . . . . . . 10 g
Bacto Tryptone . . . . . . . . . . . . . . . 10 g
Bacto L-Ornithine Hydrochloride  . . . . 5 g
Bacto Dextrose . . . . . . . . . . . . . . . 1 g
Bacto Agar . . . . . . . . . . . . . . . . . . 2 g
Bacto Brom Cresol Purple  . . . . . . 0.02 g
```

Final pH 6.5 ± 0.2 at 25°C.

One pound will make 14.6 liters of medium.

METHOD OF PREPARATION
1. Suspend 31 g in 1 liter distilled or deionized water and heat to boiling to dissolve completely.
2. Dispense in 5 ml amounts into 13 × 100 mm screw-capped test tubes.
3. Sterilize in the autoclave for 15 minutes at 15 lbs pressure (121°C).

PROCEDURE

Test cultures are inoculated into prepared tubes of Bacto MIO Medium by stabbing with a straight wire to the bottom of the medium. The tubes are then incubated for 18 – 24 hours at 35°C. Motility and ornithine decarboxylase reactions are read before adding the indole reagent for the indole test. For the indole test 3 or 4 drops of Kovacs reagent are added to the top and the tube shaken gently. The appearance of a pink to red color in the reagent is interpreted as a positive indole test.

STORAGE

Bacto MIO Medium   Below 30°C
Prepared tubes   15 – 30°C

QUALITY CONTROL

Identity Specifications

Dehydrated powder: beige, homogeneous, free-flowing
Reaction of 3.1% solution: pH 6.5 ± 0.2 at 25°C
Prepared medium: purple, clear to slightly opalescent

Typical Cultural Response in Bacto MIO Medium
After 40 – 48 Hours at 35 ± 2°C

Organism	Growth	Motility	Indole	Ornithine Decarboxylase
Enterobacter aerogenes ATCC® 13048	good	+	−	+
Escherichia coli ATCC® 25922	good	+	+	+
Klebsiella pneumoniae ATCC® 13883	good	−	−	−
Proteus mirabilis ATCC® 25933	good	±*	−	+

*Motility of *Proteus* is temperature dependent, being more pronounced at 20°C and possibly absent at 35°C.

REFERENCES

1. Appl. Micro., 20:849, 1970.
2. MacFaddin, Jean F., Biochemical Tests for Medical Bacteria, 2nd Ed., Baltimore: Williams & Wilkins, pp. 173 – 178, 1981.

PACKAGING

Bacto MIO Medium

1 lb (454 g)	0735-01-1
1/4 lb (114 g)	0735-02-0
5 lb (2.27 kg)	0735-05-7
12 tubes	0735-34-2

BACTO MR - VP MEDIUM

INTENDED USE

Bacto MR - VP Medium is used to differentiate strains of coliform bacteria by the Methyl Red and Voges-Proskauer tests.

HISTORY/PRINCIPLES

Clark and Lubs[1] first pointed out that in a suitable medium coli organisms produced a high acidity which was constant, while the aerogenes group produced a much less acid reaction, and on continued incubation became more alkaline. This difference in the acidity produced in the cultivation of coli or aerogenes could be recognized by the addition of the indicator, methyl red. This Clark and Lubs test has become known as the Methyl Red Reaction.

Closely associated with the Methyl Red Reaction is the test described by Voges and Proskauer,[2] who noted that a color reaction took place if certain cultures in a suitable medium were treated with potassium hydroxide and allowed to stand for some time. This color reaction develops particularly in that part of the medium exposed to the air, and is very similar to that of a dilute alcoholic solution of eosin. The development of the color reaction upon treatment of cultures with potassium hydroxide was found to be due to the presence of acetyl methyl carbinol, (3-hydroxy-2-butanone). Levine[3] recommended that the term "Voges-Proskauer Reaction" be restricted to designate the formation of acetyl methyl carbinol from dextrose.

Bacto MR - VP Medium was developed as a simple and reliable medium for use in the performance of the methyl red and Voges-Proskauer tests. Ruchhoft, Kallas, Chinn and Coulter[4] reported that Bacto MR - VP Medium is uniform and superior to laboratory made media for these tests.

FORMULA

BACTO MR - VP MEDIUM
DEHYDRATED

Ingredients per liter

Buffered Peptone . . . . 7 g
Dipotassium Phosphate . . . . 5 g
Bacto Dextrose . . . . 5 g

Final pH 6.9 ± 0.2 at 25°C.

One pound of Bacto MR - VP Medium will make 26.7 liters of medium.

METHOD OF PREPARATION
1. Dissolve 17 g in 1 liter distilled or deionized water.
2. Distribute into test tubes in 10 ml amounts.
3. Sterilize in the autoclave for 15 minutes at 15 lbs pressure (121°C).

PROCEDURES
For performance of the methyl red and Voges-Proskauer tests, each tube should be inoculated with a pure culture.

The methyl red test is performed after 5 days incubation at 30°C as recommended by Vaughn, Mitchell and Levine.[5] For the Voges-Proskauer test the culture should be incubated for 24 to 48 hours at 30°C.[4]

Methyl Red Test
To 5 ml of culture add 5 drops of methyl red solution. A positive reaction is indicated by a distinct red color, showing the presence of acid. A negative reaction is indicated by a yellow color. The indicator solution is prepared by dissolving 0.1 g Bacto Methyl Red in 300 ml of 95% alcohol and diluting to 500 ml with distilled water.

Voges-Proskauer Test
To 5 ml of culture add 5 ml of a 10% solution of potassium hydroxide, mix well, allow to stand exposed to the air, and observe at intervals of 2, 12, and 24 hours. A positive test is indicated by the development of an eosin pink color. This is the test as originally described.[1]

Various other tests have been suggested as being excellently adapted to the demonstration of the development of acetyl methyl carbinol, chief among which are those of Werkman,[6] O'Meara,[7] Levine, Epstein and Vaughn[8] and Vaughn, Mitchell and Levine.[5]

Werkman's test[6] consists of the addition of 2 drops of a 2% solution of ferric chloride to 5 ml of culture, followed by 5 ml of 10% sodium hydroxide, and shaking the tube well to mix. A stable copper color appearing in a few minutes is indicative of a positive test.

O'Meara[7] recommends the addition of approximately 25 mg of solid creatine to 5 ml of culture and then adding 5 ml concentrated (at least 40%) sodium hydroxide. The development of a red color in a few minutes, after thorough agitation of the tube, is a positive test. Levine, Epstein and Vaughn[8] modified the O'Meara technique by dissolving the creatine in a concentrated solution of potassium hydroxide. Vaughn, Mitchell and Levine[5] recommended the method of Barritt[9] for this test. The test is made by adding 0.6 ml of 5% alpha-naphthol in absolute ethyl alcohol and 0.2 ml of 40% potassium hydroxide to 1 ml of culture.

STORAGE

Bacto MR - VP Medium   Below 30°C
Prepared medium   15 – 30°C

QUALITY CONTROL

Identity Specifications

Dehydrated powder: light beige, homogeneous, free-flowing
Reaction of 1.7% solution: pH 6.9 ± 0.2 at 25°C
Prepared medium: light amber, clear, no significant precipitate

Typical Cultural Response on Bacto MR - VP Medium
After 48 Hours at 30°C

Organism	Growth	MR	VP
Enterobacter aerogenes ATCC® 13048	good	– (yellow)	+ (red)
Escherichia coli ATCC® 25922	good	+ (red)	– (no change)
Klebsiella pneumoniae ATCC® 23357	good	+	–

REFERENCES

1. Clark and Lubs, 1915, Journal of Infectious Diseases, 17:160.
2. Voges and Proskauer, 1898, Zeit. Hyg., 28:20.
3. Levine, 1916, Journal of Bacteriology, 1:153.
4. Kallas, Chinn and Coulter, 1931, Journal of Bacteriology, 22:125.
5. Vaughn, Mitchell, and Levine, 1939, Journal of the American Water Works Association, 31:993.
6. Werkman, 1930, Journal of Bacteriology, 20:121.
7. O'Meara, 1931, Journal of Pathogenic Bacteriology, 34:401.
8. Levine, Epstein and Vaughn, 1934, American Journal of Public Health, 24:505.
9. Barritt, 1936, Journal of Pathogenic Bacteriology, 42:441.

PACKAGING

Bacto MR - VP Medium

1 lb (454 g)	0016-01-1
1/4 lb (114 g)	0016-02-0
5 lb (2.27 kg)	0016-05-7
12 tubes	0016-34-2
144 tubes	0016-37-9

MacCONKEY MEDIA

BACTO MacCONKEY AGAR
BACTO MacCONKEY AGAR BASE
BACTO MacCONKEY AGAR CS
BACTO MacCONKEY AGAR w/o CV
BACTO MacCONKEY AGAR w/o SALT

INTENDED USE

Bacto MacConkey Agar is a differential plating medium recommended for use in the isolation and differentiation of lactose-fermenting organisms from lactose nonfermenting gram-negative enteric bacteria.

Bacto MacConkey Agar Base is Bacto MacConkey Agar prepared without carbohydrates, allowing their addition either individually or in combination, for fermentation studies of coliforms.

Bacto MacConkey Agar CS (Controlled Swarming) is Bacto MacConkey Agar which has been prepared using raw materials specially pretested for use in isolating gram-negative enteric bacilli from clinical and environmental specimens which may contain swarming strains of *Proteus*.

Bacto MacConkey Agar w/o CV is a modification of Bacto MacConkey Agar prepared without crystal violet. It is slightly less selective than the original formula and permits growth of staphylococci and enterococci.

Bacto MacConkey Agar w/o Salt is a differential medium which restricts the swarming of most *Proteus* species thereby permitting greater ease in the detection and isolation of enteric microorganisms. It is especially useful for culturing specimens, such as urines, which may contain large numbers of *Proteus* organisms as well as gram-positive organisms which may be potentially pathogenic.

HISTORY

Bacto MacConkey Agar is based on the bile salt-neutral red-lactose agar of MacConkey and has been generally used for the selective isolation of gram-negative enteric bacilli from clinical and nonclinical sources.[1] Bacto MacConkey Agar is recommended for use in the microbiological examination of foodstuffs,[2] for direct plating of water samples for coliform counts[3] and for the isolation of pathogenic bacteria from cheese and other dairy products.[4] It has been specified by such references as: *Standard Methods for the Examination of Water and Sewage, Standard Methods for the Examination of Dairy Products, Diagnostic Procedures and Reagents,*[5] and *Compendium of Methods for the Microbiological Examination of Foods*[6] of the American Public Health Association.

Over a period of years, particularly in Great Britain, the neutral red bile salt agar of MacConkey[7] has been quite generally used for differentiating strains of *Salmonella typhi* from members of the coliform group. Bacto MacConkey Agar as modified by the addition of 0.5% sodium chloride, decreasing the agar content to 1.35% and by altering the concentration of bile salts and neutral red, has the added advantage of supporting excellent growth of *Shigella* and *Salmonella*. It also gives a more clear cut differential between these enteric pathogens and the coliform group, making it easier to read than the original medium. Block and Ferguson,[8] investigating an outbreak of Shiga dysentery, found MacConkey agar satisfactory in the isolation of this fastidious strain.

The fact that this medium promotes development of these organisms, and at the same time differentiates them from lactose-fermenting gram-negative bacilli, makes it an excellent substrate for the cultural detection of dysentery, typhoid and other *Salmonella* organisms in stools and other infected material. Gram-positive bacteria are inhibited.

PRINCIPLES
Bacto MacConkey Agar
The differential action of Bacto MacConkey Agar is based on fermentation of lactose. Colonies of organisms capable of fermenting lactose produce a localized pH drop which, followed by absorption of the neutral red, imparts a red color to the colony. A zone of precipitated bile may also be present due to this localized drop in pH. Colonies of organisms which do not ferment lactose remain colorless and translucent.

When growing in proximity to coliform colonies, they have the appearance of clearing the areas of precipitated bile. On plates which are not overcrowded, the differentiation is exceptionally distinct. A plate crowded with coli will appear red and opaque, yet, if not too crowded, Salmonella or other lactose-nonfermenting organisms may easily be detected by transmitted light. On such plates they will appear as small transparent areas against the red background. A plate showing discrete colonies is to be desired for isolation purposes. Selectivity of the medium is due to the presence of crystal violet and bile salts which markedly to completely inhibit the growth of gram-positive organisms.

Bacto MacConkey Agar Base
Some members of the coliform group will ferment sucrose more readily than lactose. Since pathogenic significance is sometimes assigned to such members of the coliform group, which are reported by Holt, Harris and Teague,[9] it is desirable to have a basal medium to which carbohydrates can be added as required. Similarly, certain enteropathogenic *E. coli* can ferment sorbitol whereas others do not, but these will all ferment lactose. It is obvious that the addition of sorbitol to the base will aid in the differentiation of some enteropathogenic *E. coli* from the nonpathogenic strains. It is recommended that carbohydrates be added in a concentration of 1% to the basal medium.

Bacto MacConkey Agar CS
The controlled swarming property of Bacto MacConkey Agar CS is due to careful selection and piloting of the raw materials by actual performance testing. Only those raw materials are used which are compatible with one another and which do not contribute to the swarming of *Proteus*.

Although numerous reports have appeared in the literature regarding the influence of specific culture media additives on swarming of *Proteus*, the actual factors responsible for this swarming have not been completely identified.[10-15]

While MacConkey Agar CS is selective primarily for gram-negative enteric bacilli, biochemical and (if indicated) serological testing using pure cultures are recommended for complete identification. Consult appropriate references for further information.[16-19]

Bacto MacConkey Agar w/o CV
Bacto MacConkey Agar w/o Salt
Without the crystal violet as a selective agent, the former medium is less selective and allows the growth of gram-positive cocci as well as the *Enterobacteriaceae*. It has been widely used in the United Kingdom and in Canada for enumerating coliforms.[20,21]

Coliform bacteria produce red colonies on Bacto MacConkey Agar w/o CV and Bacto MacConkey Agar w/o Salt. Enteric organisms such as *Shigella, Salmonella,* and *Pro-*

teus which do not ferment lactose produce transparent or light opaque colonies. Surface colonies of *Escherichia coli* are usually red and not mucoid, while those of *Enterobacter aerogenes* are pinkish red and mucoid. Subsurface colonies of *E. coli* or *E. aerogenes* are red and usually have a red zone around the colonies. Staphylococci produce pale pink to red colonies. Enterococci produce compact tiny red colonies, either on or beneath the surface.

FORMULAE

	Bacto MacConkey Agar	Bacto MacConkey Agar Base	Bacto MacConkey Agar CS	Bacto MacConkey Agar w/o CV	Bacto MacConkey Agar w/o Salt
Bacto Peptone	17 g	17 g	17 g	20 g	20 g
Proteose Peptone, Difco	3 g	3 g	3 g	—	—
Bacto Lactose	10 g	—	10 g	10 g	10 g
Bacto Bile Salts No. 3	1.5 g	1.5 g	—	—	—
Bile Salts	—	—	1.5 g	5 g	5 g
Sodium Chloride	5 g	5 g	5 g	5 g	—
Bacto Agar	13.5 g	13.5 g	13.5 g	12 g	12 g
Bacto Neutral Red	0.03 g	0.03 g	0.03 g	0.05 g	0.075 g
Bacto Crystal Violet	0.001 g	0.001 g	0.001 g	—	—
Final pH ± 0.2 at 25°C	7.1	7.1	7.1	7.4	7.4
Rehydration	50 g/liter	40 g/liter	50 g/liter	52 g/liter	47 g/liter
Yield	9.08 liter/lb	12.5 liter/lb	10 liter/500 g	8.73 liter/lb	10.6 liter/500 g

METHOD OF PREPARATION
1. Suspend the appropriate amount of medium in 1 liter distilled or deionized water. Heat to boiling with gentle swirling to dissolve completely. Add 10 g of lactose or other carbohydrate to Bacto MacConkey Agar Base and swirl the flasks to dissolve completely.
2. Sterilize in the autoclave for 15 minutes at 15 lbs pressure (121°C). Avoid overheating of the medium. Bacto MacConkey Agar Media inoculated the same day as rehydrated may be used without autoclave sterilization if it is boiled for 2 – 3 minutes.
3. Cool to 45 – 50°C and dispense in approximately 20 ml amounts into sterile Petri dishes.
4. The surface of the medium should be dry when inoculated. Drying may be accomplished by allowing the medium to stand after dispensing for 1 – 2 hours with lids slightly ajar.

PROCEDURE
For the cultural detection of *S. typhi* and other *Salmonella* in infected material, it is recommended that Bacto MacConkey Agar be used in conjunction with the more selective media, Bacto SS Agar and Bacto Bismuth Sulfite Agar.

A procedure designed to show the largest number of pathogens from a specimen would be:
A. Streak or smear a large inoculum on 1 plate of SS agar and 1 plate of bismuth sulfite agar.
B. Streak or smear a light inoculum on 1 plate of MacConkey agar.
C. Prepare bismuth sulfite agar poured plates with 5 ml and 1 drop of inoculum.
D. Enrich for 12 – 18 hours in tetrathionate broth or selenite broth, followed by streaking on 1 plate of bismuth sulfite agar or brilliant green agar and 1 plate of SS agar.

Sucrose in 1% concentration may be added to isolation media, such as Bacto MacConkey Agar, to permit the detection of certain members of the coliform group which ferment sucrose more readily than lactose. This principle was described by Holt-Harris and Teague[9] and has been employed by many other bacteriologists. In some laboratories pathogenic significance is assigned to these organisms, and under such conditions, sucrose should not be added to the medium.

Inoculate the specimen directly onto the surface of a MacConkey agar medium plate and streak for isolation. Incubate inoculated plates aerobically at 35°C. Examine for growth after 18 – 24 hours incubation.

Due to the selective properties of this medium, some strains of gram-negative enteric bacilli may be encountered that fail to grow or grow poorly on this medium; similarly, some strains of gram-positive organisms may be encountered that are not inhibited or only partially inhibited on this medium. Some strains of enterococci may grow on this medium after prolonged incubation. Incubation of MacConkey agar plates under increased CO_2 has been reported to reduce the growth and recovery of a number of strains of gram-negative enteric bacilli.[19] For optimal performance, therefore, it is recommended that plates prepared from Bacto MacConkey Agar be incubated under aerobic conditions.

STORAGE

Bacto MacConkey media	Below 30°C
Prepared media	2 – 8°C

QUALITY CONTROL

Identity Specifications

	Bacto MacConkey Agar	Bacto MacConkey Agar Base	Bacto MacConkey Agar CS	Bacto MacConkey Agar w/o CV	Bacto MacConkey Agar w/o Salt
Dehydrated powder:	pinkish beige	pinkish beige	pink	beige	pinkish beige
% solution:	5%	4%	5%	5.2%	4.7%
pH ± 0.2 at 25°C:	7.1	7.1	7.1	7.4	7.4
Prepared medium:	reddish-purple	red	reddish-purple	reddish-orange	red

-------------------- homogeneous, free-flowing --------------------

-------------------- slightly opalescent --------------------

Typical Cultural Response on Bacto MacConkey Agar and Bacto MacConkey Agar Base with 1% Lactose After 18 – 24 Hours at 35°C

Organism	Growth	Color of Colony
Enterobacter aerogenes ATCC® 13048	good to excellent	pink to red
Escherichia coli ATCC® 25922	good to excellent	pink to red w/bile ppt.
Proteus vulgaris ATCC® 13315	good to excellent	colorless
Salmonella enteritidis ATCC® 13076	good to excellent	colorless
Shigella dysenteriae ATCC® 13313	fair to good	colorless
Staphylococcus aureus ATCC® 25923	inhibited	—

Typical Cultural Response on Bacto MacConkey Agar CS After 18 – 24 Hours at 25°C

Organism	Growth	Color of Colony
Escherichia coli ATCC® 25922	good	pink to red with or without a zone of precipitated bile

Organism	Growth	Color of Colony
Proteus mirabilis ATCC® 9240	good	translucent and color-less; swarming is markedly to com-pletely inhibited
Shigella flexneri ATCC® 12022	good	translucent, colorless
Staphylcoccus aureus ATCC® 25923	marked to com-plete inhibition	—

Typical Cultural Response on Bacto MacConkey Agar w/o CV
After 18 – 48 Hours at 35°C

Organism	Growth	Color of Colony
Enterobacter aerogenes ATCC® 13048	good to excellent	colorless
Escherichia coli ATCC® 25922	good to excellent	pink to red
Salmonella enteritidis ATCC® 13076	good to excellent	colorless
Shigella flexneri ATCC® 21022	fair to good	colorless
Staphylococcus aureus ATCC® 25923	poor to good	pink
Staphylococcus epidermidis ATCC® 12228	poor to good	pink

Typical Cultural Response on Bacto MacConkey Agar w/o Salt
After 18 – 48 Hours at 35°C

Organism	Growth	Color of Colony
Enterobacter aerogenes ATCC® 13048	good to excellent	sl. pink to pink
Escherichia coli ATCC® 25922	good to excellent	pink to red
Proteus mirabilis ATCC® 9240	good to excellent	colorless
Salmonella enteritidis ATCC® 13076	good to excellent	colorless
Shigella flexneri ATCC® 12022	fair to good	colorless
Staphylococcus aureus ATCC® 25923	fair to good	pink

REFERENCES

1. MacConkey, A. 1905. Lactose-fermenting bacteria in feces. J. Hyg. 5:333 – 378.
2. Leininger, H. V. 1976. Equipment, media, reagents, routine tests, p. 10 – 94. *In*: M.L. Speck (ed.), Compendium of methods for the microbiological examination of foods. American Public Health Association, Inc., Washington, D.C.
3. Standard methods for the examination of water and wastewater, 15th ed. 1980. American Public Health Association, Washington, D.C.
4. Standard methods for the examination of dairy products, 9th ed. 1948. American Public Health Association, Washington, D.C.
5. Diagnostic procedures and reagents, 3rd ed. 1950.
6. Speck, M. L. 1976. Compendium of methods for the microbiological examination of foods. American Public Health Association, Washington, D.C.
7. J. Hyg., 5:333, 1905.
8. Am. J. Pub. Health, 130:42, 1940.
9. J. Infectious Dis., 18:596, 1916.
10. Jeffries, C. D., and H. E. Rogers. 1968. Enhancing effect of agar on swarming of *Proteus*. J. Bacteriol. 95:732 – 733.
11. Sandys, G. H. 1960. A new method of preventing swarming of *Proteus* ssp. with a description of a new medium suitable for use in routine laboratory practice. J. Med. Lab. Tech. 17:224 – 233.
12. Jones, H. A., and R. W. A. Park. 1967. The influence of medium composition on the growth and swarming of *Proteus*. J. Gen. Microbiol. 47: 369 – 378.
13. Brogan, T. D., J. Nettleton and C. Reid. 1971. The swarming of *Proteus* on semisynthetic media. J. Med. Microbiol. 4:1 – 11.
14. Kopp, R., J. Mueller, and R. Lemme. 1966. Inhibition of swarming of *Proteus* by sodium tetradecyl sulfate, B-phenethyl alcohol, and p-nitrophenylglycerol. Appl. Microbiol. 14:873 – 878.
15. Williams, F. D. 1973. Abolition of swarming of *Proteus* by p-nitrophenyl glycerin: general properties. Appl. Microbiol. 25:745 – 750.

16. Cowan, S. T. and K. J. Steel. 1970. Manual for the identification of medical bacteria. Cambridge University Press, Cambridge, U.K.
17. Lennette, E. H., A. Balows, W. J. Hausler, Jr., and J. P. Truant (ed.). 1980. Manual of clinical microbiology, 3rd ed. American Society for Microbiology, Washington, D.C.
18. Speck, M. L. (ed.). 1976. Compendium of methods for the microbiological examination of foods. American Public Health Association, Washington, D.C.
19. Mazura-Reetz, G., T. R. Neblett, and J. M. Galperin. 1979. MacConkey Agar: CO_2 vs. ambient incubation. Abst. Ann. Mtg. Amer. Soc. Microbiol. C179.
20. Ministry of Health: Public Health Laboratory Service Water Committee. 1956. The bacteriological examination of water supplies, 3rd ed. HMSO, London.
21. Windle, T. E. 1958. The examination of waters and water supplies, 7th ed. Churchill Ltd., London.

PACKAGING

Bacto MacConkey Agar	1 lb (454 g)	0075-01-9
	1/4 lb (114 g)	0075-02-8
	5 lb (2.27 kg)	0075-05-5
	10 kg	0075-08-2
Bacto MacConkey Agar Base	1 lb (454 g)	0818-01-1
Bacto MacConkey Agar CS	500 g	1818-17-1
	5 lb (2.27 kg)	1818-05-5
	10 kg	1818-08-2
Bacto MacConkey Agar w/o CV	1 lb (454 g)	0470-01-0
Bacto MacConkey Agar w/o Salt	500 g	0331-17-1
	10 kg	0331-08-2

BACTO MacCONKEY BROTH

INTENDED USE

Bacto MacConkey Broth is used in the presumptive phase for the presence of coliform bacteria in water and foods. It is a selective medium that detects gram-negative lactose-fermenting bacilli.

HISTORY/PRINCIPLES

Bacto MacConkey Broth is a modification of the original bile salt broth recommended by MacConkey[1] which contained 0.5% sodium taurocholate and litmus as an indicator. In later publications[2,3] MacConkey suggested variations of this formulation, using neutral red instead of litmus as an indicator. Childs and Allen[4] demonstrated the inhibitory effect of samples of neutral red and substituted brom cresol purple which is less inhibitory. The oxgall in the medium replacing the original 0.5% sodium taurocholate inhibits the growth of gram-positive organisms.

Bacto MacConkey Broth is also used for the rapid determination of fecal contamination in oysters as determined by Quadri, Buckle and Edwards.[5]

FORMULA

BACTO MacCONKEY BROTH
DEHYDRATED

Ingredients per liter

Bacto Oxgall	5 g
Bacto Peptone	20 g
Bacto Lactose	10 g
Bacto Brom Cresol Purple	0.01 g

Final pH 7.3 ± 0.1 at 25°C.

One pound will make 12.9 liters of medium.

METHOD OF PREPARATION
1. Dissolve 35 g in 1 liter distilled or deionized water.
2. Distribute in tubes.
3. Sterilize in the autoclave for 15 minutes at 15 lbs pressure (121°C).

When inocula are larger than 1 ml, the proper concentration of ingredients must be maintained by preparing a more concentrated medium.

STORAGE
Bacto MacConkey Broth   Below 30°C
Prepared medium   15 – 30°C

QUALITY CONTROL

Identity Specifications

Dehydrated powder: light beige, homogeneous, free-flowing
Reaction of 3.5% solution: 7.3 ± 0.1 at 25°C
Prepared medium: purple, clear, no scum or precipitate

Typical Cultural Response in Bacto MacConkey Broth
After 18 – 48 Hours at 35°C

Organism	Growth	Acid	Gas
Enterobacter aerogenes ATCC® 13048	good to excellent	+	+
Escherichia coli ATCC® 25922	good to excellent	+	+
Salmonella choleraesuis ATCC® 12011	fair to good	–	–
Staphylococcus aureus ATCC® 25923	inhibited	–	–

REFERENCES
1. MacConkey, 1901, Centr. Bakt. 29:740.
2. MacConkey, 1905, Journal of Hygiene, 5:333.
3. MacConkey, 1905, Journal of Hygiene, 8:322.
4. Childs and Allen, 1953, J. Hyg. Camb. 51 (4), 468 – 477.
5. Quadri, Buckle and Edwards, 1974, Journal of Applied Bacteriology, 37:7 – 14.

PACKAGING
Bacto MacConkey Broth   1 lb (454 g)   0020-01-5
     5 lb (2.27 kg)   0020-05-1

BACTO MALONATE BROTH

INTENDED USE
Bacto Malonate Broth is recommended for the differentiation of Enterobacter and Escherichia.

HISTORY/PRINCIPLES
Bacto Malonate Broth is prepared according to the formula described by Leifson.[1] It is a liquid medium prepared with materials of known chemical composition with ammonium sulfate and sodium malonate as the only sources of nitrogen and carbon.

The ability of members of the *Enterobacter* group to utilize malonate and the inability of members of the *Escherichia* group to grow in this medium was pointed out by Leifson in his investigations. A pH indicator, brom thymol blue, was incorporated in the medium for the purpose of differentiation. *Enterobacter,* utilizing malonate, produce an alkaline reaction and change the color of the medium to blue. *Escherichia,* not capable of utilizing malonate, fail to grow, leaving the medium unchanged.

FORMULA

BACTO MALONATE BROTH
DEHYDRATED

Ingredients per liter

Ammonium Sulfate	2 g
Dipotassium Phosphate	0.6 g
Monopotassium Phosphate	0.4 g
Sodium Chloride	2 g
Sodium Malonate	3 g
Bacto Brom Thymol Blue	0.025 g

Final pH 6.7 ± 0.2 at 25°C.

One pound will make 56.7 liters of medium.

METHOD OF PREPARATION
1. To rehydrate, dissolve 8 g in 1 liter distilled water.
2. Sterilize in the autoclave for 15 minutes at 15 lbs pressure (121°C).
3. Avoid introduction of carbon and nitrogen from other sources.

STORAGE

Bacto Malonate Broth	Below 30°C
Prepared medium	2 – 8°C

QUALITY CONTROL

Identity Specifications

Dehydrated powder:	light green, homogeneous, free-flowing
Reaction of 0.8% solution:	pH 6.7 ± 0.2 at 25°C
Prepared medium:	green, clear, no significant precipitate

Typical Cultural Response in Bacto Malonate Broth
After 18 – 48 Hours at 35°C

Organism	Growth	Color of Medium
Enterobacter aerogenes ATCC® 13048	good	blue
Enterobacter cloacae ATCC® 13047	good	blue
Escherichia coli ATCC® 25922	poor to fair	green
Klebsiella pneumoniae ATCC® 13883	good	blue
Salmonella arizonae ATCC® 13314	good	blue
Salmonella typhimurium ATCC® 14028	fair to good	green

REFERENCE
1. J. Bact., 25:329, 1933.

PACKAGING

Bacto Malonate Broth	1 lb (454 g)	0395-01-2
	1/4 lb (114 g)	0395-02-1

BACTO MALONATE BROTH MODIFIED

INTENDED USE
Bacto Malonate Broth Modified is recommended for differentiating members of *Entero-bacteriaceae* on the basis of malonate utilization.

HISTORY/PRINCIPLES
Bacto Malonate Broth Modified is essentially Bacto Malonate Broth to which Bacto Yeast Extract and dextrose have been added as described by Edwards and Ewing.[1] The small amount of yeast extract and dextrose initiates growth of some organisms that otherwise fail to respond and permits observation of their reaction upon malonate.

Malonate utilization by microorganisms is indicated by an increase in alkalinity and development of a deep blue color in the medium.

Bacto Malonate Broth Modified is of considerable value in the differentiation of *Salmonella*. The majority of salmonellae do not utilize malonate whereas the Arizona group does.[2,3,4,5] Many cultures belonging to the *Klebsiella* and *Enterobacter* genera utilize malonate whereas *Serratia* and *Proteus* strains, with rare exceptions, do not, as reported by Davis and Ewing[6] and Ewing and Davis.[7] Further, this medium may be used in conjunction with the organic acid media of Kauffmann and Peterson[8] and Ellis, Edwards and Fife[9] for differentiation of *Citrobacter* as well as *Salmonella* cultures described by Ewing and Edwards.[10]

FORMULA
BACTO MALONATE BROTH MODIFIED
DEHYDRATED
Ingredients per liter

Bacto Yeast Extract	1 g	Sodium Chloride	2 g
Ammonium Sulfate	2 g	Sodium Malonate	3 g
Dipotassium Phosphate	0.6 g	Bacto Dextrose	0.25 g
Monopotassium Phosphate	0.4 g	Bacto Brom Thymol Blue	0.025 g

Final pH 6.7 ± 0.2 at 25°C.

One pound will make 49 liters of medium.

METHOD OF PREPARATION
1. To rehydrate, suspend 9.3 g in 1 liter distilled or deionized water and agitate to dissolve completely.
2. Dispense into tubes and sterilize in the autoclave for 15 minutes at 15 lbs pressure (121°C).

STORAGE
Bacto Malonate Broth Modified　　　　　　　　　　Below 30°C
Prepared medium　　　　　　　　　　　　　　　　2 – 8°C

QUALITY CONTROL
Identity Specifications

Dehydrated powder:　　　　beige, homogeneous, free-flowing
Reaction of 0.93% solution:　pH 6.7 ± 0.2 at 25°C
Prepared medium:　　　　　bluish-green, clear, no significant precipitate

Typical Cultural Response in Bacto Malonate Broth Modified
After 18 – 48 Hours at 35°C

Organism	Growth	Color of Medium
Enterobacter aerogenes ATCC® 13048	good to excellent	blue
Enterobacter cloacae ATCC® 13047	good to excellent	blue

Organism	Growth	Color of Medium
Escherichia coli ATCC® 25922	good to excellent	green
Klebsiella pneumoniae ATCC® 13883	good to excellent	blue
Salmonella arizonae ATCC® 13314	good to excellent	blue
Salmonella typhimurium ATCC® 14028	good to excellent	green

REFERENCES

1. Publ. Hlth. Serv. Bull., No. 734, pg. 19, 1962.
2. John Hopkins Bull., 83:367, 1948.
3. Int. Bull. Bact. Nomen. & Tax., 6:1, 1956.
4. Publ. Hlth. Lab., 15:89, 1957.
5. Publ. Hlth. Lab., 15:153, 1957.
6. Int. Bull. Bact. Nomen. & Tax., 7:151, 1957.
7. Studies on Serratia Group, CDC Monograph, 1959.
8. Acta. Path. et Microbiol. Scand., 88:481, 1956.
9. Publ. Hlth. Lab., 15:89, 1957.
10. Int. Bull. Bact. Nomen. & Tax., 10:1, 1960.

PACKAGING

Bacto Malonate Broth Modified   1 lb (454 g)   0569-01-2

BACTO MALT AGAR

INTENDED USE

Bacto Malt Agar is recommended for the detection and isolation of yeasts and molds from dairy products, food and other materials. It is also recommended for carrying stock cultures of yeasts and molds.

HISTORY

Malt media for yeasts and molds have been in use for many years.[1] In 1919, Reddish[2] prepared a satisfactory substitute for beer wort from malt extract. Fullmer and Grimes[3] employed a malt agar for their studies of the growth of yeasts on synthetic media. Reddish's medium was used by Thom and Church[4] in their studies of the aspergilli. Standard Methods for the Examination of Dairy Products specified the use of dehydrated Bacto Malt Agar or Bacto Potato Dextrose Agar for the mold count of dry milk.[5]

PRINCIPLES

An acidified medium inhibits the growth of bacteria, thus allowing colonies of yeasts and molds to flourish.[6]

Heating processes during the rehydration and sterilization should be completed in as short a period as possible. Excessive exposure to heat causes partial hydrolysis of the agar with resultant inability to gel properly when cooled. Normally, a medium prepared from Bacto Malt Agar is slightly soft and is ideal for plating purposes. However, if the medium is desired for streaking, use 54 g of Bacto Malt Agar per liter distilled water, or include 5 g Bacto Agar with the 45 g of Bacto Malt Agar per liter.

FORMULA

BACTO MALT AGAR
DEHYDRATED

Ingredients per liter

Malt Extract, Difco . . . . 30 g
Bacto Agar . . . . 15 g

Final pH 5.5 ± 0.2 at 25°C.

One pound will make 10 liters of medium.

METHOD OF PREPARATION
1. To rehydrate, suspend 45 g in 1 liter distilled or deionized water and heat to boiling to dissolve completely.
2. Distribute in tubes or flasks and sterilize in the autoclave for 15 minutes at 15 lbs pressure (121°C). AVOID OVERHEATING, which will result in a softer medium.
3. Acidify the medium to pH 4.5 or 3.5. On the label for each package of Bacto Malt Agar, there is specified an amount of lactic acid USP (85%) that should be added to 100 ml of the sterile liquid medium to adjust the pH.
4. Mix well and pour plates. Allow to harden, inoculate and incubate as desired. Never heat medium after the addition of acid, as heating in the acid state will hydrolyze the agar, reducing its solidifying properties so that the resulting medium will be soft.

STORAGE
Bacto Malt Agar — Below 30°C
Prepared medium — 2 – 8°C

QUALITY CONTROL
Identity Specifications
Dehydrated powder: light tan, homogeneous, free-flowing
Reaction of 4.5% solution: pH 5.5 ± 0.2 at 25°C
Prepared medium: light to medium amber, very slightly opalescent

Typical Cultural Response on Bacto Malt Agar
After 40 – 48 Hours at 30°C

Organism	Growth
Aspergillus niger ATCC® 16404	good to excellent
Candida albicans ATCC® 26790	good to excellent
Saccharomyces cerevisiae ATCC® 9763	good to excellent

REFERENCES
1. DIFCO Manual, 9th Ed., 1953, pp 65 – 67.
2. Abs. Bact., 3:6, 1919.
3. J. Bact., 8:586, 1923.
4. Thom and Church: The Aspergilli, 1926.
5. Standard Methods for the Examination of Dairy Products, 9th Ed.:181, 1948.
6. Can. Dept. Agr. Pamphlet, 92 - N.S.

PACKAGING
Bacto Malt Agar — 1 lb (454 g) — 0024-01-1
— 1/4 lb (114 g) — 0024-02-0

BACTO MALT EXTRACT

Bacto Malt Extract is a useful ingredient of culture media designed for the propagation of yeasts and molds. This product was used by Thom and Church[1] in the preparation of "wort" medium as originally described by Reddish.[2] Bacto Malt Extract Broth is a complete dehydrated medium and duplicates the medium of Thom and Church.

Bacto Malt Extract is employed in the preparation of Bacto Malt Agar, a widely used medium for the detection and isolation of yeasts and molds from dairy products, food and other materials. For a complete discussion, see Bacto Malt Agar. Partansky and McPherson[3] used Bacto Malt Extract in conjunction with Bacto Yeast Extract and Bacto Agar for the cultivation of molds in their laboratory method for testing mold resistant properties of oil paints.

REFERENCES

1. Thom and Church: The Aspergilli, 1926.
2. Abst. Bact., 3:6, 1919.
3. Ind. Eng. Chem., Anal. Edition:12:443, 1940.

PACKAGING

Bacto Malt Extract	1 lb (454 g)	0186-01-5
	1/4 lb (114 g)	0186-02-4
	10 kg	0186-08-8

BACTO MALT EXTRACT BROTH
BACTO MALT EXTRACT AGAR

INTENDED USE

Bacto Malt Extract media are recommended for the isolation, detection and enumeration of yeasts and molds. Bacto Malt Extract Broth may be employed as a sterility test medium to detect the presence of these organisms.

HISTORY/PRINCIPLES

The use of malt and malt extracts for the propagation of yeasts and molds is quite common. Reddish[1] described a culture medium prepared from malt extract which was a satisfactory substitute for wort. Thom and Church, following the formula of Reddish, used Bacto Malt Extract, as a base from which they prepared the complete media.

Bacto Malt Extract Broth is prepared from Bacto Malt Extract according to the directions given by Thom and Church.[2]

The carbohydrates present in such media are well suited to the growth requirements of fungi, particularly if the reaction of the media is somewhat acid. Comparative tests have shown that early and luxuriant growth will be initiated in Bacto Malt Extract Broth from inocula of yeasts and molds as small or smaller than those required for other media, including broths prepared with honey.

FORMULAE

BACTO MALT EXTRACT BROTH
DEHYDRATED

Ingredients per liter

Malt Extract Base	6 g
Maltose, Technical	1.8 g
Dextrose	6 g
Yeast Extract	1.2 g

Final pH 4.7 ± 0.2 at 25°C.

One pound will make 30.2 liters
of medium.
Rehydrate 15 g/liter.

BACTO MALT EXTRACT AGAR
DEHYDRATED

Ingredients per liter

Maltose, Technical	12.75 g
Dextrin, Difco	2.75 g
Glycerol	2.35 g
Bacto Peptone	0.78 g
Bacto Agar	15 g

Final pH 4.7 ± 0.2 at 25°C.

One pound will make 14.8 liters
of medium.
Rehydrate 33.6 g/liter.

METHOD OF PREPARATION

1. Rehydrate by suspending the appropriate amount in 1 liter distilled or deionized water. (A more nutritious medium can be obtained by using more Bacto Malt Extract Broth per unit of water.) Heat the agar medium to boiling to dissolve completely.
2. Sterilize in the autoclave for 15 minutes at 15 lbs pressure (121°C).
3. Avoid overheating which will result in a softer agar and increased darkening of both media.

STORAGE

Bacto Malt Extract Broth and Agar	Below 30°C
Prepared medium	2 – 8°C

QUALITY CONTROL

Identity Specifications

	Bacto Malt Extract Broth	Bacto Malt Extract Agar
Dehydrated powder:	light beige, homogeneous, free-flowing	off-white, homogeneous, free-flowing
% solution:	1.5	3.36
pH at 25°C:	4.7 ± 0.2	4.7 ± 0.2
Prepared medium:	light amber, clear	very light amber, very slightly opalescent

Typical Cultural Response
After 40 – 48 Hours at 25 – 30°C

Organism	Recovery
Aspergillus niger ATCC® 16404	good to excellent
Candida albicans ATCC® 10231	good to excellent
Saccharomyces cerevisiae ATCC® 9763	good to excellent
Saccharomyces uvarum ATCC® 9080	good to excellent

REFERENCES

1. Abst. Bact., 3:6:1919.
2. Thom and Church: The aspergilli, 1926.

PACKAGING

Bacto Malt Extract Broth	1 lb (454 g)	0113-01-3
	10 kg	0113-08-6
Bacto Malt Extract Agar	500 g	0112-17-6

BACTO MANNITOL SALT AGAR

INTENDED USE

Bacto Mannitol Salt Agar is a selective medium used for the isolation of pathogenic staphylococci.

HISTORY/PRINCIPLES

Bacto Mannitol Salt Agar is prepared according to the formula suggested by Chapman.[1] Growth of most bacteria other than staphylococci is inhibited by the high salt concentration.

Koch[2] reported that on solid media, staphylococci were not inhibited by a concentration of 7.5% sodium chloride. Chapman[1] confirmed this observation and noted that the addition of 7.5% sodium chloride to Bacto Phenol Red Mannitol Agar gave a medium on which staphylococci that coagulated rabbit plasma grew luxuriantly, producing colonies with yellow zones. Nonpathogenic staphylococci produced small colonies with no color change of the surrounding medium. Other bacteria were generally inhibited, making possible the use of a heavy inoculum without danger of overgrowth. Chapman recommended incubation for 36 hours at 37°C. In a study of the resistance of chronic staphylococcal bovine mastitis to massive penicillin therapy, McCulloch[3] stated that the staphylococci responsible for the mastitis grew well and formed acid in phenol red mannitol agar to which 7.0% sodium chloride had been added. Velilla, Faber and Pelczar[4] used Bacto Mannitol Salt Agar for the isolation of coagulase producing staphylococci from milk in bovine mastitis. They recommended the use of both Bacto Mannitol Salt Agar and Bacto Staphylococcus Medium No. 110, to insure maximum recovery of these organisms.

FORMULA

BACTO MANNITOL SALT AGAR
DEHYDRATED

Ingredients per liter

Proteose Peptone No. 3, Difco	10 g
Bacto Beef Extract	1 g
D-Mannitol	10 g
Sodium Chloride	75 g
Bacto Agar	15 g
Phenol Red	0.025 g

Final pH 7.4 ± 0.2 at 25°C.

One pound will make 4.0 liters of medium.

METHOD OF PREPARATION

1. Suspend 111 g in 1 liter distilled or deionized water and heat to boiling to dissolve completely.
2. Sterilize in the autoclave for 15 minutes at 15 lbs pressure (121°C). Cool to 45 – 50°C. (If a tubed medium is desired, dispense prior to autoclaving and cool in a slanted position.)
3. Dispense into sterile Petri dishes.

STORAGE

Bacto Mannitol Salt Agar	Below 30°
Prepared medium	2 – 8°

QUALITY CONTROL

Identity Specifications

Dehydrated powder:	pinkish beige, homogeneous, free-flowing
Reaction of 11.1% solution:	pH 7.4 ± 0.2 at 25°C
Prepared medium:	red, very slightly opalescent

Typical Cultural Response on Bacto Mannitol Salt Agar
After 18 – 40 Hours at 35°C

Organism	Growth	Color of Colony
Enterobacter aerogenes ATCC® 13048	inhibited	—
Escherichia coli ATCC® 25922	inhibited	—
Staphylococcus aureus ATCC® 25923	good to excellent	yellow
Staphylococcus epidermidis ATCC® 12228	poor to fair	red
Staphylococcus epidermidis ATCC® 14990	good to excellent	red

REFERENCES

1. J. Bact., 50:201, 1945.
2. Zentr. Bakt., I Abt. Orig., 149:122, 1942.
3. Am. J. Vet. Res., 8:173, 1947.
4. Am. J. Vet. Res., 8:275, 1947.

PACKAGING

Bacto Mannitol Salt Agar		
	1 lb (454 g)	0306-01-0
	1/4 lb (114 g)	0306-02-9
	5 lb (2.27 kg)	0306-05-6
	10 kilo	0306-08-3

BACTO MANNITOL SALT BROTH

INTENDED USE

Bacto Mannitol Salt Broth is a modified formulation for the detection and isolation of pathogenic staphylococci from foods.

PRINCIPLES

Bacto Mannitol Salt Broth contains only mannitol as a fermentable carbohydrate, the indicator phenol red for ease of detection and an unusually high concentration of salt to inhibit undesired organisms.

FORMULA

BACTO MANNITOL SALT BROTH
DEHYDRATED

Ingredients per liter

Bacto Tryptone . . . . 17 g
   Pancreatic Digest Casein
Bacto Soytone . . . . 3 g
   Papaic Digest Soy Bean Meal
Bacto Mannitol . . . . 2.5 g
Sodium Chloride . . . . .100 g
Dipotassium Phosphate . . . . 2.5 g
Phenol Red . . . . 0.025 g

Final pH 7.3 ± 0.2 at 25°C.

One pound will make 3.63 liters of final medium.

METHOD OF PREPARATION

1. Suspend 125 g in 1 liter distilled or deionized water. Heat to boiling to dissolve completely.
2. Dispense into tubes or flasks as desired.
3. Sterilize in the autoclave for 15 minutes at 15 lbs pressure (121°C).
4. Allow to cool to at least 37°C and inoculate with specimen or culture.
5. Incubate at 35°C for 18 – 48 hours. Seventy - two hours may be required for some organisms.

STORAGE

Bacto Mannitol Salt Broth	Below 30°C
Prepared medium	15 – 30°C

QUALITY CONTROL

Identity Specifications

Dehydrated powder:	pink, homogeneous, free-flowing
Reaction of 12.5% solution:	pH 7.3 ± 0.2 at 25°C
Prepared medium:	red, clear to very slightly opalescent

Typical Cultural Response in Bacto Mannitol Salt Broth After 18 – 48 Hours at 35°C

Organism	Growth	Acid
Escherichia coli ATCC® 25922	inhibited	−
Proteus vulgaris ATCC® 13315	inhibited	−
Staphylococcus aureus ATCC® 25923	good to excellent	+
Staphylococcus epidermidis ATCC® 12228	good to excellent	slight + or −

+ = positive, yellow
− = negative, no change, red

PACKAGING

Bacto Mannitol Salt Broth	1 lb (454 g)	0926-01-0

BACTO MARINE AGAR 2216
BACTO MARINE BROTH 2216

INTENDED USE

Bacto Marine Agar 2216 and Bacto Marine Broth 2216 are recommended for culturing heterotropic marine bacteria. The agar medium can also be used to isolate and enumerate these organisms.

HISTORY/PRINCIPLES

Bacto Marine Agar 2216 and Bacto Marine Broth 2216, prepared according to the formula of ZoBell[1,2] contain all of the nutrients necessary for the growth of marine bacteria. Besides including minerals which nearly duplicate the major mineral composition of sea water,[3] it contains peptone and yeast extract which were reported Jones[4] to be the best source of nutrients for marine bacteria in general.

Marine bacteria are present in nutrient sea water b... ...ions per ml and are essential to the life cycle of all marine flora and fa... ...use of their increasing importance to the food industry for the conser... ...ine life, the enumeration and activity of marine bacteria are important. ...ditional pour plate and spread plate

In the use of Bacto Marine Agar 2... ...echnique ...cause of the thermo-sensitive nature of techniques of enumeration are...han ...ion. This latter technique the agar is poured while hot and carefully cooled to 42°C be... ...ve bacteria. ...the pour method was reported by Buck most marine bacteria. In ... ...pour plate method because of the in- allowed to cool and s... and Cleverdon[6] t... creased grow...

FORMULAE

BACTO MARINE AGAR 2216
DEHYDRATED

Ingredients per liter

Bacto Peptone	5 g	Potassium Bromide	0.08 g
Bacto Yeast Extract	1 g	Strontium Chloride	0.034 g
Ferric Citrate	0.1 g	Boric Acid	0.022 g
Sodium Chloride	19.45 g	Sodium Silicate	0.004 g
Magnesium Chloride	8.8 g	Sodium Fluoride	0.0024 g
Sodium Sulfate	3.24 g	Ammonium Nitrate	0.0016 g
Calcium Chloride	1.8 g	Disodium Phosphate	0.008 g
Potassium Chloride	0.55 g	Bacto Agar	15 g
Sodium Bicarbonate	0.16 g		

Final pH 7.6 ± 0.2 at 25°C.

One pound will make 8.25 liters of medium.
Use 55.1 g per liter.

BACTO MARINE BROTH 2216
DEHYDRATED

Ingredients per liter

Bacto Peptone	5 g	Sodium Bicarbonate	0.16 g
Bacto Yeast Extract	1 g	Potassium Bromide	0.08 g
Ferric Citrate	0.1 g	Strontium Chloride	0.034 g
Sodium Chloride	19.45 g	Boric Acid	0.022 g
Magnesium Chloride Dried	5.9 g	Sodium Silicate	0.004 g
Sodium Sulfate	3.24 g	Sodium Fluoride	0.0024 g
Calcium Chloride	1.8 g	Ammonium Nitrate	0.0016 g
Potassium Chloride	0.55 g	Disodium Phosphate	0.008 g

Final pH 7.6 ± 0.2 at 25°C.

One pound will make 12.13 liters of medium.
Use 37.4 g per liter.

METHOD OF PREPARATION
1. Suspend appropriate amount in 1 liter cold distilled or deionized water. Heat to boiling to dissolve completely.
2. Distribute into tubes or flasks as desired.
3. Sterilize in the autoclave for 15 minutes at 15 lbs pressure (121°C).

STORAGE
Bacto Marine   Below 30°C
Prepared media 2216   15 – 30°C

QUALITY CONTROL

Identity Specifications

Dehydrated powder: light beige with a few dark particles, free-flowing
Reaction of appropriate so H 7.6 ± 0.2 at 25°C
Prepared medium: amber, clear to slightly opalescent, may
                            a slight precipitate

Typical Cultural Res
After 2 – 3 Days   Bacto Marine Media 2216
     Organism   quired) at 20°C
     Vibrio fischeri   Growth
     Vibrio harveyi   excellent
                                xcellent

REFERENCES

1. J. Marine Research, 4:42, 1941.
2. Personal Communication.
3. J. Marine Research, 3:134, 1940.
4. Bact. Proc., Pg. 36 (A29), 1960.
5. Ecology, 40:712, 1959.
6. Limnology and Oceanography, 5:78, 1960.

PACKAGING

Bacto Marine Agar 2216	1 lb (454 g)	0979-01-6
Bacto Marine Broth 2216	1 lb (454 g)	0791-01-2

BACTO McCLUNG TOABE AGAR BASE

INTENDED USE

Bacto McClung Toabe Agar Base is recommended for the detection and isolation of *Clostridium perfringens* in foods.

HISTORY/PRINCIPLES

Bacto McClung Toabe Agar Base, prepared according to the formulation of McClung and Toabe,[1,2,3] differentiates various species of *Clostridium* on the basis of their lecithinase production. Lecithinase lyses egg yolk lecithin, producing an opaque precipitate in the agar surrounding the slightly raised colonies.

FORMULA

BACTO McCLUNG TOABE AGAR BASE
DEHYDRATED

Ingredients per liter

Proteose Peptone, Difco	40 g
Bacto Dextrose	2 g
Sodium Phosphate Dibasic	5 g
Potassium Phosphate Monobasic	1 g
Sodium Chloride	2 g
Magnesium Sulfate	0.1 g
Bacto Agar	25 g

Final pH 7.6 ± 0.2 at 25°C.

One pound will make 6 liters of medium.

METHOD OF PREPARATION

1. Suspend 75 g in 1 liter distilled or deionized water. ___ boiling to dissolve completely.
2. Dispense into flasks in 90 ml amounts. ___ pressure (121°C).
3. Sterilize in the autoclave for 20 minutes ___ Egg Yolk Enrichment 50% to each
4. Allow to cool to 50 – 55°C and add 10 ___shes in approximately 15 ml amounts. 90 ml of medium.
5. Mix thoroughly and pour into s___ple to be tested in each of 2 tubes containing

PROCEDURE

1. Place approximately 2___ medium with fermentation vials (Durham tubes). 25 ml Bacto Fluid ___ for 4 – 6 hours. Observe for growth and gas formation.
2. Incubate inocula___

3. Streak inoculum onto prepared plates of Bacto McClung Toabe Agar.
4. Incubate anaerobically for 24 hours at 35°C or as desired.

STORAGE

Bacto McClung Toabe Agar Base	Below 30°C
Prepared plates	2 – 8°C

QUALITY CONTROL

Identity Specifications

Dehydrated powder:	very light beige, homogeneous, free-flowing
Reaction of 7.5% solution:	pH 7.6 ± 0.2 at 25°C
Prepared medium:	light amber, opalescent with a precipitate

Typical Cultural Response on Bacto McClung Toabe Agar
After 18 – 48 Hours at 35°C *anaerobically

Organism	Growth	Lecithinase Halos
Clostridium perfringens ATCC® 12919	good to excellent	+
Clostridium perfringens ATCC® 12924	good to excellent	+
Staphylococcus aureus ATCC® 25923	good to excellent	+
Staphylococcus epidermidis ATCC® 12228	good to excellent	–

REFERENCES

1. J. Bact., 53:139, 1947.
2. Public Health Service Publication No. 1142, March, 1964.
3. Laboratory Manual for Food Canners and Processors Vol. 1 p 25, 1968.

PACKAGING

Bacto McClung Toabe Agar Base	1 lb (454 g)	0941-01-1

BACTO McFARLAND BARIUM SULFATE STANDARDS

INTENDED USE

Bacto McFarland Barium Sulfate Standards are sealed tubes of barium sulfate suspensions prepared as described by McFarland[1] for use as standards in adjusting densities of bacterial suspensions and other turbid suspensions. They may also be used as standard suspensions for standardizing electrometric turbidimeters for turbidimetric measurements relating to the McFarland series.

Publications describing the McFarland barium sulfate standards are cited in references.[2,3,4,5]

PRINCIPLES

The density of bacterial and other suspensions with the comparison of the suspensions are determined by direct visual comparison of the suspensions with that of the standard barium sulfate standards. It is obvious, of course, that the diameter of the suspensions are measured must be comparable with that of the standard st... ch the suspensions are measured be thoroughly shaken to insure complete s... o essential that the standards they are used. he barium sulfate each time

Visual comparison of turbidity are easily and accurately made by viewing the test suspensions and standard against a white background with a black line running horizontally at midpoint.

Bacto McFarland Barium Sulfate Standards are prepared in a ten tube series numbered 1 through 10, and represent the approximate turbidity of most bacteria per milliliter as follows:

Standard No.	Turbidity — in millions/ml
1	300
2	600
3	900
4	1200
5	1500
6	1800
7	2100
8	2400
9	2700
10	3000

REFERENCES

1. McFarland, J., JAMA 49:1176, 1907.
2. Kolmer, J. A., Infection, Immunity and Biologic Therapy, Sanders, Philadelphia, Pa., 1923.
3. Boyd, W. C., Fundamentals of Immunology, 2nd Ed., Interscience, New York, 1947.
4. Bray, W. E., Clinical Laboratory Methods, 5th Ed., C. V. Mosby, St. Louis, Mo., 1957.
5. Diagnostic Procedures and Reagents, 4th Ed., APHA, New York, 1963.

PACKAGING

Bacto McFarland Barium Sulfate Standards   1 set   0691-32-6

BACTO METHYLENE BLUE LOEFFLER

Bacto Methylene Blue Loeffler is a basic dye used in studying the morphology of microorganisms. It is particularly helpful when used in the Quellung reaction as part of the serological identification of *Klebsiella pneumoniae*, *Streptococcus pneumoniae* and *Haemophilus influenza*. It may also be used as a counterstain with Bacto Flagella Stain. For complete discussion refer to Flagella Stain.

STORAGE

Bacto Methylene Blue Loeffler $2 - 8°C$

PACKAGING

Bacto Methylene Blue Loeffler   5 ml   3111-56-3

BACTO MICROBIAL CONTENT TEST AGAR

INTENDED USE

Bacto Microbial Content Test Agar is a Standard Methods medium[1] used for determining the efficiency of sanitization of containers, equipment, and environmental surfaces as well as for use in the microbial content test for water miscible cosmetic products.[2]

HISTORY/PRINCIPLES

The value of a highly nutritional medium containing a neutralizing agent for quaternary ammonium compounds in sanitizers was first described by Weber and Black.[3] Bacto Letheen Agar and Bacto Letheen Broth, based on the Weber and Black formulae, are also modifications of the AOAC[4] formulae and are used to determine the bactericidal activity of quaternary ammonium compounds. Letheen agar is also recommended by The National Aeronautics and Space Administration[5] for the microbiological sampling of environmental surfaces sanitized with quaternary ammonium compounds.

Bacto Microbial Content Test Agar is a modification of letheen agar. It has also been referred to as Casein Soy Peptone Agar w/Polysorbate 80 and Lecithin.

FORMULA

BACTO MICROBIAL CONTENT TEST AGAR
DEHYDRATED

Ingredients per liter

Bacto Tryptone .... 15 g
Pancreatic Digest of Casein USP
Bacto Soytone .... 5 g
Soy Peptone USP
Sodium Chloride .... 5 g
Lecithin .... 0.7 g
Sorbitan Monooleate Complex ...... 5 g
Bacto Agar .... 15 g

Final pH 7.3 ± 0.2 at 25°C.

Five hundred grams will make 11 liters of medium.

METHOD OF PREPARATION

1. Suspend 45.7 g in 1 liter distilled or deionized water and heat to boiling to dissolve.
2. Continue heating for 1 – 2 minutes with frequent gentle swirling of the medium.
3. Dispense as desired.
4. Sterilize in the autoclave for 15 minutes at 15 lbs pressure (121°C).

STORAGE

Bacto Microbial Content Test Agar   2 – 8°C
Prepared medium   2 – 8°C

QUALITY CONTROL

Identity Specifications

Dehydrated powder: tan, homogeneous, free-flowing
Reaction of 4.57% solution: 7.3 ± 0.2 at 25°C
Prepared medium: medium amber, opalescent

Typical Cultural Response on Bacto Microbial Content Test Agar
After 40 – 48 Hours at 35°C

Organism	Growth	Growth w/Disinfectant* (Roccal®)
Escherichia coli ATCC® 11229	excellent	good to excellent
Staphylococcus aureus ATCC® 6538P	excellent	good to excellent

*Depending on concentration of disinfectant

REFERENCES

1. Am. Pub. Hlth. Assoc., 1978, Standard Methods for the Examination of Dairy Products, 14th Ed., Am. Pub. Hlth. Assoc., Inc., Washington, D.C.
2. Tenebaum, 1970, TGA Cosmetic Journal, 2:24 – 29.
3. Weber and Black, 1948, Soap and Sanitary Chemicals, October.
4. AOAC, 10th Ed., 1965.
5. National Aeronautics and Space Administration, 1966, Standard Procedures for the Microbiological Examination of Space Hardware.

PACKAGING

Bacto Microbial Content Test Agar   500 g   0553-17-2

MIDDLEBROOK MEDIA

BACTO MIDDLEBROOK 7H9 BROTH
BACTO MIDDLEBROOK 7H10 AGAR
BACTO MYCOBACTERIA 7H11 AGAR
BACTO MIDDLEBROOK ADC ENRICHMENT
BACTO MIDDLEBROOK OADC ENRICHMENT
BACTO MIDDLEBROOK OADC ENRICHMENT w/WR 1339

INTENDED USE

Middlebrook media are defined culture media used for isolating, cultivating and determining the antimicrobial susceptibility of mycobacteria.

HISTORY

Cohn and Middlebrook formulated a series of defined culture media during the 1950's for experimental and clinical cultivation of mycobacteria. The Middlebrook media currently in use are, for the most part, modifications of the original formulae.

Bacto Mycobacteria 7H11 Agar is a modification of Middlebrook 7H10 agar special[1] as described by Cohn et al. in 1968.

Bacto Middlebrook 7H10 Agar is prepared according to the 1960 formulation of Middlebrook, Cohn, Dye, Russell and Levy modifying and replacing Middlebrook 7H9 agar B.

PRINCIPLES

Agar base media for culturing mycobacteria are not affected by proteolytic organisms as are egg base media, except as they may overgrow the mycobacteria. Further, agar base media are relatively clear, permitting the study of colony morphology with a stereo microscope, and may better be used as a base for added therapeutic agents for sensitivity testing since the final medium will not be inspissated.

The differing formulations provide somewhat different growth characteristics. The increased level of malachite green in 1 formulation provides for better isolation, although some workers prefer a reduced level for primary isolation.

Cohn et al.[2] demonstrated that the addition of enzymatic digest of casein stimulated growth of the more fastidious strains of *M. tuberculosis* and, as a result, improved drug susceptibility studies.

The addition of WR 1339 (Triton®), as recommended by Lorian,[4,5,6] accelerates growth during the lag phase and alters colony morphology of *M. tuberculosis* so that "cording" is detected at 100X magnification, facilitating differentiation from atypical mycobacteria.

The broth formulation is suitable for cultivating pure cultures of mycobacteria and for preparing an emulsion of tubercle bacilli for sensitivity testing.

FORMULAE

BASAL MEDIA
Ingredients per liter

	Bacto Middlebrook 7H9 Broth	Bacto Middlebrook 7H10 Agar	Bacto Mycobacteria 7H11 Agar
Pancreatic Digest of Casein	—	—	1 g
Ammonium Sulfate	0.5 g	0.5 g	0.5 g
Monopotassium Phosphate	1.0 g	1.5 g	1.5 g
Disodium Phosphate	2.5 g	1.5 g	1.5 g
Sodium Citrate	0.1 g	0.4 g	0.4 g
Magnesium Sulfate	0.05 g	0.025 g	0.05 g
Calcium Chloride	0.0005 g	0.0005 g	—
Zinc Sulfate	0.001 g	0.001 g	—
Copper Sulfate	0.001 g	0.001 g	—
L-Glutamic Acid	0.5 g	0.5 g	0.5 g
Ferric Ammonium Citrate	0.04 g	0.04 g	0.04 g
Pyridoxine	0.001 g	0.001 g	0.001 g
Biotin	0.0005 g	0.0005 g	0.0005 g
Malachite Green	—	0.00025 g	0.001 g
Bacto Agar	—	15 g	15 g
Solution g/liter (final solution)	4.7 g/liter	19 g/liter	21 g/liter
Final pH at 25°C.	6.6 ± 0.2	6.6 ± 0.2	6.6 ± 0.2

ENRICHMENTS
Ingredients per 100 ml

(For preparing 1 liter final medium)

	Bacto Middlebrook ADC Enrichment	Bacto Middlebrook OADC Enrichment	Bacto Middlebrook OADC Enrichment w/WR 1339
Oleic Acid	—	0.05 g	0.05 g
Albumin Fraction V, Bovine	5 g	5 g	5 g
Dextrose	2 g	2 g	2 g
Catalase (Beef)	0.003 g	0.004 g	0.004 g
Sodium Chloride	—	0.85 g	0.85 g
Distilled Water	100 ml	100 ml	100 ml
WR 1339, Triton®	—	—	0.25 g

METHOD OF PREPARATION
1. To prepare 1 liter of final medium, suspend the amount of dehydrated powder indicated under FORMULAE-Solution (and on the package label) in 900 ml distilled or deionized water containing 5 ml glycerol (2 ml for Bacto Middlebrook 7H9 Broth) and heat to boiling to dissolve completely.
2. Sterilize in the autoclave for 10 or 15 minutes at 15 lbs pressure (121°C), per label copy. Cool to 50 – 55°C.
3. Aseptically add 100 ml of the appropriate Bacto Middlebrook Enrichment and mix thoroughly.

4. If preparing the medium for sensitivity testing, add therapeutic agents at this point.
5. Dispense as desired.

STORAGE

Dehydrated media	Below 30°C
Enrichments	2 – 8°C
Prepared media	2 – 8°C

QUALITY CONTROL

Identity Specifications

	Bacto Middlebrook 7H9 Broth	Bacto Middlebrook 7H10 Agar	Bacto Mycobacteria 7H11 Agar
Dehydrated powder:	light beige	light beige w/slight greenish tint	light beige w/slight greenish tint
% solution:	0.47%	1.9%	2.1%
pH at 25°C:	6.6 ± 0.2	6.6 ± 0.2	6.6 ± 0.2
Prepared medium:	light amber, clear	light amber, very slightly opalescent	light amber to grey, slightly opalescent

	Bacto Middlebrook ADC Enrichment	Bacto Middlebrook OADC Enrichment	Bacto Middlebrook OADC Enrichment w/WR 1339
Appearance:	very light to light amber, clear liquid	very light to light amber, clear liquid	very light to light amber, clear liquid
Reaction at 25°C:	pH 6.8 ± 0.2	pH 6.8 ± 0.2	pH 6.8 ± 0.2

Typical Cultural Response on/in Bacto Middlebrook Media, Enriched as Required After 2 Weeks at 35°C

Organism	Growth
Mycobacterium fortuitum ATCC® 6841	good to excellent
Mycobacterium tuberculosis ATCC® 27294	good to excellent

REFERENCES

1. AJPH, 98:844, 1958.
2. Am. Rev. Resp. Dis., 98:295, 1968.
3. Acta. Tubercul. Scand., 38:66, 1960.
4. Appl. Microbiol., 14:603, 1966.
5. Appl. Microbiol., 15:1202, 1967.
6. Am. Rev. Resp. Dis., 97:1133, 1968.

PACKAGING

Bacto Middlebrook 7H9 Broth	1 lb (454 g)	0713-01-7
Bacto Middlebrook 7H10 Agar	1 lb (454 g)	0627-01-2
	12 tubes	0627-34-3
	144 tubes	0627-37-0
	12 × 1 oz	0627-44-1
	144 × 1 oz	0627-45-0
Bacto Mycobacteria 7H11 Agar	1 lb (454 g)	0838-01-7
	12 tubes	0838-34-8
	144 tubes	0838-37-5
Bacto Middlebrook ADC Enrichment	12 × 20 ml	0714-64-0
Bacto Middlebrook OADC Enrichment	12 × 20 ml	0722-64-0
	6 × 100 ml	0722-73-9
Bacto Middlebrook OADC Enrichment w/WR 1339	6 × 20 ml	0801-63-5

BACTO MinESS ANTISERA SET II

INTENDED USE

Bacto MinESS Antisera Set II is a minimal essential antiserum set used in the preliminary serological identification of *Salmonella* and *Shigella* species and serotypes as well as the complete serological identification of certain significant and/or prevalent salmonellae.

SUMMARY AND EXPLANATION OF THE TEST

The importance of preliminary serological identification of *Salmonella*, *Shigella* and other significant bacteria belonging to the family of *Enterobacteriaceae* is stressed by Edwards and Ewing.[1] The world-wide distribution of especially *Salmonella* and *Shigella*, points to the need for preliminary identification of at least the more frequently encountered species of these genera. More importantly, every laboratory, regardless of size should be equipped to completely identify the most significant salmonellae and to identify any species of *Shigella*.[1]

Such organisms are:

Salmonella typhi	*Shigella dysenteriae*
Salmonella choleraesuis	*Shigella flexneri*
Salmonella enteritidis	*Shigella boydii*
bioser Paratyphi A	*Shigella sonnei*
ser Paratyphi B	
ser Typhimurium	

For those laboratories who wish to have a minimum basic capability in the serology of *Salmonella* and *Shigella* while also having the potential of identifying the most prevalent *Salmonella*, Bacto MinESS Antisera Set II is provided.

REAGENTS

Bacto MinESS Antisera Set II contains Bacto Salmonella O Antiserum Factors 2, 4 & 5, 7, 8, 9, Poly A-I and Vi; Salmonella Vi Antiserum; Salmonella H Antisera a, b, c, d, eh, i, k, m, q, r, s, t, 2, 5; Shigella Antiserum Poly Groups A, A₁, B, C, C₁, C₂, D and Alkalescens-Dispar Poly.

Bacto Salmonella O, H, and Vi are stable, desiccated, absorbed (when necessary) antisera prepared in rabbits.

Bacto Salmonella O Antisera used in the slide agglutination technique for somatic O antigen identification are listed below:

	Identifying Factor of Group
Factor 2	A
Factors 4 & 5	B
Factor 7	C₁
Factor 8	C₂
Factor 9	D
Poly A-I Factors 1 –16, 19, 22 – 25, 34 & Vi	

Bacto Salmonella H Antisera are used following serological grouping with Salmonella O Antisera to identify the H antigens for determining the individual serotypes. The H antisera must be diluted according to instruction under the heading "Flagellar H Antigen

Analysis." For a complete discussion refer to Bacto Salmonella Antisera. The following H Antisera are included in the set:
Antisera a, b, c, d, i, 2, 5, eh, k, m, q, r, s and t

Bacto Shigella Antisera Poly are stable, desiccated, high titered, absorbed rabbit antisera for the serological grouping of the genus *Shigella* and the Alkalescens-Dispar Group. Six Shigella antisera poly groups and 1 Alkalescens-Dispar poly are available:
Bacto Shigella Antiserum Poly Group A reacts with *S. dysenteriae* types 1 – 7.
Bacto Shigella Antiserum Poly Group A_1 reacts with *S. dysenteriae* types 8ab8ac, 9 & 10.
Bacto Shigella Antiserum Poly Group B reacts with *S. flexneri* types 1 – 6.
Bacto Shigella Antiserum Poly Group C reacts with *S. boydii* types 1 – 7.
Bacto Shigella Antiserum Poly Group C_1 reacts with *S. boydii* types 8 – 11.
Bacto Shigella Antiserum Poly Group C_2 reacts with *S. boydii* types 12 – 15.
Bacto Shigella Antiserum Poly Group D reacts with *S. sonnei* types I and II.
Bacto Alkalescens-Dispar Antiserum Poly reacts with Alkalescens-Dispar types 1, 2, 3 and 4.

PROCEDURE
Complete descriptions of the use of these antisera are found under "Shigella Antisera" and "Salmonella O and H Antisera" in this Manual.

REFERENCE
1. Edwards, P. R., and Ewing, W. H., Identification of *Enterobacteriaceae*, 3rd Ed., 1972, Burgess Publ. Co.

PACKAGING
Bacto MinESS Antisera Set II   29 × 3 ml   2975-32-9
Contains 1 vial each Salmonella O Antisera Factors 2, 4 & 5, 7, 8, 9, Poly A-1 and Vi; Salmonella H Antisera a, b, c, d, eh, i, k, m, q, r, s, t, 2, 5; Salmonella Vi Antiserum; Shigella Antisera Poly Groups A, A_1, B, C, C_1, C_2, D; and Alkalescens-Dispar Antiserum Poly

BACTO MINIMAL MEDIA DAVIS
BACTO MINIMAL AGAR DAVIS
BACTO MINIMAL BROTH DAVIS
BACTO MINIMAL BROTH DAVIS w/o DEXTROSE

INTENDED USE
Bacto Minimal Agar Davis, Bacto Minimal Broth Davis and Bacto Minimal Broth Davis w/o Dextrose are recommended for use as minimal media for the isolation and characterization of nutritional mutants of *Escherichia coli*.

Bacto Minimal Broth Davis w/o Dextrose is, in addition, recommended for the isolation and characterization of nutritional mutants from wild type strains of *Bacillus subtilis* when used in conjunction with Bacto Minimal Agar Davis and Bacto Antibiotic Medium 3.

HISTORY/PRINCIPLES

Bacto Minimal Broth Davis and Bacto Minimal Agar Davis are prepared according to the formulation of Davis as described by Lederberg.[1] Bacto Minimal Broth Davis w/o Dextrose is the same as Bacto Minimal Broth Davis except that the dextrose has been omitted from the medium. The minimal media contain only the essential nutrients for the growth of wild type strains of E. coli.

Nutritional mutants derived from irradiated cultures of wild type E. coli can be isolated by random isolation described by Lederberg,[1] delayed enrichment also described by Lederberg[1] and by use of penicillin described by Davis[2] and Lederberg.[1] B. subtilis mutants can be isolated by these techniques and the modification of the penicillin technique as described by Nester, Schafer and Lederberg[3] is recommended.

FORMULAE

BACTO MINIMAL AGAR DAVIS
DEHYDRATED

Ingredients per liter

Bacto Dextrose	1 g
Dipotassium Phosphate	7 g
Monopotassium Phosphate	2 g
Sodium Citrate	0.5 g
Magnesium Sulfate	0.1 g
Ammonium Sulfate	1 g
Bacto Agar	15 g

One pound will make 17 liters
of medium.
Rehydrate with 26.6 grams/liter.

BACTO MINIMAL BROTH DAVIS
DEHYDRATED

Ingredients per liter

Bacto Dextrose	1 g
Dipotassium Phosphate	7 g
Monopotassium Phosphate	2 g
Sodium Citrate	0.5 g
Magnesium Sulfate	0.1 g
Ammonium Sulfate	1 g

One pound will make 40 liters
of medium.
Rehydrate with 11.6 grams/liter.

BACTO MINIMAL BROTH DAVIS w/o DEXTROSE
DEHYDRATED

Ingredients per liter

Dipotassium Phosphate	7 g
Monopotassium Phosphate	2 g
Sodium Citrate USP	0.5 g
Magnesium Sulfate	0.1 g
Ammonium Sulfate	1 g

One pound will make 38.8 liters of medium.
Rehydrate with 10.6 grams/liter.

Final pH 7.0 ± 0.2 at 25°C for all above formulae.

PROCEDURE

1. To rehydrate, suspend appropriate amount in 1 liter cold distilled water. Heat to boiling to dissolve completely.
2. Dispense into tubes or flasks and sterilize in the autoclave for 15 minutes at 15 lbs pressure (121°C).
3. When Bacto Minimal Broth Davis w/o Dextrose has cooled to room temperature, aseptically add 10 ml of Bacto Dextrose Solution 10%.

PROCEDURE FOR E. COLI.

In the random technique for isolating and characterizing nutritional mutants, a cell suspension of wild type E. coli is irradiated, then diluted 100 – 500 times and cultured for colony isolation on plates of a complete agar medium containing all the necessary growth substances. These cultures are incubated for 24 hours at 35°C after which the isolated colonies are aseptically picked with a wire loop and cultured in tubes of both Bacto Minimal Broth Davis and a nutritionally complete broth. After 24 hours incubation at 35°C, growth in the complete broth with no growth in the minimal broth indicates a mutant.

In the delayed enrichment method, plates of minimal agar Davis are prepared by pouring a 15 – 20 ml base layer in 95 mm sterile Petri plates followed by a 5 ml seed layer inoculated with the diluted irradiated *E. coli* suspension. A 5 – 10 ml cover layer of uninoculated minimal agar Davis is then poured over the seed layer. The plates are incubated for 24 hours or longer to allow for the growth of prototroph cells (wild type cells). At this time, the mutant cells will have grown very little, if at all. The mutant colonies are then developed by pouring a layer of complete medium over the minimal agar medium and incubating for an additional 6 – 12 hours at 35°C. Nutrients from the complete agar will diffuse down through the minimal agar and allow the mutants to grow. The small new mutant colonies can then be picked and transferred for subsequent characterization.

In the isolation of mutants using penicillin, the irradiated *E. coli* suspension is washed with sterile saline and diluted to 20 times the original volume in sterile minimal broth and dispensed into tubes in desired amounts. A freshly prepared solution of penicillin is then added to each tube to give a final concentration of 300 units per ml. The tubes are incubated for 4 – 24 hours at 35°C on a shaker. Samples of 0.1, 0.01 and 0.001 ml are then spread on to complete agar plates and incubated for 24 hours at 35°C. The isolated colonies which develop are picked at random and tested for growth in minimal broth.

The penicillin is bactericidal to the prototrophic cells and leaves only the mutants to be streaked onto the complete agar plates. After the mutants are isolated, they can be characterized biochemically by culturing them in minimal broth supplemented with growth factors or groups of growth factors. It is generally best to classify them first in groups according to their requirements for amino acids, vitamins, nucleic acids or other substances. This is done by supplementing the minimal medium with Bacto Vitamin Assay Casamino Acids plus tryptophane, or a mixture of water soluble vitamins, alkaline-hydrolyzed yeast, nucleic acid or yeast extract depending on the particular mutants desired. The supplemented minimal broth tubes are inoculated with a slightly turbid suspension of the mutant cultures and incubated for 24 hours at 35°C. Growth with Bacto Vitamin Assay Casamino Acids indicates an amino acid requirement whereas growth with the vitamin mixture indicates a vitamin requirement. When a major growth factor group response is obtained, the characterization is carried further by the same general procedure to subgroups and finally to individual growth substances.

PROCEDURE FOR *B. SUBTILIS*

In the isolation of *B. subtilis* mutants by use of penicillin, cultures of *B. subtilis* are grown for 18 hours in Bacto Antibiotic Medium 3 and centrifuged to sediment the cells. The supernatant culture fluid is decanted aseptically. The cells are washed by resuspending them in minimal medium and sedimenting them by centrifugation. The washed cells are then resuspended in minimal medium to give a cell concentration of about 2×10^8 (200 million) cells per ml. This suspension is irradiated with a low pressure mercury ultraviolet lamp for a sufficient amount of time to give a cell survival of about 10^4 (10 thousand) cells per ml. The irradiated suspension is then incubated for 4 – 18 hours at room temperature in the minimal medium with appropriate substances added to allow for the growth of desired mutants. The culture is washed in sterile minimal medium, centrifuged and resuspended in the same medium. It is then diluted 1 to 10 with sterile minimal medium and allowed to stand for 60 minutes to starve the mutants. Penicillin is added to give a concentration of 2,000 units per ml and the incubation continued for an additional 15 minutes. The culture is then plated on nutrient agar for colony isolation.

For the identification of the nutrition mutants, the colonies are transferred by replicate plating onto plates of minimal agar which has been supplemented with the appropriate nutritional substances.

STORAGE

Bacto Minimal media	Below 30°C
Prepared media	15 – 30°C

QUALITY CONTROL

Identity Specifications

	Bacto Minimal Agar Davis	Bacto Minimal Broth Davis	Bacto Minimal Broth Davis w/o Dextrose
Dehydrated medium:	light beige	white	white
		—homogeneous, free-flowing—	
Reaction			
pH at 25°C:	7.0 ± 0.2	7.0 ± 0.2	7.0 ± 0.2
Prepared medium:	Medium amber, slightly opalescent	very light amber, clear	colorless, clear

Typical Cultural Response on/in Bacto Minimal Media Davis
After 18 – 24 Hours at 35°C

Organism	Growth
Escherichia coli ATCC® 13762	good to excellent
Escherichia coli fresh isolate	good to excellent

REFERENCES

1. Methods in Med. Res., 3:5, 1950.
2. Proc. Nat'l Acad. Sci., 35:1, 1949.
3. Genetics, 48:529, 1963.

PACKAGING

Bacto Minimal Agar Davis	1 lb (454 g)	0544-01-2
Bacto Minimal Broth Davis	1 lb (454 g)	0748-01-6
Bacto Minimal Broth Davis w/o Dextrose	1 lb (454 g)	0756-01-5

BACTO MITIS SALIVARIUS AGAR

INTENDED USE

Bacto Mitis Salivarius Agar, when used with Bacto Chapman Tellurite Solution 1%, is a selective medium for the isolation of Streptococcus mitis, S. salivarius and enterococci. Bacto Chapman Tellurite Solution is a 1% solution of potassium tellurite prepared and standardized especially for use with Bacto Mitis Salivarius Agar. For a complete discussion see Bacto Chapman Tellurite Solution 1%.

HISTORY

Bacto Mitis Salivarius Agar is prepared according to the formula described by Chapman.[1,2] Some bacteriologists refer to these organisms as *Streptococcus viridans* and "nonhemolytic streptococci", respectively, because of their alpha and gamma hemolysis on blood agar prepared from Bacto Heart Infusion Agar or Bacto Tryptose Blood Agar Base. The final medium, containing Bacto Chapman Tellurite Solution 1%, is highly selective for these organisms, making possible their isolation from grossly contaminated specimens such as feces or exudates from different body cavities.

Different methods have been employed for the isolation of streptococci and enterococci from mixed cultures. Snyder and Lichstein[3] and Lichstein and Snyder[4] used sodium azide to inhibit the growth of gram-negative bacteria including *Proteus*. Chapman[5] described a tellurite medium and an azide medium for isolation of *S. salivarius* and *S. mitis*.

Chapman was able to demonstrate pathogenic streptococci in about 95% of fecal specimens from chronic invalids. Pathogenicity of these streptococci was determined culturally according to the method described by Chapman[1,5] using hexylresorcinol.

Comparative tests have shown this medium to be satisfactory for the isolation of streptococci and enterococci from grossly contaminated specimens from a variety of clinical specimens.

PRINCIPLES

Chapman[1,2] reported complete and detailed methods for the isolation and testing for pathogenicity of fecal streptococci. Decimal dilutions of the specimen are prepared and 0.01 ml amounts spread by a glass spreader, over the surface of the Mitis Salivarius Agar containing tellurite. Plates are incubated for exactly 24 hours at 37°C. *S. mitis* produces small or minute blue colonies. Some *S. mitis* colonies may be more easily distinguished with a longer incubation. *S. salivarius* produces blue, smooth or rough "gum drop" colonies 1 – 5 mm in diameter, depending on the number of colonies on the plate. Enterococci form colonies dark blue or black in color, shiny, slightly raised, 1 – 2 mm in diameter. These organisms, few of which are pathogenic, may be readily differentiated from *S. mitis* and *S. salivarius,* particularly when viewed by reflected light. Beta hemolytic streptococci resemble *S. mitis*. Other types of streptococci have not been studied on this medium. Chapman reported that *Erysipelothrix rhusiopathiae* produce colorless circular convex colonies. He also reported that when coliform organisms are not inhibited they produce brown colonies. Spreaders are rarely observed. Molds grow after 2 days incubation.

FORMULA

BACTO MITIS SALIVARIUS AGAR
DEHYDRATED

Ingredients per liter

Bacto Tryptose	10 g	Dipotassium Phosphate	4 g
Proteose Peptone No. 3, Difco	5 g	Trypan Blue	0.075 g
Proteose Peptone, Difco	5 g	Bacto Crystal Violet	0.0008 g
Bacto Dextrose	1 g	Bacto Agar	15 g
Saccharose, Difco	50 g		

Final pH 7.0 ± 0.2 at 25°C.

One pound will make 5 liters of medium.

METHOD OF PREPARATION

1. To rehydrate the medium, suspend 90 g in 1 liter distilled or deionized water.
2. Heat to boiling to dissolve completely.
3. Sterilize in the autoclave for 15 minutes at 15 lbs pressure (121°C).
4. Allow to cool to 50 – 55°C.
5. Just prior to pouring plates, add exactly 1 ml/liter of Bacto Chapman Tellurite Solution. DO NOT HEAT THE MEDIUM AFTER THE ADDITION OF THE TELLU- RITE SOLUTION.
6. Pour into 95 mm plates in approximately 25 ml amounts.

STORAGE

Bacto Mitis Salivarius Agar   Below 30°C
Prepared plates   2 – 8°C

QUALITY CONTROL

Identity Specifications

Dehydrated powder: bluish beige, homogeneous, free-flowing
Reaction of 9% solution: pH 7.0 ± 0.2 at 25°C
Prepared medium: deep royal blue, very slightly opalescent

Typical Cultural Response on Bacto Mitis Salivarius Agar with Bacto Chapman Tellurite Solution After 18 – 48 Hours at 35°C

Organism	Growth	Colony Description
Escherichia coli ATCC® 25922	inhibited	—
Staphylococcus aureus ATCC® 25923	inhibited	—
Streptococcus mitis ATCC® 9895	good to excellent	blue
Streptococcus pyogenes ATCC® 19615	good to excellent	blue
Streptococcus salivarius	good to excellent	blue, "gum drop"

REFERENCES

1. Am. J. Digestive Diseases, 13:105:1946.
2. Trans. N.Y. Acad. Sciences, 10:45:1947.
3. J. Infectious Diseases, 67:113:1940.
4. J. Bact., 42:653:1941.
5. J. Bact., 48:113:1944.

PACKAGING

Bacto Mitis Salivarius Agar	1 lb (454 g)	0298-01-0
Bacto Chapman Tellurite Solution 1%	6 × 1 ml	0299-51-8
	6 × 25 ml	0299-66-1

BACTO MOTILITY GI MEDIUM

INTENDED USE

Bacto Motility GI Medium is a semisolid gelatin heart infusion medium for demonstrating motility of microorganisms and for separating organisms in their motile phase. It is adaptable to use in both tubes and plates for motility test studies.

HISTORY/PRINCIPLES

Bacto Motility GI Medium is prepared according to the formulation of Jordan, Caldwell and Reiter.[1] Motility is evidenced by the presence of diffuse growth away from the line or spot of inoculation. Nonmotile organisms grow only along the line of inoculation.

FORMULA

BACTO MOTILITY GI MEDIUM
DEHYDRATED

Ingredients per liter

Bacto Heart Infusion Broth .... 25 g
Bacto Gelatin .... 53.4 g
Bacto Agar .... 3 g

Final pH 7.2 ± 0.2 at 25°C.

One pound will make 5.6 liters of medium.

METHOD OF PREPARATION

1. Suspend 81.4 g in 1 liter cold distilled or deionized water.
2. Heat to boiling to dissolve completely. If tubes are desired, dispense into tubes to give a medium depth of 60 – 75 mm.
3. Sterilize in the autoclave for 15 minutes at 15 lbs pressure (121°C).
4. Cool by placing tubes in cold water up to depth of medium or cool flasks to 50 – 55°C and pour into sterile Petri dishes to give a medium depth of an eighth of an inch or more and allow to solidify.

PROCEDURE

If tubes are used, inoculate by stab inoculation with the test culture. If plates are used, the inoculum is spotted on the surface or stabbed just below the medium surface. Motile bacteria will be evidenced by a raying of growth into the medium from the line or point of inoculation whereas nonmotile organisms will not.

STORAGE

Bacto Motility GI Medium | Below 30°C
Prepared medium | 15 – 30°C

QUALITY CONTROL

Identity Specifications

Dehydrated powder: light tan, homogeneous, free-flowing
Reaction of 8.14% solution: pH 7.2 ± 0.2 at 25°C
Prepared medium: medium amber, clear, may have a fine precipitate

Typical Cultural Response in Bacto Motility GI Medium
After 40 – 48 Hours at 35°C

Organism	Growth	Motility
Enterobacter aerogenes ATCC® 13048	good	+
Escherichia coli ATCC® 25922	good	+
Klebsiella pneumoniae ATCC® 13883	good	−
Proteus mirabilis ATCC® 25933	good	±*

*Motility of Proteus is temperature dependent, being more pronounced at 20°C and possibly absent at 35°C.

REFERENCE

1. J. Bact., 27:165, 1934.

PACKAGING

Bacto Motility GI Medium | 1 lb (454 g) | 0869-01-9
| 1/4 lb (114 g) | 0869-02-8

BACTO MOTILITY MEDIUM S

INTENDED USE
Bacto Motility Medium S is a basal medium to which 2,3,5-triphenyltetrazolium chloride (TTC) is added for the determination of bacterial motility.

HISTORY/PRINCIPLES
Bacto Motility Medium S is prepared according to the formula of Ball and Sellers.[1]

The composition of the medium is such that it offers no more resistance to motility during incubation than would a broth culture, yet it preserves the stab line.

Motility is evidenced by the presence of diffuse growth away from the line of inoculation. Nonmotile organisms grow along the stab line. Growth of microorganisms capable of reducing TTC will show up red along the stab line as well as in the area into which the cells have migrated.

FORMULA

BACTO MOTILITY MEDIUM S
DEHYDRATED
Ingredients per liter

Beef Heart, Infusion from	500 g
Bacto Tryptose	10 g
Bacto Gelatin	30 g
Sodium Chloride	5 g
Dipotassium Phosphate	2 g
Potassium Nitrate	2 g
Bacto Agar	1 g

Final pH 7.3 ± 0.2 at 25°C.

One pound will make 7.56 liters of medium.

METHOD OF PREPARATION
1. Suspend 60 g in 1 liter distilled or deionized water and heat to boiling to dissolve completely.
2. Sterilize in the autoclave for 15 minutes at 15 lbs pressure (121°C).
3. Cool the medium to 60°C and aseptically add 10 ml of Bacto TTC Solution 1%.
4. Dispense the medium into sterile test tubes and refrigerate at 2 – 8°C until use.

PROCEDURE
Stab inoculation is made immediately after removal from the refrigerator. Cultures are incubated overnight at 35 – 37°C.

STORAGE
Bacto Motility Medium S	Below 30°C
Prepared tubes	2 – 8°C

QUALITY CONTROL

Identity Specifications

Dehydrated powder:	tan, homogeneous, free-flowing
Reaction of 6.0% solution:	pH 7.3 ± 0.2 at 25°C
Prepared medium:	medium amber, slightly opalescent. Appearance of a red color indicates the TTC has been reduced and tubes should be discarded.

Typical Cultural Response in Bacto Motility Medium S
After 18 – 48 Hours at 35°C

Organism	Growth	Motility	TTC Reduction
Enterobacter aerogenes ATCC® 13048	good	+	+
Escherichia coli ATCC® 25922	good	+	+
Klebsiella pneumoniae ATCC® 13883	good	–	+
Proteus mirabilis ATCC® 25933	good	±*	+

*Motility of Proteus is temperature dependent, being more pronounced at 20°C and possibly absent at 35°C.

REFERENCE
1. Appl. Microbiol., 14:670, 1966.

PACKAGING

Bacto Motility Medium S	1 lb (454 g)	0761-01-8
Bacto TTC Solution 1%	30 ml	3112-67-9

BACTO MOTILITY SULFIDE MEDIUM

INTENDED USE
Bacto Motility Sulfide Medium is a semisolid medium suitable for determining motility and the production of H_2S from L-cystine. It is used in the rapid method of differentiation and identification of various groups of Enterobacteriaceae.

HISTORY
Bacto Motility Sulfide Medium is prepared according to the formula given by Hajna.[1,2] As pointed out by Hajna,[1] this medium is the semisolid agar-gelatin medium of Edwards and Bruner[3] modified to permit observation of motility and simultaneous H_2S production.

PRINCIPLES
Motility is evidenced by presence of diffuse growth away from the line of inoculation. Nonmotile organisms grow in Bacto Motility Sulfide Medium only along the line of inoculation. H_2S producing organisms show a blackening of the medium, blackening being confined to the inoculated portion of the medium with nonmotile organisms, but diffusing throughout the medium with motile organisms. H_2S reactions on this medium may differ from the reactions usually obtained by a group of organisms since it contains free L-cystine which may give a positive reaction by organisms considered negative by classical methods.

The ability of the organism to hydrolyze urea is determined following observation of motility and H_2S production. The motility sulfide medium culture is overlaid with 1 ml of Bacto Urea Broth and incubated at 35°C for not more than 6 hours. A positive urease reaction is indicated by a reddish purple color forming in the Bacto Urea Broth. Bacto F35M Hajna may also be used for determining urease activity. After reading the medium for motility and H_2S production, overlay the Bacto Motility Sulfide Medium with 1 ml Bacto F35M Hajna and incubate at 35°C for no longer than 4 hours. Readings for urease are made at the end of 30 minutes and 4 hours. A change in color from green to yellow indicates a negative urease test and deep blue to purple is a positive test.

FORMULA

BACTO MOTILITY SULFIDE MEDIUM
DEHYDRATED

Ingredients per liter

Bacto Beef Extract	3 g	Sodium Citrate	2 g
Proteose Peptone No 3	10 g	Sodium Chloride	5 g
L-Cystine	0.2 g	Bacto Gelatin	80 g
Ferrous Ammonium Citrate	0.2 g	Bacto Agar	4 g

Final pH 7.3 ± 0.2 at 25°C.

Five hundred grams will make 4.8 liters of medium.

METHOD OF PREPARATION

1. Suspend 104 g in 1 liter cold distilled or deionized water and stir until medium is thoroughly wetted.
2. Carefully heat to boiling to dissolve the medium completely. The medium requires nearly constant agitation during the heating process.
3. Dispense in 4 ml amounts in 13 × 100 mm test tubes and sterilize in the autoclave for 15 minutes at 10 lbs pressure (117°C).
4. Allow to cool to at least 37°C before inoculation.

PROCEDURE

The suspected colony from the primary plating medium is picked with a straight needle and inoculated into the center of the tubed medium by stab inoculation to a depth of about 1/4 of the tube and then into Bacto Triple Sugar Iron Agar and Bacto H Broth, without securing further inoculum. The Bacto Sulfide Motility Medium is incubated at 30°C overnight.

STORAGE

Bacto Motility Sulfide Medium
Prepared tubes

Below 30°C
15 – 30°C

QUALITY CONTROL

Identity Specifications

Dehydrated powder: light beige, homogeneous, free-flowing
Reaction of 10.4% solution: pH 7.3 ± 0.2 at 25°C
Prepared medium: light amber, slightly opalescent, may have a slight precipitate

Typical Cultural Response in Bacto Motility Sulfide Medium
After 18 – 24 Hours at 30°C

Organism	Growth	Motility	H₂S	Urease
Enterobacter aerogenes ATCC® 13048	good	+	+	–
Escherichia coli ATCC® 25922	good	+	+	–
Proteus mirabilis ATCC® 25933	good	± *	+	+
Salmonella typhimurium ATCC® 14028	good	+	+	–
Shigella sonnei ATCC® 25931	good	–	–	–
Staphylococcus aureus ATCC® 25923	good	–	–	–

*Motility of Proteus is temperature dependent, being more pronounced at 20°C and possibly absent at 35°C.

REFERENCES

1. Pub. Health Lab., 8:36, 1950.
2. ibid., 9:23, 1951.
3. Univ. Ky. Cir., 54, 1942.

PACKAGING
Bacto Motility Sulfide Medium   500 g   0450-17-6

BACTO MOTILITY TEST MEDIUM

INTENDED USE
Bacto Motility Test Medium is recommended for use in determination of bacterial motility.

HISTORY/PRINCIPLES
Bacto Motility Test Medium is a modification of the formula of Tittsler and Sandholzer,[1] as suggested by Darby.[2] Motility is manifested macroscopically by a diffuse zone of growth spreading from a line of inoculation. Certain species of motile bacteria will show diffuse growth throughout the entire medium, while others may show diffusion from one or two points only, appearing as nodular outgrowths along the stab. Tittsler and Sandholzer reported that tubes incubated for one day gave identical results with the hanging drop method and that incubation for two days permitted them to demonstrate motility in an additional 4% of the cultures tested.

FORMULA
BACTO MOTILITY TEST MEDIUM
DEHYDRATED

Ingredients per liter

Bacto Tryptose . . . . 10 g
Sodium Chloride . . . . 5 g
Bacto Agar . . . . 5 g

Final pH 7.2 ± 0.2 at 25°C.

One pound will make 22.7 liters of medium.

PROCEDURE
1. To rehydrate, suspend 20 g in 1 liter distilled or deionized water and heat to boiling to dissolve completely.
2. Dispense into tubes and sterilize in the autoclave for 15 minutes at 15 lbs pressure (121°C).
3. Allow the medium to cool quickly with tubes in an upright position. Place tubes in cold water for best results.
4. Inoculate sterile medium by stabbing through the center of the medium.
5. Incubate at the proper temperature for organism under consideration and examine at 18 and 40 hours.
6. Motility is manifested macroscopically by a diffuse zone of growth spreading from the line of inoculation. Certain species of motile bacteria will show diffuse growth throughout the entire medium, while others may show diffusion from one or two points only, appearing as nodular growths along the line of the stab.
7. A more luxuriant growth of some microorganisms may be obtained if 0.1% – 0.3% Bacto Beef Extract is added.

STORAGE
Bacto Motility Test Medium   Below 30°C
Prepared tubes   15 – 30°C

QUALITY CONTROL

Identity Specifications

Dehydrated powder: light beige, homogeneous, free-flowing
Reaction of 2.0% solution: pH 7.2 ± 0.2 at 25°C
Prepared medium: light amber, clear to slightly opalescent

Typical Cultural Response in Bacto Motility Test Medium
After 18 – 48 Hours at 35 ± 2°C

Organism	Growth	Motility
Enterobacter aerogenes ATCC® 13048	good to excellent	positive
Escherichia coli ATCC® 25922	good to excellent	positive
Klebsiella pneumoniae ATCC® 23357	good to excellent	negative
Salmonella enteritidis ATCC® 13076	good to excellent	positive
Staphylococcus aureus ATCC® 25923	good to excellent	negative

REFERENCES

1. J. Bact., 31:575, 1936.
2. Personal Communication.

PACKAGING

Bacto Motility Test Medium   1 lb (454 g)   0105-01-3
                                          1/4 lb (114 g)   0105-02-2

BACTO MUCIN BACTERIOLOGICAL

Bacto Mucin Bacteriological, derived from hog gastric mucosa, has been used to increase virulence of some pathogenic bacteria when injected into a variety of animals.

PACKAGING

Bacto Mucin Bacteriological   100 g   0709-15-7

BACTO MUELLER HINTON MEDIUM
BACTO MUELLER HINTON BROTH

INTENDED USE

Bacto Mueller Hinton media are used for testing the susceptibility of microorganisms to antimicrobial agents.

Bacto Mueller Hinton Medium is used in the disc diffusion technique described in *NCCLS Approved Standard: ASM-2, Performance Standards for Antimicrobic Disc Susceptibility Tests.*[1] Bacto Mueller Hinton Broth is used for determining minimal inhibitory concentrations (MIC's) in the technique described in NCCLS M7-T *Tentative Standard, Methods for Dilution Antimicrobial Susceptibility Tests for Bacteria That Grow Aerobically.*[2]

HISTORY

In an attempt to develop a simple transparent medium containing no heat-labile materials and capable of withstanding autoclaving, Mueller and Hinton selected the complex Gordon and Hine[3] pea meal extract agar as the most suitable complete medium available and attempted to break it into its essential components. The authors found that starch could replace the growth-promoting properties of pea extract, acting as a "protective colloid" against toxic substances present in the medium. Further, they found that the tryptic digest of meat could be replaced by casamino acids technical. Growth of gonococci and meningococci on the medium was highly satisfactory and colonies were usually large and easily recognizable, especially with the aid of oxidase reagent.

Goodale, Gould, Schwab and Winter[4] recognized the value of Mueller Hinton medium in susceptibility test methods and developed a culture technique for testing the susceptibility of *Neisseria gonorrhoeae* to sulfonamides by adding various concentrations of sulfathiazole to the medium. Later, Nelson[5] used Mueller Hinton medium in correlating sulfonamide resistance with the clinical picture in gonorrhea.

Bauer, Kirby, Sherris and Turck[6] recommended Mueller Hinton medium for performing antibiotic susceptibility tests using a single disc of high concentration.

The World Health Organization Committee on Standardization of Susceptibility Testing has accepted Mueller Hinton medium for determining the susceptibility of microorganisms because of its reproducibility and acceptability to workers in the field.[7]

PRINCIPLES
The use of a suitable medium for testing the susceptibility of microorganisms to sulfonamides and trimethoprim is essential. Antagonism to sulfonamide activity is demonstrated by para-aminobenzoic acid (PABA) and its analogs. Reduced activity of trimethoprim, resulting in smaller inhibition zones and innerzonal colonies, is demonstrated on unsuitable Mueller Hinton medium possessing high levels of thymidine. Both the PABA and thymine/thymidine content in Bacto Mueller Hinton Medium and Broth are reduced to a minimum, thus markedly reducing the inactivation of sulfonamides and trimethoprim when the media are used for testing the susceptibility of bacterial isolates to these antimicrobics.

Bacto Mueller Hinton Medium is determined by performance testing to be low in sulfonamide and trimethoprim inhibitors.

FORMULAE

	BACTO MUELLER HINTON MEDIUM	BACTO MUELLER HINTON BROTH
	Ingredients per liter	
Beef, Infusion from	300 g	300 g
Casamino Acids, Technical	17.5 g	17.5 g
Starch	1.5 g	1.5 g
Bacto Agar	17 g	—
Solution g/liter	38 g/liter	21 g/liter
Final pH at 25°C.	7.3 ± 0.1	7.4 ± 0.2
Liters per pound of dehydrated medium.	11.9 liters	21.6 liters

METHOD OF PREPARATION
1. Suspend the amount of dehydrated medium indicated under FORMULA-Solution (and on the package label) in 1 liter distilled or deionized water and heat to boiling to dissolve completely.

2. Sterilize in the autoclave for 15 minutes at 15 lbs pressure (121°C).
3. Dispense as desired.

STORAGE

Dehydrated media	Below 30°C
Prepared media	2 – 8°C

QUALITY CONTROL

Identity Specifications

	Bacto Mueller Hinton Medium	Bacto Mueller Hinton Broth
Dehydrated powder:	beige	light beige
Appropriate solution:	3.8%	2.1 %
Reaction at 25°C:	pH 7.3 ± 0.1	pH 7.4 ± 0.2
Prepared medium:	light amber, slightly opalescent	very light amber, clear with no significant precipitate

Typical Culture Response in/on
Bacto Mueller Hinton Broth/Medium

Organism	Growth
Escherichia coli ATCC® 25922	good to excellent
Neisseria meningitidis ATCC® 13090	good to excellent
Staphylococcus aureus ATCC® 25923	good to excellent
Streptococcus faecalis ATCC® 33186	good to excellent
Streptococcus pneumoniae ATCC® 6303	good to excellent
Pseudomonas aeruginosa ATCC® 27853	good to excellent

Refer to "Bacto Sensitivity Discs, Table 2," for control zone diameters to be obtained when using antimicrobial discs.

REFERENCES

1. NCCLS Approved Standard: ASM-2, Performance standards for antimicrobic disc susceptibility tests, 2nd Ed. Villanova, Pa., National Committee for Clinical Laboratory Standards, 1979.
2. Tentative Standard, Methods for dilution antimicrobial susceptibility tests for bacteria that grow aerobically. Villanova, Pa., National Committee for Clinical Laboratory Standards, 1983.
3. Brit. Med. J., 678, 1916.
4. JAMA, 123: 547, 1943.
5. Personal communication.
6. Bauer et al., Am. J. Clin. Path., 45:493, 1966.
7. Present Status and Future Work, WHO Sponsored Collaborative Study, Chicago, October, 1967.

PACKAGING

Bacto Mueller	1 lb (454 g)	0252-01-4
Hinton Medium	1/4 lb (114 g)	0252-02-3
	5 lb (2.27 kg)	0252-05-0
	10 kg	0252-08-7
	1 × 6 × 500 ml	0252-28F-0
	6 × 6 × 500 ml	0252-29F-9
	6 × 80 ml	0252-86-2
Bacto Mueller	1 lb (454 g)	0757-01-4
Hinton Broth	1/4 lb (114 g)	0757-02-3
	5 lb (2.27 kg)	0757-05-0

BACTO MUELLER TELLURITE BASE
BACTO MUELLER TELLURITE SERUM

INTENDED USE
Bacto Mueller Tellurite Base is used in combination with Bacto Mueller Tellurite Serum for isolating, differentiating and identifying *Corynebacterium diphtheriae*.

HISTORY
The basal medium and sterile enrichment are prepared according to the formula developed by Mueller and Miller[1] after four years of practical use of tellurite plate media in the detection of diphtheria carriers.

PRINCIPLES
Tellurite medium permits differentiation of the mitis, gravis and intermedius types of *C. diphtheriae*. In 48 hours, mitis colonies are 1.0 – 1.5 mm in diameter, black and convex with a glistening surface. Gravis types show flat irregular colonies with a dull surface, slate grey in color and 2 – 3 mm in diameter. Gravis types seldom show typical "daisy head" colonies. The size of the colony and degree of darkening increases with length of incubation. Intermedius colonies are pinpoint in size in 24 hours and approximately 0.2 – 0.3 mm in diameter after 48 hours incubation. The colonies show little darkening but appear brownish grey with a white background in good light. Diphtheroids grow on the medium, often resembling mitis types. Cocci, non-diphtheroid bacilli and yeast are generally inhibited. Cocci, when they do produce colonies, resemble mitis types. Colonies or growth suggestive of diphtheria bacilli should be examined microscopically to determine the presence of organisms having corynebacterial characteristics.

Mueller and Miller report that the morphology of the various type of *C. diphtheriae* is entirely uniform on tellurite medium. Further, they suggest that diphtheria organisms cultured on the medium be tested for virulence or toxin production.

FORMULA
BACTO MUELLER TELLURITE BASE
DEHYDRATED
Ingredients per liter

Casamino Acids, Technical	20 g
Casein	5 g
Monopotassium Phosphate	0.3 g
Magnesium Sulfate	0.1 g
L-Tryptophane	0.05 g
Bacto Agar	20 g

Final pH 7.4 ± 0.2 at 25°C.

Five hundred grams will make 11.1 liters of medium.

BACTO MUELLER TELLURITE SERUM

Bacto Mueller Tellurite Serum is a sterile selective enrichment used in the preparation of the tellurite plating medium described by Mueller and Miller[1] for the isolation of *Corynebacterium diphtheriae*. It is prepared according to the directions given by the authors and contains lactate, pantothenate, beef serum and potassium tellurite, as a selective agent.

METHOD OF PREPARATION

1. Suspend 22.5 g in 500 ml distilled or deionized water and heat to boiling to dissolve completely. **AVOID EXCESSIVE HEATING.**
2. Sterilize in the autoclave for 15 minutes at 15 lbs pressure (121°C). Cool promptly to 50°C.
3. Aseptically add 12.5 ml Bacto Mueller Tellurite Serum. Mix thoroughly.
4. Dispense into sterile Petri dishes. Allow the surface of the medium to dry by partially removing the covers while the medium solidifies.

STORAGE

Bacto Mueller Tellurite Base	Below 30°C
Bacto Mueller Tellurite Serum	2 – 8°C
Prepared tellurite medium	not recommended

QUALITY CONTROL

Identity Specifications

Dehydrated powder:	light tan, homogeneous, free-flowing
Bacto Mueller Tellurite Serum:	medium to dark amber, opalescent
Reaction of 4.5% solution:	pH 7.4 ± 0.2 at 25°C
Prepared medium:	light amber, opalescent with a precipitate

Typical Cultural Response on Mueller Tellurite Medium
After 18 – 48 Hours at 35°C

Organism	Growth	Color of Colony
Corynebacterium diphtheriae type gravis	good to excellent	grey
Corynebacterium diphtheriae type intermedius	good to excellent	brownish grey
Corynebacterium diphtheriae type mitis	good to excellent	black
Staphylococcus aureus ATCC® 25923	none to poor	black
Streptococcus pyogenes ATCC® 19615	none to poor	black

REFERENCE

1. J. Bact., 51:743, 1946.

PACKAGING

Bacto Mueller Tellurite Base	500 g	0264-17-2
Bacto Mueller Tellurite Serum	6 × 25 ml	0266-66-0

BACTO MYCOBIOTIC AGAR

INTENDED USE

Bacto Mycobiotic Agar is a selective medium, nearly neutral in reaction, to be used for isolating of pathogenic fungi.

HISTORY/PRINCIPLES

Numerous media, such as Sabouraud dextrose agar, Sabouraud maltose agar, Littman oxgall agar, brain heart infusion agar and Bacto Malt Agar, have been used widely in culturing pathogenic fungi. The Sabouraud media and malt agar were somewhat selective in nature due to their low pH, which suppressed or inhibited bacterial proliferation. Media have since been made more selective by the addition of cycloheximide and/or other antibiotics, such as, chloramphenicol. The latter media have proven very useful in the isolation of the dermatophytes and other pathogenic fungi.

Littman[2,3] demonstrated that reactions near neutral were more suitable for growth of pathogenic fungi. He increased the reaction to pH 7.0 and added bacteriostatic agents to obtain proper selectivity. Littman oxgall agar was considerably more selective than the Sabouraud and malt agars as the bile, crystal violet and antibiotic content restricted the growth of most bacteria. It also retarded the growth of many saprophytic fungi and according to Moss and McQuown[4] surpassed the less selective media for primary isolation of the dermatophytes.

Leach, Ford and Whitten[5] and Whitten[6] reported that cycloheximide had little effect upon bacteria while it inhibited certain fungi. The saprophytic fungi generally were more sensitive than were the pathogenic yeasts and molds.

Phillips and Hanel[7] suggested the use of cycloheximide in media for isolation of bacteria from specimens contaminated with fungi. Georg, Ajello and Gordon[8] added cycloheximide, streptomycin and penicillin to Sabouraud dextrose agar for the isolation of *Coccidioides immitis*. The cycloheximide was especially helpful in surpressing saprophytic fungi. Fuentes, Trespalacios, Baquero and Aboulafia[9] reported the use of cycloheximide in Bacto Sabouraud Dextrose Agar to be very advantageous in isolating pathogenic fungi from lesions wherein the saprophytic fungi were present. Georg[10] demonstrated that Sabouraud dextrose agar to which cycloheximide, streptomycin and penicillin were added provided an excellent medium for the primary isolation of the dermatophytes. On this medium all dermatophyte species grew luxuriantly while bacteria and most saprophytic fungi were inhibited.

Chloramphenicol subsequently was used to inhibit bacteria contamination by Cooke[11] and Robinson, Cohen, Robinson and Bereston.[12] Difco[13] employed both cycloheximide and chloramphenicol in Bacto Mycobiotic Agar for the selective culture of pathogenic fungi. Georg et al[14,15] recommended the addition of both cycloheximide and chloramphenicol to Sabouraud dextrose agar for the primary isolation of the dermatophytes. These authors also successfully employed a medium consisting of brain heart infusion agar containing 0.5 mg/ml cycloheximide and 0.05 mg/ml chloramphenicol for the isolation of fungi which cause systemic disease. A comparable formulation is available as Bacto Brain Heart CC Agar.

Media containing cycloheximide and chloramphenicol are useful in isolating both dermatophytes and the fungi which cause systemic disease. However, there are certain limitations and precautions which must be understood. Georg et al[16] have pointed out that some of the agents of systemic mycoses may be inhibited by one or the other of these antibiotics. *Cryptococcus neoformans*, for example, is completely inhibited by cycloheximide. McDonough et al[17] have shown further that the temperature of incubation affects the sensitivity of certain systemic pathogenic fungi to these antibiotics.

Georg[14] has made the following recommendations in the selection and incubation of media for isolating the pathogenic fungi:
1. In the isolation of dermatophytes, media containing cycloheximide and chloramphenicol, such as Bacto Mycobiotic Agar, may be used exclusively, since none of the dermatophytes are sensitive to these antibiotics.

2. In the isolation of the fungi which cause systemic disease, media without antibiotics must be used in parallel with antibiotic-containing media such as Bacto Mycobiotic Agar.
3. For those fungi which are fastidious in the nutritional requirements, Bacto Brain Heart Agar or Bacto Brain Heart CC Agar is recommended.
4. Antibiotic containing media such as Bacto Mycobiotic Agar and Bacto Brain Heart CC Agar should be incubated only at room temperature.

FORMULA

BACTO MYCOBIOTIC AGAR
DEHYDRATED

Ingredients per liter

```
Bacto Soytone  . . . . . . . . . . . . . . .  10 g
Bacto Dextrose  . . . . . . . . . . . . . .  10 g
Bacto Agar . . . . . . . . . . . . . . . . .  15 g
Actidione® (cycloheximide)  . . . . . . .  0.5 g
Chloromycetin® (chloramphenicol) . .  0.05 g
```

Final pH 6.5 ± 0.2 at 25°C.

One pound will make 12.8 liters of final medium.

METHOD OF PREPARATION
1. To rehydrate, suspend 35.6 g in 1 liter distilled or deionized water. Heat to boiling to dissolve completely.
2. Distribute in 100 ml or smaller amounts.
3. Sterilize in the autoclave for 10 minutes at 15 lbs pressure (121°C).
4. Remove from autoclave and cool medium promptly. Allow tubed media to solidify in desired slanted position. Media in flasks should be poured into sterile plates.

CAUTION: Overheating during dissolving and autoclaving, or holding medium in melted state will tend to destroy the selective properties of the medium.

STORAGE
Bacto Mycobiotic Agar   Below 30°C
Prepared plates   2 – 8°C

QUALITY CONTROL
Identity Specifications

Dehydrated powder: light beige, homogeneous, free-flowing
Reaction of 3.56% solution: pH 6.5 ± 0.2 at 25°C
Prepared medium: light to medium amber, slightly opalescent

Typical Cultural Response on Bacto Mycobiotic Agar
After up to 7 Days at 25 – 30°C

Organism	Growth
Aspergillus niger ATCC® 16404	inhibited
Candida albicans ATCC® 10231	good to excellent
Candida tropicalis	inhibited
Staphylococcus epidermidis ATCC® 12228	inhibited
Trichophyton equinum ATCC® 22443	good to excellent
Trichophyton verrucosum ATCC® 36058	good to excellent

REFERENCES

1. Ann. Dermatol. Syphilogy, 1892 – 1893.
2. Science, 106:109, 1947.
3. Am. J. Clin. Path., 18:409, 1948.
4. Atlas of Medical Mycology, Williams & Wilkins, 1953 & 1960.
5. J. Am. Chem. Soc., 69:474, 1947.
6. J. Bacteriology, 56:283, 1948.
7. J. Bacteriology, 60:104, 1950.
8. Science, 114:387, 1951.
9. Mycologia, 44:170, 1952.
10. Arch. Dermatol. and Syphilol., 67:355, 1953.
11. Antibiotics and Chemotherapy, 4:657, 1954.
12. J. Am. Med. Assoc., 160:537, 1956.
13. Difco Leaflet No. 161, 1960.
14. Personal Communication, 1960.
15. J. Lab. & Clin. Med., 55:116, 1960.
16. J. Lab. & Clin. Med., 44:422, 1954.
17. Mycopath. et. Mycolog. Appl., 13:113, 1960.

PACKAGING

Bacto Mycobiotic Agar	1 lb (454 g)	0689-01-7
	1/4 lb (114 g)	0689-02-6
	5 lb (2.27 kg)	0689-05-3
	12 tubes	0689-34-8
	144 tubes	0689-37-5
	12 × 1 oz	0689-44-6

MYCOLOGICAL MEDIA

BACTO MYCOLOGICAL AGAR
BACTO MYCOLOGICAL AGAR W/LOW pH
BACTO MYCOLOGICAL BROTH
BACTO MYCOLOGICAL BROTH W/LOW pH

INTENDED USE

Bacto Mycological Media are recommended for the cultivation, isolation and identification of fungi.

HISTORY

The value of selective media for the initial cultivation of pathogenic fungi particularly has been demonstrated by numerous investigators.[1,2,3] They have shown that many fungi prefer neutral or slightly alkaline rather than acid reactions for early and luxuriant growth.[4,5] Earlier media for fungi generally relied on the acid reaction to make the medium less suited for the growth of many bacteria.[6] More recently developed media and modifications use neutral or slightly alkaline reactions, antibiotics, bile salts and dyes as selective agents against bacteria without affecting the growth of fungi.

PRINCIPLES

Mycological media are excellent basal media to which antifungal agents may be added for checking their effect upon fungi. They also afford excellent basal media to which antibacterial substances can be added to render them more selective for the isolation and cultivation of the fungi. For this latter purpose it is recommended that Bacto Mycological Agar and Bacto Mycological Broth of neutral pH be employed when working with pathogenic fungi and that Bacto Mycological Agar w/Low pH and Bacto Mycological Broth w/Low pH be employed for the saprophytic fungi.

Mycological Agar is recommended for the cultivation and maintenance of fungi. It has a lower dextrose content than Bacto Sabouraud Dextrose Agar and is adjusted to neutral reaction. This medium may be adjusted to a reaction of pH 4.0 after autoclaving by adding sterile lactic or acetic acid, for making mold and yeast counts on beverages and food products.

Bacto Mycological Agar w/Low pH is provided for laboratories wishing to use this formulation as a selective agar for culturing and enumerating fungi and aciduric bacteria.

Bacto Mycological Broth is recommended as a neutral medium for culturing fungi. It like Bacto Mycological Agar can be adjusted to any desired lower pH by the addition of sterile lactic acid or acetic acid after autoclave sterilization.

Bacto Mycological Broth w/Low pH has the same composition as Bacto Mycological Broth except the reaction has been adjusted to pH 4.8 to make it somewhat selective for the fungi and to be more suitable for culturing the saprophytic species of yeasts and molds. It is also satisfactory for culturing aciduric bacteria.

FORMULAE

MYCOLOGICAL AGAR AND MYCOLOGICAL BROTH
with Neutral and Low pH Reactions

BACTO MYCOLOGICAL AGAR
DEHYDRATED

Ingredients per liter

Bacto Soytone . . . . 10 g
Bacto Dextrose . . . . 10 g
Bacto Agar . . . . 15 g

Final pH 7.0 ± 0.2 at 25°C.

BACTO MYCOLOGICAL BROTH
DEHYDRATED

Ingredients per liter

Bacto Soytone . . . . 10 g
Bacto Dextrose . . . . 40 g

Final pH 7.0 ± 0.2 at 25°C.

BACTO MYCOLOGICAL AGAR
w/LOW pH
DEHYDRATED

Same formulation as above except reaction has been adjusted to pH 4.8 ± 0.2 at 25°C

BACTO MYCOLOGICAL BROTH
w/LOW pH
DEHYDRATED

Same formulation as above except reaction has been adjusted to pH 4.8 ± 0.2 at 25°C

METHOD OF PREPARATION

1. To rehydrate, suspend specified amount in 1 liter distilled or deionized water and heat to boiling to dissolve completely.
2. Sterilize in the autoclave for 15 minutes at 15 pounds pressure at 121°C.

Rehydration and Yield

Medium	Amount/Liter	Yield/500 g
Bacto Mycological Agar	35 g	13 liters
Bacto Mycological Agar w/Low pH	35 g	13 liters
Bacto Mycological Broth	50 g	9.1 liters
Bacto Mycological Broth w/Low pH	50 g	9.1 liters

STORAGE

Bacto Mycological media	Below 30°C
Prepared media	2 – 8°C

QUALITY CONTROL

Identity Specifications

	Mycological Agar	Mycological Agar w/Low pH	Mycological Broth	Mycological Broth w/Low pH
Dehydrated powder:	light beige, homogeneous, free-flowing			
Reaction of solution at 25°C:	3.5% pH 7.0 ± 0.2	3.5% pH 4.8 ± 0.2	5% pH 7.0 ± 0.2	5% pH 4.8 ± 0.2
Prepared medium:	light to medium amber, very slightly opalescent, no significant precipitate		light amber, clear, no precipitate	

Typical Cultural Response After 40 – 72 Hours at 25 – 30°C

Organism	Recovery Mycological Agar Mycological Broth	Mycological Agar w/Low pH Mycological Broth w/Low pH
Aspergillus niger ATCC® 16404	good to excellent	excellent
Candida albicans ATCC® 10231	good to excellent	excellent
Lactobacillus acidophilus ATCC® 11506	good to excellent	excellent
Staphylococcus aureus ATCC® 25923	good to excellent	inhibited
Saccharomyces cerevisiae ATCC® 9763	good to excellent	excellent
Saccharomyces uvarum ATCC® 9080	good to excellent	excellent

REFERENCES

1. Am. J. Publ. Health, 41:292, 1951.
2. Bull. d. Inst. Sieroteropl, Melan, 5:173, 1926.
3. Am. Rev. Resp. Dis., 95:1041, 1967.
4. A. J. Clin. Path., 24:621, 1954.
5. Rev. Latinoam Microbiol., 1:125, 1958.
6. A. J. Clin. Path., 21:684, 1951.

PACKAGING

Bacto Mycological Agar	500 g	0405-17-2
Bacto Mycological Agar w/Low pH	500 g	0305-17-3
Bacto Mycological Broth	1 lb (454 g)	0406-01-9
Bacto Mycological Broth w/Low pH	1 lb (454 g)	0304-01-2

BACTO MYCOPLASMA SUPPLEMENT
BACTO MYCOPLASMA SUPPLEMENT S

INTENDED USE

Bacto Mycoplasma Supplement and Bacto Mycoplasma Supplement S are sterile desiccated enrichments for use in Bacto PPLO Agar, as described by Hayflick,[1] and in Bacto Heart Infusion Agar and Broth[2] for the isolation and propagation of *Mycoplasma*. They are prepared according to the formulations of Chanock, Hayflick and Barile[3] and Hayflick.[4]

FORMULA

Bacto Mycoplasma Supplement contains fresh yeast extract and horse serum.

Bacto Mycoplasma Supplement S is a selective enrichment prepared by adding thallium acetate and penicillin to Bacto Mycoplasma Supplement.

METHOD OF PREPARATION

Rehydrate Bacto Mycoplasma Supplements using 30 ml sterile distilled or deionized water per vial and rotate gently to dissolve. Add the contents of one vial to 70 ml sterile Bacto PPLO Agar or Broth cooled to 50°C. Dispense in tubes or plates as desired.

STORAGE

Bacto Mycoplasma Supplement	2 – 8°C
Bacto Mycoplasma Supplement S	2 – 8°C

QUALITY CONTROL
Identity Specifications

Appearance:	light straw colored button or powder
Rehydrated appearance:	light to dark straw colored, clear to slightly opalescent

Typical Cultural Response on/in Bacto PPLO Agar or Broth
After 48 Hours or Longer at 35°C Under 10% CO_2

Organism	Growth w/Mycoplasma Supplement	Growth w/Mycoplasma Supplement S
Bacillus subtilis ATCC® 6633	good to excellent	inhibited
Staphylococcus aureus ATCC® 25923	good to excellent	inhibited
Mycoplasma bovis ATCC® 25523	good to excellent	good to excellent
Mycoplasma pneumoniae ATCC® 15531	good to excellent	good to excellent

REFERENCES

1. Texas Repts. Biol. Med., 23:285, 1965.
2. SAM – 403 Biologic Services Vet. Biol. Div., USDA.
3. Proc. Nat. Acad. Sci., 48:41, 1962.
4. Personal Communication, 1968.

PACKAGING

Bacto Mycoplasma Supplement	6 × 30 ml	0836-68-9
Bacto Mycoplasma Supplement S	6 × 30 ml	0837-68-8

NIH MEDIA

BACTO NIH THIOGLYCOLLATE BROTH
BACTO NIH AGAR MEDIUM

INTENDED USE

Bacto NIH Thioglycollate Broth, prepared according to the formula for broth medium for sterility tests as specified by the USPHS, may be substituted for Bacto Fluid Thioglycollate Medium in the sterility testing of certain biological products that are turbid or otherwise cannot be cultured satisfactorily in fluid thioglycollate medium because of its viscosity. Bacto NIH Thioglycollate Broth conforms to the specifications for the alternate fluid medium for sterility tests as given in the U.S. Pharmacopeia and the National Formulary.[2]

Bacto NIH Agar Medium, prepared according to the formula for the agar medium as specified by the USPHS sterility test,[1] is recommended for use when a solid agar medium is needed for the maintenance of cultures isolated in connection with the sterility testing of biological products. It may also be used as a solid medium for sterility testing.

PRINCIPLES

Bacto NIH Thioglycollate Broth contains thioglycollate to neutralize the bacteriostatic effect of mercurial preservatives. It contains no agar, which makes it suitable for testing viscous materials and devices having tubes with small lumina. In testing solutions for sterility, Bacto NIH Thioglycollate Broth should be used in fermentation tubes and be heated in boiling water or a steam bath just prior to use to drive off dissolved oxygen. When testing the sterility of solutions containing preservatives, the preservative must be inactivated or sufficient medium used to dilute the inoculum beyond the bacteriostatic limits of the preservative. Bacto NIH Thioglycollate Broth contains thioglycollate to neutralize the bacteriostatic effect of mercurial preservatives. The following table indicates minimum quantities of medium to use in testing the sterility of solutions containing other preservatives.

Minimum Amount of Culture Media per 1 ml of Inoculum

Preservative	Concentration of Preservative	Biological Product	Minimum Volume of Culture Medium
Phenol	0.5%	Serums and vaccines	40 ml
Cresol, NF	0.35%	Serum	60 ml
Merthiolate®	1:10,000	Toxoids	
	1:35,000	Human Plasma	10 ml
Phenylmercuric Acetate	1:50,000	Human Plasma	
	1:100,000	Normal saline	10 ml
Phenylmercuric Borate	1:50,000	Human Plasma	
	1:100,000	Normal saline	10 ml
Phenylmercuric Nitrate	1:50,000	Human Plasma	
	1:100,000	Normal saline	10 ml
Merthiolate® and Phenol	1:10,000 and 0.2%	Serum	20 ml
Chlorobutanol	0.5%	. . .	40 ml
Formalin	0.4%	. . .	40 ml

Bacto NIH Agar Medium contains no thioglycollate. If the medium is to be used to test the sterility of a biological product containing a mercurial preservative, 0.05% sodium thioglycollate or 0.03% thioglycollic acid should be added.

FORMULAE

BACTO NIH
THIOGLYCOLLATE BROTH
(USP Alternate Thioglycollate Medium)
DEHYDRATED

Ingredients per liter

Bacto Casitone	15 g	Bacto Dextrose	5.5 g
Bacto Yeast Extract	5 g	Sodium Chloride	2.5 g
L-Cystine, Difco	0.5 g	Sodium Thioglycollate	0.5 g

Final pH at 25°C.	7.1 ± 0.2
Rehydration	29 g/liter
Yield	15.6 liters/lb

BACTO NIH AGAR MEDIUM
DEHYDRATED

Ingredients per liter

Bacto Casitone	15 g	Bacto Dextrose	5.5 g
Bacto Yeast Extract	5 g	Sodium Chloride	2.5 g
L-Cystine, Difco	0.05 g	Bacto Agar	15 g

Final pH at 25°C.	7.1 ± 0.2
Rehydration	43 g/liter
Yield	11.6 liters/500 g

METHOD OF PREPARATION
1. Suspend the appropriate amount of medium in 1 liter distilled or deionized water.
2. Heat to boiling to dissolve completely and distribute into suitable test tubes or flasks as desired.
3. Sterilize in the autoclave at 15 lbs pressure at 121°C, the broth for 15 minutes but the agar for 18 – 20 minutes.

STORAGE
Bacto NIH Media	Below 30°C
Bacto NIH Thioglycollate Broth, prepared	15 – 30°C
Bacto NIH Agar Medium, prepared	2 – 8°C

QUALITY CONTROL
Identity Specifications

Dehydrated powders:	light tan, homogeneous, free-flowing
Prepared tubes:	light amber, clear, may have a slight precipitate
Prepared medium:	light amber, slightly opalescent

Typical Cultural Response on Bacto NIH Thioglycollate Broth
After 18 – 48 Hours at 35°C

Organism	Growth
Bacillus subtilis ATCC® 6633	good to excellent
Bacteroides vulgatus ATCC® 8482	good to excellent
Candida albicans ATCC® 10231	good to excellent
Clostridium sporogenes ATCC® 11437	good to excellent
Micrococcus luteus ATCC® 9341	good to excellent
Staphylococcus aureus ATCC® 6538	good to excellent

Typical Cultural Response on Bacto NIH Agar Medium
After 18 – 24 Hours at 35 – 37°C

Organism	Growth
Escherichia coli ATCC® 25922	good to excellent
Staphylococcus aureus ATCC® 6538	good to excellent
Streptococcus mitis ATCC® 9895	good to excellent
Streptococcus pyogenes ATCC® 19615	good to excellent

REFERENCES
1. USPHS Reg. 73, 730: Federal Register, Vol. 35. No. 0171, Sept. 2, 1970, p. 13930.
2. The United States Pharmacopeia/National Formulary USP XX/NFXV Washington, D.C., U.S. Pharmacopeial Convention, p. 879, 1980.

PACKAGING
Bacto NIH Thioglycollate Broth   1 lb (454 g)   0257-01-9
Bacto NIH Agar Medium   500 g   0258-17-0

NEISSERIA CULTURE MEDIA

READY TO USE MEDIA FOR THE ISOLATION OF N. GONORRHOEAE OR N. MENINGITIDIS

BACTO CHOCOLATE AGAR ENRICHED
BACTO THAYER-MARTIN MEDIUM
BACTO TRANSGROW MEDIUM (MTM)

BACKGROUND

Neisseria gonorrhoeae and *Neisseria meningitidis* are aerobic, gram-negative diplo-cocci, which grow optimally in an aerobic environment with $2-10\%$ CO_2 on enriched media. Since these organisms are very fastidious in their growth requirements, en-riched media such as Bacto Chocolate Agar Enriched, Bacto Thayer-Martin Medium and Bacto Transgrow Medium (Modified Thayer-Martin Medium-MTM) are recom-mended for their isolation and cultivation. Bacto Transgrow Medium is also designed for shipping specimens suspected of containing *N. gonorrhoeae* or *N. meningitidis*. Typ-ical colonies isolated on these media which are oxidase positive and composed of gram-negative diplococci can be considered presumptive of *N. gonorrhoeae* or *N. men-ingitidis*.

The marked increase in the incidence of gonorrhea necessitates the use of a sensitive laboratory procedure for the isolation of the etiologic agent of this disease, *N. gonor-rhoeae*. At the present time the cultural technique is considered to be the most sen-sitive method for detecting this organism.[1] Cultural methods for the detection of *N. gonorrhoeae* from cervical and urethral exudates have proven more reliable and effi-cient than microscopic techniques. These procedures are particularly valuable in early, chronic or treated cases and in isolation of the organism from asymptomatic males and females.

Gonorrhea is especially difficult to detect in women as many of them are asymptomatic. Gram stains, which are often used as presumptive diagnosis of the disease in males, are considered inconclusive in the diagnosis of gonorrhea in females. Recommended procedures for presumptive identification of *N. gonorrhoeae* in females requires iso-lation of the gonococcus on Thayer-Martin medium (or the more recently described modified Thayer-Martin medium-MTM[2]), a positive oxidase test, and microscopic de-tection of a gram-negative diplococcus.[1] Recent studies have also shown that some males may be asymptomatic and in such cases the presence of *N. gonorrhoeae* may go undetected when diagnosis is based solely on gram-stain results.[3]

Interest in the cultural procedure for the diagnosis of gonococcal infections was stim-ulated by Ruys and Jens,[4] McLeod et al,[5] Thompson,[6] Leahy and Carpenter,[7] Car-penter, Leahy and Wilson[8] and Carpenter,[9] who clearly demonstrated the superiority of this method over the microscopic technique. Further studies in cooperation with Car-penter and McLeod[10] and Herrold[11] resulted in the development of chocolate agar prepared with Bacto Proteose No. 3 Agar and Bacto Hemoglobin, which proved to be satisfactory for isolating the organism from all types of gonococcal infections.

Efforts made by Langford,[12] Langford, Scott, Cox and Cooke,[13] and Langford and Snell,[14] on suitable enrichments to increase the number of isolations and growth rate of N. gonorrhoeae were instrumental in the development of Bacto Supplement A and Bacto Supplement B.

Studies were also conducted on agar bases to improve their productivity and shorten incubation time. Johnston[15] observed, in a comparative study on an enriched medium for isolation of N. gonorrhoeae, that decreasing the agar content increased the productivity of the medium allowing cultivation in 24 rather than 48 hours. Mueller and Hinton[16] and Mueller and Ley,[17] observed that the presence of starch in a medium neutralized the toxic effects of factors which tend to inhibit the growth of the gonococcus. These findings culminated in the introduction of Bacto GC Medium Base. In a subsequent comparative study of 12 different media, an enriched chocolate agar prepared with Bacto GC Medium Base, Bacto Hemoglobin and Bacto Supplement B proved superior for isolating N. gonorrhoeae.[18]

Bacto Chocolate Agar Enriched has been proven to be as effective in the number of isolations of N. gonorrhoeae as was any other nonselective medium recommended for this purpose. Further studies describing the advantages and superiority of this medium led to the acceptance of Bacto Chocolate Agar Enriched as a standard medium for the cultivation of N. gonorrhoeae.[19,20,21] Bacto Chocolate Agar Enriched, though an excellent culture medium for gonococci, is not a selective medium and; therefore, when mixed cultures are encountered, growth of contaminants is not restricted and may result in over-growth of gonococci by bacterial and fungal flora often found in specimens from urethral and cervical sites.

Attempts to formulate a selective medium for isolating N. gonorrhoeae resulted in the development of Thayer-Martin medium. Thayer and Martin[22,23] explored the use of antibiotics as inhibitors of bacterial and fungal contaminants usually found in specimens obtained from the urogenital tract. In 1966, these authors reported on a medium composed of chocolate agar enriched, as described by Christensen and Schoenlein,[24] to which vancomycin, colistimethate and nystatin were added. The Thayer-Martin medium proved to be selective for N. gonorrhoeae and N. meningitidis while maintaining the cultural value of the chocolate agar enriched. Thayer-Martin medium has been recommended as the primary isolation medium of N. gonorrhoeae[1] from specimens suspected of containing this organism.

The introduction of Thayer-Martin medium does not negate, necessarily, the use of chocolate agar enriched. It has been found that some strains of N. gonorrhoeae may be inhibited by the antibiotics incorporated in this medium. Therefore, to insure optimal recovery of gonococci, recommended procedure calls for the use of both the selective medium and chocolate agar enriched when cultivating gonococci from urogenital specimens.[25]

Bacto Thayer-Martin Medium and Bacto Chocolate Agar Enriched are valuable culture media in overall techniques for the isolation of N. gonorrhoeae. However, many cases of gonorrhea are undetected because viable organisms which may be present originally will die during transit to the hospital or public health laboratory. Therefore, studies have continued to find an adequate transport medium for N. gonorrhoeae which would maintain the viability of gonococci during normal transit periods.

Martin and Lester[26] modified the Thayer-Martin medium to meet the requirements of transporting gonococcal specimens. The modified formulation has an increased agar content to reduce spreading of *Proteus* species and permit shipping of the medium without rupturing the slant. The addition of dextrose provides a more nutritional environment for gonococci and meningococci while colistimethate, nystatin and vancomycin markedly inhibit the growth of contaminating organisms. Further, improvement of transgrow medium was effected by the addition of trimethoprim lactate[27] which markedly to completely inhibits the growth of proteus species.

Subsequent studies[2] of the performance of transgrow (modified Thayer-Martin medium - MTM) in both plates and bottles showed this modified formula to be slightly more proficient in recovery of *N. gonorrhoeae* from mixed cultures. In a study by Martin, Armstrong and Smith,[28] Thayer-Martin medium and modified Thayer-Martin medium were found to be equally sensitive and selective in the growth of stock cultures of *N. gonorrhoeae* and from male urethral specimens, but 10% more cultures from female cervical specimens were positive for gonococci on modified Thayer-Martin medium. After consideration of the various comparative studies, the Center for Disease Control recommended the use of modified Thayer-Martin medium plates or bottles of transgrow in "GC Culture Screening Programs."

Bacto Transgrow Medium is packaged in 1 oz bottles, rather than culture tubes, providing a larger suface area on which specimens may be streaked. The presence of $3 - 10\%$ CO_2 in the bottle eliminates the need for incubating the inoculated medium in a CO_2 incubator.

Use of Bacto Transgrow Medium cannot be taken as an assurance that all specimens transported will be recovered. However, it does provide increased recovery over previous methods of transport and thereby is considered a valuable addition to the techniques for isolation and control of gonorrhea.

Because some of the growth requirements of *N. meningitidis* are similar to those of *N. gonorrhoeae*, the above media can be used successfully for isolating this organism. Bacto Chocolate Agar Enriched provides good growth of *N. meningitidis* and is particularly valuable when isolating the organism from spinal fluid where it usually exists in pure culture. Bacto Thayer-Martin Medium is used for recovering *N. meningitidis* from sites where microbial flora or contaminating organisms might be present. Bacto Transgrow Medium (modified Thayer-Martin medium-MTM) allows transport of specimens suspected of containing *N. meningitidis* when facilities for the cultivation and isolation are unavailable at the time of collection.

FORMULAE

BACTO TRANSGROW MEDIUM

Bacto GC Medium Base . . . . 36 g
Bacto Hemoglobin . . . . 10 g
Bacto Supplement B . . . . 10 ml
Bacto Antimicrobic Vial CNVT . . . . . 10 ml

Bacto Dextrose . . . . 1.5 g
Bacto Agar . . . . 10 g
Distilled or Deionized
  Water . . . . 1 liter

BACTO THAYER-MARTIN MEDIUM

Bacto GC Medium Base . . . . 36 g
Bacto Hemoglobin . . . . 10 g
Bacto Supplement B . . . . 10 ml

Bacto Antimicrobic
  Vial CNVT . . . . 10 ml
Distilled or Deionized
  Water . . . . 1 liter

BACTO CHOCOLATE AGAR ENRICHED

Bacto GC Medium Base . . . . 36 g
Bacto Hemoglobin . . . . 10 g

Bacto Supplement
  B . . . . 10 ml
Distilled or Deionized
  Water . . . . 1 liter

Bacto Transgrow Medium (MTM)

Bacto Transgrow Medium (MTM) is available in 1 oz bottles which provides an increased surface area for direct inoculation of specimens suspected of containing *N. gonorrhoeae* or *N. meningitidis*. Bottles should be stored in the refrigerator at 2 – 8°C in an upright position to avoid loss of CO_2. Do not freeze or expose to temperatures above 25 – 27°C. Bacto Transgrow Medium will maintain its cultural value through the expiration date, which appears on the label of each bottle, when stored as specified above. Allow the medium to reach room temperature prior to inoculation. Do not use bottles that show signs of cracking, drying or contamination. As with all tube and bottle media, water of syneresis should be present. However, fluid in transgrow bottles should not exceed 0.2 ml. If excessive fluid is apparent, do not use.

Loss of CO_2 occasionally occurs in storage and transit of Bacto Transgrow Medium. This may arise from insufficient closure, changes in pressure and temperature or absorption of the CO_2 gases into the medium itself. CO_2 levels in transgrow bottles below 3% may adversely effect the cultural potential of this product. It is suggested that when loss of CO_2 is suspected the following check be made on random samples to determine the presence of CO_2.

STOCK SOLUTIONS
A. 0.1 N Na_2CO_3: Dissolve 0.53 g of Na_2CO_3 in 100 ml distilled water. Store at room temperature.
B. O.1% phenolpthalein: Dissolve 100 mg of phenolpthalein in 50% solution of 95% ethyl alcohol. Discard if crystalization occurs.

Reagent: Prepare when ready to use. Mix equal parts of stock solution A and B.

PERFORMANCE OF TEST
1. Hold the Bacto Transgrow bottles in an upright position; remove cap and quickly add 1 ml phenolpthalein reagent indicator. **Do not** blow through pipette into bottle.
2. Replace cap immediately and shake the bottle vigorously.
3. A change of indicator from violet to colorless in 1 minute or less indicates the presence of approximately 5% or more CO_2.
NOTE: A control bottle of transgrow void of CO_2 may be run for comparison.

Bacto Chocolate Agar Enriched, Bacto Thayer-Martin Medium

Bacto Thayer-Martin Medium and Bacto Chocolate Agar Enriched are available in screw cap test tubes and should be stored in the refrigerator at 2 – 8°C. Do not expose to high temperatures. These products will remain stable, maintaining their cultural value, through the expiration date on the label of each tube or bottle when stored as specified above.

PRECAUTIONS

Tube and bottle media showing signs of cracking, drying or contamination should not be used.

SPECIMEN COLLECTION FOR DETECTION OF *N. GONORRHOEAE*

Asymptomatic females present the most difficulty in recovering *N. gonorrhoeae*. Therefore, to assure optimal recovery of the gonococcus, the specimen should be taken by the anocervical technique. For all suspect cases of gonorrhea in females an endocervical as well as anal culture is recommended. Care should be taken not to contaminate specimens with microbial flora of the urogenital tract or from anal canal. Cultures from early morning voided urine may be used for male subjects, however, the rec-

ommended method (USPHS) is urethral culture to insure optimum recovery of gonococcus.

The following method is recommended in USPHS *"Criteria for Diagnosis of N. gonorrhoeae."*[1]

FEMALE
Cervical Culture:
1. Moisten speculum with warm water only. Do not use any other lubricant.
2. Cleanse cervical orifice carefully removing mucous. This can be done with a cotton
   ball held in ring forceps. Do not use antiseptic solution, alcohol, etc.
3. Insert sterile cotton tip swab into endocervical canal. Move swab from side to side
   allowing several seconds for absorption.

Anal Canal:
1. Cleanse rectal area with warm water only. Do not use bacteriostatic, bactericidal or
   antiseptic solution.
2. Insert sterile, cotton-tipped swab into anal canal (approximately 1″ to avoid contact
   with feces.) If swab is inadvertantly pushed into feces, use another swab to obtain
   specimen.
3. Move swab from side to side in the anal canal to sample crypts. Allow several seconds for absorption of organisms to swab.

MALE
Urine Sedimentation:
1. Aseptically obtain a few ml of urine in a sterile dry test tube.
2. Centrifuge specimen for 15 minutes at 2500 rpm.
3. Discard supernatant.
4. Absorb sediment with sterile cotton swab.

Urethral Culture:
Strip the urethra toward orifice to allow emmission of exudate. Obtain specimen by use
of a sterile loop or cotton swab.

or,

Using a sterile bacteriological wire loop, obtain specimen from anterior urethra by gently
scraping mucous.

PROCEDURE FOR DETECTION OF *N. GONORRHOEAE*
Materials Provided:
Bacto Transgrow Medium (MTM)
Bacto Thayer-Martin Medium
Bacto Chocolate Agar Enriched

Materials Required but not Provided:
1 candle jar or CO_2 incubator (not required
   when using transgrow medium)
Swabs
Wire loops
Slides
Mailing container for each inoculated
   bottle of transgrow to be shipped

Purchase Separately:
Bacto Oxidase Reagent
(p-Aminodimethyaniline Oxalate)
Bacto Differentiation Disks Oxidase
50 disks per vial
Bacto Gram Stain Set contains
   1 each Gram
Iodine, Crystal Violet, Safranin, and
   Decolorizer

Bacto Transgrow Medium may be used for transport only or as an isolation medium for presumptive identification of *N. gonorrhoeae*.

TRANSPORTING SPECIMENS TO CLINICAL OR PUBLIC HEALTH LAB
1. Bring Bacto Transgrow Medium to room temperature prior to inoculation.
2. Unscrew cap holding bottle in an upright position to prevent loss of CO_2.
3. Insert swab containing specimen and inoculate surface of medium in formation of a "Z" with a rolling motion of the swab.
4. Replace cap and tighten securely.
5. The inoculated bottle is packed for mailing. Pack bottles in such a way as to protect from adverse extremes in temperature.

Careful collection of the specimen and proper inoculation of transgrow are essential for optimal results. Whenever possible the inoculated medium should be preincubated 16 – 18 hours at 35°C before shipment to increase growth and prevent loss of organisms during transit.

PREFERRED METHOD
1. Preincubate bottles 16 – 18 hours at 35 – 37°C in an upright position.
2. Package for mailing, protecting specimens against adverse extremes in temperature.
3. Examine medium for opaque, grayish, glistening convex colonies of *N. gonorrhoeae*.
4. Select isolated colonies and prepare for Gram stain.
5. Flood remaining colonies with oxidase reagent and examine for oxidase positive colonies, i.e., a color change of the colonies to pink and finally to dark purple or black.

Cultivation and Isolation on Bacto Thayer-Martin Medium and Bacto Chocolate Agar Enriched
1. Allow medium to reach room temperature.
2. Make certain surface of agar is dry.
3. Roll swab containing specimen, inoculating medium in a "Z" pattern.
4. Cross streak with sterile wire loop to facilitate isolation of the gonococcus from mixed culture.
5. Place in a CO_2 incubator or candle jar with atmosphere of 3 – 10% CO_2.
6. Incubate at 35 – 37°C for 24 – 48 hours.
7. Examine for *N. gonorrhoeae* colonies.
8. Pick colonies and prepare Gram stain.
9. Flood surface of the medium with oxidase reagent to detect oxidase positive colonies.

Results and Interpretation
Tubes or bottles of Bacto Thayer-Martin Medium and Bacto Chocolate Agar enriched or bottles of Bacto Transgrow Medium should be examined after 24 – 48 hours for presence of small, translucent, convex and raised, grayish white colonies. Such colonies should be selected for Gram stain and microscopic examination. Remaining colonies on the plate or slant should be subjected to the oxidase test, using Bacto Oxidase Reagent or Bacto Differentiation Disks Oxidase. Positive oxidase colonies will appear pink to black.

SPECIMEN COLLECTION FOR DETECTION OF *N. MENINGITIDIS*
Specimens for the isolation of *N. meningitidis* may be obtained from cerebrospinal fluid, joint fluid, blood, petechiae of the skin, rectum or the nasopharynx.

All culture sites and specimen collection should be executed under qualified medical supervision, particularly when involving spinal taps and nasopharyngeal procedures.

Specimens should be collected under sterile conditions and forwarded to the laboratory **immediately.** Cultures suspected of containing *N. meningitidis* should not be refrigerated.

FLUID SPECIMENS
1. Centrifuge specimen for 15 minutes at 2500 rpm immediately upon receipt.
2. Aspirate off supernatant with a sterile capillary pipette.
3. Use remaining fluid and sediment to inoculate Bacto Thayer-Martin Medium, Bacto Chocolate Agar or Bacto Transgrow Medium as stated above. Proceed as stated under *N. gonorrhoeae.*

WIRE OR SWAB SPECIMENS
Streak onto surface of Bacto Thayer-Martin Medium, Bacto Chocolate Agar Enriched or Bacto Transgrow Medium (MTM). Proceed as described under *N. gonorrhoeae.*

Results and Interpretation
Cultures grown on Bacto Thayer-Martin Medium, Bacto Transgrow Medium (MTM) or Bacto Chocolate Agar Enriched should be examined for smooth, glistening, translucent, soft, bluish gray colonies, typical of *N. meningitidis.* Flood surface of the medium with Bacto Oxidase Reagent or use Bacto Differentiation Disks Oxidase as stated above under *N. gonorrhoeae.* Prepare Gram stain using Bacto Gram Stain Set and examine microscopically. Oxidase positive, gram-negative diplococci having typical morphology can be presumptively identified as *N. meningitidis.*

LIMITATIONS OF THE PROCEDURE
Although the cultural technique is considered superior for detecting the presence of *N. gonorrhoeae*, it cannot be assumed that recovery of the gonococcus is inevitable in all suspect cases. Various factors can adversely affect the recovery of *N. gonorrhoeae* from clinical specimens:
1. Improper specimen collection, permitting excessive contaminants may result in overgrowth of the gonococci.
2. Improper streaking technique of specimens containing mixed culture can contribute to overgrowth of the gonococci by bacterial or fungal flora.
3. Improper environment, temperature, CO_2 level, moisture and the pH can adversely affect the growth and viability of the organism.
4. Inactivation or deterioration of antibiotics in Thayer-Martin or transgrow medium may allow growth of nongonococcal organisms and possible overgrowth by these "contaminants."
5. Inhibition of gonococci by *C. albicans* present in vaginal specimens has been reported.[29]
At optimal performance, isolation on Bacto Chocolate Agar Enriched, Bacto Thayer-Martin Medium or Bacto Transgrow Medium, of oxidase positive, gram-negative diplococci can only be considered as a presumptive identification of *N. gonorrhoeae* or *N. meningitidis.*

Bacto Chocolate Agar Enriched, a nonselective medium, and Bacto Thayer-Martin Medium and Bacto Transgrow Medium, as selective media, will allow growth of other *Neisseria* species as well as *Moraxella osloensis (Mima polymorpha var. oxidans), Pseudomonas* and *Flavobacterium* species, (although on Thayer-Martin medium and transgrow medium these organisms, in most cases, are markedly suppressed or inhibited). These organisms are also gram-negative and oxidase positive. *Flavobacterium* and *Pseudomonas* can be distinguished microscopically as they form rods rather than diplococci. *N. lactimicus* also grows on these media but is not normally found in specimens suspect of *N. gonorrhoeae*. However, *M. osloensis* is more difficult to distinguish as it is a cocobacillus resembling *N. gonorrhoeae*. Further, *N. meningitidis* is morphologically similar to *N. gonorrhoeae* and these organisms occur in sites in which either one or both may be present, i.e., rectal cultures.

Therefore, final identification and confirmation of *N. gonorrhoeae* or *N. meningitidis* can only be determined through further biochemical and serological tests. Specifically, a carbohydrate utilization pattern should be established for the isolate. Bacto Cystine Tryptic Agar w/Carbohydrates may be used for this purpose. In addition, "The Rapid Fermentation Test" as described in CDC Current Item No. 242 is particularly useful when rapid results are essential.

Further corroborative evidence may be obtained through the direct FA test. This procedure is outlined in "Reagents for the Confirmation of *Neisseria gonorrhoeae* by the Direct Fluorescent Antibody Technique."

REFERENCES

1. U.S. Pub. Hlth. Serv. CDC, Veneral Dis. Br.: Criteria and Tech. for the Diag. of Gonorrhea, 1971.
2. Memor., Recom. to use the same medium, MTM in both plates and bottles for the GC Cul. Scr. Prog., CDC, Atlanta, GA, 1975.
3. Med. Clin. North Am., 56:1127, 1972.
4. Muench. Wochschr., 80:846, 1933.
5. J. Path. Bact., 38:221, 1934.
6. J. Infect. Dis., 61:129, 1937.
7. Am. J. Syphilis, 22:347, 1938.
8. Am. J. Syphilis, 22:55, 1938.
9. 7th Ann. Yearbook (1936 – 37), p. 133, Suppl. Am. J. Publ. Hlth., 27:3, 1937.
10. Personal Communication, 1938.
11. Personal Communication, 1938.
12. J. Bact., 44:139, 1942.
13. J. Bact., 45:321, 1943.
14. J. Bact., 45:410, 1943.
15. J. Ven. Dis. Inform., 26:239, 1945.
16. Proc. Soc. Exp. Biol. Med., 48:330, 1941.
17. J. Bact., 52:453, 1946.
18. Diagnostic Proc. and Reagents, 3rd Ed., 30 – 107, 1950.
19. Paper read at the Annual Meeting of the Canadian Publ. Hlth. Assoc., 1947.
20. Paper read at the Annual Meeting of the American Publ. Hlth. Assoc., 1947.
21. Personal Communication, 1947.
22. Publ. Hlth. Reports, 79:49, 1964.
23. Publ. Hlth. Reports, 81:599, 1966.
24. Ann. Meeting CA. Publ. Assn., 1947.
26. Proficiency Test Rep., CDC, Atlanta, GA.
26. HSMHA Hlth. Rep., 86:30, 1971.
27. Brit. J. Pharmogol., 33:72, 1968.
28. Appl. Micro., 27:802, 1974.
29. Appl. Micro., 27:192, 1974.

PACKAGING

Bacto Chocolate Agar Enriched	12 tubes	1002-34-6
	144 tubes	1002-37-3
Bacto Thayer-Martin Medium	12 tubes	1247-34-1
Bacto Transgrow Medium (MTM)	12 × 1 oz	1410-44-0
	144 × 1 oz	1410-45-9

REAGENTS FOR THE CONFIRMATION OF *NEISSERIA GONORRHOEAE* BY THE DIRECT FLUORESCENT ANTIBODY TECHNIQUE

BACTO FA N. GONORRHOEAE
BACTO FA CLOACAE, DESICCATED
BACTO FA BUFFER, DRIED
BACTO FA MOUNTING FLUID pH 9
BACTO DIFFERENTIATION DISKS OXIDASE
BACTO p-AMINODIMETHYLANILINE OXALATE

INTENDED USE

Bacto FA N. gonorrhoeae, is a fluorescein conjugated antirabbit globulin used to confirm the identity of isolated cultures of biochemically confirmed *Neisseria gonorrhoeae* by the direct FA technique. Bacto FA Cloacae, Desiccated, is used as a nonspecific staining control.

SUMMARY & EXPLANATION OF THE TEXT

Originally Deacon et al,[1] described the FA technique for use in both the direct and delayed methods for identification of this organism in females. After its original description, cross reactions with *Neisseria meningitidis* necessitated absorption of this reagent to render it specific for *N. gonorrhoeae*.

While the direct FA technique is useful to corroborate the identification of cultures presumed to be *N. gonorrhoeae* the USPHS[2] does not recommend its use for:
1. The staining of direct smears for the diagnosis of GC in women except as an adjunct to the cultures.
2. The delayed technique for the diagnosis of GC.
3. The staining of direct smears or the delayed technique as a test of cure in women.
4. The staining of direct smears of urethral exudate for the diagnosis of gonorrhoeae in males.

PRINCIPLES OF THE PROCEDURE

The direct fluorescent antibody technique, for use in the detection of microorganisms, is the simplest of the immunofluorescence methods and may be defined as follows:

The organism (antigen) is fixed to a microscope slide by heat or chemical methods. The fixed antigen is overlayed with a specific antibody which is labeled with a fluorescent marker (FITC). The resultant antigen-antibody reaction is then observed microscopically, utilizing a suitable wave length of light compatable to the fluorescent marker employed.

The current lots, as well as past lots of Bacto FA N. Gonorrhoeae, meet CDC specifications.

REAGENTS

Bacto FA N. Gonorrhoeae is a fluorescein-conjugated antiglobulin prepared according to the method of Deacon, Peacock, Freeman and Harris[1] and Peacock.[5] It is absorbed with Bacto FA Bovine Bone Marrow as well as with suspensions of *N. meningitidis* and blocked for reaction with *Staphylococcus aureus*. The absorbed conjugate is tested in accordance with the "Provisional Protocol for Check Testing Fluorescein-labeled Anti-Gonococcal Sera."[6] This reagent meets specifications as established by the USPHS.

To rehydrate the conjugate add the appropriate amount of sterile distilled or deionized water and rotate gently to dissolve the contents completely.

The working dilution of the conjugate should be determined shortly after its rehydration. The titer of a conjugate varies with the technique used, the fluorescent microscope used, the filters used and the age of the bulb. Generally, a 1:4 to 1:8 titer has been found with our equipment using the procedure as outlined here.

The conjugate should be titrated using a freshly isolated colony of *N. gonorrhoeae* following the test procedure. The conjugate is further diluted with rehydrated Bacto FA Buffer for the titration as well as to obtain the working dilution.

The working dilution of the conjugate is determined as follows:

Dilution of Conjugate	Fluorescence
undiluted	4+
1:2	4+
1:4	4+
1:8	4+
1:16	4+
1:32	3+
1:64	1+
1:128	—
1:256	—

In this example, the last 4+ fluorescence is found in a 1:16 dilution of the conjugate. One less dilution is chosen for a margin of safety. The working dilution, therefore, in this case is 1:8.

The working dilution of the conjugate should be checked with *E. cloacae* to insure absence of nonspecific fluorescence.

Aliquots of the titered conjugate should be put into small vials and frozen in the undiluted state and stored below −20°C for optimal stability.

Bacto FA Cloacae, Desiccated is a preparation of *Enterobacter cloacae* for use as the negative control in this test.

To rehydrate add 1 ml sterile distilled or deionized water to the vial and agitate vigorously to suspend uniformly. Process in the same manner as the unknown specimen commencing with step 3 of the procedure. The rehydrated suspension is stable to the expiry date on the label when stored at 2 – 8°C.

Bacto FA Buffer, Dried is a phosphate buffer yielding a pH of 7.2 upon rehydration.

To rehydrate add the contents of a 10 g vial to 1 liter or a 100 g bottle to 10 liters distilled or deionized water and stir until completely dissolved.

Rehydrated Bacto FA Buffer, Dried should be stored at 2 – 8°C. Solution showing turbidity or mold growth should be discarded.

Bacto FA Mounting Fluid pH 9 is a reagent grade glycerin adjusted to pH 9. It is stable to the expiry date on the label when stored at 15 – 30°C.

Bacto Differentiation Disks Oxidase are paper disks containing para-aminodimethylaniline oxalate to detect oxidase production by microorganisms. They are stable to the expiry date on the label when stored at 2 – 8°C.

Bacto p-Aminodimethylaniline Oxalate is in powder form and is prepared for use in the oxidase test by preparing a 1% solution in distilled or deionized water. The distilled or deionized water may be warmed gently to hasten solution of the salt. The solution is stable for up to 5 days if stored at 2 – 8°C.

SPECIMEN PREPARATION
Cultures to be tested should be freshly isolated on Bacto Thayer Martin Medium or other suitable selective and specialized media such as Bacto Transgrow Medium.

PROCEDURE
Materials Provided:
Bacto FA N. Gonorrhoeae
Bacto FA Cloacae, Desiccated
Bacto FA Buffer, Dried
Bacto FA Mounting Fluid pH 9
Bacto Differentiation Disks Oxidase
Bacto p-Aminodimethylaniline Oxalate
(Prepare 1% solution)

Materials Required but not Provided:
Forceps
Bunsen burner
Bacteriological wire loop
Microscope slide with etched circles 1 cm i.d.
Moisture chamber
Cover glasses 22 × 50 mm
Fluorescent microscope set-up
Sterile 0.85% NaCl solution

1. Determine the oxidase activity of the organism to be tested by one of the following methods:

Method A
On the surface of the medium containing isolated colonies of organisms suspected of being *N. gonorrhoeae*, i.e., small raised, convex, translucent, greyish-white colonies with undulate margins, place a Bacto Differentiation Disk Oxidase using a pair of flamed forceps. Add 1 drop of distilled or deionized water to the disk. Observe for a pink to violet color change in the colonies in approximately 10 – 20 minutes.

Method B
Flood the surface of the medium with a 1% solution of Bacto p-Aminodimethylaniline Oxalate. The surface of the medium is observed for a period of 6 – 10 minutes. Oxidase-positive colonies will turn pink, then maroon, and finally black.

Note: *N. gonorrhoeae* treated with oxidase reagent for up to 10 minutes still fluoresce strongly by the FA technique.[5]

2. Prepare a light suspension of the oxidase-positive organism to be tested in distilled or deionized water.

To insure a viable culture for subsequent biochemical studies a portion of the colony should be picked as soon as it becomes pink and transferred to a medium such as Bacto Cystine Tryptic Agar.

3. Smear a loopful of the suspension on a microscope slide using a bacteriological loop. Simultaneously prepare a smear for the Gram stain. The Gram stain may be performed during any convenient interval in the FA technique.

4. Allow the smears to air dry.
5. Cover the dried smear(s) with the working dilution of Bacto FA N. Gonorrhoeae (see Reagent Section).

Note: Include a N. gonorrhoeae positive control and an E. cloacae nonspecific staining control in each "run."

6. Incubate at room temperature for 5 minutes in a moisture chamber. Such a chamber may be conveniently prepared for a few slides by affixing a piece of moist filter paper to the inside surface of a Petri dish and inverting the dish over the slides.
7. Rinse **gently** in running distilled or deionized water.
8. Allow the slides to air dry.
9. Add a very small drop of Bacto FA Mounting Fluid pH 9 on the center of each smear and cover with a cover glass; care being taken not to include entrapped air bubbles.
10. Observe for fluorescence through a suitable fluorescent microscope set-up.

RESULTS

The section of the control slide containing E. cloacae should be negative. If not the test is invalid.

The section of the control slide containing N. gonorrhoeae should be strongly positive.

The reactivity of the specimen smear is determined by comparing its reactivity with the reactivity of the positive and negative controls. Record as positive or negative.

An oxidase-positive, gram-negative cocci exhibiting strong fluorescence in the direct FA technique for GC is strong evidence that the organism is N. gonorrhoeae.

LIMITATIONS OF THE PROCEDURE

Apart from the general limitations of the FA procedure mentioned in the introduction, it is well to remember that Thayer Martin medium or its modifications will support the growth of N. meningitidis. While it is unlikely that both N. gonorrhoeae and N. meningitidis would be isolated from the same specimen, it must be considered a possibility. In these circumstances a positive reaction with the conjugate, particularly of low grade fluorescence, could represent an unabsorbable cross reaction with N. meningitidis.

If there is any doubt about the specificity of the reaction in the direct FA test a carbohydrate utilization pattern should be established for the isolate to further corroborate its identity as N. gonorrhoeae.

REFERENCES

1. Proc. Soc. Exp. Biol. and Med., 101:322, 1959.
2. Pub. Health Lab., Vol. 28, No. 3, 1970.
3. Pub. Health Reports, 85:733, 1970.
4. CDC Release, March 27, 1970.
5. Pub. Health Report, Vol. 88, No. 4, 1968.

PACKAGING

Bacto FA N. Gonorrhoeae	5 ml	2361-56-2
Bacto FA Cloacae, Desiccated	1 ml	3116-50-4
Bacto FA Buffer, Dried	6 × 10 g	2314-33-8
	100 g	2314-15-0
Bacto FA Mounting Fluid pH 9	6 × 5 ml	3340-57-5
Bacto Differentiation Disks Oxidase	1 vial	1633-35-2
	6 vials	1633-33-4
Bacto p-Aminodimethylaniline Oxalate	25 g	0329-13-9
	100 g	0329-15-7

SEROLOGICAL IDENTIFICATION OF *NEISSERIA MENINGITIDIS*

INTENDED USE

Bacto Neisseria Meningitidis Antisera are used in detecting and serogrouping meningococci by the slide agglutination technique.

SUMMARY AND EXPLANATION OF THE TEST

In 1884, Marchiafava and Celli[1] stained gram-negative diplococci present in meningeal exudates from patients suffering from meningitis. Unfortunately, these organisms were not cultured. Three years later, Weichselbaum[2] isolated *Neisseria meningitidis* from several cases of epidemic cerebrospinal meningitis, establishing this organism as an etiologic agent of the disease.

The medical significance of *N. meningitidis* lies not only in its role in cases of meningitis, but also in its role in bacteremia, pneumonia and urethritis. *N. meningitidis* has also been isolated from a number of normal, healthy individuals, establishing a significantly high carrier rate for this disease.

Meningococci are divided into serological groups depending on the presence of either cell-wall or capsular antigens. The recognized serogroups are A, B, C, D, X, Y, Z, Z' and W135. The subcommittee on *Neisseriaceae* of the International Committee on Bacteriological Nomenclature accepted serogroups A, B, C and D.[3] Slaterus described serogroups X, Y, Z and Z'.[4,5]

More recently, Evans et al. described three additional serogroups which they labeled Bo, 29-E and 135.[6] All were isolated from cases of meningococcal infections. Bo serogroup was found to be identical to Slaterus Y strain and 29-E to be that of the Z' strain. The third group, W135, was found to be distinct from all others yet described.

Insofar as the relative importance of the various serogroups of *N. meningitidis* are concerned, investigators found that over a two year period of testing approximately 1,000 cultures, the most frequently isolated serogroups were B, followed by C, Y, X, Z, A and D, in that order. Groups A, B and C appeared to account for the greatest case and carrier rates, while group D was very rarely isolated. The actual existence of a specific group D has been questioned. Some investigators believe that the organisms classified as group D meningococci are actually rough strains of serological group C.[8] Groups X and Y have been isolated from individuals in both the case and carrier states, while group Z has only been isolated from carriers.

Craven et al.,[10] in their study of the prevalence of *N. meningitidis* serotypes in a disease-free military population, found that 13.8% of the cultures isolated and serologically classified were group Y, 13.1% were group W135, 12.4% were Z', 12.0% were group B, 4% were group X and 2.5% were group C. There were no isolates from groups A, D or Z.

PRINCIPLES OF THE PROCEDURE

Serologic grouping of *N. meningitidis* involves mixing a bacteriological loopful of a biochemically characterized culture (the antigen) with a drop of hyperimmune serum (the antibody). If the test culture contains antigens corresponding to the antiserum, a rapid clumping (agglutination) to at least a 3+ reaction will occur. Such a reaction in an individual serum identifies the serogroup of the organism. This method, the slide agglutination technique, is the method of choice for serogrouping isolates of *N. meningitidis*.

Some laboratories perform a capsular swelling (Quellung) reaction for serogrouping *N. meningitidis*. If so, the technician must be aware that capsules have not been demonstrated in strains of serogroup B organisms, so this serogroup should be defined serologically by the slide agglutination test rather than by the Quellung reaction. Further, since Bacto Neisseria Meningitidis Antisera are not assayed using the Quellung reaction, the technician is advised to examine each lot of antiserum with known positive and negative control strains when these sera are used in the Quellung reaction.

A positive serological test result combined with a biochemically defined culture of *N. meningitidis* confirms the presence of *N. meningitidis* in the specimen, permitting the prompt and specific therapeutic action necessary in an explosive disease state. It also provides invaluable data for epidemiological studies.

REAGENTS
Bacto Neisseria Meningitidis Antisera are stable, desiccated, absorbed polyvalent or group-specific antisera prepared in rabbits.

The various antisera detect the following antigenic groups:

Bacto Neisseria Meningitidis Antiserum	Antigenic Groups Detected
Poly	A, B, C, D
Poly 2	X, Y, Z*
Z′	Z, Z′*
W135	W135*
A	A*
B	B*
C	C*
D	D
X	X*
Y	Y*
Z	Z*

*Not absorbed for group D, which is rarely isolated.

When properly rehydrated, each antiserum contains approximately 1:10,000 Merthiolate®, sufficient to maintain a bacteriostatic condition at a storage of 2 – 8°C.

When used according to the suggested PROCEDURE (0.05 ml per test), each 1 ml vial is sufficient to perform approximately 20 tests.

Rehydration
To rehydrate Bacto Neisseria Meningitidis Antisera, add 1 ml distilled or deionized water and rotate the vial gently to completely dissolve the contents.

If desired, the antisera may be rehydrated with 50% glycerol solution having a pH 7.0 – 7.4 before use.

Storage
Store lyophilized Bacto Neisseria Meningitidis Antisera at 2 – 8°C.

Store rehydrated antisera at 2 – 8°C.

Prolonged exposure to room temperature or repeated freezing and thawing are detrimental to antisera.

Expiration Date
Lyophilized Bacto Neisseria Meningitidis Antisera are stable through the expiration date on the label when stored as directed.

Rehydrated antisera are stable for one year after rehydration, not exceeding the expiration date on the label, when stored as directed.

Discard any serum that is cloudy or has a precipitate after rehydration or storage unless it can be clarified by centrifugation and demonstrates proper reactivity with validated positive and negative control cultures.

SPECIMEN PREPARATION
1. Determine that the test culture has the following characteristics of *Neisseria meningitidis*.

Morphology:	smooth, butyrous, nonpigmented colony
Gram stain:	gram-negative diplococcus
Oxidase reaction:	+
ONPG reaction:	+
Nitrate reduction test:	−
Carbohydrate degradation:	
Glucose:	+
Maltose:	+
Sucrose:	−
Fructose:	−
Lactose:	−

2. Subculture the test organism on an enrichment medium (blood agar, chocolate agar) and incubate for 18 – 24 hours at 35 – 37°C in a moist, 5 – 10% CO_2 atmosphere.
3. To determine that the culture will not agglutinate spontaneously, transfer a loopful of growth from the test culture on blood agar or chocolate agar to a drop of saline or, preferably, normal rabbit serum on a clean slide and emulsify the organism. Rock the slide for one minute, then observe for spontaneous agglutination which results from a rough culture.

If agglutination occurs:	Subculture the organism successively on blood agar or chocolate agar until the culture is smooth (no longer agglutinates in this procedure).
If no agglutination occurs:	Proceed with the MACROSCOPIC SLIDE AGGLUTINATION TEST.

PROCEDURE
Materials Provided
Bacto Neisseria Meningitidis Antisera Poly, Poly 2, A, B, C, D, W135, X, Y, Z, Z'

Materials Required but not Provided
Glass plate	Burner
Marking pencil	Autoclave
Bacteriological loop	Fluorescent light source

MACROSCOPIC SLIDE AGGLUTINATION TEST
Perform the test initially with Bacto Neisseria Meningitidis Antisera Poly, Poly 2, Group Z' and Group W135. If the reaction is positive using the Poly, Poly 2 or Z' antiserum

and the identity of the particular group is desired, perform a second series of tests using monospecific sera.

Bring all test materials to room temperature before using.
1. Place a drop of rehydrated Bacto Neisseria Meningitidis Antiserum on a clean slide.
2. Transfer a loopful of growth of the test specimen (preferably from more than one colony) to the drop of antiserum and emulsify it.
3. Rock the slide for one minute to ensure complete contact between the antiserum and the microorganisms.
4. Record the reaction as follows:

4+	agglutination	= all of the cells agglutinate
3+	agglutination	= 75% of the cells agglutinate
2+	agglutination	= 50% of the cells agglutinate
1+	agglutination	= 25% of the cells
±	agglutination	= less than 25% of the cells agglutinate
−	agglutination	= none of the cells agglutinate

QUALITY CONTROL
Use known positive and negative control cultures in parallel with the test culture to ascertain the validity of test results.

RESULTS
A 3+ reaction or greater is considered a positive result.

LIMITATIONS OF THE PROCEDURE
1. Serological identification of Neisseria meningitidis is presumptive in the presence of typical results to biochemical studies. Serological identification, alone, is insufficient for identification of an etiological agent.
2. Organisms unrelated to Neisseria (yet capable of causing meningitis) and other species of Neisseria can cross react with meningococcal antisera. Cultural isolation must preceed serological examination.
3. Groups A and C meningococci may cross react due to the presence of common capsular polysaccharides.
4. Cultures of group Z' meningococci will not agglutinate group Z antiserum. Cultures of group Z meningococci will agglutinate group Z' antiserum.
5. Bacto Neisseria Meningitidis Antisera are designed for and tested in the slide agglutination test, only, using undiluted cultures taken from a blood agar or chocolate agar plate. Variant procedures must include known control cultures to ensure proper homologous and heterologous reactions under the variant test conditions.

REFERENCES
1. Gass. d. osp., Milano, 8, 1884.
2. Fortschr. d. Med., 5:573, 1887.
3. Internat. Bull. Bact. Nomen. Tax., 4:95 – 105, 1954.
4. Antonie von Leeuwenhoek J. Microbiol. Serol., 27:304 – 315, 1961.
5. ibid., 29:265 – 271, 1963.
6. Am. J. Epidemiol., 87:643 – 646, 1968.
7. J. Bacteriol., 95:1 – 4, 1968.
8. NAMRU #4 Annual Progress Report to the Commission on Acute Respiratory Disease Armed Forces Epidemiological Board, p. 20 – 28, October 1982.
9. Diagnostic Procedures, APHA, 1970.
10. J. Clin. Micro., 10:302 – 307, 1979.
11. J. Pediat., 63:76 – 83, 1963.
12. CDC, Reference Neisseria Meningitidis Antisera, August 1974.
13. J. Bacteriol., 96:563, 1968.

PACKAGING

Bacto Neisseria Meningitidis Antiserum Group A	1 ml	2228-50-1
Bacto Neisseria Meningitidis Antiserum Group B	1 ml	2229-50-0
Bacto Neisseria Meningitidis Antiserum Group C	1 ml	2230-50-7
Bacto Neisseria Meningitidis Antiserum Group D	1 ml	2231-50-6
Bacto Neisseria Meningitidis Antiserum Group X	1 ml	2880-50-0
Bacto Neisseria Meningitidis Antiserum Group Y	1 ml	2881-50-9
Bacto Neisseria Meningitidis Antiserum Group Z	1 ml	2891-50-7
Bacto Neisseria Meningitidis Antiserum Group Z'	1 ml	2252-50-0
Bacto Neisseria Meningitidis Antiserum Group W135	1 ml	2253-50-9
Bacto Neisseria Meningitidis Antiserum Poly		
Contains Groups A, B, C, D	1 ml	2232-50-5
Bacto Neisseria Meningitidis Antiserum Poly 2		
Contains Groups X, Y, Z	1 ml	2910-50-4
Bacto Neisseria Meningitidis Antisera Set A – D		
Contains 1 vial each of Groups A, B, C, D and Poly	5 × 1 ml	2233-32-7
Bacto Neisseria Meningitidis Antisera Set X, Y, Z		
Contains 1 vial each of Groups X, Y, Z and Poly 2	4 × 1 ml	2912-32-5

NEOPEPTONE, DIFCO

Neopeptone, Difco is an enzymatic protein digest especially adapted for the preparation of media for culturing bacteria usually considered to be difficult to cultivate *in vitro.* Neopeptone, Difco has been shown to be particularly well suited to the growth requirements of many fastidious bacteria.

Bacto Todd Hewitt Broth prepared with Neopeptone, Difco is excellent for growing Group A streptococci for the production of type-specific m substance.

HISTORY
Refer to *Difco Manual,* 9th Edition, p. 263 – 264.

PACKAGING

Neopeptone, Difco	1 lb (454 g)	0119-01-7
	1/4 lb (114 g)	0119-02-6
	10 kg	0119-08-0

BACTO NEOPEPTONE INFUSION AGAR

INTENDED USE
Bacto Neopeptone Infusion Agar is used with or without enrichment for cultivating fastidious microorganisms.

FORMULA

BACTO NEOPEPTONE INFUSION AGAR
DEHYDRATED
Ingredients per liter

Beef Heart, Infusion from . . . . .500 g
Neopeptone, Difco . . . . 20 g
Sodium Chloride . . . . 5 g
Bacto Agar . . . . 20 g

Final pH 7.4 ± 0.2 at 25°C.

One pound will make 8.2 liters of basal medium.

METHOD OF PREPARATION
1. Suspend 55 g in 1 liter distilled or deionized water and heat to boiling to dissolve completely.
2. Sterilize in the autoclave for 15 minutes at 15 lbs pressure (121°C).
3. Allow medium to cool to 45 – 50°C. Add 5% sterile defibrinated sheep blood.
4. Dispense into Petri dishes or tubes as desired.

STORAGE
Bacto Neopeptone Infusion Agar	Below 30°C
Prepared plates or tubes	2 – 8°C

QUALITY CONTROL
Identity Specifications

Dehydrated powder:	tan, homogeneous, free-flowing
Reaction of 5.5% solution:	pH 7.4 ± 0.2 at 25°C
Plates prepared with 5% sheep blood:	cherry red, opaque, no hemolysis

Typical Cultural Response on Bacto Neopeptone Infusion Agar Prepared with 5% Sheep Blood After 40 – 48 Hours at 35°C

Organism	Growth	Hemolysis
Neisseria meningitidis ATCC® 13090	excellent	none
Staphylococcus aureus ATCC® 25923	excellent	beta
Staphylococcus epidermidis ATCC® 12228	excellent	gamma
Streptococcus pneumoniae ATCC® 6303	excellent	alpha
Streptococcus pyogenes ATCC® 19615	excellent	beta

PACKAGING
Bacto Neopeptone Infusion Agar	1 lb (454 g)	0493-01-3

BACTO NEUROSPORA MINIMAL MEDIUM

INTENDED USE
Bacto Neurospora Minimal Medium is recommended for the detection of *Neurospora* mutants.

HISTORY
Bacto Neurospora Minimal Medium, prepared according to the formula of Beadle,[1] contains the essential nutrients for the growth of *Neurospora* except for vitamins and amino acids which may be needed by mutant strains.

FORMULA

BACTO NEUROSPORA MINIMAL MEDIUM
DEHYDRATED
Ingredients per liter

Biotin	0.005 mg	Calcium Chloride	0.1 g
Sodium Molybdate	0.05 mg	Sodium Chloride	0.1 g
Boric Acid	0.057 mg	Magnesium Sulfate	0.5 g
Manganese Sulfate	0.063 mg	Monopotassium Phosphate	1 g
Copper Sulfate	0.39 mg	Sodium Nitrate	1 g
Ferrous Sulfate	0.54 mg	Ammonium Tartrate	5 g
Zinc Sulfate	5.5 mg	Sucrose	20 g

Final pH 5.7 ± 0.2 at 25°C.

One pound will make 16.3 liters of medium.

METHOD OF PREPARATION
1. Suspend 27.7 g in 1 liter distilled or deionized water.
2. Heat to boiling for 1 – 2 minutes.
3. Dispense into tubes or flasks and sterilize in the autoclave for 15 minutes at 15 lbs pressure (121°C).

STORAGE
Bacto Neurospora Minimal Medium	Below 30°C
Prepared medium	15 – 30°C

QUALITY CONTROL
Identity Specifications

Dehydrated powder:	white, homogeneous, free-flowing
Reaction of 2.77% solution:	pH 5.7 ± 0.2 at 25°C
Prepared medium:	colorless, clear, may have a slight precipitate

Typical Cultural Response on Bacto Neurospora Minimal Medium
After 3 – 5 Days at 25 – 30°C

Organism	Growth
Neurospora crassa ATCC® 9277	good
Neurospora crassa ATCC® 9278	good
Neurospora sitophilia ATCC® 9276	good

REFERENCE
1. Physiological Reviews, 25:643 – 663, 1945.

PACKAGING
Bacto Neurospora Minimal Medium	1 lb	0817-01-2

BACTO NEUTRALIZING BUFFER

INTENDED USE
Bacto Neutralizing Buffer, a modification of the Standard Methods buffered distilled water, is designed for use in the sterility tests for dairy equipment as specified in *Standard Methods for the Examination of Dairy Products*[1] by the swab contact method and for the bacteriological examination of food utensils according to the method of the Sub-committee on Food Utensil Sanitation,[2] American Public Health Association.

PRINCIPLES

Bacto Neutralizing Buffer has the ability to inactivate the bactericidal and bacteriostatic effect of chlorine as well as quaternary ammonium compounds, as shown by laboratory and field tests using mixed flora as well as a large variety of pure cultures of organisms, including a *Sarcina* (now *Micrococcus*) particularly sensitive to quaternary ammonium compounds.[8] Bacto Neutralizing Buffer is used in conjunction with Bacto Tryptone Glucose Extract Agar for the performance of such viability, sanitation or sterility tests.

Bacto Neutralizing Buffer is not toxic for microorganisms. Plate counts made at intervals up to 5 hours on a suspension of organisms in the Bacto Neutralizing Buffer solution showed no reduction in numbers. This permits transfer of rinse water or swabs to the laboratory without danger of loss of viable organisms. Bacto Neutralizing Buffer may be safely used in a concentration of 10 times the single strength in procedures requiring multiple strength solutions without danger of toxicity.

In using Bacto Neutralizing Buffer in sterility tests for dairy equipment, it is suggested that the procedure detailed in *Standard Methods for the Examination of Dairy Products*[1] be closely followed.

Bacto Neutralizing Buffer is also recommended in swab test procedures or visual sanitation tests. Under these conditions it is often desired to examine more than one individual utensil as a unit. When so employed, the neutralizing buffer is distributed in screw cap tubes or screw cap swab bottles (cotton stoppers are not suitable) so that after autoclaving there will be 1 ml of neutralizing buffer for each surface or utensil to be examined. If four similar utensils are to be examined as a unit, 4 ml should be present in the tube. For the examination of food utensils the following procedure may be employed:
1. Dip a sterile swab into the tube of sterile neutralizing buffer.
2. Remove from the solution and squeeze the swab against the inside of the tube so as to remove the excess solution leaving the swab moist but not wet. One swab may be used for each group of four or more similar utensils.
3. Rub the swab slowly and firmly three times over the significant surfaces of the utensil to be examined, reversing the direction of the swab each time.

The significant surfaces of utensils are generally considered as the upper half inch of the inner and outer rims of glasses and cups and the entire inner and outer surfaces of the bowls of the spoons. In the examination of forks and knives, the inner and outer tines of the fork and both sides of the blade of the knife should be swabbed. Plates and bowls should be swabbed on the inner and outer surfaces.

After swabbing each utensil, return the swab to the neutralizing buffer solution, rotate well and press free from excess solution before swabbing the next utensil in the group with the moist swab.

4. When the last utensil, generally four or more, has been swabbed, replace the swab in the neutralizing buffer and shake vigorously. If separate swabs were used, break off each swab in the container under aseptic conditions.
5. Keep neutralizing buffer containing swabs at $0 - 6°C$ until plated in the laboratory.
6. For procedures requiring a plate count, break the stick of the swab just above the cotton with sterile forceps, if this has not already been done as indicated above. Shake the swab in the neutralizing buffer thoroughly to disintegrate the cotton swabs. Plate 1 ml or desired quantity using Bacto Tryptone Glucose Extract Agar as the plating medium. Incubate at 32 or 35°C for 48 hours before making the count.

If desired, a visual sanitation test may be run by following the above directions through Step 4, except that in Step 4 the swab is not broken. Following the swabbing of the last utensil, it is immersed in the neutralizing buffer, shaken vigorously, squeezed to remove excess moisture and then smeared directly on the surface of a slant of Bacto Tryptone Glucose Extract Agar. Inoculation may be made by moving the moist swab across the surface of the slant from top to bottom horizontally and then vertically. The swab is rotated while being streaked across the surface of the medium. The inoculated medium is then incubated as desired.

FORMULA

BACTO NEUTRALIZING BUFFER
DEHYDRATED

Ingredients per liter

Monopotassium Phosphate . . . . 0.0425 g
Sodium Thiosulfate . . . . 0.16 g
Aryl Sulfonate Complex . . . . 5 g
Sodium Hydroxide . . . . 0.008 g

Final pH 7.2 ± 0.2 at 25°C.

One hundred grams will make 19.2 liters of buffer.

METHOD OF PREPARATION
1. Rehydrate by dissolving 5.2 g in 1 liter distilled or deionized water.
2. Dispense into screw cap containers and sterilize in the autoclave for 15 minutes at 15 lbs pressure (121°C).

STORAGE
Bacto Neutralizing Buffer   Below 30°C
Prepared buffer   15 – 30°C

QUALITY CONTROL
Identity Specifications

Dehydrated powder: beige, homogeneous, free-flowing
Reaction of 0.52% solution: pH 7.2 ± 0.2 at 25°C
Prepared medium: almost colorless, clear

Typical Cultural Response on Bacto Tryptone Glucose Extract Agar After Bacto Neutralizing Buffer for 15 Minutes at 25 – 30°C

Organism	Growth	Growth w/disinfectant (Roccal®)
Staphylococcus aureus ATCC® 6538P	good to excellent	good to excellent

REFERENCES
1. Standard Methods for the Examination of Dairy Products, 9th Ed., 216:1948.
2. Ann. Year Book 1947 – 48, Suppl. Am. J. Pub. Health, 5:68:1948.
3. J. Milk Food Tech., 12:224:1949.

PACKAGING
Bacto Neutralizing Buffer   100 g   0362-15-5

BACTO NITRATE AGAR
BACTO NITRATE BROTH

INTENDED USE
Bacto Nitrate Agar and Bacto Nitrate Broth are recommended for the determination of nitrate reduction by bacteria.

HISTORY/PRINCIPLES
Bacto Nitrate media are prepared according to the formulae published in *Pure Culture Study of Bacteria*[1] of the Society of American Bacteriologists. Nitrate reduction by microorganisms is a valuable criterion for differentiating and identifying various types of bacteria. Certain bacteria reduce nitrates to nitrites only, while others are capable of further reduction to free nitrogen or even ammonia.

Nitrites are colorless but in an acid environment, they will react with alpha-naphthylamine generating a pink or red color. When nitrate positive organisms reduce the nitrates to nitrites, a pink color develops when the reagents are added to the broth. However, nitrate negative organisms, unable to reduce nitrates, yield no color after the reagents are added. If an organism grows rapidly and reduces nitrate actively, it is suggested that the test for nitrite be performed after an early incubation period since the reduction may be carried beyond nitrite to nitrogen.

FORMULAE

BACTO NITRATE BROTH
DEHYDRATED

Ingredients per liter

Bacto Beef Extract . . . . 3 g
Bacto Peptone . . . . 5 g
Potassium Nitrate . . . . 1 g

Final pH 7.0 ± 0.2 at 25°C.

One pound will make 50 liters
of medium.
Rehydrate with 9 g/liter.

BACTO NITRATE AGAR
DEHYDRATED

Ingredients per liter

Bacto Beef Extract . . . . 3 g
Bacto Peptone . . . . 5 g
Potassium Nitrate . . . . 1 g
Bacto Agar . . . . 12 g

Final pH 6.8 ± 0.2 at 25°C.

One pound will make 21 liters
of medium.
Rehydrate with 21 g/liter.

PROCEDURE
1. To rehydrate, suspend appropriate amount in 1 liter distilled or deionized water. Heat agar medium to boiling to dissolve completely.
2. Dispense into tubes and sterilize in the autoclave for 15 minutes at 15 lbs pressure (121°C).
3. Tubes of sterile slanted medium are inoculated by streaking over the surface of the slant and stabbing the butt.
4. Incubate at 37°C.
5. Agar slants are examined on various days for gas production, indicated by splitting of the agar.
6. Test for nitrates with sulfanilic acid and alpha-naphthylamine reagent solutions.
7. Prepare sulfanilic acid reagent by dissolving 8 g of sulfanilic acid in 1 liter 5 N acetic acid. The alpha-naphthylamine reagent consists of 5 g of alpha-naphthylamine dissolved in 1 liter 5 N acetic acid.

8. Put a few drops of each reagent into tubes to be tested. A distinct red or pink color indicates the presence of nitrite reduced from original nitrate. The test should always be controlled by comparing with an uninoculated tube of the medium which has been kept under the same conditions as the inoculated tubes. The evolution of gas in nitrate medium containing no sugar or fermentable substance is a definite indication of reduction to free nitrogen.

STORAGE

Bacto Nitrate Agar	Below 30°C
Bacto Nitrate Broth	Below 30°C
Prepared medium	2 – 8°C

QUALITY CONTROL

Identity Specifications

	Bacto Nitrate Agar	Bacto Nitrate Broth
Dehydrated powder:	tan, homogeneous, free-flowing	tan, homogeneous, free-flowing
Reaction of solution at 25°C:	2.1% pH 6.8 ± 0.2	0.9% pH 7.0 ± 0.2
Prepared medium:	light amber, slightly opalescent	light to medium amber, clear

Typical Cultural Response on/in Bacto Nitrate Agar or Broth After 18 – 24 Hours at 35 – 37°C

Organism	Recovery	Nitrate Reduction
Acinetobacter calcoaceticus ATCC® 19606	good	–
Enterobacter aerogenes ATCC® 13048	good	+
Escherichia coli ATCC® 25922	good	+
Salmonella typhimurium ATCC® 14028	good	+

REFERENCE

1. Pure Culture Study of Bacteria, 12:Leaflet II: 8:1944.

PACKAGING

Bacto Nitrate Agar	1 lb (454 g)	0106-01-2
Bacto Nitrate Broth	1 lb (454 g)	0268-01-6

BACTO NITRITE TEST STRIPS

INTENDED USE

Bacto Nitrite Test Strips are used for differentiating species of microorganisms, particularly of mycobacteria, based on their ability to reduce nitrates to nitrites. Test results are equivalent to those obtained using the classical tube test or culturing the test specimen in nitrite broth.

HISTORY

Bacto Nitrite Test Strips are a modification of those described by Kilburn.[1] Yu, Birk and Washington[2] used nitrite strips in parallel with nitrite broth to determine nitrate reduction by isolates from clinical specimens and obtained comparable results. Quigley and Elston[3] evaluated nitrite strips for differentiating mycobacteria and found them to be as sensitive as the broth method.

PRINCIPLES
When the sodium nitrate portion of the test strip is introduced into an aqueous suspension of the test organism, the nitrate will be reduced during a two hour incubation period if the test organism is reductase-positive. After incubation, the remainder of the strip is wetted with the suspension, releasing acid into the suspension to complete the reaction. If nitrate reduction occurred, the upper portion of the strip will become blue, indicating a positive test.

FORMULA
Bacto Nitrite Test Strips are absorbent paper strips impregnated with an area of sodium nitrate, an area of citric acid, an area of potassium iodide and starch, and imprinted with an arrow for proper positioning of the strip in the test tube.

STORAGE
Store Bacto Nitrite Test Strips at 2 – 8°C in the dark.

EXPIRATION DATE
A brown discoloration at the distal end of the strip indicates deterioration of the reagent and such strips should be discarded.

SPECIMEN PREPARATION
Obtain the test specimen from a pure culture growing on the following appropriate media:
Mycobacteria — Lowenstein medium or other coagulated egg medium
Enteric organisms — triple sugar iron agar
Other organisms — tryptic soy agar

Using a sterile 1 ml pipette or spade, remove two spades of growth from the culture and add it to 0.5 ml distilled water in a clean 20 × 113 mm screw cap tube. Break up the culture with the pipette or spade. A **dense** suspension should result.

PROCEDURE
1. Using sterile forceps, remove one Bacto Nitrite Test Strip from the vial and insert, arrow end first, into the tube containing the suspension of the test specimen, maintaining the tube in a vertical position. Recap tube.
2. Incubate at 37°C for 2 hours. Shake tube gently at end of first hour but do not tip tube.
3. Shake tube gently at end of second hour. Carefully tilt tube back and forth 6 times to wet the entire strip.
4. Slant tube to cover the entire strip with liquid and retain in this position for 10 minutes at room temperature.

QUALITY CONTROL
Use known positive and negative control cultures in parallel with the test culture to ascertain the validity of test results. An uninoculated tube containing reagents only may be substituted for the negative control.

RESULTS
A color change to light or dark blue at the top of the strip is a positive test for nitrate reduction.

REFERENCES
1. Personal Communication.
2. Tech. Bull. Regist. Med. Tech., 39:257, 1969.
3. Journal of South Carolina ASMT, 14:10, 1970.

PACKAGING

Bacto Nitrite Test Strips   25 strips   3183-30-7

BACTO NUTRIENT AGAR

INTENDED USE

Bacto Nutrient Agar is a general purpose medium used for the examination of water and dairy products according to *Standard Methods for the Examination of Water and Wastewater*[1] and *Standard Methods for the Examination of Dairy Products.*[2] It is also used for the cultivation of the majority of the less fastidious microorganisms.

HISTORY

Infusions of meat were first generally employed together with peptone as nutriments in culture media. Later it was found that for many routine procedures beef extract gave fully as good results and had the decided advantages of greater ease of preparation, greater uniformity, and economy. A simple medium composed of beef extract, peptone, and agar has been one of the most generally used media in bacteriological procedures. It is used for the ordinary routine examinations of water, sewage, and food products; for the carrying of stock cultures; for the preliminary cultivation of samples submitted for bacteriological examination; and for isolating organisms in pure culture.

Bacto Nutrient Agar was originally prepared to duplicate an extract agar of approved and standard formula. The American Public Health Association in its earliest reports on methods for water analysis emphasized the necessity of the universal use of a standard medium, and since the third edition of *Standard Methods of Water Analysis* in 1917[3] has recommended the use of beef extract rather than infusion of meat in the preparation of nutrient agar. In the fifth edition of "Standard Methods of Water Analysis," 1923,[4] and the fourth edition of *Standard Methods of Milk Analysis*, 1923,[5] the use of dehydrated media of *Standard Methods* composition had been permitted for the bacteriological examination of water and milk. In the fifth edition of *Standard Methods of Milk Analysis*, 1927,[6] the use of Bacto dehydrated media was approved as being on a par with laboratory-made media for the bacteriological plate count of milk, and in this connection the work of Norton and Seymour[7] was cited. Bacto Nutrient Agar is prepared in accordance with the formula specified in *Standard Methods of Water Analysis.*[8]

FORMULA

BACTO NUTRIENT AGAR
DEHYDRATED

Ingredients per liter

Bacto Beef Extract . . . . 3 g
Bacto Peptone . . . . 5 g
Bacto Agar . . . . 15 g

Final pH 6.8 ± 0.2 at 25°C.

One pound will make 19.7 liters of final medium.

METHOD OF PREPARATION
1. Suspend 23 grams in 1 liter distilled or deionized water and heat to boiling to dissolve completely.
2. Sterilize in the autoclave for 15 minutes at 15 pounds pressure (121°C).
3. Dispense as desired.

STORAGE
Bacto Nutrient Agar — Below 30°C
Prepared medium — 15 – 30°C

QUALITY CONTROL
Identity Specifications
Dehydrated powder: beige, homogeneous, free-flowing
Reaction of 2.3% solution: 6.8 ± 0.2 at 25°C
Prepared medium: light amber, clear to slightly opalescent, no significant precipitate

Typical Cultural Response on Bacto Nutrient Agar
After 18 – 24 Hours at 35°C

Organism	Recovery
Corynebacterium diphtheriae type mitis	good to excellent
Escherichia coli ATCC® 25922	good to excellent
Streptococcus pneumoniae ATCC® 6303	good to excellent

REFERENCES
1. American Public Health Association, 1980, Standard Methods for the Examination of Water and Wastewater, 15th Ed. American Public Health Association, Inc., Washington, D.C.
2. American Public Health Association, 1978, Standard Methods for the Examination of Dairy Products, 14th Ed. American Public Health Association, Inc., Washington, D.C.
3. American Public Health Association, 1917, Standard Methods of Water Analysis, 3rd Ed. American Public Health Association, Inc., Washington, D.C.
4. American Public Health Association, 1923, Standard Methods of Water Analysis, 5th Ed. American Public Health Association, Inc., Washington, D.C.
5. American Public Health Association, 1923, Standard Methods of Milk Analysis, 4th Ed. American Public Health Association, Inc., Washington, D.C.
6. American Public Health Association, 1927, Standard Methods of Milk Analysis, 5th Ed. American Public Health Association, Inc., Washington, D.C.
7. Norton and Seymour, 1926, American Journal of Public Health 16:35, American Public Health Association, Inc., Washington, D.C.
8. American Public Health Association, 1936, Standard Methods of Water Analysis. 8th Ed. American Public Health Association, Inc., Washington, D.C.

PACKAGING
Bacto Nutrient Agar

1 lb	(454 g)	0001-01-8
1/4 lb	(114 g)	0001-02-7
5 lb	(2.27 kg)	0001-05-4
12 tubes		0001-34-9
144 tubes		0001-37-6
6 × 100 ml		0001-73-1

BACTO NUTRIENT AGAR 1.5%

INTENDED USE
Bacto Nutrient Agar 1.5% is a slightly alkaline general purpose medium. Since this medium contains 0.8% sodium chloride, it can be used as a base for enrichment with blood, ascitic fluid or other similar substances for cultivating fastidious microorganisms.

FORMULA

BACTO NUTRIENT AGAR 1.5%
DEHYDRATED

Ingredients per liter

Bacto Beef Extract .... 3 g
Bacto Peptone .... 5 g
Sodium Chloride .... 8 g
Bacto Agar .... 15 g

Final pH 7.3 ± 0.1 at 25°C.

One pound will make 14.6 liters of final medium.

METHOD OF PREPARATION

1. Suspend 31 g in 1 liter distilled or deionized water and heat to boiling to dissolve completely.
2. Sterilize in the autoclave for 15 minutes at 15 lbs pressure (121°C).
3. To prepare an enriched medium, cool the sterile base to 45 – 50°C, aseptically add the desired enrichment and mix thoroughly.
4. Dispense as desired.

STORAGE

Bacto Nutrient Agar 1.5%   Below 30°C
Prepared medium   2 – 8°C

QUALITY CONTROL

Identity Specifications

Dehydrated powder: light tan, homogeneous, free-flowing
Reaction of 3.1% solution: pH 7.3 ± 0.1 at 25°C
Prepared medium: light to medium amber, slightly opalescent

Typical Cultural Response on Bacto Nutrient Agar 1.5%
After 18 – 48 Hours at 35°C

| Organism | Growth | | Hemolysis |
	Plain	With 5% Sheep Blood	
Neisseria meningitidis ATCC® 13090	good	good to excellent	none
Staphylococcus aureus ATCC® 25923	good	good to excellent	beta
Streptococcus pneumoniae ATCC® 6303	good	good to excellent	alpha
Streptococcus pyogenes ATCC® 19615	good	good to excellent	beta

PACKAGING

| Bacto Nutrient Agar 1.5% | 1 lb (454 g) | 0069-01-7 |
| | 1/4 lb (114 g) | 0069-02-6 |

BACTO NUTRIENT AGAR pH 6.0

INTENDED USE

Bacto Nutrient Agar pH 6.0 is used for microbiological procedures requiring a slightly acid nutrient agar. For a complete discussion refer to Bacto Nutrient Agar.

FORMULA

BACTO NUTRIENT AGAR pH 6.0
DEHYDRATED
Ingredients per liter

Bacto Beef Extract . . . . 3 g
Bacto Peptone . . . . 5 g
Bacto Agar . . . . 15 g

Final pH 6.0 ± 0.2 at 25°C.

One pound will make 19.7 liters of medium.

METHOD OF PREPARATION
1. Suspend 23 g in 1 liter distilled or deionized water and heat to boiling to dissolve completely.
2. Sterilize in the autoclave for 15 minutes at 15 lbs pressure (121°C).

STORAGE
Bacto Nutrient Agar pH 6.0   Below 30°C
Prepared media   2 – 8°C

QUALITY CONTROL
Identity Specifications

Dehydrated powder: tan, homogeneous, free-flowing
Reaction of 2.3% solution: pH 6.0 ± 0.2 at 25°C
Prepared medium: light to medium amber, very slightly opalescent

Typical Cultural Response on Bacto Nutrient Agar pH 6.0
After 18 – 48 Hours at 35°C

Organism	Growth
Escherichia coli ATCC® 25922	good to excellent
Staphylococcus aureus ATCC® 25923	good to excellent
Staphylococcus epidermidis ATCC® 12228	good to excellent

PACKAGING
Bacto Nutrient Agar pH 6.0   1 lb   0634-01-3

BACTO NUTRIENT BROTH

INTENDED USE
Bacto Nutrient Broth is a general purpose medium for the cultivation of microorganisms that are not exacting in their food requirements. It is the formula as specified by the American Public Health Association in *Standard Methods for the Examination of Water and Sewage*[1] and *Standard Methods for the Examination of Dairy Products*.[2]

HISTORY/PRINCIPLES
Bouillon or beef broth as suggested by Loeffler was one of the earliest media used in bacteriology. An infusion of meat and a peptone constituted the nutriments of this medium. Later it was shown that for many routine purposes beef extract could satisfactorily replace the infusion of fresh meat and had the decided advantage of ease of preparation, uniformity, and economy. The American Public Health Association rec-

ognized the advantage of beef extract in standard culture media and in 1917 discontinued the use of infusion of beef in standard media. Bacto Nutrient Broth has been prepared to duplicate the formula approved by the American Public Health Association and since 1927 our label has carried a statement to this effect.

FORMULA

BACTO NUTRIENT BROTH
DEHYDRATED

Ingredients per liter

Bacto Beef Extract .... 3 g
Bacto Peptone .... 5 g

Final pH 6.8 ± 0.2 at 25°C.

One pound will make 56.7 liters of final medium.

METHOD OF PREPARATION

1. Dissolve 8 grams in 1 liter distilled or deionized water.
2. Dispense into tubes.
3. Sterilize in the autoclave for 15 minutes at 15 pounds pressure (121°C).

STORAGE

Bacto Nutrient Broth   Below 30°C
Prepared medium   15 – 30°C

QUALITY CONTROL

Identity Specifications

Dehydrated powder: medium tan, homogeneous, free-flowing
Reaction of 8% solution: 6.8 ± 0.2 at 25°C
Prepared medium: light to medium amber, clear with no precipitate

Typical Cultural Response in Bacto Nutrient Broth
After 18 – 24 Hours at 35°C

Organism	Recovery
Corynebacterium diphtheriae type mitis	good to excellent
Enterobacter aerogenes ATCC® 13048	good to excellent
Escherichia coli ATCC® 25922	good to excellent
Salmonella typhi ATCC® 6539	good to excellent
Staphylococcus epidermidis ATCC® 14990	good to excellent

REFERENCES

1. American Public Health Association, 1946, Standard Methods for the Examination of Water and Sewage, 9th Ed. American Public Health Association, Inc., Washington, D.C.
2. American Public Health Association, 1948, Standard Methods for the Examination of Dairy Products, American Public Health Association, Inc., Washington, D.C.

PACKAGING

Bacto Nutrient Broth		
	1 lb (454 g)	0003-01-6
	1/4 lb (114 g)	0003-02-5
	5 lb (2.25 kg)	0003-05-2
	10 kg	0003-08-9
	12 tubes	0003-34-7
	144 tubes	0003-37-4

BACTO NUTRIENT GELATIN

INTENDED USE
Bacto Nutrient Gelatin is used for determining gelatin liquefaction (gelatinase production) by proteolytic microorganisms.

HISTORY/PRINCIPLES
Gelatin was the first gelling agent used to solidify liquid culture media, enabling the direct plate count to be used rather than the dilution method for determination of bacterial populations, and for the isolation of pure cultures. However, there are certain limitations to the use of gelatin in plating and isolation procedures, as incubation must be carried out at approximately 20°C, a temperature lower than the optimum for many organisms. Further, many organisms have the ability to attack and liquefy the gelatin.

In the past, Bacto Nutrient Gelatin was recommended for the 20°C plate count according to *Standard Methods for the Examination of Water and Sewage.*[1] Plating procedures now, however, generally utilize media containing agar as the solidifying agent.

FORMULA

BACTO NUTRIENT GELATIN
DEHYDRATED

Ingredients per liter

Bacto Beef Extract	3 g
Bacto Peptone	5 g
Bacto Gelatin	120 g

Final pH 6.8 ± 0.2 at 25°C.

One pound will make 3.5 liters of final medium.

PROCEDURE
To Rehydrate Medium
1. Suspend 128 grams of Bacto Nutrient Gelatin in 1 liter distilled or deionized water.
2. Warm to 50°C to dissolve completely.
3. Distribute into tubes.
4. Sterilize in the autoclave for 15 minutes at 15 lbs pressure (121°C).

To Determine Gelatin Liquefaction[2]
1. Inoculate the tube with a pure culture using a stab technique.
2. Incubate at 20 – 22°C for up to 6 weeks.
3. Observe tubes for liquefaction.

If the organism under test requires incubation at a temperature above 20 – 22°C, incubate at the desired temperature. After incubation cool the tubes to 20°C to determine if gelatin has been liquefied. An uninoculated control tube should be used simultaneously.

STORAGE

Bacto Nutrient Gelatin	Below 30°C
Prepared tubes	2 – 8°C

QUALITY CONTROL
Identity Specifications

Dehydrated powder:	tan, homogeneous, free-flowing
Reaction of 12.8% solution:	6.8 ± 0.2 at 25°C
Prepared medium:	medium amber, clear to slightly opalescent

Typical Cultural Response in Bacto Nutrient Gelatin
After 1 – 7 Days at 35°C.

Organism	Growth	Gelatinase*
Bacillus subtilis ATCC® 6633	good to excellent	+
Clostridium perfringens ATCC® 12924	good to excellent	+
Escherichia coli ATCC® 25922	good to excellent	−
Staphylococcus aureus ATCC® 25923	good to excellent	+

*For gelatinase test, cool to 20°C.

REFERENCES

1. American Public Health Association, 1946, Standard Methods for the Examination of Water and Sewage, 9th Ed. American Public Health Association, Inc., Washington, D.C.
2. American Society for Microbiology, 1981, Manual of Methods for General Bacteriology, 415, ASM., Washington, D.C.

PACKAGING

Bacto Nutrient Gelatin	1 lb (454 g)	0011-01-6
	1/4 lb (114 g)	0011-02-5

BACTO OF BASAL MEDIUM

INTENDED USE

Bacto OF Basal Medium is used to differentiate and classify gram-negative microorganisms on the basis of fermentative and oxidative metabolism of carbohydrates.

HISTORY/PRINCIPLES

Bacto OF Basal Medium is the formula of Hugh and Leifson.[1] They showed that when a gram-negative organism is inoculated into 2 tubes of a suitable medium containing a carbohydrate and then covering the medium in 1 tube with petrolatum to exclude oxygen while leaving the medium in the second tube uncovered, reactions of differential value may be observed. Fermentative organisms will produce an acid reaction in both the covered and uncovered media. Oxidative organisms will produce an acid reaction in the uncovered medium and yield slight to no growth without change in the covered media. Organisms which are not classified either as oxidative or fermentative yield no change in the covered medium and an alkaline reaction in the uncovered medium.

The authors classify *Alcaligenes faecalis* as nonoxidative and nonfermentative; *Shigella dysenteriae* and *sonnei* as fermentative (anaerogenic); *Salmonella enteritidis*, *Escherichia coli* and *Enterobacter aerogenes* as fermentative (aerogenic). They observed that some of the Paracolon bacilli are both fermentative and oxidative.

The carbohydrate metabolism of representative types of bacteria as described by the authors is tabulated below:

Organism	Glucose Open	Glucose Covered	Lactose Open	Lactose Covered	Sucrose Open	Sucrose Covered	Group
Alcaligenes faecalis	—	—	—	—	—	—	I Nonoxidizers Nonfermenters

Organism	Glucose Open	Glucose Covered	Lactose Open	Lactose Covered	Sucrose Open	Sucrose Covered	Group
Pseudomonas aeruginosa	A	–	–	–	–	–	II
Bacterium antitratum	A	–	A	–	–	–	Oxidizers
Agrobacterium tumefaciens	A	–	A	–	A	–	Nonfermenters
Malleomyces pseudomallei	A		A	–	A	–	
Shigella dysenteriae	A	A	–	–	–	–	IIIa
Shigella sonnei	A	A	A	A	–	–	Fermenters
Vibrio cholerae	A	A	–	–	A	A	(Anaerogenic)
Salmonella enteritidis	AG	AG	–	–	–	–	IIIb
Escherichia coli	AG	AG	AG	AG	–	–	Fermenters
Aeromonas liquefaciens	AG	AG	–	–	AG	AG	(Aerogenic)
Enterobacter aerogenes	AG	AG	AG	AG	AG	AG	
Unclassified species	A	A	A	–	Variable		IIIc
Some Paracolon	AG	AG	A	–			Fermenters
							Oxidizers

A = acid reaction; AG = acid and gas formation; – = no change or alkaline reaction

FORMULA

BACTO OF BASAL MEDIUM
DEHYDRATED

Ingredients per liter

Bacto Tryptone .... 2 g
Sodium Chloride .... 5 g
Dipotassium Phosphate .... 0.3 g
Bacto Brom Thymol Blue .... 0.08 g
Bacto Agar .... 2 g

Final pH is 6.8 ± 0.2 at 25°C.

One pound will make 48.3 liters of basal medium.

METHOD OF PREPARATION

1. Suspend 9.4 g in 1 liter distilled or deionized water and heat to boiling to dissolve completely.
2. Distribute in 100 ml amounts and sterilize in the autoclave for 15 minutes at 15 lbs pressure (121°C).
3. To 100 ml sterile medium add 10 ml Bacto Dextrose Solution 10%. To another 100 ml of basal medium, add 10 ml Bacto Lactose Solution 10%. To a third 100 ml of basal medium, add 10 ml Bacto Saccharose Solution 10%.
4. Mix each flask thoroughly and aseptically dispense in 5 ml amounts into sterile 13 × 100 mm culture tubes.

PROCEDURE

Two tubes of each carbohydrate medium are used for each organism tested. All tubes are inoculated by stabbing. One of the inoculated tubes of each carbohydrate medium is covered with 2 ml sterile melted petrolatum (mineral oil) and the other is left uncovered. The tubes are then incubated at 35°C for 48 hours or longer. Recordings of no change, acid or acid and gas production are made in 48 hours.

STORAGE

Bacto OF Basal Medium	Below 30°C
Prepared tubes	15 – 30°C

QUALITY CONTROL

Identity Specifications

Dehydrated powder:	light beige with green tinge, homogeneous free-flowing
Reaction of 0.94% solution:	pH 6.8 ± 0.2 at 25°C
Prepared medium:	green, clear to very slight opalescent

Typical Culture Response in Bacto OF Basal Medium
After 18 – 48 Hours at 35°C

Organism	No Additions		w/Dextrose		w/Lactose		w/Sucrose	
	Open	Closed	Open	Closed	Open	Closed	Open	Closed
Acinetobacter calcoaceticus ATCC® 15149	K	K	A	K	A	K	K	K
Alcaligenes faecalis ATCC® 8750	K	K	K	K	K	K	K	K
Enterobacter aerogenes ATCC® 13048	K	K	AG	AG	AG	AG	AG	AG
Escherichia coli ATCC® 25922	K	K	AG	AG	AG	AG	K	K
Pseudomonas aeruginosa ATCC® 27853	K	K	A	K	K	K	K	K
Salmonella enteritidis ATCC® 13076	K	K	AG	AG	K	K	K	K
Shigella flexneri ATCC® 12022	K	K	A	A	K	K	K	K
Vibrio cholerae El Tor ATCC® 15748	K	K	A	A	K	K	A	A

K = alkaline, green (no change)
A = acid, yellow
G = gas, sometimes observable

REFERENCE

1. Hugh, R. and Leifson, E., The Taxonomic Significance of Fermentative Versus Oxidative Metabolism of Carbohydrates by Various Gram-Negative Bacteria, J. Bact., 66:24 – 26, 1953.

PACKAGING

Bacto OF Basal Medium	1 lb (454 g)	0688-01-8
	1/4 lb (114 g)	0688-02-7
	12 tubes	0688-34-9
	114 tubes	0688-37-6
Bacto OF Basal Medium w/1% Dextrose	12 tubes	1404-34-0
Bacto OF Basal Medium w/1% Lactose	12 tubes	1405-34-9
Bacto OF Basal Medium w/1% Maltose	12 tubes	1406-34-8
Bacto OF Basal Medium w/1% Mannitol	12 tubes	1407-34-7
Bacto OF Basal Medium w/1% Saccharose	12 tubes	1408-34-6
Bacto OF Basal Medium w/1% Xylose	12 tubes	1409-34-5

BACTO OGYE AGAR BASE
BACTO ANTIMICROBIC VIAL OXYTETRACYCLINE

INTENDED USE

Bacto OGYE Agar Base (Oxytetracycline-Glucose-Yeast-Extract Agar Base) and Bacto Antimicrobic Vial Oxytetracycline are used in combination for selectively isolating and enumerating yeasts and molds in foods.

HISTORY

Mossel et al[1,2] described the use of oxytetracycline-glucose-yeast extract agar for selectively isolating and enumerating yeasts and molds in foods. Including oxytetracycline[3] in a medium having a neutral pH permitted increased microbial yields from a variety of foodstuffs when compared with media having a low pH as the mechanism for selecting against bacterial growth. In addition, acidophilic microorganisms, including lactobacilli which may be prevalent in foods, are inhibited on OGYE agar but produce undesired overgrowth on media with an acid pH.

FORMULA

BACTO OGYE AGAR BASE
DEHYDRATED

Ingredients per liter

Bacto Yeast Extract . . . . 5 g
Bacto Dextrose . . . . 20 g
Bacto Agar . . . . 12 g

Final pH 7.0 ± 0.2 at 25°C.

Five hundred grams will make 13 liters of medium.

Bacto Antimicrobic Vial Oxytetracycline is a lyophilized vial containing 100 mg of oxytetracycline. The vial is rehydrated with 10 ml sterile distilled or deionized water.

PREPARATION

1. Rehydrate Bacto OGYE Agar Base by suspending 37 g in 1 liter distilled or deionized water. Heat to boiling to dissolve the medium completely.
2. Sterilize in the autoclave for 15 minutes at 15 lbs pressure (121°C). Cool to 50°C.
3. Rehydrate one vial Bacto Antimicrobic Vial Oxytetracycline with 10 ml sterile distilled or deionized water.
4. Add entire contents of Bacto Antimicrobic Vial Oxytetracycline to sterile cooled Bacto OGYE Agar Base. Swirl gently to mix thoroughly.

PROCEDURE

1. Pour 15 – 20 ml complete medium into Petri dishes containing 1 ml of diluted test sample. Rotate plates to completely mix the contents. Allow the medium to solidify.
2. Invert the plates and incubate at controlled room temperature (22 ± 3°C).
3. Examine plates daily. Count colonies of yeasts and/or molds after 2 days and daily thereafter up to 5 days, at which time a maximum number of discrete colonies should be present.
4. Using the colony count which represents the maximum number of **discrete** colonies, calculate the number of yeasts/molds per ml or g of test sample by multiplying the number of colonies by the dilution of specimen used.

STORAGE

Bacto OGYE Agar Base	Below 30°C
Bacto Antimicrobic Vial Oxytetracycline	2 – 8°C
Prepared medium	2 – 8°C

QUALITY CONTROL

Identity Specifications

Dehydrated powder:	tan, homogeneous, free-flowing
Reaction of 3.7% solution:	pH 7.0 ± 0.2 at 25°C
Prepared medium:	medium amber, slightly opalescent without significant precipitate
Appearance of vial:	light yellow button or powder
Rehydrated vial:	yellow, clear

Typical Cultural Response on Bacto OGYE Agar
After 5 Days at 22 ± 3°C

Organism	Growth
Aspergillus niger ATCC® 16404	good to excellent
Escherichia coli ATCC® 25922	inhibited
Saccharomyces cerevisiae ATCC® 9763	good to excellent
Saccharomyces uvarum ATCC® 9080	good to excellent

REFERENCES
1. UNICEF, 1973.
2. J. Appl. Bact., 33:454, 1970.
3. Lab. Prac. II:109, 1962.

PACKAGING

Bacto OGYE Agar Base	500 g	1811-17-8
Bacto Antimicrobic Vial Oxytetracycline	10 ml	3267-59-2

BACTO OATMEAL AGAR

INTENDED USE
Bacto Oatmeal Agar is recommended for cultivation of fungi, particularly for macrospore formation.

FORMULA

BACTO OATMEAL AGAR
DEHYDRATED

Ingredients per liter

Oatmeal	60 g
Bacto Agar	12.5 g

Final pH 6.0 ± 0.2 at 25°C.

One pound will make 6 liters of medium.

METHOD OF PREPARATION
1. To rehydrate, suspend 72.5 g in 1 liter distilled or deionized water. Heat to boiling to dissolve.

2. Agitate the medium to obtain a uniform suspension and distribute into flasks in desired amounts.
3. Sterilize in the autoclave for 15 minutes at 15 lbs pressure (121°C).
4. Swirl the sterile medium in flasks to obtain a uniform mixture before pouring into Petri dishes.

STORAGE

Bacto Oatmeal Agar   Below 30°C
Prepared media   2 – 8°C

QUALITY CONTROL

Identity Specifications

Dehydrated powder: beige, nonhomogeneous, slightly lumpy
Reaction of 7.25% solution: pH 6.0 ± 0.2 at 25°C
Prepared medium: off-white, even opaque suspension with non-homogeneous particles

Typical Cultural Response on Bacto Oatmeal Agar
After 18 – 48 Hours at 30°C

Organism	Growth
Aspergillus niger ATCC® 16404	good to excellent
Candida albicans	good to excellent
Saccharomyces cerevisiae ATCC® 9763	good to excellent

PACKAGING

Bacto Oatmeal Agar   1 lb (454 g)   0552-01-1

ORANGE SERUM MEDIA

BACTO ORANGE SERUM AGAR
BACTO ORANGE SERUM BROTH CONCENTRATE 10X

INTENDED USE

Bacto Orange Serum Agar and Bacto Orange Serum Broth Concentrate 10X are used for the cultivation and enumeration of microorganisms associated with the spoilage of citrus products, the cultivation of lactobacilli, other aciduric organisms, and pathogenic fungi.

HISTORY/PRINCIPLES

Media prepared with clarified orange juice have been used for the control of the processing of citrus products by several investigators. Hays[1] described an orange serum agar for the enumeration and isolation of organisms causing spoilage in frozen concentrated orange juice. Murdock, Folinozzo and Troy[2] used a similar orange serum agar containing 0.5% dextrose and 2% agar in their evaluation of plating media for citrus concentrates. Their medium was based on previous field studies of Troy and Beisel.[3] They reported that orange serum agar at pH 5.4 was a suitable differential medium for *Leuconostoc*, lactobacilli and yeasts, and yielded maximum counts when mixed cultures of these organisms were compared. Hays and Riester[4] studying the control of "off odor" spoilage in frozen concentrated orange juice recommended orange serum agar at pH 5.5 as having been widely accepted as a control medium by the citrus industry since, at this reaction, the medium was most productive for the growth of spoilage organisms. Bacto Orange Serum Agar is prepared according to the formula described by Hays.[1]

FORMULAE

BACTO ORANGE SERUM AGAR

Ingredients per liter

Orange Serum	200 ml
Bacto Yeast Extract	3 g
Bacto Tryptone	10 g
Bacto Dextrose	4 g
Dipotassium Phosphate	2.5 g
Bacto Agar	17 g

Final pH 5.5 ± 0.2 at 25°C

One pound will make 10 liters of medium.

BACTO ORANGE SERUM BROTH CONCENTRATE 10X

Ingredients per liter

Orange Serum Concentrate	100 ml
Bacto Yeast Extract	3 g
Bacto Tryptone	10 g
Bacto Dextrose	4 g
Dipotassium Phosphate	2.5 g

Final pH 5.6 ± 0.2 at 25°C.

One hundred milliliters will make 1 liter of medium.

METHOD OF PREPARATION
Bacto Orange Serum Agar
1. Suspend 45.5 g in 1 liter distilled or deionized water and heat to boiling to dissolve completely.
2. Sterilize in the autoclave for 15 minutes at 15 lbs pressure (121°C).
   NOTE: AVOID OVERHEATING.

Bacto Orange Serum Broth Concentrate 10X
1. Under aseptic conditions, add the contents of one bottle (100 ml) to 900 ml of sterile cold distilled or deionized water.
2. Distribute aseptically in 10 ml amounts into sterile test tubes.

STORAGE
Bacto Orange Serum Agar and Broth Concentrate 10X	2 – 8°C
Prepared medium	2 – 8°C

QUALITY CONTROL
Identity Specifications
Bacto Orange Serum Agar
Dehydrated powder:	light tan, homogeneous, free-flowing
Reaction of 4.55% solution:	5.5 ± 0.2 at 25°C
Prepared medium:	medium to dark amber, very slightly opalescent

Bacto Orange Serum Broth Concentrate 10X
Concentrate:	dark amber, clear
Reaction of solution:	5.6 ± 0.2 at 25°C
Prepared medium:	medium amber, clear

Typical Cultural Response on/in Bacto Orange Serum Agar or Broth After 40 – 48 Hours at 35°C

Organism	Growth
Aspergillus niger ATCC® 16404	good to excellent
Lactobacillus fermentum ATCC® 9338	good to excellent
Saccharomyces cerevisiae ATCC® 9763	good to excellent

REFERENCES

1. Hays, 1952, Florida State Hort. Soc., 54:135.
2. Murdock, Folinozzo and Troy, 1951, Food Technology, 6:181.
3. Troy and Beisel, 1948, Unpublished Laboratory Data, Florida.
4. Hays and Riester, 1952, Food Technology, 6:386.

PACKAGING

Bacto Orange Serum Agar	1 lb (454 g)	0521-01-9
	10 kg	0521-08-2
Bacto Orange Serum Broth Concentrate 10X	6 × 100 ml	0518-73-7

BACTO ORCHID AGAR

INTENDED USE

Bacto Orchid Agar is recommended for use in the germination of orchid seeds.

HISTORY/PRINCIPLES

Extensive investigations on the germination of orchid seeds were conducted by Lewis Knudson[1,2,3] whereby he noted the importance of the presence of minor elements such as zinc, copper and manganese. Further studies by Somers and Shive[4] and Hopkins[5] demonstrated optimum orchid germination if the concentration of iron in the medium is 2 to 3 times greater than the concentration of manganese.

FORMULA

BACTO ORCHID AGAR
DEHYDRATED

Ingredients per liter

Calcium Nitrate	1 g	Ferrous Sulfate	0.025 g
Potassium Dihydrogen Phosphate	0.25 g	Manganese Sulfate	0.0075 g
Magnesium Sulfate	0.25 g	Saccharose, Difco	20 g
Ammonium Sulfate	0.5 g	Bacto Agar	15 g

Final pH 5.0 ± 0.1 at 25°C.

One pound will make 12 liters of final medium.

METHOD OF PREPARATION

1. To rehydrate, suspend 37 g in 1 liter distilled or deionized water and heat to boiling to dissolve completely.
2. Sterilize in the autoclave for 15 minutes at 15 lbs pressure (121°C).

STORAGE

Bacto Orchid Agar	Below 30°C
Prepared plates	2 – 8°C

QUALITY CONTROL

Identity Specifications

Dehydrated powder:	very light beige, fine, free-flowing
Reaction of 3.7% solution:	pH 5.0 ± 0.1 at 25°C
Prepared medium:	light amber, opalescent, may have a slight precipitate

REFERENCES
1. Knudson, Lewis, American Orchid Society Bulletin, 15:214, 1946.
2. Knudson, L., Nonsymbiotic germination of orchid seeds, Bot. Gaz., 73:1 – 25, 1922.
3. Knudson, L., Nutrient solutions for orchid seed germination, Amer. Orchid Soc. Bul., 12:77 – 78, 1943.
4. Somers, I. I., and Shive, J. W., The iron-manganese relation in plant metabolism, Plant Physiol., 17:582 – 602, 1942.
5. Somers, I. I., and J. W. Shive, The iron-manganese relation in plant metabolism, Plant Physiol., 17:582 – 602, 1942.

PACKAGING
Bacto Orchid Agar   500 g   0242-17-9

BACTO OXGALL

Bacto Oxgall is dehydrated fresh bile, and is prepared especially for use in selective media for differentiating groups of bile tolerant bacteria. It is manufactured from large quantities of fresh bile by rapid concentration to a uniform end product. The use of Bacto Oxgall insures a readily available source of supply, and assures a degree of uniformity impossible to obtain with fresh materials. The equivalent of fresh bile is attained in a 10% solution of Bacto Oxgall.

Bacto Oxgall is recommended for use in the preparation of media for the detection and propagation of intestinal organisms. Specifically, a large number of bile-containing media have been devised for use in the bacteriological examination of water. Bacto Oxgall is most frequently used in the preparation of culture media for water analysis. The Ninth Edition of *Standard Methods for the Examination of Water and Sewage*[1] recommended brilliant green lactose bile broth for confirming positive presumptive tests of coliform bacteria in water. In this text, brilliant green lactose bile agar is described as a selective agar medium for the direct plate count of the coliform group. Bacto Oxgall is recommended as an ingredient of both these media. Brilliant green lactose peptone bile 2% was also approved in *Standard Methods for the Examination of Dairy Products*[2] for detecting the presence of coliform bacteria in milk.

Solutions of Bacto Oxgall, added to broth cultures of pneumococci, produce lysis of the cells, and can therefore be used in bile solubility tests for differentiating pneumococci from the bile-insoluble streptococci. The general procedure of the bile solubility test is to add one part of a sterile 10% solution of Bacto Oxgall to 9 or 10 parts of culture. Evans[3] and McKinney[4] have used Bacto Oxgall for this purpose. Greey[5] reported excellent results using dry Bacto Oxgall sprinkled directly on blood agar plates. According to this procedure, *Pneumococcus* colonies are dissolved and disappear entirely, but leave evidence of their presence by means of the fixed blood cells in the clear medium. Colonies of *Streptococcus viridans* are not dissolved or otherwise altered.

Bacto Oxgall has been used in the preparation of a selective medium for fungi. Littman[6] described an agar medium for the isolation of pathogenic fungi, employing Bacto Oxgall, crystal violet and streptomycin as inhibiting agents for bacteria. Refer to Bacto Littman Oxgall Agar for a complete discussion.

REFERENCES
1. Standard Methods for the Examination of Water and Sewage, 9th Ed., 1946.
2. Standard Methods for the Examination of Dairy Products, 9th Ed., 1948.

3. J. Bact., 31:423, 1936.
4. J. Bact., 27:373, 1934.
5. J. Infectious Diseases, 64:206, 1939.
6. Science, 106:109, 1947.

PACKAGING

Bacto Oxgall	1 lb (454 g)	0128-01-6
	1/4 lb (114 g)	0128-02-5

BACTO PKU TEST KIT AND REAGENTS

Bacto PKU Test Kit is an assembly of reagents for estimating phenylalanine in blood as an aid in detecting phenylketonuria (PKU). The kit contains sufficient reagents for testing approximately 125 samples.

BACKGROUND

Phenylketonuria is a result of a congenital deficiency of phenylalanine hydroxylase, resulting in the accumulation in the body of this amino acid with subsequent brain damage and mental retardation.

The disease can be detected in the infant by determining the serum phenylalanine level or the level of phenylpyruvic acid in urine. Early detection of abnormally high phenyl-alanine levels facilitates initiation of dietary correction and prevention of severe mental retardation.

The observation that the inhibition of growth of *B. subtilis* in a minimal culture medium containing B-2-thienylalanine is prevented by phenylalanine, proline, phenylpyruvic acid or phenyllactic acid is the basis for Guthrie's inhibition test[1,2,3] for estimating the concentration of phenylalanine in blood.

Briefly, the test consists of applying filter paper disks saturated with the blood specimen to the surface of a minimal medium containing B-2-thienylalanine which has been inoculated with a suspension of *Bacillus subtilis*. Control disks which have been impregnated with blood containing 2, 4, 6, 8, 10, 12, and 20 mg% L-phenylalanine respectively are also placed on the medium. The plates are incubated for 12 – 16 hours at 35°C and the zones of growth around the test disks are compared with the zones around the control disks. A growth zone around the test disk which is comparable to the zone around a 4 mg% or higher disk is presumptive indication of a positive result and must be repeated using a duplicate test disk and a chemical or spectrofluorometric procedure.[4,5]

REAGENTS

Bacto PKU Test Agar is a minimal medium containing B-2-thienylalanine as an inhibitor of *B. subtilis*. The concentration of inhibitor is standardized to insure inhibition of growth of *B. subtilis* in the absence of phenylalanine.

BACTO PKU TEST AGAR
DEHYDRATED
Ingredients per liter

L-Glutamic Acid	0.5 g	Sodium Sulfate	0.5 g
DL-Alanine	0.5 g	Magnesium Sulfate	0.05 g
Bacto Asparagine	0.5 g	Manganese Chloride	0.005 g
Bacto Dextrose	10 g	Ferric Chloride	0.005 g
Dipotassium Phosphate	15 g	Calcium Chloride	0.0025 g
Monopotassium Phosphate	5 g	B_2 Thienylalanine	0.0033 g
Ammonium Chloride	2.5 g	Bacto Agar	15 g
Ammonium Nitrate	0.5 g		

pH 7.0 ± 0.2 at 25°C.

One pound will make 9.08 liters of medium.
Use 50 g per liter.

Bacto PKU Test Agar w/o Thienylalanine has the same formula as Bacto PKU Test Agar but with B-2-thienylalanine omitted. For a complete medium, add 1 ml 0.33% B-2-thienylalanine solution (0.15 ml per 150 ml medium) and mix thoroughly.

Bacto Subtilis Spore Suspension No. 2 is a standardized, stable suspension of *Bacillus subtilis* ATCC® 6633 spores which is inhibited by B-2-thienylalanine and which grows luxuriantly when the inhibitor is neutralized by phenylalanine.

Care should be taken in opening ampules of the *B. subtilis* culture. The emptied ampul should be autoclaved for 20 minutes at 121°C.

Bacto PKU Standard Disks are 1/4 inch disks which have been impregnated with blood to which had been added increasing amounts of phenylalanine for use as controls in the PKU test. The standard disks represent concentrations of 2, 4, 6, 8, 10, 12, and 20 mg% phenylalanine, respectively.

Control disks which have absorbed excess moisture may leach out hemoglobin and phenylalanine and are unsatisfactory.

CAUTION: Do not touch areas on which blood is to be collected. Bacto Blood Test Forms are for collecting and shipping blood samples to the laboratory for the PKU Test. A sterile lancet for puncturing the skin to obtain the specimen is provided. Directions for obtaining the blood specimen are given on each form.

STORAGE
Bacto PKU Test Kit will remain stable to the expiry date on the label when stored at 2 – 8°C in a dry area.

Exposure of the kit to high humidity may result in caking and hardening of the medium and/or deterioration of the control disks. Deteriorated reagents should not be used.

SPECIMEN COLLECTION
The sample must be taken at least 48 hours after first milk feeding.
1. Hold infant's limb in a position to increase venous pressure.
2. Cleanse heel and wipe with alcohol sponge or other available sterilizing solution.
3. Puncture heel with sterile lancet to obtain sufficient blood to fill each circle by single application of paper to drop of blood. Complete saturation of entire circle is essential for accuracy.

4. Collect blood in all 3 circles.
5. Allow blood samples to air dry.

PROCEDURE
Materials Provided:
Bacto PKU Test Agar
Bacto Subtilis Spore Suspension No. 2
Bacto PKU Standard Disks Set
Bacto Blood Test Forms w/Lancet
Disk Test Pattern for 150 mm Petri dish
150 mm Petri dish

Materials Required but not Provided:
1/4" punch
Forceps with fine points
Alcohol sponges
Autoclave
Incubator 35 – 37°C

Preparation of Culture Medium
1. Suspend the contents of 1 vial of Bacto PKU Test Agar in 50 ml distilled or deionized water and heat to boiling with gentle swirling to dissolve completely. Simmer 5 minutes.
2. Cool the medium to 50°C and add the contents of 1 vial of Bacto Subtilis Spore Suspension No. 2 (approximately 0.33 ml).
3. Twirl the flask to distribute spores uniformly throughout the medium.
4. Pour the contents of the flask into a 150 mm Petri dish resting on a flat horizontal surface or another appropriate container and allow to solidify.
5. Place Disk Test Pattern for 150 mm Petri dish or tray under the inoculated dish.

Preparation of Disks
Controls
Control disks impregnated with blood containing 2, 4, 6, 8, 10, 12 and 20 mg% L-phenylalanine are processed in the same manner as the sample and must be used on each plate.
1. Punch a 1/4" disk from 1 of the blood spots and place into a clean dry vial or place the entire specimen card on a wire rack in the autoclave.
2. Place 1 control disk of each concentration into separate clean vials in the autoclave.
3. Autoclave for exactly 3 minutes at 15 lbs pressure (121°C) and drop pressure rapidly to avoid condensation of moisture to wet the disks. Remove the disks promptly after temperature has dropped below 100°C. The disks are not to be used until they are dry.
4. Apply the test disks and the Bacto PKU Standard Disks to the inoculated Bacto PKU Test Agar with clean forceps and press down gently. Incubate plates at 35°C for 12 – 16 hours and examine for zones of growth around the control disks to obtain the approximate concentration of phenylalanine in the blood.

RESULTS
Zones of growth with diameters related to the concentration of phenylalanine will appear around the control disks. A zone of growth may or may not be present around the test disk depending on the presence or absence of phenylalanine in the specimen. The culture medium away from the zones of growth will be clear.

INTERPRETATION
The presence of a zone of growth around a test disk comparable to the zone around the standard disks containing 4 mg% phenylalanine or more is a presumptive positive and should be confirmed using a second sample. If the second sample gives a similar result, it is essential to determine the serum phenylalanine concentration by either a chemical[6] or a spectrofluorimetric procedure.

A negative test of an infant on antibiotics should be reconfirmed after antibiotic therapy is terminated. Antibiotics present in the blood sample are usually inactivated by the autoclaving procedure, but could be a source of error in that some antibiotics will inhibit the growth of *B. subtilis.*

POSSIBLE SOURCES OF ERROR

1. Collection of blood specimen must be done with care. The area of the absorbent paper on which the blood is collected must not come in contact with the collector's hands.
2. The blood sample must saturate the paper.
3. The sample and control disks must not be allowed to become soaked with water in the autoclave.
4. Autoclaved samples and controls must be dry before using.
5. PKU Test Agar must not be overheated. Bring to boiling and mix gently during heating. DO NOT AUTOCLAVE.
6. The temperature of the medium cannot be above 55°C when adding the spores.
7. It is important to uniformly distribute the spores in the medium without introducing bubbles.
8. The Petri dish must be placed on a horizontal surface for pouring the medium.
9. Leave the Petri dish cover slightly ajar to prevent condensate from forming on the surface of the medium.
10. Use the Test Pattern to locate the sample and control disks.
11. Incubate inverted plates at 35 – 37°C for no longer than 16 hours.

LIMITATIONS OF THE PROCEDURE

1. A zone of growth around the sample disk comparable to the zone around the 4 mg% or higher disk is a presumptive positive result and should be confirmed by a quantitative procedure.
2. Underestimating or overestimating of phenylalanine at the "cut off level of 4 and 6 mg%" yields questionable results. When questionable or borderline results are obtained, the test should be repeated using a second sample disk or the results confirmed by a quantitative procedure.

REFERENCES

1. Guthrie, R., and Tiechelmann, H., July 1960, London Conference on the Scientific Study of Mental Deficiency.
2. Guthrie, R., JAMA, 178:863, 1961.
3. Demain, A. L., J. Bact., 75:517, 1958.
4. Ambrose, J. A., et al., Clin. Chim. Acta., 15:493, 1967.
5. Ambrose, J. A., Clin. Chem., 15:15, 1969.
6. LaDu, B. N., and Michael, P. J., J. Lab. Clin. Med., 55:491, 1960.

PACKAGING

Bacto PKU Test Kit	125 Tests	0983-32-3
Bacto PKU Test Agar	1 lb (454 g)	0980-01-3
	1/4 lb (114 g)	0980-02-2
Bacto PKU Test Agar w/o Thienylalanine	1 lb (454 g)	0474-01-6
Bacto PKU Test Forms	1 × 50 forms	3240-30-8
	6 × 50 forms	3240-33-5
Bacto PKU Standard Disks (1 vial each 2 mg%, 4 mg%, 6 mg%, 8 mg%, 10 mg%, 12 mg%, 20 mg%, 50 disks per vial)	7 vials	1572-32-8
Bacto PKU Standard Disks 4 mg%	6 × 50 disks	1574-33-5
Bacto PKU Standard Strips (1 vial each 2 mg%, 4 mg%, 6 mg%, 8 mg%, 10 mg%, 12 mg% and 20 mg%, 10 strips per vial)	7 vials	1573-32-7
Bacto Subtilis Spore Suspension No. 2	25 × 1 ml	0981-36-1
	10 × 1 ml	0981-52-0
	100 × 1 ml	0981-84-2

BACTO PPLO AGAR

INTENDED USE
Bacto PPLO Agar, enriched with either Bacto PPLO Serum Fraction or Bacto Ascitic Fluid, is used for isolating and culturing *Mycoplasma* (pleuro-pneumonia-like organisms).

HISTORY/PRINCIPLES
PPLO agar was described by Morton, Smith and Leberman.[1]

PPLO colonies are round with a dense center and a less dense periphery, giving a "fried egg" appearance. Vacuoles, large bodies characteristic of the pleuropneumonia group, are seen in the periphery. Colonies vary in diameter from 10 to 500 microns (0.01 – 0.5 mm) and penetrate into the medium.

Colonial growth is visible microscopically after 48 or more hours. The plate may be considered negative after one week with no growth.

FORMULA

BACTO PPLO AGAR
DEHYDRATED

Ingredients per liter

Bacto Beef Heart for Infusions,
   Infusion from . . . . 50 g
Bacto Peptone . . . . 10 g
Sodium Chloride . . . . 5 g
Bacto Agar . . . . 14 g

Final pH 7.8 ± 0.2 at 25°C.

One pound will make 12.9 liters of unenriched medium.

METHOD OF PREPARATION
1. Suspend 35 g in 1 liter distilled or deionized water and heat to boiling to dissolve completely.
2. Sterilize in the autoclave for 15 minutes at 15 lbs pressure (121°C). Cool to 50 – 60°C.
3. Add 1% Bacto PPLO Serum Fraction or 25% Bacto Ascitic Fluid and mix thoroughly.
4. Aseptically dispense into sterile Petri dishes.

STORAGE
Bacto PPLO Agar	Below 30°C
Prepared medium	2 – 8°C

QUALITY CONTROL

Identity Specifications

Dehydrated powder:	beige, homogeneous, free-flowing
Reaction of 3.5% solution:	pH 7.8 ± 0.2 at 25°C
Prepared medium:	medium amber, slightly opalescent

Typical Cultural Response on Bacto PPLO Agar
After 40 – 48 Hours at 35°C Under 10% CO_2

Organism	Growth
Mycoplasma bovis ATCC® 25523	good
Mycoplasma gallinarium ATCC® 19708	good
Mycoplasma pneumoniae ATCC® 15531	good
Streptococcus pneumoniae ATCC® 6303	good

REFERENCE
1. Am. J. Syphilis Gonorrh. Venereal Diseases, 35:361, 1951.

PACAKGING

Bacto PPLO Agar	1 lb (454 g)	0412-01-1
	1/4 lb (114 g)	0412-02-0
Bacto PPLO Serum Fraction	6 × 20 ml	0441-63-1
Bacto Ascitic Fluid	6 × 10 ml	0135-60-5
	12 × 10 ml	0135-61-4

PPLO BROTH MEDIA
BACTO PPLO BROTH w/CV
BACTO PPLO BROTH w/o CV

INTENDED USE
Bacto PPLO Broth w/CV, enriched with Bacto Ascitic Fluid and supplemented with Bacto Chapman Tellurite Solution 1%, is used for isolating *Mycoplasma* from clinical specimens and mixed cultures.

Bacto PPLO Broth w/o CV, supplemented with Bacto PPLO Serum Fraction or Bacto Ascitic Fluid, is recommended as a basal broth medium for the enrichment of pleuropneumonia-like organisms (PPLO).

HISTORY/PRINCIPLES
Bacto PPLO Broth w/CV is prepared according to the formula of Morton, Smith, Williams and Eickenberg[1] for the selective enrichment of PPLO. While Bacto PPLO Broth w/o CV is prepared according to the formula described by Morton and Lecci,[2] it is identical to the formula of Bacto PPLO Broth w/CV with the exception that the crystal violet has been omitted due to the inhibitory action of the dye on some *Mycoplasma*. Crystal violet suppresses growth of both gram-negative and gram-positive organisms, facilitating isolation of *Mycoplasma* from clinical specimens or contaminated cultures.

FORMULAE

BACTO PPLO BROTH w/CV
DEHYDRATED
Ingredients per liter

Bacto Beef Heart for Infusion,
Infusion from . . . . 50 g
Bacto Peptone . . . . 10 g
Sodium Chloride . . . . 5 g
Bacto Crystal Violet . . . . 0.01 g

Final pH 7.8 ± 0.2 at 25°C.

Five hundred grams will make 23.8 liters of unenriched medium.

BACTO PPLO BROTH w/o CV
DEHYDRATED
Ingredients per liter

Bacto Beef Heart for Infusion,
Infusion from . . . . 50 g
Bacto Peptone . . . . 10 g
Sodium Chloride . . . . 5 g

Final pH 7.8 ± 0.2 at 25°C.

One pound will make 21.6 liters of unenriched medium.

METHOD OF PREPARATION
PPLO Broth w/CV
1. Dissolve 21 g in 1 liter distilled or deionized water.
2. Sterilize in the autoclave for 15 minutes at 15 lbs pressure (121°C). Allow to cool below 50°C.
3. Aseptically add 2.85 ml Bacto Chapman Tellurite Solution 1% (1% potassium tellurite) to yield a concentration of 1:35,000 and also add 25% Bacto Ascitic Fluid. Mix thoroughly.
4. Aseptically dispense into sterile tubes. Allow to cool to below 37°C.
5. Inoculate with clinical specimens, pure cultures or bacterial contaminated agar block colonies of PPLO and incubate at 35°C for 36 – 72 hours under 10% CO_2.
6. Streak incubated broth culture onto the surface of enriched Bacto PPLO Agar plates. Incubate plates for 36 – 72 hours at 35°C and examine for growth.

PPLO Broth w/o CV
1. Dissolve 21 g in 1 liter distilled or deionized water.
2. Sterilize in the autoclave for 15 minutes at 15 lbs pressure (121°C).
3. Allow to cool to below 50°C and aseptically add 25% Bacto Ascitic Fluid or 1% Bacto PPLO Serum Fraction. The medium should be enriched with Bacto Ascitic Fluid if crystal violet is used as a selective agent.
4. Add selective agents such as thallium acetate (1:2000) and penicillin as desired.
5. Mix medium and enrichments thoroughly and dispense aseptically into sterile tubes.
6. Inoculate and incubate same as for Bacto PPLO Broth w/CV, see above.

STORAGE
Bacto PPLO Broth media — Below 30°C
Prepared media — 2 – 8°C

QUALITY CONTROL
Identity Specifications

	Bacto PPLO Broth w/CV	Bacto PPLO Broth w/o CV
Dehydrated powder:	light beige w/blue-grey tinge, homogeneous, free-flowing	light beige, homogeneous, free-flowing
Reaction of 2.1% solution at 25°C	pH 7.8 ± 0.2	pH 7.8 ± 0.2
Prepared medium:	purple, clear	light amber, clear

Typical Cultural Response in Bacto PPLO Broth Supplemented As Above After 36 – 72 Hours at 35°C Under 10% CO_2

Organism	Growth
Mycoplasma bovis ATCC® 25523	good
Mycoplasma gallinarium ATCC® 19708	good
Mycoplasma pneumoniae ATCC® 15531	good
Streptococcus pneumoniae ATCC® 6303	none to good, depending on inhibitors used

REFERENCES
1. J. Dental Research, 30:415,1951.
2. J. Bact., 66:646, 1954.

PACKAGING

Bacto PPLO Broth w/CV	500 g	0410-17-5
Bacto PPLO Broth w/o CV	1 lb (454 g)	0554-01-9
	1/4 lb (114 g)	0554-02-8
Bacto PPLO Serum Fraction	6 × 20 ml	0441-63-1
Bacto Chapman Tellurite Solution 1%	6 × 1 ml	0299-51-8
	6 × 25 ml	0299-66-1
Bacto Ascitic Fluid	6 × 10 ml	0135-60-5
	12 × 10 ml	0135-61-4
Bacto PPLO Agar	1 lb (454 g)	0412-01-1
	1/4 lb (114 g)	0412-02-0

BACTO PPLO SERUM FRACTION

INTENDED USE

Bacto PPLO Serum Fraction is a sterile solution of a partially purified serum fraction and is used as an enrichment in media for the cultivation of mycoplasma (pleuro-pneumonia-like-organisms). It is employed in 1% concentration in Bacto PPLO Agar, Bacto PPLO Broth w/CV and Bacto PPLO Broth w/o CV.

Bacto PPLO Serum Fraction has also been used for the propagation, stabilization and maintenance of the treponemes *in vitro*.

HISTORY

Bacto PPLO Serum Fraction is the partially purified serum fraction required for growth by mycoplasma described by Smith and Morton.[1] It has the advantage over sera or ascitic fluid in that only 1% is required for enriching broth or solid media for culturing mycoplasma and does not exhibit the inhibitory effect which normal sera has upon some mycoplasma strains.

Bacto PPLO Serum Fraction was shown by Rose and Morton[2] to replace effectively the serum or albumin fractions commonly employed in the cultivation of avirulent treponemes.

STORAGE

Bacto PPLO Serum Fraction   2 – 8°C

QUALITY CONTROL
Identity Specifications

Appearance: medium to dark amber, clear to slightly opalescent
Reaction: pH 7.0 ± 0.2 at 25°C

For a complete discussion, refer to Bacto PPLO Agar.

REFERENCES

1. J. Bact., 61:395, 1961.
2. Am. J. Syphilis Gonorrh. Venereal Diseases, 36:1, 1952.

PACKAGING

Bacto PPLO Serum Fraction	6 × 20 ml	0441-63-1

BACTO PAGANO LEVIN BASE

INTENDED USE

Bacto Pagano Levin Base, a basal medium to which 2,3,5-triphenyltetrazolium chloride (TTC) and antibiotics are added, is used for the primary isolation and differentiation of *Candida* species.

HISTORY/PRINCIPLES

Bacto Pagano Levin Base is prepared according to the formulation described by Pagano, Levin and Trejo.[1] The complete medium is selective for the *Candida*. Its differential principle is based upon the different capacities of the *Candida* species to reduce the TTC and produce colonies with varying degrees of color.

The authors reported that, of the many tetrazolium compounds employed, the monotetrazolium compounds possessed the properties best suited to their purposes. One hundred μg TTC per ml of medium was the optimal concentration. Of the many antibiotics tried for inhibiting bacteria, neomycin gave best results. This antibiotic in a concentration of 500 μg per ml restricted growth of most bacteria without appreciably influencing the *Candida*.

FORMULA

BACTO PAGANO LEVIN BASE
DEHYDRATED

Ingredients per liter

Bacto Peptone . . . . 10 g
Bacto Yeast Extract . . . . 1 g
Bacto Dextrose . . . . 40 g
Bacto Agar . . . . 15 g

Final pH 6.0 ± 0.2 at 25°C.

One pound will make 6.9 liters of medium.

PROCEDURE

1. To rehydrate, suspend 66 g in 1 liter distilled or deionized water and heat to boiling to dissolve completely.
2. Sterilize in the autoclave for 15 minutes at 15 lbs pressure (121°C).
3. Allow the medium to cool to 50 – 55°C.
4. Aseptically add 100 μg TTC and 500 μg neomycin per ml of medium, and mix to obtain a uniform solution.
5. Dispense into sterile Petri dishes or tubes. Allow tubes to cool in the slanted position.
6. Inoculate the surface of medium with specimen and incubate for 48 – 72 hours at 25°C.

Colonies of *C. albicans* will appear cream-colored to light pink and will be smooth, round, raised, opaque and glistening. Typical *C. albicans* colonies may be confirmed readily on Bacto Chlamydospore Agar or Bacto Rice Extract Agar based on chlamydospore production.

STORAGE

Bacto Pagano Levin Base   Below 30°C
Prepared medium   2 – 8°C

QUALITY CONTROL
Identity Specifications

Dehydrated powder:	light beige, homogeneous, free-flowing
Reaction of 6.6% solution:	pH 6.0 ± 0.2 at 25°C
Prepared medium:	light amber, slightly opalescent

Typical Cultural Response on Bacto Pagano Levin Base
After 48 – 72 Hours at 25 – 30°C

Organism	Growth	Color of Colony
Candida albicans ATCC® 26790	good	cream to light pink
Candida krusei	good	white, spreading
Candida pseudotropicalis	good	pink
Candida stellatoidea ATCC® 36232	good	light red
Candida tropicalis	good	deep red
Saccharomyces cerevisiae ATCC® 9763	good	pink to red
Torulopsis glabrata	good	pink

REFERENCE
1. Antibiotics Annual, pg. 137, 1957 – 1958.

PACKAGING

Bacto Pagano Levin Base	500 g	0141-17-1
Bacto Pagano Levin Agar	12 tubes	1173-34-9

PANCREATIN NF

Pancreatin is a preparation of enzymes obtained from the pancreas of the hog, containing principally amylopsin, trypsin and steapsin. It is standardized according to the method given in the *United States Pharmacopeia/National Formulary*.[1]

REFERENCE
1. United States Pharmacopeia XX/National Formulary XV, Pages 580 – 581, 1980.

PACKAGING

Pancreatin NF	100 g	0296-15-6
	500 g	0296-17-4

BACTO PANTONE

Bacto Pantone is a peptone containing equal parts of pancreatic digest of casein and peptic digest of animal tissue for preparing microbiological culture media.

PACKAGING

Bacto Pantone	1 lb (454 g)	0906-01-4

PAPAIN NF

Papain NF is a purified proteolytic enzyme obtained from the fruit of papaya (*Carica papaya*). It is standardized according to the method given in the National Formulary.[1]

REFERENCE
1. National Formulary, VIII, 1946.

PACKAGING
Papain NF   500 g   0253-17-5

BACTO PENASE
BACTO PENASE CONCENTRATE

INTENDED USE
Bacto Penase and Bacto Penase Concentrate are beta-lactamases which hydrolyze the beta-lactam ring in penicillins, thereby, inactivating the antimicrobial properties of penicillin.

Both products are used in media for culturing microorganisms from blood and other body fluids containing penicillin and for estimating penicillin levels in such fluids. Bacto Penase Concentrate is also used for sterility testing of penicillins and for determining microbial counts of materials containing penicillin.

FORMULA
Bacto Penase and Bacto Penase Concentrate are highly purified, sterile, nontoxic, water-clear preparations of penicillinase having the potency and ability to inactivate penicillin listed below.

	Potency	Inactivation of Penicillin G[2]
Bacto Penase	2000 LU[1]/ml/min	1,000,000 IU/ml
Bacto Penase Concentrate	20,000 LU/ml/min	10,000,000 IU/ml

1. Levy unit (LU) – amount of penicillinase that inactivates 59.3 IU (international units) of sodium penicillin G per hour at pH 7.0 and 25°C.
2. Not all penicillins are comparable to penicillin G in regard to their inactivation by penicillinase. Some may require as much as 20 times more penicillinase than does penicillin G.

METHOD OF PREPARATION
1. Prepare the desired culture medium (Bacto Brain Heart Infusion, Bacto Tryptic Soy Broth, Bacto Fluid Thioglycollate Medium, etc.), dispense into 100 ml amounts, sterilize in the autoclave and cool to 50°C.
2. Aseptically add 1 ml Bacto Penase or 0.1 ml Bacto Penase Concentrate, or more as dictated by the situation.
3. Aseptically add 5 – 10 ml patient blood or 300 mg or less of test sample penicillin and proceed as required.

STORAGE
Bacto Penase   2 – 8°C
Bacto Penase Concentrate   2 – 8°C

PACKAGING

Bacto Penase	6 × 20 ml	0345-63-8
Bacto Penase Concentrate	6 × 20 ml	0346-63-7
	100 ml	0346-72-6
	6 × 100 ml	0346-73-5

See: Bacto Concentration Disks Penase

PENICILLIN IN MILK
AOAC RAPID DETECTION PROCEDURE
BACTO PM INDICATOR AGAR
BACTO THERMOSPORE SUSPENSION PM
BACTO PM POSITIVE CONTROL (PENICILLIN G)
BACTO PM NEGATIVE CONTROL

INTENDED USE
Bacto PM Indicator Agar, Bacto Thermospore Suspension PM, Bacto PM Positive Control and Bacto PM Negative Control are used for rapidly detecting trace amounts of penicillin in milk using the AOAC *Bacillus stearothermophilus* Qualitative Disc Method II.[1]

HISTORY
The AOAC procedure is similar to that published by the 18th Annual Meeting of the National Mastitis Council, Inc.[2]

PRINCIPLES
The qualitative disk test is used to determine whether or not penicillin is present in a milk sample as evidenced by growth or inhibition of growth of *B. stearothermophilus* around milk-impregnated disks placed on the surface of plates of Bacto PM Indicator Agar or Bacto Antibiotic Medium 4.

Bacto PM Indicator Agar is formulated to promote and demonstrate growth and acid formation by *B. stearothermophilus,* an organism sensitive to penicillin and beta-lactam residues. Milk containing as little as 0.005 IU penicillin/ml milk will produce a zone of inhibition, although the official definition of a positive test is a clear, well defined zone of inhibition equivalent to the zone around a positive control disk containing 0.008 IU penicillin/ml milk.

REAGENTS
BACTO PM INDICATOR AGAR
DEHYDRATED

Bacto Beef Extract	3 g	Sodium Chloride	0.5 g
Bacto Peptone	5 g	Dipotassium Phosphate	0.25 g
Bacto Tryptone	1.7 g	Sorbitan Monooleate Complex	1 g
Bacto Soytone	0.3 g	Bacto Agar	15 g
Bacto Dextrose	5.25 g	Brom Cresol Purple	0.06 g

Final pH 7.8 ± 0.2 at 25°C.

One hundred grams will make 31.25 liters of medium.

Bacto Thermospore Suspension PM is a suspension of *Bacillus stearothermophilus* standardized to provide clear, readable zones of growth under the time and temperature conditions of the AOAC milk test procedure.

Bacto PM Positive Control is a formulation containing 0.12 IU potassium penicillin in inhibitor-free non-fat dry milk. Each vial can be reconstituted to values of approximately 0.005, 0.006, 0.007, 0.008 or 0.01 IU penicillin/ml, depending on the volume of reconstituting water added.

Bacto PM Negative Control is inhibitor-free non-fat dry milk.

METHOD OF PREPARATION
1. FOR IMMEDIATE USE:
   A. Rehydrate Bacto PM Indicator Agar by suspending 3.2 g in 100 ml distilled or deionized water and heat to boiling to dissolve completely. Cool to 55°C in a water bath. Sterilization is not required.
   B. Inoculate the medium with 1 ml uniformly dispersed Bacto Thermospore Suspension PM and mix thoroughly to distribute the inoculum throughout the medium.
   C. Dispense 6 ml amounts into 100 mm flat-bottom Petri dishes. The use of flat-bottom Petri dishes is very important to obtain meaningful and reproducible results. Immediately tilt and rotate the plate to obtain a uniform layer of medium covering the bottom. Allow the medium to solidify on a flat surface, leaving the cover ajar for 15 minutes to allow the medium to dry.

   FOR LATER USE:
   A. Rehydrate as above.
   B. Sterilize the dissolved medium in the autoclave for 15 minutes at 15 lbs pressure (121°C).
   C. Inoculate and dispense medium as above.
   D. Immediately incubate solidified plates in an inverted position at 55 – 65°C for 30 minutes.
   E. Store plates inverted in plastic bags at 2 – 8°C. Use within 5 days.
2. Reconstitute Bacto PM Positive Control to the desired potency with cool distilled or deionized water as indicated.

Desired Potency IU Penicillin	Reconstituting Volume H$_2$O
0.005	24 ml
0.006	20 ml
0.007	17 ml
0.008 AOAC Test Control	15 ml
0.01	12 ml

3. Reconstitute Bacto PM Negative Control with 20 ml distilled or deionized water.

STORAGE
Bacto PM Indicator Agar	2 – 8°C
Bacto Thermospore Suspension PM	2 – 8°C
Bacto PM Positive Control	2 – 8°C
Bacto PM Negative Control	Below 30°C
Prepared, inoculated medium	2 – 8°C
Reconstituted, aliquotted Bacto PM Positive and Negative Controls	−15 to −30°C for up to 1 week

QUALITY CONTROL

Identity Specifications

Bacto PM Indicator Agar

Dehydrated powder: blue-grey, homogeneous, free-flowing
Reaction of 3.2% solution: pH 7.8 ± 0.2 at 25°C
Prepared medium: purple, very slightly to slightly opalescent

Bacto Thermospore Suspension PM

Appearance: white, opalescent, homogeneous suspension

Bacto PM Positive Control

Appearance: white crystals or powder
Rehydrated appearance: milky white, opaque solution

Bacto PM Negative Control

Appearance: white crystals or powder
Rehydrated appearance: milky white, opaque solution

SPECIMEN HANDLING

Test milk samples immediately, if possible.

Store samples not tested immediately at 5°C for no more than 10 hours or at −15 to −30°C for no more than 1 week.

Mix samples thoroughly before saturating disk.

PROCEDURE

SCREENING ASSAY

1. Using clean, flamed forceps, touch a blank disk to the surface of the milk sample and saturate by capillary action. Drain excess milk by touching disk to inner surface or rim of sample vessel. Immediately place disk on agar surface about 9 mm from rim of plate and at least 10 mm from other disks. Press disk gently to ensure good contact.
2. Using clean, flamed forceps, touch a blank disk to the surface of reconstituted PM Negative Control, saturate, and drain excess. Place on agar surface and press gently.
3. Using clean, flamed forceps, touch a blank disk to the surface of reconstituted PM Positive Control, saturate, and drain excess. Place on agar surface and press gently.
4. Invert plate and incubate until well defined, 17 – 20 mm zones of inhibition are obtained around the positive control disk.

Incubation Temperature	Approximate Time
55 ± 2°C	3 – 4 hrs
64 ± 2°C	2 – 3 hrs

5. Interpret results according to Table 1.

CONFIRMING ASSAY

1. Inactivate milk sample by heating at 82°C for 2 minutes or longer. Cool promptly to room temperature.
2. Using clean, flamed forceps, touch a disk to the surface of the inactivated milk sample, saturate, and drain excess. Place disk on agar surface and press gently.
3. Using clean, flamed forceps, touch a blank disk to the surface of reconstituted Bacto PM Positive Control, saturate, and drain excess. Place disk on agar surface and press gently.

4. Using clean, flamed forceps, touch a Bacto Concentration Disk Penase 1/2″ to the surface of the inactivated milk sample, saturate, and drain excess. Place disk on agar surface and press gently.
5. Invert plate and incubate as in SCREENING ASSAY, above.
6. Interpret results according to Table 2.

RESULTS

Table 1. Interpretation of Screening Assay

Test Sample Disk	Positive Control Disk	Negative Control Disk	Interpretation
No zone	Clear zone 17 – 20 mm	No zone	Negative test indicating that significant amounts of inhibiting substances are not present.
Zone ≤14 mm	Clear zone 17 – 20 mm	No zone	Negative test indicating that significant amounts of inhibitory substances are not present.
Clear zone >14 mm	Clear zone 17 – 20 mm	No zone	Positive test indicating that an inhibitor is present in test sample. Perform CONFIRMING ASSAY to determine if inhibitor is a beta-lactam residue.
Any reaction	Any reaction	**ANY ZONE**	Either result indicates error in the test system.
Any reaction	**NO ZONE**	Any reaction	Determine source of error.

Table 2. Interpretation of Confirming Assay

Inactivated Test Sample Disk	Positive Control Disk	Penicillinase/ Inactivated Milk Sample Disk	Interpretation
Clear zone >14 mm	Clear zone 17 – 20 mm	No zone	Positive test indicating the presence of beta-lactam residues.
Clear zone >14 mm	Clear zone 17 – 20 mm	Clear zone same size as test sample	Positive test indicating the presence of inhibitor(s) other than beta-lactam residues.
Clear zone >14 mm	Clear zone 17 – 20 mm	Clear zone substantially smaller than 14 mm	Positive test indicating presence of beta-lactam residues as well as other inhibitors.

LIMITATIONS
1. Petri dishes with warped or uneven bottoms will give rise to irregular zone responses due to variation in the depth of agar.
2. Disk slippage on the agar surface will cause distorted zones because disk content diffuses on contact.

REFERENCES
1. Penicillins in milk: *Bacillus stearothermophilus* qualitative disc method II. *In* W. Horwitz (ed.), Changes in methods, 3rd supplement to the 13th Ed.: official method of analysis, 1980, Association of Official Analytical Chemists, Washington, D.C.
2. Publ. of the 18th Annual Meeting of Natl. Mastitis Council, Inc.

PACKAGING

Bacto PM Indicator Agar	100 g	1800-15-3
Bacto PM Positive Control	6 × 1 vial	1802-33-9
Bacto PM Negative Control	6 × 20 ml	1803-63-1
Bacto Thermospore Suspension PM	25 × 1 ml	1801-36-7
	10 × 1 ml	1801-52-6
	6 × 10 ml	1801-60-6
Bacto Concentration Disks Penase 1/2″	6 × 50 disks	1595-33-0
Bacto Concentration Disks Sterile	6 × 50 disks	1571-33-8
Blanks 1/2″	50 disks	1571-35-6

PENICILLIN IN MILK
STANDARD METHODS PROCEDURE
BACTO ANTIBIOTIC MEDIUM 1
BACTO SUBTILIS SPORE SUSPENSION

INTENDED USE
Bacto Antibiotic Medium 1 and Bacto Subtilis Spore Suspension are used for detecting the presence of penicillin and other antibiotics in milk according to *Standard Methods for the Examination of Dairy Products*, 14th Edition.[1]

HISTORY
Churchill and Frank[2] used concentration disks penicillin to detect penicillin in milk. They compared the phosphatase method, growth coagulation method and the disk method and reported the superiority of the disk procedure.

Henningson, Silverman and Kosikowsky[3] and Kosikowsky, Henningson and Silverman[4] used Bacto Concentration Disks Penicillin for the detection of penicillin in milk. Silverman and Kosikowsky[3] in a systematic testing of inhibitory substances in milk reported that under optimal conditions, concentrations of penicillin as low as 0.05 unit/ml could be detected.

Standard Methods for the Examination of Dairy Products[5] first described the disk method using Bacto Concentration Disks in their APHA recommended test for residual penicillin in fluid milk. The test as described used Bacto Whey Agar, Bacto Subtilis Spore Suspension, Bacto Concentration Disks Sterile Blanks, Bacto Concentration Disks Penicillin as standards for quantitative work and Bacto Concentration Disks Penase to identify the inhibiting agent as penicillin.

Later, Arret and Kirshbaum[6] modified the disk assay method described by Difco[7] in 1953, and as published in *Standard Methods for the Examination of Dairy Products,*[8] for detecting and estimating penicillin in milk supplies after 3 – 4 hours incubation. Their modification was described in *Standard Methods for the Examination of Dairy Products.*[9] It specified the use of 1/2″ Concentration Disks and Bacto Antibiotic Medium 1. Standard methods procedure detects penicillin as low as 0.01 unit/ml milk sample.

More recently, the International Dairy Federation[10] published a disk assay procedure based on the method of Galesloot and Hassing[11] which detects penicillin in raw milk and milk products in excess of 0.0025 unit/ml.

REAGENTS

BACTO ANTIBIOTIC MEDIUM 1
DEHYDRATED

```
Bacto Peptone . . . . . . . . . . . . . . . . 6 g
Bacto Dextrose . . . . . . . . . . . . . . . . 1 g
Bacto Agar . . . . . . . . . . . . . . . . . . 15 g
Bacto Beef Extract . . . . . . . . . . . . 1.5 g
Bacto Yeast Extract . . . . . . . . . . . . 3 g
Bacto Casitone . . . . . . . . . . . . . . . 4 g
```

Final pH 6.6 ± 0.1 at 25°C.

One pound will make 14.88 liters of medium.

Bacto Subtilis Spore Suspension is a suspension of *Bacillus subtilis* ATCC® 6633 for use in the disk assay described below.

METHOD OF PREPARATION
Three variations of the test procedure are described in *Standard Methods* for disk assay of penicillin in milk. In each case the conditions of temperature and time of incubation, depth of medium and concentration of spores in the inoculum must be followed to obtain accurate and meaningful results.

	Method of Incubation*		
	Overnight	5 – 7 Hours	3 – 4 Hours
Amount of Medium	250 ml	250 ml	100 ml
Inoculation Temperature	50 – 55°C	50 – 55°C	70°C
Amount of Subtilis Spore Suspension	0.1 ml	1 ml	2 ml
Approximate Spore Concentration	1×10^5 ml	1×10^6 ml	5×10^6 ml
Incubation Temperature	32°C	35°C	37°C
Incubation Time	14 – 24 Hours	5 – 7 Hours	3 – 4 Hours

*Choose one Method of Incubation. Insert values in the PROCEDURE where appropriate.

1. Prepare and sterilize the prescribed amount of Bacto Antibiotic Medium 1 (according to label directions). Cool to the indicated temperatures as described above.
2. Prepare seed agar by adding the stated amount of Bacto Subtilis Spore Suspension to the medium, giving the approximate stated spore concentration. Swirl the flask gently to mix well, avoiding air bubbles. When performing the 3 – 4 hour procedure, hold the seeded agar for 15 minutes in a 70°C controlled temperature water bath before continuing.
3. Dispense into Petri dishes, 6 ml per 100 × 15 mm dish.

STORAGE
Bacto Antibiotic Medium 1   Below 30°C
Bacto Subtilis Spore Suspension   2 – 8°C

QUALITY CONTROL

Identity Specifications

Dehydrated powder:	beige, homogeneous, free-flowing
Reaction of 3.05% solution:	pH 6.6 $\pm$ 0.1 at 25°C
Prepared medium:	light to medium amber, slightly opalescent
Subtilis Spore Suspension appearance:	milky white, opaque suspension

PROCEDURE

1. Using sterile forceps, aseptically remove a disk from the vial of Bacto Concentration Disks Sterile Blanks. Touch the edge of the disk into milk sample and allow to become completely wet by capillary action. Place disk on surface of the agar, touching disk gently with the tip of the forceps to insure proper contact. Place sample disks so that they are at least 20 mm apart when measured from center to center to avoid overlapping zones.
2. Place a control disk containing 0.05 unit of penicillin per ml on each plate.
3. Invert plates and incubate at the specified temperature for the specified amount of time.
4. Examine the plates for zones of inhibition of the test organism.
5. Heat milk samples which give a zone of inhibition to 82°C for 2 – 5 minutes. Heating inactivates naturally occurring inhibitory substances. Retest heated milk samples. Any clear zone around the disk of a heated sample is a positive test.

IDENTIFICATION OF THE INHIBITOR

If a zone of inhibition is obtained at the end of the incubation period, it may be due to 1) penicillin, 2) an antibiotic other than penicillin, or 3) a combination of inhibitors. To determine which situation exists:

1. Place on plate one Bacto Concentration Disks Sterile Blanks wetted with the heated positive milk sample.
2. Place on plate one Bacto Concentration Disks Penase wetted with heated positive milk sample OR one Bacto Concentration Disks Sterile Blanks wetted with a thoroughly shaken mixture of 0.05 ml Bacto Penase Concentration in 5 ml heated positive milk sample.
3. Place on plate one Bacto Concentration Disks Penicillin 0.05 unit control disk.
4. Incubate inverted plates at appropriate temperature and time as in METHOD OF PREPARATION, above.

RESULTS

1. Penicillin is present if a clear zone of inhibition surrounds the test sample disk but not the penicillinase disk.
2. An inhibitor other than penicillin is present if clear zones of inhibition of equal size surround the sample and penicillinase disks or if the penicillinase disk zone is less than 5 mm smaller than the sample disk zone.
3. Both penicillin and another inhibitor are present if the clear zone of inhibition surrounding the penicillinase disk is 5 mm, or more than 5 mm, smaller than the zone surrounding the sample disk.
4. If penicillin only is present, estimate the concentration by comparing the zones surrounding the test sample disk and the 0.05 unit penicillin control disk.

REFERENCES

1. Standard Methods for the Examination of Dairy Products, APHA, Washington, D.C., 14th Ed., 1978.
2. Twenty-fifth Annual Report, New York State Assn. Milk Sanitarions, 1951.
3. J. Milk and Food Tech., 15:120, 1952.
4. J. Dairy Science, 35:533, 1952.
5. Standard Methods for Examination of Dairy Products, 10th Ed., 1953.
6. J. Milk and Food Tech., 22:329, 1959.

7. Difco Manual, 9th Ed., 1953.
8. Standard Methods for the Examination of Dairy Products, 11th Ed., 1960.
9. ibid., 12th Ed., 1967.
10. International Dairy Federation FIL-IFD, 57, 1970.
11. Neth. Milk Dairy J., 16:89 – 95, 1962.

PACKAGING

Bacto Antibiotic Medium 1	1 lb (454 g)	0263-01-1
	1/4 lb (114 g)	0263-02-0
	5 lb (2.27 kg)	0263-05-7
	10 kg	0263-08-4
Bacto Subtilis Spore Suspension	25 × 1 ml	0453-36-0
	10 × 1 ml	0453-52-9
	10 ml	0453-59-2
	6 × 10 ml	0453-60-9
Bacto Concentration Disks 1/2″		
Penase	6 × 50 disks	1595-33-0
Penicillin 0.05 unit	25 disks	1588-35-7
	6 × 25 disks	1588-33-9
Penicillin 0.1 unit	25 disks	1589-35-6
	6 × 25 disks	1589-33-8
Sterile Blanks	50 disks	1571-35-6
	6 × 50 disks	1571-33-8
Bacto Concentration Disks 1/4″		
Penicillin 0.05 unit	1 × 25 disks	1563-35-6
	6 × 25 disks	1563-33-8
Sterile Blanks	6 × 100 disks	1599-33-6
Bacto Penase Concentrate	6 × 20 ml	0346-63-7
	100 ml	0346-72-6
	6 × 100 ml	0346-73-5

PEPSIN 1:3,000
PEPSIN 1:10,000

Pepsin is employed in the preparation of culture media solely because of its ability to hydrolyze (digest) proteins to smaller molecules as the source of nitrogen. Pepsin 1:3,000 is an active preparation of the principal proteolytic enzymes of the gastric mucosa, and one part will digest 3,000 times its weight of coagulated egg albumin to proteoses and peptones. Pepsin 1:10,000 is highly active and will digest 10,000 times its weight of coagulated egg albumin.

Both Pepsin 1:3,000 and 1:10,000 are assayed according to a modification of the Food Chemical Codex method.[1]

Pepsin acts best at a reaction of pH 1.8.

REFERENCE
1. Food Chemicals Codex, 3rd Ed., p. 494.

PACKAGING

Pepsin 1:3,000	100 g	0308-15-2
	500 g	0308-17-0
Pepsin 1:10,000	100 g	0151-15-0
	500 g	0151-17-8

BACTO PEPTAMIN

Bacto Peptamin is a peptic digest of animal tissue for preparing microbiological culture media.

PACKAGING
Bacto Peptamin   1 lb (454 g)   0905-01-5

BACTO PEPTONE

Bacto Peptone was first introduced in 1914 and has long since become the universal standard peptone for the preparation of bacteriological culture media. It has continually been recommended as the peptone to be employed for culture media preparation in *Standard Methods of Water Analysis* and is at the present time included in the 15th Edition.[1] It is similarly specified for use in a number of special studies of milk and other dairy products.[2,3]

Bacto Peptone contains nitrogen in a form which is readily available for bacterial growth requirements. It is completely soluble in water and yields sparklingly clear solutions in the concentrations usually employed for culture media. Bacto Peptone has been standardized to a pH of approximately 7.0 in a 1% solution, as generally employed in culture media.

REFERENCES
1. Standard Methods for the Examination of Water and Sewage, 15th Ed., 1980.
2. Bull. 524, N.Y. Agr. Exp. Sta., 1924.
3. J. Dairy Science, 16:277:1933.

HISTORY
Refer to *Difco Manual*, 9th Edition, p. 256 – 257, 1953.

PACKAGING
Bacto Peptone	1 lb (454 g)	0118-01-8
	1/4 lb (114 g)	0118-02-7
	5 lb (2.27 kg)	0118-05-4
	10 kg	0118-08-1

PEPTONE BACTERIOLOGICAL, TECHNICAL

Peptone Bacteriological, Technical can be used as the nitrogen source in microbiological culture media when a standardized peptone is not essential. Although it has not been as carefully standardized as other peptones, certain parameters such as solubility, clarity, pH and growth supporting properties are monitored to permit its use as a nitrogen source.

PACKAGING
Peptone Bacteriological, Technical	1 lb (454 g)	0885-01-9
	5 lb (2.27 kg)	0885-05-5

BACTO PEPTONE IRON AGAR

INTENDED USE
Bacto Peptone Iron Agar is recommended for use as an indicator of hydrogen sulfide production by microorganisms.

HISTORY/PRINCIPLES
Levine and co-workers,[1,2] in their studies on the reactions in the colon group of bacteria, described a medium containing Proteose Peptone, Difco and ferric citrate as being particularly satisfactory for the detection of hydrogen sulfide. They further showed that such a medium served to differentiate the strains which were Voges-Proskauer negative, methyl red positive and citrate positive from other members of the *Enterobacteriaceae* family.

Levine reported that the ferric citrate was a much more sensitive indicator of hydrogen sulfide production than was lead acetate. Their medium gave definite clear-cut reactions within 12 hours.

Tittsler and Sandholzer[3] compared Bacto Peptone Iron Agar with lead acetate agar for the detection of hydrogen sulfide and found that the Bacto Peptone Iron Agar had the advantage of giving earlier reactions and more clear-cut results.

Bacto Peptone Iron Agar is a modification of Levine's original formula, in which Bacto Peptone has been included with Proteose Peptone, Difco, and the more soluble ferric ammonium citrate is used in place of ferric citrate.

FORMULA
BACTO PEPTONE IRON AGAR
DEHYDRATED
Ingredients per liter

Bacto Peptone	15 g
Proteose Peptone, Difco	5 g
Ferric Ammonium Citrate	0.5 g
Sodium Glycerophosphate	1 g
Sodium Thiosulfate	0.08 g
Bacto Agar	15 g

Final pH 6.7 ± 0.2 at 25°C.

Five hundred grams will make 13.8 liters of medium.

PROCEDURE
1. To rehydrate, suspend 36 g in 1 liter distilled or deionized water and heat to boiling to dissolve completely.
2. Dispense into tubes or appropriate containers and sterilize in the autoclave for 15 minutes at 15 lbs pressure (121°C).
3. Inoculate tubed medium by the stab method. Intense blackening of the medium indicates hydrogen sulfide production.
4. Plates may also be prepared and are convenient when it is desired to estimate the number of hydrogen sulfide producers.

STORAGE
Bacto Peptone Iron Agar   Below 30°C
Prepared medium   2 – 8°C

QUALITY CONTROL
Identity Specifications

Dehydrated powder: light beige, homogeneous, free-flowing
Reaction of 3.6% solution: pH 6.7 ± 0.2 at 25°C
Prepared medium: light amber, slightly opalescent

Typical Cultural Response on Bacto Peptone Iron Agar
After 18 – 48 Hours at 35 ± 2°C

Organism	Recovery	H_2S Production
Enterobacter aerogenes ATCC® 13048	good	–
Escherichia coli ATCC® 25922	good	+
Salmonella enteritidis ATCC® 13076	good	+
Salmonella typhi ATCC® 6539	good	+

REFERENCES
1. Proc. Soc. Exp. Biol. Med., 29:1022, 1932.
2. Am. J. Publ. Health, 24:505, 1934.
3. Am. J. Publ. Health, 27:1240, 1937.

PACKAGING
Bacto Peptone Iron Agar   500 g   0089-17-5

BACTO PEPTONE WATER

INTENDED USE
Bacto Peptone Water[1] is a liquid minimal medium used for culturing nonfastidious organisms, for studying carbohydrate fermentation patterns and for performing the indole test.

FORMULA
BACTO PEPTONE WATER
DEHYDRATED

Ingredients per liter

Peptone . . . . 10 g
Sodium Chloride . . . . 5 g

Final pH 7.2 ± 0.2 at 25°C.

Five hundred grams will make 33.3 liters of medium.

PROCEDURE
For General Culture Work
1. Rehydrate Bacto Peptone Water by dissolving 15 g in 1 liter distilled or deionized water.
2. Dispense as desired.
3. Sterilize in the autoclave for 15 minutes at 15 lbs pressure (121°C).

For Determining Carbohydrate Fermentation Patterns
1. Add 1.8 ml TC Phenol Red Solution 1% (1% solution of phenol red) to 1 liter rehydrated Bacto Peptone Water. Mix thoroughly.

2. Dispense into test tubes containing inverted Durham vials.
3. Sterilize in the autoclave for 15 minutes at 15 lbs pressure (121°C).
4. Aseptically add sufficient sterile carbohydrate solution to yield a 1% final concentration. Rotate each tube to thoroughly distribute the carbohydrate.

For Performing the Indole Test
1. Using aseptic technique, suspend a Bacto Indole Test Strip 10 mm above the surface of a 24 or 48 hour culture.
2. Incubate at 37°C for 5 – 30 minutes.
3. Observe for the formation of a violet color on the strip which indicates a positive test for indole production.

STORAGE

Bacto Peptone Water	Below 30°C
Prepared medium	15 – 30°C

QUALITY CONTROL

Identity Specifications

Dehydrated medium: cream white to light tan, homogeneous, free-flowing
Reaction of 1.5% solution: 7.2 ± 0.2 at 25°C
Prepared medium: light amber, clear with no precipitate

Typical Cultural Response in Bacto Peptone Water
After 18 – 24 Hours at 35°C

Organism	Recovery
Escherichia coli ATCC® 25922	excellent
Salmonella typhimurium ATCC® 14028	excellent
Staphylococcus aureus ATCC® 25923	excellent

REFERENCE

1. Cowan, S. T. 1974. Manual for the identification of medical bacteria, 2nd Ed. Cambridge University Press, London.

PACKAGING

Bacto Peptone Water	500 g	1807-17-4

BACTO PETRAGNANI MEDIUM

INTENDED USE
Bacto Petragnani Medium is a prepared coagulated-egg medium used for isolating and culturing mycobacteria.

HISTORY
Bacto Petragnani Medium is a modification of Petragnani medium[1] described by Norton, Thomor and Broom.[2]

PRINCIPLES
Bacto Petragnani Medium is slightly more selective than ATS and LJ media, inhibiting growth of commensal organisms to a slightly greater extent.

FORMULA

BACTO PETRAGNANI MEDIUM

```
Whole Milk . . . . . . . . . . . . . . . . . 900 ml
Potato Flour . . . . . . . . . . . . . . . . . 36 g
Potato . . . . . . . . . . . . . . . . . . . . 500 g
Whole eggs . . . . . . . . . . . . . . . 1200 ml
Egg yolks . . . . . . . . . . . . . . . . 115 ml
Bacto Glycerol . . . . . . . . . . . . . . . 70 ml
Bacto Malachite Green . . . . . . . . . 1.2 g
```

This product is available as a prepared tube medium.

STORAGE
Prepared medium   2 – 8°C

QUALITY CONTROL

Identity Specifications

Reaction of medium: pH 7.2 ± 0.2 at 25°C
Prepared medium: light green, opaque slants

Typical Cultural Response on Bacto Petragnani Medium
After 2 Weeks at 35°C Under 10% CO_2

Organism	Growth
Mycobacterium fortuitum ATCC® 6841	good to excellent
Mycobacterium tuberculosis H37Rv (ATCC® 25618)	good to excellent
Streptococcus pyogenes ATCC® 19615	inhibited

REFERENCES
1. Rend. d. adunanze dell, accad. med. fis. florentina sperimentale, 77:101, 1923.
2. Am. Rev. Tuberc., 25:378, 1932.

PACKAGING
Bacto Petragnani Medium	12 tubes	1010-34-6
	144 tubes	1010-37-3

BACTO PHENOL RED AGAR BASE
BACTO PHENOL RED LACTOSE AGAR
BACTO PHENOL RED MANNITOL AGAR

INTENDED USE
Bacto Phenol Red Agar Base is a basal medium to which carbohydrates are added for use in fermentation studies.

Bacto Phenol Red Lactose Agar and Bacto Phenol Red Mannitol Agar consist of Bacto Phenol Red Agar Base with the carbohydrate added to a final concentration of 1%.

PRINCIPLES
While liquid media are generally employed in studying the fermentation reactions of microorganisms, many bacteriologists prefer a solid medium for this purpose. The solid media employed usually contain 1% of the selected carbohydrate and an indicator of pH change.

The advantages claimed for a solid fermentation medium are that it permits observation of the fermentation reactions under both aerobic and anaerobic conditions, that gas formation is indicated by splitting of the agar or accumulation of gas bubbles in the base, and that deep tubes can provide sufficiently anaerobic conditions for the development of the obligately anaerobic bacilli.

Bacto Phenol Red Agar Base is particularly well adapted to the study of fermentation reactions of microorganisms. This medium supports excellent growth of many fastidious bacteria. The basal medium is free from fermentable carbohydrates which could give erroneous interpretation. With the exception of the carbohydrate, which has been omitted, it is a complete medium prepared with phenol red as an indicator of changes in reaction. Bacto Phenol Red Agar Base permits the user to prepare any quantity of medium required adding to different portions any fermentable substance desired (usually 1% of the test carbohydrate being added). An entire series of carbohydrate agars may thus be made up readily, conveniently and economically.

FORMULAE

Ingredients per liter

	Bacto Phenol Red Agar Base Dehydrated	Bacto Phenol Red Lactose Agar Dehydrated	Bacto Phenol Red Mannitol Agar Dehydrated
Proteose Peptone No. 3	10 g	10 g	10 g
Bacto Beef Extract	1 g	1 g	1 g
Bacto Lactose	–	10 g	–
Bacto Mannitol	–	–	10 g
Sodium Chloride	5 g	5 g	5 g
Bacto Agar	15 g	15 g	15 g
Bacto Phenol Red	25 mg	25 mg	25 mg
Rehydrate with	31 g/liter	41 g/liter	41 g/liter
Yield	14.6 liter/lb	11.07 liter/lb	12.19 liter/500 g

Final pH 7.4 ± 0.2 at 25°C.

METHOD OF PREPARATION
BACTO PHENOL RED AGAR BASE
1. Suspend 31 g in 1 liter distilled or deionized water and heat to boiling to dissolve completely.
2. Add 10 g of the desired carbohydrate and dispense into tubes.
3. Sterilize in the autoclave for 15 minutes at 15 lbs pressure. The minimum amount of heat required for sterilization is to be desired. By packing tubes loosely in the autoclave to allow free circulation of the steam, the time required for sterilization may be appreciably shortened, provided the temperature in the autoclave is actually 121°C.
4. Allow to cool in a slanted position to provide a slope and generous butt.

Alternatively, and preferably, the carbohydrate may be added aseptically after autoclaving. In this case, dissolve the medium in 900 ml distilled or deionized water and sterilize. After cooling the basal medium to 45 – 50°C, add 100 ml of a sterile 10% carbohydrate solution.

NOTE: The addition of some carbohydrates may result in an acid reaction. In this case, it is suggested that 0.1N sodium hydroxide be added drop by drop to restore the original color, taking care not to obtain too deep red or cerise color.

BACTO PHENOL RED LACTOSE AGAR
BACTO PHENOL RED MANNITOL AGAR
1. Suspend 41 g of medium in 1 liter cold distilled or deionized water and heat to boiling to dissolve completely.
2. Dispense in tubes and sterilize in the autoclave for 15 minutes at 15 lbs pressure (121°C).

If these media are not used within a few days after preparation, it is recommended that they be melted with minimum heat and resolidified to give a moist surface.

PROCEDURE
Tubes of the sterile medium are inoculated by smearing over the surface of the slant and stabbing into the butt. Obligately anaerobic bacteria may be inoculated into the melted medium previously cooled to 45°C and subsequently allowing it to solidify. After incubation, fermentation will be denoted by a change in the color of the medium from red to canary yellow. Gas formation is indicated by the collection of gas bubbles in the base, or by splitting of the agar.

STORAGE
Bacto Phenol Red Agar Media   Below 30°C
Prepared media   15 – 30°C

QUALITY CONTROL
Identity Specifications

Dehydrated powder: pink, homogeneous, free-flowing
Reaction of appropriate solution: pH 7.4 ± 0.2 at 25°C
Prepared media: red to orange red, slightly opalescent

Typical Cultural Response After 4 – 18 Hours* at 35°C

Organism	Growth	Bacto Phenol Red Agar Base w/Dextrose		w/Maltose		w/Sucrose		Bacto Phenol Red Lactose Agar		Bacto Phenol Red Mannitol Agar	
		Acid	Gas	Acid	Gas	Acid	Gas	Acid	Gas	Acid	Gas
Alcaligenes faecalis ATCC® 8750	good to excellent	–	–	–	–	–	–	–	–	–	–
Escherichia coli ATCC® 25922	good to excellent	–	–	+	+	+	+	–	–	+	+	+	+
Klebsiella pneumoniae ATCC® 13883	good to excellent	–	–	+	+	+	+	+	+	+	+	+	+
Proteus vulgaris ATCC® 6380	good to excellent	–	–	+	+	+	+	+	+	–	–	–	–
Salmonella typhimurium ATCC® 14028	good to excellent	–	–	+	+	+	+	–	–	–	–	+	+
Shigella flexneri ATCC® 12022	good to excellent	–	–	+	–	–	–	–	–	–	–	+	–

*longer if necessary

PACKAGING
Bacto Phenol Red Agar Base	1 lb (454 g)	0098-01-2
	1/4 lb (114 g)	0098-02-1
Bacto Phenol Red Lactose Agar	1 lb (454 g)	0100-01-8
Bacto Phenol Red Mannitol Agar	500 g	0103-17-7

PHENOL RED CARBOHYDRATE MEDIA
BACTO PHENOL RED BROTH BASE
BACTO PHENOL RED DEXTROSE BROTH
BACTO PHENOL RED LACTOSE BROTH
BACTO PHENOL RED MALTOSE BROTH
BACTO PHENOL RED MANNITOL BROTH
BACTO PHENOL RED SACCHAROSE BROTH

INTENDED USE

Bacto Phenol Red Broth Base is a basal medium to which carbohydrates are added for use in fermentation studies for the identification of pure cultures.

PRINCIPLES

The fermentative properties of bacteria are valuable criteria in their identification, and may be determined by culturing the organisms in a suitable medium containing the appropriate fermentable substance. A satisfactory basal medium for determining the fermentation reactions of microorganisms must be capable of supporting growth of the organisms under study, and free from fermentable carbohydrates which could give erroneous interpretations. It must be stable, uniform in composition, give distinct reactions and yield accurate results.

Bacto Phenol Red Broth Base is an excellent substrate for streptococci, as well as for other less fastidious bacteria. The cultural value of the medium can be greatly improved for some of the more delicate strains by the addition of a small amount (0.1 – 0.2%) of Bacto Agar. A medium containing this small quantity of agar may be used to best advantage by heating it to the boiling point to drive out the dissolved air and cooling it below 40°C, without excessive agitation, just prior to inoculation. Such a procedure also makes the medium sufficiently oxygen-free for propagation of the obligate anaerobes as well as microaerophiles. Bacto Phenol Red Broth Base with 0.5% selected carbohydrate and 0.15% agar is suggested as a satisfactory medium for the fermentation determinations as given in "Diagnostic Procedures and Reagents."[1] Some bacteriologists, determining the fermentation reactions of gonococci, may prefer to use 0.8% Bacto Agar and add 5% sterile fresh rabbit serum to the sterile Bacto Phenol Red Broth Base containing the selected carbohydrate.

For the determination of fermentative properties of members of the enteric group of bacteria, Bacto Purple Broth media are recommended. These media have the same nutrients, but have a slightly more acid reaction, and brom cresol purple is employed as an indicator.

With the exception of the carbohydrate, which has been omitted, Bacto Phenol Red Broth Base is a complete basal medium prepared with phenol red as an indicator of changes in reaction. This product makes it possible to prepare as much or as little medium as is required, adding to different portions any fermentable substance in any concentration desired. The concentration of carbohydrate generally employed for testing the fermentation reactions of bacteria is 0.5 or 1%. Some investigators prefer to use 1% rather than 0.5% to insure against reversion of the reaction due to depletion of the carbohydrate by some microorganisms. An entire series of carbohydrate broths, may thus be made up readily, conveniently, and economically.

When an organism ferments the added carbohydrate, the pH of the medium drops resulting in a change in the color of the pH indicator, phenol red, from red-orange to yellow. The production of gas can be determined by placing a small inverted fermentation tube (Durham tube) in the tubes of media at the time of preparation. To ensure accuracy of interpretation, noninoculated control tubes should be run in parallel with the fermentation tests.

FORMULAE

BACTO PHENOL RED BROTH

	BASE	DEXTROSE	LACTOSE	MALTOSE	MANNITOL	SACCHAROSE
Proteose Peptone No. 3, Difco	10 g	10 g	10 g	10 g	10 g	10 g
Bacto Beef Extract	1 g	1 g	1 g	1 g	1 g	1 g
Bacto Dextrose	—	5 g	—	—	—	—
Bacto Lactose	—	—	5 g	—	—	—
Bacto Maltose	—	—	—	5 g	—	—
Bacto Mannitol	—	—	—	—	5 g	—
Bacto Saccharose	—	—	—	—	—	5 g
Sodium Chloride	5 g	5 g	5 g	5 g	5 g	5 g
Bacto Phenol Red	0.018 g	0.018 g	0.018 g	0.018 g	0.018 g	0.018 g

Final pH 7.4 ± 0.2 at 25°C.

One pound of Bacto Phenol Red Broth Base will make 28.3 liters of medium.
One pound of Bacto Phenol Red Carbohydrate Broths will make 21.6 liters of media.

METHOD OF PREPARATION
Bacto Phenol Red Broth Base
Suspend 16 g of Bacto Phenol Red Broth Base and 5 to 10 g (0.5 to 1.0%) of the desired carbohydrate in 1 liter distilled or deionized water and stir to dissolve completely. If preferred, the base may be prepared without the carbohydrate added. A filter sterilized or autoclave sterilized carbohydrate solution may be aseptically added after the basal medium has been sterilized and cooled.

Difco supplies many carbohydrates in dehydrated form, as ampuled 10% solutions or in the form of Differentiation Disks. The addition of some carbohydrates may result in an acid reaction. In this case, it is suggested that 0.1N sodium hydroxide be added drop by drop to restore the original color, taking care not to obtain too deep red or cerise color.

Bacto Phenol Red Carbohydrate Broths
1. Suspend 21 g of the appropriate Bacto Phenol Red Carbohydrate Broth in 1 liter distilled or deionized water and stir to dissolve completely.
2. Better growth of more fastidious organisms, such as streptococci, pneumococci and gonococci, will be obtained if 1 g of Bacto Agar is dissolved in the broth prior to sterilization. If Bacto Agar is added, bring the medium to a boil before autoclaving.
3. Dispense into tubes containing fermentation tubes, if desired.
4. Sterilize in the autoclave for 15 minutes at 15 lbs pressure (121°C). The minimum amount of heat required for complete sterilization is desired. By packing the tubes loosely in the autoclave to allow free circulation of steam, the time required may be appreciably shortened, provided the temperature in the autoclave is actually 121°C.

5. If the medium is not used the same day it is sterilized, place in flowing steam or a boiling water bath for a few minutes to drive off dissolved gases and allow to cool without agitation.

PROCEDURE
1. Inoculate tubes with one drop of diluted culture.
2. Incubate for 4 – 18 hours at 35 – 37°C with caps loosened.
3. Read for growth, acid production (yellow color of medium) and gas production (if fermentation vials were used).

STORAGE
Bacto Phenol Red Carbohydrate Broths   Below 30°C
Prepared tubes   15 – 30°C

QUALITY CONTROL
Identity Specifications
Dehydrated powder: pink, homogeneous, free-flowing
Reaction of appropriate solution: pH 7.4 ± 0.2 at 25°C
Prepared media: orange-red to red, clear

Typical Cultural Response after 4 – 18 Hours at 35 – 37°C
Bacto Phenol Red Broth

Organism	Growth	Base		Dextrose		Lactose		Maltose		Mannitol		Saccharose	
		Acid	Gas	Acid	Gas	Acid	Gas	Acid	Gas	Acid	Gas	Acid	Gas
Alcaligenes faecalis ATCC® 8750	good to excellent	−	−	−	−	−	−	−	−	−	−	−	−
Escherichia coli ATCC® 25922	good to excellent	−	−	+	+	+	+	+	+	+	+	−	−
Klebsiella pneumoniae ATCC® 13883	good to excellent	−	−	+	+	+	+	+	+	+	+	+	+
Proteus vulgaris ATCC® 6380	good to excellent	−	−	+	+	−	−	+	+	−	−	+	+
Salmonella typhimurium ATCC® 14028	good to excellent	−	−	+	+	−	−	+	+	+	+	−	−
Shigella flexneri ATCC® 12022	good to excellent	−	−	+	−	−	−	−	−	+	−	−	−

REFERENCE
1. Diagnostic Procedures and Reagents, 3rd Ed., 107, 1950.

PACKAGING
Bacto Phenol Red Broth Base	1 lb (454 g)	0092-01-8
	1/4 lb (114 g)	0092-02-7
Bacto Phenol Red Dextrose Broth	1 lb (454 g)	0093-01-7
	1/4 lb (114 g)	0093-02-6
Bacto Phenol Red Lactose Broth	1 lb (454 g)	0094-01-6
	1/4 lb (114 g)	0094-02-5
Bacto Phenol Red Maltose Broth	1 lb (454 g)	0096-01-4
Bacto Phenol Red Mannitol Broth	1 lb (454 g)	0097-01-3
Bacto Phenol Red Saccharose Broth	1 lb (454 g)	0095-01-5

BACTO PHENOL RED TARTRATE AGAR

INTENDED USE
Bacto Phenol Red Tartrate Agar is recommended for the identification of gram-negative bacteria of the intestinal groups, particularly members of the *Salmonella* (paratyphoid) group.

HISTORY/PRINCIPLES
Brown, Duncan and Henry[1] observed that the members of the paratyphoid group varied in their ability to attack sodium tartrate and incorporated this principle in a medium for subdividing the group. Jordan and Harmon[2] claimed that the medium of Brown, Duncan and Henry failed to give sharp differentiation and devised a medium which possessed the advantage of being more definite in its differentiation.

Bacto Phenol Red Tartrate Agar duplicates the medium of Jordan and Harmon. On this medium, an acid reaction is produced by *Salmonella typhimurium*, *S. enteritidis*, *S. choleraesuis*, *S. abortiroequina*, *S. typhi*, *Escherichia coli* and *Proteus vulgaris* strains, while the *S. schottmuelleri* and *S. paratyphi* strains produce an alkaline reaction.

FORMULA
BACTO PHENOL RED TARTRATE AGAR
DEHYDRATED
Ingredients per liter

Bacto Peptone . . . . 10 g
Sodium Potassium Tartrate . . . . 10 g
Sodium Chloride . . . . 5 g
Bacto Agar . . . . 15 g
Bacto Phenol Red . . . . 0.024 g

Final pH 7.6 ± 0.2 at 25°C.

Five hundred grams will make 12.5 liters of medium.

PROCEDURE
1. To rehydrate, suspend 40 g in 1 liter cold distilled or deionized water and heat to boiling to dissolve completely.
2. Dispense into tubes which are stoppered with cotton plugs or loosely fitting caps and sterilize in the autoclave for 15 minutes at 15 lbs pressure (121°C).
3. Allow the medium to cool with tubes in an upright position.
4. Inoculate tubes of media by stabbing with a wire loop or needle.
5. Observations are made at 24 and 48 hour intervals. An acid reaction is indicated by the development of a distinct yellow color in the lower portion of the tube, the surface zone remaining red.

STORAGE
Bacto Phenol Red Tartrate Agar	Below 30°C
Prepared tubes	2 – 8°C

QUALITY CONTROL
Identity Specifications

Dehydrated powder:	pink, homogeneous, free-flowing
Reaction of 4% solution:	pH 7.6 ± 0.2 at 25°C
Prepared medium:	red, clear to slightly opalescent

**Typical Cultural Response in Bacto Phenol Red Tartrate Agar
After 24 – 48 Hours at 35 ± 2°C**

Organism	Recovery	Reaction
Edwardsiella tarda ATCC® 15947	good	–
Escherichia coli ATCC® 25922	good	+
Salmonella enteritidis ATCC® 13076	good	+
Salmonella schottmuelleri ATCC® 8759	good	–
Salmonella typhimurium ATCC® 14028	good	+

+ = acid, yellow
– = no acid, red

REFERENCES
1. J. Hyg., 23:1, 1924.
2. J. Infectious Diseases, 42:238, 1928.

PACKAGING
Bacto Phenol Red Tartrate Agar   500 g   0090-17-2

BACTO PHENYLALANINE AGAR

INTENDED USE
Bacto Phenylalanine Agar is recommended for the differentiation of members of the *Proteus* and *Providencia* groups from other members of the *Enterobacteriaceae*.

HISTORY/PRINCIPLES
Buttiaux, Osteux, Fresnoy and Moriamez[1] described a method based upon the formation of phenyl pyruvic acid from phenylalanine by members of the *Proteus* and *Providencia* groups. A modification of this method was made by Bynae[2] who incorporated phenylalanine in the medium used to grow the organism. Ewing, Davis and Reavis[3] improved and simplified the formulation of Bynae by omitting the proteose peptone. The production of phenyl pyruvic acid (by deaminase) was indicated by the formation of the characteristic green color upon the addition of an acidified ferric chloride solution to the culture. Bacto Phenylalanine Agar is prepared according to the formula of Ewing, Davis and Reavis.[3]

FORMULA
BACTO PHENYLALANINE AGAR
DEHYDRATED
Ingredients per liter

Bacto Yeast Extract . . . . 3 g
Dipotassium Phosphate . . . . 1 g
Sodium Chloride . . . . 5 g
DL-Phenylalanine . . . . 2 g
Bacto Agar . . . . 12 g

Final pH 7.3 ± 0.2 at 25°C.

One pound will make 19.7 liters of final medium.

PROCEDURE
1. To rehydrate, suspend 23 g in 1 liter distilled or deionized water and heat to boiling to dissolve completely.

2. Distribute in tubes and sterilize in the autoclave for 15 minutes at 15 lbs pressure (121°C).
3. Allow the medium to solidify in a slanted position.
4. Inoculate the slant with test organisms and incubate at 37°C for 18 – 24 hours.
5. Add 3 to 5 drops of an 8 – 12% ferric chloride solution and 3 to 5 drops of a 0.1N HCI solution to a 24 hour culture. Rotate the tubes to wet and loosen the growth.
6. A positive test is indicated by the formation of a characteristic green color. *Proteus* and *Providencia* groups will give a positive test in 1 to 5 minutes. Other members of the *Enterobacteriaceae* give negative reactions.

STORAGE
Bacto Phenylalanine Agar	Below 30°C
Prepared medium	2 – 8°C

QUALITY CONTROL
Identity Specifications

Dehydrated powder:	light tan, homogeneous, free-flowing
Reaction of 2.3% solution:	pH 7.3 ± 0.2 at 25°C
Prepared medium:	light amber, slightly opalescent

Typical Cultural Response on Bacto Phenylalanine Agar
After 18 – 24 Hours at 35 ± 2°C

Organism	Recovery	Reaction
Enterobacter aerogenes ATCC® 13048	good	–
Escherichia coli ATCC® 25922	good	–
Proteus mirabilis ATCC® 25933	good	+
Proteus vulgaris ATCC® 13315	good	+
Providencia alcalifaciens ATCC® 12013	good	+

REFERENCES
1. Ann. Inst. Pasteur, 87:375 – 386, 1954.
2. Personal Communication, 1956.
3. Pub. Hlth. Lab., 15:153, 1957.

PACKAGING
Bacto Phenylalanine Agar	1 lb (454 g)	0745-01-9
	1/4 lb (114 g)	0745-02-8
	5 lb (2.27 kg)	0745-05-5

BACTO PHENYLALANINE MALONATE BROTH

INTENDED USE
Bacto Phenylalanine Malonate Broth is used for the differentiation of gram-negative enteric bacteria on the basis of malonate utilization and formation of pyruvic acid from phenylalanine.

HISTORY/PRINCIPLES
Bacto Phenylalanine Malonate Broth is prepared according to the formulation given by Shaw and Clarke.[1] Ewing, Davis and Reavis[2] rather than using the Bacto Phenylalanine Malonate Broth preferred to use Bacto Malonate Broth and Bacto Phenylalanine Agar

separately. By so doing, their reactions were more easily observed and the differentiation more complete.

Klebsiella and *Salmonella arizonae*, capable of utilizing malonate, produce an alkaline reaction changing the color of the medium from light green to dark blue. Malonate negative organisms leave the medium a light greenish color. Members of the *Proteus* and *Providence* groups are capable of forming pyruvic acid from phenylalanine. The production of this material is determined by the addition of 3 to 5 drops ferric chloride reagent (8 – 12% $FeCl_3$ in acidified distilled water) to a 24 hour culture. Organisms producing pyruvic acid from phenylalanine give a deep green color; phenylalanine negative organisms show no change in color.

FORMULA
BACTO PHENYLALANINE MALONATE BROTH
DEHYDRATED

Ingredients per liter

Bacto Yeast Extract	1 g	Dipotassium Phosphate	0.6 g
Sodium Malonate	3 g	Monopotassium Phosphate	0.4 g
DL-Phenylalanine	2 g	Sodium Chloride	2 g
Ammonium Sulfate	2 g	Bacto Brom Thymol Blue	0.025 g

Final pH 6.3 ± 0.2 at 25°C.

One pound will make 41.2 liters of medium.

METHOD OF PREPARATION
1. Suspend 11 g in 1 liter distilled or deionized water and heat to boiling to dissolve completely.
2. Dispense into tubes.
3. Sterilize in the autoclave for 10 minutes at 10 lbs pressure (115°C).

STORAGE
Bacto Phenylalanine Malonate Broth   Below 30°C
Prepared medium   2 – 8°C

QUALITY CONTROL
Identity Specifications

Dehydrated powder: yellowish beige, homogeneous, free-flowing
Reaction of 1.1% solution: pH 6.3 ± 0.2 at 25°C
Prepared medium: yellowish green, clear

Typical Cultural Response in Bacto Phenylalanine Malonate Broth
After 18 – 24 Hours at 35°C

Organism	Growth	Malonate	Phenylalanine
Escherichia coli ATCC® 25922	good	–	–
Klebsiella pneumoniae ATCC® 13883	good	+	–
Proteus mirabilis ATCC® 25933	good	–	+
Providencia alcalifaciens ATCC® 9886	good	–	+
Salmonella arizonae ATCC® 13314	good	+	–
Salmonella typhimurium ATCC® 14028	good	–	–

REFERENCES
1. Shaw, C. and Clarke, Ph. H., 1955, Biochemical Classification of Proteus and Providence Cultures, J. Gen. Microbiol., 13:155 – 161.

2. Ewing, W. H., Davis, B. R. and Reavis, R. W., 1957, Phenylalanine and Malonate Media and Their Use in Enteric Bacteriology, The Public Health Laboratory, 15:153.

PACKAGING
Bacto Phenylalanine Malonate Broth   1 lb (454 g)   0806-01-5

BACTO PHENYLETHANOL AGAR

INTENDED USE
Bacto Phenylethanol Agar is a selective medium used for the isolation of staphylococci and streptococci from specimens also containing gram-negative organisms such as *Proteus* and *E. coli.*

HISTORY/PRINCIPLES
Brewer and Lilley[1] reported that the addition of phenylethanol to a nutritive medium will permit the growth of gram-positive organisms but markedly to completely inhibit the growth of gram-negative organisms found in the same specimen. When desired, blood plates of Bacto Phenylethanol Agar may be prepared. However, this blood medium is not recommended for the determination of hemolytic reactions. Only Bacto Azide Blood Agar Base to which is added 5% blood is satisfactory for the simultaneous isolation of staphylococci and streptococci and the identification of their hemolytic reactions.

FORMULA
BACTO PHENYLETHANOL AGAR
DEHYDRATED
Ingredients per liter

Bacto Tryptose	10 g
Bacto Beef Extract	3 g
Sodium Chloride	5 g
Bacto Agar	15 g
Phenylethanol	2.5 g

Final pH 7.3 ± 0.2 at 25°C.

One pound will make 12.5 liters of medium.

METHOD OF PREPARATION
1. To rehydrate, suspend 35.5 g in 1 liter distilled or deionized water. Heat to boiling to dissolve completely.
2. Sterilize in the autoclave for 15 minutes at 15 lbs pressure (121°C).
3. In the preparation of blood agar, cool the sterile medium to 45 – 50°C and add 5% sterile defibrinated blood under aseptic conditions.

STORAGE
Bacto Phenylethanol Agar   2 – 8°C
Prepared plates   2 – 8°C

QUALITY CONTROL
Identity Specifications

Dehydrated powder:	beige, homogeneous with soft clumps
Reaction of 3.55% solution:	pH 7.3 ± 0.2 at 25°C
Prepared medium:	light to medium amber, slightly opalescent

Typical Cultural Response on Bacto Phenylethanol Agar
After 18 – 48 Hours at 35 ± 2°C

Organism	Recovery (Plain)	Recovery (+5% Sheep Blood)
Escherichia coli ATCC® 25922	none to poor	none to poor
Salmonella typhi ATCC® 6539	none to fair	none to fair
Staphylococcus aureus ATCC® 25923	good to excellent	good to excellent
Streptococcus faecalis ATCC® 19433	fair to good	good to excellent

REFERENCE
1. Paper presented at the December, 1949, meeting of the Maryland Association of Medical and Public Health Laboratories.

PACKAGING
Bacto Phenylethanol Agar

1 lb (454 g)	0504-01-0
1/4 lb (114 g)	0504-02-9
5 lb (2.27 kg)	0504-05-6

BACTO PHOTOBACTERIUM BROTH

INTENDED USE
Bacto Photobacterium Broth is used for cultivating and demonstrating luminescence by photobacteria.

HISTORY
Bacto Photobacterium Broth is patterned after the formulations of Daudoroff[1] and Giese.[2]

PRINCIPLES
This medium contains the nutrients necessary for good growth and luminescence by photobacteria.

Cultures of photobacteria emit a green-blue light. The intensity of luminescence is related to the aeration of the culture: the greater the oxygen supply, the greater will be the luminescence.

Demonstrations of luminescence have effectively been used to arouse popular interest in the field of microbiology.

FORMULA
BACTO PHOTOBACTERIUM BROTH
DEHYDRATED
Ingredients per liter

Bacto Tryptone	5 g	Calcium Carbonate	1 g
Bacto Yeast Extract	2.5 g	Monopotassium Phosphate	3 g
Ammonium Chloride	0.3 g	Sodium Glycerol Phosphate	23.5 g
Magnesium Sulfate	0.3 g	Sodium Chloride	30 g
Ferric Chloride	0.01 g		

Final pH 7.0 ± 0.2 at 25°C.

One pound will make 6.8 liters of medium.

METHOD OF PREPARATION
1. Suspend 66 g in 1 liter distilled or deionized water and heat to boiling to dissolve completely.
2. Dispense into tubes or flasks to form a shallow layer of broth.
3. Sterilize in the autoclave for 15 minutes at 15 lbs pressure (121°C).

STORAGE
Bacto Photobacterium Broth   Below 30°
Prepared medium   15 – 30°

QUALITY CONTROL
Identity Specifications

Dehydrated powder: white, homogeneous, free-flowing
Reaction of 6.6% solution: pH 7.0 ± 0.2 at 25°C
Prepared medium: light amber, clear to very slightly opalescent with a heavy white precipitate

Typical Cultural Response in Bacto Photobacterium Broth
After 18 – 24 Hours at 25 – 30°C

Organism	Growth	Luminescence*
Lucibacterium harveyi ATCC® 14126	good to excellent	+
Vibrio fischeri ATCC® 7744	good to excellent	+

*best results are with fresh cultures and with aeration (shaking) during incubation.

REFERENCES
1. J. Bact., 44:451, 1942.
2. J. Bact., 46:323, 1943.

PACKAGING
Bacto Photobacterium Broth   1 lb (454 g)   0417-01-6

PHYTOHEMAGGLUTININ M/P

BACTO PHYTOHEMAGGLUTININ M
BACTO PHYTOHEMAGGLUTININ P

INTENDED USE
Bacto Phytohemagglutinin M and P are recommended for the isolation of viable lymphocytes from blood plasma by the selective agglutination and sedimentation of the erythrocytes and also for the initiation of mitosis of the lymphocytes *in vitro*.

HISTORY AND PRINCIPLES
HEMAGGLUTINATION
Bacto Phytohemagglutinin M or P is eminently suited for all hemagglutination techniques such as those described by Li and Osgood[1] and Takikawa, Ito, Kato, Yoshita, Kondo and Miyta.[2] Chen and Palmer[3] employed Bacto Phytohemagglutinin in conjunction with dextran and obtained excellent yields of morphologically and physiologically intact leukocytes in a suspension having no hemolysis. Amoeboid movement of the leukocytes so isolated were still observable 4 hours after isolation of the suspension.

Skoog and Beck[4] have obtained similar results after a preliminary sedimentation of a large fraction of the erythrocytes with dextran or fibrinogen followed by treatment with Bacto Phytohemagglutinin to remove the remaining erythrocytes. Seabright[5] found the phytohemagglutinin technique gave the highest yield of leukocytes (95%) of 6 different procedures investigated.

Bacto Phytohemagglutinin M or P agglutinates the erythrocytes of all human blood groups as well as those of many animals such as rabbit, dog, cat, chicken, duck, mouse, rat, sheep, horse, pig, frog and guinea pig. Phytohemagglutinin has been used to obtain the plasma suspension of trypanosomes from the blood of infected rats.[6]

The optimal amount of phytohemagglutinin required for the agglutination of all erythrocytes, leaving the maximal yield of leukocytes as determined by count, varies with the blood from different individuals. Usually the addition of 0.1 ml of the reconstituted Bacto Phytohemagglutinin M solution to 5 ml of blood containing an anticoagulant is optimal. Bacto Phytohemagglutinin P may be used in concentrations approximately 1/40th that of the Bacto Phytohemagglutinin M for the differential separation of the white cells and erythrocytes. After mixing the blood and phytohemagglutinin thoroughly, the tube is allowed to stand for 30 minutes and then gently centrifuged 2 minutes at 50 RCF (500 rpm) to obtain the maximal volume of plasma containing the suspension of leukocytes without the presence of a "buffy layer." The supernatant plasma containing viable leukocytes is obtained by aspiration.

MITOGENIC ACTIVITY
Bacto Phytohemagglutinin M was discovered by Nowell,[7] after related studies by Hungerford, Donnelly, Nowell and Beck,[8] to possess the remarkable ability to initiate mitosis in cultures of lymphocytes isolated from peripheral blood. Later, Bacto Phytohemagglutinin P was also shown to possess this property. The application of this technique has become of utmost importance in the characterization of chromosomes, particularly those existing in pathological conditions.

The procedure employed by Nowell[7] to observe the progress of mitosis in lymphocytes was a modification of the gradient technique described by Osgood and Krippachne.[9] Later, a simplified procedure was developed by Moorhead, Nowell, Mellman, Batipps and Hungerford[10] in which the cultures were routinely allowed to incubate for 3 days (65 – 70 hours). Their method incorporated the hypotonic treatment developed by Hughes[11] and by Hsu and Pomerat.[12] The flame drying of slides by Scherz[13] and the staining procedure by Rothfels and Siminovitch[14] were helpful contributions in this procedure. More recently, the staining of chromosomes by one of a number of methods[15,16,17,18,19,20] to produce characteristic bands has permitted their identification and karyotyping more accurately.

REAGENTS
Bacto Phytohemagglutinin M is a stable, nontoxic, desiccated mucophytohemagglutinin. Bacto Phytohemagglutinin P is a sterile, desiccated, purified, highly potent protein phytohemagglutinin from which the polysaccharide moiety has been removed.

Both Bacto Phytohemagglutinin M and P will agglutinate the erythrocytes of all human blood types, as well as those of animals. The rehydrated P-form has approximately 40 times more hemagglutinating potency than the M-form. Both forms will also stimulate the lymphocytes of peripheral blood to undergo mitosis in vitro. Each preparation of Bacto Phytohemagglutinin M and P must demonstrate a mitotic index of at least 30 with the lymphocytes from each sample of 4 different normal human bloods, when assayed according to the method outlined in the procedure, before it is approved for release.

METHOD OF PREPARATION
Bacto Phytohemagglutinin M or the sterile Bacto Phytohemagglutinin P is rehydrated by adding 5 ml of sterile distilled water, which produces a solution of approximately 1% in 0.85% saline (approximately 50 mg protein per 5 ml).

STORAGE
The desiccated Bacto Phytohemagglutinin M and P are stable to the expiry date on the label when stored at 2 – 8°C. The rehydrated solutions are stable for at least 2 weeks at – 20°C. If microbial contamination is inadvertently introduced in the rehydrated vial it will usually be evident by typical bacterial growth.

SPECIMEN COLLECTION AND PRESERVATION
For each culture, a 5 ml sample of blood is adequate. The blood is withdrawn with a sterile syringe and immediately placed in a sterile screw-capped test tube containing 30 units of sterile heparin and mixed thoroughly. It is convenient to have the heparin dissolved in 1 ml of a sterile 0.85% saline solution. The agglutination and mitotic procedures may be started without further delay or they may be postponed for at least 24 hours, if the specimen is kept at 2 – 8°C. Postponement will permit mailing the blood sample, if desired.

In conducting mitotic investigations, aseptic techniques should be observed from the collection of the blood sample until the addition of the TC Chromosome Arresting Solution.

Interferring Substances
For mitogenic investigations, the following should be avoided:
1. Anticoagulants containing oxalates or phenols;
2. Cytotoxic antibiotics, drugs or heavy metals (Penicillin and Streptomycin are permissible);
3. Hypertonic and hypotonic media, except for the intentional hypotonic swelling of the chromosomes;
4. Irradiation of the patient or culture, which can produce "breaks" in the chromosomes.

For laboratories performing chromosomal analyses only occasionally, the TC Chromosome Culture Kit (code 5842), which contains most of the essential reagents in pretested quantities, is very convenient.

For laboratories performing chromosomal analyses on only a few drops of whole blood, the TC Chromosome Microtest Kit (code 5060) is extremely useful, especially for examining the chromosomes of infants, small animals or for the screening of adults.

PROCEDURE
Materials Provided:
Bacto Phytohemagglutinin M or P

Available from Difco:
TC Medium RPMI #1640   TC Hanks Solution
TC Chromosome Arresting   Bacto Giemsa Stain
Solution

Materials Required but not Provided:
Pipette, 5 ml in 1/10 ml   Water aspirator
Pipette, 1 ml in 1/100 ml   Methanol, reagent grade, 15 ml
Pipette, 0.1 ml in 1/1000 ml   Acetic acid, reagent grade, 5 ml

Pasteur pipette with rubber bulb
Centrifuge, small
Culture bottle, screw-capped, 30 ml
Incubator at 37°C
Centrifuge tube, 15 ml

Microscope slides
Microscope (12.5X eyepiece with 10X low power, 40X high dry, and 100X oil immersion objectives)
Camera, if pictures desired

Lymphocyte Separation and Inoculation

1. Transfer 5 ml of blood containing 30 units of heparin to a sterile screw-capped test tube under aseptic conditions.
2. Add either 0.1 ml of rehydrated Bacto Phytohemagglutinin M or 0.0025 ml of Bacto Phytohemagglutinin P to the 5 ml of heparinized blood, and mix the contents by inverting several times.
3. Let the erythrocytes agglutinate at room temperature for 15 – 30 minutes.
4. Centrifuge the tube at 500 rpm (50 RCF) for 2 minutes. Excessive centrifuging **must** be avoided to prevent sedimentation of the lymphocytes.
5. Transfer the hazy plasma-lymphocyte suspension (about 2 ml) by means of a sterile Pasteur pipette and rubber bulb to 7 ml of a culture medium consisting of TC Medium RPMI #1640, 700 units of Penicillin, 700 μg of Streptomycin, and either 0.1 ml of Bacto Phytohemagglutinin M, if the erythrocytes had been agglutinated with the M-form, or 0.01 ml of Bacto Phytohemagglutinin P, if the erythrocytes had been agglutinated with the P-form. The optimal concentration of lymphocytes in the culture is $1.0 - 1.2 \times 10^6$/ml. If Bacto Phytohemagglutinin P and aseptic conditions are used, the antibiotics may be omitted. A culture medium containing all components, sterile and ready for use, is available from Difco as TC Chromosome Medium w/Reconstituting Fluid (code 5102-32-8). If only 3 or 4 drops of whole blood are to be used for culture, the TC Chromosome Microtest Medium w/Reconstituting Fluid (code 5398-32-1) is available.

Incubation of Culture

6. Incubate the culture, preferably in a vertical position, at 37°C with occasional swirling for **3 – 4 days.** Care should be taken to maintain proper incubation temperature. **A significant increase in mitotic index is often obtained by incubating 4 days instead of 3.** It is very important to maintain the proper pH range in the culture at all times. The phenol red indicator should not become more acidic than a light amber nor more alkaline than a light pink. If the indicator becomes amber, loosen the cap for an hour or so to allow the escape of CO_2. This precaution is often most necessary at the beginning and termination of the incubation.
7. Terminate the mitosis by the addition of 1 ml of 10^{-5} molar colchicine, available from Difco as TC Chromosome Arresting Solution (code 5845), and continue the incubation for another 4 – 6 hours at 37°C.
   The exposure of cells to the TC Chromsome Arresting Solution should not be less than 4 hours or more than 6.

Harvesting and Fixation of Cells

8. Transfer the entire culture to a 15 ml graduated conical centrifuge tube and centrifuge 6 – 8 minutes at 600 – 800 rpm.
9. Aspirate off the supernatant fluid.
10. Add 5 ml of TC Hanks Solution and resuspend the cells with a Pasteur pipette.
11. Centrifuge at 600 – 800 rpm for 6 – 8 minutes.
12. Aspirate off the supernatant fluid.
13. Add 1 ml of TC Hanks Solution and resuspend the cells with a Pasteur pipette.
14. Add 3 ml of distilled or deionized water in 1 ml portions, with swirling after each addition, to produce a hypotonic solution.
15. Incubate the suspension 10 minutes at 37°C. The exposure of the cells should not exceed 10 minutes.

16. Centrifuge the lymphocytes at 600 – 800 rpm for 6 – 8 minutes.
17. Aspirate off the supernatant fluid carefully to avoid loss of any lymphocytes.
18. Add slowly, without disturbing the button of cells, 4 ml of freshly prepared fixative consisting of 3 parts methanol and 1 part of glacial acetic acid (reagent grades).
19. Let the button of cells soak in the fixative 15 – 30 minutes. The cells should be treated gently while in the fixative.
   At this point, cells may be stored overnight at 2 – 8°C, if desired.
20. Resuspend the cells with the Pasteur pipette.
21. Centrifuge at 600 – 800 rpm for 6 – 8 minutes, and discard the supernatant fluid.
22. Resuspend the cells in 4 ml of fresh fixative with the Pasteur pipette, and centrifuge again at 600 – 800 rpm for 6 – 8 minutes. Repeat this step again, if necessary, to disperse clumps of cells.
23. Aspirate off the supernatant fluid.
24. Add 0.5 – 1.0 ml of fresh fixative to the button of cells and resuspend with the Pasteur pipette to get a hazy solution.

Preparation of Slides
25. Label clean microscope slides and place them in clean, chilled, distilled or deionized water.
26. In rapid succession, shake the excess water off a chilled slide, wipe the water off the underside, add 3 – 4 drops of the cell suspension by means of the Pasteur pipette, tip the slide several times to spread the suspension, and ignite the fixative by bringing it momentarily in contact with a flame as described by Scherz.[13] As soon as the fixative is burned off, wave the slide vigorously to hasten drying. The slides should not get hot, but drying should be accomplished as rapidly as possible.

Staining of Slides
Slides may be stained with Giemsa, orcein or other stains according to the method of Rothfels and Siminovitch.[14] The procedure using Giemsa is given below:

27. Dilute 1 ml of Bacto Giemsa Stain, (20 X stock), code 0846, with 19 ml of distilled or deionized water.
28. Stain the slides in a small staining dish or Petri dish for 10 – 20 minutes.
29. Rinse the slides in distilled or deionized water and air dry.
30. Examine the slides under the microscope. The mitotic spreads may be scanned at a total magnification of 125X, or examined more closely at 500X, or photographed under oil immersion at 1000X. Slides may be protected by cover slips and made permanent by conventional procedures.

Alternatively, the slides may be treated by staining procedures using Bacto Giemsa Stain to demonstrate G-banding of the chromosomes.[15,16,17,18,19,20]

RESULTS
A mitotic index of at least 30 may be expected from the lymphocytes from the heparinized peripheral blood of a healthy individual. The mitotic index is usually considerably higher. Ample mitotic spreads are produced for karyotyping the chromosomes.

Mitogenic Activity of Animals
The lymphocytes of many animals other than man have been stimulated in culture to produce blast cells and to exhibit chromosomes under the influence of Bacto Phytohemagglutinin M or P. Because of the differences in species, the cultural conditions must sometimes be altered. Also, as there are differences in the responses of the lymphocytes from two normal humans, there are sometimes differences in the responses obtained from two normal animals of the same species. Examples of animals

whose lymphocytes have been successfully stimulated by Bacto Phytohemagglutinin M or P to develop chromosomes are: cow, rabbit, rat, mouse, hamster, guinea pig, alligator, snake, frog, toad, kangaroo, salamander, trout, dog and African hedgehog.

Other Applications

Bacto Phytohemagglutinin M or P has the property of transforming lymphocytes of peripheral blood *in vitro* into "blast cells." These transformed cells are the precursors of mitosis and its concomitant chromosomes. The transformed blast cells are metabolically active and they have been used in the study of several lines of investigation. These studies include the sequence and rate of RNA and DNA syntheses; the relationship of an individual's age to mitotic response to phytohemagglutinin; the life-spans of lymphocytes; the stimulation of protein syntheses; the effects of drugs, antibiotics, and other therapeutic agents on cellular physiology; the examination of defective lymphocytes in disease; the protection of lymphocytes from the harmful effects of irradiation, nitrogen mustard and viruses; the sequence of chromosome formation in mitosis; taxonomic classification of animals, the enhancement of certain cellular enzymatic activities and other studies. An extensive bibliography on these applications is available from Difco.

LIMITATIONS OF THE PROCEDURE

The presence of a number of drugs in the culture, either inadvertently introduced by way of the individual's blood sample, or intentionally added to the culture for the study of their effects on cellular physiology, may be deleterious to the development of normal mitoses and chromosomes. The *in vitro* effects of some drugs are due to their inhibition of one of the sequential steps occurring in cell division, such as DNA-replication, RNA-synthesis, protein synthesis, deposition of messenger RNA on the ribosomes, and for other reasons. The *in vitro* determination of the step in lymphocyte cell division affected by a drug producing abnormal mitoses and chromosomes suggests a new method for the study of its mode of action. Cytotoxic effects are also shown by some plastic materials. Irradiation of patients often produces "breaks" in the chromosomes from their cultured lymphocytes.

The use of Bacto Phytohemagglutinin M or P only provides chromosomes in metaphase for subsequent studies either by banding techniques or by karotyping. Such techniques are beyond the scope of this manual. For details on these procedures refer to a standard text on cytogenetics.[21]

REFERENCES

1. Li and Osgood, Blood, 4:670, 1949.
2. Takikawa, et al., Acta Haemat., 18:179, 1957.
3. Chen and Palmer, Am. J. Clin. Path., 30:567, 1958.
4. Skoog and Beck, Blood, 11:436, 1956.
5. Seabright, J. Med. Lab. Tech., 14:85, 1957.
6. Yaeger, R. G., J. Parasitology, 46(3):288, 1960.
7. Nowell, P. C., Cancer Research, 20:462, 1960.
8. Hungerford, Donnelly, Nowell, and Beck, Amer. J. Human Genetics, 2:215, 1959.
9. Osgood and Krippachne, Exptl. Cell Res., 9:116, 1955.
10. *ibid.*, Moorehead, Nowell, Mellman, Batipps, and Hungerford, 20:613, 1960.
11. Hughes, Quart. J. Microscopic Sci., 93:207, 1952.
12. Hsu and Pomerat, J. Heredity, 44:23, 1953.
13. Scherz, Stain Tech., 37:386, 1962.
14. *ibid.*, Rothfels and Siminovitch, 33:73, 1958.
15. Sumner, Evans, and Buckland, Nature (New Biol.), 232:31, 1971.
16. *ibid.*, Schnell, 233:93, 1971.
17. Patil, Merrick, and Lubs, Science, 173:821, 1971.
18. Seabright, Lancet, ii:971, 1971.
19. Wang and Federoff, Nature (New Biol.), 235:52, 1972.
20. Dev. Warburton, and Miller, Lancet, i:1285, 1972.
21. Roberts, Fraser, J. A., An Introduction to Medical Genetics, Oxford University Press, 1963.

PACKAGING

Bacto Phytohemagglutinin M	5 ml	0528-56-6
	6 × 5 ml	0528-57-5
Bacto Phytohemagglutinin P	5 ml	3110-56-4
	6 × 5 ml	3110-57-3

BIBLIOGRAPHY OF CYTOGENETIC TECHNIQUES EMPLOYING PHYTOHEMAGGLUTININ M OR P

References to some of the investigations employing Bacto Phytohemagglutinin M or P have been arranged in the following groups:

Agglutination of erythrocytes (6, 29);

Transformation of lymphocytes (16, 20, 23, 27, 28, 31, 33, 48, 49, 50, 56, 58, 62, 74, 77, 83, 85, 88, 95, 104, 115);

Transformation of lymphocytes from pathological states (32, 43, 47, 53, 65, 66, 68, 81, 113, 116);

Effect on cellular physiology (21, 25, 29, 38, 41, 42, 45, 47, 53, 54, 67, 70, 71, 72, 92, 96, 106, 110, 111, 112, 113, 120);

Effect on virus and interferon production (17, 19, 34, 79, 84, 90, 103);

Effect on immunity (19, 43, 51, 52, 57, 61, 73, 89, 91, 97, 99, 101, 119);

Effect on transplantation (46, 89, 93, 94, 97, 99, 109, 114);

Effect in vivo (30, 36, 43, 52, 76, 78, 86, 87, 89, 90, 107, 108, 117, 118);

Effect of exposure to X-rays and radiation (15, 22, 41, 64, 69, 108);

Effect of chemicals on mitosis (26, 55, 59, 60, 102);

Effect on thymus, spleen, and Tonsillar Cells (37, 75, 80, 82, 100);

Miscellaneous Effects (24, 35, 44, 63).

REFERENCES

1. Li and Osgood, Blood, 4:670, 1949.
2. Takikawa, et al., Acta Haemat., 18:179, 1957.
3. Chen and Palmer, Am. J. Clin. Path., 30:567, 1958.
4. Skoog and Beck, Blood, 11:436, 1956.
5. Seabright, J. Med. Lab. Tech., 14:85, 1957.
6. Yaeger, R. G., J. Parasitology, 46(3):288, 1960.
7. Nowell, P. C., Cancer Research, 20:462, 1960.
8. Hungerford, Donnelly, Nowell, and Beck, Amer. J. Human Genetics, 2:215, 1959.
9. Osgood and Krippaehne, Exptl. Cell Res., 9:116, 1955.
10. Moorhead, Nowell, Mellman, Batipps, and Hungerford, ibid., 20:613, 1960.
11. Hughes, Quart. J. Microscopic Sci., 93:207, 1952.
12. Hsu and Pomerat, J. Heredity, 44:23, 1953.
13. Scherz, Stain Tech., 37:386, 1962.
14. Rothfels and Siminovitch, ibid., 33:73, 1958.
15. Schrek and Stefani, Nature, 200:482, 1963.
16. Elves, Gough, Chapman and Elves, Lancet, 1:306, 1964.
17. Nahmias, Kibrick and Rosan, J. Immun., 93:69, 1964.
18. Tien-wen Tao, Science, 146:247, 1964.
19. E. F. Wheelock, Science, 149:310, 1965.
20. Michalowski, Jasinska, Brzosko and Nowoslanski, Exp. Cell Res., 34:417, 1964, Exp. Med. Microbiol., 17:197, 1965.
21. Killander and Rigles, Exp. Cell Res., 39:701, 1965.
22. R. E. Millard, J. Clin. Path., 18:783, 1965.
23. W. O. Rieke, Leucocyte Workshop, Washington, D.C., 1965.
24. Sarkany and Caron, Brit. J. Dermatology, 77:439, 1965.

25. Tormey and Mueller, Blood, 26:569, 1965.
26. Nasjleti, Walden and Spencer, Cancer Research, 25:275, 1965.
27. Ling, Spicer, James and Williamson, Brit. J. Haemat., 11:421, 1965.
28. Joffey, Winter, Osmond and Meek, Brit. J. Haemat., 11:488, 1965.
29. A. A. MacKinney, Blood, 26:36, 1965.
30. C. N. Gamble, Blood, 28:174, 1965.
31. Sasaki and Norman, Nature, 210–913, 1966.
32. Kourilsky, Lovric and Levacher, Lancet, II:856, 1966.
33. Michalowski, Bartoszewicz and Kozubowski, Lancet, II:1130, 1966.
34. Edelman and Wheelock, Science, 154:1053, 1966.
35. H. L. Ioachim, Nature, 210:919, 1966.
36. Sarkany and Caron, Nature, 210:105, 1966.
37. Harris and Littleton, J. Exp. Med., 124:621, 1966.
38. Kleinsmith, Allfrey and Mirsky, Science, 154:780, 1966.
39. Borberg, Woodruff, Hirschhorn, Gesner, Miescher and Silber, Science, 154:1019, 1966.
40. Volante, Bussolati and Stramignoni, Pathologica, 48:1, 1966.
41. Nasjleti, Walden and Spencer, J. Nuclear Med., 7:159, 1966.
42. Z. J. Lucas, Science, 156:1237, 1967.
43. Spreadico and Lerner, J. Immun., 98:407, 1967.
44. Naspitz and Richter, Blood, 30:381, 1967.
45. Rabinowitz, Lubrano, Wilhite and Dietz, Exp. Cell Res., 48:675, 1967.
46. Pierre, Younger and Zmijewski, Proc. Soc. Exp. Biol. Med., 126:687, 1967.
47. Nadler, Monteleone and Hsia, Life Sci., 6:2003, 1967.
48. Yam, Castoldi and Mitus, J. Lab. Clin. Med., 70:699, 1967.
49. Batra and Schrek, Proc. Soc. Exp. Biol. Med., 125:871, 1967.
50. Moorhead, Connolly and McFarland, J. Immun., 99:413, 1967.
51. Gengozian and Hubner, J. Immun., 99:184, 1967.
52. Hege and Coli, J. Immun., 99:61, 1967.
53. R. Schrek, Arch. Path., 83:58, 1967.
54. Kong-oo-Goh, J. Lab. Clin. Med., 69:938, 1967.
55. Hersh and Oppenheim, Cancer Res., 27:98, 1967.
56. G. A. Caron, Brit. J. Haemat., 13:68, 1967.
57. D. E. Comings, Am. J. Obst. and Gynec., 97:213, 1967.
58. M. R. Schwarz, Proc. Soc. Exp. Biol. Med., 125:701, 1967.
59. Rawls, Olson, Melnick, Dent and Good, Sci., 158:506, 1967.
60. Laughman, Sargent and Israelstam, Sci., 158:508, 1967.
61. Golub and Weigle, J. Immun., 98:1241, 1967.
62. Mayron and Baram, J. Immun., 98:1274, 1967.
63. Holm and Perlmann, J. Exp. Med., 125:721, 1967.
64. Stefani and Oester, Transplantation, 5:317, 1967.
65. Kasajura and Lowenstein, Transplantation, 5:283, 1967.
66. Winter, McCarthy, Read and Joffe, Brit. J. Exp. Path., 48:66, 1967.
67. Hartog, Cline and Gradkey, Clin. Exp. Immun., 2:217, 1967.
68. Stefani and Fink, Gut, 8:249, 1967.
69. Roy Schmickel, Am. J. Human Genetics, 19:1, 1967.
70. Lobb, Curtain and Kidson, Nature, 214:783, 1967.
71. Parker, Wakasa and Lukes, Blood, 29:608, 1967.
72. Hirschhorn, Hirschhorn and Weissmann, Blood, 30:84, 1967.
73. Singhal, Naspitz and Richter, Int. Arch. Allergy, 31:390, 1967.
74. Kremer, Mengel, Nowlin and Nagaya, Blood, 30:62, 1967.
75. Winkelstein and Craddock, Blood, 29:594, 1967.
76. Chen-Mei Shaw, J. Immun., 102:63, 1968.
77. Naspitz and Richter, Brit. J. Haemat., 15:77, 1968.
78. Machado and Lozzio, Nature, 218:268, 1968.
79. Edelman and Wheelock, J. Virology, 2:440, 1968.
80. Claman and Brunstetter, J. Immun., 100:1127, 1968.
81. Dierks and Shepard, Proc. Soc. Exp. Biol. Med., 127:391, 1968.
82. Pick and Feldman, Proc. Soc. Exp. Biol. Med., 127:524, 1968.
83. Cooperband, Bondevik, Schmid and Mannick, Science, 159:1243, 1968.
84. Miller and Enders, J. Virology, 2:787, 1968.
85. Richter and Naspitz, Blood, 32:135, 1968.
86. Epstein and Smith, J. Immun., 100:421, 1968.
87. Richter, Naspitz, Blennerhassett and Rose, Int. Arch. Allergy, 35:417, 1969.
88. Hirschhorn, Nadler, Waithe, Broth and Hirschhorn, Sci., 166:1632, 1969.
89. Gertner, Harrah, Sample and Chretien, Surgical Forum, 20:254, 1969.
90. Wheelock and Edelman, J. Immun., 103:429, 1969.
91. Brody and Soltys, Blood, 34:765, 1969.
92. Williams and Granger, J. Immun., 103:170, 1969.
93. Markley, Thornton, Smallman and Markley, Transplantation, 8:258, 1969.

94. Hunter, Millman and Lerner, Transplantation, 8:413, 1969.
95. Brahmachary and Tapaswi, Experientia, 25:586, 1969.
96. Saillen, Jequier and Vannotti, J. Reticuloendothelial Soc., 6:175, 1969.
97. P. C. Y. Chan., Int. Arch. Allergy, 36:486, 1969.
98. Soltys and Brody, J. Lab. Clin. Med., 75:967, 1970.
99. Stefani and Moore, J. Immun., 104:780, 1970.
100. Levanthal and Talal, J. Immun., 104:918, 1970.
101. Abdou and Richter, J. Immun., 104:1009, 1970.
102. Stoltz, Khera, Bendall and Gunner, Sci., 167:1501, 1970.
103. Wheelock, Toy and Stjernholm, J. Immun., 105:1304, 1970.
104. Fanger, Hart, Wells and Nisonoff, J. Immun., 105:1484, 1970.
105. McMillan, Smith, Longmire, Reid and Craddock, J. Lab. Clin. Med., 76:333, 1970.
106. A. E. Rubin, Blood, 35:708, 1970.
107. Mohr, Beneke and Murr, Experientia, 26:1347, 1970.
108. S. Stefani, Experientia, 26:80, 1970.
109. Chen, Chung and Tsai, J. Immun., 107:601, 1971.
110. Rabinowitz, Farmer and Czbotar, Blood, 38:312, 1971.
111. Houck, Irausquin and Leikin, Science, 173:1139, 1971.
112. Riddick and Gallo, Blood, 37:282, 1971.
113. ibid., 37:293, 1971.
114. Hersh, Butler, Rossen, Morgan and Suki, J. Immun., 107:571, 1971.
115. E. M. Hersh, Science, 172:736, 1971.
116. Newberry and Sanford, J. Clin. Invest., 50:1262, 1971.
117. Folk and Pierre, Experientia, 27:444, 1971.
118. Mohr, Beneke and Murr, Experientia, 27:303, 1971.
119. Hellstrom-Zeromski and Perlman, Immun., 20:1099, 1971.
120. Marchesi and Andrews, Science, 174:1247, 1971.
121. Sumner, Evans, and Buckland, Nature (New Biol.), 232:31, 1971.
122. Schnell, ibid., 233:93, 1971.
123. Patil, Marrick and Lubs, Science, 173:821, 1971.
124. Seabright, Lancet, ii:971, 1971.
125. Wang and Federoff, Nature (New Biol.), 235:52, 1972.
126. Dev. Warburton, and Miller, Lancet, i:1285, 1972.

BACTO PIKE STREPTOCOCCAL BROTH

INTENDED USE

Bacto Pike Streptococcal Broth, prepared according to the formulation of Pike,[1] is a selective enrichment broth to which blood is added for the detection and isolation of hemolytic streptococci from throat swabs and other clinical specimens.

HISTORY/PRINCIPLES

Pike[1,2] devised this medium especially as a selective enrichment for hemolytic streptococci in throat swabbings and other specimens containing commensal microorganisms. The optimal balance of selective agents in the highly nutritive basal medium favored the growth of beta hemolytic streptococci and suppressed the growth of contaminating organisms such as Neisseria, Staphylococcus, Haemophilus, nonhemolytic streptococci and many Enterobacteriaceae. Using this medium (with 5% fresh rabbit blood) as a preliminary enrichment for throat swab specimens, Pike obtained 51 positive isolations of beta hemolytic streptococci, in contrast to 16 positives by direct streaking on blood agar. Preliminary enrichment suppressed most organisms other than beta hemolytic streptococci capable of producing hemolysis on blood agar, thereby reducing the work required in isolating and identifying hemolytic colonies of nonstreptococcal origin.

FORMULA

BACTO PIKE STREPTOCOCCAL BROTH
DEHYDRATED
Ingredients per liter

Bacto Tryptose . . . . 10 g
Bacto Casitone . . . . 10 g
Bacto Yeast Extract . . . . 10 g
Bacto Dextrose . . . . 0.2 g
Sodium Azide . . . . 0.065 g
Bacto Crystal Violet . . . . 0.002 g

Final pH 7.4 ± 0.2 at 25°C.

One pound will make 15 liters of medium.

PROCEDURE
1. Suspend 30.3 g in 1 liter distilled or deionized water. Heat to boiling to dissolve completely.
2. Dispense in 100 ml amounts in flasks.
3. Sterilize in the autoclave for 15 minutes at 15 lbs pressure (121°C).
4. Allow to cool to below 50°C and add 5% sterile defibrinated rabbit blood.
5. Mix well and dispense aseptically in 2 ml amounts in sterile tubes.
6. Place throat swabs or other clinical specimens into the enrichment broth. Incubate at 35°C for 18 – 24 hours then streak for isolation onto blood agar plates (Bacto Tryptic Soy Agar w/5% blood). Incubate plates at 35°C for 18 – 24 hours and examine for hemolytic colonies.

STORAGE
Bacto Pike Streptococcal Broth   Below 30°C
Prepared tubes   2 – 8°C

QUALITY CONTROL
Identity Specifications

Dehydrated powder: beige, homogeneous, free-flowing
Reaction of 3.03% solution: pH 7.4 ± 0.2 at 25°C
Prepared medium: purple, clear

Typical Cultural Response in Bacto Pike Streptococcal Broth
After 18 – 24 Hours at 35°C, then Plated on Bacto Tryptic Soy Agar w/5% Sheep Blood After 18 – 24 Hours at 35°C.

Organism	Growth	Hemolysis
Escherichia coli ATCC® 25922	poor to good	beta
Streptococcus faecalis ATCC® 19433	good to excellent	alpha, gamma
Streptococcus mitis ATCC® 9895	good to excellent	alpha
Streptococcus pyogenes ATCC® 19615	fair to good	beta

REFERENCES
1. Am. J. Hygiene, 41:211 – 220, 1945.
2. Proc. Soc. Exp. Biol. & Med., 57:187, 1944.

PACKAGING
Bacto Pike Streptococcal Broth   1 lb (454 g)   0898-01-4

BACTO PLATE COUNT AGAR
(STANDARD METHODS AGAR)
(TRYPTONE GLUCOSE YEAST AGAR)

INTENDED USE
Bacto Plate Count Agar is a Standard Methods medium used for enumeration of bacteria in water, waste water, dairy products and foods.[1,2,3]

HISTORY/PRINCIPLES
Buchbinder, Baris, Alff, Reynolds, Dillon, Pessin, Pincus and Strauss[4] in their studies to formulate new media for the standard plate count of dairy products, described media without added milk that gave satisfactory plate counts of raw and pasteurized milk when compared with tryptone glucose extract milk agar. A medium composed of Bacto Tryptone, Bacto Yeast Extract, Bacto Glucose and Bacto Agar was suggested as giving plate counts comparable with those obtained on the Standard Methods medium, tryptone glucose extract milk agar. The Committee on Applied Laboratory Methods of the International Association of Milk and Food Sanitarians Inc. continued this investigation. The results of this survey as reported by Pessin and Black[5] showed that media without added skim milk was superior to the standard medium with milk in that this medium was free from precipitate, gave a clearer background and produced larger colonies. A continuation of this study was carried out by the Subcommittee on Standard Methods for the Examination of Dairy Products of the American Public Health Association, the results of which were reported by Pessin and Robertson.[6] The medium containing 0.5% Bacto Tryptone, 0.1% Bacto Dextrose, 0.35% Bacto Yeast Extract and 1.5% Bacto Agar was superior to the tryptone glucose extract agar with added milk. The advantages of clarity of the new medium with accompanying ease of reading the plates was also noted. At the request of the American Public Health Association, Buchbinder, Baris and Goldstein[7] continued the study of milk-free media for the standard plate count of dairy products. Their results showed that a dehydrated milk-free medium containing 0.25% Bacto Yeast Extract, 0.5% Bacto Tryptone, 0.1% Bacto Dextrose and 1.5% Bacto Agar per liter approximated the productivity of tryptone glucose extract agar with added milk more closely than did the medium with 0.35% of Bacto Yeast Extract per liter. They recommended that a dehydrated culture medium be used in preparing the standard plate count medium rather than preparing the medium from ingredients. This formula, with 0.25% Bacto Yeast Extract is specified in Standard Methods for the Examination of Dairy Products. Its use became official in 1953.

Bacto Plate Count Agar is prepared with the same ingredients originally suggested by Buchbinder et al.[4] In this study they compared various brands of yeast extract and reported that for the preparation of the standard plate count medium only Bacto Yeast Extract gave satisfactory results. Combinations of Bacto Yeast Extract and Bacto Tryptone have been used in media for the examination of dairy products for the presence of thermophilic organisms since 1928.[8,9,10] The use of Bacto Tryptone in media for plate counts of dairy products has been specified by Standard Methods since 1939.[11]

Bacto Plate Count Agar is prepared from universally accepted, standardized ingredients[1,2,3] for microbiological culture procedures and is tested by APHA Methods. The label indicates that each control lot meets the prescribed standards of the APHA, USP and AOAC. The medium has the same productivity as the former Standard Methods medium, which was Bacto Tryptone Glucose Extract Agar with added milk.[12] The clarity of the medium and the increased size of the colonies permit the determination of bacterial counts with ease.

FORMULA

BACTO PLATE COUNT AGAR
DEHYDRATED

Ingredients per liter

```
Bacto Tryptone . . . . . . . . . . . . . . .  5 g
Bacto Yeast Extract . . . . . . . . . . . . 2.5 g
Bacto Dextrose (Glucose) . . . . . . . . .  1 g
Bacto Agar . . . . . . . . . . . . . . . . . 15 g
```

Final pH 7.0 ± 0.2 at 25°C.

One pound will make 19.3 liters of medium.

METHOD OF PREPARATION

1. Suspend 23.5 g in 1 liter of distilled or deionized water and heat to boiling to dissolve completely.
2. Sterilize in the autoclave for 15 minutes at 15 lbs pressure (121°C).

STORAGE

Bacto Plate Count Agar   Below 30°C
Prepared medium   15 – 30°C

QUALITY CONTROL

Identity Specifications

Dehydrated powder: light tan, homogeneous, free-flowing
Reaction of 2.35% solution: pH 7.0 ± 0.2 at 25°C
Prepared medium: light amber, slightly opalescent, no precipitate

Typical Cultural Response in Bacto Plate Count Agar
After 48 ± 1 Hour at 31 ± 1°C

Test Sample	Recovery
Pasteurized milk	good to excellent
Unpasteurized (raw) milk	good to excellent

REFERENCES

1. Standard Methods for the Examination of Water and Wastewater, 15th Ed., American Public Health Association, Inc., Washington, D.C., 1980.
2. Standard Methods for the Examination of Dairy Products, 14th Ed., American Public Health Association, Inc., Washington, D.C., 1978.
3. Compendium of Methods for the Microbiological Examination of Foods, American Public Health Association, Inc., Washington, D.C., 1976.
4. Buchbinder, Baris, Alff, Reynolds, Dillon, Pessin, Pincus and Strauss, 1951, Public Health Reports, 66:327.
5. Pessin and Black, 1951, Journal of Milk and Food Technology, 14:98.
6. Pessin and Robertson, 1952, Journal of Milk and Food Technology, 15:104.
7. Buchbinder, Baris and Goldstein, 1953, American Journal of Public Health 43:869.
8. Tech. Bulletin 147, NY State Agric. Exp. Station, 1928.
9. Journal of Dairy Science, 1932, 15:383.
10. Standard Methods of Milk Analysis, 6th Ed., 60:1934.
11. Standard Methods for the Examination of Dairy Products, 7th Ed. American Public Health Association, Inc., Washington, D.C., 1939.
12. Standard Methods for the Examination of Dairy Products, 9th Ed. American Public Health Association, Inc., Washington, D.C., 1948.

PACKAGING

Bacto Plate Count Agar

	1 lb (454 g)	0479-01-1
	1/4 lb (114 g)	0479-02-0
	5 lb (2.27 kg)	0479-05-7
	10 kg	0479-08-4
6 × 100 ml		0479-73-4

BACTO PLATE COUNT AGAR SPECIAL

INTENDED USE
Bacto Plate Count Agar Special, a modification of Bacto Plate Count Agar (Standard Methods Agar), is prepared according to the requirements of the Netherlands Dairy Association for determining the bacterial content of raw milk. The modified formula permits the use of the medium for performing electronic colony counts on milk and other dairy products.

FORMULA
BACTO PLATE COUNT AGAR SPECIAL
DEHYDRATED
Ingredients per liter

Bacto Yeast Extract . . . . 3.06 g
Bacto Tryptone . . . . 6.13 g
Bacto Dextrose . . . . 1.23 g
Bacto Agar . . . . 30.10 g

Final pH 7.0 ± 0.2 at 25°C.

Five hundred grams will make 12.3 liters of medium.

PREPARATION
1. Suspend 40.5 g in 1 liter distilled or deionized water. Warm flask gently with constant swirling, slowly increasing temperature to boiling. Continue boiling to completely dissolve the medium.
2. Sterilize in the autoclave for 20 minutes at 15 lbs pressure (121°C).
3. Cool slowly to 45 – 50°C in a water bath. Avoid gelling of the agar before pouring.

PROCEDURE
1. Pour medium into Petri dishes or tubes containing 0.5 – 1 ml diluted milk sample and rotate the container to disperse the sample throughout the medium. Allow the medium to solidify.
2. Invert and incubate at 32 or 35°C for 48 hours.

STORAGE
Bacto Plate Count Agar Special   Below 30°C
Prepared medium   15 – 30°C

QUALITY CONTROL
Identity Specifications
Dehydrated powder: beige, homogeneous, free-flowing
Reaction of 4.05% solution: pH 7.0 ± 0.2 at 25°C
Prepared medium: light amber, very slightly opalescent

Typical Cultural Response on Bacto Plate Count Agar Special
After 48 ± 1 Hours at 32 – 35°C

Inoculum	Recovery
Pasteurized Milk	excellent
Unpasteurized Milk	excellent

PACKAGING
Bacto Plate Count Agar Special   500 g   1812-17-7

BACTO m PLATE COUNT BROTH

INTENDED USE
Bacto m Plate Count Broth is a nonselective nutrient medium for determining bacterial counts by the membrane filter technique.

HISTORY/PRINCIPLES
Bacto m Plate Count Broth has the same formulation as Bacto Plate Count Agar except Bacto Agar has been omitted and the ingredients are employed in twice the concentration as in the solid medium.

FORMULA
BACTO m PLATE COUNT BROTH
DEHYDRATED
Ingredients per liter

Bacto Yeast Extract . . . . 5 g
Bacto Tryptone . . . . 10 g
Bacto Dextrose . . . . 2 g

Final pH 7.0 ± 0.2 at 25°C.

One pound will make 26.7 liters of medium.

METHOD OF PREPARATION
1. Suspend 17 g in 1 liter distilled or deionized water.
2. Sterilize in the autoclave for 15 minutes at 15 lbs pressure (121°C).

STORAGE
Bacto m Plate Count Broth   Below 30°C
Prepared medium   15 – 30°C

QUALITY CONTROL
Identity Specifications

Dehydrated powder: beige, homogeneous, free-flowing
Reaction of 1.7% solution: pH 7.0 ± 0.2 at 25°C
Prepared medium: light to medium amber, clear to very slightly opalescent

Typical Cultural Response on Bacto m Plate Count Broth
After 18 – 24 Hours at 35°C

Organism	Growth
Escherichia coli ATCC® 25922	good to excellent
Escherichia coli ATCC® 13762	good to excellent
Staphylococcus aureus ATCC® 25923	good to excellent
Staphylococcus epidermidis ATCC® 12228	good to excellent

PACKAGING
Bacto m Plate Count Broth   1 lb (454 g)   0751-01-0

BACTO PNEUMOCOCCUS ANTISERA

INTENDED USE
Bacto Pneumococcus Antisera Pools and Serotypes are prepared for use in the Neufeld (Quellung) reaction for the identification of the predominant serotypes of *Strepto-*

coccus (Diplococcus) pneumoniae. Pooled sera are employed for screening of suspect specimens, and individual serotype antisera provide specific serological identification.

SUMMARY AND EXPLANATION OF THE TEST

S. pneumoniae is the most common cause of bacterial (lobar) pneumonia and one of the 3 most common etiological agents of bacterial meningitis. Virulent pneumococci can be isolated from the nasopharynx of 30 – 70% of the normal population[1] making potential infection an ongoing clinical and public health concern. It is particularly detrimental for aged or debilitated individuals. The infection rate is lowest in children but progressively increases after age 40.

A capsular polysaccharide determines both the virulence and the specific serotype of pneumococcus. Serotyping is accomplished by capsular swelling (Quellung reaction) in the presence of specific immune serum. The Quellung reaction does not appear to be an actual "swelling" of the capsule. It has been reported that the apparent increase in capsular size is due to the reaction of the capsular substance and type specific serum antibody which forms an antigen-antibody complex on the cell surface and appears as a ground glass enlarged capsule when stained with methylene blue Loeffler.

The Quellung reaction may be used for rapid screening of sputum, spinal and pleural fluids in critical situations or from primary isolates to corroborate cultural and biochemical findings. It should be emphasized that direct examination of sputum, pleural and spinal fluid can only be considered presumptive evidence of *S. pneumoniae* as crossreactions may occur between these organisms and *Klebsiella pneumoniae* or *Haemophilus influenzae.*

An alternate rapid screening procedure can be performed using Bacto FA Pneumococcus Poly. For details refer to the monograph for this product.

Since the International Congress for Microbiology meeting held in Mexico in 1970, there has been a tendency to abandon the American system of nomenclature of *S. pneumoniae* and adapt the more comprehensive Danish system. Accordingly, Bacto Pneumococcus Antisera are in the process of such revision.

The currently supplied sera still follow the American system. However, in the future, Bacto Pneumococcus Antisera will be labeled according to the Danish system.

The following chart compares the two systems:

Table I Comparison of the Antigenic Schema for *Streptococcus Pneumoniae*

Danish	USA	Antigenic Formula	Danish	USA	Antigenic Formula
1	1	1a	21	21	21a
2	2	2a	22A	63	22a,22c
3	3	3a	22F	22	22a,22b
4	4	4a	23A	46	23a,23c,15c
5	5	5a	23B	64	23a,23b,23d
6A	6	6a,6b	23F	23	23a,23b,18b
6B	26	6a,6c	24A	65	24a,24c,24d
7A	7	7a,7b,7c	24B	60	24a,24b,24e,7h
7B	48	7a,7d,7e,7h	24F	24	24a,24b,24d,7h
7C	50	7a,7d,7f,7g,7h	25	25	25a,25b
7F	51	7a,7b	27	27	27a,27b
8	8	8a	28A	79	28a,28c,23d
9A	33	9a,9c,9d	28F	28	28a,28b,16b,23d

Table I (continued)

Danish	USA	Antigenic Formula	Danish	USA	Antigenic Formula
9L	49	9a,9b,9c,9f	29	29	29a,29b,13b
9N	9	9a,9b,9e	31	31	31a,20b
9V	68	9a,9c,9d,9g	32A	67	32a,32b,27b
10A	34	10a,10c,10d	32F	32	32a,27b
10F	10	10a,10b	33A	40	33a,33b,33d,20b
11A	43	11a,11c,11d,11e	33B	42	33a,33c,33d,33f
11B	76	11a,11b,11f,11g	33C	39	33a,33c,33e
11C	53	11a,11b,11c,11d,11f	33F	70	33a,33b,33d
11F	11	11a,11b,11e,11g	34	41	34a,34b
12A	83	12a,12c,12d	35A	47,62	35a,35c,20b
12F	12	12a,12b,12d	35B	66	35a,35c,29b
13L	13	13a,13b	35C	61	35a,35c,20b,42a
14L	14	14a	35F	35	35a,35b,34b
15A	30	15a,15c,15d,15g	36	36	36a,9e
15B	54	15a,15b,15d,15e,15h	37	37	37a
15C	77	15a,15d,15e	38	71	38a,25b
15F	15	15a,15b,15c,15f	39	69	39a,10d
16L	16	16a,16b,11d	40	45	40a,7g,7h
17A	78	17a,17c	41A	74	41a
17F	17	17a,17b	41F	38	41a,41b
18A	44	18a,18b,18d	42	80	42a,20b,35c
18B	55	18a,18b,18e,18g	43	75	43a,43b
18C	56	18a,18b,18c,18e	44	81	44a,44b,12b,12d
18F	18	18a,18b,18c,18f	45	72	45a
19A	57	19a,19c,19d	46	73	46a,12c,44b
19B	58	19a,19c,19e,7h	47A	84	47a,43b
19C	59	19a,19c,19f,7h	47F	52	47a,35a,35b
19F	19	19a,19b,19d	48	82	48a
20	20	20a,20b,7g			

REAGENTS

Bacto Pneumococcus Antisera are stable, desiccated antisera prepared in rabbits for use in the Neufeld (Quellung) reaction for the identification of *S. pneumoniae*. Antisera pools and specific antisera only for the more commonly encountered types of the currently known 83 types have been prepared. The pools and individual antisera are as follows:

Bacto Pneumococcus Antiserum Pool A 1,2,7
Bacto Pneumococcus Antiserum Pool B 3,4,5,6,8
Bacto Pneumococcus Antiserum Pool C 9,12,14,15,17,33
Bacto Pneumococcus Antiserum Pool D 10,11,13,20,22,24
Bacto Pneumococcus Antiserum Pool E 16,18,19,21,28
Bacto Pneumococcus Antiserum Pool F 23,25,27,29,31,32
Bacto Pneumococcus Antisera Set (Poly) Pools A,B,C,D,E,F
Bacto Pneumococcus Antiserum Type 1
Bacto Pneumococcus Antiserum Type 2
Bacto Pneumococcus Antiserum Type 3
Bacto Pneumococcus Antisera Set (Mono) Types 1,2,3,4,5,6,7,8,14,18 and 19

When properly rehydrated, each antiserum contains approximately 1:5000 Merthiolate®, sufficient to maintain a bacteriostatic condition at storage of 2 – 8°C.

When used according to the suggested procedure (0.05 ml per test) each 1 ml vial is sufficient to perform approximately 20 tests.

PRECAUTIONS
In the test all equipment and materials (slide, loop, culture) must be sterilized after use by flame (loop), autoclave (applicator sticks, slides, and culture) or by an adequate chemical method since living organisms are used as the antigen.

REHYDRATION
To rehydrate Bacto Pneumococcus Antisera add 1 ml sterile distilled or deionized water and rotate the vial to dissolve the contents completely.

STORAGE
Both unrehydrated and rehydrated Bacto Pneumococcus Antisera should be stored at 2 – 8°C.

Do not expose rehydrated serum to room temperature for prolonged periods of time. Discard any antiserum which becomes cloudy during storage. They should not be subjected to repeated freezing and thawing. Such treatment is detrimental to the antibody content.

EXPIRATION DATE
Both unrehydrated and rehydrated Bacto Pneumococcus Antisera will remain stable to the expiry date on the label when stored at 2 – 8°C.

SPECIMEN COLLECTION
Specimens for the isolation of S. pneumoniae may be obtained from cerebrospinal fluid, blood, pleural fluid, or sputum from the nasopharynx.

Specimens should be collected under sterile conditions and forwarded to the laboratory immediately.

SPINAL FLUID SPECIMENS
1. Centrifuge specimens for 15 minutes at 2,500 rpm immediately upon receipt.
2. Aspirate off supernatant with a sterile capillary pipette.
3. Use remaining fluid and sediment to inoculate appropriate medium (see Direct Screening below).

WIRE OR SWAB SPECIMENS
Streak onto surface of appropriate medium (see Direct Screening below).

CULTURAL ISOLATES
Inoculate Bacto Blood Agar Base, enriched w/5% blood with any of the specimens listed above. Incubate at 37°C for 18 – 24 hours in a candle jar or CO_2 incubator. Colonies exhibiting alpha hemolysis should be picked for serological examination and emulsified in sterile physiological saline, Bacto Todd Hewitt Broth, or Bacto Tryptose Phosphate Broth.

DIRECT SCREENING OF SPUTUM, SPINAL OR PLEURAL FLUID
A loopful of spinal or pleural fluid should be spread over an area 0.5 – 1 cm in diameter on a slide and allowed to air dry at room temperature.

A loopful of sputum emulsified in 1 – 2 ml sterile broth should be spread over an area 0.5 – 1 cm in diameter on a slide and allowed to dry at room temperature.

PROCEDURE

Materials Provided:
Bacto Pneumococcus Antisera Pools
Bacto Pneumococcus Antisera Types

Available from Difco:
Bacto Methylene Blue Loeffler
Bacto Rabbit Serum Normal
Bacto McFarland Barium Sulfate
  Standards

Materials Required but not Provided:
Burner
Microscope slides
Bacteriological loop

Applicator sticks
Microscope
Cover glass

Note: Bacto Pneumococcus Antisera are designed and tested for the Quellung procedure as outlined below. Any deviation from these techniques must be verified by the use of adequate controls.

QUELLUNG REACTION

1. Place a loopful of the sediment from the centrifuged spinal or pleural fluid or a diluted sputum or a broth culture of the organism to be tested on a microscope slide, or use air dried slide prepared for direct examination.

   Care should be taken not to employ too dense a bacterial suspension for the Quellung reaction. The optimal suspension will contain 25 – 50 organisms per microscopic field. A suspension approximating a Bacto McFarland Barium Sulfate Standard No. 1 is satisfactory. If the number of organisms is much greater than this and they appear clumped without apparent positive Quellung reaction, prepare another slide using fewer organisms before judging the slide to be negative.

2. Add a small drop of the appropriate antiserum and mix well.
3. Add a loopful of Bacto Methylene Blue Loeffler.
4. Mix thoroughly using a clean applicator stick and place a cover glass over the mixture.
5. Prepare a second microscope slide of the organisms to be tested using Bacto Rabbit Serum Normal instead of the Pneumococcus Antiserum. Add a loopful of Bacto Methylene Blue Loeffler and mix well. Place a cover glass over the mixture. This slide is a normal control slide to observe the normal capsule size of the unknown organism.

   It is necessary to run a known negative slide since the size of capsules varies considerably. A normal capsule size of one organism might be considered swelling when compared with the capsule size of another organism.

6. Even though a positive reaction should occur within 3 – 5 minutes, the specimen should not be considered negative until it has been examined 1 hour after preparation.

   Note: Avoid evaporation or drying of the test mixture during incubation. To prevent evaporation of the test mixture the slides should be incubated in a moist chamber. A Petri dish to which a piece of moist paper toweling is added is satisfactory.

7. Examine the slide microscopically using the oil immersion lens. The known negative slide is observed first to establish the normal size of the capsule before the test slide containing the immune serum is examined. A positive reaction is indicated by a definite increase in capsule size.

   If the organisms appear clumped with no swelling, prepare another slide using fewer organisms.

8. If swelling is not evident in one field, several fields should be examined since all cells do not quell.
9. A clinical specimen might contain more than one serotype of pneumococcus or a cross reaction might occur in 2 different sera by a given strain. If a culture exhibits capsular swelling in more than one immune serum, prepare dilutions of those antisera so reacting to determine the true type by differences in titer.

QUALITY CONTROL
Use known positive and negative control cultures in parallel with the test culture to ascertain the validity of test results.

RESULTS
If a positive result is obtained with one of the Bacto Pneumococcus Antisera Pools, the test should be repeated with the specific serotypes available comprised in that pool.

Further tests should be performed to corroborate the results of the Quellung reaction. The following characteristics should be confirmed to establish the organism as *S. pneumoniae.*

1. Exhibits alpha-hemolysis
2. Ferments inulin
3. Soluble in bile
4. Susceptible to optochin

LIMITATIONS OF THE PROCEDURE
While serological procedures, as applied to microorganisms, provide corroborative evidence and can be used to identify particular antigenic sites of genus or species under study, they cannot be used alone to identify the etiological agent of disease. Cultural isolation and at least preliminary biochemical differentiation must precede any serological examination, and final identification cannot be made without biochemical and serological characterization.

When interpreting the results of the Quellung reaction the following possible cross-reactions should be kept in mind:

K. pneumoniae Type 2 = *S. pneumoniae* Type 2
H. influenzae Type a = *S. pneumoniae* subgroup 6
H. influenzae Type b = *S. pneumoniae* group 15A and 35B, subgroups 6 and 29
H. influenzae Type c = *S. pneumoniae* 11

More than one organism has been known to occur in spinal fluid specimens.[2] Therefore, the combination of careful morphological, biochemical and serological evaluation is essential for positive identification of the etiological agent(s).

These antisera have been prepared for use to identify cultures which have been defined biochemically. Direct screening of sputum, spinal fluid or pleural fluid provides only presumptive evidence of *Streptococcus pneumoniae.*

REFERENCES
1. Lennette Spaulding and Truant, Manual of Clinical Micro., 3rd Ed. ASM, Ch. 8, 1980.
2. J. Pediatrics, 63:76 – 83, 1963.

PACKAGING
Bacto Pneumococcus Antiserum Pool A	1 ml	2369-50-0
Bacto Pneumococcus Antiserum Pool B	1 ml	2370-50-7
Bacto Pneumococcus Antiserum Pool C	1 ml	2371-50-6
Bacto Pneumococcus Antiserum Pool D	1 ml	2372-50-5

Bacto Pneumococcus Antiserum Pool E	1 ml	2372-50-4
Bacto Pneumococcus Antiserum Pool F	1 ml	2374-50-3
Bacto Pneumococcus Antisera Set (Poly)	6 × 1 ml	2330-32-9
Bacto Pneumococcus Antiserum Type 1	1 ml	2479-50-7
Bacto Pneumococcus Antiserum Type 2	1 ml	2480-50-4
Bacto Pneumococcus Antiserum Type 3	1 ml	2481-50-3
Bacto Pneumococcus Antisera Set (Mono)	11 × 1 ml	2446-32-0
Bacto Methylene Blue Loeffler	5 ml	3111-56-3

BACTO PORCINE HEART AGAR

INTENDED USE
Bacto Porcine Heart Agar, formerly known as Bacto Sensitivity Test Medium, is a general purpose infusion medium containing peptone for use in the disc plate technique for microbial sensitivity testing.

HISTORY/PRINCIPLES
Bacto Porcine Heart Agar is prepared with Bacto Proteose Peptone No. 3, Bacto Yeast Extract and Bacto Soytone. These three materials are rich in p-aminobenzoic (PABA), a natural by-product of protein hydrolysis. This material has the ability to inactivate sulfonamides and accordingly, Bacto Porcine Heart Agar cannot be recommended for testing the sensitivity of microorganisms to the sulfa drugs. Bacto Mueller Hinton Medium is recommended for antimicrobic sensitivity testing (including sulfonamides.) Bacto Porcine Heart Agar may be used in sensitivity tests of therapeutic agents other than sulfonamides.

This medium will also support the growth of a variety of microorganisms. Blood agar plates may be prepared but because of the dextrose content, porcine heart agar is not recommended for the determination of hemolytic reactions.

FORMULA
BACTO PORCINE HEART AGAR
DEHYDRATED
Ingredients per liter

Porcine Heart, Infusion from	375 g
Bacto Yeast Extract	3.5 g
Proteose Peptone No. 3, Difco	5 g
Bacto Soytone	6.5 g
Bacto Dextrose	5 g
Sodium Chloride	5 g
Bacto Agar	15 g

Final pH 7.2 ± 0.2 at 25°C.

One pound will make 9 liters of medium.

METHOD OF PREPARATION
1. Suspend 50 g in 1 liter distilled or deionized water. Heat to boiling to dissolve completely.
2. Sterilize in the autoclave for 15 minutes at 15 lbs pressure (121°C).
3. Allow to cool to 45 – 50°C and add 5% sterile sheep blood. Pour plates as desired.

STORAGE

Bacto Porcine Heart Agar
Prepared medium

Below 30°C
2 – 8°C

QUALITY CONTROL

Identity Specifications

Dehydrated powder: light tan, homogeneous, free-flowing
Reaction of 5.0% solution: pH 7.2 ± 0.2 at 25°C
Prepared medium: medium amber, slightly opalescent, may have a slight precipitate

Typical Cultural Response on Bacto Porcine Heart Agar (with or without 5% Sheep Blood) After 18 – 24 Hours at 35°C

Organism	Growth
Escherichia coli ATCC® 25922	good to excellent
Staphylococcus aureus ATCC® 25923	good to excellent
Streptococcus pneumoniae ATCC® 6303	good to excellent
Streptococcus pyogenes ATCC® 19615	good to excellent

PACKAGING

Bacto Porcine Heart Agar	1 lb (454 g)	0670-01-8

POTASSIUM TELLURITE, DIFCO

Potassium Tellurite, Difco is recommended for use as a selective agent in media for isolating and differentiating *Corynebacterium diphtheriae* varieties eg. *gravis*, *mitis* and *intermedius* on the basis of their colonial morphology on recommended culture media. It has also been used as a selective agent in media designed for isolating and differentiating certain species of streptococci and mycoplasmas.

Potassium Tellurite, Difco is readily water soluble and well suited for use in media requiring this chemical. Bacto Chapman Tellurite Solution 1% is prepared from potassium tellurite. A complete description of the media using this selective agent is given under Bacto PPLO Broth w/CV and Bacto Mitis Salivarius Agar.

PACKAGING

Potassium Tellurite, Difco	25 g	0384-13-1
	100 g	0384-15-9

BACTO POTATO DEXTROSE AGAR

INTENDED USE

Bacto Potato Dextrose Agar is a Standard Methods[1,2] plating medium used for culturing yeasts and molds from dairy and other food products.

HISTORY/PRINCIPLES

In a study of comparative methods and media used in the microbiological examination of creamery butter, Shadwick[3] investigated a number of media and found that Bacto

Potato Dextrose Agar gave the most consistent and highest count of yeasts and molds in salted and unsalted butter. The infusion from the potatoes encourages the luxuriant development of fungi.

It is frequently desirable in making yeast and mold counts to inhibit bacterial growth by acidifying the medium, and *Standard Methods* recommends that the reaction of the medium be reduced to pH 3.5 ± 0.1 subsequent to sterilization. The label of each package of Bacto Potato Dextrose Agar specifies the quantity of sterile tartaric acid (10% solution) that should be added to each 100 ml of the sterile melted medium to adjust the reaction to pH 3.5. After the tartaric acid has been added to the medium, it is mixed well and poured plates are prepared as usual. The medium should not be heated after the acid is added inasmuch as heating in the acid state will hydrolyze the agar, destroying its solidifying properties.

FORMULA

BACTO POTATO DEXTROSE AGAR
DEHYDRATED

Ingredients per liter

```
Potatoes, Infusion from  . . . . . . . . . 200 g
Bacto Dextrose . . . . . . . . . . . . . . .  20 g
Bacto Agar . . . . . . . . . . . . . . . . . .  15 g
```

Final pH 5.6 ± 0.2 at 25°C.

One pound will make 11.3 liters of medium.

METHOD OF PREPARATION

1. Suspend 39 g in 1 liter of distilled or deionized water and heat to boiling to dissolve completely.
2. Sterilize in the autoclave for 15 minutes at 15 lbs pressure (121°C).
3. Cool medium to 45 – 50°C and dispense into sterile Petri dishes.

The final pH of Bacto Potato Dextrose Agar is 5.6 ± 0.2 at 25°C. To obtain a reaction of pH 3.5 as required by *Standard Methods*,[1,2] acidify the medium with sterile 10% tartaric acid (U.S.P.). The amount of acid added per 100 ml of sterile cooled medium is stated on the package label. Do not heat the medium after addition of the acid.

STORAGE

Bacto Potato Dextrose Agar   Below 30°C
Prepared medium   15 – 30°C

QUALITY CONTROL

Identity Specifications

Dehydrated powder: light beige, homogeneous, free-flowing
Reaction of 3.9% solution: pH 5.6 ± 0.2 at 25°C
Prepared medium: light amber, slightly opalescent

Typical Cultural Response on Bacto Potato Dextrose Agar
After 5 Days at 21 – 25°C

Organism	Growth
Aspergillus niger ATCC® 16404	good to excellent
Candida albicans ATCC® 10231	good to excellent
Saccharomyces cerevisiae ATCC® 9763	good to excellent

REFERENCES

1. American Public Health Association, 1978, Standard Methods for the Examination of Dairy Products, 14th Ed., American Public Health Assoc., Inc., Washington, D.C.
2. American Public Health Association, 1976, Compendium of Methods for the Microbiological Examination of Foods, American Public Health Association, Inc., Washington, D.C.
3. Shadwick, 1938, Food Research, 3:287.

PACKAGING

Bacto Potato Dextrose Agar

1 lb (454 g)	0013-01-4
1/4 lb (114 g)	0013-02-3
5 lb (2.27 kg)	0013-05-0
12 tubes	0013-34-5
144 tubes	0013-37-2
6 × 100 ml	0013-73-7

BACTO POTATO DEXTROSE BROTH

INTENDED USE

Bacto Potato Dextrose Broth is a dehydrated medium recommended for culturing many yeasts and molds.

FORMULA

BACTO POTATO DEXTROSE BROTH
DEHYDRATED

Ingredients per liter

Potato, Infusion from . . . . 200 g
Bacto Dextrose . . . . 20 g

Final pH 5.1 ± 0.2 at 25°C.

One pound will make 18.9 liters of medium.

METHOD OF PREPARATION

1. To rehydrate, suspend 24 g in 1 liter distilled or deionized water and warm slightly to dissolve completely.
2. Distribute into tubes or flasks as desired and sterilize in the autoclave for 15 minutes at 15 lbs pressure (121°C).

STORAGE

Bacto Potato Dextrose Broth	Below 30°C
Prepared medium	15 – 30°C

QUALITY CONTROL

Identity Specifications

Dehydrated powder:	light beige, homogeneous, free-flowing
Reaction of 2.4% solution:	pH 5.1 ± 0.2 at 25°C
Prepared medium:	very light amber, clear without significant precipitate

Typical Cultural Response in Bacto Potato Dextrose Broth
After 40 – 48 Hours at 30°C

Organism	Growth
Aspergillus niger ATCC® 16404	good to excellent
Candida albicans ATCC® 10231	excellent
Lactobacillus casei ATCC® 7469	fair to good
Saccharomyces cerevisiae ATCC® 9763	good to excellent

PACKAGING
Bacto Potato Dextrose Broth 1 lb (454 g) 0549-01-7

POTATO FLOUR

Potato Flour is used especially for preparing infusions for culture media formulations, and as an additive to replace whole potatoes. It is also used in place of potato starch in media for isolating and culturing mycobacteria.

PACKAGING
Potato Flour 500 g 0380-17-1

BACTO POTATO INFUSION AGAR

INTENDED USE
Bacto Potato Infusion Agar is prepared according to the formula used by Stockman and MacFadyean for the isolation of *Brucella abortus*.

HISTORY/PRINCIPLES
This medium permits luxuriant growth of characteristic colonies of *B. abortus* from infected materials and may be used with excellent results in mass cultivation of *Brucella* in the preparation of vaccines and antigens. Bacto Tryptose Agar, which does not contain an infusion, is recommended as being far more satisfactory than Bacto Potato Infusion Agar for the isolation and cultivation of the *Brucella*.

FORMULA

BACTO POTATO INFUSION AGAR
DEHYDRATED

Ingredients per liter

Potatoes, Infusion from	200 g
Bacto Beef Extract	5 g
Proteose Peptone, Difco	10 g
Bacto Dextrose	10 g
Sodium Chloride	5 g
Bacto Agar	15 g

Final pH 6.8 ± 0.2 at 25°C.

Five hundred grams will make 10.2 liters of medium.

METHOD OF PREPARATION
1. To rehydrate, suspend 49 g in 1 liter of a 2% solution of Bacto Glycerol in distilled or deionized water and bring to a boil to dissolve the medium completely.
2. Distribute in tubes or flasks and sterilize in the autoclave for 15 minutes at 15 lbs pressure (121°C).

The final medium will contain a slight precipitate which settles rapidly. The presence of this precipitate in no way interferes with the use of the medium.

Best results are obtained on freshly prepared media with a moist surface. It is suggested, if the medium is not used the day it is prepared, that the agar be melted and allowed to resolidify in order to provide most satisfactory conditions for growth.

STORAGE

Bacto Potato Infusion Agar	Below 30°C
Prepared medium	2 – 8°C

QUALITY CONTROL

Identity Specifications

Dehydrated powder:	medium tan, homogeneous, free-flowing
Reaction of 4.9% solution	
in 2% glycerol:	pH 6.8 ± 0.2 at 25°C
Prepared medium:	medium amber, opalescent, with a precipitate

**Typical Cultural Response on Bacto Potato Infusion Agar
After 24 – 72 Hours at 35°C**

Organism	Growth
Brucella abortus ATCC® 4315	good to excellent
Brucella melitensis ATCC® 4309	good to excellent
Brucella suis ATCC® 6597	good to excellent
Streptococcus pneumoniae ATCC® 6303	good to excellent

PACKAGING

Bacto Potato Infusion Agar	500 g	0051-17-9

BACTO POTATO MALT AGAR

INTENDED USE

Bacto Potato Malt Agar is a potato infusion medium recommended for the cultivation and maintenance of fungi, and other aciduric microorganisms.

HISTORY

Bacto Potato Malt Agar is prepared according to the formulation given by Fischer.[1] In addition to the cultivation of smut fungi, this medium is suggested for the cultivation of other plant disease organisms and aciduric microorganisms requiring a high carbohydrate content and neutral or slightly alkaline reaction for optimum growth.

FORMULA

**BACTO POTATO MALT AGAR
DEHYDRATED**

Ingredients per liter

Potatoes, Infusion from	200 g
Bacto Malt Extract	20 g
Bacto Peptone	1 g
Saccharose, Difco	60 g
Bacto Agar	20 g

Final pH 7.4 ± 0.2 at 25°C.

One pound will make 4.3 liters of final medium.

METHOD OF PREPARATION
1. To rehydrate, suspend 105 g in 1 liter distilled or deionized water. Heat to boiling to dissolve completely.
2. Sterilize in the autoclave for 15 minutes at 15 lbs pressure (121°C).

STORAGE
Bacto Potato Malt Agar   Below 30°C
Prepared plates   2 – 8°C

QUALITY CONTROL
Identity Specifications
Dehydrated powder: tan, homogeneous, free-flowing
Reaction of 10.5% solution: pH 7.4 ± 0.2 at 25°C
Prepared medium: medium amber, slightly opalescent

Typical Cultural Response on Bacto Potato Malt Agar
After 40 – 48 Hours at 30°C

Organism	Growth
Aspergillus niger ATCC® 16404	good to excellent
Lactobacillus fermentum ATCC® 9338	good to excellent
Saccharomyces cerevisiae ATCC® 9763	good to excellent

REFERENCE
1. Personal Communication, 1955.

PACKAGING
Bacto Potato Malt Agar   1 lb (454 g)   0625-01-4

POTATO STARCH

Potato Starch is prepared and recommended for use in microbiological and other laboratory procedures requiring potato starch. It is used especially in culture media for isolating and propagating mycobacteria, yeasts and molds.

PACKAGING
Potato Starch   500 g   0381-17-0

BACTO PROTEOSE NO. 3 AGAR

INTENDED USE
Bacto Proteose No. 3 Agar is a basal medium used with enrichments for isolating and cultivating *Neisseria* and *Haemophilus*.

HISTORY
Interest in the cultural procedure for diagnosis of gonococcal infections was stimulated by Ruys and Jens,[1] McLeod and co-workers,[2] Thompson,[3] Leahy and Carpenter,[4] Carpenter, Leahy and Wilson[5] and Carpenter[6] who demonstrated the superiority of the

cultural procedure over the microscopic technique. However, lack of a simple, readily available and easily prepared culture medium delayed immediate, widespread acceptance of the new technique.

Difco Laboratories joined the effort and in 1938 introduced Bacto Proteose No. 3 Agar for use in preparing an enriched chocolate agar.

In a controlled study of twelve media used for isolating gonococci,[7] Bacto Proteose No. 3 Agar enriched with Bacto Hemoglobin and Bacto Supplement A yielded results comparable to more complex media, ranking just slightly less than Bacto GC Medium Base at 24 hours.

A complete historical discussion of the development of the technique for isolating and cultivating *Neisseria gonorrhoeae* is contained in the *Difco Manual, Ninth Edition*.

PRINCIPLES
Chocolate agar prepared from Bacto Proteose No. 3 Agar and Bacto Hemoglobin supports growth of the gonococcus not only in pure culture but also from the mixed cultures encountered in chronic gonococcal infections where other organisms often overgrow the gonococcus.

The growth rate of the gonococcus is increased by the addition of Bacto Supplement A, B or VX, all of which provide the growth factors glutamine and cocarboxylase.

FORMULA
BACTO PROTEOSE NO. 3 AGAR
DEHYDRATED
Ingredients per liter

Proteose Peptone No. 3, Difco	20 g
Bacto Dextrose	0.5 g
Sodium Chloride	5 g
Disodium Phosphate	5 g
Bacto Agar	15 g

Final pH 7.3 ± 0.2 at 25°C.

One pound will make 10 liters of single-strength medium.

METHOD OF PREPARATION
1. Suspend 45 g in 500 ml distilled or deionized water and heat to boiling to dissolve completely.
2. Sterilize in the autoclave for 15 minutes at 15 lbs pressure (121°C). Cool to 50 – 60°C.
3. Aseptically add 500 ml sterile 2% solution of Bacto Hemoglobin or Bacto Hemoglobin Solution 2%. Mix well.
4. Add 10 ml of Bacto Supplement A, B or VX. Mix well.
5. Dispense into sterile Petri dishes.

STORAGE
Bacto Proteose No. 3 Agar	Below 30°C
Prepared medium	2 – 8°C

QUALITY CONTROL
Identity Specifications

Dehydrated powder:	beige, homogeneous, free-flowing
Reaction of 4.5% solution:	pH 7.3 ± 0.2 at 25°C.
Prepared medium:	light to medium amber, opalescent with slight flocculent precipitate

Typical Cultural Response on Chocolate Agar Prepared with Bacto Proteose No. 3 Agar After 18 – 48 Hours at 35°C

Organism	Recovery
Haemophilus influenzae ATCC® 19418	good to excellent
Neisseria gonorrhoeae ATCC® 19424	good to excellent
Neisseria meningitidis ATCC® 13102	good to excellent
Neisseria sicca ATCC® 9913	good to excellent

REFERENCES

1. Muench. Wochschr., 80:846, 1933.
2. J. Path. Bact., 39:221, 1934.
3. J. Infectious Diseases, 61:129, 1937.
4. Am. J. Syphilis, 20:347, 1936.
5. Am. J. Syphilis, 22:55, 1938.
6. Seventh Annual Yearbook (1936 – 37), p. 133, Suppl., Am. J. Pub. Health, 27: No. 3, 1937.
7. Am. J. Syphilis Gonorrh. Venereal Diseases, 33:164, 1949.

PACKAGING

Bacto Proteose No. 3 Agar	1 lb (454 g)	0065-01-1
	1/4 lb (114 g)	0065-02-0
	5 lb (2.27 kg)	0065-05-7

PROTEOSE PEPTONE

PROTEOSE PEPTONE DIFCO
PROTEOSE PEPTONE NO. 2 DIFCO
PROTEOSE PEPTONE NO. 3 DIFCO

The development of Proteose Peptone Difco, Proteose Peptone No. 2 Difco, and Proteose Peptone No. 3 Difco was the result of accumulated information in our laboratories that no single, previously developed peptone was the most suitable nitrogen source for use in all media for culturing nutritionally fastidious bacteria. Extensive investigations were undertaken at Difco, using peptic digests of animal tissue prepared under varying digestion parameters. These were evaluated for provision of growth requirements of many fastidious pathogenic and nonpathogenic bacteria, and the initial premise was proven to be correct. The final outcome of the studies of the nutritional values of these peptones indicated that Proteose Peptone No. 3 Difco was the most superior of the 3 peptones for nutritional purposes in most media formulations for culturing fastidious microorganisms. This peptone can also replace the meat infusion-peptone combination in infusion media. It is thus well-suited for the growth of a large variety of organisms including streptococci, staphylococci, meningococci, pneumococci, gonococci and other microorganisms requiring a highly nutritious substrate.

Proteose Peptone No. 2 Difco is used for producing bacterial toxins and is suitable for preparing culture media for nutritionally less-demanding bacteria.

Proteose Peptone Difco quickly became established as the peptone of choice for use in the production of diphtheria toxin and is now employed extensively in laboratories throughout the world engaged in the production of this and other bacterial toxins.

PACKAGING

Proteose Peptone Difco	1 lb (454 g)	0120-01-4
	1/4 lb (114 g)	0120-02-3
	10 kg	0120-08-7
Proteose Peptone No. 2 Difco	1 lb (454 g)	0121-01-3
	10 kb	0121-08-6
Proteose Peptone No. 3 Difco	1 lb (454 g)	0122-01-2
	1/4 lb (114 g)	0122-02-1
	5 lb (2.27 kg)	0122-05-8
	10 kg	0122-08-5

BACTO PROTEUS ANTIGENS
BACTO PROTEUS ANTISERA

PRODUCT DESCRIPTION AND INTENDED USE

Bacto Proteus Antigens are chemically inactivated suspensions of internationally recognized strains of this genus for use in the slide and tube agglutination procedures to detect antibodies (agglutinins).

The rapid slide procedure is a screening test designed to detect agglutinins whereas the tube test is a presumptive test procedure designed to quantitate agglutinin compositions. It is, therefore, necessary that any positive results obtained in the screening (slide test) of specimens be confirmed by a tube test.

Bacto Proteus Antisera are hyperimmune antisera prepared in rabbits designed to demonstrate agglutination for use as a positive control employing these test procedures.

These reagents are for use only in the test procedure described herein. They are not recommended for the direct diagnosis of a disease but rather only for the detection of the presence of agglutinins, particularly in cases of seroconversion.

SUMMARY AND BACKGROUND

"Febrile Antigen" is a term which has been accepted generally as referring to bacterial suspensions representative of a number of species of microorganisms pathogenic to man and which are accompanied by a fever in the host. There are a number of such organisms but, by classical definition, the most prominent include the genera *Salmonella, Brucella, Francisella* (*Pasteurella*), *Leptospira* and some members of the *Rickettsia*.[1,2,3,4]

The *Rickettsiaceae* are obligate parasites primarily regarded as bacteria. Microorganisms within this family require natural hosts and an arthropod vector to maintain the disease producing cycle and serve as the mode of transmission. According to the 8th Edition of *Bergey's Manual*[5] there are 3 distinct biotypes of the genus *Rickettsia*, the typhus group, the Rocky Mountain spotted fever group and the scrub typhus group as outlined in Table I. Two other genera within this family produce rickettsioses in man, *Rochalimaea quintana* and *Coxiella burnetii* causing trench fever and Q fever, respectively. The latter 2 etiological agents do not elicit agglutinins to Proteus antigens in the host.

Ideally, diagnosis of rickettsial infections is made by 1) isolation and identification of the causative agent and 2) demonstration of a specific rising antibody titer. Generally, serological procedures are the primary requisite of identification since rickettsiae require host cells for cultivation *in vitro* and because of the danger involved in handling live pathogenic *Rickettsia*. Serological tests for the diagnosis of rickettsial disease include the complement fixation test, immunofluorescence, microagglutination technique and nonspecific agglutination tests.

One of the most widely used serological procedures is a nonspecific agglutination test, the Weil-Felix reaction.

Weil and Felix discovered a fortuitous cross reaction between the agglutinins produced by the causative agent of typhus fever, *R. prowazekii*, and certain types of *Proteus vulgaris* isolated from patients suffering from typhus fever. These *Proteus* strains designated OX19 and OX2 seem to share an alkali stable polysaccharide with *R. prowazekii* and, therefore, produce agglutinins identical to it.[6,7]

The well known Weil-Felix reaction utilizing the Proteus antigen was extended when, in 1923, Kingsbury isolated another Proteus strain, OXK, which gave a positive reaction with serum from patients convalescing from scrub typhus.

In the rapid slide test, varying amounts of the sera to be tested are distributed to the corresponding squares of a previously marked glass slide. A Proteus slide antigen is added and mixed with the sera. Positive or negative results are read within 1 minute after mixing.

The macroscopic tube test should be used to confirm the presumed presence of rickettsial agglutinins demonstrated by the slide technique and to quantitate their titer in suspected sera. In this test, the serum is serially diluted in test tubes, a constant amount of the Proteus tube antigen is added to each tube, the resultant mixture is incubated according to directions and the agglutination pattern is read and recorded.

Because the Weil-Felix reaction is a nonspecific agglutination test, in some cases there will be no demonstrable titer to the Proteus antigens. Patients with typhus fever do not always develop an appreciable titer to these antigens. Yet, patients who have been immunized or those with a *Proteus* infection may show significant titers. Finally, species differentiation cannot be accomplished with these nonspecific antigens. In spite of this, the sensitivity of the Weil-Felix reaction is such that the test is a valuable tool in screening specimens for typhus, scrub typhus, or spotted fever, although it will not differentiate between the typhus group and the spotted fever group. The test provides a simple, rapid screening method of acceptable reliability in laboratories which are not fully equipped to isolate and identify these pathogens or for those in which the occurrence of rickettsial disease is infrequent.

Table I Weil-Felix Reactions

Biotype	Rickettsial Agent	Disease	OX19	OX2	OXK
Typhus group	*R. prowazekii*	epidemic typhus	+	+	−
		Brill Zinnser	−(+)*	−(+)	−
	R. typhi	murine typhus	+	+	−

Table I (continued)

| Biotype | Rickettsial Agent | Disease | Reaction with | | |
			OX19	OX2	OXK
Spotted fever group	R. rickettsii	Rocky Mountain spotted fever	+	+	–
	R. sibirica	Siberian tick typhus	+	+	–
	R. conorii	fievre boutonneuse	+	+	–
	R. australis	queensland tick typhus	+	+	–
	R. akari	rickettsialpox	–	–	–
Scrub typhus group	R. tsutsugamushi	scrub typhus	–	–	+**

* = generally negative   ** = positive in approximately 40 – 60% of cases

REAGENTS
Antigens
Bacto Proteus OX19 Antigens, slide and tube, are prepared from nonmotile strains of *Proteus vulgaris* OX19.

Bacto Proteus OX2 Antigens, slide and tube, are prepared from nonmotile strains of *Proteus vulgaris* OX2.

Bacto Proteus OXK Antigens, slide and tube, are prepared from nonmotile strains of *Proteus mirabilis* OXK.

Concentration of Antigens
Bacto Proteus Antigens, slide are adjusted to a concentration of approximately 5% v/v.

Bacto Proteus Antigens, tube are adjusted to approximate a McFarland Barium Sulfate Standard No. 3 (9×10^8 organisms/ml).

Note: The density of the antigens may vary from the above values. They are adjusted to perform at their optimum when standardized with hyperimmune sera obtained from laboratory animals. In some antigens glycerin to a final volume of 20% is added when necessary to adjust the suspensions' sensitivity.

Preservatives
Bacto Proteus Antigens, slide contain a final concentration of 0.25% formaldehyde, crystal violet in approximately 0.005% final concentration and brilliant green in approximately 0.003% final concentration.

Bacto Proteus Antigens, tube contain a final concentration of 0.25% formaldehyde. They contain no dye.

Note: Due to various factors, including the genera of the organisms involved, the density of each lot of suspension, etc., the color intensity of the dye in each lot of slide antigen may vary and is normal. Such differences in color will not affect the outcome of the test.

Bacto Proteus Antigens are not to be used for immunization of man or animals.

Control Sera

Bacto Proteus Antisera are unabsorbed antisera prepared in rabbits designed to demonstrate agglutination employing these test procedures to a "significant" titer or greater with its homologous antigen, significant titer being defined as a 1:160 final test dilution or greater.

Merthiolate® is added as a preservative in the amount of 1:5,000 (w/v) final concentration.

Bacto Febrile Negative Control is a standardized protein solution containing Merthiolate® as a preservative in the amount of 1:5,000 (w/v) final concentration.

INSTRUCTIONS FOR REHYDRATION AND STORAGE

1. The antigens for both tests are in suspension and ready for use.
2. The Bacto Proteus Antisera are rehydrated by adding 3 ml sterile 0.85% NaCl solution at room temperature.
3. Bacto Febrile Negative Control is rehydrated by adding 5 ml sterile distilled or deionized water.
4. Store all reagents at 2 – 8°C. These reagents are stable to the expiry date on the label when stored at 2 – 8°C provided they have not been exposed to temperature extremes (excessive heat or freezing) and have not become contaminated chemically or bacterially during routine usage.

When used according to the recommended procedure, the reagents will yield the following:

Reagent	Vial	Number of Tests
Bacto Proteus Antigens (slide)	5 ml	33 slide
Bacto Proteus Antigens (tube)	25 ml	6 tube
Bacto Proteus Antisera	3 ml	19 slide or 30 tube
Bacto Febrile Negative Control	5 ml	32 slide or 50 tube

SPECIMEN COLLECTION

1. Collect 5 – 10 ml whole blood aseptically from the patient.
2. Allow the blood to clot and obtain the syneresed serum with a Pasteur pipette. If the serum is not free of erythrocytes, clarify by centrifugation. Note: Do not heat the serum, since significant antibodies may be thermolabile.

PROCEDURE

Materials Provided:

Bacto Proteus OX2 Antigen (slide)
Bacto Proteus OX2 Antigen (tube)
Bacto Proteus OX2 Antiserum
Bacto Proteus OX19 Antigen (slide)
Bacto Proteus OX19 Antigen (tube)
Bacto Proteus OX19 Antiserum
Bacto Proteus OXK Antigen (slide)
Bacto Proteus OXK Antigen (tube)
Bacto Proteus OXK Antiserum
Droppers for the above antigens (supplied droppers deliver approximately 0.03 ml per drop)

Materials Required but not Provided:
Rapid Slide Test
  Wax marking pencils
  Ruler
  Glass plate 8″ × 8″
  0.2 ml serological pipettes
  Applicator sticks
  Suitable light source
  Isotonic saline (0.85 g NaCl per 100 ml distilled or deionized water)

Purchase Separately:
Bacto Febrile Negative
  Control, 2239-56-0

Tube Test
  Isotonic saline (0.85 g NaCl per 100 ml distilled or deionized water)
  Sterile distilled or deionized water
  Kahn tubes 2 × 75 mm
  Kahn type tube supports
  Serological pipettes 1 ml and 5 ml capacity
  Water bath at 37°C
  Refrigerator 2 – 8°C
  Suitable light source

Because a change or no change in titer over a period of time are the best indicators of active infection or noninfection, respectively, and because the accuracy and precision of the tests can be affected not only by test conditions but also by the subjectivity of the person in reading the end point, the following protocol is recommended.

A preliminary test using either rapid slide test and/or the macroscopic tube test may be performed on the initial serum specimen and reported to the physician at that time. An aliquot of the serum should be transferred to a sterile test tube, sealed tightly, and kept in the freezer. When the second serum is obtained, it should be run in parallel with the original specimen. In this manner the original serum will serve as a control and any difference in titer will be more credible since the bias associated with the performance of the test as well as determining the end point will be minimized. Obviously, it would be best if the person who performed the original test also performed the second test.

Rapid Slide Test
The rapid slide test should be used for screening only. Any positive results in the slide procedure should be quantitated by the presumptive tube test.
1. Prepare the glass plate 8″ × 8″ by making a series of 1-1/2″ ruled squares with the ruler and wax marking pencil. Each row of 5 squares is sufficient to test one antigen against the sera diluted to 1:320. Note: The glass plate should be thoroughly cleaned and dried after each use.
2. Pipette 0.08, 0.04, 0.02, 0.01, and 0.005 ml serum onto a row of squares on the ruled glass plate using a 0.2 ml serological pipette.
3. Place a drop of the appropriate antigen for the slide test on each drop of serum. Note: Shake the antigen well before using.
4. Mix each serum-antigen composite with an applicator stick starting with the 0.005 ml serum dilution and working towards the 0.08 ml dilution. The final dilutions are correlated approximately to the macroscopic tube dilutions and are 1:20, 1:40, 1:80, 1:160 and 1:320, respectively.
5. Hold the glass plate in both hands and gently "rotate" it 15 – 20 times. Observe the agglutination within 1 minute over a suitable light source. Reactions occuring later may be due to the reactants drying on the slide.

Exposure to heat (from external sources, light source, burner flame, etc.) may result in evaporation of the test mixture and may result in false positive interpretations. Avoid such evaporation of the test mixture.

Macroscopic Tube Test

Prepare serial dilutions of serum to be tested and the control sera in 0.5 ml amounts in Kahn tubes in the following manner:

1. Place 8 Kahn tubes in a rack for each serum to be tested.
2. Pipette 0.9 ml of 0.85% NaCl into the first tube of each row and 0.5 ml into each of the remaining tubes.
3. Add 0.1 ml of the serum to the first tube containing 0.9 ml of NaCl solution.
4. Mix well with a pipette and transfer 0.5 ml to the second tube. Mix thoroughly.
5. Continue carrying the 0.5 ml of the serum dilution through tube 7. Discard 0.5 ml from tube 7 after mixing thoroughly. Tube 8 is the antigen control tube.
6. Add 0.5 ml of the desired antigen to each of the 8 tubes. Shake the racks to mix the antigen and antiserum. The resultant dilutions are 1:20 through 1:1280, respectively.
7. Incubate in a water bath at 37°C for 2 hours followed by overnight incubation at 2 – 8°C. It is important in this test to use the recommended time and temperature of incubation and to make certain the water bath is in a location free of mechanical vibration.

CALIBRATION AND QUALITY CONTROL

For greater proficiency in test interpretation always include a positive and negative serum control in each test protocol.

Both positive and negative controls are diluted in the same proportion as the patient's serum and processed in exactly the same manner following procedure above for the rapid slide test or for the macroscopic tube test. Bacto Febrile Negative Control may be used as the negative control in both tests.

It is optimum, if acute and convalescent specimens of known titer of each disease are available, that they be used as positive controls in the test protocols.

An antigen is considered to be satisfactory if it does not clump with the negative control and it reacts to a titer of 1:160 or more with the positive control.

DISCARD

1. Any control serum in which a precipitate forms or which does not dissolve completely upon rehydration.
2. Any antigen when clumps appear on the cap or dropper or which agglutinates spontaneously or which agglutinates in the negative control.
3. All antigens and control sera when past the end of the month of the expiry date.

RESULTS

Rapid Slide Test

Record results as follows:

4+	complete agglutination
3+	approximately 75% of the cells are clumped
2+	approximately 50% of the cells are clumped
1+	approximately 25% of the cells are clumped
±	trace agglutination
−	no agglutination

The titer of the serum is recorded as that dilution of the specimen in which at least 2+ (50%) agglutination occurs. See Tables II and III.

Macroscopic Tube Test
Using a fluorescent lamp against a black background, record the degree of agglutination as follows:

4+ 100% of the organisms are agglutinated and the supernatant fluid is clear

3+ approximately 75% of the organisms are agglutinated and the supernatant fluid is slightly cloudy

2+ approximately 50% of the organisms are agglutinated and the supernatant fluid is moderately cloudy

1+ approximately 25% of the organisms are agglutinated and the supernatant fluid is cloudy

− no agglutination is visible and the organisms remain as a cloudy suspension. This reaction should be observed in tube 8.

INTERPRETATION
1. Report the serum titer as the reciprocal of the highest dilution showing a 2+ reaction. See Tables II and III.

SAMPLE CALCULATIONS

Table II Rapid Slide Test					Table III Macroscopic Tube Test			
ml Serum	Correlated Dilution	Spec. 1	Reactions Spec. 2	Spec. 3	Serum Dilution	Spec. 1	Reactions Spec. 2	Spec. 3
					1:20	4+	3+	4+
0.08	1:20	3+	4+	4+	1:40	4+	2+	4+
0.04	1:40	2+	4+	3+	1:80	3+	1+	4+
0.02	1:80	1;	3+	2+	1:160	2+	−	4+
0.01	1:160	−	3+	+	1:320	1+	−	3+
0.005	1:320	−	1+	−	1:640	−	−	2+
					1:1280	−	−	1+
Serum Titer		1:40	1:160	1:80	Serum Titer	1:160	1:40	1:640

2. A difference in titer of 1 dilution, plus or minus, between replicate samples or between several samples drawn 1 – 2 weeks apart and run in parallel cannot be considered significant. A 1 dilution, plus or minus, deviation is within the limits of laboratory error.[10,11]

3. Past history in the use of *Proteus* suspensions has resulted in a pattern of titers which has been considered "significant." A 1:160 titer or greater is considered positive.

4. A large percentage of the sera of the normal population contains low titers to *Proteus* antigens. A titer of less than 1:160 should not be considered significant.

5. *Proteus* agglutinins to *Rickettsia* generally appear 7 – 15 days after the onset of the disease. The titer peaks during the 3rd week and begins to decline rapidly over the next several months.

6. The information supplied from a single serum specimen may be completely misleading. It is necessary to demonstrate a fourfold or greater rise in agglutinin content between 2 serum specimens taken at the acute and convalescent stage.

7. Absence of agglutinins in a patient's serum does not exclude presence of the disease.

PRECAUTIONS

All reagents and equipment used in the slide test must be at room temperature prior to use.

Specimen

1. The specimen must be clear and free of visible fat. It must be free of excessive hemolysis and not bacterially contaminated.
2. The specimen must be used in the unheated state. If inactivated by heat, some thermolabile agglutinins may be destroyed.
3. In mixing the tubes in the tube test, the specimen should not be mixed too vigorously since foaming may result in agglutinin denaturization.
4. In order to establish a pattern of reactivity (see Table I), all 3 Proteus antigens should be tested concurrently with the same serum specimen.

Glassware

1. All glassware that is employed in the preparation, testing and storage of these reagents must be free of detergents or other harmful residues.
2. Pipettes employed must be clean (acid washed, if necessary) to deliver proper volumes.

Antigens

1. Shake antigen vial well before use to insure a smooth, uniform suspension.
2. Discard any antigen demonstrating positive reactions in the negative control or any demonstrating spontaneous agglutination in the vial. Febrile antigens will demonstrate autoagglutination if at any time during shipment or storage they are subjected to freezing temperatures. Do not allow to freeze.
3. Discard any antigen which does not react properly in the positive control serum.
4. Variations in the color intensity of the slide antigen is normal and will not affect the outcome of the test.

Control Sera

1. Discard any control serum in which a precipitate forms. If bacterial contamination is suspected, autoclave the reagent or use some suitable chemical procedure to inactivate the contaminant before discarding the reagent.

LIMITATIONS

1. The major limitation of the febrile tests is that of interpretation. Again, it must be stressed that the antigens are to be used for the detection and quantitation of agglutinin content of the specimen. They are not to be used singly for a definite diagnosis. A definitive diagnosis must be made by the physician, taking into consideration the history and physical state of the patient as well as data obtained from other laboratory tests.
2. It is necessary to run several serum specimens taken at different times, from the same patient, in parallel, to detect quantitative differences in agglutinin content.
3. The slide test is for screening only. Any quantitation must be done using the tube test.
4. Since the Weil-Felix test is nonspecific, any positive results obtained should be considered presumptive and should be confirmed by the use of specific serological tests or isolation procedures. In this case, specimens should be sent to a specialized reference laboratory performing such tests, care being taken to ship the specimens under adequate conditions.[9]
5. There are many known antigenic similarities and cross reactions. Cognizance should be made when interpreting data obtained in serological reaction patterns that cross

reactions may and do exist between *Proteus* and other genera including *Brucella*, *Francisella, Vibrio* and *Yersinia* and possibly others.

6. Antibiotic therapy may result in a suppression of the production of *Proteus* agglutinins.
7. The Weil-Felix test does not differentiate between epidemic and murine typhus.

REFERENCES

1. Schubert, J. H., Holdeman, L. and Martin, D. S., J. Lab. Clin. Med., 44:194, 1964.
2. Welch, H. and Stuart, C. A., J. Lab. and Clin. Med., 21:411, 1936.
3. Welch, H. and Mickle, F. L., Amer. J. Publ. Hlth., 26:248, 1936.
4. Diamond, B. E., Pub. Hlth. Lab., 6:74, 1948.
5. Bergey's Manual of Determinative Bacteriology, 8th Ed., Williams and Wilkins Co., 1974.
6. Eyer and Grutzner, Z., Hyg. Infekt. Kr., 122:589, 1940.
7. Steuer, Z., Immun. Forsch., 101:102, 1942.
8. Vinson, J. W., Manual of Clinical Immunology, ASM 1st Ed., 1976.
9. Lennette, Spaulding and Truant, Manual of Clinical Microbiology, ASM 2nd Ed., 1974.
10. Syphilis A Synopsis, HEW, PHS, Publ., No. 1660, Jan. 1968.
11. Quality Control in Clinical Micro., Revised, ASCP Comm. on Continuing Education, Coun. on Micro., 1960.

PACKAGING

Bacto Proteus OX2 Antigen (slide)	5 ml	2243-56-6
Bacto Proteus OX2 Antigen (tube)	25 ml	2248-65-0
Bacto Proteus OX2 Antiserum	3 ml	2245-47-6
Bacto Proteus OX19 Antigen (slide)	5 ml	2234-56-7
Bacto Proteus OX19 Antigen (tube)	25 ml	2247-65-1
Bacto Proteus OX19 Antiserum	3 ml	2235-47-8
Bacto Proteus OXK Antigen (slide)	5 ml	2244-56-5
Bacto Proteus OXK Antigen (tube)	25 ml	2249-65-9
Bacto Proteus OXK Antiserum	3 ml	2246-47-5

BACTO PSEUDOMONAS AERUGINOSA ANTISERA
BACTO PSEUDOMONAS AERUGINOSA ANTIGENS

INTENDED USE
Bacto Pseudomonas Aeruginosa Antisera are recommended for use in the rapid slide agglutination technique,[1] for the serotyping of *Pseudomonas aeruginosa* cultures.

Bacto Pseudomonas Aeruginosa Antigens are recommended for use as control antigens in the serotyping of cultures of *P. aeruginosa*.

SUMMARY AND EXPLANATION OF THE TEST
Bacto Pseudomonas Aeruginosa Antisera have been prepared in cooperation with Liu, coordinator of an international panel of the subcommittee on *Pseudomonadaceae* of the International Committee on Systemic Bacteriology, which included Bergan, Duncan, Homma, Kusama, Lanyi, Matsumoto, Meitert, Mikkelsen, Parker, Sakazaki and Veron.[2]

REAGENTS
Bacto Pseudomonas Aeruginosa Antisera have been prepared in rabbits from autoclaved antigens representing 17 tentative serotypes. Types 1 – 12 were prepared using Hab's cultures 1 – 12, type 13 with Veron's 013 (Sandvik's type II), type 14 with Verder and Evan's 05, type 15 with Lanyi's 012, type 16 with Homma's 13 and type 17 with

Meitert's type X strain. When properly rehydrated, these antisera contain approximately 1:10,000 Merthiolate®, sufficient to maintain a bacteriostatic condition at 2 – 8°C.

Bacto Pseudomonas Aeruginosa Antigens are heated suspensions of the current standard 17 strains of the organisms preserved in Merthiolate® 1:10,000.

When used according to the suggested procedure, each vial of Bacto Pseudomonas Aeruginosa Antiserum, after dilution of 1:10, is sufficient to perform approximately 200 tests. Each vial of Bacto Pseudomonas Aeruginosa Antigen is sufficient to perform approximately 100 tests.

REHYDRATION
To rehydrate Bacto Pseudomonas Aeruginosa Antisera, add 1 ml sterile distilled or deionized water per vial and rotate gently to dissolve contents completely. Check purity (bacterial) and pH of the water used in rehydration if the antiserum rehydrates cloudy. Discard any serum which is cloudy and/or has a precipitate unless it has been clarified and shown to react properly with known control cultures.

EXPIRATION DATE AND STORAGE
Bacto Pseudomonas Aeruginosa Antisera, unrehydrated and rehydrated but not diluted 1:10, as well as Bacto Pseudmonas Aeruginosa Antigens are stable through the expiry date on the label when stored at 2 – 8°C.

Preliminary results indicated that the 1:10 dilution of Bacto Pseudomonas Aeruginosa Antisera, if properly stored at 2 – 8°C, are stable for at least 4 – 5 months. However, it is recommended that since various cultures, especially those of type 2, may vary considerably in their antigenic moiety the diluted sera be discarded at the end of a working week unless routinely tested with the control antigens.

PROCEDURE
Materials Provided:
Bacto Pseudomonas Aeruginosa Antisera
Bacto Pseudomonas Aeruginosa Antigens

Available from Difco:
Bacto Veal Infusion Agar
Bacto Rabbit Serum Normal

Materials Required but not Provided:
Distilled or deionized water
0.85% NaCl solution
Merthiolate®
Petri dishes 100 × 15 mm
Nylon cloth, 200 mesh
Wax marking pencils
Ruler

Glass plates 8″ × 8″
3 mm bacteriological loop
Suitable light source
Applicator sticks
Centrifuge
Autoclave
Incubator
Pasteur pipettes

NOTE: Many investigators prefer the use of viable cells taken directly from a plate of a suitable medium as an antigen. Work in our laboratories, however, has been based entirely upon autoclaved antigens. There are merits to both procedures, but it is recommended that, if difficulties are encountered in the typing of a culture of known *P. aeruginosa* by one method, the other method should be explored. There are occasions when a strain of *P. aeruginosa* may agglutinate in a given serum in an unheated state, but will not do so in a heated state and vice versa. In such cases, it is recommended that, if discrepancies exist between viable antigens and heated antigens, reactions with heated cultures be the criterion in the determination of the serotype since these sera were prepared with a somatic antigen not a whole cell antigen.

DILUTION OF SERA
For use it is recommended that a 1:10 dilution of the rehydrated antiserum be prepared using 1 part of the rehydrated antiserum to 9 parts of 0.85% NaCl solution containing 1:10,000 Merthiolate®.

SPECIMEN PREPARATION
The genus and species of the organism *P. aeruginosa* must be identified and confirmed biochemically before serological methods are performed. Many species of the *Pseudomonas* resemble *P. aeruginosa* and must be identified prior to serology.

SLIDE AGGLUTINATION TECHNIQUE
In the slide test, all materials and equipment must be at room temperature at the time of test performance.

Unheated Cells (Viable Cultures) as an Antigen
1. Streak the pure, biochemically authenticated test culture of *P. aeruginosa* on a Petri dish of Bacto Veal Infusion Agar (or other appropriate medium) using a bacteriological loop in such a manner as to obtain individual isolated colonies.
2. Incubate the plates in an inverted position for 18 – 24 hours at 35°C.
3. Using a Pasteur pipette, place droplets of Bacto Pseudomonas Aeruginosa Antisera (diluted 1:10) on appropriate squares of the glass plate.
4. Using a bacteriological loop, suspend part of an isolated colony (step 2 above) into the droplets of antisera. Care must be taken not to cross mix the antiserum droplets.
5. Rotate the plate by hand for 1 minute.
6. Observe for agglutination.

Note: Exposure of the organism or plate to heat from external sources (a hot bacteriological loop, burner flame, light source, etc.) may result in evaporation and/ or precipitation of the test mixture which may result in false positive reactions.

Heated Cells as an Antigen
1. Streak the pure, biochemically authenticated test culture of *P. aeruginosa* on a Petri dish of Bacto Veal Infusion Agar (or other appropriate medium) using a bacteriological loop or sterile applicator in such a manner as to obtain confluent (mass) growth.
2. Incubate the plates in an inverted position for 18 – 24 hours at 35°C.
3. Wash and suspend the bacterial growth from the agar plate using 10 ml 0.85% NaCl solution.
4. Autoclave the bacterial suspension for 30 minutes at 121°C. If necessary, filter the cool autoclaved antigen through nylon or cheese cloth.
5. Centrifuge at 1,000 – 2,000 rpm for 5 – 10 minutes, discard the supernatant fluid and resuspend the bacterial mass in 0.75 ml Merthiolate® saline (1:10,000 Merthiolate® in 0.85% NaCl solution). This constitutes the stable heated antigen and may be used in the slide agglutination technique as outlined in 3 – 6 above.

Table 1, below, indicates slide titers and antigenic relationships of the 17 tentative serotypes of *P. aeruginosa* as determined in this laboratory, Lot 1. Note the antigenic similarity of types 7 and 8. Also a reciprocal antigen relationship is observed between types 13 and 14.

PSEUDOMONAS AERUGINOSA ANTIGENIC RELATIONSHIPS
(Lot 1, Difco Antisera)

SLIDE TEST

ANTIGENS

ANTISERA	1	2	3	4	5	6	7	8	9	10	11	12	13	14	15	16	17
1	320*								10								
2		40														20	
3			160														
4				80													
5					40											20	
6						40											
7							160	80	10								
8							160	160									
9									160								
10									40	160							
11											320						
12												40					
13													80	10			
14													40	320			
15															160		
16																40	
17																	80

*Reciprocal of dilution.

QUALITY CONTROL
Use known positive and negative control cultures in parallel with the test culture to ascertain the validity of test results. This may be accomplished using Bacto Pseudomonas Aeruginosa Antigens.

It is advisable to check the test culture and/or control antigens in a 1:10 dilution of normal rabbit serum to detect the possibility of autoagglutination.

If the control antigens tend to autoagglutinate, as evidenced by reactivity in all sera, discard the antigen.

If the control antigen appears smooth in normal rabbit serum and the sera do not react with the control antigens properly, discard the sera.

RESULTS
A 4+ reaction is considered a positive result.

- 4+ = all of the cells agglutinate
- 3+ = 75% of the cells agglutinate
- 2+ = 50% of the cells agglutinate
- 1+ = 25% of the cells agglutinate
- ± = less than 25% of the cells agglutinate
- − = none of the cells agglutinate

LIMITATIONS OF THE PROCEDURE
1. The schema outlined in the procedure is tentative. More serotypes may be added in the future or the current tentative schema could be altered. It is recommended that correlative studies be adapted by all persons interested in *Pseudomonas* serlogy and reports be made to the international panel coordinator or to our laboratory so that it might be determined which serotypes predominate in the various disease states.
2. It should be remembered that some strains of *P. aeruginosa* are autoagglutinable and will react in many or even all the typing sera. Such strains should be reported either as nontypable or as rough cultures.
3. Due to the relationship in the antigen-antibody response, it may be necessary to increase or decrease the amount of antigen required to induce a proper reaction.

REFERENCES
1. "Identification of *Enterobacteriaceae*," P. R. Edwards, W. H. Ewing, 3rd Ed., Burgess Publishing Co.
2. P. V. Liu, University of Louisville, Louisville, Kentucky, USA.
   T. Bergan, University of Tromso, Tromso, Norway.
   I. B. R. Duncan, Sunnybrook Medical Center, Toronto, Canada.
   J. Y. Homma, University of Tokyo, Tokyo, Japan.
   H. Kusama, New York Dept. of Health, Albany, New York, USA.
   B. Lanyi, State Institute of Hygiene, Budapest, Hungary.
   H. Matsumoto, Shinshu University, Matsumoto, Japan.
   E. Meitert, Institutul De Microbiologie, Bucurasti, Roumania.
   S. Mikkelsen, State Serum Institute, Aarhus, Denmark.
   M. T. Parker, Central Public Health Laboratory, London, England.
   R. Sakazaki, National Institute of Health, Tokyo, Japan.
   M. Veron, Laboratoire de Bacteriologie et Verologie, Paris, France.

PACKAGING

Bacto Pseudomonas Aeruginosa Antisera Set	17 × 1 ml	3081-32-8
contains 1 vial each type 1 − 17		
Bacto Pseudomonas Aeruginosa Antigen Set	17 × 0.5 ml	3082-32-7

PSEUDOMONAS AGAR MEDIA

BACTO PSEUDOMONAS AGAR F
BACTO PSEUDOMONAS AGAR P

INTENDED USE
Bacto Pseudomonas Agar Media are recommended for the detection and differentiation of *Pseudomonas aeruginosa* on the basis of pigment production.

HISTORY/PRINCIPLES
Bacto Pseudomonas Agar F and Bacto Pseudomonas Agar P are patterned after the formulations described by King, Ward and Raney[1] and as modified in the USP XIX specifications.[2]

Bacto Pseudomonas Agar F enhances the elaboration of fluorescein by cultures of *Pseudomonas* and inhibits the formation of pyocyanin whereas Bacto Pseudomonas Agar P enhances the elaboration of pyocyanin by these cultures and inhibits the formation of fluorescein. The pigments diffuse from the colonies of *Pseudomonas* into the agar. Fluorescein elaborated on Bacto Pseudomonas Agar F is a yellow fluorescent color and pyocyanin elaborated on Bacto Pseudomonas Agar P is a blue color.

Occasionally, a *Pseudomonas* culture is encountered which will produce small amounts of the inhibited pigment in the medium. When this happens a yellow-green color will appear on Bacto Pseudomonas Agar F or a blue-green color will appear on Bacto Pseudomonas Agar P.

Some *Pseudomonas* organisms elaborate fluorescein with no pyocyanin, others elaborate both pigments, while still others elaborate pyocyanin with no fluorescein. When Bacto Pseudomonas Agar F and Bacto Pseudomonas Agar P are used together, they provide for the easy and rapid identification of most *Pseudomonas* organisms.

FORMULAE

BACTO PSEUDOMONAS AGAR F
DEHYDRATED

Ingredients per liter

Bacto Tryptone	10 g
Proteose Peptone No. 3, Difco	10 g
Dipotassium Phosphate	1.5 g
Magnesium Sulfate	1.5 g
Bacto Agar	15 g

Final pH 7.0 ± 0.2 at 25°C.

One pound will make 11.9 liters of medium.
Rehydrate with 38 g/liter.

BACTO PSEUDOMONAS AGAR P
DEHYDRATED

Ingredients per liter

Bacto Peptone	20 g
Potassium Sulfate	10 g
Magnesium Chloride	1.4 g
Bacto Agar	15 g

Final pH 7.0 ± 0.2 at 25°C.

One pound will make 9.8 liters of medium.
Rehydrate with 46.4 g/liter.

METHOD OF PREPARATION

1. To rehydrate, suspend the appropriate amount in 1 liter distilled or deionized water and add 10 g of Bacto Glycerol.
2. Heat the medium to boiling to dissolve completely.
3. Distribute into tubes or flasks and autoclave for 15 minutes at 15 lbs pressure (121°C).
4. Allow tubes to cool in the slanted position. Dispense sterile medium into sterile Petri dishes and allow to solidify.
5. Inoculate both tubes and plates by surface streaking.

STORAGE

Bacto Pseudomonas Agar Media	Below 30°C
Prepared media	2 – 8°C

QUALITY CONTROL
Identity Specifications
Dehydrated powder: Light beige, homogeneous, free-flowing
Reaction of solution: pH 7.0 ± 0.2 at 25°C
Prepared medium: light to medium amber, clear to slightly opalescent

Typical Cultural Response on Bacto Pseudomonas Agar Media
After 18 – 24 Hours at 35°C

Organism	Recovery	Pseudomonas Agar F	Pseudomonas Agar P
Pseudomonas aeruginosa ATCC® 9027	good	greenish yellow	blue
Pseudomonas aeruginosa ATCC® 10145	good	greenish yellow	—
Pseudomonas aeruginosa ATCC® 17934	good	—	—
Pseudomonas aeruginosa ATCC® 25619	good	greenish yellow	blue-green
Pseudomonas aeruginosa ATCC® 27853	good	greenish yellow	blue

REFERENCES
1. J. Lab. and Clin. Med., 44:301, 1954.
2. USP XIX.

PACKAGING

Bacto Pseudomonas Agar F	1 lb (454 g)	0448-01-9
	1/4 lb (114 g)	0448-02-8
Bacto Pseudomonas Agar P	1 lb (454 g)	0449-01-8
	1/4 lb (114 g)	0449-02-7

BACTO PSEUDOMONAS ISOLATION AGAR

INTENDED USE
Bacto Pseudomonas Isolation Agar is a selective medium for the isolation of *Pseudomonas*.

HISTORY/PRINCIPLES
Bacto Pseudomonas Isolation Agar is prepared according to a slight modification of the Medium A formulation of King, Ward and Raney.[1] It is especially useful for isolating *Pseudomonas* from industrial materials such as cosmetics and lotions. Bacto Pseudomonas Isolation Agar is modified to include Irgasan® CH3565, which was described by Furia and Schenkel[2] to be a potent broad spectrum antimicrobial, not active against *Pseudomonas*. As well as being selective, Bacto Pseudomonas Isolation Agar is formulated to enhance the formation of the blue or blue-green pyocyanin pigment of *Pseudomonas aeruginosa*.

FORMULA
BACTO PSEUDOMONAS ISOLATION AGAR
DEHYDRATED
Ingredients per liter

Bacto Peptone . . . . 20 g
Magnesium Chloride . . . . 1.4 g
Potassium Sulfate . . . . 10 g
Irgasan® . . . . 0.025 g
Bacto Agar . . . . 13.6 g

Final pH 7.0 ± 0.2 at 25°C.

One pound will make 10 liters of medium.

METHOD OF PREPARATION
1. To rehydrate, suspend 45 g in 980 ml distilled or deionized water.
2. Add 20 ml of Bacto Glycerol.
3. Heat to boiling to dissolve completely.
4. Dispense into flasks and sterilize in the autoclave for 15 minutes at 15 lbs pressure (121°C).

STORAGE
Bacto Pseudomonas Isolation Agar   Below 30°C
Prepared medium   2 – 8°C

QUALITY CONTROL
Identity Specifications
Dehydrated powder: very light beige, homogeneous, free-flowing
Reaction of 4.5% solution: pH 7.0 ± 0.2 at 25°C
Prepared medium: light amber, slightly opalescent.

Typical Cultural Response on Bacto Pseudomonas Isolation Agar
After 18 – 48 Hours at 35°C

Organism	Growth	Color of Colonies
Escherichia coli ATCC® 25922	inhibited	—
Proteus mirabilis ATCC® 25933	inhibited	—
Pseudomonas aeruginosa ATCC® 10145	good to excellent	green
Pseudomonas aeruginosa ATCC® 27853	good to excellent	blue-green

REFERENCES
1. J. Lab. & Clin. Med., 44:301, 1954.
2. Soap & Chemical specialties, January 1968.

PACKAGING
Bacto Pseudomonas Isolation Agar	1 lb (454 g)	0927-01-9
Bacto Glycerol	100 g	0282-15-2
	500 g	0282-17-0

PURPLE AGAR/BROTH BASE

BACTO PURPLE AGAR BASE
BACTO PURPLE BROTH BASE
BACTO PURPLE BROTH w/CARBOHYDRATES

INTENDED USE
Bacto Purple Agar Base and Bacto Purple Broth Base are recommended for the preparation of carbohydrate media used in fermentation studies for the cultural identification of pure cultures of microorganisms, particularly members of the enteric group.

PRINCIPLES
Bacto Purple Agar Base and Bacto Purple Broth Base are prepared for those bacteriologists preferring fermentation media with a slightly acid reaction (pH 6.8). They contain a sensitive sulfonphthalein indicator capable of demonstrating minute changes in reaction. The media are free from fermentable carbohydrates.

The concentration of carbohydrate generally employed for testing the fermentation reactions of bacteria is 0.5 or 1%. Some investigators prefer to use 1% rather than 0.5% to insure against reversion of the reaction due to depletion of the carbohydrate by some microorganisms. Tubes of purple lactose broth and purple saccharose broth should be tightly stoppered during the incubation period for fermentation studies of the enteric group to avoid reversion of reaction.

Bacto Purple Broth w/carbohydrates added are 16 × 125 mm tubes containing Bacto Purple Broth Base with 0.5 or 1% of the appropriate carbohydrate added (see chart). These tubes, ready for inoculation, contain Durham tubes for determination of gas production.

Carbohydrate	% Solution
Dextrose	1%
Dulcitol	0.5%
Lactose	1%
Maltose	0.5%
Mannitol	1%
Rhamnose	0.5%
Saccharose	1%
Salicin	0.5%
Xylose	0.5%

FORMULAE

BACTO PURPLE BROTH BASE
DEHYDRATED

Indredients per liter

Proteose Peptone No. 3, Difco .... 10 g
Bacto Beef Extract .... 1 g
Sodium Chloride .... 5 g
Bacto Brom Cresol Purple ..... 0.015 g

Final pH 6.8 ± 0.2 at 25°C.

Rehydration   16 g/liter
Yield   28.3 liters/lb

BACTO PURPLE AGAR BASE
DEHYDRATED

Ingredients per liter

Proteose Peptone No. 3, Difco .... 10 g
Bacto Beef Extract .... 1 g
Sodium Chloride .... 5 g
Bacto Agar .... 15 g
Bacto Brom Cresol Purple ...... 0.02 g

Final pH 6.8 ± 0.2 at 25°C.

31 g/liter
12.9 liters/500 g

METHOD OF PREPARATION

1. Suspend the appropriate amount in 1 liter distilled or deionized water and heat to boiling to dissolve completely.
2. If desired, add 5 to 10 g of the carbohydrate to be tested.
3. Distribute into tubes as desired and sterilize in the autoclave for 15 minutes at 15 lbs pressure (121°C). The minimum amount of heat required for sterilization is to be desired so as to avoid hydrolysis of the carbohydrate. By packing the tubes loosely in the autoclave to allow free circulation of steam, the time required may be appreciably shortened, provided the temperature in the autoclave is actually 121°C.

   Alternately, the basal media can be prepared using 900 ml distilled or deionized water and sterilized. One hundred ml of a sterile 5 – 10% carbohydrate solution can then be added aseptically after cooling the basal medium to 45 – 50°C.

   NOTE: The addition of some carbohydrates may result in an acid reaction. In this case, it is suggested that the proper pH be restored by adding sterile 0.1N sodium hydroxide dropwise.
4. Cool Bacto Purple Agar Base tubes in a slanted position to provide a slope and generous butt.

5. If the media are not used the same day as they are sterilized, place in flowing steam or boiling water bath for a few minutes to drive off dissolved gases and to provide a fresh, moist slant on the agar media. Allow agar media tubes to cool on a slant and broth tubes to cool without agitation.

STORAGE

Bacto Purple Base media   Below 30°C
Prepared media   2 – 8°C

QUALITY CONTROL

Identity Specifications

Dehydrated powder: light tan with greyish cast, homogeneous, free-flowing
Reaction of appropriate solution: pH 6.8 ± 0.2 at 25°C
Prepared medium: purple, clear to slightly opalescent

Typical Cultural Response After 18 – 48 Hours at 35°C

Organism	w/o Carbohydrates			w/1% Dextrose		
	Growth	Acid	Gas	Growth	Acid	Gas
Escherichia coli ATCC® 25922	excellent	–	–	excellent	+	+
Neisseria meningitidis ATCC® 13090	good to excellent	–	–	good to excellent	+	–
Staphylococcus aureus ATCC® 25923	excellent	–	–	excellent	+	–

PACKAGING

Bacto Purple Agar Base	500 g	0228-17-7
Bacto Purple Broth Base	1 lb (454 g)	0227-01-6
	12 tubes	0227-95-3
	144 tubes	0227-96-2
Bacto Purple Broth w/Dextrose	12 tubes	1084-95-3
	144 tubes	1084-96-2
Bacto Purple Broth w/Dulcitol	12 tubes	1090-95-5
Bacto Purple Broth w/Lactose	12 tubes	1087-95-0
Bacto Purple Broth w/Maltose	12 tubes	1085-95-2
Bacto Purple Broth w/Mannitol	12 tubes	1089-95-8
Bacto Purple Broth w/Rhamnose	12 tubes	1088-95-9
Bacto Purple Broth w/Saccharose	12 tubes	1086-95-1
Bacto Purple Broth w/Salicin	12 tubes	1091-95-4
Bacto Purple Broth w/Xylose	12 tubes	1083-95-4

BACTO PURPLE LACTOSE AGAR

INTENDED USE

Bacto Purple Lactose Agar is used for detecting coliform organisms and for performing differential studies.

HISTORY

Bacto Purple Lactose Agar is patterned after litmus lactose agar, as described by Wurtz in 1892,[1] except that brom cresol purple has been substituted for litmus which is less sensitive and less stable.

PRINCIPLES

Colonies of lactose-fermenting organisms are differentiated from lactose-nonfermenters by a color change of the indicator from blue-purple (alkaline) to yellow (acid).

FORMULA

BACTO PURPLE LACTOSE AGAR
DEHYDRATED

Ingredients per liter

Bacto Beef Extract	3 g
Bacto Peptone	5 g
Bacto Lactose	10 g
Bacto Agar	10 g
Bacto Brom Cresol Purple	0.025 g

Final pH 6.8 ± 0.1 at 25°C.

One pound will make 16.2 liters of medium.

METHOD OF PREPARATION

1. Suspend 28 g in 1 liter distilled or deionized water and heat to boiling to dissolve completely.
2. Sterilize in the autoclave for 15 minutes at 15 lbs pressure (121°C).
3. Dispense as desired.
4. If tubes are used, allow to solidify in a slanted position.
5. Inoculate by stabbing the butt and streaking over the surface of the slant.

STORAGE

Bacto Purple Lactose Agar	Below 30°C
Prepared medium	2 – 8°C

QUALITY CONTROL

Identity Specifications

Dehydrated powder:	light amber with greyish-purple tinge, homogeneous, free-flowing
Reaction of 2.8% solution:	pH 6.8 ± 0.1 at 25°C
Prepared medium:	purple, slightly opalescent

Typical Cultural Response on Bacto Purple Lactose Agar
After 18 – 24 Hours at 35°C

Organism	Growth	Acid	Gas
Enterobacter aerogenes ATCC® 13048	good	+	+
Escherichia coli ATCC® 25922	good	+	+
Salmonella typhi ATCC® 19430	good	−	−
Staphylococcus aureus ATCC® 25923	good	+	−

+ = positive, yellow
− = negative, no change

PACKAGING

Bacto Purple Lactose Agar	500 g	0082-17-2

LIMULUS AMEBOCYTE LYSATE
PYROTEST™ - LAL TEST
License No. 761

Individuals performing the Limulus Amebocyte Lysate (LAL) test should be familiar with
the theoretical aspects of the test, as well as all the details of the methodology. For
these reasons, it is urged that the contents of this product description be read carefully
prior to the initial performance of the LAL test.

INTRODUCTION

Limulus Amebocyte Lysate (LAL), Pyrotest™ can be used to detect and quantitate bacterial endotoxins in investigational samples and in a variety of pharmaceutical and clinical solutions including body fluids, such as cerebrospinal fluids and urines, parenteral
solutions and the starting materials from which they are prepared, medical device rinse
waters, and some biologicals such as vaccines and serum albumins. However, this
statement should not be misconstrued as recommending that the LAL test be used to
detect bacterial endotoxins in these materials. Data must be generated to substantiate
the validity of results obtained using the LAL procedures.

The Limulus Amebocyte Lysate (LAL) test is the most sensitive test available for detecting endotoxins produced by gram-negative bacteria. The procedure is simple, specific, rapid and inexpensive compared to the USP rabbit test for pyrogens and is as
much as ten times more sensitive.

The method described in this product description is based on the classical gel clot end
point method.[1] However, other procedures based on the same reaction using radioiso-
topes,[2] spectrophotometers,[3] or chromogenic substrates[4] have been described. In addition, various micromethods conserving the lysate reagent have been used.[5-8]

Briefly, the Pyrotest™ Limulus Amebocyte Lysate (LAL) test consists of adding 0.1 ml
Pyrotest™ (from the Pyrotest™ 50 Test vial) to 0.1 ml of test solution or 0.2 ml of the
test solution directly into the Pyrotest™ 1 Test vial. The contents of the vials are mixed
gently and incubated in a 37°C water bath undisturbed for 1 hour. The test vials are
carefully inverted one at a time. The presence of endotoxin (pyrogen) in the test solution is detected by the formation of a firm gel clot which remains intact when the
tubes are inverted through 180°.

EXPLANATION OF THE LAL TEST
HISTORY

Although reference to blood lymph cells and the inflammatory process in *Limulus
polyphemus* was made first by Loeb[9] in 1902, it was not until 1964, that Levin and
Bang[10] demonstrated that the blood from this animal clotted in the presence of gram-
negative bacterial endotoxins. They[1] further reported that the endotoxin sensitive portion was found within the amebocyte and not in the liquid fraction of the blood. These
reports, as well as those of Rojas-Corona et al,[11] Levin et al,[12] and Reinhold and Fine[13]
are primarily responsible for the increased interest in the use of LAL for detecting endotoxins in the starting materials used for preparing parenteral and biological solutions,
and for in-process monitoring of these materials. Additional impetus was given to the
use of this test by Cooper et al[14] who reported on its superiority over the widely used
USP rabbit test for pyrogens, and by Jorgensen et al[15] in the report on the use of the
LAL test for detecting endotoxins in radiopharmaceuticals and biologics.

BIOLOGICAL PRINCIPLES OF THE LAL TEST

Although the exact mechanism of the LAL-endotoxin reaction has not been determined, it appears that lysate gelation is an enzymatic coagulation reaction similar to the blood clotting system in mammals. A schematic representation of the postulated reaction is shown in FIGURE 1. A large molecular weight proclotting enzyme (or enzymes) in the lysate reacts with endotoxin in the presence of divalent metal cations to form an activated clotting enzyme(s) with properties similar to a serine proteinase. The activated clotting enzyme(s) cleaves a soluble small molecular weight, protein (coagulogen) to form an insoluble protein gel (coagulin).[16-20]

Figure 1.

Proclotting Enzyme(s)   + Endotoxin $\xrightarrow{Mg^{++}}$ Activated Clotting enzyme(s)

Activated Clotting Enzyme(s)   + Coagulogen $\xrightarrow{Ca^{++}}$ Coagulin (Gel Formation)

PRECAUTIONS

Strict adherence to aseptic technique and the quality control procedures outlined in this manual are essential to assure reliability of the test.

All glassware, reagents and water used in performing the test must be pyrogen free. Clean glassware, i.e., flasks, test tubes, pipettes, etc., can be depyrogenated by placing them in a dry hot air oven at 180°C for three hours or longer. Materials which cannot be depyrogenated by dry heat should be autoclaved at 121.5°C for at least 4 hours.

The pH of the test samples should be adjusted (if necessary) to a range of between 6.0 and 8.0. See method outlined below under SPECIMEN COLLECTION AND PREPARATION.

All reagents supplied in the Pyrotest kits should be prepared and handled as outlined in the reagents section of this product description.

Care must be taken in performing the LAL test. The gel formed in the test is susceptible to irreversible damage due to shaking during the incubation period. Once the test sample and lysate reagent are mixed, they are to be left undisturbed for 1 hour in a 37°C water bath. The reaction is both time and temperature dependent; therefore, even minor deviation from the prescribed method may yield aberrant results. If a large number of samples are to be tested, it is suggested that appropriate intervals between samples or groups of samples be allowed to permit for the final reading of all samples after 1 hour incubation.

THE LAL TEST IS NOT RECOMMENDED FOR DETECTING ENDOTOXEMIA IN MAN

McAuley et al[21] have found certain drugs to interfere with the LAL reaction. High concentrations of certain salts have also been shown to be inhibitory. Elin,[22] Elin et al[23] and Wildfeuer et al[24] observed false positive reactions with some polynucleotides, proteins, and high concentrations of certain fractions of the cell wall of gram-positive bacteria. Therefore, it is recommended that when a product is tested for the first time with Pyrotest, the product should be analyzed for its compatibility with LAL. The procedure is described under the COMPATIBILITY TEST using Pyrotest. This test will determine what effect the product may have on the sensitivity (reactivity) of the LAL reagent.

Picogram quantities of endotoxin isolated from gram-negative bacilli when added to Pyrotest will result in gelation of the latter when the reaction mixture is incubated in a water bath at 37°C for 1 hour. Since the test has been shown to be up to 10 times more sensitive than the USP rabbit test for pyrogens, these lesser concentrations of endotoxin may not elicit a fever in rabbits. Concentrations of endotoxin in the nanogram range and indeed even in picogram amounts, depending on the preparation, when present in parenteral solutions, may result in a fever in patients receiving these solutions, as was reported by Jorgensen et al.[25] They also demonstrated that these low concentrations of endotoxin can be detected with LAL, thus providing a potentially safer detection level. Nandan et al[26] also reported that contamination of intravenous fluids with picogram amounts of endotoxin can be detected with LAL.

Unless a manufacturer of licensed parenteral solutions, biological solutions, and/or medical devices has received FDA approval for final product testing, Pyrotest is permitted for use in detecting endotoxins only in their starting materials and in-process test solutions.

An LAL test procedure has now been adopted by the United States Pharmacopeial Convention (see Appendix A, Attachment B) and the FDA has recently given its approval for the LAL test as an alternative to the USP rabbit test for pyrogens for end product pyrogen testing of parenteral solutions and medical devices. Manufacturers of such products wishing to use the LAL test in place of the USP rabbit test for pyrogens may use the information reprinted in Appendices A and B of this product description as a guide for the development of data to support submission of their license amendments for approval from the appropriate FDA bureau. **Only licensed LAL, such as Pyrotest™, may be used for the official USP LAL test for pyrogens of final products.**

REPORTED CLINICAL APPLICATIONS FOR THE LAL TEST
Although the use of the LAL test to detect gram-negative septicemia and gram-negative pyrogenic arthritis is limited due to the inhibitory factors found in blood and synovial fluid,[27] it has been found useful in detecting the presence of gram-negative bacteria or the endotoxins they produce in other body fluids. Procedures for detecting gram-negative meningitis,[28,29] gram-negative bacteriuria,[30,31,32] and gonococcal urethritis[33] have been described. Recent advancements in LAL technology have indicated that its future clinical value for detecting gram-negative organisms and the endotoxins they produce has only begun to be realized.[27]

SENSITIVITY CONSIDERATIONS FOR LAL, PYROTEST™
All manufacturers of LAL have been instructed by the FDA that effective June 1, 1982, the sensitivity of their LAL preparations must be designated in Endotoxin Units/ml (EU/ml), in place of the classical nanogram/ml (ng/ml). This change was a result of observed discrepancies in LAL gel clot end points with various endotoxins in relationship to the US Standard Endotoxin when expressed in concentrations (ng/ml). In an attempt to standardize the LAL test, the FDA established an Endotoxin Unit as the potency unit of a given endotoxin preparation relative to the US Standard Endotoxin EC2. The US Standard Endotoxin EC2 was assigned an arbitrary value of 5 EU/ng and was specified to be used for all subsequent calibration of endotoxin potency for the LAL test.

US Standard Endotoxin EC5 is currently available through the United States Pharmacopeial Convention as USP Endotoxin Reference Standard and contains 10,000 EU/vial which is to be used for the calibration of the USP LAL test for pyrogens. The USP Convention recommends that an in-house control standard endotoxin (CSE), such as that supplied by Difco as Pyrotrol™ Positive Control be used in routine daily LAL testing to conserve on the supply of the USP Endotoxin Reference Standard.

Since Pyrotrol Positive Control is dispensed by weight (100 ng/vial) and different lysate preparations have been found to react differently with various endotoxins,[34,35] it is recommended that the relationship in EU/ng of the in-house control standard endotoxin Pyrotrol Positive Control to the US Standard Endotoxin must be determined every time a new lot of Pyrotest or Pyrotrol Positive Control is used. For "unofficial" use, Pyrotrol Positive Control can be used uncalibrated.

METHOD FOR DETERMINING THE RELATIONSHIP BETWEEN PYROTROL POSITIVE CONTROL AND US STANDARD ENDOTOXIN

Prior to performing this or any LAL assay, initial assessment of the testing laboratory's performance variability must be done. The procedure for this can be found below. It is recommended that at least 4 vials of Pyrotrol Positive Control (control standard endotoxin [CSE]) and US Standard Endotoxin be tested in parallel, each endotoxin being tested with a minimum of four replicate test rows. The Pyrotrol Positive Control (CSE) and the US Standard Endotoxin should be rehydrated according to directions. Serial two-fold dilutions in pyrogen-free water are made with each vial according to the methods described in the PREPARATION OF ENDOTOXIN DILUTIONS section below. To insure observation of the titration end points, bracket the labeled end point by plus or minus two tubes. The values obtained should be expressed as the geometric mean in EU/ml for the US Standard Endotoxin and ng/ml for the Pyrotrol Positive Control (CSE). See the example below:

4 vials US Standard Endotoxin yielding 16 end points at 0.25 EU/ml
4 vials Pyrotrol Positive Control yielding 16 end points at 0.1 ng/ml

The potency of the Pyrotrol Positive Control in EU/ng can be determined by dividing the geometric mean of the US Standard Endotoxin by the geometric mean of the Pyrotrol Positive Control.

$$0.25 \text{ EU/ml} \div 0.1 \text{ ng/ml} = 2.5 \text{ EU/ng}$$

Thus the titration using the Pyrotrol Positive Control can be converted to EU/ml and can now be used in place of the US Standard Endotoxin with a given lysate. An example of this relationship is illustrated in TABLE 1. The designation of the sensitivity of the LAL reagent in EU/ml standardizes the end point titration irrespective of the endotoxin used. Note the equivalent end points in EU/ml with variable end points in ng/ml.

REAGENTS

Limulus Amebocyte Lysate, Pyrotest™ is available from Difco Laboratories in kits with one 50 test vial or 100 single test vials of LAL reagent. Each kit also contains five vials of Pyrotrol Positive Control and five vials of Pyrotrol Negative Control. The positive and negative controls can be purchased separately and are to be used as LAL test performance controls only (see below).

Pyrotest reagent is a heat-labile lyophilized preparation of the intracellular fluid obtained from lysed blood cells (amebocytes) of the horseshoe crab, *Limulus polyphemus*.

The sensitivity to endotoxins of each Pyrotest lot is based on the geometric mean of a 20-row test calibrated in Endotoxin Units/ml with the US Standard Endotoxin and appears on the package and vial labels of each kit.

Pyrotest is present in the vial as a white pellet which may be broken into several pieces during shipment and which does not affect its activity. When exposed to heat or if permitted to absorb moisture, Pyrotest will become light brown and should not be used. Pyrotest is soluble in pyrogen-free water and yields a clear to slightly opalescent so-

Table 1. Example of LAL Reactivity Comparing Sensitivity in EU/ml and ng/ml.

Endotoxins	US Reference Limulus Amebocyte Lysate Lot 8 (0.0625 EU/ml)a	US Reference Limulus Amebocyte Lysate Lot 9 (0.25 EU/ml)b	Difco PyrotestTM 1 Test Limulus Amebocyte Lysate Lot HR0032 (0.25 EU/ml)a	Difco PyrotestTM 50 Test Limulus Amebocyte Lysate Lot HK9314 (0.5 EU/ml)a
US Standard Endotoxin EC2 (5 EU/ng)b	0.0125 ng/ml	0.05 ng/ml	0.05 ng/ml	0.1 ng/ml
US Standard Endotoxin EC5 (10 EU/ng)b	ND	0.025 ng/ml	0.025 ng/ml	0.05 ng/ml
Difco E. coli 0127:B8 Endotoxin (2.5 EU/ng)c	0.025 ng/ml	0.1 ng/ml	0.1 ng/ml	0.2 ng/ml
Difco E. coli 055:B5 Endotoxin (10 EU/ng)c	0.006 ng/ml	0.025 ng/ml	0.025 ng/ml	0.05 ng/ml

[a]Estimated sensitivity designations in EU/ml as observed in Difco LAL Laboratory.
[b]Potency designation set by FDA for Reference Standard Endotoxins.
[c]The potency designations were observed in the Difco LAL Laboratory and are meant for the above example comparisons only. Specific performance of all LAL reagents must be verified in the user's laboratory.

lution. Gentle shaking may aid in solubilizing the lysate. Occasionally, the larger size package may contain some finely suspended particles. This does not indicate a change in sensitivity. If so desired, the solution can be centrifuged gently in a pyrogen-free test tube for about 5 minutes and the supernatant fluid can be used as indicated.

Pyrotest is stable through the expiry date on the label when stored at 2 – 8°C. Do not expose Pyrotest to room temperature longer than is necessary to reconstitute and to obtain sufficient material to perform the test. The 1 test vials should be allowed to come to room temperature before using. The tubes containing the Pyrotest and test solutions or endotoxin dilutions are shaken individually and gently to obtain a homogeneous mixture before incubation and must not be disturbed during the incubation period. Pyrotest 1 Test is rehydrated with 0.2 ml of the test solution. **NOTE: No additional Pyrotest need be added when using the Pyrotest 1 Test vials.** Pyrotest 50 Test is rehydrated with 5.2 ml Pyrotrol Negative Control or the equivalent pyrogen-free distilled water (see below). The rehydrated Pyrotest 50 Test reagent is stable for one week at 2 – 8°C. It can also be stored frozen at −20°C or below for two weeks. Do not freeze lysate reagent more than once. Repeated freezing and thawing may reduce the reagent's sensitivity.

Each lot of Pyrotest must be tested for its sensitivity and compatibility with the test samples before beginning use in routine assay work. Calibration of the sensitivity and analysis of compatibility assays are outlined below. In addition, these procedures are outlined in the appendices.

Pyrotrol Positive Control is a heat-stable lyophilized preparation of endotoxin isolated from E. coli. Each vial contains 100 ng of endotoxin with sufficient quantities of filler for stabilization during freeze drying. It has been observed in our laboratory to contain approximately 2.5 EU/ng when assayed with the reference lysate and Pyrotest lysates. However, it is recommended that this potency value be used only as a reference point. The relationship of the Pyrotrol Positive Control and the US Standard Endotoxin should be determined every time a new lot of lysate or Pyrotrol Positive Control is used (see above).

Since Pyrotrol Positive Control contains endotoxin, it is pyrogenic and should be handled with care. It is not to be used for any purpose other than those described in this manual.

Pyrotrol Positive Control is stable through the expiry date on the label when stored at 2 – 8°C. Pyrotrol Positive Control is rehydrated with 4 ml Pyrotrol Negative Control or equivalent pyrogen-free distilled water (see below). Thorough mixing and preferably, vortexing for five minutes of the reconstituted stock solution and subsequent dilutions is essential to obtain full solubilization and accurate dilutions. The rehydrated Pyrotrol Positive Control stock solution is stable for one month at 2 – 8°C if free from contamination. For optimum results, further dilutions of the endotoxin should be discarded at the end of each working day. TABLE 2 outlines the preparation and recommended dilution scheme for the US Standard Endotoxin and Pyrotrol Positive Control for use in all LAL assays requiring titrated endotoxins.

Pyrotrol Negative Control is pyrogen-free distilled water. It is to be used for reconstituting Pyrotest 50 Test vial and Pyrotrol Positive Control, for making dilutions of the positive control and sample material, and for a negative control in the sensitivity calibration and the compatibility tests described below. If the vial of Pyrotrol Negative Control has been previously opened or if pyrogen-free water other than that supplied by Difco is used, such as Sterile Water for Irrigation USP, it must be tested for pyrogenicity prior to use. Pyrotrol Negative Control is stable through the expiry date on the label.

PROCEDURES
The procedures outlined in this manual are meant to be used as a guide for the proper use of the LAL test. These procedures consist of:
  I. Preparation of Endotoxin Dilutions
 II. Performance of the LAL Test
III. Initial Quality Control Assessment of Testing Laboratory Proficiency in Performing the LAL Test
IV. Determination of the Relationship between Pyrotrol Positive Control (Control Standard Endotoxin [CSE]) and the US Standard Endotoxin
 V. Calibration of Pyrotest Sensitivity
VI. Compatibility Test
VII. Analysis of an Unknown Test Solution for Endotoxins
STRICT ADHERENCE TO DIRECTIONS IS ESSENTIAL

I. PREPARATION OF ENDOTOXIN DILUTIONS
A. Dilution of the US Standard Endotoxin
1. Reconstitute the US Standard Endotoxin with 5 ml Pyrotrol Negative Control (or equivalent).
2. Vortex the stock solution for 30 minutes to thoroughly solubilize endotoxin.
3. Label nine depyrogenated tubes 1 – 9 (15 ml conical disposable tubes such as Falcon 2095 or equivalent can be used).
4. Add 9 ml Pyrotrol Negative Control to the first three tubes and 5 ml to each of the remaining tubes.

5. Transfer 1 ml of the stock solution to the first tube. Mix thoroughly.
6. Transfer 1 ml from the first dilution tube to the second. Mix thoroughly.
7. Transfer 1 ml from the second dilution tube to the third. Mix thoroughly.
8. Serially transfer 5 ml from the third through the ninth tube. Mix each dilution thoroughly.
9. The resultant dilution scheme is shown in TABLE 2a.

B. Dilution of the Pyrotrol Positive Control (In-House Control Standard Endotoxin [CSE])

As indicated above in the section entitled METHOD FOR DETERMINING THE RELATIONSHIP BETWEEN PYROTROL POSITIVE CONTROL AND US STANDARD ENDOTOXIN, Pyrotrol Positive Control (CSE) can be used in the daily routine of LAL tests, once its potency with a given lysate is established (see above procedure). The method for dilution of the Pyrotrol Positive Control which will yield titration end points with most lots of Pyrotest using the gel clot end point method is as follows:

1. Reconstitute 1 vial of Pyrotrol Positive Control with 4 ml Pyrotrol Negative Control (or equivalent).
2. Vortex the stock solution for five minutes to thoroughly solubilize endotoxins.
3. Label nine depyrogenated tubes 1 – 9.
4. Add 9 ml Pyrotrol Negative Control to the first tube, 6 ml to the second and third tubes and 5 ml to each of the remaining tubes.
5. Transfer 1 ml of the stock solution to the first tube. Mix thoroughly.
6. Transfer 4 ml from the first dilution tube to the second. Mix thoroughly.
7. Transfer 4 ml from the second dilution tube to the third. Mix thoroughly.
8. Serially transfer 5 ml from the third through the ninth tube. Mix each dilution thoroughly.
9. The resultant dilution scheme is shown in TABLE 2a. – 2b.

The above dilution schemes are recommended for all test procedures in which a titration end point is desired. To ensure obtaining an end point, bracket the labeled LAL sensitivity as specified in EU/ml by plus and minus 2 tubes. When using the Pyrotrol Positive Control, determine the potency in EU/ng by using 2.5 EU/ng as an estimate of its potency and the equivalent dilution titration calibrated in EU/ml (see example in TABLE 1).

II. PERFORMANCE OF THE LAL TEST

Before the LAL test can be performed on a test solution, the pH of the test solution must be in the range of 6.0 to 8.0 (see SPECIMEN COLLECTION AND PREPARATION method below) and a compatibility test should be run (see COMPATIBILITY TEST below). Consult PRECAUTION section above.

When performing the LAL test using Pyrotest 1 Test vial, the lysate reagent within the vial is rehydrated with 0.2 ml of the test solution or endotoxin dilution and the test is run in the 1 test vial. Conversely, the Pyrotest 50 Test vial is rehydrated with 5.2 ml of Pyrotrol Negative Control, and 0.1 ml of the rehydrated lysate reagent is mixed with 0.1 ml of the solution to be tested. The test should be done in a depyrogenated 10 × 75 mm (or 12 × 75 mm) glass round bottom tube (not supplied). All other aspects of the tests are the same.

Allow all reagents and test solutions to come to room temperature. Remove a sufficient number of vials from the Pyrotest 1 Test Kit to perform the number of tests necessary for the desired test procedure (consult individual assay procedure

Table 2. Dilution Titration Scheme for Endotoxins.
2a. US Standard Endotoxin.

Dilution (Tube#)	US Standard Endotoxin 10,000 EU/vial Reconstituted with 5 ml Pyrotrol Negative Control (2000 EU/ml) Stock Solution	Pyrotrol Negative Control	Dilution Factor	Concentration of Endotoxin (EU/ml)	Example of gel end point of an LAL w/ a labeled sensitivity of 0.25 EU/ml
1	1 ml of stock solution +	9 ml	1:10	200	ND
2	1 ml of dilution #1 +	9 ml	1:10	20	ND
3	1 ml of dilution #2 +	9 ml	1:10	2	ND
4	5 ml of dilution #3 +	5 ml	1:2	1	+
5	5 ml of dilution #4 +	5 ml	1:2	0.5	+
6	5 ml of dilution #5 +	5 ml	1:2	0.25	+
7	5 ml of dilution #6 +	5 ml	1:2	0.125	–
8	5 ml of dilution #7 +	5 ml	1:2	0.062	–
9	5 ml of dilution #8 +	5 ml	1:2	0.031	ND
	Negative Control	5 ml	–	–	–

2b. Pyrotrol Positive Control

Dilution (Tube#)	Pyrotrol Positive Control (CSE) 100 ng/vial Reconstituted with 4 ml Pyrotrol Negative Control (25 ng/ml) Stock Solution	Pyrotrol Negative Control	Dilution Factor	Concentration of Endotoxin (ng/ml)	Example of gel end point of an LAL w/ a labeled sensitivity of 0.25 EU/ml
1	1 ml of stock solution +	9 ml	1:10	2.5	ND
2	4 ml of dilution #1 +	6 ml	1:25	1	ND
3	4 ml of dilution #2 +	6 ml	1:25	0.4	+
4	5 ml of dilution #3 +	5 ml	1:2	0.2	+
5	5 ml of dilution #4 +	5 ml	1:2	0.1	+
6	5 ml of dilution #5 +	5 ml	1:2	0.05	–
7	5 ml of dilution #6 +	5 ml	1:2	0.025	–
8	5 ml of dilution #7 +	5 ml	1:2	0.0125	ND
9	5 ml of dilution #8 +	5 ml	1:2	0.006	ND
	Negative Control	5 ml	–	–	–

As shown in the above example, 0.25 EU/ml US Standard Endotoxin yields an equivalence end point with Pyrotrol Positive Control (CSE) at 0.1 ng/ml. The potency of the Pyrotrol Positive Control in the example would be 0.25 EU/ml ÷ 0.1 ng/ml = 2.5 EU/ng. See METHOD FOR DETERMINATION OF THE RELATIONSHIP BETWEEN THE US STANDARD ENDOTOXIN AND PYROTROL POSITIVE CONTROL (CSE) above.

sections for the required number of tests). Aseptically remove the aluminum seals
and rubber stoppers.

Begin with the lowest endotoxin dilution (dilution 9, TABLE 2a or 2b) when titrating
endotoxins in applicable procedures (i.e., Procedures III – VI). It is also recommended that the endotoxin dilutions should be added to the predispensed Pyrotest
50 Test reagent to avoid potential contamination of the Pyrotest 50 Test stock vial.
1. Adjust water bath to 37 ± 0.5°C.
2. Arrange a sufficient number of assay tubes (or Pyrotest 1 Test tubes) in a test
   tube rack to perform the desired assay.
3. When using Pyrotest 50 Test, add 0.1 ml of the lysate reagent to the vials.
4. Add the solutions to be tested to the lysate reagent.
   a. Pyrotest 1 Test vials are rehydrated with 0.2 ml of the solutions to be tested.
   b. Pyrotest 50 Test (predispensed) are mixed with 0.1 ml of the solutions to
      be tested.
      **NOTE: If a large number of tests are to be assayed at one time, ap-
      propriate intervals between groups of samples to be tested should be
      taken to allow for the reading of all samples after 1 hour.**
5. Gently mix the lysate-test sample thoroughly.
6. Place the test tube rack in a 37°C water bath for 1 hour. It is extremely important that the tubes are not disturbed during this incubation time (see PRE-
   CAUTIONS above).
7. At the end of the incubation time the test tube rack is **carefully** removed from
   the water bath and gently placed on a lab bench.
8. The tubes are carefully removed, one at a time, and inverted through 180°. A
   positive test is defined as the presence of a firm gel which remains intact.
9. The end point (in a titration assay) is the highest dilution of endotoxin showing
   a firm gel

III. INITIAL QUALITY CONTROL
PROCEDURE FOR A TESTING LABORATORY
Before any official tests are performed, the variability of the testing laboratory should
be assessed. A minimum of 4 and a maximum of 8 tests should be performed by
each technician using a single lot of LAL and a single lot of endotoxin. For each
test, a group of 4 replicate tests on a single series of dilutions of the endotoxin
should be prepared. A two-fold dilution series should be used with a range suitable
for yielding an end point with the LAL being tested (see TABLES 2a and 2b). After
incubation, the reaction in each tube should be recorded as either positive or negative. The end points should be expressed as EU/ml, therefore, the US Standard
Endotoxin should be used in the initial assay. Pyrotrol Positive Control can be used
if its potency (EU/ml) relative to the US Standard Endotoxin is determined with
each lysate. The end points are converted to logs and the standard deviation (S.D.)
is calculated.

If the S.D. is less than or equal to the value for the 99% limit on the S.D. for 4
samples (see TABLE 3), the test is in control. If the S.D. is greater than the tabulated value, the test may be expanded to a maximum of 8 tests and a new S.D.
calculated using data from all tests. If, after doing 8 tests, the S.D. exceeds the
99% upper limit value corresponding to 8 tests, the tests are invalid due to excessive variability. If after any set of 4 replicates the S.D. is equal to or less than
the tabulated value for that number of tests, it can be concluded that the variability
is under control and no more tests are necessary. At that point the geometric mean
(G.M.) of the endpoints from all tests performed should be calculated. If the labeled

value for the lysate used falls within plus or minus one two-fold dilution of the geometric mean, the tests have confirmed the label claim. If it is not, the source of variation should be determined before doing additional testing.

Table 3. 99% Upper Limit on Standard Deviation.

Size	LOG_2	LOG_{10}
4	1.214	0.365
8	1.015	0.306

IV. **DETERMINATION OF THE RELATIONSHIP OF THE PYROTROL POSITIVE CONTROL (CSE) TO THE US STANDARD ENDOTOXIN**
This procedure is described above. It should be noted that this procedure as well as the above procedure are similar and can be used to calibrate (confirm) the Pyrotest label sensitivity.

V. **CONFIRMATION OF THE PYROTEST LABEL SENSITIVITY**
The procedure is outlined above under DETERMINATION OF THE RELATION-SHIP OF THE PYROTROL POSITIVE CONTROL (CSE) TO THE US STANDARD ENDOTOXIN and INITIAL QUALITY CONTROL PROCEDURE. The test consists of running a 4 row titration assay of an endotoxin calibrated in EU/ml, preferably in parallel with a reference lysate of known sensitivity.

VI. **COMPATIBILITY TEST**
Before using Pyrotest for assaying endotoxins in test solutions, the compatibility of the test solution in the LAL test must be established. Certain drugs, increased concentrations of salt solutions and biologicals have been observed to inhibit or enhance the LAL's sensitivity to endotoxin. All compatibility tests shall be performed on the undiluted test sample or the Maximum Valid Dilution (MVD). For further information consult the attached FDA appendix.

Before running the compatibility test, the sample must have a pH in the range of 6.0 to 8.0. Check SPECIMEN COLLECTION AND PREPARATION section below.

The compatibility test consists of preparing a two-fold serial dilution of endotoxin in the test solution. Use the pattern in TABLE 2a or 2b as a guide. Test the sample material spiked with endotoxins in concentrations which bracket the labeled sensitivity (in EU/ml) by plus or minus two tubes. As a control the same endotoxin equivalently titrated in Pyrotrol Negative Control is to be assayed in parallel. The end point titration determination of the control endotoxin series and the spiked test solution series should not differ by more than plus or minus 1 two-fold dilution. See TABLE 4. If the spiked test solution displays more than a plus or minus 1 two-fold dilution of inhibition, or enhancement, then the USP rabbit test for pyrogens will still be the appropriate pyrogen test. Exceptions to this would be if the user can by some method neutralize the factor in the test solution which causes the incompatibility. Consult the FDA appendix or call Difco Laboratories Technical Services Department for assistance.

Lysate preparations are crude extracts that can vary from lot to lot. Therefore, it is possible that one lot may be compatible with the test sample, while a future lot may not. Although it is not necessary to run the complete compatibility test every time a new Pyrotest lot is used, it is recommended that a test sample be spiked with a concentration of endotoxin (in EU/ml) equal to twice the labeled sensitivity

Table 4. Example of the Compatibility Test.
US Standard Endotoxin or Pyrotrol Positive Control(a).

Titration in Pyrotrol Negative Control	Dilution Pattern EU/ml(b)	Titration in Test Solution(c)				
		Acceptable Range			Unacceptable Range(d)	
+	1	+	+	+	+ or (−)	+
+	0.5	+	+	+	−	+
+	0.25	−	+	+	−	+
−	0.125	−	−	+	−	+
−	0.06	−	−	−	−	+
−	Negative Control	−	−	−	−	−
Control Titration		Compatible			Inhibition	Enhancement

a. In order to perform the compatibility test with Pyrotrol Positive Control or any endotoxin preparation other than the US Standard Endotoxin, its potency in EU/ng should be determined (see METHOD FOR DETERMINATION OF THE RELATIONSHIP OF THE PYROTROL POSITIVE CONTROL [CSE] AND THE US STANDARD ENDOTOXIN).
b. Select a dilution pattern (in EU/ml) which brackets the observed end point by ±2 tubes (the above table uses 0.25 EU/ml).
c. The test solution should be prepared for use in the LAL test as indicated in the SPECIMEN COLLECTION AND PREPARATION section below.
d. Based on the observed end point of the control titration, if the test solution is found to inhibit or enhance the sensitivity of the lysate, repeat the test with fresh dilutions of endotoxin. If it is still found to be incompatible, check the discussion on compatibility in this manual and the appendices or call Difco Laboratories for technical service.

NOTE: The control titration should come out as expected ±1 tube and the acceptable range of the test will vary accordingly (i.e., ±1 tube of the control). If unusual results persist call Difco Laboratories for technical service.

and assayed in parallel with an equivalent endotoxin dilution in Pyrotrol Negative Control. For a complete discussion on this topic consult the attached appendices.

VII. ANALYSIS OF AN UNKNOWN TEST SOLUTION FOR ENDOTOXIN

A. Specimen Collection and Preparation

Specimens must be handled aseptically and in pyrogen-free glassware. Using a pyrogen-free pipette, aseptically remove a small volume of the product to be tested and determine the pH. Calculate the volume of pyrogen-free 0.1N NaOH or pyrogen-free 0.1N HCl required to adjust the pH of the remaining sample to a range of 6 – 8. Adjustment must be done using pyrogen-free glassware. Samples which are in a pH range of 6 – 8 need not be adjusted. If a dry chemical is being tested, prepare a working concentration in pyrogen-free water and adjust the pH accordingly.

B. Compatibility of Test Solution with Pyrotest

Results of tests for pyrogen of a variety of parenteral and biological products to which have been added endotoxin, i.e., "spiked" with endotoxin, when performed in parallel with the untreated specimen indicate that some of these products may contain substances which decrease the sensitivity of the LAL test. It has been demonstrated also that increased concentrations of a variety of salts may adversely affect the activity of the lysate. For these reasons, it is necessary that possible incompatibility be determined prior to routine testing of a product for the first time, by "spiking" with known amounts of endotoxin. Prior to testing of unknown solutions, a compatibility test must be run (see above). Specimens to be tested are stored at 2 – 8°C and allowed to come to room temperature before testing.

Each lot of Pyrotest must be tested for compatibility with the test solution before beginning the assay.

Table 5. Interpretation of the LAL Assay on Test Solutions.

Sample	Test Solution	Test Solution	Positive Control	Positive Control	Negative Control	
Tube#	1	2	3	4	5	Interpretation
	−	−	+	+	−	Test solution contains less endotoxin than can be detected with the Pyrotest lot and can be considered pyrogen free if the testing procedures for compatibility and sample preparations are met.
	+	+	+	+	−	Test solution contains more endotoxin than the Pyrotest labeled sensitivity if the testing procedures for compatibility and sample preparations are met. In order to quantitate the amount of endotoxin in the test solution, prepare a serial dilution of the test solution in Pyrotrol Negative Control and multiply the dilution factor at the end point by the label sensitivity in EU/ml.
Possible Reactions Patterns	soft gel or + or −	soft gel	+	+	−	A soft gel is observed when amount of endotoxin in the test solution is equal to the limit of the sensitivity of the Pyrotest. Variable results (+ and −) in duplicate assays can also be observed. Soft gels are also observed if the tubes are disturbed during incubation or if the test solution's pH is outside the 6.0 - 8.0 range. Repeat assay. Call Difco Technical Services.
	any reaction		−	+ or −	−	Repeat endotoxin dilution. The addition of 4 times the labeled sensitivity of endotoxin to the positive control may be necessary to insure a positive reaction with the positive control. If problem persists, call Difco Technical Services.
	any reaction	any reaction			+	If a positive reaction is observed with the negative control, repeat entire test with fresh vial of Pyrotrol Negative Control. If problem persists contact Difco Technical Services.

C. The Test

Using the procedure outlined above under the LAL test procedure, a minimum of 5 tubes are required.

1. Tubes 1 and 2 contain the test solution in duplicate.
2. Tubes 3 and 4 contain positive control (a solution of endotoxin in Pyrotrol Negative Control at a concentration of twice the labeled sensitivity (EU/ml) in duplicate.
3. Tube 5 contains Pyrotrol Negative Control. TABLE 5 illustrates the test procedure and interpretation.

STABILITY OF FINAL REACTION

The LAL test has been standardized on the basis of incubating the reaction mixture at 37°C for 1 hour. Increasing the incubation period may result in gelation of lysate with lower concentration of endotoxin than is observed after 1 hour. The results obtained with increased incubation time may not be equivalent to the sensitivity of the test based on a 1 hour incubation period.

LIMITATIONS OF THE TEST PROCEDURE

As indicated above there have been some reports of false positive reactions with some polynucleotides, proteins, and fractions of the cell wall of gram-positive bacteria. Whether these false positive reactions are indeed the result of an interaction between these substances and LAL or possible contamination of these materials with endotoxins from gram-negative bacteria has not been proven.

Similarly, false negative reactions, i.e., partial loss or complete loss of reactivity of LAL, may result from certain components of parenteral and biological solutions.

For the above reasons, it is essential that a compatibility test be performed with the product being tested for pyrogen with Pyrotest by "spiking" with a known concentration of endotoxin. This will determine if this product will affect the activity of the Pyrotest. See COMPATIBILITY TEST.

EXPECTED VALUES

The sensitivity of Pyrotest is calibrated with US Standard Endotoxin and is stated on the vial label. Pyrotest is available in a range of sensitivities.

SPECIFIC PERFORMANCE CHARACTERISTICS

Pyrotest™, 1 Test vial and 50 Test vials have been tested in parallel with US Reference Limulus Amebocyte Lysate using US Standard Endotoxin and yielded entirely satisfactory results. The specifications by the FDA for LAL, Pyrotest, require that its sensitivity be in a range ±1 doubling dilution of that specified on the label (see above). The results obtained in our laboratories with US Standard Endotoxin Lot EC2 show that the sensitivities of various lots of Pyrotest™, 1 Test vials and 50 Test vials are in a range of 1.0 – 0.0625 EU/ml. LAL tests with the Difco endotoxin isolated from a strain of *E. coli,* Pyrotrol Positive Control, were observed to yield potencies of one-fourth that of the US Standard Endotoxin EC5 (see above).

Tests performed in our laboratories with endotoxins isolated from a variety of gram-negative bacilli indicate that Pyrotest is able to detect them all, but at different concentrations (ng/ml). Different lots of lysate having the same sensitivity to the US Standard Endotoxin have been observed to vary in sensitivity to other endotoxins by four two-fold dilutions. As indicated above, this observation necessitates internal standardization of the lysate every time a new batch is used, and the designation of its sensitivity to endotoxins in EU/ml.

REFERENCES

1. Levin, J., and F. B. Bang, Thrombos, Diathes, Haemorrh., 19:186, 1968.
2. Munsford, R. S., Anal. Biochem., 91:509, 1978.
3. Hollandar, V. P., and W. C. Harding, Biochem. Med., 15:28, 1976.
4. Harada, T., T. Morita, S. Iwanaga, S. Noakamura, and M. Niwa, In E. Cohen (ed), Biomedical Application of the Horseshoe Crab (Limulidae) 209,220 Alan R. Liss, Inc., N.Y., 1979.
5. Flowers, D. J., Med. Lab. Sci., 36:171, 1979.
6. Frauch, P., J. Pharm. Sci., 63:808, 1974.
7. Goto, H., and S. Nakamura, Jpn. J. Exp. Md., 49:19, 1979.
8. Melvaer, K. L., and D. Frystro, Appl. Env. Micro., 43:493, 1982.
9. Loeb, L., J. Med. Res., 7:145, 1902.
10. Levin, J., and F. B. Bang, Bull. John Hopkins Hosp., 11:265, 1964.
11. Rojas-Corona, R. R., R. Sharkes, S. Tomakuma and J. Fine, Proc. Soc. Exp. Biol. Med., 132:599, 1969.
12. Levin, J., T. E. Poore, N. P. Zauber and R. S. Oser, N. E. J. Med., 283:1313, 1970.
13. Reinhold, R., and J. Fine, Proc. Soc. Exp. Biol. and Med., 137:334, 1970.
14. Cooper, J. F., J. Levin, and N. N. Wagner, Jr., J. Lab. Clin. Med., 78:138, 1971.
15. Jorgensen, J. H., H. F. Carrajal, B. E. Chipps, and R. F. Smith, Applied Micro., 26:38, 1973.
16. Solumn, N. O., Thromb. Res., 2:55, 1973.
17. Levin, J., In E. Cohen (ed), Biomedical Applications of the Horseshoe Crab (Limulidae), p. 131 – 146, Alan R. Liss, Inc., N.Y., 1979.
18. Liu, T., et al., ibid, p. 147 – 158.
19. Takagi, T., et al., ibid, p. 169 – 184.
20. Mosesson, N. W., et al., ibid, p. 159 – 168.
21. McAuley, R. J., R. D. Ice and E. G. Curtis, Amer. J. Hosp. Pharm., 31:688, 1974.
22. Elin, R. J., Personal Communication, 1973.
23. Elin, R. J., and S. Wolff, J. Infec. Diseases, 128:349, 1973.
24. Wildfeuer, A., B. Haymer, K. H. Schleijer and O. Hajerkamp, Applied Micro., 28:867, 1974.
25. Jorgensen, J. H., R. F. Smith, Applied Micro., 26:521, 1973.
26. Nandan, R., and D. R. Brown, J. Lab. Clin. Med., 89:910, 1977.
27. Elin, R. J., Biomedical Applications of the Horseshoe Crab (Limulidae), p. 279 – 292, Alan R. Liss, Inc., N.Y., 1979.
28. Dyson, D., and Cassady, G., Pediat., 58:105, 1976.
29. Nachum, R., A. Lipsev and S. E. Siegel, N. Eng. J. Med., 289:931, 1973.
30. Jorgensen, J. H., J. C. Lee and P. M. Jones, Texas Rep. Bio. Med., 33:475, 1975.
31. Jorgensen, J. H., Biomedical Applications of the Horseshoe Crab (Limulidae), p. 255 – 264, Alan R. Liss, Inc., N.Y., 1979.
32. Jorgensen, J. H., and R. M. Jones, Am. J. Clin. Path., 63:142, 1975.
33. Spagna, V. A., R. B. Prior and R. L. Perkins, British J. Vener. Dis., 55:179, 1979.
34. Difco Laboratories, Unpublished Data.
35. Weary, M. E., G. Donohue, F. C. Pearson and K. Story, Applied and Environmental Microbiology, 40:1148, 1980.

PACKAGING

Pyrotest™	100 × 1 test	3375-32-3
Contains:		
3362 Pyrotest™, 1 Test Vial	100 vials	
3364 Pyrotrol™ Positive Control 100 ng	5 vials	
3365 Pyrotrol™ Negative Control	5 vials	
Pyrotest™	50 tests	3378-32-0
Contains:		
3381 Pyrotest™, 50 Test Vial	1 vial	
3364 Pyrotrol™ Positive Control 100 ng	5 vials	
3365 Pyrotrol™ Negative Control	5 vials	
Pyrotrol™ Positive and Negative Controls	10 vials	3366-32-4
Contains:		
3364 Pyrotrol™ Positive Control 100 ng	5 vials	
3365 Pyrotrol™ Negative Control	5 vials	

APPENDIX A

FDA GUIDELINE FOR VALIDATION OF THE LIMULUS AMEBOCYTE LYSATE TEST AS AN ALTERNATE END-PRODUCT PYROGEN TEST FOR HUMAN AND VETERINARY INJECTABLE DRUGS OTHER THAN LICENSED BIOLOGICALS

In the Federal Register of November 4, 1977, (42 FR 57749), the Food and Drug Administration (FDA) issued a notice stating the conditions under which manufacturers can use the Limulus Amebocyte Lysate (LAL) test for end-product testing of biologicals and medical devices for endotoxins. The notice stated that the conditions for use of the LAL test as an end-product endotoxin test for human drugs other than biologicals would be published at a later date. The following guideline has been developed to inform manufacturers of human and veterinary drugs, other than licensed biologicals as to what is necessary to validate the use of the LAL test as an alternate to the USP rabbit end-product pyrogen test. While a manufacturer who complies with the following would be considered as complying with Current Good Manufacturing Practices and New Drug Application requirements, utilization of methods and techniques not presented does not necessarily represent a violation, provided the manufacturer is able to adequately assure through validation the nonpyrogenicity of the product.

Validation of the LAL test as an alternate pyrogen release test for such drugs requires the determination of LAL sensitivity and the performance of inhibition and activation testing.

A. LYSATE SENSITIVITY

Various methodologies for the detection of endotoxin, using limulus amebocyte lysate, have been described. Currently, all commercially available, licensed lysate, employ either the gel clot or turbidimetric methods. Other methods have been reported and some show potential to further increase the sensitivity of the LAL method. Manufacturers shall use a LAL reagent licensed by the FDA Bureau of Biologics. See Attachment A for a method of assessing the capability of the laboratory and analysis.

Based on a review of the literature by the Bureau of Drugs, it appears that the lower 95% confidence limit for the PD_{50}[1] for endotoxin in the rabbit test is 5.0 EU/Kg[2] (0.5 EU/ml at a dose of 10 ml/kg. While levels of endotoxin down to 2.5 EU/Kg are indeed pyrogenic, these levels cannot be **reliably** detected in this test. A review of data submitted by manufacturers to the Bureau of Drugs and information from the Bureau of Biologics the LAL test can reliably detect less than 0.25 EU/ml of endotoxin.

The sensitivity of the LAL reagent should be checked using a series of reference endotoxin concentrations bracketing the labeled sensitivity. All endotoxin testing shall be performed using equipment and diluent demonstrated to be negative for the lysate used. Positive and negative controls shall be included in each test run. Two-fold dilutions of the reference standard endotoxin or control standard endotoxin (see Attachments B and C) made with nonpyrogenic, LAL negative diluent are combined with specified amounts of LAL preparation. The tubes are incubated and the endpoint is determined according to the technique selected. Confirmation of the LAL sensitivity should be reproducible.

---

[1] PD_{50} is that dose of pyrogen which is pyrogenic (0.6°C rise to 50% of the rabbits.)

[2] EU = Endotoxin Unit. An Endotoxin Unit is the activity in a defined weight of the U.S. Standard Endotoxin. The Bureau of Biologics established this standard and maintains continuity of the Endotoxin Unit with successive lots.

The following shall be considered the tolerance limit that all parenteral drugs, except biologicals, must meet if the LAL test is to be used as an end-product endotoxin test: 5 EU/kg

Non Detectable: for parenteral drugs that have an intrathecal route of administration. A negative response at the lowest detectable limit using a lysate with a sensitivity of 0.2 EU/ml or one more sensitive.

Drugs exempted from the above tolerance limits are:
a. Compendial drugs for which endotoxin limits have been established.
b. Drugs covered by New Drug applications where different limits have been approved.
c. Drugs which are not amenable to the LAL and rabbit test.

If a manufacturer uses **either** the rabbit or the LAL test and if **either** test fails, the batch shall be rejected. A batch which fails the LAL test shall not be retested in the rabbit and released if it passes.

B. INHIBITION TESTING

Before using the LAL test for pyrogenicity of drugs, lack of product inhibition or activation of the test shall be shown for each drug formulation (see Attachment B, Preparatory Testing). If the drug is manufactured in various concentrations of the active ingredient and the rest of the formulation remain constant, then only the highest and lowest concentrations need be tested. If either concentrations shows inhibition or activation then each concentration should be tested.

All inhibition tests shall be performed on the undiluted drug product or the Maximum Valid Dilution (See Attachment D). If the drug product shows inhibition or activation and is amenable to rabbit testing, then the rabbit test will still be the appropriate pyrogen test for that drug. Exceptions to the above will be accepted if the inhibitor or activator can be neutralized without affecting the sensitivity of the test or if the LAL Test is more sensitive than the rabbit test even though it cannot meet the 5 EU/kg limit. For those drugs not amenable to rabbit testing the manufacturer shall demonstrate the smallest quantity of endotoxin that can be detected. At least three (3) production batches of each finished product shall be tested for inhibition.

The sampling technique selected should result in a statistically acceptable sampling of a finished production batch. If the source of a particular ingredient of the drug formulation is changed, three (3) batches of the finished product shall be tested for the presence of inhibitors or activators. Also, if the manufacturing procedure is changed, the inhibition test shall be repeated.

For inhibition testing, the drug product should be "spiked" with endotoxin and then diluted with drug product or appropriate dilution of drug product to bracket the sensitivity of the LAL. The diluent used in the inhibition tests must be the same as that used for release testing. Positive and negative controls shall be included. Results of endotoxin determinations in the standard series and drug shall not differ by more than plus or minus a two-fold dilution.

Lysates are crude limulus amebocyte extracts that have not been characterized as to protein, lipid, carbohydate, or lipoprotein content. Therefore, it is possible that one lysate or lot of lysate may contain a component which, through a synergistic interaction with a particular drug formulation could inhibit or activate the assay, while another lysate or lot of lysate, void of this component, would not. Consequently, it is crucial that a positive product control should be included in the release test protocol. The positive product control would consist of the drug product or diluted drug product which has been spiked with endotoxin at a concentration of 2 λ (λ = Labeled Lysate Sensitivity).

ATTACHMENT A
INITIAL QUALITY CONTROL PROCEDURE FOR A TESTING LABORATORY
Before any official tests are performed the variability of the testing laboratory should be assessed. A minimum of 4 and a maximum of 8 tests should be performed by each technician using a single lot of LAL and a single lot of endotoxin. For each test, a group of 4 replicate tests, a single series of dilutions of the endotoxin should be prepared. A two-fold dilution series should be used with a range suitable for yielding an endpoint with the LAL being tested. After incubation, the reaction in each tube should be recorded as either positive or negative. The endpoints should be expressed as EU/ml and then converted to logs and the standard deviation (S.D.) calculated.

If the S.D. is less than or equal to the value for the 99% limit on the S.D. for 4 samples (see Table 1), the test is in control. If the S.D. is greater than the tabulated value, the test may be expanded to a maximum of 8 tests and a new S.D. calculated using data from all tests. If after doing 8 tests, the S.D. exceeds the 99% upper limit value corresponding to 8 tests, the tests are invalid due to excessive variability. If after any set of 4 replicates the S.D. is equal to or less than the tabulated value for that number of tests, it can be concluded that the variability is under control and no more tests are necessary. At that point the geometric mean (G.M.) of the endpoints from all tests performed should be calculated. If the labeled value for the lysate used falls within plus or minus one two-fold dilution of the geometric mean, the tests have confirmed the label claim. If it is not, the source of variation should be determined before doing additional testing.

Table 1. 99% Upper Limit on Standard Deviation.

Size	LOG$_2$	LOG$_{10}$
4	1.214	0.365
8	1.015	0.306

ATTACHMENT B
USP XX Page 888 – 889 (1980)
(85) BACTERIAL ENDOTOXINS TEST
This chapter provides a test for estimating the concentration of bacterial endotoxins that may be present in or on the sample of the article(s) to which the test is applied. These endotoxins are lipopolysaccharides that are pyrogens. The reagent employed for detecting and measuring such endotoxins is Limulus Amebocyte Lysate (LAL) obtained from aqueous extracts of the circulating amebocytes of the horseshoe crab, *Limulus polyphemus*. In the presence of minute concentrations of endotoxins (of the order of pg/ml in the case of potent endotoxin preparations made from *Escherichia coli*), LAL reagent exhibits a protein coagulation reaction (gel, turbidity or precipitate) under prescribed in-vitro test conditions and in the absence of interfering substances. Biochemical studies indicate that, in addition to endotoxin, three other factors are necessary to produce the protein coagulation reaction: a proclotting enzyme, clottable protein (coagulogen), and certain divalent cations. The constituted LAL reagent provides these factors.

In preparing for and applying the test, observe precautions in handling the specimens in order to avoid gross microbial contamination, and treat any containers or utensils employed to destroy extraneous endotoxins that may be present on their surfaces, such as by heating in an oven at 250° or above for sufficient time to render their surfaces negative to the test, but not less than 30 minutes. Where the test is conducted as a limit test, the specimen is determined to be positive or negative to the test judged against some pre-established endotoxin concentration. Positive and negative controls are essential.

For validating the test for an article, or for special purposes where so specified, testing of specimens is conducted quantitatively to determine response endpoints. Usually graded strengths of the specimen and endotoxin control are made by multifold dilutions. Dilutions selected so that they correspond to a geometric series in which each step is greater than the next lower by a constant ratio, or to an arithmetic series with constant intervals in concentrations are not precluded. Do not store diluted endotoxin, because of loss of activity by absorption. As a general rule and in the absence of supporting data to the contrary, negative and positive controls are incorporated in the test. Replication within dilution series is essential for assessing the error and confidence in the results obtained and for comparing the geometric means of end-point values.

REFERENCE STANDARD AND CONTROL STANDARD ENDOTOXINS
The reference standard endotoxin (RSE) is the U.S. Reference Endotoxin designated EC-2 by the Bureau of Biologics of the Federal Food and Drug Administration. It has a defined potency of 5.0 Endotoxin Units per ng. A control standard endotoxin (CSE) is an endotoxin preparation other than the RSE that has been standardized against the RSE.

PREPARATORY TESTING
The validity of test results for bacterial endotoxins requires an adequate demonstration that specimens of the article, or of solutions, washings, or extracts thereof to which the test is to be applied do not of themselves inhibit or enhance the reaction of otherwise interfere with the test. Validation is accomplished by testing untreated specimens and specimens to which known and demonstrable amounts of RSE or a CSE have been added, and comparing the results with those obtained with a control consisting of the same RSE or CSE dissolved and serially diluted in LAL reagent water or saline (Water for Injection or Sodium Chloride Injection that does not give a positive response with the LAL reagent used in this test). Appropriate negative controls are included. Use a LAL reagent of confirmed label potency. Confirm the label potency (sensitivity) of the particular LAL reagent with the RSE (or CSE) using not less than 4 replicates, under conditions shown to achieve an acceptable variability of the test, e.g., where the standard deviation of the geometric mean lot lysate sensitivity is such that the 99% upper limit of the result does not exceed the 99% upper limit specified for licensing the LAL reagent (see Proposed Additional Standards, Limulus Amebocyte Lysate, 43 Fed. Reg. 156, August 11, 1978, applicable to 21 CFR Sec. 660.102(8)(d)). Conduct replicated assays of the endotoxin controls and of the specimens to which endotoxin was added as directed under Test Procedure. Record the endpoints (E, in Units or Milliunits per ml) observed in the replicated assays. Take the logarithms (e) of the endpoints of the dilution series used, and compute the geometric means of the log endpoints for the control and for specimens containing endotoxin by the formula antilog Σ e/f, in which Σe is the sum of the log endpoints of the dilution series used and f is the number of replicate endpoints in each case. The test is valid for the article if the geometric mean value obtained for the specimen is not significantly different from that value for the control, i.e., the geometric mean endpoints are separated by not more than a 5-fold difference in the parallel dilution series and have a statistical probability (determined from the logarithmic values of the observed endpoints for the calculation) of 0.05 or less of a true difference between the means.

TEST PROCEDURE

Use not less than 4 replicates at each level of the dilution series for each specimen under test. Whether the test is employed as a limit test or as a quantitative assay, an endotoxin dilution series involving not less than 4 replicates is conducted in parallel as a control measure. A single set of endotoxin dilution series may serve as a control for several blocks of limit tests or of assays, provided the environmental conditions within test blocks are uniform. Where there is to be a change in the lot of LAL or of standard endotoxins, test a prior satisfactory lot simultaneously on changeover.

Preparation

Since the amounts per container of standard endotoxin and of LAL reagent may vary, constitution of contents and storage conditions should be as directed in the labeling. The pH of the specimen during test is in the range 6.0 to 7.5 unless specifically directed otherwise in the individual monograph. The pH may be adjusted by the addition of sterile, endotoxin-free 0.1N sodium hydroxide or 0.1N hydrochloric acid or suitable buffers.

Procedure

To 10 × 75 mm test tubes add aliquots of the appropriately constituted LAL reagent, and the specified volumes of specimens, endotoxin standard, positive controls, and negative controls. Swirl gently to mix, and place in an incubating device such as a water bath or heating block, recording the exact time at which the tubes are so placed. Incubate each tube, undisturbed, for 60 ± 2 minutes at 37 ± 1°, and carefully remove it for observation.

CALCULATION AND INTERPRETATION

Gel Clot Readings

A positive reaction is characterized by the formation of a firm gel that remains firm when inverted through 180°. Record such a result as positive (+). A negative result is characterized by the absence of such a gel or by the formation of a viscous gel that does not maintain its integrity. Record such a result as negative (−). The fine gel structure is susceptible to irreversible breakdown. Handle the tubes with care, and avoid subjecting them to unwanted vibrations, or false negative observations may result in the gel clot test.

Turbidimetric Readings

Compare absorbances of the test specimens with a standard curve obtained from an endotoxin dilution series when using LAL reagents that have been modified to yield reactions that can be measured turbidimetrically (transmittance), or after using a process for developing a color reaction with the centrifuged precipitate (see Spectrophotometry and Light-scattering [851] and Color — Instrumental Measurement [1061]), in order to obtain the concentration of endotoxin present in the specimen.

Calculations for Limit Test

Calculate the concentration of endotoxin (in Units or Milliunits per ml or in Units or Milliunits per g) in or on the article under test by the formula $(\rho\lambda)(f/\Sigma E)$, in which ΣE is the sum of endpoint dilution factors expressed as decimal fraction, f is the number of replicate endpoints, λ is the labeled lysate sensitivity in Units or Milliunits per ml appropriate to the standard endotoxin employed in the assay, and ρ is the correction factor for those cases where a specimen of the article cannot be taken directly into test but is processed as an extract, solution, or washing.

Calculations for Assay

Calculate the concentration of endotoxin in or on the article under test from the antilog of the mean log endpoint values and the confidence limits from the logarithmic values of the observed endpoints, or by using any other suitable statistical method (see Calculation of Potency from a Single Assay [111]).

Interpretation

The article meets the requirements of the test if there is no formation of a firm gel at the level of endotoxin specified in the individual monograph or, where an assay of endotoxin content has been made, such result is not more than the maximum permissible amount specified in the individual monograph, and the confidence limits of the assay do not exceed those specified.

ATTACHMENT C
DETERMINATION OF THE RELATIONSHIP BETWEEN THE CONTROL STANDARD ENDOTOXIN (CSE) AND THE REFERENCE STANDARD ENDOTOXIN (RSE)

If a manufacturer chooses to use an endotoxin preparation (CSE) other than the RSE Bureau of Biologics Standard Endotoxins, the CSE will have to be standardized against the RSE. If the CSE is not a commercial preparation which has been adequately characterized, it should be studied and fully characterized as to reaction in the rabbit, uniformity, stability of the preparation, etc. Whenever a new lot, sensitivity, or manufacturer of the LAL, or a new lot, source or manufacturer of the CSE is obtained, the relationship of the CSE to the RSE should be determined.

The following is an example of a procedure to obtain the relationship of the CSE to the RSE:

At least 4 samples (vials) for the lot of CSE and RSE should be assayed. State in ng/ml the endpoint for the CSE and in EU/ml for the RSE. The values obtained should be the geometric mean of the endpoints using a minimum 4 replicates.

$$\text{Example CSE} = 0.018 \text{ ng/ml}$$
$$\text{RSE} = 0.3 \text{ EU/ml}$$

This indicates that 0.018 ng of the CSE is equal to 0.3 EU of the RSE. Thus, the CSE contains 16.6 EU/ng.

Until we have more data on how different CSE endotoxins react to the lysate the range of the EU/ng for the CSE shall be between 0.1 and 10 times that of the RSE.

ATTACHMENT D
MAXIMUM VALID DILUTION

To determine how much the product can be diluted and still be able to detect the limit endotoxin concentration, the following formula will determine the Maximum Valid Dilution:

$$\text{MVD} = \frac{\lambda\theta}{K}$$

where,

λ = Sensitivity of LAL reagent in Endotoxin Units (EU) per ml.

θ = Rabbit or Maximum Human dose/kg, whichever is more stringent. Use 70 kg as the weight of the average human when calculating the maximum human dose per kg. Also, if the pediatric dose/kg is higher than the adult dose then it shall be the dose used in the formula.

K = Tolerance Limit: 5 EU/kg

APPENDIX B

BUREAU OF MEDICAL DEVICES GUIDELINES FOR ADOPTION OF THE LIMULUS AMEBOCYTE LYSATE TEST FOR PYROGENICITY OF DEVICES
In the Federal Register of November 4, 1977, (42FR 57749), the Commissioner of Food and Drugs issued a notice stating that manufacturers of medical devices must submit data establishing that the LAL test is at least equivalent to the USP rabbit pyrogen test and obtain written approval to use the LAL test from the Director of the Bureau of Medical Devices. The following guidelines have been developed to assist manufacturers converting from the rabbit to the LAL test.

Submissions of data should include the following:
1. Data demonstrating the sensitivity and reproducibility of the LAL test (if the firm will pass/fail lots on the results of the LAL test); or
2. Data demonstrating the sensitivity and reproducibility of the LAL test and the USP pyrogen test (if the firm will retest LAL test failures with the rabbit test).
3. Inhibition Testing: Each product line of devices utilizing different materials or methods of manufacture must be checked for inhibition or activation of the LAL test; it is only necessary to demonstrate this at one plant for each product line. However, consideration should be given to the potential for impurities present in different sources of raw materials that may inhibit the LAL test.
4. Parallel Testing: Testing of device production lots by both rabbit and LAL test.

The Bureau has reviewed the results of a HIMA collaborative study undertaken to determine a sensitivity endpoint for the "average" rabbit colony. The sensitivity endpoint determined by this study, 0.1 ng/ml of *E. coli* 055:B5 endotoxin from Difco Laboratories can be referenced by manufacturers who will pass/fail lots solely on the results of the LAL test. Manufacturers who will retest LAL test failures with a USP rabbit test must submit data establishing the sensitivity and reproducibility of the rabbit colony used for retesting.

These data and the results of inhibition and parallel testing should be submitted to the Bureau for review. Data should be expressed graphically or in tabular form when possible.

If Bureau review of the data submission does not reveal any problems, the Bureau will send an official letter giving the manufacturer permission to release the devices included in their submission using the LAL test alone for pyrogenicity. For devices not included in the original submission, manufacturers should proceed with inhibition and parallel testing before switching to the LAL test. This data should be submitted to the Bureau for review before using LAL for final product testing of these devices.

1. DATA AND DEMONSTRATING THE SENSITIVITY AND REPRODUCIBILITY OF THE LAL TEST
The LAL test must be shown to be at least as sensitive as the USP rabbit pyrogen test. A manufacturer must be able to demonstrate a test failure rate of 90% or greater at the sensitivity endpoint determined for the "average" rabbit colony, 0.1 ng/ml of *E. coli* 055:B5 endotoxin. The level of endotoxin selected as the pass/fail point for evaluating pyrogenicity of products using the LAL test must be equivalent to or below this endpoint sensitivity level.

The sensitivity of the LAL should be determined using a series of reference endotoxin concentrations bracketing the endpoint of the lysate. Two-fold dilutions of the endotoxin reference made with non-pyrogenic water or saline should bracket the lysate endpoint sensitivity. Aliquots of reference endotoxin are added to tubes containing specified amounts of lysate preparation. Tubes are incubated and the endpoint is determined according to the technique selected. Results should be reproducible and differences between runs of each test should not be statistically significant.

Routine performance of the LAL test should include an endotoxin standard series (run in duplicate) and a negative control. Running an endotoxin series can be useful for checking lysate sensitivity, the competence of the technician, and for identifying other problems such as contamination of glassware. Consideration should also be given to the stability of endotoxin standards and appropriate storage conditions; weak endotoxin solutions may not be as stable as more concentrated solutions under certain conditions.

2. DATA DEMONSTRATING THE SENSITIVITY AND REPRODUCIBILITY OF THE LAL TEST AND THE USP PYROGEN TEST

The Bureau will accept a provision for retest of device lots in the USP rabbit test if a manufacturer demonstrates that its rabbit colony is at least as sensitive as the sensitivity endpoint of the "average" rabbit colony (i.e., 0.1 ng/ml of *E. coli* 055:B5 endotoxin from Difco Laboratories).

The sensitivity of the rabbit colony and of the LAL test should be determined using a series of reference endotoxin concentrations bracketing the endpoint for both procedures. The sensitivity of the LAL test should be determined as described above.

The procedure for performing the USP rabbit pyrogen test is explained in the USP XIX, 1975. Sufficiently small dilutions (five-fold or less) of endotoxin bracketing the endpoint should be used to determine the sensitivity of the rabbit colony.

For manufacturers contracting for comparison of the sensitivities of the USP rabbit pyrogen and LAL tests who intend eventually to do LAL tests on devices, in-house sensitivity of the LAL test in-house should not be significantly different from the sensitivity of the LAL test done by the contractor using the same lots of lysate and endotoxin.

3. INHIBITION TESTING

Before using the LAL test for pyrogenicity of devices, lack of product inhibition or activation of the LAL test must be shown for each type of device. Consideration should be given to possible inhibition of different chemical components of otherwise similar devices.

Manufacturers can logically divide their device products into groups of products according to common chemical formulation; it should then only be necessary to qualify a representative product from each such group. Ideally, the product chosen from each group would be the one with the largest surface area exposed to the fluid path.

At least three production lots of each product type should be tested for inhibition. In general, use of the sampling technique selected should result in a random sampling of a finished production lot. The number of devices in any finished production lot may vary and individual devices may differ in size, configuration, composition, expense, and their criticality in terms of pyrogenicity. It is recommended that a sample of at least ten test units per production lot be used. In special cases, other sampling schedules can be adapted in consultation with the Bureau.

The process of preparing an eluate/extract for pyrogen or inhibition testing may vary for each device. Some medical devices can be flushed, some may have to be immersed in the non-pyrogenic rinse solution (water or saline), while others may be best tested by disassembling or by cutting the device up into pieces prior to extraction by immersion. In general, for devices being flushed, the non-pyrogenic rinse solution should be held in the fluid pathway for 1 hour at room temperature (above 65°F); effluents should be combined. If a device is to undergo extraction, a minimum extraction time should be 15 minutes at 37°C, 1 hour at room temperature (above 65°F) or other demonstrated equivalent conditions.

Guidelines for rinse volumes include the following:
1. Each of the 10 test units should be rinsed with 20 – 40 mls of non-pyrogenic saline or water depending on the size of the device.
2. For unusually small or large devices, the surface area of the device which comes in contact with the patient may be used as an adjustment factor in selecting the rinsing or extracting volume.

The rinsing dilution scheme should not result in a greater dilution of endotoxin than used previously in USP rabbit testing.

For inhibition testing, both the rinsing/extraction solution and the device extract should be spiked with the same endotoxin levels used in standard series bracketing the sensitivity of the lysate, negative controls for both solutions should be included. Results of endotoxin determination in the standard series and device extracts should not differ significantly.

For each group of devices protocols and test results from inhibition studies including the end-points of the standard series, inhibition series and results of negative controls should be compiled and submitted to the Bureau for review.

4. PARALLEL TESTING OF DEVICES BY LAL AND USP RABBIT PYROGEN TEST.

The general procedure for parallel testing by LAL and USP rabbit pyrogen testing will include the following aspects. After obtaining a suitable eluate/extract pool from a finished production lot, this pooled extract is kept under conditions appropriate for endotoxin stability until it is tested. A smaller portion of the pool is used for LAL testing, while the major portion is used for rabbit pyrogen testing.

Parallel testing should be done on 3 finished product lots for each group of devices using rinsing/eluting and sampling techniques similar to those described for inhibition testing. As in inhibition testing sampling can be adjusted for special situations.

Special attention should be given to any inconsistent results such as false negatives or false positives. For each product group, protocols and test results from parallel testing including the endpoint of the standard series, results of negative controls, LAL and rabbit device tests should be compiled and submitted to the Bureau for review.

BACTO REINFORCED CLOSTRIDIAL MEDIUM
BACTO REINFORCED CLOSTRIDIAL AGAR

INTENDED USE
Bacto Reinforced Clostridial Medium and Bacto Reinforced Clostridial Agar are used for culturing anaerobes, particularly Clostridia, and other microorganisms from foods and clinical specimens.

HISTORY/PRINCIPLES

Hirsch and Grinstead[1] formulated reinforced clostridial medium to initiate growth from small inocula and to obtain the highest viable count of Clostridia. Barnes and Ingram[2] used the medium for diluting an inoculum of vegetative cells of *Clostridium perfringens*.

Barnes et al,[3] used reinforced clostridial agar, a solid version of reinforced clostridial medium, for estimating Clostridia in food. Attenborough and Scarr[4] used the membrane filter technique to count *Clostridium saccharolyticum* in sugar. The culture medium contains no inhibitors and uses cysteine as the reducing agent.

FORMULAE

BACTO REINFORCED CLOSTRIDIAL MEDIUM
DEHYDRATED

Ingredients per liter

Bacto Tryptose	10 g
Bacto Beef Extract	10 g
Bacto Yeast Extract	3 g
Bacto Dextrose	5 g
Sodium Chloride	5 g
Soluble Starch	1 g
Cysteine Hydrochloride	0.5 g
Sodium Acetate	3 g
Bacto Agar	0.5 g

Final pH 6.8 ± 0.2 at 25°C.

Five hundred grams will make 13.1 liters of medium. Rehydrate with 38 g/liter.

BACTO REINFORCED CLOSTRIDIAL AGAR
DEHYDRATED

Ingredients per liter

Bacto Tryptose	10 g
Bacto Beef Extract	10 g
Bacto Yeast Extract	3 g
Bacto Dextrose	5 g
Sodium Chloride	5 g
Soluble Starch	1 g
Cysteine Hydrochloride	0.5 g
Sodium Acetate	3 g
Bacto Agar	15 g

Final pH 6.8 ± 0.2 at 25°C.

Five hundred grams will make 9.5 liters of medium. Rehydrate with 52.5 g/liter.

METHOD OF PREPARATION

1. Suspend the appropriate amount of medium in 1 liter distilled or deionized water. Heat to boiling to dissolve completely.
2. Dispense into tubes or flasks as desired.
3. Sterilize in the autoclave for 15 minutes at 15 lbs pressure (121°C).

STORAGE

Bacto Reinforced Clostridial media	Below 30°C
Prepared medium	15 – 30°C

QUALITY CONTROL
Identity Specifications

Dehydrated powder:	light tan, homogeneous, free-flowing
Reaction of appropriate solution:	pH 6.8 ± 0.2 at 25°C
Prepared medium:	medium amber, slightly opalescent

Typical Cultural Response on Bacto Reinforced Clostridial Media
After 40 – 48 Hours or Longer if Required at 35°C

Organism	Growth
Bacteroides fragilis ATCC® 23745	good to excellent
Bacteroides vulgatus ATCC® 8482	good to excellent
Clostridium butyricum ATCC® 9690	good to excellent
Clostridium perfringens ATCC® 13124	good to excellent

REFERENCES
1. J. Dairy Res., 21:101, 1954.
2. J. Appl. Bact., 19:117, 1956.
3. J. Appl. Bact., 26:415, 1963.
4. J. Appl. Bact., 20:460, 1957.

PACKAGING

Bacto Reinforced Clostridial Medium	500 g	1808-17-3
Bacto Reinforced Clostridial Agar	500 g	1813-17-6

RENNIN NF XI

Rennin NF XI is a desiccated enzyme obtained from the glandular layer of the stomach of the calf and is used for the enzymatic coagulation of milk. It is standardized according to the method given in the National Formulary, 11th edition.

PACKAGING

Rennin NF XI	100 g	0287-15-7
	500 g	0287-17-5

BACTO RHEUMATOID SLIDE TEST
(Eosin Latex Test)

INTENDED USE
Bacto Rheumatoid Slide Test containing Bacto Latex 0.81 and Bacto Eosin Solution is used in a qualitative test[1] to rapidly screen sera for macroglobulins which are closely associated with the rheumatoid factor (RF)[2,3] and also with certain nonrheumatic diseases.

TEST SUMMARY AND HISTORY
The value of the rheumatoid slide test was first demonstrated by Singer and Plotz[1] who also concluded that this screening procedure is highly sensitive for detecting macroglobulins associated with rheumatoid arthritis and suggested its suitability in the clinical laboratory and physician's office. Bianchi and Keech[4] used whole blood from the finger tip to perform the rheumatoid slide test for rheumatoid arthritis and obtained good agreement with another latex fixation test and the clinical diagnosis. They stressed the advantages of this slide test using whole blood, as being less expensive and more easily performed than other slide tests in which serum is used.

The mechanism of the eosin latex test is not clear. It is not considered an immune reaction like that between an agglutinogen and its specific agglutinin. Rather it is probably more related to the reaction between cardiolipin "antigen" and "reagin" and their subsequent flocculation as found in a nontreponemal antigen test for syphilis.

Essentially, the eosin latex test is a reaction between macroglobulins in the serum and latex 0.81 with agglutination of the latex particles as evidence of the reaction. The eosin solution inhibits nonspecific agglutination that may occur in some sera lacking macroglobulins. For this reason the eosin solution must be added to a serum before the latex is added. The eosin solution also provides contrast to make the agglutination easier to observe.

Despite the ambiguity of the reaction mechanism of this test, Lane and Decker[5] using serum, compared the rheumatoid slide test with other slide tests and obtained comparable results. The further usefulness and practicality of the test in the laboratory for the detection of rheumatoid arthritis and certain nonrheumatoid conditions were described by Dresner and Trombly,[6] Caplan,[7] Howell, Malcolm and Pike,[8] Bonomo, Lo Spalluto and Ziff,[9] and Kunkel, Simon and Fudenberg.[10]

Slide tests are used regularly for the detection of RF. Singer[11] has stressed that when using slide tests all positive screening results, and those negative results which are at variance with clinical findings, must be repeated using a tube dilution test for RF. Bacto Rheumatoid Titration Test, code 3300 is recommended for this purpose.

REAGENTS

Bacto Latex 0.81 is a standardized suspension of latex particles having an average diameter of 0.81 micrometers. Thoroughly resuspend contents with gentle end-over-end motion before use. Always avoid shaking latex with dropper cap on the bottle. Use regular plastic cap. The latex suspension is stable to the expiry date on the label when stored at 2 – 8°C. **Do not freeze.**

Bacto Eosin Solution is a 1% solution of Bacto Eosin Y. The reagent is stable to the expiry date on the label when stored at 2 – 8°C.

Bacto Rheumatoid Slide Positive Control will yield a strong positive reaction with Bacto Latex 0.81 when used according to the described technique. It is not for use as a positive control in other latex fixation tests for RF. The reagent is stable to the expiry date on the label when stored at 2 – 8°C.

Bacto Rheumatoid Slide Negative Control will not agglutinate Bacto Latex 0.81 when used according to the described technique. The reagent is stable to the expiry date on the label when stored at 2 – 8°C.

Bacto Rheumatoid Slide is a glass slide divided into 3 sections and is used for the performance of the slide test. The slide should be thoroughly cleaned and dried before each use.

SPECIMEN PREPARATION

Fingertip blood or serum are equally satisfactory. Blood must not be mixed with an anticoagulant. Sera should not be diluted prior to testing nor do they require inactivation.

PROCEDURE

Materials Provided:
Bacto Eosin Solution
Bacto Latex 0.81
Bacto Rheumatoid Slide Positive Control
Bacto Rheumatoid Slide Negative Control
Bacto Rheumatoid Slide

Materials Required but not Provided:
Toothpicks
Light viewer

The rheumatoid slide test is performed either with serum or with whole blood as follows:
1. Place 1 drop of patient's serum (or 3 drops of whole blood) on the center section of the Bacto Rheumatoid Slide.

2. Place 1 drop of Bacto Rheumatoid Slide Positive Control on the left hand section of the slide. Care should be taken not to use the dropper from the positive control vial for dispensing the eosin solution or latex 0.81.
3. Place 1 drop of Bacto Rheumatoid Slide Negative Control on the right hand section of the slide.
4. Add 1 drop of Bacto Eosin Solution to the test material and to each control on the slide.
5. Mix each thoroughly with toothpicks.
6. Invert vial of Bacto Latex 0.81 to obtain a uniform suspension.
7. Add 2 drops of latex suspension to each of the above mixtures.

Note: Nonspecific agglutination may occur if latex is added before the eosin solution.

8. Mix again, spreading the mixtures over an area of about 2 square cm.
9. Tilt the slide back and forth over a light viewer for 3 minutes and observe for agglutination. This is best seen in the thin section of the mixture as the slide is tipped. Avoid attempting to read results from dried out edge of mixture.

CONTROLS

The positive control serves as a reading standard for agglutination. The negative control assures that agglutination does not occur with normal globulins. As a further control on the reagents it is suggested that when the latex and eosin are first removed from the refrigerator a spot check be made by placing a drop of Bacto Eosin Solution on a slide, adding 2 drops of uniformly suspended Bacto Latex 0.81 to the eosin droplet, mixing with a toothpick and spreading the mixture over about a 2 square cm area. Tilt the slide back and forth over a light viewer for 3 minutes. Absence of agglutination of the latex indicates the reagents to be satisfactory.

RESULTS

Any clumping of the latex is considered a positive test.

LIMITATIONS OF THE PROCEDURE

Bacto Rheumatoid Slide Test is a screening procedure only and does not by itself prove or rule out a diagnosis of rheumatoid arthritis. Sera found positive in the rapid slide screening test should be confirmed and quantitated by the Bacto Rheumatoid Titration Test. Negative results by the rapid slide screening test which are at variance with clinical findings should be retested using the Bacto Rheumatoid Titration Test.

REFERENCES

1. JAMA, 168:180, 1958.
2. Acta. Med. Scand. Supp., 341:111, 1958.
3. JAMA, 177:50, 1961.
4. J. Lab and Clin. Med., 50:113, 1957.
5. J. Am. Med. Assoc., 164:1999, 1957.
6. New Engl. J. Med., 251:981, 1958.
7. Tufts New Engl. Med. Ctr. Bull., 4:136, 1958.
8. Am. J. Med., 29:662, 1960.
9. Arth. & Rheum., 6:104, 1963.
10. ibid., 1:289, 1958.
11. Personal Communication, 1974.

PACKAGING

Bacto Rheumatoid Slide Test	100 test set	3237-32-1
Contains:		
Bacto Eosin Solution	5 ml	
Bacto Latex 0.81 (2)	5 ml	
Bacto Rheumatoid Slide Positive Control	2 ml	
Bacto Rheumatoid Slide Negative Control	2 ml	
Bacto Rheumatoid Slide	1 slide	

Bacto Latex 0.81	5 ml	3102-56-4
	6 × 5 ml	3102-57-3
	25 ml	3102-65-3
	1 liter	3102-81-3

RHEUMATOID FACTOR REAGENTS

BACTO RHEUMATOID TITRATION TEST
BACTO RA BUFFER DRIED
BACTO RA PLASMA FRACTION II
BACTO RHEUMATOID NEGATIVE CONTROL
BACTO RHEUMATOID POSITIVE CONTROL

TEST SUMMARY
Bacto Rheumatoid Titration Test Set contains all the reagents required for determining the titer of the rheumatoid factor (RF). Note: Singer has recommended that the antibody designation AHGG (antihuman gamma globulin) should replace RF in this test because these antibodies are directed to the HGG antigen and because they are found in conditions other than RA.

Singer and Plotz[1] first described the use of polystyrene latex particles sensitized with human gamma globulin in a serological test for rheumatoid arthritis. They also described a slide test in which the patient's own gamma globulin was used to sensitize the latex and which resulted in aggregation when exposed to the RF. Svartz[2,3] demonstrated that RF is closely associated with macroglobulins and, in a personal communication[4] stressed that a positive result in any system in which the carrier is sensitized with gamma globulin indicates only that macroglobulins are present in the serum.

Slide tests are used regularly for the detection of RF. Singer[5] has stressed that when using slide tests all positive screening results, and those negative results which are at variance with clinical findings, must be repeated using a tube dilution test for RF.

REAGENTS
Bacto Latex 0.81 is a standardized suspension of latex particles having an average diameter of 0.81 microns. It is particularly applicable to the RA serological procedures described by Singer and Plotz[1] and as modified by Rheins et al,[6] Rothermich and Philips,[7] Goften et al[8] and Singer and Plotz.[9] Thoroughly resuspend contents with gentle end-over-end motion before use. The latex suspension is stable to the expiry date on the label when stored at 2 – 8°C. **Do not freeze.**

Bacto RA Plasma Fraction II is a purified human plasma gamma globulin for use in the RA latex fixation tests of Singer and Plotz[4] and in other procedures. The rehydrated Bacto RA plasma Fraction II is used for sensitizing Bacto Latex 0.81. The entire contents of 1 vial of Bacto RA Plasma Fraction II (150 mg) is added to a vial for rehydrating Bacto RA Plasma Fraction II. Fifteen ml of Bacto RA Buffer Solution is added and the vial gently shaken to dissolve the plasma. If a Bacto RA Plasma Fraction II vial containing 250 mg (available separately) is being reconstituted, then 25 ml of Bacto RA Buffer Solution is added to the reconstitution vial. It is recommended that the vial remain at 2 – 8°C overnight and only the supernatant fluid be used to sensitize the latex. Care should be taken to avoid contaminating the solution.

Bacto Rheumatoid Positive Control is rehydrated by adding 3 ml distilled or deionized water and gently shaking the vial. The rehydrated positive control serum when used in the tube test will agglutinate the sensitized latex particles.

Bacto Rheumatoid Negative Control is rehydrated by adding 3 ml distilled or deionized water and gently shaking the vial. The rehydrated negative control serum will not agglutinate the sensitized latex particles.

Bacto RA Buffer, Dried is an isotonic amino. acetic acid buffer having a pH of 8.2 for rehydrating Bacto RA Plasma Fraction II and for preparing dilution of the test and control sera. To rehydrate, dissolve 17.5 g in 1 liter distilled or deionized water. RA buffer solution should be stored at 2 – 8°C for maximum stability and may be used indefinitely as long as it remains sparklingly clear.

SPECIMEN COLLECTION
Five ml whole blood is collected from a patient. A fasting specimen is best. The blood is allowed to clot and the syneresed serum is transferred to a clean dry test tube. Store serum in the refrigerator at 2 – 8°C until ready to use.

PROCEDURE
Materials Provided:
Bacto Latex 0.81
Bacto RA Plasma Fraction II
Bacto Rheumatoid Positive Control
Bacto Rheumatoid Negative Control
Bacto RA Buffer, Dried
Vial for reconstituting RA Plasma
  Fraction II

Materials Required but not Provided:
Test tubes 12 × 75 mm
Test tube rack
1 ml pipette
Water bath at 56°C
Centrifuge
Interval timer
Refrigerator 2 – 8°C

1. Invert the vial of Bacto Latex 0.81 to obtain a uniform suspension.
2. Transfer exactly 0.1 ml of the latex suspension to 10 ml rehydrated Bacto RA Buffer, Dried to which has been added 0.5 ml rehydrated Bacto RA Plasma Fraction II. Unused portions of the sensitized latex particles should be discarded if not used within several hours after preparation.
3. Place 10 tubes (12 × 75 mm) in a rack and label.
4. Add 0.95 ml RA buffer solution to tube 1 and 0.5 ml buffer to tubes 2 – 10.
5. Add 0.05 ml of test serum to tube 1 and mix thoroughly.
6. Transfer 0.5 ml of solution, tube 1 to tube 2 and mix thoroughly.
7. Transfer 0.5 ml of solution, tube 2 to tube 3 and mix, then continue dilutions through tube 9.
8. Remove 0.5 ml from tube 9 and discard. Each tube now contains a total of 0.5 ml solution and with serum dilutions 1:20 to 1:5120 (tube 9). Tube 10 is a buffer control tube.
9. Prepare positive and negative control series of dilution using Bacto Rheumatoid Positive Control and Bacto Rheumatoid Negative Control instead of test serum as described in steps 1 – 8.
10. Add 0.5 ml sensitized latex from step 2 to each tube. To obtain the final dilution multiply the initial serum dilution by 2.
11. Shake tubes vigorously to obtain an even suspension and incubate in a 56°C water bath for 1 hour.
12. Place tubes in the refrigerator at 2 – 8°C overnight.
13. Centrifuge the tubes at 2300 rpm for 3 minutes.
14. Gently shake the tubes and read the agglutination with the unaided eye rating them – to 4+.

15. The highest serum dilution with 1+ agglutination is reported as the titer.
16. The negative control should not exhibit any granularity.

RESULTS AND INTERPRETATION
A titer of 1:80 (final dilution or greater) is considered a positive test.

LIMITATIONS OF THE PROCEDURE
A positive test indicates only that the RF is present in the patient's blood. It is not evidence that the patient has rheumatoid arthritis, but is only one of the criteria used by the rheumatologist in making a diagnosis.

EXPECTED VALUES
The rheumatoid factor (RF) is not usually found except in cases of rheumatoid arthritis or certain other pathological conditions as reported by Bartfield.[10]

SPECIFIC PERFORMANCE CHARACTERISTICS
The RF is detected in 85 – 95% of RA cases.[1,6,7,8,9] It is also found in 4 – 8% of apparently healthy people as reported by Valkenberg.[11]

REFERENCES
1. Am. J. Med., 21:888, 1956.
2. Acta. Med. Scand. Supp., 341:111, 1958.
3. JAMA, 177:50, 1961.
4. Personal Communication, 1965.
5. Personal Communication, 1974.
6. J. Lab. and Clin. Med., 50:113, 1957.
7. J. Am. Med. Assoc., 164:1999, 1957.
8. Can. J. Med. Assoc., 77:1098, 1957.
9. Arth. and Rheum., 1:142, 1958.
10. Ann. N.Y. Acad. Sciences, May 1960.
11. Ann. Rheum. Dis., 26, 1966.

PACKAGING

Bacto Rheumatoid Titration Test	50 – 65 tests	3300-32-3
Bacto Latex 0.81	5 ml	
Bacto RA Plasma Fraction II	2 × 150 mg	
Bacto RA Buffer, Dried	1 liter	
Bacto Rheumatoid Positive Control	3 ml	
Bacto Rheumatoid Negative Control	3 ml	
Vial for rehydrating RA Plasma Fraction II	2	
Bacto Latex 0.81	5 ml	3102-56-4
	6 × 5 ml	3102-57-3
	25 ml	3102-65-3
	1 liter	3102-81-3
Bacto RA Buffer, Dried	10 × 1 liter	3103-37-7
Bacto RA Plasma Fraction II	6 × 250 mg	3104-33-0
Bacto Rheumatoid Positive Control	3 ml	3301-47-5
Bacto Rheumatoid Negative Control	3 ml	3302-47-4

BACTO RICE EXTRACT AGAR

INTENDED USE
Bacto Rice Extract Agar is recommended for the identification of *Candida albicans* by chlamydospore formation.

HISTORY/PRINCIPLES

Bacto Rice Extract Agar is prepared according to the formulation of Taschdjian.[1]

Chlamydospores were observed consistently and in abundance upon this medium 17 – 24 hours after inoculation with *C. albicans*. The morphology of both the pathogenic, and nonpathogenic *Candida* agreed in every respect with that of the cultures grown on corn meal agar. Recently it has been shown by Kelly and Funigiello,[2] and Walker and Huppert[3] that the addition of Tween® 80 to Bacto Rice Extract Agar enhanced chlamydospore formation by *C. albicans*. Since Tween® 80 also favored chlamydospore formation in other *Candida* species, its use necessitated employing further media for species identification.

The addition of 2% dextrose to Bacto Rice Extract Agar enhanced pigment production by *Trichophyton rubrum* permitting the differentiation between this dermatophyte and *Trichophyton mentagrophytes*.

Taubert and Smith recommended rice extract agar for use in diagnosis of vulvovaginal candidiasis. A cotton-tipped applicator was used for obtaining a specimen and then rolled on the surface of a rice extract agar plate; a cover glass was then applied to the agar, covering most of the inoculum.[4,5] The lack of nutrients together with the glass-covered, oxygen-deficient culture conditions create a deficient environment inducing the formation of specific morphological forms (chlamydospores and pseudomycelia in particular) in some yeasts.

FORMULA

BACTO RICE EXTRACT AGAR
DEHYDRATED

Ingredients per liter

White Rice, Extract from . . . . 20 g
Bacto Agar . . . . 20 g

Final pH 7.1 ± 0.2 at 25°C.

One pound will make 18.1 liters of medium.

PROCEDURE

1. To rehydrate, suspend 25 g in 1 liter cold distilled water and heat to boiling to dissolve the medium completely.
2. Distribute into tubes or flasks and sterilize in the autoclave for 15 minutes at 15 lbs pressure (121°C).
3. Allow the tubes to cool in a slanted position. Medium in flasks should be aseptically dispensed into sterile Petri dishes.
4. Inoculate by cutting through the surface of the agar with an inoculating wire.
5. The application of a sterile cover slip on a portion of the inoculated plated medium will stimulate chlamydospore production.
6. Incubate the inoculated medium at 23 – 25°C for 18 – 72 hours.
7. Examine for chlamydospores microscopically using approximately 100X magnification and by focusing upon the line of inoculation.

STORAGE

Bacto Rice Extract Agar   Below 30°C
Prepared medium   2 – 8°C

QUALITY CONTROL
Identity Specifications

Dehydrated power: beige, homogeneous, free-flowing
Reaction of 2.5% solution: pH 7.1 ± 0.2 at 25°C
Prepared medium: light amber, opalescent

Typical Cultural Response on Bacto Rice Extract Agar
After 40 – 48 hours at 23 – 25°C

Organism	Growth	Chlamydospores
Candida albicans ATCC® 10231	good to excellent	+
Candida albicans ATCC® 26790	good to excellent	+

REFERENCES

1. Mycologia, 45:474, 1953.
2. J. Lab. and Clin. Med., 53:807, 1959.
3. Tech. Bull. Reg. of Med. Tech., 30:10, 1960.
4. J. Lab. and Clin. Med., 55:820, 1960.
5. Mycopath. et Myc. App., 17:269, 1962.

PACKAGING

Bacto Rice Extract Agar	1 lb (454 g)	0899-01-3
	1/4 lb (114 g)	0899-02-2

BACTO RICE POWDER

INTENDED USE
Bacto Rice Powder is recommended for use in the cultivation of *Endamoeba histolytica*.

HISTORY
Bacto Rice Powder was prepared at the suggestion of L. R. Cleveland and is recommended for use in the propagation of *Endamoeba histolytica*. Cleveland and his co-workers[1,2,3] have given detailed descriptions of the methods they employed for the cultivation of this organism on a medium prepared from Bacto Liver Infusion Agar with rice powder. Their method is discussed under Bacto Endamoeba Medium. Essentially, the procedure is to cover each slant of the medium with fresh horse serum-saline solution and to add sterile rice powder to each tube prior to inoculation with the organism.

Bacto Rice Powder is also used with balamuth medium tubes, where a loopful of sterile powder is added along with the inoculum. For a complete discussion refer to Bacto Balamuth Medium.

PREPARATION
Bacto Rice Powder is sterilized in a dry heat oven at 160°F for one hour. Scorching must be avoided.

REFERENCES

1 Arch. Protistenkunde, 70:223, 1930.
2. Science, 72:149, 1930.
3. Am. J. Hyg., 12:606, 1930.

PACKAGING
Bacto Rice Powder   100 g   0146-15-8

BACTO ROGOSA SL BROTH
BACTO ROGOSA SL AGAR

INTENDED USE
Bacto Rogosa SL Broth and Bacto Rogosa SL Agar are selective media for the cultivation of oral, vaginal and fecal lactobacilli.

PRINCIPLES
The media have been prepared according to a modification of the formula of Rogosa, Mitchell and Wisemann[1,2] as recommended by Rogosa.[3] The low pH and high acetate concentration effectively suppress other bacterial flora and allow the lactobacilli to flourish.

FORMULAE

BACTO ROGOSA SL BROTH
DEHYDRATED
Ingredients per liter

Bacto Tryptone	10 g
Bacto Yeast Extract	5 g
Bacto Dextrose	10 g
Bacto Arabinose	5 g
Bacto Saccharose	5 g
Sodium Acetate	15 g
Ammonium Citrate	2 g
Monopotassium Phosphate	6 g
Magnesium Sulfate	0.57 g
Manganese Sulfate	0.12 g
Ferrous Sulfate	0.03 g
Sorbitan Monooleate	1 g

BACTO ROGOSA SL AGAR
DEHYDRATED
Ingredients per liter

Bacto Tryptose	10 g
Bacto Yeast Extract	5 g
Bacto Dextrose	10 g
Bacto Arabinose	5 g
Bacto Saccharose	5 g
Sodium Acetate	15 g
Ammonium Citrate	2 g
Monopotassium Phosphate	6 g
Magnesium Sulfate	0.57 g
Manganese Sulfate	0.12 g
Ferrous Sulfate	0.03 g
Sorbitan Monooleate	1 g
Bacto Agar	15 g

Final pH 5.4 ± 0.2 at 25°C

One pound will make 7.5 liters of final medium. Rehydrate with 60 grams/liter.

One pound will make 6 liters of final medium. Rehydrate with 75 grams/liter.

METHOD OF PREPARATION
1. To rehydrate, suspend the appropriate amount in 1 liter cold distilled water and heat to boiling to dissolve completely.
2. Add 1.32 ml glacial acetic acid and mix thoroughly.
3. Heat to 90 – 100°C for 2 – 3 minutes. Do not autoclave.
4. Distribute into sterile culture tubes, Petri dishes, or flasks.
5. With the agar, cool to 45°C for plate counts. For direct inoculation, cool medium below 40°C.

STORAGE
Bacto Rogosa SL Media   2 – 8°C
Prepared media   2 – 8°C

QUALITY CONTROL

Identity Specifications

	BACTO ROGOSA SL BROTH	BACTO ROGOSA SL AGAR
Dehydrated powder:	beige, slightly lumpy	beige, homogeneous with soft clumps
Reaction of solution: (after addition of 1.32 ml glacial acetic acid)	6% solution pH 5.4 ± 0.2 at 25°C	7.5% solution pH 5.4 ± 0.2 at 25°C
Prepared medium:	light amber, clear to slightly opalescent	light amber, slightly opalescent, may have slight precipitate

Typical Cultural Response After 40 – 48 Hours at 35 ± 2°C

Organism	Recovery
Lactobacillus casei ATCC® 9595	good to excellent
Lactobacillus fermentum ATCC® 9338	good to excellent
Lactobacillus leichmannii ATCC® 4797	good to excellent
Lactobacillus plantarum ATCC® 8014	good to excellent
Staphylococcus aureus ATCC® 25923	inhibited

REFERENCES
1. J. Bact., 62:132:1951.
2. J. Dental Research, 30:682:1951.
3. Personal Communication.

PACKAGING

Bacto Rogosa SL Broth	1 lb (454 g)	0478-01-2
Bacto Rogosa SL Agar	1 lb (454 g)	0480-01-8

ROSE BENGAL AGAR

BACTO ROSE BENGAL AGAR BASE
BACTO ROSE BENGAL ANTIMICROBIC SUPPLEMENT C

INTENDED USE
Bacto Rose Bengal Agar Base and Bacto Rose Bengal Antimicrobic Supplement C are used in the preparation of Bacto Rose Bengal Chloramphenicol Agar. The complete medium is recommended in the selective isolation and enumeration of yeasts and molds from environmental materials and foodstuffs.

HISTORY OF THE TEST
A number of methods and media have been described for the selective isolation of fungi from environmental materials and foodstuffs containing mixed populations of fungi and bacteria. The use of media with an acid pH to selectively inhibit the growth of bacteria and thereby promote the growth of fungi is widely recognized.[1-3] However, a number of workers have reported that acidified media tend to be inhibitory to the growth of fungi,[4,5] do not completely inhibit the growth of bacteria[5] and have little effect in restricting the size of mold colonies.[6] In 1944, Smith and Dawson reported on the use of rose bengal in neutral pH media for the selective isolation of fungi from soil samples.[7] Martin in 1950 reported that the selectivity of rose bengal-containing media for

fungi could be further improved by the incorporation of streptomycin.[4] Since that time, a number of workers have reported on the use of rose bengal in combination with various antimicrobics including chloramphenicol, streptomycin, oxytetracycline and chlortetracycline for the improved, selective isolation and enumeration of yeasts and molds from soil, sewage and foodstuffs.[8-11] Rose bengal chloramphenicol agar from Difco Laboratories is a modification of the rose bengal chlortetracycline agar formula of Jarvis.[11] Chloramphenicol is employed in this medium as a selective supplement rather than the chlortetracycline originally recommended by Jarvis.

PRINCIPLES OF THE PROCEDURE

Bacto Rose Bengal Agar Base is a selective basal medium which supports good growth of yeasts and molds. The pH of the medium is near neutrality for improved growth and recovery of acid sensitive strains.[4,5,11] The presence of rose bengal in the base suppresses the growth of bacteria and restricts the size and height of mold colonies.[8,10,11] This restriction in growth of molds aids in the isolation of slow-growing fungi by preventing their overgrowth by more rapidly growing species. In addition, rose bengal is taken up by yeast and mold colonies thereby facilitating their recognition and enumeration. Bacto Rose Bengal Antimicrobic Supplement C is a lyophilized antimicrobic supplement containing chloramphenicol. Supplementation of Bacto Rose Bengal Agar Base with chloramphenicol provides for improved inhibition of bacteria present in environmental materials and foodstuffs.

FORMULAE

BACTO ROSE BENGAL AGAR BASE
DEHYDRATED

Ingredients per liter

Bacto Soytone	5 g
Bacto Dextrose	10 g
Monopotassium Phosphate	1 g
Magnesium Sulfate	0.5 g
Rose Bengal	0.05 g
Bacto Agar	15 g

BACTO ROSE BENGAL ANTIMICROBIC SUPPLEMENT C

Ingredient per 2 ml vial

Chloramphenicol	0.05 g

PRECAUTIONS

Use aseptic technique in rehydrating Bacto Rose Bengal Antimicrobic Supplement C and adding the supplement to the base. Follow proper, established laboratory procedures in handling and disposing of infectious materials. Avoid the formation of aerosols.

STORAGE INSTRUCTIONS FOR BACTO ROSE BENGAL AGAR BASE

Store below 30°C. The powder is very hygroscopic. Keep container tightly closed. The expiration date applies to the product in its intact container when stored as directed.

STORAGE AND REHYDRATION INSTRUCTIONS FOR BACTO ROSE BENGAL ANTIMICROBIC SUPPLEMENT C

Store at 2 – 8°C. Protect from light. Use rehydrated vials within 24 hours. Store rehydrated vial at 2 – 8°C. To rehydrate, aseptically add 2 ml of ethanol and invert several times to dissolve the powder. The expiration date applies to the product in its intact container when stored as directed. Do not open or rehydrate vials until ready to use.

PRODUCT DETERIORATION
Do not use the rehydrated supplement if it is contaminated, partially or completely evaporated or shows other signs of deterioration. Do not use the agar base if it is caked, discolored or shows other signs of deterioration.

SPECIMEN COLLECTION
Specimens should be collected in sterile containers and transported immediately to the laboratory in accordance with recommended guidelines.[12-14]

PROCEDURE
Materials Provided:
Bacto Rose Bengal Agar Base
Bacto Rose Bengal Antimicrobic Supplement C

Materials Not Provided:
Specimen container
Inoculating loop
Sterile dilution blanks
Bunsen burner or incinerator
Incubator (25°C)

Sterile pipettes
User quality control cultures
Sterile Petri dishes
Ethanol

Preparation of Rose Bengal Chloramphenicol Agar
1. To rehydrate the base, suspend 16 g Bacto Rose Bengal Agar Base in 500 ml of distilled or deionized water and heat to boiling to dissolve completely.
2. Sterilize in the autoclave for 15 minutes at 15 lbs pressure (121°C).
3. Cool to 43 – 46°C.
4. Aseptically add 2 ml of rehydrated Bacto Rose Bengal Antimicrobic Supplement C to the molten agar. Mix thoroughly.

Inoculation and Incubation
Prepare the sample for pour plate inoculation in accordance with recommended guidelines.[12-14] Aseptically add 0.1 or 1 ml of each dilution to a sterile 90 mm Petri dish. Add 10 – 20 ml of sterile molten agar, mix by gently rocking and tilting the plate. and allow the medium to solidify. Incubate plates aerobically at 20 – 25°C. Examine plates for growth after 5 to 7 days incubation.

RESULTS
Colonies of molds and yeasts should be apparent within 5 days of incubation. Colonies of yeasts appear pink due to uptake of rose bengal. Determine the yeast or mold count by counting the number of colonies appearing on those plates containing 30 to 300 colony forming units and multiplying that number by the dilution factor. Results may be reported as yeast or mold count per gram or milliliter of sample, as applicable. Appropriate references may be consulted for further information.[12-14]

USER QUALITY CONTROL
1. Examine the agar base for color and texture. The powder should be beige to faint pink, free-flowing and homogeneous.
2. Determine the pH of the base after preparation and cooling to 25°C. The pH should be 7.2 ± 0.2.
3. Examine the lyophilized and rehydrated supplement for evidence of deterioration as described under PRODUCT DETERIORATION.
4. Check the performance of the base and supplement by testing in the complete medium. Plates should be inoculated with approximately 100 colony forming units of

the cultures listed below and incubated aerobically at 20 – 25°C. Examine plates for growth after 5 days incubation. Results should be as stated below:

Organism	Results After 5 Days Incubation Growth	Colonial Morphology
Candida albicans ATCC® 10231	good	colonies appear pink, smooth, pasty and raised.
Aspergillus niger ATCC® 1015	good	colonies appear white and filamentous becoming salt and pepper and eventually black.
Escherichia coli ATCC® 25922	marked to complete inhibition	
Micrococcus luteus ATCC® 10240	marked to complete inhibition	

LIMITATIONS OF THE PROCEDURE

1. Bacto Rose Bengal Agar Base and Bacto Rose Bengal Antimicrobic Supplement C are intended for use in the preparation of rose bengal chloramphenicol agar. Although this medium is selective primarily for fungi, microscopic examination is recommended for presumptive identification. Biochemical testing using pure cultures is required for complete identification. Consult appropriate references for further information.[12-14]

2. Due to the selective properties of this medium and the type of specimen being cultured, some strains of fungi may be encountered that fail to grow or grow poorly on the complete medium; similarly some strains of bacteria may be encountered that are not inhibited or only partially inhibited.

REFERENCES

1. Waksman, S. A. 1922. A method for counting the number of fungi in the soil. J. Bact. 7:339 – 341.
2. Koburger, J. A. 1976. Yeasts and molds, p. 225 – 229. *In* M. L. Speck (ed.), Compendium of methods for the microbiological examination of foods. American Public Health Association, Inc., Washington, D.C.
3. Mossel, D. A. A., M. Visser, and W. H. J. Mengerink. 1962. A comparison of media for the enumeration of moulds and yeasts in foods and beverages. Lab Practice 11:109 – 112.
4. Martin, J. P. 1950. Use of acid, rose bengal and streptomycin in the plate method for estimating soil fungi. Soil Sci. 69:215 – 232.
5. Koburger, J. A. 1972. Fungi in foods. IV. Effect of plating medium pH on counts. J. Milk Food Technol. 35:659 – 660.
6. Tyner, L. E. 1944. Effect of media compositions on the numbers of bacterial and fungal colonies developing in Petri plates. Soil Sci. 57:271 – 274.
7. Smith, N. R., and V. T. Dawson. 1944. The bacteriostatic action of rose bengal in media used for the plate counts of soil fungi. Soil Sci. 58:467 – 471.
8. Cooke, W. B. 1954. The use of antibiotics in media for the isolation of fungi from polluted water. Antibiotics and Chemotherapy 4:657 – 662.
9. Papavizas, G. C., and C. B. Davey. 1959. Evaluation of various media and antimicrobial agents for isolation of soil fungi. Soil Sci. 88:112 – 117.
10. Overcast, W. W., and D. J. Weakley. 1969. An aureomycin-rose bengal agar for enumeration of yeast and mold in cottage cheese. J. Milk Technol. 32:442 – 445.
11. Jarvis, B. 1973. Comparison of an improved rose bengal-chlortetracycline agar with other media for the selective isolation and enumeration of molds and yeasts in foods. J. Appl. Bact. 36:723 – 727.
12. Speck, M. L. (ed.). 1976. Compendium of methods for the microbiological examination of foods. American Public Health Association, Washington, D.C.
13. Marth, E. H. (ed.). 1978. Standard methods for the examination of dairy products, 14th ed. American Public Health Association, Washington, D.C.
14. Greenberg, A. E., J. J. Connors, and D. Jenkins (ed.). 1981. Standard methods for the examination of water and wastewater, 15th ed. American Public Health Association - American Water Works Association - Water Pollution Control Federation, Washington, D.C.

PACKAGING

Bacto Rose Bengal Agar Base	500 g	1831-17-4
	10 kg	1831-08-5
Bacto Rose Bengal Antimicrobic Supplement C	6 × 2 ml	3352-54-3

BACTO ROSOLIC ACID

Bacto Rosolic Acid is a powdered rosolic acid standardized for use in microbiological culture media. It is particularly recommended for use in Bacto *m* FC Broth Base prepared according to the formulation of Geldreich, Clark and Kabler[1] for detecting and enumerating fecal coliforms in raw waters.

To prepare a 1% solution, dissolve 1 g in 100 ml 0.2N NaOH. The rosolic acid solution is stable for 2 weeks at refrigerator temperature.

REFERENCE

1. J. Am. Water Works Assoc., 57:208, 1965.

PACKAGING

Bacto Rosolic Acid	6 × 1 g	3328-09-1

BACTO RUSSELL DOUBLE SUGAR AGAR

INTENDED USE

Bacto Russell Double Sugar Agar is recommended for the identification of gram-negative enteric bacilli, particularly the colon-typhoid-salmonellae-dysentery groups, based on the fermentation of dextrose and lactose.

HISTORY/PRINCIPLES

Bacto Russell Double Sugar Agar conforms to the original formula of Russell[1] except that phenol red replaces litmus, and Proteose Peptone No. 3, Difco is utilized in place of Bacto Peptone. The phenol red indicator gives exceptionally clear-cut reactions on both sides of the neutral point. Alkaline reactions turn the indicator red, and acid reactions change it to yellow. Investigations have also demonstrated that faster and clearer reactions are secured in the medium prepared with Proteose Peptone No. 3, Difco.

This medium may be used to aid in the identification of pure cultures of colonies picked from primary plating media such as Bacto MacConkey Agar, Bacto SS Agar, Bacto Bismuth Sulfite Agar and others.

Bacto Russell Double Sugar Agar is used in tubes which are slanted in order to provide a deep butt. Inoculation is made from isolated colonies or pure cultures by smearing over the surface of the slant and stabbing the butt. After suitable incubation, the production of acid under aerobic conditions (on the slant and under anaerobic conditions in the butt) can be detected by changes in color of the indicator. Gaseous fermentation is indicated by splitting of the agar or formation of bubbles in the butt.

FORMULA
BACTO RUSSELL DOUBLE SUGAR AGAR
DEHYDRATED

Ingredients per liter

Bacto Beef Extract	1 g
Proteose Peptone No. 3, Difco	12 g
Bacto Lactose	10 g
Bacto Dextrose	1 g
Sodium Chloride	5 g
Bacto Agar	15 g
Bacto Phenol Red	0.025 g

Final pH 7.5 ± 0.2 at 25°C.

One pound will make 10.3 liters of medium.

METHOD OF PREPARATION

1. To rehydrate, suspend 44 g in 1 liter distilled or deionized water and heat to boiling to dissolve completely.
2. Distribute into tubes and sterilize in the autoclave for 15 minutes at 15 lbs pressure (121°C).
3. Allow the tubes to solidify in a manner which will give a generous butt.
4. After 24 – 48 hours incubation, a properly inoculated tube showing a red or cerise slope and a yellow butt with or without gas formation indicates fermentation of the dextrose. Some strains of *S. typhi* may require as long as 30 – 40 hours to produce a characteristic alkaline slant.

A tube showing a yellow slant and butt with or without gas formation indicates fermentation of the lactose.

A tube showing no change indicates that neither dextrose nor lactose has been fermented.

Organisms capable of fermenting dextrose but not lactose, *Salmonella typhi*, for example, will show an initial acid slant in short incubation periods. As the dextrose is utilized, the reaction under aerobic conditions reverts and becomes alkaline. Under anaerobic conditions, in the butt of the tube, these same organisms are not capable of causing a reversion of the reaction, and remain acid.

STORAGE

Bacto Russell Double Sugar Agar	Below 30°C
Prepared medium	2 – 8°C

QUALITY CONTROL
Identity Specifications

Dehydrated powder:	pink, homogeneous, free-flowing
Reaction of 4.4% solution:	pH 7.4 ± 0.2 at 25°C
Prepared medium:	orange-red, clear to slightly opalescent

Typical Cultural Response on Bacto Russell Double Sugar Agar
After 18 – 40 Hours at 35°C

Organism	Recovery	Slant/Butt	Gas
Escherichia coli ATCC® 25922	good	A/A	+
Proteus vulgaris ATCC® 6380	good	K/A	+
Pseudomonas aeruginosa ATCC® 27853	good	K/K	−
Salmonella typhimurium ATCC® 14028	good	K/A	+
Shigella dysenteriae ATCC® 13313	good	K/A	−

A = acid, yellow   K = alkaline, red

REFERENCE
1. J. Med. Research, 25:217, 1911.

PACKAGING

Bacto Russell Double Sugar Agar   500 g   0084-17-0

BACTO SABHI AGAR BASE

INTENDED USE

Bacto SABHI (Sabouraud Dextrose & Brain Heart Infusion) Agar Base is recommended for the cultivation and isolation of pathogenic fungi.

HISTORY/PRINCIPLES

Bacto SABHI Agar Base to which chloromycetin is added is prepared according to the formulation of Gorman.[1] It was particularly useful for the maximum recovery of *Blastomyces dermatitidis* and *Histoplasma capsulatum* from body tissues and fluids. It may be used with or without blood. The addition of blood to the medium has been shown by Gorman[1] to increase the recovery of *H. capsulatum*. Blood is also added to the medium when it is desired to convert *H. capsulatum* and *B. dermatitidis* to the yeast phase.

FORMULA

BACTO SABHI AGAR BASE
DEHYDRATED

Ingredients per liter

Calf Brains, Infusion from	100 g	Bacto Dextrose	21 g
Beef Heart, Infusion from	125 g	Sodium Chloride	2.5 g
Proteose Peptone, Difco	5 g	Disodium Phosphate	1.25 g
Bacto Neopeptone	5 g	Bacto Agar	15 g

Final pH 7.0 ± 0.2 at 25°C.

One pound will make 7.7 liters of final medium.

METHOD OF PREPARATION

1. To rehydrate, suspend 59 g in 1 liter distilled or deionized water and heat to boiling to dissolve completely.
2. Sterilize in the autoclave for 15 minutes at 15 lbs pressure (121°C).
3. Cool to 50 – 55°C and add 1 ml sterile 100 mg/ml chloromycetin solution.
4. Dispense in 5 ml amounts into sterile 20 × 110 mm screwcap tubes or plates, and allow to solidify. Let the tubes solidify in a slanted position.
5. To prepare blood agar, add sterile sheep or human blood to a concentration of 10% before dispensing into sterile tubes or plates.
6. To convert to the yeast phase, incubate the cultures at 37°C. For maximum recovery incubate at room temperature and observe frequently. Keep for two months before final reporting.

STORAGE

Bacto SABHI Agar Base   Below 30°C
Prepared plates   2 – 8°C

QUALITY CONTROL
Identity Specifications

Dehydrated powder:	light beige, homogeneous, free-flowing
Reaction of 5.9% solution:	pH 7.0 ± 0.2 at 25°C
Prepared medium:	medium amber, slightly opalescent

Typical Cultural Response on Bacto SABHI Agar Base With 100 mg/ml Chloromycetin
After 40 – 48 Hours at 25 – 30°C

Organism	Plain	w/10% Sheep Blood
Aspergillus niger ATCC® 16404	good	good
Candida albicans ATCC® 26790	good to excellent	good to excellent
Escherichia coli ATCC® 25922	inhibited	inhibited
Saccharomyces cerevisiae ATCC® 9763	good to excellent	good to excellent
Saccharomyces uvarum ATCC® 9080	good to excellent	good to excellent
Staphylococcus aureus ATCC® 25923	inhibited	inhibited

REFERENCE
1. Am. J. Med. Tech., 33:151, 1967.

PACKAGING
Bacto SABHI Agar Base	1 lb (454 g)	0797-01-6
	5 lb (2.27 kg)	0797-05-2

BACTO SBG ENRICHMENT
BACTO SBG SULFA ENRICHMENT

INTENDED USE
Bacto SBG (Selenite Brilliant Green) Enrichment and Bacto SBG Sulfa Enrichment are selective enrichments for the isolation of *Salmonella* from egg products.

HISTORY/PRINCIPLES
Bacto SBG Enrichments are prepared according to the formulae described by Stokes and Osborne.[1] They contain brilliant green, selenite, taurocholate and sulfapyridine as selective agents to inhibit growth of gram-positive organisms and enteric organisms other than *Salmonella*. Mannitol, Bacto Yeast Extract and Bacto Peptone supply the required nutrients.

The presence of sulfapyridine in the formula stabilizes the selectivity in the presence of egg products. It was shown by Osborne and Stokes[2] that whole egg and egg yolk reduced the selective properties of selenite brilliant green enrichment and that the addition of sulfapyridine restored these selective properties, even in the presence of as much as 20% liquid whole egg. *Proteus* and *Escherichia coli* were inhibited while growth of *Salmonella* was obtained from small inocula. They reported recovery of *Salmonella* from a naturally contaminated dried egg containing 1 viable *Salmonella* per 100 g dry egg. These authors also reported that the addition of 0.1% sulfapyridine to brilliant green agar (Bacto BG Sulfa Agar) enhanced the selectivity of this medium for *Salmonella*.

Bacto MacConkey Agar, Bacto Bismuth Sulfite Agar, Bacto Brilliant Green Agar and Bacto BG Sulfa Agar are suggested as plating media following enrichment with Bacto SBG Enrichment for recovery of *Salmonella*.

FORMULAE

BACTO SBG ENRICHMENT
DEHYDRATED
Ingredients per liter

Bacto Yeast Extract . . . . 5 g
Bacto Peptone . . . . 5 g
D-Mannitol, Difco . . . . 5 g
Sodium Taurocholate . . . . 1 g
Sodium Selenite . . . . 4 g
Dipotassium Phosphate . . . . 2.65 g
Monopotassium Phosphate . . . . . . 1.02 g
Bacto Brilliant Green . . . . 0.005 g

Final pH 7.2 ± 0.2 at 25°C.

One pound will make
19.1 liters of medium.
Rehydrate with 23.7 g/liter.

BACTO SBG SULFA ENRICHMENT
DEHYDRATED
Ingredients per liter

Bacto Yeast Extract . . . . 5 g
Bacto Peptone . . . . 5 g
D-Mannitol, Difco . . . . 5 g
Sodium Taurocholate . . . . 1 g
Sodium Sulfapyridine . . . . 0.5 g
Sodium Selenite . . . . 4 g
Dipotassium Phosphate . . . . 2.65 g
Monopotassium Phosphate . . . . . . 1.02 g
Bacto Brilliant Green . . . . 0.005 g

Final pH 7.2 ± 0.2 at 25°C.

One pound will make
18.7 liters of medium.
Rehydrate with 24.2 g/liter.

METHOD OF PREPARATION
1. Suspend appropriate amount in 1 liter distilled or deionized water. Heat to boiling to dissolve completely.
2. Sterilize by holding at boiling temperature for 5 – 10 minutes. DO NOT AUTO-CLAVE. Avoid excessive heating.
3. Dispense into sterile tubes and allow to cool to room temperature.

STORAGE
Bacto SBG Enrichments   Below 30°C
Prepared tubes   2 – 8°C

QUALITY CONTROL
Identity Specifications

Dehydrated powder: light beige, homogeneous, free-flowing
Reaction of appropriate solution: pH 7.2 ± 0.2 at 25°C
Prepared medium: green, opalescent

Typical Cultural Response on Bacto MacConkey Agar After 18 – 24 Hours at 35°C Transferred from Bacto SBG or SBG Sulfa Enrichment After 18 – 24 Hours at 35°C

Organism	Recovery	Color of Colony
Enterobacter aerogenes ATCC® 13048	little or no increase in numbers	pink to colorless, bile precipitate
Escherichia coli ATCC® 25922	little or no increase in numbers	pink to colorless, bile precipitate
Salmonella choleraesuis ATCC® 12011	good to excellent	colorless
Salmonella typhi ATCC® 6539	good to excellent	colorless

REFERENCES

1. Appl. Microbiol., 3:217, 1955.
2. *ibid.*, 3:295, 1955.

PACKAGING

Bacto SBG Enrichment	1 lb (454 g)	0661-01-9
Bacto SBG Sulfa Enrichment	1 lb (454 g)	0715-01-5

BACTO SF MEDIUM

INTENDED USE

Bacto SF Medium is a selective medium for the detection of fecal streptococci from milk, water, sewage and feces.

HISTORY/PRINCIPLES

Bacto SF Medium (Streptococcus Faecalis medium) is prepared according to the formula of Hajna and Perry.[1] The medium contains sodium azide, an inhibitor of the cytochrome oxidase enzyme in the electron transport chain, as the selective agent. Sodium azide has been employed as a selective agent in media for the isolation of streptococci by a number of investigators. Snyder and Lichstein[2] and Lichstein and Snyder[3] reported that 0.01% sodium azide in blood agar prevented swarming of *Proteus* and permitted the isolation of streptococci from stools and other infected material without overgrowth by gram-negative organisms. Mallmann[4] used 1:5000 sodium azide in a buffered Tryptose broth for the examination of sewage and reported that streptococci grew while coliform organisms were inhibited. This method made possible an easy procedure for the routine laboratory detection of streptococci. Parker[5,6] used sodium azide and crystal violet in the preparation of a selective blood agar for streptococci and *Erysipelothrix rhusiopathiae*.

Hajna and Perry[1] devised a selective medium containing 0.05% sodium azide, dextrose and the indicator, brom cresol purple, for the detection of fecal streptococci in swimming pools, water samples and milk. They specified 45.5°C as the incubation temperature, and reported that growth accompanied with an acid reaction, as shown by the change in color of the medium to yellow, was almost complete evidence of the presence of fecal streptococci.

FORMULA

BACTO SF MEDIUM
DEHYDRATED

Ingredients per liter

Bacto Tryptone	20 g
Bacto Dextrose	5 g
Dipotassium Phosphate	4 g
Monopotassium Phosphate	1.5 g
Sodium Chloride	5 g
Sodium Azide	0.5 g
Bacto Brom Cresol Purple	0.032 g

Final pH 6.9 ± 0.2 at 25°C.

One pound will make 12.6 liters of medium.

METHOD OF PREPARATION

1. Dissolve 36 g in 1 liter distilled or deionized water.
2. Dispense into tubes as desired.
3. Sterilize in the autoclave for 15 minutes at 15 lbs pressure (121°C).

When the inoculum is greater than 1 ml, particular care must be taken to preserve the correct concentration of the ingredients after dilution with the sample. For example, if 10 ml of water are to be added to 10 ml of medium, the medium should be prepared in double strength.

STORAGE

Bacto SF Medium	Below 30°C
Prepared medium	15 – 30°C

QUALITY CONTROL

Identity Specifications

Dehydrated powder:	beige, homogeneous, free-flowing
Reaction of 3.6% solution:	pH 6.9 ± 0.2 at 25°C
Prepared medium:	purple, clear, no significant precipitate

Typical Cultural Response in Bacto SF Medium
After 18 – 48 Hours at 45.5 – 46°C

Organism	Growth	Acid
Escherichia coli ATCC® 25922	inhibited	–
Streptococcus bovis ATCC® 33317	poor	–
Streptococcus faecalis ATCC® 19433	good	+
Streptococcus faecium ATCC® 27270	good	+

REFERENCES

1. Hajna and Perry, 1943, American Journal of Public Health, 33:550.
2. Snyder and Lichstein, 1940, Journal of Infectious Diseases, 67:113.
3. Lichstein and Synder, 1941, Journal of Bacteriology, 42:653.
4. Mallman, 1940, Sewage Works Journal, 12:875.
5. Parker, 1941, Journal of Bacteriology, 42:137.
6. Parker, 1943, Journal of Bacteriology, 46:343.

PACKAGING

Bacto SF Medium		
	1 lb (454 g)	0315-01-9
	1/4 lb (114 g)	0315-02-8
	12 tubes	0315-34-0

BACTO SFP AGAR BASE

INTENDED USE

Bacto SFP Agar Base is a basal medium which, after supplementation with egg yolk emulsion, polymyxin B sulfate and kanamycin sulfate, is used for the detection and enumeration of *Clostridium perfringens* in foods.

HISTORY/PRINCIPLES

Bacto SFP Agar Base is prepared according to the formulation of Shahidi and Ferguson[1,2] for the maximum growth of clostridia. With the addition of 50% egg yolk emulsion, both the lecithinase and sulfite reaction can be utilized to identify *Clostridium perfringens*. The complete supplemented medium in plates must be incubated anaerobically. In using Bacto SFP Agar Base, a base layer and a cover layer are prepared. The medium for the base layer is supplemented with egg yolk emulsion, polymyxin B sulfate and kanamycin sulfate, whereas the medium for the cover layer is supplemented with only polymyxin B sulfate and kanamycin sulfate.

FORMULA

BACTO SFP AGAR BASE
DEHYDRATED

Ingredients per liter

Bacto Yeast Extract . . . . 5 g
Bacto Tryptose . . . . 15 g
Bacto Soytone . . . . 5 g
Ferric Ammonium Citrate . . . . 1 g
Sodium Bisulfite . . . . 1 g
Bacto Agar . . . . 20 g

Final pH 7.6 ± 0.2 at 25°C.

One pound will make 9.6 liters of either the complete medium or the cover layer.

METHOD OF PREPARATION
BASAL LAYER

1. To rehydrate the medium, suspend 47 g in 900 ml distilled or deionized water. Heat to boiling to dissolve the medium completely.
2. Dispense into flasks and sterilize in the autoclave for 15 minutes at 15 lbs pressure (121°C).
3. Add 100 ml of Bacto Egg Yolk Enrichment 50%, warmed to room temperature, to 900 ml of sterile rehydrated SFP agar base which has been cooled to about 50°C in a water bath.
4. Rehydrate 1 Bacto Antimicrobic Vial P (30,000 units polymycin B sulfate) with 10 ml sterile distilled or deionized water and add the entire contents.
5. Rehydrate 1 Bacto Antimicrobic Vial K (25,000 μg kanamycin), and add 4.8 ml to the base.
6. Swirl to mix thoroughly and pour plates as desired.

COVER LAYER

1. To rehydrate the medium, suspend 47 g in 1 liter distilled or deionized water. Heat to boiling to dissolve completely.
2. Sterilize and cool as above.
3. Add contents of 1 rehydrated Bacto Antimicrobic Vial P and the remaining rehydrated Bacto Antimicrobic Vial K from above.

PROCEDURE

Aseptically dispense 10 – 12 ml of the completed base layer medium into sterile Petri dishes and incubate overnight to test for sterility and to dry the plates. Then inoculate the plates by using dilutions of the test samples applied in 0.1 ml amounts to the center of the base layer plates. Spread the inoculum over the entire surface of the medium using sterile L shaped spreading rods. Dispense 10 ml of cover layer medium on each inoculated plate. Incubate the plates in anaerobic jars under an atmosphere of 90% nitrogen and 10% carbon dioxide. This is done by evacuating the jars to a reading of 20 – 26 inches of vacuum on a mercury manometer and subsequently filling the jar to atmospheric pressure with the nitrogen-carbon dioxide gas mixture. Repeat the procedure two or more times to remove residual oxygen. Incubate the plates in the anaerobic jars at 37°C for 24 hours. Observe black colonies surrounded by a zone of precipitate. These colonies are counted and enumerated according to standard procedure to obtain good statistical results.

STORAGE

Bacto SFP Agar Base   Below 30°C
Prepared plates   2 – 8°C

QUALITY CONTROL

Identity Specifications

Dehydrated powder: light cream, homogeneous, free-flowing
Reaction of 4.7% solution: pH 7.6 ± 0.2 at 25°C
Prepared medium: medium to dark amber, slightly opalescent

Typical Cultural Response on Bacto SFP Agar
After 40 – 48 Hours at 35°C, Anaerobically

Organism	Growth	Color of Colony	Lecithinase (Halos)
Clostridium perfringens ATCC® 12919	good to excellent	black	+
Clostridium perfringens ATCC® 12924	good to excellent	black	+

REFERENCES

1. Private Communication.
2. Applied Microbiology 21:500, 1971.

PACKAGING

Bacto SFP Agar Base	1 lb (454 g)	0811-01-8
Bacto Egg Yolk Enrichment 50%	12 × 10 ml	3347-61-2
	6 × 100 ml	3347-73-8
Bacto Antimicrobic Vial K	6 × 10 ml	3339-60-3
Bacto Antimicrobic Vial P	6 × 10 ml	3268-60-8

BACTO SIM MEDIUM

INTENDED USE

Bacto SIM Medium is recommended for use in the determination of hydrogen sulfide production, indole formation and motility of members of the *Salmonella* and *Shigella* groups.

HISTORY

Orlowski[1] noted that *Salmonella typhosa (S. typhi)* could be distinguished from the coliform organisms by culturing them in a medium containing lead acetate, an indicator of hydrogen sulfide production. Jordan and Victorson[2] showed further that *S. paratyphi* (Paratyphoid A) and *S. schottmuelleri* (Paratyphoid B) could be distinguished on the basis of hydrogen sulfide production by growing them in a lead acetate medium.

Semisolid media, as described by Hiss,[3] Hesse,[4] Jackson and Melia,[5] Tittsler and Sandholzer[6] and others, have been employed quite extensively in the determination of motility by bacteria. Sulkin and Willett,[7] in Bacto Triple Sugar Iron Agar, used 1% agar to demonstrate motility or lack of motility in addition to hydrogen sulfide production and carbohydrate fermentation by members of the *Salmonella* and *Shigella* groups. They called attention to the "brush-like" growth or motility of the typhoid organisms. Green and co-workers[8] used Bacto SIM Medium in 1 – 2 ml amounts in 10 × 75 mm tubes and reported the detection of motility by a large series of cultures following incubation at 37°C for 90 to 120 minutes. Friewer and Shaughnessy[9] used the fermentation of lactose, hydrogen sulfide production and motility in a lead acetate semisolid agar as a screening medium in the isolation of enteric pathogens from stool cultures. Sosa[10] described a peptone medium with a low agar concentration to determine motility, and stated that indole determination could be made using the Ehrlich reagent in this medium.

PRINCIPLES

In the development of Bacto SIM Medium it was determined that 0.0025% sodium thiosulfate added to the semisolid 3% Bacto Peptone medium produced results comparable to those obtained when hydrogen sulfide was determined in a 1% Bacto Peptone solution using lead acetate paper strips as an indicator of hydrogen sulfide. Smaller quantities of sodium thiosulfate did not give a satisfactory response in the semisolid medium, while larger quantities obscured motility and also permitted some false reactions. Any blackening along the line of inoculation is considered as a positive hydrogen sulfide reaction. Hydrogen sulfide producing organisms generally give a positive reaction in 18 – 24 hours. Motile cultures show diffuse growth or turbidity away from the line of inoculation and intensify hydrogen sulfide reactions. Strains of *Proteus* and *Salmonella* often show diffuse growth throughout the entire medium, while *S. typhi* is not as actively motile. In Bacto SIM Medium motility of typical members of the enteric group is demonstrated in 18 – 24 hours, or less, incubation at 37°C.

Bacto Tryptone has been used universally in the test for indole production. It has been demonstrated that Bacto SIM Medium gives parallel indole production in comparison with a 1% Bacto Tryptone solution after 18 – 24 hours incubation at 37°C.

Bacto SIM Medium is especially recommended as an aid in the routine confirmation of members of the *Salmonella* and *Shigella* groups following presumptive evidence as obtained on the differential tube media (Bacto Russell Double Sugar Agar and Bacto Triple Sugar Iron Agar). As with other differential media, it is necessary to use pure cultures for inoculation. Generally, an incubation period of 18 – 24 hours or less is sufficient to give reactions by typical organisms in this group. Occasionally an atypical culture may be encountered which will fail to produce hydrogen sulfide, indole or motility in 18 – 24 hours incubation. Typical cultural reactions of the pathogenic gram-negative enteric bacteria are:

	S. typhi	Salmonella	Shigella
Motility	+	+	−
Hydrogen sulfide	+	±	−
Indole	−	−	±

FORMULA
BACTO SIM MEDIUM
DEHYDRATED
Ingredients per liter

Bacto Peptone .... 30 g
Bacto Beef Extract .... 3 g
Peptonized Iron, Difco .... 0.2 g
Sodium Thiosulfate .... 0.025 g
Bacto Agar .... 3 g

Final pH 7.3 ± 0.2 at 25°C.

One pound will make 12.6 liters of medium.

PROCEDURE
1. To rehydrate, suspend 36 g in 1 liter distilled water and heat to boiling to dissolve completely.
2. Dispense the medium in tubes to a depth of approximately 3 inches (or approximately 15 ml for standard sized tubes).

3. Sterilize in the autoclave for 15 minutes at 15 lbs pressure (121°C).
4. Allow the medium to solidify in a vertical position.
5. Inoculate with a single stab using a straight needle through the center, to a depth of about two-thirds of the medium. Place a Bacto Indole Test Strip over the lip of each inoculated tube. Secure in a position at least 1 cm from the surface of the medium using the cap to hold strip in place. Alternative methods of indol test are listed below.
6. Incubate for 18 – 24 hours at 37°C.
7. Read tubes for growth, H_2S production (blackening), motility (diffuse growth from line of inoculum) and indole production (pink area on strip).

Kovacs' Test Method
1. Overlay the medium with 2 ml chloroform without agitation.
2. Add 2 ml Kovacs' reagent.

KOVACS' REAGENT

Amyl alcohol	75 ml
Hydrochloric acid, concentrated	25 ml
p-dimethylaminobenzaldehyde	5 g

A pink to deep red color is formed in the chloroform layer if indole is present. In a negative test, no color is formed in the chloroform layer.

NOTE: When using this test, hydrogen sulfide and motility readings should be made **BEFORE** testing for indole.

Ehrlich Test Method
The indole test may also be performed in accordance with the *Manual of Methods of Pure Culture Study of Bacteria*[11] of the Society of American Bacteriologists.
1. Add approximately 2 ml of ethyl ether to each tube.
2. Shake gently but do not break the agar.
3. Let stand several minutes.
4. Add approximately 1 ml Ehrlich reagent by dropping it down the side of the tube so that it spreads out as a layer between the ether and medium.

EHRLICH REAGENT

p-dimethylaminobenzaldehyde	1 g
Ethyl alcohol (95%)	95 ml
Hydrochloric acid	20 ml

The formation of a purplish red color at the interface of the two liquids **WITHIN 5 MINUTES** indicates indole production. A slight pink color on the surface of the medium **AFTER 5** minutes indicates a negative reaction.

NOTE: When using this test, hydrogen sulfide and motility readings should be made **BEFORE** testing for indole.

STORAGE
Bacto SIM Medium	Below 30°C
Prepared tubes	15 – 30°C

QUALITY CONTROL
Identity Specifications

Dehydrated powder:	beige, homogeneous, free-flowing
Reaction of 3.6% solution:	pH 7.3 ± 0.2 at 25°C
Prepared medium:	medium amber, slightly opalescent

Typical Cultural Response in Bacto SIM Medium
After 18 – 24 Hours at 37°C

Organism	Growth	H$_2$S	Motility	Indole
Escherichia coli ATCC® 25922	good	–	+	+
Salmonella typhimurium ATCC® 14028	good	+	+	–
Shigella flexneri ATCC® 12022	good	–	–	–

REFERENCES

1. Dissert. St. Petersburg, 1897.
2. J. Infectious Diseases, 21:554, 1917.
3. J. Exp. Med., 2:677, 1897.
4. Zeit. Hyg. Infektionskrank., 58:441, 1908.
5. J. Infectious Diseases, 6:194, 1909.
6. J. Bact., 31:575, 1936.
7. J. Lab. Clin. Med., 25:649, 1940.
8. J. Bact., 62:347, 1951.
9. Tech. Bull. Reg. Med. Tech., 5:1, 1944.
10. Rev. Inst. Bact., "Dr. Carlos G. Malbran," 11:286, 1943.
11. Manual of Methods of Pure Culture Study of Bacteria, Leaflet V, p. 6, 7th Ed., 1939.

PACKAGING

Bacto SIM Medium	1 lb (454 g)	0271-01-1
	1/4 lb (114 g)	0271-02-0
	12 tubes	0271-34-2
	144 tubes	0271-37-9
Bacto Indole Test Strips	1 vial	1627-30-5

BACTO SPS AGAR

INTENDED USE

Bacto SPS Agar (Sulfite-Polymyxin-Sulfadiazine) is a selective medium for detecting and enumerating *Clostridium perfringens* in foods.

HISTORY/PRINCIPLES

Bacto SPS Agar is a slight modification of the formulae of Mossel[1] and of Angelotti, Hall, Foster and Lewis.[2] Most clostridia reduce the sulfite to sulfide which then reacts with the ferric citrate to produce black colonies. Other sulfite-reducing organisms are suppressed by polymyxin and sulfadiazine.

FORMULA

BACTO SPS AGAR
DEHYDRATED
Ingredients per liter

Bacto Tryptone	15 g
Bacto Yeast Extract	10 g
Ferric Citrate	0.5 g
Sodium Sulfite	0.5 g
Bacto Sodium Thioglycollate	0.1 g
Sorbitan Monooleate	0.05 g
Sulfadiazine	0.12 g
Polymyxin B Sulfate	0.01 g
Bacto Agar	15 g

Final pH 7.0 ± 0.2 at 25°C.

One pound will make 11 liters of final medium.

METHOD OF PREPARATION

1. Suspend 41 g in 1 liter distilled or deionized water. Heat to boiling to dissolve completely.
2. Sterilize in the autoclave for 15 minutes at 15 lbs pressure (121°C). Allow to cool to 50 – 55°C.
3. Dispense into sterile Petri dishes to which inoculum has been added. Gently but thoroughly mix inoculum and medium. Allow to solidify on a flat surface.
4. If desired, pour cover layers using approximately 5 ml of medium.
5. Incubate anaerobically at 35 – 37°C for 24 – 48 hours or as desired.
6. Enumerate black colonies according to standard procedures.

STORAGE

Bacto SPS Agar   2 – 8°C
Prepared plates (not recommended)

QUALITY CONTROL

Identity Specifications

Dehydrated powder:	beige, homogeneous, free-flowing
Reaction of 4.1% solution:	pH 7.0 ± 0.2 at 25°C
Prepared medium:	light to medium amber, slightly opalescent

Typical Cultural Response on Bacto SPS Agar
After 18 – 48 Hours at 35°C Anaerobically

Organism	Growth	Color of Colony
Clostridium perfringens ATCC® 12919	good to excellent	black
Clostridium sporogenes ATCC® 11437	poor to good	black
Escherichia coli ATCC® 25922	inhibited	—
Staphylococcus aureus ATCC® 25923	inhibited to good	white

REFERENCES

1. J. Sci. Food Agric., 10, December, 1959.
2. Appl. Microbiol., 10:193,1962.

PACKAGING

Bacto SPS Agar	1 lb (454 g)	0845-01-8
	1/4 lb (114 g)	0845-02-7

BACTO SS AGAR

INTENDED USE

Bacto SS Agar is a highly selective plating medium used for isolating *Salmonella* and some *Shigella*.

HISTORY

Hormaeche and Surraco,[1] Hardy and co-workers[2] and Rose and Kolodny,[3] have reported SS Agar as superior to other media that have been recommended for isolation of *Shigella* and *Salmonella* organisms. Mayfield and Goeber[4] compared media for isolation of *Shigella* organisms and found Bacto SS Agar to yield the greatest number of positive isolations. Pots[5] and Caudill[6] have reported on the satisfactory use of SS Agar in isolation of *Shigella* organisms. In their studies of the acute diarrheal diseases, Mosher, Wheeler, Chant and Hardy[7] first enriched their specimens in selenite broth and then plated on bismuth sulfite agar, SS agar, MacConkey agar and desoxycholate agar, while Watt and Cummings[8] plated directly upon SS agar. Hormaeche and his co-workers[9] used SS agar in conjunction with others for isolation of *Shigella* as the causative agent of infantile summer diarrhea. Vacarro et al.[10] employed SS agar in conjunction with other plating media in the isolation of *Salmonella* and *Shigella* from healthy carriers, and Neter[11] used it similarly in his study of the *Proteus* and *Paracolobactrum* (paracolon bacilli) in feces of healthy infants. McClure and Crossley,[12] using SS agar and other media, isolated *S. newport* in an epidemic of food poisoning, and Cordy and Davis[13] isolated *S. morbificans* from horses and mules in an outbreak of salmonellosis. Watt, DeCapito and Morgan[14] isolated *S. texas* on SS agar after enrichment in tetrathionate broth. For the isolation and typing of *Salmonella* and *Shigella,* Borman, Wheeler and Mickle[15] and Nelson et al.[16] indicate the desirability of plating specimens on SS agar and other substrates before and after enrichment. Neter and Clark[17] have reported on the effectiveness of culture media in isolation of enteric organisms, the usefulness of SS agar, especially with other media, being clearly demonstrated by their results.

Schaub used SS agar as slants in tubes and found that this was a satisfactory method of demonstrating hydrogen sulfide production. The medium at the base of the slant became markedly blackened when the surface of the slant was streaked with a hydrogen sulfide producing organism.

Newman,[18] in a study of the detection of food poisoning attributable to dairy products, used direct streaking on SS agar as well as enrichment in tetrathionate broth followed by streaking on SS agar and bismuth sulfite agar for the isolation of salmonellae. In a study of methods to be used as a standard for the bacterial examination of pullorum reactors, Jungherr, Hall and Pomeroy,[19] in a committee report showed that in a comparative study of media and enrichments from October, 1946 to February, 1950, bismuth sulfite agar and SS agar permitted the highest number of specific isolations of *S. pullorum* and *S. gallinarum.* These selective media suppressed the growth of coliform organisms. Following enrichment of the specimens in selenite broth, streaking on bismuth sulfite agar gave the largest number of positive isolations, followed by SS agar and then MacConkey agar. Selenite broth yielded a higher number of successful isolations on follow-up media than did tetrathionate broth. The highest percentage of organisms were isolated from the ovary, followed by gall bladder, peritoneum, oviduct, intestines and pericardial sac in the order listed.

SS agar is specified in *Compendium of Methods for the Microbiological Examination of Foods*[20] and *Official Methods of Analysis of the Association of Official Analytical Chemists*[21] as a selective plating medium for examining food specimens.

PRINCIPLES
Bacto SS Agar was formulated to differentiate lactose-fermenters from lactose-nonfermenters and to provide maximum inhibition of coliform organisms without restricting growth of pathogenic gram-negative bacilli. *Shigella, Salmonella* and other lactose-nonfermenters produce opaque, translucent or transparent colonies that are, generally, smooth. The few lactose-fermenting organisms that develop are readily differentiated by their reddish, mucoid or black-centered colonies.

FORMULA
BACTO SS AGAR
DEHYDRATED
Ingredients per liter

Bacto Beef Extract .... 5 g	Sodium Thiosulfate .... 8.5 g
Proteose Peptone, Difco .... 5 g	Ferric Citrate .... 1 g
Bacto Lactose .... 10 g	Bacto Agar .... 13.5 g
Bacto Bile Salts No. 3 .... 8.5 g	Bacto Brilliant Green .... 0.33 mg
Sodium Citrate .... 8.5 g	Bacto Neutral Red .... 0.025 g

Final pH 7.0 ± 0.2 at 25°C.

One pound will make 7.5 liters of medium.

METHOD OF PREPARATION
1. Suspend 60 g in 1 liter distilled or deionized water and heat to boiling.
2. Boil for 2 – 3 minutes with frequent and careful swirling to dissolve completely. Avoid overheating. **Do not autoclave.** Cool to 55 – 60°C.
3. Dispense into sterile Petri dishes. Allow the surface of the medium to become quite dry by partially removing the covers while the medium solidifies (about 2 hours).

STORAGE
Bacto SS Agar   Below 30°C
Prepared medium   2 – 8°C

QUALITY CONTROL
Identity Specifications

Dehydrated powder: very light buff to pink, homogeneous, free-flowing
Reaction of 6% solution: pH 7.0 ± 0.2 at 25°C
Prepared medium: red-orange, very slightly opalescent

Typical Cultural Response on Bacto SS Agar
After 18 – 24 Hours at 35°C

Organism	Growth	Color of Colony
Enterobacter aerogenes ATCC® 13048	poor to good	cream-pink
Proteus mirabilis ATCC® 25933	poor to good	colorless
Salmonella enteritidis ATCC® 13076	good	colorless
Shigella flexneri ATCC® 12022	fair to good	colorless
Streptococcus faecalis ATCC® 19433	none to poor	colorless

REFERENCES
1. Apartado De Los Archivos Uruguayos De Medicina, 18:485, 1941.
2. Public Health Reports, 57:521, 524, 1942.
3. J. Lab. Clin. Med., 27:1081, 1942.
4. Am. J. Pub. Health, 31:363, 1941.
5. The Lancet, Vol. I, (XXIII):677, 1942.
6. J. Am. Med. Assoc., 119:1402, 1942.
7. Public Health Reports, 56:2415, 1944.
8. Public Health Reports, 60:1355, 1945.
9. Am. J. Diseases Children, 66:539, 1943.
10. Rev. Chileana Hyg. Med. Prev., 4:353, 1942.
11. J. Pediatrics, 26:39, 1945.
12. Can. J. Pub. Health, 36:401, 1945.
13. J. Am. Vet. Med. Assoc., 58:20, 1946.
14. Public Health Reports, 62:806, 1947.
15. Am. J. Pub. Health, 33:127, 1943.
16. Am. J. Pub. Health, 36:51, 1946.
17. Am. J. Digestive Diseases, 7:229, 1944.
18. J. Milk and Food Tech., 13:226, 1950.
19. Proc. 22nd Ann. Meet. Northeastern Conf. Lab. Workers in Pullorum Disease Control, Burlington, Vermont, June 20 – 21, 1950.
20. Speck, M. L. (ed.). 1976. Compendium of methods for the microbiological examination of foods. American Public Health Association, Washington, D.C.
21. Horwitz, W. (ed.). 1980. Official methods of analysis of the Association of Official Analytical Chemists. Association of Official Analytical Chemists, Washington, D.C.

PACKAGING
Bacto SS Agar	1 lb (454 g)	0074-01-0
	1/4 lb (114 g)	0074-02-9
	5 lb (2.27 kg)	0074-05-6

SABOURAUD CULTURE MEDIA
BACTO SABOURAUD DEXTROSE AGAR
BACTO SABOURAUD DEXTROSE BROTH
BACTO SABOURAUD AGAR MODIFIED
BACTO SABOURAUD MALTOSE AGAR
BACTO SABOURAUD MALTOSE BROTH
BACTO FLUID SABOURAUD MEDIUM

INTENDED USE
Bacto Sabouraud culture media are used for culturing yeasts, molds and aciduric microorganisms.

HISTORY/PRINCIPLES
Bacto Sabouraud Dextrose Agar and Bacto Sabouraud Dextrose Broth are modifications of the dextrose agar described by Sabouraud.[1] They are recommended for the cultivation and growth of fungi, particularly those associated with skin infections.

The majority of molds are not pathogenic, but some are true parasites, producing a number of common diseases such as ringworm, favus and various other hair and skin lesions. Internal infections of the lung and lymphatics may also be traced to molds.

For the primary isolation of fungi from scales and crusts, Ch'in[2] suggests the addition of 0.015% potassium tellurite or 0.05% copper sulfate to this medium in order to suppress the growth of bacteria. Emmons and Ashburn[3] used a Sabouraud dextrose agar containing 1% neopeptone and 2% dextrose for the isolation of *Histoplasma capsulatum* from rats. Emmons and Hollaender[4] used Sabouraud dextrose agar prepared with neopeptone for growing *Trichophyton gypseum asteroides* in their studies on mutation of the dermatophytes induced by ultraviolet irradiation. These comparative tests have shown that Neopeptone, Difco is a most satisfactory source of nitrogen for the development of fungi.

Robinson and Kotcher[5] used Sabouraud dextrose agar containing 20 units penicillin and 40 units dihydrostreptomycin hydrochloride per ml of medium for the isolation of *Histoplasma* from dogs. Kotcher, Robinson and Miller[6] in a study of media for the recovery of *H. capsulatum* from tissues of artificially infected rats reported that the highest percentage recovery of the organism was from spleen on Sabouraud dextrose agar. Serowy and Jung[7] used Bacto Sabouraud Dextrose Agar in their study of the *Microspora* and called attention to the suitability of this medium for the cultivation of *Microspora* and other pathogenic fungi, as well as the ease with which this medium may be prepared and used.

The addition of antibiotics to acid as well as neutral media for the isolation of pathogenic fungi has proven especially satisfactory.

Bacto Sabouraud Agar Modified is a modification of Bacto Sabouraud Dextrose Agar with the dextrose reduced to 20 grams per liter and the reaction adjusted to pH 7.0. This formulation has been widely used by Emmons for the cultivation of pathogenic fungi and is listed in *Diagnostic Procedures and Reagents,* 4th Edition of the APHA, 1963[8] for this purpose. It serves as an excellent basal medium to which antibiotics and other inhibitors may be added for the selective cultivation of various groups of microorganisms.

Bacto Sabouraud Maltose Agar, like Bacto Sabouraud Dextrose Agar, contains no selective agent. It depends entirely on the acid reaction, pH 5.6, for the selective growth of fungi over bacteria. In the initial cultivation of fungi from specimens many investigators prefer to use a selective medium such as Bacto Littman Oxgall Agar or Bacto Brain Heart Infusion Agar.

The use of maltose or the extractives of malt in media designed for the cultivation of molds and other fungi is quite universal. Maltose is well adapted to the nutritional requirements of these organisms.

Frank[9] has used Bacto Sabouraud Maltose Agar successfully in cultivating the causative organisms of perleche. Davidson, Dawding and Buller[10] reported that Bacto Sabouraud Maltose Agar was a most satisfactory medium in their studies of the infections caused by *Microsporon audouini*, *M. lanosum* and *Trichophyton gypseum*. Davidson and Dawding[11] also used this medium in isolating *T. gypseum* from a case of tinea barbae. Serowy and Jung[12] used Bacto Sabouraud Maltose Agar in their study of the *Microspora* and called attention to the suitability of this medium for the cultivation of *Microspora* and other pathogenic fungi, as well as the ease with which this medium may be prepared and used. A. W. Bengston[13] observed that Sabouraud maltose agar could be used to advantage in the isolation and differentiation of *Pseudomonas*. On this medium the blue pyocyanin pigment is enhanced making it easy to determine pigment production, thereby detecting *Pseudomonas* organisms in mixed infections. Sabouraud dextrose agar on the other hand, tends to elicit the production of the pink fluorescein pigment with suppression of the pyocyanin.

Chapman[14] modified Sabouraud maltose agar in the preparation of a selective medium for the isolation and identification of *Monilia* and other fungi. The medium was prepared by adding 0.1 ml of Tergitol 7 and 0.0025% brom cresol purple to Sabouraud maltose agar at pH 5.6. The medium was sterilized in the autoclave for 10 minutes and when cooled to 45 – 55°C, 0.3 ml of Bacto Chapman Tellurite Solution and 3 ml of 2,3,5-triphenyltetrazolium chloride (TTC), was added. Surfaces of the plates were inoculated followed by incubation at 37°C for 48 hours. Chapman reported that the Tergitol 7 inhibited all bacteria except members of the coliform group, while the potassium tellurite inhibited these organisms. *Candida albicans* produced "off white" circular smooth entire convex to pulvinate colonies about 4 mm in diameter. Other *Candida* produce colored colonies ranging from orange to tan to lilac, often discoloring the medium. *Saccharomyces* grew in 48 hours on the medium producing colonies somewhat resembling those of *Candida*.

Bacto Sabouraud Maltose Broth is acid in reaction and has the same formula as Bacto Sabouraud Maltose Agar except agar is omitted. Bacto Sabouraud Maltose Broth is also well suited for the detection of fungi in sterility test procedures.

Bacto Fluid Sabouraud Medium is prepared according to the formula of the NIH, FDA, USP and NF for sterility testing. It is used for detecting molds and yeasts by official sterility test procedures.

FORMULAE

BACTO SABOURAUD DEXTROSE AGAR
DEHYDRATED

Ingredients per liter

Neopeptone, Difco	10 g
Bacto Dextrose	40 g
Bacto Agar	15 g

Final pH 5.6 ± 0.2 at 25°C.

BACTO SABOURAUD MALTOSE AGAR
DEHYDRATED

Ingredients per liter

Neopeptone, Difco	10 g
Bacto Maltose	40 g
Bacto Agar	15 g

Final pH 5.6 ± 0.2 at 25°C.

BACTO SABOURAUD DEXTROSE BROTH
DEHYDRATED

Ingredients per liter

Neopeptone, Difco . . . . 10 g
Bacto Dextrose . . . . 20 g

Final pH 5.6 ± 0.2 at 25°C.

BACTO SABOURAUD MALTOSE BROTH
DEHYDRATED

Ingredients per liter

Neopeptone, Difco . . . . 10 g
Bacto Maltose . . . . 40 g

Final pH 5.6 ± 0.2 at 25°C.

BACTO SABOURAUD AGAR MODIFIED
DEHYDRATED

Ingredients per liter

Neopeptone, Difco . . . . 10 g
Bacto Dextrose . . . . 20 g
Bacto Agar . . . . 20 g

Final pH 7.0 ± 0.2 at 25°C.

BACTO FLUID SABOURAUD MEDIUM
DEHYDRATED

Ingredients per liter

Bacto Casitone . . . . 5 g
Bacto Peptamin . . . . 5 g
Bacto Dextrose . . . . 20 g

Final pH 5.7 ± 0.2 at 25°C.

Rehydration and Yield

	Grams / Liter	Liters / Pound
Bacto Sabouraud Dextrose Agar	65 g	6.9 liter
Bacto Sabouraud Dextrose Broth	30 g	15.1 liter
Bacto Sabouraud Agar Modified	50 g	9 liter
Bacto Sabouraud Maltose Agar	65 g	6.9 liter
Bacto Sabouraud Maltose Broth	50 g	9 liter
Bacto Fluid Sabouraud Medium	30 g	15.1 liter

METHOD OF PREPARATION
1. To rehydrate, suspend appropriate amount in 1 liter distilled or deionized water. Mix thoroughly. When the medium contains agar, heat to boiling to dissolve completely.
2. Dispense into appropriate containers.
3. Sterilize in the autoclave for 15 minutes at 15 pounds pressure at 121°C. Avoid overheating which could result in a softer agar medium, especially those with a low pH.

ADDITION OF ANTIBIOTICS
4. Under aseptic conditions, antibiotics can be added to the sterile melted medium at 45 – 50°C.

Generally 20 units of penicillin and 40 micrograms of streptomycin or dihydrostreptomycin per ml of medium can be added. These desired concentrations of penicillin may be readily obtained by dissolving the contents of one vial of penicillin containing 100,000 units penicillin in 10 ml. sterile distilled water. Two ml of this solution are added to one liter of sterile melted medium at 45 – 50°C (0.2 ml per 100 ml of medium). To obtain the desired concentration of streptomycin in the same medium, dissolve the contents of a one gram vial of streptomycin (one million micrograms) in 10 ml sterile distilled water. One ml of this solution is added to 9 ml sterile distilled water to give a solution containing 10,000 micrograms streptomycin per ml. To each liter of medium are added 4 ml of this solution to obtain 40 micrograms per ml (0.4 ml for 100 ml medium).

STORAGE
Bacto Sabouraud media	Below 30°C
Prepared media without antibiotics	15 – 30°C
Prepared media with antibiotics	2 – 8°C

QUALITY CONTROL

Identity Specifications

	Sabouraud Dextrose Agar	Sabouraud Dextrose Broth	Sabouraud Agar Modified
Dehydrated powder:	————light beige, homogeneous, free-flowing————		
Reaction of solution at 25°C:	6.5% pH 5.6 ± 0.2	3% pH 5.6 ± 0.2	5% pH 7.0 ± 0.2
Prepared medium:	light to medium amber, slightly opalescent with no significant precipitate	light amber, clear	light amber, slightly opalescent

	Sabouraud Maltose Agar	Sabouraud Maltose Broth	Fluid Sabouraud Medium
Dehydrated powder:	light beige	white	off white
	————homogeneous, free-flowing————		
Reaction of solution at 25°C:	6.5% pH 5.6 ± 0.2	5% pH 5.6 ± 0.2	3% pH 5.7 ± 0.2
Prepared medium:	light amber, slightly opalescent	light amber, clear to slightly opalescent	light amber, clear

Typical Cultural Response on/in any of the above Sabouraud Media After 40 – 72 Hours at 30 ± 2°C

Organism	Recovery
Aspergillus niger ATCC® 16404	good to excellent
Candida albicans ATCC® 26790	good to excellent
Escherichia coli ATCC® 25922	good to excellent*
Lactobacillus casei ATCC® 9595	good to excellent
Saccharomyces cerevisiae ATCC® 9763	good to excellent

*inhibited on media with lower pH.

REFERENCES

1. Ann. dermatol. syphilol., 1892–1893.
2. Proc. Soc. Exp. Biol. Med., 38:700:1938.
3. Pub. Health Reports, 63:1416:1948.
4. Arch. Dermatol Syphilol., 52:257:1945.
5. Pub. Health Reports, 66:1533:1951.
6. J. Bact., 62:613:1951.
7. Derm. Wschr., 124:665:1951.
8. Diagnostic Procedures APHA, 4th Ed. 175, 1963.
9. Arch. Dermatol. Syphilol., 26:451:1932.
10. Can. J. Research, 6:1:1932.
11. Arch. Dermatol. Syphilol., 26:660:1932.
12. Derm. Wschr., 124:665:1951.
13. Personal Communication, 1951.
14. Trans. New York Acad. Sci Series II. 14:254:1952.

PACKAGING

Bacto Sabouraud Dextrose Agar		
	5 lb (2.27 kg)	0109-05-5
	10 kg	0109-08-2
	100 g	0109-15-3
	500 g	0109-17-1
	12 tubes	0109-34-0
	144 tubes	0109-37-7
	6 × 100 ml	0109-73-2

Bacto Sabouraud Dextrose Broth	5 lb (2.27 kg)	0382-05-3
	100 g	0382-15-1
	500 g	0382-17-9
Bacto Sabouraud Agar Modified	1 lb (454 g)	0747-01-7
	5 lb (2.27 kg)	0747-05-3
Bacto Sabouraud Maltose Agar	1 lb (454 g)	0110-01-6
	1/4 lb (114 g)	0110-02-5
	5 lb (2.27 kg)	0110-05-2
Bacto Sabouraud Maltose Broth	1 lb (454 g)	0429-01-2
Bacto Fluid Sabouraud Medium	1 lb (454 g)	0642-01-3
	5 lb (2.27 kg)	0642-05-9

SALMONELLA ANTIGENS

BACTO SALMONELLA H ANTIGENS
BACTO SALMONELLA O ANTIGENS
BACTO WIDAL ANTIGEN SET

PRODUCT DESCRIPTION AND INTENDED USE

Bacto Salmonella Antigens are chemically inactivated, dyed suspensions of internationally recognized strains of commonly encountered *Salmonella* for use in the slide and tube agglutination procedures to detect antibodies (agglutinins) implicated in certain febrile disease states.

The rapid slide procedure is a screening test for detecting agglutinins whereas the tube test is a confirmatory procedure for quantitating agglutinin compositions. It is therefore necessary that any serum yielding a positive result in the screening (slide test) be retested using the tube test to establish the titer.

Bacto Febrile Positive Control Polyvalent is a hyperimmune antiserum prepared in goats, designed to demonstrate agglutination employing these test procedures, to a "significant titer"* (or greater) with all the bacterial antigens provided in this set.

Bacto Febrile Negative Control is a serum preparation of animal origin designed to demonstrate a negative reaction with the antigens included in this set.

These reagents are for use only in the test procedure described herein. They are not recommended for the direct diagnosis of a disease but only for the detection of agglutinins particularly in cases of seroconversion.

*Past history in the use of bacterial suspension for the detection of febrile agglutinins, has resulted in a pattern of titers which have been considered "significant." These titers may be found in Table I. It must be emphasized that although the information obtained is not absolute, a four-fold increase in titer between two specimens is regarded clinically significant. The value of febrile antigens is in the detection of an increase in the titer of agglutinins in sera from the same patient in successive bleedings when the test is performed at the same time using the same reagents by the same technician.

SUMMARY AND BACKGROUND

Salmonella is representative of a number of species of pathogenic microorganisms which upon invasion produce a fever in its host. Consequently it is often referred to as a "Febrile Antigen."

For a complete discussion refer to Febrile Antigens.

Grunbaum and Widal[2] introduced techniques which were to be the first practical application of the new science of immunology. Their tests quantitatively measured the agglutinins in the sera of patients with Typhoid Fever. This test has become known universally as the Widal Test. It has been extended to include *Salmonella* other than *S. typhi* by the use of a variety of *Salmonella* O and H antigens.

Bacto Widal Antigen Set employs four *Salmonella* antigens commonly found as etiological agents causing fever in the host.

The Rapid Slide Test is the most widely used procedure employing Widal antigens because of the simplicity with which the results may be reported. Negative reactions with this test can usually be reported as such if all 5 serum dilutions have been used. Note: Even though the slide test is not quantitative, it is necessary to run the series of dilutions in order to detect agglutinin control of a serum which might be overlooked in the case of a "prozone phenomenon." This often occurs in the sera containing a high titer of typhoid agglutinins where higher concentrations of the serum may yield negative results but a dilution of the serum is positive.

In this test, varying amounts of the sera to be tested are distributed to the corresponding square of a previously marked glass slide. Widal antigen is added and mixed with the sera. Positive or negative results are read 1 minute after mixing. Any positive reaction in the slide test must be confirmed using the tube test.

The Macroscopic Tube Test[3] should be used to confirm the presence of antibodies demonstrated by the slide technique and to quantitate their titer in suspected sera. In this test, the patient's serum is serially diluted in test tubes, a constant amount of the appropriate dilution of Widal antigen is added to each tube, the resultant mixture is incubated according to directions and the agglutination pattern is read and recorded.

REAGENTS
Antigens
Bacto Salmonella O Antigen Group D is prepared from *Salmonella typhi* 0901.
Bacto Salmonella H Antigen a — *Salmonella enteritidis* bioser Paratyphi A.
Bacto Salmonella H Antigen b — *Salmonella enteritidis* ser Paratyphi B.
Bacto Salmonella H Antigen d — *Salmonella typhi* H901.

When used according to the suggested PROCEDURE, each 5 ml vial of Bacto Salmonella O or H Antigen is sufficient to perform approximately 20 slide tests or approximately 25 tube tests.

Concentration of Antigen[4,5]
Bacto Salmonella Antigens (both O and H) have a density of approximately 20X a Bacto McFarland Barium Sulfate Standard No. 3 (1.8×10^{10} organisms per ml) for use in the undiluted state employing the rapid slide screening test and diluted 1:20 for use in the tube test.

Note: The density of the antigens may vary from the above values. They are adjusted to perform at their optimum when standardized with hyperimmune sera obtained from laboratory animals. In some antigens glycerin to a final volume of 20% is added when necessary to adjust the suspension sensitivity.

Preservatives
Bacto Salmonella H Antigens contain:
A final concentration of 0.3% formaldehyde.
Crystal Violet in approximately 0.002% final concentration.*
Brilliant Green in approximately 0.005% final concentration.

Bacto Salmonella O Antigen Group D contains:
A final concentration of 0.5% phenol.
Crystal Violet in approximately 0.002% final concentration.*
Brilliant Green in approximately 0.005% final concentration.

*Due to various factors, including the genera of the organism involved, the density of each lot of suspension, etc., the color intensity of the dye in each lot of antigen may vary and is normal. Such differences in color will not affect the outcome of the test.

Control Sera
Bacto Febrile Positive Control Polyvalent contains agglutinins for the Salmonella Antigens in the Bacto Widal Antigen Set to a titer of 1:80 or greater as determined in our laboratories.

Merthiolate® is added as a preservative in the amount of 1:5,000 (w/v) final concentration.

Bacto Febrile Negative Control is a standardized protein solution containing Merthiolate® as a preservative in the amount of 1:5,000 (w/v) final concentration.

When used according to the suggested PROCEDURE each 5 ml vial of Bacto Febrile Positive Control Polyvalent and Bacto Febrile Negative Control is sufficient to perform approximately 32 slide tests or 50 tube tests using the four antigens in the Widal Kit or 20 slide or 25 tube tests using a single Salmonella Antigen.

INSTRUCTIONS FOR REHYDRATION AND STORAGE
1. The antigens are liquid and ready for use in the slide test. The *Salmonella* antigens may be used in the tube test also, provided they are first diluted 1:20 with formalized saline.
2. The control sera are rehydrated by adding 5 ml sterile distilled or deionized water at room temperature.
3. Store all reagents at 2 – 8°C. These reagents are stable to the expiry date on the label when stored at 2 – 8°C provided they have not been exposed to temperature extremes (excessive heat or freezing) and have not become contaminated chemically or bacterially during routine usage.

Bacto Salmonella Antigens are not to be used for immunization of man or animals.

SPECIMEN COLLECTION
1. Collect 5 – 10 ml whole blood aseptically from the patient.
2. Allow the blood to clot and obtain the syneresed serum with a Pasteur pipette. If the serum is not free of erythrocytes, clarify by centrifugation. Note: Do not heat the serum, since significant antibodies may be thermolabile.

PROCEDURE
Materials Provided:
Bacto Widal Antigen Set contains:
  Bacto Salmonella O Antigen Group D
  Bacto Salmonella H Antigen a
  Bacto Salmonella H Antigen b
  Bacto Salmonella H Antigen d
  Bacto Febrile Positive Control Polyvalent
  Bacto Febrile Negative Control
Droppers for the above antigens (supplied droppers deliver approximately 0.03 ml/drop)
Individual Salmonella Antigens are listed in the packaging section.

Materials Required but not Provided:
Rapid Slide Test
  Wax marking pencils
  Ruler
  Glass plate 8 × 8″
  0.2 ml serological pipettes

  Distilled or deionized water
  Applicator sticks
  Suitable light source

Tube Test
  Formalinized saline (0.5 ml formalde-
    hyde and 0.85g NaCl per 100 ml
    distilled or deionized water)
  Isotonic saline (0.85g NaCl per 100
    ml distilled or deionized water)
  Erlenmeyer flask
  Distilled or deionized water

  Kahn tubes 12 × 75 mm
  Kahn type tube supports
  Serological pipettes, 1 ml and 5 ml capacity
  Water bath at 50°C
  Refrigerator at 2 – 8°C
  Suitable light source

PRECAUTIONS
All reagents and equipment used in the slide test must be at room temperature prior to use.

Specimen
1. The specimen must be clear and free of visible fat. It must be free of excessive hemolysis and not bacterially contaminated.
2. The specimen must be used in the unheated state. If inactivated by heat, some thermolabile agglutinins may be destroyed.
3. In mixing the tubes in the tube test, the specimen should not be mixed too vigorously since foaming may result in agglutinin denaturization.

Glassware
1. All glassware that is employed in the preparation, testing and storage of these reagents must be free of detergents or other harmful residues.
2. Pipettes employed must be clean (acid washed if necessary) to deliver proper volumes.

Antigens
1. Shake antigen vial well before use to insure a smooth, uniform suspension.
2. Discard any antigen demonstrating positive reactions in the negative control or any demonstrating spontaneous agglutination in the vial. Widal antigens will demonstrate autoagglutination if at any time during shipment or storage they are subjected to freezing temperatures. Do not allow antigens to freeze.
3. Discard any antigen which does not react properly in the positive control serum.
4. Variation in the color intensity of the antigen is normal and will not affect the outcome of the test.

Control Sera
Discard any control serum in which a precipitate forms. If bacterial contamination is suspected, autoclave the reagent or use some suitable chemical procedure to inactivate the contaminant before discarding the reagent.

METHODOLOGY
Special Note: Because a change or no change in titer over a period of time are the best indicators of active infection or noninfection respectively, and because the accuracy and precision of the tests can be affected not only by test conditions, but also by the subjectivity of the person in reading the end-point, the following protocol is recommended always.

A preliminary test using either rapid slide test and/or the macroscopic tube test may be performed on the initial serum specimen and reported to the physician at that time. An aliquot of the serum should be transferred to a sterile test tube, sealed tightly, and kept in the freezer. When the second serum is obtained, it should be run in parallel with the original specimen. In this manner the original serum will serve as a control and any difference in titer will be more credible since the bias associated with the performance of the test as well as determining the end-point will be minimized. Obviously, it would be best if the person who performed the original test also performed the second test.

Exposure to heat (from external source, i.e., burner flame, etc.) may result in evaporation of the test mixture and may result in false-positive interpretation.

Rapid Slide Test
Note: The rapid slide test should be used for screening only. Any positive results in the slide procedure should be confirmed by the tube test.
1. Prepare the glass plate 8 × 8″ by making a series of 1-1/2″ ruled squares with the ruler and wax pencil. Each row of 5 squares is sufficient to test one antigen against the sera diluted to 1:320.
   Note: The glass plate should be thoroughly cleaned and dried after each use.
2. Pipette 0.08, 0.04, 0.02, 0.01 and 0.005 ml serum onto a row of squares on the ruled glass plate using a 0.2 ml serological pipette.
3. Place a drop of the appropriate antigen for the slide test on each drop of serum.
   Note: Shake the antigen well before using.
4. Mix each serum-antigen composite with an applicator stick starting with the 0.005 ml serum dilution and working towards the 0.08 ml dilution. The final dilutions are correlated approximately to the macroscopic tube dilutions and are 1:20, 1:40, 1:80, 1:160 and 1:320, respectively.
5. Hold the glass plate in both hands and gently "rotate" it 15 – 20 times. Observe the agglutination within 1 minute over a suitable light source. Reactions occurring later may be due to the reactants drying on the slide.

Macroscopic Tube Test
Bacto Salmonella Antigens may also be used as the tube antigens by diluting 1 part of antigen with 19 parts formalinized 0.85% NaCl (0.5 ml formaldehyde per 100 ml saline).

Prepare serial dilutions of serum to be tested and the control sera in 0.5 ml amounts in Kahn tubes in the following manner:
1. Place 8 Kahn tubes in a rack for each serum to be tested.
2. Pipette 0.9 ml of 0.85% NaCl into the first tube of each row and 0.5 ml into each of the remaining tubes.
3. Add 0.1 ml of the serum to the first tube containing 0.9 ml of NaCl solution.
4. Mix well with a pipette and transfer 0.5 ml to the second tube. Mix thoroughly.
5. Continue carrying the 0.5 ml of the serum dilution through tube 7. Discard 0.5 ml from tube 7 after mixing thoroughly. Tube 8 is the antigen control tube.
6. Add 0.5 ml of the desired antigen to each of the 8 tubes. Shake the racks to mix the antigen and antiserum. The resultant dilutions are 1:20 through 1:1280, respectively.
7. Incubate in a water bath as follows: Salmonella O Antigen, 16 – 18 hours at 50°C; Salmonella H Antigen, 1 hour at 50°C. It is important in this test to use the recommended time and temperature of incubation and to make certain the water bath is in a location free of mechanical vibration.

CALIBRATION AND QUALITY CONTROL

For greater proficiency in test interpretation always include a positive and negative serum control in each test protocol.

Both positive and negative controls are diluted in the same proportion as the patient's serum and processed in exactly the same manner following procedure above for the rapid slide test or for the macroscopic tube test.

An antigen is considered to be satisfactory if it does not clump with the negative control and it reacts to a titer of 1:80 or more with the positive control.

RESULTS
Rapid Slide Test
Record results as follows:

4+	complete agglutination
3+	approximately 75% of the cells are clumped
2+	approximately 50% of the cells are clumped
1+	approximately 25% of the cells are clumped
±	trace agglutination
−	no agglutination

The titer of serum is recorded as that dilution of the specimen in which at least 2+ (50%) agglutination occurs. See Tables I and II.

Macroscopic Tube Test
Using a fluorescent lamp against a black background, record the degree of agglutination as follows:

4+	100% of the organisms are agglutinated and the supernatant fluid is clear
3+	approximately 75% of the organisms are agglutinated and the supernatant fluid is slightly cloudy
2+	approximately 50% of the organisms are agglutinated and the supernatant fluid is moderately cloudy
1+	approximately 25% of the organisms are agglutinated and the supernatant fluid is cloudy
−	no agglutination is visible and the organisms remain as a cloudy suspension. This reaction should be observed in tube 8.

The following illustration and explanation is provided to help interpret the macroscopic tube reactions.

The genus *Salmonella* is composed mainly of 2 kinds of antigens, the "O" or heat stable somatic antigen and the "H" or heat labile flagellar antigen. Since some febrile sera, particularly those of typhoid fever cases, may react with both O and H antigens, particular attention should be given to the kind of agglutination obtained especially in the tube titration of such sera.

A somatic "O" reaction is characterized by coarse, compact agglutination which tends to be difficult to disperse, while the flagellar "H" has a characteristic loose, flocculent agglutination. Care must be taken not to shake an "H" antigen reaction too vigorously while examining. Characteristic O and H agglutination is indicated in the following diagrams.

SOMATIC "O" AGGLUTINATION

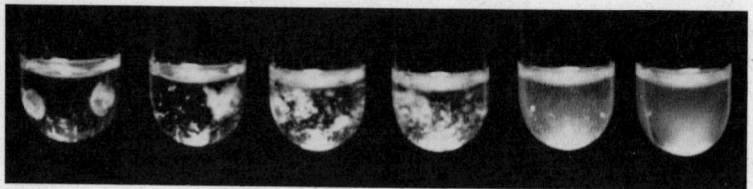

FLAGELLAR "H" AGGLUTINATION

Courtesy of New York State Health Dept., Albany, N.Y.

SAMPLE CALCULATIONS
Table I Rapid Slide Test

ml Serum	Correlated Dilution	Reaction Specimen 1	Reaction Specimen 2	Reaction Specimen 3
0.08	1:20	3+	4+	4+
0.04	1:40	2+	4+	3+
0.02	1:80	1+	3+	2+
0.01	1:160	−	3+	+
0.005	1:320	−	1+	−
Serum Titer		1:40	1:160	1:80

Table II Macroscopic Tube Test

Serum Dilution	Reaction Specimen 1	Reaction Specimen 2	Reaction Specimen 3
1:20	4+	3+	4+
1:40	4+	2+	4+
1:80	3+	1+	4+
1:160	2+	−	4+
1:320	1+	−	3+
1:640	−	−	2+
1:1280	−	−	1+
Serum Titer	1:160	1:40	1:640

INTERPRETATION
Report the serum titer as the reciprocal of the highest dilution showing a 2+ reaction. See Tables I and II.

A difference in titer of 1 dilution, plus or minus, between replicate samples or between several samples drawn 1–2 weeks apart and run in parallel cannot be considered significant. A 1 dilution, plus or minus, deviation is within the limits of laboratory error.[6,7]

LIMITATIONS

1. The major limitation of the febrile tests is that of interpretation. Again, it must be stressed that the antigens are to be used for the detection and quantitation of agglutinin content in sera. They are not to be used singly for a definite diagnosis. A definitive diagnosis must be made by the physician, taking into consideration the history and physical state of the patient as well as data obtained from other laboratory tests. Refer to Bacto Blood Culture Bottles for isolation procedures of etiological agents from blood; and to Bacto Salmonella Antisera for additional isolation and serological procedures.

   Although the Widal antigens are useful for screening purposes, this technique should not be a complete substitute for the conventional isolation and serological identification of the etiological agent.
2. It is necessary to run several serum specimens taken at different times, from the same patient, in parallel, to detect quantitative differences in agglutinin content.
3. The slide test is for screening only. Any quantitation must be done using the tube test.
4. There are many known antigenic similarities and cross-reactions.
5. Significant titers may be obtained in specimens from individuals immunized with TAT and typhoid vaccines.
6. Nonspecific agglutination has been noted with Salmonella O Group D antigen in sera of patients with influenza.
7. Sera from narcotic addicts appear to contain broad nonspecific activity to the Widal antigen.[8]
8. Sera from patients with chronic active liver disease may demonstrate high agglutinin titers.[9]
9. Serum from cases of typhoid fever often agglutinate with Salmonella O Antigen Group B (Paratyphi-B) but rarely with Salmonella O Antigen Group A.
10. In some cases of typhoid fever, sera may demonstrate a prozone. (Ability of an antigen to react in higher serum concentrations). It is advised that all five serum dilutions be run in the rapid slide test rather than just one dilution to eliminate the possibility of missing positive reactions due to the prozone.

EXPECTED VALUES

Table III presents data that will be helpful in interpreting serological tests with the Salmonella antigens. The values tabulated will vary somewhat in certain cases.

Table III

Bacto Antigen	Pathology	Time of Maximum Titer	Significant Titer
B. Abortus	Brucellosis	3 – 5 weeks	1:80 over 1:160 indicative
Proteus OX19	Rocky Mt. Spotted Fever	2 – 3 weeks	1:160 – 1:320
Proteus OX19	Typhus	2 – 3 weeks	1:160 and higher
Salmonella H Antigen d (Typhoid H)	Typhoid Fever	4 – 5 weeks	1:80
Salmonella O Antigen Group D (Typhoid O)	Typhoid Fever	3 – 5 weeks	1:80* over 1:160 indicative
Salmonella H Antigen a (Para A)	Paratyphoid Fever	3 – 5 weeks	1:80*
Salmonella H Antigen b (Para B)	Paratyphoid Fever	3 – 5 weeks	1:80*
Leptospira	Leptospirosis	1 – 2 weeks	1:100
F. Tularensis	Tularemia	4 – 8 weeks	1:160

*Significant in nonvaccinated individuals.

REFERENCES

1. Grunbaum, A. S., Lancet, 2:806, 1896.
2. Widal, F., Bull. Soc. Med. Hop. de Paris, 13, 1896.
3. Spink, W. W., McCullough, N. B., Hutchings, L. H., Amer. J. Clin. Path., 24:486, 1954.
4. Identification of *Enterobacteriaceae*, Edwards, P. R., and Ewing, W. H., 3rd Ed., Burgess Publishing Co.
5. Huddleson, J. F., and Abell, E., J. Infect. Dis., 42:242, 1928.
6. Syphilis, A Synopsis, U.S. Dept. of Hlth., Ed. and Welf., Publ. Hlth. Serv. Publ. No. 1660, Jan. 1968.
7. Quality Control in Clinical Micro., Revised, ASCP Comm. on Continuing Education, Coun. on Micro., 1968.
8. Vogel, H., Cherubin, C. E., and Millian, S. J., Amer. J. Clin. Path., 53:932, 1970.
9. Protell, R. L., Soloway, R. D., Martin, W. J., Schoenfield, L. J., and Summerskill, W. H. J., Lancet ii:330, 1971.

PACKAGING

Bacto Widal Antigen Set	6 × 5 ml	2642-32-2
Contains: (also available separately)		
Bacto Salmonella O Antigen Group D (Typhoid O)	5 ml	2842-56-1
Contains factors 9,12		
Bacto Salmonella H Antigen a (Paratyphoid A)	5 ml	2844-56-9
Bacto Salmonella H Antigen b (Paratyphoid B)	5 ml	2845-56-8
Bacto Salmonella H Antigen d (Typhoid H)	5 ml	2847-56-6
Bacto Febrile Positive Control Polyvalent	5 ml	3238-56-1
Bacto Febrile Negative Control	5 ml	3239-56-0
Bacto Salmonella Antigens		
Bacto Salmonella H Antigen c	5 ml	2846-56-7
Contains factor c		
Bacto Salmonella H Antigen eh	5 ml	2514-56-8
Contains factors e,h		
Bacto Salmonella H Antigen g	5 ml	2513-56-9
Contains factor g		
Bacto Salmonella H Antigen i	5 ml	2515-56-7
Contains factor i		
Bacto Salmonella O Antigen Group A	5 ml	2839-56-6
Contains factors 1,2,12		
Bacto Salmonella O Antigen Group B	5 ml	2840-56-3
Contains factors 1,4,5,12		
Bacto Salmonella O Antigen Group C	5 ml	2841-56-2
Contains factors 6,7,(8),20		
Bacto Salmonella O Antigen Group E	5 ml	2843-56-0
Contains factors 1,3,10,15,19,34		
Bacto Salmonella O Antigen Group F	5 ml	2385-56-4
Contains factor 11		
Bacto Salmonella O Antigen Group G	5 ml	2386-56-3
Contains factors 13,22,23,[36],[37]		
Bacto Salmonella O Antigen Group H	5 ml	2387-56-2
Contains factors 6,14,24,25		
Bacto Salmonella O Antigen Group I	5 ml	2388-56-1
Contains factor 16		
Bacto Salmonella Vi Antigen	5 ml	2953-56-6

QUALITY CONTROL ANTIGENS FOR *SALMONELLA* SOMATIC SEROLOGY

PRODUCT DESCRIPTION AND INTENDED USE

Bacto QC Antigens Salmonella are chemically stabilized and inactivated suspensions of known strains of the genus *Salmonella* for use in quality control of *Salmonella* somatic antisera, employing the slide agglutination technique.

These reagents are designed as homologous controls for testing the efficacy of the *Salmonella* grouping antisera employed in routine laboratory procedures. They are not recommended for the direct diagnosis of a disease, but rather for detection of agglutinins in the hyperimmune serum employed.

SUMMARY AND BACKGROUND

Many clinical laboratories desiring homologous control suspensions for quality control purposes do not have them available on a routine basis due to several factors among which are lack of an adequate culture collection, lack of time to maintain the cultures in their proper antigenic state and the possible hazard of handling such large numbers of potentially pathogenic organisms. The use of Bacto QC Antigens eliminates these problems.

To determine that nothing deleterious has occurred to a polyvalent antiserum that has been approved by a manufacturer, it should only be necessary to obtain proper reactivity with one major antibody component and its homologous antigen.

REAGENTS

Antigen	Organism Used for Antigen Preparation	Homologous Control for the Following Identifying Antigen(s)
Bacto QC Antigen Salmonella O Group A	*Salmonella enteritidis* bioser Paratyphi A factors 1,2,12	2
Bacto QC Antigen Salmonella O Group B	*Salmonella enteritidis* ser Typhimurium factors 1,4,5,12	4,5
Bacto QC Antigen Salmonella O Group C$_1$	*Salmonella choleraesuis* factors 6,7	7
Bacto QC Antigen Salmonella O Group C$_2$	*Salmonella enteritidis* ser Newport factors 6,8	8
Bacto QC Antigen Salmonella O Group D	*Salmonella enteritidis* bioser Gallinarum factors 1,9,12	9
Bacto QC Antigen Salmonella O Group E$_1$	*Salmonella enteritidis* ser Anatum factors 3,10	10
Bacto QC Antigen Salmonella O Group E$_2$	*Salmonella enteritidis* ser Newington factors 3,15	15
Bacto QC Antigen Salmonella O Group E$_4$	*Salmonella enteritidis* ser Senftenberg factors 1,3,19	19
Bacto QC Antigen Salmonella O Group F	*Salmonella enteritidis* ser Rubislaw factor 11	11
Bacto QC Antigen Salmonella O Group G$_1$	*Salmonella enteritidis* ser Poona factors [1], 13,22,[36]	22
Bacto QC Antigen Salmonella O Group H	*Salmonella enteritidis* ser Carrau factors 6,14,24	14
Bacto QC Antigen Salmonella O Group I	*Salmonella enteritidis* ser Hvittingfoss factor 16	16
Bacto QC Antigen Salmonella Vi	Citrobacter ballerup 029	Vi

CONCENTRATION OF ANTIGENS

The antigens are adjusted to 70 – 100 International Units of Opacity. They are ready to use without further dilution.

PRESERVATIVES

Bacto QC Antigens Salmonella are preserved with 0.5% phenol USP (v/w), except for the Vi Antigen which contains 0.01% Merthiolate® (v/w).

STORAGE AND STABILITY

Store all Bacto QC Antigens Salmonella at 2 – 8°C. Do not allow to freeze at any time. Bring to room temperature prior to use. Return to the refrigerator after use. Prolonged exposure to room or higher temperature may, over a period of time, prove to be detrimental to the antigen.

These reagents are stable to the expiry date on the label when stored at 2 – 8°C provided they have not been exposed to temperature extremes and have not become contaminated with chemicals or bacteria during routine use.

Bacto QC Antigens are not to be used for immunization of man or animals.

SPECIMEN

The specimen to be tested with a Bacto QC Antigen Salmonella is the homologous antiserum used in the grouping of *Salmonella* organisms. The antiserum must be within its expiry dating, be used at its recommended routine test dilution (RTD), be at room temperature at the time of testing and clear. (If the antiserum is cloudy, it may be interpreted as a false-positive reaction.) Often, due to lipid content, antisera will become cloudy upon refrigeration but still maintain proper reactivity. If cloudy, clarify the antiserum by centrifugation and test with the homologous QC antigen.

Bacto QC Antigens may be used also as a negative control system by using a heterologous antigen (possessing no common antigen) with a given test serum. For example, use a QC Antigen Salmonella O Group I to test a Salmonella O Antiserum Group A factors 1,2,12.

One should be cognizant of existing cross reactions. For further details on cross-reactivity, consult references.

TEST PRECAUTIONS

1. If autoagglutination of the QC Antigen is apparent, discard the antigen. The most likely cause of such autoagglutination is exposure of the antigen to freezing conditions. Avoid freezing.
2. Allow the QC Antigens, the antisera to be tested and all equipment used to be at room temperature at the time of testing. The test reagents, if cold, may cause false-negative reactions.
3. Shake the antigen well before use to suspend the organisms.
4. A glass plate left under or near a burner may be too warm for the test, resulting in a rapid evaporation of the test mixture (false-positive).
5. The presence of artifacts on the glass plate (dust, etc.) or particulate matter in the antiserum may be interpreted as a false reaction.
6. The QC Antigens Salmonella will react with their homologous polyvalent, homologous grouping, single factor antisera to the identifying antigen(s), and other antigenic components. Some of the latter, however, may give reactions somewhat weaker than the grouping sera due to the specificity of a single factor sera to the identifying antigen(s).
7. Generally, because of the density of Bacto QC Antigens Salmonella, they are negative in heterologous grouping antisera possessing a common antigen. However, if a given lot of an antiserum is particularly avid for a given factor, it may agglutinate a QC antigen possessing the common factor; i.e., Salmonella O Group E_1 Antiserum factors 3,10 will agglutinate a QC Antigen Salmonella E_1 (3,10) but may also agglutinate a QC Antigen Salmonella E_2 (3,15) if the serum is of high titer for factor 3.
8. Dispel all antigen from the dropper after use and before storage.
9. Avoid contact of the antigen into the rubber bulb.
10. Keep upright when stored at 2 - 8°C.

PROCEDURE
Materials Provided:
Bacto QC Antigens Salmonella
Droppers

Materials Required but not Provided:
Bacto Salmonella O Antisera (available from Difco) Plain applicator sticks
Bacto Febrile Negative Control (available from Difco) Wax marking pencil
Glass plate 8 × 8″ or equivalent Ruler
Droppers for the antisera Suitable light source

TECHNIQUE
1. Mark off a glass plate into 1″ square sections with a wax marking pencil.
2. Place a drop of the Salmonella Antiserum to be tested in one square.
3. Place a drop of Bacto Febrile Negative Control in the square next to the antiserum. This is the negative control for the antigen. A heterologous Bacto Salmonella Antiserum may also be used as a negative control. (See chart in RESULTS section for proper selection of the heterologous control.)
4. Add one drop each of the QC Antigen Salmonella designated as the homologous control to one of the drops of the test antiserum and to the Bacto Febrile Negative Control or heterologous antiserum control.
5. Mix the contents of each square with a fresh end of an applicator stick. Spread each mixture uniformly throughout the square.
6. Rotate the plate by hand continuously for one minute.
7. Read and record the results as follows:

4+	Agglutination of all the cells (clear background)
3+	Agglutination of approximately 75% of the cells
2+	Agglutination of approximately 50% of the cells
1+	Agglutination of approximately 25% of the cells
±	Very slight roughness
−	None of the cells agglutinate

INTERPRETATION
An antiserum is satisfactory for use if all of the following conditions are met:
1. A 3+ or greater reaction is obtained with its recommended homologous control antigen(s).
2. No more than a 1+ reaction is obtained with its recommended heterologous control antigen(s).
3. A negative or, rarely, a ± reaction is obtained in the Bacto Febrile Negative Control.

RECOMMENDED QC ANTIGENS FOR MOST COMMONLY USED SALMONELLA ANTISERA

Code	Salmonella O Antiserum Designation	Recommended Bacto QC Antigen Salmonella for Homologous Control	Recommended Bacto QC Antigen Salmonella for Heterologous Control
2814-47-7	factor 2	A	I or F
2659-47-5	factor 4	B	I or F
2815-47-6	factor 4,5	B	I or F
2660-47-2	factor 5	B	I or F
2947-47-7	Group A factors 1,2,12	A	I or F
2264-47-2	Poly A-I & Vi	$A,B,C_1,C_2,D,E_1,E_2,$ E_4,F,G_1,H,I,Vi	*
2534-47-6	Poly A	A,B,E_1,E_2,E_4	I or F
2816-47-5	factor 7	C_1	I or F
2949-47-5	Group C_1, factors 6,7	C_1	I or F
2535-47-5	Poly B	C_1,C_2,F,G_1,H	I or E_1
2817-47-4	factor 8	C_2	I or F
2950-47-1	Group C_2, factors 6,8	C_2	I or F
2818-47-3	factor 9	D	I or F
2951-47-0	Group D_1, factors 1,9,12	D	I or F
2257-47-1	factor 10	E_1	I or F
2819-47-2	Group E, factors 1,3,10,15,19,34	E_1,E_2,E_4	I or F
2260-47-6	Group F, factor 11	F	I or E_1
2661-47-1	factor 14	H	I or F
2262-47-4	Group H, factors 1,6,14,24,25	H	I or F

2258-47-0	factor 15	E_2	I or F
2954-47-7	Group E_2, factors 3,15	E_2	I or F
2259-47-9	factor 19	E_4	I or F
3019-47-8	Group E_4, factors 1,3,19	E_4	I or F
2663-47-9	factor 22	G_1	I or F
3029-47-6	Group G, factors 13,22,23,(36),(37)	G_1	I or F
2261-47-5	Group G_1, factors 13,22,(36)	G_1	I or F
2263-47-3	Group I, factor 16	I	F or E_1
2536-47-4	Poly C	I	F or E_1
2827-47-2	Vi	Vi	I or F

*A heterologous control is not recommended for Poly A-I and Vi. It is an unabsorbed serum and is likely to react not only with some of the higher O groups of *Salmonella* but possibly with other members of *Enterobacteriaceae* as well. Since it is the major screening reagent in *Salmonella* serology, if it reacts with all identifying antigens of Groups A-I and Vi to a 3+ or greater reaction under the proper test conditions, it is suitable for use.

REFERENCES

1. Department of Health, Education, and Welfare, 1975, Specifications and evaluation methods for laboratory, immunological and microbiological reagents, Volume 1, 4th Ed. Centers for Disease Control, Atlanta.
2. Edwards, P. R., and W. H. Ewing, 1972, Identification of *Enterobacteriaceae*, 3rd Ed., Burgess Publishing Co., Minneapolis.
3. Lennette, E. H., A. Balows, W. J. Hausler, Jr., and J. P. Truant, 1980, Manual of clinical microbiology, 3rd Ed., American Society for Microbiology, Washington, D.C.
4. Wright, D. N., D. R. Welch, and J. M. Matsen, 1980, J. Clin. Microbiol., 11:305.

PACKAGING

Bacto QC Antigen Salmonella O Group A	1 ml	2130-50-8
Bacto QC Antigen Salmonella O Group B	1 ml	2131-50-7
Bacto QC Antigen Salmonella O Group C_1	1 ml	2132-50-6
Bacto QC Antigen Salmonella O Group C_2	1 ml	2133-50-5
Bacto QC Antigen Salmonella O Group D	1 ml	2134-50-4
Bacto QC Antigen Salmonella O Group E_1	1 ml	2135-50-3
Bacto QC Antigen Salmonella O Group E_2	1 ml	2136-50-2
Bacto QC Antigen Salmonella O Group E_4	1 ml	2137-50-1
Bacto QC Antigen Salmonella O Group F	1 ml	2138-50-0
Bacto QC Antigen Salmonella O Group G_1	1 ml	2139-50-9
Bacto QC Antigen Salmonella O Group H	1 ml	2140-50-6
Bacto QC Antigen Salmonella O Group I	1 ml	2141-50-5
Bacto QC Antigen Salmonella Vi	1 ml	2142-50-4
Bacto Febrile Negative Control	5 ml	3239-56-0

SALMONELLA SEROLOGY

BACTO SALMONELLA O ANTISERA
BACTO SALMONELLA H ANTISERA
BACTO SALMONELLA H ANTISERA SPICER-EDWARDS SET

INTENDED USE

Bacto Salmonella Antisera are prepared for use in the serological identification of members of the genus *Salmonella*. Bacto Salmonella O Antisera are employed in the identification of the somatic antigens by the slide agglutination technique, while Bacto Salmonella H Antisera are employed in the identification of flagellar antigens by the tube agglutination technique.

SUMMARY AND EXPLANATION OF THE TEST

Salmonella typhi was the first member of the *Salmonella* to be recognized as a pathogen. It was first seen in 1880, by Eberth and isolated by Gaffky in 1884.[1] Later, other *Salmonella* associated with disease processes were isolated. Salmon and Smith isolated *S. choleraesuis* in 1885.[2] *S. enteritidis* was isolated by Gaertner[3] and, in 1892, Loeffler isolated *S. typhimurium*.[4] Studies continued on the identification and classification of *Salmonella* culminating in the adoption of the terminology and antigenic schematization of Kauffmann and White. There are currently over 1,000 specific antigenic types delineated in the Kauffmann-White Schema.

Members of the genus *Salmonella* are widely distributed in the environment and, as enteric pathogens, cause a wide variety of diseases in man and animals.

They have been responsible for gastroenteritis, bacteremia, septicemia, and enteric fevers of which typhoid fever is the most notorious. The mode of transmission of *Salmonella* including the typhoid bacillus — via food, water, poultry and human carriers — makes eradication difficult and of great public concern. Serological identification aids in locating the source of the infection and thereby controlling or preventing its spread.

Before proceeding, as a matter of expediency in taxonomy, Ewing[5] recommends the use of a "three species concept" in *Salmonella* nomenclature. According to this concept, only 3 species of *Salmonella* are recognized. They are *Salmonella typhi, Salmonella choleraesuis,* and *Salmonella enteritidis.* As an example, *Salmonella typhimurium* as formerly known, would become *Salmonella enteritidis* ser Typhimurium, while *Salmonella pullorum,* which is an aberrant strain, would become *Salmonella enteritidis* bioser Pullorum.

Complete identification of *Salmonella* requires cultural isolation, biochemical characterization and serotyping. However well defined the serology of *Salmonella,* the use of serological procedures do not supersede cultural isolation and biochemical characterization. The intergeneric similarities of the *Enterobacteriaceae* as well as the antigenic relationships of its members dictate reliance on all the above factors for complete and final identification of *Salmonella.* Table I outlines the schema for complete identification of *Salmonella.* The significant biochemical reactions for this genus are also listed. Any serological results obtained prior to biochemical identification must be considered as presumptive identification only.

Edwards and Ewing[6] and Ewing[7] stressed that every laboratory regardless of size should be equipped to completely identify the most significant *Salmonella* namely, *Salmonella typhi, Salmonella enteritidis* bioser Paratyphi A, *Salmonella enteritidis* bioser Paratyphi B, *Salmonella choleraesuis* and *Salmonella enteritidis* ser Typhimurium.

Identification of these species and serotypes requires few serological reagents. Bacto Salmonella O Antiserum Poly A – I and Vi should be used to screen suspect specimens; followed by the use of Bacto Salmonella O Antisera Factors 2, 4 & 5, 7, and 9, Bacto Salmonella Vi Antiserum and Salmonella H Antisera a – d, i, 2, and 5, for serotyping and species identification.

Beyond the complete serological identification of the above species and serotypes of *Salmonella,* the extent to which *Salmonella* serology should be performed depends upon the kind of information the laboratory wishes to obtain and the facilities and amount of time which can be allowed for serological procedures.

Some laboratories may perform complete serological identification of the more commonly occurring *Salmonella* serotypes in addition to the significant serotypes mentioned previously. Reference laboratories and laboratories interested in epidemiology perform complete identification of all known serotypes. Difco offers antisera to satisfy the needs of most of these laboratories.

For those laboratories performing complete identification of the significant *Salmonella* as well as many of the more common serotypes, the assortment of Salmonella Antisera found in Bacto MinESS Antisera Set II is recommended. This assortment includes Bacto Salmonella O Antisera Poly A – I and Vi, Factors 2, 4 & 5, 7, 8, 9, Bacto Salmonella Vi Antiserum, Salmonella H Antisera a, b, c, d, eh, i, k, m, q, r, s, t, 2, and 5.

For reference laboratories or those interested in complete characterization and identification of *Salmonella,* the 7 polyvalent O antisera supplemented by sera containing O groups A – Z, 51 – 64 and Vi are recommended. For H antigen analysis the use of Salmonella H Antisera Polys a – z, A, B, C, D, and E supplemented by the individual H specific sera a – z_{52}, will identify all except a very few rare strains. In addition, single factor O antisera have been prepared which makes it possible to identify most somatic antigens listed in the Antigenic Schema for *Salmonella* (Kaufmann-White Schema, Modified).

It is recognized that not all clinical laboratories will be equipped to completely identify *Salmonella* beyond the 5 significant serotypes stated above, nor will the incidence of specimens justify such procedures. However, the world wide distribution of *Salmonella* points to the need for preliminary identification of at least the more frequently encountered types of this genus. Edwards and Ewing[6] and Ewing and Martin[8] have recognized the importance of preliminary serological identification of *Salmonella* and other significant members of the *Enterobacteriaceae*.

As outlined in Table I preliminary serological identification can be accomplished after differential biochemical tests have been performed. Cultures which have been presumptively identified as "*Salmonella*" by results obtained from reactions on triple sugar iron agar (or Kligler iron agar), lysine iron agar, MIO medium and urea agar should be examined serologically with Bacto Salmonella Polyvalent Antisera. Reactions on above media are listed in Table II. The polyvalent antisera are prepared for use as screening reagents. They are useful adjuncts when used in conjunction with differential media in the preliminary characterization of the *Salmonella*.

Kauffman described the rapid identification of *Salmonella* by the use of 5 polyvalent O and 5 polyvalent H antisera which reduces the time necessary for identification of a given species. More recently numerous isolates from various sources made it necessary to enlarge the antigenic schema. Therefore, to make available antisera to detect most organisms, the battery of 8 polyvalent O and 6 polyvalent H antisera just described have been prepared.

SALMONELLA ANTIGENIC ANALYSIS
Serological identification of *Salmonella* is based on the detection of specific antigenic components present. First, the organism is placed into "O" group according to its somatic composition and then confirmed by the identification of its flagellar "H" antigen.

Somatic "O" Antigen
The somatic antigen (O = ohne Hauch) is a heat-stable, polysaccharide associated with the body of the cell. It is the antigen first determined in *Salmonella* serology using the slide agglutination technique to group the organism.

Briefly, somatic "O" antigens have been assigned Arabic numbers, in consecutive order, from 1 – 64. Some somatic antigens were deleted from this schema for one reason or another, therefore there is not a complete continuity of the numbers.

For convenience, the *Salmonella,* using Arabic numeral system, are placed into "serogroups" depending upon their somatic content. Serogroups A through Z exist, the somatic serogroup Z organism possessing antigen No. 50. Since additional antigens were delineated and the alphabet exhausted, the next antigen (No. 51) is also called Group 51, etc.

Further, numerous organisms possess more than one antigen. Also a number of organisms contain antigens in common which will cause cross-reactions in an unabsorbed or "partially absorbed" antiserum. There is at least one somatic antigen which is shared in common by all members of a given serogroup, which identifies that group. Thus, serogroup A is represented by three members, *Salmonella enteritidis* bioser Paratyphi A, *Salmonella enteritidis* ser Kiel, both of which contain somatic antigens 1, 2, 12; while *Salmonella enteritidis* ser Nitra contains antigens 2, 12. As is noted all three members of this serogroup contain antigens 2 and 12 in common.

Serogroup B is represented by a fairly large number of organisms consisting of somatic antigens:

1, 4, 5, 12	4, 5, 12
1, 4, 12	4, 12
1, 4, 12, 27	4, 12, 27, etc.

while serogroup D organisms contain somatic antigens 1, 9, 12; 9, 12, etc.

In the above example, all three serogroups A, B, and D, contain in common the antigens 1 and 12. A serum prepared from a 1, 2, 12 culture, if not absorbed, will react with cultures of serogroups B and D in varying degrees depending upon the concentration of the commonly shared 1 and 12 factors. This must be taken into consideration when one chooses a serum to be used in the serological examination of the *Salmonella,* since several different sera are available (i.e., a serum is available which covers all three antigens represented in the group, while a single factor serum containing only the 2 factor is also prepared). If a serum is desired to be one that is specific for an identifiable antigen in a given serogroup, **a single factor serum should be used.** Such a single factor serum is not called a "group" serum, even though it contains the group identifiable agglutinin, since it has been recommended by CDC that the term "group" be applied only to those sera possessing all the major agglutinins found in that group.

It is also evident, in view of the fact that 1 and 12 antigens are found in common in some members of these three serogroups (1 and 9 being found in still higher O groups), that, in reality, only somatic antigen 2 is common for serogroup A organisms; somatic antigen 4 is the only common antigen for serogroup B organisms, while antigen 9 specifically identifies the serogroup D organisms.

Even though a single factor serum is recommended to be used for serogrouping of the *Salmonella* due to its specificity, one note of precaution must be made. A 1, 2, 12 serum absorbed of its 1 and 12 agglutinins — because of its specificity — might react weakly or might fail to agglutinate some strains of a 1, 2, 12 culture. A 1, 2, 12 culture will not at all times possess 1/3 antigenic component of somatic antigen No. 1; 1/3 antigen No. 2 and 1/3 antigen No. 12. These will vary in all combinations from time to time. Even isolated colonies from the same agar plate, derived from the same parent colony, might or will possess the three somatic antigens in varying proportions. One must take this into consideration in the selection of an antiserum — the sacrifice of sensitivity to specificity or vice versa. This is why it is recommended that several isolated colonies be checked in the slide agglutination procedure rather than just a single colony.

Flagellar "H" Antigen

The flagellar antigen (H = Hauch) is heat-labile. It is a protein located in the flagella of the organism. It is identified using a tube agglutination technique to determine the serotype of the organism. This antigen generally is associated with the motility of the organism although some nonmotile cultures are known to possess flagella.[5] In the main, most *Salmonella* have two different "H" antigens. A very few (*S. enteritidis* bioser Gallinarum and Pullorum) have none — they are nonmotile. Others possess three or even four "H" antigens.

Practically, a pure "H" serum cannot be prepared without containing some somatic composition; however, since "H" antigens are highly antigenic, the serum derived from motile cultures may be used at a dilution which eliminates somatic agglutination. The "H" antigens are designated by the lower case elements of the alphabet, a, b, c, d, etc.

There are several situations in which a complex antigen will share a common antigen with another (i.e., l;v, l,w; l,z_{13} − e,h; e,n; e,n,z_{15}; etc.) Also the Arabic numeral system has been used to identify some H antigens, particularly in the 1 complex system, 1,2; 1,5; 1,6; etc., all of which contain a common 1 H antigen.

The 2 different "H" antigens possessed by most *Salmonella* occur in phases. An organism may exist in one or all of its phases when isolated. For more detailed information, refer to *Identification of Enterobacteriaceae*.[5]

Occasionally, a third type antigen called the Vi antigen, is present. It is a heat-labile, envelope antigen, surrounding the cell wall which masks the somatic antigen rendering the organism inagglutinable in O sera.

The principle of the serological identification of *Salmonella* involves the intimate mixing of the organism (the antigen) with the immune serum (the antibody). If the serum contains agglutinins for the antigen present on the organism, a rapid and at least a 3+ clumping (agglutination) of the organism will occur. This is known as a homologous reaction. In some cases a reaction in the immune serum with another species of *Salmonella* might occur. This reaction is due to the presence of one or more commonly shared antigens between various species and is called a heterologous reaction. The homologous reaction is characterized by the rapidity and avidity of reaction, as compared to the heterologous reaction which is slow or weak. The agglutination of the somatic antigen in the rapid slide test appears as a firm, granular clumping; while that of the flagellar antigen (tube test) is a loose floccular reaction, in which the agglutination tends to be easily resuspended.

REAGENTS

Bacto Salmonella O, H and Vi Antisera are stable, desiccated, absorbed (when necessary), single factor or whole group antisera prepared in rabbits.

Bacto Salmonella O Antiserum Poly A − I and Vi has been prepared with representative members of those somatic groups and has not been absorbed. It is obvious that this serum may and will react with higher O groups of *Salmonella*.

Bacto Salmonella O Group Antisera contain all factors present in the named group and Bacto Salmonella O Factor Antisera contain only the factors indicative of the individual groups. When using whole group antisera, one must be aware of the presence of common antigens with other groups, which will cause cross-reactions. With the Salmonella O Factor Antisera these cross-reactions have been absorbed.

Bacto Salmonella O Antisera **Poly A** through **Poly G** should not be confused with Salmonella O Antisera **Groups** A through G. Bacto Salmonella O Antisera **Poly A** through G are polyvalent antisera containing a number of different groups to help to narrow the number of individual antisera which must be tested. The contents of the Salmonella O Antisera Poly A through G are as follows:

Salmonella Poly Groups	Somatic Groups Present
Bacto Salmonella O Antiserum Poly A	A, B, D, E_1, E_2, E_3, E_4, L
Bacto Salmonella O Antiserum Poly B	C_1, C_2, F, G, H
Bacto Salmonella O Antiserum Poly C	I, J, K, M, N, O
Bacto Salmonella O Antiserum Poly D	P, Q, R, S, T, U
Bacto Salmonella O Antiserum Poly E	V, W, X, Y, Z
Bacto Salmonella O Antiserum Poly F	51 – 55
Bacto Salmonella O Antiserum Poly G	56 – 61

When properly rehydrated, each antiserum contains in most cases approximately 1:10,000 Merthiolate®, sufficient to maintain a bacteriostatic condition at storage of 2 – 8°C.

The protein content of Bacto Salmonella O Antisera approximates that of a glycerinated serum and therefore should be considered as a 1:2 dilution. The somatic (O) sera at this dilution are to be considered at the RTD (Routine Test Dilution). This RTD is based on the state of the antiserum at the time of testing under our laboratory conditions, using antigenically defined and recognized reference cultures. Bacto Salmonella H Antisera must be diluted further for use. (These recommended dilutions are found under "Flagellar (H) Antigen Analysis" below.)

When used according to the suggested procedure each 3 ml vial of Bacto Salmonella O or Vi Antisera is sufficient to perform approximately 60 tests. Each 3 ml vial of Bacto Salmonella H Antisera will perform between 150 and 1500 tests, depending upon which sera are employed.

REHYDRATION
To rehydrate the Salmonella antisera add 3 ml of 0.85% NaCl solution and rotate gently to dissolve completely.

Check purity (bacterial) and pH of the saline used in rehydration if the antiserum rehydrates cloudy.

STORAGE
Store both desiccated and rehydrated antisera at 2 – 8°C.

Prolonged exposure to room temperature or repeated freezing and thawing are detrimental to antisera. **Do not expose rehydrated serum to room temperature for prolonged periods of time.**

EXPIRATION DATE
Both desiccated and rehydrated, but undiluted, Bacto Salmonella Antisera are stable through the expiration date on the label when stored as directed. Diluted Salmonella H Antisera should be discarded at the end of the day.

Discard any serum which is cloudy and/or has a precipitate unless it has been clarified and shown to react properly with known control cultures.

SPECIMEN COLLECTION

Fecal specimens should be collected early in the course of an enteric infection, preferably during the acute stage when pathogenic microorganisms are present in maximal numbers, and if possible prior to antibiotic therapy. Freshly passed stools are the specimen of choice for examination of the presence of enteric pathogens. Rectal swabs are acceptable when stool specimens cannot be obtained but these should not be expected to contain an optimal number of *Salmonella* organisms.

Although *S. typhi, S. enteritidis* bioser Paratyphi A and *S. enteritidis* ser Paratyphi B are the prime causes of enteric fever, almost any serotype of *Salmonella* can produce enteric fever. For this reason a blood culture should also be considered, particularly in the first week of the disease.

SPECIMEN PREPARATION

Again, it is emphasized that biochemical testing must also be employed to characterize and confirm the suspected pathogen as a member of the genus *Salmonella* prior to applying serological techniques.

Suspect fecal specimens should be immediately plated to appropriate media or enrichments as soon as possible. When specimens cannot be cultured soon after collection they should be placed in a transport or preservative medium until proper cultural methods can be undertaken. Table I shows a schema for the examination of suspect specimens.

It should be remembered that some strains of *Salmonella* may not grow well on highly selective media and particularly that *S. typhi* does not grow on brilliant green agar. Therefore, both highly selective and slightly selective differential plating media should be employed in cultural procedures for optimal recovery of *Salmonella*.

Table I

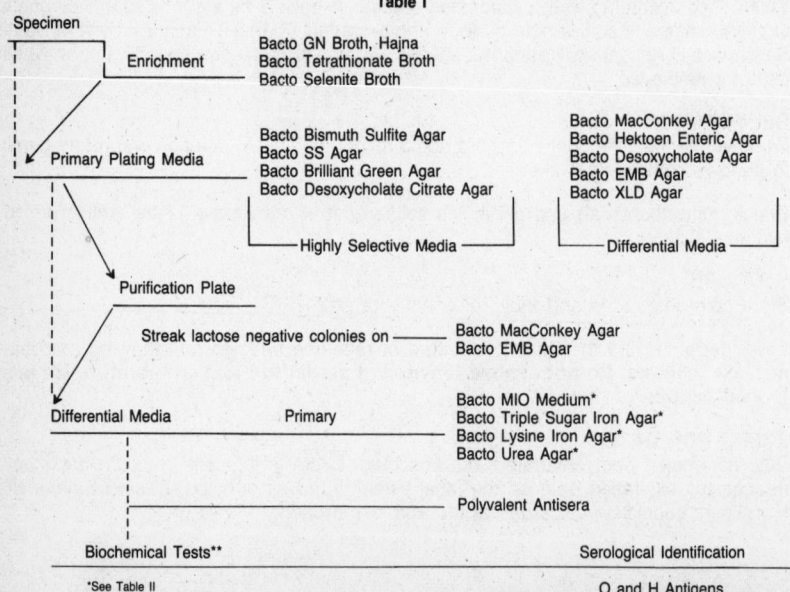

Table II Differential Tests for Salmonella

Medium	Salmonella
Triple Sugar Iron Agar	
Slant	K
Butt	A
Gas	+(−)
H₂S	+
Lysine Iron Agar	
Slant	K
Butt (Decarboxylase)	K or N
Gas	−
H₂S	+(−)
MIO Medium	
Motility	+
Indole	−
Ornithine	+
Urea Agar*	
Urease	−

KEY

*The urease reaction insures differentiation of *Proteus* (which is urease positive) from *Salmonella*

( ) Signs in parenthesis indicate occasional reactions.

K = Alkaline
A = Acid
N = Neutral

Table III Additional Biochemical Reactions for Identification of Salmonella

dulcitol	d	alginate	−	
salicin	−	Simmons citrate	d	
adonitol	−	phenylalanine	−	d = delayed
inositol	d	sodium malonate	−	(+) = variable
raffinose	−	KCN	−	(−) = S. enteritidis bioser
erythritol	−	arginine	+ or (+)	Paratyphi A, S. choleraesuis
esculin	−	ornithine	+	(diphasic), some cultures of
urease	−	lysine	+	S. typhi, S. enteritidis bioser
MR	+	VP	−	Typhisuis, S. enteritidis bioser
nitrate reduction	+	indole	−	Sendai, S. enteritidis ser Berta
glucose	+	gelatin	−	and a few rare types fail to
lactose	−	H₂S	+ (−)	produce H₂S (TSI).
sucrose	−			
oxidase	−			

There exists in the genus *Salmonella* some species which do not conform completely to all the biochemical tests ascribed to that genus. The "aberrant" strains occur occasionally, differing slightly from the general schema. The rationale as to placement of a given organism within a given genus, specifically *Salmonella,* depends upon the sum total of all the organisms biochemical characteristics not just in 1 or 2 reactions.

PROCEDURE

Depending upon the Salmonella O Antisera selected, follow Schema I or Schema II for the complete preliminary or final serological identification of a given isolate.

SCHEMA I

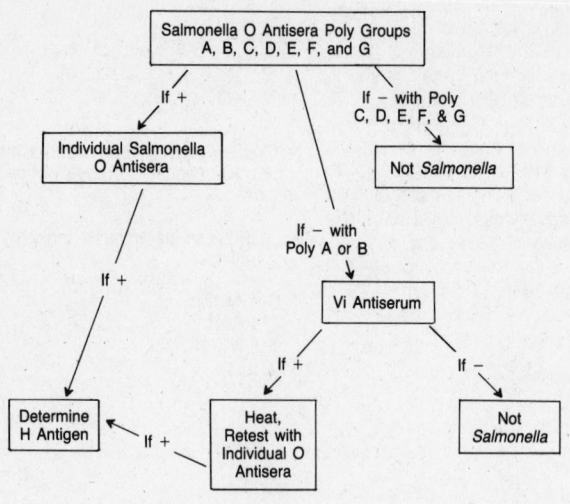

SCHEMA II

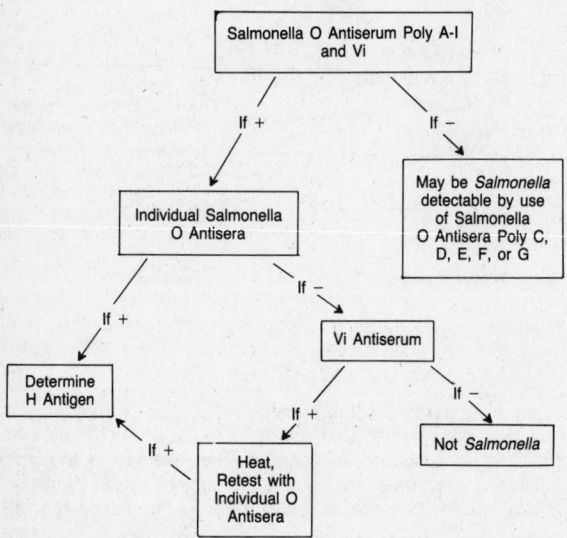

GENERAL INSTRUCTIONS

1. In the slide test, all materials and equipment must be at room temperature at the time of test performance.
2. In the tube test make certain the proper dilution is prepared for a given antiserum. Various dilutions are used for various sera. This information is given below under

"H" analysis. Also it is important in this test to use the recommended time and temperature of incubation and to make certain the water bath is in a location free of mechanical vibration.

3. Exposure of the organism or plate to heat from external sources (a hot bacteriological loop, burner flame, light source, etc.) may result in either a culture which cannot be suspended readily or evaporation and/or precipitation of the test mixture which may result in false-positive reactions.

4. The test culture must be checked in a saline control (both in the tube and slide technique) for smoothness, or preferably in normal rabbit serum in the slide test. Often stock cultures and sometimes isolated cultures are rough and will agglutinate spontaneously in a normal serum. Therefore, it is necessary to select smooth colonies for serological testing. It is recommended that more than one colony be tested from both assay and control cultures.

5. In the slide test, all equipment and materials (glass plates, loop, culture) must be sterilized after use by flame (loop) or autoclave (glass plate and culture) or by an adequate chemical method since living organisms are used as the antigen.

Materials Provided:
Bacto Salmonella O Antisera
Bacto Salmonella H Antisera
Bacto Salmonella Vi Antiserum

Materials Required but not Provided:

Slide Test
0.85% NaCl solution
Bacteriological loop
Burner
Glass plate ruled into 1″ squares
Fluorescent desk lamp

Tube Test
0.85% NaCl solution
Kahn type tube racks
Culture tubes w/o lip (12 × 75 mm)
1 ml serological pipettes
50°C water bath
Timer

SOMATIC O ANTIGEN ANALYSIS

Cultures of suspected *Salmonella* (as indicated by results from differential tests outlined in Table II) should be subjected to serological procedures using Bacto Salmonella O Antisera Poly A – I and Vi (Schema II) or Bacto Salmonella O Antisera Poly A, B, C, D, E, F and G (Schema I).

For preliminary serological identification growth may be taken directly from the triple sugar iron agar slant and used in the procedure outlined below. Results of these serological procedures must be considered tentative. Confirmatory biochemical tests should be made and the H antigens of such cultures must be determined before final reports can be made. If the tentative serological results deviate from what might be expected from the confirmatory biochemical reactions the antisera used for the preliminary identification should be rerun using the procedure outlined below.

When it is necessary to perform final serological identification, streak colonies which produce an alkaline slant, acid butt and blacken Bacto Triple Sugar Iron Agar (except those strains stated in Table III) on Bacto Veal Infusion Agar plates in such a manner as to obtain isolated colonies. Incubate for 18 – 24 hours at 37°C.

The following procedure may be used for preliminary and final identification of *Salmonella*.
1. Mark off a glass plate into 1″ square sections with a wax pencil.
2. Place a small drop of the appropriate Bacto Salmonella O Antiserum on the ruled section of the plate using the supplied dropper.

3. To the square next to the one containing the antiserum, place one drop of 0.85% NaCl solution or a drop of normal rabbit serum. This will serve as a negative control of the bacterial suspension.
4. Transfer a portion of a loopful of growth from the appropriate medium to the section containing the NaCl solution or normal rabbit serum, and suspend thoroughly.
5. Similarly, transfer a portion of a loopful of growth to the square containing the antiserum and suspend thoroughly.

Note: It is recommended that more than one colony be tested. The antisera have not been tested employing antigen suspension in saline or alcohol-treated cultures. If variations in the recommended procedures are to be used, the investigator is advised to test each lot of antiserum with known positive control cultures to insure its proper homologous and heterologous reactions under their test conditions.

6. Rock the slide for 1 minute and avoid excessive evaporation.
7. Record agglutination as follows:

$++++$ agglutination = all of the cells agglutinate
$+++$ agglutination = 75% of the cells agglutinate
$++$ agglutination = 50% of the cells agglutinate
$+$ agglutination = 25% of the cells agglutinate
$\pm$ agglutination = <25% of the cells agglutinate
$-$ agglutination = none of the cells agglutinate

INTERPRETATION OF RESULTS

1. A $+++$ reaction or greater should be considered as the end-point at the RTD.
2. Occasionally clinical isolates of *Salmonella* will possess M agglutinins which are O inagglutinable. Ewing and Martin[8] have reported that suspensions prepared from cultures of this type should be heated in a boiling water bath for 15 minutes, cooled, and retested with O antisera.
3. If positive agglutination occurs, proceed further to identify the group to which the organism belongs by using the desired individual Salmonella O Antiserum groups in the same manner as described above. In general serogroup B organisms are found most frequently followed by serogroup D and serogroup C_1, respectively. Using the somatic antisera for these groups first will usually be the most efficient procedure when Salmonella Poly A – I and Vi, Poly A or Poly B are positive initially.
4. When Salmonella O Poly Groups are employed in the above procedure and a negative reaction is obtained with Poly C, D, E, F and G the organism can be considered presumptively negative for *Salmonella*. Biochemical tests should be performed to confirm this negative result. If a negative result is obtained with Poly A or Poly B, it should be tested with Bacto Salmonella Vi Antiserum by the above procedure.
5. When Salmonella O Antiserum Poly A – I and Vi is employed in the above procedure and a negative reaction is obtained, the organisms can only be considered presumptively negative for *Salmonella* belonging to serogroups A – I. Biochemical tests should be performed to confirm this negative result. If biochemical tests prove the organism to be a *Salmonella,* a serogroup beyond serogroup A – I is probably involved.

If the organism reacts with Poly A – I and Vi but does not react with the specific somatic antisera, it should be checked with Bacto Salmonella Vi Antiserum by the procedure mentioned previously.

Note: One must be aware of the existance of common antigens between various "O" serogroups of the *Salmonella*. As an example, Salmonella O Antiserum Poly A contains — among others — agglutinins for factor 1, since cultures possessing a factor 1 was used in immunization. It may be expected, therefore, that this polyvalent antiserum will react with cultures other than those contained in "O" serogroups A, B, D, E, and L due to the common 1 antigen (those organisms in groups G_1, G_2, H, R, T, etc., which contain factor 1).

6. If no agglutination is observed in the test for Vi antigen the culture may be regarded as not of the *Salmonella* genus. This must be confirmed biochemically. If the culture reacts with the Salmonella Vi Antiserum, a dense suspension of the culture prepared by suspending a loopful of growth (from TSI or veal infusion agar as applicable) in 0.5 ml 0.85% NaCl is heated in a boiling water bath for 10 minutes and cooled. After cooling, the heated culture should be retested with Salmonella O Antiserum Factors 9 and 7 and Salmonella Vi Antiserum.

If the heated organism does not react with the Vi antiserum, but reacts with Bacto Salmonella O Antiserum Factor 9, it is most likely *Salmonella typhi* and should be confirmed using Bacto Salmonella H Antiserum d and a motile suspension of the culture.

If the organism does not react with the Vi or factor 9 antisera after heating, but reacts with Bacto Salmonella O Antiserum Factor 7, it is most likely *Salmonella enteritidis* ser Paratyphi C and should be confirmed using Bacto Salmonella H Antiserum c for Phase 1 and Bacto Salmonella H Antiserum 5 for Phase 2.

7. If the heated culture continues to react with Vi antiserum and does not react with the Salmonella O Antisera it is most likely a member of the *Citrobacter*.

This result must be confirmed by biochemical testing with particular attention given to lysine decarboxylase and KCN reactions.

Organism	KCN	Lysine Decarboxylase Moeller Method
Salmonella	−	+
Citrobacter	+	−

Organisms giving positive agglutination with Salmonella O Antiserum Groups should be analyzed further for their antigens using the appropriate Bacto Salmonella H Antisera.

Note: Refer to the Kauffmann-White Schema to determine which H antigens are associated with the somatic antigen(s) in question.

FLAGELLAR H ANTIGEN ANALYSIS
Bacto Salmonella H Antisera (flagellar) were prepared and assayed for use in the tube agglutination technique. The slide technique is not recommended for H antigen analysis since cross-reactions may occur with somatic agglutinins. No attempt has been made to absorb or to test for O agglutinins in H sera.

TUBE TEST
For final identification of the *Salmonella* serotypes within a group as determined by the Salmonella O Antisera it is necessary to determine the H antigens and the phase of the organism. The tube test of Edwards and Bruner[9] is recommended. It is necessary to have a motile organism when testing for H antigens. Usually broth cultures of fresh isolates are satisfactory for use as antigens for this purpose.

Occasionally, it is necessary to increase the motility of the test organisms. This is accomplished by making several consecutive transfers in Bacto Motility GI Medium. Inoculate the tube slightly below the surface of the medium by the stab method. Incubate the tubes at 37°C for 18 – 20 hours. Transfer only those organisms that have migrated to the bottom of the tube when making successive cultures. This may be accomplished by sterilizing a hollow glass tube, approximately 12 × 120 mm, with a rubber stopper in the bottom and cotton plug in the top, aseptically dispensing in sterile Bacto Motility GI Medium, cooling, then inoculating. After several transfers if the motility of the culture is such that they travel 50 – 60 mm through the medium in 18 – 20 hours, it is ready for use.

1. Inoculate a Bacto Veal Infusion Broth tube with the motile organism from the last transfer to motility medium and incubate at 37°C overnight.
2. Inactivate the culture using equal volumes of 0.6% formalized physiological saline solution (6 ml formalin + 8.5 g NaCl in 1 liter distilled water). This is the antigen to be determined.

Note: It is recommended that the final concentration of the test antigen approximate the density of Bacto McFarland Barium Sulfate Standard No. 2 or No. 3.

3. The dilutions recommended using Bacto Salmonella H Antisera differ depending upon which sera are to be employed. Generally, a 1:1,000 final serum dilution is prepared when using the majority of the H sera. This is done by diluting the rehydrated antiserum in a ratio of 0.1 ml antiserum to 25 ml of 0.85% NaCl solution. A few of the specific single factor sera must be used at a 1:500 dilution since extensive adsorption is necessary to render them specific resulting in a reduced titer of the serum. The 1:500 dilution is recommended when using Bacto Salmonella H Antisera x, z_{13}, z_{15}, and z_{28}. To prepare a 1:500 dilution add 0.1 ml of the rehydrated antiserum to 12.5 ml of 0.85% NaCl solution. When using Bacto Salmonella H Antiserum Poly a – z use a dilution of 1:100. To obtain this dilution add 1 part of the rehydrated polyvalent antiserum to 25 parts of 0.85% NaCl solution. Bacto Salmonella H Antisera Poly A, B, C, D, E, and F, however, are used at 1:1,000 dilution as prepared above.

It is recommended that only that amount of diluted Salmonella H sera be prepared which is to be used in any given day. All excess should be discarded.

4. Add 0.5 ml of the appropriate serum dilution to Kahn type serological tubes.
5. Add 0.5 ml of the antigen and incubate in a water bath at 50°C for 1 hour.
6. Read for presence or absence of agglutination and record.

Often it is necessary to isolate the second phase of a diphasic organism for complete serological identification. This is done by using Bacto Motility GI Medium either in tubes or Petri dishes according to the following procedure:

1. Prepare a 1% solution of Bacto Sodium Thioglycollate in 0.85% NaCl solution and sterilize by autoclaving.
2. Add an equal amount of the sterilized thioglycollate solution to the antiserum opposite of the phase desired.
3. Prepare a 1:10 dilution of the antiserum sodium thioglycollate mixture using 0.85% NaCl as diluent employing aseptic procedures.
4. Sterilize the serum dilution by passing through a "Swinny" filter using aseptic techniques and procedures.
5. Add 1 ml of the 1:10 serum dilution to 25 ml of sterile motility medium and mix well. Add the medium containing the serum to a sterile tube or Petri dish.
6. Inoculate the medium either slightly below the surface of the medium (if a tube is used) or on one side of the Petri dish. Incubate at 37°C overnight to permit the organism to migrate to the bottom of the tube or to the opposite side of the Petri dish. Several transfers in this medium containing serum might be necessary to change the phase.

Note: Since O agglutination may be present in H sera and since heterologous H agglutinins may interfere — due to the fact that the H serum is used in comparatively heavy concentrations in phase change — it might be necessary to vary (increase or decrease) the amount of flagellar antiserum. This might be determined:
a. If migration is interfered with (serum too concentrated)
b. If the migrated organisms still occur in the original phase (serum too dilute).

SPICER-EDWARDS RAPID H ANTIGEN IDENTIFICATION TECHNIQUE

Spicer[10] described a simplified rapid method for screening and identifying the most commonly encountered *Salmonella*. He used 4 main polyvalent Salmonella H Antisera and 2 adjunctive antisera for identifying more than 17 of the more commonly occurring H antigens. Edwards modified the Spicer antisera by omitting the L Complex from the main antisera to minimize cross-reactions and suggested it be used independently. He also modified the antisera components so that no single H antigen reacted with all 4 Salmonella H Antisera Spicer-Edwards.

1. For preparation of the antigen see directions under FLAGELLAR H ANTIGEN ANALYSIS.
2. Dilute the rehydrated antisera in the ratio of 0.1 ml antiserum to 25 ml 0.85% NaCl solution. It is recommended that only that amount of diluted Salmonella H Antisera be prepared which is to be used in any given day. All excess should be discarded.
3. Add 0.5 ml of the required groups of diluted Salmonella H Antisera to Kahn type serological tubes. Add 0.5 ml of the test organism suspension and incubate in a water bath at 50°C for 1 hour. Read for presence or absence of agglutination and record.

The reactions of the Salmonella H Antisera Spicer-Edwards with H antigens are tabulated:

H Antigens	Salmonella H Antisera Spicer-Edwards				H Antigens	Salmonella H Antisera Spicer-Edwards			
	1	2	3	4		1	2	3	4
a	+	+	+	−	k	−	+	+	+
b	+	+	−	+	r	−	+	−	+
c	+	+	−	−	y	−	+	−	−
d	+	−	+	+	z	−	−	+	+
e,h	+	−	+	−	Z_4 Complex**	−	−	+	−
G Complex*	+	−	−	+	z_{10}	−	−	−	+
i	+	−	−	−	z_{29}	−	+	+	−

H Antigens	Salmonella H Antisera
e, n, x; e, n, z_{15};	EN Complex
l, v; l, w; l, z_{13}; l, z_{28};	L Complex
1, 2; 1, 5; 1, 6; 1, 7	1 Complex

*The G Complex component of Salmonella H Antisera Spicer-Edwards 1 and 4 reacts with antigens f,g; g,g,s; f,g,t; g,m; g,m,q; g,m,s; g,m,s,t; g,m,t; g,p; g,p,s; g,p,u; g,q; g,s,t; g,t; m,p,t,u, and m,t.
**The Z_4 Complex component reacts with z_4, z_{23}; z_4, z_{24} and z_4, z_{32}.

It may be noted in the 4 Bacto Spicer-Edwards H Antisera no antigen is positive with all 4 sera. Any antigen thus reacting should be checked for smoothness.

FINAL RESULTS

A serological reaction pattern of a biochemically confirmed *Salmonella* should now be available. For example, an organism that reacted in Poly A − I and Vi or Poly A, Somatic Factors 4, 5, Flagellar b (Phase 1) and Flagellar 2 (Phase 2) would be the very common *S. enteritidis* ser Typhimurium as shown in the Kauffmann-White Schema. Notice that somatic single factor 5 is required if it is deemed necessary to distinguish between ser Typhimurium and its variant Copenhagen.

LIMITATIONS OF THE PROCEDURE

While serological procedures, as applied to microorganisms, provide corroborative evidence and can be used to identify particular antigenic sites of genus or species under study, they cannot be used alone to identify the etiological agent of disease. Cultural isolation and at least preliminary biochemical differentiation must preceed any serological examination, and final identification cannot be made without biochemical and serological characterization.

This is particularly true of members of *Enterobacteriaceae*. Members of this family are antigenically related to *Citrobacter* and *Arizona* (now *Salmonella arizonae*). Antigens of these microorganisms are identical with, or similar to, each other. Further, species within the *Salmonella* may have similar or identical antigenic formulae. Following are a few of these antigenic relationships which could cause cross-reactions when serological procedures are employed.

Arizona O 1,2	= *Salmonella* O 51	*Citrobacter* O 42	≠ *Salmonella* O 54	
Arizona O 1,3	= *Salmonella* O 44	*S. choleraesuis*	= *S. enteritidis*	
Arizona 7a, 7b	= *Salmonella* O 18	6,7:c:1,5	bioser Decatur	
Arizona H1	= *Salmonella* H z_4		6,7:c:1,5	
Arizona H 25	= *Salmonella* H z_{53}	*S. choleraesuis*	= *S. enteritidis*	
Arizona H 39	= *Salmonella* H z	bioser Kunzendorf	bioser Typhisuis	
Citrobacter O 37	≠ *Salmonella* O 48	6,7:[c]:1,5	6,7:[c]:1,5	
Citrobacter O 40	≠ *Salmonella* O 57			

= Identical　　≠ Similar

For further explanation regarding antigenic relationships refer to *Identification of Enterobacteriaceae*.[5]

QUALITY CONTROL

In *Salmonella* serology, as in any serological test, known positive and negative control cultures should be employed to ascertain validity of test results. For complete discussion refer to Salmonella QC Antigens.

REFERENCES

1. Gaffky, G., Mitt. a.d. Kaiserl. Gsndhtsamte., 2:372, 1884.
2. Salmon, D. E., and Smith, T., U.S. Bur. Animal Industry, 2nd Ann. Rep., p. 184, 1885.
3. Gaertner. Korresp. Be. artze. ver Thuringen, 17:233, 573, 1888.
4. Loeffler, F., Zentralbl. f. Bakt., 11:129, 1892.
5. Edwards, P. R. and Ewing, W. H., Identification of *Enterobacteriaceae*, 3rd Ed., p. 146, 1972.
6. ibid.
7. Bodily, H. L., Updyke, E. L., and Mason, J. O., Diagnostic Procedures, 5th Ed., APHA, Ch. 11, 1970.
8. Lennette, Spaulding, and Truant, Manual of Clinical Microbiology, 2nd Ed., ASM Ch. 18, 1974.
9. Edwards and Bruner, Am. J. Hyg., 45:19, 1947.
10. Spicer, C. C., J. Clin. Path., 9:378, 1956.

PACKAGING

Bacto Salmonella H Antiserum　　　　　　　　　　　　　　　　　3 ml

Antiserum a	2820-47-9
Antiserum b	2821-47-8
Antiserum c	2822-47-7
Antiserum d	2823-47-6
Antiserum eh	2273-47-1
Antiserum i	2824-47-5
Antiserum k	2274-47-0
Antiserum r	2275-47-9
Antiserum y	2276-47-8
Antiserum z	2277-47-7
Antiserum z_6	2473-47-9
Antiserum z_{10}	2279-47-5
Antiserum z_{27}	2560-47-3

Antiserum z_{29}	2280-47-2
Antiserum EN Complex	2270-47-4
Antiserum G Complex	2269-47-7
Antiserum L Complex	2271-47-3
Antiserum Z_4 Complex	2278-47-6
Antiserum 1 Complex	2272-47-2
Antiserum Poly a – z	2406-47-1
Antiserum Poly A (a,b,c,d,i,z_{10},z_{29})	2539-47-1
Antiserum Poly B (e,h;e,n;e,n,x;e,n,z_{15};G)	2540-47-8
Antiserum Poly C (k,l,r,y,z,z_4)	2541-47-7
Antiserum Poly D (z_{35},z_{36},z_{37},z_{38},z_{39},z_{41},z_{42})	2542-47-6
Antiserum Poly E (l,z_6)	2543-47-5
Antiserum single factor f	2544-47-4
Antiserum single factor h	2545-47-3
Antiserum single factor m	2546-47-2
Antiserum single factor p	2548-47-0
Antiserum single factor s	2550-47-5
Antiserum single factor t	2551-47-4
Antiserum single factor u	2552-47-3
Antiserum single factor w	2554-47-1
Antiserum single factor x	2555-47-0
Antiserum single factor z_{13}	2556-47-9
Antiserum single factor z_{15}	2557-47-8
Antiserum single factor z_{23}	2558-47-7
Antiserum single factor z_{28}	2561-47-2
Antiserum single factor z_{32}	2562-47-1
Antiserum single factor 2	2474-47-8
Antiserum single factor 5	2475-47-7
Antiserum single factor 6	2476-47-6
Antiserum single factor 7	2477-47-5
Antiserum Spicer-Edwards 1	2265-47-1
Antiserum Spicer-Edwards 2	2266-47-0
Antiserum Spicer-Edwards 3	2267-47-9
Antiserum Spicer-Edwards 4	2268-47-8
Bacto Salmonella O Antiserum	3 ml
Antiserum Factor 1	2658-47-6
Antiserum Factor 2 (Group A)†	2814-47-7
Antiserum Factor 4 (Group B)†	2659-47-5
Antiserum Factor 5 (Group B)†	2660-47-2
Antiserum Factor 4,5 (Group B)†	2815-47-6
Antiserum Factor 7 (Group C_1)†	2816-47-5
Antiserum Factor 8 (Group C_2)†	2817-47-4
Antiserum Factor 9 (Group D)†	2818-47-3
Antiserum Factor 10	2257-47-1
Antiserum Factor 12	2779-47-0
Antiserum Factor 14	2661-47-1
Antiserum Factor 15	2258-47-0
Antiserum Factor 19	2259-47-9
Antiserum Factor 20 (Group C_3)†	2662-47-0
Antiserum Factor 22	2663-47-9
Antiserum Factor 23	2664-47-8
Antiserum Factor 25	2666-47-6

Antiserum Factor 27	2667-47-5
Antiserum Factor 34	2512-47-2
Antiserum Group A Factors 1,2,12	2947-47-7
Antiserum Group B Factors 1,4,5,12	2948-47-6
Antiserum Group B Factors 1,4,12,27	2973-47-4
Antiserum Group C_1 Factors 6,7	2949-47-5
Antiserum Group C_2 Factors 6,8	2950-47-1
Antiserum Group C_3 Factors (8),20	3016-47-1
Antiserum Group D_1 Factors 1,9,12	2951-47-0
Antiserum Group D_2 Factors (9),46	3017-47-0
Antiserum Group E Factors 1,3,10,15,19,34	2819-47-2
Antiserum Group E_1 Factors 3,10	2952-47-9
Antiserum Group E_2 Factors 3,15	2954-47-7
Antiserum Group E_3 Factors (3),(15),34	3018-47-9
Antiserum Group E_4 Factors 1,3,19	3019-47-8
Antiserum Group F Factor 11	2260-47-6
Antiserum Group G Factors 13,22,23,(36),(37)	3029-47-6
Antiserum Group G_1 Factors 13,22,(36)	2261-47-5
Antiserum Group G_2 Factors 1,13,23,(37)	3020-47-5
Antiserum Group H Factors 1,6,14,24,25	2262-47-4
Antiserum Group I Factor 16	2263-47-3
Antiserum Group K Factor 18	2518-47-6
Antiserum Group L Factor 21	2519-47-5
Antiserum Group N Factor 30	2521-47-1
Poly A – I and Vi Factors 1 – 16,19,22 – 25,34 and Vi	2264-47-2
Poly A (somatic groups A,B,D,E_1,E_2,E_3,E_4, & L)+	2534-47-6
Poly B (somatic groups C_1,C_2,F,G & H)+	2535-47-5
Poly C (somatic groups I,J,K,M,N & O)+	2536-47-4
Poly D (somatic groups P,Q,R,S,T & U)+	2537-47-3
Poly E (somatic groups V,W,X,Y & Z)+	2538-47-2
Poly F (somatic groups 51 – 55)+	2645-47-2
Poly G (somatic groups 56 – 61)+	2646-47-1
Vi Antiserum	2827-47-2

†**Special note:** As will be noted there has been a change in some of the product names of Bacto Salmonella O Antisera.

It has been recommended by CDC that the term "Group" be applied only to those sera possessing all the major agglutinins found in that group, i.e., Salmonella O Antiserum Group A Factors 1, 2, 12, as opposed to Salmonella Antiserum Factor 2. The latter is a single factor serum and does not contain all of the major factors of somatic group A organisms (1, 2, 12).

+These polyvalent sera do not contain Vi agglutinins. Salmonella Vi Antiserum should be used in all negative reactions or when *Salmonella enteritidis* bioser Paratyphi C or *Salmonella typhi* is suspected.

Bacto Salmonella O Antisera Set A – I	14 × 3 ml	2892-32-9
Contains Factors 2, 4 & 5, 7, 8, 9, 10, 15, 19;		
Group F Factor 11; G Factors 13, 22, (36);		
H Factors 1, 6, 14, 24, 25; I Factor 16;		
Poly A – I and Vi; and Salmonella Vi		
Bacto Salmonella O Antisera Set A	10 × 3 ml	2897-32-4
Contains Factors 2, 4 & 5, 9, 10, 15, 19, 34;		
Group L Factor 21; Poly A and Salmonella Vi		
Bacto Salmonella O Antisera Set B	6 × 3 ml	2898-32-3
Contains Factors 7, 8; Group F Factor 11;		
Group G Factors 13, 22, 23, (36), (37);		
H Factors 1, 6, 14, 24, 25; and Poly B		

Bacto Salmonella O Antisera Set C   7 × 3 ml   2899-32-2
Contains Salmonella O Antisera Groups
I, J, K, M, N, O and Poly C
Bacto Salmonella H Antisera Set a – z   19 × 3 ml   2893-32-8
Contains Groups a, b, c, d, eh, i, k, r, y,
z_6, z_{10}, z_{29} and Complexes 1, EN, G, L,
Z_4 and Poly a – z
Bacto Salmonella H Antisera Spicer-Edwards Set   7 × 3 ml   2328-32-3
Contains Complexes 1, EN and L, and
Spicer-Edwards 1, 2, 3, and 4
Bacto MinESS Antisera Set II   29 × 3 ml   2975-32-9
Contains Salmonella O Antisera Factors
2, 4 & 5, 7, 8, 9, Poly A – I and Vi; Salmonella
Vi Antiserum; Salmonella H Antisera a, b,
c, d, eh, i, k, m, q, r, s, t, 2, 5; Shigella
Antisera Poly Groups A, A_1, B, C, C_1, C_2,
D, and Alkalescens-Dispar Poly

ANTIGENIC SCHEMA FOR *SALMONELLA* (KAUFFMANN-WHITE SCHEMA, MODIFIED)

The antigenic schema that follows is the modification of the Kauffmann-White Schema[1] as proposed by Ewing.[2] It is to be used in conjuction with Difco Salmonella Antisera as an aid in the serological identification of *Salmonella*.

Subgenus III *S. arizonae* (formerly *Arizona* group) is not included in the outline. Refer to *Bergey's Manual of Determinative Bacteriology*, 8th Edition, for an expanded Kauffmann-White Schema which includes subgenus III organisms.

Ewing recommends the use of a "three species concept" in *Salmonella* nomenclature. According to this concept, only 3 species of *Salmonella* are recognized. They are *Salmonella typhi*, *Salmonella choleraesuis*, and *Salmonella enteritidis*. The species *S. enteritidis* includes all the salmonellae other than *S. typhi* and *S. choleraesuis*. As an example, *Salmonella typhimurium* as formerly known, would become *Salmonella enteritidis* ser Typhimurium while *Salmonella pullorum*, which is a biochemically aberrant strain, would become *Salmonella enteritidis* bioser Pullorum.

Because there is still no universal agreement on the nomenclature of *Salmonella* a laboratory may wish to use the schema as outlined here for antigenic analysis only and continue to report the organisms identity in the nomenclature as outlined in the unmodified Kauffmann-White Schema. The Ewing nomenclature is easily converted by dropping *enteritidis* from *Salmonella enteritidis* and adding the Roman lettered "species" name after it has been converted to italics using accepted rules of nomenclature. Variants of serotypes or bioserotypes only appear in the modified schema and are not convertible.

Ewing Nomenclature	Examples Serogroup	Unmodified Kauffmann-White Nomenclature
Salmonella enteritidis ser Newport	C_2	*Salmonella newport*
Salmonella enteritidis ser Enteritidis	D_1	*Salmonella enteritidis*
Salmonella enteritidis bioser Sendai	D_1	*Salmonella sendai*
Salmonella enteritidis ser Manila	E_2	*Salmonella manila*

ORGANISM	SOMATIC (O) ANTIGEN	FLAGELLAR (H) ANTIGEN Phase 1	Phase 2
Salmonella enteritidis			
bioser Paratyphi-A	Serogroup A 1,2,12	a	—
variant Durazzo	2,12	a	—
ser Nitra	2,12	g,m	—
ser Kiel	1,2,12	g,p	—
S. enteritidis	Serogroup B		
*ser Makoma	4,12	a	—
ser Kisangani	1,4,5,12	a	1,2
ser Hessarek	4,12,[27]	a	1,5
ser Fulica	4,5,12	a	1,5
ser Arechavaleta	4,5,12	a	[1,7]
ser Bispebjerg	1,4,5,12	a	e,n,x
ser Abortusequi	4,12	—	e,n,x
ser Tinda	1,4,12,27	a	e,n,z_{15}
ser Nakura	1,4,12,27	a	z_6
ser Paratyphi-B	1,4,5,12	b	1,2
variant Odense	1,4,12	b	1,2
bioser Java	1,4,5,12	b	[1,2]
*ser Sofia	4,12,[27]	b	[e,n,x]
ser Limete	1,4,12,27	b	1,5
ser Canada	4,12	b	1,6
ser Uppsala	4,12,27	b	1,7
ser Schleissheim	4,12,27	b,z_{12}	—
ser Abony	1,4,5,12	b	e,n,x
variant Haifa	4,12	b	e,n,x
ser Abortusbovis	1,4,12,27	b	e,n,x
ser Sladun	1,4,12,27	b	e,n,x
ser Wagenia	1,4,12,27	b	e,n,z_{15}
ser Wien	1,4,12,[27]	b	l,w
ser Abortuscanis	4,5,12	b	z_5
ser Legon	[1],4,12,[27]	c	1,5
ser Abortusovis	4,12	c	1,6
ser Altendorf	4,12	c	1,7
ser Womba	4,12,27	c	1,7
ser Jericho	1,4,12,27	c	e,n,z_{15}
ser Bury	4,12,27	c	z_6
ser Stanley	4,5,12	d	1,2
ser Cairo	1,4,12,27	d	1,2
ser Eppendorf	[1],4,12,[27]	d	1,5
ser Schwarzengrund	1,4,12,27	d	1,7
*ser Kluetjenfelde	4,12	d	e,n,x
ser Sarajane	4,12,27	d	e,n,x
ser Duisburg	[1],4,12,[27]	d	e,n,z_{15}
ser Mons	1,4,12,[27]	d	l,w
ser Ayinde	4,12,27	d	z_6
ser Salinatis	4,12	d,e,h	d,e,n,z_{15}
ser Saintpaul	1,4,[5],12	e,h	1,2
ser Reading	4,[5],12	e,h	1,5
ser Kaapstad	4,12	e,h	1,7
ser Chester	4,5,12	e,h	e,n,x
ser Sandiego	4,5,12	e,h	e,n,z_{15}
*ser Makumira	[1],4,12,[27]	e,n,x	1,7
*ser Derby	4,12	(f),g	[1,2]
ser Agona	1,4,5,12	f,g,s	—
ser Essen	4,12	f,g,s	—
*ser Caledon	4,12	g,m	e,n,x
ser Hato	4,5,12	g,m,s	—
ser Joenkoeping	4,5,12	g,s,t	—
ser Kingston	[1],4,12,[27]	g,s,t	—
variant Copenhagen	4,12	g,s,t	—
ser Budapest	1,4,12	g,t	—
*ser Bechuana	4,12,27	g,t	—
ser Travis	4,5,12	g,z_{51}	1,7
ser California	4,5,12	m,t	—
ser Typhimurium	1,4,5,12	i	1,2
variant Copenhagen	1,4,12	i	1,2
variant binns	1,4,5,12	—	1,2
ser Lagos	1,4,12	i	1,5

ORGANISM	SOMATIC (O) ANTIGEN	FLAGELLAR (H) ANTIGEN Phase 1	Phase 2
ser Agama	4,12	i	1,6
ser Gloucester	1,4,12,(27)	i	l,w
ser Massenya	1,4,12,27	k	1,5
* ser Neumuenster variant	1,4,12,27	k	1,6
ser Ljubljana	1,4,12,27	k	e,n,x
ser Texas	4,5,12	k	e,n,z_{15}
ser Azteca	4,5,12	l,v	1,5
ser Bredeney	1,4,12,27	l,v	1,7
ser Kimuenza	1,4,12,27	l,v	e,n,x
ser Brandenburg	4,12	l,v	e,n,z_{15}
ser Clackamas	4,12	$l,v,(z_{13})$	1,6
* ser Kilwa	4,12	l,w	e,n,x
ser Togo	4,12	l,w	1,6
ser Ayton	1,4,12,27	l,w	z_6
ser Vom	4,12,27	l,z_{13},z_{28}	e,n,z_{15}
ser Kunduchi	1,4,[5],12,27	l,z_{28}	1,2
ser Heidelberg	[1],4,5,[12]	r	1,2
ser Bradford	4,12,27	r	1,5
ser Remo	1,4,12,27	r	1,7
ser Bochum	4,5,12	r	l,w
ser Africana	4,12	r(i)	l,w
ser Coeln	4,5,12	y	1,2
ser Trachau	4,12,27	y	1,5
ser Teddington	4,12,27	y	1,7
ser Ruki	4,5,12	y	e,n,x
ser Ball	1,4,12,27	y	e,n,x
ser Jos	1,4,12,27	y	e,n,z_{15}
ser Kamoru	4,12,27	y	z_6
ser Shubra	4,5,12	z	1,2
ser Kiambu	4,12	z	1,5
ser Indiana	1,4,12	z	1,7

ORGANISM	SOMATIC (O) ANTIGEN	FLAGELLAR (H) ANTIGEN Phase 1	Phase 2
*ser Nordenham	1,4,12,27	z	e,n,x
ser Preston	1,4,12	z	l,w
ser Entebbe	1,4,12,27	z	z_6
ser Stanleyville	1,4,5,12	z_4,z_{23}	[1,2]
ser Jaja	4,12,27	z_4,z_{23}	—
ser Kalamu	(1),4,12	z_4,z_{24}	[1,5]
ser Haifa	1,4,5,12	z_{10}	1,2
variant Afula, 01 & 05—			
ser Ituri	4,12	z_{10}	1,2
ser Tudu	1,4,12	z_{10}	1,5
ser Albert	4,12	z_{10}	1,6
ser Tokoin	4,12	z_{10}	e,n,x
ser Mura	1,4,12	z_{10}	e,n,z_{15}
ser Fortune	4,12,27	z_{10}	l,w
ser Vellore	1,4,12,27	z_{10}	z_{35}
ser Brancaster	1,4,12,27	z_{29}	—
*ser Helsinki	1,4,12	z_{29}	[e,n,x]
ser Tafo	1,4,12,27	z_{35}	1,7
ser Sloterdijk	1,4,12,27	z_{35}	z_6
ser Tejas	4,12	z_{36}	—
ser Wilhelmsburg	4,12,27	z_{38}	—
variant Teufels-brueck	1,4,12		—
*ser Durbanville	[1],4,12,[27]	$[z_{39}]$	1,[5],7
*ser _____	4,12	—	1,6
	Serogroup C₁		

S. enteritidis

| ser Sanjuan | 6,7 | a | 1,5 |

ORGANISM	SOMATIC (O) ANTIGEN	FLAGELLAR (H) ANTIGEN Phase 1	Phase 2
ser Umhlali	6,7	a	1,6
ser Austin	6,7	a	1,7
ser Oslo	6,7	a	e,n,x
ser Denver	6,7	a	e,n,z_{15}
ser Coleypark	6,7	a	l,w
*ser	6,7	a	z_6
*ser Calvinia	6,7	a	z_{42}
ser Nissii	6,7,14	b	—
ser Brazzaville	6,7	b	1,2
ser Edinburg	6,7	b	1,5
ser Koumra	6,7	b	1,7
ser Georgia	6,7	b	e,n,z_{15}
ser Ohio	6,7	b	l,w
ser Leopoldville	6,7	b	z_6
ser Kotte	6,7	b	z_{35}
*ser Bloemfontein	6,7	b	[e,n,x]:z_{42}
Salmonella choleraesuis			
bioser Kunzendorf	6,7	c	1,5
S. enteritidis	6,7	[c]	1,5
bioser Paratyphi-C	6,7[Vi]	c	1,5
bioser Decatur	6,7	c	1,5
bioser Typhisuis	6,7	[c]	1,5
ser Birkenhead	6,7	c	1,6
ser Mission	6,7	d	1,5
ser Kivu	6,7	d	1,6
ser Amersfoort	6,7	d	e,n,x
ser Gombe	6,7	d	e,n,z_{15}
ser Livingstone	6,7	d	l,w
ser Wil	6,7	d	l,z_{13},z_{28}
ser Nieukerk variant Zollenspicker	6,7	d	z_6
ser Larochelle	6,7	e,h	1,2

ORGANISM	SOMATIC (O) ANTIGEN	FLAGELLAR (H) ANTIGEN Phase 1	Phase 2
ser Lomita	6,7	e,h	1,5
ser Norwich	6,7	e,h	1,6
ser Braenderup	6,7	e,h	e,n,z_{15}
ser Rissen	6,7	f,g	—
ser Montevideo	6,7	g,m,s	—
ser Othmarschen	6,7	g,m,[t]	—
ser Menston	6,7	g,s,t	—
*ser	6,7	g,t	e,n,x:z_{42}
ser Riggil	6,7	g,t	—
ser Alamo	6,7	g,z_{51}	1,5
ser Haelsingborg	6,7	m,p,t,[u]	—
ser Oranienburg	6,7	m,t	—
ser Augustenborg	6,7	i	1,2
ser Oritamerin	6,7	i	1,5
ser Garoli	6,7	i	1,6
ser Norton	6,7	i	l,w
ser Galiema	6,7	k	1,2
ser Thompson	6,7	k	1,5
variant 14	6,7,14	k	1,5
ser Daytona	6,7	k	1,6
ser Baiboukoum	6,7	k	1,7
ser Cardiff	6,7	k	1,10
ser Singapore	6,7	k	e,n,x
ser Escanaba	6,7	k	e,n,z_{15}
*ser	6,7	k	[z_6]
ser Concord	6,7	l,v	1,2
ser Irumu	6,7	l,v	1,5
ser Bonn	6,7	l,v	e,n,x
ser Potsdam	6,7	l,v	e,n,z_{15}
ser Gdansk	6,7	l,v	z_6
ser Gabon	6,7	l,w	1,2
ser Colorado	6,7	l,w	1,5

ORGANISM	SOMATIC (O) ANTIGEN	FLAGELLAR (H) ANTIGEN Phase 1	Phase 2
ser Nessziona	6,7	$1,z_{13}$	1,5
ser Kenya	6,7	$1,z_{13}$	e,n,x
ser Neukoelln	6,7	$1,z_{13},z_{28}$	e,n,z_{15}
ser Makiso	6,7	$1,z_{13},z_{28}$	z_6
*ser Heilbron	6,7	$1,z_{28}$	1,5:[z_{42}]
ser Virchow	6,7	r	1,2
ser Infantis	6,7,[14]	r	1,5
ser Nigeria	6,7	r	1,6
ser Colindale	6,7	r	1,7
ser Papuana	6,7	r	e,n,z_{15}
ser Richmond	6,7	y	1,2
ser Bareilly	6,7	y	1,5
ser Gatow	6,7	y	1,7
ser Hartford	6,7	y	e,n,x
ser Mikawasima	6,7	y	e,n,z_{15}
*ser Tosamanga	6,7	z	1,5
ser Oakland	6,7	z	1,6,(7)
ser Businga	6,7	z	e,n,z_{15}
*ser	6,7	z	z_6
*ser Oysterbeds	6,7	z	z_{42}
*ser Roterberg	6,7	z_4,z_{23}	—
ser Obogu	6,7	z_4,z_{23}	1,5
ser Aequatoria	6,7	z_4,z_{23}	e,n,z_{15}
*ser Kralendyk	6,7	z_4,z_{24}	—
*ser Cape	6,7	z_6	1,7
ser Menden	6,7	z_{10}	1,2
ser Inganda	6,7	z_{10}	1,5
ser Eschweiler	6,7	z_{10}	1,6
ser Ngili	6,7	z_{10}	1,7
ser Djugu	6,7	z_{10}	e,n,x

ORGANISM	SOMATIC (O) ANTIGEN	FLAGELLAR (H) ANTIGEN Phase 1	Phase 2
ser Mbandaka	[1],6,7,[25]	z_{10}	e,n,z_{15}
ser Jerusalem var. 014—	6,7	z_{10}	1,w
*ser	6,7	z_{10}	z_{35}
ser Tennessee	6,7	z_{29}	—
*ser	6,7	z_{29}	—
*ser Argentina	6,7	z_{36}	—
*ser Bacongo	6,7	z_{36}	z_{42}
ser Lille	6,7	z_{38}	—
*ser Gilbert	6,7	z_{39}	1,7
ser Hillsborough	6,7	z_{41}	1,w
*ser	6,7	z_{42}	e,n,x:1,6
*ser Sullivan	6,7	z_{42}	1,7
ser Thompson variant Berlin	6,7	—	1,5
ser Nienstedten	6,(7),(14)	b	1,w
ser Kaduna	6,(7),(14)	c	e,n,z_{15}
ser Omderman	6,(7),(14)	d	e,n,x
ser Eimsbuettel	6,(7),(14)	d	1,w
ser Nieukerk	6,(7),(14)	d	z_6
ser Ardwick	6,(7),(14)	f,g	—
ser Thielallee	6,(7),(14)	m,t	—
ser Thompson variant 014+	6,(7),(14)	k	1,5
ser Gelsenkirchen	6,(7),(14)	l,v	z_6
ser Bareilly var. 014+	6,(7),(14)	y	1,5
ser Jerusalem	6,(7),(14)	z_{10}	1,w
ser Bornum	6,(7),(14)	z_{38}	—

S. enteritidis

ORGANISM	SOMATIC (O) ANTIGEN Serogroup C_2	FLAGELLAR (H) ANTIGEN Phase 1	Phase 2
ser Doncaster	6,8	a	1,5
ser Curacao	6,8	a	1,6
ser Nordufer	6,8	a	1,7
ser Narashino	6,8	a	e,n,x
ser Leith	6,8	a	e,n,z_{15}
*ser Tulear	6,8	a	z_{52}
ser Nagoya	6,8	b	1,5
ser Stourbridge	6,8	b	1,6
ser Gatuni	6,8	b	e,n,x
ser Presov	6,8	b	e,n,z_{15}
ser Bukuru	6,8	b	l,w
ser Banalia	6,8	b	z_6
ser Wingrove	6,8	c	1,2
ser Utah	6,8	c	1,5
ser Bronx	6,8	c	1,6
ser Belem	6,8	c	e,n,x
ser Quiniela	6,8	c	e,n,z_{15}
ser Muenchen	6,8	d	1,2
ser Manhattan	6,8	d	1,5
ser Sterrenbos	6,8	d	e,n,x
ser Herston	6,8	d	e,n,z_{15}
ser Labadi	6,8	d	z_6
ser Newport	6,8	e,h	1,2
ser Kottbus	6,8	e,h	1,5
ser Tshiongwe	6,8	e,h	e,r,z_{15}
ser Sandow	6,8	f,g	e,n,z_{15}
ser Chincol	6,8	g,m,s	e,n,x
*ser	6,8	g,(m),t	e,n,x
*ser Baragwanath	6,8	m,t	1,5

ORGANISM	SOMATIC (O) ANTIGEN	FLAGELLAR (H) ANTIGEN Phase 1	Phase 2
*ser Germiston	6,8	m,t	e,n,x
ser Lindenburg	6,8	i	1,2
ser Takoradi	6,8	i	1,5
ser Warnow	6,8	i	1,6
ser Bonariensis	6,8	i	e,n,x
ser Aba	6,8	i	e,n,z_{15}
ser Blockley	6,8	k	1,5
ser Schwerin	6,8	k	e,n,x
ser Litchfield	6,8	l,v	1,2
ser Loanda	6,8	l,v	1,5
ser Manchester	6,8	l,v	1,7
ser Holcomb	6,8	l,v	e,n,x
ser Edmonton	6,8	l,v	e,n,z_{15}
ser Fayed	6,8	l,w	1,2
ser Breukelen	6,8	l,z_{13}	e,n,z_{15}
ser Bovismorbificans	6,8	r	1,5
ser Akanji	6,8	r	1,7
ser Hidalgo	6,8	r	e,n,z_{15}
ser Goldcoast	6,8	r	l,w
ser Tananarive	6,8	y	1,5
ser Alagbon	6,8	y	1,7
ser Praha	6,8	y	e,n,z_{15}
ser Mowanjium	6,8	z	1,5
ser Kuru	6,8	z	l,w
ser Lezennes	6,8	z_4,z_{23}	1,7
ser Chailey	6,8	z_4,z_{23}	[e,n,z_{15}]
ser Duesseldorf	6,8	z_4,z_{24}	—
ser Tallahassee	6,8	z_4,z_{32}	—
ser Mapo	6,8	z_{10}	1,5
ser Cleveland	6,8	z_{10}	1,7
ser Hadar	6,8	z_{10}	e,n,x

ORGANISM	SOMATIC (O) ANTIGEN	FLAGELLAR (H) ANTIGEN Phase 1	Phase 2
ser Glostrup	6,8	z_{10}	e,n,z_{15}
ser Wippra	6,8	z_{10}	z_6
ser Uno	6,8	z_{29}	—
ser Yarm	6,8	z_{35}	1,2
Serogroup C_3			
ser Korbol	(8),20	b	1,5,(6)
ser Sanga	(8)	b	1,7
ser Shipley	(8),20	b	e,n,z_{15}
ser Alexanderpolder	(8)	c	l,w
ser Virginia	(8)	d	[1,2]
ser Bardo	(8)	e,h	1,2
ser Rechovot	(8),20	e,h	z_6
ser Emek	(8),20	g,m,s	—
ser Kentucky variant Jerusalem	(8),20	i	z_6
	(8)	i	z_6
ser Haardt	(8)	k	1,5
ser Pakistan	(8)	l,v	1,2
ser Amherstiana	(8)	l,(v)	1,6
ser Hindmarsh	(8),20	r	1,5
ser Pikine	(8),20	r	z_6
ser Cocody	(8),20	r(i)	e,n,z_{15}
ser Altona	(8),20	r(i)	z_6
ser Brunei	(8),20	y	1,5
ser Sunnycove	(8),20	y	e,n,x
ser Kralingen	(8),20	y	z_6
ser Corvallis	(8),20	z_4,z_{23}	—
ser Albany	(8),20	z_4,z_{24}	—

ORGANISM	SOMATIC (O) ANTIGEN	FLAGELLAR (H) ANTIGEN Phase 1	Phase 2
ser Paris	(8),20	z_{10}	1,5
ser Molade	(8),20	z_{10}	z_6
ser Tamale	(8),20	z_{29}	—
Serogroup D_1			
S. enteritidis			
ser Miami	1,9,12	a	1,5
bioser Sendai	1,9,12	a	1,5
ser Os	9,12	a	1,6
ser Saarbruecken	[1],9,12	a	1,7
ser Lomalinda	9,12	a	e,n,x
ser Durban	9,12	a	e,n,z_{15}
ser Onarimon	1,9,12	b	1,2
ser Frintrop	1,9,12	b	1,5
*ser Mjimwema	1,9,12	b	e,n,x
*ser Blankenese	1,9,12	b	z_6
*ser Suederelbe	1,9,12	b	z_{39}
ser Goeteborg	9,12	c	1,5
ser Ipeko	9,12	c	1,6
ser Alabama	9,12	c	e,n,z_{15}
ser Ridge	9,12	c	z_6
*ser Zuerich	1,9,12	c	z_{39}
Salmonella typhi	9,12,[Vi]	d	—
S. enteritidis			
ser Ndolo	[1],9,12	d	1,5
ser Tarshyne	9,12	d	1,6
*ser Rhodesiense	9,12	d	e,n,x
ser Zega	9,12	d	z_6
ser Jaffna	1,9,12	d	z_{35}
ser Bournemouth	9,12	e,h	1,2
ser Eastbourne	1,9,12	e,h	1,5

ORGANISM	SOMATIC (O) ANTIGEN	FLAGELLAR (H) ANTIGEN Phase 1	Phase 2
ser Israel	9,12	e,h	e,n,z_{15}
*ser Lindrick	9,12	e,n,x	1,[5],7
*ser	9,12	e,n,x	1,6
ser Berta	9,12	f,g,t	—
ser Enteritidis	1,9,12	g,m	—
ser Blegdam	9,12	g,m,q	—
ser Muizenberg	9,12	g,m,s,t	1,5
*ser Kuilsrivier	1,9,12	g,m,s,t	e,n,x
*ser Manica	1,9,12	g,m,s,t	z_{42}
ser Dublin	1,9,12	g,p	—
ser Naestved	1,9,12	g,p,s	—
ser Rostock	1,9,12	g,p,u	—
ser Moscow	9,12	g,q	—
*ser Neasden	1,9,12	g,s,t	e,n,x
*ser Hamburg	1,9,12	g,t	—
ser Newmexico	9,12	g,z_{51}	1,5
ser Pensacola	9,12	m,t	—
ser Seremban	9,12	i	1,5
ser Claibornei	1,9,12	k	1,5
ser Goverdhan	9,12	k	1,6
ser Mendoza	9,12	l,v	1,2
ser Panama	1,9,12	l,v	1,5
ser Kapemba	9,12	l,v	1,7
ser Italiana	9,12	l,v	1,11
ser Goettingen	9,12	l,v	e,n,z_{15}
ser Victoria	1,9,12	l,w	1,5
*ser Daressalaam	1,9,12	l,w	e,n,x
ser Miyazaki	9,12	l,z_{13}	1,7
ser Napoli	1,9,12	l,z_{13}	e,n,x
ser Javiana	1,9,12	l,z_{28}	1,5
ser Jamaica	9,12	r	1,5
ser Lome	9,12	r	z_6

ORGANISM	SOMATIC (O) ANTIGEN	FLAGELLAR (H) ANTIGEN Phase 1	Phase 2
ser Lawndale	1,9,12	z	1,5
*ser Stellenbosch	1,9,12	z	1,7
*ser Angola	1,9,12	z	z_6
*ser Hueningen	9,12	z	z_{39}
ser Wangata	9,12	z_4,z_{23}	[1,7]
ser Portland	9,12	z_{10}	1,5
*ser Canastel	9,12	z_{29}	1,5
ser Penarth	9,12	z_{35}	z_6
ser Elomrane	1,9,12	z_{38}	—
*ser Wynberg	1,9,12	z_{39}	1,7
bioser Gallinarum	1,9,12	—	—
bioser Pullorum	9,12	—	—
Serogroup D_2			
ser Baildon	(9),46	a	e,n,x
ser Zadar	(9),46	b	1,6
*ser Lundby	(9),46	b	e,n,x
ser Bamboye	(9),46	b	l,w
ser Itutaba	(9),46	c	z_6
ser Strasbourg	(9),46	d	1,7
ser Plymouth	(9),46	d	z_6
ser Bergedorf	(9),46	e,h	1,2
ser Wernigerode	(9),46	f,g	—
*ser Duivenhoks	(9),46	g,m,s,t	e,n,x
ser Gateshead	(9),46	g,s,t	—
ser Mathura	(9),46	i	e,n,z_{15}
ser Marylebone	(9),46	k	1,2

ORGANISM	SOMATIC (O) ANTIGEN	FLAGELLAR (H) Phase 1	Phase 2
• ser Ceyco	(9),46	k	z_{35}
ser India	(9),46	l,v	1,5
ser Shoreditch	(9),46	r	e,n,z_{15}
ser Mayday	(9),46	y	z_6
*ser Haarlem	(9),46	z	e,n,x
ser Ekotedo	(9),46	z_4,z_{23}	—
*ser Maarssen	(9),46	z_4,z_{24}	z_{39},z_{42}
ser Lishabi	(9),46	z_{10}	1,7
ser Inglis	(9),46	z_{10}	e,n,x
*ser ___	(9),46	z_{10}	z_6
ser Ouakam	(9),[12],[34],46	z_{29}	—
ser Hillegersberg	(9),46	z_{35}	1,5
ser Fresno variant	(9),46	z_{38}	—
*	(9),46	z_{38}	—
ser Potto	(9),12,46	i	z_6
*ser ___	1,9,12,(46),27	y	z_{39}

Serogroup E₁

S. enteritidis

ORGANISM	SOMATIC (O) ANTIGEN	FLAGELLAR (H) Phase 1	Phase 2
ser Aminatu	3,10	a	1,2
ser Goelzau	3,10	a	1,5
ser Oxford	3,10	a	1,7
*ser Matroosfontein	3,10	a	e,n,x
ser Galil	3,10	a	e,n,z_{15}
ser Kalina	3,10	b	1,2
ser Butantan	3,10	b	1,5
ser Allerton	3,10	b	1,6
ser Huvudsta	3,10	b	1,7

ORGANISM	SOMATIC (O) ANTIGEN	FLAGELLAR (H) Phase 1	Phase 2
ser Benfica	3,10	b	e,n,x
ser Yaba	3,10	b	e,n,z_{15}
ser Epicrates	3,10	b	l,w
ser Pramiso	3,10	c	1,7
ser Agege	3,10	c	e,n,z_{15}
ser Anderlecht	3,10	c	l,w
ser Okefoko	3,10	c	z_6
ser Stormont	3,10	d	1,2
ser Shangani	3,10	d	1,5
ser Lekke	3,10	d	1,6
ser Onireke	3,10	d	1,7
ser Souza	3,10	d	e,n,x
ser Madjorio	3,10	d	e,n,z_{15}
ser Birmingham	3,10	d	l,w
ser Weybridge	3,10	d	z_6
ser Maron	3,10	d	z_{35}
ser Vejle	3,10	e,h	1,2
ser Muenster	3,10	e,h	1,5
ser Anatum	3,10	e,h	1,6
ser Nyborg	3,10	e,h	1,7
ser Newlands	3,10	e,h	e,n,x
ser Meleagridis	3,10	e,h	l,w
ser Sekondi	3,10	e,h	z_6
*ser Chudleigh	3,10	e,n,x	1,7
ser Regent	3,10	f,g	—
ser Suberu	3,10	g,m	—
ser Amsterdam	3,10	g,m,s	—
ser Westhampton	3,10	g,s,t	—
*ser Islington	3,10	g,t	—
ser Southbank	3,10	m,t	—
*ser Stikland	3,10	m,t	e,n,x
ser Amounderness	3,10	i	1,5

ORGANISM	SOMATIC (O) ANTIGEN	FLAGELLAR (H) ANTIGEN Phase 1	Phase 2
ser Falkensee	3,10	i	e,n,z_{15}
ser Yeerongpilly	3,10	i	z_6
ser Wimborne	3,10	k	1,2
ser Zanzibar	3,10	k	1,5
ser Marienthal	3,10	k	e,n,z_{15}
ser Newrochelle	3,10	k	l,w
ser Nchanga	3,10	l,v	1,2
ser Sinstorf	3,10	l,v	1,5
ser London	3,10	l,v	1,6
ser Give	3,10	l,v	1,7
ser Ruzizi	3,10	l,v	e,n,z_{15}
*ser Fuhlsbuettel	3,10	l,v	z_6
ser Uganda	3,10	l,z_{13},z_{28}	1,5
ser Fallowfield	3,10	l,z_{13},z_{28}	e,n,z_{15}
*ser Westpark	3,10	l,z_{28}	e,n,x
*ser	3,10	l,z_{28}	z_{39}
ser Rutgers	3,10	l,z_{40}	1,7
ser Seegefeld	3,10	r(i)	1,2
ser Ughelli	3,10	r	1,5
ser Elisabethville	3,10	r	1,7
ser Simi	3,10	r	e,n,z_{15}
ser Weltevreden	3,10	r	z_6
ser Amager	3,10	y	1,2
ser Orion	3,10	y	1,5
ser Mokola	3,10	y	1,7
ser Ohlstedt	3,10	y	e,n,x
ser Bolton	3,10	y	e,n,z_{15}
ser Langensalza	3,10	y	l,w
ser Stockholm	3,10	y	z_6
*ser Alexander	3,10	z	1,5
*ser Finchley	3,10	z	e,n,x
ser Clerkenwell	3,10	z	l,w

ORGANISM	SOMATIC (O) ANTIGEN	FLAGELLAR (H) ANTIGEN Phase 1	Phase 2
*ser Tafelbaai	3,10	z	z_{39}
ser Adabraka	3,10	z_4,z_{23}	[1,7]
ser Okerara	3,10	z_{10}	1,2
ser Lexington	3,10	z_{10}	1,5
ser Coquilhatville	3,10	z_{10}	1,7
ser Kristianstad	3,10	z_{10}	e,n,z_{15}
ser Biafra	3,10	z_{10}	z_6
ser Jedburgh	3,10	z_{29}	—
ser Cairina	3,10	z_{35}	z_6
ser Macallen	3,10	z_{36}	—
ser Bolombo	3,10	z_{38}	—
*ser Mpila	3,10	z_{38}	z_{42}
*ser Winchester	3,10	z_{39}	1,7

Serogroup E_2

S. enteritidis

ORGANISM	SOMATIC (O) ANTIGEN	FLAGELLAR (H) ANTIGEN Phase 1	Phase 2
ser Rosenthal	3,15	b	1,5
ser Pankow	3,15	d	1,5
ser Eschersheim	3,15	d	e,n,x
ser Goerlitz	3,15	e,h	1,2
ser Newhaw	3,15	e,h	1,5
ser Newington	3,15	e,h	1,6
ser Selandia	3,15	e,h	1,7
ser Cambridge	3,15	e,h	l,w
ser Drypool	3,15	g,m,s	—
*ser Parow	3,15	g,m,s,t	—
ser Halmstad	3,15	g,s,t	—
ser Portsmouth	3,15	l,v	1,6
ser Newbrunswick	3,15	l,v	1,7

ORGANISM	SOMATIC (O) ANTIGEN	FLAGELLAR (H) ANTIGEN Phase 1	Phase 2
ser Kinshasa	3,15	l,z_{13}	1,5
ser Lanka	3,15	r	z_6
ser Tuebingen	3,15	y	1,2
ser Binza	3,15	y	1,5
ser Tournai	3,15	y	z_6
ser Manila	3,15	z_{10}	1,5
ser Hamilton	3,15	—	z_{27}

Serogroup E$_3$

S. enteritidis

ORGANISM	SOMATIC (O) ANTIGEN	FLAGELLAR (H) ANTIGEN Phase 1	Phase 2
ser Khartoum	(3),(15),34	a	1,7
ser Arkansas	(3),(15),34	e,h	1,5
ser Minneapolis	(3),(15),34	e,h	1,6
ser Wildwood	(3),(15),34	e,h	1,w
ser Canoga	(3),(15),34	g,s,t	—
ser Menhaden	(3),(15),34	l,v	1,7
ser Thomasville	(3),(15),34	y	1,5
ser Illinois	(3),(15),34	z_{10}	1,5
ser Harrisonburg	(3),(15),34	z_{10}	1,6

Serogroup E$_4$

S. enteritidis

ORGANISM	SOMATIC (O) ANTIGEN	FLAGELLAR (H) ANTIGEN Phase 1	Phase 2
ser Gwoza	1,3,19	a	e,n,z_{15}
ser Gnesta	1,3,19	b	1,5
ser Visby	1,3,19	b	1,6
ser Broughton	1,3,19	b	1,w
ser Accra	1,3,19	b	z_6
ser Madiago	1,3,19	c	1,7

ORGANISM	SOMATIC (O) ANTIGEN	FLAGELLAR (H) ANTIGEN Phase 1	Phase 2
ser Ahamdi	1,3,19	d	1,5
ser Liverpool	1,3,19	d	e,n,z_{15}
ser Tilburg	1,3,19	d	1,w
ser Nilose	1,3,19	d	z_6
ser Sanktmarx	1,3,19	e,h	1,7
ser Sao	1,3,19	e,h	e,n,z_{15}
ser Calabar	1,3,19	e,h	1,w
ser Rideau	1,3,19	f,g	—
ser Maiduguri	1,3,19	f,g,t	e,n,z_{15}
ser Senftenberg	1,3,19	g,s,t	—
ser Cannstatt	1,3,19	m,t	—
ser Stratford	1,3,19	i	1,2
ser Machaga	1,3,19	i	e,n,x
ser Avonmouth	1,3,19	i	e,n,z_{15}
ser Zuilen	1,3,19	i	1,w
ser Taksony	1,3,19	i	z_6
ser Ngor	1,3,19	l,v	1,5
ser Westerstede	1,3,19	l,z_{13}	—
ser Lokstedt	1,3,19	l,z_{13},z_{28}	1,2
ser Bedford	1,3,19	l,z_{13},z_{28}	e,n,z_{15}
ser Yalding	1,3,19	r	e,n,z_{15}
ser Krefeld	1,3,19	y	1,w
ser Korlebu	1,3,19	z	1,5
ser Schoeneberg	1,3,19	z	e,n,z_{15}
ser Carno	1,3,19	z	1,w
ser Dallgow	1,3,19	z_{10}	e,n,z_{15}
ser Simsbury	1,3,19	—	z_{27}
ser Llandoff	1,3,19	z_{29}	—
ser Chittagong	(1),3,10,(19)	b	z_{35}

ORGANISM	SOMATIC (O) ANTIGEN	FLAGELLAR (H) ANTIGEN Phase 1	Phase 2
ser Bilu	(1),3,10,(19)	f,g,t	1,(2),7
ser Ilugun	(1),3,10,(19)	z_4,z_{23}	z_6
ser Dessau	(1),3,15,(19)	g,s,t	—

Serogroup F

S. enteritidis

ORGANISM	SOMATIC (O) ANTIGEN	FLAGELLAR (H) ANTIGEN Phase 1	Phase 2
ser Marseille	11	a	1,5
ser Luciana	11	a	e,n,z_{15}
*ser Glencairn	11	a	$z_4;z_{42}$
ser Leeuwarden	11	b	1,5
*ser	11	b	1,7
*ser Srinagar	11	b	e,n,x
ser Pharr	11	b	e,n,z_{15}
ser Chandans	11	d	e,n,x
*ser Montgomery	11	d,a	d,e,n,z_{15}
ser Findorff	11	d	z_6
ser Chingola	11	e,h	1,2
ser Adamstua	11	e,h	1,6
ser Redhill	11	e,h	l,z_{13},z_{28}
*ser Grabouw	11	g,(m),s,t	$[z_{39}]$
ser Missouri	11	g,s,t	—
*ser Mundsburg	11	g,z_{51}	—
*ser Lincoln	11	m,t	e,n,x
ser Aberdeen	11	i	1,2
ser Brijbhumi	11	i	1,5
ser Heerlen	11	i	1,6
ser Veneziana	11	i	e,n,x
ser Pretoria	11	k	1,2
ser Abaetetuba	11	k	1,5
ser Sharon	11	k	1,6

ORGANISM	SOMATIC (O) ANTIGEN	FLAGELLAR (H) ANTIGEN Phase 1	Phase 2
ser Kisarawe	11	k	e,n,x
ser Amba	11	k	l,z_{13},z_{28}
ser Stendal	11	l,v	1,2
ser Maracaibo	11	l,v	1,5
ser Fann	11	l,v	e,n,x
ser Bulbay	11	l,v	e,n,z_{15}
ser Osnabrueck	11	l,z_{13},z_{28}	e,n,x
ser Huila	11	l,z_{28}	e,n,x
ser Senegal	11	r	1,5
*ser Rubislaw	11	[d],r	[d],e,n,x
ser Volta	11	r	l,z_{13},z_{28}
ser Solt	11	y	1,5
ser Herzliya	11	y	e,n,x
ser Nyanza	11	z	z_6
*ser Soutpan	11	z	z_{39}
*ser Parera	11	z_4,z_{23}	—
ser Etterbeek	11	z_4,z_{23}	e,n,z_{15}
*ser	11	z_4,z_{23}	—
ser Wentworth	11	z_{10}	1,2
ser Straengnaes	11	z_{10}	1,5
ser Telhashomer	11	z_{10}	e,n,x
ser Maastricht	11	z_{41}	1,2

Serogroup G_1

S. enteritidis

ORGANISM	SOMATIC (O) ANTIGEN	FLAGELLAR (H) ANTIGEN Phase 1	Phase 2
ser Mim	13,22	a	1,6
ser Ibadan	13,22	b	1,5
ser Vaertan	13,22	b	e,n,x
ser Bahati	13,22	b	e,n,z_{15}

ORGANISM	SOMATIC (O) ANTIGEN	FLAGELLAR (H) ANTIGEN Phase 1	Phase 2
ser Haouaria	13,22	c	e,n,x,z_{15}
ser Friedenau	13,22	d	1,6
ser Diguel	1,13,22	d	e,n,z_{15}
ser Willemstad	1,13,22	e,h	1,6
ser Raus	13,22	f,g	e,n,x
*ser Bron	13,22	g,m	$[e,n,z_{15}]$
*ser Limbe	1,13,22	g,m,t	[1,5]
*ser Rotterdam	1,13,22	g,t	[1,5]
ser Lovelace	13,22	l,v	1,5
ser Borbeck	13,22	l,v	1,6
ser Tanger	1,13,22	y	1,6
ser Poona	[1],13,22	z	1,6
variant 37	1,13,22,36,37	z	1,6
ser Bristol	13,22	z	1,7
ser Roodepoort	[1],13,22	z_{10}	1,5
*ser Clifton	13,22	z_{29}	1,5
*ser Goodwood	13,22	z_{29}	e,n,x
ser Mampong	13,22	z_{35}	1,6
ser Leiden	13,22	z_{38}	—

Serogroup G_2

S. enteritidis

ORGANISM	SOMATIC (O) ANTIGEN	FLAGELLAR (H) ANTIGEN Phase 1	Phase 2
ser Chagoua	1,13,23	a	1,5
*ser Tygerberg	1,13,23	a	z_{42}
ser Mississippi(Atlanta)	1,13,23	b	1,5
ser Bracknell	13,23	b	1,6
ser Ullevi	13,23	b	e,n,x
ser Durham	13,23	b	e,n,z_{15}
*ser Acres	1,13,23	b	z_{42}:[1,5]

ORGANISM	SOMATIC (O) ANTIGEN	FLAGELLAR (H) ANTIGEN Phase 1	Phase 2
ser Mishmarhaemek	1,13,23	d	1,5
ser Grumpensis	13,23	d	1,7
ser Telelkebir	13,23	d	e,n,z_{15}
ser Putten	13,23	d	1,w
ser Isuge	13,23	d	z_6
ser Wichita	1,13,23	d	z_{37}
*ser Epping	1,13,23	e,n,x	1,7
ser Havana	1,13,23	f,g,[s]	—
ser Agbeni	13,23	g,m	—
*ser Kraaifontein	1,13,23	g,(m),t	[e,n,x]
*ser Luanshya	13,23	g,s,(t)	—
ser Okatie	13,23	g,s,t	—
ser Congo	13,23	g,t	—
*ser Gojenberg	1,13,23	g,t	1,5
ser Kintambo	13,23	m,t	—
*ser Katesgrove	1,13,23	m,t	1,5
*ser Worcester	1,13,23	m,t	e,n,x
*ser Boulders	13,23	m,t	z_{42}
ser Idikan	13,23	i	1,5
ser Jukestown	13,23	i	e,n,z_{15}
*ser	13,23	$1,z_{28}$	z_6
*ser Vredelust	1,13,23	$1,z_{28}$	z_{42}
ser Adjame	13,23	r	1,6
ser Linton	13,23	r	e,n,z_{15}
ser Yarrabah	13,23	y	1,7
ser Ordonez	1,13,23	y	1,w
ser Tunis	1,13,23	y	z_6
*ser Nachshonim	1,13,23	z	1,5
*ser Farmsen	13,23	z	1,6
ser Worthington	1,13,23	z	1,w
ser Ajiobo	13,23	z_4,z_{23}	—

ORGANISM	SOMATIC (O) ANTIGEN	FLAGELLAR (H) ANTIGEN Phase 1	Phase 2
ser Romanby	13,23	z_4,z_{24}	–
ser Demerara	13,23	z_{10}	1,w
ser Cubana	1,13,23	z_{29}	–
ser Fanti	13,23	z_{38}	–
ser Stevenage	1,13,23	[z_{42}]	1,7

Serogroup H

S. enteritidis

ORGANISM	SOMATIC (O) ANTIGEN	FLAGELLAR (H) ANTIGEN Phase 1	Phase 2
ser Garba	1,6,14,25	a	1,5
ser Ferlac	1,6,14,25	a	e,n,x
ser Tucson	1,6,14,25	b	[1,7]
ser Blijdorp	1,6,14,25	c	1,5
ser Kassberg	1,6,14,25	c	1,6
ser Heves	6,14,24	d	1,5
ser Finkenwerder	1,6,14,25	d	1,5
ser Florida	1,6,14,25	d	1,7
ser Lindern	6,14,24	d	e,n,x
ser Charity	1,6,14,25	d	e,n,x
ser Teko	1,6,14,25	d	$1,z_{13},z_{28}$
ser Encino	1,6,14,25	d	e,n,z_{15}
ser Albuquerque	6,14,24	d	z_6
ser Bahrenfeld	6,14,24	e,h	1,5
ser Onderstepoort	1,6,14,25	e,h	1,5
ser Magumeri	1,6,14,25	e,h	1,6
ser Warragul	1,6,14,25	g,m	–
ser Caracas	1,6,14,25	g,m,s	–
ser Kaitaan	1,6,14,25	m,t	–
*ser Rooikrantz	1,6,14	m,t	1,5
*ser Emmerich	6,14	[m,t]	e,n,x
ser Mampeza	1,6,14,25	i	1,5

ORGANISM	SOMATIC (O) ANTIGEN	FLAGELLAR (H) ANTIGEN Phase 1	Phase 2
ser Buzu	1,6,14,25	i	1,7
ser Schalkwijk	6,14,(24)	i	e,n,......
*ser	(6),14	k	[e,n,x]
ser Harburg	1,6,14,25	k	1,5
*ser	1,(6),14	k	$z_6:z_{42}$
ser Boecker	[1],6,14,[25]	l,v	1,7
ser Horsham	1,6,14,25	l,v	e,n,x
ser Surat	1,6,14,25	r(i)	e,n,z_{15}
variant Hr-	1,6,14,25	i	e,n,z_{15}
ser Carrau	6,14,24	y	1,7
ser Madelia	1,6,14,25	y	1,7
ser Fischerkietz	1,6,14,25	y	e,n,x
ser Homosassa	1,6,14,25	z	1,5
ser Soahamina	6,14,24	z	e,n,x
ser Sundsvall	1,6,14,25	z	e,n,x
ser Poano	1,6,14,25	z	$1,z_3,z_{28}$
ser Bousso	1,6,14,25	z_4,z_{23}	–
ser Uzaramo	1,6,14,25	z_4,z_{24}	–
ser Nessa	1,6,14,25	z_{10}	1,2
*ser Bornheim	1,6,14,25	z_{10}	1,(2),7
*ser Simonstown	1,6,14	z_{10}	1,5
*ser Slangkop	1,6,14	z_{10}	$z_6:z_{42}$
ser Sara	1,6,14,25	z_{38}	[e,n,x]
ser	1,(6),14	z_{42}	1,6

Serogroup I

S. enteritidis

ORGANISM	SOMATIC (O) ANTIGEN	FLAGELLAR (H) ANTIGEN Phase 1	Phase 2
ser Hannover	16	a	1,2
ser Brazil	16	a	1,5
ser Amunigun	16	a	1,6

ORGANISM	SOMATIC (O) ANTIGEN	FLAGELLAR (H) ANTIGEN	
		Phase 1	Phase 2
ser Fischerhuette	16	a	e,n,z_{15}
ser Heron	16	a	z_6
ser Hull	16	b	1,2
ser Wa	16	b	1,5
ser Glasgow	16	b	1,6
ser Hvittingfoss	16	b	e,n,x
ser Malstatt	16	b	z_6
*ser	16	b	z_{42}
ser Vancouver	16	c	1,5
ser Shamba	16	c	e,n,x
ser Oldenburg	16	d	1,2
ser Gaminara	16	d	1,7
ser Barranguilla	16	d	e,n,x
ser Nottingham	16	d	e,n,z_{15}
ser Barmbek	16	d	z_6
ser Malakal	16	e,h	1,2
ser Weston	16	e,h	z_6
*ser Bellville	16	e,n,x	1,7
ser Tees	16	f,g	—
ser Adeoyo	16	g,m	—
ser Nikolaifleet	16	g,m,s	—
*ser Mobeni	16	g,m,s,t	—
*ser Merseyside	16	g,t	1,5
*ser Rowbarton	16	m,t	—
ser Amina	16	i	1,5
ser Frankfurt	16	i	e,n,z_{15}
ser Szentes	16	k	1,2
ser Nuatja	16	k	e,n,x
ser Orientalis	16	k	e,n,z_{15}
ser Shanghai	16	l,v	1,6

ORGANISM	SOMATIC (O) ANTIGEN	FLAGELLAR (H) ANTIGEN	
		Phase 1	Phase 2
ser Welikade	16	l,v	1,7
ser Salford	16	l,v	e,n,x
ser Burgas	16	l,v	e,n,z_{15}
ser Losangeles	16	l,v	z_6
*ser Noordhoek	16	l,w	z_6
ser Mandera	16	l,z_{13}	e,n,z_{15}
ser Enugu	16	l,z_{13},z_{28}	—
*ser Sarepta	16	l,z_{28}	z_{42}
*ser	16	l,z_{40}	—
ser Annedal	16	r(i)	e,n,x
ser Zwickau	16	r(i)	e,n,z_{15}
ser Saphra	16	y	1,5
ser Akuafo	16	y	1,6
ser Kikoma	16	y	e,n,x
ser Lingivala	16	z	1,7
*ser Louwbester	16	z	e,n,x
ser Kibi	16	z_4,z_{23}	—
*ser Haddon	16	z_4,z_{23}	—
*ser Ochsenzoll	16	z_4,z_{23}	—
*ser Chameleon	16	z_4,z_{32}	—
ser Lisboa	16	z_{10}	1,6
ser Redlands	16	z_{10}	e,n,z_{15}
*ser Jacksonville	16	z_{29}	—
*ser Woodstock	16	z_{42}	1,(5),7
*ser Elsiesivier	16	z_{42}	1,6

Serogroup J

S. enteritidis

ORGANISM	SOMATIC (O) ANTIGEN	FLAGELLAR (H) ANTIGEN	
		Phase 1	Phase 2
ser Bonames	17	a	1,2
ser Jangwani	17	a	1,5

ORGANISM	SOMATIC (O) ANTIGEN	FLAGELLAR (H) ANTIGEN Phase 1	Phase 2
ser Kinondoni	17	a	e,n,x
ser Kirkee	17	b	1,2
*ser Hillbrow	17	b	e,n,x,z_{15}
ser Victoriaborg	17	c	1,6
*ser Woerden	17	c	z_{39}
ser Berlin	17	d	1,5
ser Niamey	17	d	l,w
*ser Verity	17	e,n,x,z_{15}	1,6
*ser Bleadon	17	(f),g,t	[e,n,x,z_{15}]
*ser ____	17	k	—
ser Irenea	17	k	1,5
ser Matadi	17	k	e,n,x
ser Morotai	17	l,v	1,2
ser Michigan	17	l,v	1,5
ser Carmel	17	l,v	e,n,x
ser Gori	17	z	1,2
*ser Constantia	17	z	l,w:z_{42}
ser Kandla	17	z_{29}	—
Serogroup K			
ser Fluntern	6,14,18	b	1,5
ser ____	18	b	1,5
ser Usumbura	18	d	1,7
ser Langenhorn	18	m,t	—
*ser ____	18	m,t	1,5
ser Memphis	18	k	1,5
*ser ____	18	y	e,n,x,z_{15}

ORGANISM	SOMATIC (O) ANTIGEN	FLAGELLAR (H) ANTIGEN Phase 1	Phase 2
ser Siegburg	6,14,18	z_4,z_{23}	[1,5]
ser Cerro	18	z_4,z_{23}	[z_{45}]
ser Blukwa	18	z_4,z_{24}	—
*ser Shomron	18	z_4,z_{32}	—
*ser Zeist	18	z_{10}	z_6
*ser Beloha	18	z_{36}	—
ser Sinthia	18	z_{38}	—
Serogroup L			
ser Assen	21	a	—
ser Ghana	21	b	1,6
ser Minnesota	21	b	e,n,x
ser Rhone	21	c	e,n,x
ser Spartel	21	d	1,5
ser Magwa	21	d	e,n,x
ser Good	21	f,g	e,n,x
ser Diourbel	21	i	1,2
ser Ruiru	21	y	e,n,x
*ser Soesterberg	21	z_4,z_{23}	—
*ser Gwaai	21	z_4,z_{24}	—
*ser Wandsbek	21	z_{10}	z_6
ser Gambaga	21	z_{35}	e,n,z_{15}
Serogroup M			
ser Solna	28	a	1,5
ser Dakar	28	a	1,6
ser Seattle	28	a	e,n,x

ORGANISM	SOMATIC (O) ANTIGEN	FLAGELLAR (H) ANTIGEN Phase 1	Phase 2
ser Honelis	28	a	e,n,z_{15}
ser Moëro	28	b	1,5
ser Ashanti	28	b	1,6
ser Bokanjac	28	b	1,7
ser Langford	28	b	e,n,z_{15}
*ser Kaltenhausen	28	b	z_6
ser Hermannswerder	28	c	1,5
ser Eberswalde	28	c	1,6
ser Halle	28a,28c	c	1,7
variant vidin	28a,28b	c	1,7
ser Dresden	28	c	e,n,x
ser Wedding	28	c	e,n,z_{15}
ser Techimani	28	c	z_6
ser Amoutive	28	d	1,5
ser Mundonobo	28	d	1,7
ser Mocamedes	28	d	e,n,x
ser Patience	28	d	e,n,z_{15}
*ser ___	28	e,n,x	1,7
ser Friedrichsfelde	28	f,g	–
ser Abadina	28	g,m	$[e,n,z_{15}]$
ser Croft	28	g,m,s	–
ser Ona	28	g,s,t	–
*ser Llandudno	28	g,s,t	1,5
ser Vinohrady	28	m,t	–
*ser variant	28	m,t	–
ser Cotham	28	i	1,5
ser Volkmarsdorf	28	i	1,6
ser Kuessel	28	i	e,n,z_{15}
ser Guildford	28	k	1,2
ser Ilala	28	k	1,5
ser Adamstown	28	k	1,6
ser Taunton	28	k	e,n,x

ORGANISM	SOMATIC (O) ANTIGEN	FLAGELLAR (H) ANTIGEN Phase 1	Phase 2
ser Ank	28	k	e,n,z_{15}
ser Leoben	28	l,v	1,5
ser Vitkin	28	l,v	e,n,x
ser Nashua	28	l,v	e,n,z_{15}
ser Chicago	28	r	1,5
ser Kibusi	28	r	e,n,x
*ser Oevelgoenne	28	r	e,n,z_{15}
ser Sanktgeorg	28	r(i)	e,n,z_{15}
ser Oskarshamn	28	y	1,2
ser Nima	28	y	1,5
ser Pomona	28	y	1,7
ser Kitenge	28	y	e,n,x
ser Telaviv	28	y	e,n,z_{15}
ser Shomolu	28	y	l,w
ser Ezra	28	z	1,7
ser Brisbane	28	z	e,n,z_{15}
*ser Ceres	28	z	z_{39}
ser Babelsberg	28	z_4,z_{23}	e,n,z_{15}
ser Teltow	28	z_4,z_{23}	1,6
ser Rosy	28	z_{10}	1,2
ser Umbilo	28	z_{10}	e,n,x
ser Luckenwalde	28	z_{10}	e,n,z_{15}
ser Moroto	28	z_{10}	l,w
ser Djermaia	28	z_{29}	–
ser Aderike	28	z_{38}	–
Serogroup N			
ser Overvecht	30	a	1,2
ser Zehlendorf	30	a	1,5
*ser Odijk	30	a	z_{39}

ORGANISM	SOMATIC (O) ANTIGEN	FLAGELLAR (H) ANTIGEN Phase 1	Phase 2
ser Louga	30	b	1,2
ser Aschersleben	30	b	1,5
ser Urbana	30	b	e,n,x
ser Messina	30	d	1,5
*ser Slatograd	30	f,g,(p),t	–
ser Godesberg	30	g,m	–
ser Giessen	30	g,m,s	–
ser Sternchanze	30	g,s,t	–
ser Wayne	30	g,z_{51}	–
ser Landau	30	i	1,2
ser Morehead	30	i	1,5
ser Soerenga	30	i	l,w
ser Hilversum	30	k	1,2
ser Ramatgan	30	k	1,5
ser Aqua	30	k	1,6
ser Angoda	30	k	e,n,x
ser Odozi	30	k	e,n,x,z_{15}
* variant	30	k	e,n,x,z_{15}
ser Ligeo	30	l,v	1,2
ser Donna	30	l,v	1,5
ser Morocco	30	l,z_{13},z_{28}	e,n,z_{15}
ser Gege	30	r	1,5
ser Matopeni	30	y	1,2
ser Steinplatz	30	y	1,6
ser Baguirmi	30	y	e,n,x
ser Bodjonegoro	30	z_4,z_{24}	–
ser Kumasi	30	z_{10}	e,n,z_{15}
ser Ago	30	z_{38}	–
*ser	30	z_{39}	1,(7)

ORGANISM	SOMATIC (O) ANTIGEN	FLAGELLAR (H) ANTIGEN Phase 1	Phase 2
	Serogroup O		
ser Umhlatazana	35	a	e,n,z_{15}
ser Tchad	35	b	–
ser Yolo	35	c	–
ser Dembe	35	d	l,w
ser Gassi	35	e,h	z_6
ser Adelaide	35	f,g	–
ser Ealing	35	g,m,s	–
ser Ebrie	35	g,m,t	–
*ser	35	g,m,s,t	–
ser Agodi	35	g,t	–
ser Monschaui	35	m,t	–
ser Ganbia	35	i	e,n,z_{15}
ser Bandia	35	i	l,w
ser	35	l,z_{28}	–
ser Massakory	35	r	l,w
ser Alachua	35	z_4,z_{23}	–
ser Westphalia	35	z_4,z_{24}	–
ser Camberene	35	z_{10}	1,5
ser Enschede	35	z_{10}	l,w
ser Ligna	35	z_{10}	z_6
*ser Utbremen	35	z_{29}	e,n,x
	Serogroup P		
ser Sheffield	38	c	1,5
ser Kidderminster	38	c	1,6
*ser Carletonville	38	d	[1,5]
ser Thiaroye	38	e,h	1,2

Salmonella antigenic scheme — continued

(O:38 group)

ORGANISM	SOMATIC (O) ANTIGEN	FLAGELLAR (H) ANTIGEN Phase 1	Phase 2
ser Kasenyi	38	e,h	1,5
ser Korovi	38	g,m,s	—
*ser Foulpointe	38	g,t	—
ser Mgulani	38	i	1,2
ser Lansing	38	i	1,5
ser Echa	38	k	1,2
ser Inverness	38	k	1,6
ser Alger	38	l,v	1,2
ser Kimberly	38	l,v	1,5
ser Roan	38	l,v	e,n,x
ser Lindi	38	r	1,5
ser Emmastad	38	r	1,6
ser Freetown	38	y	1,5
ser Colombo	38	y	1,6
ser Perth	38	y	e,n,x
ser Yoff	38	z_4,z_{23}	1,2

Serogroup Q

ORGANISM	SOMATIC (O) ANTIGEN	FLAGELLAR (H) ANTIGEN Phase 1	Phase 2
ser Wandsworth	39	b	1,2
ser Logone	39	d	1,5
ser Mara	39	e,h	[1,5]
ser Hofit	39	i	1,5
ser Champaign	39	k	1,5
ser Kokomlemle	39	l,v	e,n,x
*ser Mondeor	39	l,z_{28}	e,n,x
ser Anfo	39	y	1,2
ser Windermere	39	y	1,5
ser Cook	39	z_{48}	1,5

Serogroup R

ORGANISM	SOMATIC (O) ANTIGEN	FLAGELLAR (H) ANTIGEN Phase 1	Phase 2
ser Shikmonah	40	a	1,5
ser Greiz	40	a	z_6
*ser Springs	40	a	z_{39}
*ser	40	b	1,5
ser Riogrande	40	b	e,n,x
ser Johannesburg	1,40	b	e,n,z_{15}
ser Duval	1,40	b	z_6
ser Benguella	40	c	e,n,x,z_{15}
*ser Suarez	1,40	d	1,5
*ser Ottershaw	40	d	1,2
ser Driffield	1,40	e,h	—
ser Tilene	1,40	g,m,t	—
*ser Alsterdorf	1,40	g,p	e,n,x,z_{15}
ser Maartensdyijk	40	g,s	—
*ser Boksburg	1,40a,40b	g,z_{51}	—
*ser Seminole	1,40a,40c	g,z_{51}	—
*ser	1,40	m,t	z_{42}
ser Goulfy	1,40	k	1,5,(6)
ser Allandale	1,40	k	1,6
*ser Sunnydale	1,40	k	e,n,x,z_{15}
ser Millesi	1,40	l,v	1,2
ser Bukavu	1,40	l,z_{28}	1,5
ser Santhiaba	40	l,z_{28}	1,6
*ser Bulawayo	(1),40	z	1,5
ser Nowawes	40	z	z_6
*ser Sachsenwald	1,40	z_4,z_{23}	—
*ser Degania	40	z_4,z_{24}	—
*ser Bern	1,40	z_4,z_{32}	—

Serogroup S

ORGANISM	SOMATIC (O) ANTIGEN	FLAGELLAR (H) ANTIGEN Phase 1	Phase 2
*ser ___	1,40	z_6	1,5
ser Trotha	40	z_{10}	z_6
ser Omifisan	40	z_{29}	—
*ser Fandran	1,40	z_{35}	e,n,x,z_{15}
*ser Grunty	1,40	z_{39}	1,6
ser Karamoja	40	z_{41}	1,2
*ser ___	40	—	1,7
ser Vietnam	41	b	—
* variant	41	b	—
ser Egusi	41	d	—
*ser Hennepin	41	d	z_6
*ser Lethe	41	g,t	—
*ser ___	41	k	—
*ser Dubrovnik	41	z	1,5
ser Waycross	41	z_4,z_{23}	—
* variant	41	z_4,z_{23}	—
ser Ipswich	41	z_4,z_{24}	—
*ser Negev	41	z_{10}	1,2
ser Leipzig	41	z_{10}	1,5
ser Landala	41	z_{10}	1,6
ser Inpraw	41	z_{10}	e,n,x
*ser Lurup	41	z_{10}	e,n,x,z_{15}
*ser Lichtenberg	41	z_{10}	$[z_6]$
ser Offa	41	z_{38}	—

Serogroup T

ORGANISM	SOMATIC (O) ANTIGEN	FLAGELLAR (H) ANTIGEN Phase 1	Phase 2
ser Faji	1,42	a	e,n,z_{15}
*ser Chinovum	42	b	1,5
*ser Uphill	42	b	e,n,x,z_{15}
ser Egusitoo	1,42	b	z_6
ser Kampala	1,42	c	z_6
*ser Fremantle	42	(f),g,t	—
ser Maricopa	1,42	g,z_{51}	1,5
*ser ___	42	m,t	e,n,x,z_{15}
ser Kaneshie	1,42	i	1,w
ser Middlesbrough	1,42	i	z_6
ser Haferbreite	42	k	[1,6]
*ser Portbech	42	l,v	e,n,x,z_{15}
*ser Nairobi	42	r	—
ser Harvestehude	1,42	y	z_6
*ser Detroit	42	z	1,5
*ser Ursenbach	1,42	z	1,6
*ser Rand	42	z	e,n,x,z_{15}
*ser Nuernberg	42	z	z_6
ser Gera	1,42	z_4,z_{23}	1,6
ser Loenga	1,42	z_{10}	z_6
ser Kahla	1,42	z_{35}	1,6
ser Weslaco	42	z_{36}	—
*ser ___	42	—	1,6

ORGANISM	SOMATIC (O) ANTIGEN	FLAGELLAR (H) ANTIGEN Phase 1	Phase 2
Serogroup U			
ser Graz	43	a	1,2
ser Berkeley	43	a	1,5
*ser Kommetje	43	b	z_{42}
*ser _____	43	e,n,x,z_{15}	1,(5),7
*ser _____	43	e,n,z_{15}	1,6
*ser Milwaukee	43	f,g	–
*ser Mosselbay	43	g,s,(t)	z_{42}
*ser Veddel	43	g,t	–
ser Mbao	43	i	1,2
ser Ahuza	43	k	1,5
ser Farcha	43	y	1,2
ser Kingabwa	43	y	1,5
*ser _____	43	z	1,5
*ser Houten	43	z_4,z_{23}	–
*ser Tuindorp	43	z_4,z_{32}	–
*ser Volksdorf	43	z_{36},z_{38}	–
ser Irigny	43	z_{38}	–
*ser Bunnik	43	z_{42}	[1,5]
Serogroup V			
ser Niarembe	44	a	1,w
ser Sedgwick	44	b	e,n,z_{15}
ser Madigan	44	c	1,5
ser Bobo	44	d	1,5
ser Fischerstrasse	44	d	e,n,x,z_{15}

ORGANISM	SOMATIC (O) ANTIGEN	FLAGELLAR (H) ANTIGEN Phase 1	Phase 2
ser Vleuten	44	f,g	–
ser Gamaba	44	g,m,s	–
*ser Muguga	44	g,z_{51}	–
ser Lawra	44	k	e,n,z_{15}
ser Uhlenhorst	44	z	1,w
*ser _____	44	z_4,z_{23}	–
*ser Christiansborg variant	44	z_4,z_{24}	–
*ser _____	44	z_4,z_{24}	–
*ser Lohbruegge	44	z_4,z_{32}	–
ser Guinea	44	z_{10}	[1,7]
*ser _____	44	z_{36},z_{38}	–
*ser Clovelly	1,44	z_{39}	$[e,n,x,z_{15}]$
Serogroup W			
*ser Vrindaban	45	a	e,n,x
*ser Ejeda	45	a	z_{10}
ser Riverside	45	b	1,5
ser Deversoir	45	c	e,n,x
ser Dugbe	45	d	1,6
ser Karachi	45	d	e,n,x
ser Suelldorf	45	f,g	–
ser Tornow	45	g,m	e,n,x
*ser Bremen	45	g,m,s,t	e,n,x
*ser Windhoek	45	g,t	1,5
*ser _____	45	g,z_{51}	–
ser Apapa	45	m,t	1,5
*ser Perinet	45	m,t	e,n,x,z_{15}

ORGANISM	SOMATIC (O) ANTIGEN	FLAGELLAR (H) ANTIGEN Phase 1	Phase 2
ser Casablanca	45	k	1,7
ser Cairns	45	k	e,n,z_{15}
*ser Klapmuts	45	z	z_{39}
ser Jodhpur	45	z_{29}	—
ser Lattenkamp	45	z_{35}	1,5
Serogroup X			
*ser Bilthoven	47	a	[1,5]
ser Saka	47	b	—
*ser Phoenix	47	b	1,5
*ser Khami	47	b	e,n,x,z_{15}
ser Stellingen	47	d	e,n,x
*ser Quimbamba	47	d	z_{39}
ser Sljeme	1,47	f,g	—
ser Luke	1,47	g,m	—
ser Mesbit	47	m,t	e,n,z_{15}
ser Bergen	47	i	e,n,z_{15}
ser Bootle	47	k	1,5
ser Lyon	47	k	e,n,z_{15}
ser Teshie	1,47	l,z_{13},z_{28}	e,n,z_{15}
ser Moualine	47	y	1,6
ser Mountpleasant	47	z	1,5
ser Kaolack	47	z	1,6
*ser Chersina	47	z	z_6
ser Bere	47	z_4,z_{23}	z_6
*ser _____	47	z_6	1,6

ORGANISM	SOMATIC (O) ANTIGEN	FLAGELLAR (H) ANTIGEN Phase 1	Phase 2
ser Alexanderplatz	47	z_{38}	—
ser Quinhon	47	z_{44}	—
Serogroup Y			
ser Hisingen	48	a	1,5,7
*ser Pumila	48	a	z_6
*ser Hagenbeck	48	d	z_6
ser Fitzroy	48	e,h	1,5
*ser Hammonia	48	e,n,x,z_{15}	z_6
*ser Erlangen	48	g,m,t	—
*ser Marina	48	g,z_{51}	—
*ser Sydney	48	i	z
ser Dahlem	48	k	e,n,z_{15}
*ser Sakaraha	48	[k]	z_{39}
ser Djakarta	48	z_4,z_{24}	—
*ser	48	z_4,z_{32}	—
*ser Ngozi	48	z_{10}	[1,5]
ser Bongor	48	z_{35}	—
Serogroup Z			
ser Rochdale	50	b	e,n,x
*ser Krugersdorp	50	e,n,x	1,7
*ser Namib	50	g,m,s,t	1,5
*ser Wassenaar	50	g,z_{51}	—
ser Atra	50	m,t	z_6:z_{42}
*ser Seaforth	50	k	z_6

ORGANISM	SOMATIC (O) ANTIGEN	FLAGELLAR (H) ANTIGEN Phase 1	Phase 2
*ser _____	50	l,w	$e,n,x,z_{15}:z_{42}$
*ser _____	50	l,z	z_{42}
ser Dougi	50	y	1,6
*ser Greenside	50	z	e,n,x
*ser Flint	50	z_4,z_{23}	—
*ser _____	50	z_4,z_{24}	—
*ser Bonaire	50	z_4,z_{32}	—
*ser Hooggraven	50	z_{10}	$z_6:z_{42}$
*ser Faure	50	z_{42}	1,7
Serogroup 51			
ser Tione	51	a	e,n,x
ser Gokul	1,51	d	—
ser Meskin	51	e,h	1,2
ser Dan	51	k	e,n,z_{15}
ser Overschie	51	l,v	1,5
*ser Askraal	51	l,z_{28}	—
ser Antsalova	51	z	1,5
ser Treforest	1,51	z	1,6
*ser Harnelen	51	z_4,z_{23}	—
*ser Roggeveld	51	—	1,7
Serogroup 52			
ser Flottbek	52	b	—
ser Utrecht	52	d	1,5

ORGANISM	SOMATIC (O) ANTIGEN	FLAGELLAR (H) ANTIGEN Phase 1	Phase 2
*ser _____	52	d	e,n,x,z_{15}
ser Saintemarie	52	g,t	—
*ser Wilhelmstrasse	52	z_{44}	1,5,7
*ser Lobatsi	52	—	1,5,7
Serogroup 53			
*ser Midhurst	53	l,z_{28}	z_{39}
*ser _____	53	z	z_6
*ser Humber	53	z_4,z_{24}	—
*ser Bockenheim	1,53	z_{36},z_{38}	—
Serogroup 54			
ser Tonev	54	b	e,n,x
ser Rossleben	54	e,h	1,6
ser Uccle	54	g,s,t	—
ser Poeseldorf	54	i	z_6
ser Ochsenwerder	54	k	1,5
ser Yerba	54	z_4,z_{23}	—
Serogroup 55			
*ser Tranoroa	55	k	z_{39}
Serogroup 56			
*ser Artis	56	b	—
*ser _____	56	e,n,x	1,7

ORGANISM	SOMATIC (O) ANTIGEN	FLAGELLAR (H) ANTIGEN Phase 1	Phase 2
	Serogroup 57		
ser Antonio	57	a	z_6
*ser _____	57	g,m,s,t	z_{42}
*ser Locarno	57	z_{29}	z_{42}
*ser Manombo	57	z_{39}	e,n,x,z_{15}
*ser Tokai	57	z_{42}	1,6:z_{53}
	Serogroup 58		
*ser _____	58	a	1,5
*ser _____	58	a	—
*ser Basel	58	1,z_{13},z_{28}	1,5
	Serogroup 59		
*ser Betioky	59	k	(z)

ORGANISM	SOMATIC (O) ANTIGEN	FLAGELLAR (H) ANTIGEN Phase 1	Phase 2
	Serogroup 60		
*ser Setubal	60	g,m,t	z_6
*ser Luton	60	z	e,n,x
	Serogroup 61		
*ser Eilbek	61	i	z
	Serogroup 64		
*ser _____	64	k	e,n,x,z_{15}
*ser _____	64	z_{29}	—

* = biochemically aberrant   ( ) = antigen incomplete

[ ] = antigen may be present or absent

ALPHABETICAL LIST OF THE SEROTYPES OF *SALMONELLA*

ORGANISM	O GROUP
Salmonella enteritidis	
ser Aba 6,8:i:e,n,z_{15}	C_2
ser Abadina 28:g,m:[e,n,z_{15}]	M
ser Abaetetuba 11:k:1,5	F
ser Aberdeen 11:i:1,2	F
ser Abony 1,4,5,12:b:e,n,x	B
variant Haifa 4,12:b:e,n,x	B
ser Abortusbovis 1,4,12,27:b:	
e,n,x	B
ser Abortuscanis 4,5,12:b:z_5	B
ser Abortusequi 4,12:-:e,n,x	B
ser Abortusovis 4,12:c:1,6	B
ser Accra 1,3,19:b:z_6	E_4
*ser Acres 1,13,23:b:z_{42}:[1,5]	G_2
ser Adabraka 3,10:z_4,z_{23}:[1,7]	E_1
ser Adamstown 28:k:1,6	M
ser Adamstua 11:e,h:1,6	F
ser Adelaide 35:f,g:-	O
ser Adeoyo 16:g,m:-	I
ser Aderike 28:z_{38}:-	M
ser Adjame 13,23:r:1,6	G_2
ser Aequatoria 6,7:z_4,z_{23}:e,n,z_{15}	C_1
ser Aertrycke = ser Typhimurium	
ser Aflao 1,6,14,25:l,z_{28}:e,n,x	H
ser Africana 4,12:r,(i):l,w	B
ser Agama 4,12:i:1,6	B
ser Agbeni 13,23:g,m:-	G_2
ser Agege 3,10:c:e,n,z_{15}	E_1
ser Ago 30:z_{38}:-	N
ser Agodi 35:g,t:-	O
ser Agona 4,12:f,g,s:-	B
ser Ahmadi 1,3,19:d:1,5	E_4
ser Ahuza 43:k:1,5	U
ser Ajiobo 13,23:z_4,z_{23}:-	G_2
ser Akanji 6,8:r:1,7	C_2
ser Akuafo 16:y:1,6	I
ser Alabama 9,12:c:e,n,z_{15}	D_1
ser Alachua 35:z_4,z_{23}	O
ser Alagbon 6,8:y:1,7	C_2
ser Alamo 6,7:g,z_{51}:1,5	C_1
ser Albany (8),20:z_4,z_{24}:-	C_3
ser Albert 4,12:z_{10}:e,n,x	B
ser Albuquerque 6,14,24:d:z_6	H
*ser Alexander 3,10:z:1,5	E_1
ser Alexanderplatz 47:z_{38}:-	X
ser Alexanderpolder (8):c:l,w	C_3
ser Alger 38:l,v:1,2	P
ser Allandale 1,40:k:1,6	R
ser Allerton 3,10:b:1,6	E_1
*ser Alsterdorf 1,40:g,m,t:-	R
ser Altendorf 4,12:c:1,7	B
ser Altona (8),20:r(i):z_6	C_3
ser Amager 3,10:y:1,2	E_1
ser Amba 11:k:l,z_{13},z_{28}	F

ORGANISM	O GROUP
ser Amersfoort 6,7:d:e,n,x	C_1
ser Amherstiana (8):l,(v):1,6	C_3
ser Amina 16:i:1,5	I
ser Aminatu 3,10:a:1,2	E_1
ser Amounderness 3,10:i:1,5	E_1
ser Amoutive 28:d:1,5	M
ser Amsterdam 3,10:g,m,s:-	E_1
ser Amunigun 16:a:1,6	I
ser Anatum 3,10:e,h:1,6	E_1
ser Anderlecht 3,10:c:l,w	E_1
ser Anfo 39:y:1,2	Q
ser Angoda 30:k:e,n,x	N
*ser Angola 1,9,12:z:z_6	D_1
ser Ank 28:k:e,n,z_{15}	M
ser Annedal 16:r(i):e,n,x	I
ser Antonio 57:a:z_6	57
ser Antsalova 51:z:1,5	51
ser Apapa 45:m,t:-	W
ser Aqua 30:k:1,6	N
ser Ardwick 6,(7),(14):f,g:-	C_1
ser Arechavaleta 4,5,12:a:[1,7]	B
*ser Argentina 6,7:z_{36}:-	C_1
ser Arkansas (3),(15),34:e,h:1,5	E_3
*ser Artis 56:b:-	56
ser Aschersleben 30:b:1,5	N
ser Ashanti 28:b:1,6	M
*ser Askraal 51:l,z_{28}:-	51
ser Assen 21:a:-	L
ser Atherton = ser Waycross	
ser Atlanta (Mississippi)	
[1],13,23:b:-	G_2
*ser Atra 50:m,t:z_6:z_{42}	Z
ser Augustenborg 6,7:i:1,2	C_1
ser Austin 6,7:a:1,7	C_1
ser Avonmouth 1,3,19:i:e,n,z_{15}	E_4
ser Ayinde 4,12,27:d:z_6	B
ser Ayton 1,4,12,27:l,w:z_6	B
ser Azteca 4,5,12:l,v:1,5	B
ser Babelsberg 28:z_4,z_{23}:e,n,z_{15}	M
*ser Bacongo 6,7:z_{36}:z_{42}	C_1
ser Baguirmi 30:y:e,n,x	N
ser Bahati 13,22:b:e,n,z_{15}	G_1
ser Bahrenfeld 6,14,24:e,h:1,5	H
ser Baiboukoum 6,7:k:1,7	C_1
ser Baildon (9),46:a:e,n,x	D_2
ser Ball 1,4,12,27:y:e,n,x	B
ser Bambesa, joined w/ser Miami	
ser Bamboye (9),46:b:l,w	D_2
ser Banalia 6,8:b:z_6	C_2
ser Banana, joined w/ser California	
ser Bandia 35:i:l,w	O
ser Bantam = ser Meleagridis	

ORGANISM	O GROUP
*ser Baragwanath $6,8:m,t:1,5$	C_2
ser Bardo $(8):e,h:1,2$	C_3
ser Bareilly $6,7,[14]:y:1,5$	C_1
ser Barmbek $16:d:z_6$	I
ser Barranguilla $16:d:e,n,x$	I
*ser Basel $58:l,z_{13},z_{28}:1,5$	58
ser Batavia = ser Lexington	
*ser Bechuana $4,12,27:g,t:-$	B
ser Bedford $1,3,19:l,z_{13},z_{28}:e,n,z_{15}$	E_4
ser Belem $6,8:c:e,n,x$	C_2
*ser Bellville $16:e,n,x:1,7$	I
*ser Beloha $18:z_{36}:-$	K
ser Benfica $3,10:b:e,n,x$	E_1
variant T_1 $T_1:b:e,n,x$	
ser Benguella $40:b:z_6$	R
ser Bere $47:z_4,z_{23}:z_6$	X
ser Bergedorf $(9),46:e,h:1,2$	D_2
ser Bergen $47:i:e,n,z_{15}$	X
ser Berkeley $43:a:1,5$	U
ser Berlin $17:d:1,5$	J
*ser Bern $1,40:z_4,z_{32}:-$	R
ser Berta $9,12:f,g,t:-$	D_1
*ser Betioky $59:k:(z)$	59
ser Biafra $3,10:z_{10}:z_6$	E_1
*ser Bilthoven $47:a:[1,5]$	X
ser Bilu $(1),3,10,(19):f,g,t:1,(2),7$	E_4
ser Binza $3,15:y:1,5$	E_2
ser Birkenhead $6,7:c:1,6$	C_1
ser Birmingham $3,10:d:l,w$	E_1
ser Bispebjerg $1,4,5,12:a:e,n,x$	B
*ser Blankenese $1,9,12:b:z_6$	D_1
*ser Bleadon $17:(f),g,t:[e,n,x,z_{15}]$	J
ser Bledgam $9,12:g,m,q:-$	D_1
ser Blijdorp $1,6,14,25:c:1,5$	H
ser Blockley $6,8:k:1,5$	C_2
*ser Bloemfontein $6,7:b:[e,n,x]:z_{42}$	C_1
ser Blukwa $18:z_4,z_{24}:-$	K
ser Bobo $44:d:1,5$	V
ser Bochum $4,5,12:r:l,w$	B
*ser Bockenheim $1,53:z_{36},z_{38}:-$	53
ser Bodjonegoro $30:z_4,z_{24}:-$	N
ser Boecker $[1],6,14[25]:l,v:1,7$	H
ser Bokanjac $28:b:1,7$	M
*ser Boksburg $40:g,s:e,n,x,z_{15}$	R
ser Bolombo $3,10:z_{38}:-$	E_1
ser Bolton $3,10:y:e,n,z_{15}$	E_1
ser Bombay, not confirmed	
*ser Bonaire $50:z_4,z_{32}:-$	Z
ser Bonames $17:a:1,2$	J
ser Bonariensis $6,8:i:e,n,x$	C_2
ser Bongor $48:z_{35}:-$	Y
ser Bonn $6,7:l,v:e,n,x$	C_1
ser Bootle $47:k:1,5$	X
ser Borbeck $13,22:l,v:1,6$	G_1
*ser Bornheim $1,6,14,25:z_{10}:1,(2),7$	H
ser Bornum $6,(7),(14):z_{38}:-$	C_1
*ser Boulders $13,23:m,t:z_{42}$	G_2
ser Bournemouth $9,12:e,h:1,2$	D_1
ser Bousso $1,6,14,25:z_4,z_{23}:-$	H

ORGANISM	O GROUP
ser Bovismorbificans $6,8:r:1,5$	C_2
ser Bracknell $13,23:b:1,6$	G_2
ser Bradford $4,12,27:r:1,5$	B
ser Braenderup $6,7:e,h:e,n,z_{15}$	C_1
ser Brancaster $1,4,12,27:z_{29}:-$	B
ser Brandenburg $4,12:l,v:e,n,z_{15}$	B
ser Brazil $16:a:1,5$	I
ser Brazzaville $6,7:b:1,2$	C_1
ser Bredeney $1,4,12,27:l,v:1,7$	B
*ser Bremen $45:g,m,s,t:e,n,x$	W
ser Breukelen $6,8:l,z_{13}:e,n,z_{15}$	C_2
ser Brijbhumi $11:i:1,5$	F
ser Brisbane $28:z:e,n,z_{15}$	M
ser Bristol $13,22:z:1,7$	G_1
ser Bron $13,22:g,m:[e,n,z_{15}]$	G_1
ser Bronx $6,8:c:1,6$	C_2
ser Broughton $1,3,19:b:l,w$	E_4
ser Broxbourne = ser Wien	
ser Brunei $(8),20:y:1,5$	C_3
ser Budapest $1,4,12:g,t:-$	B
ser Buenosaires = ser Bonariensis	
ser Bukavu $1,40:l,z_{28}:1,5$	R
ser Bukuru $6,8:b:l,w$	C_2
*ser Bulawayo $1,40:z:1,5$	R
ser Bulbay $11:l,v:e,n,z_{15}$	F
*ser Bunnik $43:z_{42}:[1,5]$	U
ser Burgas $16:l,v:e,n,z_{15}$	I
ser Bury $4,12,27:c:z_6$	C_1
ser Businga $6,7:z:e,n,z_{15}$	C_1
ser Butantan $3,10:b:1,5$	E_1
ser Buzu $1,6,14,25:i:1,7$	H
ser Cairina $3,10:z_{35}:z_6$	E_1
ser Cairns $45:k:e,n,z_{15}$	W
ser Cairo $1,4,12,27:d:1,2$	B
ser Calabar $1,3,19:e,h:l,w$	E_4
*ser Caledon $4,12:g,m:e,n,x$	B
ser California $4,5,12:m,t:-$	B
*ser Calvinia $6,7:a:z_{42}$	C_1
ser Camberene $35:z_{10}:1,5$	O
ser Cambridge $3,15:e,h:l,w$	E_2
ser Canada $4,12:b:1,6$	B
*ser Canastel $9,12:z_{29}:1,5$	D_1
ser Canoga $(3),(15),34:g,s,t:-$	E_3
ser Cannstatt $1,3,19:m,t:-$	E_4
*ser Cape $6,7:z_6:1,7$	C_1
ser Caracas $1,6,14,25:g,m,s:-$	H
ser Cardiff $6,7:k:1,10$	C_1
*ser Carletonville $38:d:[1,5]$	P
ser Carmel $17:l,v:e,n,x$	J
ser Carno $1,3,19:z:l,w$	E_4
ser Carrau $1,6,14,24:y:1,7$	H
ser Casablanca $45:k:1,7$	W
*ser Ceres $28:z:z_{39}$	M
ser Cerro $18:z_4,z_{23}:[z_{45}]$	K
ser Ceyco $(9),46:k:z_{35}$	D_2
ser Chagoua $1,13,23:a:1,5$	G_2
ser Chailey $6,8:z_4,z_{23}:[e,n,z_{15}]$	C_2

ORGANISM	O GROUP
*ser Chameleon 16:z_4,z_{32}:-	I
ser Champaign 39:k:1,5	Q
ser Chandans 11:d:e,n,x	F
ser Charity 1,6,14,25:d:e,n,x	H
*ser Chersina 47:z:z_6	X
ser Chester 4,5,12:e,h:e,n,x	B
ser Chiba, not a Salmonella	
ser Chicago 28:r:1,5	M
ser Chincol 6,8:g,m,s:e,n,x	C_2
ser Chingola 11:e,h:1,2	F
*ser Chinovum 42:b:1,5	T
ser Chittagong (1),3,10,(19):b:z_{35}	E_4
Salmonella cholerae-suis 6,7:c:1,5	C_1
bioser Kunzendorf 6,7:[c]:1,5	C_1
S. enteritidis	
ser Christiansborg 44:z_4,z_{24}:-	V
* variant 44:z_4,z_{24}:-	V
*ser Chudleigh 3,10:e,n,x:1,7	E_1
ser Clackamas 4,12:l,v,(z_{13}):1,6	B
ser Claibornei 1,9,12:k:1,5	D_1
ser Clerkenwell 3,10:z:l,w	E_1
ser Cleveland 6,8:z_{10}:1,7	C_2
*ser Clifton 13,22:z_{29}:1,5	G_1
*ser Clovelly 1,44:z_{39}:[e,n,x,z_{15}]	V
ser Cocody (8),20:r,(i):e,n,z_{15}	C_3
ser Coeln 4,5,12:y:1,2	B
ser Coleypark 6,7:a:l,w	C_1
ser Colindale 6,7:r:1,7	C_1
ser Colombo 38:y:1,6	P
ser Colorado 6,7:l,w:1,5	C_1
ser Concord 6,7:l,v:1,2	C_1
ser Congo 13,23:g,t:-	G_2
*ser Constantia 17:z:l,w:z_{42}	J
ser Cook 39:z_{48}:1,5	Q
ser Coquilhatville 3,10:z_{10}:1,7	E_1
ser Corvallis (8),20:z_4,z_{23}:-	C_3
ser Cotham 28:i:1,5	M
ser Croft 28:g,m,s:-	M
ser Cuba = ser Cubana	
ser Cubana 1,13,23:z_{29}:-	G_2
ser Curacao 6,8:a:1,6	C_2
ser Dahlem 48:k:e,n,z_{15}	Y
ser Dakar 28:a:1,6	M
ser Dalat, joined w/ser Ball	
ser Dallgow 1,3,19:z_{10}:e,n,z_{15}	E_4
ser Dan 51:k:e,n,z_{15}	51
*ser Daressalaam 1,9,12:l,w:e,n,x	D_1
ser Daytona 6,7:k:1,6	C_1
bioser Decatur 6,7:c:1,5	C_1
*ser Degania 40:z_4,z_{24}:-	R
ser Dembe 35:d:l,w	O
ser Demerara 13,23:z_{10}:l,w	G_2
ser Denver 6,7:a:e,n,z_{15}	C_1
ser Derby 1,4,5,12:f,g:[1,2]	B
ser Dessau (1),3,15,(19):g,s,t:-	E_4
*ser Detroit 42:z;1,5	T

ORGANISM	O GROUP
ser Deversoir 45:c:e,n,x	W
ser Diguel 1,13,22:d:e,n,z_{15}	G_1
ser Diourbel 21:i:1,2	L
ser Djakarta 48:z_4,z_{24}:-	Y
ser Djermaia 28:z_{29}:-	M
ser Djugu 6,7:z_{10}:e,n,x	C_1
ser Doncaster 6,8:a:1,5	C_2
ser Donna 30:l,v:1,5	N
ser Dougi 50:y:1,6	Z
ser Dresden 28:c:e,n,x	M
ser Driffield 1,40:d:1,5	R
ser Drypool 3,15:g,m,s:-	E_2
ser Dublin 1,9,12:g,p:-	D_1
variant Vi + 1,9,12,Vi:g,p:-	D_1
*ser Dubrovnik 41:z:1,5	S
ser Duesseldorf 6,8:z_4,z_{24}:-	C_2
ser Dugbe 45:d:1,6	W
ser Duisburg [1],4,12,[27]:d:e,n,z_{15}	B
*ser Duivenhoks (9),46:g,m,s,t:e,n,x	D_2
ser Durban 9,12:a:e,n,z_{15}	D_1
*ser Durbanville [1],4,12,[27]:	
[z_{39}]:1,5,7	B
ser Durham 13,23:b:e,n,z_{15}	G_2
ser Duval 1,40:b:e,n,z_{15}	R
ser Ealing 35:g,m,s:-	O
ser Eastbourne 1,9,12:e,h:1,5	D_1
ser Eberswalde 28:c:1,6	M
ser Ebrie 35:g,m,t:-	O
ser Echa 38:k:1,2	P
ser Edinburg 6,7:b:1,5	C_1
ser Edmonton 6,8:l,v:e,n,z_{15}	C_2
ser Egusi 41:d:-	S
ser Egusitoo 1,42:b:z_6	T
*ser Eilbek 61:i:z	61
*ser Eimsbuettel 6,(7),(14),:d:l,w	C_1
*ser Ejeda 45:a:z_{10}	W
ser Ekotedo (9),46:z_4,z_{23}:-	D_2
ser Elizabethville 3,10:r:1,7	E_1
ser Elomrane 1,9,12:z_{38}:-	D_1
*ser Elsiesivier 16:z_{42}:1,6	I
ser Emek (8),20:g,m,s:-	C_3
ser Emmastad 38:r:1,6	P
*ser Emmerich 6,14:[m,t]:e,n,x	H
ser Encino 1,6,14,25:d:l,z_{13},z_{28}	H
ser Enschede 35:z_{10}:l,w	O
ser Entebbe 1,4,12,27:z:z_6	B
ser Enteritidis 1,9,12:g,m:-	D_1
ser Enugu 16:l,z_{13},z_{28}:-	I
ser Epicrates 3,10:b:l,w	E_1
ser Eppendorf [1],4,12,[27]:d:1,5	B
*ser Epping 13,23:e,n,x:1,7	G_2
*ser Erlangen 48:g,m,t:-	Y
ser Escanaba 6,7:k:e,n,z_{15}	C_1
ser Eschersheim 3,15:d:e,n,x	E_2
ser Eschweiler 6,7:z_{10}:1,6	C_1
ser Essen 4,12:g,m:-	B
ser Etterbeek 11:z_4,z_{23}:e,n,z_{15}	F
ser Ezra 28:z:1,7	M

ORGANISM	O GROUP
ser Faji 1,42:a:e,n,z_{15}	T
ser Falkensee 3,10:i:e,n,z_{15}	E_1
ser Fallowfield 3,10:l,z_{13},z_{28}:e,n,z_{15}	E_1
*ser Fandran 1,40:z_{35}:e,n,x,z_{15}	R
ser Fann 11:l,v:e,n,x	F
ser Fanti 13,23:z_{38}:-	G_2
ser Farcha 43:y:1,2	U
*ser Farmsen 13,23:z:1,6	G_2
*ser Faure 50:z_{42}:1,7	Z
ser Fayed 6,8:l,w:1,2	C_2
ser Ferlac 1,6,14,25:a:e,n,x	H
*ser Finchley 3,10:z:e,n,x	E_1
ser Findorff 11:d:z_6	F
ser Finkenwerder 1,6,14,25:d:1,5	H
ser Fischerhuette 16:a:e,n,z_{15}	I
ser Fischerkietz 1,6,14,25:y:e,n,x	H
ser Fischerstrasse 44:d:e,n,z_{15}	V
ser Fitzroy 48:e,h:1,5	Y
*ser Flint 50:z_4,z_{23}:-	Z
ser Florida 1,6,14,25:d:1,7	H
ser Flottbek 52:b:-	52
ser Fluntern 6,14,18:b:1,5	H
ser Fortune 4,12,27:z_{10}:z_6	B
*ser Foulpointe 38:g,t:-	P
ser Frankfurt 16:i:e,n,z_{15}	I
ser Freetown 38:y:1,5	P
*ser Fremantle 42:(f),g,t:-	T
ser Fresno (9),46:z_{38}:-	D_2
* variant (9),46:z_{38}:-	D_2
ser Friedenau 13,22:d:1,6	G_1
ser Friedrichsfelde 28:f,g:-	M
ser Frintrop 1,9,12:b:1,5	D_1
*ser Fuhlsbuettel 3,10:l,v:z_6	E_1
ser Fulica 4,5,12:a:1,5	B
ser Gabon 6,7:l,w:1,2	C_2
ser Galiema 6,7:k:1,2	C_1
ser Galil 3,10:a:e,n,z_{15}	E_1
bioser Gallinarum 1,9,12:-:-	D_1
ser Gamaba 44:g,m,s:-	V
ser Gambaga 21:z_{35}:e,n,z_{15}	L
ser Gambia 35:i:e,n,z_{15}	O
ser Gaminara 16:d:1,7	I
ser Garba 1,6,14,25:a:1,5	H
ser Garoli 6,7:i:1,6	C_1
ser Gassi 35:e,h;z_6	O
ser Gateshead (9),46:g,s,t:-	D_2
ser Gatow 6,7:y:1,7	C_1
ser Gatuni 6,8:b:e,n,x	C_2
ser Gdansk 6,7:l,v:z_6	C_1
ser Gege 30:r:1,5	N
ser Gelsenkirchen 6,(7),(14):l,v:z_6	C_1
ser Georgia 6,7:b:e,n,z_{15}	C_1
ser Gera 1,42:z_4,z_{23}:1,6	T
*ser Germiston 6,8:m,t:e,n,x	C_2
ser Ghana 21:b:1,6	L
ser Giessen 30:g,m,s:-	N
*ser Gilbert 6,7:z_{39}:1,7	C_1

ORGANISM	O GROUP
ser Give 3,10:l,v:1,7	E_1
ser Glasgow 16:b:1,6	I
*ser Glencairn 11:a:z_6:z_{42}	F
ser Glostrup 6,8:z_{10}:e,n,z_{15}	C_2
ser Gloucester 1,4,12,(27):i:l,w	B
ser Gnesta 1,3,19:b:1,5	E_4
ser Godesberg 30:g,m:-	N
ser Goelzau 3,10:a:1,5	E_1
ser Goerlitz 3,15:e,h:1,2	E_2
ser Goeteborg 9,12:c:1,5	D_1
ser Goettingen 9,12:l,v:e,n,z_{15}	D_1
*ser Gojenberg 1,13,23:g,t:1,5	G_2
ser Gokul 1,51:d:-	51
ser Goldcoast 6,8:r:l,w	C_2
ser Gombe 6,7:d:e,n,z_{15}	C_1
ser Good 21:f,g:e,n,x	L
*ser Goodwood 13,22:z_{29}:e,n,x	G_1
ser Gori 17:z:1,2	J
ser Goulfy 1,40:k:1,5,(6)	R
ser Goverdhan 9,12:k:1,6	D_1
*ser Grabouw 11:g,m,s,t:[z_{39}]	F
ser Graz 43:a:1,2	U
*ser Greenside 50:z:e,n,x	Z
ser Greiz 40:a:z_6	R
ser Grumpensis 13,23:d:l,7	G_2
*ser Grunty 1,40:z_{39}:1,6	R
ser Guildford 28:k:1,2	M
ser Guinea 44:z_{10}:[1,7]	V
*ser Gwaai 21:z_4,z_{24}:-	L
ser Gwoza 1,3,19:a:e,n,z_{15}	E_4
ser Haardt (8):k:1,5	C_3
*ser Haarlem (9),46:z:e,n,x	D_2
ser Habana = ser Havana	
ser Habar 6,8:z_{10}:e,n,x	C_2
*ser Haddon 16:z_4,z_{23}:-	I
ser Haelsingborg 6,7:m,p,t,[u]:-	C_1
ser Haferbreite 42:k:[1,6]	T
*ser Hagenbeck 48:d:z_6	Y
ser Haifa 1,4,5,12:z_{10}:1,2	B
variant afula 01 & 05- 4,12:z_{10}:1,2	B
ser Halle 28a,28c:c:1,7	M
variant Vidin 28a,28b:c:1,7	M
ser Halmstad 3,15:g,s,t:-	E_2
*ser Hamburg 1,9,12:g,t:-	D_1
ser Hamilton 3,15:z_{27}:-	
= R phase of ser Goerlitz	E_2
*ser Hammonia 48:e,n,x,z_{15}:z_6	Y
ser Hannover 16:a:1,2	I
ser Haouaria 13,22:c:e,n,x,z_{15}	G_1
ser Harburg 1,6,14,25:k:1,5	H
*ser Harmelen 51:z_4,z_{23}:-	51
ser Harrisonburg (3),(15),34:z_{10}:1,6	E_3
ser Hartford 6,7:y:e,n,x	C_1
ser Harvestehude 1,42:y:z_6	T
ser Hato 4,5,12:g,m,s:-	B.
ser Havana 1,13,23:f,g,[s]:-	G_2
ser Heerlen 11:i:1,6	F

ORGANISM	O GROUP
ser Heidelberg [1],4,[5],12:r:1,2	B
*ser Heilbron 6,7:l,z_{28}:1,5:[z_{42}]	C_1
*ser Helsinki 1,4,12:z_{29}:[e,n,x]	B
*ser Hennepin 41:d:z_6	S
ser Hermannswerder 28:c:1,5	M
ser Heron 16:a:z_6	I
ser Herston 6,8:d:e,n,z_{15}	C_2
ser Herzliya 11:y:e,n,x	F
ser Hessarek 4,12,[27]:a:1,5	B
ser Heves 6,14,24:d:1,5	H
ser Hidalgo 6,8:r:e,n,z_{15}	C_2
*ser Hillbrow 17:b:e,n,x,z_{15}	J
ser Hillegersberg (9),46:z_{35}:1,5	D_2
ser Hillsborough 6,7:z_{41}:l,w	C_1
ser Hilversum 30:k:1,2	N
ser Hindmarsh (8):r:1,5	C_3
ser Hirschfeldii = ser Paratyphi C	
ser Hisingen 48:a:1,5,7	Y
ser Hofit 39:i:1,5	Q
ser Holcomb 6,8:l,v:e,n,x	C_2
ser Holstein, not a salmonella	
ser Homosassa 1,6,14,25:z:1,5	H
ser Honelis 28:a:e,n,z_{15}	M
*ser Hooggraven 50:z_{10}:z_6:z_{42}	Z
ser Horsham 1,6,14,25:l,v:e,n,x	H
*ser Houten 43:z_4,z_{23}:-	U
*ser Hueningen 9,12:z:z_{39}	D_1
*ser Huila 11:l,z_{28}:e,n,x	F
ser Hull 16:b:1,2	H
*ser Humber 53:z_4,z_{24}:-	53
ser Huvudsta 3,10:b:1,7	E_1
ser Hvittingfoss 16:b:e,n,x	I
ser Ibadan 13:22:b:1,5	G_1
ser Idikan 13,23:i:1,5	G_2
ser Ilala 28:k:1,5	M
ser Illinois (3),(15),34:z_{10}:1,5	E_3
ser Ilugun (1),3,10,(19):z_4,z_{23}:z_6	E_4
ser Indiana 1,4,12:z:1,7	B
ser Infantis 6,7,[14]:r:1,5	C_1
ser Inganda 6,7:z_{10}:1,5	C_1
ser Inglis (9),46:z_{10}:e,n,x	D_2
ser Inpraw 41:z_{10}:e,n,x	S
ser Inverness 38:k:1,6	P
ser Ipeko 9,12:c:1,6	D_1
ser Ipswich 41:z_4,z_{24}:-	S
ser Irenea 17:k:1,5	I
ser Irigny 43:z_{38}:-	U
ser Irumu 6,7:l,v:1,5	C_1
*ser Islington 3,10:g,t:-	E_1
ser Israel 9,12:e,h:e,n,z_{15}	D_1
ser Isuge 13,23:d:z_6	G_2
ser Italiana 9,12:l,v:1,11	D_1
ser Ituri 1,4,12:z_{10}:1,5	B
ser Itutaba (9),46:c:z_6	D_2
ser Iwojima = ser Kentucky	

ORGANISM	O GROUP
*ser Jacksonville 16:z_{29}:-	I
ser Jaffna 1,9,12:d:z_{35}	D_1
ser Jaja 4,12,27:z_4,z_{23}:-	B
ser Jamaica 9,12:r:1,5	D_1
ser Jangwani 17:a:1,5	J
bioser Java 1,4,5,12:b:[1,2]	B
ser Javiana 1,9,12:l,z_{28}:1,5	D_1
ser Jedburgh 3,10:z_{29}:-	E_1
ser Jericho 1,4,12,27:c:e,n,z_{15}	B
ser Jerusalem 6,(7),[14]:z_{10}:l,w	C_1
ser Jodhpur 45:z_{29}:-	W
ser Joenkoeping 4,5,12:g,s,t:-	B
ser Johannesburg 1,40:b:e,n,x	R
ser Jos 1,4,12,27:y:e,n,z_{15}	B
ser Jukestown 13,23:i:e,n,z_{15}	G_2
ser Kaapstad 4,12:e,h:1,7	B
ser Kaduna 6,(7),(14):c:e,n,z_{15}	C_1
ser Kahla 1,42:z_{35}:1,6	T
ser Kaitaan 1,6,14,25:m,t:-	H
ser Kalamu 1,4,12:z_4,z_{24}:[1,5]	B
ser Kalina 3,10:b:1,2	E_1
*ser Kaltenhausen 28:b:z_6	M
ser Kamoru 4,12,27:y:z_6	B
ser Kampala 1,42:c:z_6	T
ser Kanda = ser Meleagridis	
ser Kandla 17:z_{29}:-	J
ser Kaneshie 1,42:i:l,w	T
ser Kaolack 47:z:1,6	X
ser Kapemba 9,12:l,v:1,7·	D_1
ser Kaposvar, combined w/ser Reading	
ser Karachi 45:d:e,n,x	W
ser Karamoja 40:z_{41}:1,2	R
ser Kasenyi 38:e,h:1,5	P
ser Kassberg 1,6,14,25:c:1,6	H
*ser Katesgrove 1,13,23:m,t:1,5	G_2
ser Kentucky (8),20:i:z_6	C_3
variant Jerusalem (8):i:z_6	C_2
ser Kenya 6,7:l,z_{13}:e,n,x	C_1
*ser Khami 47:b:e,n,x,z_{15}	X
ser Khartoum (3),(15),34:a:1,7	E_3
ser Kiambu 4,12:z:1,5	B
ser Kibi 16:z_4,z_{23}:-	I
ser Kibusi 28:r:e,n,x	M
ser Kidderminster 38:c:1,6	P
ser Kiel 1,2,12:g,p:-	A
ser Kikoma 16:y:e,n,x	I
*ser Kilwa 4,12:l,w:e,n,x	B
ser Kimberly 38:l,v:1,5	P
ser Kimuenza 1,4,12,27:l,v:e,n,x	B
ser Kingabwa 43:y:1,5	U
ser Kingston 1,4,12,27:g,s,t:-	B
variant Copenhagen 4,12:g,s,t:-	B
ser Kinondoni 17:a:e,n,x	J
ser Kinshasa 3,15:l,z_{13}:1,5	E_2
ser Kintambo 13,23:m,t:-	G_2
ser Kirkee 17:b:1,2	I
ser Kisangani 1,4,5,12:a:1,2	B

ORGANISM	O GROUP
ser Kisarawe 11:k:e,n,x	F
ser Kitenge 28:y:e,n,x	M
ser Kivu 6,7:d:1,6	C_1
*ser Klapmuts 45:z:z_{39}	W
*ser Kluetjenfelde 4,12:d:e,n,x	B
ser Kokemlemle 39:l,v:e,n,x	Q
*ser Kommetje 43:b:z_{42}	U
ser Korbol (8),20:b:1,5,(6)	C_3
ser Korelbu 1,3,19:z:1,5	E_4
ser Korovi 38: g,m,s:-	P
ser Kottbus 6,8:e,h:1,5	C_2
ser Kotte 6,7:b:z_{35}	C_1
ser Koumra 6,7:b:1,7	C_1
ser Kralendyk 6,7:z_4,z_{24}	C_1
*ser Kraaifontein 1,13,23:g,(m),t: [e,n,x]	G_2
ser Kralingen (8),20:y:z_6	C_3
ser Krefeld 1,3,19:y:l,w	E_4
ser Kristianstad 3,10:z_{10}:e,n,z_{15}	E_1
*ser Krugersdorp 50:e,n,x:1,7	Z
ser Kuessel 28:i:e,n,z_{15}	M
*ser Kuilsrivier 1,9,12:g,m,s,t:e,n,x	D_1
ser Kumasi 30:z_{10}:e,n,z_{15}	N
ser Kunduchi 1,4,[5],12,27:l,z_{28}:1,2	B
ser Kuru 6,8:z:l,w	C_2
ser Labadi 6,8:d:z_6	C_2
ser Lagos 1,4,12:i:1,5	B
ser Landala 41:z_{10}:1,6	S
ser Landau 30:i:1,2	N
ser Langenhorn 18:m,t:-	K
ser Langensalza 3,10:y:l,w	E_1
ser Langford 28:b:e,n,z_{15}	M
ser Lanka 3,15:r:z_6	E_2
ser Lansing 38:i:1,5	P
ser Larochelle 6,7:e,h:1,2	C_1
ser Lattenkamp 45:z_{35}:1,5	W
ser Lawndale 1,9,12:z:1,5	D_1
ser Lawra 44:k:e,n,z_{15}	V
ser Leeuwarden 11:b:1,5	F
ser Legon [1],4,12,[27]:c:1,5	B
ser Leiden 13,22:z_{38}:-	G_1
ser Leipzig 41:z_{10}:1,5	S
ser Leith 6,8:a:e,n,z_{15}	C_2
ser Lekke 3,10:d:1,6	E_1
ser Leoben 28:l,v:1,5	M
ser Leopoldville 6,7:b:z_6	C_1
*ser Lethe 41:g,t:-	S
ser Lexington 3,10:z_{10}:1,5	E_1
ser Lezennes 6,8:z_4,z_{23}:1,7	C_2
*ser Lichtenberg 41:z_{10}:[z_6]	S
ser Ligeo 30:l,v:1,2	N
ser Ligna 35:z_{10}:z_6	O
ser Lille 6,7:z_{38}:-	C_1
*ser Limbe 1,13,22:g,m,t:[1,5]	G_1
ser Limete 1,4,12,27:b:1,5	B
*ser Lincoln 11:m,t:e,n,x	F
ser Lindenburg 6,8:i:1,2	C_2

ORGANISM	O GROUP
ser Lindern 6,14,24:d:e,n,x	H
ser Lindi 38:r:1,5	P
*ser Lindrick 9,12:e,n,x:1,[5],7	D_1
ser Lingivala 16:z:1,7	I
ser Linton 13,23:r:e,n,z_{15}	G_2
ser Lisboa 16:z_{10}:1,6	I
ser Lishabi (9),46:z_{10}:1,7	D_2
ser Litchfield 6,8:l,v:1,2	C_2
ser Liverpool 1,3,19:d:e,n,z_{15}	E_4
ser Livingstone 6,7:d:l,w	C_1
ser Ljubljana 4,12,27:k:e,n,x	B
ser Llandoff 1,3,19:z_{29}:-	E_4
*ser Llandudno 28:g,s,t:1,5	M
ser Loanda 6,8:l,v:1,5	C_2
*ser Lobatsi 52:-:1,5,7	52
*ser Locarno 57:z_{29}:z_{42}	57
ser Loenga 1,42:z_{10}:z_6	T
ser Logone 39:d:1,5	Q
*ser Lohbruegge 44:z_4,z_{32}:-	V
ser Lokstedt 1,3,19:l,z_{13},z_{28}:2	E_4
ser Lomalinda 9,12:a:e,n,x	D_1
ser Lome 9,12:r:z_6	D_1
ser Lomita 6,7:e,h:1,5	C_1
ser London 3,10:l,v:1,6	E_1
ser Losangeles 16:l,v:z_6	C_2
ser Louga 30:b:1,2	N
*ser Louwbester 16:z:e,n,x	I
*ser Lovelace 13,22:l,v:1,5	G_1
*ser Luanshya 13,23:g,s,(t):-	G_2
ser Luciana 11:a:e,n,z_{15}	F
ser Luckenwalde 28:z_{10}:e,n,z_{15}	M
ser Luke 1,47:g,m:-	X
*ser Lundby (9),46:b:e,n,x	D_2
*ser Lurup 41:z_{10}:e,n,x,z_{15}	S
*ser Luton 60:z:e,n,x	60
ser Lyon 47:k:e,n,z_{15}	
*ser Maarssen (9),46:z_4,z_{24}:z_{39}:z_{42}	D_2
ser Maartensdyijk 40:g,p:-	R
ser Maastricht 11:z_{41}:1,2	F
ser Macallen 3,10:z_{36}:-	E_1
ser Machaga 1,3,19:i:e,n,x	E_4
ser Madelia 1,6,14,25:y:1,7	H
ser Madiago 1,3,19:c:1,7	E_4
ser Madigan 44:c:1,5	V
ser Madjorio 3,10:d:e,n,z_{15}	E_1
ser Magumeri 1,6,14,25:e,h:1,6	H
ser Magwa 21:d:e,n,x	L
ser Maiduguri 1,3,19:f,g,t:e,n,z_{15}	E_4
ser Makiso 6,7:l,z_{13},z_{28}:z_6	C_1
*ser Makoma 4,12:a:-	B
*ser Makumira [1],4,12,[27]:e,n,x:1,7	B
ser Malakal 16:e,h:1,2	I
ser Malstatt 16:b:z_6	I
ser Mampeza 1,6,14,25:i:1,5	H
ser Mampong 13,22:z_{35}:1,6	G_1
ser Manchester 6,8:l,v:1,7	C_2
ser Mandera 16:l,z_{13}:e,n,z_{15}	I

ORGANISM	O GROUP
ser Manhattan 6,8:d:1,5	C_2
*ser Manica 1,9,12:g,m,s,t:z_{42}	D_1
ser Manila 3,15:z_{10}:1,5	E_2
*ser Manombo 57:z_{39}:e,n,x,z_{15}	57
ser Mapo 6,8:z_{10}:1,5	C_2
ser Mara 39:e,h:[1,5]	Q
ser Maracaibo 11:l,v:1,5	F
ser Maricopa 1,42:g,z_{51}:1,5	T
ser Marienthal 3,10:k:e,n,z_{15}	E_1
*ser Marina 48:g,z_{51}:-	Y
ser Maritza = ser Salford var. Maritza	
ser Maron 3,10:d:z_{35}	E_1
ser Marseille 11:a:1,5	F
ser Marylebone (9),46:k:1,2	D_2
ser Massakory 35:r:l,w	O
ser Massenya 1,4,12,27:k:1,5	B
ser Matadi 17:k:e,n,x	J
ser Mathura (9),46:i:e,n,z_{15}	D_2
ser Matopeni 30:y:1,2	N
*ser Matroosfontein 3,10:a:e,n,x	E_1
ser Mayday (9),46:y:z_6	D_2
ser Mbandaka [1],6,7,[25]:z_{10}: e,n,z_{15}	C_1
ser Mbao 43:i:1,2	U
ser Meleagridis 3,10:e,h:l,w	E_1
ser Memphis 18:k:1,5	K
ser Menden 6,7:z_{10}:1,2	C_1
ser Mendoza 9,12:l,v:1,2	D_1
ser Menhaden (3),(15),34:l,v:1,7	E_3
ser Menston 6,7:g,s,t:-	C_1
*ser Merseyside 16:g,t:1,5	I
ser Mesbit 47:m,t:e,n,z_{15}	X
ser Meskin 51:e,h:1,2	51
ser Messina 30:d:1,5	N
ser Mexicana, combined w/ser Muenchen	
ser Mgulani 38:i:1,2	P
ser Miami 1,9,12:a:1,5	D_1
ser Michigan 17:l,v:1,5	F
ser Middlesbrough 1,42:i:z_6	T
*ser Midhurst 53:l,z_{28}:z_{39}	53
ser Mikawasima 6,7:y:e,n,z_{15}	C_1
ser Millesi 1,40:l,v:1,2	R
ser Milwaukee 43:f,g:-	U
ser Mim 13,22:a:1,6	G_1
ser Minneapolis (3),(15),34:e,h:1,6	E_3
ser Minnesota 21:b:e.n.x	L
ser Mishmarhaemek 1,13,23:d:1,5	G_2
ser Mission 6,7:d:1,5	C_1
ser Mississippi (Atlanta) 1,13,23:b:1,5	G_2
ser Missouri 11:g,s,t:-	F
ser Miyazaki 9,12:l,z_{13}:1,7	D_1
*ser Mjimwema 1,9,12:b:e,n,x	D_1
*ser Mobeni 16:g,m,s,t:-	I
ser Mocamedes 28:d:e,n,x	M
ser Moero 28:b:1,5	M
ser Mokola 3,10:y:1,7	E_1
ser Molade (8),20:z_{10}:z_6	C_3
*ser Mondeor 39:l,z_{28}:e,n,x	Q

ORGANISM	O GROUP
ser Mons 1,4,12,[27]:d:l,w	B
ser Monschaui 35:m,t:-	O
ser Montevideo 6,7:g,m,s:-	C_1
*ser Montgomery 11:d,a:d,e,n,z_{15}	F
ser Montreal = ser Wein	
ser Morehead 30:i:1,5	N
ser Morocco 30:l,z_{13},z_{28}:e,n,z_{15}	N
ser Morotai 17:l,v:1,2	J
ser Moroto 28:z_{10}:l,w	M
ser Moscow 9,12:g,q:-	D_1
*ser Mosselbay 43:g,s,(t):z_{42}	U
ser Moualine 47:y:1,6	X
ser Mountpleasant 47:z:1,5	X
ser Mowanjum 6,8:z:1,5	C_2
*ser Mpila 3,10:z_{38}:z_{42}	E_1
ser Muenchen 6,8:d:1,2	C_2
ser Muenster 3,10:e,h:1,5	E_1
ser Muguga 44:m,t:-	V
*ser Muizenberg 9,12:g,m,s,t:1,5	D
ser Mundonobo 28:d:1,7	M
*ser Mundsburg 11:g,z_{51}:-	F
*ser Mura 1,4,12:z_{10}:i,w	B
*ser Nachshonim 1,13,23:z:1,5	G_2
ser Naestved 1,9,12:g,p,s:-	D_1
ser Nagoya 6,8:b:1,5	C_2
*ser Nairobi 42:r:-	T
ser Nakura 1,4,12,27:a:z_6	B
*ser Namib 50:g,m,s,t:1,5	Z
ser Napoli 1,9,12:l,z_{13}:e,n,x	D_1
ser Narashino 6,8:a:e,n,x	C_2
ser Nashua 28:l,v:n,z_{15}	M
ser Nchanga 3,10:l,v:1,2	E_1
ser Ndolo [1],9,12:d:1,5	D_1
*ser Neasden 9,12:g,s,t:e,n,x	D_1
*ser Negev 41:z_{10}:1,2	S
ser Nessa 1,6,14,25:z_{10}:1,5	H
ser Nessziona 6,7:l,z_{13}:1,5	C_1
ser Neukoelln 6,7:l,z_{13},z_{28}:e,n,z_{15}	C_1
ser Neumuenster 1,4,12,27:k:1,6	B
* variant 1,4,12,27:k:1,6	B
ser Newbrunswick 3,15:l,v:1,7	E_2
ser Newhaw 3,15:e,h:1,5	E_2
ser Newington 3,15:e,h:1,6	E_2
ser Newlands 3,10:e,h:e,n,x	E_1
ser Newmexico 9,12:g,z_{51}:1,5	D_1
ser Newport 6,8:e,h:1,2	C_2
ser Newrochelle 3,10:k:l,w	E_1
ser Newyork = ser Javiana	
ser Ngili 6,7:z_{10}:1,7	C_1
ser Ngor 1,3,19:l,v:1,5	E_4
*ser Ngozi 48:z_{10}:[1,5]	Y
ser Niamey 17:d:l,w	J
ser Niarembe 44:a:l,w	V
ser Nienstedten 6,(7),(14):b:[l,w]	C_1
ser Nieukerk 6,(7),(14):d:z_6	C_1
variant zollenspicker 6,7:d:z_6	C_1
ser Nigeria 6,7:r:1,6	C_1

ORGANISM	O GROUP
ser Nikolaifleet 16:g,m,s:-	I
ser Niloese 1,3,19:d:z_6	E_4
ser Nima 28:y:1,5	M
ser Nipponbasi, not confirmed	
ser Nissii 6,7,14:b:-	C_1
ser Nitra 2,12:g,m:-	A
*ser Noordhoek 16:l,w:z_6	I
*ser Nordenham 1,4,12,27:z:e,n,x	B
ser Nordufer 6,8:a:1,7	C_2
ser Norton 6,7:i:l,w	C_1
ser Norwich 6,7:e,h:1,6	C_1
ser Nottingham 16:d:e,n,z_{15}	I
ser Nowawes 40:z:z_6	R
ser Nuatja 16:k:e,n,x	I
*ser Nuernberg 42:z:z_6	T
ser Nyanza 11:z:z_6	F
ser Nyborg 3,10:e,h:1,7	E_1
ser Oahu, not confirmed	
ser Oakland 6,7:z:1,6,(7)	C_1
ser Obogu 6,7:z_4,z_{23}:1,5	C_1
ser Ochsenwerder 54:k:1,5	54
*ser Ochsenzoll 16:z_4,z_{23}:-	I
*ser Odijk 30:a:z_{39}	N
ser Odozi 30:k:e,n,x,z_{15}	N
* variant 30:k:e,n,x,z_{15}	N
*ser Oevelgoenne 28:r:e,n,z_{15}	M
ser Offa 41:z_{38}:-	S
ser Ohio 6,7:b:l,w	C_1
ser Ohlstedt 3,10:y:e,n,x	E_1
ser Okatie 13,23:g,s,t:-	G_2
ser Okefoko 3,10:c:z_6	E_1
ser Okerara 3,10:z_{10}:1,2	E_1
ser Oldenburg 16:d:1,2	I
ser Omderman 6,(7),(14):d:e,n,x	C_1
ser Omifisan 40:z_{29}:-	R
ser Ona 28:g,s,t:-	M
ser Onarimon 1,9,12:b:1,2	E_1
ser Onderstepoort 1,6,14,25:e,h:1,5	H
ser Onireke 3,10:d:1,7	E_1
ser Oranienburg 6,7:m,t:-	C_1
ser Ordonez 1,13,23,37:y:l,w	G_2
ser Oregon, combined w/ser Muenchen	
ser Orientalis 16:k:e,n,z_{15}	I
ser Orion 3,10:y:1,5	E_1
ser Oritamerin 6,7:i:1,5	C_1
ser Os 9,12:a:1,6	D_1
ser Oskarshamn 28:y:1,2	M
ser Oslo 6,7:a:e,n,x	C_1
ser Osnabrueck 11:l,z_{13},z_{28}:e,n,x	F
ser Othmarschen 6,7:g,m,[t] :-	C_1
*ser Ottershaw 40:d:-	R
ser Ouakam (9),[12],[34]:46:z_{29}:-	D_2
ser Overschie 51:l,v:1,5	51
ser Overvecht 30:a:1,2	N
ser Oxford 3,10:a:1,7	E_1
*ser Oysterbeds 6,7:z:z_{42}	C_1

ORGANISM	O GROUP
ser Pakistan (8):l,v:1,2	C_3
ser Panama 1,9,12:l,v:1,5	D_1
ser Pankow 3,15:d:1,5	E_2
ser Papuana 6,7:r:e,n,z_{15}	C_1
ser Paratyphi = ser Paratyphi A	
bioser Paratyphi A 1,2,12:a:-	A
variant Durazzo 2,12:a:-	A
ser Paratyphi B java = ser Java	
1,4,5,12:b:[1,2]	B
ser Paratyphi B 1,4,5,12,:b:1,2	B
variant Odense 1,4,12:b:1,2	B
bioser Paratyphi C 6,7,[Vi]:c:1,5	C_1
*ser Parera 11:z_4,z_{23}:-	F
*ser Parow 3,15:g,m,s,t:-	E_2
ser Paris (8):20,z_{10}:1,5	C_3
ser Patience 28:d:e,n,z_{15}	M
ser Penarth 9,12:z_{35}:z_6	D_1
ser Pensacola 9,12:m,t:-	D_1
*ser Perinet 45:m,t:e,n,x,z_{15}	W
ser Perth 38:y:e,n,x	P
ser Pharr 11:b:e,n,z_{15}	F
*ser Phoenix 47:b:1,5	X
ser Pikine (8),20:r:z_6	C_3
ser Plymouth (9),46:d:z_6	D_2
ser Poano 1,6,14,25:z:l,z_{13},z_{28}	H
ser Poeseldorf 54:i:z_6	54
ser Pomona 28:y:1,7	M
ser Poona [1],13,22,[37]:z:1,6	G_1
variant 37 1,13,22,36,37:z:1,6	G_1
*ser Portbech 42:l,v:e,n,x,z_{15}	T
ser Portland 9,12:z_{10}:1,5	D_1
ser Portsmouth 3,15:l,v:1,6	E_2
ser Potsdam 6,7:l,v:e,n,z_{15}	C_1
ser Potto (9),12,46:i:z_6	D_2
ser Praha 6,8:y:e,n,z_{15}	C_2
ser Pramiso 3,10:c:1,7	E_1
ser Presov 6,8:b:e,n,z_{15}	C_2
ser Preston 1,4,12:z:l,w	B
ser Pretoria 11:k:1,2	B
ser Pueris, combined w/ser Newport	
bioser Pullorum 9,12:-:-	D_1
ser Pumila 48:a:z_6	Y
ser Putten 13,23:d:l,w	G_2
*ser Quimbamba 47:d:z_{39}	X
ser Quinhon 47:z_{44}:-	X
ser Quiniela 6,8:c:e,n,z_{15}	C_2
ser Ramatgen 30:k:1,5	N
*ser Rand 42:z:e,n,x,z_{15}	T
ser Raus 13,22:f,g:e,n,x	G_1
ser Reading 4,[5],12:e,h:1,5	B
ser Rechovot (8),20:e,h:z_6	C_3
ser Redhill 11:e,h:l,z_{13},z_{28}	F
ser Redlands 16:z_{10}:e,n,z_{15}	I
ser Regent 3,10:f,g:-	E_1
ser Remo 1,4,12,27:r:1,7	B

ORGANISM	O GROUP	ORGANISM	O GROUP
*ser Rhodensiense 9,12:d:e,n,x	D_1	ser Sedgwick 44:b:e,n,z_{15}	V
ser Rhone 21:c:e,n,x	L	ser Seegefeld 3,10:r(i):1,2	E_1
ser Richmond 6,7:y:1,2	C_1	ser Sekondi 3,10:e,h:z_6	E_1
ser Rideau 1,3,19:f,g:-	E_4	ser Selandia 3,15:e,h:1,7	E_2
ser Ridge 9,12:c:z_6	D_1	*ser Seminole 1,40a,40b:g:z_{51}	R
ser Riggil 6,7:g,t:-	C_1	bioser Sendai 1,9,12:a:1,5	D_1
ser Riogrande 40:b:1,5	R	ser Senegal 11:r:1,5	F
ser Rissen 6,7:f,g:-	C_1	ser Senftenberg 1,3,19:g,s,t:-	E_4
ser Riverside 45:b:1,5	W	ser Seremban 9,12:i:1,5	D_1
ser Roan 38:l,v:e,n,x	P	*ser Setubal 60:g,m,t:z_6	60
ser Rochdale 50:b:e,n,x	Z	ser Shamba 16:c:e,n,x	I
*ser Roggeveld 51:-:1,7	51	ser Shangani 3,10:d:1,5	E_1
ser Rogy 28:z_{10}:1,2	M	ser Shanghai 16:l,v:1,6	I
ser Romanby 13,23:z_4,z_{24}:-	G_2	ser Sharon 11:k:1,6	F
ser Roodepoort [1],13,22,[37]:z_{10}:1,5	G_1	ser Sheffield 38:c:1,5	P
*ser Rooikrantz 1,6,14:m,t:1,5	H	ser Shikmonah 40:a:1,5	R
ser Rosenthal 3,15:b:1,5	E_2	ser Shipley (8),20:b:e,n,z_{15}	C_3
ser Rossleben 54:e,h:1,6	54	ser Shomolu 28:y:l,w	M
ser Rostock 1,9,12:g,p,u:-	D_1	*ser Shomron 18:z_4,z_{32}:-	K
*ser Roterberg 6,7:z_4,z_{23}:-	C_1	ser Shoreditch (9),46:r:e,n,z_{15}	D_2
*ser Rotterdam 1,13,22:g,t:1,5	G_1	ser Shubra 4,5,12:z:1,2	B
*ser Rowbarton 16:m,t:-	I	ser Siegburg 6,14,18:z_4,z_{23}:[1,5]	K
ser Rubislaw 11:[d],r:[d],e,n,x	F	ser Simi 3,10:r:e,n,z_{15}	E_1
ser Ruiru 21:y:e,n,x	L	*ser Simonstown 1,6,14:z_{10}:1,5	H
ser Ruki 4,5,12:y:e,n,x	B	ser Simsbury 1,3,19:z_{27}:-	E_4
ser Rutgers 3,10:l,z_{40}:1,7	E_1	ser Singapore 6,7:k:e,n,x	C_1
ser Ruzizi 3,10:l,v:e,n,z_{15}	E_1	ser Sinstorf 3,10:l,v:1,5	E_1
		ser Sinthia 18:z_{38}:-	K
		ser Sladun 1,4,12,27:b:e,n,x	B
ser Saarbruecken [1],9,12:a:1,7	D_1	*ser Slangkop 1,6,14:z_{10}:z_6:z_{42}	H
*ser Sachsenwald 1,40:z_4,z_{23}:-	R	*ser Slatograd 30:f,g,(p),t:-	N
ser Saintmarie 52:g,t:-	52	ser Sljeme 1,47:f,g:-	X
ser Saintpaul 1,4,[5],12:e,h:1,2	B	ser Sloterdijk 1,4,12,27:z_{35}:z_6	B
ser Saipam, not confirmed		ser Soahamina 6,14,24:z:e,n,x	H
ser Saka 47:b:-	X	ser Soerenga 30:i:l,w	H
ser Sakai = ser Potsdam		*ser Soesterberg 21:z_4,z_{23}:-	L
*ser Sakaraha 48:[k]:z_{39}	Y	*ser Sofia 4,12,[27]:b:[e,n,x]	B
ser Salford 16:l,v:e,n,x	I	ser Solna 28:a:1,5	M
ser Salinatis 4,12:d,e,h:d,e,n,z_{15}	B	ser Solt 11:y:1,5	F
ser Sandiego 4,12:e,h:e,n,z_{15}	B	ser Southbank 3,10:m,t:-	E_1
ser Sandow 6,8:f,g:e,n,z_{15}	C_2	*ser Soutpan 11:z:z_{39}	F
ser Sanga (8):b:1,7	C_3	ser Souza 3,10:d:e,n,x	E_1
ser Sanjuan 6,7:a:1,5	C_1	ser Spartel 21:d:1,5	L
ser Sanktgeorg 28:r,(i):e,n,z_{15}	M	*ser Springs 40:a:z_{39}	R
ser Sanktmarx 1,3,19:e,h:1,7	E_4	*ser Srinagar 11:b:e,n,x	F
ser Santhiaba 40:l,z_{28}:1,6	R	ser Stanley 4,5,12:d:1,2	B
ser Sao 1,3,19:e,h:e,n,z_{15}	E_4	ser Stanleyville 1,4,5,12:z_4,z_{23}:[1,2]	B
ser Saphra 16:y:1,5	I	ser Steinplatz 30:y:1,6	N
ser Sara 1,6,14,25:z_{38}:[e,n,x]	H	*ser Stellenbosch 1,9,12:z:1,7	D_1
ser Sarajane 4,12,27:d:e,n,x	B	ser Stellingen 47:d:e,n,x	X
*ser Sarepta 11:l,z_{28}:z_{42}	I	ser Stendal 11:l,v:1,2	F
ser Schalkwijk 6,14,(24):i:e,n......	H	ser Sternchanze 30:g,s,t:-	N
ser Schleissheim 4,12,27:b,z_{12}:-	B	ser Sterrenbos 6,8:d:e,n,x	C_2
ser Schoeneberg 1,3,19:z:e,n,z_{15}	E_4	*ser Stevenage 1,13,23:[z_{42}]:1,7	G_2
ser Schottmuelleri = ser Paratyphi B		*ser Stikland 3,10:m,t:e,n,x	E_1
ser Schwarzengrund 1,4,12,27:d:1,7	B	ser Stockholm 3,10:y:z_6	E_1
ser Schwerin 6,8:k:e,n,x	C_2	ser Stormont 3,10:d:1,2	E_1
*ser Seaforth 50:k:z_6	Z	ser Stourbridge 6,8:b:1,6	C_2
ser Seattle 28:a:e,n,x	M	ser Straengnaes 11:z_{10}:1,5	F

ORGANISM	O GROUP
ser Strasbourg (9),46:d:1,7	D_2
ser Stratford 1,3,19:i:1,2	E_4
*ser Suarez 1,40:c:e,n,x,z_{15}	R
ser Suberu 3,10:g,m:-	E
*ser Suederelbe 1,9,12:b:z_{39}	D_1
ser Suelldorf 45:f,g:-	W
ser Suez = ser Shubra	
ser Suipestifer = ser Choleraesuis	
*ser Sullivan 6,7:z_{42}:1,7	C_1
ser Sundsvall 1,6,14,25:z:e,n,x	H
ser Sunnycove (8):y:e,n,x	C_3
*ser Sunnydale 1,40:k:e,n,x,z_{15}	R
ser Surat 1,6,14,25:r(i):e,n,z_{15}	H
variant Hr- 1,6,14,25:i:e,n,z_{15}	H
*ser Sydney 48:i:z	Y
ser Szentes 16:k:1,2	I
*ser Tafelbaai 3,10:z:z_{39}	E_1
ser Tafo 1,4,12,27:z_{35}:1,7	B
ser Taihoku = ser Meleagridis	
ser Takoradi 6,8:i:1,5	C_2
ser Taksony 1,3,19:i:z_6	E_4
ser Tallahassee 6,8:z_4,z_{32}:-	C_2
ser Tamale (8),20:z_{29}:-	C_3
ser Tananarive 6,8:y:1,5	C_2
ser Tanger 1,13,22:y:1,6	G_1
ser Tarshyne 9,12:d:1,6	D_1
ser Taunton 28:k:e,n,x	M
ser Tchad 35:b:-	O
ser Techimani 28:c:z_6	M
ser Teddington 4,12,27:y:1,7	B
ser Tees 16:f,g:-	I
ser Tejas 4,12:z_{36}:-	B
ser Teko 1,6,14,25:d:e,n,z_{15}	H
ser Telaviv 28:y:e,n,z_{15}	M
ser Telekebir 13,23:d:e,n,z_{15}	G_2
ser Telhashomer 11:z_{10}:e,n,x	F
ser Teltow 28:z_4,z_{23}:1,6	M
ser Tennessee 6,7:z_{29}:-	C_1
ser Teshie 1,47:l,z_{13},z_{28}:e,n,z_{15}	X
ser Texas 4,5,12:k:e,n,z_{15}	B
ser Thiaroye 38:e,h:1,2	P
ser Thielallee 6,(7),(14):m,t:-	C_1
ser Thomasville (3),(15),34:y:1,5	E_3
ser Thompson 6,7,[14]:k:1,5	C_1
variant Berlin 6,7:-:1,5	C_1
variant 14 6,7,14:k:1,5	C_1
ser Tilburg 1,3,19:d:l,w	E_4
ser Tilene 1,40:e,h:1,2	R
ser Tim = ser Newington var tim	
ser Tinda 1,4,12,27:a:e,n,z_{15}	B
ser Tione 51:a:e,n,x	51
ser Togo 4,12:l,w:1,6	B
*ser Tokai 57:z_{42}:1,6:z_{53}	57
ser Tokoin 4,12:z_{10}:e,n,z_{15}	B
ser Tokyo, not confirmed	
ser Tonev 54:b:e,n,x	54
ser Tornow 45:g,m:-	W

ORGANISM	O GROUP
*ser Tosamanga 6,7:z:1,5	C_1
ser Tournai 3,15:y:z_6	E_2
ser Trachau 4,12,27:y:1,5	B
*ser Tranoroa 55:k:z_{39}	55
ser Travis 4,5,12:g,z_{51}:1,7	B
ser Treforest 1,51:z:1,6	51
ser Trotha 40:z_{10}:z_6	R
ser Tshiongwe 6,8:e,h:e,n,z_{15}	C_2
ser Tucson 1,6,14,25:b:[1,7]	H
ser Tudu 4,12:z_{10}:1,6	B
ser Tuebingen 3,15:y:1,2	E_2
*ser Tuindorp 43:z_4,z_{32}:-	U
*ser Tulear 6,8:a:z_{52}	C_2
ser Tunis 1,13,23:y:z_6	G_2
*ser Tygerberg 1,13,23:a:z_{42}	G_2
Salmonella typhi 9,12,[Vi]:d:-	D_1

S. enteritidis

ser Typhimurium 1,4,5,12:i:1,2	B
variant binns 1,4,5,12:-:1,2	B
variant Copenhagen 1,4,12:i:1,2	B
bioser Typhisuis 6,7:[c]:1,5	C_1
ser Uccle 54:g,s,t:-	54
ser Uganda 3,10:l,z_{13}:1,5	E_1
ser Ughelli 3,10:r:1,5	E_1
ser Uhlenhorst 44:z:l,w	V
ser Ullevi 1,13,23,27:b:e,n,x	G_2
ser Umbilo 28:z_{10}:e,n,x	M
ser Umhlali 6,7:a:1,6	C_1
ser Umhlatazana 35:a:e,n,z_{15}	O

*Unnamed serotypes

ser _____ 4,12:(f),g:-	B
ser _____ 4,12:-:1,6	B
ser _____ 6,7:a:z_6	C_1
ser _____ 6,7:g,t:e,n,x:z_{42}	C_1
ser _____ 6,7:k:[z_6]	C_1
ser _____ 6,7:z:z_6	C_1
ser _____ 6,7:z_{10}:z_{35}	C_1
ser _____ 6,7:z_{29}:-	C_1
ser _____ 6,7:z_{42}:e,n,x:1,6	C_1
ser _____ 6,8:g,(m),t:e,n,x	C_2
ser _____ 9,12:e,n,x:1,6	D_1
ser _____ (9),46:z_{10}:z_6	D_2
ser _____ 1,9,12,(46),27:y:z_{39}	D_2
ser _____ 3,10:l,z_{28},z_{39}	E_1
ser _____ 11:b:1,7	F
ser _____ 11:z_4,z_{23}:-	F
ser _____ 13,23:l,z_{28}:z_6	G_2
ser _____ 14:k:[e,n,x]	H
ser _____ 1,(6),14:k:z_6:z_{42}	H
ser _____ 1,(6),14:z_{42}:1,6	H
ser _____ 16:b:z_{42}	I
ser _____ 16:l,z_{40}:-	I
ser _____ 17:k:-	J
ser _____ 18:b:1,5	K
ser _____ 18:m,t:1,5	K

ORGANISM	O GROUP	ORGANISM	O GROUP
ser_____ 18:y:e,n,x,z_{15}	K	* variant 28:m,t:-	M
ser_____ 28:e,n,x:1,7	M	ser Virchow 6,7:r:1,2	C_1
ser_____ 30:z_{39}:1,(7)	N	ser Virginia (8):d:[1,2]	C_3
ser_____ 35:g,m,s,t:-	O	ser Visby 1,3,19:b:1,6	E_4
ser_____ 35:l,z_{28}:-	O	ser Vitkin 28:l,v:e,n,x	M
ser_____ 40:b:-	R	ser Vleuten 44:f,g:-	V
ser_____ 1,40a,40c:g,z_{51}:-	R	ser Volkmarsdorf 28:i:1,6	M
ser_____ 1,40:m,t:z_{42}	R	*ser Volksdorf 43:z_{36},z_{38}:-	U
ser_____ 1,40:z_6:1,5	R	ser Volta 11:r:l,z_{13},z_{28}	F
ser_____ 1,40:-:1,7	R	ser Vom 4,12,27:l,z_{13},z_{28}:e,n,z_{15}	B
ser_____ 41:k:-	S	*ser Vredelust 1,13,23:l,z_{28}:z_{42}	G_2
ser_____ 42:m,t:e,n,x,z_{15}	T	*ser Vrindaban 45:a:e,n,x	W
ser_____ 42:-:1,6	T		
ser_____ 43:e,n,x,z_{15}:1,(5),7	U		
ser_____ 43:e,n,z_{15}:1,6	U	ser Wa 16:b:1,5	I
ser_____ 43:z:1,5	U	ser Wagenia 1,4,12,27:b:e,n,z_{15}	B
ser_____ 44:g,z_{51}:-	V	*ser Wandsbek 21:z_{10}:z_6	L
ser_____ 44:z_4,z_{23}:-	V	ser Wandsworth 39:b:1,2	Q
ser_____ 44:z_{36},z_{38}:-	V	ser Wangata 9,12:z_4,z_{23}:[1,7]	D_1
ser_____ 45:g,z_{51}:-	W	ser Warnow 6,8:i:1,6	C_2
ser_____ 47:z_6:1,6	X	ser Warragul 1,6,14,25:g,m:-	H
ser_____ 48:z_4,z_{32}:-	Y	*ser Wassenaar 50:g,z_{51}:-	Z
ser_____ 50:l,w:e,n,x,z_{15}:z_{42}	Z	ser Waycross 41:z_4,z_{23}:-	S
ser_____ 50:l,z_{28}:z_{42}	Z	* variant 41:z_4,z_{23}:-	S
ser_____ 50:z_4,z_{24}:-	Z	ser Wayne 30:g,z_{51}:-	N
ser_____ 52:d:e,n,x,z_{15}	52	ser Wedding 28:c:e,n,z_{15}	M
ser_____ 53:z:z_6	53	ser Welikade 16:l,v:1,7	I
ser_____ 56:e,n,x:1,7	56	ser Weltevreden 3,10:r:z_6	E_1
ser_____ 57:g,m,s,t:z_{42}	57	ser Wentworth 11:z_{10}:1,2	F
ser_____ 58:a:-	58	ser Wernigerode (9),46:f,g:-	D_2
ser_____ 58:a:1,5	58	ser Weslaco 42:z_{36}:-	T
ser_____ 64:k:e,n,x,z_{15}	64	ser Westerstede 1,3,19:l,z_{13}:-	E_4
ser_____ 64:z_{29}:-	64	ser Westhampton 3,10:g,s,t:-	E_1
ser Uno 6,8:z_{29}:-	C_2	ser Weston 16:e,h:z_6	I
*ser Uphill 42:b:e,n,x,z_{15}	T	*ser Westpark 3,10:l,z_{28}:e,n,x	E_1
ser Uppsala 4,12,27:b:1,7	B	ser Westphalia 35:z_4,z_{24}:-	O
ser Urbana 30:b:e,n,x	N	ser Weybridge 3,10:d:z_6	E_1
ser Ursenbach 1,42:z:1,6	T	ser Wichita 1,13,23:d:z_{37}	G_2
ser Usumbura 18:d:1,7	K	ser Wien 1,4,12,[27]:b:l,w	B
ser Utah 6,8:c:1,5	C_2	ser Wil 6,7:d:l,z_{13},z_{28}	C_1
*ser Utbremen 35:z_{29}:e,n,x	O	ser Wildwood (3),(15),34:e,h:l,w	E_3
ser Utrecht 52:d:1,5	52	ser Wilhelmsburg 4,[5],12,[27]:z_{38}:-	B
ser Uzaramo 1,6,14,25:z_4,z_{24}:-	H	variant Teufelsbrueck 1,4,12:z_{38}:-	B
		*ser Wilhelmstrasse 52:z_{44}:1,5,7	52
ser Vaertan 13,22:b:e,n,x	G_1	ser Willemstad 1,13,22:e,h:1,6	G_1
ser Vancouver 16:c:1,5	I	ser Wimborne 3,10:k:1,2	E_1
*ser Veddel 43:g,t:-	U	*ser Winchester 3,10:z_{39}:1,7	E_1
ser Vejle 3,10:e,h:1,2	E_1	ser Windermere 39:y:1,5	Q
ser Vellore 1,4,12,27:z_{10}:z_{35}	B	*ser Windhoek 45:g,t:1,5	W
ser Venusberg = ser Nchanga variant venusberg		ser Wingrove 6,8:c:1,2	C_2
ser Veneziana 11:i:e,n,x	F	ser Wippra 6,8:z_{10}:z_6	C_2
*ser Verity 17:e,n,x,z_{15}:1,6	J	*ser Woerden 17:c:z_{39}	J
ser Victoria 1,9,12:l,w:1,5	D_1	ser Womba 4,12,27:c:1,7	B
ser Victoriaborg 17:c:1,6	J	*ser Woodstock 16:z_{42}:1,(5),7	I
ser Vietnam 41:b:-	S	*ser Worcester 1,13,23:m,t:e,n,x	G_2
* variant 41:b:-	S	ser Worthington 1,13,23:z:l,w	G_2
ser Vinohrady 28:m,t:-	M	*ser Wynberg 1,9,12:z_{39}:1,7	D_1

DIFFERENTIATION OF ENTEROBACTERIACEAE BY BIOCHEMICAL TESTS

Test	Escherichia coli	Shigella sonnei	Shigella Other	Edwardsiella tarda	Salmonella typhi	Salmonella typhi	Arizona	Citrobacter freundii	Citrobacter diversus	Citrobacter amalonaticus	Klebsiella pneumoniae	Klebsiella oxytoca	Enterobacter cloacae	Enterobacter aerogenes	Enterobacter agglomerans	Enterobacter sakazakii	Enterobacter gergoviae	Hafnia alvei	Serratia marcescens	Serratia liquefaciens	Serratia rubidaea	Proteus vulgaris	Proteus mirabilis	Morganella morganii	Providencia rettgeri	Providencia alcalifaciens	Providencia stuartii	Yersinia enterocolitica	Yersinia pseudotuberculosis	Yersinia pestis
Indole	+	-	-	+	-	-	-	-	+	+	-	+	-	-	>	>	-	-	-	-	-	+	-	+	+	+	+	>	-	-
Methyl Red	+	+	+	+	+	+	>	+	+	+	-	>	-	-	>	>	>	>	>	>	>	+	+	+	+	+	+	+	+	+
Voges-Proskauer	-	-	-	-	-	-	-	-	-	-	+	+	+	+	>	+	>	>	+	+	+	>	>	-	-	-	-	-	-	-
Simmons' Citrate	-	-	-	-	-	>	+	+	+	+	+	+	+	+	>	+	+	>	+	+	+	(V)	+	-	+	+	+	-	-	-
Hydrogen Sulfide (TSI)	-	-	-	+	+ʷ	-	+	+	-	-	-	-	-	-	-	-	-	-	-	-	-	+	+	-	-	-	-	-	-	-
Urea	-	-	-	-	(V)	(V)	-	vʷ	vʷ	+	+	+	-	-	(+)	-	+	-	vʷ	-	-	+	+	+	+	-	-	+	+	(V)
KCN	-	-	-	-	-	-	(+)	+	-	+	+	-	+	+	vʷ	+	-	+	+	+	vʷ	+	+	+	+	+	+	-	-	-
Motility	+	-	-	+	+	+	+	+	+	+	-	-	+	+	>	+	+	+	+	+	+	+	+	+	+	+	+	-37C +22C	-37C +22C	-37C -22C
Gelatin (22°C)	-	-	-	+	-	-	(+)	-	-	-	-	-	-	-	>	-	-	-	+	+	+	+	>	-	-	-	-	-	-	-
Lysine Decarboxylase	+	-	-	+	+	+	+	-	-	-	+	+	-	+	>	-	+	+	+	+	+	-	+	-	-	-	-	-	-	-
Arginine Dihydrolase	(V)	-	-	-	-	(V)	+	(V)	(V)	+	-	-	+	-	>	+	-	-	-	+	(V)	-	-	-	-	-	-	-	-	-
Ornithine Decarboxylase	+	+	-	+	+	+	+	+	+	+	-	-	+	+	>	+	+	+	+	+	-	-	+	+	-	-	-	+	-	-
Phenylalanine Deaminase	-	-	-	-	-	-	-	-	-	-	-	-	-	-	>	-	-	-	-	-	-	+	+	+	+	+	+	-	-	-
Malonate	-	-	-	-	-	-	+	>	+	-	+	+	>	+	>	+	+	+	-	>	-	-	-	-	-	-	-	-	-	-
Gas from D-Glucose	+	-	-ᵇ	-	-	+	+	+	+	+	+	+	+	+	(V)	+	+	+	+	+	+	+	+	+	-	+	-	-	-	-
Lactose	+	-ᵃ	-	-	-	-	>	>	+	+	+	+	+	+	>	+	-	-	-	>	+	-	-	-	-	-	-	>	-	-
Sucrose	>	-ᵃ	-	-	-	-	-	>	+	+	+	+	+	+	+	+	+	-	+	+	+	+	-	-	-	-	>	+	+	-
D-Mannitol	+	+	+	-	+	+	+	+	+	+	+	+	+	+	+	+	+	+	+	+	+	-	-	-	+	+	>	+	+	+
Dulcitol	>	-	>	-	+ᶜ	(V)	+	>	>	+	>	>	>	-	>	-	-	-	-	-	-	-	-	-	-	-	-	-	-	-
Salicin	>	-	-	-	-	-	-	>	+	+	+	+	(V)	+	(V)	+	+	-	>	+	(V)	+	-	-	+	-	-	+	+	-
Adonitol	-	-	-	-	-	-	-	-	+	-	+	+	-	+	-	-	-	-	-	-	+	-	-	-	+	+	-	-	-	-
i (meso) Inositol	-	-	-	-	-	-	>	-	-	-	+	+	>	+	>	-	+	-	+	(V)	(V)	-	>	-	+	-	+	(V)	-	-
D-Sorbitol	+	-	>	-	+	+	+	+	+	+	+	+	+	+	+	-	+	-	+	+	+	-	+	-	-	+	-	+	>	>
L-Arabinose	+	+	+	>	+ᶜ	+	+	+	+	+	+	+	+	+	+	+	+	+	-	+	+	+	-	-	-	+	-	+	+	+
Raffinose	-	-	>	-	-	-	-	+	+	+	+	+	+	+	+	+	+	-	-	+	+	-	-	-	-	-	-	(V)	+	-
L-Rhamnose	+	(+)	>	-	+	+	+	+	+	+	+	+	+	+	(V)	+	+	+	-	>	-	-	-	-	>	-	-	+	+	+

+ = 90% or more positive within 48 h
- = less than 10% positive within 48 h
(+) = 10% to 89.9% positive within 48 h
> = 90% or more positive between 3 and 7 days
(V) = more than 50% positive within 48 h, and more than 90% positive in 3 to 7 days

w = weak reaction
a = Most S. sonnei strains are delayed positive in reactions for lactose (88%) and sucrose (85%).
b = Some biogroups of S. flexneri produce gas from glucose
c = A few serotypes including S. cholerae-suis, S. paratyphi A and S. pullorum do not ferment dulcitol within 48 h; S. cholerae-suis does not ferment arabinose.

This chart is designed to be a brief guide to the reactions of the more clinically important species of *Enterobacteriaceae*. Only 26 of the 60 or more tests used to distinguish between species are listed. Specific biotypes (H_2S^+, *E. coli*, lactose⁺ and raffinose⁺ *Y. enterocolitica*, etc.), fastidious strains and atypical strains are not addressed. For a more sophisticated treatment of these and other species of *Enterobacteriaceae* the reader should consult special publications that give the above information and percentages.

Enteric Section and Bacteriology Training Branch, CDC

ORGANISM	O GROUP	ORGANISM	O GROUP
ser Yaba 3,10:b:e,n,z_{15}	E_1	ser Zega 9,12:d:z_6	D_1
ser Yalding 1,3,19:r:e,n,z_{15}	E_4	ser Zehlendorf 30:a:1,5	N
ser Yarm 6,8:z_{35}:1,2	C_2	*ser Zeist 18:z_{10}:z_6	K
ser Yarrabah 13,23:y:1,7	G_2	*ser Zuerich 1,9,12:c:z_{39}	D_1
ser Yeerongpilly 3,10:i:z_6	E_1	ser Zuilen 1,3,19:i:l,w	E_4
ser Yerba 54:z_4,z_{23}:-	54	ser Zwickau 16:r(i):e,n,z_{15}	I
ser Yodabasi, not confirmed			
ser Yoff 38:z_4,z_{23}:1,2	P		
ser Yolo 35:c:-	O		

* = Biochemically aberrant

( ) = Antigen incomplete

ser Zadar (9),46:b:1,6	D_2
ser Zagreb, combined w/ser Saintpaul	
ser Zanzibar 3,10:k:1,5	E_1

[ ] = Antigen may be present or absent.

REFERENCES

1. Kauffmann, F., *Enterobacteriaceae*, 2nd Ed., Munksgaard, Copenhagen, 1969.
2. Edwards, P. R., and Ewing, W. H., Identification of *Enterobacteriaceae*, 3rd Ed., Burgess Publishing Co., 1972.

BACTO SCHAEDLER AGAR
BACTO SCHAEDLER BROTH

INTENDED USE

Bacto Schaedler Agar and Broth are nonselective media for cultivating and enumerating anaerobic and aerobic microorganisms. They are excellent basal media to which blood or other enrichments may be added to enhance recovery of fastidious anaerobic and microaerophilic organisms. Bacto Schaedler Broth is the same formula as Bacto Schaedler Agar except the agar has been omitted.

HISTORY

As a result of their studies of the bacterial flora of the gastrointestinal tract of mice, Schaedler, Dubos and Costello[1] formulated an agar medium that proved successful in recovering both aerobic and anaerobic microorganisms. In a subsequent study to determine and enumerate the organisms comprising the human fecal microflora, Mata et al[2] employed a slight modification of the Schaedler formula. In both studies, the authors successfully cultured and enumerated anaerobic lactobacilli, anaerobic streptococci, clostridia, *Bacteriodes*, bifidiobactria and *Viellonella*. In these studies, sodium chloride and neomycin were added to the base for selective cultivation.

Starr, Killgore and Dowell[3] used Schaedler agar supplemented with 5% rabbit blood and 0.5 mg Menadione in a comparative study of quantity, rate and quality of anaerobic bacteria. They found that colony size of *Clostridium cadaveris*, *C. haemolyticum*, *C. novyii* A and *C. perfringens* was larger on Schaedler agar than on similarly enriched tryptic soy yeast extract agar. They considered Schaedler agar superior to TSYEA for cultivating some of the most fastidious anaerobes, with the exception that some strains of clostridia produced rough colonies.

Stalons, Mornsberry and Dowell[4] evaluated nine broth media in varied carbon dioxide atmospheres for their ability to support the growth of anaerobic bacteria commonly associated with human diseases. Schaedler broth in an atmosphere of 5% CO_2, 10% hydrogen and 85% nitrogen exhibited the most rapid rate of growth and highest growth response of the nine media tested.

PRINCIPLES
The ingredients of Bacto Schaedler media will readily support the growth of most organisms, although, they are basically employed in anaerobic microbiology. In order to meet the specific requirements of some anaerobes, hemin and L-cystine are added. Blood or other enrichments may be added to further enhance growth capabilities.

FORMULAE

BACTO SCHAEDLER AGAR
DEHYDRATED
Ingredients per liter

Bacto Tryptic Soy Broth . . . . 10 g
Proteose Peptone No. 3, Difco . . . . . 5 g
Bacto Yeast Extract . . . . 5 g
Bacto Dextrose . . . . 5 g
Tris (Hydroxymethyl)
  Amino Methane . . . . 3 g
L-Cystine . . . . 0.4 g
Hemin . . . . 0.01 g
Bacto Agar . . . . 13.5 g

Final pH 7.6 ± 0.2 at 25°C.

Use 41.9 g/liter.
Five hundred grams will make
11.9 liters.

BACTO SCHAEDLER BROTH
DEHYDRATED
Ingredients per liter

Bacto Tryptic Soy Broth . . . . 10 g
Proteose Peptone No. 3, Difco . . . . . 5 g
Bacto Yeast Extract . . . . 5 g
Bacto Dextrose . . . . 5 g
Tris (Hydroxymethyl)
  Amino Methane . . . . 3 g
L-Cystine . . . . 0.4 g
Hemin . . . . 0.01 g

Final pH 7.6 ± 0.2 at 25°C.

Use 28.4 g/liter.
Five hundred grams will make
17.6 liters.

METHOD OF PREPARATION
1. Suspend appropriate amount in 1 liter distilled or deionized water and heat to boiling to dissolve completely.
2. Sterilize in the autoclave for 15 minutes at 15 lbs pressure (121°C).
3. If an enriched medium is desired, cool medium to 45 – 50°C, aseptically add 5% sterile, defibrinated blood or other enrichment and mix thoroughly.
4. Dispense as desired.

STORAGE
Bacto Schaedler media   Below 30°C
Prepared media   2 – 8°C

QUALITY CONTROL
Identity Specifications
Dehydrated powder: light tan, homogeneous, free-flowing
Reaction of appropriate solution: pH 7.6 ± 0.2 at 25°C
Prepared medium: light to medium amber, clear to slightly opalescent, may have a fine precipitate

Typical Cultural Response on/in Bacto Schaedler Agar or Broth
After 18 – 48 Hours at 35°C

Organism	Growth
Bacteriodes fragilis ATCC® 25285	good to excellent
Clostridium butyricum ATCC® 9690 (formerly *C. novyii*)	good to excellent
Clostridium perfringens ATCC® 12924 (formerly *C. welchii*)	good to excellent
Peptococcus magnus ATCC® 14956 (formerly *P. variabiles*)	good to excellent
Streptococcus pyogenes ATCC® 19615	good to excellent

*incubated anaerobically

REFERENCES
1. J. Exp. Med., 122:59, 1965.
2. Applied Micro., 17:596, 1969.
3. Applied Micro., 22:655, 1971.
4. Applied Micro., 27:1098, 1974.

PACKAGING
Bacto Schaedler Agar	500 g	0403-17-4
Bacto Schaedler Broth	500 g	0534-17-6

BACTO SELENITE BROTH

INTENDED USE

Bacto Selenite Broth is used for enriching members of the *Salmonella* group when isolating these organisms from infectious materials.

HISTORY/PRINCIPLES

The formula of this medium is essentially the same as that of selenite F broth described by Leifson.[1]

Handel and Theodorascu according to Guth[2] observed that *Escherichia coli* was much more susceptible to the toxicity of sodium selenite than was *S. typhosa* (*S. typhi*). Guth confirmed the observations of these authors and employed sodium selenite as a selective agent in an agar medium and in an enrichment broth for the isolation of *S. typhosa* from feces. Leifson[1] extended Guth's observations and developed a selenite agar and a selenite broth for use in the isolation of typhoid and paratyphoid bacilli from feces and urine and found the broth enrichment to offer the greater promise.

Leifson showed that the selenite broth was not sufficiently toxic to inhibit fecal coli and enterococci completely. However, the colon bacilli were reduced in numbers during the first 8 – 12 hours and thereafter increased rapidly. The typhoid bacilli on the other hand multiplied fairly rapidly from the start. *Proteus* and pyocyaneus were not inhibited. Dysentery and alcaligenes were inhibited. In the selenite broth the growth behavior of coli and typhoid in the presence of feces or urine was similar to that of pure cultures.

Leifson observed that the selenite medium functioned most efficiently under reduced oxygen tension. To provide optimal conditions, the broth was distributed in tubes to give a depth of 2 inches or more. Using the enrichment under optimal conditions Leifson was able to isolate many more typhoid and paratyphoid organisms than by direct plating without primary enrichment.

In a survey of methods used for the collection and preservation of stool specimens for the isolation and identification of *Salmonella, Shigella* and intestinal protozoa, Felsenfeld[3] reported an increasing number of laboratories using the selenite broth as an enrichment. In a study of methods to be used as a standard for the bacterial examination of pullorum reactors, Jungherr, Hall and Pomeroy[4] in a committee report showed that in a comparative study of media and enrichments, from October 1946 to February 1950, bismuth sulfite agar and SS agar permitted the highest number of specific isolations of *S. pullorum* and *S. gallinarum*. These favored selective media suppressed the growth of coliform organisms. Following enrichment of the specimens in selenite broth streaking on bismuth sulfite agar gave the largest number of positive isolations, followed by SS agar and then MacConkey agar. Selenite broth yielded a higher number of successful isolations on follow-up media than did tetrathionate broth. The highest percentage of organisms were isolated from the ovary, followed by gall bladder, peritoneum, oviduct, intestines and pericardial sac in the order listed.

Bacto Selenite Broth is identical to selenite F broth, recommended by *Compendium of Methods for the Microbiological Examination of Foods* for enriching *Yersinia enterocolitica.*[5]

FORMULA
BACTO SELENITE BROTH
DEHYDRATED
Ingredients per liter

Bacto Tryptone . . . . 5 g
Bacto Lactose . . . . 4 g
Sodium Selenite . . . . 4 g
Sodium Phosphate . . . . 10 g

Final pH 7.0 ± 0.2 at 25°C.

One pound will make 19.7 liters of medium.

METHOD OF PREPARATION
1. Dissolve 23 g in 1 liter distilled or deionized water and heat to boiling to pasteurize. Avoid excessive heating. **Do not sterilize in the autoclave.**
2. Dispense into sterile tubes to a depth of at least 5 cm.

PROCEDURE
1. Inoculate the medium as follows:
   Fecal specimen: Add 1 – 2 ml of fecal suspension to 10 – 15 ml selenite broth and mix thoroughly.
   Infected tissue: Mascerate 1 – 2 g of specimen in 10 – 15 ml selenite broth using a sterile pipette or stirring rod.
   Urine: Add 5 – 7.5 ml urine to an equal amount of double-strength selenite broth and mix thoroughly.
2. Incubate at 37°C for 18 – 24 hours.
3. Streak a loopful of enriched broth culture on a plate of Bacto MacConkey Agar, Bacto Bismuth Sulfite Agar, Bacto SS Agar or other medium, as desired.

STORAGE
Bacto Selenite Broth — Below 30°C
Prepared medium — 2 – 8°C

QUALITY CONTROL
Identity Specifications

Dehydrated powder: off-white, homogeneous, free-flowing
Reaction of 2.3% solution: pH 7.0 ± 0.2 at 25°C
Prepared medium: very light amber, clear to very slightly opalescent, may have a slight precipitate

Typical Cultural Response on Bacto MacConkey Agar
After Enrichment in Bacto Selenite Broth for 18 – 24 Hours at 35°C

Organism	Recovery	Color of Colony
Escherichia coli ATCC® 25922	little or no increase in numbers	white to pink with bile ppt
Salmonella choleraesius ATCC® 12011	good to excellent	colorless
Salmonella typhi ATCC® 6539	good to excellent	colorless
Salmonella typhimurium ATCC® 14028	good to excellent	colorless

REFERENCES

1. Am. J. Hyg., 24:423, 1936.
2. Centr. Bakt. I Abt. Orig., 77:487, 1916.
3. Public Health Reports, 63:1075, 1950.
4. Proc 22nd Ann. Mtg. N.E. Conf. Lab. Workers Pullorum Disease Control, Burlington, Vt., 1950.
5. Speck, M.L. 1976. Compendium of methods for the microbiological examination of foods. American Public Health Association, Washington, D.C.

PACKAGING

Bacto Selenite Broth		
	1 lb (454 g)	0275-01-7
	1/4 lb (114 g)	0275-02-6
	10 kg	0275-08-0
	12 tubes	0275-34-8
	144 tubes	0275-37-5

BACTO SELENITE CYSTINE BROTH

INTENDED USE

Bacto Selenite Cystine Broth is a selective enrichment broth recommended by AOAC Methods[1] and USP[2] for use in detecting and identifying *Salmonella* in foods, dairy products and other materials of sanitary importance.

HISTORY/PRINCIPLES

Bacto Selenite Cystine Broth is a modification of Leifson's[3] formula for selenite broth. Leifson found selenite to be useful in permitting *Salmonella* species to grow while reducing growth of fecal coli and enterococci. The formulation with cystine was proposed by the Food and Drug Administration for use as an enrichment medium for detecting *Salmonella* in food materials. It was also included among the Standard Methods media of the American Public Health Association (1960).

FORMULA

BACTO SELENITE CYSTINE BROTH
DEHYDRATED

Ingredients per liter

Bacto Tryptone . . . . 5 g
Bacto Lactose . . . . 4 g
Disodium Phosphate . . . . 10 g
Sodium Acid Selenite . . . . 4 g
L-Cystine . . . . 0.01 g

Final pH 7.0 ± 0.2 at 25°C.

Five hundred grams will make 21.7 liters of medium.

METHOD OF PREPARATION

1. Suspend 23 g in 1 liter distilled or deionized water. Heat to boiling to dissolve.
2. Dispense in sterile tubes or bottles to a depth of at least 60 mm.
3. If desired, heat for 10 minutes in flowing steam. DO NOT AUTOCLAVE. Overheating will destroy the selective properties. Medium is not sterile and should be used the same day as prepared.
4. Allow to cool to room temperature and inoculate as desired.
5. Incubate at 35°C for 12 – 18 hours.
6. Streak for isolation on differential plating media, such as Bacto MacConkey Agar, Bacto Bismuth Sulfite Agar, Bacto Hektoen Enteric Agar, etc.

STORAGE

Bacto Selenite Cystine Broth
Prepared medium

Below 30°C
not recommended

QUALITY CONTROL

Identity Specifications

Dehydrated powder:
Reaction of 2.3% solution:
Prepared medium:

off-white, homogeneous, free-flowing
pH 7.0 ± 0.2 at 25°C
very light amber, clear to very slightly opalescent, may
have a slight precipitate

NOTE: Medium has been overheated if the precipitate is brick red in color.

Typical Cultural Response on Bacto MacConkey Agar
After 18 – 24 Hours at 35°C in Bacto Selenite Cystine Broth

Organism	Recovery	Color of Colony
Escherichia coli ATCC® 25922	no increase in numbers	white to pink with bile precipitate
Salmonella choleraesuis ATCC® 12011	good to excellent	colorless
Salmonella sp. serotype Pullorum ATCC® 9120	good to excellent	colorless
Salmonella typhi ATCC® 6539	excellent	colorless

REFERENCES

1. Methods of Analysis AOAC, 11th Ed., 1980.
2. US Pharmacopeia XX, 1980.
3. Leifson, Am. J. Hyg., 24:423, 1936.

PACKAGING

Bacto Selenite Cystine Broth	5 lb (2.27 kg)	0687-05-5
	10 kg	0687-08-2
	500 g	0687-17-1
	100 g	0687-15-3

BACTO SELLERS DIFFERENTIAL AGAR

INTENDED USE

Bacto Sellers Differential Agar is recommended for the differentiation and identification of nonfermentative gram-negative bacilli, particularly Pseudomonas aeruginosa, Herellea vaginicola (Acinetobacter calcoaceticus), Mima polymorpha (Acinetobacter lwoffii), Alcaligenes facealis and Bacterium anitratum (Acinetobacter calcoaceticus).

HISTORY

Bacto Sellers Differential Agar is prepared according to the formulation of Sellers.[1,2] He designed this special medium to distinguish organisms which fail to produce acid reactions on triple sugar iron agar. Bhagwat and King[3] evaluated Sellers' medium and found the results to be in agreement with Sellers. They felt, however, the addition of dextrose did not seem to be an advantage.

PRINCIPLES

Pseudomonas aeruginosa growing on Bacto Sellers Differential Agar produces nitrogen gas and when placed under ultraviolet light, it fluoresces. Most *Pseudomonas* cultures produce an alkaline butt or an alkaline band at the base of the slant. *Acinetobacter calcoaceticus*, produces an alkaline slant and an acid band. *Acinetobacter lwoffii* produces an alkaline slant with no band or gas formation. *A. faecalis* produces an alkaline slant and most cultures produce nitrogen gas and/or alkaline butts.

Differentiation of gram-negative bacilli is on the basis of fluorescence under ultraviolet light, oxidation of dextrose to acid, production of nitrogen gas and growth under anaerobic conditions in the presence of nitrate, changing the pH. The production of nitrogen gas is stimulated by the dipotassium phosphate in the medium. It is demonstrated in the butt section of the medium as separations of the agar from the sides and bottom of the tube.

The magnesium sulfate and mannitol incorporated in the medium stimulate fluorescein production by some organisms. Brom thymol blue and phenol red serve as indicators. Oxidation of dextrose is demonstrated by a yellow band at the junction of the slant and butt. The dextrose added prior to inoculation diffuses into the medium during the incubation period. *Pseudomonas aeruginosa* which normally produces an acid reaction from the dextrose, does not do so in Sellers differential agar because of the arginine and high peptone concentration.

The reactions produced by the nonfermentative gram-negative bacilli are given in the following table:

Organism	Slant Color	Butt Color	Band Color	Fluorescent Slant	Nitrogen Gas
P. aeruginosa	green	blue or no change	sometimes blue	yellow-green	produced
A. calcoaceticus	blue	no change	yellow	absent	absent
A. lwoffii	blue	no change	absent	absent	absent
A. faecalis	blue	blue or no change	absent	absent	produced

FORMULA

BACTO SELLERS DIFFERENTIAL AGAR
DEHYDRATED

Ingredients per liter

Bacto Yeast Extract	1 g	Sodium Chloride	2 g
Bacto Peptone	20 g	Sodium Nitrate	1 g
L-Arginine	1 g	Sodium Nitrite	0.35 g
D-Mannitol	2 g	Magnesium Sulfate	1.5 g
Brom Thymol Blue	0.04 g	Dipotassium Phosphate	1 g
Phenol Red	0.008 g	Bacto Agar	15 g

Final pH 6.7 ± 0.2 at 25°C.

One pound will make 10.1 liters of medium.

PROCEDURE

1. To rehydrate, suspend 45 g in 1 liter distilled water and heat to boiling to dissolve the medium completely.
2. Dispense into test tubes and stopper with cotton plugs or loose fitting caps and sterilize in the autoclave for 10 minutes at 15 lbs pressure (121°C).

3. Allow the tubes to cool in a slanted position to give approximately 1-1/2 inch butts and 3 inch slants.
4. Immediately before inoculating add 0.15 ml or 2 drops of sterile Bacto Dextrose Solution 50% to each tube by letting it run down the side of the tube opposite the slant.
5. Inoculate with a deep stab into the butt and streak the slant from side to side, then incubate for 24 hours at 37°C. A 48 hour incubation period may be required by some organisms.

STORAGE
Bacto Sellers Differential Agar   Below 30°C
Prepared tubes   2 – 8°C

QUALITY CONTROL
Identity Specifications
Dehydrated powder: beige, homogeneous, free-flowing
Reaction of 4.5% solution: pH 6.7 ± 0.2 at 25°C
Prepared medium: green, slightly opalescent

Typical Cultural Response on Bacto Sellers Differential Agar
After 18 – 24 Hours at 37°C

Organism	Growth	Slant	Butt	Band	Fluorescence	Nitrogen Gas
Acinetobacter calcoaceticus ATCC® 19606	good	blue	green	yellow	–	–
Acinetobacter calcoaceticus ATCC® 17961	good	blue	blue	yellow	–	–
Acinetobacter lwoffii ATCC® 9957	good	blue	blue	—	–	–
Alcaligenes faecalis ATCC® 8750	good	blue-green	blue-green	—	–	sl+
Pseudomonas aeruginosa ATCC® 27853	good	blue-green	blue-green	blue	+	+

REFERENCES
1. Personal Communication, 1962.
2. Bact. Proc., p. 65, 1963.
3. Cleveland Clinic Quarterly, 33:137, 1966.

PACKAGING
Bacto Sellers Differential Agar   1 lb (454 g)   0895-01-7

BACTO SENSITIVITY DISCS

INTENDED USE
Bacto Sensitivity Discs are used for determining the in vitro susceptibility of microorganisms to a variety of antimicrobial agents.

SUMMARY AND EXPLANATION OF THE TEST

Variations of the disc diffusion technique have been in use since the late 1940's. The need for standardization of this procedure was long recognized when Bauer, Kirby, Sherris and Turck[1] suggested a standardized single-disc method in 1966. In 1971, Ericsson and Sherris[2] published the results of an extensive series of international collaborative studies of testing techniques using broth dilution, agar dilution and agar diffusion.The National Committee for Clinical Laboratory Standards (NCCLS)[3] issued approved performance standards in 1976, followed by World Health Organization (WHO)[4] requirements in 1977.

The procedure described herein is based on the modified test currently recommended by NCCLS.[5]

This test is routinely used to determine the susceptibility or resistance of a pathogenic organism to a particular antimicrobic agent.

REAGENTS

Each of the antimicrobial agents listed below is available in self-dispensing magazines of 50 discs each. Magazines can be used individually or can be included in the 6 - or 12 - magazine dispenser for convenient, properly spaced placement on a 100 or 150 mm Petri dish, respectively. Each magazine is individually packaged in a protective stay-dry pack that includes a desiccant.

Amikacin 30 μg
Ampicillin 10 μg
    25 μg
Azlocillin 75 μg
Bacitracin 10 units
Carbenicillin 100 μg
Cefamandole 30 μg
Cefoperazone 75 μg
Cefotaxime 30 μg
Cefoxitin 30 μg
Cephaloridine 30 μg
Cephalothin 30 μg
Chloramphenicol 30 μg
Chlortetracycline 30 μg
Cinoxacin 100 μg
Clindamycin 2 μg
Cloxacillin 1 μg
Colistin 10 μg
Dibekacin 10 μg
    30 μg
Dicloxacillin 1 μg
Doxycycline 30 μg

Erythromycin 15 μg
Fosfomycin 50 μg
Fusidic Acid 10 μg
Gentamicin 10 μg
Kanamycin 30 μg
Lincomycin 2 μg
Methenamine Mandelate 3 mg
Methicillin 5 μg
Mezlocillin 75 μg
Minocycline 30 μg
Moxalactam 30 μg
Nafcillin 1 μg
Nalidixic Acid 30 μg
Neomycin 5 μg
    30 μg
Netilmicin 30 μg
Nitrofurantoin 300 μg
Novobiocin 5 μg
    30 μg
Nystatin 100 units
Oxacillin 1 μg
Oxytetracycline 30 μg

Penicillin G 2 units
    10 units
Piperacillin 100 μg
Polymyxin B 300 units
Rifampin 5 μg
SxT 25 μg
Sisomicin 10 μg
    30 μg
Streptomycin 10 μg
Sulfadiazine 300 μg
Sulfamethoxazole 23.75 μg
Sulfathiazole 300 μg
Sulfisoxazole 150 μg
    300 μg
Tetracycline 30 μg
Ticarcillin 75 μg
Tobramycin 10 μg
Trimethoprim 1.25 μg
    5 μg
Triple Sulfa 300 μg
Vancomycin 30 μg

STORAGE

Bacto Sensitivity Discs   Below 8°C

PREPARATION OF REAGENTS

1. Prepare and sterilize Bacto Mueller Hinton Medium according to package directions or remelt with minimum heating previously prepared and sterilized medium. Bring to 45 – 50°C. Pour the melted medium into sterile Petri dishes (approximately 25 ml for 100 mm plates and 60 ml for 150 mm plates) on a level surface to a uniform depth of 4 mm and allow medium to harden. Avoid moisture on the surface of the medium by pouring medium cooled to or below 50°C. Leave top of plate slightly

ajar until medium solidifies to minimize moisture collection on the surface of the medium. If moisture does collect, incubate plates at 35 – 37°C for 30 minutes or at room temperature for 1 hour prior to use. Final reaction of the medium is pH 7.3 ± 0.1 at 25°C.

2. Remove Bacto Sensitivity Discs from (freezer to) refrigerator to room temperature 1 – 2 hours before use to minimize possibility of condensation.

3. Prepare a turbidity standard to measure inoculum density. Add 0.5 ml of 0.048 M $BaCl_2$ (or 0.5 ml of 1.175% $BaCl_2 \cdot 2\ H_2O$) to 99.5 ml of 0.36 N H_2SO_4. Distribute in 4 – 6 ml amounts into screw cap tubes of the same size as those being used to grow the broth culture inoculum. Seal tubes tightly and store at room temperature in the dark.

Vigorously agitate this turbidity standard on a mechanical vortex mixer just before each use. Prepare new standards at least every 6 months.

PROCEDURE
ANTIMICROBIAL DISC SUSCEPTIBILITY TEST FOR NONFASTIDIOUS BACTERIA

1. Inoculate test plates by standard method or by the alternate agar overlay method.

Standard Method
   a. Transfer 4 – 5 similar colonies from the primary isolation medium into approximately 5 ml Bacto Tryptic Soy Broth (Soybean-Casein Digest Broth, USP) by touching the top of the colonies with a flamed and cooled wire loop or needle.
   b. If the resulting culture suspension is less turbid than the turbidity standard when read against a white background with a contrasting black line, incubate the culture at 35 – 37°C for 2 – 8 hours until suspension turbidity equals or exceeds that of the standard. Dilute the culture suspension with sterile saline if necessary. If the turbidity of the just-inoculated suspension is comparable to that of the standard, proceed without incubation.
   c. Within 15 minutes of standardization, dip a sterile nontoxic cotton swab on a wooden applicator into the standardized culture suspension and express excess fluid by rotating the swab firmly against the inside wall of the test tube.
   d. Inoculate the entire agar surface, streaking in 3 different directions by rotating the plate at 60° angles after each streaking. If the plate is satisfactorily streaked, there will be a confluent lawn of growth after 16 – 18 hours incubation. Avoid extremes of inoculum density.
   e. Allow inoculum to dry 5 – 15 minutes with top in place.

Agar Overlay Method
The alternate agar overlay method of Barry et al[6] which incorporates inoculum in a specified volume of agar may be used providing the resulting diameters of zones of inhibition around approved antibiotic discs are equivalent to those obtained with the standard method.

2. Apply appropriate susceptibility discs manually or with a Dispens-O-Disc Dispenser onto the inoculated surface. Gently press each disc down with a sterile forceps, flamed and cooled between each disc, to insure complete contact with the agar. Observe a minimum spacing of 24 mm from disc center to center to avoid overlapping inhibition zones. Do not move a disc once it has contacted the agar because the drug diffuses almost instantaneously.

3. Within 15 minutes of applying discs, aerobically incubate the inverted plates at 35°C for 16 – 18 hours.

4. With the unaided eye, examine plates illuminated with reflected light against a black nonreflecting background. Measure diameters of zones of complete inhibition to the nearest whole mm using the Microbial Disc Zone Chart marked with appropriate

zone measuring tabs (Code 0345 and Code 0347, available on request from Difco) or use calipers or a ruler. On blood agar, remove the cover and measure inhibition zones from the surface illuminated with reflected light. Observe the following precautions:

a. Ignore swarming of *Proteus* if zones of inhibition are clearly outlined.
b. Measure sulfonamide zones at the margin of heavy growth. Sulfonamides may not inhibit organisms for several generations so slight growth within the inhibition zones may be disregarded.
c. Define the zone end point as that area showing no visible growth when observed with the unaided eye as described above. An occasional control culture may produce a mutant colony which is visible within the inhibition zone with transmitted light or assisted vision. Disregard such growth.
d. Subculture, reidentify and retest any large colonies growing within a clear zone of inhibition.

ANTIMICROBIAL DISC SUSCEPTIBILITY TEST FOR FASTIDIOUS BACTERIA
Haemophilus influenzae
1. Prepare Bacto Mueller Hinton Medium supplemented with 1% Bacto Hemoglobin and 1% Bacto Supplement VX.
2. Emulsify growth from an overnight agar plate in Bacto Mueller Hinton Broth. Dilute with more broth until the turbidity of the suspension is comparable to that of the turbidity standard.
3. Inoculate the medium, apply discs, incubate the plates and examine resulting growth according to the principles set forth in the above procedure. Observe special references to *H. influenzae* on Table 1.
4. In critical *H. influenzae* infections, test for β - lactamase production from growth on isolation plates to determine if the organism is susceptible to ampicillin. Strains having zones less than 20 mm with ampicillin or penicillin G discs produce β - lactamase and are not considered susceptible to ampicillin.

β - lactamase-producing *Neisseria gonorrhoeae*.
1. Prepare Bacto GC Medium Base supplemented with 1% Bacto Supplement VX.
2. Emulsify growth from an overnight agar plate in Bacto Mueller Hinton Broth. Dilute with more broth until the turbidity of the suspension is comparable to that of the turbidity standard.
3. According to the principles set forth in the above procedure inoculate the medium, apply **only the 10 unit penicillin disc**, incubate at **35°C in CO_2** and examine resulting growth.
4. Assume that strains with penicillin zones less than 20 mm produce β - lactamase since susceptible strains generally have zones of 30 – 60 mm.

RESULTS
Interpret the sizes of zones of inhibition by referring to TABLE 1. Report the reaction of the test organism as resistant, intermediate or susceptible.

TABLE 1. Interpretation of Zone Diameters of Test Cultures

| Antimicrobic Agent | Disc Content | Zone Diameter to Nearest Whole mm | | |
		Resistant mm or Less	Intermediate[15] mm Range	Susceptible mm or More
Amikacin[1]	30 µg	14	15 – 16	17
Ampicillin when testing gram-negative microorganisms and enterococci[2]	10 µg	11	12 – 13	14

TABLE 1 (continued)

| Antimicrobic Agent | Disc Content | Zone Diameter to Nearest Whole mm | | |
		Resistant mm or Less	Intermediate[15] mm Range	Susceptible mm or More
Ampicillin when testing staphylococci and penicillin G - susceptible microorganisms[2,3]	10 µg	20	21 – 28	29
Ampicillin when testing *Haemophilus* species[2,4]	10 µg	19		20
Bacitracin	10 units	8	9 – 12	13
Carbenicillin when testing *Enterobacteriaceae*	100 µg	17	18 – 22	23
Carbenicillin when testing *Pseudomonas aeruginosa*	100 µg	13	14 – 16	17
Cefamandole[5]	30 µg	14	15 – 17	18
Cefotaxime[5]	30 µg	14	15 – 22	23
Cefoxitin[5]	30 µg	14	15 – 17	18
Cephalothin[6]	30 µg	14	15 – 17	18
Cephalothin when reporting susceptibility to cephaloglycin[6]	30 µg	14		15
Chloramphenicol	30 µg	12	13 – 17	18
Cinoxacin	100 µg	14	15 – 18	19
Clindamycin[7]	2 µg	14	15 – 16	17
Colistin[8]	10 µg	8	9 – 10	11
Erythromycin	15 µg	13	14 – 17	18
Gentamicin[1]	10 µg	12	13 – 14	15
Kanamycin	30 µg	13	14 – 17	18
Methicillin[9]	5 µg	9	10 – 13	14
Nafcillin[9]	1 µg	10	11 – 12	13
Nalidixic Acid[10]	30 µg	13	14 – 18	19
Neomycin	30 µg	12	13 – 16	17
Nitrofurantoin[10]	300 µg	14	15 – 16	17
Novobiocin	30 µg	17	18 – 21	22
Oxacillin[9]	1 µg	10	11 – 12	13
Penicillin G when testing staphylococci[11]	10 units	20	21 – 28	29
Penicillin G when testing other microorganisms[12]	10 units	11	12 – 21	22
Polymyxin B[8]	300 units	8	9 – 11	12
Rifampin when testing *Neisseria meningitidis* susceptibility only	5 µg	24		25
Streptomycin	10 µg	11	12 – 14	15
Sulfonamides[10,13]	300 µg	12	13 – 16	17
Tetracycline[14]	30 µg	14	15 – 18	19
Ticarcillin when testing *P. aeruginosa*	75 µg	11	12 – 14	15
Tobramycin	10 µg	12	13 – 14	15
Trimethoprim[10,13]	5 µg	10	11 – 15	16
Trimethoprim/ Sulfamethoxazole[13]	1.25 µg/ 23.75 µg	10	11 – 15	16
Vancomycin	30 µg	9	10 – 11	12

1. The zone sizes obtained with aminoglycosides, particularly when testing *Pseudomonas aeruginosa*, are very medium dependent because of variations in divalent cation content. The zone diameter interpretive standards for amikacin, gentamicin, and tobramycin are shown in Table 1. These interpretive standards are to be used only with Mueller-Hinton medium that has yielded zone diameters within the correct range shown in Table 2 when performance tests were done with *P. aeruginosa* ATCC® 27853. In addition, the amikacin disc must be 30 µg rather than the 10 µg disc used previously. Organisms in the intermediate category may be either susceptible or resistant when tested by dilution methods and should therefore more properly be classified as "indeterminate" in their susceptibility to aminoglycosides.

2. Class disc for ampicillin (amoxicillin), cyclacillin, and hetacillin.
3. Resistant strains of *S. aureus* produce β - lactamase and the testing of the 10 unit penicillin disc is preferred.
4. For testing *Haemophilus* use Mueller-Hinton agar supplemented with 1% hemoglobin and 1% Supplement VX. Adjust pH to 7.2. Prepare the inoculum by suspending growth from a 24 - hour chocolate agar plate in Mueller-Hinton broth to the density of a turbidity standard. The vast majority of ampicillin-resistant strains of *Haemophilus* produce detectable β - lactamase.
5. Cefamandole, cefoxitin and cefotaxime are recently released cephalosporins having a wider spectrum of activity against gram-negative bacilli than do other previously approved cephalosporins. Therefore, the cephalothin disc cannot be used as the class disc for these three drugs.
6. The cephalothin disc is used for testing susceptibility to cephalothin, cefaclor, cefadroxil, cefazolin, cephalexin, cephaloridine, cephapirin, and cephradine. Cefamandole, cefoxitin and cefotaxime must be tested separately. *Staphylococcus aureus* exhibiting resistance to methicillin, nafcillin or oxacillin discs should be reported as resistant to cephalosporin-type antimicrobics, regardless of zone diameter, because in most cases infections caused by these organisms are clinically resistant to cephalosporins. Methicillin-resistant *S. epidermidis* infections also may not respond to cephalosporins.
7. The clindamycin disc is used for testing susceptibility to both clindamycin and lincomycin.
8. Colistin and polymyxin B diffuse poorly in agar, and the diffusion method is thus less accurate than with other antimicrobics. Resistance is always significant, but when treatment of systemic infections caused by susceptible strains is being considered, results of a diffusion test should be confirmed with those of a dilution method.
9. Of the antistaphylococcal β - lactamase resistant penicillins, either oxacillin, nafcillin, or methicillin may be tested, and results can be applied to the other two of these drugs and to cloxacillin and dicloxacillin. Oxacillin and nafcillin are more resistant to degradation in storage. Cloxacillin discs should not be used because they may not detect methicillin-resistant *S. aureus*. When an intermediate result is obtained with *S. aureus*, the strains should be further investigated to determine if they are heteroresistant.
10. Susceptibility data for nalidixic acid, nitrofurantoin, sulfonamides and trimethoprim apply only to organisms isolated from urinary-tract infections.
11. Penicillin G should be used to test the susceptibility of all penicillinase-sensitive penicillins, such as ampicillin, amoxicillin, hetacillin, carbenicillin and ticarcillin. Results may also be applied to phenoxymethyl penicillin or phenethicillin. The intermediate category usually contains penicillinase producing isolates and should be considered resistant to therapy.
12. Intermediate category includes some microorganisms, such as enterococci, and certain gram-negative bacilli that may cause systemic infections treatable with high parenteral dosages of benzyl penicillin but not of orally administered phenoxymethyl penicillin or phenethicillin.
13. The 300 μg sulfisoxazole discs can be used for any of the commercially available sulfonamides. Blood-containing media, except media containing lysed horse blood, are not satisfactory for testing sulfonamides. The Mueller-Hinton agar should be as thymidine-free as possible for sulfonamide and/or trimethoprim testing.
14. Tetracycline is the class disc for all tetracyclines, and the results can be applied to chlortetracycline, demeclocycline, doxycycline, methacycline, minocycline, and oxytetracycline. However, some in vitro data show that certain organisms may be more susceptible to doxycycline and minocycline than to tetracycline.
15. The category "Intermediate" should be reported. Infections with bacteria of "intermediate" susceptibility may be considered moderately susceptible and may respond to antimicrobial agents with a wide safe dosage range.

QUALITY CONTROL

1. Using the methods described above in PROCEDURE, inoculate plates individually with *Staphylococcus aureus* ATCC® 25923, *Escherichia coli* ATCC® 25922, *Pseudomonas aeruginosa* ATCC® 27853 and *Escherichia coli* ATCC®35218 from Bactrol™ Disks or stock cultures.
2. Apply antimicrobic discs appropriate for each test organism and incubate.
3. Record results and compare with Table 2, monitoring for significant change in mean zone diameter that cannot be explained by a change in methodology.

TABLE 2. Range of Zone Diameters (mm) Obtained with Control Cultures.

Antimicrobial Agent	Disc Content	*S. aureus* ATCC® 25923	*E. coli* ATCC® 25922	*P. aeruginosa* ATCC® 27853	*E. coli* ATCC® 35218
Amikacin	30 μg	20 – 26	19 – 26	18 – 26	
Ampicillin	10 μg	27 – 35	16 – 22		no zones
Augmentin	20/10 μg				8 – 22
Carbenicillin	100 μg		23 – 29	18 – 24	
Cefamandole	30 μg	26 – 34	24 – 30		
Cefoperazone	75 μg	24 – 33	28 – 34	23 – 29	
Cefotaxime	30 μg	25 – 31	29 – 35	18 – 22	
Cefoxitin	30 μg	23 – 29	23 – 29		

TABLE 2 (continued)

Antimicrobial Agent	Disc Content	*S. aureus* ATCC® 25923	*E. coli* ATCC® 25922	*P. aeruginosa* ATCC® 27853	*E. coli* ATCC® 35218
Cephalothin	30 μg	29 – 37	17 – 21		
Chloramphenicol	30 μg	19 – 26	21 – 27		
Clindamycin	2 μg	24 – 30			
Colistin	10 μg		11 – 15		
Doxycycline	30 μg	23 – 29	18 – 24		
Erythromycin	15 μg	22 – 30			
Gentamicin	10 μg	19 – 27	19 – 26	16 – 21	
Kanamycin	30 μg	19 – 26	17 – 25		
Methicillin	5 μg	17 – 22			
Mezlocillin	75 μg		23 – 29	19 – 25	
Minocycline	30 μg	25 – 30	19 – 25		
Moxalactam	30 μg	18 – 24	28 – 35	17 – 25	
Nafcillin	1 μg	16 – 22			
Nalidixic Acid	30 μg		22 – 28		
Neomycin	30 μg	18 – 26	17 – 23		
Netilmicin	30 μg	22 – 31	22 – 30	17 – 23	
Nitrofurantoin	300 μg		21 – 26		
Oxacillin	1 μg	18 – 24			
Penicillin G	10 units	26 – 37			
Piperacillin	100 μg		24 – 30	25 – 33	
Polymyxin B	300 units	7 – 13	12 – 16		
Streptomycin	10 μg	14 – 22	12 – 20		
Sulfisoxazole	250 or 300 μg	24 – 34	18 – 26		
Tetracycline	30 μg	19 – 28	18 – 25		
Ticarcillin	75 μg		24 – 30	21 – 27	
Tobramycin	10 μg	19 – 29	18 – 26	19 – 25	
Trimethoprim	5 μg	21 – 28	21 – 28		
Trimethoprim/ Sulfamethoxazole	1.25/23.75 μg	24 – 32	24 – 32		
Vancomycin	30 μg	15 – 19			

LIMITATIONS

1. The procedure described herein applies only when used with a pure culture. The results of using a mixed culture may be grossly misleading.
2. Any deviation from standard methods may cause false or erroneous results.

REFERENCES

1. Amer. J. Clin. Path., 45:493,1966.
2. Acta. Pathologica et Microbiologica Scandinavica, Section B. Supp. 217, 1971.
3. Performance Standards for Antimicrobic Disc Susceptibility Tests, NCCLS, January 1976.
4. World Health Organization Technical Report Series, No. 610, Twenty-Eighth Report, 1977.
5. Performance Standards for Antimicrobic Disc Susceptibility Tests, Second Ed., NCCLS, October 1979.
6. Amer. J. Clin. Path., 53:149, 1970.

PACKAGING

Bacto Sensitivity Discs are packaged as 10 magazines of 50 discs each. Please consult Difco PRODUCT LIST III for current antimicrobial agent availability and product code.

Dispens-O-Disc® Dispenser	6 mag 100 mm w/storage container	6003-30-8
Dispens-O-Disc® Dispenser	12 mag 150 mm w/storage container	6002-30-9
Dispens-O-Rack	w/storage container, holds 18 magazines	6005-30-6
Desiccant w/Indicator	10 × 1 unit	1821-31-8

SHIGELLA ANTISERA/ALKALESCENS-DISPAR ANTISERUM

BACTO SHIGELLA ANTISERA
BACTO ALKALESCENS-DISPAR ANTISERUM POLY

INTENDED USE
Bacto Shigella Antisera are prepared for use in the serological detection and identification of *Shigella* by the slide agglutination technique. Bacto Shigella Polyvalent Antisera are employed in the screening of *Shigella* cultures for the presumptive evidence of the presence of that species of *Shigella* while Bacto Shigella Antiserum Types are used for serotyping the organisms.

Bacto Alkalescens-Dispar Antiserum Poly likewise is prepared for the serological detection and identification of this group of organisms using the slide agglutination technique.

SUMMARY AND EXPLANATION OF THE TEST
Chantemesse and Widal[1] first reported a bacillus isolated from the feces of 5 cases of acute dysentery. However, these authors failed to establish this organism as the etiological agent of the infection. Shiga isolated and reported a bacterium identical to that described by Chantemesse and Widal and established its etiological significance by agglutinating the bacterium isolated from the stools of patients with dysentery with serums of convalescing patients.[2] Subsequent studies of Kruse, Flexner, Strong and Musgrave corroborated the findings of Shiga.[3,4,5]

The genus *Shigella* contains 4 recognized species, *S. dysenteriae, S. flexneri, S. boydii* and *S. sonnei*, all of which are pathogenic to man. *Shigella* have been implicated in summer diarrhea of children and have been incriminated in cases of epidemic infant diarrhea. They have become the most frequent cause of clinical dysentery.[6] Indeed, in recent years there have been more reported cases of shigellosis (bacillary dysentery) in the United States than cases of typhoid and amoebic dysentery combined. As recently as 1973, 10,216 cases of shigellosis were reported to the Center for Disease Control, Atlanta, Georgia, covering a period of 6 months.[7] Further, it has been estimated that for every clinical case of shigellosis, there exists in our population eight to ten asymptomatic carriers. This offers a large reservoir of potential infection, which is often triggered by the hygienic condition of a given population segment, indicating the necessity of a constant and adequate surveillance program.

Because of the clinical and public health significance of *Shigella*, and its apparent increasing occurrence throughout the population, fecal specimens submitted to the laboratory, from patients with dysentery-like symptoms, should be examined for the presence of this organism.

Those laboratories which are interested in the genus *Shigella*, should be prepared to detect — to at least the species level — those organisms serologically. Better, it is recommended that those laboratories so interested, be equipped serologically to recognize the most common members within the 4 species.[8]

To determine the most efficient use of antisera, it might be in order to review some of the literature documenting incidence of *Shigella* in the U.S.[7]

Table I

No. of Types in Species	Species			
	S. flexneri	*S. sonnei*	*S. boydii*	*S. dysenteriae*
	6	1	15	10
1967 studies				
Isolates (%)	70	22	5	3
Types	2	—	1,2,4	2 predom.
Predominate				1 – 5 common
1973 studies				
Isolates (%)	15.5	83.6	0.55	0.34
Types				
Predominate	2,3,5	—	1,7	1,2

From the above studies, it is evident that using 2 polyvalent antisera and 7 individual sera over 90% of the *Shigella* cultures might be identified. However, the above statements are included to indicate prevalence of serotypes — not necessarily severity of disease. Awareness of *S. dysenteriae* must be constant since it is of primary pathogenic importance due to its rarity and the severity of the disease state produced.

Shigellosis may occur more often than it is actually reported; many cases may go unrecognized or undetected. Two notable reasons may account for this situation. First, some selective and differential media traditionally employed in the examination of fecal specimens for enteric pathogens have been found to be too inhibitory for less hearty *Shigella* organisms. The use of more appropriate enrichments and less inhibitory plating media results in increased isolation of these organisms. Secondly, the *Shigella* are closely related and similar to *E. coli*, and Alkalescens-Dispar, (anaerogenic nonmotile biotypes of *E. coli*) resembling shigellae on TSI agar. These relationships could lead to confusing results or misidentification since all three of these groups may occur in fecal specimens and might cause dysentery-like symptoms. *Shigella* may also be confused with some nonpathogens which fail to ferment lactose.

Biochemical characterization is necessary to distinguish *Shigella* from other related members of the *Enterobacteriaceae*, particularly *E. coli*. It is known that although these 2 groups are biochemically different in a few instances, they are antigenically related. More specifically the somatic antigens of most serotypes of *Shigella* are either identical to, or related to, *E. coli*.

On the other hand some members of the Alkalescens-Dispar Group (*S. alkalescens*) are lactose nonfermentors which, like *Shigella*, do not produce gas from carbohydrates; they are best differentiated from *Shigella* by serological as well as biochemical methods.

Shigella cultures should not react in Alkalescens-Dispar Antisera but Alkalescens-Dispar cultures may at times, react in Shigella Antisera. The cross reactions of biochemically and serologically defined Alkalescens-Dispar organism in a Shigella Antisera should be considered comparable to an intergeneric cross reaction between *Escherichia* and *Salmonella*. Such cross reactions are, of course, possible among related organisms, particularly the *Enterobacteriaceae*. The Alkalescens-Dispar group is serologically related to both *Shigella* and *Escherichia*. The complexity of these relationships requires reliance on both biochemical and serological findings to accurately identify members of the Alkalescens-Dispar group. Ewing[9] worded it so emphatically and eloquently that it was felt necessary to quote:

"In the identification of an unknown enteric bacterium, one cannot rely entirely on serology, nor can complete characterization be made on the basis of biochemical reactions alone. Both methods must be employed for accurate identification. Biochemical reactions are used to determine the genus and species to which a culture belongs, and serology is employed to determine the serotype. Quite often anaerogenic and microaerogenic cultures of *E. coli* are isolated which do not utilize lactose or sucrose promptly and hence given the appearance of shigellae on triple sugar iron agar and similar media. If such cultures are examined only by serological methods, the chance of error is great."

SHIGELLA ANTIGENIC ANALYSIS

Since the *Shigella* are nonmotile, serological identification of this group of organisms is based solely on the detection of their somatic agglutinogens. However, it is often necessary to heat the suspension at 100°C for 15 – 60 minutes to destroy an envelope antigen which inhibits agglutination of the living cell.

The genus *Shigella* is divided into 4 serogroups, each group consisting of a species which contains distinctive type antigens. The type antigens within the group demonstrate significant variations. The first group — Group A contains *S. dysenteriae* with 10 serotypes, 1 – 10, and possibly 3 provisional serotypes. These organisms are mannitol negative and are unrelated serologically to members of groups B, C, and D *Shigella*.

Note: Bacto Shigella Poly A and individual *S. dysenteriae* serotypes are not tested for or absorbed with the 3 provisional serotypes.

S. dysenteriae type 1 was formerly known as "Shiga," *S. shigae* or *S. krusei*. The strains are homogeneous but some possess envelope antigens and therefore will have to be heated before serotyping. Type 1 is the type species.

S. dysenteriae type 2 was formerly known as *S. ambiqua* or *S. schmitzii*. The strains produce acid from rhamnose and dextrose, are serologically homogeneous and are generally encapsulated.

S. dysenteriae types 3 – 7 were formerly of the Large and Sankaran and Sacks group. All 5 types are well defined and are serologically distinct from types 1 and 2 even though minor cross-reactions might occur. Envelope capsules are frequently present.

S. dysenteriae types 8 – 10 were added to the schema more recently. They were described by Ewing,[10] Cox and Wallace[11] and Ewing.[12]

The second group — Group B contains *S. flexneri* with 6 serotypes and possibly 2 variants (X and Y). They generally produce acid from mannitol. Each serotype possesses a distinctive main or type antigen — however, cross-reactions will occur in unabsorbed sera due to the interrelationships of minor or subgroup antigens (see Table 2). Members of this group were formerly known as *S. paradysenteriae*, Flexner.

Table II
Shigella flexneri

Serotype	Subserotype	Abbreviated Subserotype component
1	1a	I:4
1	1b	I:4,6
2	2a	II:4
2	2b	II:7,8
3	3a	III:6,7,8
3	3b	III:3,4,6
3	3c	III:6
4	4a	IV:3,4
4	4b	IV:6
5	5	V:7,8
6	6	VI:—

Note: Bacto Shigella Poly B is not prepared with or tested for variants X and Y cultures since they are not recognized as valid types, i.e., do not possess distinctive antigens.

The third group — Group C contains *S. boydii* with 15 serotypes. This group is made up of organisms that are biochemically similar to group B organisms but are not related significantly to them serologically. The first 6 types are those of Boyd with the remaining types being described subsequently in the literature.

The last group — Group D contains *S. sonnei* with 1 serotype. The serotype is represented by both smooth forms (I) and rough forms (II) of the organism. The form II is not the classical form of an R antigen, since it is stable in saline. This group is biochemically similar to Groups B and C *Shigella* but members ferment lactose on prolonged incubation. The O antigen of *S. sonnei* is not related to the other *Shigella* and members of this group do not develop envelope capsules. There is one exception however, that must be noted. The rough antigens of *S. boydii* type 6 are identical to the rough antigen of *S. sonnei* II. If an *S. sonnei* I, II serum is absorbed with *S. boydii* 6(R), the resultant serum can be weakened to such a degree as to render it useless diagnostically therefore, this absorption is not done.

Bacto Shigella Antisera Poly and individual types are absorbed, when necessary to render each lot of serum as specific as practical without reducing the homologous reactions to an unsatisfactory level. They have been absorbed inter and intra specifically as recommended except as in the cases noted above — i.e., Shigella Poly D is not absorbed with *S. boydii* type 6; the shigella antisera in general are not prepared from or tested for *S. dysenteriae* provisional serotypes or *S. flexneri* X and Y variants nor with the Alkalescens-Dispar Groups above somatic group 05.

ALKALESCENS-DISPAR ANTIGENIC ANALYSIS
The term "Alkalescens-Dispar" (AD) is applied to a group of organisms intermediate between the genus *Shigella* and genus *Escherichia*.

These organisms are anaerogenic and nonmotile and even though some of its members demonstrate a lactose negative fermentation pattern, they are regarded as biotypes of *E. coli*, not members of the genus *Shigella*. Serological detection and grouping of these organisms are done on the somatic antigen content, using the slide agglutination technique. As in the case with some of the *Shigella* strains, envelope capsules are frequently observed in strains of the A – D group rendering the culture inagglutinable in an O antiserum therefore, it is often necessary to boil these strains in order to render the antigen agglutinable in its homologous serum.

Table III indicates the major O antigenic relationships between Alkalescens-Dispar and *E. coli* groups.

Table III

A – D O Groups	Previous Designation	*E. coli* O Relationship
01	S. alkalescens type I	01 (identical)
02	S. alkalescens type II and S. tiete	025 (strong) plus others
03	S. alkalescens type III	025 (strong) plus others
	S. dispar type II	
	S. ceylonensis B	
04	S. dispar type I and S. madampensis	04 (strong)
05	— — — — — — — — —	02a (identical)
06	S. dispar type III	09 (identical)
07	— — — — — — — — —	07 (identical)
08	— — — — — — — — —	081 (identical)

The principle of the serological use of these sera involves the intimate mixing of the organism (the antigen) with the immune serum (the antibody). If the serum contains agglutinins for the antigen present on the organism a rapid and at least 3+ clumping (agglutination) of the organism will occur. This is known as a homologous reaction. In some cases a reaction in the immune serum with another species might occur. This reaction is due to the presence of one or more commonly shared antigens between various species and is called a heterologous reaction. The homologous reaction is characterized by the rapidity and avidity of reaction, as compared to the heterologous reaction which is slow or weak. The agglutination of the somatic antigen in the rapid slide test appears as a firm, granular clumping.

REAGENTS
Bacto Shigella and Alkalescens-Dispar are stable, desiccated and absorbed (when necessary) antisera prepared in rabbits.

Bacto Shigella Antiserum Poly Group A reacts with *S. dysenteriae* types 1 – 7.
Bacto Shigella Antiserum Poly Group A_1 reacts with *S. dysenteriae* type 8ab8ac, 9, 10.
Bacto Shigella Antiserum Poly Group B reacts with *S. flexneri* types 1 – 6.
Bacto Shigella Antiserum Poly Group C reacts with *S. boydii* types 1 – 7.
Bacto Shigella Antiserum Poly Group C_1 reacts with *S. boydii* types 8 – 11.
Bacto Shigella Antiserum Poly Group C_2 reacts with *S. boydii* types 12 – 15.
Bacto Shigella Antiserum Poly Group D reacts with *S. sonnei* I and II.
Bacto Alkalescens-Dispar Antiserum Poly reacts with Alkalescens-Dispar Groups 1, 2, 3 and 4.

When rehydrated, the protein content of these antisera approximates that of a glycerinated serum and therefore should be considered as a 1:2 dilution. The sera at this dilution are to be considered at the RTD (Routine Test Dilution). This RTD is based on the state of the antiserum at the time of testing under our laboratory conditions, using antigenicially defined and recognized reference cultures.

When properly rehydrated each antiserum contains approximately 1:10,000 Merthiolate®, sufficient to maintain a bacteriostatic condition at storage of 2 – 8°C.

When used according to the suggested procedure each 3 ml vial is sufficient to perform approximately 60 tests.

REHYDRATION
To rehydrate, add 3 ml of 0.85% NaCl solution to the vial and rotate gently to dissolve contents completely.

Check purity (bacterial) and pH of the saline used in rehydration if the antiserum rehydrates cloudy. Discard any serum which is cloudy and/or has a precipitate unless it has been clarified and shown to react properly with known control cultures.

STORAGE
Store both unrehydrated and rehydrated antisera at 2 – 8°C.

Prolonged exposure to room temperature or repeated freezing and thawing are detrimental to antisera.

EXPIRATION DATE
These reagents are stable to the expiry date on the label when stored at 2 – 8°C.

Discard any antiserum which becomes cloudy during storage.

The slide agglutination antisera have been prepared for use to identify cultures which have been defined biochemically. Such cultures are taken from an isolated state from an agar medium using a bacterial loop. A portion of the isolated colony is picked from the plate or slant and emulsified in a droplet of the antiserum. The sera have been tested and absorbed using this method. It is recommended that more than one colony be tested.

These antisera have not been tested employing antigen suspensions in saline or chemically treated antigen suspensions. If variations in the recommended procedures are to be used, the investigator is advised to test each lot of antiserum with known control cultures to insure its proper homologous and heterologous reactions under their test conditions.

SPECIMEN COLLECTION
Shigella organisms occur in greatest number in the feces of patients with dysentery during the acute stage of infection (usually first 3 days) and specimens should be collected at this point whenever possible. Freshly passed specimens should be used for inoculation of appropriate media. In chronic cases fecal specimens afford the best method of obtaining possible enteric pathogens particularly *Shigella*. However, when this is not possible rectal swabs may be employed. It should be noted that *Shigella* organisms die off rapidly, and therefore samples should be inoculated to appropriate media as soon as possible after collection. *Shigella* organisms are also more difficult to recover than *Salmonella* even on differential or selective media designed for this purpose. For this reason, multiple specimens should be examined and several plates should be employed.

SPECIMEN PREPARATION
As with all bacterial serology it is important that the appropriate cultural and differential procedures be carried out before serological techniques are applied. Further, before the organism can be assumed to be a member of the genus *Shigella*, minimal biochemicals should be run and grouping sera employed for presumptive identification. Final identification is dependent upon biochemicals and serotyping. A cultural and biochemical schema for the *Shigella* and Alkalescens-Dispar is given in Table IV. Special note should be given to the biochemical reactions of this genus and group.

The choice of enrichments and primary plating media is extremely important for the optimal recovery of *Shigella*. It is important to note that early (traditional) cultural methods for the isolation of this organism assumed that the environmental conditions suit-

Table IV

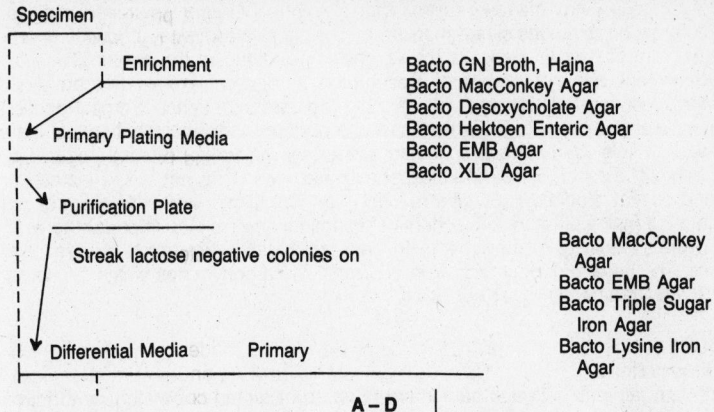

Specimen

Enrichment — Bacto GN Broth, Hajna

Primary Plating Media — Bacto MacConkey Agar / Bacto Desoxycholate Agar / Bacto Hektoen Enteric Agar / Bacto EMB Agar / Bacto XLD Agar

Purification Plate

Streak lactose negative colonies on — Bacto MacConkey Agar / Bacto EMB Agar / Bacto Triple Sugar Iron Agar / Bacto Lysine Iron Agar

Differential Media   Primary

Biochem.	Shigella	A – D Group	Biochem.	Shigella	A – D Group
H₂S	–	–	Sucrose	–(1)	d
Urease	–	–	Mannitol	+ or –	+
Indol	– or +	+	Dulcitol	d	d
Methyl Red	+	+	Salicin	–	–
Voges Proskauer	–	–	Adonitol	–	–
Simmons Citrate	–	–	Inositol	–	–
KCN	–	–	Arabinose	d	+
Motility	–	–	Raffinose	d	d
Gelatin (22°C)	–	–	Rhamnose	d	d
Lysine Decarb	–	d	Mucate	–	d
Arginine Dehyd	d	d	Christensen		
Ornithine Decarb	d	d	Citrate	–	d
Phenylalanine	–	–	Sodium		
Malonate	–	–	Acetate	–	+(+)
Gas from			Malonate	–	–
Glucose	–(1)	–	Maltose	d	+
Lactose	–(1)	d	Xylose	d	+
Serology					

(1) Certain biotypes of *S. flexneri* produce gas; cultures of *S. sonnei* ferment lactose and sucrose slowly and decarboxylate ornithine.
d = different reactions
(+) = delayed 2 – 5 days

able for *Salmonella* would also be conducive to the growth of *Shigella*. However, as has been stated, *Shigella* is more closely related to the genus *Escherichia*, which highly selective media were designed to inhibit. Thus many *Shigella* were inadvertently inhibited. Therefore, in cultural methods employed for recovery of *Shigella*, highly selective media should be avoided. They are suitable for isolation of *Salmonella* but may not allow optimal recovery of *Shigella*.

Particular attention should be paid to the purification procedure. Lactose-negative colonies are picked to MacConkey or EMB agar; green to blue-green colonies from Hektoen enteric agar, or red colonies from XLD agar are picked to MacConkey or EMB agar. Pure cultures are then obtained. Omitting this step may result in mixed cultures which will give erroneous results of TSI agar and may also be misidentified serologically.

Ewing summarizes the minimal criteria in determining whether the organism under study is indeed a *Shigella*. "Any microorganisms that blackens TSI agar, is urease positive, or is able to grow on Simmons citrate medium or decarboxylates lysine, produces phenyl pyruvic acid from phenylalanine or is motile does not belong to the genus *Shigella*." When these criteria are met, results of presumptive serological examination are likely to be confirmed during final identification. Should the user experience a cross reaction in these polyvalent antisera it will be necessary to perform additional biochemical tests (as outlined in Table IV) before proceeding further. As stated earlier anaerogenic or microaerogenic strains of *E. coli* (Alkalescens-Dispar) which do not utilize lactose or sucrose and do not produce H_2S, give reactions on TSI slants similar to *Shigella* will require more extensive differential biochemical examination. It is also important to evaluate the overall reactivity of the isolate. In general shigellae are much less reactive than *E. coli*. The latter produces acid from a variety of carbohydrates within 24 hours, while the former reacts slowly or not at all.

PROCEDURE
General Instructions
In the slide test, all materials and equipment must be at room temperature at the time of test performance.

Exposure of the organism or plate to heat from external sources (a hot bacteriological loop, burner flame, light source, etc.) may result in either a culture which cannot be suspended readily or evaporation and/or precipitation of the test mixture which may result in false-positive reactions.

The test culture must be checked in a saline control or preferably in normal rabbit serum for smoothness. Often stock cultures and sometimes isolated cultures are rough and will agglutinate spontaneously in a normal serum. Therefore, it is necessary to select smooth colonies for serological testing. It is recommended that more than one colony be tested for assay and control procedures.

In the slide test, all equipment and materials (glass plates, loop, culture) must be sterilized after use by flame (loop) or autoclave (glass plate and culture) or by an adequate chemical method since living organisms are used as the antigen.

Materials Provided:
Bacto Shigella Antisera Poly Groups
   A – D and Alkalescens-Dispar
   Antiserum Poly
Bacto Shigella Antisera Types
Droppers

Materials Required but not Provided:
Slide Test:
   0.85% NaCl solution   Fluorescent desk lamp
   Bacteriological loop   Autoclave
   Burner   Bacto Rabbit Serum Normal
   Glass plate ruled into 1″ squares

1. Prepare a microscope slide or a glass plate for use in the test by marking off 1″ squares with a wax pencil.
2. Place a drop of the appropriate Bacto Shigella Antiserum on the ruled section of the slide or plate. (Starting with Group D Antisera is the most efficient procedure.)

3. To the square next to the one containing the antiserum, place a drop of 0.85% NaCl solution or Bacto Rabbit Serum Normal. This will serve as a negative control of the bacterial culture.
4. Use a clean inoculation loop and transfer a portion of a loopful of the culture taken from a solid medium to a section containing the NaCl solution as prepared in step 3 and emulsify thoroughly.
5. Transfer a portion of a second loopful of the bacterial mass to a section containing the antiserum. Mix thoroughly to emulsify mixture.
6. Rock the glass plate for 1 minute to enhance agglutination.
7. Positive agglutination will be rapid and complete.
8. Record agglutination as follows:

  $++++$ agglutination = all of the cells agglutinate
  $+++$ agglutination = 75% of the cells agglutinate
  $++$ agglutination = 50% of the cells agglutinate
  $+$ agglutination = 25% of the cells agglutinate
  $\pm$ agglutination = <25% of the cells agglutinate
  $-$ agglutination = none of the cells agglutinate

  A $+++$ reaction or greater should be considered as the end-point at the RTD.
9. If agglutination does not occur using Bacto Shigella Antiserum Poly Group D, proceed to Poly Groups A, A_1, B, C, C_1, C_2, and to the Alkalescens-Dispar Antiserum Poly.

Note: According to Edwards and Ewing "Cultures that appear to be *Shigella* but which agglutinate poorly or not at all, should be tested for the presence of heat labile inhibitory substances. This may be done by heating a suspension in a water bath at 100°C for 15 – 60 minutes. After such treatment, the suspension is cooled and retested on a slide for agglutination. Substances that inhibit O agglutination are present commonly in cultures of *E. coli* but also occur in serotypes of *Shigella*."

This may be done as follows:
1. Wash a portion of the growth from a Petri dish or slant in 3 – 5 ml 0.5% saline.
2. Boil in a water bath for 15 – 60 minutes.
3. Discard any culture which precipitates upon heating.
4. Centrifuge at 1,000 rpm for 3 – 5 minutes.
5. Aspirate and discard supernatant solution.
6. Using a portion of a loopful of the sedimented bacterial mass, perform the slide agglutination test as outlined above.

QUALITY CONTROL
Use known positive and negative control cultures in parallel with the test culture to ascertain the validity of the test results. Difco provides these quality control cultures in the form of QC Antigens Shigella Groups and QC Antigen Alkalescens-Dispar. The complete description of these antigens is described separately under QC ANTIGENS SHIGELLA AND ALKALESCENS-DISPAR.

RESULTS
A positive reaction is characterized by a 3+ agglutination after 1 minute. Any partial or delayed reaction should be considered negative. There should be no agglutination of the suspension with 0.85% NaCl or Bacto Rabbit Serum Normal. If autoagglutination occurs streak presumptive *Shigella* culture on Bacto SS Agar several times, then transfer back onto Bacto Blood Agar Base. Recheck culture again for roughness. A positive reaction with an antiserum poly group identifies the organism as a member of that poly group.

SEROTYPING

Final identification of the specific type is accomplished by repeating steps 1 – 7 above using individual types of specific sera within the group.

Positive results with one of the specific antisera of the poly group with typical biochemical reactions of the species (see Table IV) may be considered final identification of the *Shigella* species and serotype.

LIMITATIONS OF THE PROCEDURE

While serological procedures, as applied to microorganisms, provide corroborative evidence and can be used to identify particular antigenic sites of genus or species under study, they cannot be used alone to identify the etiological agent of disease. Cultural methods must precede any serological examination, and final identification cannot be made without biochemical characterization. This is particularly true of *Enterobacteriaceae*. Members of this family are antigenically related and cross-reactions do occur, therefore, identification of genera within the family cannot be based on serological methods alone.

Some members of the genus *Shigella* are antigenically related to *E. coli*. Antigens of these microorganisms are identical or similar to each other. Following are a few of these antigenic relationships which could cross-react when serological procedures are employed without prior biochemical characterization:

Shigella-E. coli Antigenic Relationships

S. dysenteriae 1= E. coli O1	S. boydii 2 = E. coli O87ab
S. dysenteriae 2= E. coli O112ac	S. boydii 4 = E. coli O53
S. dysenteriae 3= E. coli O124	S. boydii 5 = E. coli O79
S. dysenteriae 5= E. coli O58	S. boydii 8 = E. coli O143
S. flexneri 2b = E. coli O147ab	S. boydii 11 = E. coli O105ab
S. flexneri 4b = E. coli O135	S. boydii 14 = E. coli O32
S. flexneri 5 = E. coli O129	S. boydii 15 = E. coli O112ab

It must be stressed that the above listing is that of identities (identical antigenic relationships existing between the 2 genera). It is an incomplete list — there are many more identical, partial, reciprocal and unilateral relationships. For more indepth information on cross-reactions of *Shigella, E. coli* (including Alkalescens-Dispar) refer to Edwards and Ewing, *Identification of Enterobacteriaceae*.[9]

REFERENCES

1. Chantemesse, and Widal, F., Bull. Acad. Nat. Med., 19:522, 1888.
2. Shiga, K., Zentralb, f., Bakt., 1 Abt., 23:599, 1898.
3. Kruse, W., Deutsche Med., Wchshr., 26:637, 1900.
4. Flexner, S. J., Exper. Med., 8:514, 1906.
5. Strong, R. P., and Musgrave, W. E., JAMA, 35:498, 1900.
6. Smith, D. T., Conant, N. F., Overman, J. R., Zinsser Microbiology, 13th Ed., Meredith Publ. Co., New York, N.Y., 1964.
7. Report No. 35, Center for Disease Control, Shigella Surveillance, Nov. 1974.
8. Blair, J. E., Lenette, E. H., Truant, J. P., Manual of Clinical Microbiology, ASM, Bethesda, Md., 1970.
9. Edwards, P. R., and Ewing, W. H., Identification of Enterobacteriaceae, 3rd Ed., Burgess Publ. Co., Minneapolis, 1970.
10. Ewing, W. H., Hucks, M. C., and Taylor, M. W., J. Bact., 63:319, 1952.
11. Cox, C. D., and Wallace, G. I., J. Immun., 60:465, 1948.
12. Ewing, W. H., Reavis, R. W., and Davis, B. R., Can. J. Microbiol., 4:89, 1958.

PACKAGING

Bacto Shigella Antiserum Poly Group A	3 ml	2834-47-3
Bacto Shigella Antiserum Poly Group A₁	3 ml	2776-47-3
Bacto Shigella Antiserum Poly Group B	3 ml	2835-47-2
Bacto Shigella Antiserum Poly Group C	3 ml	2836-47-1
Bacto Shigella Antiserum Poly Group C₁	3 ml	2777-47-2
Bacto Shigella Antiserum Poly Group C₂	3 ml	2778-47-1
Bacto Shigella Antiserum Poly Group D	3 ml	2837-47-0
Bacto Alkalescens-Dispar Antiserum Poly	3 ml	2838-47-9
Bacto Shigella Antisera Set*	8 × 3 ml	2796-32-6
Contains Poly Groups A, A₁, B, C, C₁, C₂, D and Alkalescens-Dispar Poly		
Bacto Shigella Dysenteriae Antisera Set A	7 × 3 ml	2313-32-0
Contains 1 vial each Types 1, 2, 3, 4, 5, 6 and 7		
Bacto Shigella Dysenteriae Antisera Set B	3 × 3 ml	2794-32-8
Contains 1 vial each Types 8, 9 and 10		
Bacto Shigella Flexneri Antisera Set	6 × 3 ml	2896-32-5
Contains 1 vial each Types 1, 2, 3, 4, 5 and 6		
Bacto Shigella Boydii Antisera Set A	7 × 3 ml	2298-32-9
Contains 1 vial each Types 1, 2, 3, 4, 5, 6 and 7		
Bacto Shigella Boydii Antisera Set B	8 × 3 ml	2795-32-7
Contains 1 vial each Types 8, 9, 10, 11, 12, 13, 14 and 15		

*Shigella Antisera comprising the Bacto Shigella Antisera Set are also included in Bacto MinESS Antisera Set II.

QUALITY CONTROL ANTIGENS *SHIGELLA* SEROLOGY

PRODUCT DESCRIPTION AND INTENDED USE

Bacto QC Antigens Shigella are chemically stabilized and inactivated suspensions of known recommended strains of the genus *Shigella* for use in quality control of *Shigella* grouping antisera commonly employed in the slide agglutination technique.

These reagents are designed as positive controls for testing the efficacy of the *Shigella* grouping antisera employed in routine laboratory procedures. They are not recommended for the direct diagnosis of a disease, but rather for detection of agglutinins in the hyperimmune serum employed.

SUMMARY AND BACKGROUND

Many clinical laboratories desiring positive control suspensions for quality control purposes do not have them available on a routine basis due to several factors, among which are lack of an adequate culture collection, lack of time to maintain the cultures in their proper antigenic state and the possible laboratory hazard of handling such a large number of potentially pathogenic organisms. The use of Bacto QC Antigens eliminates these problems.

To determine that nothing deleterious has occurred to a polyvalent antiserum that has been approved by a manufacturer, it should only be necessary to obtain proper reactivity with one major antibody component and its homologous antigen.

Reagents	Organism Identity
Bacto QC Antigen Shigella Group A	*Shigella dysenteriae* type 1
Bacto QC Antigen Shigella Group A_1	*Shigella dysenteriae* type 8
Bacto QC Antigen Shigella Group B	*Shigella flexneri* type 6
Bacto QC Antigen Shigella Group C	*Shigella boydii* type 3
Bacto QC Antigen Shigella Group C_1	*Shigella boydii* type 8
Bacto QC Antigen Shigella Group C_2	*Shigella boydii* type 12
Bacto QC Antigen Shigella Group D	*Shigella sonnei*
Bacto QC Antigen Alkalescens-Dispar Group 1	Alkalescens-Dispar Group 1

CONCENTRATIONS OF ANTIGENS

The antigens are adjusted to 70 – 100 International Units of Opacity. They are ready to use without further dilution.

PRESERVATIVES

Bacto QC Antigens Shigella are preserved with 0.5% formaldehyde solution reagent (v/v).

STORAGE AND STABILITY

Store all Bacto QC Antigens Shigella at 2 – 8°C. Do not allow to freeze at any time. Bring to room temperature prior to use. Return to the refrigerator after use.

These reagents are stable to the expiry date on the label when stored at 2 – 8°C.

Bacto QC Antigens are not to be used for immunization of man or animals.

SPECIMEN

The specimen tested with Bacto QC Antigens Shigella is the homologous grouping antiserum used in the grouping of *Shigella* organisms. The antiserum must be within its expiry dating, be used at its recommended routine test dilution (RTD), be at room temperature at the time of testing and be completely clear. (If the antiserum is cloudy, it may be interpreted as a false-positive reaction.) Often, due to lipid content, antisera will become cloudy upon refrigeration but still maintain proper reactivity. If cloudy, clarify the antiserum by centrifugation and test with the homologous QC antigen.

Bacto QC Antigens may also be used as negative controls by using a heterologous antigen (possessing no common antigen) with a given test serum. For example, use a QC Antigen Shigella Group A (*Shigella dysenteriae* type 1) with all Shigella grouping antisera except Group A. One should be cognizant of existing cross reactions, i.e., QC Antigen Shigella Group A with Alkalescens-Dispar 1 = *Shigella flexneri* 6 with *Shigella boydii* 5 and 12 as well as Alkalescens-Dispar 2. For further details on cross reactivity, consult references.

TEST PRECAUTIONS

1. If autoagglutination of the QC Antigen is apparent, discard the antigen. The most likely cause of such autoagglutination is in the subjection of the antigen to freezing conditions. Avoid freezing.
2. Allow the QC Antigens, the antisera to be tested and all equipment used to be at room temperature at the time of testing. The test reagents, if cold, may cause false-negative reactions.
3. Shake the antigen well before use to suspend the organisms.
4. A glass plate left under or near a burner may be too warm for the test, resulting in a rapid evaporation of the test mixture (false-positive).

5. The presence of artifacts on the glass plate (dust, etc.) or particulate matter in the antiserum may be interpreted as a false reaction.
6. Dispel all antigen from the dropper after use and before storage.
7. Avoid contact of the antigen into the rubber bulb.
8. Keep upright when stored at 2 – 8°C.

PROCEDURE
Materials Provided:
Bacto QC Antigens Shigella
Droppers

Materials Necessary but not Provided:
Shigella Antisera (available from Difco)   0.85% NaCl solution
Glass plate 8 × 8″ or equivalent   Wax marking pencil
Droppers for the antisera   Ruler
Plain applicator sticks   Suitable light source

Technique:
All reagents must be at room temperature.
1. Mark off a scrupulously clean glass plate into 1″ square sections with a wax pencil.
2. Place a drop of the Shigella Antiserum to be tested in one square. (See chart under SPECIFIC PERFORMANCE CHARACTERISTICS for appropriate matching of antigen to antiserum.)
3. Place a drop of 0.85% NaCl in the square next to the antiserum. This is a negative control for the antigen. A heterologous Bacto Shigella Antiserum may also be used as a negative control.
4. Add 1 drop of the appropriate QC Antigen Shigella to both drops.
5. Mix the droplets, using an applicator stick (broken into small pieces) for each droplet. Spread the mixture in such a manner as to fill the 1″ square.
6. Rotate the plate by hand for 1 minute.
7. Read and record the results as follows:
   - 4+   agglutination of all the cells (clear background)
   - 3+   agglutination of approximately 75% of the cells
   - 2+   agglutination of approximately 50% of the cells
   - 1+   agglutination of approximately 25% of the cells
   - ±   very slight roughness
   - −   none of the cells agglutinate

Interpretation:
1. A 3+ or greater reaction is considered positive.
2. The antigen should not react in the saline droplet or heterologous antiserum control.
3. If 3+ or greater reaction occurs in the homologous system with a negative reaction in saline, the serum is satisfactory for use at its RTD.

SPECIFIC PERFORMANCE CHARACTERISTICS

Reagent	Reactive 3+ or Greater with:
QC Antigen Shigella Group A	Bacto Shigella Antiserum Poly Group A, 2834-47-3 Bacto Shigella Dysenteriae Antiserum Type 1, 2283-47, contained in Bacto Shigella Dysenteriae Antiserum Set A, 2313-32-0
QC Antigen Shigella Group A_1	Bacto Shigella Antiserum Poly Group A_1, 2776-47-3 Bacto Shigella Dysenteriae Type 8, 2764-47, contained in Bacto Shigella Dysenteriae Antiserum Set B, 2794-32-8

QC Antigen Shigella Group B	Bacto Shigella Antiserum Poly Group B, 2835-47-2 Bacto Shigella Flexneri Type 6, 2887-47, contained in Bacto Shigella Flexneri Antiserum Set, 2896-32-5
QC Antigen Shigella Group C	Bacto Shigella Antiserum Poly Group C, 2836-47-1 Bacto Shigella Boydii Antiserum Type 3, 2293-47, contained in Bacto Shigella Boydii Antiserum Set A, 2298-32-9
QC Antigen Shigella Group C_1	Bacto Shigella Antiserum Poly Group C_1, 2777-47-2 Bacto Shigella Boydii Antiserum Type 8, 2768-47, contained in Bacto Shigella Boydii Antiserum Set B, 2795-32-7
QC Antigen Shigella Group C_2	Bacto Shigella Antiserum Poly Group C_2, 2778-47-1 Bacto Shigella Boydii Antiserum Type 12, 2772-47, contained in Bacto Shigella Boydii Antiserum Set B, 2795-32-7
QC Antigen Shigella Group D	Bacto Shigella Antiserum Poly Group D, 2837-47-0
QC Antigen Alkalescens-Dispar Group 1	Bacto Alkalescens-Dispar Poly, 2838-47-9

REFERENCES

1. Department of Health, Education, and Welfare, 1975, Specifications and evaluation methods for laboratory, immunological and microbiological reagents, Volume 1, 4th Ed., Centers for Disease Control, Atlanta.
2. Edwards, P. R., and W. H. Ewing, 1972, Identification of *Enterobacteriaceae*, 3rd Ed., Burgess Publishing Co., Minneapolis.
3. Lennette, E. H., A. Balows, W. J. Hausler, Jr., and J. P. Truant, 1980, Manual of clinical microbiology, 3rd Ed., American Society for Microbiology, Washington, D.C.
4. Wright, D. N., D. R. Welch, and J. M. Matsen, 1980, J. Clin. Microbiol., 11:305.

PACKAGING

Bacto QC Antigen Shigella Group A	1 ml	2100-50-4
Bacto QC Antigen Shigella Group A_1	1 ml	2101-50-3
Bacto QC Antigen Shigella Group B	1 ml	2102-50-2
Bacto QC Antigen Shigella Group C	1 ml	2103-50-1
Bacto QC Antigen Shigella Group C_1	1 ml	2104-50-0
Bacto QC Antigen Shigella Group C_2	1 ml	2105-50-9
Bacto QC Antigen Shigella Group D	1 ml	2106-50-8
Bacto QC Antigen Alkalescens-Dispar Group 1	1 ml	2116-50-6

BACTO SIMMONS CITRATE AGAR

INTENDED USE

Bacto Simmons Citrate Agar is recommended for the differentiation and identification of *Enterobacteriaceae* on the basis of citrate utilization, citrate being the sole carbon source.

HISTORY/PRINCIPLE

Koser[1] developed a liquid medium based upon the principle that the difference between fecal coli (negative) and aerogenes (positive) was the inability of fecal coli to develop in a medium containing inorganic ammonium salts as the only source of nitrogen, with citrate as the sole source of carbon; whereas strains of aerogenes grew quite unrestrictedly. Likewise, *Salmonella typhi, S. paratyphi* and *Shigella* could be differentiated from *S. schottmuelleri, S. enteritidis* and *S. typhimurium,* the latter group being able to utilize citrate in such a medium while the former could not.

This principle was first employed in a liquid medium by Koser. The liquid medium had the disadvantage of appearing turbid when large inocula were used, even when no growth ensued. This observation led Simmons[2] to devise a solid medium which obviated the disadvantage of turbidity as a criterion for growth.

Bacto Simmons Citrate Agar is essentially Koser's medium to which brom thymol blue and 1.5% agar have been added. Organisms able to metabolize the citrate, grow luxuriantly. The medium is alkalinized and changes from its initial green to deep blue in 24 – 48 hours. Fecal coli either do not grow at all on this medium or grow so sparsely that no change in reaction is apparent.

Simmons also recommended the use of this medium for isolating and identifying certain fungi and fungi imperfecti.

FORMULA
BACTO SIMMONS CITRATE AGAR
DEHYDRATED
Ingredients per liter

Magnesium Sulfate . . . . 0.2 g
Ammonium Dihydrogen Phosphate . . . 1 g
Dipotassium Phosphate . . . . 1 g
Sodium Citrate . . . . 2 g
Sodium Chloride . . . . 5 g
Bacto Agar . . . . 15 g
Bacto Brom Thymol Blue . . . . 0.08 g

Final pH 6.8 ± 0.2 at 25°C.
One pound will make 18.7 liters of final medium.

METHOD OF PREPARATION
1. To rehydrate, suspend 24.2 g in 1 liter distilled or deionized water and heat to boiling to dissolve completely.
2. Sterilize in the autoclave for 15 minutes at 15 lbs pressure (121°C).
3. Medium is usually prepared as agar slants, inoculated by stab and streak. Incubate at 37°C.
4. Growth is indicated very clearly by colony formation and is usually accompanied by a color change of the indicator due to acid or alkali production.

STORAGE
Bacto Simmons Citrate Agar   Below 30°C
Prepared medium   15 – 30°C

QUALITY CONTROL
Identity Specifications

Dehydrated powder: yellow, may have green tinge, homogeneous, free-flowing
Reaction of 2.42% solution: pH 6.8 ± 0.2 at 25°C
Prepared medium: forest green, slightly opalescent, may have a slight precipitate

Typical Cultural Response on Bacto Simmons Citrate Agar
After 18 – 24 Hours at 35°C

Organism	Growth	Color of Medium
Enterobacter aerogenes ATCC® 13048	good	blue
Escherichia coli ATCC® 25922	inhibited	green
Salmonella enteritidis ATCC® 13076	good	blue
Salmonella typhimurium ATCC® 14028	good	blue
Salmonella typhi ATCC® 19430	good	green
Shigella dysenteriae ATCC® 13313	inhibited	green

REFERENCES
1. J. Bact., 8:493, 1923.
2. J. Infectious Diseases, 39:209, 1926.

PACKAGING
Bacto Simmons Citrate Agar

1 lb (454 g)	0091-01-9
1/4 lb (114 g)	0091-02-8
12 tubes	0091-34-0
144 tubes	0091-37-7

BACTO SKIM MILK

INTENDED USE
Bacto Skim Milk is used as a complete medium or as an ingredient in other media for the propagation of organisms occuring in milk products and for demonstration of coagulation and proteolysis of casein.[1]

HISTORY/PRINCIPLES
Bacto Skim Milk is a spray dried powder of large volumes of skim milk. It is readily soluble and easily prepared. Bacto Skim Milk is recommended for use with Bacto Plate Count Agar in the preparation of skim milk agar according to Compendium of Methods for the Microbiological Examination of Foods.[2]

Bacto Skim Milk may be used in media for the detection of proteolytic enzymes, and as an ingredient in the preparation of media for the cultivation of such organisms as Mycobacterium tuberculosis and Corynebacterium diphtheria. A solution of Bacto Skim Milk containing litmus or other indicator is a widely used medium for determining acid production in milk and also the ability to peptonize or coagulate milk in the identification of microorganisms. Skim milk is used to detect the so-called "stormy fermentation" of Clostridium perfringens.

Nungester and Ellingson[3] have suggested the addition of 0.1% agar and iron, either in the form of filings or freshly pickled strips, to adjust the oxygen tension of the medium and permit more rapid development of organisms in the "stormy fermentation" test.

FORMULA
BACTO SKIM MILK
DEHYDRATED
Bacto Skim Milk is a standardized product recommended for culture media. When prepared in 10% solution it is equivalent to fresh skim milk.

METHOD OF PREPARATION
1. Dissolve 100 g in 1 liter distilled or deionized water.
2. Sterilize in the autoclave for 15 minutes at 15 lbs pressure (121°C). Avoid over-heating as caramelization will occur.

STORAGE
Bacto Skim Milk   15 – 30°C
Prepared medium   2 – 8°C

QUALITY CONTROL
Identity Specifications
Dehydrated powder: white to off-white, homogeneous, free-flowing
Reaction of 10% solution: pH 6.3 ± 0.2 at 25°C
Prepared medium: white, opaque

Typical Cultural Response of Bacto Skim Milk as Litmus Milk
After 40 – 72 Hours at 35°C

Organism	Growth	Reaction
Bacillus subtilis ATCC® 6633	good	none, digested curd
Clostridium perfringens ATCC® 12919	good	stormy fermentation
Escherichia coli ATCC® 25922	good	acid, reduction, curd
Lactobacillus casei ATCC® 9595	good	acid, reduction, curd
Salmonella typhi ATCC® 6539	good	reduction

REFERENCES
1. American Public Health Association, 1948, Standard Methods for the Examination of Dairy Products, 9th Ed., Am. Pub. Hlth. Assoc., Inc., Washington, D.C.
2. American Public Health Association, 1976, Compendium of Methods for the Microbiological Examination of Foods, Am. Pub. Hlth. Assoc., Inc., Washington, D.C.
3. Nungester and Ellingson, Personal Communication.

PACKAGING
Bacto Skim Milk   1 lb (454 g)   0032-01-1
                                     1/4 lb (114 g)   0032-02-0

BACTO SNYDER TEST AGAR

INTENDED USE
Bacto Snyder Test Agar is used for the colorimetric estimation of the relative number of lactobacilli in saliva.

HISTORY
Bacto Snyder Test Agar is prepared for the colorimetric determination of the rate and amount of acid production by organisms in saliva as described by Snyder.[1,2] Alban[3] modified the Snyder test, and after five years of experience reported that the modified test is superior in its simplicity and accuracy to the original procedure.

PRINCIPLES
The method is based on acid production in a carbohydrate medium by acidogenic microorganisms obtained from the buccal cavity. The reaction is evidenced by a change in color of the indicator, Bacto Brom Cresol Green, from a blue-green to a yellow color. The test gives excellent correlation with the *Lactobacillus* plate count.

FORMULA

BACTO SNYDER TEST AGAR
DEHYDRATED

Ingredients per liter

Bacto Tryptose . . . . 20 g
Bacto Dextrose . . . . 20 g
Sodium Chloride . . . . 5 g
Bacto Agar . . . . 20 g
Bacto Brom Cresol Green . . . . . . 0.02 g

Final pH 4.8 ± 0.2 at 25°C.

One pound will make 7 liters of final medium.

METHOD OF PREPARATION
1. Suspend 65 g in 1 liter distilled or deionized water and heat to boiling to dissolve completely.
2. Distribute in 10 ml amounts into test tubes with screw caps. (If using the Alban Modification of the test, use 100 × 17 mm test tubes.)
3. Sterilize in the autoclave for 15 minutes at 15 lbs pressure (121°C).
4. Allow tubes to cool in an upright position.

PROCEDURE AND INTERPRETATION
The time for collecting specimens is preferably before breakfast and before the teeth are brushed, otherwise just before lunch or dinner.

Snyder Procedure[1,2]
SPECIMEN COLLECTION AND INCUBATION
1. Collect specimens of saliva in a sterile flask or bottle while patient is chewing paraffin for 3 minutes.
2. Shake specimens thoroughly and transfer 0.2 ml to a tube of sterile Bacto Snyder Test Agar melted and cooled at 45°C. (Prepared medium in tubes is heated in a boiling water bath for 10 minutes and cooled to 45°C.)
3. Rotate the inoculated tubes to mix the inoculum uniformly with the medium and allow to solidify in the upright position.
4. Incubate at 37°C for 72 hours and observe color at 24, 48 and 72 hours.

OBSERVATIONS
Examine the tubes daily for three days and record changes in color compared with an uninoculated control tube. Observation of the color changes is facilitated by means of reflected light with the tubes held against a white background. The color will change from the bluish-green of the control to yellow.

Positive: Change in color so that green is no longer dominant.
    Record as ++ to ++++

Negative: No change in color or only slight deviation, but green still dominant.
    Record 0 to +.

Interpretation

Caries Activity	Hours Incubation		
	24	48	72
Marked	Positive	—	—
Moderate	Negative	Positive	—
Slight	Negative	Negative	Positive
Negative	Negative	Negative	Negative

The interpretation of laboratory data as given above with clinical activity depends upon experience and understanding of several factors:
1. The data indicate only what is happening at the time the specimen was collected.
2. At least two specimens collected within 2 – 4 days must be obtained to establish a base-line or reference point.
3. Only when two or more specimens have been cultured can any reliability or prediction be obtained.
4. The clinician must study enough cases by use of periodic laboratory data to establish in his own mind the value of significance for the purpose intended.

Data summarizing the correlation between the Snyder colorimetric test and *Lactobacillus* counts on specimens of saliva collected routinely are tabulated.

Colorimetric Change (hours)

No. Lactobacilli per ml Saliva	No. Spec.	24 Pos	24 % Pos.	48 Pos.	48 % Pos.	72 Pos.	72 % Pos.
0	348	2	0.6	22	6.3	85	24.5
0 – 100	59	0	0.0	7	11.9	32	54.3
100 – 1000	157	5	3.2	47	30.0	111	70.0
1,000 – 10,000	203	6	5.7	69	65.7	99	94.3
10,000 – 20,000	138	18	13.0	105	71.2	131	95.0
20,000 – 50,000	264	59	22.2	229	86.8	260	98.7
50,000 – 100,000	245	72	29.4	221	90.3	243	99.3
100,000 –	497	231	46.6	476	95.5	494	99.3

Alban Modification of the Snyder Test
Note: The medium is not remelted prior to use.
1. Collect sufficient unstimulated saliva directly into the tube to just cover the medium. When specimen collection is difficult, dip a sterile cotton swab into the saliva under the tongue or rub on tooth surfaces and place the swab just below the surface of the medium.
2. Incubate the inoculated tubes and an uninoculated control at 37°C and examine daily for four days for any color change in the inoculated tubes compared to the uninoculated control.
3. Record the daily color change as follows:
   a. No color change
   b. Color beginning to change to yellow from top of medium down   +
   c. One half of medium yellow   + +
   d. Three fourths of medium yellow   + + +
   e. The entire medium is yellow   + + + +
4. The final report is a composite of the daily readings, for example; − + + + + + +. This reading indicates the rapidity of and amount of acid production.

STORAGE
Bacto Snyder Test Agar   Below 30°C
Prepared tubes   2 – 8°C

QUALITY CONTROL
Identity Specifications
Dehydrated powder: light green, homogeneous, free-flowing
Reaction of 6.5% solution: pH 4.8 ± 0.2 at 25°C
Prepared medium: dark emerald green, slightly opalescent

Typical Cultural Response on Bacto Snyder Test Agar
After 18 – 24 Hours at 35°C

Organism	Growth	Acid Production
Lactobacillus casei ATCC® 9595	good to excellent	+
Lactobacillus fermentum ATCC® 9338	good to excellent	+
Saliva sample	variable, depending on flora	

REFERENCES
1. J. Dent. Res., 20:189, 1941.
2. J. Am. Dent. Assoc., 28:44, 1941.
3. J. Dent. Res., 49:641, 1970.

PACKAGING
Bacto Snyder Test Agar	1 lb (454 g)	0247-01-2
	12 tubes	0247-34-3

SODIUM AZIDE

Sodium Azide is used as a selective agent in culture media for isolating and cultivating streptococci and staphylococci from clinical and nonclinical sources. Most enteric bacilli are markedly to completely inhibited by sodium azide when added in a concentration of 0.02%. Refer to the discussions on Bacto Azide Blood Agar Base and Bacto Azide Dextrose Broth for a complete description of the uses of these media.

PRECAUTIONS
1. Poison.
2. Avoid inhaling and contact with skin, eyes or clothing. For contact with skin or clothing, promptly wash with copious amounts of water. For contact with eyes, immediately irrigate with copious amounts of water and call a physician. For spillage on bench or floor, flood area with water and dry.
3. Sodium azide solutions, when discarded into a sink, must be washed into the drain with large amounts of water to prevent the possible formation of lead azide (if lead plumbing is used), a highly explosive compound. The use of copious flooding with water will prevent the reaction between sodium azide and lead and the accumulation of lead azide.

PACKAGING
Sodium Azide	25 g	0601-13-8

SODIUM CASEINATE, DIFCO

Sodium Caseinate, Difco is a valuable ingredient of many culture media, and has frequently been referred to as "Nutrose". Sodium Caseinate, Difco is a purified product recommended for use in all media in which "Nutrose" or sodium caseinate has been employed.

PACKAGING
Sodium Caseinate, Difco	500 g	0187-17-6

BACTO SODIUM DESOXYCHOLATE

Bacto Sodium Desoxycholate is the sodium salt of desoxycholic acid. It is used in bacteriological culture media and in the pneumococcus bile solubility test. It is readily soluble in distilled water, giving a clear, colorless solution that is neutral in reaction.

Bacto Sodium Desoxycholate, like bile and other bile salts, when used in bacteriological culture media, inhibits gram-positive cocci and spore forming organisms, but not gram-negative enteric bacilli. Since sodium desoxycholate is a salt of a highly purified bile acid, it is used in culture media in lower concentrations than is naturally occurring bile.

Leifson[1] described media for the enumeration of coliform organisms from milk, water, sewage, etc., and for the isolation of enteric pathogens using sodium desoxycholate to inhibit gram-positive organisms. He also prepared a selective medium using sodium desoxycholate and sodium citrate for the isolation of intestinal pathogens. Coliform as well as gram-positive organisms are inhibited on desoxycholate citrate agar. Bacto Desoxycholate Agar and Bacto Desoxycholate Citrate Agar, are prepared with sodium desoxycholate.

Sodium desoxycholate is also used in 10% concentration in the bile solubility test for identifying pneumococci.

REFERENCE
1. J. Path. Bact., 40:581, 1935.

PACKAGING

Bacto Sodium Desoxycholate	25 g	0248-13-7
	100 g	0248-15-5

BACTO SODIUM HIPPURATE

Bacto Sodium Hippurate is the sodium salt of hippuric acid and is used in the hippurate hydrolysis test for the differentiation of group B streptococci from other beta-hemolytic streptococci.

PACKAGING

Bacto Sodium Hippurate	10 g	0330-12-7
	25 g	0330-13-6

BACTO SODIUM SELENITE

Bacto Sodium Selenite is recommended for use as a selective agent in media for enriching the growth of salmonellae and markedly inhibiting the growth of coliforms from specimens such as feces, urine and other natural specimens which may contain these bacteria.

Refer to the discussion on Bacto Selenite Broth and Bacto Selenite Cystine Broth for a complete description of the uses of these media.

WARNING:
Avoid inhaling and contact with skin, eyes, and clothing. For contact with skin or clothing, promptly wash with copious amounts of water. In case of contact with eyes, promptly flood with water and immediately obtain medical attention. For spillage on bench or floor, flood area with water.

PACKAGING

Bacto Sodium Selenite	100 g	0608-15-9
	500 g	0608-17-7

BACTO SODIUM TAUROCHOLATE

Bacto Sodium Taurocholate is the sodium salt of a conjugated bile acid used generally in bacteriological culture media for the isolation and cultivation of gram-negative enteric bacilli. Like whole bile, it has the cultural characteristic of inhibiting the growth of gram-positive organisms and spore-forming bacteria, without inhibiting the development of gram-negative bacilli.

Bacto Sodium Taurocholate contains about 75% sodium taurocholate in addition to other naturally occurring salts of bile acids. It is readily soluble in distilled water, forming a clear amber solution, neutral in reaction.

PACKAGING

Bacto Sodium Taurocholate	1 lb (454 g)	0278-01-4
	1/4 lb (114 g)	0278-02-3

SODIUM THIOGLYCOLLATE

Sodium thioglycollate or thioglycollic acid is recommended for use in liquid culture media for testing the sterility of biological and other materials containing heavy metal compound preservatives, such as the mercurials. The active sulfhydryl group neutralizes the toxicity of the metallic preservative, thus permitting the development of any viable organisms present and, further, behaves similarly to glutathione, cysteine and the alkali sulfides in being able to lower the oxidation-reduction potential of the medium. This principle was first described by Trenkmann[1] who showed that the presence of an alkali sulfide induced "aerobic growth of anaerobic organisms."

Quastel and Stephenson[2] showed likewise that cysteine and thioglycollic acid made possible the growth of anaerobes through the lowering of the oxidation-reduction potential of the medium. Brewer[3,4] combined the principle of lowering the oxidation reduction potential with Hitchen's[5] method of adding 0.05 to 0.1% agar to the medium for the growth of anaerobes. Such a combination was found particularly useful as a sterility test medium for biologicals containing mercurial preservatives, since the sodium thioglycollate neutralizes the toxic effect of any mercurial carried over with the inoculum, while the agar provides anaerobic conditions necessary for growth.

Since sodium thioglycollate is toxic for some organisms, especially with inocula containing few organisms, it is recommended that no more of the sodium thioglycollate be added to the medium than is required for mercurial neutralization. Sodium thioglycollate in 0.05% concentration is specified in the thioglycollate media of the National Institute of Health[6] for sterility testing. These media are discussed under Bacto Fluid Thioglycollate Medium and Bacto NIH Thioglycollate Broth. Bacto Brewer Anaerobic Agar contains sodium thioglycollate, and is employed as a solid medium for the cultivation of anaerobes.

REFERENCES

1. Centr. Bakt. I. Abt., 23:1038, 1898.
2. Biochem. J., 20:1125, 1926.
3. J. Bact., 39:10, 1940.
4. J. Am. Med. Assoc., 115:598, 1940.
5. J. Infectious Diseases, 29:390, 1921.
6. National Instutute of Health Circular: Culture Media for Sterility Test, 2nd Revision, Feb. 5, 1946.

WARNING:
Causes irritation. Avoid contact with skin, eyes and clothing. Wash thoroughly after handling.

PACKAGING

Sodium Thioglycollate	25 g	0233-13-4
	100 g	0233-15-2

SOLUBLE STARCH

Soluble Starch is a polysaccharide, soluble in water with gentle warming. It may be added in a 1% concentration to culture media substrates to differentiate certain bacteria on their ability to hydrolyze this carbohydrate. It is extremely useful as an additive to culture media for improving the growth response and protective properties for some bacteria.

PACKAGING

Soluble Starch	100 g	0178-15-9
	500 g	0178-17-7

BACTO SOYTONE

Bacto Soytone is an enzymatic hydrolysate of soybean meal prepared under controlled conditions especially for use in microbiological procedures. It is recommended for use in media for the cultivation of a large variety of organisms, including fungi, and is also used in media for microbiological assay. This nitrogen source contains the naturally occurring carbohydrate of the soybean and is suitable in all microbiological procedures wherein carbohydrates are not objectionable. Bacto Soytone is completely soluble in distilled or deionized water in concentrations generally employed in bacteriological culture media and yields clear solutions with an almost neutral reaction after autoclaving.

PACKAGING

Bacto Soytone	1 lb (454 g)	0436-01-3

BACTO SPIRIT BLUE AGAR
BACTO LIPASE REAGENT

INTENDED USE

Bacto Spirit Blue Agar is a basal medium to which is added lipoidal substrates for the detection, enumeration and study of lipolytic microorganisms. Bacto Lipase Reagent is recommended as the lipid source.

HISTORY/PRINCIPLES

Bacto Spirit Blue Agar is prepared according to the formulation of Starr.[1] He reported that previous formulations containing dyes as indicators of lipolysis exerted a toxic effect upon many microorganisms. His studies demonstrated spirit blue to be inert bacteriologically and ideally suited for use as an indicator in his medium for detecting fat-splitting microorganisms in food products and other materials. Using the spirit blue dye in his relatively simple basal medium to which lipoidal emulsions were added, Starr obtained accurate counts of lipolytic microorganisms. He also could obtain total microbial counts on this same medium.

Bacto Lipase Reagent is recommended as the lipid source. Other lipoidal emulsions may be prepared from a variety of lipids. Cotton seed meal, cream, Wesson® oil, olive oil and the higher fatty acids have been used successfully. A satisfactory emulsion can be prepared by dissolving 10 g gum acacia or one ml Tween 80® in 400 ml warm distilled water, adding 100 ml cotton seed or olive oil and agitating vigorously to emulsify.

FORMULA

BACTO SPIRIT BLUE AGAR
DEHYDRATED

Ingredients per liter

Bacto Tryptone	10 g
Bacto Yeast Extract	5 g
Bacto Agar	20 g
Spirit Blue	0.15 g

Final pH 6.8 ± 0.2 at 25°C.

One pound will make 12.9 liters of final medium.

METHOD OF PREPARATION

1. Suspend 36 g in 1 liter distilled or deionized water. Heat to boiling to dissolve completely.
2. Sterilize in the autoclave for 15 minutes at 15 lbs pressure (121°C).
3. Allow to cool to 50 – 55°C and add 30 ml Bacto Lipase Reagent slowly while agitating medium in the flask to obtain an even distribution.
4. Use either the pour plate or streak plate technique.

STORAGE

Bacto Spirit Blue Agar	Below 30°C
Bacto Lipase Reagent	15 – 30°C
Prepared plates	2 – 8°C

QUALTITY CONTROL
Identity Specifications

Dehydrated powder: light beige, homogeneous, free-flowing
Reaction of 3.5% solution: pH 6.8 ± 0.2 at 25°C
Prepared medium (plain):   royal blue, slightly opalescent
Complete medium: pale lavender, opalescent

Typical Cultural Response on Bacto Spirit Blue Agar
After 40 – 72 hours at 35°C

Organism	Growth	Halo/Lipolysis
Proteus mirabilis ATCC® 25933	good	−
Staphylococcus aureus ATCC® 25923	good	+
Staphylococcus aureus ATCC® 6538	good	+
Staphylococcus epidermidis ATCC® 12228	good	+

REFERENCE
1. Science, 93:333, 1941.

PACKAGING

Bacto Spirit Blue Agar	1 lb (454 g)	0950-01-9
	1/4 lb (114 g)	0950-02-8
Bacto Lipase Reagent	6 × 20 ml	0431-63-3

BACTO STANDARD II NUTRIENT AGAR

INTENDED USE
Bacto Standard II Nutrient Agar is a general purpose medium for culturing nonfastidious microorganisms (such as the Enterobacteriaceae and other gram-negative and gram-positive bacteria) and for carrying stock cultures. The medium may be used as a base for blood and other enrichments for culturing fastidious microorganisms. In addition, indole production can be detected on the medium.

HISTORY
Standard II Nutrient Agar is based on the reports of Kuczynski and Ferner.[1] The medium contains a high proportion of essential amino acids, lower and higher peptides, and the necessary essential and accessory growth-promoting requirements of nonfastidious microorganisms.

Levetzow[2] suggested using Standard II Nutrient Agar in detecting inhibitory agents in the bacteriological examination of meat.

FORMULA
BACTO STANDARD II NUTRIENT AGAR
DEHYDRATED

Ingredients per liter

Bacto Tryptose . . . . 7 g
Sodium Chloride . . . . 5 g
Bacto Agar . . . . 13 g

Final pH 7.5 ± 0.2 at 25°C.

Five hundred grams will make 20 liters of medium.

PREPARATION

1. Rehydrate Bacto Standard II Nutrient Agar by suspending 25 g in 1 liter distilled or deionized water. Heat to boiling to dissolve completely.
2. Sterilize in the autoclave for 15 minutes at 15 lbs pressure (121°C).
3. When adding blood or other enrichments, cool sterile medium to 45 – 50°C, aseptically add the enrichment and gently swirl the flask to mix thoroughly.
4. Dispense as desired.

STORAGE

Bacto Standard II Nutrient Agar   Below 30°C
Prepared medium   15 – 30°C

QUALITY CONTROL

Identity Specifications

Dehydrated powder: beige, homogeneous, free-flowing
Reaction of 2.5% solution: pH 7.5 ± 0.2 at 25°C
Prepared medium: light amber, clear

Typical Cultural Response on Bacto Standard II Nutrient Agar
After 18 – 48 Hours at 35°C

Organism	Growth	Indole	Acetyl Methyl Carbinol
Enterobacter aerogenes ATCC® 13048	good	–	+
Escherichia coli ATCC® 25922	good	+	–

REFERENCES

1. Klin. Wschr., 2:826, 1923.
2. Bundesgesundheitsblatt, 14:211, 1971.

PACKAGING

Bacto Standard II Nutrient Agar   500 g   1816-17-3

BACTO m STAPHYLOCOCCUS BROTH

INTENDED USE

Bacto m Staphylococcus Broth is a selective medium used in the membrane filter technique for isolating pathogenic and enterotoxigenic staphylococci.

HISTORY/PRINCIPLES

Bacto m Staphylococcus Broth is patterned after the formulation of Bacto Staphylococcus Medium 110 with the gelatin and agar omitted. The high selectivity of the medium will restrict the growth of most contaminating bacteria, therefore, autoclaving is optional.

FORMULA

BACTO m STAPHYLOCOCCUS BROTH
DEHYDRATED

Ingredients per liter

Bacto Tryptone . . . . 10 g
Bacto Yeast Extract . . . . 2.5 g
Bacto Lactose . . . . 2 g
Bacto Mannitol . . . . 10 g
Dipotassium Phosphate . . . . 5 g
Sodium Chloride . . . . 75 g

Final pH 7.0 ± 0.2 at 25°C.

One pound will make 4.3 liters of medium.

METHOD OF PREPARATION
1. Suspend 104 g in 1 liter of distilled or deionized water and heat to boiling to dissolve completely.
2. Sterilize in the autoclave for 15 minutes at 15 lbs pressure (121°C).

   NOTE: For field studies where autoclaving is not practical, boil the medium for 5 minutes.

APPLICATION
Use 2 – 2.5 ml of medium to saturate the paper pads on which the inoculated membrane is placed in this technique.

STORAGE
Bacto m Staphylococcus Broth	Below 30°C
Prepared medium	15 – 30 °C

QUALITY CONTROL
Identity Specifications
Dehydrated powder: light beige, homogeneous, free-flowing
Reaction of 10.4% solution: 7.0 ± 0.2 at 25°C
Prepared medium: light amber, clear to slightly opalescent, may have a slight precipitate

Typical Cultural Response on Pads Soaked with Bacto m Staphylococcus Broth After 40 – 48 Hours at 35°C in Humid Atmosphere

Organism	Growth
Staphylococcus aureus ATCC® 25923	good to excellent
Staphylococcus aureus ATCC® 19095	good to excellent
Staphylococcus epidermidis ATCC® 12228	good to excellent

PACKAGING
Bacto m Staphylococcus Broth	1 lb (454 g)	0649-01-6

BACTO STAPHYLOCOCCUS MEDIUM 110

INTENDED USE
Bacto Staphylococcus Medium 110 due to its high concentration of sodium chloride, is a selective culture medium for the isolation of pathogenic strains Staphylococcus. It is also well suited for pigment formation, may be used for the determination of mannitol, for the Stone type gelatinase test, and gives a growth satisfactory for the coagulase test.

HISTORY
Stone[2] described a culture medium on which food poisoning staphylococci gave a gelatinase test; while Chapman, Lieb and Curcio reported that, in addition, typical food poisoning staphylococci must produce pigment, coagulate plasma, hemolyze rabbit blood and ferment mannitol. Because most bacteria other than staphylococci were inhibited on a medium with a concentration of 7.5% NaCl, Chapman[3] suggested that 7.5% NaCl be added to Bacto Phenol Red Mannitol Agar to achieve a selective isolation medium for staphylococci. Further studies by Chapman[4] led to the development of staphylococcus medium 110.

PRINCIPLES

The high NaCl content of Bacto Staphylococcus Medium 110 makes it selective for pathogenic staphylococci. According to Chapman, and Chapman and Domingo,[5] staphylococci incriminated in food poisoning produce an orange pigment, coagulate plasma, ferment mannitol and give a positive Stone reaction or gelatinase test on staphylococcus medium 110 when tested at the time of isolation.

FORMULA

BACTO STAPHYLOCOCCUS MEDIUM 110
DEHYDRATED

Ingredients per liter

Bacto Tryptone	10 g
Bacto Yeast Extract	2.5 g
Bacto Gelatin	30 g
Bacto Lactose	2 g
D-Mannitol	10 g
Sodium Chloride	75 g
Dipotassium Phosphate	5 g
Bacto Agar	15 g

Final pH 7.0 ± 0.2 at 25°C.

One pound will make 3 liters of medium.

PROCEDURE

1. Suspend 149 grams in 1 liter distilled or deionized water and heat to boiling to dissolve the medium completely.
2. Sterilize in the autoclave for 10 minutes at 15 lbs pressure (121°C).
3. Gently agitate to evenly dispense precipitate and dispense into sterile Petri dishes.
4. Following streaking or smearing of the specimens on plates, the medium is incubated at 35°C for exactly 48 hours.
5. Colonies are observed for signs of any orange or yellow pigment production. If pigment production is detected, colonies are picked from the surface of the medium and emulsified in 0.1 – 0.2 ml of Bacto Brain Heart Infusion or Bacto Tryptose Phosphate Buffer for the coagulase test.
6. A drop of brom cresol purple indicator is added to several areas from which pigmented colonies were removed. Any change in color of indicator compared with that of uninoculated medium is indicative of fermentation of mannitol.
7. Plate is flooded with 5 ml saturated solution of $(NH_4)_2 SO_4$, kept in the incubator, and let stand 10 minutes. Any clear zones around areas from which colonies have been picked, or around colonies, are gelatinase positive.
8. To determine the coagulative power, use a 16 – 24 hour culture of the suspected organism in Bacto Brain Heart Infusion or a Bacto Brain Heart Infusion suspension of a 16 – 24 hour culture on a slant of Bacto Heart Infusion Agar.
9. Two drops of the culture or suspension of the organism are added to 0.5 ml of Bacto Coagulase Plasma Solution. Incubate at 35°C.
10. Most coagulase positive staphylococci will clot the solution within 1 hour. A positive culture will show a definite clot often within 20 – 30 minutes. Readings, however, should be made after 2 and 3 hours of incubation.

STORAGE

Bacto Staphylococcus Medium 110	Below 30°C
Staphylococcus Medium 110 plates	2 – 8°C

QUALITY CONTROL

Identity Specifications

Dehydrated Medium: beige, free flowing, homogeneous powder
pH of final medium: 7.0 ± 0.2 at 25°C
Prepared plates: light amber, opalescent with heavy precipitate

Cultural Response

Organism	Growth	Pigment Production	Gelatinase Production*
Bacillus subtilis ATCC® 6633	good-excellent	negative	positive
Escherichia coli ATCC® 25922	inhibited	negative	negative
Staphylococcus aureus ATCC® 25923	good-excellent	positive	positive
Staphylococcus aureus ATCC® 6538	good-excellent	positive	positive
Staphylococcus epidermidis ATCC® 12228	good-excellent	negative	positive

*After 48 hours incubation at 35°C, flood plate with saturated ammonium sulfate solution to determine reaction.

REFERENCES

1. Proc. Soc. Exp. Biol. Med., 33:185:1935.
2. Food Research, 2:349:1937.
3. J. Bact., 50:201:1945.
4. J. Bact., 51:409:1946.
5. J. Bact., 51:405:1946.

PACKAGING

Staphylococcus Medium 110	1 lb (454 g)	0297-01-1
	1/4 lb (114 g)	0297-02-0
	10 kg	0297-08-4

BACTO STARCH AGAR

INTENDED USE

Bacto Starch Agar is used for cultivating organisms being tested for starch hydrolysis.

HISTORY/PRINCIPLES

Starch agar, described by Vedder in 1915,[1] was originally formulated for cultivating *Neisseria*. Other media have been developed that are superior to starch agar for that purpose, specifically Bacto GC Medium Base, Bacto Proteose No. 3 Agar and Bacto Dextrose Starch Agar.

Bacto Starch Agar is made available for those who wish to use it in comparative studies or for performing the starch hydrolysis test.

FORMULA

BACTO STARCH AGAR
DEHYDRATED

Ingredients per liter

Bacto Beef Extract . . . . 3 g
Soluble Starch, Difco . . . . 10 g
Bacto Agar . . . . 12 g

Final pH 7.5 ± 0.2 at 25°C.

One pound will make 18.1 liters of medium.

METHOD OF PREPARATION

1. Suspend 25 g in 1 liter distilled or deionized water and heat to boiling to dissolve completely.
2. Sterilize in the autoclave for 15 minutes at 15 lbs pressure (121°C).
3. Dispense as desired.
4. Starch hydrolysis test:
   Flood the surface of a 48 hour culture on Bacto Starch Agar with Bacto Gram Iodine. Starch hydrolysis (+) is indicated by a colorless zone surrounding colonies. A blue or purple zone indicates that starch has not been hydrolyzed (−).

STORAGE

Bacto Starch Agar	Below 30°C
Prepared medium	2 – 8°C

QUALITY CONTROL

Identity Specifications

Dehydrated powder:	light beige, homogeneous, free-flowing
Reaction of 2.5% solution:	pH 7.5 ± 0.2 at 25°C
Prepared medium:	light amber, slightly opalescent

Typical Cultural Response on Bacto Starch Agar
After 18 – 48 Hours at 35°C

Organism	Growth	Hydrolysis
Bacillus subtilis ATCC® 6633	good to excellent	+
Escherichia coli ATCC® 25922	good to excellent	−
Staphylococcus aureus ATCC® 25923	good to excellent	−
Streptococcus pyogenes ATCC® 19615	good to excellent	−

REFERENCE

1. J. Inf. Dis., 16:385, 1915.

PACKAGING

Bacto Starch Agar	1 lb (454 g)	0072-01-2
	1/4 lb (114 g)	0072-02-1

BACTO STOCK CULTURE AGAR

INTENDED USE

Bacto Stock Culture Agar is a soft, semisolid medium used for the maintenance of pathogenic and nonpathogenic bacteria, especially streptococci.

HISTORY/PRINCIPLES

Ayers and Johnson[1] described a medium that gave luxuriant growth and long life of streptococci. Their medium was developed from a formula originally prepared by Supplee. The success of their medium probably lies in the fact that the medium has a semisolid consistency, contains casein, is well buffered and contains a small quantity of dextrose which serves as a readily available source of energy. They reported that pathogenic streptococci remained viable for at least four months at room temperature (24°C) in their medium. Organisms other than streptococci such as pneumococci, human tubercle bacilli and others grew well on their stock culture agar.

Bacto Stock Culture Agar is prepared to duplicate the medium described by Ayers and Johnson. This medium, likewise, will support luxuriant growth of many pathogenic bacteria and preserve their viability over a long period of time.

FORMULA
BACTO STOCK CULTURE AGAR
DEHYDRATED
Ingredients per liter

Beef Heart Infusion from	500 g	Bacto Dextrose	0.5 g
Proteose Peptone, Difco	10 g	Disodium Phosphate	4 g
Bacto Gelatin	10 g	Sodium Citrate	3 g
Bacto Isoelectric Casein	5 g	Bacto Agar	7.5 g

Final pH 7.5 ± 0.2 at 25°C.

Five hundred grams will make 10 liters of medium.

METHOD OF PREPARATION
1. To rehydrate, suspend 50 g in 1 liter cold distilled or deionized water.
2. Heat to boiling to dissolve completely.
3. Sterilize in the autoclave for 15 minutes at 15 lbs pressure (121°C).

STORAGE
Bacto Stock Culture Agar   Below 30°C
Prepared medium   2 – 8°C

QUALITY CONTROL
Identity Specifications

Dehydrated powder: beige, homogeneous, free-flowing
Reaction of 5.0% solution: pH 7.5 ± 0.2 at 25°C
Prepared medium: medium amber, opalescent

Typical Cultural Response on Bacto Stock Culture Agar
After 18 – 48 Hours at 35°C

Organism	Growth
Neisseria meningitidis ATCC® 13090	excellent
Staphylococcus aureus ATCC® 25923	excellent
Streptococcus pneumoniae ATCC® 6303	excellent
Streptococcus pyogenes ATCC® 19615	excellent

REFERENCE
1. J. Bact., 9:111, 1924.

PACKAGING
Bacto Stock Culture Agar   500 g   0054-17-6

BACTO STREPTOCOCCUS ANTISERA
BACTO STREPTOCOCCUS ANTIGENS

INTENDED USE
Bacto Streptococcus Antisera are prepared for use in the serological grouping of some members of the genus *Streptococcus*, using the capillary tube precipitin technique.

Bacto Streptococcus Antigens are prepared for use as control antigens in this procedure.

Bacto Streptomyces Albus Enzyme is prepared for use in extracting the polysaccharide of the streptococci especially groups A, B, C, F and G for use as antigens in the capillary precipitin test.

SUMMARY AND EXPLANATION OF THE TEST

The genus *Streptococcus* contains organisms, some of which are found in great numbers in the normal flora of the mouth, pharynx and intestine of man and animals. Most of the members are saprophytic. However, some species are capable of causing a large variety of clinical infections, including respiratory infections, pharyngitis, tonsillitis, scarlet fever, meningitis, endocarditis, arthritis, osteomyelitis, puerperal sepsis and erysipelas in humans and numerous diseases in animals which include mastitis, pharyngitis, genital tract infections, arthritis and meningitis.

A very useful phenomenon which assists in the identification of the *Streptococcus* is its action on erythrocytes as described by Brown.[1]
1. Alpha (α) Hemolysis — characterized by a greenish or brownish-green discoloration around the colony on blood agar.
2. Beta (β) Hemolysis — a demonstration of a clear, colorless zone around the colony on blood agar.
3. Gamma (γ) Reaction — no change in the erythrocytes around the colony on blood agar.

Most streptococcal strains that are associated with an acute disease state are beta-hemolytic, while those strains that are involved in chronic cases or that are nonpathogens are alpha or gamma in reaction.

Lancefield[2,3,4] divided the *Streptococcus* into serological groups according to the group specific somatic carbohydrate they possess. The Lancefield groups differ in their clinical significance as well as in their biochemical and hemolytic reactions within the same serological group. Therefore, all streptococci exhibiting beta-hemolysis and which are involved in an infection are grouped.[5]

Maxted[6] first used the *S. albus* enzyme to group hemolytic streptococci and demonstrated that the lysates containing the group specific polysaccharides could be prepared easily and rapidly by this process.

Reactivity on blood agar, as well as other biochemical reactions are important in the tentative identification of the *Streptococcus*, but, for diagnosis, serological definition of the antigen by specific serological grouping is necessary for final identification.[7]

Table 1 displays some of the *Streptococcus* groups associated with infections in man and animals together with their hemolytic and serological characteristics.

PRINCIPLES

The serology of the streptococci is based on the reaction of precipitins prepared against the "C" carbohydrate which is elaborated by the organisms with extracts of the test organism.[2,3,4] Currently there are 20 generally accepted serogroups, A – V, but only 6 of these A, B, C, D, F and G are of clinical importance.

Table 1

Serogroup	Species of Streptococcus	Type of Hemolysis	Host Association
A	S. pyogenes	β	Human
B	S. agalactiae	α,β,γ	Human and Bovine
C	S. dysgalactiae	α	Animals
C	S. equi	β	Equine
C	S. equisimilis	β	Human and Animal
C	S. zooepidemicus	β	Animals
D	S. bovis	α,γ	Human and Animal
D	S. equinus	α(weak)	Equine
D	S. faecalis	γ	Human and Animal
D	S. faecium	α	Human and Animal
E	S. species	β	Porcine and Dairy
F*	S. anginosus	α,γ	Human
G	S. species	β	Human

Key: α = alpha hemolysis
β = beta hemolysis
γ = gamma hemolysis
* = Extracts from Streptococcus MG were considered at one time to be Lancefield's group F.

Beyond serogrouping, further antigenic delineation of some of the Lancefield serogroups, may be accomplished by serotyping using antisera prepared against the "M" and "T" proteins of the streptococci. However, serotyping is seldom performed outside a comparatively few reference centers.

Note: Streptococcus MG is not a Lancefield group but this organism was found to be antigenically identical to Streptococcus group F — type III.[9] This organism is possibly associated with primary atypical pneumonia.

This product description of "Serological Identification of Streptococcus" omits the species "pneumoniae" of the genus Streptococcus. Information on this species is described separately under "Pneumococcus Antisera."

REAGENTS
Bacto Streptococcus Antisera are stable, desiccated, absorbed (when necessary), group specific antisera prepared in rabbits for use in the capillary tube precipitin technique. A vial of 100 capillary tubes is provided with each vial of Bacto Streptococcus Antiserum.

Bacto Streptococcus Antigens are stable, clear, liquid extracts of streptococci representative of the various serogroups prepared according to the Rantz and Randall[10] method, designed for use as positive controls.

Bacto Streptomyces Albus Enzyme is a desiccated soluble enzyme of Streptomyces albus recommended for preparing lysates of streptococci for the Streptococcus group precipitin test.

Bacto Streptococcus reagents contain sufficient Merthiolate® to maintain a bacteriostatic condition at a storage of 2 – 8°C. In most cases, the final concentration of Merthiolate® is 1:5000. Bacto Streptomyces Albus Enzyme, when properly rehydrated, contains approximately 1:10,000 Merthiolate®.

REHYDRATION

To rehydrate Bacto Streptococcus Antisera, add 1 ml sterile distilled or deionized water and rotate the vial gently to dissolve the contents completely.

To rehydrate the Bacto Streptomyces Albus Enzyme, add 5 ml sterile distilled or deionized water and rotate the vial to dissolve the contents.

STORAGE

Store unrehydrated Bacto Streptococcus Antisera at 2 – 8°C. Once rehydrated the antisera may be stored at 2 – 8°C or for **optimal stability, stored in convenient aliquot amounts at –60°C.** The sera should not be subjected to repeated freezing and thawing as such treatment is detrimental to antibody content.

Bacto Streptococcus Antigens should be stored at 2 – 8°C. Do not subject to prolonged exposure to room or elevated temperatures.

Bacto Streptomyces Albus Enzyme unrehydrated should be stored at 2 – 8°C.

EXPIRATION DATE

Desiccated Bacto Streptococcus Antisera are stable through the expiration date on the label when properly stored. Once rehydrated, the rehydration date should be written on the original vial and/or aliquot vials. Rehydrated antisera are stable for one year after rehydration, not to exceed the expiration date on the label when stored according to directions and if not contaminated with chemicals or bacteria.

Should the antiserum be stored (after rehydration) at 2 – 8°C and the serum becomes "cloudy," it is absolutely necessary to clarify the serum by centrifugation and also to test the clarified serum with known positive control antigens before use. Do not expose the rehydrated antiserum to room temperature for prolonged periods of time. Discard any serum which cannot be clarified or does not react properly with a positive control antigen. **It is imperative in this technique that sparklingly clear antigens and antisera be used.**

Bacto Streptococcus Antigens, if stored at 2 – 8°C and not contaminated with chemicals or bacteria, are stable to the expiry date which appears on the label. Should a precipitate develop, clarify the antigen by centrifugation and test using a known homologous antiserum. Discard any antigen which cannot be clarified or does not demonstrate proper reactivity.

Note: If at any time an antigen or antiserum becomes turbid and it is proven by a Gram stain to be bacterially contaminated, the reagent must be discarded.

Bacto Streptomyces Albus Enzyme, unrehydrated, is stable to the expiry date on the label. After rehydration, store at 2 – 8°C, and it should be used within one week. Should rehydrated enzyme become cloudy upon storage, it should be discarded.

SPECIMEN COLLECTION

Specimens submitted to the laboratory for the possibility of containing streptococci must be obtained under proper medical guidance from nasal, nasopharyngeal or throat areas, from skin, wounds, pus, blood, cerebrospinal fluid or urine. It is imperative, especially when material is obtained from the throat to culture, that it be done properly to obtain an adequate specimen. Improperly obtained specimens may give rise to a grossly contaminated culture with few significant bacteria.

The isolation medium recommended for this species of organisms is any one of several blood agar bases containing 5% sterile, defibrinated sheep blood. Hemolytic reactions should be determined from a pure culture prior to serological examination. Sheep blood plates are recommended as they exhibit clear-cut reactions for streptococci.

Note: It has been found that hemolytic reactions of *Streptococcus* groups A, B, C, and G strains were identical when sheep, rabbit, horse and human blood were used. The only difference was in the group D streptococci which demonstrated alpha-hemolysis in sheep blood and beta-hemolysis in rabbit, horse and human blood.[11]

SPECIMEN PREPARATION
General Directions
1. For antigen preparation, Bacto Todd Hewitt Broth is recommended.
2. *Streptococcus* group D organisms should be cultured in a CO_2 jar for antigen extract preparation.
3. If the antigen is to be used in serotyping as well as serogrouping, it must be prepared only by Lancefield method.
4. In preparing the antigen extract, avoid too low a pH for the actual extraction process.
5. The antisera and antigen extracts must be absolutely clear. It may be necessary to clarify them by centrifugation prior to their use in the test.
6. All glassware that is employed in the preparation, testing and storage of these reagents must be free of detergents or other harmful residues.

A water-clear antigen extract of the test organism is prepared by one of the following methods:

Autoclave Method of Rantz and Randall[10]
1. Culture the *Streptococcus* to be grouped in 40 ml of Bacto Todd Hewitt Broth for 18 – 24 hours at 37°C.
2. Centrifuge to sediment the growth completely.
3. Discard the supernatant fluid.
4. Add 0.5 ml of 0.85% NaCl solution.
5. Suspend the organisms by gently shaking the tube.
6. Autoclave the tube for 15 – 20 minutes at 15 lbs pressure (121°C).
7. Centrifuge to sediment the cellular debris.
8. Use the supernatant solution for the antigen in the precipitin test.

Hot HCl Method of Lancefield[3]
Note: The HCl and NaOH solutions used in the preparation of antigen extracts by the Lancefield method, must be sparklingly clear. It is advisable to prepare fresh solutions every month or two.
1. Culture the *Streptococcus* to be grouped in 40 ml of Bacto Todd Hewitt Broth for 18 – 24 hours at 37°C
2. Centrifuge to sediment the growth completely.
3. Discard the supernatant fluid.
4. Add a small amount of N/5 HCl (prepared in 0.85% NaCl solution) and suspend the bacterial cells. Generally 0.4 ml is sufficient.
5. Adjust the pH of the suspension to 2.0 by the further addition of N/5 HCl, if necessary, using a drop of 0.01% Bacto Phenol Red solution or a pH meter.
6. Place the tube in a boiling water bath for 10 minutes.
7. Cool, centrifuge and decant the supernatant fluid into a clean tube.
8. Adjust the pH of the extract to a pH of 7.4 – 7.8 with N/5 NaOH prepared in distilled or deionized water. If the pH of the antigen extract prepared by the Lancefield method is too high at the time of testing, false-positive reactions may occur.

9. Centrifuge and aspirate the supernatant extract. This extract — the antigen — is ready for use in the precipitin test. If it is to be used over a period of time it should be merthiolated to a final dilution of 1:5,000 to 1:10,000.

Enzyme Method of Maxted[6]

Note: The enzyme method of extraction is especially useful for lysing streptococci belonging to Lancefield's Groups A, C, and G. Members of other groups are less rapidly lysed or may not be lysed at all.

1. Culture the *Streptococcus* to be grouped in 5 ml of Bacto Todd Hewitt Broth or Bacto Brain Heart Infusion for 18 – 24 hours at 37°C.
2. Centrifuge to sediment the growth completely.
3. Discard the supernatant fluid.
4. Add 0.5 ml of the rehydrated Bacto Streptomyces Albus Enzyme and suspend the bacterial cells.
5. Incubate the suspension at 50°C for 90 minutes.
6. Centrifuge to sediment the cellular debris. Complete lysis (complete clearing) of the enzyme — *Streptococcus* suspension may not occur, however, the centrifuged clear suspension is satisfactory for the test antigen.
7. Use the supernatant solution for the antigen in the precipitin test.
8. Follow the STREPTOCOCCUS GROUP PRECIPITIN TEST steps 1 – 7 below except incubate the test at 37°C for 15 minutes before reading.

The enzyme method of Maxted employs the use of living cultures. It is absolutely necessary to sterilize all tubes, pipettes and other equipment after use.

PROCEDURE
Materials Provided:

		Available from Difco:
Bacto Streptococcus Antisera	Bacto Streptomyces Albus Enzyme	Bacto Phenol Red Bacto Todd Hewitt Broth
Bacto Streptococcus Antigens	Bacto Capillary Tubes	Bacto Brain Heart Infusion

Materials Required but not Provided:

Culture Media	Centrifuge	N/5 HCl and N/5 NaOH	Plasticine block
Incubator at 37°C	Glassware	0.85% NaCl solution	pH meter
Autoclave	Burner	pH indicators	Adequate light source

STREPTOCOCCUS GROUP PRECIPITIN TEST

Note: To conserve antisera only the antiserum related to the hemolytic type of the isolate should be used initially. For example, of the beta-hemolytic streptococci, over 90% belong to the group A and this antiserum therefore should be used first with an extract of an isolate from this hemolytic group. (See Table 1 for additional relationships.)

Capillary Tube Method:

1. Dip a capillary tube, which has been provided, into the rehydrated antiserum and allow a column of 2 – 3 cm to rise into the tube, the antiserum should be added to the capillary tube first so that it is layered above the extract.
2. With the forefinger on the end of the capillary tube, remove it from the serum holding the finger in place. The tip of the capillary should be cleaned with lint-free tissue to avoid contamination of the extract with any excess antiserum on the tube.

3. Insert the tube into the prepared antigen extract. Allow an equal amount (2 – 3 cm) to enter it by removing the finger. The antiserum and antigen must be in contact with each other in the capillary tube. If an air bubble separates them, discard the tube and repeat procedure.
4. Remove the tube from the extract and allow the column of antiserum to rise to the center of the tube by inverting it, leaving air spaces at both ends.
5. Wipe the excess fluid off the tube and insert it into a plasticine block. After placing the capillary tube in the plasticine block, or prior to reading the test, wipe the capillary tube free of serum, fingerprints or any other material which might obscure the reading or result in a false-positive reaction.
6. Bacto Streptococcus Antigens should be used as positive and negative controls.
7. Incubate the test at room temperature and observe for precipitation, using a fluorescent light together with a dark background after 5 minutes and at intervals up to 30 minutes. Discard the test at the end of 30 minutes, since false-positive or negative reactions might occur after this time.

The test must be read at the end of 5 minutes as well as at the end of 30 minutes since a prozone has been described in which the precipitate has been formed, may rapidly redissolve, resulting in a false-negative reaction.

For use of the enzyme treated culture, the test conditions are incubation at 37°C for 15 minutes.

RESULTS
A precipitation in a tube indicates that the extract tested is homologous to the *Streptococcus* antiserum group used.

LIMITATIONS OF THE PROCEDURE
While serological procedures, as applied to microorganisms, provide corroborative evidence and can be used to identify particular antigenic sites of the genus or species under study, they cannot be used alone to identify the etiological agent of a disease. Cultural isolation and at least preliminary biochemical differentiation must precede any serological examination and final identification cannot be made without biochemical and serological characterization.

Because of the prozone phenomenon, it is possible for the precipitate to redissolve when the capillary tube test is employed. Close attention must be given to follow the course of the reaction during the first 5 – 15 minutes of incubation.

Nonspecific reactions may occur. Causes for such reactions are dealt with in the test procedure.

REFERENCES
1. Rockefeller Inst. Med. Res. Mono. No. 9, 1919.
2. J. Exp. Med., 57:571, 1933.
3. *ibid.,* 47:91, 1928.
4. Proc. Soc. Exp. Biol. & Med., 38:473, 1938.
5. Clinical Bact., 3rd Ed., E Arnold Ltd., London, 1970.
6. Lancet, 225:255, 1948.
7. Manual of Clin. Micro., ASM, 1970.
8. Int. J. Sys. Bacteriol., 24:434, 1974.
9. J. Gen. Micro., 37:425, 1964.
10. Stanford Med. Bull., 13:290, 1955.
11. Manual of Clin. Micro., ASM, 2nd Ed., 1974.

PACKAGING

Bacto Streptococcus Antisera and Antigens		Antisera Code	Antigen Code
Group A	1 ml		2978-50-3
B	1 ml	2741-50-9	2979-50-2
C	1 ml	2742-50-8	2980-50-9
D	1 ml	2743-50-7	2981-50-8
E	1 ml	2810-50-5	
F	1 ml		2983-50-6
G	1 ml	2744-50-6	2984-50-5
Bacto Streptococcus Antigen Set	7 × 1 ml	2368-32-4	
Contains Groups A, B, C, D, E, F, G			
Bacto Streptococcus Antisera Set	7 × 1 ml	2974-32-0	
Contains Groups A, B, C, D, E, F, G			
Bacto Streptococcus Antigen MG	1 ml	2242-50-3	
Bacto Streptococcus MG Suspension	5 ml	2367-56-6	
Bacto Streptococcus Antiserum MG	1 ml	2239-50-8	
Bacto Streptomyces Albus Enzyme	5 ml	3134-56-6	
Bacto Todd Hewitt Broth	1/4 lb (114 g)	0492-02-3	
	1 lb (454 g)	0492-01-4	
Bacto Phenol Red	5 g	0203-11-2	

BACTO STREPTOCOCCUS MG (GROUP F) SUSPENSION

INTENDED USE

Bacto Streptococcus MG Suspension now known as *Streptococcus anginosus* and in the Group F of the Lancefield classification, is a cell suspension for use as antigen in the detection and quantitation of *Streptococcus* MG antibodies in serous fluids by the tube agglutination procedure. It should be stored at 2 – 8°C when not in use.

PROCEDURE

Materials Provided:
Bacto Streptococcus MG (Group F) Suspension
Bacto Streptococcus MG (Group F) Antiserum

Materials Required but not Provided:

0.85% NaCl solution	Incubator at 37°C
Kahn tubes	Refrigerator at 2 – 8°C
Serological pipettes	Suitable light source

MACROSCOPIC TUBE AGGLUTINATION TEST

1. Dilute sufficient Bacto Streptococcus MG (Group F) Suspension for one day's use 1:10 with 0.85% NaCl solution. This diluted antigen should be discarded at the end of the working day.
2. Place 8 Kahn tubes in a rack for each serum to be tested.
3. Pipette 0.9 ml of 0.85% NaCl into the first tube of each row and 0.5 ml into each of the remaining tubes.
4. Add 0.1 ml of the serum to the first tube containing 0.9 ml of NaCl solution.
5. Mix well with a pipette and transfer 0.5 ml to the second tube. Mix thoroughly.
6. Continue carrying the 0.5 ml of the serum dilution through tube 7. Discard 0.5 ml from tube 7 after mixing thoroughly. Tube 8 is the antigen control tube. Add 0.5 ml of the diluted Bacto Streptococcus MG Antigen to each of the 8 tubes. Shake the

racks to mix the antigen and antiserum. The resultant dilutions are 1:20 through 1:1280, respectively.

7. Incubate tubes at 37°C for 2 hours.
8. Place tubes in the refrigerator at 2 – 8°C overnight.
9. Reincubate at 37°C for 2 hours and read for agglutination.

CALIBRATION AND QUALITY CONTROL

For greater proficiency in test interpretation always include a positive and negative serum control in each test protocol. Bacto Streptococcus MG (Group F) Antiserum is recommended as a positive control in this test.

Both positive and negative controls are diluted in the same proportion as the patient's serum and processed in exactly the same manner following procedure above for the macroscopic tube test.

An antigen is considered to be satisfactory if it does not clump with the negative control and it reacts to a titer of 1:80 or more with the positive control.

RESULTS

Tube Agglutination Test

Record results as follows:

4+	complete agglutination
3+	approximately 75% of the cells are clumped
2+	approximately 50% of the cells are clumped
1+	approximately 25% of the cells are clumped
±	trace agglutination
−	no agglutination

The titer of serum is recorded as that dilution of the specimen in which at least 2+ (50%) agglutination occurs.

LIMITATIONS

The major limitation of this test is that of interpretation. The test result is not to be used singly for a definite diagnosis. A definitive diagnosis must be made by the physician, taking into consideration the history and physical state of the patient as well as data obtained from other laboratory tests.

PACKAGING

Bacto Streptococcus MG (Group F) Suspension	5 ml	2367-56-6
Bacto Streptococcus MG (Group F) Antiserum	1 ml	2239-50-8

STREPTOLYSIN O REAGENTS

BACTO STREPTOLYSIN O REAGENT
BACTO ASO STANDARD
BACTO STREPTOLYSIN O BUFFER, DRIED

INTENDED USE

Bacto Streptolysin O Reagent, Bacto Streptolysin O Buffer, Dried, and Bacto ASO Standard, are standardized preparations for use in the quantitative test procedures for the determination of Antistreptolysin O titers (ASO) of sera as described by Massel and

Miller,[1] Rammelkamp,[2] and by Rantz and Randall[3] or in any modification of these techniques. The Rantz and Randall method is described in detail in these instructions.

HISTORY

Todd,[4] in 1928, demonstrated the liberation of a lysin for red cells by group A *Streptococcus* and proved its antigenic nature which results in the production of antistreptolysin O after infection with these organisms. He later differentiated the streptolysins into 2 serologically identifiable lysins, streptolysin O and streptolysin S. The greater interest in streptolysin O and its antibody, antistreptolysin O, has resulted in a more thorough investigation of the significance of increased titers of antistreptolysin O in sera from group A *Streptococcus* infections and their relation to rheumatic fever, acute glomerulonephritis, rheumatoid arthritis and erythema nodosum.

Todd,[5] standardized the streptolysin O so that 0.5 ml just completely hemolyzed 0.5 ml of a 5% rabbit blood cell suspension at 37°C in 1 hour. He then defined 1 unit of antistreptolysin as the amount of serum just neutralizing 2.5 minimum hemolytic doses of streptolysin O.

PRINCIPLES

The Rantz and Randall technique consists essentially of combining a constant volume of rehydrated Bacto Streptolysin O Reagent with various dilutions of the patient's serum, incubating for 15 minutes at 37°C, adding a constant volume of 5% erythrocyte (RBC) suspension, reincubating, centrifuging, and observing the end point indicated by the highest dilution showing absence of hemolysis. The ASO titer is the reciprocal of the highest serum dilution showing no sign of hemolysis.

NOTE: It is recognized by rheumatologists that a single determination is of minimal significance, and that serial titrations at bi-weekly intervals 4 – 6 weeks following the streptococcal infection, yield information which is more pertinent. For example, a single ASO titer of 166 units does not, in itself, prove the presence or absence of streptococcal disease, but if this designates a rise from a base line of 100 ASO, it is significant, and points to a present streptococcal infection or a recent infection. Also a drop from a titer of 250 units to 125 units is strongly indicative of recovery. Robinson[6] recommends serial ASO determination during the illness and into convalescence. He further states[7] that ASO determinations provide an excellent tool for the detection of host exposure to most streptococcal infections and that this information is especially useful in epidemiological studies where immunological confirmation of beta streptococcal disease is desired.

REAGENTS

Bacto Streptolysin O Reagent is a desiccated, standardized filtrate of streptolysin O, in reduced form, prepared from group A *Streptococcus*. Bacto Streptolysin O Reagent is so standardized that 0.5 ml will combine with 1 unit of antistreptolysin O as defined by Todd[8] and modified by Hodge and Swift.[9] The reagent can, therefore, be used with any serum dilution schema. Bacto Streptolysin O Reagent is standardized with both Todd's original standard and with the more recently prepared International (WHO) Standard.[10]

Bacto Streptolysin O Buffer, Dried, contains 7.40 g NaCl, 3.17 g of potassium dihydrogen phosphate, 1.81 g of disodium phosphate per liter and has a reaction of pH 6.5 – 6.7 when reconstituted.

Bacto ASO Standard is a control serum having an end point of 166 Todd units when used as described in the test procedure.

RECONSTITUTION

1. Bacto Streptolysin O Reagent is rehydrated with 10 ml distilled water. Mixing should be done gently. Shaking the vial during mixing can result in the oxidation of the reagent making it unsuitable for the test. **Do not rehydrate the reagent until ready to add to the tubes for the performance of the test.**
2. Bacto Streptolysin O Buffer, Dried is reconstituted by dissolving 1 vial (12.4 g) in 1 liter distilled or deionized water.
3. Bacto ASO Standard is rehydrated by adding 10 ml distilled water to each vial. Solution is hastened by a gentle end-over-end rotation. Avoid vigorous shaking resulting in foam formation.

STORAGE

1. Bacto Streptolysin O Reagent, unreconstituted should be stored at 2 – 8°C. Unused portions of the reconstituted reagent should be discarded within 15 minutes after rehydration. Solutions showing turbidity should be discarded.
2. Bacto Streptolysin O Buffer, Dried is stable at temperatures below 30°C. The reconstituted buffer should be stored at 2 – 8°C. Solutions showing turbidity or mold growth should be discarded.
3. Bacto ASO Standard, unreconstituted, should be stored at 2 – 8°C. Unused portions of reconstituted standard may be distributed into clean, dry test tubes in 3.3 ml amounts, tightly sealed, and stored in the deep freeze at below 0°C. Each aliquot thus prepared is sufficient for one control test.

SPECIMEN COLLECTION

Five ml whole blood is collected aseptically from a patient. A fasting specimen is best. The blood is allowed to clot and the tube is centrifuged. The clear serum is transferred to a clean, dry test tube. Store serum in the refrigerator until ready to use. Fresh or inactivated sera are equally satisfactory for the test.

NOTE: Contaminated sera or those that contain increased beta lipoprotein inhibit streptolysin O, thereby yielding a falsely high ASO titer.

PROCEDURE

Materials Provided:
  Bacto ASO Standard
  Bacto Streptolysin O Reagent
  Bacto Streptolysin O Buffer, Dried

Materials Required but not Provided:
  Erythrocyte (RBC) suspension
  Chemically clean pipettes (1,5,10 ml)
  Chemically clean test tubes
  Centrifuge
  Test tube racks
  Distilled water
  Water bath at 37°C

1. Red cell suspension — defibrinated rabbit cells, human cells from the blood bank or freshly drawn and defibrinated human blood cells are satisfactory for the test. Blood obtained from a human donor suspected of having a streptococcal infection should not be used. Wash cells in normal saline and centrifuge at 2,000 rpm for 3 – 5 minutes. The red tinged supernatant is syphoned off and the process repeated until the supernatant is colorless. It is essential that the last centrifugation yield a colorless supernatant. The final centrifugation should be at 2,000 rpm for 15 minutes. Three washings are usually required. Cells requiring more than 5 washings should not be used. Suspend 5 ml of the washed, packed cells in 95 ml of Bacto Streptolysin O Buffer solution.
2. Serum dilutions — prepare the following serum dilutions using Bacto Streptolysin O Buffer solution as the diluent:

A. 1:10 = 0.5 ml of serum + 4.5 ml Bacto Streptolysin O Buffer
B. 1:100 = 1 ml of 1:10 + 9.0 ml Bacto Streptolysin O Buffer
C. 1:500 = 2 ml of 1:100 + 8.0 ml Bacto Streptolysin O Buffer
Various amounts of these dilutions are pipetted into tubes according to the table.

NOTE: Glassware used in the test must be chemically clean; lipids, detergents and other contaminants will cause inaccurate titers.

The entire range from 12 – 2500 Todd units as outlined in the table is seldom used for a serum titration. Generally, for the initial titration, the first 7 dilutions covering the range of 12 – 333 Todd units are employed. This range will indicate normal or elevated titers, and subsequent titration dilutions may be based on the previous titers.

CONTROLS
1. Red Cell Control — 1.5 ml of the Bacto Streptolysin O Buffer solution used as diluent plus 0.5 ml cell suspension. The supernatant after centrifugation or 2 hours standing should be without hemolysis. This tube may be used to aid in determining slight amounts of hemolysis in other tubes.
2. Streptolysin Control — 1.0 ml of Bacto Streptolysin O Buffer solution, 0.5 ml of reconstituted lysin and 0.5 ml cell suspension. This tube should show marked to complete hemolysis.
3. Bacto ASO Standard, having a titer of 166 Todd units, is recommended as a control of the conditions of the test. Rehydrated Bacto ASO Standard is used without further dilution as the 1:100 serum dilution in the range of 100 through 333 Todd units (tubes 3 – 7) as indicated in the table. If desired, sera with known low or high titers may also be included as controls. These additional control sera are treated in the same manner as are unknown sera. The person performing the test should consistently reproduce a stated titer within ±1 dilution.

RESULTS
The antistreptolysin O titer, expressed in Todd units, is the reciprocal of the highest dilution of serum showing no hemolysis. For example, the ASO titer of a serum showing no hemolysis in tube 1 – 4, a trace of hemolysis in tube 5, and complete hemolysis in other tubes would be 125.

LIMITATIONS OF THE PROCEDURE
Some patients with a group A streptococcal infection do not produce a significant ASO titer. If repeated ASO tests do not demonstrate a significant titer and the clinician is still unable to prove or disprove a diagnosis of a group A *Streptococcus* infection, an AHT test is recommended, using Bacto AHT Kit, Code 0934-32-3.

EXPECTED VALUES
Klein, Baker and Jones[10] determined the "Upper Limits of Normal" ASO titers on serum specimens from various age groups of pediatric patients (425 sera) with no history of a recent streptococcal infection and on healthy adult hospital employees (220 sera). The upper limit of normal ASO titer was defined as that level of antibody titer exceeded by no more than 15% of the total subjects in each age group. The upper limit of normal ASO titer for the preschool and adult group was 85 (100) and for the school age group it was 170 (166). The study was done using a microtitration method.

The figures in parenthesis represent the upper limit of normal titer using the Rantz and Randall method.

Methodology for the Titration of Antistreptolysin O Titers

Tube	Serum Dilution												Red Cell Control 13	Streptolysin Control 14
	1:10		1:100						1:500					
	1	2	3	4	5	6	7	8	9	10	11	12		
ml Diluted Serum	0.8	0.2	1.0	0.8	0.6	0.4	0.3	1.0	0.8	0.6	0.4	0.2	—	—
ml Bacto Streptolysin O Buffer ..... Shake gently to mix.	0.2	0.8	0.0	0.2	0.4	0.6	0.7	0.0	0.2	0.4	0.6	0.8	1.5	1.0
ml Bacto Streptolysin O Reagent ... Shake gently to mix. Incubate at 37°C for 15 min.	0.5	0.5	0.5	0.5	0.5	0.5	0.5	0.5	0.5	0.5	0.5	0.5	—	0.5
ml of 5% Red Cell Suspension .... Shake gently. Incubate at 37°C for 45 min, shaking after first 15 min. Centrifuge tubes for 1 min at 1000 – 1500 rpm.	0.5	0.5	0.5	0.5	0.5	0.5	0.5	0.5	0.5	0.5	0.5	0.5	0.5	0.5
Unit Value	12	50	100	125	166	250	333	500	625	833	1250	2500	Controls	

Rantz, DiCaprio and Randall[11] in their ASO studies in various diseases and in healthy controls suggest that very low ASO titers (50 units or less) are strong evidence against the presence of either rheumatic fever or glomerulonephritis, and that increased ASO titers, 500 units or more, are present in individuals who have recently had a streptococcal infection. They consider both the low and increased titers helpful in "building a complete picture in certain pathologies." It is generally agreed that a low ASO titer is of greater value as an aid in the detection of rheumatic fever and glomerulonephritis than a single high titer; also, that changes in titers of serial specimens offer more information than isolated examinations.

McCarty[12] stated that laboratory tests are often necessary as an aid in the detection of rheumatic fever and that the most valuable and commonly employed of these tests is the ASO determination. He emphasized that quantitative variations between patients are marked and that the rapid rise in ASO is noted mainly during the month after the streptococcal infection. He noted that the ASO titer in rheumatic fever had a mean titer of 500 units per ml and that 150 units would be considered in the normal range.[15] This is in agreement with other investigators such as Massell and Miller,[13] Warren,[14] Bernheimer[15] and Bunim.[16]

SPECIFIC PERFORMANCE CHARACTERISTICS

Klein et al[17] found the ASO titer test to be reproducible well within the usual ±1 dilution standard. This is emphasized by the fact that "less than 2% of the titers of duplicate sera would differ by more than 1 dilution, even if processed by different technicians."

Cravitz,[18] reporting the results of a 5 year evaluation of routine antistreptolysin O titrations in a hospital laboratory, concludes that the antistreptolysin O test constitutes a single, reliable serologic procedure that can be used as a laboratory aid for the detection of group A *Streptococcus* infection including acute or early rheumatic fever and acute glomerulonephritis.

While Stallerman et al[19] agree with others[20,21,22,23,24] that between 80 – 90% of group A *Streptococcus* infections can be detected by the ASO titers if the test was performed within the first month of an acute rheumatic fever attack.

REFERENCES

1. Personal Communication, 1952.
2. Am. J. Med., 10:673, 1951.
3. Personal Communication, 1952.
4. J. Exp. Med., 48:493, 1928.
5. J. Exp. Med., 55:267, 1932.
6. Naval Med. Research Rpt., 1950.
7. Am. J. Clin. Path., 22:237, 1952.
8. J. Exp. Med., 58:277, 1933.
9. Applied Microbiology, 21:999 – 1001, 1971.
10. Bull. Wld. Hlth. Org., 24:271 – 279, 1961.
11. Am. J. Med. Sc., 224:194, 1952.
12. Annals Int. Med., 37:1027, 1952.
13. Personal Communication, 1952.
14. *ibid.*, 1952.
15. *ibid.*, 1952.
16. *ibid.*, 1952.
17. Am. J. Clin. Path., 53:159 – 162, 1970.
18. Proc. N.Y State Assoc. of Publ. Hlth. Lab., 33:6, 1953.
19. Am. J. Med., 20:163 – 169, 1956.
20. J. Immunol., 41:35, 1941.
21. *ibid.*, 41:61, 1941.
22. *ibid.*, 41:87, 1941.
23. Am. J. Med., 5:3, 1948.
24. McCarty, M., Rheumatic Fever, edited by Thomas, L., Univ. of Minn. Press, p. 136, 1952.

PACKAGING

Bacto ASO Standard		6 × 10 ml	0602-60-9
Bacto Streptolysin O Reagent	12 tests	6 × 10 ml	0482-60-4
	36 tests	6 × 25 ml	0482-66-8
	168 tests	6 × 100 ml	0482-73-9
Bacto Streptolysin O Buffer, Dried		6 × 12.4 g	0516-33-8

BACTO STREPTOMYCES ALBUS ENZYME

INTENDED USE
Bacto Streptomyces Albus Enzyme is a desiccated soluble enzyme of *Streptomyces albus* recommended for preparing lysates of streptococci for the *Streptococcus* group precipitin test.

HISTORY
Maxted[1] first used the *S. albus* enzyme to lyse hemolytic streptococci for classifying them into appropriate serological groups. He demonstrated that the lysate containing the group specific polysaccharides could be prepared easily and rapidly by this process. The method is especially useful for lysing streptococci belonging to groups A, C and G. Members of other groups are less readily lysed.

PREPARATION
To rehydrate the enzyme add 5 ml sterile distilled water and rotate the vial to dissolve the contents.

Either of 2 methods for preparing the *Streptococcus* extract using Bacto Streptomyces Albus Enzyme may be employed. In the first method a bead or loopful of growth from a blood agar slant or plate is suspended in 0.25 ml of the enzyme in a small bore test tube and placed in the water bath at 45 – 50°C until the solution is clear. Usually, 60 – 90 minutes will be required for complete clearing. The tube is then centrifuged to sediment cell walls and the supernatant tested with streptococcal antisera.

In the second method for preparing the extract the sediment from 5 ml of overnight growth in 5 ml of Bacto Todd Hewitt Broth or Bacto Brain Heart Infusion is suspended in 0.5 ml of the enzyme and the suspension treated as above.

TECHNIQUES OF THE GROUP PRECIPITIN TEST
Capillary Tube Method
1. Dip a capillary tube, 0.7 – 1 mm i.d., and 75 – 90 mm long, into the rehydrated antiserum and allow a column 2 – 3 cm to rise into the tube.
2. With the forefinger on the end of the capillary tube, remove it from the serum holding the finger in place.
3. Wipe excess serum off the tube and insert into the prepared antigen extract. Allow an equal amount (2 – 3 cm) to enter the tube by removing the finger.
4. Remove the tube from the extract and allow the column of antiserum and extract to rise to the center of the tube leaving about 1 cm air space at both ends. The extract and antiserum must be in contact. Repeat the test if an air bubble appears between them.
5. Wipe excess fluid off the tube and insert into a plastic block.
6. Normal serum controls and Bacto Streptococcus Antigen controls should be included.
7. The antigen extract, derived from an organism belonging to the corresponding *Streptococcus* group, will produce a precipitate in 10 – 15 minutes.

Small Tube Method
1. Use small tubes, 3 – 4 mm i.d.
2. Place 0.05 ml of the clear *Streptococcus* extract in the tube, using a Pasteur pipette.
3. Add an equal amount (0.05 ml) of Bacto Streptococcus Antisera. Normal serum controls and Bacto Streptococcus Antigen controls should be included. Repeat test is an air bubble is present between the extract and antiserum.
4. The antigen extract derived from an organism belonging to the corresponding *Streptococcus* group will produce a precipitate in 10 – 15 minutes.

STORAGE
Bacto Streptomyces Albus Enzyme is stable when stored at 2 – 8°C. After rehydration it should be stored at 2 – 8°C and used within 1 week.

REFERENCE
1. Lancet, 225:255, 1948.

PACKAGING
Bacto Streptomyces Albus Enzyme   5 ml   3134-56-6

BACTO STUART MEDIUM BASE

INTENDED USE
Bacto Stuart Medium Base is recommended for the cultivation of *Leptospira*.

HISTORY/PRINCIPLES
Bacto Stuart Medium Base, prepared according to the formulation of Stuart,[1] has been demonstrated by Galton, et al.[2] to yield *Leptospira* cultures of excellent antigenic properties for the detection of *Leptospira* antibodies in both humans and animals. Heat inactivation of this medium is not required.

FORMULA

BACTO STUART MEDIUM BASE
DEHYDRATED

Ingredients per liter

Bacto Asparagine	0.132 g
Ammonium Chloride	0.268 g
Magnesium Chloride · 6H$_2$O	0.406 g
Sodium Chloride	1.808 g
Disodium Phosphate	0.666 g
Monopotassium Phosphate	0.087 g
Bacto Phenol Red	0.01 g

Final pH 7.6 ± 0.2 at 25°C.

One pound will make 133.5 liters of final medium.

METHOD OF PREPARATION
1. To rehydrate, dissolve 3.4 g in 995 ml distilled or deionized water. Add 5 ml Bacto Glycerol.
2. Sterilize in the autoclave for 15 minutes at 15 lbs pressure (121°C).
3. Allow the medium to cool to room temperature.

4. Aseptically add 8 – 10% Bacto Leptospira Enrichment. Mix well.
5. Dispense aseptically into sterile screw capped test tubes in 8 – 10 ml amounts.

STORAGE
Bacto Stuart Medium Base	Below 30°C
Prepared medium	15 – 30°C

QUALITY CONTROL
Identity Specifications

Dehydrated powder:	pink, homogeneous, free-flowing
Reaction of 0.34% solution:	pH 7.6 ± 0.2 at 25°C
Prepared medium:	red, clear

Typical Cultural Response in Bacto Stuart Medium Base with Bacto Leptospira Enrichment After up to 5 Days at 30°C

Organism	Growth
Leptospira interrogans serotype *australis*	good to excellent
Leptospira interrogans serotype *canicola*	good to excellent
Leptospira interrogans serotype *grippotyphosa*	good to excellent

REFERENCES
1. J. Path. and Bact., 58:343, 1946.
2. Am. J. Vet. Res., 19:505, 1958.

PACKAGING
Bacto Stuart Medium Base	1 lb (454 g)	0988-01-5
Bacto Leptospira Enrichment	6 × 10 ml	0452-60-0
Bacto Glycerol	100 g	0282-15-2
	500 g	0282-17-0

BACTO SULFATE API BROTH

INTENDED USE
Bacto Sulfate API Broth is used for the detection, differentiation and estimation of sulfate-reducing bacteria.

HISTORY/PRINCIPLES
Bacto Sulfate API Broth is prepared according to the formulation given in the "American Petroleum Institute Recommended Practice."[1]

Sulfate-reducing bacteria convert the sulfate to sulfide which reacts with the ferrous ion to give a black color. The detection and estimation of these bacteria depend upon their ability to grow and produce sulfide in the medium. Sulfate-reducing bacteria cause corrosion in oil well flood systems resulting in perforations in the pipes. The formation of insoluble sulfide results in plugging, necessitating increased injection pressures.

Bacto Sulfate API Broth vials, sterile and evacuated for maintaining anaerobic conditions, are ready for inoculation and dilution of water flood samples.

FORMULA

BACTO SULFATE API BROTH
DIFCO CERTIFIED

Ingredients per liter

Bacto Yeast Extract . . . . 1 g
Ascorbic Acid . . . . 0.1 g
Sodium Lactate . . . . 5.2 g
Magnesium Sulfate . . . . 0.2 g
Dipotassium Phosphate . . . . 0.01 g
Ferrous Ammonium Sulfate . . . . 0.1 g
Sodium Chloride . . . . 10 g

Final pH 7.5 ± 0.2 at 25°C.

One pound will make 31.3 liters of medium.

PROCEDURE

1. Suspend 14.5 g in 1 liter distilled or deionized water.
2. Warm gently to facilitate solution of the medium, if necessary.
3. Dispense in 9 ml amounts into test tubes and autoclave for 10 minutes at 15 lbs pressure (121°C).
4. Inoculate the sample by diluting serially in a series of Bacto Sulfate API Broth tubes as follows:

   Dispense 1 ml of the water sample into the first tube. Stopper the tube and invert several times to mix the inoculum and medium. Transfer 1 ml from this tube to a second tube and mix as before. Continue this serial transfer until a dilution of 1 to 100,000 is reached.

5. Incubate within 5°C of the temperature of the injection water under anaerobic conditions for a minimum of 3 weeks. Examine daily for blackening of the medium. If prepared vials are used, the sample is inoculated using a syringe with needle, or similar system, and then incubated aerobically.

STORAGE

Bacto Sulfate API Broth   2 – 8°C
Prepared medium   15 – 30°C

QUALITY CONTROL

Identity Specifications

Dehydrated powder: light beige, homogeneous, free-flowing
Reaction of 1.45% solution: pH 7.5 ± 0.2 at 25°C
Prepared medium: light amber, clear to very slightly opalescent

Typical Cultural Response in Bacto Sulfate API Broth
After up to 1 Week at 30°C

Organism	Growth
Desulfovibrio desulfuricans	good to excellent

REFERENCE

1. American Petroleum Institute Recommended Practice 38, First Ed., May 1959.

PACKAGING

Bacto Sulfate API Broth   1 lb (454 g)   0500-01-4
                                     6 × 9 ml   0500-86-2

BACTO SULFITE AGAR

INTENDED USE
Bacto Sulfite Agar is used for detecting thermophilic anaerobes producing H_2S.

HISTORY
Bacto Sulfite Agar is prepared according to the formula of Clark and Tanner.[1] It has been recommended by APHA[2] *Standard Methods for the Examination of Dairy Products*.

FORMULA
BACTO SULFITE AGAR
DEHYDRATED
Ingredients per liter

Bacto Tryptone . . . . 10 g
Sodium Sulfite . . . . 1 g
Bacto Agar . . . . 20 g

Final pH 7.6 ± 0.2 at 25°C.

One pound will make 14.6 liters of medium.

METHOD OF PREPARATION
1. To rehydrate suspend 31 g in 1 liter distilled water and heat to boiling to dissolve completely.
2. Distribute into tubes in 15 ml amounts.
3. Add a clean iron nail to each tube and cap with screw caps.
4. Sterilize in the autoclave for 15 minutes at 15 lbs pressure (121°C).
5. Allow to cool, keeping medium at 45 – 50°C until ready to inoculate.
6. Inoculate tubes and incubate aerobically at 35°C for 18 – 48 hours.
7. *Clostridia* (sulfite reducers) will turn the medium black. Other organisms will grow but not turn black.

STORAGE
Bacto Sulfite Agar	Below 30°C
Prepared medium	15 – 30°C

Quality Control
Identity Specifications

Dehydrated powder:	light beige, homogeneous, free-flowing
Reaction of 3.1% solution:	pH 7.6 ± 0.2 at 25°C
Prepared medium:	light amber, very slightly opalescent

Typical Cultural Response in Bacto Sulfite Agar
After 18 – 48 Hours at 35°C

Organism	Growth	Blackening of medium
Clostridium botulinum ATCC® 25763	good to excellent	+
Clostridium butyricum ATCC® 9690	good to excellent	+
Clostridium sporogenes ATCC® 11437	good to excellent	+
Escherichia coli ATCC® 25922	good to excellent	−
Salmonella enteritidis ATCC® 13076	good to excellent	−

REFERENCES
1. Food Research, 2:27, 1937.
2. Standard Methods for the Examination of Dairy Products, 11th Ed., APHA, Inc., New York, 1960.

PACKAGING
Bacto Sulfite Agar   1 lb (454 g)   0972-01-3

BACTO SUPPLEMENT A

INTENDED USE
Bacto Supplement A is a sterile yeast concentrate containing crystal violet. It is used as an enrichment in chocolate agar media for the detection and isolation of *Neisseria gonorrhoeae* from specimens suspected of containing other bacteria, especially the gram-positive organisms. It is particularly recommended for use in chocolate agar prepared from Bacto Proteose No. 3 Agar or Bacto GC Medium Base and Bacto Hemoglobin.

HISTORY/PRINCIPLES
The marked increase in the incidence of gonorrhea requires the use of a sensitive laboratory procedure for the isolation of the etiologic agent of this disease, *N. gonorrhoeae*. At the present time the cultural technique is considered to be the most sensitive method for detecting this organism.[1] Cultural methods for the detection of *N. gonorrhoeae* from cervical and urethral exudates have proven more reliable and efficient than microscopic techniques. These procedures are particularly valuable in early, chronic or treated cases and in isolation of the organism from asymptomatic males and females.

Interest in the cultural procedure for the diagnosis of gonococcal infections was stimulated by Ruys and Jens,[2] McLeod et al,[3] Thompson,[4] Leahy and Carpenter.[5] Carpenter, Leahy and Wilson[6] and Carpenter[7] clearly demonstrated the superiority of this method over the microscopic technique. Further studies in cooperation with Carpenter, and McLeod[8] and Herrold[9] resulted in the development of chocolate agar prepared with Bacto Proteose No. 3 Agar and Bacto Hemoglobin, which proved to be satisfactory for isolating the organism from all types of gonococcal infections.

Efforts made by Langford,[10] Langford, Scott, Cox and Cooke,[11] and Langford and Snell,[12] on suitable enrichments to increase the number of isolations and growth rate of *N. gonorrhoeae* were instrumental in the development of Bacto Supplement A and Bacto Supplement B.

Bacto Supplement A is processed to preserve both the thermolabile and thermostable growth accessory factors, including glutamine, coenzyme (V factor), cocarboxylase, and others for the most exacting strains of *Neisseria gonorrhoeae* and *Haemophilus influenzae*. It also contains the hematin (X factor) required by *H. influenzae*.

In addition, this enrichment contains crystal violet which suppresses growth of many of the common contaminants, especially *Staphylococcus*.

METHOD OF PREPARATION
Use this enrichment in media as discussed in detail under Bacto GC Medium Base.

STORAGE
Bacto Supplement A   2 – 8°C

QUALITY CONTROL
Identity Specifications
Appearance: purple, clear to very slightly opalescent

Typical Cultural Response
See Bacto GC Medium Base

REFERENCES
1. U.S. Public Health Service, CDC, Venereal Disease Branch: Criteria and Techniques for the Diagnosis of Gonorrhea, 1971.
2. Muench Wochschr, 80:846, 1933.
3. J. Path. Bact., 38:221, 1934.
4. J. Infect. Dis., 61:129, 1937.
5. Am. J. Syphilis, 22:347, 1938.
6. Am. J. Syphilis, 22:55, 1938.
7. Seventh Annual Yearbook (1936 – 37) p. 133, Suppl. Am. J. Publ. Health, 27:3, 1937.
8. Personal Communication, 1938.
9. Personal Communication, 1938.
10. J. Bact., 44:139, 1942.
11. J. Bact., 45:321, 1943.
12. J. Bact., 45:410, 1943.

PACKAGING
Bacto Supplement A   6 × 5 ml   0246-57-6

BACTO SUPPLEMENT B

INTENDED USE
Bacto Supplement B (desiccated) is a sterile enrichment for use in supplementing media for culturing fastidious microorganisms with exacting growth requirements. It is particularly recommended for use in the preparation of chocolate agar as described by Christensen and Schoenlein,[1] prepared from Bacto GC Medium Base or Bacto Proteose No. 3 Agar and Bacto Hemoglobin employed in the detection and isolation of *Neisseria gonorrhoeae*. It is also recommended for use as an enrichment in both solid and liquid media for the isolation and propagation of *Haemophilus influenzae* as described by Neter.[2,3]

PRINCIPLES
Bacto Supplement B is processed to preserve both the thermolabile and thermostable growth accessory factors of fresh yeast, and also contains glutamine, coenzyme (V factor), cocarboxylase and other growth factors for the most exacting strains of *N. gonorrhoeae* and *H. influenzae*. It also contains hematin (X factor) required by *H. influenzae*.

Bacto Supplement B is used in a final 1% concentration in either solid or liquid media. It is added aseptically after the medium has been sterilized in the autoclave and cooled below 50°C.

METHOD OF PREPARATION
1. Aseptically add the contents of 1 vial Bacto Supplement B Reconstituting Fluid to 1 vial Bacto Supplement B.
2. Rotate the vial to dissolve the contents completely.

3. Use this enrichment in media as discussed in detail under Bacto GC Medium Base. It may also be used in place of Bacto Supplement VX for modified Thayer Martin medium. See Bacto Supplement VX for preparation.

STORAGE
Bacto Supplement B with Bacto
  Supplement B Reconstituting Fluid $2 - 8°C$
Rehydrated $2 - 8°C$

QUALITY CONTROL
Identity Specifications
Appearance: tan button or powder
Rehydrated appearance: light to medium amber, clear to slightly opalescent

Typical Cultural Response
See Bacto GC Medium Base

REFERENCES
1. Ann. Meeting CA Publ. Hlth. Assn., 1947.
2. Science, 106:350, 1947.
3. J. Bact., 54:70, 1947.

PACKAGING
Bacto Supplement B   6 × 10 ml   0276-60-4
                                    100 ml   0276-72-0

BACTO SUPPLEMENT C

INTENDED USE
Bacto Supplement C is a sterile desiccated yeast concentrate for use in supplementing media for the cultivation of fastidious microorganisms with exacting growth requirements. It is particularly recommended for enriching media for the cultivation of *Haemophilus influenzae* and *Neisseria gonorrhoeae.* Bacto Supplement C can also be used for the preparation of chocolate agar as described by Christensen and Schoenlein.[1]

HISTORY/PRINCIPLES
Mattman[2] used blood agar with Supplement C for culturing and studying hemolytic reactions of fastidious streptococci and pneumococci. More recently Supplement C has been used with Bacto Mueller Hinton Medium in a standardized disk-diffusion susceptibility test for *Haemophilus influenzae* by Jorgensen and Lee.[3]

Bacto Supplement C contains the thermolabile and thermostable growth accessory factors of fresh yeast, including glutamine, coenzyme (V factor), hematin (X factor), cocarboxylase and other growth factors. The rehydrated Bacto Supplement C is used in 1% concentration in solid or liquid media. It is added aseptically after the medium has been sterilized in the autoclave and cooled below 50°C.

METHOD OF PREPARATION
1. Rehydrate by adding 5 ml sterile distilled or deionized water to one vial Bacto Supplement C.
2. Rotate the vial to dissolve the contents.

3. Use this enrichment in media as discussed in detail under Bacto GC Medium Base same as for Bacto Supplement A or B, as desired or as below for sensitivity testing.
4. Prepare and sterilize 2 × 50 ml Bacto Mueller Hinton Medium per label instructions for each organism to be tested. Allow to cool to 50 – 55°C.
5. Add 2.5 ml Bacto Penase to one flask (for each organism) and mix well.
6. Add 2.5 ml rehydrated Bacto Supplement C to each flask and mix well.
7. Pour entire contents of each flask into separate 150 × 15 mm Petri plates and allow to solidify.
8. Prepare inocula in 0.85% sterile saline. Adjust density to Bacto McFarland Standard 0.5.
9. Spread inoculum onto surface of plates (one with and one without Bacto Penase) using a sterile cotton tipped applicator. Inoculum must evenly cover the entire surface. Smear in three directions.
10. Allow to dry slightly then apply Dispens-O-Disc Ampicillin 10 μg and Chloramphenicol 30 μg. Be sure to place sufficiently far apart to allow for large zones.
11. Incubate at 35 – 37°C for 18 – 24 hours.
12. Record zone diameters. The organism is considered susceptible to ampicillin if the zone is greater than 20 mm or resistant if less than 20. The organism is considered susceptible to chloramphenicol if the zone is greater than 30 mm or resistant if less than 30. There should be no zones on plates with Bacto Penase.

STORAGE
Bacto Supplement C	2 – 8°C
Rehydrated	2 – 8°C

QUALITY CONTROL
Identity Specifications

Appearance:	tan button or powder
Rehydrated appearance:	light to medium amber, clear to slightly opalescent

Typical Cultural Response
See Bacto GC Medium Base

REFERENCES
1. Ann. Meeting CA Publ. Hlth. Assn., 1947.
2. J. Lab. & Clin. Med., 42:485, 1953.
3. A.J.C.P., 67:3, March 1977.

PACKAGING
Bacto Supplement C	6 × 5 ml	0527-57-6
Bacto Mueller Hinton Medium	available in various forms	
Bacto Penase	6 × 20 ml	0345-63-8
Bacto McFarland Barium Sulfate Standards	1 set	0691-32-6
Dispens-O-Disc Magazine		
Ampicillin 10 μg	10 mag	6363-89-2
Chloramphenicol 30 μg	10 mag	6133-89-1

BACTO SUPPLEMENT VX

INTENDED USE
Bacto Supplement VX is recommended as an enrichment for the cultivation of *Neisseria gonorrhoeae, Haemophilus influenzae* and other fastidious organisms.

PRINCIPLES

Bacto Supplement VX is a sterile lyophilized concentrate of essential growth factors for supplementing media for culturing microorganisms with exacting growth requirements. It is used for the preparation of chocolate agar from Bacto GC Medium Base and Bacto Hemoglobin for isolation and cultivation of *N. gonorrhoeae.* It is also recommended for enriching Bacto Brain Heart Infusion Agar and Bacto Brain Heart Infusion. More recently supplement VX has been recommended for use in modified Thayer Martin medium. Bacto Supplement VX is used in a 1% concentration in either solid or liquid media. Bacto GC Medium Base has sufficient buffering capacity to offset the very low pH of the small amount of supplement VX added, however, the pH of some media may have to be readjusted (with 1% NaOH) after the addition of Bacto Supplement VX.

METHOD OF PREPARATION

1. Aseptically add the contents of one vial Bacto Reconstituting Fluid VX to one vial Bacto Supplement VX.
2. Rotate the vial to dissolve the contents completely.
3. Use this enrichment in media as discussed in detail under Bacto GC Medium Base.
4. One liter Bacto Thayer Martin Medium Modified can be prepared by the following technique:
   a. Suspend 36 g Bacto GC Medium Base plus 1.5 g Bacto Dextrose in 500 ml distilled or deionized water. Heat to boiling to dissolve completely. Sterilize per label instructions.
   b. Prepare 500 ml Bacto Hemoglobin per label instructions or use Bacto Hemoglobin Solution 2%.
   c. Allow medium to cool (or warm if using Bacto Hemoglobin Solution 2%) to 50 – 55°C and aseptically add 1% (10 ml) rehydrated Bacto Supplement VX to the GC medium base. Mix well.
   d. Combine the supplemented base with the hemoglobin, mix well. Aseptically add 10 ml rehydrated Bacto Antimicrobic Vial CNVT. Mix thoroughly.
   e. Dispense into sterile Petri dishes, tubes or as desired. If using tubes, allow to solidify with a long slant to maximize surface area.
   f. Inoculate and incubate as described under Bacto GC Medium Base. The antimicrobics added above will suppress growth of contaminating flora while allowing most pathogenic *Neisseria* or *Haemophilus* to grow.

STORAGE

Bacto Supplement VX	2 – 8°C
Rehydrated	2 – 8°C

QUALITY CONTROL

Identity Specifications

Appearance:	pink button or powder
Rehydrated appearance:	pink, clear solution

REFERENCE

1. Memorandum, Recommendation to use the same medium, MTM in both plates and bottles for the GC Culture Screening Program, CDC, Atlanta, GA, 1975.

PACKAGING

Bacto Supplement VX	6 × 2 ml	3354-54-1
	6 × 10 ml	3354-60-3
	100 ml	3354-72-9

Bacto GC Medium Base	1 lb (454 g)	0289-01-1
	1/4 lb (114 g)	0289-02-0
Bacto Hemoglobin	1 lb (454 g)	0136-01-6
	1/4 lb (114 g)	0136-02-5
Bacto Hemoglobin Solution 2%	100 ml	3248-72-9
	6 × 100 ml	3248-73-8
Bacto Antimicrobic Vial CNVT	6 × 10 ml	3198-60-3
	100 ml	3198-72-9
Bacto Dextrose	500 g	0155-17-4

SYPHILIS SEROLOGY

The exacting technique of serological diagnostic methods requires carefully prepared and standardized materials. Difco offers antigens and ancillary reagents in the following test classifications.

Nontreponemal Antigen
Bacto USR Antigen —
　Agglutination Test
Bacto VDRL Antigen —
　Flocculation Test

Treponemal Antigen
Bacto FTA Antigen —
　Immunofluorescence Test
Bacto HATTS Kit —
　Hemagglutination Test

The reagents are prepared in strict accordance with the published methods. Before distribution each lot of antigen as well as applicable ancillary reagents are tested by CDC for acceptability. For details on Bacto HATTS Kit refer to Technical Information 3400, available from Difco upon request.

VDRL TEST REAGENTS
BACTO VDRL ANTIGEN
BACTO VDRL TEST CONTROL SERUM SET

Bacto VDRL Antigen is used for the VDRL slide (qualitative, quantitative) test for syphilis as described in the *Manual of Tests for Syphilis 1969*.

VDRL antigen is a nontreponemal antigen.[1,2] As discussed in *Syphilis a Synopsis*, reactive nontreponemal tests confirm the diagnosis in the presence of early or late lesion syphilis, offer a diagnostic clue in latent subclinical syphilis, are an effective tool for detecting cases in epidemiologic investigations and are superior to the treponemal test for following the response to therapy.[3]

Nontreponemal antigen tests are not entirely specific for syphilis nor do they have satisfactory sensitivity in all stages of syphilis. Whenever the results of a nontreponemal antigen test disagree with the clinical impression, a treponemal antigen test such as the FTA-ABS[2,3] or HATTS should be performed.

TEST SUMMARY
VDRL antigen is a cardiolipin-lecithin antigen. Spinal fluid or inactivated serum and VDRL antigen emulsion, prepared from VDRL antigen and VDRL buffered saline, are mixed with the aid of a rotating machine for a prescribed period of time. The cardiolipin-lecithin-coated cholesterol particles flocculate in the presence of an antibody-like substance (reagin) that is present in sera from syphilitic persons and, occasionally, in sera of persons with other acute and chronic diseases.[2,4]

Qualitative testing is performed first on all specimens. Each serum that produces a reactive, weakly reactive or "rough" nonreactive result in the qualitative VDRL slide test is retested, quantitatively, to an end point titer.

REAGENTS
Bacto VDRL Antigen contains 0.03% cardiolipin and 0.9% cholesterol dissolved in absolute alcohol with sufficient purified lecithin (0.21 ± 0.01%) to produce standard reactivity. It is prepared according to the directions given by Harris, Rosenberg and Riedel.[5] Cardiolipin and lecithin are prepared according to directions given by Pangborn.[6,7,8] Cardiolipin is prepared under license from the New York State Department of Health.

Bacto VDRL Buffered Saline (pH 6.0 ± 0.1) for the preparation of the VDRL Antigen emulsion contains:
  0.037 g Disodium Phosphate, Anhydrous, (A.C.S.)
  0.170 g Monopotassium Phosphate, (A.C.S.)
  10.0 g Sodium Chloride (A.C.S.)
  1.0 liter distilled water

Bacto VDRL Buffered Saline is a slight modification of the formulation given by Harris, Rosenberg and Riedel.[5] The formaldehyde has been omitted.

Bacto Nontreponemal Antigen Reactive Serum is a human serum standardized to provide a reactive reading when tested according to the USR or VDRL test procedures.

Bacto VDRL Weakly Reactive Serum is a human serum standardized to provide a weakly reactive reading when tested according to the VDRL test procedure.

Bacto Nontreponemal Antigen Nonreactive Serum is a human serum standardized to provide a nonreactive reading when tested according to the USR or VDRL test procedure.

Precautions
Each human serum used in preparing the control serum contained in Bacto VDRL Test Control Serum Set has been screened for the presence of hepatitis B surface antigen. Only sera exhibiting a negative reaction are used in preparing these reagents. As an additional precaution, the user should employ proper and aseptic techniques in handling all human sera.

Before performing this test the user should be familiar with the contents of sections "Introduction" and "Appendix" of the *Manual of Tests for Syphilis 1969.*

Rehydration and Storage
On receipt, store the package of Bacto VDRL Antigen and Bacto VDRL Buffered Saline at room temperature. When ready to use the Bacto VDRL Antigen, pour the entire contents of the ampule into the Bacto VDRL Antigen Storage Vial. Cap the bottle tightly and store it in the dark at 15 – 30°C. Withdraw antigen as required. After the bottle of Bacto VDRL Buffered Saline is opened, it should be stored at 2 – 8°C between the periodic withdrawals of saline for preparing the emulsion. Discard the buffered saline if mold or turbidity appears.

The three control sera contained in Bacto VDRL Test Control Serum Set are stored at 2 – 8°C. Rehydrate the sera with 3 ml sterile distilled or deionized water. Transfer rehydrated sera that are in excess of the first day's use into an aliquant vial and maintain at −20°C for up to one month. Do not thaw and refreeze. The expiry date of the aliquant is not to exceed the expiry date of the original container.

SPECIMEN COLLECTION AND PREPARATION
1. Obtain blood specimen and allow to clot. Separate serum from the clot by centrifugation (1500 – 2000 rpm for 5 minutes).
2. Inactivate serum specimen at 56°C for 30 minutes.
3. Centrifuge sera containing particulate debris prior to testing.
4. Reinactivate, at 56°C for 10 minutes, all sera to be tested subsequent to 4 hours after original inactivation.
5. Cool all sera to room temperature (23 – 29°C) prior to testing.

PROCEDURE
Procedures as described here are essentially those described in the *Manual of Tests for Syphilis*, US Department of Health, Education and Welfare.[1,2]

Materials Provided
Bacto VDRL Antigen with Bacto VDRL Buffered Saline
Bacto VDRL Test Control Serum Set

Materials Required But Not Provided
Rotating machine, adjustable to 180 rpm circumscribing a circle 3/4″ in diameter on a horizontal plane
Hypodermic needles, without bevels — for serum test, 18 gauge; for spinal fluid test, 21 or 22 gauge
2 × 3″ glass slide with 12 paraffin or ceramic rings approximately 14 mm in diameter for serum test
Agglutination slide approximately 2 × 3″ with concavities measuring 16 mm in diameter and 1.75 mm in depth for spinal fluid
Syringe, Luer-type, 1 or 2 ml
VDRL antigen emulsion bottles, 30 ml round, glass-stoppered, narrow-mouth, approximately 35 mm in diameter with flat inner bottom surface
Serological pipettes — 5.0 ml, 1.0 ml and 0.2 ml
Water bath 56°C
Light microscope with 10X ocular and 10X objective
Distilled water
0.9% saline solution
10% saline solution
Absolute alcohol and acetone
Interval timer
pH meter
Aliquant vials

Preparation of Specific Glassware
Syringes with needles and emulsion bottles. Wash by hand in the following manner:
1. Prerinse with tap water.
2. Soak and wash thoroughly in a glassware detergent solution.
3. Rinse with tap water 6 – 8 times.
4. Rinse with unused distilled or demineralized water.
5. Rinse with absolute alcohol.
6. Rinse with acetone.
7. Air dry until acetone odor is competely eliminated.
8. Remove needles from syringes for storage.

Ceramic-ringed slides:
1. Prerinse with tap water.
2. Wash with a glassware detergent solution.
3. Rinse with tap water 3 – 4 times.
4. Rinse with unused distilled or deionized water.
5. Wipe dry with clean lint-free cloth. If cleaned slides do not allow serum to spread evenly within inner surface of circle, treat the slides as follows.
6. Scrub the slides with a nonscratching cleanser.
7. Dry and polish with a clean lint-free cloth.

NOTE: Avoid prolonged soaking of ceramic-ringed slides in detergent solution since the ceramic rings will become brittle and flake off.

Preparation of VDRL Antigen Emulsion
1. Pipette 0.4 ml of VDRL buffered saline directly to the bottom of a VDRL antigen emulsion bottle.
2. Gently tilt bottle so that VDRL buffered saline will cover entire inner surface of bottom.
3. From the lower half of a 1.0 ml pipette graduated to the tip, add 0.5 ml of VDRL antigen as follows:
   a. Keep the pipette in the upper 1/3 area of the bottle. Do not let it touch the saline.
   b. While rotating the bottle around a circle approximately 2″ in diameter, allow the antigen to be added dropwise to the VDRL buffered saline.
   c. Allow approximately 6 seconds to add antigen, then blow out remaining antigen from pipette.
4. Continue rotation of bottle for 10 seconds.
5. Add 4.1 ml of VDRL buffered saline to bottle, allowing the VDRL buffered saline to flow down the side of the bottle.
6. Place the glass stopper in the bottle and shake the bottle from bottom to top and back approximately 30 times in 10 seconds.
7. Let VDRL antigen emulsion stand without further disturbance for 10 minutes.
8. VDRL antigen emulsion is ready to use. Swirl gently prior to using. Antigen emulsion may be used during one day.

NOTE: Check pH of VDRL buffered saline prior to preparing VDRL antigen emulsion. VDRL buffered saline outside the range of pH 6.0 ± 0.1 should be discarded.

Allow VDRL antigen and VDRL buffered saline to reach 23 – 29°C before preparing VDRL antigen emulsion.

Use only emulsion bottles with flat inner bottom surfaces that allow the initial VDRL buffered saline to evenly cover the inner bottom surface of the bottle. If the VDRL buffered saline beads or does not spread evenly to cover the bottom of bottle, rewash bottle.

For reproducible results the VDRL antigen emulsion must be checked daily for proper reactivity by testing with Bacto VDRL Test Control Serum Set. Only those VDRL emulsions producing the established reactivity pattern of the control serum should be used.

NOTE: Mix VDRL emulsion prior to using in test by gentle swirling of emulsion bottle. Do not mix emulsion by forcing back and forth through the syringe and needle.

Calibration of Delivery Needles without Bevels
1. With 18 gauge needle attached to 1 – 2 ml Luer-type syringe, fill syringe with antigen emulsion.
2. Hold the syringe with needle in a vertical position and expel the emulsion dropwise into the emulsion bottle.
3. Count the number of drops delivered per ml of emulsion. The needle should deliver 60 drops ± 2 drops per ml of antigen emulsion.
4. Adjust drops per ml by either narrowing the open end of the needle to allow more drops per ml to be delivered or opening the end of the needle to allow fewer drops per ml to be formed.

Standardization of VDRL Antigen Emulsion
1. Rehydrate each control serum with 3 ml distilled water and rotate in an end-over-end motion to dissolve completely.
2. Inactivate serum dilutions at 56°C for 30 minutes.
3. Cool to room temperature.
4. Perform qualitative VDRL slide test on all serum dilutions with VDRL antigen emulsion. Test serum dilutions within 1 hour after inactivation.
5. Read test immediately as follows:

Reactive Serum	Medium or large clumps
Weakly Reactive Serum	Small clumps
Nonreactive Serum	No clumping with complete dispersion of emulsion particles

Select a VDRL antigen emulsion which produces the specified reactivity pattern with each control serum.

Perform the VDRL slide test procedures within the temperature range 23 – 29°C since lower temperatures decrease test reactivity and higher temperatures increase test reactivity.

VDRL SLIDE QUALITATIVE TEST ON SERUM
1. With 1.0 ml pipette, mix inactivated serum several times, then using same pipette add 0.05 ml of serum into 1 ring of the paraffin-ringed or ceramic-ringed glass slide. Do not use glass slides with concavities, wells, or glass rings for the VDRL qualitative or quantitative test procedures on serum. Do not use glass slides with ceramic rings that have begun to flake off or that permit spillage of the antigen serum mixture when slides are rotated at the prescribed speed.
2. Spread the serum with a circular motion of the pipette tip so that the serum covers the entire inner surface of the paraffin or ceramic ring. Use only clean slides that allow serum to evenly cover entire surface within ceramic or paraffin ring.
3. While holding the syringe with an 18 gauge needle in a vertical position, carefully add 1 drop of antigen (1/60 ml) to the serum. Do not allow needle to touch the serum.
4. Place the paraffin- or ceramic-ringed slides containing the serum and antigen on a rotator and rotate for 4 minutes at 180 rpm circumscribing a 3/4″ diameter circle.
5. Prevent evaporation of slides during rotation in a dry climate by covering slides with a moisture chamber (box lid containing a moistened blotter).

6. Observe ringed slides microscopically with a 10X ocular and a 10X objective immediately after rotation.
7. Read test as follows:

Medium and large clumps	R	Reactive
Small clumps	W	Weakly Reactive
No clumping or very slight roughness	N	Nonreactive

NOTE: Quantitatively retest each serum producing weakly reactive or rough nonreactive results when tested qualitatively since prozone reactions are encountered occasionally.

VDRL SLIDE QUANTITATIVE TEST ON SERUM

1. Prepare a two-fold serial dilution of inactivated serum in 0.9% saline as follows: 1:2, 1:4, 1:8, 1:16, 1:32.
2. Test each serum dilution using qualitative test procedure.
3. Report results in terms of the greatest serum dilution that produces a reactive (not weakly reactive) result in accordance with the following examples:

Undiluted Serum (1:1)	1:2	1:4	1:8	1:16	1:32	Report
R	W	N	N	N	N	Reactive, undiluted only, or 1 dilution
R	R	W	N	N	N	Reactive, 1:2 dilution, or 2 dilutions
R	R	R	W	N	N	Reactive, 1:4 dilution, or 4 dilutions
W	W	R	R	W	N	Reactive, 1:8 dilution, or 8 dilutions
N (rough)	W	R	R	R	N	Reactive, 1:16 dilution, or 16 dilutions
W	N	N	N	N	N	Weakly Reactive, undiluted only or 0 dilutions

If reactive results are obtained through dilution 1:32, prepare further two-fold serial dilutions in 0.9% saline (1:64, 1:128 and 1:256) and retest using qualitative test procedure.

VDRL SLIDE TESTS ON SPINAL FLUID

VDRL antigen and VDRL buffered saline may be used for spinal fluid testing in accordance with the VDRL slide tests on spinal fluids as stated on pages 40, 41 and 42 in the *Manual of Tests for Syphilis 1969.*[2]

LIMITATIONS OF THE PROCEDURE

VDRL antigen slide tests are nonspecific tests for syphilis. All sera and spinal fluids expressing a reactive or weakly reactive VDRL test in the absence of clinical evidence of syphilis should be subjected to other syphilis serology tests such as FTA-ABS[2,3] or HATTS.

Reactive or weakly reactive VDRL test results should not be considered as conclusive evidence that the patient is syphilitic. Conversely a nonreactive VDRL test by itself does not rule out the diagnosis of syphilis.

SPECIFIC PERFORMANCE CHARACTERISTICS

In various stages of untreated syphilis, the likelihood of obtaining a reactive VDRL is as follows:

Stage of Untreated Syphilis	% Reactive
Primary	76
Secondary	100
Early Latent	95
Late Latent	72
Late (Tertiary)	70

REFERENCES

1. U.S. Department of Health, Education, and Welfare, National Communicable Disease Center, Venereal Disease Program, 1964, Manual of tests for syphilis, U.S. Government Printing Office, Washington, D.C.
2. U.S. Department of Health, Education, and Welfare, National Communicable Disease Center, Venereal Disease Program, 1969, Manual of tests for syphilis, U.S. Government Printing Office, Washington, D.C.
3. U.S. Department of Health, Education, and Welfare, National Communicable Disease Center, Venereal Disease Program, 1968, Syphilis a synopsis, U.S. Government Printing Office, Washington, D.C.
4. Kraus, S. J., J. R. Haserick and M. A. Lantz, 1970, Fluorescent treponemal antibody-absorption test reactions in lupus erythematosis: atypical beading pattern and probable false-positive reactions, N. Engl. J. Med., 282:1287 – 1290.
5. Harris, A., A. A. Rosenberg and L. M. Riedel, 1946, A microflocculation test for syphilis using cardiolipin antigen, J. Ven. Dis. Infor., 27:169 – 174.
6. Pangborn, M. C., 1941, A new serologically active phospholipid from beef heart, Proc. Soc. Exp. Biol. and Med., 48:484 – 486.
7. Pangborn, M. C., 1944, Acid cardiolipin and an improved method for the preparation of cadiolipin from beef heart, J. Biol. Chem., 153:343 – 348.
8. Pangborn, M. C., 1945, A simplified preparation of cardiolipin, with a note on purification of lecithin for serologic use, J. Biol. Chem., 161:71 – 82.

PACKAGING

Bacto VDRL Antigen with	5 ml	0388-56-5
Bacto VDRL Buffered Saline	10 × 0.5 ml	0388-49-5
Bacto VDRL Test Control Serum Set	1 set	3520-32-7

USR TEST REAGENTS

BACTO USR ANTIGEN
BACTO USR TEST CONTROL SERUM SET

Bacto USR Antigen is used for the unheated serum reagin test on serum as described in the *Manual of Tests for Syphilis 1969* and its supplement, January 1982. The antigen is applicable for qualitative as well as quantitative determinations.

Test results with Bacto USR Antigen provide only presumptive evidence for syphilis. Further testing may be necessary as discussed under test limitations.

Essentially, the USR test consists of the following steps. Unheated serum and Bacto USR Antigen are placed on a slide and are mixed with the aid of a rotating machine for a prescribed period of time. The cardiolipin-lecithin-coated cholesterol particles flocculate in the presence of an antibody-like substance, reagin, that is present in syphilitic serum and occasionally in serum of persons with other acute or chronic diseases.

REAGENTS
Bacto USR Antigen contains the cardiolipin-lecithin-coated particles of a VDRL antigen emulsion in USR resuspending solution. VDRL antigen emulsion is prepared from VDRL

antigen and VDRL buffered saline as described under VDRL Test Reagents. USR resuspending solution contains the following:
    1.25 ml EDTA (0.1M) (ethylenedinitrilo) tetraacetic acid disodium salt
    2.5 ml choline chloride (40%)
    5.0 ml Phosphate (0.02M), Merthiolate® (0.2%)
    1.25 ml distilled water

Bacto Nontreponemal Antigen Reactive Serum is a human serum standardized to provide a reactive reading when tested according to the USR or VDRL test procedures.

Bacto USR Weakly Reactive Serum is a human serum standardized to provide a weakly
reactive reading when tested according to the USR test procedure.

Bacto Nontreponemal Antigen Nonreactive Serum is a human serum standardized to
provide a nonreactive reading when tested according to the USR or VDRL test procedure.

Precautions
Each human serum used in preparing the control serum contained in Bacto USR Test
Control Serum Set has been screened for the presence of hepatitis B surface antigen.
Only sera exhibiting a negative reaction are used in preparing these reagents. As an
additional precaution, the user should employ proper and aseptic techniques in han-
dling all human sera.

Before performing this test the user should be familiar with the contents of sections
"Introduction" and "Appendix" of the *Manual of Tests for Syphilis* 1969.

Rehydration and Storage
Bacto USR Antigen is stored at 2 – 8°C. If the original 3 ml quantity is in excess of
that needed for one testing period, transfer the remainder from the first day's use to
one or more aliquant vials and store at 2 – 8°C. The expiry date of the aliquant is not
to exceed the expiry date of the original container. However, do not use the antigen
suspension if at any time the expected results are not obtained with Bacto USR Test
Control Serum Set. Allow cold antigen to remain at room temperature for at least 30
minutes. Unused, warm antigen should be discarded each day.

The three control sera contained in Bacto USR Test Control Serum Set are stored at
2 – 8°C. Rehydrate the sera with 3 ml sterile distilled or deionized water. Transfer the
rehydrated sera that are in excess of the first day's use into aliquant vials and maintain
at −20°C for up to one month. Do not thaw and refreeze. The expiry date of the aliquant is not to exceed the expiry date of the original container.

SPECIMEN COLLECTION AND PREPARATION
Obtain blood specimen and allow to clot. Separate serum from the clot by centrifugation
(1500 – 2000 rpm for 5 minutes).

Sera to be tested in the USR test for syphilis, if refrigerated, should be allowed to
remain at room temperature for at least 30 minutes after removing them from the refrigerator before being used in the tests.

PROCEDURE
Materials Provided:
Bacto USR Antigen
Bacto USR Test Control Serum Set

Materials Not Provided:
Rotating machine, adjustable to 180 rpm circumscribing a circle 3/4" in diameter on a
    horizontal plane
Hypodermic needles without bevels, 18 gauge adjusted to drop 1/45 ml per drop
2 × 3" glass slide with 12 paraffin or ceramic rings approximately 14 mm in diameter
Serological pipettes, 5.0 ml and 1.0 ml
Syringe, Luer-type, 1 – 2 ml
Light microscope with 10X ocular and 10X objective
0.9% saline
Absolute alcohol and acetone
Distilled water
Interval timer
Aliquant vials

Preparation of Specific Glassware
Syringes with needles, wash by hand in the following manner:
1. Prerinse with tap water.
2. Soak and wash thoroughly in a glassware detergent solution.
3. Rinse with tap water 6 – 8 times.
4. Rinse with unused distilled or demineralized water.
5. Rinse with absolute alcohol.
6. Rinse with acetone.
7. Air dry until acetone odor is completely eliminated.
8. Remove needles from syringes for storage.

Ceramic-ringed slides:
1. Prerinse with tap water.
2. Wash with a glassware detergent solution.
3. Rinse with tap water 3 – 4 times.
4. Rinse with unused distilled or demineralized water.
5. Wipe dry with clean lint-free cloth. If slides do not allow serum to spread evenly
    within inner surface of circle, treat the slides as follows.
6. Scrub the slides with a nonscratching cleanser.
7. Dry and polish with a clean lint-free cloth.

NOTE: Avoid prolonged soaking of ceramic-ringed slides in detergent solution since
the ceramic rings will become brittle and flake off. All such slides should be discarded.

Calibration of Delivery Needles without Bevels
1. With 18 gauge needle attached to 1 – 2 ml Luer-type syringe, fill syringe with an-
    tigen suspension.
2. Hold the syringe with needle in a vertical position and expel the suspension drop-
    wise into the suspension bottle.
3. Count the number of drops delivered per ml of USR antigen. The needle should
    deliver 45 drops ± 1.
4. Adjust drops per ml by either narrowing the open end of the needle to allow more
    drops per ml to be delivered or opening the end of the needle to allow fewer drops
    per ml to be formed.

Preliminary Testing of Antigen

Sufficient antigen for a day's use should be removed from the refrigerated stock Bacto USR Antigen vial, transferred to a clean container and allowed to remain at room temperature for at least 30 minutes before use. Unused warm antigen should be discarded each day.

Bacto USR Antigen has been standardized for use in the USR test. It is essential that it be rechecked daily using reactive, weakly reactive and nonreactive control sera prepared from Bacto USR Test Control Serum Set to insure proper reactivity. The antigen should be reactive, weakly reactive and nonreactive with the specific serum controls.

USR SLIDE QUALITATIVE TEST ON UNHEATED SERUM

The USR test is performed as described in the USPHS *Manual of Tests for Syphilis 1969*. The procedure is essentially as follows:
1. Pipette 0.05 ml of the unheated serum directly into a ring of a 14 mm diameter paraffin-ringed glass slide.
2. Add 1 drop of Bacto USR Antigen. It is recommended that an 18 gauge pointless needle delivering 45 drops per ml be used.
3. Rotate the slide for 4 minutes at 180 rpm.
4. Read the tests immediately upon completion of rotation using a microscope with 100X magnification.
5. Report as follows:

Medium and large clumps	R	Reactive
Small clumps	W	Weakly Reactive
No clumping or a slight roughness	N	Nonreactive

USR SLIDE QUANTITATIVE TEST ON UNHEATED SERUM

Test quantitatively, **to an endpoint titer,** each unheated serum that produces a reactive, weakly reactive or "rough" nonreactive result in the USR slide qualitative test. The dilutions to be tested are: Undiluted (1:1), 1:2, 1:4, etc. Three serum quantitative tests through 1:8 dilutions may be performed on one slide.
1. Measure 0.05 ml of 0.9% saline onto the second, third and fourth paraffin or ceramic rings in a row on the slide. Do not spread saline. Saline may be delivered from an 18 gauge needle without point (0.025 ml per drop — use 2 drops) or a large needle or calibrated dropper that delivers 0.05 ml in a single drop. These should be checked daily for accuracy of delivery.
2. Using a safety pipettor device with disposable tip that delivers 0.05 ml or 50 µl, measure 0.05 ml of serum to the first and second rings. Avoid contamination of the instrument with serum.
3. Use the same pipettor and tip to prepare serial 2-fold dilutions by drawing the serum/saline mixture up and down in the tip five or six times. Avoid excess bubbles. Use a clean tip for each serum tested.
4. Mix the serum and saline in ring #2 (1:2 dil); transfer 0.05 ml of the 1:2 dilution to ring #3 and mix (1:4 dil); transfer 0.05 ml of the 1:4 dilution to ring #4 and mix (1:8 dil); discard 0.05 ml. Additional serial dilutions may be set up for strongly reactive sera. (If the 0.05 ml of serum dilution has not spread within the entire area of a paraffin or glass ring, spread with the pipettor tip before proceeding to the next ring.)
5. Repeat steps 2, 3, 4 and 5 of USR qualitative test on unheated serum.
6. Report titer in terms of the greatest serum dilution that produces a reactive (not weakly reactive) result, in accordance with the following examples:

Undiluted Serum (1:1)	1:2	1:4	1:8	1:16	1:32	Report
R	W	N	N	N	N	Reactive, undiluted only, or 1 dilution
R	R	W	N	N	N	Reactive, 1:2 dilution, or 2 dilutions
R	R	R	W	N	N	Reactive, 1:4 dilution, or 4 dilutions
W	W	R	R	W	N	Reactive, 1:8 dilution, or 8 dilutions
N (rough)	W	R	R	R	N	Reactive, 1:16 dilution, or 16 dilutions
W	N	N	N	N	N	Weakly reactive, undiluted only or 0 dilutions

CHECK LIST

Points to observe for best test results:

1. Adjust the speed of rotating machine to meet the prescribed speed then periodically check during testing.
2. Check the accuracy of the needles prior to performing the USR test procedure. Adjust all needles that do not meet the required number of drops of emulsion per ml.
3. Always hold syringe and needle in a vertical position when testing accuracy and applying USR antigen to serum. Exercise care to obtain drops of uniform size.
4. Mix USR antigen prior to using in test by gently inverting bottle end-over-end several times. Do not mix antigen by forcing back and forth through the syringe and needle.
5. Perform USR antigen slide test procedures within the temperature range 23 – 29°C.
6. Use only clean slides that allow serum to evenly cover entire surface within ceramic or paraffin ring. Serum may be spread with the aid of a pipette tip following deliverance of the serum.
7. Prevent evaporation of slides during rotation in a dry climate by covering slides with a moisture chamber (box lid containing a moistened blotter).

TEST LIMITATIONS

USR antigen slide tests are nonspecific tests for syphilis. Each serum reported as positive in the USR qualitative test should be subjected to further serologic study, including quantitation and, if indicated, to other syphilis serology tests such as FTA - ABS or HATTS. Positive tests obtained by using USR antigen should not be considered as conclusive evidence that the patient is syphilitic. Conversely, a nonreactive USR test, by itself, does not rule out the diagnosis of syphilis.

REFERENCES

1. Portnoy, J., and W. Garson, 1960, New and improved antigen suspension for rapid reagin tests for syphilis, Pub. Health Rep., 75:985 – 988.
2. Portnoy, J., H. N. Bossak, V. H. Falcone and A. Harris, Rapid reagin test with unheated serum and new improved antigen suspension, Pub. Health Rep., 76:933 – 935.
3. Pettit, D. E., S. A. Larsen, V. Pope, M. W. Perryman and M. R. Adams, The unheated serum reagin (USR) test as a quantitative test for syphilis, Submitted for publication.

PACKAGING

Bacto USR Antigen	6 × 3 ml (810 tests)	2405-46-3
Bacto USR Test Control Serum Set	1 set	3516-32-3

FLUORESCENT TREPONEMAL ANTIBODY ABSORBED (FTA-ABS) TEST REAGENTS

BACTO FTA ANTIGEN
BACTO FTA SERUM REACTIVE
BACTO FTA SORBENT CONTROL
BACTO FTA SORBENT
BACTO FA HUMAN GLOBULIN ANTIGLOBULIN (RABBIT)
BACTO FA MOUNTING FLUID
BACTO FA BUFFER, DRIED
TWEEN 80®

The fluorescent treponemal antibody absorbed test performed with the above listed reagents is an indirect FA procedure for detecting human antibody against *Treponema pallidum*, the causative agent of syphilis. Since its introduction in 1957, by Deacon, Falcone and Harris,[1] the FTA Test has increased in popularity. However, certain difficulties with respect to sensitivity versus specificity have been encountered.

In its original form the test utilized a 1:5 dilution of patient's serum. At this dilution, a large number of false-positive reactions were encountered. There seemed to be a cross reaction of the treponemal antigen with antibodies to group antigens which are common to all treponemes. The titer of serums containing the nonspecific group antibodies ranged from 1:5 – 1:100.

In 1960, Deacon, Freeman and Harris[2] introduced a modification of the FTA test. This procedure was called the FTA-200 test and utilized a 1:200 dilution of patient's serum. By increasing the dilution of the serum, the nonspecific antibodies were diluted beyond their titer and could no longer interfere with the test. However, by testing a high dilution of serum, the sensitivity of the test is decreased in that the low titered primary cases of syphilis are not detected. This has been substantiated by Eng, Nielsen, Wercide and Wilkinson.[3,4]

It has been the trend since the introduction of the FTA-200 test to return to the 1:5 dilution and eliminate the nonspecific antibodies in the serum. Deacon and Hunter[5] demonstrated that by using appropriate absorption procedures, the reactivity of non-specific antibodies can be eliminated or blocked. This absorptive procedure has developed into an improved test, the FTA-ABS test.[6]

Treponemal antigen tests such as the FTA-ABS test are used primarily as confirmatory tests in diagnostic problem cases; e.g., patients in which the clinical, historical or epidemiological evidence of syphilis is equivocal. The FTA-ABS test is more sensitive than the VDRL in primary syphilis and late latent and tertiary syphilis.

The persistant reactivity of the FTA-ABS test to a treated case of syphilis, sometimes for life, minimizes its use for following the response to therapy as well as making it unreliable for detecting new untreated cases in epidemiologic investigations.

REAGENTS
Bacto FTA Antigen is a desiccated, standardized killed suspension of the Nichols strain of *T. pallidum*. Unreconstituted antigen is stored at 2 – 8°C. To rehydrate, add 1 ml

sterile distilled water per vial and rotate to effect complete solution. The antigen suspension should be thoroughly mixed with a disposable pipette and rubber bulb. Drawing the suspension into and expelling from the pipette 8 – 10 times will break the treponemal clumps and ensure an even distribution of treponemes.

Treponemal antigen smears may be prepared in the following manner:
1. On clean grease-free slides inscribe 2 circles 1 cm in diameter with a diamond point stylus. Remove any free glass particles by wiping slides with a piece of clean gauze.
2. Within the inscribed circles, smear 1 loopful of Bacto FTA Antigen using a standard 2 mm, 26 gauge, platinum wire loop. Allow to dry at least 15 minutes at room temperature.
3. Immerse the dry slides into acetone for 10 minutes. This fixes the treponemes to the slide so they will not wash off during the rinsing procedures later in the test. Not more than 50 slides should be fixed with 200 ml of acetone.
4. It is recommended that a freshly rehydrated bottle of Bacto FTA Antigen be used in its entirety for making treponemal antigen smears on the same day. These slides may then be stored at −20°C or lower after acetone fixation. The frozen slides are usable as long as satisfactory results are obtained with the controls. Thawed slides should not be refrozen.

Bacto FTA Serum Reactive is a standardized, desiccated human serum for use as a positive control in the FTA-ABS test. To rehydrate this serum, add 5 ml distilled water as indicated by the label. When diluted 1:5 in FA buffer or sorbent, a 4+ fluorescence will be obtained. To obtain a 1+ fluorescence, this serum must be diluted further in Bacto FA Buffer. This dilution will vary with each lot and is provided in the instructions received with each vial. The undiluted serum may be dispensed in 0.4 ml aliquants and frozen until needed. Do not refreeze thawed aliquants. This serum should be heated at 56°C for 30 minutes before using.

Bacto FTA Sorbent is a standardized extract of the nonpathogenic Reiters treponeme containing group treponemal antigens. It is designed to be used in the FTA-ABS test for blocking the action of nonspecific antibodies in the serum. The sorbent comes ready to use and is supplied in vials of 5 ml and boxes of 6 × 5 ml. It is recommended that Bacto FTA Sorbent be stored either at 2 – 8°C or in the frozen state at −20°C after it is received in the laboratory. A 1:5 dilution of the serum is made by placing 0.05 ml of heated serum into 0.2 ml sorbent and mixing thoroughly.

Bacto FTA Sorbent Control is a desiccated, standardized human serum demonstrating at least a 2+ fluorescence in the FTA-ABS test at a dilution in Bacto FA Buffer of 1:5. At a dilution of 1:5 in Bacto FTA Sorbent no reactivity will be observed. This serum is supplied in 6 × 0.5 ml vials and is good for 1 week after rehydration if refrigerated at 2 – 8°C or longer if stored in the frozen state.

Tween 80®
A 2% solution of Tween 80 may be prepared as follows:
1. Heat the bottle of Tween 80 and 98 ml Bacto FA Buffer in a 56°C water bath until both solutions are at 56°C.
2. Using a 2 ml pipette graduated to the tip, add 2 ml Tween 80 to the Bacto FA Buffer and rinse the pipette thoroughly.
3. The 2% solution should have a pH of 7.0 – 7.2. The solution keeps well in the refrigerator at 2 – 8°C, but should be discarded if the pH changes or if a precipitate is noticed.

Bacto FA Human Globulin Antiglobulin (Rabbit) is a fluorescein conjugated antihuman globulin designed for use in the FTA - ABS test. It is used as an indicator system to detect the presence of human syphilitic antibodies on the treponemal antigen. It is a desiccated product and is supplied in 5 ml and 1 ml quantities. In rehydration, use the amount of good quality distilled water indicated by the package size. The rehydrated conjugate may be stored in the refrigerator at 2 – 8°C or in 0.5 ml aliquants at −20 to −40°C.

Each vial is supplied with a dilution titer. When a new lot of conjugate is used, the titer should be confirmed with the fluorescent assembly in use at the time. This may be done in the following manner:

1. Prepare serial dilutions of the conjugate with the 2% Tween 80 solution. The serial dilutions should include the titer specified on the vial.
2. Each conjugate dilution is tested with the Bacto FTA Serum Reactive diluted 1:5 in Bacto FA Buffer and a nonspecific staining control (conjugate + antigen).
3. A known conjugate is set up with the Bacto FTA Serum Reactive (4+), a minimally reactive (1+) serum, and a nonspecific staining control to control reagents and test conditions.
4. The dilution to be used in the test is taken as 1 doubling dilution lower than the 4+ end point (the highest dilution of conjugate giving a 4+ fluorescence with the 4+ control serum).

PREPARATION OF THE SPECIMEN
Test and control sera should be heated at 56°C for 30 minutes before being tested. Previously heated serum should be reheated for 10 minutes on the day of testing.

PROCEDURE
Procedure steps are taken essentially from Pub. Hlth. Serv., Pub. No. 411, 1969.[8]

Materials Provided:

Bacto FTA Antigen
Bacto FTA Serum Reactive
Bacto FTA Sorbent Control
Bacto FTA Sorbent

Tween 80®
Bacto FA Human Globulin
  Antiglobulin (Rabbit)
Bacto FA Mounting Fluid
Bacto FA Buffer, Dried

Materials Required but not Provided:
Diamond point stylus (not needed if pre-etched slides are used)
Plain slides or 1 × 3 inch frosted and approximately 1 mm thick or slides with circles
  already inscribed
Cover slips, No. 1, 22 mm square
2 mm, 26 gauge, platinum wire loop
Acetone
Distilled water
Water bath at 56°C
Interval timer
Staining dish with removable slide carriers
Moisture chamber
0.2 ml serological pipette
5 ml serological pipette
1 ml serological pipette
Bibulous paper
Applicator sticks
Fluorescent microscope assembly

1. Identify previously prepared slides to correspond to the sera and control sera to be tested.
2. Before using, slides taken from the freezer must be completely thawed and thoroughly dry.
3. Prepare the proper dilutions for the 4+ control, the 1+ control, the nonspecific serum controls and for the unknown sera to be tested. The 1:5 dilutions may be prepared by adding 0.05 ml serum (using a 0.2 ml pipette graduated to the tip and measuring from the bottom) to 0.2 ml sorbent or Bacto FA Buffer, whichever is required. Mix the dilution no less than 8 times. All unknown sera are diluted in FTA sorbent only.
4. Do not dilute the control and unknown sera more than 30 minutes before they are to be placed on the antigen smears.

Controls
   a. Reactive 4+ Control
      Bacto FTA Serum Reactive diluted 1:5 in Bacto FA Buffer and in Bacto FTA Sorbent. A 4+ fluorescence should be observed with both dilutions.
   b. Minimally Reactive (1+) Control
      A dilution of Bacto FTA Serum Reactive to give a 1+ fluorescence. This control is used as a reading standard to indicate the minimal degree of fluorescence that can be reported as reactive.
   c. Nonspecific Serum Control
      1. Antigen smear treated with 0.03 ml of Bacto FA Buffer.
      2. Antigen smear treated with 0.03 ml of Bacto FTA Sorbent.

   Test runs in which the proper control results were not obtained are considered unsatisfactory and should not be reported.
5. Cover each identified antigen smear with the corresponding serum dilution and the proper nonspecific staining controls with Bacto FA Buffer or Bacto FTA Sorbent. Make certain the entire smear is covered. The shaft of an applicator stick held parallel to the surface of the smear is an effective tool for this purpose.
6. Place the slides into a moist chamber to prevent evaporation and incubate at 35 – 37°C for 30 minutes.
7. Place the slides in a slide carrier and rinse them in running Bacto FA Buffer for 5 seconds and then soak in Bacto FA Buffer for 5 minutes. Agitate the slides by dipping them in and out of the buffer 10 times. Repeat the soaking and agitation process for another 5 minutes in fresh Bacto FA Buffer.
8. Rinse the slides in running distilled water for 5 seconds and blot dry with bibulous paper.
9. Dilute the conjugate to its working titer using FA buffer containing 2% Tween 80.
10. Cover each smear with approximately 0.03 ml diluted conjugate. Make certain the entire smear is covered.
11. Repeat steps 6, 7 and 8.
12. Mount slides immediately using a small drop of Bacto FA Mounting Fluid and a cover slip. Be careful not to entrap air bubbles in the mounting fluid between the slide and cover slip.
13. Examine slides microscopically as soon as possible using mercury arc illumination and a high power dry objective. A combination of BG 12 exciting filter, not greater than 3 mm thickness, and OG 1 barrier filters or their equivalents have been found to be satisfactory. Slides should be stored in a darkened room and read within 4 hours.
14. Nonreactive smears should be checked using illumination from a tungsten light source in order to verify the presence of treponemes.

RESULTS

Using the 1+ control serum as a reading standard, record the intensity of fluorescence and report as follows:

Reading	Intensity of Fluorescence	Report
2+ to 4+	Moderate to strong	Reactive
1+	Equivalent to 1+ control	Reactive
less than 1+	Weak but definite	Borderline
−	None or vaguely visible	Nonreactive

Retest all specimens with a fluorescence of 1+ or less. When a specimen initially read as 1+ gives a 1+ or better fluorescence, it is reported as reactive. All other results are reported as borderline. It is not necessary to retest nonreactive specimens.

LIMITATIONS OF THE PROCEDURE

There is a degree of difficulty in determining whether a reading is weak or vaguely visible. The ability to make this distinction is critical since a nonreactive (vaguely visible to none) serum is not retested.

The test cannot be used to follow the response to therapy nor can it be relied upon to detect new, untreated cases in epidemiological investigations.

SPECIFIC PERFORMANCE CHARACTERISTICS

In various stages of untreated syphilis the likelihood of obtaining a reactive FTA - ABS is as follows[7]:

Stage of Untreated Syphilis	% Reactive
Primary	86
Secondary	100
Early Latent	99
Late Latent	96
Late (Tertiary)	97

REFERENCES

1. Proc. Soc. Exp. Biol. and Med., 96:477, 1957.
2. ibid., 103:827, 1960.
3. Wld. Hlth. Org. Document WHO/VDT 314 WHO/VDT/RES/29 March 1963.
4. Proc. Royal Soc. Med., 56:478,1963.
5. Proc. Soc. Exp. Biol. and Med., 110:352, 1962.
6. Publ. Hlth. RPTS., 79:410, 1964.
7. Publ. Hlth. Service Publ. No. 1660, 1968.
8. Publ. Hlth. Service Publ. No. 411, 1969.

PACKAGING

Bacto FTA Antigen	1 ml	2344-50-0
Bacto FTA Serum Reactive	5 ml	2439-56-0
Bacto FTA Sorbent	5 ml	3259-56-5
	6 × 5 ml	3259-57-4
Bacto FTA Sorbent Control	6 × 0.5 ml	3266-49-6
Bacto FA Human Globulin Antiglobulin (Rabbit)	1 ml	2449-50-4
	5 ml	2449-56-8
Bacto FA Mounting Fluid pH 7.2	6 × 5 ml	2329-57-2
Bacto FA Buffer, Dried	6 × 10 g	2314-33-8
	100 g	2314-15-0
Tween 80®	100 g	3118-15-6

BACTO TAT BROTH BASE

INTENDED USE
Bacto TAT Broth Base with the addition of Tween® 20 is recommended for sterility testing of highly viscous or gelatinous substances such as salves or ointments. It is especially adapted to the sterility testing of cosmetics.

HISTORY
Bacto TAT (Tryptone-Azolectin-Tween) Broth Base is prepared according to the formula obtained from the United States Food and Drug Administration.[1]

FORMULA
BACTO TAT BROTH BASE
DEHYDRATED
Ingredients per liter

Bacto Tryptone . . . . 20 g
Azolectin . . . . 5 g

Final pH 7.2 ± 0.2 at 25°C.

One pound will make 18 liters of medium.

METHOD OF PREPARATION
1. Suspend 25 g in 960 ml of distilled or deionized water and add 40 ml of Tween® 20.
2. Heat to 50 – 60°C and let stand for 15 – 30 minutes with occasional agitation to dissolve the medium completely.
3. Sterilize in the autoclave for 15 minutes at 15 lbs pressure (121°C).

PROCEDURE
One gram of test material can be added to 40 ml of complete medium and agitated to obtain an even suspension.

See the FDA *Bacteriological Analytical Manual*[1] for additional test procedures.

STORAGE
Bacto TAT Broth Base   Below 30°C
Prepared medium   15 – 30°C

QUALITY CONTROL
Identity Specifications

Dehydrated powder: off-white, homogeneous, free-flowing
Reaction of 2.5% solution
  (with 4% Tween® 20):   pH 7.2 ± 0.2 at 25°C
Prepared medium: light amber, clear to very slightly opalescent, may have a very slight precipitate.

Typical Cultural Response in Bacto TAT Broth
After 18 – 48 Hours at 35°C

Organism	Growth
Bacillus subtilis ATCC® 6633	good to excellent
Candida albicans ATCC® 26790	fair to excellent
Pseudomonas aeruginosa ATCC® 27853	good to excellent
Salmonella typhi ATCC® 6539	good to excellent
Staphylococcus aureus ATCC® 25923	good to excellent

REFERENCE

1. FDA Bacteriological Analytical Manual, 5th Ed., Washington, D.C.: Association of Official Analytical Chemists, pp. XXIII-6-XXIII-7, April 1978.

PACKAGING

Bacto TAT Broth Base 1 lb (454 g) 0984-01-9

BACTO TB HYDROLYSIS REAGENT

INTENDED USE

Bacto TB Hydrolysis Reagent is used for differentiating mycobacteria based on their ability to hydrolyze Polysorbate 80 (Tween® 80).

HISTORY

Wayne, Doubek and Russell[1] differentiated various species and subgroups of acid-fast bacilli using the Tween 80 hydrolysis test. Kubica and Dye[2] used the test to differentiate clinically significant from clinically insignificant mycobacteria. In 1970, Runyan, Kubica, Morse, Smith and Wayne[3] defined a positive reaction as one that occurs in less than five days, a doubtful reaction as one that occurs in five to ten days, and a negative reaction as one occurring after ten days.

In 1973, Kubica[4] recognized the greater reliability of a 10-day Tween hydrolysis test for separation of clinically insignificant from clinically significant members of both the scotochromogenic and nonphotochromogenic mycobacteria. These observations were later confirmed in an international cooperative study by Wayne et al.[5] and the 10-day reading is now regarded as a better cutoff for the Tween hydrolysis test.

PRINCIPLES

In this test, polysorbate 80 acts as a lipid, binding the neutral red indicator, causing the solution to be amber colored. If the mycobacterial lipase splits the polysorbate 80, it can no longer complex with the neutral red indicator which then exhibits its normal red color at pH 7. The intensity of the red depends upon how much polysorbate 80 is actually split.

Some mycobacteria possess a lipase capable of splitting polysorbate 80 into oleic acid and polyoxyethylated sorbitol, modifying the solution from yellow to pink. The differential criterion of this test is based on the relative time necessary for a particular species or subgroup to hydrolyze the compound. The majority of M. kansasii[2] strains and clinically insignificant species are positive in five days[4,5] or less, while clinically significant species may be negative even after three weeks. Mycobacterium tuberculosis generally yields a positive reaction in 10 – 20 days.

REAGENT

Bacto TB Hydrolysis Reagent is a sterile, phosphate-buffered solution of Tween® 80 and neutral red prepared according to the formulation of Kilburn.[4]

Each vial contains sufficient reagent to perform 50 tests.

STORAGE

Bacto TB Hydrolysis Reagent 2 – 8°C

EXPIRATION DATE

The reagent is stable through the expiry date on the label when stored as directed.

A change of color of the reagent from its original amber to red indicates deterioration of the reagent. Such reagent should be discarded.

SPECIMEN PREPARATION
The test organism is a pure, actively metabolizing 3 – 4 week old culture of *Mycobacterium*. Carefully exclude underlying culture medium.

PROCEDURE
1. Prepare and sterilize 13 × 75 mm screw cap test tubes containing 1 ml distilled or deionized water. Cool to room temperature.
2. Add 2 drops Bacto TB Hydrolysis Reagent, taking care to not touch the glass dropper, which could contaminate the reagent and cause aberrant test results.
3. Transfer 1 loopful of test culture to the tube.
4. Incubate at 37°C for up to 21 days.
5. Visually inspect the color of the reagent daily in a strong light against a white background.
6. Record results.
7. Upon completion of the test, sterilize all tubes in the autoclave for 20 minutes at 15 lbs pressure (121°C).

QUALITY CONTROL
Use known positive (*M. kansasii*) and negative (uninoculated tube and *M. scrofulaceum* or *M. avium* complex) controls in parallel with the test culture to ascertain the validity of test results.

RESULTS
A positive reaction is indicated by a change of the color of the solution from amber to pink or red.
Report reactions as follows:

Color change in:	Report:
10 days or less	Positive for Tween hydrolysis
11 days or more	Negative for Tween hydrolysis

The positive control should exhibit a color change in less than 5 days.

The negative control should exhibit no color change during 21 days.

REFERENCES
1. Amer. Rev. Resp. Dis., 80:585 – 597, 1964.
2. Lab. Methods for Clin. and Public Health, Mycobacteriology, UPHS Pub. No. 1547, 44, 1967.
3. Manual of Clin. Micro., ASM, 128, 1970.
4. Amer. Rev. Resp. Dis. 107: 9 – 21, 1973.
5. Inter. J. Syst. Bacteriol. 24: 412 – 419, 1974.
6. Kilburn, J.O., Personal Communication.

PACKAGING
Bacto TB Hydrolysis Reagent   5 ml   3192-56-5

TB NIACIN TEST REAGENTS

BACTO TB NIACIN TEST STRIPS
BACTO TB NIACIN TEST CONTROL

INTENDED USE
Bacto TB Niacin Test Strips are used for differentiating species of mycobacteria on the basis of niacin production.

Bacto TB Niacin Test Control is used as a positive control in the niacin test.

HISTORY

Konno[1] devised the standard niacin test which was later modified by Runyon.[2] Kilburn and Kubica[3] demonstrated that the Runyon test could be modified to permit use of a single paper strip impregnated with test reagents. Morse, Blair, Weiser and Sproat[4] also modified the Runyon procedure, using a slight variation of the single strip method. Studies carried out by Kilburn and Kubica and Morse et al. indicated that the strip test was as sensitive and specific as the classical test.

PRINCIPLES

Mycobacterium tuberculosis, growing on egg-base media or agar media containing 0.25% L-asparagine,[5] releases large quantities of niacin into the medium. *M. simae* and *M. chelonei* also produce niacin but are not common clinical isolates and are, when encountered, readily differentiated from *M. tuberculosis.*

Niacin thus released by a culture can be detected because it reacts with cyanogen bromide and a primary or secondary amine (ie., aniline) to produce a yellow color. While this test is reliable, laboratories are hesitant to use it because of the toxicity and relative instability of cyanogen bromide.

Paper strips impregnated with acidified potassium thiocyanate and chloramine T will release cyanogen chloride which reacts with PAS to produce a yellow color in the presence of niacin. No color is produced if niacin is not present.

A positive niacin test on a non-chromogenic mycobacteria culture isolated from a clinical specimen is strongly indicative of *M. tuberculosis.* The intensity of the reaction depends on the amount of culture growth present. Niacin-negative strains of *M. tuberculosis* are rare, while niacin-positive strains of mycobacteria other than *M. tuberculosis* are also uncommon.

FORMULA

Bacto TB Niacin Test Strips are absorbent paper strips impregnated with potassium thiocyanate, chloramine T, citric acid and sodium aminosalicylate according to the method of Kilburn and Kubica with some modification.

Bacto TB Niacin Test Control is a disk impregnated with nicotinomide that will yield a yellow solution equivalent to approximately 5 μg niacin when used according to test procedure.

STORAGE

Bacto TB Niacin Test Strips	2 – 8°C
Bacto TB Niacin Test Control	2 – 8°C

SPECIMEN PREPARATION

1. Prepare an extract of a pure, well-developed 3 – 4 week old *Mycobacterium* culture using one of the following methods:
   A. Overlay the culture with sterile distilled water or sterile 0.85% NaCl in the following amount: 1.5 ml per tube or Petri dish quadrant, or 6 ml per Petri dish. If culture growth is confluent, stab the pipette through the medium or lift part of the culture to permit the water to contact and extract niacin from the medium.
   B. ROOM TEMPERATURE EXTRACTION
      Position the tube or plate so the liquid covers the entire culture and keep it in this position 20 – 30 minutes to extract the niacin, or
      AUTOCLAVE EXTRACTION (FROM EGG MEDIA, ONLY)
      Place tube in autoclave so liquid covers entire culture. Autoclave for 15 minutes

at 15 lbs pressure (121°C). Permit tube to cool, then stand upright for 10 minutes to allow particles of medium to settle.
C. Using a sterile capillary pipette and bulb, transfer 0.6 ml of the extract to a 13 × 75 mm test tube.
2. Prepare the positive control as follows:
A. Place 0.6 ml distilled water in a 13 × 75 mm test tube.
B. Add one Bacto TB Niacin Test Control disk, stopper tube and shake gently 3 times during 15 minutes. Retain tube at room temperature.
3. Prepare the negative control by placing 0.6 ml distilled water or 0.85% NaCl (obtained from the same source as was used for extracting niacin) into a 13 × 75 mm test tube.

PROCEDURE
1. Using sterile forceps, aseptically place one Bacto TB Niacin Test Strip, arrow downward, in each tube: test culture, negative control and positive control. Stopper tubes immediately.
2. Shake tubes gently but do not tilt. Repeat gentle shaking after 5 – 10 minutes.
3. Compare the color of each extract after 12 – 15 minutes, again at 20 – 25 minutes but no later than 30 minutes. Record results.
4. Neutralize test reagents at completion of test by adding 1 – 2 drops of 10% NaOH to each tube. Restopper tubes and sterilize in the autoclave.

QUALITY CONTROL
Use the positive and negative controls described above or known positive and negative control cultures in parallel with the test culture to ascertain the validity of test results.

RESULTS
A yellow color in the extract of the test culture constitutes a positive test for niacin production. The positive control should be yellow. The negative control should remain colorless.

REFERENCES
1. Science, 124:985, 1956.
2. Amer. Rev. Tuberc., 79:663 – 665, 1959.
3. Trans. 26th VA Armed Forces Pulmonary Dis., Res. Conf., pg. 24; January 1967.
4. Mycobacteriol. Lab. Methods Fitzsimons Gen. Hosp. Rpt. No. 317; pg. 6, 1968.
5. Trans. 26th VA Armed Forces Pulmonary Dis., Res. Conf., pg. 26; January 1967.

PACKAGING
Bacto TB Niacin Test Strips	25 strips	3182-30-8
Bacto TB Niacin Test Control	50 disks	3184-90-3

TB STAINS

BACTO TB STAIN SET K
BACTO TB STAIN SET ZN
BACTO TB FLUORESCENT STAIN SET M
BACTO TB FLUORESCENT STAIN SET T

TEST SUMMARY
One of the earliest methods devised for the detection of the tubercle bacilli and which presently is a standard procedure in every laboratory is the microscopic staining tech-

nique. The unparalleled acid-fastness of mycobacteria renders this method valuable for early presumptive diagnosis of mycobacterial infections and gives information as to the number of acid-fast bacilli. Sputa, direct smear or concentrate, spinal or other body fluids, including catheterized urine, and homogenized tissue material or impression slides are satisfactory specimens for smears.

The use of acid-fast staining techniques – conventional or fluorescent procedures – do not differentiate pathogenic from nonpathogenic mycobacteria nor can they be used for species identification, i.e., *M. tuberculosis*. Conversely, the absence of acid-fast bacilli in the smear does not rule out the presence of mycobacteria. An improperly prepared specimen may adversely influence the outcome of the test. Runyon, Kubica and Wayne[1] report that a carefully selected portion of a specimen may yield a better preparation than a concentrate. Further, insufficient destaining may also interfere with detection of acid-fast bacilli. Finally, the low sensitivity of the microscopic technique may account for negative smears from sputa or other specimens. Therefore, findings obtained from acid-fast microscopic staining procedures must be regarded solely as presumptive.

As pointed out by Willis and Cummings,[2] the use of phenol as a mordant, of fuchsin as the dye, and a mineral acid as a decolorizing agent according to the method introduced by Ehrlich[3] and later modified by Ziehl[4] and Neelsen,[5] is the classical method. This method as outlined by Willis and Cummings[2] is but one of the many satisfactory staining procedures described and used successfully in many laboratories.

Fluorescent microscopy for the detection of mycobacteria offers numerous advantages over the classical methods in the speed and simplicity of staining as well as in the case of examining the slide as pointed out by McClure.[6] The use of low power and/ or high dry objectives yielding magnifications of approximately 100X – 650X permits examination of a larger slide area with less effort. The reliability and superiority of this method has been described by Weiser, Sproat, Hakes and Morse.[7] Previously, Truant, Brett and Thomas[8] published on the theory and use of the auramine and rhodamine dyes for the fluorochrome staining of clinical material for mycobacteria. Morse et al[9] stressed the greater simplicity of their procedure since an ordinary blue light with suitable filters may be used in place of the ultra violet light. They described in detail the minimum equipment required to yield optimum results and pointed out that this method eliminates the need of oil immersion and a dark room for examining slides. Joseph and Houk[10] and Joseph, Vaichulis and Houk[11] have reported that most of the Runyon Group IV strains do not fluoresce after staining with auramine-rhodamine dyes. Since these are rapid growers they can be easily differentiated from the other mycobacteria. Somlo, Black and Somlo[12] found the auramine-rhodamine method to be more sensitive than the Ziehl-Neelsen technique for the demonstration of acid-fast bacilli in sputum smears.

Morse et al[9] used a microscope fitted with an incandescent bulb, a KG 1 heat filter, a 3 – 4 mm thick BG 12 excitation filter, an ordinary substage condenser and a No. 51 bright field or GG barrier filter. The above, or comparable, combination of light source, filter, lamp housing and condenser used with most simple laboratory microscopes will give a light green background with brilliant light-yellow stained mycobacteria.

Difco offers a choice of 2 stain sets that utilize the principals of the classical staining methods and 2 stain sets that utilize the newer fluorescent procedures. Whichever method is chosen, satisfactory results can be expected if the recommended procedure is adhered to.

REAGENTS

Bacto TB Stain Set K contains the reagents used to stain smears prepared from specimens suspected of containing mycobacteria according to the procedure described in the *Handbook of Tuberculosis Laboratory Methods*.[13] The use of the Bacto TB Carbolfuchsin KF prepared according to the Kinyoun formula obviates the need for heating the carbolfuchsin.

Bacto TB Stain Set K contains:

> Bacto TB Carbolfuchsin KF ..... 250 ml
> Bacto TB Decolorizer .... 250 ml
> Bacto TB Brilliant Green K ...... 250 ml

Bacto TB Stain Set ZN contains the reagents used to stain smears prepared from specimens suspected of containing mycobacteria according to the procedure described by Kubica and Dye[14] and in the *Handbook of Tuberculosis Laboratory Methods*.[13]

Bacto TB Stain Set ZN contains:

> Bacto TB Carbolfuchsin ZN ..... 250 ml
> Bacto TB Decolorizer .... 250 ml
> Bacto TB Methylene Blue .... 250 ml

Bacto TB Fluorescent Stain Set M contains the reagents used to stain smears prepared from specimens suspected of containing mycobacteria as described by Morse, Blair, Weiser and Sproat.[9]

Bacto TB Fluorescent Stain Set M contains:

> Bacto TB Auramine M .... 250 ml
> Bacto TB Decolorizer TM .... 250 ml
> Bacto TB Potassium
> Permanganate .... 250 ml

It is characteristic for Bacto TB Auramine M to show some separation from solution. This in no way interferes with the effectiveness of the stain. Shake the bottle of Bacto TB Auramine M thoroughly before using.

Bacto TB Fluorescent Stain Set T contains the reagents used to stain smears prepared from specimens suspected of containing mycobacteria as described by Truant, Brett and Thomas.[8]

Bacto TB Fluorescent Stain Set T contains:

> Bacto TB Auramine-
> Rhodamine T .... 250 ml
> Bacto TB Decolorizer TM .... 250 ml
> Bacto TB Potassium
> Permanganate .... 250 ml

It is characteristic of Bacto TB Auramine-Rhodamine T to show some separation from solution. This in no way interferes with the effectiveness of the stain. Shake the bottle of Bacto TB Auramine-Rhodamine T thoroughly before using.

STORAGE

The stain sets should be stored in the dark at 15 – 30°C. TB Decolorizer should be kept away from any open flame. These reagents are stable to the expiry date on the label when stored properly.

SPECIMEN PREPARATION

The specimen to be examined may be smeared directly on a clean unused slide or first concentrated by any of the recommended techniques and the sediment examined for acid-fast bacilli. The smear is allowed to dry in the air and the slide passed through a low flame 2 or 3 times to fix the smear, avoiding the use of excessive heat. Thick smears are to be avoided.

Clean slides are important in detecting tubercle bacilli since bacilli from previous smears may persist. New slides are recommended, however, slides may be cleansed satisfactorily in an acid dichromate solution prepared by dissolving 75 g potassium dichromate in 200 ml distilled water with heat. Allow solution to cool and carefully add very slowly 400 ml of concentrated sulfuric acid with gentle and frequent stirring.

PROCEDURE

Materials Provided:
Bacto TB Stain Set K
Bacto TB Stain Set ZN
Bacto TB Fluorescent Stain Set M
Bacto TB Fluorescent Stain Set T

Materials Required but not Provided:
Microscope slides
Staining rack
Fluorescent microscope assembly
Ordinary microscope with oil immersion lens

TB STAIN SET K

1. Place slides on a staining rack and flood with Bacto TB Carbolfuchsin KF for 4 minutes. Do not heat.
2. Wash gently in running water.
3. Decolorize with Bacto TB Decolorizer.
4. Wash gently in running water.
5. Counterstain with Bacto TB Brilliant Green K for 30 seconds.
6. Wash gently in running water and air dry.

An alternate method for staining mycobacteria with Bacto TB Carbolfuchsin KF may be used:
1. Place slides on a staining rack and flood with Bacto TB Carbolfuchsin KF for 4 minutes. Do not heat.
2. Wash gently in running water.
3. Decolorize with Bacto TB Decolorizer.
4. Wash gently in running water.
5. Counterstain for 30 seconds with Bacto TB Methylene Blue.
6. Wash gently with tap water and dry over gentle heat.

TB STAIN SET ZN

1. Place slides on a staining rack and flood with Bacto TB Carbolfuchsin ZN and heat gently to steaming. Allow to steam for 5 minutes.
2. Wash gently with tap water.
3. Decolorize with Bacto TB Decolorizer with 2 changes of reagent for 1 – 2 minutes or until no more red color appears in washing.
4. Wash gently with tap water.
5. Counterstain for 30 seconds with Bacto TB Methylene Blue.
6. Wash gently with tap water and dry over gentle heat.

TB FLUORESCENT STAIN SET M

1. Place slides on a staining rack and flood with Bacto TB Auramine M for 15 minutes.
2. Wash gently in running tap water.
3. Decolorize for 30 – 60 seconds with Bacto TB Decolorizer TM.

4. Wash slides gently in running water.
5. Apply Bacto TB Potassium Permanganate for 2 minutes.
6. Wash gently in running water.
7. Air dry.
8. Examine the slides, using a microscope as described below.

Morse et al, used a microscope fitted with an incandescent bulb, a KG 1 heat filter, a 3 – 4 mm thick BG excitation filter, an ordinary substage condenser, and a No. 51 bright field or GG barrier filter.

TB FLUORESCENT STAIN SET T
1. Place slides on a staining rack.
2. Shake bottle of Bacto TB Auramine-Rhodamine T solution and cover slides. Leave for 20 – 25 minutes at room temperature.
3. Wash gently in running tap water.
4. Decolorize for 2 – 3 minutes with Bacto TB Decolorizer TM.
5. Wash gently in running tap water.
6. Flood smears for 4 – 5 minutes with Bacto TB Potassium Permanganate.
7. Wash slides gently in running tap water.
8. Blot lightly and dry in air or very gently over a flame.
9. Smears are examined under a laboratory microscope fitted with suitable attachment for fluorescent microscopy as described by Truant et al.[1] The following or comparable combination of attachments can be used successfully: a 25X objective, an HBO L2 bulb heat filter, a BG 12 primary filter and OG 1 barrier filter.

CONTROLS
It is recommended that controls be run daily using known acid-fast and nonacid-fast organisms.

Positive organism: *Mycobacterium tuberculosis* H37 Ra or known positive culture
Negative organism: *Bacillus subtilis* ATCC® 6633.
or
Bactrol™ TB Slide

A suggested technique is as follows:
1. Divide a microscope slide into 3 equal segments with a wax marking pencil.
2. In the center segment make a smear of the specimen.
3. On the left-hand segment place a loopful of a suspension of *B. subtilis.*
4. Repeat step 3 using *M. tuberculosis* H37 Ra, desiccated.
5. Fix slides and proceed with the staining procedure.

RESULTS
With Bacto TB Stain Set K the acid-fast bacilli are stained dark pink to red; nonacid fast microorganisms are stained green.

With Bacto TB Stain Set ZN and Bacto TB Stain Set K (alternate method) the acid-fast bacilli are stained dark pink to red; the nonacid-fast microorganisms are stained blue.

With Bacto TB Fluorescent Stain Set M, *Mycobacteria* will have a bright yellow-green fluorescence.

With Bacto TB Fluorescent Stain Set T, *Mycobacteria* will have a reddish-orange fluorescence.

LIMITATIONS OF THE PROCEDURE

A positive staining reaction is presumptive information concerning the presence of *M. tuberculosis* in the specimen. Conversely, a negative finding does not necessarily indicate that the specimen will not be positive for *M. tuberculosis* by culture. For positive identification of *M. tuberculosis* cultural methods must be employed. For complete discussion on media for the primary isolation and identification of *Mycobacteria,* refer to Middlebrook and/or Lowenstein media.

REFERENCES

1. Lennette, Spaulding & Truant, Manual of Clinical Microbiology, 2nd Ed., 1974, ASM.
2. Diagnostic and Experimental Methods in Tuberculosis, 2nd Ed., 78:1952.
3. Deutsche Med. Wchnschr., 8:269, 1882.
4. Deutsche Med. Wchnschr., 8:451, 1882.
5. Cited by Johne, fortsche. d. Med., 3:198, 1885.
6. J. Clin. Path., 6:273, 1957.
7. Am. J. Clin. Path., 46:587, 1966.
8. Henry Ford Hospital. Med. Bull., 10:287, 1962.
9. Mycobact. Lab. Methods, Fitzsimans Gen. Hospital., Rept. No. 17, May, 1968.
10. Am. Rev. Resp. Dis., 98:1044, 1968.
11. *ibid.,* 95:114, 1967.
12. Tech. Bull., Reg. Med. Tech., 39:51, 1969.
13. Vet. Adm. Arm. Forces Coop. Study on the Chemo. of Tuber. Nov. 1962, VA Dept. of Med. & Sur., Wash., D.C.
14. Lab. Methods Clin. and Publ. Hlth. Mycobact., USPHS, 1968.

PACKAGING

Bacto TB Stain Set K	3 × 250 ml	3326-32-3
Contains TB Carbolfuchsin KF, TB Decolorizer, TB Brilliant Green K		
Bacto TB Stain Set ZN	3 × 250 ml	3324-32-5
Contains TB Carbolfuchsin ZN, TB Decolorizer, TB Methylene Blue		
Bacto TB Fluorescent Stain Set M	3 × 250 ml	3323-32-6
Contains TB Auramine M, TB Decolorizer TM, TB Potassium Permanganate		
Bacto TB Fluorescent Stain Set T	3 × 250 ml	3325-32-4
Contains TB Auramine-Rhodamine T, TB Decolorizer TM, TB Potassium Permanganate		
Components available individually:		
Bacto TB Auramine M	6 × 250 ml	3316-76-2
Bacto TB Auramine-Rhodamine T	6 × 250 ml	3317-76-1
Bacto TB Brilliant Green K	6 × 250 ml	3327-76-9
Bacto TB Carbolfuchsin KF	6 × 250 ml	3321-76-5
Bacto TB Carbolfuchsin ZN	6 × 250 ml	3313-76-5
Bacto TB Decolorizer	6 × 250 ml	3318-76-0
Bacto TB Decolorizer TM	6 × 250 ml	3314-76-4
Bacto TB Methylene Blue	6 × 250 ml	3319-76-9
Bacto TB Potassium Permanganate	6 × 250 ml	3315-76-3
Bactrol™ TB Slide	50 slides	3139-26-8

BACTO TCBS AGAR

INTENDED USE

Bacto TCBS Agar (thiosulfate-citrate-bile-sucrose agar) is used for the selective isolation and cultivation of *Vibrio cholerae* and other enteropathogenic vibrios.

HISTORY/PRINCIPLES

Bacto TCBS Agar is prepared according to the formula of Kobayashi, Enomoto, Sakazaki and Kuwabara.[1] Fishbein, Mehlman and Pitcher[2] reported that *Vibrio parahaemolyticus* will grow on this medium as light bluish colonies. Enteric bacteria such as coliforms and *Proteus* which are found to contaminate fecal materials are inhibited. Some strains of *Proteus* and enterococci may grow and form small colorless colonies but they are easily distinguished. Colonies of *Vibrio cholerae* (*V. comma*) and Type El tor grow to a readily recognizable size in 18 – 24 hours and are easily distinguished by their yellow color.

West, Ressek, Brayton and Colwell[3] reported on the usefulness of TCBS agar for recovering the relatively new pathogens, *V. fluvialis* and *V. vulnificus*.

FORMULA

BACTO TCBS AGAR
DEHYDRATED

Ingredients per liter

Bacto Yeast Extract	5 g	Sodium Chloride	10 g
Proteose Peptone No. 3, Difco	10 g	Ferric Citrate	1 g
Sodium Citrate	10 g	Bacto Brom Thymol Blue	0.04 g
Sodium Thiosulfate	10 g	Bacto Thymol Blue	0.04 g
Bacto Oxgall	8 g	Bacto Agar	15 g
Bacto Saccharose	20 g		

Final pH 8.6 ± 0.2 at 25°C.

One pound will make 5.1 liters of medium.

METHOD OF PREPARATION

1. Suspend 89 g in 1 liter distilled or deionized water.
2. Heat to boiling to dissolve completely. DO NOT AUTOCLAVE.
3. Dispense into sterile Petri dishes as desired.

STORAGE

Bacto TCBS Agar	Below 30°C
Prepared medium	2 – 8°C

QUALITY CONTROL

Identity Specifications

Dehydrated powder:	light tan with green tinge, homogeneous, free-flowing
Reaction of 8.9% solution:	pH 8.6 ± 0.2 at 25°C
Prepared medium:	forest green, slightly opalescent without precipitation or scum

Typical Cultural Response on Bacto TCBS Agar
After 18 – 24 Hours at 35°C

Organism	Growth	Color of Colony
Escherichia coli ATCC® 25922	none to poor	translucent
Proteus mirabilis ATCC® 25933	none to poor	small, yellow
Streptococcus faecalis ATCC® 19433	none to poor	small, yellow
Vibrio cholerae	good	yellow
Vibrio fluvialis	good	yellow
Vibrio parahaemolyticus	good	blue
Vibrio vulnificus	fair to good	yellow or translucent

REFERENCES
1. Jap. J. Bact., 18:387, 1963.
2. Applied Microbiology, 20:176, 1970.
3. J. Clin. Micro., 16:1110 – 1116, 1982.

PACKAGING

| Bacto TCBS Agar | 1 lb (454 g) | 0650-01-2 |
| | 1/4 lb (114 g) | 0650-02-1 |

BACTO m TGE BROTH

INTENDED USE

Bacto m TGE Broth is a nonselective nutrient medium for the determination of bacterial counts by the membrane filter method.

HISTORY/PRINCIPLES

Bacto m TGE Broth has the same formulation as Bacto Tryptone Glucose Extract Agar except it contains no agar and the ingredients are employed in twice the concentration as in the solid medium.

FORMULA

BACTO m TGE BROTH
DEHYDRATED

Ingredients per liter

Bacto Beef Extract . . . . 6 g
Bacto Tryptone . . . . 10 g
Bacto Dextrose . . . . 2 g

Final pH 7.0 ± 0.2 at 25°C.

One pound will make 25.2 liters of medium.

METHOD OF PREPARATION

1. Suspend 18 g in 1 liter distilled or deionized water.
2. Sterilize in the autoclave for 15 minutes at 15 lbs pressure (121°C).

STORAGE

| Bacto m TGE Broth | Below 30°C |
| Prepared medium | 15 – 30°C |

QUALITY CONTROL

Identity Specifications

Dehydrated powder:	light tan, homogeneous, free-flowing
Reaction of 1.8% solution:	pH 7.0 ± 0.2 at 25°C
Prepared medium:	medium amber, clear

Typical Cultural Response on Membrane Filters on Pads Soaked with Bacto m TGE Broth After 18 – 24 Hours at 35°C

Organism	Growth
Escherichia coli ATCC® 25922	good to excellent
Escherichia coli ATCC® 13762	good to excellent
Staphylococcus aureus ATCC® 25923	good to excellent

PACKAGING
Bacto m TGE Broth

1 lb (454 g)   0750-01-1
1/4 lb (114 g)   0750-02-0

TPEY MEDIA

BACTO TPEY AGAR BASE
BACTO TPEY ENRICHMENT

INTENDED USE
Bacto TPEY Agar Base, after supplementation with Bacto TPEY Enrichment, is recommended for the detection and enumeration of coagulase-positive staphylococci in foods and other materials.

Bacto TPEY Enrichment is a tellurite, egg yolk emulsion prepared for use with Bacto TPEY Agar Base for the detection and enumeration of coagulase-positive staphylococci.

HISTORY/PRINCIPLES
Bacto TPEY Agar Base is prepared according to the formulation of Crisley.[1,2] The complete medium is prepared by adding aseptically Bacto TPEY Enrichment and Bacto Antimicrobic Vial P (polymyxin B 30,000 units) to the sterile Bacto TPEY Agar Base. This medium permits the isolation and enumeration of coagulase-positive staphylococci, after 24 hours incubation, from a variety of specimens such as food products, air, dust, and soil. Coagulase-negative staphylococci and other organisms are markedly to complete inhibited.

The coagulase-positive staphylococci are distinguished by their formation of jet black or dark gray colonies and the formation of a zone of precipitated egg yolk around the colonies or a clear zone around the colonies and precipitation beneath the colonies or precipitation beneath the colonies only. Generally, other organisms do not produce growth during the 24 hour incubation period with the exception of an occasional coagulase-negative strain which may produce small black pinpoint colonies. These colonies do not produce egg yolk precipitation or clearing around the colonies.

FORMULAE

BACTO TPEY AGAR BASE
DEHYDRATED

Ingredients per liter

Bacto Tryptone	10 g
Bacto Yeast Extract	5 g
Bacto Mannitol	5 g
Sodium Chloride	20 g
Lithium Chloride	2 g
Bacto Agar	18 g

Final pH 7.2 ± 0.2 at 25°C.

One pound will make
7.5 liters of medium.

BACTO TPEY ENRICHMENT

Egg Yolk	30 ml
Sodium Chloride	0.6 g
Potassium Tellurite	0.1 g
Distilled Water	70 ml

Final pH 6.5 ± 0.5.

Use one hundred ml
per 900 ml base.

METHOD OF PREPARATION
1. Suspend 60 g basal medium in 900 ml distilled or deionized water. Heat to boiling to dissolve completely.

2. Dispense into appropriate sized flasks.
3. Sterilize in the autoclave for 15 minutes at 15 lbs pressure (121°C).
4. Allow basal medium to cool to 50 – 55°C in a water bath and add aseptically 100 ml Bacto TPEY Enrichment warmed to room temperature and 10 ml (one vial rehydrated per label instructions) Bacto Antimicrobic Vial P. Alternatively, 100 ml of a 30% egg yolk emulsion plus 10 ml Bacto Chapman Tellurite Solution 1% plus 0.4 ml of filter sterilized 1% polymyxin B solution may be used.
5. Mix thoroughly and pour in 15 – 17 ml amounts in sterile Petri dishes.
6. In making counts, the procedure to be used is described in the U.S. Public Health Service Publication No. 1142, 1964.

In food samples, where large numbers of staphylococci are expected, the plate count method is recommended. In making plate counts, spread 0.1 ml of each dilution of food homogenate, prepared according to the method on page 84 of the above publication, evenly and completely over the agar surface with a sterile glass spreading rod. Spread the 0.1 ml inoculum of each dilution onto plates labeled with the next higher dilution. Use serial dilutions through 10^{-6} or 1:1,000,000. Incubate the plates for 24 hours at 37°C and observe for jet black circular, convex colonies of 1.0 – 1.5 mm diameter. Specifically, look for colonies which show a zone of precipitation around the colony, a clear zone around the colony and precipitation beneath the colony or precipitation beneath the colony only. Count the number of these colonies on plates which contain 30 – 300 colonies. Record the presumptive members of coagulase-positive staphylococci per gram of sample.

The coagulase test is used as an aid to identify coagulase-positive staphylococci. Aseptically touch each of 5 colonies with a wire needle and inoculate onto prepared slants of Bacto Tryptic Soy Agar. Incubate for 24 hours at 35 – 37°C. Pick the remainder of the colony with a wire loop and inoculate 0.2 ml of prepared Bacto Tryptic Soy Broth in an agglutination tube. Add 0.5 ml of Bacto Coagulase Plasma, mix and incubate in a water bath for 4 hours at 35°C. Observe for coagulation of the plasma. Any amount of coagulation is positive. Those colonies which give a negative reaction are further checked by performing a coagulase test from the agar slant cultures.

In food samples where small numbers of staphylococci are expected, the most probable number method of enumeration is recommended. For this procedure see the U.S. Public Health Service Publication No. 1142, page 98, 1964.

STORAGE

Bacto TPEY Agar Base	Below 30°C
Bacto TPEY Enrichment	2 – 8°C
Prepared medium	2 – 8°C

QUALITY CONTROL
Identity Specifications

Dehydrated powder:	light tan, homogeneous, free-flowing
Reaction of 6% solution:	pH 7.2 ± 0.2 at 25°C
Bacto TPEY Enrichment:	canary yellow, opaque solution with resuspendable precipitate
Prepared medium:	light to medium amber, opalescent
Prepared plates:	yellow, opaque

Typical Cultural Response on Bacto TPEY Agar
After 18 – 48 Hours at 35°C

Organism	Growth	Color of Colony
Bacillus subtilis ATCC® 6633	poor to fair	brown
Escherichia coli ATCC® 25922	inhibited	—
Proteus mirabilis ATCC® 25933	poor to fair	brown
Staphylococcus aureus ATCC® 25923	good to excellent	black with halos
Staphylococcus epidermidis ATCC® 12228	poor to fair	small, black, no halo
Streptococcus pyogenes ATCC® 19615	inhibited	—

REFERENCES

1. Personal Communication.
2. Publ. Health Serv. Publ., No. 1142, 1964.

PACKAGING

Bacto TPEY Agar Base	1 lb (454 g)	0556-01-7
Bacto TPEY Enrichment	6 × 100 ml	0354-73-4
Bacto Antimicrobic Vial P	6 × 10 ml	3268-60-8

BACTO TSA BLOOD AGAR BASE

INTENDED USE

Bacto TSA Blood Agar Base is Bacto Tryptic Soy Agar which has been prepared using specially selected raw materials for improved hemolytic reactions of beta-hemolytic streptococci. It is intended for use with blood in the preparation of TSA Blood Agar and is recommended for use in the isolation and cultivation of fastidious microorganisms from specimens where clear and distinct hemolytic reactions are of prime importance.[1,2]

HISTORY

Bacto Tryptic Soy Agar (TSA) is based on the soybean casein digest agar formula as described in The U.S. Pharmacopeia[3] and is widely used as a general purpose medium for the isolation and cultivation of fastidious and nonfastidious microorganisms.[2] This medium has been supplemented with 5 – 10% blood for improved growth of the more fastidious microorganisms and for the determination of hemolytic reactions.[1,2]

Since good growth of fastidious microorganisms and correct, distinct hemolytic reactions are of major concern to the public health and clinical microbiologist, it was considered of utmost importance that blood agar perform optimally in these requirements. For these reasons, Bacto TSA Blood Agar Base was developed at Difco Laboratories with the purpose of improving beta-hemolytic reactions obtained on TSA Blood Agar, particularly with the group A streptococci, while continuing to maintain good growth of fastidious microorganisms.

PRINCIPLES

Bacto TSA Blood Agar Base is a nutritionally rich medium which supports good growth of a wide variety of fastidious and nonfastidious microorganisms. The medium contains two peptones, a pancreatic digest of casein and a papaic digest of soybean meal, agar and sodium chloride.

Supplementation of this medium with 5 – 10% blood provides for improved growth of the more fastidious microorganisms as well as the capability of determining hemolytic reactions. The medium is relatively free of reducing sugars, which have been reported to adversely influence the hemolytic reactions of the beta-hemolytic streptococci.[4] The improved beta-hemolytic reaction obtained on Bacto TSA Blood Agar Base is due to careful selection and piloting of the raw materials by actual performance testing. Only those raw materials are used which are complementary to one another and which do not adversely affect hemolytic reactions, particularly with the streptococci.

FORMULA

BACTO TSA BLOOD AGAR BASE
DEHYDRATED

Ingredients per liter

Pancreatic Digest of Casein	15 g
Papaic Digest of Soybean Meal	5 g
Sodium Chloride	5 g
Agar	15 g

Final pH 7.3 ± 0.2 at 25°C.

Five hundred grams will make 12.5 liters of medium.

METHOD OF PREPARATION

1. Suspend 40 g in 1 liter of distilled or deionized water and heat to boiling with gentle swirling to dissolve completely.
2. Sterilize in the autoclave for 15 minutes at 15 lbs pressure (121°C).
3. Cool to 45 – 50°C and aseptically add 5% sterile defibrinated blood.
4. Mix thoroughly, avoiding the formation of air bubbles and dispense into sterile Petri dishes.

SPECIMEN COLLECTION

Specimens should be collected in sterile containers or with sterile swabs and transported immediately to the laboratory in accordance with recommended guidelines.[5,6]

PROCEDURE

1. Process specimen as appropriate for that specimen and inoculate directly onto the surface of a Bacto TSA Blood Agar plate. Streak for isolation.
2. Incubate plates aerobically, anaerobically or under conditions of increased CO_2 (5 – 10%) in accordance with established laboratory procedures.
3. For further information on processing and inoculation of specimens and incubation conditions consult appropriate references.[1,2,6]
4. Examine plates for growth and hemolytic reactions after 18 – 24 hours and after 48 hours incubation.

LIMITATIONS OF THE PROCEDURE

1. Bacto TSA Blood Agar Base is intended for use with blood in the preparation of TSA Blood Agar. Although certain diagnostic tests may be performed directly on this medium, biochemical and, if indicated, immunological testing using pure cultures are recommended for complete identification. Appropriate references should be consulted for further information.[1,2]
2. Since the nutritional requirements of the organisms vary, some strains of microorganisms may be encountered that fail to grow or grow poorly on this medium.
3. Hemolytic reactions of some strains of group D streptococci have been shown to

be affected by differences in animal blood. These strains are beta-hemolytic on horse, human and rabbit blood agar and alpha-hemolytic on sheep blood agar.[4]

4. Colonies of *Haemophilus haemolyticus* are beta-hemolytic on horse and rabbit blood agar and must be distinguished from colonies of beta-hemolytic streptococci, which they resemble, using other criteria. The use of sheep blood has been suggested to obviate this problem since sheep blood is deficient in pyridine nucleotides and does not support growth of *H. haemolyticus*.[4]

5. Atmosphere of incubation has been shown to influence hemolytic reactions of beta-hemolytic streptococci.[4] For optimal performance, incubation of TSA Blood Agar under increased CO_2 or under anaerobic conditions is recommended.

STORAGE

Bacto TSA Blood Agar Base   Below 30°C
Prepared medium   2 – 8°C

QUALITY CONTROL

Identity Specifications

Dehydrated powder:	light beige, homogeneous, free-flowing
Reaction of 4% solution:	pH 7.3 ± 0.2 at 25°C
Prepared medium:	medium amber, slightly opalescent
Plates prepared with 5% sheep blood:	cherry red, opaque

Typical Cultural Response on Bacto TSA Blood Agar Base w/5% Sheep Blood After 18 – 24 Hours at 35°C with 5 – 10% CO_2

Organism	Growth	Hemolytic Reaction	Colonial Morphology
Escherichia coli ATCC® 25922	good	beta	smooth, gray, slightly convex, moist with an entire edge
Streptococcus faecalis ATCC® 29212	good	nonhemolytic	small, smooth, off-white with an entire edge
Streptococcus pneumoniae ATCC® 6303	good	alpha	small, round, smooth, glistening, dome shaped
Streptococcus pyogenes ATCC® 19615	good	beta	small, white to off-white, smooth with an entire edge

REFERENCES

1. Finegold, S. M., and Martin, W. J. 1982. Diagnostic Microbiology, 6th Ed., C. V. Mosby Company, St. Louis.
2. Lennette, E. H., Balows, A., Hausler, Jr., W. J., and Truant, J. P. (ed.). 1980. Manual of Clinical Microbiology, 3rd Ed. American Society for Microbiology, Washington, D.C.
3. The United States Pharmacopeial Convention, Inc. 1980. The United States Pharmacopeia XX and National Formulary XV. Mack Publishing Company, Easton, PA.
4. Facklam, R. R. 1980. Streptococci and aerococci, p. 88 – 110. In Lennette, E. H., Balows, A., Hausler, Jr., W. J., and Truant, J. P. (ed.). Manual of Clinical Microbiology, 3rd Ed. American Society for Microbiology, Washington, D.C.
5. Isenberg, H. D, Washington II, J. A., Balows, A., and Sonnenwirth, A. C., 1980. Collection, handling, and processing of specimens, p. 52 – 80. In Lennette, E. H., Balows, A., Hausler, Jr., W. J., and Truant, J. P. (ed.). Manual of Clinical Microbiology, 3rd Ed. American Society for Microbiology, Washington, D.C.
6. Isenberg, H. D., Schoenknecht, F. D., and VonGravenitz, A. 1979. Cumitech 9, Collection and processing of bacteriological specimens. Coordinating ed., Rubin, S. J., American Society for Microbiology, Washington, D.C.

PACKAGING

Bacto TSA Blood Agar Base	500 g	0026-17-1
	5 lb (2.27 kg)	0026-05-5
	10 kg	0026-08-2

BACTO TT BROTH BASE, HAJNA

INTENDED USE
Bacto TT Broth Base, Hajna (Tetrathionate Broth Base, Hajna) is a selective medium for the enrichment of members of the *Salmonella* group.

HISTORY/PRINCIPLES
Bacto TT Broth Base conforms to the formulation of Hajna and Damon,[1] being a modification of the enrichment as described by Kauffmann[2] and Knox.[3] Bacto Yeast Extract has been added as an additional nutriment and fermentation of dextrose and mannitol aids in the selective properties of the medium by tetrathionate decomposition. Grampositive organisms are inhibited by brilliant green and sodium desoxycholate.

FORMULA
BACTO TT BROTH BASE, HAJNA
DEHYDRATED
Ingredients per liter

Bacto Yeast Extract	2 g
Bacto Tryptose	18 g
Bacto Dextrose	0.5 g
D-Mannitol, Difco	2.5 g
Sodium Desoxycholate	0.5 g
Sodium Chloride	5 g
Sodium Thiosulfate	38 g
Calcium Carbonate	25 g
Bacto Brilliant Green	0.01 g

Final pH 7.6 ± 0.2 at 25°C.

One pound will make 5 liters of final medium.

METHOD OF PREPARATION
1. Suspend 91.5 g in 1 liter distilled or deionized water. Heat to boiling. DO NOT AUTOCLAVE.
2. Allow to cool to below 50°C and add 40 ml iodine solution.

   Prepare iodine solution by dissolving 8 g of potassium iodine in 20 ml distilled or deionized water. When solution is complete, add and dissolve 5 g of iodine crystals. Bring the volume of the complete solution to 40 ml with distilled or deionized water. Use 40 ml of this iodine solution per liter of Bacto TT Broth Base, Hajna.
3. Keep suspension well mixed while distributing into sterile tubes. Do not reheat after the addition of iodine.
4. Specimens are incubated for 18, 24 or 48 hours at 35°C in Bacto TT Broth Base, Hajna and then streaked on the desired isolation media such as Bacto BCP-D Agar, Bacto SS Agar, Bacto MacConkey Agar or Bacto Bismuth Sulfite Agar.

STORAGE
Bacto TT Broth Base, Hajna   Below 30°C

The complete medium, containing iodine, should be used the day it is prepared; the base medium can be stored indefinitely at 2 – 8°C.

QUALITY CONTROL

Identity Specifications

Dehydrated powder: very light green, homogeneous, free-flowing
Reaction of 9.15% solution
(with iodine): pH 7.6 ± 0.2 at 25°C
Prepared medium: light green, slightly opalescent with heavy white precipitate

Typical Cultural Response on Bacto MacConkey Agar
After 18 – 24 Hours at 35°C in Bacto TT Broth Base, Hajna

Organism	Recovery	Color of Colony on MacConkey Agar
Escherichia coli ATCC® 25922	fair to good	pink
Salmonella arizonae ATCC® 13314	good to excellent	colorless
Salmonella typhimurium ATCC® 14028	good to excellent	colorless
Shigella dysenteriae ATCC® 13313	good to excellent	colorless

REFERENCES

1. Appl. Microbiol., 4:341, 1956.
2. Zentr. Bakt., I Abt., Orig., 113:148, 1930.
3. J. Pathol. Bacteriol., 54:469, 1952.

PACKAGING

Bacto TT Broth Base, Hajna   1 lb (454 g)   0491-01-5

BACTO TTC SOLUTION 1%

INTENDED USE

Bacto TTC Solution 1% is prepared from microbiologically tested 2, 3, 5 triphenyltetrazolium chloride (TTC) for use in all microbiological methods utilizing this compound. It is used as a redox indicator in culture media for differentiating bacteria. It is colorless in the oxidized form and is reduced to insoluble red triphenylformazan by certain actively metabolizing bacteria, i.e., bacterial reducing systems. The formation of the insoluble formazan is an irreversible reaction. Once a microorganism reduces the indicator the red color persists.

HISTORY

Chapman[1] reported that the addition of TTC to tergitol 7 agar permitted the detection and confirmation of E. coli after only 10 hours incubation. Chapman also reported that tergitol 7 agar with added TTC gave a selective medium for the detection and isolation of Candida albicans. Chapman[2] added TTC, tergitol 7 and brom cresol purple to Sabouraud maltose agar for the isolation and identification of C. albicans and other fungi. Pagano, Levin and Trejo[3] incorporated TTC in Pagano Levin base for the detection and isolation of Candida species.

Slanetz and Bartley[4] used TTC in m enterococcus agar for the enumeration of enterococci. On this medium the colonies appear as pink to dark maroon in color. Kenner, Clark and Kabler[5] used TTC with KF Streptococcus agar and broth to detect and enumerate streptococci in surface waters.

PROCEDURE

TTC is generally used in a concentration of 0.01 g per 100 ml of broth or agar medium. This concentration is achieved by adding 1 ml Bacto TTC Solution 1% to 100 ml sterile medium at 50°C.

STORAGE

Bacto TTC Solution 1%   2 – 8°C, in the dark

REFERENCES

1. Am. J. Pub. Hlth., 41:1381, 1951.
2. Tran. N.Y. Acad. Sci. Series II, 14:254, 1952.
3. Antibiotics Annual, p. 137, 1957 – 1958.
4. J. Bact., 74:591, 1957.
5. Appl. Microbiol., 9:15, 1961.

PACKAGING

Bacto TTC Solution 1%   30 ml   3112-67-9

BACTO TELLURITE BLOOD SOLUTION

INTENDED USE

Bacto Tellurite Blood Solution is a mixture of defibrinated blood to which is added potassium tellurite to markedly reduce growth of commensal organisms encountered when isolating *Corynebacterium diphtheriae*. The enrichment is used in combination with Bacto Dextrose Proteose No. 3 Agar.

PROCEDURE

See entry for Bacto Dextrose Proteose No. 3 Agar.

STORAGE

Bacto Tellurite Blood Solution   2 – 8°C

QUALITY CONTROL

Appearance: dark red-brown, opaque

Determine CULTURAL RESPONSE in prepared tellurite medium. See Bacto Dextrose Proteose No. 3 Agar.

PACKAGING

Bacto Tellurite Blood Solution   6 × 25 ml   0139-66-5

BACTO TELLURITE GLYCINE AGAR

INTENDED USE

Bacto Tellurite Glycine Agar is a basal selective medium for the isolation of coagulase-positive staphylococci.

HISTORY/PRINCIPLES

Bacto Tellurite Glycine Agar is prepared according to the formula of Zebovitz, Evans and Niven.[1] The complete medium is prepared by adding Bacto Chapman Tellurite

Solution 1% to the sterile melted Bacto Tellurite Glycine Agar. The medium permits the isolation of coagulase-positive staphylococci after 24 hours incubation from a variety of specimens such as food products, air, dust, soil, fecal specimens and from skin and mucous membranes, while coagulase-negative staphylococci and other organisms are markedly to completely inhibited.

Ludlam[2] described a selective medium for the isolation of staphylococci. This medium was alkaline in reaction, contained mannitol and lithium chloride with potassium tellurite as a selective agent. Chapman[3] compared Ludlam's medium with Bacto Staphylococcus Medium 110, a selective medium for pathogenic staphylococci. He reported complete agreement in 83% of 232 specimens and suggested that, for highest recovery, both media be employed. If a single medium were to be used, Bacto Staphylococcus Medium No. 110 was preferred since some coagulase-positive staphylococci failed to grow on the Ludlam medium and, furthermore, growth from Bacto Staphylococcus Medium No. 110 could be employed for application of confirmatory tests.

Zebovitz, Evans and Niven[1] modified Ludlam's medium by including glycine as a selective agent and having the reaction of the basal medium pH 7.2 instead of pH 9.6, prior to sterilization. They obtained a medium that permitted the development of coagulase-positive staphylococci, while coagulase-negative and other organisms were inhibited. These authors found it necessary to use a surface or streak inoculation for better results. A careful evaluation of the Bacto Tellurite Glycine Medium, using a variety of specimens including those with many organisms other than staphylococci, showed it to be of decided practical value in the isolation of coagulase-positive staphylococci even though a quantitative comparison with known coagulase-positive strains indicated a slight inhibition of some strains and not complete inhibition of all coagulase-negative strains. In contrast, Bacto Staphylococcus Medium 110 and Bacto Mannitol Salt Agar permitted unrestricted development of all staphylococci without differentiation between coagulase-positive and coagulase-negative strains.

Coagulase-positive staphylococci produce black colonies within 24 hours incubation at 35°C. Generally, other organisms produce no growth during this incubation period with the exception of an occasional coagulase-negative strain that may produce small gray colonies, not readily confused with the black coagulase-positive colonies.

FORMULA
BACTO TELLURITE GLYCINE AGAR
DEHYDRATED
Ingredients per liter

Bacto Yeast Extract	6.5 g	D-Mannitol, Difco	5 g
Bacto Soytone	3.5 g	Dipotassium Phosphate	5 g
Bacto Tryptone	10 g	Lithium Chloride	5 g
Glycine	10 g	Bacto Agar	17.5 g

Final pH 7.2 ± 0.2 at 25°C.

One pound will make 7.2 liters of medium.

METHOD OF PREPARATION
1. Suspend 62.5 g in 1 liter distilled or deionized water. Heat to boiling to dissolve completely.
2. Sterilize in the autoclave for 15 minutes at 15 lbs pressure (121°C).
3. Allow to cool to 50 – 55°C and add aseptically 1 ml of Bacto Chapman Tellurite

Solution 1% for each 100 ml sterile medium. Do not heat after tellurite has been added.
4. Mix thoroughly and pour plates as desired. Allow agar surface to dry slightly before inoculating.

STORAGE
Bacto Tellurite Glycine Agar   Below 30°C
Prepared medium   2 – 8°C

QUALITY CONTROL
Identity Specifications
Dehydrated powder: light beige, homogeneous, free-flowing
Reaction of 6.25% solution: pH 7.2 ± 0.2 at 25°C
prepared medium: medium amber, opalescent with precipitate

Typical Cultural Response on Bacto Tellurite Glycine Agar
After 18 – 24 Hours at 35°C

Organism	Growth	Color of Colony
Escherichia coli ATCC® 25922	inhibited	—
Staphylococcus aureus ATCC® 25923	good to excellent	black
Staphylococcus epidermidis ATCC® 12228	none to fair	gray
Salmonella typhimurium ATCC® 14028	inhibited	—

REFERENCES
1. J. Bact., 70:686:1955.
2. Monthly Bull. of the Ministry of Hlth., 8:15:1949.
3. J. Bact., 58:823:1949.

PACKAGING
Bacto Tellurite Glycine Agar	1 lb (454 g)	0617-01-4
Bacto Chapman Tellurite Solution 1%	6 × 1 ml	0299-51-8
	6 × 25 ml	0299-66-1

BACTO TERGITOL 7 AGAR H

INTENDED USE
Bacto Tergitol 7 Agar H is used as a selective differential medium for isolating enteric bacteria from urine.

HISTORY/PRINCIPLES
Bacto Tergitol 7 Agar H, Hinds' modification of Bacto Tergitol 7 Agar[1] contains an indicator for detecting H_2S producing bacteria and lactose for detecting lactose fermentation. H_2S producing organisms appear as black colonies or colonies with black centers. Lactose fermentors produce yellow colonies surrounded by yellow zones. Lactose nonfermenting organisms usually produce colonies surrounded by blue zones.

FORMULA

BACTO TERGITOL 7 AGAR H
DEHYDRATED
Ingredients per liter

Proteose Peptone No. 3, Difco	5 g	Sodium Thiosulfate	0.5 g
Bacto Yeast Extract	3 g	Bacto Agar	15 g
Bacto Lactose	10 g	Brom Thymol Blue	0.025 g
Ferric Ammonium Citrate	0.5 g	Tergitol 7	0.1 g

Final pH 7.2 ± 0.2 at 25°C.

One pound will make 13.3 liters of medium.

METHOD OF PREPARATION
1. Suspend 34 g in 1 liter distilled or deionized water and heat to boiling to dissolve completely.
2. Dispense into tubes or flasks.
3. Sterilize in the autoclave for 15 minutes at 15 lbs pressure (121°C).

PROCEDURE
1. Cool medium to 50 – 55°C and pour into Petri dishes.
2. Inoculate by smearing with a sterile bent glass rod 0.1 ml of the test organisms or sample.
3. Incubate for 18 – 48 hours at 35°C.
4. Examine for lactose fermentation (yellow colonies and zones) and hydrogen sulfide production (black centers). Lactose nonfermenting organisms usually produce colonies surrounded by blue zones.

STORAGE
Bacto Tergitol 7 Agar H — Below 30°C
Prepared medium — 2 – 8°C

QUALITY CONTROL
Identity Specifications

Dehydrated powder: beige, homogeneous, free-flowing
Reaction of 3.4% solution: pH 7.2 ± 0.2 at 25°C
Prepared medium: yellowish green, very slightly opalescent

Typical Cultural Response on Bacto Tergitol 7 Agar H
After 18 – 48 Hours at 35°C

Organism	Growth	Color of Colony	H₂S
Escherichia coli ATCC® 25922	good to excellent	yellow	–
Klebsiella pneumoniae ATCC® 13883	fair to good	yellow	–
Proteus mirabilis ATCC® 25933	good to excellent	blue	+
Providencia alcalifaciens ATCC® 9886	good to excellent	blue	+
Salmonella enteritidis ATCC® 13076	good to excellent	blue	+
Streptococcus faecalis ATCC® 19433	inhibited	—	—

REFERENCE
1. Hinds, D.B., and Howard, A.T. 1974. Applied Micro. 28:521 – 522.

PACKAGING

Bacto Tergitol 7 Agar H	1 lb (454 g)	0800-01-1

BACTO TERGITOL 7 AGAR
BACTO TERGITOL 7 BROTH

INTENDED USE
Bacto Tergitol 7 Agar and Bacto Tergitol 7 Broth are selective media for the enumeration and identification of members of the coliform group.

HISTORY/PRINCIPLES
Bacto Tergitol 7 Broth is the same formula as Bacto Tergitol 7 Agar except the agar has been omitted.

Bacto Tergitol 7 Agar is a selective medium for *Escherichia coli* and members of the coliform group, prepared according to the formula published by Chapman.[1] Chapman[2] reported that the addition of triphenyltetrazolium chloride (TTC) to this medium permitted the confirmation of *E. coli* after 10 hours incubation, and also that this medium gave excellent results in the cultivation of *Candida* and other fungi.

Chapman[1] reported that the addition of Tergitol 7 to an agar medium consisting of Proteose Peptone No. 3, Bacto Yeast Extract, lactose and brom thymol blue permitted unrestricted development of all coliform organisms and inhibited the development of gram-negative spore formers as well as gram-positive microorganisms. *Escherichia* produce yellow colonies surrounded by yellow zones; *Enterobacter* produce large mucoid colonies, usually surrounded by yellow zones; *Salmonella, Shigella* and other lactose nonfermenting organisms produce colonies usually surrounded by blue zones, on this medium. *Proteus* and other organisms have little tendency to form spreading colonies. Counts of coliform organisms on Bacto Tergitol 7 Agar were found to be 30% higher than on other selective media for members of this group. Tergitol 7 agar is inoculated by smearing the surface with the specimen using a bent glass rod. As much as 0.1 ml of inoculum may be used per plate if the surface of the medium is dry. Pour plates do not give satisfactory results.

The addition of 40 mg of TTC to a liter of sterile tergitol 7 agar permitting the confirmation of *E. coli* after 10 hours incubation was described by Chapman.[2] *E. coli* does not reduce the dye while other coliform organisms rarely fail to do so. Surface colonies of *E. coli* on this medium are greenish yellow surrounded by a yellow halo while other coliform surface colonies are dark red. Readings can be made following incubation at 35°C for 10 hours. Chapman also reported that tergitol 7 agar with added TTC (40 mg per liter) gave a selective medium for the isolation of *Candida* and other fungi. *Candida* growing on this medium produce white circular convex entire colonies about 1 mm in diameter in 24 – 48 hours. The colonies may appear pale blue because of the color of the medium. Yeasts produce red colonies. Since the medium permits the development of coliform organisms this fact must be taken into consideration in the isolation of *Candida* from specimens. Chapman[3] also used tergitol 7 in a modified Sabouraud maltose agar, for the isolation and differentiation of *Candida* and other fungi.

FORMULAE

BACTO TERGITOL 7 AGAR
DEHYDRATED

Ingredients per liter

Proteose Peptone No. 3, Difco ..... 5 g
Bacto Yeast Extract .... 3 g
Bacto Lactose .... 10 g
Bacto Agar .... 15 g
Tergitol 7 .... 0.1 g
Bacto Brom Thymol Blue ...... 0.025 g

Final pH 6.9 ± 0.2 at 25°C.

One pound will make 13.7 liters
of medium.
Rehydrate with 33 g/liter.

BACTO TERGITOL 7 BROTH

Ingredients per liter

Proteose Peptone No. 3, Difco ..... 5 g
Bacto Yeast Extract .... 3 g
Bacto Lactose .... 10 g
Tergitol 7 .... 0.1 g
Bacto Brom Thymol Blue ...... 0.025 g

Final pH 6.9 ± 0.2 at 25°C.

One pound will make 25.2 liters
of medium.
Rehydrate with 18 g/liter.

METHOD OF PREPARATION

1. To rehydrate, suspend appropriate amount in 1 liter distilled water and heat to boiling to dissolve completely.
2. Sterilize in the autoclave for 15 minutes at 15 lbs pressure (121°C).

STORAGE

Bacto Tergitol 7 media	Below 30°C
Prepared media	15 – 30°C

QUALITY CONTROL

Identity Specifications

Dehydrated powder:	beige, homogeneous, free-flowing
Reaction of appropriate solution:	pH 6.9 ± 0.2 at 25°C
Prepared medium:	green, clear to slightly opalescent

Typical Cultural Response in/on Bacto Tergitol 7 Broth/Agar
After 18 – 48 Hours at 35°C

Organism	Growth	Acid
Enterobacter aerogenes ATCC® 13048	good to excellent	+
Escherichia coli ATCC® 25922	good to excellent	+
Salmonella typhimurium ATCC® 14028	good to excellent	−
Shigella flexneri ATCC® 12022	good to excellent	−
Staphylococcus aureus ATCC® 25923	inhibited	−

+ = positive, yellow colony or medium
− = negative, blue colony or medium

REFERENCES

1. J. Bact., 53:504, 1947.
2. Am. J. Pub. Health, 41:1381, 1951.
3. Trans. New York Acad. Sci. Series II, 14:254, 1952.

PACKAGING

Bacto Tergitol 7 Agar	1 lb (454 g)	0455-01-9
	5 lb (2.27 kg)	0455-05-5
Bacto Tergitol 7 Broth	1 lb (454 g)	0912-01-6

BACTO TETRATHIONATE BROTH BASE

INTENDED USE
Bacto Tetrathionate Broth Base is used for enriching members of the *Salmonella* group in the isolation of these organisms from infectious material.

HISTORY
The credit for discovering the usefulness of a tetrathionate broth for enriching typhoid and the paratyphoid group is ascribed to Mueller[1] who demonstrated clearly that it inhibited or killed the coliform organisms and permitted typhoid and the paratyphoids to grow almost unrestrictedly. Mueller obtained pure cultures of typhoid by incubating mixtures containing few typhoid organisms and large numbers of coli in his tetrathionate enrichment.

Kauffmann,[2,3] using a modified Mueller's broth claimed to have increased his positive isolations of *S. typhosa* (*S. typhi*) over 30% and of other members of the *Salmonella* group from 100 to 700% over that possible by direct streaking onto plate media. Schaeffer,[4] using tetrathionate broth also demonstrated the greater efficiency of tetrathionate enrichment by detecting four times as many typhoid and paratyphoid positive fecal specimens as could be found by direct plating. Further demonstrations of the usefulness of tetrathionate enrichment were made by Jones,[5] Ruys[6] and by Szper,[7] who developed a technique for its use in examining large volumes of water and sewage material. Newman[8] in a study of the detection of food poisoning attributable to dairy products, used tetrathionate broth followed by streaking to bismuth sulfite agar and SS agar for the isolation of *Salmonella*.

Standard Methods for the Examination of Water and Wastewater, Fifteenth Edition,[9] and *Compendium of Methods for the Microbiological Examination of Foods*[10] both specify tetrathionate broth for enriching specimens containing *Salmonella*.

FORMULA
BACTO TETRATHIONATE BROTH BASE
DEHYDRATED
Ingredients per liter

Proteose Peptone, Difco	5 g
Bacto Bile Salts	1 g
Sodium Thiosulfate	30 g
Calcium Carbonate	10 g

Final pH 8.4 ± 0.2 at 25°C.

Five hundred grams will make 10.8 liters of medium.

METHOD OF PREPARATION
1. Suspend 4.6 g in 100 ml distilled or deionized water and heat to boiling. Cool below 60°C.
2. Add 2 ml iodine solution (prepared by dissolving 6 g iodine crystals and 5 g potassium iodide in 20 ml distilled or deionized water) to medium. Do not heat medium after adding iodine.
3. Dispense 10 – 12 ml quantities into sterile test tubes. Use medium the same day it is prepared.

STORAGE
Bacto Tetrathionate Broth Base   Below 30°C
Basal medium without iodine   2 – 8°C
Prepared medium — Use the same day as iodine is added.

QUALITY CONTROL

Identity Specifications

Dehydrated powder: white, homogeneous, free-flowing
Reaction of 4.6% solution: pH 8.4 ± 0.2 at 25°C
Prepared medium: milky white opaque suspension; on standing yields a nearly colorless liquid over a heavy white precipitate.

Typical Cultural Response on Bacto MacConkey Agar
After Enrichment in Tetrathionate Broth for 18 – 24 Hours at 35°C

Organism	Recovery	Color of Colony
Escherichia coli ATCC® 25922	little or no increase in numbers	white to pink w/bile ppt
Salmonella choleraesuis ATCC® 12011	good to excellent	colorless
Salmonella typhi ATCC® 6539	good to excellent	colorless
Salmonella typhimurium ATCC® 14028	good to excellent	colorless

REFERENCES

1. Compt. Rend. Soc. Biol., 89:434, 1923.
2. Zentr. Bakt., I Abt., Orig., 113:148, 1930 – 31.
3. Zeit. Hyg., 117:26, 1935 – 36.
4. Zentr. Bakt. I Abt., Orig., 133:458, 1935.
5. J. Path. Bact., 42:455, 1936.
6. Brit. Med. J., I:606, 1940.
7. Comp. Rend. Soc. Biol., 118:1675, 1935.
8. J. Milk and Food Tech., 13:226, 1950.
9. Franson, M. A. 1980. Standard methods for the examination of water and wastewater, 15th ed. American Public Health Association, Washington, D.C.
10. Speck, M. L. 1976. Compendium of methods for the microbiological examination of foods. American Public Health Association, Washington, D.C.

PACKAGING

Bacto Tetrathionate Broth Base

	500 g	0104-17-6
	100 g	0104-15-8
	5 lb (2.27 kg)	0104-05-0

BACTO m TETRATHIONATE BROTH BASE

INTENDED USE

Bacto m Tetrathionate Broth Base is a selective medium used in the membrane filter technique for the preliminary enrichment of Salmonella prior to placing the filter on selective media.

HISTORY/PRINCIPLES

Bacto m Tetrathionate Broth has the same formulation as Bacto Tetrathionate Broth Base except that it contains no calcium carbonate, as suggested by Kabler and Clark.[1]

Tetrathionate broth, in single strength, without added calcium carbonate, was used by Kabler and Clark[1] for the preliminary enrichment of Salmonella other than S. typhi. They reported that, in mixed cultures, about 80% of the Salmonella organisms were recovered while most coliforms were suppressed. The presence of calcium carbonate in the medium tended to give poor, erratic results. These authors reported good enrichment of S. typhimurium, in the membrane filter technique, using a 3 hour preliminary incubation on pads saturated with tetrathionate broth followed by 15 hours incubation on Bacto m Brilliant Green Broth. *Shigella* recovery was reported using the tetrathionate broth.

FORMULA

BACTO m TETRATHIONATE BROTH BASE
DEHYDRATED
Ingredients per liter

Proteose Peptone, Difco . . . . 5 g
Bacto Bile Salts . . . . 1 g
Sodium Thiosulfate . . . . 30 g

Final pH 8.0 ± 0.2 at 25°C.

One pound will make 12.6 liters of medium.

PROCEDURE
1. Suspend 3.6 g in 100 ml distilled or deionized water and heat to boiling.
2. Cool to below 60°C and add 2 ml iodine solution.* Do not heat after the addition of iodine.

    NOTE: The complete medium containing iodine should be used the day it is prepared; the base medium can be stored indefinitely.
3. Place absorbent pads in 50 – 60 mm Petri dishes and dispense 2 ml of medium onto each.
4. Prepare membrane filter with inoculum, place on pad and incubate at 35°C for 3 hours.
5. Transfer membranes aseptically to pads soaked with 2 ml Bacto m Brilliant Green Broth in another set of 50 – 60 mm Petri dishes.
6. Incubate at 35°C for 15 – 21 more hours (total incubation of 18 – 24 hours) in humidified atmosphere. Examine for growth.

*Iodine Solution - dissolve 6 g iodine crystals and 5 g potassium iodide in 20 ml distilled water.

STORAGE
Bacto m Tetrathionate Broth Base — Below 30°C
Prepared medium (w/o iodine) — 15 – 30°C

QUALITY CONTROL
Identity Specifications

Dehydrated powder: — beige, homogeneous, free-flowing
Reaction of 3.6% solution
 w/o iodine: — pH 8.0 ± 0.2 at 25°C
Prepared medium: — light amber, clear, may have some precipitate

Typical Cultural Response Using Bacto m Tetrathionate Broth Base and Bacto *m* Brilliant Green Broth After 18 – 24 Hours at 35°C

Organism	Growth	Color of Colony
Escherichia coli ATCC® 25922	fair to good	yellow-green
Salmonella typhimurium ATCC® 14028	good to excellent	pink to red

REFERENCE
1. Kabler and Clark, American Journal of Public Health, 42:390, 1952.

PACKAGING

Bacto m Tetrathionate Broth Base	1 lb (454 g)	0580-01-7
Bacto m Brilliant Green Broth	1 lb (454 g)	0494-01-2

BACTO THERMOACIDURANS AGAR

INTENDED USE

Bacto Thermoacidurans Agar, a Standard Methods medium[1], is used for the isolation and cultivation of *Bacillus thermoacidurans* from food products.

HISTORY/PRINCIPLES

Bacto Thermoacidurans Agar is prepared according to the formula described by Stern, Hegarty and Williams[2] for the isolation of *B. thermoacidurans* (*Bacillus coagulans*) the causative organism of "flat sour" spoilage in tomato juice.

For the detection of *B. thermoacidurans* Stern, Hegarty and Williams[2] recommended the plating of 1 ml of tomato juice per 20 ml of agar medium. They observed that larger quantities of tomato juice exhibited an inhibitory effect on the growth of the organisms. Plates were poured with the sterile melted agar at 45 – 55°C and, following solidification, incubated at 55°C for 48 hours.

FORMULA

BACTO THERMOACIDURANS AGAR
DEHYDRATED

Ingredients per liter

Bacto Yeast Extract . . . . 5 g
Proteose Peptone, Difco . . . . 5 g
Bacto Dextrose . . . . 5 g
Dipotassium Phosphate . . . . 4 g
Bacto Agar . . . . 20 g

Final pH 5.0 ± 0.2 at 25°C.

Five hundred grams will make 12.8 liters of medium.

METHOD OF PREPARATION

1. Suspend 39 g in 1 liter distilled or deionized water and heat to boiling to dissolve completely.
2. Sterilize in the autoclave for 15 minutes at 15 lbs pressure (121°C). Overheating during the sterilization period or prolonged holding of the melted medium should be avoided or a soft medium will result.
3. Consult *The Compendium of Methods for the Microbiological Examination of Foods*[1] for food testing procedure.

STORAGE

Bacto Thermoacidurans Agar	Below 30°C
Prepared medium	15 – 30°C

QUALITY CONTROL
Identity Specifications

Dehydrated powder:	light tan, homogeneous, free-flowing
Reaction of 3.9% solution:	pH 5.0 ± 0.2 at 25°C
Prepared medium:	light amber, opalescent

Typical Cultural Response on Bacto Thermoacidurans Agar
After 18 – 48 Hours at 55°C

Organism	Growth
Bacillus coagulans ATCC® 7050	fair to good
Staphylococcus aureus ATCC® 25923	none to poor

REFERENCES

1. American Public Health Association, 1976, Compendium of Methods for the Microbiological Examination of Foods, American Public Health Association, Inc., Washington, D.C.
2. Stern, Hegarty and Williams, 1942, Food Research, 7:186.

PACKAGING

Bacto Thermoacidurans Agar	500 g	0303-17-5

BACTO THIOGLYCOLLATE GELATIN MEDIUM

INTENDED USE

Bacto Thioglycollate Gelatin Medium is used for the determination of gelatin liquefaction by aerobes, microaerophiles and anaerobes without special incubation.

PRINCIPLES

Bacto Thioglycollate Gelatin Medium will inactivate mercurial salts with the thioglycollate in the medium. The presence of 0.2% dextrose gives a readily available source of carbohydrate, permitting early growth of the inoculum. The medium maintains a low redox potential, when incubated at temperatures above the melting point of the gelatin, due to the presence of the agar, thus assuring optimal conditions for the development of anaerobes without special seal.

FORMULA

BACTO THIOGLYCOLLATE GELATIN MEDIUM
DEHYDRATED

Ingredients per liter

Bacto Casitone	15 g
Bacto Yeast Extract	5 g
Bacto Dextrose	2 g
Sodium Chloride	2.5 g
L-Cystine, Difco	0.25 g
Sodium Sulfite	0.1 g
Thioglycollic Acid	0.3 ml
Bacto Agar	0.75 g
Bacto Gelatin	50 g

Final pH 7.0 ± 0.2 at 25°C.

One pound will make 6.5 liters of medium.

METHOD OF PREPARATION

1. Suspend 76 g in 1 liter of cold distilled or deionized water and swirl to wet all of the dehydrated medium.
2. Warm to 50°C in a water bath, then heat to boiling to dissolve the medium completely.
3. Dispense into test tubes if desired.
4. Sterilize in the autoclave for 15 minutes at 15 lbs pressure (121°C).

PROCEDURE

Gelatin liquefaction may readily be determined in tubed Bacto Thioglycollate Gelatin Medium. At incubation temperatures of 20°C or less, the medium will be solid, and liquefaction of the gelatin may readily be noted. At incubation temperatures above 30°C, the medium will be liquid. Following the incubation period the inoculated tubes of medium are cooled to 20°C or less. If gelatin has been liquefied, the medium will remain liquid. If the microorganism failed to hydrolyze the gelatin, the medium will solidify.

STORAGE

Bacto Thioglycollate Gelatin Medium   Below 30°C
Prepared medium   2 – 8°C

QUALITY CONTROL

Identity Specifications

Dehydrated powder: light amber, homogeneous, free-flowing
Reaction of 7.6% solution: pH 7.0 ± 0.2 at 25°C
Prepared medium: light to medium amber, very slightly opalescent, may have a very slight precipitate

Typical Cultural Response in Bacto Thioglycollate Gelatin Medium
After 18 – 48* Hours at 35°C

*Some gelatinase reactions require longer incubation.

Organism	Growth	Gelatinase
Clostridium butyricum ATCC® 9690	good to excellent	+
Clostridium sporogenes ATCC® 11437	good to excellent	+
Escherichia coli ATCC® 25922	good to excellent	−
Salmonella enteritidis ATCC® 13076	good to excellent	−
Staphylococcus aureus ATCC® 25923	good to excellent	+

PACKAGING

Bacto Thioglycollate Gelatin Medium	1 lb (454 g)	0530-01-8

THIOGLYCOLLATE MEDIA

BACTO FLUID THIOGLYCOLLATE MEDIUM
BACTO FLUID THIOGLYCOLLATE MEDIUM w/BEEF EXTRACT
BACTO FLUID THIOGLYCOLLATE MEDIUM w/K AGAR
BACTO THIOGLYCOLLATE MEDIUM w/o DEXTROSE
BACTO THIOGLYCOLLATE MEDIUM w/o INDICATOR
BACTO THIOGLYCOLLATE MEDIUM w/o DEXTROSE OR INDICATOR

INTENDED USE

Bacto Thioglycollate media are used for cultivating anaerobic, microaerophilic and aerobic microorganisms and for detecting the presence of bacteria in normally sterile materials.

Bacto Fluid Thioglycollate Medium conforms to the requirements of The United States Pharmacopeia[1] and the Association of Official Analytical Chemists[2] for testing the sterility of antibiotics, biologics and foods and for determining the phenol coefficient and sporicidal effect of disinfectants.

The five variations of Bacto Fluid Thioglycollate Medium each offer alternate performance characteristics based on formula modification, as discussed in PRINCIPLES.

HISTORY

In 1898, Trenkmann[3] reported growth of anaerobes under aerobic conditions in the presence of alkaline sulfide. Quastel and Stephenson[4] found that the presence of a small amount of a compound containing an −SH group (cysteine, thioglycollic acid, glutathione) permitted "aerobic" growth of Clostridium sporogenes in tryptic digest broth.

Falk, Bucca and Simmons[5] pointed out the advantages of using small quantities of agar (0.06 − 0.25%) in detecting contaminants during sterility testing of biologicals. Brewer[6] demonstrated the value of combining a small amount of agar and a reducing substance. In a liquid medium containing 0.05% agar, anaerobes grew equally well in the presence or absence of sodium thioglycollate. Marshall, Gunnison and Luxen[7] reported satisfactory cultivation of anaerobes in Brewer's thioglycollate medium in the presence of a mercurial preservative. Nungester, Hood and Warren[8] and Portwood[9] confirmed the neutralization of the bacteriostatic effect of mercurial compounds by sodium thioglycollate. Malin and Finn[10] observed that thioglycollate is inhibitory to some organisms in the presence of carbohydrate.

PRINCIPLES

Thioglycollate media, formulated with Bacto Yeast Extract and Bacto Casitone or pancreatic digest of casein, support growth of a wide variety of fastidious microorganisms having a range of growth requirements. Sodium thioglycollate lowers the oxidation-reduction potential of the media and neutralizes the antibacterial effect of mercurial preservatives in the specimen. Resazurin and methylene blue indicate the status of oxidation or aerobiosis. Since methylene blue is toxic to some bacteria, resazurin has replaced it as the OR indicator. Dextrose is included in four of the formulations because most organisms show earlier and more vigorous growth in the presence of a carbohydrate. Dextrose is omitted from two formulations, making them suitable for use in fermentation studies with added carbohydrates.

Substituting K Agar and potassium chloride for Bacto Agar and sodium chloride produces a medium with greater clarity to facilitate earlier visual recognition of growth.

Beef extract is added to one formulation as recommended by Agricultural Research Service CFR Amendment 70-253, 35FR16038, 10, 13, 70.

METHOD OF PREPARATION

1. Suspend the amount of dehydrated medium indicated under FORMULA — Solution (and on the package label) in 1 liter distilled or deionized water and heat to boiling to dissolve completely.
2. Dispense as desired.
3. Sterilize in the autoclave for 15 minutes at 15 lbs pressure (121°C). Cool to 25°C.
4. After storage of prepared medium, if more than 30% of the medium demonstrates oxidation (color change of indicator), reheat medium once in a boiling water or steam bath to drive off absorbed oxygen.

FORMULAE

	Bacto Fluid Thioglycollate Medium	Bacto Fluid Thioglycollate Medium w/Beef Extract	Bacto Fluid Thioglycollate Medium w/K Agar	Bacto Thioglycollate Medium w/o Dextrose	Bacto Thioglycollate Medium w/o Indicator	Bacto Thioglycollate Medium w/o Dextrose or Indicator
Bacto Beef Extract		5 g				
Bacto Yeast Extract	5 g	5 g	5 g	5 g	5 g	5 g
Bacto Casitone	15 g	15 g	15 g	15 g	15 g	15 g
Pancreatic Digest of Casein						
Bacto Dextrose	5.5 g	5.5 g	5 g		5 g	
Sodium Chloride	2.5 g	2.5 g	2.5 g	2.5 g	2.5 g	2.5 g
Potassium Chloride			0.5 g			
L-Cystine, Difco	0.5 g	0.5 g		0.25 g	0.25 g	0.25 g
Sodium Thioglycollate	0.5 g	0.5 g			0.5 g	
Thioglycollic Acid			0.3 ml	0.3 ml		0.3 ml
Bacto Agar	0.75 g	0.75 g		0.75 g	0.75 g	0.75 g
K Agar			0.45 g			
Resazurin	0.001 g	0.001 g	0.001 g			
Bacto Methylene Blue				0.002 g		
Final pH at 25°C	7.1 ± 0.2	7.2 ± 0.2	7.2 ± 0.2	7.2 ± 0.2	7.2 ± 0.2	7.2 ± 0.2
One pound will make:	15.23 liters	13.08 liters	15.65 liters	18.91 liters	15.65 liters	18.91 liters
Solution (g/liter)	29.8 g/liter	34.7 g/liter	29 g/liter	24 g/liter	29 g/liter	24 g/liter

QUALITY CONTROL

Identity Specifications

	Bacto Fluid Thioglycollate Medium	Bacto Fluid Thioglycollate Medium w/Beef Extract	Bacto Fluid Thioglycollate Medium w/K Agar	Bacto Thioglycollate Medium w/o Dextrose	Bacto Thioglycollate Medium w/o Indicator	Bacto Thioglycollate Medium w/o Dextrose or Indicator
Dehydrated powder:	light beige	beige	light beige	light tan	light beige	light beige
Reaction of solution at 25°C: Prepared medium:	pH 7.1 ± 0.2 light straw color, very slightly opalescent with upper 10% or less medium pink on standing	pH 7.2 ± 0.2 medium amber, very slightly opalescent with upper 10% or less medium pink on standing	pH 7.2 ± 0.2 light straw color, clear with upper 10% or less medium pink on standing	pH 7.2 ± 0.2 light amber, very slightly opalescent with upper 10% or less medium green on standing	pH 7.2 ± 0.2 light amber, very slightly opalescent	pH 7.2 ± 0.2 light to medium amber, very slightly opalescent

(Dehydrated powder: —————— homogeneous, free-flowing ——————)

Typical Cultural Response After* 18 – 48 Hours at 35°C

Growth

Organism	Bacto Fluid Thioglycollate Medium	Bacto Fluid Thioglycollate Medium w/Beef Extract	Bacto Fluid Thioglycollate Medium w/K Agar	Bacto Thioglycollate Medium w/o Dextrose	Bacto Thioglycollate Medium w/o Indicator	Bacto Thioglycollate Medium w/o Dextrose on Indicator
Bacillus subtilis ATCC® 6633	good to excellent	good to excellent	good to excellent	good	good to excellent	good
Bacteroides vulgatus ATCC® 8482	poor to fair	poor to fair	poor	poor	poor to fair	poor
Candida albicans ATCC® 10231	good to excellent	good to excellent	good to excellent	good	good to excellent	good
Clostridium sporogenes ATCC® 11437	good to excellent	good to excellent	good to excellent	good to excellent	good to excellent	good to excellent
Micrococcus luteus ATCC® 9341	good to excellent	good to excellent	good to excellent	good	good to excellent	good
Neisseria meningitidis ATCC® 13090	good to excellent	good to excellent	good	good	good to excellent	good
Streptococcus pyogenes ATCC® 19615	good to excellent	good to excellent	good to excellent	good to excellent	good to excellent	good to excellent

*these cultures may be incubated at 30 – 35°C for 2 – 7 days, if preferred, per USP.[1]

STORAGE

Dehydrated media   Below 30°C
Prepared media   15 – 30°C*

*if for prolonged storage, store at 2 – 8°C

REFERENCES

1. The United States Pharmacopeia, 20th Ed., Rockville, Md., United States Pharmacopeial Convention, Inc., 1980.
2. Horwitz, W. (ed.): Official Methods of Analysis of the Association of Official Analytical Chemists, 13th Ed., Washington, D.C., Association of Official Analytical Chemists, 1980.
3. Centr. Bakt. I., Abt., 23:1038, 1898.
4. Biochem. J., 20:1125, 1926.
5. J. Bact., 37:121, 1939.
6. J. Am. Med. Assoc., 115:598, 1940.
7. Proc. Soc. Exp. Biol. Med., 43:672, 1940.
8. Proc. Soc. Exp. Biol. Med., 52:287, 1943.
9. J. Bact., 48:255, 1944.
10. J. Bact., 62:349, 1957.

PACKAGING

Bacto Fluid Thioglycollate Medium	1 lb (454 g)	0256-01-0
	1/4 lb (114 g)	0256-02-9
	5 lb (2.27 kg)	0256-05-6
	10 kg	0256-08-3
	12 tubes	0256-34-1
	144 tubes	0256-37-8
	6 × 100 ml	0256-73-3
Bacto Fluid Thioglycollate Medium w/Beef Extract	1 lb (454 g)	0697-01-7
	10 kg	0697-08-0
Bacto Fluid Thioglycollate Medium w/K Agar	1 lb (454 g)	0607-01-6
Bacto Thioglycollate Medium w/o Dextrose	1 lb (454 g)	0363-01-0
Bacto Thioglycollate Medium w/o Indicator	500 g	0430-17-1
	12 tubes	0430-34-0
	144 tubes	0430-37-7
Bacto Thioglycollate Medium w/o Dextrose or Indicator	1 lb (454 g)	0432-01-7
	1/4 lb (114 g)	0432-02-6

THIOGLYCOLLIC ACID

Thioglycollic Acid is recommended for use in media designed for isolating or culturing bacteria in the presence of mercurial salt preservative. It may also be used for the preparation of media of a low Eh. In the preparation of media for the sterility testing of biologicals, thioglycollic acid or sodium thioglycollate may be employed. A complete discussion of these media is given under Bacto Fluid Thioglycollate Medium and Bacto Thioglycollate Broth, NIH.

WARNING:

Causes irritation. Avoid contact with the skin, eyes and clothing. Wash thoroughly after handling.

PACKAGING

Thioglycollic Acid	10 g	0250-12-3

BACTO THIOL MEDIUM
BACTO THIOL BROTH

INTENDED USE
Bacto Thiol Medium and Bacto Thiol Broth are used for culturing organisms from body fluids and other materials containing penicillin, streptomycin or sulfonamides. Bacto Thiol Broth is prepared for laboratories requiring a medium having the properties of Bacto Thiol Medium except without agar.

HISTORY
Broom, in a personal communication, reported excellent results with enriched Bacto Thiol Medium in isolating *Haemophilus influenzae* and meningococci. Huddleson[2] found the medium satisfactory for isolating, cultivating and maintaining stock cultures of *Vibrio fetus*. The cultures remained viable in Bacto Thiol Medium for at least 150 days without transfer.

PRINCIPLES
Bacto Thiol Medium contains 0.1% agar to maintain an Eh potential conducive to growth of aerobic, microaerophilic and strictly anaerobic microorganisms. Ten ml of the medium has the ability to inactivate up to 100 units of penicillin or up to 1000 units of streptomycin, producing luxuriant growth of staphylococci and other test organisms from even dilute inocula in 24 hours.

FORMULAE

BACTO THIOL MEDIUM
DEHYDRATED
Ingredients per liter

Proteose Peptone No. 3, Difco	10 g
Bacto Yeast Extract	5 g
Bacto Dextrose	1 g
Sodium Chloride	5 g
Thiol Complex	8 g
Bacto Agar	1 g
p-Aminobenzoic Acid	0.05 g

Final pH 7.1 ± 0.2 at 25°C.

Five hundred grams will make
16.6 liters of medium.
Rehydrate with 30 g/liter.

BACTO THIOL BROTH
DEHYDRATED
Ingredients per liter

Proteose Peptone No. 3, Difco	10 g
Bacto Yeast Extract	5 g
Bacto Dextrose	1 g
Sodium Chloride	5 g
Thiol Complex	8 g
p-Aminobenzoic Acid	0.05 g

Final pH 7.1 ± 0.2 at 25°C.

Five hundred grams will make
17.2 liters of medium.
Rehydrate with 29 g/liter.

METHOD OF PREPARATION
1. Suspend appropriate amount in 1 liter distilled or deionized water and heat to boiling to dissolve completely.
2. Dispense medium into tubes or flasks to a depth of 60 mm for neutralization of penicillin or in shallow layers for neutralization of streptomycin. Dispense thiol broth into tubes as desired.
3. Sterilize in the autoclave for 15 minutes at 15 lbs pressure (121°C).
4. Use within 4 days.

STORAGE
Bacto Thiol media	Below 30°C
Prepared media	15 – 30°C*

*No longer than 4 days

QUALITY CONTROL

Identity Specifications

Dehydrated powder:	light beige, homogeneous, free-flowing
Reaction of appropriate solution:	pH 7.1 ± 0.2 at 25°C
Prepared medium:	very light amber, clear to very slightly opalescent

Typical Cultural Response in Bacto Thiol Medium or Broth
After 18 – 48 Hours at 35°C

Organism	Growth*
Neisseria meningitidis ATCC® 13090	fair to good
Staphylococcus aureus ATCC® 6538	good to excellent
Streptococcus pneumoniae ATCC® 6303	good to excellent
Streptococcus pyrogenes ATCC® 19615	good to excellent

*Growth may be less if antibiotics are present in high concentrations.

REFERENCE
1. J. Bact., 56:508, 1948.

PACKAGING

Bacto Thiol Medium	500 g	0307-17-1
Bacto Thiol Broth	500 g	0434-17-7

PROTHROMBIN REAGENTS

BACTO THROMBOPLASTIN
BACTO CALCIUM CHLORIDE 0.02M
BACTO SODIUM CHLORIDE 0.85%

INTENDED USE

Bacto Thromboplastin, Bacto Calcium Chloride 0.02M and Bacto Sodium Chloride 0.85% are prepared for use in the determination of prothrombin times according to the methods described by Quick.[1]

SUMMARY AND EXPLANATION

Bacto Thromboplastin is prepared from rabbit brain by acetone dehydration according to the methods first described by Quick[1] and is carefully standardized to give prothrombin times comparable to Quick's classical thromboplastin.

Its freedom from extraneous tissue extractives makes it applicable to all tests requiring a potent, sensitive thromboplastin. In conjunction with anciliary reagents available from Difco and some easily prepared in the user's laboratory, they are used for:
1. Control of patients on coumarin derivatives or allied anticoagulant therapy.
2. Prothrombin consumption tests in the detection and differentiation of hemophilias.
3. Determination of the presence or absence of Factors II, V, VII, VIII and IX.

For uniformity and brevity the international nomenclature using Roman numerals corresponding to the various blood coagulation factors and their synonyms as shown in figure 1, will be used in the discussions.

Figure 1

Factor	Synonym
I	Fibrinogen
II	Prothrombin
III	Tissue Thromboplastin
IV	Calcium
V	Labile factors, Proaccelerin, Accelerin, Ac Globulin
VII	Stable factors, Proconvertin, Serum Prothrombin Conversion Accelerator (SPCA)
VIII	Antihemophilic Globulin (AHG), Antihemophilic factor (AHF)
IX	Plasma Thromboplastin Component (PTC), Christmas factor
X	Stuart-Prower factor
XI	Plasma Thromboplastin Antecedent (PTA)
XII	Hageman factor
XIII	Fibrin stabilizing factor

It is generally accepted that the process by which blood clots occurs in 3 stages. These stages are (1) the formation of a prothrombin activator, (2) the conversion of prothrombin to thrombin, (3) the conversion of fibrinogen to fibrin.

Stage 1 can occur either when factor III is involved or when plasma constituents alone are involved. When factor III is involved it will activate prothrombin to thrombin in the presence of factors V, X and IV. Factor VII is also necessary as a cofactor for factor III. Because factor III is not normally found in circulating blood its involvement in stage 1 is termed extrinsic.

When the plasma constituents alone are involved in stage 1, the system is termed intrinsic. In this system, factors VIII, IX, XI, XII as well as platelets are involved with factors V, X and IV for the prothrombin activator.

Figure 2 is a diagramatic schema of the blood clotting mechanism just discussed.

Figure 2

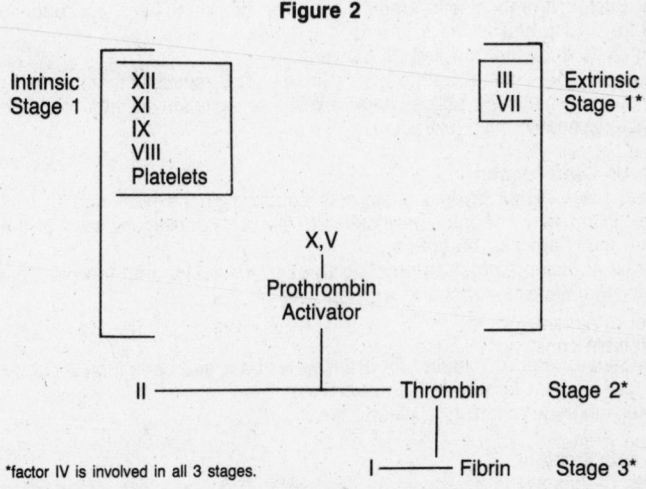

*factor IV is involved in all 3 stages.

The Quick 1-Stage Prothrombin Time Determination will be described in this product description. For more detailed information on blood coagulation factors refer to the many standard texts on the subject.

The Quick 1-Stage Prothrombin Time Determination is a misnomer because more than just prothrombin activity is included in the measurement. Despite this, the test remains one of the best tests for the control of patients on coumarin derivatives or allied anticoagulant therapy.

Principle of the 1-Stage Prothrombin Time
In the presence of excess factor III (extract of Bacto Thromboplastin) and factor IV (Bacto Calcium Chloride 0.02M) plasma will clot rapidly. A prolonged clotting time may result from one or more of the following: Factors I, II, V, VII or X. It is also possible that a circulating anticoagulant inhibiting one or more of these factors may be present.

Any plasma with an abnormal, unexplained 1-stage prothrombin time should be tested for a deficiency in Factor II as well as other factors which might be causing the abnormal prothrombin time. An excellent reference for such test procedures is "Laboratory Methods in Blood Coagulation."[2]

REAGENTS
Bacto Thromboplastin is prepared from rabbit brain by acetone dehydration according to the methods first described by Quick.[1]

Preparation of Thromboplastin
The preparation of the extract of Bacto Thromboplastin can be done simultaneously with the centrifuging of the blood specimens under test without any loss of time.

Reagents Required
Bacto Thromboplastin
Bacto Sodium Chloride 0.85% (saline)

Extraction
1. Empty Bacto Thromboplastin ampule contents into clean, dry test tube. Add 4 ml saline for 0.15 g and 28 ml saline for 1 g.
2. Mix lightly to suspend and wet all particles.
3. Place in 45 – 48°C water bath for 10 minutes. At 3 minute intervals, twirl the tube gently to resuspend the solids. Avoid excessive agitation as this will yield an unsatisfactory, heavy, milky extract.

Filtration or Centrifugation
1. To filter, place a thin layer of absorbent cotton over the wide end of a clean dry pipette. Insert this end into the incubated mixture keeping the end of the pipette pressed firmly against the cotton.
2. Withdraw as much extract as possible through the cotton into the pipette.
3. Transfer the filtered extract into a clean tube.

Alternative Method
1. To centrifuge, remove extract tube from water bath and centrifuge for 3 minutes at 1500 rpm (500 G) to sediment the particles.
2. Transfer the supernatant to a clean tube.

Storage and Stability
The extract of thromboplastin should be used within 6 hours unless frozen below −10°C for future use. Properly frozen extracts will retain potency for at least 3 months.

Thromboplastin extracts that are deeply opaque or milky may give prolonged times. Deeply opaque or dense thromboplastin extracts result through excessive agitation of the suspension during the extraction period and may give unsatisfactory results. Unfiltered extracts which are not deeply opaque even though they contain gross particles of brain tissue generally give satisfactory results.

Since repeated freezing and thawing will damage the extract, distribute it in tubes holding approximately 1 day's usage or in single test dose (0.1 ml) amounts. Freeze it quickly and solidly, cover and store in freezer. Before using in the test, thaw and reincubate for 5 minutes at 37°C.

Bacto Thromboplastin in the ampule is stable to the expiry date on the label when stored at 2 – 8°C.

Bacto Calcium Chloride 0.2M is prepared especially for use with Bacto Thromboplastin and Bacto Thromboplastin Extract in testing the prothrombin activity of blood plasma by the Quick test and related techniques. It is packaged in sealed bottles to avoid any change in composition due to evaporation or contamination.

Bacto Sodium Chloride 0.85% is prepared especially for use with Bacto Thromboplastin and Bacto Thromboplastin Extract in prothrombin activity determination of blood or plasma. It is recommended as an extraction fluid in preparing thromboplastin extract from Bacto Thromboplastin and for diluting test plasmas for determining prothrombin activity by the dilute plasma technique.

Bacto Sodium Chloride 0.85% is packaged in sealed bottles with a diaphragm stopper so as to avoid any change in composition due to evaporation, or contamination with other chemicals or microorganisms. It is recommended that the solution be removed with a sterile, clean dry syringe.

SPECIMEN PREPARATION
Blood plasma samples must be carefully prepared. Blood must be withdrawn with a clean venipuncture. Immediately 9 parts of blood should be mixed with 1 part sodium oxalate 0.1M, reagent grade. 4.5 ml blood to 0.5 ml sodium oxalate 0.1M is usual. No other anticoagulant is satisfactory. Within a half hour, centrifuge for 10 minutes at 2000 rpm (900 G), and immediately decant the plasma. Use within 1/2 hour at 37°C or within 3 hours at 2 – 8°C. Plasma allowed to stand at room temperature or refrigerator temperature for more than 2 – 3 hours may give clotting times slightly higher than does fresh plasma. A plasma contaminated with hemoglobin must not be used as results will be in error. Blood drawn for prothrombin determinations should be taken previous to meal time and not directly after meals. High lipid content of plasma following ingestion of high fat food materials may prolong the clotting time. Aspirin and other salicylates, as well as caffeine when ingested may influence the clotting mechanism by prolonging the prothrombin time.

Other factors in patients' plasma that may influence results independently of the technique and reagents are disease-induced inhibitors; liver damage, reaction variability of various anticoagulants and patients' reliability in taking anticoagulant.

TEST PROCEDURE for the Quick 1-Stage Prothrombin Time Test
Before performing test see section "Factors Affecting Test Results."

Materials Provided:
Bacto Thromboplastin
Bacto Calcium Chloride 0.02M

Materials Required but not Provided:
Electronic clot timer set-up (see mfg. instructions for test performance) or water bath at 37°C
Stop watch or other timing device
Nichrome or platinum wire with a terminal loop about 1/8″ in diameter
Control plasma with known prothrombin time
Test tubes
Serological pipettes

1. Prepare oxalated patient's plasma according to standard directions.
2. Place tubes with sufficient amounts of thromboplastin extract or extract prepared from Bacto Thromboplastin and Bacto Calcium Chloride 0.02M for the test series in 37°C water bath to bring them to working temperature.

    NOTE: Gently rotate thromboplastin extract end-over-end to assure a uniform suspension.

3. Transfer 0.1 ml plasma to bottom of clean dry tube in water bath. Add 0.1 ml thromboplastin extract and twirl tube to mix contents. Allow 30 seconds to come to working temperature. The clotting time is shortened in comparison with that obtained if the calcium chloride is added immediately following the addition of the extract. Either procedure is satisfactory if adhered to consistently. Our curve is based upon the addition of calcium chloride a few seconds after the addition of the thromboplastin extract.
4. Forcibly blow 0.1 ml calcium chloride vertically and directly into the plasma-thromboplastin extract mixture and simultaneously start your timer.
5. Quickly shake the tube, and then hold it in the bath without agitation until 2 or 3 seconds before the clot is expected to form. Remove tube from bath, touching the bottom to a towel to absorb water droplets.
    a. Tilt to the horizontal position at 1/4 second intervals and observe for formation of fibrin web which is the endpoint. Stop timer at this point, or
    b. Insert a clean iron, nichrome or platinum wire with a terminal loop about 1/8″ in diameter into the tube and stir thoroughly. Draw the loop from the back of the tube to the front through the plasma-thromboplastin extract calcium chloride mixture at the rate of 2 sweeps per second, and observe the time at which the clot forms. When the clot forms it will usually adhere to the loop and is easily recognizable. Stop the watch at the first evidence of the clot and record the time required for the plasma to clot after adding the calcium chloride to the plasma-thromboplastin mixture. This is the prothrombin time.
6. Record time to the tenth of a second.
7. Run each plasma in triplicate to obtain an average time for reporting.

Alternative Method-Single Reagent
1. Mix equal quantities of Bacto Thromboplastin Extract and Bacto Calcium Chloride 0.02M in a test tube to the volume desired for the immediate series of tests. Best results will be obtained when the mixture is used within an hour. Do not incubate in bath. NOTE: When Bacto Thromboplastin Extract and Bacto Calcium Chloride 0.02M are mixed in equal quantities and immediately refrigerated at 2 – 8°C the mixture may be used within a period of 48 hours or longer. Stored mixtures must be checked with control plasma to assure proper reactivity.
2. Transfer 0.2 ml mixture to the bottom of a tube in the 37°C water bath and allow to stand for about 3 minutes to attain temperature.
3. Forcibly blow oxalated patient's plasma vertically and directly into the mixture and simultaneously start timer.
4. Follow steps 5, 6 and 7 above.

Factors Affecting Test Results

Apart from the factors associated with the specimen (see Specimen Collection) the following factors can influence the test adversely.

1. Reagents must be of highest purity and reasonably fresh. For example, the Bacto Calcium Chloride 0.02M must not be exposed to air for long periods of time as the calcium may be precipitated. Stored Bacto Thromboplastin should be tested regularly to assure that potency is maintained. Prolonged clotting times obtained with Bacto Thromboplastin have been traced most frequently to a faulty calcium chloride reagent. A calcium chloride solution either more or less concentrated than that indicated for use in the Quick technique may prolong the clotting time. Anhydrous calcium chloride is hygroscopic, and absorbs moisture from the atmosphere. Compensation for the moisture content must be taken into consideration when preparing the 0.02M solution.

2. Glassware used in the Prothrombin Time Test should be washed and stored separately from all other glassware. It must be chemically clean. Assure this with acid cleaning and thorough rinsing in distilled water. Soaps and detergents, when used, must be thoroughly rinsed away. Frequently glassware appears optically clean but remains chemically contaminated with adsorbed soap or detergent. These materials interfere with prothrombin procedures. Results will still be irregular if the container in which the blood has been collected is not chemically clean.

3. Etched, scratched or chipped items should be discarded. Uniform size pipettes and test tubes are mandatory for optimum results.

4. Metallic ions present in excess in the distilled water used may lead to prolonged clotting times.

5. Temperature of the reaction mixture is critical. The water bath or dry heat incubator must be kept at $37 \pm 0.5°C$. A circulator may be necessary to assure even temperature throughout a water bath. When dry heat is used, incubation times should be checked as it usually takes longer for the reaction mixture to attain 37°C.

6. The pH of the reaction mixture should be within 7.0 – 7.5.

7. The reagents and plasma must be mixed thoroughly at the time of addition or a prolonged clotting time may result.

8. Lighting must provide a constant source without glare for consistent determination of the end point. A black background with indirect lighting is ideal for the tilt-tube method.

9. Timing, properly done, is one of the keys to reproducible tests. Both the reaction time of the operator and the accuracy of the clock mechanism used are involved.

Unusual stress (mental or organic), tiredness, stimulants or depressants, all will affect technologist reaction time in starting, or stopping, the clock-timer used in the test and the variation may be significant.

The clock itself, whether stop-watch, other hand or foot-controlled electrical timer, or electronic instrument timer mechanism may not function properly. These should be checked for accuracy periodically. The sweep second hand of a wrist watch should never be used. All results should be recorded to the tenth of a second.

CONTROLS

Control of reagents and techniques is particularly necessary in view of the many influencing factors that can cause error. Therefore, a constant control plasma that reacts in the same manner as normal human plasma is essential.

Because of variations in technique each laboratory should establish its own Prothrombin Time Activity Curve and then check before each series of tests to assure that all factors are constant and results of the tests adhere to the established pattern.

The standard curve to which Bacto Thromboplastin is controlled in our laboratories is reproduced (Figure 3). If values check within 1/2 second of those given on the standard curve (use at least 3 plasma dilutions including 20% plasma) this curve may be used to interpolate the % prothrombin activity. If, on the other hand, a variation greater than 1/2 second of the standard curve values is produced, a new standard curve should be constructed.

Plotting a Standard Curve

Use the same reagents as in the Prothrombin Time Test, plus your control plasma and 0.85% Sodium Chloride (saline). The simplest and most economical method is to prepare the plasma dilutions and immediately (after mixing and warming) run triplicate prothrombin times on each dilution, recording the results as you proceed. It is theoretically possible to use only 1 ml of control plasma but accuracy may be sacrificed, so rehydrate 2 ml control plasma as directed, and decant to a tube marked No. 1 placed in the water bath. Other tubes similarly marked are placed in the bath. Dilutions are made as follows:

% Plasma	Tube no.	Mixture
100%	1	Undiluted
60	2	Mix 0.6 ml from tube 1 with 0.4 ml saline
40	3	Mix 0.4 ml from tube 1 with 0.6 ml saline
30	4	Mix 0.5 ml from tube 2 with 0.5 ml saline
20	5	Mix 0.5 ml from tube 3 with 0.5 ml saline
15	6	Mix 0.4 ml from tube 4 with 0.4 ml saline
10	7	Mix 0.4 ml from tube 5 with 0.4 ml saline

Figure 3

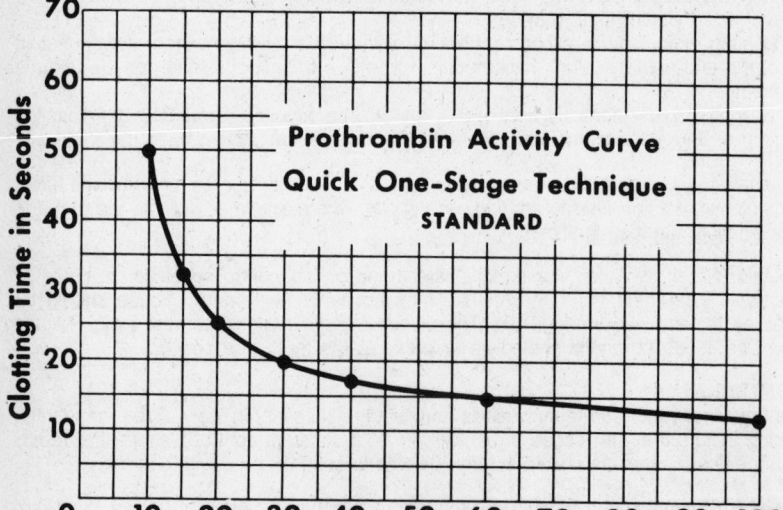

Prothrombin Activity Curve
Quick One-Stage Technique
STANDARD

Use the average of the 3 times as a point on the graph, which is plotted on 4-square graph paper. Plot the Time in Seconds on the vertical axis and the % Plasma on the horizontal axis. The % Plasma and the % Prothrombin Activity become identical when using the graph for interpretation of results. Connect the points with a smooth curving line and it becomes your Standard Activity Curve ready for use as required.

To use the standard curve as a check on technique and reagents follow the regular prothrombin time procedure substituting Bacto Prothrombin Control Plasma for the patients plasma at the 100% level. Then run at least 2 more dilutions including the 20% level to check other points on the curve. When these results are within 1/2 second of the value on the curve you are assured that technique and reagents are working properly.

RESULTS
The clotting time in seconds is plotted on the standard curve at the point where the curve intersects the time value in seconds (vertical axis). The % Prothrombin Activity is then read from the % value on the horizontal axis. The % Prothrombin and the % of concentration of the control plasma diluted with saline becomes identical in use on this chart.

It is desirable to report both in terms of seconds to the 1/10 (for multiple tests report average) and % Prothrombin activity, i.e., 25 seconds, 20%. The whole (100%) clotting time of the control plasma also may be given.

LIMITATIONS OF THE PROCEDURE
The 1-Stage Prothrombin Time Determination does not measure the activity of Prothrombin alone. A prolonged prothrombin time can be the result of a deficiency in the plasma of one or more clotting factors or the presence of a circulating anticoagulant. Additional tests are required to identify the deficient factor(s) in unexplainable prolonged prothrombin time. "Laboratory Methods in Blood Coagulation" referenced previously discusses these additional tests in detail.

REFERENCES
1. Quick, A. J., Hemorrhagic Diseases, Philadelphia, Lea and Febiger, 1957.
2. Eichelberger, J. W., Laboratory Methods in Blood Coagulation, New York, Harper and Row, 1965.

PACKAGING
Bacto Thromboplastin	210 Tests	6 × 150 mg	0226-33-9
	1380 Tests	6 × 1 g	0226-09-9
Bacto Sodium Chloride 0.85%		6 × 100 ml	0379-73-5
Bacto Calcium Chloride 0.02M		6 × 100 ml	0378-73-6

BACTO TINSDALE BASE
BACTO TINSDALE ENRICHMENT, DESICCATED

INTENDED USE
Bacto Tinsdale Base enriched with Bacto Tinsdale Enrichment is an excellent differential plating medium for the primary isolation of *Corynebacterium diphtheriae*.

HISTORY/PRINCIPLES

Bacto Tinsdale Base and Enrichment are prepared according to Billings'[1] modification of the Tinsdale serum-cystine-thiosulfate-tellurite enrichment.

Bacto Tinsdale Enrichment, Desiccated contains bovine serum, sodium hydroxide, L-cystine, sodium thiosulfate and potassium tellurite in the quantity and proportion described by Billings[1] for use with Bacto Tinsdale Base.

Tinsdale[2] developed a serum-cystine-thiosulfate-tellurite-agar medium for the primary isolation and differentiation of *C. diphtheriae.* His formulation distinguished between the *C. diphtheriae* and diphtheroids which exhibited similar pleomorphism as well as between the *C. diphtheriae* and other microorganisms of the respiratory tract. The differential principle was based upon the capacity of *C. diphtheriae* to produce a brown or black halo around the colonies whereas other microorganisms of the respiratory tract did not. Billings[1] investigated the Tinsdale medium to determine its usefulness for routine diagnostic use and to elucidate the mechanism of its differential principle. He simplified the Tinsdale basal medium by using Proteose Peptone No. 3, Difco as a nutrient and improved the differential qualities as well as the reproducibility of the medium. Moore and Parsons[3] using Billings'[1] simplified formulation for the Tinsdale medium obtained excellent reproducible results. They demonstrated the characteristic, sharply outlined dark halos around *C. diptheriae* colonies to be specific for these organisms with but one exception. *C. ulcerans,* occasionally encountered in the nasopharynx, gave colonies similar to *C. diphtheriae* and require further biochemical differentiation. They investigated the reactions of numerous genera on the nodified Tinsdale medium and declared it to be an excellent medium for routine use in the primary isolation of *C. diphtheriae.*

FORMULA

BACTO TINSDALE BASE
DEHYDRATED

Ingredients per liter

Proteose Peptone No. 3, Difco . . . . 20 g
Sodium Chloride . . . . 5 g
Bacto Agar . . . . 20 g

Final pH 7.4 ± 0.2 at 25°C.

One pound will make 10 liters basal medium.

PROCEDURE

1. Suspend 45 g in 1 liter distilled or deionized water. Heat to boiling to dissolve completely.
2. Distribute in 100 ml amounts and sterilize in the autoclave for 15 minutes at 15 lbs pressure (121°C).
3. Allow to cool to 50 – 55°C.
4. Rehydrate Bacto Tinsdale Enrichment with sterile distilled or deionized water using the amount stated on label. Rotate in an end-over-end motion to dissolve completely.
5. Add 15 ml rehydrated enrichment per 100 ml of medium. Swirl to mix thoroughly. If single plates are required, mix 3 ml enrichment and 20 ml rehydrated Bacto Tinsdale Base.
6. Dispense into sterile Petri plates as desired and allow to solidify.
7. Inoculate the medium in a manner to obtain discrete colonies. Also stab the agar several times using a wire inoculating loop or straight wire.

STORAGE

Bacto Tinsdale Base	Below 30°C
Bacto Tinsdale Enrichment, Desiccated	2 – 8°C
Prepared plates	2 – 8°C

QUALITY CONTROL

Identity Specifications

Dehydrated powder:	light beige, homogeneous, free-flowing
Reaction of 4.5% solution:	pH 7.4 ± 0.2 at 25°C
Prepared medium:	light amber, slightly opalescent
Desiccated Enrichment:	tan button
Rehydrated Enrichment:	light to dark amber, slightly opalescent, may have a slight precipitate.

Typical Cultural Response on Bacto Tinsdale Medium
(Bacto Tinsdale Base with Bacto Tinsdale Enrichment)
After 40 – 48 Hours at 35°C

Organism	Growth	Colony Description
Corynebacterium diphtheriae Type gravis	good to excellent	dark brown to black w/halos
Corynebacterium diphtheriae Type intermedius	good to excellent	dark brown to black w/halos
Corynebacterium diphtheriae Type mitis	good to excellent	dark brown to black w/halos
Klebsiella pneumoniae ATCC® 13883	inhibited	—
Streptococcus pyogenes ATCC® 19615	good	black, pinpoint w/o halos

REFERENCES

1. An Investigation of Tinsdale-Tellurite Medium: Its Usefulness and Mechanism of Halo Formation. MS Thesis, University of Michigan, 1955.
2. J. Path. & Bact., 59:461, 1947.
3. J. Inf. Dis., 102:88, 1958.

PACKAGING

Bacto Tinsdale Base	1 lb (454 g)	0786-01-9
	1/4 lb (114 g)	0786-02-8
Bacto Tinsdale Enrichment, Desiccated	6 × 15 ml	0342-33-8
	6 × 3 ml	0342-46-3

TISSUE CULTURE

INTRODUCTION

As in the field of culture media and reagents for bacteriology and mycology, Difco Laboratories has pioneered in the preparation of reagents for in vitro propagation and maintenance of tissue cells and viruses. These reagents are tested and certified for use in tissue culture procedures. They are applicable to the slide, roller tube and flask culture techniques commonly employed for propagation and study of tissue cells and for virus studies.

TC media and reagents are prepared in the dried state and in liquid or solution form. Many of the chemically defined media are available in 10-fold concentration (10X), and are prepared for use by diluting 1 part of the 10X media with 9 parts TC Distilled Water, DE. Stability of some reagents and media is increased by drying. Sterile reagents and media, dried from the frozen state, to preserve unaltered the properties of the original material, are designated by the term Desiccated in the title. For the convenience of laboratories using large quantities of media and balanced salt solutions we have prepared them in dehydrated form. Dehydrated media, requiring sterilization following rehydration, are designated as Dried. Solutions and desiccated materials are packaged in bottles with a rubber closure permitting aseptic entry, unless otherwise specified. Dried formulations are supplied in screw capped bottles.

The reagents currently available are those commonly employed for culturing tissue cells, maintaining tissue banks and for the propagation of viruses. Additional reagents will be made available as required for these and other procedures related to tissue culture.

Cytogenetic reagents for a special application of the tissue culture method, i.e., the propagation and examination of chromosomes from human and animal lymphocytes, are available from Difco. For details refer to TC CHROMOSOME MEDIA and PHY-TOHEMAGGLUTININ M and P.

METHODS OF TISSUE CULTURE
Brief outlines of the more commonly used tissue culture techniques employing Difco Tissue Culture Media and Reagents are given below. These directions apply generally for any type of animal tissue; however, special investigation may be required to determine the optimal conditions for culturing a given tissue. For more detailed tissue culture techniques refer to a standard text on the subject.

Proportions of the various reagents given in the discussion may be varied to meet individual requirements.

PLASMA CULTURES
Slide Cultures
The sterile tissue to be cultured should be placed on a clean sterile glass surface and cut into appropriate sized explants while immersed in sterile balanced salt solution (pH 7.4). The following dimensions are usually most suitable for the special tissue.

minced tissues	0.5 – 1 mm
embryonic tissues	1 mm
adult tissues	1.5 mm
adult skin	2 mm

Further rinsing of the explants in balanced salt solution may be done if desired. Slide cultures are prepared by adding 1 drop of TC Chicken Plasma to a cover glass, followed by 1 drop of TC Embryo Extract EE_{20}. Larger amounts may be used in the same ratio if desired. Mix the 2 with a spatula and spread over an area having a diameter of 10 mm. Add 2 or 3 explants and separate them by about equal distances. Cover with a well-slide and set aside to clot. Later, seal the cover glass to the well-slide with melted paraffin or a mixture of 5 parts paraffin and 1 part yellow petroleum jelly.

A double cover glass technique permitting greater facility of handling and feeding of slide cultures may be employed. In this double cover glass method, a large sterile cover glass or mica cover slip is placed on a clean surface, a drop of sterile Balanced Salt

Solution placed in the center of the mica slide and a sterile cover glass placed on the drop of salt solution. The TC Chicken Plasma, Embryo Extract and the explants are then added as described above. A small amount of petroleum jelly is placed on the 4 corners of the mica cover slip and a sterile well-slide inverted over the 2 cover slips centering the inoculated cover glass. The edges of the mica cover slip are then cemented down with paraffin or the mixture of paraffin and petroleum jelly.

More diffuse and extensive cell migration may be encouraged by diluting the plasma before using with an equal or double volume of a Balanced Salt Solution. Incubate the slide culture at 37°C, observe and continue culture in accordance with standard procedures.

Roller Tube Cultures
Prepare explants as for slide cultures and place them in a small volume of nutrient solution in a well-slide or Petri dish. The nutrient may consist of various proportions of a Balanced Salt Solution, TC Medium 199, TC Horse Serum, or TC Human Serum and TC Embryo Extract EE_{20} (5 parts of Balanced Salt Solution, 3 parts Serum and 2 parts Embryo Extract is satisfactory for general purposes). Place 2 drops of whole TC Chicken Plasma (50% Plasma is also satisfactory) in a 16 × 150 mm tube and rotate until bottom third or half of tube is coated. With a pipette, place 4 or 5 explants on the uncoated area near the top of the tube and remove excess nutrient from around explant with a Pasteur pipette. Then with pipette, transfer individual explants to plasma coated surface. After about 10 minutes, stopper tubes and place them in the roller drum in the 37°C incubator for 2 hours. The nutrient medium (mentioned above) should now be added in 0.5 ml volumes to each tube and the tubes returned to roller drum in the incubator. Incubate at 37°C, observe and continue culture in the usual manner.

Flask or Bottle Cultures
Prepare explants in the desired balanced salt solution. Explants may be made considerably larger for culture in flasks than for slide or roller tube cultures. While the conventional size of the explants has approximated 1 × 1 × 1 mm, Earle has used explants ranging from 0.5 × 1 × 2 mm to 0.5 × 3 × 2 mm. In other cases he has substituted cell suspensions for the explants.

To a 35 mm flask (Carrel type), add 0.3 ml of TC Chicken Plasma, either whole or diluted to 50 or even 25% with the balanced salt solution and spread over bottom of flask. This is then increased in volume to 1 ml by introducing 0.6 ml of the balanced salt solution and 0.1 ml of TC Embryo Extract EE_{20}. Add explants and orient them. Stopper flasks and set aside to clot. Incubate for 1 – 2 hours to give firm clot, then add 0.5 ml of nutrient mixture as described under Roller Tube Culture. Where optical perfection is not important, flat sided prescription bottles make convenient culture vessels.

Watch Glass Cultures
This technique is used by experimental embryologists to study organized growth *in vitro*. Rudimentary tissues and organs grow and maintain their histological and often their anatomical development during cultivation. The method can be used to cultivate many kinds of undifferentiated tissues.

Prepare explants as for slide cultures but carry out the dissection on the stage of a dissecting microscope. Free the desired rudiment from surrounding tissues of the embryo with a dissecting needle and a cataract knife. Wash the rudiment several times in a sterile balanced salt solution and transfer to the watch glass of a prepared culture vessel.

To prepare a culture vessel cut a circle of absorbent cotton to fit an 80 mm Petri dish. A 25 mm disc is removed from the center of the cotton. A 40 mm watch glass is located over the central hole in the cotton. After sterilization by dry heat, 10 ml of TC Distilled Water DE is pipetted into the Petri dish. The water moistens the cotton to form a moist chamber. The medium may consist of 4 drops each of reconstituted TC Chicken Plasma and TC Embryo Extract EE_{20}. Mix and spread with a glass rod. Several explants can be transferred in a drop of sterile balanced salt solution to the plasma clot in the prepared culture vessel. Finally the excess saline is aspirated from the tissue with a capillary pipette. Cover the Petri dish and place in the 37°C incubator. Every 2 – 3 days the explants are freed from the old clot, washed in saline and transferred to freshly prepared culture vessels.

SERUM CULTURES, STATIONARY

Many cells and tissues may be cultivated without the support of a plasma or fibrin clot in fluid media. Fresh tissue explants, minces or suspensions and certain strain cells left undisturbed will attach to a substratum which may be perforated cellophane, agar or the glass wall of the culture vessel. For example, fresh adult monkey kidney, chick embryo tissues, the HeLa cell strain of Gey, and the murine L strain cell of Earle will grow on glass of culture vessels containing suitably diluted serum.

Thin sheets of cells or cell monolayers adhering to surfaces in culture vessels provide excellent means for studying the cytopathogenic effect of viral agents upon the cells and for observing the effect of other additives to the culture. The cellular response can be viewed microscopically through the vessel wall. Massive sized cultures of minced kidney are used for the routine production of poliomyelitis vaccine.

Roller Tube Cultures

Explants for roller tube cultures can be prepared as for plasma cultures. After introducing the explants, a short drying period is necessary to allow the tissue to adhere to the glass of the roller tubes before introducing the fluid medium. Media may consist of various preparations of TC Sera in a balanced salt solution containing 2 – 3% TC Embryo Extract EE_{100}. Alternatively, TC Sera in combination with certain TC Media will eliminate the embryo extract requirement. The cultures should be incubated at 37°C in stationary racks or in a roller drum at 10 – 20 revolutions per hour with fluid changes twice a week.

In some experimental situations it is desirable to prepare uniform cell suspensions from fresh tissues or established cultures. For example, suspensions of cells are used to inoculate replicate cultures, to estimate cell populations by direct counting and to isolate single cells from a mixed population. Cell suspensions can be readily obtained by the use of Trypsin 1:250, Difco as described by Syverton and Scherer[1] and Swim.[2] The concentration and duration of activity of the enzyme varies somewhat for different cells and tissues.

The procedure for the preparation of a cell suspension of fresh monkey kidney will be described. The kidney capsule is carefully removed aseptically and the kidney minced into small cubes 1 – 4 mm. The fragments are transferred to a small flask containing about 30 ml of 0.25% Trypsin 1:250, Difco in TC Dulbecco Solution. The flask is incubated at 37°C for 10 minutes and the supernatant decanted and replaced by 15 ml of fresh trypsin solution at 37°C. The fragments are then rapidly pipetted back and forth to disrupt the fragments into cell clumps and single cells. Allow the remaining cell fragments to settle and decant the supernatant into a sterile centrifuge tube. The remaining cell fragments are again trypsinized, pipetted and the supernatant decanted as before. This procedure is repeated 7 or 8 times. The supernatants are all centrifuged at 600 rpm for 2 minutes, and resuspended in 1/2 the original volume of salt solution. Centrifugation and resuspension is repeated twice more in half the previous volume of salt solution. Finally the cells are resuspended in 10 – 20 volumes of TC Earle Solution. After vigorous pipetting the suspension is ready to inoculate directly into prepared culture vessels. Nutrient medium together with 0.25 ml of suspension will provide luxuriant growth in a 16 × 150 mm roller tube.

Flask or Bottle Cultures
Flask or bottle cultures may be prepared using explants as for plasma cultures or they may be inoculated with trypsinized cell suspensions as described above.

Petri Dish Cultures
This technique as developed by Dulbecco[3] is widely used in experimental biology and medicine to study animal virus-host cell relations. Essentially the method consists of growing cells on agar in conventional Petri dishes until a monolayer of cells is formed. At this stage, the cultures are inoculated with a virus suspension and incubated briefly to permit adsorption of the virus. The monolayer is then washed, overlaid with a thin layer of agar and incubated at 37°C. As incubation proceeds the infected cells produce macroscopically visible plaques, characteristic of the kind and type of virus. The number of plaques per plate varies inversely with the dilution of virus thus providing a method of virus titration. Also, pure virus lines can be isolated from the viruses produced in a single plaque.

SERUM CULTURES, AGITATED, SPINNER
Roller Tube Cultures
In contrast to the aforementioned techniques, certain strains of animal cells will multiply in suspension in suitable nutrient media provided that the cultures are continuously agitated as described by Gey and Earle and their co-workers.[4,5,6,7] Suspended cell cultures can also be propagated in conventional roller drums rotated at relatively high speeds, for example 40 – 50 rpm. The minimal inoculum should be of the order of 10^5 to 3×10^5 cells per ml, as determined by light staining with 1% methylene blue followed by enumeration in a bacterial counting chamber or hemocytometer. A suitable suspension medium consists of 40% TC Serum in a TC Balanced Salt Solution with or without 2% TC Chick Embryo Extract.

Cultures may be inoculated into any size conventional test tube with a suitable closure, but to ensure adequate oxygen supply the total culture volume should not occupy more than two-thirds the volume of the containing tube. Cultures of such cells may be kept growing in the logarithmic phase indefinitely and a 100-fold increase in cell population can be achieved without an early lag. Portions of the culture can be withdrawn from time to time for biochemical, cell enumeration or virus infectivity tests.

Flask Cultures
Earle, Bryant and Schilling[5] grew the L cell in suspension in Erlenmeyer flasks kept agitated on a Brunswick type shaker. The inoculum size and the medium used are similar to conditions used for the rapidly rotating roller tubes described above. Successful applications of agitated flask culture methods were reported by Kuchler and Merchant,[9] Cherry and Hull,[10] and McLimans, Davis, Glover and Rake.[11] The Spinner culture method of McLimans et al[11] utilizing a suspended magnetic stirrer in a closed system permits large scale growth of cells in suspension.

CHEMICALLY DEFINED CULTURE MEDIA
A selection of chemically defined TC Media are now available for the maintenance and growth of a variety of animal and plant cells in tissue culture. These media were de-

vised for the purpose of studying cell nutrution, virus propagation, virus isolation and other applications where reproducible nutrient fluids are required that are free of antibodies, non-antigenic and economical.

Roller Tube, Flask or Bottle Cultures

The techniques for preparing roller tube, flask or bottle cultures in chemically defined TC Media are exactly as mentioned above with the exception that the agitated or rapidly rotating cultures usually do not give good growth in chemically defined media without added serum or tissue extract.

AGAR SLANT TISSUE CULTURE

Make a 2-fold concentration of Bacto Noble Agar in distilled or deionized water so that when diluted with an equal quantity of tissue culture medium, the agar concentration will be 1 – 2% as desired. Sterilize the agar solution for 15 minutes at 15 lbs pressure (121°C).

Prepare a sterile two-fold concentration of the liquid tissue culture medium to be employed.

Cool the agar to about 50°C. Prewarm the sterile tissue culture medium to the same temperature. Combine equal quantities of each, add antibiotics if desired, and adjust the pH with TC Bicarbonate Solution 10% or 0.3 N HCl to 7.2 – 7.4. The easiest way to determine the pH is to remove a 1 ml sample with a sterile pipette to a small glass vial and compare this to a vial containing fluid adjusted to the desired pH value.

Dispense the combined agar-tissue culture medium in 4 ml amounts into glass tubes fitted with rubber-lined screw caps and slant to give a long surface. When solidified, add 0.3 ml of the sterile single strength liquid tissue culture medium to the tube at the base of the slant. Store the slants at 2 – 8°C in an upright position. Such tubes are satisfactory for periods in excess of 6 months.

To prepare the inoculum, centrifuge harvested cells from monolayer or suspension culture at 165 × G for 5 minutes. Decant the supernatant fluid, and disperse the cells in the small volume of fluid remaining. Use an ordinary 2 mm bacteriological loop to transfer the inoculum cells to slants. One loopful per slant is sufficient. Touch the loopful to several places on the agar slant rather than streaking the entire surface. Incubate the slants in an upright position at 35°C. Transplants of colonies to fresh slants are made with the loop in the same manner described above.

Colonies appear within 48 hours when incubated in an air atmosphere supplemented with 5 – 10% CO_2. In ordinary atmospheric air, colonies appear in about 4 days. The colonies develop to 5 – 10 mm in diameter within a week to 10 days. The nutrient solution should be changed once a month, or more often if desired.

BALANCED SALT SOLUTIONS

Balanced Salt Solutions prepared in accordance with accepted formulae are supplied for all tissue culture and virus procedures. Most of these formulations are available in sterile ten-fold (10X) concentration and in the dried state, as well as in single strength solution.

REFERENCES

1. Ann. N.Y. Acad. Sci., 58:1056, 1954.
2. Proceeding Joint Meeting Michigan-Ohio S.A.B., May 1953.
3. J. Exp. Med., 99:167, 1954.

4. Proc. Am. Assoc. Cancer Res., 4:1, 1953.
5. Ann. N.Y. Acad. Sci., 58:1000, 1954.
6. J. Nat. Cancer Inst., 14:1159, 1954.
7. Ann. N.Y. Acad. Sci., 58:1039, 1954.
8. Proc. Soc. Exp. Biol. Med., 89:326, 1955.
9. ibid., 92:803, 1956.
10. Abstract Tissue Culture Assoc., Milwaukee, 1956.
11. J. Immun., 79:428, 1957.

TISSUE CULTURE MEDIA AND REAGENTS

TC AMINO ACIDS HeLa, 100X

INTENDED USE
TC Amino Acids HeLa, 100X is a sterile 100-fold concentration, certified for use in the preparation of Eagle HeLa Medium for tissue culture procedures and virus studies. It provides a ready source of amino acids in the ratio recommended by Eagle[1] for the preparation of Eagle HeLa Medium.

FORMULA
TC AMINO ACIDS HeLa, 100X
Ingredients per liter

L-Arginine	17.4 mg	L-Methionine	7.5 mg
L-Cystine	12 mg	L-Phenylalanine	16.5 mg
L-Histidine	7.8 mg	L-Threonine	23.8 mg
L-Isoleucine	26.2 mg	L-Tryptophane	4.1 mg
L-Leucine	26.2 mg	L-Tyrosine	18.1 mg
L-Lysine	29.2 mg	L-Valine	23.4 mg

STORAGE
Store at 2 – 8°C.

REFERENCE
1. Science, 122:501, 1955.

PACKAGING
TC Amino Acids HeLa, 100X   100 ml   5790-72-6

TC AMINO ACIDS MINIMAL EAGLE, 50X

INTENDED USE
TC Amino Acids Minimal Eagle, 50X is a sterile 50-fold concentrate, certified for use in the preparation of TC Minimal Medium Eagle for tissue culture procedures and virus studies. It provides a ready source of amino acids in the ratio recommended by Eagle[1] for the preparation of TC Minimal Medium Eagle.

FORMULA
TC AMINO ACIDS MINIMAL EAGLE, 50X
Ingredients per liter

L-Arginine	105 mg	L-Methionine	15 mg
L-Cystine	24 mg	L-Phenylalanine	32 mg
L-Histidine	31 mg	L-Threonine	48 mg
L-Isoleucine	52 mg	L-Tryptophane	10 mg
L-Leucine	52 mg	L-Tyrosine	36 mg
L-Lysine	58 mg	L-Valine	46 mg

PROCEDURE

To prepare 1 liter of TC Minimal Medium Eagle, aseptically combine the amounts indicated:

TC Amino Acids Minimal
Eagle 50X . . . . 20 ml
TC Vitamins Minimal Eagle 100X . . 10 ml
TC Glutamine 5% . . . . 6 ml
TC Sodium Bicarbonate 10% . . . . . 20 ml

Sterile Salt Solution containing the following:
Bacto Dextrose . . . . 1000 mg
Sodium Chloride . . . . 6800 mg
Potassium Chloride . . . . 400 mg
Calcium Chloride . . . . 200 mg
Magnesium Chloride . . . . 200 mg
Monosodium Phosphate . . . . . 150 mg
TC Distilled Water, DE . . . . . . 944 ml

STORAGE

Store at 2 – 8°C.

REFERENCE

1. Science, 130:432, 1959.

PACKAGING

TC Amino Acids Minimal Eagle, 50X   100 ml   5831-72-7

TC BICARBONATE SOLUTION 10%

INTENDED USE

TC Bicarbonate Solution 10% is a sterile concentrated solution certified for use in adjusting the reaction of media and balanced salt solutions used in tissue culture procedures.

STORAGE

Store at 15 – 30°C.

PACKAGING

TC Bicarbonate Solution 10%   6 × 10 ml   5788-60-4
                                      100 ml   5788-72-0

TC BOVINE EMBRYO EXTRACT EE$_{100}$, DESICCATED

INTENDED USE

TC Bovine Embryo Extract EE$_{100}$, Desiccated is sterile, desiccated, undiluted bovine embryo extract certified for use in tissue culture procedures requiring bovine embryo extract. It is prepared from fresh eviscerated bovine embryos 60 – 90 days of age.

PROCEDURE

TC Bovine Embryo Extract EE$_{100}$, Desiccated is rehydrated to its original volume by adding aseptically 2 ml TC Reconstituting Fluid or TC Distilled Water DE to each bottle. When the contents of the bottle have uniformly combined with the liquid they are diluted further, as desired, with TC Earle Solution. A 20% embryo extract EE$_{20}$ has been found useful in tissue culture procedures. This concentration is obtained by adding 4 volumes (8 ml) of TC Earle Solution to the rehydrated TC Bovine Embryo Extract EE$_{100}$. The reconstituting fluid and balanced salt solution may be added conveniently by means of a sterile syringe and needle.

The diluted embryo extract EE_{20}, or other desired concentration, is allowed to stand at $35 - 37°C$ for $30 - 60$ minutes to obtain maximum solution of the nutrients. Solution of the desiccated material is accelerated by gentle end-over-end rotation of the bottle. Do not shake. A small portion of the rehydrated bovine embryo extract may remain in suspension. This insoluble residue should be removed by centrifugation of the bottle for 10 minutes at $2500 - 3000$ rpm. The supernatant liquid is now ready for use.

STORAGE
Store at $2 - 8°C$.

Unused portions of the diluted embryo extract should be stored in tightly stoppered containers in the refrigerator at $2 - 8°C$. Storage of the rehydrated TC Bovine Embryo Extract for periods longer than 1 week, even though refrigerated, is not recommended.

PACKAGING
TC Bovine Embryo Extract EE_{100}, Desiccated   2 ml   5396-53-7

TC BOVINE SERUM, DESICCATED

INTENDED USE
TC Bovine Serum, Desiccated is sterile, desiccated, whole, pooled, normal calf serum certified for use in all tissue culture and other microbiological procedures requiring bovine serum. It is obtained from fasted healthy calves.

PROCEDURES
TC Bovine Serum, Desiccated is rehydrated to its original volume by adding aseptically the quantity of TC Reconstituting Fluid or TC Distilled Water DE specified on the label. Solution of the material is accelerated by gentle agitation of the bottle.

The rehydrated TC Bovine Serum is alkaline in reaction due to loss of CO_2 in the desiccation process and, if desired, may be adjusted to pH $7.2 - 7.4$ in an atmosphere of CO_2 as described for the adjustment of TC Chicken Plasma, Desiccated.

STORAGE
TC Bovine Serum, Desiccated is stable when stored below $8°C$. The rehydrated TC Bovine Serum should be kept in tightly stoppered containers in the refrigerator at $2 - 8°C$. Storage of the rehydrated TC Bovine Serum for periods longer than 1 week, even though refrigerated, is not recommended.

PACKAGING

TC Bovine Serum, Desiccated		
	6×5 ml	5565-57-8
	30 ml	5565-67-6
	100 ml	5565-72-9

TC CHICK EMBRYO EXTRACT EE_{100}, DESICCATED

INTENDED USE
TC Chick Embryo Extract EE_{100}, Desiccated is sterile, desiccated, whole, undiluted chick embryo extract certified for use in tissue culture procedures requiring chick embryo extract. It is prepared from selected 11 day old chick embryos.

HISTORY

Duncan, Silverthorne, McNaughton, Johnson and Rhodes[1] embedded monkey testes fragments in rehydrated TC Chicken Plasma and clotted the mixture with reconstituted TC Chick Embryo Extract for the cultivation of the tissue in their method of poliomyelitis diagnosis.

PROCEDURE

TC Chick Embryo Extract EE_{100}, Desiccated is rehydrated by adding aseptically 2 ml of TC Reconstituting Fluid or TC Distilled Water DE to the 2 ml bottle. When the contents of the bottle have uniformly combined with the liquid, they are diluted further as desired with TC Earle Solution. A 20% embryo extract EE_{20} has been found to be useful in tissue culture procedures. This concentration is obtained by adding 4 volumes (8 ml) of TC Earle Solution to the rehydrated embryo extract EE_{100}. The TC Reconstituting Fluid and the TC Earle Solution may be added conveniently by means of a sterile syringe and needle.

The diluted embryo extract EE_{20}, or other desired concentration, is allowed to stand at $35 - 37°C$ for $30 - 60$ minutes to obtain maximum solution of the nutrients. Solution of the desiccated material is accelerated by gentle end-over-end rotation of the bottle. Do not shake. A small portion of the desiccated embryo extract may remain in suspension. This insoluble residue should be removed by centrifugation of the bottle for 10 minutes at $2500 - 3000$ rpm. The supernatant liquid is now ready for use.

It has been our observation that embryo extracts diluted with balanced salt solutions containing calcium in levels comparable with TC Earle Solution give a faster clotting rate and a firmer coagulum with TC Chicken Plasma than do extracts prepared with a lower calcium content.

STORAGE

Store at $2 - 8°C$.

Unused portions of the diluted embryo extract should be stored in tightly stoppered containers in the refrigerator at $2 - 8°C$. Storage of the rehydrated TC Chick Embryo Extract for periods longer than 1 week, even though refrigerated, is not recommended.

REFERENCE

1. Canad. J. Pub. Health, 45:55, 1954.

PACKAGING

TC Chick Embryo Extract EE_{100}, Desiccated　　　　　2 ml　　　　　5355-53-6

TC CHICKEN PLASMA, DESICCATED

INTENDED USE

TC Chicken Plasma, Desiccated is sterile, desiccated, whole, chicken plasma certified for use in culturing tissue cells *in vitro*. It is prepared from selected disease-free cockerels $4 - 8$ months old.

HISTORY

Duncan, Silverthorne, McNaughton, Johnson and Rhodes[1] used rehydrated TC Chicken Plasma, Desiccated and TC Chick Embryo Extract EE_{100}, Desiccated for the cultivation of monkey testes fragments in their method of diagnosis of poliomyelitis. They reported that this method could be used routinely in hospital laboratories for poliomyelitis diagnosis.

PROCEDURES

TC Chicken Plasma, Desiccated is rehydrated to its original volume by adding aseptically 5.0 ml sterile TC Reconstituting Fluid or TC Distilled Water DE to each bottle. The reconstituting fluid may be added conveniently with a sterile 5 ml syringe. Solution of the material is accelerated by gentle end-over-end rotation of the bottle. Do not shake. Allow 30 minutes for complete solution of the desiccated plasma.

The rehydrated plasma is alkaline in reaction due to loss of CO_2 in the desiccation process and requires adjustment with CO_2 to pH 7.2 – 7.4 if a normal clotting is desired. Adjustment of the reaction with CO_2 to pH 7.2 – 7.4 accelerates the clotting time when embryo extract and tissue explants are added. The adjustment of the reaction of the reconstituted plasma is best accomplished by first adding TC Phenol Red Solution 1% to the plasma to give a final phenol red concentration of 0.002%. (Dilute 0.1 ml of TC Phenol Red 1% to 50 ml, and use 5 ml of the 1:500 dilution to rehydrate the 5 ml of TC Chicken Plasma.) Introduce CO_2 to bring the reaction to pH 7.2 – 7.4.

Three methods of introducing CO_2 that have given good results include:
1. Transfer the rehydrated TC Chicken Plasma containing 0.002% phenol red to a clean, dry, sterile cotton-stoppered or loosely capped test tube. Place the tube in a beaker or container of sufficient depth so that the walls extend 1 inch or more above the mouth of the tube. Introduce CO_2 or a mixture of CO_2 and nitrogen into the beaker to displace the air and then cover the beaker with paper or suitable cover. Gently agitate the plasma in the tube occasionally to bring all parts of the solution in contact with the CO_2. Observe color change in the indicator from cerise alkaline reaction to reddish-orange almost neutral reaction. Reasonable care should be exercised to prevent over acidification as this will tend to coagulate the plasma spontaneously. When the indicator color change begins to turn from cerise to reddish-orange, double check the pH by removing a drop of the plasma aseptically and transferring it to a few drops of distilled or deionized water to which indicator has been added on a spot plate or in a test tube. The recheck of the reaction by the test tube or spot plate method is essential because indicators do not always give true color changes in undiluted protein solutions.

   Failure to adjust the reaction of TC Chicken Plasma may give prolonged clotting times. For example, undiluted chicken plasma containing 0.002% phenol red may give a color reaction indicative of pH 7.2 whereas the true reaction as determined by dilution in a test tube or spot plate may be pH 8.0 or greater. Also, color reactions with phenol red in undiluted plasma indicative of pH 7.2 reaction may actually be more acid by the test tube dilution and spot plate methods. When the reaction has been adjusted to pH 7.2 – 7.4, stopper tightly to prevent escape of CO_2. Store in refrigerator at 2 – 8°C if not used immediately. Storage for periods longer than 1 week is not recommended.
2. Transfer the rehydrated plasma containing phenol red to a clean dry sterile tube. Bubble sterile CO_2 or a mixture of sterile CO_2 and nitrogen through the solution until the cerise color changes to reddish-orange. Recheck pH by test tube or spot plate as outlined above. Adjust final reaction to pH 7.2 – 7.4.
3. Same as method 2 above, except blow sterile alveolar air through the plasma solution to give pH 7.2 – 7.4 using precautions stated above. It has been our observation that embryo extracts diluted with balanced salt solutions containing calcium in levels comparable with TC Earle Solution give a faster clotting rate and a firmer

coagulum with TC Chicken Plasma than do extracts prepared with a lower calcium content.

STORAGE
TC Chicken Plasma, Desiccated   2 – 8°C

REFERENCE
1. Canad. J. Pub. Health, 45:55, 1954.

PACKAGING
TC Chicken Plasma, Desiccated   6 × 5 ml   5354-57-3

TC DISTILLED WATER, DE

INTENDED USE
TC Distilled Water, DE is sterile and certified for use in tissue culture procedures. It is recommended as a reconstituting fluid for tissue culture reagents and for the preparation of balanced salt solutions and other media.

STORAGE
Store at 15 – 30°C.

PACKAGING
TC Distilled Water, DE   100 ml   5353-72-5
                                         1 liter   5353-81-4

TC DULBECCO SOLUTION

INTENDED USE
TC Dulbecco Solution is a sterile balanced salt solution certified for use in tissue culture and virus procedures. It is prepared according to the formulation of Dulbecco and Vogt.[1]

FORMULA
TC DULBECCO SOLUTION
Ingredients per liter

Sodium Chloride	8000 mg	Monopotassium Phosphate	200 mg
Potassium Chloride	200 mg	Calcium Chloride	100 mg
Disodium Phosphate	1150 mg	Magnesium Chloride	100 mg

STORAGE
Store at 15 – 30°C.

PACKAGING
TC Dulbecco Solution   100 ml   5785-72-3

TC DULBECCO SOLUTION, DRIED

INTENDED USE

TC Dulbecco Solution, Dried is the dried formulation of the balanced salt solution described by Dulbecco and Vogt.[1] It provides a ready stable source medium for the preparation of fresh liquid solution whenever required.

PROCEDURE

To rehydrate, suspend 9.8 g in 1 liter TC Distilled Water, DE by adding the powder slowly to the water while stirring. Continue stirring until dissolved as completely as possible. (A slight precipitate may be present, which will be removed by filtration.) Filter sterilize by vacuum or pressure filtration.

STORAGE

Store at 15 – 30°C.

REFERENCE

1. J. Exp. Med., 99:167, 1954.

PACKAGING

TC Dulbecco Solution, Dried   10 × 1 liter   5787-37-5

TC EARLE SOLUTION

INTENDED USE

TC Earle Solution is a balanced salt solution certified for use in tissue culture procedures. It is prepared according to the formula of Earle.[1]

FORMULA

TC EARLE SOLUTION

Ingredients per liter

Bacto Dextrose	1000 mg	Magnesium Sulfate	100 mg
Sodium Chloride	6800 mg	Monosodium Phosphate	125 mg
Potassium Chloride	400 mg	Sodium Bicarbonate	2200 mg
Calcium Chloride	200 mg		

PROCEDURE

TC Phenol Red Solution 1% may be added in 0.002% concentration as an indicator of reaction by adding 0.2 ml TC Phenol Red Solution 1% to 100 ml TC Earle Solution.

STORAGE

Store at 2 – 8°C.

PACKAGING

TC Earle Solution   100 ml   5351-72-7
                                           400 ml   5351-78-1

TC EARLE SOLUTION, DRIED

INTENDED USE
TC Earle, Solution, Dried is the dried formulation of the balanced salt solution of Earle[1] without sodium bicarbonate. It provides a ready stable source for the preparation of fresh liquid solution whenever required.

PROCEDURE
To rehydrate, suspend 8.7 g in 1 liter TC Distilled Water, DE by adding the powder slowly to the water while stirring. Continue stirring until completely dissolved.

Adjust to the desired pH by adding TC Bicarbonate Solution 10%.

Filter sterilize the solution preferably using pressure, and stopper tightly to prevent loss of CO_2. Filtration or storage can result in a loss of CO_2, which causes increased alkalinity. The reaction of the sterile filtrate should be readjusted with sterile CO_2 if it is too alkaline.

STORAGE
Store at $2-8°C$.

REFERENCE
1. National Cancer Inst., 4:167, 1943.

PACKAGING
TC Earle Solution, Dried   10 × 1 liter   5773-37-1

TC FETAL CALF SERUM, DESICCATED

INTENDED USE
TC Fetal Calf Serum, Desiccated is a desiccated, sterile, whole, pooled, bovine embryonic serum for use as an enrichment in tissue culture media. It is obtained from bovine fetuses less than 7 months of age and processed carefully to preserve all growth factors.

PROCEDURE
TC Fetal Calf Serum, Desiccated is rehydrated for use by adding aseptically the quantity of TC Reconstituting Fluid or TC Distilled Water DE specified on the label.

Solution of the desiccated material is accelerated by gentle agitation of the bottle. The rehydrated serum will have an elevated pH and may require adjustment with sterile CO_2 to $7.2-7.5$ before use.

STORAGE
The desiccated serum is stable when refrigerated below 8°C. The rehydrated serum should be stored at $2-8°C$ and used within a week.

PACKAGING

TC Fetal Calf Serum, Desiccated	30 ml	5065-67-1
	100 ml	5065-72-4
	500 ml	5065-88-6

TC GLUTAMINE 5%

INTENDED USE

TC Glutamine 5% is a ready stable source of sterile L-glutamine in 0.85% NaCl solution, certified for use in the preparation of culture media for the maintenance and propagation of tissue cells and for virus studies.

PROCEDURE

TC Glutamine 5% is rehydrated by adding sterile TC Distilled Water, DE in the amount indicated on the label. This provides a 5% L-glutamine solution (340 mMolar), or 50 mg L-glutamine per ml, in saline. The rehydrated solutions are stable for one week when stored at 2 – 8°C.

STORAGE

Store at 2 – 8°C.

PACKAGING

TC Glutamine 5%	10 ml	5789-59-6
	30 ml	5789-67-6
	60 ml	5789-83-6

TC HANKS SOLUTION

INTENDED USE

TC Hanks Solution is a sterile balanced salt solution certified for use in tissue culture procedures. It is prepared according to the formula of Hanks.[1]

FORMULA

TC HANKS SOLUTION

Ingredients per liter

Bacto Dextrose	1000 mg	Magnesium Chloride	100 mg
Sodium Chloride	8000 mg	Monopotassium Phosphate	60 mg
Potassium Chloride	400 mg	Disodium Phosphate	60 mg
Calcium Chloride	140 mg	Sodium Bicarbonate	350 mg
Magnesium Sulfate	100 mg	Phenol Red	20 mg

STORAGE

Store at 2 – 8°C.

PACKAGING

TC Hanks Solution	6 × 12 ml	5508-33-7
	100 ml	5508-72-9
	400 ml	5508-78-3

TC HANKS SOLUTION, DRIED

INTENDED USE
TC Hanks Solution, Dried is the dried formulation of the balanced salt solution of Hanks[1] without sodium bicarbonate. It provides a ready stable source for the preparation of fresh liquid solution whenever required.

PROCEDURE
To rehydrate, suspend 9.9 g in 1 liter TC Distilled Water, DE by adding the powder slowly to the water while stirring. Continue stirring until the solution is completely dissolved.

Adjust to the desired pH by adding TC Bicarbonate Solution 10%.

Filter sterilize the medium preferably using pressure, and stopper tightly to prevent loss of CO_2. Filtration or storage can result in a loss of CO_2, which causes increased alkalinity. The reaction of the sterile filtrate should be readjusted with sterile CO_2 if it is too alkaline. (Alkalinity may be indicated by a change in color towards cerise.)

STORAGE
Store at 15 – 30°C.

REFERENCE
1. Proc. Soc. Exp. Biol. Med., 71:196, 1949.

PACKAGING
TC Hanks Solution, Dried	10 × 1 liter	5775-37-9

TC HANKS SOLUTION, 10X

INTENDED USE
TC Hanks Solution, 10X is a sterile 10-fold concentration of Hanks Solution without sodium bicarbonate. It is recommended for use in preparing single strength TC Hanks Solution for tissue culture procedures and other uses requiring an isotonic balanced salt solution.

PROCEDURE
One liter of single strength Hanks Solution is prepared by adding 100 ml TC Hanks Solution, 10X to 900 ml sterile TC Distilled Water, DE. Adjust to the desired pH using TC Bicarbonate Solution 10%.

STORAGE
Store at 2 – 8°C.

PACKAGING
TC Hanks Solution, 10X	100 ml	5774-72-6
	400 ml	5774-78-0

TC HORSE SERUM, DESICCATED

INTENDED USE
TC Horse Serum, Desiccated is sterile, desiccated, whole, pooled horse serum certified for use in culturing tissue cells *in vitro*. It is obtained from the blood of fasted, healthy, lean geldings.

PROCEDURE
TC Horse Serum, Desiccated is rehydrated to its original volume by adding aseptically the quantity of TC Reconstituting Fluid or TC Distilled Water DE specified on the label. Solution of the material is accelerated by end-over-end rotation of the bottle. Do not shake.

The rehydrated TC Horse Serum, Desiccated is alkaline in reaction due to loss of CO_2 in the desiccation process and, if desired, may be adjusted to pH 7.2 – 7.4 in an atmosphere of CO_2 as described for the adjustment of TC Chicken Plasma, Desiccated.

STORAGE
TC Horse Serum, Desiccated is stable when stored below 8°C. The rehydrated serum should be kept tightly stoppered, stored at 2 – 8°C and used within a week.

PACKAGING
TC Horse Serum, Desiccated	30 ml	5357-67-8
	100 ml	5357-72-1

TC HUMAN SERUM, DESICCATED

INTENDED USE
TC Human Serum, Desiccated is sterile, fresh, whole, human serum certified for use in culturing tissue cells *in vitro*.

PROCEDURE
TC Human Serum, Desiccated is rehydrated to its original volume by adding aseptically the quantity of TC Reconstituting Fluid or TC Distilled Water DE specified on the label. Solution of the material is accelerated by gentle end-over-end rotation of the bottle. Do not shake.

The rehydrated TC Human Serum, Desiccated is alkaline in reaction due to loss of CO_2 in the desiccation process and, if desired, may be adjusted to pH 7.2 – 7.4 in an atmosphere of CO_2 as described for the adjustment of TC Chicken Plasma, Desiccated.

STORAGE
TC Human Serum, Desiccated is stable when stored below 8°C. Unused portions of the rehydrated TC Human Serum should be kept in tightly stoppered containers in the refrigerator at 2 – 8°C. Storage of the rehydrated TC Human Serum for periods longer than 1 week, even though refrigerated, is not recommended.

PACKAGING
TC Human Serum, Desiccated	6 × 5 ml	5578-57-3
	100 ml	5578-72-4

TC LACTALBUMIN HYDROLYSATE

INTENDED USE
TC Lactalbumin Hydrolysate is an enzymatic digest of lactalbumin for use as an enrichment in tissue culture media. It is usually employed in concentrations of 0.25 – 0.5% as described by Melnick,[1] Merchant, Kahn and Murphy,[2] and Takaoka, Katsuta, Kaneko, Kauana and Farukawa.[3]

STORAGE
Store at 15 – 30°C.

REFERENCES
1. Annals N.Y. Acad. Science, 61:75, 1955.
2. Handbook of Cell and Organ Culture, Burgess Publ. Co., 1960.
3. Japanese J. of Exp. Med., 30:391, 1960.

PACKAGING
TC Lactalbumin Hydrolysate	1 lb (454 g)	5996-01-4
	10 kg	5996-08-7

TC MEDIUM EAGLE, EARLE BSS

INTENDED USE
TC Medium Eagle, Earle BSS is the sterile chemically defined basal medium of Eagle[1] prepared with Earle Balanced Salt Solution for the cultivation of HeLa, KB and other cell lines.

FORMULA
TC MEDIUM EAGLE, EARLE BSS
BME, EARLE
Ingredients per liter

L-Arginine	17.4 mg	Nicotinamide	1 mg
L-Cystine	12 mg	Calcium Pantothenate	1 mg
L-Tyrosine	18 mg	Pyridoxal Hydrochloride	1 mg
L-Histidine	8 mg	Thiamine Hydrochloride	1 mg
L-Isoleucine	26 mg	Riboflavin	0.1 mg
L-Leucine	26 mg	Inositol	1.8 mg
L-Lysine	26 mg	TC Phenol Red	5 mg
L-Methionine	7.5 mg	Bacto-Dextrose	1000 mg
L-Phenylalanine	16.5 mg	Sodium Chloride	6800 mg
L-Threonine	24 mg	Potassium Chloride	400 mg
L-Tryptophane	4 mg	Monosodium Phosphate	125 mg
L-Valine	23.5 mg	Magnesium Sulfate	100 mg
Biotin	1 mg	Calcium Chloride	200 mg
Folic Acid	1 mg	Sodium Bicarbonate	2200 mg
Choline Chloride	1 mg		

PROCEDURE
The medium is completed by adding 0.6 ml rehydrated TC Glutamine 5% per 100 ml of basal medium, plus 5 – 10% TC Human Serum, TC Fetal Calf Serum, TC Horse Serum, TC Bovine Serum, or other enrichments as desired.

STORAGE
Store at 2 – 8°C.

PACKAGING
TC Medium Eagle, Earle BSS   100 ml   5068-72-1

TC MEDIUM EAGLE, EARLE BSS, DRIED
(BME, EARLE, DRIED)

INTENDED USE
TC Medium Eagle, Earle BSS, Dried is the dried chemically defined formulation of Eagle[1] with Earle Balanced Salt Solution and without glutamine and sodium bicarbonate. This standardized formulation provides a ready stable source of medium from which fresh liquid medium can be prepared as required.

PROCEDURE
To rehydrate, suspend 8.9 g in 1 liter TC Distilled Water, DE by adding the medium slowly to the water while stirring. Continue stirring until the medium is completely dissolved.

Adjust to the desired pH by adding TC Bicarbonate Solution 10%.

Filter sterilize the medium preferably using pressure, and stopper tightly to prevent loss of CO_2. Filtration or storage can result in a loss of CO_2, which causes increased alkalinity. The reaction of the sterile filtrate should be readjusted with sterile CO_2 if it is too alkaline. (Alkalinity may be indicated by a change in color towards cerise).

Add 6 ml rehydrated TC Glutamine 5%, and 5 – 10% TC Human Serum, TC Fetal Calf Serum, TC Horse Serum, TC Bovine Serum, or other enrichments as desired.

STORAGE
Store at 2 – 8°C.

REFERENCE
1. Proc. So. Exp. Biol. & Med., 89:362, 1955.

PACKAGING
TC Medium Eagle, Earle BSS, Dried   10 × 1 liter   5069-37-4

TC MEDIUM EAGLE, HANKS BSS, DRIED
(BME, HANKS, DRIED)

INTENDED USE
TC Medium Eagle, Hanks BSS, Dried is the chemically defined formulation of Eagle[1] with Hanks Balanced Salt Solution and without glutamine or sodium bicarbonate. This standardized formulation provides a ready stable source medium from which fresh liquid medium can be prepared as required.

FORMULA

TC MEDIUM EAGLE, HANKS BSS, DRIED
BME, HANKS, DRIED
Ingredients per liter

L-Arginine	17.4 mg	Nicotinamide	1 mg
L-Cystine	12 mg	Calcium Pantothenate	1 mg
L-Tyrosine	18 mg	Pyridoxal Hydrochloride	1 mg
L-Histidine	8 mg	Thiamine Hydrochloride	1 mg
L-Isoleucine	26 mg	Riboflavin	0.1 mg
L-Leucine	26 mg	Inositol	1.8 mg
L-Lysine	26 mg	TC Phenol Red	5 mg
L-Methionine	7.5 mg	Bacto Dextrose	1000 mg
L-Phenylalanine	16.5 mg	Sodium Chloride	8000 mg
L-Threonine	24 mg	Potassium Chloride	400 mg
L-Tryptophane	4 mg	Monopotassium Phosphate	60 mg
L-Valine	23.5 mg	Disodium Phosphate	50 mg
Biotin	1 mg	Magnesium Sulfate	100 mg
Folic Acid	1 mg	Calcium Chloride	140 mg
Choline Chloride	1 mg		

PROCEDURE

To rehydrate, suspend 10 g in 1 liter TC Distilled Water, DE by adding the medium slowly to the water while stirring. Continue stirring until the medium is completely dissolved.

Adjust to the desired pH by adding TC Bicarbonate Solution 10%.

Filter sterilize the medium preferably using pressure, and stopper tightly to prevent loss of CO_2. Filtration or storage can result in a loss of CO_2, which causes increased alkalinity. The reaction of the sterile filtrate should be readjusted with sterile CO_2 if it is too alkaline. (Alkalinity may be indicated by a change in color towards cerise).

Add 6 ml rehydrated TC Glutamine 5%, and 5 – 10% TC Human Serum, TC Fetal Calf Serum, TC Horse Serum, TC Bovine Serum, or other enrichments as desired.

STORAGE
Store at 2 – 8°C.

REFERENCE
1. Pro. Soc. Exp. Biol. & Med., 89:362, 1955.

PACKAGING
TC Medium Eagle, Hanks BSS, Dried   10 × 1 liter   5071-37-0

TC MEDIUM EAGLE HeLa

INTENDED USE
When glutamine and serous enrichments are added, TC Medium Eagle HeLa is used for the propagation and maintenance of the HeLa and other cell lines, and for studying the cytopathogenicity of viral agents.

HISTORY
TC Medium Eagle HeLa is the sterile original chemically defined basal HeLa medium of Eagle,[1] who used the medium containing glutamine and 1% dialyzed horse serum or 5% dialyzed human serum for nutritional studies.

FORMULA

TC MEDIUM EAGLE HeLa

Ingredients per liter

L-Arginine	17.4 mg	Folic Acid	0.44 mg
L-Cystine	12 mg	Pantothenic Acid	0.22 mg
L-Histidine	7.8 mg	Pyridoxal Hydrochloride	0.2 mg
L-Isoleucine	26.2 mg	Thiamine Hydrochloride	0.34 mg
L-Leucine	26.2 mg	Nicotinamide	0.12 mg
L-Lysine	29.2 mg	Riboflavin	0.04 mg
L-Methionine	7.5 mg	Bacto Dextrose	900 mg
L-Phenylalanine	16.5 mg	Sodium Chloride	5850 mg
L-Threonine	23.8 mg	Potassium Chloride	373 mg
L-Tryptophane	4.1 mg	Calcium Chloride	111 mg
L-Tyrosine	18.1 mg	Magnesium Chloride	102 mg
L-Valine	23.4 mg	Monosodium Phosphate	138 mg
Biotin	0.24 mg	Sodium Bicarbonate	1680 mg
Choline Chloride	0.14 mg	Phenol Red	5 mg

PROCEDURE

The basal medium is completed by adding 0.6 ml rehydrated TC Glutamine 5% per 100 ml, and 5 – 10% TC Horse Serum, TC Bovine Serum, TC Human Serum, TC Fetal Calf Serum, or other enrichments as desired.

STORAGE
Store at 2 – 8°C.

PACKAGING
TC Medium Eagle HeLa   100 ml   5651-72-4

TC MEDIUM EAGLE HeLa, DRIED

INTENDED USE
TC Medium Eagle HeLa, Dried is the dried formulation of the original chemically defined basal HeLa medium of Eagle[1] without glutamine and sodium bicarbonate. This standardized medium provides a ready stable source medium for the preparation of fresh liquid medium whenever required.

PROCEDURE
To rehydrate, suspend 7.7 g in 1 liter TC Distilled Water, DE by adding the medium slowly to the water while stirring. Continue stirring until the medium is completely dissolved.

Adjust to the desired pH by adding TC Bicarbonate Solution 10%.

Filter sterilize the medium preferably using pressure, and stopper tightly to prevent loss of CO_2. Filtration or storage can result in a loss of CO_2, which causes increased alkalinity. The reaction of the sterile filtrate should be readjusted with sterile CO_2 if it is too alkaline. (Alkalinity may be indicated by a change in color towards cerise.)

Add 6 ml rehydrated TC Glutamine 5%, and 5 – 10% TC Human Serum, TC Fetal Calf Serum, TC Horse Serum, TC Bovine Serum, or other enrichments as desired.

STORAGE
Store at 2 – 8°C.

REFERENCE
1. Science, 122:501, 1955.

PACKAGING

TC Medium Eagle HeLa, Dried   10 × 1 liter   5765-37-1

TC MEDIUM EAGLE HeLa, 10X

INTENDED USE

TC Medium Eagle HeLa, 10X is a sterile 10-fold concentration of the original chemically defined basal HeLa medium of Eagle[1] without glutamine and sodium bicarbonate.

PROCEDURE

One liter of single strength medium is prepared by adding 100 ml TC Medium Eagle HeLa, 10X to 900 ml TC Distilled Water, DE under aseptic conditions. Adjust to the desired pH by adding TC Bicarbonate Solution 10%. Add 5 – 10% TC Human Serum, TC Fetal Calf Serum, TC Horse Serum, TC Bovine Serum, or other enrichments as desired.

STORAGE

Store at 2 – 8°C.

PACKAGING

TC Medium Eagle HeLa, 10X   100 ml   5764-72-8

TC MEDIUM HAM F10, DRIED

INTENDED USE

When supplemented with sera, TC Medium Ham F10, Dried affords an excellent medium for culturing most cell lines. This standardized medium provides a ready stable source medium for the preparation of fresh liquid medium whenever required.

HISTORY

TC Medium Ham F10, Dried is the dried chemically defined medium formulation of Ham[1] without sodium bicarbonate.

Purified serum protein fractions are added for supporting growth of diploid Chinese hamster cell lines. TC Medium Ham F10, Dried contains each nutrient at the optimum level suggested by the author who reported it to be superior to the previous F7 formulation.

FORMULA

TC MEDIUM HAM F10, DRIED

Ingredients per liter

L-Arginine	211 mg	Biotin	0.024 mg
L-Histidine	21 mg	Calcium Pantothenate	0.7 mg
L-Lysine	29.3 mg	Choline Chloride	0.69 mg
L-Methionine	4.48 mg	i - Inositol	0.54 mg
L-Phenylalanine	4.96 mg	Niacinamide	0.6 mg

Ingredients per liter (continued)

L-Tryptophane	0.6 mg	Pyridoxine Hydrochloride	0.2 mg
L-Tyrosine	1.81 mg	Riboflavin	0.37 mg
L-Alanine	8.91 mg	Thymidine	0.7 mg
Glycine	7.51 mg	Cyanocobalamin	1.3 mg
L-Serine	10.5 mg	Sodium Pyruvate	110 mg
L-Threonine	3.57 mg	Lipoic Acid	0.2 mg
L-Aspartic Acid	13.3 mg	Calcium Chloride	44 mg
L-Glutamic Acid	14.7 mg	Magnesium Sulfate	153 mg
L-Asparagine	15 mg	Bacto Dextrose	1100 mg
L-Glutamine	146.2 mg	Sodium Chloride	7400 mg
L-Isoleucine	2.6 mg	Potassium Chloride	285 mg
L-Leucine	13.1 mg	Disodium Phosphate	290 mg
L-Proline	11.5 mg	Monopotassium Phosphate	83 mg
L-Valine	3.5 mg	Phenol Red	1.2 mg
L-Cysteine	31.5 mg	Ferrous Sulfate	0.83 mg
Thiamine Hydrochloride	1 mg	Cupric Sulfate	0.0025 mg
Hypoxanthine	4 mg	Zinc Sulfate	0.028 mg
Folic Acid	1.3 mg		

PROCEDURE

To rehydrate, suspend 9.84 g in 1 liter TC Distilled Water, DE by adding the medium slowly to the water while stirring. Continue stirring until the medium is completely dissolved.

Adjust to the desired pH by adding TC Bicarbonate Solution 10%.

Filter sterilize the medium preferably using pressure, and stopper tightly to prevent loss of CO_2. Filtration or storage can result in a loss of CO_2, which causes increased alkalinity. The reaction of the sterile filtrate should be readjusted with sterile CO_2 if it is too alkaline. (Alkalinity may be indicated by a change in color towards cerise). The medium is completed by the addition of albumin for serum-free growth or by addition of TC Fetal Calf Serum.

STORAGE
Store at $2 - 8°C$.

REFERENCE
1. Exp. Cell Res., 29:515, 1963.

PACKAGING
TC Medium Ham F10, Dried   10 × 1 liter   5825-37-9

TC MEDIUM NCTC 109

INTENDED USE
TC Medium NCTC 109 is a standardized chemically defined medium. It provides an excellent medium for the rapid proliferation of many cell lines when supplemented with tissue extracts and serous enrichments such as TC Bovine Serum, TC Horse Serum, TC Human Serum, or TC Fetal Calf Serum.

HISTORY
TC Medium NCTC 109 is prepared according to the formulation of Evans, Bryant, Fioramonti, McQuilkin, Sanford, and Earle.[1] This medium was used by the authors as a background medium for nutritional studies of Clone L 929 cells. Although the produc-

tivity of the chemically defined medium was not as great as media containing tissue extracts and serous enrichments, it supported continued proliferation of Clone L 929 cells.

FORMULA

TC MEDIUM NCTC 109

Ingredients per liter

L-Alanine	31.48 mg	Folic Acid	0.025 mg
L-Arginine	25.76 mg	i-Inositol	0.125 mg
L-Aspartic Acid	9.91 mg	p-Aminobenzoic Acid	0.125 mg
L-Glutamic Acid	8.26 mg	Vitamin A	0.25 mg
L-Histidine	19.73 mg	Calciferol	0.25 mg
L-Hydroxyproline	4.09 mg	Menadione	0.025 mg
L-Proline	6.13 mg	α-Tocopherol Phosphate	0.025 mg
Glycine	13.51 mg	Ascorbic Acid	49.9 mg
L-Valine	25.0 mg	Glutathione	10.1 mg
L-Isoleucine	18.04 mg	Choline Chloride	1.25 mg
L-Leucine	20.44 mg	Diphosphopyridine Nucleotide	7 mg
L-Lysine	30.75 mg	Triphosphopyridine Nucleotide	1 mg
L-Phenylalanine	16.53 mg	5-Methyl Cytosine	0.1 mg
L-Tyrosine	16.44 mg	Cyanocobalamin (Vit. B_{12})	10 mg
L-Threonine	18.93 mg	Coenzyme A	2.5 mg
L-Tryptophane	17.5 mg	Cocarboxylase	1 mg
L-Serine	10.75 mg	Flavin Adenine Dinucleotide	1 mg
L-Taurine	4.18 mg	Uridine Triphosphate	1 mg
L-Ornithine	7.38 mg	Desoxyadenosine	10 mg
L-Methionine	4.44 mg	Desoxycytidine	10 mg
L-Cystine	10.49 mg	Desoxyguanosine	10 mg
L-Cysteine	259.9 mg	Thymidine	10 mg
L-Glutamine	135.73 mg	Glucuronolactone	1.8 mg
D-Glucosamine	3.2 mg	Sodium Glucuronate	1.8 mg
L-Asparagine	8.09 mg	Bacto-Dextrose	1000 mg
L-α-N-Butyric Acid	5.51 mg	Sodium Acetate	50 mg
Thiamine Hydrochloride	0.025 mg	Calcium Chloride	200 mg
Riboflavin	0.025 mg	Magnesium Sulfate	100 mg
Pyridoxine Hydrochloride	0.0625 mg	Monosodium Phosphate	140 mg
Pyridoxal Hydrochloride	0.0625 mg	Sodium Bicarbonate	2200 mg
Niacin	0.0625 mg	Sodium Chloride	6800 mg
Niacinamide	0.0625 mg	Potassium Chloride	400 mg
Pantothenate	0.025 mg	Tween 80	12.5 mg
Biotin	0.025 mg	Phenol Red	20 mg

PROCEDURE

Add 5 – 10% TC Human Serum, TC Fetal Calf Serum, TC Horse Serum, or other enrichments as desired.

STORAGE

Store at 2 – 8°C.

PACKAGING

TC Medium NCTC 109   100 ml   5926-72-3

TC MEDIUM NCTC 109, DRIED

INTENDED USE

TC Medium NCTC 109, Dried is the dried formulation of Evans et al[1] without sodium bicarbonate. It provides a ready stable source medium for the preparation of fresh liquid medium whenever required. When supplemented with tissue extracts or serous enrichments, it provides an excellent medium for the rapid proliferation of many cell lines.

PROCEDURE

To rehydrate, suspend 9.7 g in 1 liter TC Distilled Water, DE by adding the medium slowly to the water while stirring. Continue stirring until the medium is completely dissolved.

Adjust to the desired pH by adding TC Bicarbonate Solution 10%.

Filter sterilize the medium preferably using pressure, and stopper tightly to prevent loss of CO_2. Filtration or storage can result in a loss of CO_2, which causes increased alkalinity. The reaction of the sterile filtrate should be readjusted with sterile CO_2 if it is too alkaline. (Alkalinity may be indicated by a change in color towards cerise).

Add 5 – 10% TC Human Serum, TC Fetal Calf Serum, TC Horse Serum, TC Bovine Serum, or other enrichments as desired.

STORAGE

Store at 2 – 8°C.

REFERENCE

1. Cancer Res., 16:77, 1956.

PACKAGING

TC Medium NCTC 109, Dried   10 × 1 liter   5927-37-6

TC MEDIUM 199

INTENDED USE

TC Medium 199 is a chemically defined medium certified for use in tissue culture procedures and virus studies, especially recommended for virus culture. It is a nutritive basal medium to which additional growth factors may be added for the propagation of tissue cells for morphological, histochemical and physiological studies. TC Medium 199 is recommended for the diagnosis and study of viral infections including their detection, titering and typing. It is also especially adapted for the preservation of tissue and cells in the viable state in tissue banks and cultures.

FORMULA

TC MEDIUM 199
(Morgan & Morton Medium 150)
Ingredients per liter

L-Arginine	70 mg	Calcium Pantothenate	0.01 mg
L-Histidine	20 mg	Biotin	0.01 mg
L-Lysine	70 mg	Folic Acid	0.01 mg
L-Tyrosine	40 mg	Choline	0.5 mg
DL-Tryptophane	20 mg	Inositol	0.05 mg
DL-Phenylalanine	50 mg	p-Aminobenzoic Acid	0.05 mg
L-Cystine	20 mg	Vitamin A	0.1 mg
DL-Methionine	30 mg	Calciferol	0.1 mg
DL-Serine	50 mg	Menadione	0.01 mg
DL-Threonine	60 mg	α-Tocopherol Phosphate	0.01 mg
DL-Leucine	120 mg	Ascorbic Acid	0.05 mg
DL-Isoleucine	40 mg	Glutathione	0.05 mg
DL-Valine	50 mg	Cholesterol	0.2 mg
DL-Glutamic Acid	150 mg	L-Glutamine	100 mg
DL-Aspartic Acid	60 mg	Adenosinetriphosphate	1 mg

Ingredients per liter (continued)

DL-Alanine	50 mg	Adenylic Acid	0.2 mg
L-Proline	40 mg	Ribose	0.5 mg
L-Hydroxyproline	10 mg	Desoxyribose	0.5 mg
Glycine	50 mg	Bacto-Dextrose	1000 mg
L-Cysteine	0.1 mg	Tween 80	5 mg
Adenine	10 mg	Sodium Acetate	50 mg
Guanine	0.3 mg	Iron (as Ferric Nitrate)	0.1 mg
Xanthine	0.3 mg	Sodium Chloride	8000 mg
Hypoxanthine	0.3 mg	Potassium Chloride	400 mg
Thymine	0.3 mg	Calcium Chloride	140 mg
Uracil	0.3 mg	Magnesium Sulfate	100 mg
Thiamine Hydrochloride	0.01 mg	Disodium Phosphate	60 mg
Riboflavin	0.01 mg	Monopotassium Phosphate	60 mg
Pyridoxine Hydrochloride	0.025 mg	Sodium Bicarbonate	350 mg
Pyridoxal Hydrochloride	0.025 mg	Bacto-Phenol Red	20 mg
Niacin	0.025 mg	Carbon Dioxide	to pH 7.2
Niacinamide	0.025 mg		

PROCEDURES

Add 5 – 10% TC Human Serum, TC Fetal Calf Serum, TC Horse Serum, TC Bovine Serum or other enrichments as desired.

Antibiotics to discourage development of bacteria may be added to TC Medium 199 when ready for use if desired. Generally, 500 units of penicillin and 500 µg of streptomycin are added per ml of medium. These concentrations for 100 ml of TC Medium 199 may readily be obtained as follows: Penicillin - Add 10 ml sterile distilled or deionized water to a bottle containing 1 million units of penicillin, and add 0.5 ml to 100 ml of TC Medium 199, (or use 2.5 ml penicillin solution prepared by dissolving 100,000 units of penicillin in 5 ml sterile distilled or deionized water). Streptomycin — Add 10 ml sterile distilled or deionized water to a bottle containing 1 g (1 million µg) streptomycin, and add 0.5 ml to 100 ml of TC Medium 199 (or use 2.5 ml streptomycin solution prepared by dissolving 0.1 g streptomycin in 5 ml sterile distilled or deionized water). Farrell, Wood, MacMorine, Shimada and Graham[1] used Medium 199 prepared in Earle's Solution to which was added 500 units per ml penicillin and 250 µg/ml streptomycin in the preparation of the poliomyelitis virus for the production of the vaccine employed in the 1954 field trials.

STORAGE
Store at 2 – 8°C.

PACKAGING
TC Medium 199

100 ml	5477-72-6
400 ml	5477-78-0
1 liter	5477-81-5

TC MEDIUM 199, DRIED

INTENDED USE

TC Medium 199, Dried is the dried formulation of the chemically defined medium of Morgan, Morton and Parker[2] prepared with Hanks balanced salt solution without sodium bicarbonate. This medium has been standardized for use in tissue culture procedures and provides a ready stable source medium for the preparation of fresh liquid media whenever required.

PROCEDURE
To rehydrate, suspend 11 g in 1 liter TC Distilled Water, DE by adding the medium slowly to the water while stirring. Continue stirring until the medium is completely dissolved.

Adjust to the desired pH by adding TC Bicarbonate Solution 10%.

Filter sterilize the medium preferably using pressure, and stopper tightly to prevent loss of CO_2. Filtration or storage can result in a loss of CO_2, which causes increased alkalinity. The reaction of the sterile filtrate should be readjusted with sterile CO_2 if it is too alkaline. (Alkalinity may be indicated by a change in color towards cerise).

Add 5 – 10% TC Human Serum, TC Fetal Calf Serum, TC Horse Serum, TC Bovine Serum, or other enrichments as desired.

STORAGE
Store at 2 – 8°C.

PACKAGING

TC Medium 199, Dried	10 × 1 liter	5701-37-8
	10 liters	5701-24-3

TC MEDIUM 199, 10X

INTENDED USE
TC Medium 199, 10X is the sterile 10-fold concentration of the formulation of Morgan, Morton and Parker[2] using Hanks balanced salt solution without sodium bicarbonate. It is certified for use in the preparation of single strength TC Medium 199 for tissue culture procedures and virus studies.

PROCEDURE
One liter of single strength medium is prepared by adding 100 ml TC Medium 199, 10X to 900 ml sterile TC Distilled Water, DE under aseptic conditions. Adjust to the desired pH by adding TC Bicarbonate Solution 10%. Add 5 – 10% TC Horse Serum, TC Fetal Calf Serum, TC Human Serum, TC Bovine Serum, or other enrichments as desired.

STORAGE
Store at 2 – 8°C.

PACKAGING

TC Medium 199, 10X	100 ml	5696-72-1
	400 ml	5696-78-5
	1 liter	5696-81-0

TC MEDIUM 199 w/o BICARBONATE

INTENDED USE
TC Medium 199 w/o Bicarbonate is the same as TC Medium 199 except sodium bicarbonate has not been added. It is used in tissue culture procedures and virus studies.

PROCEDURE
Adjust to the desired pH by adding TC Bicarbonate Solution 10%. Add 5 – 10% TC Human Serum, TC Fetal Calf Serum, TC Horse Serum, TC Bovine Serum, or other enrichments as desired.

STORAGE
Store at 2 – 8°C.

PACKAGING
TC Medium 199 w/o Bicarbonate   100 ml   5529-72-4

TC MEDIUM 199 w/o PHENOL RED

INTENDED USE
TC Medium 199 w/o Phenol Red is the same as TC Medium 199 except that the phenol red indicator has been omitted. It is recommended for use in tissue culture and other procedures where the medium without pH indicator is desired.

STORAGE
Store at 2 – 8°C.

PACKAGING
TC Medium 199 w/o Phenol Red   100 ml   5702-72-3

TC MEDIUM 199 w/o PHENOL RED, DRIED

INTENDED USE
TC Medium 199 w/o Phenol Red, Dried is the dried formulation of the chemically defined medium of Morgan, Morton and Parker[2] prepared with Hanks balanced salt solution and without phenol red or sodium bicarbonate. This medium has been standardized for use in tissue culture procedures and provides a ready stable source medium for the preparation of fresh liquid media whenever required.

PROCEDURE
To rehydrate, suspend 11 g in 1 liter TC Distilled Water, DE by adding the medium slowly to the water while stirring. Continue stirring until the medium is completely dissolved.

Adjust to the desired pH by adding TC Bicarbonate Solution 10%. The reaction should be checked carefully electrometrically or with phenol red indicator.

Filter sterilize the medium preferably using pressure, and stopper tightly to prevent loss of CO_2. Filtration or storage can result in a loss of CO_2, which causes increased alkalinity. The reaction of the sterile filtrate should be readjusted with sterile CO_2 if it is too alkaline.

Add 5 – 10% TC Human Serum, TC Fetal Calf Serum, TC Horse Serum, TC Bovine Serum, or other enrichments as desired.

STORAGE
Store at 2 – 8°C.

PACKAGING
TC Medium 199 w/o Phenol Red, Dried	10 × 1 liter	5915-37-0
	10 liters	5915-24-5
	100 liters	5915-20-9

REFERENCES
1. Canad. J. Pub. Health, 46:265, 1955.
2. Proc. Soc. Exp. Biol. Med., 73:1, 1950.

TC MEDIUM RPMI #1640

INTENDED USE
When supplemented with serous enrichments, TC Medium RPMI #1640 is excellent for the rapid proliferation of many cell lines.

HISTORY
TC Medium RPMI #1640 is the sterile, chemically defined medium developed at the Roswell Park Memorial Institute by Moore, Gerner, and Franklin.[1] It was formulated for the cultivation of human and mouse leukemic cells in tissue culture.

FORMULA
TC MEDIUM RPMI #1640
Ingredients per liter

L-Arginine	200 mg	p-Aminobenzoic Acid	1 mg
L-Asparagine	50 mg	Biotin	0.20 mg
L-Aspartic Acid	20 mg	Calcium Pantothenate	0.25 mg
L-Cystine	50 mg	Choline Chloride	3 mg
L-Glutamic Acid	20 mg	Folic Acid	1 mg
L-Glutamine	300 mg	Inositol	35 mg
Glycine	10 mg	Nicotinamide	1 mg
L-Histidine	15 mg	Pyridoxine HCl	1 mg
Hydroxy-L-Proline	20 mg	Riboflavin	0.20 mg
L-Isoleucine	50 mg	Thiamine HCl	1 mg
L-Leucine	50 mg	Vitamin B_{12}	0.005 mg
L-Lysine HCl	40 mg	Dextrose	2000 mg
L-Methionine	15 mg	Glutathione	1 mg
L-Phenylalanine	15 mg	Calcium Nitrate	100 mg
L-Proline	20 mg	Potassium Chloride	400 mg
L-Serine	30 mg	Magnesium Sulfate	100 mg
L-Threonine	20 mg	Sodium Chloride	6460 mg
L-Tryptophane	5 mg	Sodium Bicarbonate	2000 mg
L-Tyrosine	20 mg	Monosodium Phosphate	1512 mg
L-Valine	20 mg	Phenol Red	5 mg

PROCEDURE
The basal medium is completed by the addition of 10% TC Human Serum, TC Fetal Calf Serum, TC Horse Serum, TC Bovine Serum, or other enrichments as desired.

STORAGE
Store at 2 – 8°C.

PACKAGING
TC Medium RPMI #1640   100 ml   5087-72-8

TC MEDIUM RPMI #1640, DRIED

INTENDED USE
When supplemented with serous enrichments, TC Medium RPMI #1640 is excellent for the rapid proliferation of many cell lines. It does not contain sodium bicarbonate. This standardized medium provides a ready stable source medium for the preparation of fresh liquid medium whenever required.

TC Medium RPMI #1640, Dried is the dried chemically defined formulation developed by Moore, Gerner, and Franklin.[1] It was formulated especially for the cultivation of mouse and human leukemic cells in tissue culture.

PROCEDURE
To rehydrate, suspend 10.4 g in 1 liter TC Distilled Water, DE by adding the medium slowly to the water while stirring. Continue stirring until the medium is completely dissolved.

Adjust to the desired pH by adding TC Bicarbonate Solution 10%.

Filter sterilize the medium preferably using pressure. Loss of CO_2 during filtration or storage results in increased alkalinity as indicated by a change in color towards cerise. The reaction of the sterile filtrate should be readjusted with sterile CO_2 if it is too alkaline.

The basal medium is completed by the addition of 10% TC Human Serum, TC Fetal Calf Serum, TC Horse Serum, TC Bovine Serum, or other enrichments as desired.

STORAGE
Store at 2 – 8°C.

REFERENCE
1. JAMA, 199:519, 1967.

PACKAGING
TC Medium RPMI #1640, Dried   10 × 1 liter   5085-37-4

TC MINIMAL MEDIUM EAGLE, EARLE BSS, DRIED
(MEM, EARLE DRIED)

INTENDED USE
TC Minimal Medium Eagle, Earle BSS, Dried is used for preparing Eagle's Minimal Medium for the cultivation of mammalian cells in monolayer or suspension. It provides

a flexible ready stable source medium for the preparation of fresh liquid media whenever required.

HISTORY
TC Minimal Medium Eagle, Earle BSS, Dried is the dried minimal essential medium of Eagle[1] without glutamine, sodium bicarbonate, and calcium chloride.

FORMULA
TC MINIMAL MEDIUM EAGLE, EARLE BSS, DRIED
MEM, EARLE DRIED
Ingredients per liter

L-Arginine	105 mg	Folic Acid	1 mg
L-Cystine	24 mg	i-Inositol	2 mg
L-Histidine	31 mg	Nicotinamide	1 mg
L-Isoleucine	52 mg	Calcium Pantothenate	1 mg
L-Leucine	52 mg	Pyridoxal	1 mg
L-Lysine	58 mg	Riboflavin	0.1 mg
L-Methionine	15 mg	Thiamine Hydrochloride	1 mg
L-Phenylalanine	32 mg	Bacto Dextrose	1 g
L-Threonine	48 mg	Sodium Chloride	6.8 g
L-Tryptophane	10 mg	Potassium Chloride	0.4 g
L-Tyrosine	36 mg	Calcium Chloride	0.2 g
L-Valine	46 mg	Magnesium Chloride	0.2 g
Choline Chloride	1 mg	Monosodium Phosphate	0.15 g

PROCEDURE
To rehydrate, suspend 9.06 g in 1 liter TC Distilled Water, DE by adding the medium slowly to the water while stirring. Continue stirring until the medium is completely dissolved.

For growing cells in monolayer or other stationary culture, add 2 ml of a 10% solution of calcium chloride. For suspension cultures, greatly reduce the calcium chloride or omit it altogether.

Adjust to the desired pH by adding TC Bicarbonate Solution 10%.

Filter sterilize the medium preferably using pressure, and stopper tightly to prevent loss of CO_2. Filtration or storage can result in a loss of CO_2, which causes increased alkalinity.

The reaction of the sterile filtrate should be readjusted with sterile CO_2 if it is too alkaline.

Add 6 ml rehydrated TC Glutamine 5%, and 5 – 10% TC Human Serum, TC Fetal Calf Serum, TC Horse Serum, TC Bovine Serum, or other enrichments as desired.

STORAGE
Store at 2 – 8°C.

REFERENCE
1. Science, 130: 432, 1959.

PACKAGING
TC Minimal Medium Eagle Earle BSS, Dried   10 × 1 liter   5675-37-0

TC MINIMAL MEDIUM EAGLE, HANKS BSS, DRIED
(MEM, HANKS, DRIED)

INTENDED USE
TC Minimal Medium Eagle, Hanks BSS, Dried is the dried basal minimal essential medium of Eagle[1] with Hanks Balanced Salt Solution, but without glutamine or sodium bicarbonate. It provides a ready stable source medium for the preparation of fresh liquid basal medium for propagating mammalian cells in monolayer culture. Sodium bicarbonate and glutamine are added at the time of preparing the liquid medium.

PROCEDURE
To rehydrate, suspend 10.4 g in 1 liter TC Distilled Water, DE by adding the medium slowly to the water while stirring. Continue stirring until the medium is completely dissolved.

Adjust to the desired pH by adding TC Bicarbonate Solution 10%.

Filter sterilize the medium preferably using pressure, and stopper tightly to prevent loss of CO_2. Filtration or storage can result in a loss of CO_2, which causes increased alkalinity. The reaction of the sterile filtrate should be readjusted with sterile CO_2 if it is too alkaline. (Alkalinity may be indicated by a change in color towards cerise.)

Add 6 ml rehydrated TC Glutamine 5%, and 5 – 10% TC Human Serum, TC Fetal Calf Serum, TC Horse Serum, TC Bovine Serum, or other enrichments as desired.

STORAGE
Store at 2 – 8°C.

REFERENCE
1. Science, 130:432, 1959.

PACKAGING
TC Minimal Medium Eagle, Hanks BSS, Dried   10 × 1 liter   5073-37-8

TC MINIMAL MEDIUM EAGLE SPINNER MODIFIED, DRIED
(MEM-S, DRIED)

INTENDED USE
When glutamine and serous enrichments are added to TC Minimal Medium Eagle Spinner Modified, Dried, it is used for the cultivation of mammalian cells in suspension. This standarized medium provides a ready stable source medium for the preparation of fresh liquid medium whenever required.

HISTORY
TC Minimal Medium Eagle Spinner Modified, Dried is the dried formulation of Eagle[1] without glutamine, calcium, and sodium bicarbonate.

FORMULA
TC MINIMAL MEDIUM EAGLE SPINNER MODIFIED, DRIED
MEM-S, DRIED

Ingredients per liter

TC Minimal Medium Eagle, Earle BSS, Dried	9.06 g
Monosodium Phosphate	1.35 g

PROCEDURE
To rehydrate, suspend 10.4 g in 1 liter TC Distilled Water, DE by adding the medium slowly to the water while stirring. Continue stirring until the medium is completely dissolved.

Adjust to the desired pH by adding TC Bicarbonate Solution 10%.

Filter sterilize the medium preferably using pressure, and stopper tightly to prevent loss of CO_2. Filtration or storage can result in a loss of CO_2, which causes increased alkalinity. The reaction of the sterile filtrate should be readjusted with sterile CO_2 if it is too alkaline.

Add 6 ml rehydrated TC Glutamine 5%, and 5 – 10% TC Human Serum, TC Fetal Calf Serum, TC Horse Serum, TC Bovine Serum, or other enrichments as desired.

STORAGE
Store at 2 – 8°C.

REFERENCE
1. Science, 130:432, 1959.

PACKAGING
TC Minimal Medium Eagle Spinner Modified, Dried   10 × 1 liter   5839-37-3

TC PENICILLIN-STREPTOMYCIN, DESICCATED

INTENDED USE
TC Penicillin-Streptomycin, Desiccated is a sterile stable preparation of penicillin G and streptomycin sulfate ready for use in tissue culture or other media to inhibit certain organisms.

FORMULA
Each vial contains 100,000 units Penicillin G and
         100,000 μg Streptomycin Sulfate

PROCEDURE
The rehydrated TC Penicillin-Streptomycin solution may be added aseptically in the desired concentration directly to the growth medium or diluted with balanced salt solution and used to sterilize tissues or cells prior to inoculating the growth medium.

To rehydrate, add 10 ml sterile distilled or deionized water and rotate the vial end-over-end to dissolve the contents. Each ml then contains 10,000 units penicillin and

10,000 μg streptomycin. To obtain a higher concentration of antibiotics than the 10,000 units per ml, use less sterile distilled or deionized water with which to rehydrate each vial.

The rehydrated material should be used when prepared or stored in the frozen state. The frozen solution will retain its potency for 1 week.

STORAGE
Store at 2 – 8°C.

PACKAGING
TC Penicillin-Streptomycin, Desiccated   6 × 10 ml   5854-60-3

BACTO PEPTONE
(For complete discussion refer to Bacto Peptone.)

Bacto Peptone has been shown to be a satisfactory enrichment, replacing serum, in chemically defined media, for the continuous propagation of strain mouse cells by Waymouth.[1] Her experience has been confirmed by Merchant and others.[2] A concentration of 0.5% Bacto Peptone in the final medium has been employed successfully. Sterilization of Bacto Peptone solutions used as enrichments in media may be by filtration or in the autoclave.

REFERENCES
1. J. Nat. Cancer Inst., 17:315, 1956.
2. Personal Communication, 1956.

PACKAGING
Bacto Peptone

1/4 lb (114 g)	0118-02-7
1 lb (454 g)	0118-01-8
5 lb (2.27 kg)	0118-05-4
10 kg	0118-08-1

TC PHENOL RED SOLUTION 1%

INTENDED USE
TC Phenol Red Solution 1% is sterile and certified for use as an indicator of pH in tissue culture procedures. It is employed in tissue culture media in concentrations of 0.001 – 0.002%. For judging the pH of very small volumes of media (1 – 2 ml), the concentration of phenol red should be 3 – 5 times that given above, as suggested under the discussion of TC Chicken Plasma.

STORAGE
Store at 2 – 8°C.

PACKAGING
TC Phenol Red Solution 1%

10 ml	5358-59-7
6 × 1 ml	5358-51-5

TC RECONSTITUTING FLUID

INTENDED USE
TC Reconstituting Fluid is sterile TC Distilled Water, DE saturated with CO_2 and is certified for use in tissue culture procedures. It is used for reconstituting desiccated tissue culture reagents where a lower pH is desired than obtainable through use of TC Distilled Water, DE.

STORAGE
Store at 15 – 30°C.

PACKAGING
TC Reconstituting Fluid   100 ml   5352-72-6

BACTO TRYPTOSE PHOSPHATE BROTH
For complete discussion see Bacto Tryptose Phosphate Broth.

Bacto Tryptose Phosphate Broth has decided value in tissue culture procedures. Ginsberg, Gold and Jordan[1] maintained HeLa cells for 10 days in a combination consisting of 15 – 25% Bacto Tryptose Phosphate Broth, 67.5 – 77.5% Scherer's Maintenance Solution and 7.5% chicken serum. They reported a 3 – 5 fold cell increase during this period. HeLa cells maintained in maintenance solution supplemented with 15% Bacto Tryptose Phosphate Broth, yielded substantially greater amounts of ARD-AD and poliomyelitis type I virus than did the unsupplemented medium. Furthermore, the greater viral cytopathogenic sensitivity of the HeLa cells growing in the TPB Supplemental Medium permitted the detection of small quantities of virus in clinical specimens. The proteose content of the Bacto Tryptose Phosphate Broth was considered to be the stimulating factor for the cells.

To rehydrate Bacto Tryptose Phosphate Broth, suspend 29.5 g in 1 liter cold TC Distilled Water, DE. Heat to boiling to dissolve completely. Sterilize in the autoclave for 15 minutes at 15 lbs pressure (121°C).

REFERENCE
1. Proc. Soc. Exp. Biol. & Med., 89:66, 1955.

PACKAGING
Bacto Tryptose Phosphate Broth	1 lb (454 g)	0060-01-6
	1/4 lb (114 g)	0060-02-5
	5 lb (2.27 kg)	0060-05-2
	10 kg	0060-08-9

TC TYRODE SOLUTION

INTENDED USE
TC Tyrode Solution is a sterile balanced salt solution certified for use in tissue culture procedures. It is prepared according to the formula of Tyrode.[1]

FORMULA

TC TYRODE SOLUTION
Ingredients per liter

Bacto Dextrose	1000 mg	Magnesium Chloride	100 mg
Sodium Chloride	8000 mg	Monosodium Phosphate	50 mg
Potassium Chloride	200 mg	Sodium Bicarbonate	1000 mg
Calcium Chloride	200 mg		

PROCEDURE
TC Phenol Red Solution 1% may be added in a 0.002% concentration as an indicator of reaction by adding 0.2 ml TC Phenol Red Solution 1% to 100 ml TC Tyrode Solution.

STORAGE
Store at 2 – 8°C.

REFERENCE
1. Arch. internat. de pharmacodyn, ét de therup., 20:205, 1910.

PACKAGING
TC Tyrode Solution   100 ml   5556-72-0

TC TYRODE SOLUTION w/o CALCIUM AND MAGNESIUM

INTENDED USE
TC Tyrode Solution w/o Calcium and Magnesium is prepared according to the formulation of Tyrode except the calcium chloride and magnesium chloride have been omitted. It is recommended for specialized uses in tissue culture and other procedures requiring a calcium-free and magnesium-free basal substrate.

STORAGE
Store at 2 – 8°C.

PACKAGING
TC Tyrode Solution w/o Calcium and Magnesium   100 ml   5662-72-1

TC VITAMINS MINIMAL EAGLE, 100X

INTENDED USE
TC Vitamins Minimal Eagle, 100X is a sterile 100-fold concentration of vitamins prepared according to the formulation of Eagle[1] for the preparation of TC Minimal Medium Eagle.

FORMULA

TC VITAMINS MINIMAL EAGLE, 100X
Ingredients per liter

Choline Chloride	1 mg	Calcium Pantothenate	1 mg
Folic Acid	1 mg	Pyridoxal	1 mg
Inositol	2 mg	Riboflavin	0.1 mg
Nicotinamide	1 mg	Thiamine Hydrochloride	1 mg

PROCEDURE
One liter of TC Minimal Medium Eagle is prepared as described under TC Amino Acids Minimal Eagle, 50X.

STORAGE
Store at 2 – 8°C.

REFERENCE
1. Science, 130:432, 1959.

PACKAGING
TC Vitamins Minimal Eagle, 100X   100 ml   5833-72-5

TC YEASTOLATE

TC Yeastolate is a desiccated, clarified, soluble portion of autolyzed fresh yeast prepared for use in tissue culture procedures. It may be used as a source of vitamin B complex and enrichment in bacteriological culture media.

TC Yeastolate has been reported of decided value in tissue culture procedures. Morgan[1] reported that TC Yeastolate in 0.05 – 0.1% concentration in TC Medium 199 markedly increased the longevity of tissue cells and showed promise of being a favorable substitute for serous fluids in tissue culture. Syverton and Robertson[2] also reported that the addition of TC Yeastolate to their medium for the cultivation of HeLa cells permitted a marked saving in the serum requirements. TC Yeastolate in 0.1 – 0.2% concentration enhanced the growth of HeLa cells and maintained them with no evidence of toxicity. Westwood, MacPherson and Titmuss[3] employed this product in 0.5% concentration in their medium for studying cell transformations.

REFERENCES
1. Personal Communication, April, 1954.
2. Proc. Soc. Exp. Biol. and Med., 88:119, 1955.
3. Brit. J. Exp. Path., 38:138, 1957.

PACKAGING
TC Yeastolate   100 g   5577-15-5
                           5 lb (2.27 kg)   5577-05-7

BACTO TODD HEWITT BROTH

INTENDED USE
Bacto Todd Hewitt Broth is recommended for the cultivation of group A hemolytic streptococci used in serological typing, as a general culture medium for the cultivation of pathogenic microorganisms and for blood culture.

HISTORY/PRINCIPLES
Bacto Todd Hewitt Broth is prepared according to the formula suggested by Updyke[1] and Updyke and Nickle.[2] Elliott[3] reported that Todd Hewitt broth prepared with neopeptone was excellent for growing group A streptococci for the production of type specific M substance since proteinase was not produced in this medium. Updyke and Nickle[2] compared the suitability of various dehydrated media for the production of type specific extracts of group A streptococci, using a Todd Hewitt broth prepared with an infusion of fresh beef heart as a control. Their findings showed that Bacto Todd Hewitt Broth gave results comparable with those obtained with the control medium, and that this medium was particularly satisfactory for growth of group A streptococci for serological typing.

FORMULA
BACTO TODD HEWITT BROTH
DEHYDRATED
Ingredients per liter

Beef Heart, Infusion from . . . . .500 g
Neopeptone, Difco . . . . 20 g
Bacto Dextrose . . . . 2 g
Sodium Chloride . . . . 2 g
Disodium Phosphate . . . . 0.4 g
Sodium Carbonate . . . . 2.5 g

Final pH 7.8 ± 0.2 at 25°C.

One pound will make 15 liters of medium.

METHOD OF PREPARATION
1. Dissolve 30 g in 1 liter distilled or deionized water.
2. Distribute as desired in tubes, flasks or bottles and sterilize in the autoclave for 15 minutes at 15 lbs pressure (121°C).

STORAGE
Bacto Todd Hewitt Broth Below 30°C
Prepared medium 15 – 30°C

QUALITY CONTROL
Identity Specifications
Dehydrated powder: light beige, homogeneous, free-flowing
Reaction of 3% solution: pH 7.8 ± 0.2 at 25°C
Prepared medium: light to medium amber, clear

Typical Cultural Response in Bacto Todd Hewitt Broth
After 18 – 48 Hours at 35°C

Organism	Growth
Neisseria meningitidis ATCC® 13090	good to excellent
Streptococcus mitis ATCC® 9895	good to excellent
Streptococcus pneumoniae ATCC® 6303	good to excellent
Streptococcus pyogenes ATCC® 19615	good to excellent

REFERENCES
1. Personal Communication, 1951.
2. Applied Microbiol., 2:177, 1954.
3. J. Exp. Med., 81:573, 1945.

PACKAGING

Bacto Todd Hewitt Broth	1 lb (454 g)	0492-01-4
	1/4 lb (114 g)	0492-02-3
	5 lb (2.27 kg)	0492-05-0
	10 kg	0492-08-7

BACTO TOMATO JUICE AGAR
BACTO TOMATO JUICE AGAR SPECIAL

INTENDED USE
Bacto Tomato Juice Agar and Bacto Tomato Juice Agar Special are used for the cultivation and enumeration of lactobacilli.

HISTORY/PRINCIPLES
The lactobacilli grow poorly on ordinary cluture media and require special nutrients. Mickle and Breed[1] reported the use of tomato juice in culture media for lactobacilli. Kulp,[2] while investigating the use of tomato juice on bacterial development, found that the growth of L. acidophilus was enhanced in media containing this material. Colonies on plates of this new medium were large and more characteristic than on other media. Later Kulp and White[3] described a modification of the original medium which yielded relatively high quantitative counts.

Bacto Tomato Juice Agar is prepared according to Kulp and White's modification and contains Bacto Peptone and Bacto Peptonized Milk. The dehydrated medium is excellently suited for the cultivation of members of the Lactobacillus group.

Bacto Tomato Juice Agar Special is recommended for the direct plate count of the lactobacilli from saliva and for the cultivation of other acidophilic microorganisms. It is prepared according to the formula suggested by Jay.[4,5] The number of lactobacilli in saliva is an index of predisposition to dental caries as described by Jay.[4,5] Many dentists prefer to use the direct count of lactobacilli for the diagnosis of caries.

The reaction of Bacto Tomato Juice Agar Special is adjusted to pH 5.0 so as to encourage the growth of lactobacilli and at the same time inhibit the growth of many commensal bacteria that may be encountered in saliva. This is a slightly more selective medium for lactobacilli than is Bacto Tomato Juice Agar.

FORMULAE

BACTO TOMATO JUICE AGAR

DEHYDRATED

Ingredients per liter

Tomato Juice (400 ml)	20 g
Bacto Peptone	10 g
Bacto Peptonized Milk	10 g
Bacto Agar	11 g

Final pH 6.1 ± 0.2 at 25°C.

One pound will make 8.9 liters
of medium.

BACTO TOMATO JUICE AGAR SPECIAL

DEHYDRATED

Ingredients per liter

Tomato Juice (400 ml)	20 g
Bacto Peptone	10 g
Bacto Peptonized Milk	10 g
Bacto Agar	20 g

Final pH 5.0 ± 0.2 at 25°C.

Five hundred grams will make
8.3 liters of medium.

METHOD OF PREPARATION
CAUTION: Allow media to come to room temperature before opening.

Bacto Tomato Juice Agar
1. Suspend 51 g in 1 liter distilled or deionized water and heat to boiling to dissolve completely.
2. Sterilize in the autoclave for 15 minutes at 15 lbs pressure (121°C).

Bacto Tomato Juice Agar Special
1. Suspend 60 g in 1 liter distilled or deionized water and heat to boiling to dissolve completely.
2. Sterilize in the autoclave for 15 minutes at 15 lbs pressure (121°C).

Avoid overheating of these media to prevent hydrolysis of agar resulting in a soft medium.

STORAGE
Bacto Tomato Juice Agar	2 – 8°C
Bacto Tomato Juice Agar Special	2 – 8°C
Prepared media	2 – 8°C

QUALITY CONTROL
Identity Specifications
Dehydrated powder:	tan, homogeneous, free-flowing
Prepared medium:	medium to dark amber, slightly opalescent

Typical Cultural Response
After 40 – 48 Hours at 35°C

Organism	Bacto Tomato Juice Agar Growth	Bacto Tomato Juice Agar Special Growth
Lactobacillus casei ATCC® 9595	good to excellent	good to excellent
Lactobacillus leichmannii ATCC® 4797	good to excellent	good to excellent
Lactobacillus sp ATCC® 11506	good to excellent	good to excellent
Staphylococcus aureus ATCC® 25923	good to excellent	inhibited

REFERENCES
1. Mickle and Breed, 1925, Technical Bulletin 110, NY State Agriculture Exp. Station.
2. Kulp, 1927, Science, 66:512.
3. Kulp and White, 1932, 76:17.
4. Jay, 1938, Bacteriology and Immunology of Dental Caries and Dental Science and Dental Art.
5. Jay, 1949, Dentistry in Public Health.

PACKAGING
Bacto Tomato Juice Agar	1 lb (454 g)	0031-01-2
	1/4 lb (114 g)	0031-02-1
Bacto Tomato Juice Agar Special	500 g	0389-17-2

BACTO TOMATO JUICE BROTH

INTENDED USE
Bacto Tomato Juice Broth is a liquid medium for the cultivation of yeast and other aciduric microorganisms.

FORMULA

BACTO TOMATO JUICE BROTH
DEHYDRATED

Ingredients per liter

Tomato Juice, Desiccated	20 g
Bacto Yeast Extract	10 g
Bacto Dextrose	10 g
Dipotassium Phosphate	0.5 g
Monopotassium Phosphate	0.5 g
Magnesium Sulfate	0.2 g
Sodium Chloride	0.01 g
Ferrous Sulfate	0.01 g
Manganese Sulfate	0.01 g

Final pH 6.7 ± 0.2 at 25°C.

One pound will make 11 liters of medium.

METHOD OF PREPARATION
1. To rehydrate, suspend 41 g in 1 liter distilled or deionized water.
2. Distribute as desired and sterilize in the autoclave for 15 minutes at 15 lbs pressure (121°C).

STORAGE
Bacto Tomato Juice Broth	2 – 8°C
Prepared medium	2 – 8°C

QUALITY CONTROL
Identity Specifications

Dehydrated powder:	tan, homogeneous, free-flowing
Reaction of 4.1% solution:	pH 6.7 ± 0.2 at 25°C
Prepared medium:	dark amber, clear, no significant precipitate

Typical Cultural Response in Bacto Tomato Juice Broth
After 40 – 48 Hours at 35°C

Organism	Recovery
Lactobacillus casei ATCC® 9595	good to excellent
Lactobacillus leichmannii ATCC® 4797	good to excellent
Saccharomyces cerevisiae ATCC® 9763	good to excellent
Saccharomyces uvarum ATCC® 9080	good to excellent

PACKAGING
Bacto Tomato Juice Broth	1 lb (454 g)	0517-01-5

TRANSPORT MEDIA
FOR MICROBIOLOGICAL SPECIMENS

SWAB TRANSPORT PACK AMIES MEDIUM w/o CHARCOAL
SWAB TRANSPORT PACK AMIES MEDIUM
SWAB TRANSPORT PACK STUART'S MEDIUM MODIFIED
BACTO TRANSPORT MEDIUM AMIES
BACTO TRANSPORT MEDIUM AMIES w/o CHARCOAL
BACTO TRANSPORT MEDIUM STUART
BACTO TRANSPORT MEDIUM w/CHARCOAL
BACTO SWABS STUART

INTENDED USE
Swab Transport Packs and Bacto Transport Media are employed in the safe collection, transport and preservation of microbiological specimens.

HISTORY
Several methods are employed in the transport and preservation of specimens. The method used in dependent upon the kind of specimen under study. A common means of transport and preservation incorporates the use of a transport medium.

In 1948, Moffett, Young and Stuart[1] described a medium for transporting gonococcal specimens to the laboratory. Stuart, Toschach and Patsula[2] elaborated upon the rationale of their transport method and presented the formulation of the Stuart transport medium which they used successfully for the routine shipment of gonococcal specimens to the laboratory. It is essentially a nonnutrient, semisolid, highly reductive medium which inhibits self-destructive enzymatic reactions within the cells as well as preventing the lethal effects of oxidation, thus maintaining the organisms within the specimen at "status quo."

Crookes and Stuart[3] observed that coliform organisms were occasionally encountered in gonorrhoeal specimens and that they were able to propagate in the transport medium and over-grow the gonococci. They added Aerosporin® to the Stuart transport medium to suppress the contaminants. This antibiotic seems to restrict the growth of coliforms satisfactorily without appreciably affecting the gonococci.

Although Stuart's medium was originally designed for the transport of gonococci, its suitability in maintaining the viability of this fastidious organism led other researchers to explore the use of this medium in transporting a variety of specimens. Cooper[4] found that Stuart's methodology, utilizing the transport medium Stuart and charcoal-impregnated swabs, proved successful in maintaining the viability of enteric pathogens as well as those isolated from the upper respiratory tract.

Microorganisms which have been successfully transported in this medium include *Neisseria gonorrhoeae, N. intracellularis, Haemophilus influenzae, Bordetella pertussis,* streptococci, staphylococci, pneumococci, *Shigella, Salmonella, Escherichia, Enterobacter* and *Trichomonas vaginalis.* As a rule, these organisms show little or no reduction in viability in this medium in transport within 24 hours. Thereafter, there is a gradual diminution in viable cells. The more hardy microorganisms may remain viable for periods in excess of 72 hours. It is recommended that the transported specimens be cultured without delay after their receipt in the laboratory.

Cary and Blair[5], using the Stuart transport medium in transporting fecal specimens containing *Shigella*, also experienced difficulty with overgrowth by contaminants in many specimens. These authors took cognizance of the contaminating organisms (*E. coli, E. freundii* and *E. aerogenes*) in relation to the composition of the transport medium and concluded that the contaminants derived their energy from the glycerophosphate. Cary and Blair[5] observed that Verkatraman and Ramakrishnan[6] had maintained *Vibrio cholerae* viable for more than 92 days in a nutrient-free, sea salt medium adjusted to pH 9.2. They, therefore, substituted inorganic phosphates for the glycerophosphate in Stuart transport medium and eliminated overgrowth by contaminants.

Amies[7,8] confirmed Cary and Blair's[5] observations that the inorganic salt buffer was superior to the glycerophosphate. He further modified the formulation of Cary and Blair[5] by using a balanced salt solution containing the inorganic phosphate buffer and omitting the methylene blue. The medium so modified according to the author yielded a significantly higher percentage of positive cultures than did the former transport medium of Stuart.

PRINCIPLES OF THE PROCEDURE

Swab Transport Packs are a complete ready-to-use system to collect, preserve and transport a specimen to the laboratory. The swab is a sterile, nonreactive vehicle to collect the specimen. It can be removed from the transport pack without handling or contaminating the swab tip. The medium Amies, Amies w/o charcoal or Stuart's medium modified provide a moist, inert protective environment to preserve the specimen. The specimen is taken via the swab, inserted into the medium of choice and transported in the plastic casing to the laboratory.

Bacto Transport Medium Amies is recommended for transporting microbial specimens by mail or otherwise to the laboratory for culturing. It is patterned after Cary and Blair's modification of the Stuart transport medium.

Bacto Transport Medium Amies w/o Charcoal is formulated similarly to the Bacto Transport Medium Amies except the charcoal has been omitted. It can be used as a transport medium without charcoal or can be used as the basal transport medium to which charcoal is added to make to make the complete Amies formulation.

Bacto Transport Medium Stuart is recommended as a substrate in which to transport clinical specimens containing bacteria, fungi or parasites in public health laboratory practice. It is prepared according to the formulation of Stuart and possesses the capacity to maintain fastidious microorganisms in the viable state during shipment.

Bacto Transport Medium w/Charcoal is designed for use as a transport medium for specimens containing bacteria, yeasts or molds. It is Bacto Transport Medium Stuart to which 1% charcoal has been added. Some laboratory workers prefer to use plain swabs and the medium containing charcoal rather than the Transport Medium Stuart and Swabs Stuart which are impregnated with charcoal.

FORMULAE

SWAB TRANSPORT PACK AMIES MEDIUM
and plain swab in plastic tube

Sodium Chloride	3 g	Disodium Phosphate	1.15 g
Potassium Chloride	0.2 g	Sodium Thioglycollate	1 g
Calcium Chloride	0.1 g	Charcoal	10 g
Magnesium Chloride	0.1 g	Bacto Agar	7.5 g
Monopotassium Phosphate	0.2 g	Distilled Water	1 liter

Final pH 7.3 ± 0.2 at 25°C.

SWAB TRANSPORT PACK AMIES MEDIUM w/o CHARCOAL
and plain swab in plastic tube

Sodium Chloride	3 g	Disodium Phosphate	1.15 g
Potassium Chloride	0.2 g	Sodium Thioglycollate	1 g
Calcium Chloride	0.1 g	Bacto Agar	7.5 g
Magnesium Chloride	0.1 g	Distilled Water	1 liter
Monopotassium Phosphate	0.2 g		

Final pH 7.3 ± 0.2 at 25°C.

SWAB TRANSPORT PACK STUART'S MEDIUM MODIFIED
and plain swab in plastic tube

Sodium Glycerophosphate	10 g	Bacto Agar	7.5 g
Calcium Chloride	0.1 g	Distilled Water	1 liter
Mercaptoacetic Acid	1 g		

Final pH 7.4 ± 0.1 at 25°C.

BACTO SWABS STUART

Bacto Swabs Stuart are sterile, buffered, charcoal-impregnated swabs for use with Bacto Transport Medium Stuart in taking and transporting clinical specimens containing bacteria, fungi or parasites. They are prepared according to the procedure described by Stuart.

BACTO TRANSPORT MEDIUM AMIES

Sodium Chloride	3 g	Disodium Phosphate	1.15 g
Potassium Chloride	0.2 g	Sodium Thioglycollate	1 g
Calcium Chloride	0.1 g	Charcoal	10 g
Magnesium Chloride	0.1 g	Bacto Agar	4 g
Monopotassium Phosphate	0.2 g	Distilled or Deionized Water	1 liter

Final pH 7.3 ± 0.2 at 25°C.

BACTO TRANSPORT MEDIUM AMIES w/o CHARCOAL

Bacto Transport Medium Amies w/o Charcoal is the same formulation as Bacto Transport Medium Amies except the charcoal has been omitted.

Final pH 7.3 ± 0.2 at 25°C.

BACTO TRANSPORT MEDIUM STUART

Sodium Thioglycollate	0.9 g	Methylene Blue	0.002 g
Sodium Glycerophosphate	10 g	Bacto Agar	3 g
Calcium Chloride	0.1 g	Distilled or Deionized Water	1 liter

Final pH 7.4 ± 0.1 at 25°C.

Vials of Transport Medium Stuart will undergo a slight degree of oxidation at the mouth of the vial. This is indicated by the blue color at the upper periphery of the medium; however, vials which demonstrate a distinct blue color throughout should be discarded.

BACTO TRANSPORT MEDIUM w/CHARCOAL

Bacto Transport Medium w/Charcoal is the same formulation as Bacto Transport Medium Stuart except 1% charcoal has been added.

Final pH 7.4 ± 0.1 at 25°C.

Note: To prepare ready-to-use medium from the dehydrated base, see Dehydrated Media — **Section DM.**

PRECAUTIONS
Follow established laboratory procedure in handling and disposing of infectious materials.

STORAGE INSTRUCTIONS
Swab Transport Packs and Transport Medium vials should be stored at 15 – 30°C. Avoid freezing. See outer package for expiry. The expiration date applies to the product in its intact container when stored as directed.

Bacto Transport Medium (dehydrated) storage is below 30°C. The powder is very hygroscopic. Keep container tightly closed. The expiration date applies to the product in its intact container when stored as directed.

PROCEDURE
Materials Provided:

Swab Transport Pack Amies Medium w/o Charcoal
Swab Transport Pack Amies Medium
Swab Transport Pack Stuart's Medium Modified
Bacto Transport Medium Amies
Bacto Transport Medium Amies w/o Charcoal

Bacto Transport Medium Stuart
Bacto Transport Medium
w/Charcoal
Bacto Swabs Stuart
(to be used with Bacto
Transport Medium Stuart)

Materials Not Provided:

Long-shanked forceps
Culture medium plates

Broth tubes
37°C incubator

SPECIMEN COLLECTION
Ready-to-Use Swab Transport Packs
1. Open, peel pack to point indicated "to open" until swab cap is visible.
2. Remove sterile swab and take sample.
3. Remove media tube from package.
4. Remove cap from media tube, place swab into media tube and seal cap.
5. Complete the label and send to the laboratory with minimum delay.

Transport Medium and Transport Medium Vials
1. Take specimen swab or swabs and insert into the upper third of the medium in the transport container.
2. Cut or break off the protruding portion of the swab stick and screw the lid on the bottle or vial tightly.
3. Label the bottle or vial and send to the laboratory with minimum delay.

INOCULATION AND INCUBATION
1. Grasp swab cap and remove from Swab Transport Pack. For inoculation from transport vials, grasp end of the short swab stick firmly with long-shanked forceps.
2. Apply swab to suitable culture medium plate and inoculate in the usual manner.
3. If broth medium is employed, submerge swab and aseptically sever the swab shaft below the cap. For transport medium vials, use long-shanked forceps to remove and submerge the swab.
4. Incubate broth tubes or agar plates at 37°C for 18 – 24 hours. Other incubation conditions may be used if organism sought or suspected requires it.
5. Examine plates or tubes for growth.

RESULTS

Acceptable growth and typical morphology should be apparent after inoculation to a suitable culture medium and incubation for 18 – 24 hours. For slow-growing organisms, it may be necessary to incubate up to 48 hours.

LIMITATIONS OF THE PROCEDURE

Specimens taken from a transport medium will not exhibit the optimal or comparative growth expected from direct inoculation and cultivation. These media do, however, provide an adequate degree of preservation for those specimens which cannot be forwarded immediately to the laboratory for prompt evaluation. Viability of cells will diminish over time and some degree of multiplication or growth of contaminants can occur during prolonged periods of transit. This is particularly true of fecal specimens which contain substantial numbers of coliform organisms.

The condition of the specimen received by the laboratory for culture is a significant variable in recovery and final identification of the suspect pathogen. An unsatisfactory specimen — one which has been overgrown by contaminants, contains nonviable organisms, or in which the number of pathogens is greatly diminished — can lead to erroneous or inconclusive results.

Although Stuart's medium and the later formulation of Amies have been used in transporting specimens suspected of containing *N. gonorrhoeae* and *N. meningitidis,* it is generally accepted and recommended that Bacto Transgrow Medium be used for this purpose.

Section DM

Preparation of those transport media listed under FORMULAE from dehydrated media formulations may be accomplished by the methods stated below:

To rehydrate Bacto Transport Medium Amies, suspend 20 g in 1 liter distilled or deionized water and heat to boiling to dissolve completely. Dispense the hot medium into small screw cap vials of 6 – 8 ml capacity to within 5 mm of the top. Screw caps on snugly and autoclave vials for 15 minutes at 15 lbs pressure (121°C). Invert the vials just before the medium gels to distribute the charcoal uniformly. Retighten caps, if necessary, when vials are cool enough to handle. Final reaction of the medium will be pH 7.3 ± 0.2 at 25°C. One pound will make 22.7 liters of medium.

To rehydrate Bacto Transport Medium Amies w/o Charcoal, suspend 10 g in 1 liter distilled or deionized water and heat to boiling to dissolve completely. Dispense hot medium into small screw cap vials of 6 – 8 ml capacity to within 5 mm of the top. Screw caps snugly and autoclave for 15 minutes at 15 lbs pressure (121°C). Retighten caps, if necessary, when vials are cool enough to handle. Final reaction of the medium will be 7.3 ± 0.2 at 25°C. One pound will make 45.4 liters of medium.

To rehydrate Bacto Transport Medium Stuart, suspend 14.1 g in 1 liter distilled or deionized water and heat to boiling to dissolve completely. Dispense hot medium into small screw cap vials of 6 – 8 ml capacity to within 5 mm of the top. Screw caps snugly and autoclave for 15 minutes at 15 lbs pressure (121°C). Retighten caps, if necessary, when vials are cool enough to handle. Final reaction of the medium will be pH 7.4 ± 0.1 at 25°C. One pound will make 32 liters of medium.

To rehydrate Bacto Transport Medium w/Charcoal, suspend 24 g in 1 liter distilled or deionized water and heat to boiling to dissolve completely. Dispense the hot medium into screw cap vials of 6 – 8 ml capacity to within 5 mm of the top. Screw caps on snugly and autoclave vials for 15 minutes at 15 lbs pressure (121°C). Invert the vials just before the medium gels to distribute charcoal uniformly. Retighten caps, if necessary, when vials are cool enough to handle. Final reaction of the medium will be pH 7.4 ± 0.1 at 25°C. One pound will make 19 liters of medium.

REFERENCES

1. Brit. Med. J., 2:241, 1948.
2. Can. J. Publ. Hlth., 45:73, 1954.
3. J. Path & Bact., 78:283, 1959.
4. J. Clin. Path., 10:226 – 230, 1957.
5. J. Bact., 88:96, 1964.
6. Indian J. Med. Res., 29:681, 1941.
7. Paper presented at Can. Publ. Hlth. Meetings, Dec. 1964.
8. Can. J. Publ. Hlth., 58:296, 1967.

PACKAGING

Ready-to-Use Swab Transport Packs

Amies Medium w/o Charcoal	1 × 50 packs	9343-26-7
and plain swab in plastic tube	10 × 50 packs	9343-27-6
Amies Medium	1 × 50 packs	9345-26-5
and plain swab in plastic tube	10 × 50 packs	9345-27-4
Stuart's Medium Modified	1 × 50 packs	9340-26-0
and plain swab in plastic tube	10 × 50 packs	9340-27-9

Ready-to-Use Media

Bacto Transport Medium Amies	12 vials	0996-34-6
	144 vials	0996-37-3
Bacto Transport Medium Amies w/o Charcoal	12 vials	0832-34-4
Bacto Transport Medium Stuart	12 vials	0621-34-9
	144 vials	0621-37-6
Bacto Transport Medium w/Charcoal	12 vials	0744-34-1
	144 vials	0744-37-8
Bacto Swabs Stuart	50 × 2 swabs	3108-30-9

Dehydrated Media

Bacto Transport Medium Amies	1 lb (454 g)	0996-01-5
	5 lb (2.27 kg)	0996-05-1
Bacto Transport Medium Amies w/o Charcoal	1 lb (454 g)	0832-01-3
Bacto Transport Medium Stuart	1/4 lb (114 g)	0621-02-7
	1 lb (454 g)	0621-01-8
	5 lb (2.27 kg)	0621-05-4
Bacto Transport Medium w/Charcoal	1 lb (454 g)	0744-01-0

BACTO TRICHINELLA ANTIGEN
BACTO TRICHINELLA ANTISERUM

INTENDED USE

Bacto Trichinella Antigen is a stable, specific, sensitive extract of *Trichinella spiralis* recommended for use with Bacto Latex in the Latex Flocculation Test for trichinosis. The procedure is easily performed and the results parallel the sensitivity of the complement fixation test and the clinical course of the disease.

Bacto Trichinella Antiserum is a stable, specific serum obtained from animals with trichinosis. It is recommended as a positive control in the latex procedure as well as in the complement fixation test for trichinosis.

SUMMARY AND BACKGROUND
The significance of trichinosis in man was stressed by Gould,[1] who estimated that 16% of the American population has this disease in varying degrees, and by Kagan,[2] who reported that approximately 15% of the sera received at the National Communicable Disease Center for testing were positive for trichinosis. Because of the high incidence and protean nature of trichinosis and the difficulty in making clinical diagnosis by demonstrating the larvae in muscle tissue, a reliable, rapid serological test fills an important need. The complexity of the complement fixation test and the instability of the reagents preclude its wide-spread usage.

The use of uniform latex particles as carriers of trichinella antigen in a tube test for trichinosis was described by Innella and Redner.[3] They demonstrated the specificity, sensitivity and ease of performance of the latex test as compared with the complement fixation procedure.

REAGENTS
When used according to the suggested procedure, each 1 ml vial of Bacto Trichinella Antigen is sufficient to perform approximately 10 tube tests or 16 slide tests. Each 1 ml vial of Bacto Trichinella Antiserum is sufficient to perform approximately 2 tube tests or 2 slide tests.

REHYDRATION
Rehydrate Bacto Trichinella Antiserum with 1 ml distilled or deionized water. Rotate gently to dissolve completely.

Rehydrate Bacto Trichinella Antigen with 1 ml distilled or deionized water. If necessary, centrifuge at 1000 rpm for 3 minutes to obtain a sparklingly clear solution. Decant the clear solution. Add the Bacto Trichinella Antigen to 4 ml Bacto RA Buffer solution.

STORAGE
Bacto Trichinella Antiserum and Bacto Trichinella Antigen in their desiccated state should be stored at 2 – 8°C

Bacto Trichinella Antigen rehydrated and diluted with 4 ml Bacto RA Buffer may be aliquoted into useable amounts and stored below −20°C. Antigen should only be frozen and thawed once.

Rehydrated Bacto Trichinella Antiserum should be stored at 2 – 8°C.

EXPIRATION DATE
These reagents in their desiccated form are stable through the expiration date on the label.

Rehydrated Bacto Trichinella Antiserum is stable for one month from date of rehydration.

Rehydrated and frozen Bacto Trichinella Antigen is stable for one month from date of rehydration.

PROCEDURE
Materials Provided:
Bacto Trichinella Antigen
Bacto Trichinella Antiserum

Materials Required but not Provided:

56°C water bath	Mirror
Kahn tubes	Suitable light source
Serological pipettes	Serological flocculation test slide
Incubator	Boerner-type shaker
Centrifuge	Trichinella-negative serum
Bacto Latex 0.81	
Bacto RA Buffer, Dried	

PERFORMANCE OF THE LATEX TUBE TEST
1. Inactivate Bacto Trichinella Antiserum along with a known *Trichinella*-negative serum and the serum to be tested at 56°C for 30 minutes.
2. Prepare dilutions 1:5, 1:10 and 1:20 of each of the inactivated sera using Bacto RA Buffer as the diluent.
3. Add 0.1 ml of each dilution as well as the undiluted serum to unscratched Kahn tubes.
4. Add 0.1 ml Bacto Latex 0.81 suspension to every 2 ml of Bacto Trichinella Antigen diluted with Bacto RA Buffer according to the rehydration instructions. Mix well.
5. Add 0.1 ml sensitized latex to each serum dilution and buffer control. Shake.
6. Incubate at 37°C for 1 hour.
7. Centrifuge for 3 minutes at 2000 rpm.
8. Gently shake each tube to resuspend the latex particles and examine for the degree of flocculation by holding the tube horizontally over the concave side of the microscope mirror. Report as follows:

4+	all latex agglutinated into large clumps with clear supernatant fluid
3+	all latex agglutinated into medium clumps with clear supernatant fluid
2+	all latex agglutinated into smaller clumps with clear supernatant fluid
1+	all latex agglutinated into small clumps with slightly cloudy supernatant fluid
±	all latex agglutinated into very small clumps with slightly cloudy supernatant fluid
−	tube is cloudy without agglutination

RESULTS
Any agglutination is considered positive. Sera showing agglutination in 1:20 dilution should be retested in higher dilutions until the end point is reached.

PERFORMANCE OF THE LATEX SLIDE TEST
The latex slide test for trichinosis should be used only as a screening test. A positive test with undiluted serum or in the 1:5 dilution should be repeated using the latex tube test.
1. Prepare serum dilutions as in steps 1 and 2 of the latex tube test.
2. Prepare Trichinella Antigen as in step 4 of the latex tube test.
3. Add 1 drop of each serum dilution from a dropper (20 − 30 drops/ml) to wells on a serological flocculation test slide.
4. Add 1 drop prepared antigen to each well.
5. Rotate on Boerner-type shaking machine for 10 minutes.
6. Examine for clumping of latex particles by holding the slide over a strong light. The RA Buffer and negative serum controls should be examined at the same time.

RESULTS
Bacto Trichinella Antiserum and Antigen should give a titer of 1:16 to 1:32.

REFERENCES
1. Gould, S.E., Trichinosis, 1945.
2. J. Infec. Dis., 107:65, 1960.
3. J. Am. Med. Assoc., 171:885, 1959.

PACKAGING

Bacto Trichinella Antigen	1 ml	2375-50-2
Bacto Trichinella Antiserum	1 ml	2376-50-1
Bacto Latex 0.81	5 ml	3102-56-4
	6 × 5 ml	3102-57-3
	25 ml	3102-65-3
Bacto RA Buffer Dried	10 × 1 liter	3103-37-7

TRICHOPHYTON MEDIA

BACTO TRICHOPHYTON AGAR 1
BACTO TRICHOPHYTON AGAR 2
BACTO TRICHOPHYTON AGAR 3
BACTO TRICHOPHYTON AGAR 4
BACTO TRICHOPHYTON AGAR 5
BACTO TRICHOPHYTON AGAR 6
BACTO TRICHOPHYTON AGAR 7

INTENDED USE
Bacto Trichophyton Media are recommended for the differentiation of the *Trichophyton* species.

HISTORY/PRINCIPLES
Bacto Trichophyton Media are prepared according to the formulations of Georg and Camp[1] and are based upon the nutritional requirements of the various species.

The authors grouped the dermatophytes into four groups. Group 1 consisted of *T. verrucosum, T. schoenleinii* and *T. concentricum.* Organisms within this group seldom produce spores or distinctive pigments and their colonies resemble one another so closely that they cannot be identified by morphological criteria. These species are differentiated as indicated in the table by their behavior toward inositol, thiamine, or a combination. The fact that *T. verrucosum* grows faster at 37°C than at room temperature is also helpful in separation from these two species; as *T. schoenleinii* and *T. concentricum* are not stimulated by incubation at 37°C.

Group 2 consists of *T. tonsurans, T. mentagrophytes* and *T. rubrum.* These are species which usually produce microconidia but only occasionally produce macroconidia. Their colonial forms and pigments are so variable, one cannot differentiate them by these means. Ajello and Georg[2] based differentiation of the species within this group upon their nutritional behavior on Bacto Trichophyton Agars 1 and 4 and by in vitro hair cultures. *T. mentagrophytes* forms perforating organs in human hair samples suspended in water within 2 – 3 weeks whereas *T. rubrum* does not.

Group 3 consists of *T. violaceum.* It seldom produces microconidia but develops characteristically pigmented colonies. *T. violaceum* has a similar nutritional pattern to that of *T. tonsurans,* however it grows very slowly even in the presence of thiamine and produces a glabrous colony without spores. *T. tonsurans* grows rapidly in presence of thiamine and shows numerous microconidia.

Group 4 includes *T. megninii* and *T. equinum* which can be identified solely on basis of nutritional requirements. *T. megninii* requires histidine as indicated on Bacto Trichophyton Agar 7 in the table. *T. equinum* requires nicotinic acid as indicated in the table.

FORMULAE

BACTO TRICHOPHYTON AGAR 1
DEHYDRATED
Ingredients per liter

Bacto Vitamin Assay
Casamino Acids . . . . 2.5 g
Bacto Dextrose . . . . 40 g
Monopotassium Phosphate . . . . 1.8 g
Magnesium Sulfate . . . . 0.1 g
Bacto Agar . . . . 15 g

BACTO TRICHOPHYTON AGAR 2
DEHYDRATED
Ingredients per liter

Bacto Vitamin Assay
Casamino Acids . . . . 2.5 g
Bacto Inositol . . . . 50 g
Bacto Dextrose . . . . 40 g
Monopotassium Phosphate . . . . . . 1.8 g
Magnesium Sulfate . . . . 0.1 g
Bacto Agar . . . . 15 g

BACTO TRICHOPHYTON AGAR 3
DEHYDRATED
Ingredients per liter

Bacto Vitamin Assay
Casamino Acids . . . . 2.5 g
Bacto Inositol . . . . 50 mg
Thiamine Hydrochloride USP . . . 200 µg
Bacto Dextrose . . . . 40 g
Monopotassium Phosphate . . . . . . 1.8 g
Magnesium Sulfate . . . . 0.1 g
Bacto Agar . . . . 15 g

BACTO TRICHOPHYTON AGAR 4
DEHYDRATED
Ingredients per liter

Bacto Vitamin Assay
Casamino Acids . . . . 2.5 g
Thiamine Hydrochloride USP . . . 200 µg
Bacto Dextrose . . . . 40 g
Monopotassium
Phosphate . . . . 1.8 g
Magnesium Sulfate . . . . 0.1 g
Bacto Agar . . . . 15 g

BACTO TRICHOPHYTON AGAR 5
DEHYDRATED
Ingredients per liter

Bacto Vitamin Assay
Casamino Acids . . . . 2.5 g
Nicotinic Acid . . . . 2.0 mg
Bacto Dextrose . . . . 40 g
Monopotassium Phosphate . . . . . . 1.8 g
Magnesium Sulfate . . . . 0.1 g
Bacto Agar . . . . 15 g

BACTO TRICHOPHYTON AGAR 6
DEHYDRATED
Ingredients per liter

Ammonium Nitrate . . . . 1.5 g
Bacto Dextrose . . . . 40 g
Monopotassium Phosphate . . . . 1.8 g
Magnesium Sulfate . . . . 0.1 g
Bacto Agar . . . . 15 g

BACTO TRICHOPHYTON AGAR 7
DEHYDRATED
Ingredients per liter

Ammonium Nitrate . . . . 1.5 g
Histidine Hydrochloride . . . . 30 mg
Bacto Dextrose . . . . 40 g
Monopotassium Phosphate . . . . . . 1.8 g
Magnesium Sulfate . . . . 0.1 g
Bacto Agar . . . . 15 g

Final pH 6.8 ± 0.2 at 25°C.

One pound will make 7.7 liters of final medium.

PROCEDURE

1. Suspend 59 g in 1 liter distilled or deionized water and heat to boiling to dissolve completely.
2. Distribute into tubes and sterilize in the autoclave for 12 minutes at 15 lbs pressure (121°C).
3. Allow tubes to cool in the slanted position.

INOCULATION

4. It is important that pure cultures from a non-vitamin enriched medium such as Bacto Sabouraud Dextrose Agar or Bacto Mycobiotic Agar be used for inoculum.
5. If cultures are contaminated with bacteria, they should be grown on a medium containing antibiotics such as Bacto Mycobiotic Agar or Bacto Brain Heart CC Agar for several generations to eliminate the bacteria. Many bacteria synthesize vitamins which may invalidate the test. For complete information regarding these methods, see Technical Information 0161, "Ready to Use Media for the Primary Isolation of Fungi."
6. In inoculating Bacto Trichophyton Agars, care must be exercised, so as not to carry over growth substances from primary cultures onto tube media used in the differential tests. Inocula to the nutrition tubes should be very small.
7. Nutritional tests are generally incubated at room temperature and read after 1 or 2 weeks. The tube in a particular test which shows maximum growth is recorded as 4+. Other tubes are read by comparison.

STORAGE

Bacto Trichophyton Agars 1 – 7　　　　　　　Below 30°C
Prepared media　　　　　　　　　　　　　　2 – 8°C

QUALITY CONTROL

Identity Specifications

Dehydrated powder:　　　　white to off-white, homogeneous, free-flowing
Reaction of 5.9% solution:　pH 6.8 ± 0.2 at 25°C
Prepared medium:　　　　　light amber, slightly opalescent

Typical Cultural Response of Certain Dermatophyte Species on Trichophyton Agars 1 – 7

Species	Trichophyton Agars No.						
	1	2	3	4	5	6	7
T. verrucosum, 84%	0	±	4+	0			
T. verrucosum, 16%	0	0	4+	4+			
T. schoenleinii	4+	4+	4+	4+			
T. concentricum, 50%	4+	4+	4+	4+			
T. concentricum, 50%	2+	2+	4+	4+			
T. violaceum	±			4+			
T. tonsurans	±			4+			
T. rubrum	4+			4+			
T. mentagrophytes	4+			4+			
T. equinum	0				4+		
T. megninii						0	4+
M. gallinae						4+	4+

REFERENCES

1. J. Bact., 74:113:1957.
2. Mycopath. ef. Mycol. Appl., 7:3:1957.

BACTO TRIPLE SUGAR IRON AGAR

INTENDED USE

Bacto Triple Sugar Iron Agar is recommended for the identification of gram-negative enteric bacilli based on the fermentation of dextrose, lactose, and sucrose and for hydrogen sulfide production.

HISTORY/PRINCIPLES

In 1911, Russell[1] described the use of two sugars in an agar medium to differentiate gram-negative organisms of intestinal origin. The ability of some members of this group to produce hydrogen sulfide was also recognized as a valuable characteristic. To detect its presence, lead or iron salts were added to the Russell medium by many investigators. Kliger[2,3] reported that adding lead acetate to Russell double sugar agar, a medium capable of differentiating typhoid, paratyphoid and dysentery could be obtained. A modification of this medium, Bacto Kligler Iron Agar, using phenol red as an indicator, and iron salts to detect hydrogen sulfide production was developed.

Krumweide and Kohn[4] modified Russell double sugar agar by the addition of sucrose to the medium. This permitted an earlier detection of those coliform organisms which ferment lactose slowly, since many of these organisms attack sucrose more readily than lactose. The added sucrose also permitted the exclusion of certain coliform and *Proteus* organisms which have the ability to attack sucrose, but not lactose, in a 24 – 48 hour incubation period.

In 1940, Sulkin and Willet[5] described a triple sugar ferrous sulfate medium for use in the identification of enteric organisms. Difco Laboratories concurrently developed a similar medium by adding 1% sucrose to Bacto Kligler Iron Agar with phenol red as the indicator. Hajna[6] described a similar medium for the identification of bacteria of the intestinal group.

Bacto Triple Sugar Iron Agar is essentially the formula originally described by Sulkin and Willet.[5] The Bacto Tryptone has been replaced by a combination of Bacto Peptone and Proteose Peptone, Difco, Bacto Yeast Extract has been added, and phenol red used as an indicator instead of brom thymol blue. Fermentation of the sugars, resulting in acid production is detected by the phenol red indicator. The color changes — yellow for acid production and red for alkalinization are striking. Sodium thiosulfate is reduced to hydrogen sulfide which then reacts with an iron salt yielding the typical black iron sulfide.

This medium is of special value when used in conjunction with MacConkey agar, SS agar, bismuth sulfite agar, brilliant green agar, EMB agar and Endo agar. For additional discussion, refer to the *Difco Manual*, 9th Edition.[7]

FORMULA

BACTO TRIPLE SUGAR IRON AGAR
DEHYDRATED
Ingredients per liter

Bacto Beef Extract	3 g	Bacto Sucrose	10 g
Bacto Yeast Extract	3 g	Ferrous Sulfate	0.2 g
Bacto Peptone	15 g	Sodium Chloride	5 g
Proteose Peptone, Difco	5 g	Sodium Thiosulfate	0.3 g
Bacto Dextrose	1 g	Bacto Agar	12 g
Bacto Lactose	10 g	Bacto Phenol Red	0.024 g

Final pH 7.4 ± 0.2 at 25°C.

One pound will make 6.6 liters of final medium.

PROCEDURE

1. To rehydrate, suspend 65 g in 1 liter distilled or deionized water and heat to boiling to dissolve completely.
2. Dispense into tubes and sterilize in the autoclave for 15 minutes at 15 lbs pressure (121°C).
3. Allow the tubes to solidify in a slanting position so that a generous butt is obtained.
4. For typical cultural reactions in 18 hours, it is recommended that the medium be inoculated heavily with growth from a solid culture medium. If inoculated from a suspension of organisms, or from broth culture, typical reactions of hydrogen sulfide production, and reversion, may not be obtained until 36 – 40 hours at 37°C.
5. To obtain true differential culture reactions on this medium, it is necessary to have a pure culture. In inoculating directly from isolation media such as MacConkey agar, SS agar or bismuth sulfite agar plates, select well isolated colonies and pick only the very center of the colony.
6. If there is any question as to the ability to obtain a pure culture from a certain colony, it is recommended that a suspicious colony be purified by streaking on MacConkey agar before inoculating into triple sugar iron agar. This procedure is always recommended to insure culture purity when picking from poured plates of bismuth sulfite agar. It is often possible to detect contaminated cultures on triple sugar iron agar slants, and when this is the case it is necessary to isolate the organism in pure culture before its typical cultural reactions can be determined.
7. Cultural reactions on Bacto Triple Sugar Iron Agar are similar to those obtained on Bacto Kligler Iron Agar, but to it is added the information of the ability of an organism to ferment sucrose. See table of typical reactions on this and other differential media.
8. Results obtained on triple sugar iron agar, as on Kligler iron agar, constitute presumptive evidence only, and must be confirmed biochemically and serologically.

STORAGE
Bacto Triple Sugar Iron Agar   Below 30°C
Prepared tubes   2 – 8°C

QUALITY CONTROL
Identity Specifications

Dehydrated powder: pink, homogeneous, free-flowing
Reaction of 6.5% solution: pH 7.4 ± 0.2 at 25°C
Prepared medium: red, very slightly opalescent, may have a slight precipitate

Typical Cultural Response in Bacto Triple Sugar Iron Agar
After 18 – 24 Hours at 35°C

	Recovery	Slant	Butt	Gas	H₂S
Escherichia coli ATCC® 25922	good	A	A	+	−
Proteus vulgaris ATCC® 13315	good	A	A	+	+
Pseudomonas aeruginosa ATCC® 9027	good	K	K	−	−
Salmonella enteritidis ATCC® 13076	good	K	A	+	+
Shigella flexneri ATCC® 12022	good	K	A	−	−

REFERENCES

1. J. Med. Research, 25:217, 1911.
2. Am. J. Pub. Health, 7:1042, 1917.
3. J. Exp. Med., 28:319, 1918.
4. J. Med. Research, 37:225, 1917.
5. J. Lab. Clin. Med., 25:649, 1940.
6. J. Bact., 49:516, 1945.
7. Difco Manual, 9th Ed., p. 166 – 168, 1953.

PACKAGING

Bacto Triple Sugar Iron Agar	1 lb (454 g)	0265-01-9
	1/4 lb (114 g)	0265-02-8
	5 lb (2.27 kg)	0265-05-5
	12 tubes	0265-34-0
	144 tubes	0265-37-7

BACTO TRYPTIC AGAR BASE

INTENDED USE

Bacto Tryptic Agar Base is a semisolid medium for use in determining motility and, with added carbohydrates, fermentation reactions of microorganisms.

FORMULA

BACTO TRYPTIC AGAR BASE
DEHYDRATED

Ingredients per liter

Bacto Tryptose . . . . 20 g
Bacto Agar . . . . 3.5 g
Bacto Phenol Red . . . . 0.02 g

Final pH 7.4 ± 0.2 25°C.

One pound will make 19.3 liters of medium.

METHOD OF PREPARATION

1. Suspend 23.5 g in 1 liter distilled or deionized water and heat to boiling to dissolve completely.
2. Sterilize in the autoclave for 15 minutes at 15 lbs pressure (121°C).

   Carbohydrate is added to the medium to a final concentration of 0.5 – 1.0%. The addition is made before or after sterilization depending on the lability of the carbohydrate. (See PREPARATION OF CULTURE MEDIA.)
3. If placed in tubes and motility reading is desired, do not slant.

STORAGE

Bacto Tryptic Agar Base	Below 30°C
Prepared medium	15 – 30°C

QUALITY CONTROL

Identity Specifications

Dehydrated powder:	pink, homogeneous, free-flowing
Reaction of 2.35% solution:	pH 7.4 ± 0.2 at 25°C
Prepared medium:	red, slightly opalescent

Typical Cultural Response in Bacto Tryptic Agar Base with 0.5% Dextrose After 18 – 24 Hours at 35°C

Organism	Growth	Acid	Gas	Motility
Escherichia coli ATCC® 25922	good to excellent	+	+	+
Salmonella typhi ATCC® 6539	good to excellent	+	–	–
Streptococcus pneumoniae ATCC® 6303	good to excellent	+	–	–

PACKAGING

Bacto Tryptic Agar Base	1 lb (454 g)	0364-01-9

BACTO TRYPTIC DIGEST BROTH

INTENDED USE

Bacto Tryptic Digest Broth is recommended for culturing fastidious microorganisms.

HISTORY/PRINCIPLES

Bacto Tryptic Digest Broth, formulated by Fields,[1] when supplemented with serum, blood or ascitic fluid, was found to be superior to some infusion media for maintenance of *Diplococcus pneumoniae (Streptococcus pneumoniae)* and *Neisseria meningitidis* in the encapsulated state. Fields demonstrated this medium to be suited for the cultivation and maintenance of *Haemophilus influenzae* and *H. pertussis (Bordetella pertussis)*. Dienes observed this medium to be satisfactory for the isolation and maintenance of PPLO (pleuro-pneumonia-like organisms). Other workers including Shepard, Murray, Foley and Wheeler declared it to be suitable for the culturing of *Streptococcus pyogenes*, *Streptobacillus moniliformis* and *Actinomycetes*.

More recently Bacto Tryptic Digest Broth has been recommended by Shepard and Lunceford[2] in the preparation of urease color test medium U9. This medium is for the detection and identification of T strain mycoplasmas. These same authors[3] later described a modification of this medium (with the addition of 0.01% L-cysteine hydrochloride) termed U9B.

FORMULA

BACTO TRYPTIC DIGEST BROTH
DEHYDRATED

Ingredients per liter

Tryptic Digest of Beef Heart	10 g
Sodium Chloride	5 g
Bacto Dextrose	1 g

Final pH 7.6 ± 0.2 at 25°C.

Five hundred grams will make 31.25 liters of medium.

METHOD OF PREPARATION
1. Suspend 16 g in 1 liter distilled or deionized water. Warm to dissolve completely, if necessary.
2. Dispense as desired and sterilize in the autoclave for 15 minutes at 15 lbs pressure (121°C).
3. Cool below 50°C and add sterile blood, serous fluids, Bacto PPLO Serum Fraction or other enrichments, if desired.

STORAGE
Bacto Tryptic Digest Broth   Below 30°C
Prepared medium   2 – 8°C

QUALITY CONTROL
Identity Specifications

Dehydrated powder: beige, homogeneous, free-flowing
Reaction of 1.6% solution: pH 7.6 ± 0.2 at 25°C
Prepared medium: light amber, clear with no precipitate

Typical Cultural Response in Bacto Tryptic Digest Broth
After 18 – 24 Hours at 35°C

Organism	Growth
Staphylococcus aureus ATCC® 25923	good to excellent
Streptococcus pneumoniae ATCC® 6303	good to excellent
Streptococcus pyogenes ATCC® 19615	good to excellent
Neisseria meningitidis ATCC® 13090	good to excellent

Typical Cultural Response in Bacto Tryptic Digest Broth with 1% Bacto PPLO
Serum Fraction After 40 – 72 Hours at 35°C under 10% CO_2

Organism	Growth
Mycoplasma bovis ATCC® 25523	good
Mycoplasma pneumoniae ATCC® 15531	good
Ureaplasma urealyticum ATCC® 27618	good

REFERENCES
1. SAB Newsletter, 22:8, 1956.
2. Applied Microbiology, 20:4:539 – 543, 1970.
3. Journal of Clinical Microbiology, 3:6:613 – 625, 1976.

PACKAGING
Bacto Tryptic Digest Broth   500 g   1829-17-8

BACTO TRYPTIC NITRATE MEDIUM

INTENDED USE
Bacto Tryptic Nitrate Medium is a semisolid medium recommended for the detection of nitrate reduction to nitrite.

PRINCIPLES
Potassium nitrate in the formula can be reduced by various microorganisms to either the nitrite or nitrogen gas forms. The presence of nitrites can be detected by several

test methods.[1,2] The method presented below was adapted from the FDA's *Bacteriological Analytical Manual* (BAM).[1]

This medium with its low agar content affords varying degrees of anaerobiosis to allow organisms with various oxygen requirements to grow.

FORMULA
BACTO TRYPTIC NITRATE MEDIUM
DEHYDRATED
Ingredients per liter

Bacto Tryptose	20 g
Bacto Dextrose	1 g
Disodium Phosphate	2 g
Potassium Nitrate	1 g
Bacto Agar	1 g

Final pH 7.2 ± 0.2 at 25°C.

One pound will make 18.1 liters of medium.

METHOD OF PREPARATION
1. Suspend 25 g in 1 liter distilled or deionized water. Heat to boiling to dissolve completely.
2. Dispense in 9 – 10 ml amounts in standard size test tubes. Apply caps.
3. Sterilize in the autoclave for 15 minutes at 15 lbs pressure (121°C).
4. Allow to cool to below 37°C and inoculate as desired.
5. Incubate for 18 – 24 hours at 35°C.
6. Read tubes for growth and prepare nitrate test reagents.

Nitrate Test Reagents
(Acetic acid, 5 Normal: add 1 part glacial acetic acid to 2.5 parts distilled water.)

Reagent A

Sulfanilic Acid	1 g
Acetic Acid, 5N	125 ml

Reagent B

Alpha-Naphthol (dimethyl-α-naphthylamine)	1 g
Acetic Acid, 5N	200 ml

7. Perform the nitrate test by shaking the contents of each tube and adding 0.1 to 0.5 ml each of Reagent A and Reagent B. The development of a red-violet color indicates that nitrate has been reduced to nitrite. If no color develops, add a pinch of zinc dust to the tube. The formation of red-violet color indicates that nitrate is still present and has not been reduced.

STORAGE
Bacto Tryptic Nitrate Medium	Below 30°C
Prepared medium	15 – 30°C

QUALITY CONTROL
Identity Specifications

Dehydrated powder:	beige, homogeneous, free-flowing
Reaction of 2.5% solution:	pH 7.2 ± 0.2 at 25°C
Prepared medium:	light amber, very slightly opalescent

**Typical Cultural Response in Bacto Tryptic Nitrate Medium
After 18 – 24 Hours at 35°C**

Organism	Growth	Nitrate Reduction
Clostridium sporogenes ATCC® 11437	good to excellent	−
Escherichia coli ATCC® 25922	good to excellent	+
Proteus mirabilis ATCC® 25933	good to excellent	+
Staphylococcus aureus ATCC® 25923	good to excellent	+
Streptococcus mitis ATCC® 9895	good to excellent	−

+ = positive, red-violet color after the addition of reagent
− = no change in color after the addition of reagents.

REFERENCES
1. FDA Bacteriological Analytical Manual, 5th Ed., P. B-1, 1978.
2. Lennette, Balows, Hansler and Truant, Manual of Clinical Microbiology, 3rd Ed., ASM, 1980.

PACKAGING
Bacto Tryptic Nitrate Medium

<table>
<tr><td></td><td>1 lb (454 g)</td><td>0367-01-6</td></tr>
<tr><td></td><td>12 tubes</td><td>0367-34-7</td></tr>
</table>

BACTO TRYPTIC SOY AGAR

INTENDED USE
Bacto Tryptic Soy Agar is a general purpose medium used with or without blood or other enrichment for isolating and cultivating a variety of fastidious microorganisms.

PRINCIPLES
When prepared without blood, Bacto Tryptic Soy Agar conforms to standards set for Soybean-Casein Digest Agar Medium by *The United States Pharmacopeia, Twentieth Revision*[1] for use in Microbiological Tests, [51] Antimicrobial Preservatives-Effectiveness and [61] Microbial Limits Test.

When blood is added, the medium is suitable for determining hemolytic reactions. When prepared in chocolate agar, the medium supports good growth of *Neisseria, Haemophilus influenzae* and related organisms. Under anaerobic conditions, clostridia and non-sporulating anaerobes grow luxuriantly.

FORMULA
BACTO TRYPTIC SOY AGAR
DEHYDRATED

Ingredients per liter

Bacto Tryptone . . . . 15 g
  Pancreatic Digest of Casein
Bacto Soytone . . . . 5 g
  Papaic Digest of Soybean Meal
Sodium Chloride . . . . 5 g
Bacto Agar . . . . 15 g

Final pH 7.3 ± 0.2 at 25°C.

One pound will make 11.4 liters of single-strength medium.

METHOD OF PREPARATION

1. Suspend 40 g in 1 liter distilled or deionized water (40 g/500 ml when preparing chocolate agar) and heat to boiling to dissolve completely.
2. Sterilize in the autoclave for 15 minutes at 15 lbs pressure (121°C). Cool to 45 – 50°C.
3. If desired, aseptically prepare one of the following variations:
   BLOOD AGAR: Add 5% sterile defibrinated blood and mix well.
   CHOCOLATE AGAR: Add 500 ml of a sterile, uniform 2% solution of Bacto Hemoglobin to 500 ml double-strength Bacto Tryptic Soy Agar and mix well. Add 10 ml (1%) Bacto Supplement B and mix well.
4. Dispense as desired.

STORAGE

Bacto Tryptic Soy Agar	Below 30°C
Prepared plates or tubes	15 – 30°C
Prepared blood plates or tubes	2 – 8°C

QUALITY CONTROL

Identity Specifications

Dehydrated powder:	light beige, free-flowing, homogeneous
Reaction of 4% solution:	pH 7.3 ± 0.2 at 25°C
Plates prepared without blood:	light amber, slightly opalescent, firmly solid
Plates prepared with 5% blood:	cherry red, opaque, firmly solid

Typical Cultural Response on Bacto Tryptic Soy Agar
After 18 – 24 Hours at 35°C

Organism	Growth w/o Blood	Growth w/5% Sheep Blood	Hemolysis
Neisseria meningitidis ATCC® 13090	excellent	excellent	none
Staphylococcus aureus ATCC® 25923	excellent	excellent	beta
Staphylococcus epidermidis ATCC® 12228	excellent	excellent	gamma
Streptococcus pneumoniae ATCC® 6303	good	excellent	alpha
Streptococcus pyogenes ATCC® 19615	good	excellent	beta

REFERENCE

1. The United States Pharmacopeia, Twentieth Edition, 1980.

PACKAGING

Bacto Tryptic Soy Agar		
	1 lb (454 g)	0369-01-4
	1/4 lb (114 g)	0369-02-3
	5 lb (2.27 kg)	0369-05-0
	10 kg	0369-08-7
	6 × 500 ml	0369-28F-0
	6 × 6 × 500 ml	0369-29F-9
	12 tubes	0369-34-5
	144 tubes	0369-37-2
	6 × 100 ml	0369-73-7

TRYPTIC SOY BROTH

BACTO TRYPTIC SOY BROTH
BACTO TRYPTIC SOY BROTH w/o DEXTROSE

INTENDED USE

Bacto Tryptic Soy Broth is a general purpose medium for cultivating fastidious and nonfastidious microorganisms. It conforms to the *United States Pharmacopeia XX* (1980) formula for soybean-casein digest medium.

Bacto Tryptic Soy Broth w/o Dextrose is a low-carbohydrate formulation of Bacto Tryptic Soy Broth.

PRINCIPLES

Bacto Tryptic Soy Broth, enriched with 1% Bacto Supplement B or C, supports excellent growth of *Neisseria, Haemophilus influenzae* and related organisms.

Clostridia and nonsporulating anaerobes grow luxuriantly when incubated under anaerobic conditions.

Bacto Tryptic Soy Broth w/o Dextrose may be used with added carbohydrates, 0.05% or more agar, or Bacto Phenol Red or other indicators.

FORMULAE

BACTO TRYPTIC SOY BROTH
Soybean-Casein Digest Medium, USP
DEHYDRATED

Ingredients per liter

Bacto Tryptone . . . . 17 g
  Pancreatic Digest of Casein
Bacto Soytone . . . . 3 g
  Papaic Digest of Soybean Meal
Bacto Dextrose . . . . 2.5 g
Sodium Chloride . . . . 5 g
Dipotassium Phosphate . . . . 2.5 g

Final pH 7.3 ± 0.2 at 25°C.

One pound will make 15.1 liters
of medium.
Rehydrate with 30 grams/liter.

BACTO TRYPTIC SOY BROTH
w/o DEXTROSE
DEHYDRATED

Ingredients per liter

Bacto Tryptone . . . . 17 g
Bacto Soytone . . . . 3 g
Sodium Chloride . . . . 5 g
Dipotassium Phosphate . . . . 2.5 g

Final pH 7.3 ± 0.2 at 25°C.

One pound will make 16.5 liters
of medium.
Rehydrate with 27.5 grams/liter.

METHOD OF PREPARATION

1. Dissolve the appropriate amount of dehydrated medium in 1 liter distilled or deionized water. If necessary, warm slightly to dissolve completely.
2. Dispense as desired.
3. Sterilize in the autoclave for 15 minutes at 15 lbs pressure (121°C).
4. Prepare variations, when desired, prior to or following sterilization, as appropriate.

STORAGE

Bacto Tryptic Soy Broth media　　　　　Below 30°C
Prepared media　　　　　　　　　　　　15 – 30°C

QUALITY CONTROL

Identity Specifications

Dehydrated powders: light beige, homogeneous, free-flowing
Reaction of solutions: pH 7.3 ± 0.2 at 25°C
Prepared media: very light amber, clear

Typical Cultural Response After 18 – 48 Hours at 35°C

Organism	Growth in Bacto Tryptic Soy Broth	Growth in Bacto Tryptic Soy Broth w/o Dextrose
Neisseria meningitidis ATCC® 13090	good to excellent	good
Streptococcus pneumoniae ATCC® 6303	good to excellent	good to excellent
Streptococcus pyogenes ATCC® 19615	good to excellent	good to excellent
Staphylococcus epidermidis ATCC® 12228	excellent	excellent

PACKAGING

Bacto Tryptic Soy Broth	1 lb (454 g)	0370-01-1
	1/4 lb (114 g)	0370-02-0
	5 lb (2.27 kg)	0370-05-7
	10 kg	0370-08-4
	12 tubes	0370-34-2
	144 tubes	0370-37-9
	12 × 1 ml	0370-52-9
	12 × 5 ml	0370-86-9
	144 × 5 ml	0370-87-8
Bacto Tryptic Soy Broth w/o Dextrose	1 lb (454 g)	0862-01-6

BACTO TRYPTONE

Bacto Tryptone is a pancreatic digest of casein used as a nitrogen source in culture media formulated for isolating and cultivating fastidious and nonfastidious bacteria and fungi.

The absence of detectable levels of carbohydrates in Bacto Tryptone makes it a suitable peptone for use in media for differentiating bacteria on the basis of their ability to ferment various carbohydrates. The high tryptophane content of Bacto Tryptone makes it valuable for use in detecting indole production[1,2,3] when differentiating bacteria that, otherwise, have similar biochemical characteristics.

This casein digest is recommended for use in culture media for detecting bacteria of sanitary significance in dairy products and treated and untreated waters, for sterility testing and for assaying antimicrobics as described in Standard Methods for the Examination of Dairy Products, Standard Methods for the Examination of Water and Wastewater, U.S. Pharmacopeia, Code of Federal Regulations and other official publications from such agencies as the Food and Drug Administration and the National Institutes of Health.

HISTORY
Refer to *Difco Manual*, 9th Edition, p. 260 – 261, 1953.

REFERENCES
1. J. Bact., 25:623, 1933.
2. Pure Culture Study of Bacteria, 5:No. 3, 1947.
3. Centr. Bakt. I Abt., 76:1, 1915.

PACKAGING
Bacto Tryptone	1 lb (454 g)	0123-01-1
	1/4 lb (114 g)	0123-02-0
	5 lb (2.27 g)	0123-05-7
	10 kg	0123-08-4

BACTO TRYPTONE GLUCOSE EXTRACT AGAR

INTENDED USE
Bacto Tryptone Glucose Extract Agar is used for the Standard Plate Count according to *Standard Methods for the Examination of Water and Wastewater*, 15th Edition.[1] It has also been used for cultivation and enumeration of bacteria in milk and other dairy products.[2]

HISTORY
Bacto Tryptone Glucose Extract Agar is a modification of the Tryptone Glucose Skim Milk Agar of Bowers and Hucker.[3] Extensive investigations in many widely separated laboratories established the superiority of their medium over nutrient agar for estimations of bacteria in milk and other dairy products. This work has been ably summarized by Yale[4] in his report on the use of the medium. Robertson[5] has employed his medium in a study of the bacterial count of ice cream, and Dennis and Weiser[6] employed it in their study of the influence of the incubation temperature on bacterial counts of milk. Prickett[7] used a glucose agar containing Bacto Tryptone in his study of the thermophilic bacteria in milk which was described in the Sixth Edition of *Standard Methods of Milk Analysis*[8] and was prepared in the dehydrated form as Bacto Yeast Dextrose Agar. *Standard Methods for the Examination of Dairy Products*[9] presently recommends the use of tryptone glucose extract agar for the detection of thermophilic bacteria. Media similar to that used by Prickett were recommended by Downs, Hammer, Cordes, and Macy[10] in their report on the bacteriological methods for the analysis of dairy products.

A committee on Standard Methods for the Examination of Dairy Products evaluated, comparatively, the modified medium of Bowers and Hucker to determine whether it was superior to the then standard nutrient agar for the plate count of milk. As a result of these studies, the committee[11,12] recommended the adoption of tryptone glucose extract milk agar as the standard medium for the bacteriological plate count of milk. Details of the *Standard Methods* committee's studies are reported by Abele.[13]

FORMULA

BACTO TRYPTONE GLUCOSE EXTRACT AGAR
DEHYDRATED
Ingredients per liter

Bacto Beef Extract . . . . 3 g
Bacto Tryptone . . . . 5 g
Bacto Dextrose (Glucose) . . . . 1 g
Bacto Agar . . . . 15 g

Final pH 7.0 ± 0.2 at 25°C.

One pound will make 18.9 liters of medium.

METHOD OF PREPARATION

1. Rehydrate medium by suspending 24 g in 1 liter distilled or deionized water and heat to boiling to dissolve completely.
2. Sterilize in the autoclave for 15 minutes at 15 lbs pressure (121°C).

For additional information concerning use with dairy products refer to the *Difco Manual*, 9th Ed.[14]

STORAGE

Bacto Tryptone Glucose Extract Agar   Below 30°C
Prepared medium   15 – 30°C

QUALITY CONTROL
Identity Specifications

Dehydrated medium: tan, homogeneous, free-flowing.
Reaction of 2.4% solution: pH 7.0 ± 0.2 at 25°C
Prepared medium: light amber, clear to slightly opalescent

Typical Cultural Response on Bacto Tryptone Glucose Extract Agar
After 47 – 49 Hours at 31 – 33°C

Pasteurized Milk ⎫
Unpasteurized Milk ⎬ excellent recovery of organisms present
Water Samples ⎭

REFERENCES

1. Standard Methods for the Examination of Water and Wastewater, 15th Ed., APHA, 1980.
2. Standard Methods for the Examination of Dairy Products, 13th Ed., APHA, 1972.
3. Tech. Bull. 228, N.Y. State Agr. Exp. Sta., 1935.
4. Am. J. Pub. Health, 28:148:1938.
5. Proc. 36th Cong. Intern. Assoc. Ice Cream Manufacturers, 2:132:1936.
6. J. Dairy Science, 20:445:1937.
7. Tech. Bull. 147, N.Y. State Agr. Exp. Sta., 1928.
8. Standard Methods of Milk Analysis, 6th Edition:60:1934.
9. Standard Methods for the Examination of Dairy Products, 9th Ed.: 343:1948.
10. J. Dairy Science, 18:647:1935.
11. Am. J. Pub. Health, 28:1447:1938.
12. Ninth Annual Year Book (1938 – 39) p. 79, Suppl., Am. J. Pub. Health, 29:No. 2:1939.
13. Am. J. Pub. Health, 29:821:1939.
14. Difco Manual, 9th Ed., 58 – 59, 1953.

PACKAGING

Bacto Tryptone Glucose Extract Agar

1 lb (454 g)	0002-01-7
1/4 lb (114 g)	0002-01-7
5 lb (2.27 kg)	0002-05-3
12 tubes	0002-34-8
144 tubes	0002-37-5
6 × 100 ml	0002-73-0

BACTO TRYPTOSE

Bacto Tryptose is the enzymatic hydrolysate of protein which, when added to media formulations as the only nitrogen source, easily demonstrates its equivalence and often its superiority to the meat infusion media used for culturing many of the nutritionally fastidious bacteria.

The problems involved in preparing meat infusions for use in culture media formulations for the nutritionally fastidious bacteria encouraged us to attempt the development of a protein hydrolysate which was as culturally productive in a complete medium as were the meat infusions. The superiority of the infusion media over other available media had been demonstrated and reported.

Bacto Tryptose is the recommended nitrogen source in media for culturing *Brucella*, streptococci, meningococci and other nutritionally demanding organisms.

HISTORY
Refer to *Difco Manual*, 9th Ed., p. 262 – 263, 1953.

PACKAGING

Bacto Tryptose	1 lb (454 g)	0124-01-0
	1/4 lb (114 g)	0124-02-9
	5 lb (2.27 kg)	0124-05-6
	10 kg	0124-08-3

BACTO TRYPTOSE BLOOD AGAR BASE

INTENDED USE
Bacto Tryptose Blood Agar Base is a nutritious, infusion-free basal medium designed especially for use with blood in isolating, cultivating and determining the hemolytic reactions of fastidious microorganisms.

HISTORY
Casman[1,2] reported that a medium consisting of 2% Bacto Tryptose, 0.3% Bacto Beef Extract, 0.5% sodium chloride, 1.5% Bacto Agar and 0.03% dextrose equaled fresh beef infusion base with respect to growth of microorganisms. The small amount of carbohydrate, however, interfered with hemolytic reactions unless the medium was incubated in an atmosphere of carbon dioxide.

PRINCIPLES
Blood agar base prepared with Bacto Tryptose rather than an infusion maintains blood cells in an excellent state of preservation, insuring distinct and typical hemolytic reactions.

FORMULA

BACTO TRYPTOSE BLOOD AGAR BASE
DEHYDRATED
Ingredients per liter

Bacto Tryptose . . . . 10 g
Bacto Beef Extract . . . . 3 g
Sodium Chloride . . . . 5 g
Bacto Agar . . . . 15 g

Final pH 7.2 ± 0.2 at 25°C.

One pound will make 13.7 liters of single-strength medium.

METHOD OF PREPARATION

1. Suspend 33 g in 1 liter distilled or deionized water and heat to boiling to dissolve completely.
2. Sterilize in the autoclave for 15 minutes at 15 lbs pressure (121°C). Cool to 45 – 50°C.
3. Aseptically add 5% sterile, defibrinated, room temperature blood. Mix well.
4. Dispense into Petri dishes or tubes, as desired.

STORAGE

Bacto Tryptose Blood Agar Base   Below 30°C
Prepared medium   2 – 8°C

QUALITY CONTROL

Identity Specifications

Dehydrated powder: tan, free-flowing, homogeneous
Reaction of 3.3% solution: pH 7.3 ± 0.2 at 25°C
Plates prepared with 5% sheep blood: cherry red, opaque, firmly solid

Typical Cultural Response on Bacto Tryptose Blood Agar Base Prepared with 5% Sheep Blood After 18 – 24 Hours at 35°C

Organism	Growth w/o Blood	Growth w/Blood	Hemolysis
Neisseria meningitidis ATCC® 13090	good to excellent	excellent	none
Staphylococcus aureus ATCC® 25923	good to excellent	excellent	beta/gamma
Staphylococcus epidermidis ATCC® 12228	good to excellent	excellent	gamma
Streptococcus pneumoniae ATCC® 6303	fair to good	good to excellent	alpha
Streptococcus pyogenes ATCC® 19615	fair to good	good to excellent	beta

REFERENCES
1. J. Bact., 43:33, 1942.
2. Am. J. Clin. Path., 17:281, 1947.

PACKAGING

Bacto Tryptose Blood Agar Base	1 lb (454 g)	0232-01-9
	5 lb (2.27 kg)	0232-05-5
	10 kg	0232-08-2

BACTO TRYPTOSE BLOOD AGAR BASE W/YEAST EXTRACT

INTENDED USE
Bacto Tryptose Blood Agar Base w/Yeast Extract is used with or without blood for culturing microorganisms.

FORMULA
BACTO TRYPTOSE BLOOD AGAR BASE W/YEAST EXTRACT
DEHYDRATED
Ingredients per liter

Bacto Tryptose Blood Agar Base . . . 33 g
Bacto Yeast Extract . . . . 1 g

Final pH 7.3 ± 0.1 at 25°C.

One pound will make 13.3 liters of basal medium.

METHOD OF PREPARATION
1. Suspend 34 g in 1 liter distilled or deionized water and heat to boiling to dissolve completely.
2. Sterilize in the autoclave for 15 minutes at 15 lbs pressure (121°C). Cool to 45 – 50°C.
3. Aseptically add 5% sterile, defibrinated, room-temperature blood. Mix well.
4. Dispense into Petri dishes or tubes, as desired.

STORAGE
Bacto Tryptose Blood Agar Base w/Yeast Extract	Below 30°C
Prepared medium	2 – 8°C

QUALITY CONTROL
Identity Specifications

Dehydrated powder:	beige, homogeneous, free-flowing
Reaction of 3.4% solution:	pH 7.3 ± 0.1 at 25°C
Plates prepared with 5% sheep blood:	cherry red, opaque, no hemolysis

Typical Cultural Response on Bacto Tryptose Blood Agar Base w/Yeast Extract Prepared with 5% Sheep Blood After 40 – 48 Hours at 35°C

Organism	Growth w/o Blood	Growth w/Blood	Hemolysis
Neisseria meningitidis ATCC® 13090	excellent	excellent	none
Staphylococcus aureus ATCC® 25923	excellent	excellent	beta
Staphylococcus epidermidis ATCC® 12228	excellent	excellent	gamma
Streptococcus pneumoniae ATCC® 6303	good to excellent	excellent	alpha
Streptococcus pyogenes ATCC® 19615	excellent	excellent	beta

PACKAGING
Bacto Tryptose Blood Agar Base w/Yeast Extract	1 lb (454 g)	0662-01-8
	10 kg	0662-08-1

TRYPTOSE MEDIA

BACTO TRYPTOSE AGAR
BACTO TRYPTOSE AGAR W/THIAMINE
BACTO TRYPTOSE BROTH
BACTO TRYPTOSE BROTH W/THIAMINE

INTENDED USE

Bacto Tryptose media are recommended for the isolation, cultivation and differentiation of *Brucella*. *Standard Methods for the Examination of Dairy Products*[1,2] and *Diagnostic Procedures and Reagents*[3,4] of the American Public Health Association list the medium with thiamine for this purpose and also for the cultivation of a large variety of pathogenic microorganisms, including streptococci.

HISTORY/PRINCIPLES

Bacto Tryptose Media are prepared without extract or infusion of meat and are recommended for the cultivation of discriminating pathogenic as well as saprophytic bacteria. Huddleson[5] used a broth containing 2% Bacto Tryptose as an enrichment medium in the isolation of *Brucella* from man. McCullough, Mills, Herbst, Roessler and Brewer[6] reported that the addition of thiamine, dextrose and iron salts increased the growth of *B. suis*. Sanders and Huddleson[3] showed that the addition of dextrose and thiamine hydrochloride to the medium resulted in the stimulation of the growth of some species of *Brucella*. Bacto Tryptose Broth w/Thiamine is prepared according to the formula of tryptose dextrose vitamin B broth as given in the *Diagnostic Procedures and Reagents*[3] of the American Public Health Association.

In the past, it has been considered necessary to include meat extract or meat infusions in culture media for the cultivation of bacteria, except possibly for some of the more easily cultivated strains. It has been shown that many fastidious pathogenic organisms can be isolated and cultivated in media prepared without meat extract, infusions of meat or other enrichment, if a suitable peptone is employed. Bacto Tryptose, in 2% concentration, satisfactorily replaces the usual infusion peptone portion of many media. Huddleson[5] pointed out that the isolation of *Brucella* from human blood is more rapid and improved if the blood is incubated in Bacto Tryptose Broth. Sodium citrate in 1% concentration added to the medium serves as an anticoagulant and assists in fixing the complement of the blood specimen.

The procedure for isolating and culturing *Brucella* from human blood is given in detail in the discussion of Bacto Tryptose Agar in the *Difco Manual*, 9th Ed.[7]

The addition of 0.1% agar to tryptose broth is highly recommended, if the use of the small amount of agar is not objectionable. *Diagnostic Procedures and Reagents*[3] prefers the use of the tryptose broth w/thiamine with the addition of 0.05 – 0.1% agar for culturing *Brucella* from whole blood. Growth of aerobes and anaerobes in liquid media is greatly increased by the addition of 0.1% agar.

Bacto Tryptose Agar and Bacto Tryptose Agar w/Thiamine are also recommended for the cultivation of pathogenic organisms without enrichment, for streptococci, pneumococci, meningococci and other fastidious bacteria. The high productivity of Bacto Tryptose Agar and Bacto Tryptose used clinically for the isolation and cultivation of *Brucella* attests to its value for the primary cultivation of *Brucella* as well as other fastidious organisms.

Blood agar may be prepared by adding 5% sterile defibrinated blood to the melted sterile medium at 50°C. Bacto Tryptose Agar contains 0.1% dextrose, probably slightly more than is present in the average meat infusion medium in the form of muscle sugar.

FORMULAE

BACTO TRYPTOSE AGAR
DEHYDRATED

Ingredients per liter

Bacto Tryptose	20 g
Bacto Dextrose	1 g
Sodium Chloride	5 g
Bacto Agar	15 g

BACTO TRYPTOSE AGAR
W/THIAMINE
DEHYDRATED

Ingredients per liter

Bacto Tryptose	20 g
Bacto Dextrose	1 g
Sodium Chloride	5 g
Bacto Agar	15 g
Thiamine Hydrochloride	0.005 g

FinaL pH 7.2 ± 0.2 at 25°C.

One pound will make 11 liters of medium.
Use 41 g/liter.

BACTO TRYPTOSE BROTH
DEHYDRATED

Ingredients per liter

Bacto Tryptose	20 g
Sodium Chloride	5 g
Bacto Dextrose	1 g

BACTO TRYPTOSE BROTH
W/THIAMINE
DEHYDRATED

Ingredients per liter

Bacto Tryptose	20 g
Sodium Chloride	5 g
Bacto Dextrose	1 g
Thiamine Hydrochloride	0.005 g

Final pH 7.2 ± 0.2 at 25°C.

One pound will make 18.1 liters of medium.
Use 25 g/liter.

METHOD OF PREPARATION
1. Dissolve the appropriate amount in 1 liter distilled or deionized water. If desired, add 0.5 – 1% agar to the broth media. Heat to boiling to dissolve completely.
2. Distribute in tubes, bottles or flasks and sterilize in the autoclave for 15 minutes at 15 lbs pressure (121°C).

   For best results the medium should be freshly prepared. If not used the same day as sterilized, heat in boiling water or flowing steam to remove absorbed oxygen and cool quickly without agitation just prior to inoculation.

STORAGE
Bacto Tryptose media   Below 30°C
Prepared media   2 – 8°C

QUALITY CONTROL
Identity Specifications

Dehydrated powder:	beige, homogeneous, free-flowing
Reaction of solution:	pH 7.2 ± 0.2 at 25°C
Prepared media:	light amber, very slightly opalescent

**Typical Cultural Response on Bacto Tryptose Media
After 40 – 48 Hours at 35°C Under 10% CO_2**

Organism	Growth
Brucella abortus ATCC® 4315	good to excellent
Brucella melitensis ATCC® 4309	good to excellent
Brucella suis ATCC® 6597	good to excellent
Streptococcus pneumoniae ATCC® 6303	good to excellent
Streptococcus pyogenes ATCC® 19615	good to excellent

REFERENCES

1. Standard Methods for the Examination of Dairy Products, 9th Ed., 1948.
2. ibid., 12th Ed., p. 93, 1967.
3. Diagnostic Procedures and Reagents, 3rd Ed., 1950.
4. ibid., 4th Ed., 1963.
5. Huddleson: Brucellosis in Man and Animals, 14, 1939.
6. J. Bact., 53:5, 1947.
7. Difco Manual, 9th Ed., p. 111 – 115, 1953.

PACKAGING

Bacto Tryptose Agar	1 lb (454 g)	0064-01-2
	1/4 lb (114 g)	0064-02-1
	5 lb (2.27 kg)	0064-05-8
	10 kg	0064-08-5
Bacto Tryptose Agar w/Thiamine	1 lb (454 g)	0633-01-4
Bacto Tryptose Broth	1 lb (454 g)	0062-01-4
	10 kg	0062-08-7
Bacto Tryptose Broth w/Thiamine	1 lb (454 g)	0623-01-6

BACTO TRYPTOSE PHOSPHATE BROTH

INTENDED USE

Bacto Tryptose Phosphate Broth is a buffered medium used for cultivating fastidious microorganisms.

HISTORY

Standard Methods for the Examination of Dairy Products, 9th Edition,[1] described tryptose phosphate agar broth (0.1% agar added) for use in isolating pathogenic bacteria from cheese. Diagnostic Procedures and Reagents[2] (APHA) also recommended adding 0.1 to 0.2% agar to the formulation.

While studying the detection of food poisoning attributable to dairy products, Newman[3] used tryptose phosphate broth with 0.1% agar and 1:2500 sodium azide added for cultivating streptococci at 37°C.

Tryptose phosphate broth has been used for testing the sensitivity of microorganisms to antibiotics by the tube dilution method. Waisbren, Carr and Dunnett[4] demonstrated that a soy bean peptone medium inhibited the action of neomycin, Aureomycin®, terramycin and polymyxin against the test organism while tryptose phosphate broth, dextrose broth and nutrient broth were suitable for the test.

PRINCIPLES

The addition of 0.1 to 0.2% agar to a broth medium improves the productivity of the medium for most purposes. The low agar concentration provides suitable conditions for both aerobic growth in the clear upper zone and for microaerophilic and anaerobic growth in the lower, flocculent agar zones.

FORMULA

BACTO TRYPTOSE PHOSPHATE BROTH
DEHYDRATED

Ingredients per liter

Bacto Tryptose	20 g
Bacto Dextrose	2 g
Sodium Chloride	5 g
Disodium Phosphate	2.5 g

Final pH 7.3 ± 0.2 at 25°C.

One pound will make 15.3 liters of final medium.

METHOD OF PREPARATION

1. Dissolve 29.5 g in 1 liter distilled or deionized water.
2. If a medium containing 0.1% agar is desired, suspend 1 g agar with above ingredients and heat to boiling to dissolve completely.
3. Dispense as desired.
4. Sterilize in the autoclave for 15 minutes at 15 lbs pressure (121°C).

STORAGE

Bacto Tryptose Phosphate Broth   Below 30°C
Prepared medium   15 – 30°C

QUALITY CONTROL
Identity Specifications

Dehydrated powder: beige, homogeneous, free-flowing
Reaction of 2.95% solution: pH 7.3 ± 0.2 at 25°C
Prepared medium: light amber, may be very slightly opalescent with very
                               slight precipitate

Typical Cultural Response in Bacto Tryptose Phosphate Broth
After 18 – 24 Hours at 35°C

Organism	Recovery
Neisseria meningitidis ATCC® 13090	good
Staphylococcus epidermidis ATCC® 12228	excellent
Streptococcus pneumoniae ATCC® 6303	excellent
Streptococcus pyogenes ATCC® 19615	excellent

REFERENCES

1. Standard Methods for the Examination of Dairy Products, 9th Ed.: 165, 1948.
2. Diagnostic Procedures and Reagents, 3rd Ed.: 16, 1950.
3. J. Milk and Food Tech., 13:226, 1950.
4. Am. J. Clin. Path., 21:884, 1951.

PACKAGING

Bacto Tryptose Phosphate Broth	1 lb (454 g)	0060-01-6
	1/4 lb (114 g)	0060-02-5
	5 lb (2.27 kg)	0060-05-2
	10 kg	0060-08-9

TWEEN 80®

Tween 80® (Polysorbate 80, USP) is a surface active agent which when included in a medium, lowers the interfacial tension around bacteria suspended in the medium, thereby permitting more rapid entry of desired compounds into the bacterial cell. This may result in more rapid growth or other activity between the bacteria and the reactive compounds in the medium. It is used in certain culture media, in FA diluents and in numerous other reagents to improve their productivity or reactivity. It is used in Bacto TB Hydrolysis Reagent for differential identification of mycobacteria.

PACKAGING
Tween 80®   100 g   3118-15-6

BACTO UBA MEDIUM

INTENDED USE
Bacto UBA Medium (Universal Beer Agar) is a basal medium to which beer or beer and Actidione® are added for culturing microorganisms of significance in the brewing industry. It is prepared according to the formulation described by Kozulis and Page.[1]

HISTORY/PRINCIPLES
The authors compared UBA medium with other media commonly used in breweries for controlling microbial contamination and found it to be superior to the others for this purpose. It supported growth of more varieties of lactic acid bacteria and yielded larger colonies of these bacteria in a shorter period of time than did the conventional media. The UBA medium was demonstrated to have other advantages among which are:
1. It is a selective medium because of the presence of beer in the medium. Growth of microorganisms that have adapted themselves to existent conditions in the brewery is readily obtained. The presence of hop constituents and alcohol eliminates the growth of many airborne microorganisms and thus gives a true picture of bacterial contamination in the plant.
2. The completed UBA medium supports the growth of lactobacilli, pediococci, acetobacter and yeast strains better than the traditional brewer's media tested.
3. The characteristics of the UBA medium are closer to the natural environmental conditions found in the typical brewery than other media studied.

Actidione® at a level of 1 ppm (1 µg per ml) is incorporated when samples containing yeast are cultured.

FORMULA
BACTO UBA MEDIUM
DEHYDRATED
Ingredients per liter

Bacto Yeast Extract	6.1 g	Magnesium Sulfate	0.12 g
Bacto Peptonized Milk	15 g	Sodium Chloride	0.006 g
Tomato Juice Desc.	12.2 g	Ferrous Sulfate	0.006 g
Bacto Dextrose	16.1 g	Manganese Sulfate	0.006 g
Dipotassium Phosphate	0.31 g	Bacto Agar	12 g
Monopotassium Phosphate	0.31 g		

Final pH 6.3 ± 0.2 at 25°C.

One pound will make 7.3 liters of final medium.

METHOD OF PREPARATION
1. To rehydrate, suspend 62 g in 750 ml cold distilled or deionized water (halogen free tap water may be used). Heat to boiling to dissolve medium completely.
2. While still hot, add 250 ml commercial beer without degassing and mix by gently swirling or stirring.
3. Dispense immediately into sterile proper sized containers and sterilize in the auto-clave for 10 minutes at 15 lbs pressure (121°C).
4. When using Actidione,® aseptically incorporate into the medium shortly before pouring into plates according to standard procedure.

STORAGE

Bacto UBA Medium	Below 30°C
Prepared medium	2 – 8°C

QUALITY CONTROL

Identity Specifications

Dehydrated powder:	medium beige, homogeneous, free-flowing
Reaction of 6.2% solution:	pH 6.3 ± 0.2 at 25°C
Prepared medium:	medium to dark amber, clear to very slightly opalescent, no significant precipitate

Typical Cultural Response on Bacto UBA Medium After 40 – 48 Hours at 35°C

Organism	Recovery
Acinetobacter calcoaceticus ATCC® 19606	good to excellent
Lactobacillus fermentum ATCC® 9338	good to excellent
Lactobacillus sp 11506	good to excellent
Proteus vulgaris ATCC® 13315	fair to good

REFERENCE
1. Am. Soc. Brewing Chemists Proc., p. 52, 1968.

PACKAGING

Bacto UBA Medium	1 lb (454 lb)	0856-01-4

UREA, DIFCO

Urea, Difco is purified urea recommended for use in bacteriological culture media. The ability of an organism to hydrolyze urea is often a salient characteristic in its differentiation from other bacteria which have identical biochemical reactivities except in their urease producing characteristic. Refer to the discussion on Bacto Urea Broth and Bacto Urea Agar for a description of the use of these products as aids for differentiating *Enterobacteriaceae*.

PACKAGING

Urea, Difco	100 g	0190-15-3
	500 g	0190-17-1
Differentiation Disks Urea	1 vial	1625-35-2
	6 vials	1625-33-4

BACTO UREA AGAR
BACTO UREA AGAR BASE
BACTO UREA AGAR BASE CONCENTRATE 10X

INTENDED USE

Bacto Urea Agar Base is a basal medium, to which Bacto Agar is added, for detecting urease producing bacteria. Bacto Urea Agar is the prepared medium ready to use.

Bacto Urea Agar Base Concentrate is a sterile 10X solution of Bacto Urea Agar Base ready to use as recommended. It is suggested for laboratories that require only small amounts of medium.

HISTORY

Christensen[1] considered the well buffered urea medium described by Rustigian and Stuart,[2] as being suited only for the identification of *Proteus*. However, other gram-negative intestinal bacteria which are capable of splitting urea are unable to do so in this medium because of the small amount of nutritive material and increased amount of buffer present. Christensen[1] devised the urea agar medium in which he included Bacto Peptone and Bacto Dextrose and reduced the buffer content. His medium supported a more vigorous growth of many of the gram-negative enteric bacilli and readily permitted observation of urease production by *Proteus* and members of the paracolon aerobacter and paracolon intermediate groups.

Ewing[3] used the urea agar as a differential medium in the examination of many cultures from stool specimens and confirmed the findings of Christensen. Ewing and Bruner[4] utilized the urease reaction as a screening medium for the selection of *Salmonella* and *Shigella* cultures for serologic classification. Typical colonies suspected of being pathogens were picked from primary plating media onto triple sugar iron agar. All tubes showing acid and gas in the slant and butt were discarded. Transfers were made from the tubes showing an alkaline slant and acid or acid and gas butt onto the urea agar medium of Christensen. A preliminary reading was made at the end of 2 – 4 hours at 37°C and all tubes showing alkaline reactions were discarded. The tubes were reincubated and reactions were generally complete in 24 hours. All cultures producing an alkaline reaction were *Proteus* or members of the *Enterobacter* group. *Salmonella* and *Shigella* cultures failed to produce any increase in alkalinity in the medium. Hydrogen sulfide positive cultures, as determined on the previously inoculated triple sugar iron agar, were tested with Salmonella polyvalent antisera, while hydrogen sulfide negative cultures were tested with Shigella or Salmonella polyvalent antisera.

Urea agar may also be used to show urease production by other organisms. Some gram-positive cocci, certain diphtheroids, as well as certain pigment producing members of *Pseudomonas* give a positive reaction on this medium.

PRINCIPLES

Bacto Urea Agar Base is a highly nutritive medium because of the addition of Bacto Peptone and Bacto Dextrose to the medium originally described by Rustigian et al.[2] The buffer content is reduced thereby shortening the time of incubation. Bacto Urea Agar Base supports more vigorous growth of many of the enteric organisms than does the medium for the identification of *Proteus* described under Bacto Urea Broth. On this solid medium, *Proteus* attacks the urea very rapidly and after 6 hours incubation the color change, due to ammonia production, has penetrated deep into the medium. After 24 hours incubation, the entire butt of the tube is alkaline in reaction.

Urease positive *Enterobacter, Citrobacter,* or *Klebsiella,* in contrast, hydrolyse urea much more slowly, showing only slight penetration of the alkaline reaction into the butt of the medium in 6 hours and requiring 3 to 5 days to change the reaction of the entire butt. According to Christensen[1] most *Enterobacter hafniae* (*Hafnia alvei*) and *Citrobacter freundii* cultures are urease positive and *Escherichia coli* cultures are urease negative. *Salmonella* and *Shigella* species fail to produce a tract of alkalinity on the medium.

FORMULA

BACTO UREA AGAR BASE
DEHYDRATED

Ingredients per liter

Bacto Peptone	1 g
Bacto Dextrose	1 g
Sodium Chloride	5 g
Potassium Phosphate, Monobasic	2 g
Urea, Difco	20 g
Bacto Phenol Red	0.012 g

Final pH 6.8 ± 0.1 at 25°C.

One pound will make 15.6 liters of final medium.

PROCEDURE
Urea Agar Base
Allow medium to come to room temperature before opening bottle.
1. To rehydrate, suspend 29 g Bacto Urea Agar Base in 100 ml distilled or deionized water and mix thoroughly to dissolve completely.
2. Filter sterilize this concentrated base. DO NOT BOIL OR AUTOCLAVE CONCENTRATED BASE.
3. Dissolve 15 g of Bacto Agar in 900 ml distilled or deionized water by boiling and sterilize in the autoclave for 15 minutes at 15 lbs pressure (121°C).
4. Allow to cool to 50 – 55°C and aseptically add 100 ml of the filter sterilized concentrated Bacto Urea Agar Base to the cooled Bacto Agar.
5. Mix thoroughly and distribute in sterile tubes. Slant the tubes so as to have a butt about 2 cm in depth and a slant about 3 cm in length.
6. Inoculate slants by spreading growth from an agar culture over the entire surface. Do not inoculate into the butt.

Urea Agar Base Concentrate 10X
1. To prepare the medium, dissolve 1.5 g of Bacto Agar in 90 ml distilled or deionized water by boiling.
2. Sterilize in the autoclave for 15 minutes at 15 lbs pressure (121°C).
3. Allow to cool to 50 – 55°C and aseptically add the contents of one 10 ml tube Bacto Urea Agar Base Concentrate.
4. Mix thoroughly, dispense into tubes and slant.

Bacto Urea Agar Base cannot be heated above 50°C during the preparation or sterilization process. The tubes of complete urea agar medium must be slanted before the medium solidifies to avoid the necessity of remelting the agar medium, causing the hydrolysis of the urea.

Should crystals form in the concentrate before preparing the final medium, place the tubes of sterile concentrate in a water bath at 40 – 50° for a few moments with agitation to effect complete solution of the crystals.

STORAGE

Bacto Urea Agar Base and Concentrate　　　　　2 – 8°C
Prepared medium　　　　　　　　　　　　　　　　2 – 8°C

QUALITY CONTROL
Identity Specifications

Dehydrated powder:　　　　orange-red, homogeneous, inherently lumpy
Reaction of 2.9% solution:　pH 6.8 ± 0.1 at 25°C
Prepared solution:　　　　　orange, clear
Prepared medium:　　　　　　reddish orange, very slightly opalescent slants

Typical Cultural Response on Bacto Urea Agar
After 18 – 24 Hours at 35°C

Organism	Growth	Urease
Enterobacter aerogenes ATCC® 13048	good to excellent	–
Escherichia coli ATCC® 25922	good to excellent	–
Klebsiella pneumoniae ATCC® 13883	good to excellent	sl +
Proteus vulgaris ATCC® 13315	good to excellent	+
Salmonella typhimurium ATCC® 14028	good to excellent	–

+ = positive, cerise
– = negative, no change

REFERENCES
1. J. Bact., 52:461, 1946.
2. ibid., 49:437, 1945.
3. ibid., 51:433, 1946.
4. Am. J. Clin. Path., 17:1, 1947.

PACKAGING

Bacto Urea Agar	12 tubes	0283-34-8
	144 tubes	0283-37-5
Bacto Urea Agar Base	1 lb (454 g)	0283-01-7
	1/4 lb (114 g)	0283-02-6
Bacto Urea Agar Base Concentrate 10X	12 × 10 ml	0284-61-3

BACTO UREA BROTH
BACTO UREA BROTH CONCENTRATE

INTENDED USE
Bacto Urea Broth is a complete medium for detecting urease producing bacteria, particularly members of the genus Proteus.

Bacto Urea Broth Concentrate is a sterile 10X solution of Bacto Urea Broth ready to use as recommended. It is suggested for laboratories that require only small amounts of medium.

HISTORY
Rustigian and Stuart[1] and later Stuart, Van Stratum and Rustigian[2] described a highly buffered urea medium by means of which members of the genus Proteus could be

distinguished from other gram-negative enteric bacilli. *Proteus* possessed the ability to produce sufficient ammonia to give a reaction more alkaline than pH 8.1 after 12 – 48 hours incubation at 37°C. These investigators pointed out that a large number of organisms other than *Proteus* were capable of urease production and would give a positive test in weakly buffered media. It was also noted that by decreasing the amount of buffer in their standard medium to one-tenth or one-hundredth that of the original concentration, the time of incubation for identification of *Proteus* could be decreased from 12 – 48 hours to 2 – 4 hours. Rustigian and Stuart[2] used urea decomposition as a limiting characteristic for the identification of *Proteus* strains from other members of the family *Enterobacteriaceae*.

Ferguson and Hook[3] also reported that urease production, as indicated by the Rustigian and Stuart Method, was an excellent means of differentiating between members of the *Proteus* and *Salmonella* groups.

PRINCIPLES
Bacto Urea Broth is prepared according to the formula of Stuart, Van Stratum and Rustigian[1] and provides all essential growth requirements for *Proteus*. The decomposition of urea by members of the *Proteus* group has been especially helpful to differentiate these organisms from other gram-negative intestinal bacteria and to eliminate them from further study in various schema for the identification of the gram-negative intestinal pathogens.

FORMULA

BACTO UREA BROTH
DEHYDRATED

Ingredients per liter

Bacto Yeast Extract . . . . 0.1 g
Potassium Phosphate, Monobasic . . 9.1 g
Potassium Phosphate, Dibasic . . . . 9.5 g
Urea, Difco . . . . 20 g
Bacto Phenol Red . . . . 0.01 g

Final pH 6.8 ± 0.1 at 25°C.

One pound will make 11.7 liters of final medium.

PROCEDURE
Urea Broth
1. Allow medium to come to room temperature before opening bottle.
2. To rehydrate, suspend 38.7 g in 1 liter distilled or deionized water and mix thoroughly to dissolve completely.
3. Filter sterilize and aseptically distribute 3 ml amounts into small sterile test tubes (14 × 125 mm or equivalent). DO NOT BOIL OR AUTOCLAVE COMPLETE MEDIUM.
4. Inoculate with a straight needle from 18 – 24 hour agar slant cultures.
5. Incubate at 37°C.
6. Reactions are recorded after 8, 12, 24 and 48 hours incubation.
7. A positive urease reaction (hydrolysis of urea) is indicated by a change in color from yellow (pH 6.8) to a red or cerise color (pH 8.1 or more alkaline).

Urea Broth Concentrate
1. To prepare 100 ml of final medium, sterilize 90 ml distilled or deionized water in autoclave for 15 minutes at 15 lbs pressure (121°C).

2. Cool to 50 – 50°C and aseptically add the contents of one 10 ml tube of Bacto Urea Broth Concentrate. Mix thoroughly.
3. Distribute in 3 ml amounts in small sterile tubes (14 × 125 mm).
4. Bacto Urea Broth cannot be heated above 50°C during the preparation or sterilization process.
5. Should crystals form in the concentrate before preparing the final medium, place the tubes of sterile concentrate in a water bath at 40 – 50°C for a few moments with agitation to effect complete solution of the crystals.

STORAGE
Bacto Urea Broth and Concentrate   2 – 8°C
Prepared medium   2 – 8°C

QUALITY CONTROL
Identity Specifications
Dehydrated powder: light pink, homogeneous, inherently lumpy
Reaction of 3.87% solution: pH 6.8 ± 0.1 at 25°C
Prepared medium: reddish orange, clear

Typical Cultural Response in Bacto Urea Broth
After 12 – 24 Hours at 37°C

Organism	Growth	Urease
Enterobacter aerogenes ATCC® 13048	good to excellent	−
Escherichia coli ATCC® 25922	good to excellent	−
Klebsiella pneumoniae ATCC® 13883	good to excellent	+
Proteus vulgaris ATCC® 13315	good to excellent	+
Salmonella typhimurium ATCC® 14028	good to excellent	−

+ = positive, cerise
− = negative, no change

REFERENCES
1. J. Bact., 49:437, 1945.
2. Proc. Soc. Exp. Biol. Med., 53:241, 1943.
3. J. Lab. Clin. Med., 28:1715, 1943.

PACKAGING
Bacto Urea Broth	1 lb (454 g)	0272-01-0
	1/4 lb (114 g)	0272-02-9
	12 tubes	0272-34-1
	144 tubes	0272-37-8
Bacto Urea Broth Concentrate	6 × 10 ml	0280-60-8
	12 × 10 ml	0280-61-7

BACTO UREA R BROTH

INTENDED USE
Bacto Urea R (Rapid) Broth is a sterile desiccated medium which rapidly detects urease production by microorganisms.

HISTORY / PRINCIPLES
Bacto Urea R Broth, prepared according to the formulation of Ewing,[1] is suitable for the rapid detection of urease, by nature of its reduced buffer content.

FORMULA

BACTO UREA R BROTH
Ingredients per liter

Bacto Yeast Extract . . . . 0.1 g
Monopotassium Phosphate . . . . . 0.091 g
Disodium Phosphate . . . . 0.095 g
Urea, Difco . . . . 20 g
Bacto Phenol Red . . . . 0.01 g

Final pH 6.9 ± 0.2 at 25°C.

PROCEDURE
1. Rehydrate each vial with 3 ml sterile distilled or deionized water. If urease test is to be performed immediately, nonsterile water may be used.
2. Heavily inoculate the rehydrated broth with suspected colonies isolated on solid media or from a pure culture.
3. Incubate at 37°C for up to 4 hours.
4. Examine after 10, 30, 60 minutes and 4 hours. A color change from orange to cerise indicates a positive test for urease.

STORAGE
Bacto Urea R Broth   15 – 30°C

QUALITY CONTROL
Identity Specifications

Appearance: beige button or powder
Reaction of 2.03% solution: pH 6.9 ± 0.2 at 25°C
Prepared medium: orange, clear

Typical Cultural Response in Bacto Urea R Broth
After 4 Hours at 35°C

Organism	Urease
Enterobacter aerogenes ATCC® 13048	−
Escherichia coli ATCC® 25922	−
Klebsiella pneumoniae ATCC® 13883	+
Proteus mirabilis ATCC® 25933	+

REFERENCE
1. USPHS Publ. No. 174, 1962.

PACKAGING
Bacto Urea R Broth   12 × 3 ml   0978-34-8
                                  144 × 3 ml   0978-37-5

UREASE

Urease, prepared from jack bean *Canavalia ensiformis*,[1] is a highly specific enzyme which hydrolyzes urea to ammonium carbonate. Since urease, like other enzymes, is extremely sensitive to mercury, care must be taken to use laboratory equipment which has been thoroughly washed.

REFERENCE
1. *The Merck Index*, 10th Ed., 1983.

PACKAGING
Urease   10 g   0615-12-3

BACTO VJ AGAR

INTENDED USE

Bacto VJ (Vogel-Johnson) Agar is prepared according to the formula of Vogel and Johnson[1] for isolating coagulase-positive, mannitol fermenting *Staphylococcus aureus* from clinic specimens and foods. This culture medium complies with the formula in *USP XX*[2] and in *AOAC*.[3]

PRINCIPLES

Coagulase-positive strains of *Staphylococcus aureus* form characteristic black colonies surrounded by a yellow zone. Mannitol fermentation indicated by the yellow zone is a key differentiating characteristic of most strains of *S. aureus*. The growth of other bacteria is almost completely inhibited by lithium chloride, a high glycine concentration and tellurite. The tellurite when reduced by pathogenic staphylococci yields a black precipitate in the colonies.

FORMULA

BACTO VJ AGAR
DEHYDRATED

Ingredients per liter

Bacto Tryptone	10 g	Lithium Chloride	5 g
Bacto Yeast Extract	5 g	Glycine	10 g
Bacto Mannitol	10 g	Bacto Agar	15 g
Dipotassium Phosphate	5 g	Phenol Red	0.025 g

Final pH 7.2 ± 0.2 at 25°C.

One pound will make 7.58 liters of medium.

METHOD OF PREPARATION

1. Suspend 60 g in 1 liter distilled or deionized water. Heat to boiling to dissolve completely.
2. Sterilize in the autoclave for 15 minutes at 15 lbs pressure (121°C).
3. Allow to cool to 45 – 50°C and add 20 ml Bacto Chapman Tellurite Solution 1%. Mix thoroughly and dispense as desired. A less selective medium may be prepared by adding only 10 ml of the tellurite solution.
4. Inoculate directly with the clinic specimen or foods suspected of containing coagulase-positive, mannitol fermenting staphylococci.

STORAGE

Bacto VJ Agar	Below 30°C
Prepared medium	2 – 8°C

QUALITY CONTROL

Identity Specifications

Dehydrated powder:	pink, homogeneous, free-flowing
Reaction of 6.0% solution:	pH 7.2 ± 0.1
Prepared medium:	red, slightly opalescent, may have a slight white precipitate

Typical Cultural Response on Bacto VJ Agar
After 18 – 48 hours at 35°C

Organism	Growth	Color of Colony
Escherichia coli ATCC® 25922	inhibited	—
Proteus mirabilis ATCC® 25933	none to poor	black
Staphylococcus aureus ATCC® 25923	good to excellent	black colonies w/yellow halos
Staphylococcus epidermidis ATCC® 12228	fair to good	translucent to black

REFERENCES
1. Pub. Hlth. Lab., 18:131, 1960.
2. United States Pharmacopeia XX, 1980.
3. Methods of Analysis, AOAC, 11th Ed., 1970.

PACKAGING

Bacto VJ Agar	1 lb (454 g)	0562-01-9
	1/4 lb (114 g)	0562-02-8
Bacto Chapman Tellurite Solution 1%	6 × 1 ml	0299-51-8
	6 × 25 ml	0299-66-1

BACTO VEAL INFUSION AGAR
BACTO VEAL INFUSION BROTH

INTENDED USE
Bacto Veal Infusion Agar and Bacto Veal Infusion Broth are used for cultivating fastidious microorganisms. The agar formulation may be used as a base for enrichment with blood, ascitic fluid, serum or other enrichments.

FORMULAE

BACTO VEAL INFUSION AGAR DEHYDRATED

Lean Veal, Infusion from . . . . 500 g
Proteose Peptone No. 3, Difco . . . . 10 g
Sodium Chloride . . . . 5 g
Bacto Agar . . . . 15 g

Final pH 7.4 ± 0.2 at 25°C.

One pound will make 11.3 liters of medium.
Rehydrate with 40 grams/liter.

BACTO VEAL INFUSION BROTH DEHYDRATED

Lean Veal, Infusion from . . . . 500 g
Proteose Peptone No. 3, Difco . . . . 10 g
Sodium Chloride . . . . 5 g

Final ph 7.4 ± 0.2 at 25°C.

One pound will make 18.1 liters of medium.
Rehydrate with 25 grams/liter.

METHOD OF PREPARATION
1. Suspend the appropriate amount of dehydrated medium in 1 liter distilled or deionized water. When preparing the agar, heat to boiling to dissolve completely.
2. Dispense as desired.
3. Sterilize in the autoclave for 15 minutes at 15 lbs pressure (121°C).
4. If an enriched medium is desired, aseptically add the enrichment to the sterile medium after cooling to 45 – 50°C.

STORAGE

Bacto Veal Infusion media   Below 30°C
Prepared media   2 – 8°C

QUALITY CONTROL

Identity Specifications

	Bacto Veal Infusion Agar	Bacto Veal Infusion Broth
Dehydrated powder:	very light beige, homogeneous, free-flowing	very light beige, homogeneous, free-flowing
% solution:	4.0%	2.5%
pH of solution:	7.4 ± 0.2 at 25°C	7.4 ± 0.2 at 25°C
Prepared medium:	light to medium amber, slightly opalescent	very light amber, clear

Typical Cultural Response After 18 – 48 Hours at 35°C

Organism	Growth
Neisseria meningitidis ATCC® 13090	good to excellent
Staphylococcus epidermidis ATCC® 12228	good to excellent
Streptococcus mitis ATCC® 9895	good to excellent
Streptococcus pneumoniae ATCC® 6303	good to excellent

PACKAGING

Bacto Veal Infusion Agar	500 g	0343-17-7
Bacto Veal Infusion Broth	500 g	0344-17-6
	10 kg	0344-08-7

BACTO VEILLONELLA AGAR

INTENDED USE

Bacto Veillonella Agar, patterned after the medium devised by Rogosa[1,2] and as modified by Rogosa, Fitzgerald, MacKintosh and Beaman,[3] is a selective medium recommended for the isolation of *Veillonella*.

HISTORY/PRINCIPLES

Organisms of the genus *Veillonella* have been isolated from the oral cavity as well as from the gastrointestinal tract. They are anaerobic, gram-negative diplococci and appear as clumps of diplococci on staining when cultivated in vitro.

Rogosa,[2] comparing the nutritive qualities of Veillonella agar with other media previously used for isolation of these organisms, clearly demonstrated the superiority of his medium. His studies with oral specimens from man and rats demonstrated the medium to be highly selective for the *Veillonella*. The few colonies of streptococci and diphtheroids which appeared in the medium could easily be differentiated from the *Veillonella* colonies. Streptomycin was originally employed as the selective agent; however, in later work Rogosa et al.[3] demonstrated vancomycin to be superior to streptomycin in reducing growth of extraneous organisms without restricting the growth of *Veillonella*.

VEILLONELLA AGAR 1049

FORMULA

BACTO VEILLONELLA AGAR
DEHYDRATED
Ingredients per liter

Bacto Tryptone	5 g
Bacto Yeast Extract	3 g
Sodium Thioglycollate	0.75 g
Bacto Basic Fuchsin	0.002 g
Sodium Lactate, 60%	21 ml
Bacto Agar	15 g

Final pH 7.5 ± 0.2 at 25°C.

One pound will make 12.3 liters of medium.

METHOD OF PREPARATION
1. To rehydrate the medium, suspend 36 g in 1 liter distilled or deionized water and heat to boiling to dissolve completely. (One g of Tween® 80 may be added at this point if desired.)
2. Sterilize in the autoclave for 15 minutes at 15 lbs pressure (121°C). Cool to 50 – 55°C and aseptically add 7500 µg vancomycin.

PROCEDURE
Rogosa recommends that one ml of the diluted specimen be added to a sterile Petri dish, approximately 20 ml of medium poured into the dish and mixed with the inoculum. The plates are incubated anaerobically at 35°C for 48 hours.

STORAGE
Bacto Veillonella Agar	2 – 8°C
Prepared medium	2 – 8°C

QUALITY CONTROL
Identity Specifications

Dehydrated powder:	beige, homogeneous, free-flowing
Reaction of 3.6% solution:	pH 7.5 ± 0.2 at 25°C
Prepared medium:	pink, slightly opalescent

Typical Cultural Response on Bacto Veillonella Agar After 40 – 48 Hours at 35°C Anaerobically

Organism	Growth
Veillonella criceti ATCC® 17747	good to excellent
Veillonella dispar ATCC® 17748	good to excellent
Veillonella ratti ATCC® 17746	good to excellent
Veillonella rodentium ATCC® 17743	good to excellent
Saliva sample	variable depending on organisms present

REFERENCES
1. J. Dent. Res., 34:721, 1955.
2. J. Bact., 72:533, 1956.
3. ibid., 76:455, 1958.

PACKAGING
Bacto Veillonella Agar   1 lb (454 g)   0917-01-1

VIBRIO CHOLERAE SEROLOGY

BACTO VIBRIO CHOLERAE ANTISERUM POLY
BACTO VIBRIO CHOLERAE ANTISERUM OGAWA
BACTO VIBRIO CHOLERAE ANTISERUM INABA

INTENDED USE

Bacto Vibrio Cholerae Antisera are used in identifying *Vibrio cholerae* by the rapid slide agglutination test.

SUMMARY AND EXPLANATION OF THE TEST

In 1935, Gardner and Ventkatraman[1] defined the identification of *Vibrio cholerae* as the recognition of the somatic antigen 01 (0 group 1) as evidenced by agglutination with 01 (polyvalent) antiserum.

Cultures that agglutinate in polyvalent antiserum can be further differentiated as to serotype by use of specific Ogawa and Inaba antisera. While this latter procedure may be of little clinical or laboratory significance, it is of interest to the epidemiologist.[2]

PRINCIPLES OF THE PROCEDURE

Vibrio cholerae 0 group 1 has three known serotypes, each possessing the antigenic factors listed in Table 1.

Table 1. Antigen Factors of *Vibrio cholerae* Serotypes.

Serotype	0 Antigen Factors
Ogawa	AB
Inaba	AC
Hikojima	ABC

Bacto Vibrio Cholerae Antiserum Poly and the absorbed Bacto Vibrio Cholerae Antisera Ogawa and Inaba possess the antibody factors listed in Table 2.

Table 2. Antibody Factors Contained in Bacto Vibrio Cholerae Antisera.

Antiserum	0 Antibody Factors
Bacto Vibrio Cholerae Antiserum Poly	ABC
Bacto Vibrio Cholerae Antiserum Ogawa	B
Bacto Vibrio Cholerae Antiserum Inaba	C

Positive agglutination using a polyvalent antiserum constitutes presumptive identification of *Vibrio cholerae* in the presence of typical results to biochemical studies.

Cultures exhibiting positive agglutination in the polyvalent antiserum may be typed for epidemiologic purposes by using the absorbed Ogawa and Inaba antisera. Positive agglutination in *both* Ogawa *and* Inaba is rare and, when it occurs, is usually interpreted as identifying the Hikojima serotype.[3,4]

REAGENTS

Bacto Vibrio Cholerae Antiserum Poly, Bacto Vibrio Cholerae Antiserum Ogawa and Bacto Vibrio Cholerae Antiserum Inaba are desiccated rabbit antisera prepared from smooth strains of *Vibrio cholerae* according to the method of Finkelstein and LaBrec[5] and Burrows and Pollitzer.[6] Bacto Vibrio Cholerae Antisera Ogawa and Inaba are monospecific absorbed antisera. All Bacto Vibrio Cholerae Antisera contain 1:10,000 Merthiolate® as a preservative.

To rehydrate Bacto Vibrio Cholerae Antisera, add 3 ml 0.85% NaCl and rotate gently to dissolve completely.

Store desiccated and rehydrated Bacto Vibrio Cholerae Antisera at 2 – 8°C.

SPECIMEN COLLECTION AND PREPARATION
The test specimen is the suspected colony of *Vibrio cholerae*, preferably cultured on nonselective media such as Bacto Nutrient Agar or Bacto Kligler Iron Agar.

PROCEDURE
Materials Provided:
Bacto Vibrio Cholerae Antisera
Dropper

Materials Required but not Provided:
Microscope slides
0.85% NaCl
Bacteriological loop
Applicator sticks
Bacto Rabbit Serum Normal

RAPID SLIDE TEST
1. Place one drop of rehydrated Bacto Vibrio Cholerae Antiserum Poly at one end of a microscope slide.
2. Place one drop of 0.85% NaCl or Bacto Rabbit Serum Normal at the other end.
3. Add to each drop a loopful of growth from the test culture and emulsify the organism in each drop with separate applicator sticks. Do not allow the drops to intermix.
4. Gently rock the slide for one minute.

RESULTS
Positive agglutination, when it occurs, should be immediate and complete.

Agglutination of the negative control indicates that the culture is unsatisfactory for testing in the rapid slide agglutination technique.

Agglutination of the polyvalent antiserum constitutes presumptive identification of *Vibrio cholerae* in the presence of typical results to biochemical studies. If identification of the serotype is desired, repeat the Rapid Slide Test using the monospecific Bacto Vibrio Cholerae Antisera Ogawa and Inaba.

Agglutination of the monospecific antiserum, when used, provides preliminary presumptive identification of the serotype, as outlined in Table 3.

LIMITATION OF THE PROCEDURE
Serological identification of *Vibrio cholerae* is presumptive in the presence of typical results to biochemical studies.

REFERENCES
1. Gardner, A. D., and K. V. Ventkatraman, 1935, The antigens of the cholera group of vibrios, J. Hyg. (Camb), 35: 262 – 282.
2. Smith, H. L., Jr., 1981, Vibrio infections, p. 718, *In* A. Balows and W. J. Hausler, Jr. (Ed.), Bacterial, mycotic and parasitic infections, 6th Ed., American Public Health Association, Inc. Washington, D.C.

Table 3. Summary of Possible Results of the Rapid Slide Test for *Vibrio Cholerae.*

Result of Rapid Slide Test

0.85% NaCl Control	Bacto Vibrio Cholerae Antiserum			Interpretation
	Poly	Ogawa	Inaba	
−	−			Test organism not *Vibrio cholerae*
−	+			Test organism is presumptively *Vibrio cholerae*
−	+	+	−	*Vibrio cholerae* serotype Ogawa
−	+	−	+	*Vibrio cholerae* serotype Inaba
−	+	+	+	*Vibrio cholerae* serotype Hikojima
+	Any result	Any result	Any result	False results. Test culture is unsuitable for use in this procedure

\+ = agglutination
− = no agglutination

3. Shewan, J. M., and M. Véron, 1974, Genus I. *Vibrio* Pacini 1854, 411, p. 343, *In* R. E. Buchanan and N. E. Gibbons (ed.), Bergey's manual of determinative bacteriology, 8th Ed., The Williams & Wilkins Co., Baltimore.
4. Wachsmuth, D. K., G. K. Morris and J. C. Feeley, 1980, *Vibrio*, p. 228, *In* E. H. Lennette, A. Balows, W. J. Hausler, Jr., and J. P. Truant (ed.), Manual of clinical microbiology, American Society for Microbiology, Washington, D.C.
5. Finkelstein, R. A., and E. H. LaBrec, 1959, Rapid identification of cholera vibrios with fluorescent antibody, J. Bact., 78: 886–891.
6. Burrows, W., and R. Pollitzer, 1958, Laboratory diagnosis of cholera, Bull. Wld. Hlth. Org., 18: 275–290.

PACKAGING

Bacto Vibrio Cholerae Antiserum Poly	3 ml	2432-47-9
Bacto Vibrio Cholerae Antiserum Ogawa	3 ml	2431-47-0
Bacto Vibrio Cholerae Antiserum Inaba	3 ml	2430-47-1
Bacto Rabbit Serum Normal	1 ml	2423-50-4

BACTO VIOLET RED BILE AGAR

INTENDED USE

Bacto Violet Red Bile Agar is a Standard Methods[1,2] selective plating medium used for the presumptive phase for enumerating coliform bacteria in foods and dairy products.

HISTORY

The original work by Difco Laboratories on the development of a solid selective medium for the quantitative estimation of the number of viable coliform organisms in water, milk, dairy and other food products was undertaken in 1932 in cooperation with M. H. McCrady. He was at that time sub-referee of the committee on Standards Methods of Milk Analysis of the American Public Health Association. The formula adopted had exhaustive trials in our laboratories and was also used extensively by others in the determination of the presence and numbers of coliform organisms in milk and dairy products. The use of Bacto Bile Salts No. 3 in place of Bacto Bile Salts originally employed resulted in an improved medium.

Bacto Violet Red Bile Agar is especially applicable to the control of pasteurization of milk and cream. The use of this medium makes it possible to determine quantitatively the coliform count of the milk at any stage of the process. Therefore, it is possible to determine the cleanliness of apparatus and the efficiency of the process as well as to detect sources of contamination at various stages.

Bartram and Black[3] in an investigation of media for the isolation of the coliform group from raw, pasteurized, and certified milk, found Bacto Violet Red Bile Agar to be the most satisfactory solid medium for this work. Babel and Parfitt[4] studying media for the detection of *Escherichia-Aerobacter* in ice cream, used Bacto Violet Red Bile Agar and reported its superiority over other solid media. Yale[5,7] reported the use of Bacto Violet Red Bile Agar for making coli counts of ice cream and pasteurized milk. Fabian and Hook[6] used Bacto Violet Red Bile Agar in establishing the count of *Escherichia-Aerobacter* in a study of sanitary conditions of ice cream as served at the fountain. Miller and Prickett[8] published a note on the use of this medium in a practical case concerning recontamination of milk. The violet red bile agar counts were completed within 24 hours after plating. That was a considerable saving in time in comparison with the confirmed brilliant green bile procedure which required a minimum period of 48 hours. Quinn and Garnatz[9] used Bacto Violet Red Bile Agar for the coli-aerogenes count of frozen eggs.

PRINCIPLES

Bacto Violet Red Bile Agar, a poured plate medium demonstrates very clearly the presence of lactose fermenting enteric organisms. Those organisms of the coliform group, due to their ability to ferment lactose, form purplish red subsurface colonies, 1 to 2 mm in diameter and are generally surrounded by a reddish zone of precipitated bile. The neutral red in the medium indicates acid production, while the bile salts and crystal violet inhibit the growth of gram-positive organisms.

In using this selective medium, best results will be obtained if it is not subjected to autoclave sterilization, since organisms not killed by the boiling required to dissolve the medium, will not form colonies during the 24 hour incubation period. Following boiling to dissolve the medium completely, it is ready for use. The plates should not be incubated longer than 24 hours, inasmuch as the organisms whose growth has been suppressed, may develop and confuse the count. Best results are obtained if plates are not too heavily seeded.

FORMULA
BACTO VIOLET RED BILE AGAR
DEHYDRATED
Ingredients per liter

Bacto Yeast Extract	3 g	Sodium Chloride	5 g
Bacto Peptone	7 g	Bacto Agar	15 g
Bacto Bile Salts No. 3	1.5 g	Neutral Red	0.03 g
Bacto Lactose	10 g	Bacto Crystal Violet	0.002 g

Final pH 7.4 ± 0.2 at 25°C.

One pound will make 11 liters of medium.

METHOD OF PREPARATION
1. Suspend 41.5 g in 1 liter distilled or deionized water and heat to boiling to dissolve completely. Do not boil for more than 2 minutes. Do not autoclave.
2. Cool to 45°C and dispense 15 – 20 ml of medium into 100 mm Petri dishes containing inoculum. Rotate the plate carefully to distribute the inoculum.

3. Allow the inoculated plate to solidify.
4. Evenly distribute 4 ml of medium on the surface of the plate to form an overlay.

Consult the appropriate Standard Methods reference for the proper procedure.[1,2]

STORAGE
Bacto Violet Red Bile Agar   Below 30°C
Prepared medium   2 – 8°C

QUALITY CONTROL
Identity Specifications
Dehydrated powder: reddish beige, homogeneous, free-flowing
Reaction of 4.15% solution: 7.4 ± 0.2 at 25°C
Prepared medium: reddish purple, very slightly opalescent

Typical Cultural Response on Bacto Violet Red Bile Agar
After 18 – 24 Hours at 35°C

Organism	Growth	Color of Colony
Enterobacter aerogenes ATCC® 13048	good to excellent	pink
Escherichia coli ATCC® 25922	good to excellent	purplish-red w/bile ppt
Salmonella enteritidis ATCC® 13076	good to excellent	colorless
Staphylococcus aureus ATCC® 25923	inhibited	—

REFERENCES
1. American Public Health Association, 1976, Compendium of methods for the microbiological examination of foods, American Public Health Association, Inc., Washington, D.C.
2. American Public Health Association, 1978, Standard methods for the examination of dairy products, 14th ed., American Public Health Association, Inc., Washington, D.C.
3. Bartram and Black, 1936, Food Research, 1:551.
4. Babel and Parfitt, 1936, Journal of Dairy Science, 19:497.
5. Yale, 1936, Proc. Intern. Association of Ice Cream Manufacturers, 2:17.
6. Fabian and Hook, 1936, Proc. Intern. Association of Ice Cream Manufacturers, 2:30.
7. Yale, 1937, American Journal of Public Health, 27:564.
8. Miller and Prickett, 1936, Journal of Dairy Science, 21:559.
9. Quinn and Garnatz, 1943, Journal of Bacteriology, 45:49.

PACKAGING
Bacto Violet Red Bile Agar

1 lb (454 g)	0012-01-5
1/4 lb (114 g)	0012-02-4
5 lb (2.27 kg)	0012-05-1
6 × 500 ml (6's unipack)	0012-29F-0

BACTO VITAMIN ASSAY CASAMINO ACIDS

Bacto Vitamin Assay Casamino Acids is an acid digest of casein specially treated to markedly reduce or eliminate certain vitamins. It is recommended for use in micro-biological assay media and in studies of the growth requirements of microorganisms. Sarett[1] used Bacto Vitamin Free Casamino Acids (previous name) as the acid hydro-lyzed casein in his studies on p-aminobenzoic acid and p-teroylglutamic acid as growth factors for lactobacilli.

Bacto Vitamin Assay Casamino Acids is readily soluble in distilled or deionized water, yielding a clear, colorless solution.

REFERENCE
1. J. Biol. Chem., 171:265, 1947.

PACKAGING

Bacto Vitamin Assay Casamino Acids	1 lb (454 g)	0288-01-2
	1/4 lb (114 g)	0288-02-1

BACTO VITAMIN ASSAY CASITONE

Bacto Vitamin Assay Casitone is an enzymatic digest of casein that is essentially free from vitamins and recommended for use in preparing media for vitamin assays.

PACKAGING

Bacto Vitamin Assay Casitone	100 g	0646-15-3

BACTO VITAMIN TEST CASEIN

Bacto Vitamin Test Casein is a source of high-quality protein with minimal to no vitamin content.

PACKAGING

Bacto Vitamin Test Casein	500 g	0579-17-2

MEDIA FOR THE MICROBIOLOGICAL ASSAY OF VITAMINS AND AMINO ACIDS

INTENDED USE
Bacto Vitamin and Amino Acid Assay Media are prepared for use in the microbiological assay of vitamins and amino acids. Three types of media are used for this purpose:
1. Maintenance Media: For carrying the stock culture to preserve the viability and sensitivity of the test organism for its intended purpose.
2. Inoculum Media: To condition the test culture for immediate use.
3. Assay Media: To permit quantitation of the vitamin or amino acid under test. They contain all the factors necessary for optimal growth of the test organism except the single essential vitamin or amino acid to be determined.

PRINCIPLES
Each assay medium is free from the essential vitamin or amino acid for which the medium is recommended. When the required vitamin or amino acid is added in specified increasing concentrations, the growth of the particular test organism will show a corresponding increase in growth response. This response is measured acidimetrically, turbidimetrically, gravimetrically or linearly.

In the performance of the various tests, the basal assay medium is usually prepared in double strength and tubed in 5 ml amounts. To 1 series of tubes containing the assay medium are added graduated quantities of the standard known essential vitamin or amino acid. To the second series are added varying amounts of the material under assay. Included in each series of tubes is 1 containing only distilled water and the basal assay medium. This tube is a blank control. The tubes of each series are then made up to 10 ml with distilled water, sterilized, cooled, inoculated, incubated and observed. The growth response is determined turbidimetrically or acidimetrically.

Assays employing *Neurospora* as test organism use the weight of mycelium as a measure of growth response. These determinations are usually conducted in small flasks using 10 ml of basal medium per flask.

Plate assay procedures depend upon diffusion of a required growth factor from impregnated surface paper disks or from penicillin cylinders into an agar medium inoculated with the test organism. The growth response is determined by measuring the diameter of the zone of growth around the disks.

Bacto Test Cultures for the Microbiological Assay of Vitamins and Amino Acids

Bacteria

Acetobacter suboxydans ATCC® 621H	
(*Gluconobacter oxydans subsp. suboxydans*)	3209-30-7
Lactobacillus casei ATCC® 7469	3210-30-4
Lactobacillus plantarum ATCC® 8014	3211-30-3
Lactobacillus fermenti ATCC® 9338	
(*Lactobacillus fermentum*)	3212-30-2
Lactobacillus lactis ATCC® 8000	3213-30-1
Lactobacillus leichmannii ATCC® 4797	3214-30-0
Lactobacillus leichmannii ATCC® 7830	3215-30-9
Lactobacillus viridescens ATCC® 12706	3216-30-8
Pediococcus cerevisiae ATCC® 8081	
(*Pediococcus acidilactici*)	3217-30-7
Pediococcus cerevisiae ATCC® 8042	
(*Pediococcus acidilactici*)	3218-30-6
Streptococcus faecalis ATCC® 8043	
(*Streptococcus faecium*)	3219-30-5

Yeasts and Molds

Kloeckera apiculata (*brevis*) ATCC® 9774	3220-30-2
Saccharomyces carlsbergensis ATCC® 9080	
(*Saccharomyces uvarum*)	3221-30-1
Neurospora crassa ATCC® 9277	3224-30-8
Neurospora sitophilia ATCC® 9276	3225-30-7

Bacto Assay Cultures are made from cultures which have been properly cultured and maintained to produce good growth response curves in the proper ranges of sensitivity. Each culture is transferred into maintenance medium or inoculum broth and used according to the vitamin or amino acid assay procedures for which the culture is used. Bacto Assay Cultures are prepared to order and will remain viable to the expiry date on the label when stored at 2 – 8°C.

GUIDE TO SELECTION OF MEDIA FOR VITAMIN ASSAYS

Vitamin	Assay Medium	Test Cultures	Maintenance Media	Inoculum Media	Assay Method
Vitamin B12	B12 Assay Medium USP	Lactobacillus leichmannii ATCC® 7830	B12 Culture Agar USP	B12 Inoculum Broth USP	Turbidimetric or acidimetric
	Vitamin B12 Assay Medium	Lactobacillus leichmannii ATCC® 4797	Lactobacilli Agar AOAC or B12 Culture Agar USP	Lactobacilli Broth AOAC or B12 Inoculum Broth USP	Turbidimetric or acidimetric
	Euglena B12 Medium	Euglena gracilis ATCC® 12716	Euglena B12 Medium (single strength) w/50 pg per ml cyanocobalamin added	Euglena B12 Medium (single strength) w/50 pg per ml cyanocobalamin added	Turbidimetric
	CS Vitamin B12 Agar	Lactobacillus leichmannii ATCC® 4797	Lactobacilli Agar AOAC or B12 Culture Agar	Lactobacilli Broth AOAC or B12 Inoculum Broth USP	Cup, plate or disk
Biotin	Biotin Assay Medium	Lactobacillus plantarum ATCC® 8014	Lactobacilli Agar AOAC or Micro Assay Culture Agar	Lactobacilli Broth AOAC or Micro Inoculum Broth	Acidimetric or turbidimetric
Citrovorum Factor	CF Assay Medium	Pediococcus acidilactici ATCC® 8081 (P. cerevisiae)	Lactobacilli Agar AOAC or Micro Assay Culture Agar	Lactobacilli Broth AOAC or Micro Inoculum Broth	Turbidimetric or acidimetric
Choline	Choline Assay Medium	Neurospora crassa ATCC® 9277	Neurospora Culture Agar	None required	Gravimetric
Folic Acid	Folic Acid Assay Medium	Streptococcus faecium ATCC® 8043 (S. faecalis)	Lactobacilli Agar AOAC or Micro Assay Culture Agar	Lactobacilli Broth AOAC or Micro Inoculum Broth	Turbidimetric

GUIDE TO SELECTION OF MEDIA FOR VITAMIN ASSAYS (continued)

Vitamin	Assay Medium	Test Cultures	Maintenance Media	Inoculum Media	Assay Method
Folic Acid	Folic AOAC Medium	Streptococcus faecium ATCC® 8043 (S. faecalis)	Lactobacilli Agar AOAC	Lactobacilli Broth AOAC	Turbidimetric or acidimetric
	Folic Acid Casei Medium	Lactobacillus casei ATCC® 7469	Lactobacilli Agar AOAC	Micro Inoculum Broth	Turbidimetric
	Folic TE Medium	Streptococcus faecium ATCC® 8043 (S. faecalis)	Lactobacilli Agar AOAC or Micro Assay Culture Agar	Lactobacilli Broth AOAC or Micro Inoculum Broth	Turbidimetric or acidimetric
Inositol	Inositol Assay Medium	Saccharomyces uvarum ATCC® 9080 (S. carlsbergensis)	Neurospora Culture Agar	Inositol Assay Medium w/10 μg inositol per 10 ml added	Turbidimetric
	Inositol Assay Medium KB	Kloeckera apiculata (brevis) ATCC® 97774	Lactobacilli Agar AOAC	None required	Turbidimetric
Nicotinic Acid or Nicotinamide (Niacin)	Niacin Assay Medium	Lactobacillus plantarum ATCC® 8014	Lactobacilli Agar AOAC or Micro Assay Culture Agar	Lactobacilli Broth AOAC or Micro Inoculum Broth	Turbidimetric or acidimetric
Panthenol	Panthenol Assay Medium w/Panthenol Supplement	Gluconobacter oxydans suboxydans ATCC® 621H (Acetobacter suboxydans)	Lactobacilli Agar AOAC or Micro Assay Culture Agar	Panthenol Assay Medium w/Panthenol Supplement and .5 g Liver Concentrate NF per liter added	Turbidimetric
Pantothenic Acid	Pantothenate Medium AOAC USP	Lactobacillus plantarum ATCC® 8014	Lactobacilli Agar AOAC or Pantothenate Culture Agar USP	Pantothenate Assay Medium (single strength) w/.02 μg/ml pantothenic acid or pantothenate	Turbidimetric or acidimetric

GUIDE TO SELECTION OF MEDIA FOR VITAMIN ASSAYS (continued)

Vitamin	Assay Medium	Test Cultures	Maintenance Media	Inoculum Media	Assay Method
Pantothenic Acid	Pantothenate Assay Medium	*Lactobacillus plantarum* ATCC® 8014	Pantothenate Culture Agar	Pantothenate Assay Medium (single strength) w/.02 µg/ml pantothenate added	Turbidimetric or acidimetric
Pyridoxine	Pyridoxine Assay Medium	*Neurospora sitophilia* ATCC® 9276	Neurospora Culture Agar	None required	Gravimetric
	Pyridoxine Y Medium	*Saccharomyces uvarum* ATCC® 9080 (*S. carlsbergensis*)	Lactobacilli Agar AOAC	Pyridoxine Y Medium (single strength) w/1 ng per ml each: pyridoxal hydrochloride, pyridoxamine hydrochloride, pyridoxine hydrochloride.	Turbidimetric
Riboflavin	Riboflavin Assay Medium	*Lactobacillus casei* ATCC® 7469	Lactobacilli Agar AOAC or Micro Assay Culture Agar	Lactobacilli Broth AOAC or Micro Inoculum Broth	Turbidimetric or acidimetric
Thiamine	Thiamine Assay Medium	*Lactobacillus fermentum* ATCC® 9338 (*L. fermenti*)	Lactobacilli Agar AOAC or Micro Assay Culture Agar	Lactobacilli Broth AOAC or Micro Inoculum Broth	Turbidimetric or acidimetric
	Thiamine Assay Medium LV	*Lactobacillus viridescens* ATCC® 12706	APT Agar or Lactobacilli Agar AOAC	APT Broth or Lactobacilli Broth AOAC	Turbidimetric

GUIDE TO SELECTION OF MEDIA FOR AMINO ACID ASSAYS

Amino Acid	Assay Medium	Test Culture	Maintanance Media	Inoculum Media	Assay Method
Arginine	Arginine Assay Medium				
Cystine	Cystine Assay Medium				
Glutamic Acid	Glutamic Acid Assay Medium				
Histidine	Histidine Assay Medium	*Pediococcus acidilactici* ATCC® 8042 (*P. cerevesiae*)			
Isoleucine	Isoleucine Assay Medium				
Leucine	Leucine Assay Medium				
Lysine	Lysine Assay Medium		Lactobacilli Agar AOAC or Micro Assay Culture Agar	Lactobacilli Broth AOAC or Micro Inoculum Broth	Turbidimetric or acidimetric
Methionine	Methionine Assay Medium				
Phenylalanine	Phenylalanine Assay Medium				
Threonine	Threonine Assay Medium	*Streptococcus faecium* ATCC® 8043 (*S. faecalis*)			
Tryptophan	Tryptophan Assay Medium	*Lactobacillus plantarum* ATCC® 8014			
Tyrosine	Tyrosine Assay Medium	*Pediococcus acidilactici* ATCC® 8042 (*P. cerevisiae*)			
Valine	Valine Assay Medium				

GENERAL PROCEDURES FOR ASSAY PERFORMANCE

Great care to avoid contamination of media or glassware must be taken in microbiological assay procedures. Extremely small amounts of foreign material may be sufficient to give erroneous results. Scrupulously clean glassware free from detergents and other chemicals must be used. For example, in the assay of Vitamin B_{12}, as small a quantity as 0.01 ng will give rise to a definite growth response. The importance of clean glassware in the assay of this vitamin was stressed by the U.S. Pharmacopeia Vitamin B_{12} Study Panel, wherein they observed that glassware for this purpose required special handling. As many as 12 rinses were necessary to remove interfering substances and give satisfactory results.

The test organism used for inoculating an assay medium must be cultured and maintained on media recommended for this purpose. Aseptic technique should be used throughout. The use of altered or deficient media may give rise to mutants having different nutritional requirements and which will not give a satisfactory response.

For the successful execution of these procedures, all conditions of the assay must be adhered to meticulously. Points of prime importance for successful assays have been stressed in the description of each medium.

ASSAY SAMPLES

Assay samples are prepared for assay according to references given in the specific assay procedures. For assay, the samples should be diluted to approximately the same concentration as the standard solution.

MAINTENANCE MEDIA

Suspend the appropriate amount of dehydrated agar medium in 100 or 1000 ml of distilled or deionized water and heat to boiling to dissolve completely. Sterilize in the autoclave for 15 minutes at 15 lbs pressure (121°C).

INOCULUM MEDIA

Suspend the appropriate amount of dehydrated medium in 100 or 1000 ml of distilled or deionized water and heat to boiling for 1 – 2 minutes. Evenly distribute the slight precipitate which may form by swirling. Prepare for use and sterilize as indicated in specific medium description.

STOCK CULTURES

Bacterial and Yeast Cultures
1. Prepare stock cultures in one or more tubes of sterile maintenance medium.
2. Inoculate the tubes using a straight wire inoculating needle.
3. Incubate the tubes at 30 – 37°C for 18 – 24 hours.
4. Store at 2 – 8°C.
5. Transfer at weekly or twice monthly intervals.

Neurospora Cultures
Prepare stock cultures in a similar manner to bacteria and yeast with the following exceptions:
1. Inoculate by streaking only.
2. Incubate at 25 – 30°C in the light for 3 – 5 days.
3. Transfer at monthly intervals.

Euglena Cultures
Prepare stock cultures in a similar manner to bacteria and yeast with the following exceptions:

1. Inoculate tubes using an inoculating loop or pipette.
2. Incubate at 27 – 29°C in the light for 5 – 7 days.
3. Transfer at monthly intervals.

INOCULUM
Bacterial and Yeast Cultures
1. Subculture from 16 – 24 hour stock culture into 10 ml inoculum medium.
2. Incubate at 35 – 37°C for 16 – 24 hours or as specified in specific assay procedures.
3. Centrifuge the culture and decant the supernatant.
4. Resuspend the cells in 10 ml of sterile 0.9% NaCl solution or sterile single strength basal assay medium.
5. Wash the cells by centrifuging and decanting the supernatant two additional times unless otherwise indicated.
6. Dilute the washed suspension 1:100 with sterile 0.9% NaCl or sterile single strength basal assay medium or as indicated. Where applicable, inoculum concentration should be adjusted according to limits specified in AOAC or USP.

Neurospora Cultures
1. Remove a loop of spores from a 48 hour stock culture and suspend in 100 ml of 0.9% sterile NaCl solution.
2. This suspension is the inoculum.

Euglena Cultures
1. Subculture from a stock culture into a tube of single strength Euglena B_{12} Medium containing 50 pg vitamin B_{12} per ml.
2. Incubate in the light at 27 – 29°C for 1 week.
3. Centrifuge the culture and decant the supernatant.
4. Resuspend the cells in 10 ml sterile single strength Bacto Euglena B_{12} Medium.
5. Wash the cells by centrifuging and decanting the supernatant two additional times.
6. Resuspend the cells in 10 ml sterile single strength Bacto Euglena B_{12} Medium.
7. This suspension is the inoculum.

ASSAY PROCEDURES
TUBE METHOD
1. Preparation of tubes for standard curve:
   a. Pipette working standard solution at 0.0, 0.5, 1, 2, 3, 4 and 5 ml or amount indicated in specific assay procedure into duplicate tubes (or flasks).
   b. Add 5 or 10 ml of assay medium, as specified, to each tube or flask and sufficient distilled or deionized water to make specified volume.
   c. Sterilize the tubes or flasks by autoclaving for 5 minutes for turbidimetric assays and 10 minutes for acidimetric assays or as specified at 15 lbs pressure (121°C).
   d. Cool the tubes (flasks) quickly in cold water.
2. Preparation of tubes for test sample curve:
   a. Prepare tubes in a similar manner to standard curve tubes except add sample solution at appropriate volumes in place of standard solution.
   b. Inoculation, incubation and results are the same as the procedure for standard curve. Sample tubes and standard tubes must be sterilized, inoculated, and incubated at the same time and under the same conditions.
3. Inoculation:
   a. Place one drop of inoculum into each tube (flask) except for the duplicate tubes which contain no standard, which are uninoculated blanks.
   b. Shake the tubes (flasks) to mix the inoculum.

4. Incubation:
  a. Incubate at 25 – 37°C as indicated on specific assay procedure. As specified in individual assay procedures, incubate for 16 – 24 hours, if turbidimetric assay, or 69 – 72 hours, if acidimetric assay. Gravimetric assays are incubated as specified for 3 – 5 days.
5. Results:
  a. In turbidimetric assays, measure absorbance or light transmission using a spectrophotometer at 540 – 660 nm wavelength.
  b. For acidimetric assays, titrate the contents of each tube with 0.1N NaOH using brom thymol blue indicator or electrometrically to pH 6.8.
  c. In gravimetric determinations, recover the mycelial growth on a stiff wire needle or glass rod. Remove the excess moisture between absorbant papers, dry under vacuum for 2 hours and weigh the mycelium. Weigh on analytical balance to the nearest 0.5 mg.

AGAR PLATE METHOD (Vitamin B_{12} Assay)
1. Preparation of plates for assay:
  a. Aseptically add 1 ml of washed inoculum to 100 ml of sterile melted vitamin assay agar cooled to 45 – 50°C.
  b. Distribute 50 ml into each of 4 sterile flat bottom Petri dishes (150 mm).
  c. Allow the agar to solidify and dry with the plate covers slightly raised. It is necessary for the agar surface to be dry before placing assay cups or disks on the medium.
  d. Place 9 sterile assay cups or 9 filter paper disks (13 mm) evenly spaced on the agar surface of each plate. Use 6 cups or disks for the standard and 3 for the assay solution on each of 4 replicate plates.
2. Preparation and application of standard and assay solutions:
  a. Prepare sterile solutions of vitamin B_{12} containing 0.05, 0.1, 0.2, 0.5, 1 and 2 μg per ml by making dilutions from vitamin B_{12} stock solution.
  b. To prepare assay solution see AOAC 13th Edition, p. 762.
  c. Apply 0.1 ml standard on assay solution to appropriate disks or 0.2 – 0.3 ml to appropriate cups.
3. Incubation:
  Incubate plates for 16 – 24 hours at 35 – 37°C as specified.
4. Results:
  a. Measure the diameter of growth zones of each standard and sample on each set of 4 replicate plates.
  b. The zone diameters indicate the amount of vitamin B_{12} present.

CALCULATIONS
1. Prepare a standard concentration response curve by plotting the response readings against the amount of standard in each tube, disk or cup.
2. Determine the amount of vitamin at each level of assay solution by interpolation from the standard curve.
3. Calculate the concentration of vitamin in the sample from the average of these volumes. Use only those values which do not vary more than ±10% from the average and use the results only if two thirds of the values do not vary more than ±10%.

Sample Curve
Volume (ml) of Standard Solution or Assay Solution

CALCULATIONS
Concentration of standard solution   40 pg/ml
Dilution of assay solution   1:10

One ml of assay solutions shows the same growth as 0.5 ml of standard solution.

The sample contains $0.5/1 \times 40 \times 10 = 200$ pg/ml
Two ml assay solution shows $1.0/2 \times 40 \times 10 = 200$ pg/ml
Three ml assay solution shows $1.4/3 \times 40 \times 10 = 187$ pg/ml

Results $\dfrac{200 + 200 + 187}{3} = 197$ pg/ml

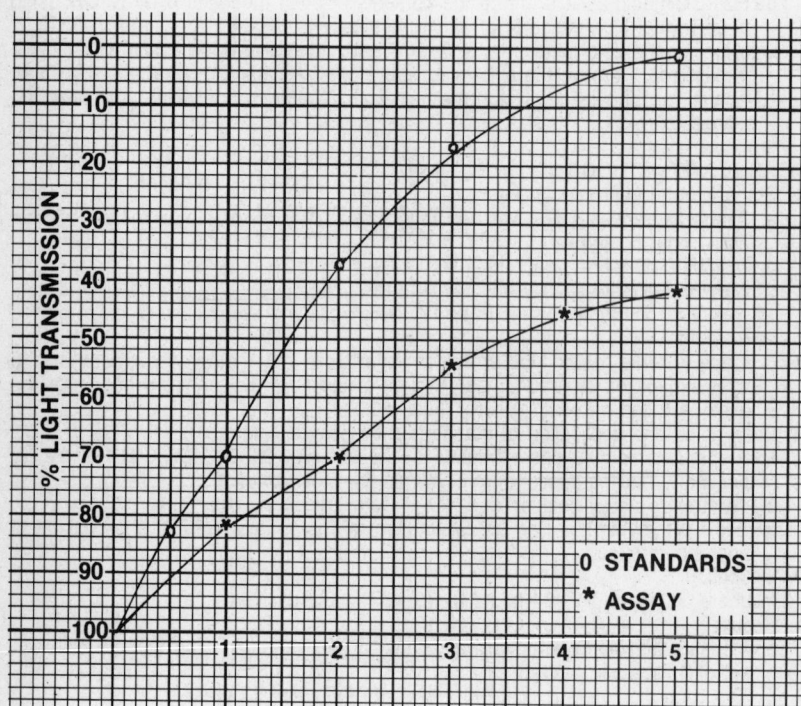

VITAMIN ASSAY-MAINTENANCE MEDIA
FOR TEST CULTURES
BACTO APT AGAR

INTENDED USE
Bacto APT Agar, patterned after the formulation of Evans and Niven[1] and Deibel, Evans and Niven,[2] is a solid medium for the cultivation of heterofermentative lactobacilli and other organisms requiring high thiamine content. It is also recommended for carrying stock cultures of *Lactobacillus veridescens* ATCC® 12706 for the microbiological assay of thiamine using Bacto Thiamine Assay Medium LV.

FORMULA

BACTO APT AGAR
DEHYDRATED

Ingredients per liter

Bacto Yeast Extract	7.5 g	Dipotassium Phosphate	5 g
Bacto Tryptone	12.5 g	Manganese Chloride	0.14 g
Bacto Dextrose	10 g	Magnesium Sulfate	0.8 g
Sodium Citrate	5 g	Ferrous Sulfate	0.04 g
Thiamine Hydrochloride	0.001 g	Sorbitan Monooleate Complex	0.2 g
Sodium Chloride	5 g	Bacto Agar	15 g

Final pH 6.7 ± 0.2 at 25°C.

One pound will make 7.4 liters of medium.

METHOD OF PREPARATION

1. Suspend 61.2 g in 1 liter distilled or deionized water and heat to boiling to dissolve completely.
2. Dispense in 10 ml amounts into tubes of 16 – 20 mm diameter.
3. Sterilize in the autoclave for 15 minutes at 15 lbs pressure (121°C). Excessive heating of the medium is to be avoided.
4. Allow tubes to cool in an upright position.

PROCEDURE

Stock cultures of the test organism are prepared by stab inoculation. It is desirable to prepare stock cultures in triplicate at monthly intervals. One of the transfers is saved for the preparation of stock cultures and the others used to prepare inoculum in Bacto APT Broth for assay as needed. Stock cultures following incubation at 35 – 37°C for 24 – 48 hours should be kept in the refrigerator at 2 – 8°C.

STORAGE

Bacto APT Agar   Below 30°C

QUALITY CONTROL

Identity Specifications

Dehydrated powder: light beige, homogeneous, free-flowing
Reaction of 6.12% solution: pH 6.7 ± 0.2 at 25°C
Prepared medium: medium amber, clear to slightly opalescent, may have
                          a slight precipitate

REFERENCES

1. J. Bact., 62:599, 1951.
2. Bact. Proc., 1955.

PACKAGING

Bacto APT Agar   1 lb (454 g)   0654-01-8

BACTO B$_{12}$ CULTURE AGAR USP

INTENDED USE

Bacto B$_{12}$ Culture Agar USP is recommended for maintaining stock cultures of *Lactobacillus leichmannii* ATCC® 7830 for use in the Vitamin B$_{12}$ Activity Assay according to the *USP XX*.[1]

FORMULA

BACTO B₁₂ CULTURE AGAR USP
DEHYDRATED

Ingredients per liter

Tomato Juice	100 ml
Proteose Peptone No. 3, Difco	7.5 g
Bacto Yeast Extract	7.5 g
Bacto Dextrose	10 g
Monopotassium Phosphate	2 g
Sorbitan Monooleate Complex	0.1 g
Bacto Agar	15 g

Final pH 6.8 ± 0.1 at 25°C.

One hundred grams will make 2.1 liters of medium.

METHOD OF PREPARATION

1. Suspend 47 g in 1 liter distilled or deionized water and heat to boiling to dissolve completely.
2. Dispense in 10 ml quantities in tubes of 16 – 20 mm diameter.
3. Sterilize in the autoclave for 15 minutes at 15 lbs pressure (121°C).
4. Allow tubes to cool in an upright position.

PROCEDURE

Stock cultures are prepared by stab inoculation of at least three tubes. Before using a fresh culture for assay, make no fewer than 10 successive transfers of the culture in a 2 week period. Incubate cultures for 16 – 24 hours at any temperature between 30 – 40°C but held constant within ±0.5°C. Store at 2 – 8°C.

Prepare stab cultures at least three times each week and do not use culture for preparing assay inoculum if over 4 days old.

STORAGE

Bacto B₁₂ Culture Agar USP   2 – 8°C

QUALITY CONTROL

Identity Specifications

Dehydrated powder:	beige, homogeneous, free-flowing
Reaction of 4.7% solution:	pH 6.8 ± 0.1 at 25°C
Prepared medium:	light-medium amber, slightly opalescent

REFERENCE

1. U.S. Pharmacopeia/National Formulary, USP XX/NF XV, Rockville: U.S. Pharmacopeial Convention, p. 904:1980.

PACKAGING

Bacto B₁₂ Culture Agar USP   100 g   0541-15-9

BACTO LACTOBACILLI AGAR AOAC

INTENDED USE

Bacto Lactobacilli Agar AOAC,[1] prepared according to the formula recommended by Loy,[2] is used for carrying stock cultures of *Lactobacillus leichmannii* ATCC® 7830,

Streptococcus faecium ATCC® 8043 (*S. faecalis* ATCC® 8043), *Lactobacillus plantarum* ATCC® 8014, *Lactobacillus casei* ATCC® 7469 and others used in the microbiological assay of the B vitamins.

FORMULA
BACTO LACTOBACILLI AGAR AOAC
DEHYDRATED
Ingredients per liter

```
Bacto Peptonized Milk . . . . . . . . . . 15 g
Bacto Yeast Extract . . . . . . . . . . . . . 5 g
Bacto Dextrose . . . . . . . . . . . . . . . 10 g
Tomato Juice (100 ml) . . . . . . . . . . . 5 g
Monobasic Potassium Phosphate . . . . 2 g
Sorbitan Monooleate Complex . . . . . . 1 g
Bacto Agar . . . . . . . . . . . . . . . . . . 10 g
```

Final pH 6.8 ± 0.2 at 25°C.

One hundred grams will make 2 liters of medium.

METHOD OF PREPARATION
1. Suspend 48 g in 1 liter distilled or deionized water and heat to boiling 2 – 3 minutes to dissolve completely.
2. Dispense in 10 ml quantities into tubes of 16 – 20 mm diameter.
3. Sterilize in the autoclave for 15 minutes at 15 lbs pressure (121°C).
4. Allow tubes to cool in an upright position.

PROCEDURE
Stock cultures are prepared by stab inoculation of one or more tubes of sterile medium. Incubate cultures for 16 – 24 hours at any temperature between 30 – 40°C but held constant within ±0.5°C, then store at 2 – 8°C.

Before using a culture in an assay it is essential to make at least 2 successive transfers during a 1 – 2 week period. No culture should be used which is more than one week old. A culture is considered satisfactory when turbidity is seen in the inoculum medium 2 – 4 hours after inoculation.

STORAGE
Bacto Lactobacilli Agar AOAC   2 – 8°C

QUALITY CONTROL
Identity Specifications

Dehydrated powder: tan, homogeneous, free-flowing
Reaction of 4.8% solution: pH 6.8 ± 0.2 at 25°C
Prepared medium: medium amber, opalescent

REFERENCES
1. AOAC Methods, 13th Ed., Washington, D.C.:AOAC, p. 760, 1980.
2. J. AOAC, 4:61, 1958.

PACKAGING
Bacto Lactobacilli Agar AOAC   100 g   0900-15-4

BACTO MICRO ASSAY CULTURE AGAR

INTENDED USE

Bacto Micro Assay Culture Agar is used for carrying stock cultures of lactobacilli and other test microorganisms used in microbiological assays. It is also used for general cultivation of lactobacilli.

FORMULA

BACTO MICRO ASSAY CULTURE AGAR
DEHYDRATED

Ingredients per liter

Proteose Peptone No. 3, Difco . . . . . 5 g
Bacto Yeast Extract . . . . 20 g
Bacto Dextrose . . . . 10 g
Monopotassium Phosphate . . . . 2 g
Sorbitan Monooleate Complex . . . . . 0.1 g
Bacto Agar . . . . 10 g

Final pH 6.7 ± 0.2 at 25°C.

One pound will make 9.6 liters of medium.

METHOD OF PREPARATION

1. Suspend 47 g in 1 liter distilled or deionized water and heat to boiling to dissolve completely.
2. Dispense in 10 ml quantities in tubes of 16 – 20 mm diameter.
3. Sterilize in the autoclave for 15 minutes at 15 lbs pressure (121°C).
4. Allow tubes to cool in an upright position. Evenly disperse the characteristic flocculent precipitate by gently twirling the tube just prior to solidification.

PROCEDURE

Stock cultures are prepared by stab inoculation in triplicate. One transfer is used for the preparation of stock cultures and the others used to prepare inoculum for assays. Incubate cultures for 24 – 48 hours at 35 – 37°C, then store in the refrigerator at 2 – 8°C. Transfers of the culture should be made at weekly or twice monthly intervals.

STORAGE

Bacto Micro Assay Culture Agar   Below 30°C

QUALITY CONTROL

Identity Specifications

Dehydrated powder: light tan, homogeneous, free-flowing
Reaction of 4.7% solution: pH 6.7 ± 0.2 at 25°C
Prepared medium: light amber, slightly opalescent with flocculent
                            precipitate

PACKAGING

Bacto Micro Assay Culture Agar   1 lb (454 g)   0319-01-5
                                        1/4 lb (114 g)   0319-02-4

BACTO NEUROSPORA CULTURE AGAR

INTENDED USE

Bacto Neurospora Culture Agar is used for the cultivation of *Neurospora sitophilia* ATCC® 9276 in the microbiological assay of pyridoxine. This medium is also used as a nearly neutral medium with 4% maltose for the cultivation of other fungi.

FORMULA

BACTO NEUROSPORA CULTURE AGAR
DEHYDRATED
Ingredients per liter

Proteose Peptone No. 3, Difco . . . . . 5 g
Bacto Yeast Extract . . . . 5 g
Bacto Maltose . . . . 40 g
Bacto Agar . . . . 15 g

Final pH 6.7 ± 0.2 at 25°C.

One hundred grams will make 1.5 liters of medium.

METHOD OF PREPARATION

1. Suspend 65 g in 1 liter distilled or deionized water and heat to boiling to dissolve completely.
2. Dispense in 8 ml quantities per tube.
3. Sterilize in the autoclave for 15 minutes at 15 lbs pressure (121°C).
4. Allow tubes to cool in a slanted position.

PROCEDURE

Stock cultures are prepared by streaking the surface of the agar slant. Incubate the cultures at 25 – 30°C in the light for 3 – 5 days. Transfers of the culture should be made at monthly intervals. Keep cultures at 2 – 8°C.

STORAGE

Bacto Neurospora Culture Agar

Below 30°C, preferably in a container with calcium chloride or other desiccant.

QUALITY CONTROL

Identity Specifications

Dehydrated powder: off white, homogeneous, free-flowing
Reaction of 6.5% solution: pH 6.7 ± 0.2 at 25°C
Prepared medium: light amber, clear to very slightly opalescent

PACKAGING

Bacto Neurospora Culture Agar   100 g   0321-15-5

BACTO PANTOTHENATE CULTURE AGAR USP

INTENDED USE

Bacto Pantothenate Culture Agar USP, prepared according to USP[1] and AOAC[2] procedures, is used for carrying cultures of *Lactobacillus plantarum* ATCC® 8014 in the microbiological assay of pantothenic acid or pantothenate. It is also used for the culture of other lactobacilli.

FORMULA

BACTO PANTOTHENATE CULTURE AGAR USP
DEHYDRATED

Ingredients per liter

Bacto Yeast Extract . . . . 20 g
Bacto Dextrose . . . . 5 g
Sodium Acetate . . . . 5 g
Bacto Agar . . . . 15 g

Final pH 6.8 ± 0.2 at 25°C.

One hundred grams will make 2.2 liters of medium.

METHOD OF PREPARATION

1. Suspend 45 g in 1 liter distilled or deionized water and heat to boiling to dissolve completely.
2. Dispense in 10 ml quantities per tube.
3. Sterilize in the autoclave for 15 minutes at 15 lbs pressure (121°C).
4. Allow tubes to cool in an upright position.

PROCEDURE

Stock cultures are prepared in triplicate by stab inoculation. Incubate cultures for 16 – 24 hours at any temperature between 30 – 37°C but held constant within ±0.5°C, then store in refrigerator at 2 – 8°C.

Prepare stab cultures each week and do not use culture for preparing assay inoculum if more than one week old.

STORAGE

Bacto Pantothenate Culture Agar USP   below 30°C, preferably in a container with calcium chloride or other desiccant

QUALITY CONTROL

Identity Specifications

Dehydrated powder: light beige, homogeneous, free-flowing
Reaction of 4.5% solution: pH 6.8 ± 0.2 at 25°C
Prepared medium: light to medium amber, clear to slightly opalescent

REFERENCES

1. U.S. Pharmacopeia/National Formulary, USP XX/NF XV, Rockville: U.S. Pharmacopeial Convention, p. 890, 1980.
2. AOAC, 8th Ed., 1955, or AOAC, 11th Ed., 1970.

PACKAGING

Bacto Pantothenate Culture Agar USP   100 g   0603-15-4

VITAMIN ASSAY-MEDIA FOR PREPARING INOCULA
BACTO APT BROTH

INTENDED USE

Bacto APT Broth, patterned after the formulation of Evans and Niven[1] and Deibel, Evans and Niven,[2] is used for the cultivation of *Lactobacillus viridescens* ATCC® 12706 in preparing the inoculum in the microbiological assay of thiamine using Bacto Thiamine Assay Medium LV. It is also used as a broth for the cultivation of heterofermentative lactobacilli and other organisms requiring high thiamine.

FORMULA

BACTO APT BROTH
DEHYDRATED
Ingredients per liter

Bacto Yeast Extract	7.5 g	Dipotassium Phosphate	5 g
Bacto Tryptone	12.5 g	Manganese Chloride	0.14 g
Bacto Dextrose	10 g	Magnesium Sulfate	0.8 g
Sodium Citrate	5 g	Ferrous Sulfate	0.04 g
Thiamine Hydrochloride	0.0001 g	Sorbitan Monooleate Complex	0.2 g
Sodium Chloride	5 g		

Final pH 6.7 ± 0.2 at 25°C.

One pound will make 9.8 liters of medium.

METHOD OF PREPARATION

1. Dissolve 46.2 g Bacto APT Broth in 1 liter distilled or deionized water.
2. Dispense in 10 ml amounts in tubes 16 – 20 mm diameter.
3. Sterilize in the autoclave for 15 minutes at 15 lbs pressure (121°C). Excessive heating of the medium is to be avoided.
4. Subcultures of *Lactobacillus viridescens* ATCC® 12706 are made from Bacto APT Agar into 10 ml tubes of Bacto APT Broth and incubated for 24 hours at 37°C. The culture is then centrifuged, washed with sterile 0.85% NaCl solution and diluted before use as inoculum.

STORAGE

Bacto APT Broth   Below 30°C

QUALITY CONTROL

Identity Specifications

Dehydrated powder: light tan, homogeneous, free-flowing
Reaction of 4.62% solution: pH 6.7 ± 0.2 at 25°C
Prepared medium: light-medium amber, clear

REFERENCES

1. J. Bact., 62:599, 1951.
2. Bact. Proc., 1955.

PACKAGING

Bacto APT Broth   1 lb (454 g)   0655-01-7

BACTO B$_{12}$ INOCULUM BROTH USP

INTENDED USE

Bacto B$_{12}$ Inoculum Broth USP is used for preparing the inoculum of *Lactobacillus leichmannii* ATCC® 7830 in the microbiological assay of vitamin B$_{12}$ according to the *USP XX.*[1]

FORMULA

BACTO B$_{12}$ INOCULUM BROTH USP
DEHYDRATED

Ingredients per liter

Tomato Juice	100 ml
Proteose Peptone No. 3, Difco	7.5 g
Bacto Yeast Extract	7.5 g
Bacto Dextrose	10 g
Sorbitan Monooleate Complex	0.1 g
Potassium Phosphate Dibasic	2 g

Final pH 6.8 ± 0.1 at 25°C.

One hundred grams will make 3.1 liters of medium.

METHOD OF PREPARATION

1. Dissolve 32 g Bacto B$_{12}$ Inoculum Broth USP in 1 liter distilled or deionized water.
2. Dispense in 10 ml amounts in tubes of 16 – 20 mm diameter.
3. Sterilize in the autoclave for 15 minutes at 15 lbs pressure (121°C).
4. Prepare inoculum as described in *USP* XX.

STORAGE

Bacto B$_{12}$ Inoculum Broth USP

2 – 8°C, preferably in a
container with calcium
chloride or other desiccant

QUALITY CONTROL

Identity Specifications

Dehydrated powder:	tan, homogeneous, free-flowing
Reaction of 3.2% solution:	pH 6.8 ± 0.1 at 25°C
Prepared medium:	medium-dark amber, clear

REFERENCE

1. U.S. Pharmacopeia/National Formulary, USP XX/NF XV, Rockville: U.S. Pharmacopeial Convention, p. 904, 1980.

PACKAGING

Bacto B$_{12}$ Inoculum Broth USP   100 g   0542-15-8

BACTO LACTOBACILLI BROTH AOAC

INTENDED USE

Bacto Lactobacilli Broth AOAC, prepared according to the formulation of Loy,[1] is used to prepare inocula of *Lactobacillus leichmannii* ATCC® 7830, *Streptococcus faecium*

ATCC® 8043 (*S. faecalis*) *Lactobacillus plantarum* ATCC® 8014, *Lactobacillus casei* ATCC® 7469 and others in the microbiological assay of B vitamins as described in *AOAC Methods* (1980).[2]

FORMULA

BACTO LACTOBACILLI BROTH AOAC
DEHYDRATED

Ingredients per liter

Bacto Peptonized Milk	15 g
Bacto Yeast Extract	5 g
Bacto Dextrose	10 g
Tomato Juice (100 ml)	5 g
Monobasic Potassium Phosphate	2 g
Sorbitan Monooleate Complex	1 g

Final pH 6.8 ± 0.2 at 25°C.

One hundred grams will make 2.6 liters of medium.

METHOD OF PREPARATION
1. Suspend 38 g Bacto Lactobacilli Broth AOAC in 1 liter distilled or deionized water and heat to boiling for 2 – 3 minutes.
2. Dispense in 10 ml amounts in tubes of 16 – 20 mm diameter.
3. Sterilize in the autoclave for 15 minutes at 15 lbs pressure (121°C).
4. Prepare inocula as described in individual monographs for vitamin assays in *AOAC Methods*.

STORAGE
Bacto Lactobacilli Broth AOAC   2 – 8°C

QUALITY CONTROL
Identity Specifications

Dehydrated powder:	tan, homogeneous, free-flowing
Reaction of 3.8% solution:	pH 6.8 ± 0.2 at 25°C
Prepared medium:	medium amber, clear, may have a slight precipitate

REFERENCES
1. J. AOAC, 41:61, 1958.
2. AOAC Methods, 13th Ed., Washington, D.C.: AOAC, p. 759 – 760, 1980.

PACKAGING
Bacto Lactobacilli Broth AOAC   100 g   0901-15-3

BACTO MICRO INOCULUM BROTH

INTENDED USE
Bacto Micro Inoculum Broth is used for the cultivation of lactobacilli used in microbiological assays. It is of particular value in the preparation of the inoculum for these tests.

FORMULA

BACTO MICRO INOCULUM BROTH
DEHYDRATED
Ingredients per liter

Proteose Peptone No. 3, Difco	5 g
Bacto Yeast Extract	20 g
Bacto Dextrose	10 g
Monopotassium Phosphate	2 g
Sorbitan Monooleate Complex	0.1 g

Final pH 6.7 ± 0.2 at 25°C.

One pound will make 12.2 liters of medium.

METHOD OF PREPARATION
1. Dissolve 37 g Bacto Micro Inoculum Broth in 1 liter distilled or deionized water.
2. Dispense in 10 ml amounts in tubes of 16 – 20 mm diameter.
3. Sterilize in the autoclave for 15 minutes at 15 lbs pressure (121°C).

PROCEDURE
Subcultures of lactobacilli are made from Bacto Micro Assay Culture Agar into 10 ml tubes of Bacto Micro Inoculum Broth and incubated for 24 hours at 37°C. The culture is then centrifuged, washed with sterile isotonic NaCl solution, and diluted before use as inoculum. It is essential that the directions given in the discussion of each assay medium be followed in minute detail, since the age, preparation and size of the inoculum are most important factors in obtaining a satisfactory assay. Although other media and methods may be used successfully for carrying the cultures and preparing the inocula, uniformly good results will be obtained if the methods described under each medium are followed.

STORAGE
Bacto Micro Inoculum Broth

below 30°C, preferably in a container with calcium chloride or other desiccant

QUALITY CONTROL
Identity Specifications
Dehydrated powder:	beige, homogeneous, free-flowing
Reaction of 3.7% solution:	pH 6.7 ± 0.2 at 25°C
Prepared medium:	light-medium amber, clear to very slightly opalescent

PACKAGING
Bacto Micro Inoculum Broth	1 lb (454 g)	0320-01-2
	1/4 lb (114 g)	0320-02-1

VITAMIN ASSAY-MEDIA FOR THE MICROBIOLOGICAL ASSAY OF VITAMINS
BACTO B$_{12}$ ASSAY MEDIUM USP

INTENDED USE
Bacto B$_{12}$ Assay Medium USP is prepared for use in the microbiological assay of vitamin B$_{12}$ according to the procedures of the Vitamin B$_{12}$ Activity Assay in the *USP XX*

and Cobalamin (Vitamin B_{12} Activity) Assay in *AOAC Methods*, 13th Edition,[2] using *Lactobacillus leichmannii* ATCC® 7830 as the test organism.

PRINCIPLES
Bacto B_{12} Assay Medium is a vitamin B_{12}-free dehydrated medium containing all other nutrients and vitamins essential for the cultivation of *Lactobacillus leichmannii* ATCC® 7830. To obtain a standard curve USP Cyanocobalamin Reference is added in specified increasing concentrations giving a growth response that can be measured acidimetrically or turbidimetrically.

FORMULA
BACTO B₁₂ ASSAY MEDIUM USP
DEHYDRATED

Ingredients per liter

Bacto Vitamin Assay Casamino Acids	15 g	Niacin	2 mg
Bacto Dextrose	40 g	p-Aminobenzoic Acid	2 mg
Bacto Asparagine	0.2 g	Calcium Pantothenate	1 mg
Sodium Acetate	20 g	Pyridoxine Hydrochloride	4 mg
Ascorbic Acid	4 g	Pyridoxal Hydrochloride	4 mg
L-Cystine	0.4 g	Pyridoxamine Hydrochloride	800 µg
DL-Tryptophane	0.4 g	Folic Acid	200 µg
Adenine Sulfate	20 mg	Monopotassium Phosphate	1 g
Guanine Hydrochloride	20 mg	Dipotassium Phosphate	1 g
Uracil	20 mg	Magnesium Sulfate	0.4 g
Xanthine	20 mg	Sodium Chloride	20 mg
Riboflavin	1 mg	Ferrous Sulfate	20 mg
Thiamine Hydrochloride	1 mg	Manganese Sulfate	20 mg
Biotin	10 µg	Sorbitan Monooleate Complex	2 g

Final pH 6.0 ± 0.2 at 25°C.

One hundred grams will make 2.35 liters of single strength medium.

METHOD OF PREPARATION
1. Suspend 8.5 g Bacto B_{12} Assay Medium USP in 100 ml distilled or deionized water and heat to boiling 2 – 3 minutes.
2. Agitate to disperse the slight precipitate that forms and distribute in 5 ml amounts into test tubes.
3. After addition of standard and test solutions according to the recommended concentrations, adjust the total volume of all tubes to 10 ml with distilled or deionized water.
4. Sterilize in the autoclave for 5 minutes at 15 lbs pressure (121°C).

PROCEDURE
Follow assay procedures as outlined in *USP XX* or *AOAC Methods*. Levels of B_{12} used in the preparations of the standard curve should be according to these references. Generally satisfactory results are obtained with B_{12} at the following levels: 0.0, 0.025, 0.05, 0.075, 0.1, 0.125, 0.15, 0.2 and 0.25 millimicrograms per assay tube (10 ml).

STORAGE
Bacto B_{12} Assay Medium USP   2 – 8°C, preferably in a container with calcium chloride or other desiccant

QUALITY CONTROL

Identity Specifications

Dehydrated powder:	light yellow or pink, homogeneous, free-flowing
Reaction of 4.25% solution:	pH 6.0 ± 0.2 at 25°C
Prepared medium:	light amber, clear, may have a slight precipitate

REFERENCES

1. US Pharmacopoeia/National Formulary, USP XX/NF XV, Rockville: US Pharmacopeial Convention, p. 903 – 905, 1980.
2. AOAC Methods, 13th Ed., Washington, D.C.: AOAC, p. 759 – 762, 1980.

PACKAGING

Bacto B$_{12}$ Assay Medium USP	100 g	0457-15-1

BACTO BIOTIN ASSAY MEDIUM

INTENDED USE

Bacto Biotin Assay Medium is prepared for use in the microbiological assay of biotin using *Lactobacillus plantarum* ATCC® 8014 as the test organism.

PRINCIPLES

Bacto Biotin Assay Medium is a biotin-free dehydrated medium containing all other nutrients and vitamins essential for the cultivation of *Lactobacillus plantarum* ATCC® 8014. The addition of biotin standard in specified increasing concentrations gives a growth response by this organism which may be measured acidimetrically and turbidimetrically.

FORMULA

BACTO BIOTIN ASSAY MEDIUM
DEHYDRATED

Ingredients per liter

Bacto Vitamin Assay		Niacin	2 mg
Casamino Acids	12 g	Calcium Pantothenate	2 mg
Bacto Dextrose	40 g	Pyridoxine Hydrochloride	2 mg
Sodium Acetate	20 g	p-Aminobenzoic Acid	200 µg
L-Cystine	0.2 g	Dipotassium Phosphate	1 g
DL-Tryptophane	0.2 g	Monopotassium Phosphate	1 g
Adenine Sulfate	20 mg	Magnesium Sulfate	0.4 g
Guanine Hydrochloride	20 mg	Sodium Chloride	20 mg
Uracil	20 mg	Ferrous Sulfate	20 mg
Thiamine Hydrochloride	2 mg	Manganese Sulfate	20 mg
Riboflavin	2 mg		

Final pH 6.7 ± 0.1 at 25°C.

One hundred grams will make 2.6 liters of single strength medium.

METHOD OF PREPARATION

1. Suspend 7.5 g Bacto Biotin Assay Medium in 100 ml distilled or deionized water and heat to boiling 2 – 3 minutes.
2. Agitate to disperse the slight precipitate that forms and distribute in 5 ml amounts into test tubes.
3. After addition of standard and test solutions according to the recommended concentrations, adjust the total volume of all tubes to 10 ml with distilled or deionized water.
4. Sterilize in the autoclave for 10 minutes at 15 lbs pressure (121°C).

PROCEDURE

Stock cultures of the test organism, *L. plantarum* ATCC® 8014, are prepared by stab inoculation of Bacto Lactobacilli Agar AOAC or Bacto Micro Assay Culture Agar. After 24 – 48 hours incubation at 35 – 37°C, the tubes are stored in the refrigerator. Transplants are made at monthly intervals in triplicate.

Inoculum for assay is prepared by subculturing from a stock culture of *L. plantarum* ATCC® 8014 to 10 ml of Bacto Lactobacilli Broth AOAC or Bacto Micro Inoculum Broth. After 16 – 24 hours incubation at 35 – 37°C, the cells are centrifuged under aseptic conditions and the supernatant liquid decanted. The cells are resuspended in 10 ml sterile 0.85% NaCl solution and again centrifuged under aseptic conditions. The final inoculum suspension is prepared by suspending these cells in 10 ml sterile 0.85% NaCl solution and diluting this suspension 1:100 with sterile 0.85% NaCl solution. One drop of this suspension is used to inoculate each of the assay tubes.

It is essential that a standard curve be constructed each time an assay is run, since conditions of autoclaving, temperature of incubation, etc., which influence the standard curve readings, cannot be duplicated exactly from time to time. A standard curve is obtained by using biotin at levels of 0.0, 0.025, 0.05, 0.1, 0.2, 0.3, 0.4, 0.5, and 1 ng per assay tube (10 ml).

The concentration of biotin required for the preparation of the standard curve may be prepared by dissolving 0.1 g of d-Biotin or equivalent, in 1000 ml of distilled water (100 µg per ml). Dilute this stock solution by adding 2 ml to 98 ml of distilled water. This solution is diluted by adding 1 ml of 999 ml distilled water giving a solution of 2 ng of biotin per ml. This solution is further diluted by adding 5 ml to 95 ml distilled water giving final solution of 0.1 ng of biotin per ml. Use 0.0, 0.25, 0.5, 1, 2, 3, 4 and 5 ml of this final solution. The last tube is prepared by adding 0.5 ml of the standard solution containing 2 ng of biotin per liter. The stock solution is prepared fresh daily.

Bacto Biotin Assay Medium may be used for both turbidimetric and acidimetric analysis. Before reading, the tubes are refrigerated for 15 – 30 minutes to stop growth. Turbidimetric readings should be made after 16 – 20 hours at 35 – 37°C. Acidimetric determinations are made after 72 hours incubation at 35 – 37°C. The most effective assay range, using Bacto Biotin Assay Medium, has been found to be between 0.025 ng and 0.5 ng biotin.

STORAGE

Bacto Biotin Assay Medium

2 – 8°C, preferably in a container with calcium chloride or other desiccant.

QUALITY CONTROL

Identity Specifications

Dehydrated powder: white to cream, homogeneous, free-flowing, may be slightly packed

Reaction of 3.75% solution: pH 6.7 ± 0.1 at 25°C
Prepared medium: light amber, clear, may have a slight precipitate

PACKAGING

Bacto Biotin Assay Medium   100 g   0419-15-8

BACTO CF ASSAY MEDIUM

INTENDED USE

Bacto CF Assay Medium, a modification of the formula given by Sauberlich and Baumann[1] and William F. Faloon,[2] is used in the microbiological assay of citrovorum factor using *Pediococcus acidilactici* ATCC® 8081 (formerly *P. cerevisiae* ATCC® 8081) as the test organism.

PRINCIPLES

Bacto CF Assay Medium is a citrovorum factor-free dehydrated medium containing all other nutrients and vitamins essential to the cultivation of *Pediococcus acidilactici* ATCC® 8081 (*P. cerevisiae* ATCC® 8081). The addition of citrovorum factor standard in specified increasing concentrations gives a growth response by this organism which can be measured acidimetrically or turbidimetrically.

FORMULA

BACTO CF ASSAY MEDIUM
DEHYDRATED

Ingredients per liter

Bacto Vitamin Assay		Pyridoxamine Hydrochloride	6 mg
Casamino Acids	10 g	Pyridoxal Hydrochloride	600 μg
Bacto Dextrose	50 g	Calcium Pantothenate	1 mg
Sodium Acetate	40 g	Riboflavin	1 mg
Ammonium Chloride	6 g	Nicotinic Acid	2 mg
L-Cystine	0.2 g	p-Aminobenzoic Acid	200 μg
L-Cysteine Hydrochloride	0.2 g	Biotin	2 μg
DL-Tryptophane	0.2 g	Folic Acid	20 μg
DL-Alanine	0.2 g	Monopotassium Phosphate	1.2 g
Glycine	20 mg	Dipotassium Phosphate	1.2 g
Adenine Sulfate	20 mg	Magnesium Sulfate	0.4 g
Guanine Hydrochloride	20 mg	Sodium Chloride	20 mg
Uracil	20 mg	Ferrous Sulfate	20 mg
Xanthine	20 mg	Manganese Sulfate	40 mg
Thiamine Hydrochloride	1 mg		
Pyridoxine Hydrochloride	2 mg		

Final pH 6.7 ± 0.2 at 25°C.

One hundred grams will make 1.8 liters of single strength medium.

METHOD OF PREPARATION

1. Suspend 11 g of Bacto CF Assay Medium in 100 ml distilled or deionized water and heat to boiling 2 – 3 minutes.
2. Agitate to disperse the slight precipitate that forms and distribute in 5 ml amounts into test tubes.
3. After addition of standard and test solutions according to the recommended concentrations, adjust the total volume of all tubes to 10 ml with distilled or deionized water.
4. Sterilize in the autoclave for 10 minutes at 15 lbs pressure (121°C).

PROCEDURE

Stock cultures of the test organism *P. acidilactici* ATCC® 8081 are prepared by stab inoculation of Bacto Lactobacilli Agar AOAC or Bacto Micro Assay Culture Agar. After 24 – 48 hours incubation at 35 – 37°C, the tubes are kept in the refrigerator at 2 – 8°C. Transplants are made at monthly intervals, in triplicate.

Inoculum for assay is prepared by subculturing from a stock culture to 10 ml of Bacto Lactobacilli Broth AOAC or Bacto Micro Inoculum Broth. After 16 – 24 hours incubation at 35 – 37°C, the cells are centrifuged under aseptic conditions, and the supernatant liquid decanted. The cells are resuspended in 10 ml of sterile 0.85% NaCl solution. The cell suspension is then diluted 1:100 with sterile 0.85% NaCl. One drop of this latter suspension is used to inoculate the assay tubes (10 ml).

It is essential that a standard curve be constructed each time an assay is run, since conditions of autoclaving, temperature of incubation, etc., which influence the standard curve readings, cannot be duplicated exactly from time to time. A standard curve is obtained by using citrovorum factor at levels of 0.0, 0.6, 1.2, 2.4, 3.6, 4.8 and 6 ng per assay tube (10 ml).

The concentration of citrovorum factor required for the construction of the standard curve may be prepared by dissolving the contents of 1 ampule leucovorin (citrovorum factor) in 1000 ml distilled water. This is the stock solution (3 μg per ml). Dilute the stock solution by adding 4 ml to 96 ml distilled water. Dilute this solution further by adding 1 ml to 99 ml distilled water. Use 0.0, 0.5, 1, 2, 3, 4, and 5 ml per tube. The stock solution is prepared fresh daily.

Bacto CF Assay Medium may be used for both acidimetric and turbidimetric analyses. Tubes should be refrigerated for 15 – 30 minutes to stop growth before reading. Acidimetric determinations are made after 72 hours incubation at 35 – 37°C. Turbidimetric readings should be made after 20 – 24 hours incubation at 35 – 37°C. The most effective assay range, using Bacto CF Assay Medium, has been found to be between 0.6 and 4.8 ng of citrovorum factor.

STORAGE
Bacto CF Assay Medium

2 – 8°C, preferably in a
container with calcium
chloride or other desiccant

QUALITY CONTROL
Identity Specifications

Dehydrated powder: white, homogeneous, free-flowing
Reaction of 5.5% solution: pH 6.7 ± 0.2 at 25°C
Prepared medium: light amber, clear, may have a slight precipitate

REFERENCES
1. J. Biol. Chem., 176:165, 1948.
2. Personal Communication, 1951.

PACKAGING
Bacto CF Assay Medium   100 g   0456-15-2

BACTO CHOLINE ASSAY MEDIUM

INTENDED USE
Bacto Choline Assay Medium, a slight modification of the medium described by Horowitz and Beadle,[1] is used in the microbiological assay of choline using *Neurospora crassa* ATCC® 9277 as the test organism.

PRINCIPLES

Bacto Choline Assay medium is a choline-free dehydrated medium containing all other nutrients and vitamins essential for the cultivation of *Neurospora crassa* ATCC® 9277. The addition of choline standard in specified increasing concentrations gives a growth response by this organism which can be measured gravimetrically.

FORMULA

BACTO CHOLINE ASSAY MEDIUM
DEHYDRATED

Ingredients per liter

Bacto Sucrose	40 g	Calcium Chloride	0.2 g
Ammonium Nitrate	2 g	Sodium Borate	700 µg
Biotin	10 µg	Ammonium Molybdate	500 µg
Potassium Sodium Tartrate	11.4 g	Ferrous Sulfate	1.1 mg
Monopotassium Phosphate	2 g	Cuprous Chloride	300 µg
Magnesium Sulfate	1 g	Manganese Sulfate	110 µg
Sodium Chloride	0.2 g	Zinc Sulfate	17.6 mg

Final pH 5.5 ± 0.2 at 25°C.

One hundred grams will make 3.5 liters of single strength medium.

METHOD OF PREPARATION

1. Dissolve 5.7 g Bacto Choline Assay Medium in 100 ml distilled or deionized water and heat to boiling 2 – 3 minutes.
2. Agitate to disperse the slight precipitate that forms and distribute in 10 ml amounts into 125 ml flasks.
3. After addition of standard and test solutions according to the recommended concentrations, adjust the total volume of all flasks to 20 ml with distilled or deionized water.
4. Sterilize in the autoclave for 10 minutes at 15 lbs pressure (121°C).

PROCEDURE

Remove 1 loop of spores from a 48 hour culture of *N. crassa* ATCC® 9277 on Bacto Neurospora Culture Agar and suspend in 100 ml sterile saline. Add 1 drop of this spore suspension to each flask. Incubate at 25°C for 3 days. At the end of the incubation period, steam the flask at 100°C for 5 minutes. Remove all the mycelium from the flask using a stiff wire needle or glass rod, press dry between paper towels, and roll into a small pellet. Dry the pellets at 100°C in a vacuum oven for 2 hours. A glazed porcelain spot plate is convenient for handling the mycelium during drying and weighing. Weigh to the nearest 0.5 mg. A standard curve is then constructed from the weights obtained and the unknown determined by interpolation. In the assay for choline, 125 ml Erlenmeyer flasks containing a total volume of 20 ml medium each are used.

It is essential that a standard curve be constructed each time an assay is run since conditions of autoclaving, temperature of incubation, etc., which influence the standard curve readings, cannot be duplicated exactly from time to time. The standard curve is obtained by using choline at levels of 0.0, 2.5, 5, 10, 15, 20 and 25 µg per assay flask (20 ml). Using Bacto Choline Assay Medium the most effective assay range has proved to be between 2.5 and 30 µg choline.

The concentration of choline required for the preparation of the standard curve may be prepared by dissolving 0.5 g choline chloride in 1000 ml distilled water. This is the stock solution (500 µg per ml). Dilute the stock solution by adding 1 ml to 99 ml distilled water. Use 0.0, 0.5, 1, 2, 3, 4 and 5 ml per tube. The stock solution is prepared fresh daily.

STORAGE

Bacto Choline Assay Medium

2 – 8°C, preferably in a container with calcium chloride or other desiccant

QUALITY CONTROL

Identity Specifications

Dehydrated powder:
Reaction of 2.85% solution:
Prepared medium:

white, homogeneous, free-flowing
pH 5.5 ± 0.2 at 25°C
colorless, clear, may have a slight precipitate

REFERENCE

1. J. Biol. Chem., 150:325, 1943.

PACKAGING

Bacto Choline Assay Medium   100 g   0460-15-6

BACTO CS VITAMIN B₁₂ AGAR

INTENDED USE

Bacto CS Vitamin B_{12} Agar is used in the microbiological assay of vitamin B_{12} by the cup plate or disk method using *Lactobacillus leichmannii* ATCC® 4797 as the test organism. It is prepared according to the formula given by Cohen and Bennett at the 1950 spring meeting of the American Chemical Society held in Philadelphia.

PRINCIPLES

Bacto CS Vitamin B_{12} Agar is a vitamin B_{12}-free dehydrated medium containing all other nutrients and vitamins essential for the cultivation of *Lactobacillus leichmannii* ATCC® 4797. The addition of vitamin B_{12} in specified increasing amounts gives a growth response that can be measured by the diameter of the growth zone around the cup or disk containing vitamin B_{12}.

FORMULA

BACTO CS VITAMIN B₁₂ AGAR
DEHYDRATED

Ingredients per liter

Bacto Vitamin Assay Casamino Acids . . . . 10 g	L-Cystine . . . . 0.2 g
Bacto Soytone Vitamin Free . . . . . . 5 g	Adenine Sulfate . . . . 17.6 mg
Bacto Dextrose . . . . 20 g	Guanine Hydrochloride . . . . 12.4 mg
Sodium Acetate . . . . 12 g	Uracil . . . . 10 mg
Sorbitan Monooleate Complex . . . . . 1 g	Xanthine . . . . 10 mg
Potassium Sulfate . . . . 20 g	Folic Acid . . . . 1 mg
Monopotassium Phosphate . . . . 1 g	Riboflavin . . . . 2 mg
Dipotassium Phosphate . . . . 1 g	Thiamine Hydrochloride . . . . 2 mg
Magnesium Sulfate . . . . 0.4 g	Calcium Pantothenate . . . . 2 mg
Sodium Chloride . . . . 20 mg	Niacin . . . . 2 mg
Ferrous Sulfate . . . . 20 mg	Pyridoxine . . . . 4 mg
Manganese Sulfate . . . . 20 mg	Pyridoxal . . . . 4 mg
Ribose Nucleic Acid . . . . 1 g	Biotin . . . . 1 µg
Sodium Thioglycollate . . . . 1.7 g	DL-Tryptophane . . . . 0.2 g
	Bacto Agar . . . . 15 g

Final pH 6.2 ± 0.1 at 25°C.

One hundred grams will make 1.1 liters of medium.

METHOD OF PREPARATION

1. Suspend 9 g Bacto CS Vitamin B_{12} Agar in 100 ml distilled or deionized water and heat to boiling to dissolve the medium completely.
2. Sterilize in the autoclave for 15 minutes at 15 lbs pressure (121°C).
3. Cool sterile medium to 45 – 50°C and inoculate with 1 ml of a properly washed and centrifuged cell suspension of *L. leichmannii* ATCC® 4797.
4. Distribute 50 ml into each of 4 sterile flat bottom Petri dishes (150 mm) and allow to solidify and dry with covers slightly raised. It is necessary for the agar surface to be dry before placing on the assay cups or disks.

PROCEDURE

Stock cultures of the test organism, *L. leichmannii* ATCC® 4797, are prepared by stab inoculation of Bacto Lactobacilli Agar AOAC or Bacto B_{12} Culture Agar USP. Following incubation at 37°C for 24 – 48 hours, the tubes are kept in the refrigerator. Transplants are made at 2 week intervals. Inoculum for assay is prepared by subculturing from a stock culture of *L. leichmannii* ATCC® 4797 into a tube containing 10 ml of Bacto Lactobacilli Broth AOAC or Bacto B_{12} Inoculum Broth USP. Incubate for 16 – 24 hours at 37°C, centrifuge the cells under aseptic conditions and decant the supernatant. Resuspend the cells in 10 ml sterile 0.85% NaCl, centrifuge, discard the supernatant and resuspend in 10 ml sterile 0.85% NaCl. Aseptically add 1 ml of washed cell suspension to sterile melted Bacto CS Vitamin B_{12} Agar as described under Method of Preparation. Place 9 sterile assay cups or 9 sterile filter paper disks (13 mm) on the surface of the agar of each plate. Six cups or disks are to be used for the standard and 3 for the assay solution.

For the preparation of the standard, prepare sterile solutions of Vitamin B_{12} USP (Cyanocobalamin Reference Standard or equivalent) containing 0.05, 0.1, 0.2, 0.5, 1 and 3 µg per ml. When metal or porcelain cups are employed, use 0.2 – 0.3 ml of the standard B_{12} solution per cup. When paper disks are employed, aseptically transfer 0.1 ml of the above solutions to the disks using an appropriately graduated pipette.

In determining the vitamin B_{12} content of unknown materials, the assay samples are appropriately diluted and similarly applied. In the assay of some preparations, better defined, clearer cut zones may be obtained by the addition of 0.5 to 1.0% of NaCl to Bacto CS Vitamin B_{12} Agar. NaCl, under these conditions, must be added to the medium in preparing the standard curve.

Incubate inoculated plates at 37°C for 24 hours and measure size of zone growth produced. The diameter of the growth zone is indicative of the amount of vitamin B_{12} present.

STORAGE

Bacto CS Vitamin B_{12} Agar   2 – 8°C, preferably in a container with calcium chloride or other desiccant.

QUALITY CONTROL

Identity Specifications

Dehydrated powder: off-white, homogeneous, free-flowing
Reaction of 4.5% solution: pH 6.2 ± 0.1 at 25°C
Prepared medium: light amber, slightly opalescent

PACKAGING

Bacto CS Vitamin B_{12} Agar   100 g   0399-15-2

BACTO EUGLENA B₁₂ MEDIUM

INTENDED USE

Bacto Euglena B_{12} Medium is patterned after the formulation of Hutner and Bach[1] and modified according to the work of Guttman,[2] and is used in the microbiological assay of vitamin B_{12} using *Euglena gracilis* ATCC® 12716 as the test organism.

PRINCIPLES

Bacto Euglena B_{12} Medium is a vitamin B_{12}-free dehydrated medium containing all other nutrients and vitamins essential for the cultivation of *Euglena gracilus* ATCC® 12716. To prepare a standard curve, USP Cyanocobalamin Reference Standard is added to the medium in specified increasing concentrations, giving a growth response that can be measured turbidimetrically.

FORMULA

BACTO EUGLENA B₁₂ MEDIUM
DEHYDRATED
Ingredients per liter

L-Glutamic Acid	6 g	Magnesium Sulfate	0.8 g
DL-Aspartic Acid	4 g	Calcium Carbonate	0.16 g
Glycine	5 g	Ammonium Carbonate	0.72 g
Sucrose	30 g	Ferric Chloride	60 mg
DL-Malic Acid	2 g	Zinc Sulfate	40 mg
Succinic Acid	1.04 g	Manganese Sulfate	6 mg
Boric Acid	1.14 mg	Copper Sulfate	620 µg
Thiamine Hydrochloride	12 mg	Cobalt Sulfate	5 mg
Monopotassium Phosphate	0.6 g	Ammonium Molybdate	1.34 mg

Final pH 3.5 ± 0.1 at 25°C.

One hundred grams will make 3.92 liters of single strength medium.

METHOD OF PREPARATION

1. Suspend 5.1 g Bacto Euglena B_{12} Medium in 100 ml distilled or deionized water and heat to dissolve completely.
2. Distribute in 2 ml amounts in test tubes.
3. After addition of standard and test solutions according to the recommended concentrations, adjust the total volume of each tube to a total volume of 4 ml per tube with distilled or deionized water.
4. Heat the tubes in flowing steam for 15 minutes.

PROCEDURE

E. gracilis is a photosynthetic protist which grows best when cultured in the light. Sunlight, bright diffuse daylight and light from fluorescent lamps are satisfactory. Stock cultures of the organism are carried on single strength Bacto Euglena B_{12} Medium containing 50 pg vitamin B_{12}. The inoculated tubes should be incubated at 27 – 29°C in the light for a week to permit good growth and pigmentation. The stock cultures may then be maintained at room temperature for as long as 3 months without transfer.

Inocula for assay procedures are prepared by subculturing from the stock cultures to tubes of single strength Bacto Euglena B_{12} Medium containing 50 pg vitamin B_{12} per ml and incubating in the light at 27 – 29°C for 1 week to obtain good growth and chlorophyll pigmentation. The broth cultures are then centrifuged under aseptic conditions to sediment the cells. The supernatant fluid is removed and the cells washed by re-suspending them in 10 ml sterile single strength Bacto Euglena B_{12} Medium, centrifuging to sediment the cells and decanting the supernatant. The washed cells are resuspended in 10 ml sterile Bacto Euglena B_{12} Medium. One drop of this cell suspension is used as inoculum for each tube of medium in the assay procedure.

It is essential that a standard curve be set up for each separate assay since cultural conditions which influence the standard curve readings cannot be duplicated exactly from time to time. The standard curve is obtained by using vitamin B_{12} at levels of 0.0, 1.25, 2.5, 5, 10, 25 and 50 pg per tube (1 picogram = 0.001 nanogram).

Concentrations of vitamin B_{12} required for the preparation of the standard curve may be prepared as follows: Dissolve the equivalent of 50 µg vitamin B_{12} (USP Cyanocobalamin Reference Standard or equivalent) dried to constant weight, in 25% ethyl alcohol and dilute with additional 25% ethyl alcohol to make a stock solution containing exactly 100 ng per ml vitamin B_{12}. Dilute 5 ml of this stock vitamin B_{12} solution with 95 ml 25% ethyl alcohol to make a stock solution containing 5 ng per ml. The stock solutions described above are prepared fresh daily.

Prepare 3 standard B_{12} solutions from the most dilute stock solution, described above, as follows: Add 5 ml of the 5 ng per ml stock vitamin B_{12} solution to 995 ml distilled water to make a 25 pg per ml standard B_{12} solution. Then add 20 ml of the 25 pg per ml standard B_{12} solution to 80 ml of distilled water to make a 5 pg per ml standard B_{12} solution. Finally add 5 ml of the 25 pg per ml standard solution to 95 ml distilled water to make a 1.25 pg per ml standard B_{12} solution.

The standards are added to 15 × 150 mm test tubes as indicated in the following table.

Tube No.	Standard Added
1	0
2	1 ml 1.25 pg per ml
3	2 ml 1.25 pg per ml
4	1 ml 5 pg per ml
5	2 ml 5 pg per ml
6	1 ml 25 pg per ml
7	2 ml 25 pg per ml

Add 2 ml distilled water to tube 1 and 1 ml distilled water to tubes 2, 4 and 6. Then add 2 ml double strength Bacto Euglena B_{12} Medium to each tube to make a total volume of 4 ml per tube. Appropriate graduated amounts of the unknown are added to duplicate sets of tubes and diluted with distilled water to a final volume of 2 ml per tube. Two ml of double strength Bacto Euglena B_{12} Medium is then added to each tube to make a final volume of 4 ml. The tubes are then heated in flowing steam at atmospheric pressure for 15 minutes and cooled quickly to room temperature. They are then inoculated and incubated for 5 – 6 days at 27 – 29°C being constantly and uniformly illuminated with 40 watt white fluorescent lamps.

Results are determined turbidimetrically using the assay procedure tube method. For further discussion on calculations, refer to general section and USP XX or AOAC 13th Edition.

STORAGE
Bacto Euglena B_{12} Medium   2 – 8°C

QUALITY CONTROL

Identity Specifications

Dehydrated powder: white, free-flowing with some clumps
Reaction of 2.55% solution: pH 3.5 ± 0.1 at 25°C
Prepared medium: almost colorless, clear with no significant precipitate

REFERENCES
1. J. Protozology, 3:101, 1956.
2. Anal. Microbiol., p. 537, 1963.

PACKAGING
Bacto Euglena B$_{12}$ Medium   100 g   0532-15-0

BACTO FOLIC AOAC MEDIUM

INTENDED USE
Bacto Folic AOAC Medium is prepared for use in the microbiological assay of folic acid according to the procedures of the Folic Acid Assay in *AOAC Methods*, 13th edition,[1] using *Streptococcus faecium* ATCC® 8043 (formerly *S. faecalis* ATCC® 8043), as the test organism.

PRINCIPLES
Bacto Folic Acid AOAC Medium is a folic acid-free dehydrated medium containing all other nutrients and vitamins essential for the cultivation of *S. faecium* ATCC® 8043 (*S. faecalis* ATCC® 8043). The addition of folic acid in specified increasing concentrations gives a growth response that can be measured turbidimetrically or acidimetrically.

FORMULA
BACTO FOLIC AOAC MEDIUM
DEHYDRATED
Ingredients per liter

Bacto Vitamin Assay		Calcium Pantothenate	800 µg
Casamino Acids	10 g	Nicotinic Acid	800 µg
L-Asparagine	0.6 g	Biotin	20 µg
L-Tryptophane	0.2 g	Riboflavin	1 mg
L-Cysteine Hydrochloride	0.76 g	Glutathione	5.2 mg
Bacto Dextrose	40 g	Sorbitan Monooleate Complex	0.1 g
Adenine Sulfate	10 mg	Sodium Citrate	52 g
Guanine Hydrochloride	10 mg	Dipotassium Phosphate	6.4 g
Uracil	10 mg	Magnesium Sulfate	0.4 g
Xanthine	20 mg	Manganese Sulfate	20 mg
p-Aminobenzoic Acid	1 mg	Sodium Chloride	20 mg
Pyridoxine Hydrochloride	4 mg	Ferrous Sulfate	20 mg
Thiamine Hydrochloride	400 µg		

Final pH 6.7 ± 0.1 at 25°C.

One hundred grams will make 2.6 liters of single strength medium.

METHOD OF PREPARATION
1. Suspend 11 g Bacto Folic AOAC Medium in 100 ml distilled or deionized water and heat to boiling 2 – 3 minutes.
2. Agitate to disperse the slight precipitate that forms and distribute in 5 ml amounts into test tubes.

3. After addition of standard and test solutions according to the recommended concentrations, adjust the total volume of all tubes to 10 ml with distilled or deionized water.
4. Sterilize in the autoclave for 5 minutes at 15 lbs pressure (121°C).

PROCEDURE

Follow assay procedures as outlined in *AOAC Methods*, 13th Edition. It is essential that a standard curve be set up for each separate assay since conditions by autoclaving, temperature of incubation, etc., which influence the standard curve readings, cannot be duplicated exactly from time to time. The standard curve is obtained by using folic acid at levels of 0.0, 1, 2, 4, 6, 8 and 10 ng per assay tube (10 ml). Bacto Folic AOAC Medium may be used for both turbidimetric and acidimetric analysis. Turbidimetric readings should be made after 16 – 18 hours incubation at 35 – 37°C. Acidimetric determinations are best made following 72 hours incubation at 35 – 37°C.

The concentration of folic acid required for the preparation of the standard curve may be prepared as follows: Dissolve 50 mg dried folic acid in about 30 ml 0.01N NaOH and 300 ml distilled water. Adjust the reaction to pH 7.5 ± 0.5 with diluted HCl solution and add distilled water to give a volume of 500 ml. Add 10 ml of the solution to 900 ml distilled water, adjusting reaction to pH 7.5 ± 0.5 with HCl solution and diluting to 1 liter with distilled water. Add 100 ml to 900 ml distilled water, adjusting the reaction to pH 7.5 ± 0.5 and diluting to 1 liter with distilled water to give a stock solution containing 100 ng folic acid per ml. The stock solution is prepared fresh daily.

The standard solution for the assay is made by diluting 10 ml of this stock solution to 500 ml with distilled water to give a solution containing 2 ng folic acid per ml. Use 0.0, 0.5, 1, 2, 3, 4, and 5 ml per assay tube.

Some laboratories may wish to alter the concentration of folic acid recommended above for the standard curve. This is permissible if the concentration used is within the limits specified by the AOAC.

STORAGE

Bacto Folic AOAC Medium

         2 – 8°C, preferably in a
         container with calcium
         chloride or other desiccant

QUALITY CONTROL

Identity Specifications

Dehydrated powder: white, homogeneous, free-flowing
Reaction of 5.5% solution: pH 6.7 ± 0.1 at 25°C
Prepared medium: light amber, clear, may have a slight precipitate

REFERENCE

1. AOAC Methods, 13th Ed., Washington, D.C.: AOAC, p. 763 – 764, 1980.

PACKAGING

Bacto Folic AOAC Medium   100 g   0967-15-4

BACTO FOLIC TE MEDIUM

INTENDED USE

Bacto Folic TE Medium, prepared according to the formulation of Teply and Elvehjem,[1] is used in the microbiological assay of folic acid using *Streptococcus faecium* ATCC® 8043 (*S. faecalis* ATCC® 8043) as the test organism.

PRINCIPLES

Bacto Folic TE Medium is a folic acid-free dehydrated medium containing all other nutrients and vitamins essential for the cultivation of *S. faecium* ATCC® 8043 (*S. faecalis* ATCC® 8043). The addition of folic acid in specified increasing concentrations gives a growth response that can be measured turbidimetrically or acidimetrically.

FORMULA

BACTO FOLIC TE MEDIUM
DEHYDRATED

Ingredients per liter

Bacto Vitamin Assay Casamino Acids	10 g	Pyridoxine Hydrochloride ....2.4 mg
DL-Alanine	0.4 g	Calcium Pantothenate .... 800 μg
L-Cystine	0.4 g	Biotin ....0.8 μg
DL-Tryptophane	0.4 g	p-Aminobenzoic Acid .... 20 μg
Bacto Dextrose	40 g	Sodium Citrate .... 50 g
Adenine Sulfate	20 mg	Charcoal Treated Peptone .....0.6 g
Guanine Hydrochloride	20 mg	Asparagine ....0.2 g
Uracil	20 mg	Magnesium Sulfate ....0.4 g
Xanthine	20 mg	Sodium Chloride .... 20 mg
Thiamine Hydrochloride	400 μg	Ferrous Sulfate .... 20 mg
Riboflavin	400 μg	Manganese Sulfate .... 14 mg
Nicotinic Acid	1.2 mg	Dipotassium Phosphate ....5 g

Final pH 6.8 ± 0.2 at 25°C.

One hundred grams will make 1.8 liters of single strength medium.

METHOD OF PREPARATION

1. Suspend 10.8 g Bacto Folic TE Medium in 100 ml distilled or deionized water and heat to boiling 2–3 minutes.
2. Agitate to disperse the slight precipitate that forms and distribute in 5 ml amounts into test tubes.
3. After addition of standard and test solutions according to the recommended concentrations, adjust the total volume of all tubes to 10 ml, with distilled or deionized water.
4. Sterilize in the autoclave for 5 minutes at 15 lbs pressure (121°C).

PROCEDURE

Stock cultures of *S. faecium* ATCC® 8043 are prepared by stab inoculation of Bacto Lactobacilli Agar AOAC or Bacto Micro Assay Culture Agar. Following incubation at 35–37°C for 18–24 hours, the tubes are stored in the refrigerator at 2–8°C. Fresh stab cultures are prepared every week and are not used for preparing the inoculum, if more than a week old. Inoculum for assay is prepared by subculturing from a stock culture of *S. faecium* ATCC® 8043 into a tube containing 10 ml Bacto Lactobacilli Broth AOAC or Bacto Micro Inoculum Broth. After 18–24 hours incubation at 35–37°C the cells are centrifuged under aseptic conditions and the supernatant liquid decanted. The cells are resuspended in 10 ml sterile 0.85% NaCl. The cell suspension is then diluted

1:100 in sterile 0.85% NaCl. One drop of this suspension is then used to inoculate each of the assay tubes.

It is essential that a standard curve be set up for each separate assay since conditions of autoclaving, temperature of incubation, etc., which influence the standard curve readings cannot be duplicated exactly from time to time. The standard curve is obtained by using folic acid at levels of 0.0, 2, 4, 6, 8 and 10 ng per assay tube (10 ml).

The concentration of folic acid required for the preparation of the standard curve may be prepared as follows: Dissolve 50 mg dried folic acid in about 30 ml 0.01N NaOH and 300 ml distilled water. Adjust the reaction to pH 7.5 ± 0.5 with HCl and dilute to 500 ml with distilled water. Dilute 10 ml of this solution to 1000 ml with distilled water. Then dilute 200 ml to 500 ml with distilled water, adjust the reaction to pH 7.5 ± 0.5 and dilute to 1 liter with distilled water to make a stock solution containing 200 ng folic acid per ml. The stock solution is prepared fresh daily.

To make the standard solution dilute 5 ml of the stock solution to 500 ml with distilled water to give a solution containing 2 ng folic acid per ml. Use 0.0, 1, 2, 3, 4 and 5 ml per assay tube.

Following inoculation, tubes are incubated at 37°C for 18 – 24 hours for turbidimetric determination and 37°C for 72 hours for acidimetric determinations.

STORAGE
Bacto Folic TE Medium

2 – 8°C, preferably in a
container with calcium
chloride or other desiccant

QUALITY CONTROL
Identity Specifications

Dehydrated powder: white, homogeneous, free-flowing
Reaction of 5.4% solution: pH 6.8 ± 0.2 at 25°C
Prepared medium: light amber, clear, may have a slight precipitate

REFERENCE
1. J. Biol. Chem., 157:303, 1945.

PACKAGING
Bacto Folic TE Medium   100 g   0968-15-3

BACTO FOLIC ACID ASSAY MEDIUM

INTENDED USE
Bacto Folic Acid Assay Medium is used in the microbiological assay of folic acid using *Streptococcus faecium* ATCC® 8043 (formerly *S. faecalis* ATCC® 8043) as the test organism. It is prepared according to the formula described by Capps, Hobbs and Fox[1] modified by use of sodium citrate instead of sodium acetate.

PRINCIPLES
Bacto Folic Acid Assay Medium is a folic acid-free dehydrated medium containing all other nutrients and vitamins essential for the cultivation of *S. faecium* ATCC® 8043 (*S.*

faecalis ATCC® 8043). The addition of folic acid in specified increasing concentrations gives a growth response that can be measured turbidimetrically.

FORMULA
BACTO FOLIC ACID ASSAY MEDIUM
DEHYDRATED
Ingredients per liter

Bacto Vitamin Assay		Riboflavin	2 mg
Casamino Acids	12 g	Niacin	2 mg
Bacto Dextrose	40 g	p-Aminobenzoic Acid	200 µg
Sodium Citrate	20 g	Biotin	0.8 µg
L-Cystine	0.2 g	Calcium Pantothenate	400 µg
DL-Tryptophane	0.2 g	Dipotassium Phosphate	1 g
Adenine Sulfate	20 mg	Monopotassium Phophate	1 g
Guanine Hydrochloride	20 mg	Magnesium Sulfate	0.4 g
Uracil	20 mg	Sodium Chloride	20 mg
Thiamine Hydrochloride	2 mg	Ferrous Sulfate	20 mg
Pyridoxine Hydrochloride	4 mg	Manganese Sulfate	20 mg

Final pH 6.8 ± 0.2 at 25°C.

One hundred grams will make 2.6 liters of single strength medium.

METHOD OF PREPARATION
1. Suspend 7.5 g Bacto Folic Acid Assay Medium in 100 ml distilled or deionized water and heat to boiling 2 – 3 minutes.
2. Agitate to disperse the slight precipitate that forms and distribute in 5 ml amounts into test tubes.
3. After addition of standard and test solutions according to the recommended concentrations, adjust the total volume of all tubes to 10 ml with distilled or deionized water.
4. Sterilize in the autoclave for 10 minutes at 15 lbs pressure (121°C).

PROCEDURE
Stock cultures of *S. faecium* ATCC® 8043 are prepared by stab inoculation of Bacto Lactobacilli Agar AOAC or Bacto Micro Assay Culture Agar. Following incubation at 35 – 37°C for 24 – 48 hours, the tubes are stored in the refrigerator. Transplants are made at monthly intervals. Inoculum for assay is prepared by subculturing from a stock culture of *S. faecium* ATCC® 8043 into a tube containing 10 ml of Bacto Lactobacilli Broth AOAC or Bacto Micro Inoculum Broth. After 24 hours incubation at 35 – 37°C, the cells are centrifuged, under aseptic conditions, and the supernatant liquid decanted. The cells are resuspended in 10 ml of sterile 0.85% NaCl. The cell suspension is then diluted 1:100 with sterile 0.85% NaCl. One drop of this latter suspension is used to inoculate each of the assay tubes.

It is essential that a standard curve be set up for each separate assay since conditions of autoclaving, temperature of incubation, etc., which influence the standard curve readings cannot be duplicated exactly from time to time. The standard curve is obtained by using folic acid at levels of 0.0, 2, 4, 6, 8 and 10 ng per assay tube (10 ml). Tubes are refrigerated for 15 – 30 minutes to stop growth before reading. Turbidimetric readings should be made at 16 – 18 hours incubation at 35 – 37°C.

The concentration of the folic acid required for the preparation of the standard curve may be prepared as follows: Dissolve 50 mg dried Folic Acid USP Reference Standard or equivalent, in about 30 ml of 0.01N NaOH and 300 ml distilled water. Adjust the reaction to pH 7.5 ± 0.5 with diluted HCl solution and add distilled water to give a volume of 500 ml. Add 10 ml to 500 ml distilled water, adjusting the reaction to pH 7.5 ± 0.5 with HCl solution and dilute to 1 liter with distilled water. Then add 100 ml to 500 ml distilled water, adjusting the reaction to pH 7.5 ± 0.5 and dilute to 1 liter with distilled water to give a stock solution containing 100 ng folic acid per ml. The stock solution is prepared fresh daily.

The standard solution for the assay is made by diluting 10 ml of this stock solution to 500 ml with distilled water to give a solution containing 2 ng folic acid per ml. Use 0.0, 1, 2, 3, 4 and 5 ml per assay tube.

Following incubation, the tubes are placed in the refrigerator for 15 – 30 minutes in order to stop growth. The growth can then be measured by any suitable nephelometric method, and the curve constructed from the values obtained. Using Bacto Folic Acid Assay Medium, we have found the most effective assay range to be between the levels of 2 and 10 ng folic acid per tube (10 ml).

STORAGE
Bacto Folic Acid Assay Medium   2 – 8°C, preferably in a container with calcium chloride or other desiccant

QUALITY CONTROL
Identity Specifications

Dehydrated powder:	white to off-white, homogeneous, free-flowing
Reaction of 3.75% solution:	pH 6.8 ± 0.2 at 25°C
Prepared medium:	light amber, clear, may have a slight precipitate

REFERENCE
1. J. Bact., 55:869, 1948.

PACKAGING
Bacto Folic Acid Assay Medium   100 g   0318-15-0

BACTO FOLIC ACID CASEI MEDIUM

INTENDED USE
Bacto Folic Acid Casei Medium is prepared for the micorbiological assay of folic acid, particularly folic acid in blood serum, using *Lactobacillus casei* ATCC® 7469 as the test organism. It is prepared according to the formulation described by Flynn, Williams, O'Dell and Hogan[1] and as modified by Baker, Herbert, Frank, Pasher, Hunter, Wasserman and Sobotka[2] and by Waters and Mollin.[3]

PRINCIPLES
Total serum folic acid activity varies in accordance with the amount of a labile component, which is low in the case of megaloblastic anemia. It has been reported that normal subjects have a mean serum folic acid level of 9.9 ng per ml, patients with uncomplicated pernicious anemia have a mean serum folic acid level of 16.6 ng per ml, whereas patients with megaloblastic anemia have serum folic acid levels of less than 4.0 ng per ml.

Bacto Folic Acid Casei Medium is a folic acid-free dehydrated medium containing all other nutrients and vitamins essential for the cultivation of *L. casei* ATCC® 7469. The addition of folic acid in specified increasing concentration gives a growth response that can be measured turbidimetrically.

FORMULA

BACTO FOLIC ACID CASEI MEDIUM
DEHYDRATED

Ingredients per liter

Charcoal Treated Casitone	10 g	Glutathione (reduced)	5 mg
Bacto Dextrose	40 g	Magnesium Sulfate	0.4 g
Sodium Acetate	40 g	Sodium Chloride	20 mg
Dipotassium Phosphate	1 g	Ferrous Sulfate	20 mg
Monopotassium Phosphate	1 g	Manganese Sulfate	15 mg
DL-Tryptophane	0.2 g	Riboflavin	1 mg
L-Asparagine	0.6 g	p-Aminobenzoic Acid	2 mg
L-Cysteine Hydrochloride	0.5 g	Pyridoxine Hydrochloride	4 mg
Adenine Sulfate	10 mg	Thiamine Hydrochloride	400 µg
Guanine Hydrochloride	10 mg	Calcium Pantothenate	800 µg
Uracil	10 mg	Nicotinic Acid	800 µg
Xanthine	20 mg	Biotin	20 µg
Sorbitan Monooleate Complex	0.1 g		

Final pH 6.7 ± 0.1 at 25°C.

One hundred grams will make 2.13 liters of single strength medium.

METHOD OF PREPARATION

1. Suspend 9.4 g Bacto Folic Acid Casei Medium in 100 ml distilled or deionized water and add 50 mg ascorbic acid. Heat to boiling 1 – 2 minutes.
2. Distribute in 5 ml amounts into test tubes.
3. After addition of standard and test solutions according to recommended concentrations, adjust the total volume of all tubes to 10 ml with distilled or deionized water.
4. Sterilize in the autoclave for 5 minutes at 15 lbs pressure (121°C).

PROCEDURE
Preparation of Stock Cultures and Inoculum
Prepare stock cultures of test organism *L. casei* ATCC® 7469 by stab inoculation into prepared tubes of Bacto Lactobacilli Agar AOAC. Incubate the cultures at 35 – 37°C for 18 – 24 hours and store them in the refrigerator at 2 – 8°C. Make stock transfers at monthly intervals.

Prepare the inoculum for assay by subculturing from a stock culture of *L. casei* ATCC® 7469 into a tube containing 10 ml prepared Bacto Micro Inoculum Broth. Incubate for 16 – 18 hours at 35 – 37°C and centrifuge the tubes to sediment the cells under aseptic conditions. Decant the supernatant liquid and resuspend the cells in 10 ml sterile single strength Bacto Folic Acid Casei Medium. Resediment the cells by centrifuging aseptically and decant the supernatant liquid. Finally, resuspend the cells in 10 ml sterile single strength medium and dilute 1 ml with 99 ml of the same medium. One drop of this suspension is used to inoculate each of the assay tubes. The growth response of the assay tubes should be read turbidimetrically after 18 – 24 hours incubation at 37°C.

Some laboratories prefer to use 0.85% NaCl instead of the single strength basal medium to wash and dilute the inoculum.

Preparation of the Standard

It is essential that a standard curve be constructed for each separate assay since conditions of autoclaving, temperature of incubation, etc., which influence the standard curve readings, cannot be duplicated exactly from time to time. The standard curve may be obtained by using folic acid at levels of 0.0, 0.1, 0.2, 0.4, 0.6, 0.8, and 1 ng per assay tube (10 ml).

The folic acid required for preparation of the standard curve may be prepared as follows: Dissolve 20 mg dried folic acid in 100 ml of a solution containing 20 ml ethanol and 80 ml distilled or deionized water.

Adjust the reaction to pH 10.0 with 0.1N NaOH to dissolve the folic acid and then adjust the reaction to pH 7.0 with 0.05N HCl. This solution contains 200 µg per ml folic acid. Dilute 1 ml of this solution with 999 ml of distilled or deionized water to make a solution containing 200 ng per ml and finally dilute 1 ml of the 200 ng per ml solution with 999 ml of Bacto Folic Buffer A solution to make a standard solution containing 0.2 ng per ml folic acid. Use 0.0, 0.5, 1, 2, 3, 4 and 5 ml per assay tube.

Preservation of Serum Specimens

1. Allow blood specimens to clot and the serum to separate from the clot.
2. Aspirate the serum into a clean dry tube and centrifuge it to remove any cells which may be present. Care should be taken to avoid hemolysis of the erythrocytes. Dispense 5 ml of each serum sample into clean dry test tubes and add 25 mg ascorbic acid to each tube.
3. If the test is not begun immediately, place tubes in a freezer and hold below −20°C.

Preparation of Serum Specimen

1. Thaw the serum containing ascorbic acid.
2. Add 5 ml of the uniform sample to 45 ml rehydrated Bacto Folic Buffer A, Dried.
3. Incubate the serum-buffer solution at 37°C for 90 minutes and then autoclave the incubated mixture for 2.5 minutes at 15 lbs pressure (121°C).
4. Remove the coagulated protein by centrifuging and transfer the clear supernatant to a clean dry tube. The clear solution is the sample to use in the folic acid assay.

Procedure for Total Folic Acid

1. Use 0.5, 1.0, 1.5 ml or other volumes of the prepared serum extracts as described above.
2. Fill each assay tube with 5 ml of rehydrated Bacto Folic Acid Casei Medium and sufficient distilled or deionized water to give a total volume of 10 ml per tube.
3. Autoclave tubes for 5 minutes at 15 lbs pressure (121°C).
4. Add 1 drop of inoculum described under PREPARATION OF STOCK CULTURE AND INOCULUM to each assay tube.
5. Incubate tubes for 18 – 24 hours at 37 ± 1°C and read turbidimetrically. Tubes are refrigerated for 15 – 30 minutes to stop growth before reading.

Determination of Folic Acid in Test Samples

The amount of folic acid in the test samples can be determined by interpolating the results with the values obtained on the standard curve, taking into consideration the dilutions of the samples.

STORAGE

Bacto Folic Acid Casei Medium   2 – 8°C, preferably in a
                                         container with calcium
                                         chloride or other desiccant

QUALITY CONTROL

Identity Specifications

Dehydrated powder: white to off-white, homogeneous, free-flowing
Reaction of 4.7% solution: pH 6.7 ± 0.1 at 25°C
Prepared medium: light amber, clear, may have a slight precipitate

REFERENCES
1. Anal. Chem., 23:180, 1951.
2. Clin. Chem., 5:275, 1959.
3. J. Clin. Path., 14:335, 1961.

PACKAGING
Bacto Folic Acid Casei Medium   100 g   0822-15-9

BACTO FOLIC BUFFER A

INTENDED USE
Bacto Folic Buffer A is prepared for use with Bacto Folic Acid Casei Medium in the microbiological assay of serum folic acid. It is used for preparing both the standard and the serum specimen.

FORMULA
BACTO FOLIC BUFFER A, DRIED
Ingredients per liter

Monopotassium Phosphate ..... 10.88 g
Dipotassium Phosphate .... 3.48 g
Ascorbic Acid .... 1 g

Final pH 6.1 ± 0.05 at 25°C.

Each 15.4 gram package will make 1 liter of 0.1 M buffer solution.

METHOD OF PREPARATION
Suspend 15.4 g in 1 liter distilled water.

STORAGE
Bacto Folic Acid Buffer A   Below 30°C

QUALITY CONTROL
Identity Specifications

Dehydrated powder: white, homogeneous, free-flowing
Reaction of 1.54% solution: pH 6.1 ± 0.05 at 25°C
Prepared medium: colorless, clear, no precipitate

PACKAGING
Bacto Folic Buffer A   6 × 15.4 g   3246-33-9

BACTO INOSITOL ASSAY MEDIUM

INTENDED USE
Bacto Inositol Assay Medium, a slight modification of the formula of Atkin, Schultz, Williams and Frey,[1] is used in the microbiological assay of inositol using *Saccharo-*

myces uvarum ATCC® 9080 (formerly *Saccharomyces carlsbergensis* ATCC® 9080) as the test organism.

PRINCIPLES

Bacto Inositol Assay Medium is an inositol-free dehydrated medium containing all other nutrients and vitamins essential for the cultivation of *S. uvarum* ATCC® 9080 (*S. carlsbergensis* ATCC® 9080). The addition of inositol in specified increasing concentrations gives a growth response that can be measured turbidimetrically.

FORMULA

BACTO INOSITOL ASSAY MEDIUM
DEHYDRATED

Ingredients per liter

Bacto Dextrose	100 g	L-Tyrosine	0.2 g
Potassium Citrate	10 g	L-Asparagine	0.8 g
Citric Acid	2 g	DL-Aspartic Acid	0.2 g
Monopotassium Phosphate	1.1 g	DL-Serine	0.1 g
Potassium Chloride	0.85 g	Glycine	0.2 g
Magnesium Sulfate	0.25 g	DL-Threonine	0.4 g
Calcium Chloride	0.25 g	L-Valine	0.5 g
Manganese Sulfate	50 mg	L-Histidine	0.124 g
Ferric Chloride	50 mg	L-Proline	0.2 g
DL-Tryptophane	80 mg	DL-Alanine	0.4 g
L-Cystine	0.1 g	L-Glutamic Acid	0.6 g
L-Isoleucine	0.5 g	L-Arginine	0.48 g
L-Leucine	0.5 g	Thiamine Hydrochloride	500 μg
L-Lysine	0.5 g	Biotin	16 μg
L-Methionine	0.2 g	Calcium Pantothenate	5 mg
DL-Phenylalanine	0.2 g	Pyridoxine Hydrochloride	1 mg

Final pH 5.2 ± 0.2 at 25°C.

One hundred grams will make 1.6 liters single strength medium.

METHOD OF PREPARATION

1. Suspend 12.2 g Bacto Inositol Assay Medium in 100 ml distilled or deionized water and heat to boiling.
2. Distribute in 5 ml amounts into 50 ml flasks.
3. After addition of standard and test solution according to recommended concentrations adjust the total volume of all flasks to 10 ml with distilled or deionized water.
4. Sterilize in the autoclave for 5 minutes at 15 lbs pressure (121°C).

PROCEDURE

Inoculate 10 ml sterile single strength Bacto Inositol Assay Medium containing 10 μg inositol in a 50 ml flask with *S. uvarum* ATCC® 9080 carried on Bacto Neurospora Culture Agar. Incubate at 30°C for 18 – 24 hours. Transfer the culture aseptically into a sterile centrifuge tube, centrifuge and decant the supernatant liquid. Suspend the sedimented organisms into 10 ml sterile 0.85% NaCl, mix and dilute 1 ml in 99 ml of 0.85% NaCl. This diluted suspension is the inoculum. Use 1 drop of inoculum suspension to inoculate each assay flask.

It is essential that a standard curve be constructed each time an assay is run since conditions of autoclaving, temperature of incubation, etc., which influence the standard curve readings, cannot be duplicated exactly from time to time. The standard curve is obtained by using inositol at levels of 0.0, 1, 2, 4, 6, 8 and 10 μg per assay flask (10 ml).

The concentrations of inositol required for the preparation of the standard curve may be prepared by dissolving 200 mg inositol in 100 ml distilled water, mix thoroughly, then dilute 1 ml of this solution with 999 ml distilled water to make a final solution containing 2 μg inositol per ml. Use 0.0, 0.5, 1, 2, 3, 4 and 5 ml per flask. This stock solution is prepared fresh daily.

Following inoculation, flasks are incubated at 30 ± 1°C for 18 – 24 hours and placed in the refrigerator for 15 – 30 minutes to stop growth. The growth is then measured turbidimetrically using any suitable turbidimeter.

STORAGE

Bacto Inositol Assay Medium	2 – 8°C, preferably in a container with calcium chloride or other desiccant.

QUALITY CONTROL

Identity Specifications

Dehydrated powder:	white, homogeneous, free-flowing
Reaction of 6.1% solution:	pH 5.2 ± 0.2 at 25°C
Prepared medium:	very light amber, clear, no precipitate should be present

REFERENCE

1. Ind. & Eng. Chem., Ann. Ed., 15:141, 1943.

PACKAGING

Bacto Inositol Assay Medium	100 g	0995-15-0

BACTO INOSITOL ASSAY MEDIUM KB

INTENDED USE

Bacto Inositol Assay Medium KB, a slight modification of the formula of Campling and Nixon,[1] is used in the microbiological assay of inositol using *Kloeckera apiculata* ATCC® 9774 (formerly *Kloeckera brevis* ATCC® 9774) as the test organism.

PRINCIPLES

Bacto Inositol Assay Medium KB is an inositol-free dehydrated medium containing all other nutrients and vitamins essential for the cultivation of *K. apiculata* ATCC® 9774 (*K. brevis* ATCC® 9774). The addition of inositol in specified increasing concentrations gives a growth response that can be measured turbidimetrically.

FORMULA

BACTO INOSITOL ASSAY MEDIUM KB
DEHYDRATED

Ingredients per liter

Bacto Dextrose	40 g	Manganese Sulfate	80 μg
Monopotassium Phosphate	3 g	Copper Sulfate	90 μg
DL-Asparagine	4 g	Ferrous Sulfate	500 μg
Calcium Chloride	0.98 g	Thiamine	400 μg
Magnesium Sulfate	1 g	Pyridoxine	400 μg
Potassium Iodide	200 μg	Calcium Pantothenate	400 μg
Ammonium Sulfate	4 g	Nicotinic Acid	400 μg
Boric Acid	200 μg	Biotin	0.4 μg
Zinc Sulfate	80 μg	Riboflavin	20 μg
Ammonium Molybdate	40 μg		

Final pH 5.0 ± 0.2 at 25°C.

One hundred grams will make 3.7 liters single strength medium.

METHOD OF PREPARATION

1. Suspend 5.3 g Bacto Inositol Assay Medium KB in 100 ml distilled or deionized water and heat to boiling.
2. Distribute in 5 ml amounts into 50 ml flasks.
3. After addition of standard and test solutions according to recommended concentrations, adjust the volume of all flasks to 10 ml with distilled or deionized water.
4. Sterilize in the autoclave for 5 minutes at 15 lbs pressure (121°C).

PROCEDURE

Transfer K. apiculata ATCC® 9774 from a stock culture carried on Bacto Lactobacilli Agar AOAC to a fresh slant of the same medium and incubate 24 hours at 30°C. Suspend the growth from a 24 hour culture in a 10 ml sterile 0.85% NaCl, then dilute 2 ml of this suspension with 98 ml sterile 0.85% NaCl. This suspension is the inoculum. To inoculate the assay flasks add 1 drop of inoculum suspension per flask.

It is essential that a standard curve be constructed each time an assay is run since conditions of autoclaving, temperature of incubation, etc., which influence the standard curve readings, cannot be duplicated exactly from time to time. The standard curve is obtained by using inositol at levels of 0.0, 0.5, 1, 2, 3, 4 and 5 µg per assay flask.

The concentrations of inositol required for the preparation of the standard curve may be prepared by dissolving 100 mg inositol into 100 ml distilled water, mixing thoroughly, then diluting 1 ml of the solution with 999 ml distilled water to make a final solution containing 1 µg inositol per ml. Use 0.0, 0.5, 1, 2, 3, 4 and 5 ml per flask (50 ml).

Following inoculation, the flasks are incubated on a suitable shaker (about 100 strokes per minute) at 30 ± 1°C for 18 – 24 hours and placed in the refrigerator for 15 – 30 minutes to stop growth. The growth is then measured turbidimetrically using any suitable turbidimeter.

STORAGE

Bacto Inositol Assay Medium KB   2 – 8°C, preferably in a container with calcium chloride or other desiccant.

QUALITY CONTROL

Identity Specifications

Dehydrated powder: white, homogeneous, free-flowing
Reaction of 2.65% solution: pH 5.0 ± 0.2 at 25°C
Prepared medium: colorless, clear to very slightly opalescent, may have a slight precipitate

REFERENCE

1. J. Physiol., 126:71, 1954.

PACKAGING

Bacto Inositol Assay Medium KB   100 g   0966-15-5

BACTO NIACIN ASSAY MEDIUM

INTENDED USE

Bacto Niacin Assay Medium, prepared according to the formula described by Snell and Wright[1] and modified by Krehl, Strong and Elvehjem[2] and Barton-Wright,[3] is used in the microbiological assay of nicotinic acid or nicotinamide (niacin) using *Lactobacillus plantarum* ATCC® 8014 as the test organism. The medium complies with the recommendations of the *USP XX*[4] and *AOAC*.[5]

PRINCIPLES

Bacto Niacin Assay Medium is a medium free from nicotinic acid and its analogs but containing all other nutrients and vitamins essential for the cultivation of *L. plantarum* ATCC® 8014. The addition of nicotinic acid or its analogs in specified increasing concentrations gives a growth response that can be measured turbidimetrically or acidimetrically.

FORMULA

BACTO NIACIN ASSAY MEDIUM
DEHYDRATED

Ingredients per liter

Bacto Vitamin Assay		Pyridoxine Hydrochloride	400 µg
Casamino Acids	12 g	Riboflavin	400 µg
Bacto Dextrose	40 g	p-Aminobenzoic Acid	100 µg
Sodium Acetate	20 g	Biotin	0.8 µg
L-Cystine	0.4 g	Dipotassium Phosphate	1 g
DL-Tryptophane	0.2 g	Monopotassium Phosphate	1 g
Adenine Sulfate	20 mg	Magnesium Sulfate	0.4 g
Guanine Hydrochloride	20 mg	Sodium Chloride	20 mg
Uracil	20 mg	Ferrous Sulfate	20 mg
Thiamine Hydrochloride	200 µg	Manganese Sulfate	20 mg
Calcium Pantothenate	200 µg		

Final pH 6.7 ± 0.2 at 25°C.

One hundred grams will make 2.6 liters single strength medium.

METHOD OF PREPARATION

1. Suspend 7.5 g Bacto Niacin Assay Medium in 100 ml distilled or deionized water and heat to boiling 2 – 3 minutes.
2. Agitate to disperse the slight precipitate that forms and distribute in 5 ml amounts into test tubes.
3. After addition of standard and test solutions according to the recommended concentrations, adjust the total volume of all tubes to 10 ml with distilled or deionized water.
4. Sterilize in the autoclave for 10 minutes at 15 lbs pressure (121°C).

PROCEDURE

Assay procedures can be followed as outlined in *USP XX*[6] or *AOAC Methods*.[7] Best results, using Bacto Niacin Assay Medium, are obtained through the use of the following procedure:

Stock cultures of the test organism *L. plantarum* ATCC® 8014 are prepared by stab inoculation of Bacto Lactobacilli Agar AOAC or Bacto Micro Assay Culture Agar. After 24 – 48 hours incubation at 35 – 37°C, the tubes are stored in the refrigerator. Transplants are made at monthly intervals in triplicate.

Inoculum for assay is prepared by subculturing from a stock culture of *L. plantarum* ATCC® 8014 to 10 ml of Bacto Lactobacilli Broth AOAC or Bacto Micro Inoculum Broth. After 18 – 24 hours incubation at 35 – 37°C, the cells are centrifuged under aseptic conditions and the supernatant liquid decanted. The cells are resuspended in 10 ml sterile 0.85% NaCl solution and again centrifuged. The cells are resuspended in 10 ml sterile 0.85% NaCl and finally diluted 1:100 with 0.85% sterile NaCl. One drop of this latter suspension is used to inoculate each of the assay tubes (10 ml).

It is essential that a standard curve be set up for each separate assay since conditions of autoclaving, temperature of incubation, etc., which influence the standard curve readings, cannot be duplicated exactly from time to time. The standard curve is obtained by using niacin at levels of 0.0, 0.025, 0.05, 0.1, 0.15, 0.2 and 0.25 μg niacin per assay tube (10 ml). Bacto Niacin Assay Medium may be used for both turbidimetric and acidimetric analyses. Turbidimetric readings should be made after 16 – 18 hours incubation at 35 – 37°C. Acidimetric determinations are best made following 72 hours incubation at 35 – 37°C.

The concentrations of niacin required for the preparation of the standard curve may be prepared by dissolving 0.05 g of niacin in 1000 ml of distilled water, giving a stock solution of 50 μg per ml. Dilute the stock solution by adding 1 ml to 999 ml distilled water. Use 0.0, 0.5, 1, 2, 3, 4 and 5 ml per tube. Other standard concentrations may be used provided the standard falls within the limits specified by the AOAC.

STORAGE
Bacto Niacin Assay Medium

2 – 8°C, preferably in a container with calcium chloride or other desiccant.

QUALITY CONTROL
Identity Specifications

Dehydrated powder: light yellow, homogeneous, free-flowing
Reaction of 3.75% solution: pH 6.7 ± 0.2 at 25°C
Prepared medium: light amber, clear, may have a slight precipitate

REFERENCES
1. J. Biol. Chem., 13:675, 1941.
2. Ind. & Eng. Chem., Ann. Ed., 15:471, 1943.
3. Biochem. J., 38:314, 1944.
4. U.S. Pharmacopeia/National Formulary, USPXX/NFXV, Rockville: U.S. Pharmocopeial Convention, p. 923 – 925, 1980.
5. AOAC Methods, 13th Ed., Washington, D.C.: AOAC, p. 760, 1980.

PACKAGING
Bacto Niacin Assay Medium   100 g   0322-15-4

BACTO PANTHENOL ASSAY MEDIUM
BACTO PANTHENOL SUPPLEMENT

INTENDED USE
Bacto Panthenol Assay Medium and Bacto Panthenol Supplement, slight modifications of the formulae of DeRitter and Ruben,[1] are used in the microbiological assay of panthenol using *Gluconobacter oxydans* subsp. *suboxydans* ATCC® 621H (formerly *Acetobacter suboxydans* ATCC® 621H) as the test organism.

PRINCIPLES
Bacto Panthenol Assay Medium with Bacto Panthenol Supplement added is a medium free from panthenol or its analogs but containing all other nutrients and vitamins essential for the cultivation of *Gluconobacter oxydans* subsp. *suboxydans* ATCC® 621H (*Acetobacter suboxydans* ATCC® 621H). The addition of pantoic acid in increasing specified concentrations gives a growth response that can be measured turbidimetrically.

FORMULA

BACTO PANTHENOL ASSAY MEDIUM
DEHYDRATED
Ingredients per liter

Bacto Dextrose	15 g	p-Aminobenzoic Acid	2 mg
Bacto Vitamin Assay		Thiamine Hydrochloride	2 mg
Casamino Acids	2 g	Riboflavin	2 mg
Charcoal Treated Casitone	10 g	Pyridoxine Hydrochloride	2 mg
Sodium Citrate	2 g	Folic Acid	20 µg
L-Tryptophane	0.2 g	Biotin	16 µg
L-Cystine	0.15 g	Magnesium Sulfate	0.8 g
Adenine Sulfate	10 mg	Sodium Chloride	40 mg
Guanine Hydrochloride	10 mg	Ferrous Sulfate	40 mg
Uracil	10 mg	Manganous Sulfate	0.16 g
Beta-Alanine	2 mg	Monopotassium Phosphate	1 g
Liver Concentrate	40 mg	Dipotassium Phosphate	1 g
Nicotinic Acid	2 mg		

Final pH 6.0 ± 0.2 at 25°C.

One hundred grams will make 6.06 liters single strength medium.

BACTO PANTHENOL SUPPLEMENT

Bacto Glycerol	33 g
Sorbitan Monooleate Complex	2 g
Lactic Acid U.S.P.	0.68 g
Distilled or Deionized Water	71.5 ml

Each vial will make 0.4 liters of complete single strength medium.

METHOD OF PREPARATION
1. Suspend 33 g Bacto Panthenol Assay Medium in 900 ml distilled or deionized water and heat to boiling.
2. Distribute in 4.5 ml amounts into 50 ml flasks.
3. After addition of standard and test solutions according to the recommended concentrations, adjust the total volume in all flasks to 9.5 ml with distilled or deionized water.
4. Sterilize in the autoclave for 10 minutes at 15 lbs pressure (121°C).
5. Cool flasks to room temperature and aseptically add 0.5 ml Bacto Panthenol Supplement to each flask.

PROCEDURE
Inoculate 2 flasks of 50 ml each containing 10 ml sterile single strength Bacto Panthenol Assay Medium with Bacto Panthenol Supplement, and 4 µg/ml pantoic acid added with *Gluconobacter oxydans* subsp *suboxydans* ATCC® 621H carried on Bacto Lactobacilli Agar AOAC. Incubate on a suitable shaker (160 – 300 rpm) for 20 – 24 hours at 30°C. Transfer the two cultures into a sterile test tube and centrifuge 30 minutes. Decant the supernatant liquid, resuspend the sedimented organisms in 10 ml sterile 0.85% NaCl. Use 1 drop of this suspension to inoculate each assay flask.

It is essential that a standard curve be constructed each time an assay is run since conditions of autoclaving, temperature of incubation, etc., which influence the growth response, cannot be duplicated exactly from time to time. The standard curve is obtained by using pantoic acid at levels of 0.0, 0.2, 0.4, 0.6, 0.8, 1.0, 1.2, 1.6 and 2 µg per assay flask.

The concentration of pantoic acid required for the preparation of the standard curve may be prepared by the following procedure: Dissolve 69.2 mg of pure panthenol in distilled water, adjust to pH 6.0, and dilute to 1 liter (1 ml contains the equivalent of 50 µg of pantoic acid). Autoclave 8 ml of this solution with 8 ml 0.1N NaOH for 30 minutes at 15 lbs pressure (121°C). Cool, add distilled water, adjust reaction to pH 6.0 with 0.1N HCl and dilute to 100 ml. This stock solution contains 4 µg pantoic acid per ml. To make the standard solution dilute 10 ml of the stock solution with 90 ml distilled water. This standard solution contains 0.4 µg of pantoic acid per ml. Use 0.0, 0.5, 1, 1.5, 2, 2.5, 3, 4 and 5 ml per flask (50 ml).

Following inoculation the cultures are incubated on a suitable shaker at 30°C for 18 – 24 hours at 160 to 300 rpm and placed in the refrigerator to stop growth. The growth is then measured turbidimetrically using any suitable turbidimeter.

STORAGE

Bacto Panthenol Assay Medium	2 – 8°C, preferably in a container with calcium chloride or other desiccant.
Bacto Panthenol Supplement	2 – 8°C

QUALITY CONTROL

Identity Specifications

Dehydrated powder:	light yellow, homogeneous, free-flowing
Reaction of 1.65% solution:	pH 6.0 ± 0.2 at 25°C
Prepared medium:	colorless, clear, may have a slight precipitate
Appearance of supplement:	colorless, clear liquid

REFERENCE

1. Anal. Chem., 21:823, 1949.

PACKAGING

Bacto Panthenol Assay Medium	100 g	0994-15-1
Bacto Panthenol Supplement	12 × 20 ml	0212-64-7

BACTO PANTOTHENATE ASSAY MEDIUM

INTENDED USE

Bacto Pantothenate Assay Medium, prepared according to the formula given in the *U.S. Pharmocopeia XV*,[1] is used in the microbiological assay of pantothenic acid and its salts using *Lactobacillus plantarum* ATCC® 8014 as the test organism. It differs from Bacto Pantothenate Medium AOAC USP, in that it does not contain Tween® 80 (Sorbitan Monooleate Complex).

FORMULA

BACTO PANTOTHENATE ASSAY MEDIUM
DEHYDRATED

Ingredients per liter

Bacto Vitamin Assay Casamino Acids	10 g	Niacin	1 mg
Bacto Dextrose	40 g	Pyridoxine	800 µg
Sodium Acetate	20 g	p-Aminobenzoic Acid	200 µg
L-Cystine	0.4 g	Biotin	0.8 µg
DL-Tryptophane	0.2 g	Monopotassium Phosphate	1 g
Adenine Sulfate	20 mg	Dipotassium Phosphate	1 g
Guanine Hydrochloride	20 mg	Magnesium Sulfate	0.4 g
Uracil	20 mg	Sodium Chloride	20 mg
Thiamine Hydrochloride	200 µg	Ferrous Sulfate	20 mg
Riboflavin	400 µg	Manganese Sulfate	20 mg

Final pH 6.7 ± 0.1 at 25°C.

One hundred grams will make 2.74 liters single strength medium.

METHOD OF PREPARATION

1. Suspend 7.3 g Bacto Pantothenate Assay Medium in 100 ml distilled or deionized water and heat to boiling 2 – 3 minutes.
2. Agitate to disperse the slight precipitate that forms and distribute in 5 ml amounts into test tubes.
3. After addition of standard solutions and test solutions according to the recommended concentrations, adjust the total volume of all tubes to 10 ml with distilled or deionized water.
4. Sterilize in the autoclave for 10 minutes at 15 lbs pressure (121°C).

PROCEDURE

Stock cultures of the test organism, *L. plantarum* ATCC® 8014, are prepared in triplicate or more by stab inoculation of Bacto Pantothenate Culture Agar USP. Following incubation for 16 – 24 hours at any selected temperature between 30°C and 37°C, but held constant to within ±0.5°C, the tubes are stored at 2 – 8°C. Prepare a fresh stab of stock culture every week and do not use a culture older than 1 week for transferring to broth for inoculation.

Inoculum for the assay is prepared by subculturing from a suitable stock culture of *L. plantarum* ATCC® 8014 on Bacto Pantothenate Culture Agar USP into a tube containing 10 ml of sterile single strength Bacto Pantothenate Assay Medium supplemented with pantothenate. The medium is prepared by dissolving 36.5 g of the dehydrated medium and 20 μg of pantothenate in 1000 ml of distilled water, distributing in tubes, and sterilizing in the autoclave for 15 minutes at 15 lbs pressure (121°C). After 18 – 24 hours incubation at 30 – 37°C the cells are centrifuged under aseptic conditions and the supernatant liquid decanted. The cells are resuspended in 10 ml sterile 0.85% NaCl. The cell suspension is then diluted 1:100 with sterile 0.85% NaCl. The cell suspension so obtained is the inoculum. Inoculate each tube aseptically with 1 drop of the inoculum.

It is essential that a standard curve be set up for each assay since conditions of autoclaving, temperature of incubation, etc., which influence the standard curve readings, cannot be duplicated exactly from time to time. The standard curve is obtained by using calcium pantothenate solution at levels to give 0.0, 0.01, 0.02, 0.03, 0.04, 0.05, 0.06, 0.07, 0.08, 0.09 and 0.1 μg pantothenic acid per assay tube (10 ml). Turbidimetric determinations are made after 16 – 24 hours incubation at any selected temperature between 30°C and 37°C, but held constant to within ±0.5°C. Acidimetric determinations are made after 72 hours incubation at 30 – 37°C. A standard curve is then constructed and the unknown determined by interpolation.

The concentration of pantothenic acid required for the preparation of the standard curve may be prepared by dissolving 50 mg dried calcium pantothenate in about 500 ml distilled water, 10 ml 0.2N acetic acid and 100 ml 0.2N sodium acetate. Dilute with additional water to make calcium pantothenate concentration 43.47 μg per ml. One ml equals 40 μg pantothenic acid.

This solution is then diluted further by adding 25 ml to 500 ml distilled water, 10 ml 0.2N acetic acid and 100 ml 0.2N sodium acetate, then dilute to 1 liter with distilled water to make a stock solution containing 1.0 μg pantothenic acid per ml. The standard solution is made by diluting 2 ml of the stock solution to 100 ml with distilled water to make a solution containing 0.02 μg pantothenic acid per ml. Use 0.0, 0.5, 1.0, 1.5, 2.0, 2.5, 3.0, 3.5, 4.0, 4.5 and 5.0 ml per assay tube. The stock solution is prepared fresh daily.

STORAGE

Bacto Pantothenate Assay Medium   2 – 8°C, preferably in a
container with calcium
chloride or other desiccant.

QUALITY CONTROL

Identity Specifications

Dehydrated powder: white to light yellow, homogeneous, free-flowing
Reaction of 3.65% solution: pH 6.7 ± 0.1 at 25°C
Prepared medium: light amber, clear, may have a slight precipitate

REFERENCE

1. USP XV, 1955.

PACKAGING

Bacto Pantothenate Assay Medium   100 g   0604-15-3

BACTO PANTOTHENATE MEDIUM AOAC USP

INTENDED USE

Bacto Pantothenate Medium AOAC USP is prepared for use in the microbiological assay of pantothenic acid or pantothenate according to the procedures of the Calcium Pantothenate Assay in the *USP XX*[1] and Pantothenate Acid Assay in *AOAC Methods*, 13th Ed.[2] using *Lactobacillus plantarum* ATCC® 8014 as the test organism. The formula is a slight modification of the medium of Loy.

PRINCIPLES

Bacto Pantothenate Medium AOAC USP is a medium free from pantothenic acid or pantothenate, but containing all other nutrients and vitamins essential for the cultivation of *Lactobacillus plantarum* ATCC® 8014. The addition of calcium pantothenate in specified increasing concentrations gives a growth response that can be measured acidimetrically or turbidimetrically.

FORMULA

BACTO PANTOTHENATE MEDIUM AOAC USP
DEHYDRATED

Ingredients per liter

Bacto Dextrose	40 g	Manganese Sulfate	20 mg
Sodium Acetate	20 g	Adenine Sulfate	20 mg
Bacto Vitamin Assay		Guanine Hydrochloride	20 mg
Casamino Acids	10 g	Uracil	20 mg
Dipotassium Phosphate	1 g	Riboflavin	400 µg
Monopotassium Phosphate	1 g	Thiamine Hydrochloride	200 µg
L-Cystine	0.4 g	Biotin	0.8 µg
L-Tryptophane	0.1 g	p-Aminobenzoic Acid	200 µg
Magnesium Sulfate	0.4 g	Nicotinic Acid	1 mg
Sodium Chloride	20 mg	Pyridoxine Hydrochloride	800 µg
Ferrous Sulfate	20 mg	Sorbitan Monooleate Complex	0.1 g

Final pH 6.7 ± 0.1 at 25°C.

One hundred grams will make 2.7 liters single strength medium.

METHOD OF PREPARATION

1. Suspend 7.3 g Bacto Pantothenate Medium AOAC USP in 100 ml distilled or deionized water and heat to boiling 2 – 3 minutes.
2. Agitate to disperse the slight precipitate that forms and distribute in 5 ml amounts into test tubes.
3. After addition of standard solutions and test solutions according to the recommended concentrations, adjust the total volume of all tubes to 10 ml with distilled or deionized water.
4. Sterilize in the autoclave for 10 minutes at 15 lbs pressure (121°C).

PROCEDURE

Follow the assay procedures as outlined in *USP XX* or *AOAC Methods,* 13th Ed.

Stock cultures of *L. plantarum* ATCC® 8014 are prepared by stab inoculation of Bacto Lactobacilli Agar AOAC for the AOAC procedure or Bacto Pantothenate Culture Agar USP for the USP procedure. Following incubation at 30 – 37°C (held constant to within ±0.5°C) for 16 – 24 hours, the tubes are kept in the refrigerator. Fresh stab cultures are prepared every week and cultures are not used for preparing the inoculum if more than a week old. Inoculum for assay is prepared by culturing from a stock culture of *L. plantarum* ATCC® 8014 into a tube of sterile single strength Bacto Pantothenate Medium AOAC USP (10 ml) containing 0.02 μg per ml pantothenic acid or pantothenate. After 16 – 24 hours incubation at 30 – 37°C (held constant to within ±0.5°C) the cell suspension obtained is the inoculum. One drop of this suspension is used to inoculate each assay tube.

Solutions of Calcium Pantothenate USP Reference Standard or equivalent or pantothenic acid are prepared according to *USP XX* or *AOAC Methods,* 13th Ed. Generally, satisfactory results are obtained with the standard curve by using pantothenic acid at levels of 0.0, 0.005, 0.01, 0.015, 0.02, and 0.025 μg per assay tube (10 ml) for the AOAC procedure or calcium pantothenate may be used at standard levels of 0.0, 0.01, 0.02, 0.03, 0.04, and 0.05 μg per assay tube for the USP procedure. Bacto Pantothenate Medium AOAC USP may be used for both turbidimetric and acidimetric analysis in the AOAC procedure and for turbidimetric analysis only in the USP procedure. Turbidimetric readings should be made after 16 – 24 hours incubation at 30 – 37°C (held constant to within ±0.5°C). Acidimetric determinations are made following 72 hours incubation at 30 – 37°C (held constant to within ±0.5°C).

The concentration of pantothenic acid or calcium pantothenate required for the preparation of the standard curve may be prepared by dissolving 50 mg dried calcium pantothenate in about 500 ml distilled water, 10 ml 0.2N acetic acid and 100 ml 0.2N sodium acetate. Dilute with additional water to make calcium pantothenate concentration 43.47 μg per ml for the AOAC procedure or dilute to 50 μg per ml for the USP procedure. At 43.47 μg per ml one ml should equal 40 μg pantothenic acid.

This solution is then diluted further by adding 25 ml to 500 ml distilled water, 10 ml 0.2N acetic acid and 100 ml 0.2N sodium acetate. This solution is then diluted to 1 liter with distilled water which makes a stock solution containing 1 μg pantothenic acid per ml. The standard solution is made by diluting 5 ml of the stock solution to 1000 ml distilled water to obtain solution containing 0.005 μg pantothenic acid per ml. Use 0.0, 1, 2, 3, 4, and 5 ml per assay tube. For the USP procedure dilute the 50 μg per ml solution with distilled water to make a standard concentration of 0.01 μg per ml. Other standard concentrations may be used provided the standard falls within the limits specified by the USP or AOAC.

STORAGE
Bacto Pantothenate Medium AOAC USP

2 – 8°C, preferably in a
container with calcium
chloride or other desiccant.

QUALITY CONTROL
Identity Specifications
Dehydrated powder: white, homogeneous, free-flowing
Reaction of 3.65% solution: pH 6.7 ± 0.1 at 25°C
Prepared medium: light amber, clear, may have a slight precipitate

REFERENCES
1. U.S. Pharmacopeia/National Formulary, USP XX/NF XV, Rockville: U.S. Pharmacopeial Convention, p. 889 – 890, 1980.
2. AOAC Methods, 13th Ed., Washington, D.C.: AOAC, p. 764 – 765, 1980.

PACKAGING
Bacto Pantothenate Medium AOAC USP   100 g   0816-15-7

BACTO PYRIDOXINE Y MEDIUM

INTENDED USE
Bacto Pyridoxine Y Medium, patterned after the formulation of Campling and Nixon[1] and modified according to the work of Hurley[2] and Parrish, Loy and Kline,[3] is used in the microbiological assay of pyridoxine using *Saccharomyces uvarum* ATCC® 9080 (formerly *Saccharomyces carlsbergensis* ATCC® 9080) as the test organism.

PRINCIPLES
Bacto Pyridoxine Y Medium is free from pyridoxine but contains all other nutrients and vitamins essential for the growth of *Saccharomyces uvarum* ATCC® 9080 (*S. carlsbergensis* ATCC® 9080). The addition of pyridoxine in specified increasing concentrations gives a growth response that can be measured turbidimetrically or acidimetrically.

FORMULA
BACTO PYRIDOXINE Y MEDIUM
DEHYDRATED
Ingredients per liter

L-Asparagine	4 g	Inositol	5 mg
L-Histidine Hydrochloride	20 mg	Boric Acid	200 µg
DL-Methionine	40 mg	Monopotassium Phosphate	3 g
DL-Tryptophane	40 mg	Magnesium Sulfate	1 g
DL-Isolecuine	40 mg	Ammonium Sulfate	4 g
DL-Valine	40 mg	Calcium Chloride	0.49 g
Bacto Dextrose	40 g	Potassium Iodide	200 µg
Thiamine Hydrochloride	400 µg	Ammonium Molybdate	40 µg
Calcium Pantothenate	400 µg	Manganese Sulfate	80 µg
Nicotinic Acid	400 µg	Copper Sulfate	90 µg
Biotin Salt	8 mg	Zinc Sulfate	80 µg
Riboflavin	20 mg	Ferrous Sulfate	500 µg

Final pH 4.4 ± 0.2 at 25°C.

One hundred grams will make 3.77 liters of single strength medium.

METHOD OF PREPARATION

1. Suspend 5.3 g Bacto Pyridoxine Y Medium in 100 ml distilled or deionized water and heat to boiling 2 – 3 minutes.
2. Agitate to disperse the slight precipitate that forms and distribute in 5 ml amounts into flasks.
3. After addition of standard solutions and test solutions according to the recommended concentrations, adjust the total volume of all flasks to 10 ml with distilled or deionized water.
4. Steam in the autoclave for 10 minutes at 100°C.

PROCEDURE

Stock cultures of *S. uvarum* ATCC® 9080 are carried on slants of Bacto Lactobacilli Agar AOAC. Following incubation at 25 – 30°C (held constant to within ±0.5°C) for 18 – 24 hours, the cultures are stored in the dark at 2 – 8°C. Fresh slant cultures are prepared every week and are not used for preparing the inoculum if more than a week old. Inoculum for assay is prepared by subculturing from a stock culture of *S. uvarum* ATCC® 9080 into a tube (10 ml) of single strength Bacto Pyridoxine Y Medium containing 1 ng per ml each of pyridoxal hydrochloride, pyridoxamine hydrochloride and pyridoxine hydrochloride. After 18 – 24 hours incubation at 25 – 30°C (held constant within ±0.5°C) the cells are centrifuged under aseptic conditions and the supernatant liquid decanted. The cells are resuspended in 10 ml sterile single strength Bacto Pyridoxine Y Medium. One drop of this suspension is then used to inoculate each of the assay flasks.

It is essential that a standard curve be set up for each separate assay since conditions of steaming, temperature of incubation, etc., which influence the standard curve readings, cannot be duplicated exactly from time to time. The standard curve is obtained by using pyridoxine hydrochloride at levels of 0, 2, 4, 6, 8 and 10 ng per flask (10 ml).

The concentrations of pyridoxine hydrochloride required for the preparation of the standard curve may be prepared as follows: Dissolve 50 mg dried pyridoxine hydrochloride in about 100 ml 25% ethyl alcohol and dilute with additional 25% alcohol to make 500 ml, then dilute by adding 2 ml to 998 ml 25% ethyl alcohol to make a stock solution containing 200 ng pyridoxine hydrochloride per ml. The stock solution is prepared fresh daily.

To make the standard solution dilute 1 ml of stock solution with 99 ml distilled water to make a solution containing 2 ng pyridoxine hydrochloride per ml. Use 0.0, 1, 2, 3, 4 and 5 ml per assay tube.

Following inoculation the tubes are incubated on a suitable shaker (about 100 strokes/minute) at 25 – 30°C for 22 hours and then steamed in the autoclave for 5 minutes to stop growth. The growth is then measured turbidimetrically using a suitable spectrophotometer at any specific wavelength between 540 and 660 nm.

STORAGE

Bacto Pyridoxine Y Medium   2 – 8°C

QUALITY CONTROL

Identity Specifications

Dehydrated powder: white to off-white, fine, free-flowing
Reaction of 2.65% solution: pH 4.4 ± 0.2 at 25°C
Prepared medium: almost colorless, clear, may have a slight precipitate

REFERENCES

1. J. Physiol., 126:71, 1954.
2. J. AOAC, 43:43, 1960.
3. ibid., 39:157, 1956.

PACKAGING

Bacto Pyridoxine Y Medium   100 g   0951-15-2

BACTO PYRIDOXINE ASSAY MEDIUM

INTENDED USE

Bacto Pyridoxine Assay Medium, patterned after the medium described by Stokes, Larsen, Woodward and Foster[1] and modified by Barton-Wright,[2] is used in the microbiological assay of pyridoxine using *Neurospora sitophilia* ATCC® 9276 as the test organism.

PRINCIPLES

Bacto Pyridoxine Assay Medium is free from pyridoxine but contains all other nutrients and vitamins essential for the growth of *Neurospora sitophila* ATCC® 9276. The addition of pyridoxine in specified increasing concentrations gives a growth response that can be measured gravimetrically.

FORMULA

BACTO PYRIDOXINE ASSAY MEDIUM
DEHYDRATED

Ingredients per liter

Bacto Sucrose	30 g	Ammonium Tartrate	10 g
Choline Chloride	10 mg	Sodium Dihydrogen Citrate	4 g
Nicotinic Acid	2 mg	Monopotassium Phosphate	5 g
Riboflavin	1 mg	Magnesium Sulfate	1 g
Thiamine Hydrochloride	10 mg	Sodium Chloride	0.2 g
Calcium Pantothenate	1 mg	Calcium Chloride	0.2 g
p-Aminobenzoic Acid	200 µg	Ferric Chloride	10 mg
Biotin	8 µg	Zinc Sulfate	4 mg

Final pH 4.5 ± 0.2 at 25°C.

One hundred grams will make 4 liters single strength medium.

METHOD OF PREPARATION

1. Suspend 5.0 g Bacto Pyridoxine Assay Medium in 100 ml distilled or deionized water and heat to boiling 2 – 3 minutes.
2. Agitate to disperse the slight precipitate that forms and distribute in 5 ml amounts into 50 ml flasks.
3. After addition of standard solutions or test solutions according to the recommended concentrations, adjust the total volume of all flasks to 10 ml with distilled or deionized water.
4. Sterilize in the autoclave for 15 minutes at 15 lbs pressure (121°C).

PROCEDURE

The test is run according to the method described by Stokes, Larsen, Woodward and Foster and modified by Barton-Wright. Remove 1 loop of spores from a 48 hour culture of *N. sitophilia* ATCC® 9276 on Bacto Neurospora Culture Agar and suspend in 100 ml sterile saline. Add 1 drop of this spore suspension to each flask. Incubate at 30°C for 5 days. At the end of the incubation period steam the flasks at 100°C for 5 minutes. Remove all the mycelium from the flask using a stiff wire needle or glass rod, press dry between paper towels, and roll into a small pellet. Dry the pellets at 100°C for 2 hours in a vacuum. A glazed porcelain spot plate is convenient for handling the mycelium during drying and weighing. A standard curve is then constructed from the weights obtained and the unknown determined by interpolation.

It is essential that a standard curve be constructed each time an assay is run since conditions of autoclaving, temperature of incubation, etc., which influence the standard curve readings, cannot be duplicated exactly from time to time. The standard curve is obtained by using pyridoxine at levels of 0.0, 0.2, 0.4, 0.6, 0.8 and 1 μg per assay flask (10 ml). Using Bacto Pyridoxine Assay Medium the most effective assay range has proved to be between 0.2 and 0.8 μg pyridoxine.

The concentrations of pyridoxine required for the preparation of the standard curve may be prepared by dissolving 0.1 g of pyridoxine in 1000 ml of distilled water, giving a stock solution of 100 μg per ml. Dilute the stock solution by adding 1 ml to 499 ml distilled water. Use 0.0, 1, 2, 3, 4 and 5 ml per flask. This stock solution is prepared fresh daily.

STORAGE
Bacto Pyridoxine Assay Medium   2 – 8°C

QUALITY CONTROL
Identity Specifications

Dehydrated powder: white to off-white, homogeneous, free-flowing
Reaction of 2.5% solution: pH 4.5 ± 0.2 at 25°C
Prepared medium: very light yellow, clear

REFERENCES
1. J. Biol. Chem., 150:17, 1943.
2. Analyst, 70:283, 1945.

PACKAGING
Bacto Pyridoxine Assay Medium   100 g   0324-15-2

BACTO RIBOFLAVIN ASSAY MEDIUM

INTENDED USE
Bacto Riboflavin Assay Medium, a slight modification of the medium described by Snell and Strong,[1] prepared according to the specifications given in the USP,[2] National Formulary[3] and AOAC,[4] except that it contains 2% anhydrous dextrose in the basal medium instead of 6%, is used in the microbiological assay of riboflavin using Lactobacillus casei ATCC® 7469 as the test organism.

PRINCIPLES
Bacto Riboflavin Assay Medium is free from riboflavin but contains all other nutrients and vitamins essential for the growth of Lactobacillus casei ATCC® 7469. The addition of riboflavin in specified increasing concentrations gives a growth response that can be measured turbidimetrically or acidimetrically.

Additional dextrose may be added if desired. However, in our experience this causes carmelization during the preparation of the medium with adverse effects on the assay procedure. The lower concentration of dextrose has given parallel results.

FORMULA

BACTO RIBOFLAVIN ASSAY MEDIUM
DEHYDRATED
Ingredients per liter

Photolyzed Peptone	22 g	Monopotassium Phosphate	1 g
Yeast Supplement	2 g	Magnesium Sulfate	0.4 g
Bacto Dextrose	20 g	Sodium Chloride	20 mg
Sodium Acetate	1.8 g	Ferrous Sulfate	20 mg
L-Cystine	0.2 g	Manganese Sulfate	20 mg
Dipotassium Phosphate	1 g		

Final pH 6.8 ± 0.2 at 25°C.

One hundred grams will make 4 liters single strength medium.

METHOD OF PREPARATION

1. Suspend 4.8 g Bacto Riboflavin Assay Medium in 100 ml distilled or deionized water and heat to boiling 2 – 3 minutes.
2. Agitate to disperse the slight precipitate that forms and distribute in 5 ml amounts into test tubes.
3. After addition of standard solutions and test solutions according to the recommended concentrations, adjust the total volume of all tubes to 10 ml with distilled or deionized water.
4. Sterilize in the autoclave for 10 minutes at 15 lbs pressure (121°C).

PROCEDURE

Follow assay procedures as outlined in *AOAC Methods,* 13th Ed. Levels of riboflavin used in the determination of the standard curve should be prepared according to this reference or according to the following procedure.

Stock cultures of *L. casei* ATCC® 7469 are prepared by stab inoculation into 10 ml of Bacto Lactobacilli Agar AOAC or Bacto Micro Assay Culture Agar. After 24 – 48 hours incubation at 35 – 37°C, the stock cultures are kept in the refrigerator. Transplants are made at monthly intervals in triplicate. Inoculum for assay is prepared by subculturing from a stock culture of *L. casei* ATCC® 7469 into 10 ml of Bacto Lactobacilli Broth AOAC or Bacto Micro Inoculum Broth. Following incubation for 24 hours at 35 – 37°C, the culture is centrifuged under aseptic conditions and the supernatant liquid decanted. The cells are resuspended in 10 ml sterile 0.85% NaCl. The cell suspension is then diluted 1:20 with sterile 0.85% NaCl. One drop of this latter suspension is then used to inoculate each of the assay tubes.

Bacto Riboflavin Assay Medium may be used for both turbidimetric and acidimetric determinations. Turbidimetric readings should be made after 18 – 24 hours incubation at 35 – 37°C, whereas acidimetric determinations are best made after 72 hours incubation at 35 – 37°C. We have found the most effective assay range, using Bacto Riboflavin Assay Medium, to be between 0.025 and 0.15 µg riboflavin.

It is essential that a standard curve be constructed each time an assay is run, since conditions of autoclaving, temperature of incubation, etc., which influence the standard curve readings, cannot be duplicated exactly from time to time. The standard curve is obtained by using Riboflavin USP Reference Standard or equivalent at levels of 0.0, 0.025, 0.05, 0.075, 0.1, 0.15, 0.2 and 0.3 µg riboflavin per assay tube (10 ml).

The concentration of riboflavin required for the preparation of the standard curve may be prepared by dissolving 0.1 g of Riboflavin USP Reference Standard or equivalent in 1000 ml of distilled water by heating, giving a stock solution of 100 μg per ml. Dilute the stock solution by adding 1 ml to 999 ml distilled water. Use 0.0, 0.25, 0.5, 0.75, 1, 1.5, 2 and 3 ml of the diluted stock solution per tube. The stock solution is prepared fresh daily.

STORAGE
Bacto Riboflavin Assay Medium   2 – 8°C, preferably in a
                                             container with calcium
                                             chloride or other desiccant.

QUALITY CONTROL
Identity Specifications

Dehydrated powder: light tan, homogeneous, free-flowing
Reaction of 2.4% solution: pH 6.8 ± 0.2 at 25°C
Prepared medium: light amber, clear, may have a slight precipitate

REFERENCES
1. Ind. & Eng. Chem. Ann. Ed., 11:346, 1939.
2. USP XV, 1955.
3. Natl. Formulary, 9th Ed., 746, 1950.
4. AOAC Methods, 13th Ed., Washington, D.C.: AOAC, p. 760, 765, 766, 1980.

PACKAGING
Bacto Riboflavin Assay Medium   100 g   0325-15-1

BACTO THIAMINE ASSAY MEDIUM

INTENDED USE
Bacto Thiamine Assay Medium, prepared according to the formula given by Sarett and Cheldelin,[1] is used in the microbiological assay of thiamine using *Lactobacillus fermentum* ATCC® 9338 (*Lactobacillus fermenti* ATCC® 9338) as the test organism.

PRINCIPLES
Bacto Thiamine Assay Medium is free from thiamine but contains all other nutrients and vitamins essential for the growth of *Lactobacillus fermentum* ATCC® 9338 (*L. fermenti* ATCC® 9338). The addition of thiamine in specified increasing concentrations gives a growth response that can be measured turbidimetrically.

FORMULA
BACTO THIAMINE ASSAY MEDIUM
DEHYDRATED
Ingredients per liter

Photolyzed Peptone	22 g	Niacin	200 μg
Bacto Vitamin Assay		Pyridoxine Hydrochloride	200 μg
Casamino Acids	5 g	p-Aminobenzoic Acid	200 μg
Bacto Dextrose	40 g	Folic Acid	0.5 μg
Sodium Acetate	15 g	Biotin	0.8 μg
L-Cystine	0.2 mg	Dipotassium Phosphate	1 g
Adenine Sulfate	20 mg	Monopotassium Phosphate	1 g
Guanine Hydrochloride	20 mg	Magnesium Sulfate	0.4 g
Uracil	20 mg	Sodium Chloride	20 mg
Riboflavin	200 μg	Ferrous Sulfate	20 mg
Calcium Pantothenate	200 μg	Manganese Sulfate	20 mg

Final pH 6.5 ± 0.2 at 25°C.

One hundred grams will make 2.35 liters single strength medium.

METHOD OF PREPARATION

1. Suspend 8.5 g Bacto Thiamine Assay Medium in 100 ml distilled or deionized water and heat to boiling 2 – 3 minutes.
2. Agitate to disperse the slight precipitate that forms and distribute in 5 ml amounts into test tubes.
3. After addition of standard solutions and test solutions according to the recommended concentrations, adjust the total volume of all tubes to 10 ml with distilled or deionized water.
4. Sterilize in the autoclave for 5 minutes at 15 lbs pressure (121°C).

PROCEDURE

Stock cultures of the test organism, *L. fermentum* ATCC® 9338, are prepared by stab inoculation of Bacto Lactobacilli Agar AOAC or Bacto Micro Assay Culture Agar. After 24 – 48 hours incubation at 35 – 37°C, the tubes are kept in the refrigerator. Transplants are made at monthly intervals in triplicate.

Inoculum for assay is prepared by subculturing from a stock culture to 10 ml of Bacto Lactobacilli Broth AOAC or Bacto Micro Inoculum Broth. After 16 – 18 hours incubation 35 – 37°C, the cells are centrifuged under aseptic conditions, and the supernatant liquid is decanted. The cells are resuspended in 10 ml sterile 0.85% NaCl. One half (0.5) ml of this suspension is then added to 100 ml of sterile 0.85% NaCl. One drop of this suspension is then used to inoculate the assay tubes.

The tubes for the standard curve, which should be set up each time an assay is run, contain 0.0, 0.005, 0.01, 0.015, 0.02, 0.03, 0.04 and 0.05 µg of thiamine hydrochloride per tube (10 ml). It is essential that a standard curve be run with each assay since conditions of heating, temperature of incubation, etc., which influence the standard curve readings, cannot be duplicated exactly from time to time. Using Bacto Thiamine Assay Medium we have found the most effective assay range to be between 0.005 and 0.03 µg thiamine.

The concentration of thiamine required for the preparation of the standard curve may be prepared by dissolving 0.1 g of thiamine in 1000 ml of distilled water, giving a stock solution of 100 µg per ml. Dilute the stock solution by adding 1 ml to 99 ml of distilled water. This solution is further diluted by adding 1 ml to 99 ml distilled water to give the final solution. Use 0.0, 0.5, 1, 1.5, 2, 3, 4 and 5 ml of this final solution per tube. The stock solution is prepared fresh daily.

After 20 – 24 hours incubation at 35 – 37°C, *L. fermentum* ATCC® 9338 is capable of utilizing the pyrimidine and thiazole moieties of the thiamine molecule. It is essential, therefore, that the growth response be measured turbidimetrically prior to this time. The tubes should be incubated at 35 – 37°C for 16 – 18 hours, and then placed in the refrigerator for 15 – 30 minutes in order to stop growth. The growth can then be measured by any suitable nephelometric method.

STORAGE

Bacto Thiamine Assay Medium   2 – 8°C

QUALITY CONTROL

Identity Specifications

Dehydrated powder: light tan, homogeneous, free-flowing
Reaction of 4.25% solution: pH 6.5 ± 0.2 at 25°C
Prepared medium: light amber, clear, may have a slight precipitate

REFERENCE
1. J. Biol. Chem., 155:153, 1944.

PACKAGING
Bacto Thiamine Assay Medium   100g   0326-15-0

BACTO THIAMINE ASSAY MEDIUM LV

INTENDED USE
Bacto Thiamine Assay Medium LV, patterned after APT medium, is prepared according to the formula of Deibel, Evans and Niven,[1] is used in the microbiological assay of thiamine using *Lactobacillus viridescens* ATCC® 12706 as the test organism.

HISTORY
Nutritional studies by Evans and Niven[2] on the heterofermentative lactobacilli causing greening in cured meat products, indicated that thiamine was one of the essential vitamins for the growth of these organisms. Deibel, Evans and Niven[3] described APT medium for their cultivation. They reported that these lactobacilli required at least 10 ng per ml for growth in contrast to 0.2 to 3 ng per ml for thiamine-requiring streptococci, leuconostocs and staphylococci. They suggested that these lactobacilli requiring large amounts of thiamine, might be employed in microbiological assay procedures. Using a thiamine-free yeast extract, these authors in 1957, described a medium for the microbiological assay of thiamine using *Lactobacillus viridescens* ATCC® 12706 as the test organism. A quantitative growth response, which can be measured turbidimetrically, was obtained by the addition of thiamine ranging from 1 – 15 ng per ml of medium, and followed by incubation at 30°C for 16 – 20 hours. Adjustment of the reaction to pH 6.0 permitted autoclaving of the medium for 5 minutes at 15 lbs pressure (121°C) without loss of thiamine. Thiamine assays using Bacto Thiamine Assay Medium LV have proven to be accurate and readily reproducible.

PRINCIPLES
Bacto Thiamine Assay Medium LV is free from thiamine but contains all other nutrients and vitamins essential for the growth of *Lactobacillus viridescens* ATCC® 12706. The addition of thiamine in specified increasing concentrations gives a growth response that can be measured turbidimetrically.

FORMULA
BACTO THIAMINE ASSAY MEDIUM LV
DEHYDRATED
Ingredients per liter

Thiamine-Free Yeast Extract	10 g	Sodium Chloride	10 g
Bacto Tryptone	20 g	Magnesium Sulfate	1.6 g
Bacto Dextrose	20 g	Manganese Chloride	0.28 g
Sodium Citrate	10 g	Ferrous Sulfate	80 mg
Dipotassium Phosphate	10 g	Sorbitan Monoleate Complex	2 g

Final pH 6.0 ± 0.2 at 25°C.

One hundred grams will make 2.4 liters of single strength medium.

METHOD OF PREPARATION
1. Suspend 8.4 g Bacto Thiamine Assay Medium LV in 100 ml distilled or deionized water and heat to boiling 2 – 3 minutes.

2. Agitate to disperse the slight precipitate that forms and distribute in 5 ml amounts into test tubes.
3. After addition of standard solutions and test solutions according to the recommended concentrations, adjust the total volume of all tubes to 10 ml with distilled or deionized water.
4. Sterilize in the autoclave for 5 minutes at 15 lbs pressure (121°C).

PROCEDURE

Stock cultures of the test organism, *Lactobacillus viridescens* ATCC® 12706, are prepared by stab inoculation of Bacto APT Agar or Bacto Lactobacilli Agar AOAC. After 24 – 48 hours incubation at 30°C, the tubes are kept in the refrigerator. Transplants are made at monthly intervals in triplicate.

Inoculum for assay is prepared by subculturing from a stock culture to 10 ml Bacto APT Broth or Bacto Lactobacilli Broth AOAC. After 16 – 20 hours incubation at 30°C, the cells are centrifuged under aseptic conditions, and the supernatant liquid decanted. The cells are resuspended in 10 ml sterile 0.85% NaCl. One ml of this cell suspension is then added to 100 ml sterile 0.85% NaCl. One drop of this suspension is then used to inoculate the assay tube.

It is essential that a standard curve be set up for each separate assay since conditions of autoclaving, temperature of incubation, etc., which influence the standard curve readings, cannot be duplicated exactly. The standard curve is obtained by using thiamine at levels of 0.0, 1, 2.5. 5, 7.5, 10, 15, 20, and 25 ng of thiamine hydrochloride per tube (10 ml). This is obtained by using 0.0, 0.2, 0.5, 1, 1.5, 2, 3, 4 and 5 ml of the standard solution, which contains 5 ng (0.005 µg) thiamine hydrochloride per ml.

The solution for preparing the standard curve may be prepared as follows:
A. Dissolve 50 mg of thiamine hydrochloride in 500 ml distilled water (100 µg/ml).
B. Add 1 ml of A to 99 ml distilled water (1 µg/ml).
C. Add 1 ml of B to 199 ml distilled water to give a final concentration of 5 ng (0.005 µg) per ml.

Following incubation at 30°C for 16 – 20 hours the growth response is measured turbidimetrically. We have found the most effective assay range to be between 2.5 and 20 ng per tube.

STORAGE

Bacto Thiamine Assay Medium LV   2 – 8°C

QUALITY CONTROL

Identity Specifications

Dehydrated powder: white, homogeneous, free-flowing
Reaction of 4.2% solution: pH 6.0 ± 0.2 at 25°C
Prepared medium: light amber, clear, may have a slight precipitate

REFERENCES

1. Paper Presented 57th Gen. Meeting Soc. Am. Bact., Det., 1957.
2. J. Bact., 62:599, 1951.
3. Bact. Proc., 1955.

PACKAGING

Bacto Thiamine Assay Medium LV   100 g   0808-15-7

BACTO VITAMIN B₁₂ ASSAY MEDIUM

INTENDED USE

Bacto Vitamin B_{12} Assay Medium, prepared according to the formula described by Capp, Hobbs and Fox,[1] is used in the microbiological assay of vitamin B_{12} using *Lactobacillus leichmannii* ATCC® 4797 as the test organism.

PRINCIPLES

Bacto Vitamin B_{12} Assay Medium is a vitamin B_{12}-free medium containing all other nutrients and vitamins essential for the cultivation of *Lactobacillus leichmannii* ATCC® 4797. To obtain a standard curve, USP Cyanocobalamin Reference is added in specified increasing concentrations giving a growth response that can be measured acidimetrically or turbidimetrically.

FORMULA

BACTO VITAMIN B₁₂ ASSAY MEDIUM
DEHYDRATED

Ingredients per liter

Bacto Vitamin Assay Casamino Acids	12 g	Calcium Pantothenate	200 µg
Bacto Dextrose	40 g	Pyridoxine Hydrochloride	4 mg
Sodium Acetate	20 g	p-Aminobenzoic Acid	200 µg
L-Cystine	0.2 g	Biotin	10 µg
DL-Tryptophane	0.2 g	Folic Acid	100 µg
Adenine	20 mg	Sorbitan Monooleate Complex	2 g
Guanine	20 mg	Dipotassium Phosphate	1 g
Uracil	20 mg	Monopotassium Phosphate	1 g
Xanthine	1 mg	Magnesium Sulfate	0.4 g
Thiamine Hydrochloride	2 mg	Sodium Chloride	20 mg
Riboflavin	2 mg	Ferrous Sulfate	20 mg
Niacin	2 mg	Manganese Sulfate	20 mg

Final pH 6.3 ± 0.2 at 25°C.

One hundred grams will make 2.6 liters of single strength medium.

METHOD OF PREPARATION

1. Dissolve 7.6 g Bacto Vitamin B_{12} Assay Medium in 100 ml distilled or deionized water and heat to boiling 2 – 3 minutes.
2. Agitate to disperse the slight precipitate that forms and distribute in 5 ml amounts into test tubes.
3. After addition of standard and test solutions according to the recommended concentrations, adjust the total volume of all tubes to 10 ml with distilled or deionized water.
4. Sterilize in the autoclave for 5 minutes at 15 lbs pressure (121°C).

PROCEDURE

A stock culture of *L. leichmannii* ATCC® 4797 is prepared by stab inoculation of Bacto Lactobacilli Agar AOAC or Bacto B_{12} Culture Agar USP. Following incubation at 37°C for 24 – 48 hours, the tubes are stored in the refrigerator. Transplants are made at 2 week intervals. Inoculum for the assay is prepared by subculturing from a stock culture of *L. leichmannii* ATCC® 4797 into a tube containing 10 ml of Bacto Lactobacilli Broth AOAC or Bacto B_{12} Inoculum Broth USP. After 24 hours incubation at 37°C, the cells are centrifuged under aseptic conditions and the supernatant liquid decanted. The cells are washed by resuspending in 10 ml of sterile 0.85% NaCl solution and centrifuging.

The washing is repeated for a total of 3 times. Finally the cells are resuspended in 10 ml of sterile 0.85% NaCl. The cell suspension is then diluted 1:100 with sterile 0.85% NaCl. One drop is used to inoculate each assay tube.

It is essential that a standard curve be constructed each time an assay is run, since conditions of autoclaving, temperature of incubation, etc., which influence the standard curve readings cannot be duplicated exactly from time to time.

The concentrations required for the prepartion of the standard curve are obtained by adding sufficient 25% ethanol to an accurately weighed amount of USP Cyanocobalamin Reference Standard to make a solution containing 1.0 μg of cyanocobalamin per ml. This stock solution is stored in the refrigerator and should be used within 60 days. In the preparation of the standard curve further dilutions of this stock solution (1 μg/ml) are made as follows:
A. Add 1 ml stock solution to 99 ml distilled water (1 ml = 10 ng)
B. Add 1 ml of A to 199 ml distilled water (1 ml = 0.05 ng).

An acceptable standard curve can be obtained by using the USP Cyanocobalamin Reference Standard at levels of 0.0, 0.025, 0.05, 0.1, 0.15, 0.2, and 0.25 ng per assay tube. This is accomplished by adding 0, 0.5, 1, 2, 3, 4, and 5 ml of the 0.05 ng/ml solution per assay tube and sufficient distilled or deionized water to make 10 ml volume per tube.

A standard concentration should be used which after incubation gives a transmittance value at the 5 ml level of not less than that which corresponds to a dry cell weight of 1.25 mg (see *USP XX* for method of calibration of spectrophotometer and determination of dry cell weight). For the acidimetric method a standard concentration should be used which after incubation will give a titration at the 5 ml level of 8 – 12 ml 0.1N sodium hydroxide.

Following inoculation, tubes are incubated at 37°C for 24 – 30 hours for turbidimetric determinations and then placed in a refrigerator at 2 – 8°C for 15 – 20 minutes in order to stop growth. The growth can then be measured by any suitable nephelometric method. Acidimetric determinations of growth are made after 72 hours of incubation at 37°C. The curve is then constructed from the values obtained.

STORAGE
Bacto Vitamin B$_{12}$ Assay Medium   2 – 8°C, preferably in a
                                                       container with calcium
                                                       chloride or other desiccant

QUALITY CONTROL
Identity Specifications
Dehydrated powder: off-white to yellow, homogeneous, free-flowing
Reaction of 3.8% solution: pH 6.3 ± 0.2 at 25°C
Prepared medium: light amber, clear, may have a slight precipitate

REFERENCE
1. Capps, Hobbs and Fox, J. Biol. Chem., 178:517, 1949.

PACKAGING
Bacto Vitamin B$_{12}$ Assay Medium   100 g   0360-15-7

AMINO ACID ASSAY MEDIA

INTENDED USE
Bacto Amino Acid Assay media, prepared according to the formulation of Steel, Sauberlich, Reynolds and Baumann,[1] are used in the microbiological assay of amino acids using *Pediococcus acidilacti* ATCC® 8042 (*Pediococcus cerevisiae* ATCC® 8042) and/or *Streptococcus faecium* ATCC® 8043 (*Streptococcus faecalis* ATCC® 8043).

Bacto Tryptophan Assay Medium is the only exception. It is prepared according to the formula of Greene and Black[2] and uses *Lactobacillus plantarum* ATCC® 8014 as the test organism.

PRINCIPLES
Each amino acid assay medium contains all factors essential for the growth of the test organism except for the amino acid under assay. The addition of the amino acid in specified increasing concentrations gives a growth response by the test organism.

FORMULAE
All amino acid assay media except Bacto Tryptophan Assay Medium contains the following general formulation. Omit the particular amino acid to be assayed from the medium.

Bacto Dextrose	50 g	Biotin	2 µg
Sodium acetate	40 g	Folic acid	20 µg
Ammonium chloride	6 g	Glycine	0.2 g
Monopotassium phosphate	1.2 g	DL-Alanine	0.4 g
Dipotassium phosphate	1.2 g	Bacto Asparagine	0.8 g
Magnesium sulfate	0.4 g	L-Aspartic Acid	0.2 g
Ferrous sulfate	20 mg	L-Proline	0.2 g
Manganese sulfate	40 mg	DL-Serine	0.1 g
Sodium chloride	20 mg	DL-Tryptophane	80 mg
Adenine sulfate	20 mg	L-Cystine	0.1 g
Guanine hydrochloride	20 mg	L-Glutamic acid	0.6 g
Uracil	20 mg	L-Histidine hydrochloride	0.124 g
Xanthine	20 mg	DL-Phenylalanine	0.2 g
Thiamine hydrochloride	1 mg	DL-Threonine	0.4 g
Pyrodoxine hydrochloride	2 mg	L-Tyrosine	0.2 g
Pyridoxamine hydrochloride	600 mg	DL-Valine	0.5 g
Pyridoxal hydrochloride	600 mg	L-Lysine hydrochloride	0.5 g
Calcium pantothenate	1 mg	DL-Methionine	0.2 g
Riboflavin	1 mg	DL-Isoleucine	0.5 g
Nicotinic acid	2 mg	DL-Leucine	0.5 g
p-Aminobenzoic acid	200 µg	L-Arginine hydrochloride	0.484 g

Final pH at 25°C is 6.7 ± 0.2 for all formulations.

One hundred grams will make 1.9 liters of single strength medium.

BACTO TRYPTOPHAN ASSAY MEDIUM
DEHYDRATED

Ingredients per liter

Bacto Vitamin Assay		Riboflavin	400 µg
Casamino Acids	12 g	p-Aminobenzoic Acid	200 µg
Bacto Dextrose	40 g	Niacin	200 µg
Sodium Acetate	20 g	Biotin	0.8 µg
L-Cystine	0.2 g	Dipotassium Phosphate	1 g
Adenine Sulfate	20 mg	Monopotassium Phosphate	1 g
Guanine Hydrochloride	20 mg	Magnesium Sulfate	0.4 g
Uracil	20 mg	Sodium Chloride	20 mg
Thiamine Hydrochloride	200 µg	Ferrous Sulfate	20 mg
Calcium Pantothenate	200 µg	Manganese Sulfate	20 mg
Pyridoxine Hydrochloride	400 µg		

Final pH 6.7 ± 0.2 at 25°C.

One hundred grams will make 2.6 liters single strength medium.

METHOD OF PREPARATION

1. Suspend 10.5 g in 100 ml distilled or deionized water. If Bacto Tryptophan Assay Medium is being prepared, use 7.5 g.
2. Heat to boiling 2 – 3 minutes. Agitate to disperse the slight precipitate that forms and distribute in 5 ml amounts into test tubes.
3. Add standard solutions and test solutions according to recommended procedures listed below and adjust the total volume of all tubes to 10 ml with distilled or deionized water.
   Bacto Cystine Assay Medium is an exception. The L-cystine solution, as well as the solution under assay, is sterilized **separately** in a similar manner and then added aseptically.
4. Sterilize in the autoclave for 10 minutes at 15 lbs pressure (121°C). **Note: Oversterilization of any of these media will give unsatisfactory results.**

PROCEDURE

Stock Culture and Inoculum

Stock cultures of the test organisms are prepared by stab inoculation into tubes of prepared Bacto Lactobacilli Agar AOAC or Bacto Micro Assay Culture Agar. After 24 hours incubation at 35 – 37°C, the tubes are stored in the refrigerator. The transplants are made at monthly intervals in triplicate.

The inoculum for assay is prepared by subculturing the test organism into 10 ml Bacto Lactobacilli Broth AOAC or Bacto Micro Inoculum Broth. After 16 – 24 hours incubation at 35 – 37°C, the cells are centrifuged under aseptic conditions and the supernatant liquid decanted. The cells are resuspended in 10 ml sterile 0.85% NaCl solution. This 10 ml cell suspension is further diluted with the appropriate amount of sterile 0.85% NaCl solution. (See table.) One drop of the diluted inoculum suspension is used to inoculate each of the assay tubes.

Amino Acid Solution

Prepare stock solutions of each amino acid as described in the table. If the DL form is used, twice the concentration of the amino acid is required. The stock solutions should be prepared fresh for each assay.

Increasing amounts of the standard or the unknown and sufficient distilled or deionized water to give a total volume of 10 ml per tube are added to the tubes containing 5 ml of the rehydrated medium. The appropriate volumes of the standards and their final concentrations are listed in the table.

The growth response can be measured acidimetrically or turbidimetrically. The turbidimetric readings should be made after 16 – 20 hours incubation at 35 – 37°C while the acidimetric determinations are made after 72 hours incubation at 35 – 37°C.

It is essential that a standard curve be constructed each time an assay is run, since conditions of autoclaving, temperature of incubation, etc., which influence the standard curve readings, cannot be duplicated exactly from time to time.

STORAGE

Dehydrated Media

These products are very hygroscopic. Keep bottle tightly stoppered. It is recommended that the bottle be stored in a container with calcium chloride or other dehydrating material at 2 – 8°C.

Assay Medium	Test Culture	Preparation of Inoculum Dilution (cell suspension) + (sterile 0.85% NaCl)	Preparation of Amino Acid Stock Solution (amino acid) + (distilled H₂O)		Standard Working Solution (stock solution) + (distilled H₂O)	Volume of Standard Working Solution (ml/10 ml tube)	Final Amino Acid Concentration µg/10 ml
Arginine Assay Medium	Pediococcus acidilacti ATCC® 8042	5 ml + 95 ml	L-arginine hydrochloride	100 mg + 100 ml	1 ml + 99 ml	0, 0.5, 1, 1.5, 2, 2.5, 3, 4, 5	0.0, 5, 10, 15, 20, 25, 30, 40, 50
				200 mg* + 100 ml	1 ml + 99 ml		0, 10, 20, 30, 40, 50, 60, 80, 100
Cystine Assay Medium	Pediococcus acidilacti ATCC® 8042	1 ml + 19 ml	L-cystine	1 g + 100 ml + 1 ml HCl heated, then cooled, add up to 1000 ml	1 ml + 99 ml	0, 0.5, 1, 1.5, 2, 2.5, 3, 4, 5	0.0, 5, 10, 15, 20, 25, 30, 40, 50
Glutamic Acid Assay Medium	Pediococcus acidilacti ATCC® 8042	5 ml + 100 ml	L-glutamic acid	200 mg + 100 ml	5 ml + 100 ml	0, 0.5, 1, 1.5, 2, 2.5, 3, 3.5, 4	0.0, 50, 100, 150, 200, 250, 300, 350, 400
Histidine Assay Medium	Pediococcus acidilacti ATCC® 8042	5 ml + 100 ml	L-histidine hydrochloride	50 mg + 100 ml	1 ml + 100 ml	0, 0.5, 1, 2, 3, 4, 5	0.0, 2.5, 5, 10, 15, 20, 25
			DL-histidine hydrochloride	100 mg + 100 ml	1 ml + 100 ml		0.0, 5, 10, 20, 30, 40, 50
Isoleucine Assay Medium	Pediococcus acidilacti ATCC® 8042	5 ml + 95 ml	DL-isoleucine	2.4 g + 1000 ml	1 ml + 99 ml	0, 0.5, 1, 1.5, 2, 2.5, 3, 4, 5	0.0, 12, 24, 36, 48, 60, 72, 96, 120

Assay Medium	Test Culture	Preparation of Inoculum Dilution (cell suspension) + (sterile 0.85% NaCl)	Preparation of Amino Acid Stock Solution (amino acid) + (distilled H$_2$O)		Standard Working Solution (stock solution) + (distilled H$_2$O)	Volume of Standard Working Solution (ml/10 ml tube)	Final Amino Acid Concentration µg/10 ml
Leucine Assay Medium	Pediococcus acidilacti ATCC® 8042	5 ml + 95 ml	L-leucine	2 g +1000 ml	1 ml + 99 ml	0, 0.5, 1, 1.5, 2, 2.5, 3, 4, 5	0.0, 10, 20, 30, 40, 50, 60, 80, 100
			DL-leucine	4 g +1000 ml	1 ml + 99 ml		0.0, 20, 40, 60, 80, 100, 120, 160, 200
Lysine Assay Medium	Pediococcus acidilacti ATCC® 8042	1 ml + 19 ml	L-lysine	6 g +1000 ml	1 ml + 99 ml	0, 0.5, 1, 1.5, 2, 2.5, 3, 4, 5	0.0, 30, 60, 90, 120, 150, 180, 240, 300
Methionine Assay Medium	Pediococcus acidilacti ATCC® 8042	1 ml + 19 ml	DL-methionine	1.2 g +1000 ml	1 ml + 99 ml	0, 0.5, 1, 2, 2.5, 3, 4, 5	0.0, 6, 12, 18, 24, 30, 36, 48, 60
Phenylalanine Assay Medium	Pediococcus acidilacti ATCC® 8042	5 ml + 95 ml	L-phenylalanine	1 g +1000 ml	1 ml + 99 ml	0, 0.5, 1, 1.5, 2, 2.5, 3, 4, 5	0.0, 5, 10, 15, 20, 25, 30, 40, 50
			DL-phenylalanine	2 g +1000 ml	1 ml + 99 ml		0.0, 10, 20, 30, 40, 50, 60, 80, 100

Assay Medium	Test Culture	Preparation of Inoculum Dilution (cell suspension) + (sterile 0.85% NaCl)	Preparation of Amino Acid Stock Solution (amino acid) + (distilled H2O)		Standard Working Solution (stock solution) + (distilled H2O)	Volume of Standard Working Solution (ml/10 ml tube)	Final Amino Acid Concentration µg/10 ml
Threonine Assay Medium	*Streptococcus faecium* ATCC® 8043	1 ml + 99 ml	L-threonine	50 mg + 100 ml	1 ml + 99 ml	0, 0.5, 1, 2, 3, 4, 5	0.0, 2.5, 5, 10, 15, 20, 25
			DL-threonine	100 mg + 100 ml	1 ml + 99 ml		0.0, 5, 10, 20, 30, 40, 50
Tryptophane Assay Medium	*Lactobacillus plantarum* ATCC® 8014	1 ml + 99 ml	L-tryptophane	0.1 g + 1000 ml	4 ml + 96 ml	0, 0.5, 1, 1.5, 2, 3, 5	0.0, 2, 4, 6, 8, 10, 12
			DL-tryptophane	0.2 g + 1000 ml	4 ml + 96 ml		0.0, 4, 8, 12, 16, 20, 24
Tyrosine Assay Medium	*Pediococcus acidilacti* ATCC® 8042	1 ml + 19 ml	L-tyrosine	2 g + 1000 ml	1 ml + 99 ml	0, 0.5, 1, 1.5, 2, 2.5, 3, 4, 5	0.0, 10, 20, 30, 40, 50, 60, 80, 100
			DL-tyrosine	4 g + 1000 ml	1 ml + 99 ml		0.0, 20, 40, 60, 80, 100, 120, 160, 200
Valine Assay Medium	*Pediococcus acidilacti* ATCC® 8042	5 ml + 95 ml	L-valine	100 mg + 100 ml	1 ml + 100 ml	0, 0.5, 1, 2, 3, 4, 5	0.05, 5, 10, 20, 30, 40, 50
			DL-valine	200 mg + 100 ml	1 ml + 100 ml		0.0, 10, 20, 40, 60, 80, 100

*For acidimetric assay stock solution should be twice the concentration for optimum growth curves.

Because of its formulation, this material has a tendency to lump in the bottle. It may be removed easily by breaking with a spatula or other sharp instrument. The cultural value of the medium is in no way impaired by this lumping.

QUALITY CONTROL

Identity Specifications

Dehydrated powder: white to off-white, homogeneous with a tendency to clump

Reaction of 5.25%* solution: pH 6.7 ± 0.2 at 25°C

Prepared medium: light to light-medium amber, clear, may have a slight precipitate

*3.75% for Bacto Tryptophan Assay Medium

REFERENCES

1. J. Biol. Chem., 177:533, 1949.
2. J. Biol. Chem., 155:1, 1944.

PACKAGING

Bacto Arginine Assay Medium	100 g	0466-15-0
Bacto Cystine Assay Medium	100 g	0467-15-9
Bacto Glutamic Acid Assay Medium	100 g	0961-15-0
Bacto Histidine Assay Medium	100 g	0992-15-3
Bacto Isoleucine Assay Medium	100 g	0437-15-6
Bacto Leucine Assay Medium	100 g	0421-15-4
Bacto Lysine Assay Medium	100 g	0422-15-3
Bacto Methionine Assay Medium	100 g	0423-15-2
Bacto Phenylalanine Assay Medium	100 g	0469-15-7
Bacto Threonine Assay Medium	100 g	0323-15-3
Bacto Tryptophan Assay Medium	100 g	0327-15-9
Bacto Tyrosine Assay Medium	100 g	0468-15-8
Bacto Valine Assay Medium	100 g	0991-15-4

WL NUTRIENT AND DIFFERENTIAL MEDIA

INTENDED USE

Bacto WL Nutrient Broth and Bacto WL Nutrient Medium are recommended for the cultivation of yeasts, molds and bacteria encountered in brewing and industrial fermentation processes.

Bacto WL Differential Medium, also used in the microbiological control processes in the fermentation industry, permits the unrestricted growth of bacteria and inhibits development of yeasts and molds.

HISTORY/PRINCIPLES

Bacto WL Nutrient media are prepared according to the formulae described by Green and Gray.[1,2] In their study of various fermentation processes, Green and Gray pointed out the inadequacy of the microscopic count in fermentation control procedures. An exhaustive study of the method of examination of worts, beers, and liquid yeast and similar fermentation products led to the development of two media; one containing no selective agent and the other, a differential medium containing the antibiotic Actidione® (cycloheximide) as a selective agent.

Bacto WL Nutrient media permit the development of yeast. In those instances in which the number of yeast cells is comparatively small, certain bacteria can be detected. Green and Gray[2] reported that counts of viable bakers' yeast may be made on the WL nutrient medium at pH 5.5.. If the reaction is adjusted to pH 6.5, the count of bakers' and distillers' yeast may be made. In making microbial counts using these media, the temperature and time of incubation will vary depending on the various materials under investigation. Temperatures of 25°C are generally employed with brewing materials and 30°C for bakers' yeast and alcohol fermentation mash analyses. Incubating periods run from 2 to 7 days, depending on the flora encountered. Incubation periods of 10 to 14 days may be used in some cases.

Bacto WL Differential Medium has the same formula as Bacto WL Nutrient Medium, with the addition of 0.004 g of Actidione® per liter. This inhibits the development of yeasts without interfering with the development of bacteria generally encountered in beers.

A reliable count of bacteria can be obtained at pH 5.5. To obtain estimations of beer cocci and lactic rods, plates should be incubated under anaerobic conditions. For estimation of acetic acid rods and termobacteria (very small rods occuring in wort as described by Linder in about 1900 as *Termobacterium lutescens, iridescens* and *erythrimum*) incubate under aerobic conditions. To analyze bakers' yeast and alcohol fermentation mashes, the reaction is adjusted to pH 6.5. Plates containing dilutions of bakers' yeast are incubated aerobically, while those from alcoholic fermentation mashes are incubated anaerobically.

FORMULAE

BACTO WL NUTRIENT BROTH
DEHYDRATED

Ingredients per liter

Bacto Yeast Extract	4 g	Calcium Chloride	0.125 g
Bacto Casitone	5 g	Magnesium Sulfate	0.125 g
Bacto Dextrose	50 g	Ferric Chloride	0.0025 g
Monopotassium Phosphate	0.55 g	Manganese Sulfate	0.0025 g
Potassium Chloride	0.425 g	Bacto Brom Cresol Green	0.022 g

Final pH 5.5 ± 0.2 at 25°C.

One pound will make 7.5 liters of final medium.
Rehydrate with 60 grams/liter.

BACTO WL NUTRIENT MEDIUM
DEHYDRATED

Ingredients per liter

Bacto Yeast Extract	4 g	Magnesium Sulfate	0.125 g
Bacto Casitone	5 g	Ferric Chloride	0.0025 g
Bacto Dextrose	50 g	Manganese Sulfate	0.0025 g
Monopotassium Phosphate	0.55 g	Bacto Agar	20 g
Potassium Chloride	0.425 g	Bacto Brom Cresol Green	0.022 g
Calcium Chloride	0.125 g		

Final pH 5.5 ± 0.2 at 25°C.

One pound will make 5.6 liters of final medium.
Rehydrate with 80 grams/liter.

BACTO WL DIFFERENTIAL MEDIUM
DEHYDRATED
Ingredients per liter

Bacto Yeast Extract	4 g		Magnesium Sulfate	0.125 g
Bacto Casitone	5 g		Ferric Chloride	0.0025 g
Bacto Dextrose	50 g		Manganese Sulfate	0.0025 g
Monopotassium Phosphate	0.55 g		Bacto Agar	20 g
Potassium Chloride	0.425 g		Bacto Brom Cresol Green	0.022 g
Calcium Chloride	0.125 g		Actidione® (cycloheximide)	0.004 g

Final pH 5.5 ± 0.2 at 25°C.

One pound will make 5.6 liters of final medium.
Rehydrate with 80 grams/liter.

METHOD OF PREPARATION
1. To rehydrate suspend appropriate amount in 1 liter cold distilled water and heat to boiling to dissolve completely.
2. Sterilize in the autoclave for 15 minutes at 15 lbs pressure (121°C).
3. To obtain a final reaction of pH 6.5 add the amount specified on the product label of a 1% solution of sodium carbonate per liter distilled water used for rehydration; dissolve and sterilize as indicated above.

STORAGE
Bacto WL Nutrient and Differential media	Below 30°C
Prepared media	2 – 8°C

QUALITY CONTROL
Identity Specifications

	WL Nutrient Broth	WL Nutrient Medium	WL Differential Medium
Dehydrated powder:	light beige w/blue tint, homogeneous, free-flowing	light tan w/blue tint, homogeneous, free-flowing	beige w/blue tint, homogeneous, free-flowing
Solution:	6% solution	8% solution	8% solution
Reaction:	pH 5.5 ± 0.2 at 25°C	pH 5.5 ± 0.2 at 25°C	pH 5.5 ± 0.2 at 25°C
Prepared medium:	blue, clear	blue-green, very slightly opalescent	greenish-blue, slightly opalescent

Typical Cultural Response in/on Bacto WL Media
After 40 – 48 Hours at 30°C (Bacteria at 35°C)

Organism	Growth	
	WL Nutrient Media	WL Differential Medium
Escherichia coli ATCC® 25922	fair to good	good
Lactobacillus fermentum ATCC® 9338	fair to good	good
Proteus mirabilis ATCC® 25933	fair to good	good
Saccharomyces cerevisiae ATCC® 9763	good	inhibited
Saccharomyces uvarum ATCC® 9080	good	inhibited

REFERENCES

1. Paper read at Am. Soc. of Brewing Chemists Meeting, Detroit, May, 1950.
2. Wallerstein Lab. Comm., 13:357, 1950.
3. Ibid., 14:169, 1951.

PACKAGING

Bacto WL Nutrient Broth	1 lb (454 g)	0471-01-9
Bacto WL Nutrient Medium	1 lb (454 g)	0424-01-7
Bacto WL Differential Medium	500 g	0425-17-8

BACTO WILKINS CHALGREN AGAR

INTENDED USE

Bacto Wilkins Chalgren Agar is used for susceptibility testing of anaerobic bacteria by the agar dilution procedure, and for isolating and culturing anaerobic bacteria.

HISTORY/PRINCIPLES

Wilkins Chalgren agar was designed by Wilkins and Chalgren[1] for use in determining the minimal inhibitory concentrations (MIC's) of antibiotics for anaerobic bacteria by the agar dilution procedure. This medium is preferred over other previously used media, since it does not require the addition of blood for satisfactory growth of anaerobic bacteria of clinical significance. Both the medium and the procedure are described in the protocol used in the NCCLS Collaborative Study of the Proposed Reference Dilution Method of Antimicrobial Susceptibility Testing of Anaerobic Bacteria.[2] This procedure is essentially the same as described by Sutter and Washington[3] except that the Wilkins Chalgren agar is used in place of the other media which require enrichment.

FORMULA

BACTO WILKINS CHALGREN AGAR
DEHYDRATED

Ingredients per liter

Bacto Tryptone	10 g	L-Arginine	1 g
Bacto Peptone	10 g	Sodium Pyruvate	1 g
Bacto Yeast Extract	5 g	Hemin	5.0 mg
Bacto Dextrose	1 g	Vitamin K_1	0.5 mg
Sodium Chloride	5 g	Bacto Agar	15 g

Final pH 7.1 ± 0.1 at 25°C.

Five hundred grams will make 10.4 liters of medium.

METHOD OF PREPARATION

1. Suspend 48 g in 1 liter distilled or deionized water. Heat to boiling to dissolve completely.
2. Dispense as desired into tubes or flasks and autoclave for 15 minutes at 15 lbs pressure (121°C).
3. Cool in waterbath to 50 – 55°C before adding antibiotics.
4. Mix gently and dispense into Petri dishes.

STORAGE

Bacto Wilkins Chalgren Agar	Below 30°C
Prepared medium	2 – 8°C

QUALITY CONTROL

Identity Specifications

Dehydrated powder: light beige, homogeneous, free-flowing
Reaction of 4.8% solution: pH 7.1 ± 0.1 at 25°C
Prepared medium: light to medium amber, slightly opalescent

Typical Cultural Response on Bacto Wilkins Chalgren Agar After 40 – 48 Hours at 35°C, Anaerobically

Organism	Growth
Bacteroides fragilis ATCC® 25285	good to excellent
Bacteroides melanogenicus ATCC® 25611	good to excellent
Clostridium perfringens ATCC® 13123	good to excellent

REFERENCES

1. Wilkins, T. D., and Chalgren, S., Antimicrob. Agents Chemother. 14:384–399, 1978.
2. NCCLS Tenative Standard Reference Agar Dilution Procedure for Susceptibility Testing of Anaerobic Bacteria Vol. 2, No. 3, 1982.
3. Sutter, V. L., and Washington, II, J. A., Manual of Clinical Microbiology, Second Ed., 1974, Chapter 49.

PACKAGING

Bacto Wilkins Chalgren Agar   500 g   1805-17-6

BACTO WILSON BLAIR BASE

INTENDED USE

Bacto Wilson Blair Base is used with added selective agents for isolating *Salmonella*, particularly *Salmonella typhi*.

HISTORY/PRINCIPLES

Bacto Wilson Blair Base is the formulation recommended by the Illinois State Health Department for preparing Bacto Wilson Blair Agar.

The selective reagent as prepared by the Illinois State Health Department,[1] for use with Bacto Wilson Blair Base, is a modification of the bismuth sulfite reagent described by Hajna and Perry.[2] The advantage of the modification is that it is more easily prepared.

FORMULAE

BACTO WILSON BLAIR BASE
DEHYDRATED

Ingredients per liter

Bacto Beef Extract . . . . 5 g
Proteose Peptone No. 3, Difco . . . . 10 g
Bacto Dextrose . . . . 10 g
Sodium Chloride . . . . 5 g
Bacto Agar . . . . 30 g

Final pH 7.3 ± 0.2 at 25°C.

Five hundred grams will make 8.3 liters of medium.

Selective Reagent for Wilson Blair Base

Solution 1 — 40 g sodium sulfite anhydrous in 100 ml distilled water.
Solution 2 — 21 g sodium phosphate dibasic anhydrous/100 ml distilled water.

Solution 3 — 12.5 g bismuth ammonium citrate granular/100 ml distilled water.
Solution 4 — 0.96 g ferrous sulfate dried/20 ml distilled water with 2 drops hydrochloric acid.

METHOD OF PREPARATION
Basal Medium
1. Suspend 60 g in 1 liter distilled or deionized water and heat to boiling to dissolve completely.
2. Sterilize in the autoclave for 15 minutes at 15 lbs pressure (121°C).
3. To prepare the complete medium, aseptically add 70 ml of selective reagent and 4 ml of a 1% titrated solution of brilliant green. Mix thoroughly.
4. Dispense as desired.

Selective Reagent
The selective reagent consists of a combination of solutions 1 through 4. Each is made up separately, dissolved, then combined. Heat combined solution to boiling until a slate grey color develops. Allow to cool and store at room temperature in a closed rubber stoppered container. It is stable for up to one month.

The optimal amount of brilliant green to be added is based on the size, number and typical characteristics of colonies. If less than 4 ml is used, *Escherichia coli* colonies will be larger and more plentiful. With an additional amount (greater than 4 ml), *E. coli* colonies are inhibited, but recovery of pathogens may also be suppressed. A titration is recommended when organisms to be recovered are unknown.

STORAGE
Bacto Wilson Blair Base	Below 30°C
Prepared medium	2 – 8°C

QUALITY CONTROL
Identity Specifications

Dehydrated powder:	tan, homogeneous, free-flowing
Reaction of 6% solution:	pH 7.3 ± 0.2 at 25°C
Prepared medium:	medium amber, opalescent, may have a fine precipitate

Typical Cultural Response on Wilson Blair Agar
After 18 – 48 Hours at 35°C

Organism	Growth	Color of Colony
Escherichia coli ATCC® 25922	inhibited	—
Proteus mirabilis ATCC® 25933	good to excellent	green
Salmonella typhi ATCC® 19430	good to excellent	black with sheen
Salmonella typhimurium ATCC® 14028	good to excellent	black with sheen
Fecal specimen	variable	variable, *Salmonella* appear black if present

REFERENCES
1. Personal Communication, 1948.
2. J. Lab. Clin. Med., 23:1:85, 1938.

PACKAGING
Bacto Wilson Blair Base	500 g	0312-17-4

BACTO WORFEL-FERGUSON AGAR

INTENDED USE

Bacto Worfel-Ferguson Agar is recommended for the production of capsules by *Klebsiella* for use in the Neufeld (Quellung) reaction for the serological identification of these organisms.

HISTORY/PRINCIPLES

Bacto Worfel-Ferguson Agar is prepared according to the formula given by Edwards and Ewing.[1] The *Klebsiella* culture is inoculated onto an agar surface and incubated 18 – 24 hours at 37°C. In order to assure optimal capsule formation, cultures of klebsiellae should be grown on Bacto Worfel-Ferguson Agar. They should be incubated 18 – 24 hours at 37°C. The culture is examined for amount and size of capsule production in Bacto Loeffler Methylene Blue wet-mounts.

Only cultures possessing capsules of moderate size should be used in the Quellung reaction. Cultures with very small capsules render the Quellung reaction difficult to observe while extremely large capsules tend to obscure positive reactions. After the size of the capsules has been determined, the Quellung reaction is performed.

A complete discussion on the Quellung reaction is given under Bacto Klebsiella Pneumoniae antisera in this manual.

FORMULA

BACTO WORFEL-FERGUSON AGAR
DEHYDRATED

Ingredients per liter

Bacto Yeast Extract	2 g
Magnesium Sulfate	0.25 g
Potassium Sulfate	1 g
Sodium Chloride	2 g
Bacto Saccharose	20 g
Bacto Agar	15 g

Final pH 6.5 ± 0.2 at 25°C.

One pound will make 11.2 liters of medium.

METHOD OF PREPARATION

1. Suspend 40.3 g in 1 liter cold distilled or deionized water. Heat to boiling to dissolve completely.
2. Sterilize in the autoclave for 15 minutes at 15 lbs pressure (121°C).

STORAGE

Bacto Worfel-Ferguson Agar	Below 30°C
Prepared plates	2 – 8°C

QUALITY CONTROL

Identity Specifications

Dehydrated powder:	light beige, homogeneous, free-flowing
Reaction of 4.03% solution:	pH 6.5 ± 0.2 at 25°C
Prepared medium:	light to medium amber, slightly opalescent

Typical Cultural Response on Bacto Worfel-Ferguson Agar
After 18 – 24 Hours at 35°C

Organism	Growth
Enterobacter aerogenes ATCC® 15038	good to excellent
Escherichia coli ATCC® 25922	good to excellent
Klebsiella pneumoniae ATCC® 13883	good to excellent
Klebsiella pneumoniae ATCC® 23357	good to excellent

REFERENCES

1. Edwards, P. R., and Ewing, W. H., Identification of Enterobacteriaceae, Burgess Publ. Co., 1955.
2. J. Inf. Dis., 91:92, 1952.

PACKAGING

Bacto Worfel-Ferguson Agar	1 lb (454 g)	0924-01-2

BACTO WORT AGAR

INTENDED USE
Bacto Wort Agar is recommended for the cultivation and enumeration of yeasts.

HISTORY
Parfitt[1] reported satisfactory results with Bacto Wort Agar in his study of the influence of culture media on the mold and yeast counts of butter. In order to procure comparative yeast and mold counts, he suggested that dehydrated Bacto Whey Agar, Bacto Malt Agar or Bacto Wort Agar should be used.[2]

PRINCIPLES
Yeasts grow well upon culture media containing dextrose or maltose, particularly if the reaction is acid. The formula of Bacto Wort Agar closely duplicates the composition of wort, which is a favorable medium for the cultivation of yeasts. Growth promoting qualities have been increased by the addition of salts and other nutriments. The reaction of the medium is adjusted so that after sterilization, it will be pH 4.8, which is near the optimum for most yeasts. At the same time, it will inhibit most bacterial growth.

FORMULA

BACTO WORT AGAR
DEHYDRATED

Ingredients per liter

Bacto Malt Extract	15 g
Bacto Peptone	0.78 g
Maltose, Technical	12.75 g
Dextrin, Difco	2.75 g
Glycerol	2.35 g
Dipotassium Phosphate	1 g
Ammonium Chloride	1 g
Bacto Agar	15 g

Final pH 4.8 ± 0.1 at 25°C.

One pound will make 9 liters of final medium.

METHOD OF PREPARATION

1. Suspend 48 g in one liter distilled or deionized water and heat to boiling to dissolve completely.
2. Distribute in tubes or flasks and sterilize in the autoclave for 15 minutes at 15 lbs pressure (121°C).
3. Due to the high acidity of the medium, the heating process should be completed in as short a period of time as possible. Excessive heating causes a breaking down of the agar, resulting in an inability to solidify properly when cooled.
4. If a medium of solidity satisfactory for streaking is desired, it can be prepared by using 60 g of Bacto Wort Agar per 1 liter of water or by adding an extra 5 g of Bacto Agar per liter of medium.

STORAGE
Bacto Wort Agar   Below 30°C
Prepared plates   2 – 8°C

QUALITY CONTROL
Identity Specifications
Dehydrated powder: medium beige, homogeneous, free-flowing
Reaction of 4.8% solution: pH 4.8 ± 0.1 at 25°C
Prepared medium: medium to dark amber, opalescent, may have a slight flocculent precipitate

Typical Cultural Response on Bacto Wort Agar
After 40 – 48 Hours at 30°C

Organism	Growth
Aspergillus niger ATCC® 16404	good to excellent
Saccharomyces cerevisiae ATCC® 9763	good to excellent
Saccharomyces uvarum ATCC® 9080	good to excellent

REFERENCES
1. J. Dairy Science, 16:141, 1933.
2. Difco Manual Ninth Ed., 1953, p. 244.

PACKAGING
Bacto Wort Agar   500 g   0111-17-7

XL/XLD MEDIA
BACTO XL AGAR BASE
BACTO XLD AGAR

INTENDED USE
Bacto XL Agar Base is the xylose lysine agar base of Taylor[1,2,3] for the isolation, differentiation and enumeration of enteric bacteria.

Bacto XLD Agar is the xylose lysine agar base of Taylor[1,2,3] to which the sodium desoxycholate, sodium thiosulfate and ferric ammonium citrate have been added to make a complete medium. This selective differential plating medium is designed for direct isolation of *Shigella* and *Providencia* from stool specimens.

PRINCIPLES

Bacto XL Agar Base, the basal medium, is completed by supplementation with sodium thiosulfate and ferric ammonium citrate. The completed medium may be supplemented further with Bacto Sodium Desoxycholate, Bacto Brilliant Green or a combination of the two to make it selective for specific groups of enteric bacteria.

Bacto XL Agar Base, not being a selective medium, permits growth of most enteric bacteria. The author recommends the medium for use in making counts of enteric bacteria.

Bacto XLD Agar is prepared from the base by aseptically adding to the prepared Bacto XL Agar Base containing the thiosulfate and ferric ammonium citrate, 25 ml of a sterile 10% sodium desoxycholate solution per liter of liquid medium and mixing just before dispensing it into Petri plates.

Published literature describing the preparation of XLD agar necessitated the adjustment of the reaction to pH 6.9 after all ingredients were added. Experience with the XLD agar has shown that a pH of 7.4 rather than pH 6.9 gives best results. At the higher pH a smoother medium for use in plates is obtained.

Bacto XLD Agar is recommended by the author for general laboratory usage and is preferred for direct isolation of *Shigella* and *Providencia* from stool specimens.

XLBG agar can be prepared by adding 1.25 ml 1% aqueous solution brilliant green per liter Bacto XL Agar Base prior to autoclaving. Sterilize and cool to 55 – 60°C. Add 20 ml solution containing 34% sodium thiosulfate and 4% ferric ammonium citrate and dispense into sterile Petri plates.

XLBG agar is recommended for use in isolating *Salmonella* after selenite or tetrathionate enrichment. Coliforms and *Shigella* are inhibited on this medium.

Sodium desoxycholate in the formula provides inhibition of gram-positive bacteria. It also markedly inhibits the swarming of *Proteus* strains.

Differentiation of *Shigella* and *Salmonella* from nonpathogenic bacteria is accomplished by three reactions: xylose fermentation, lysine decarboxylation and hydrogen sulfide production. Xylose as a carbohydrate source in the formula differentiates *Shigella* and *Providencia*, which ferment xylose slowly or not at all, from the other enterics, which ferment xylose rapidly. *Salmonella* are further differentiated from nonpathogenic xylose fermenters by the lysine decarboxylase reaction. As the organisms rapidly exhaust the xylose and decarboxylate the lysine, a reversion to alkaline conditions stimulates the *Shigella* reaction. Lactose and sucrose, added in excess, prevent the lysine-positive coliforms from similarly reverting. The production of hydrogen sulfide under alkaline conditions results in formation of colonies with black centers, whereas under acidic conditions, this black precipitation is inhibited.

FORMULAE

BACTO XL AGAR BASE
DEHYDRATED

Ingredients per liter

Bacto Yeast Extract	3 g
L-Lysine	5 g
Bacto Xylose	3.75 g
Bacto Lactose	7.5 g
Bacto Saccharose	7.5 g
Sodium Chloride	5 g
Bacto Agar	15 g
Bacto Phenol Red	0.08 g

Final pH 7.4 ± 0.2 at 25°C.

Rehydrate with 47 g/liter
One pound will make 9.45 liters.

BACTO XLD AGAR
DEHYDRATED

Ingredients per liter

Bacto Yeast Extract	3 g
L-Lysine	5 g
Bacto Xylose	3.75 g
Bacto Lactose	7.5 g
Bacto Saccharose	7.5 g
Sodium Desoxycholate	2.5 g
Ferric Ammonium Citrate	0.8 g
Sodium Thiosulfate	6.8 g
Sodium Chloride	5 g
Bacto Agar	15 g
Bacto Phenol Red	0.08 g

Final pH 7.4 ± 0.2 at 25°C

Rehydrate with 57 g/liter
One pound will make 8 liters.

METHOD OF PREPARATION

BACTO XL AGAR BASE
1. Suspend 47 g Bacto XL Agar Base in 1 liter distilled or deionized water and heat to boiling to dissolve completely.
2. Sterilize in the autoclave for 15 minutes at 15 lbs pressure (121°C).
3. Cool to 55 – 60°C and aseptically add a 20 ml sterile aqueous solution containing 34% sodium thiosulfate and 4% ferric ammonium citrate.
4. Swirl flask to obtain a uniform solution and dispense into sterile Petri dishes.

BACTO XLD AGAR
1. Suspend 57 g Bacto XLD Agar in 1 liter distilled or deionized water and heat to boiling to dissolve completely. AVOID OVERHEATING. DO NOT AUTOCLAVE.
2. Cool to 55 – 60°C and dispense as desired.

PROCEDURE AND RESULTS

Plates of Bacto XLD Agar inoculated with specimens should be incubated at 35°C for 18 – 24 hours. Longer incubation may lead to false-positive results.

Differentiation of Enteric Bacteria

Bacto XLD Agar

Group or Genus	Colonies
Salmonella*	red, black centers
Shigella, Providencia	red
Some Pseudomonas	falsely positive red
Escherichia, Citrobacter,	yellow
Klebsiella, Proteus**,	
Enterobacter (Aerobacter)	
Alkalescens-Dispar	

*S. paratyphi A, S. choleraesuis, S. pullorum and S. gallinarum — May give red colonits without black centers.
**Some Proteus strains will give black-centered colonies on Bacto XLD Agar.

STORAGE

Bacto XL Agar Base and XLD Agar	Below 30°C
Prepared medium	2 – 8°C

QUALITY CONTROL

Identity Specifications

	XL Agar Base	XLD Agar
Dehydrated powder:	pink, homogeneous, free-flowing	pink, homogeneous, free-flowing
Reaction of appropriate solution:	pH 7.4 ± 0.2	pH 7.4 ± 0.2
Prepared medium:	red, very slightly opalescent	red, very slightly opalescent

Typical Cultural Response on Bacto XLD Agar After 18 – 24 Hours at 35°C

Organism	Growth	Color of Colony
Enterobacter aerogenes ATCC® 13048	good to excellent	yellow w/bile precipitate
Escherichia coli ATCC® 25922	poor to fair	yellow
Proteus mirabilis ATCC® 25933	good to excellent	yellow, may have black center
Providencia alcalifaciens ATCC® 9886	good to excellent	red
Salmonella arizonae ATCC® 13314	good to excellent	red w/black centers
Salmonella typhimurium ATCC® 14028	good to excellent	red w/black centers
Shigella sonnei ATCC® 25931	good to excellent	red
Staphylococcus aureus ATCC® 25923	inhibited	—

REFERENCES

1. Bact. Proc. Abst., ASM, Washington, D.C., 1964.
2. Tech. Bull. Regist. Med. Techs., 35:161, 1965.
3. Am. J. Clin. Path., 44:471, 1965.

PACKAGING

Bacto XL Agar Base	1 lb (454 g)	0555-01-8
Bacto XLD Agar	1 lb (454 g)	0788-01-7
	1/4 lb (114 g)	0788-02-6
	5 lb (2.27 kg)	0788-05-3
	10 kg	0788-08-0

BACTO YM AGAR
BACTO YM BROTH

INTENDED USE

Bacto YM media are recommended for the isolation and cultivation of yeasts, molds and other aciduric microorganisms.

HISTORY/PRINCIPLES

Bacto YM Agar and Bacto YM Broth are prepared according to the formulae published by Wickerham.[1,2]

Selective media may be prepared from either Bacto YM Agar or Bacto YM Broth by acidifying the media to between pH 3.0 to 4.0 or by the addition of antibiotics. YM broth may be acidified prior to sterilization. YM agar should be sterilized without adjustment of reaction, and sterile acid added to sterile melted medium at 45°C to 55°C. Acidified agar should not be heated. Antibiotics generally are added to the sterile media under aseptic conditions.

Other fungistatic materials, such as sodium propionate and diphenyl may be added to YM agar to eliminate molds and permit the enumeration of yeasts to mixed populations.

Wickerham[1,2] suggested that YM broth adjusted to pH 3.0 to 4.0 might be used as an enrichment medium for yeast from populations consisting of bacteria and molds, by adding a layer of sterile paraffin oil 1 cm deep on the surface of the inoculated broth. The culture is incubated until growth appears and then streaked to YM agar to obtain isolated yeast colonies. This technique favors the development of fermentative species. A method favoring the development of fermentative as well as oxidative strains is to place acidified inoculated YM broth on a rotary shaker for 1 or 2 days. This greatly favors the development of yeast cells while the sporulation of molds is prevented so that yeast may be isolated readily by streaking on YM agar.

FORMULAE

BACTO YM AGAR
DEHYDRATED

Ingredients per liter

Bacto Yeast Extract . . . . 3 g
Malt Extract, Difco . . . . 3 g
Bacto Peptone . . . . 5 g
Bacto Dextrose . . . . 10 g
Bacto Agar . . . . 20 g

Final pH 6.2 ± 0.2 at 25°C.

One pound will make
11 liters of final medium.
Rehydrate with 41 g/liter.

BACTO YM BROTH
DEHYDRATED

Ingredients per liter

Bacto Yeast Extract . . . . 3 g
Malt Extract, Difco . . . . 3 g
Bacto Peptone . . . . 5 g
Bacto Dextrose . . . . 10 g

Final pH 6.2 ± 0.2 at 25°C.

One pound will make
21.6 liters of final medium.
Rehydrate with 21 g/liter.

METHOD OF PREPARATION

1. To rehydrate, suspend the appropriate amount in 1 liter distilled or deionized water. When preparing the agar, heat to boiling to dissolve completely.
2. Sterilize in the autoclave for 15 minutes at 15 lbs pressure (121°C).
3. Selective media may be prepared by acidifying the sterile melted medium to pH 3.0 – 4.0 at 45 – 55°C, or by the addition of antibiotics at 45 – 50°C or below.

   These media are not to be reheated after the addition of antibiotics or acid.

STORAGE

Bacto YM Agar/Bacto YM Broth   Below 30°C
Bacto YM Agar plates   2 – 8°C
Bacto YM Broth tubes   15 – 30°C

QUALITY CONTROL

Identity Specifications

	Bacto YM Agar	Bacto YM Broth
Dehydrated powder:	beige, homogeneous, free-flowing	beige, homogeneous, free-flowing
Solution:	4.1%	2.1%
pH ± 0.2 at 25°C:	6.2	6.2
Prepared medium:	light to medium amber, very slightly opalescent	light amber, clear

Typical Cultural Response on/in Bacto YM Media
After 40 – 72 Hours at 25 – 30°C

Organism	Recovery pH 3 – 4	pH 6.2
Aspergillus niger ATCC® 16404	good to excellent	good to excellent
Candida albicans ATCC® 10231	good to excellent	good to excellent
Escherichia coli ATCC® 25922	inhibited	good to excellent
Lactobacillus leichmannii ATCC® 7830	poor	good to excellent
Saccharomyces cerevisiae ATCC® 9763	good to excellent	good to excellent

REFERENCES

1. U.S. Dept. Agric. Tech. Bull. No. 1029, 1951.
2. J. Tropical Med. Hyg., 42:176, 1939.

PACKAGING

Bacto YM Agar	1 lb (454 g)	0712-01-8
Bacto YM Broth	1 lb (454 g)	0711-01-9
	10 kg	0711-08-2

BACTO YEAST EXTRACT

Bacto Yeast Extract is the water soluble portion of autolyzed yeast. The autolysis is carefully controlled to preserve the naturally occurring B-complex vitamins. Bacto Yeast Extract is prepared and standardized for bacteriological use. It is an excellent stimulator of bacterial growth and is used in culture media in place of, or as an adjuvant to, beef extract. In concentrations of 0.3 – 0.5%, as it is generally employed, it forms sparklingly clear solutions with a reaction of approximately pH 6.6.

Bacto Yeast Extract has been used advantageously in culture media for studies of the bacteria in milk and other dairy products. Its usefulness for this purpose is attested in such reports as those of Prickett[1] on the thermophilic and thermoduric bacteria of milk. Since publication of this work, Bacto Yeast Extract has been used with increasing frequency in the study of the bacterial flora of milk. Among the references to its use for this purpose are those of Hucker and Hucker[2] on the number and type of bacteria in commercially prepared infant foods, Breed et. al.[3] on methods for the examination of thermophiles in dry milk, and Downs and his associates[4] on methods for the study of these organisms in evaporated and condensed milk. Bowers and Hucker[5] have also studied the effect of Bacto Yeast Extract in media employed for counting bacteria in milk. An increasing number of references to Bacto Yeast Extract is to be found in bacteriological literature. Hutner[6] used this product in a stock broth for the streptococci. Nelson and Werkman[7] have used Bacto Yeast Extract as an ingredient of the medium they employed for cultivation of Lactobacillus brevis. Werkman and his associates[8,9,10,11] have used Bacto Yeast Extract media for propagating organisms of the propionic acid group. Partansky and McPherson[12] used Bacto Yeast Extract in combination with Bacto Malt Extract, and Bacto Agar for the growth of molds in their laboratory method for testing the mold resistant properties of oil paints.

Pringsheim and Robinow[13] in their studies on the cultivation of Caryophanon latum reported very good growth of the organism in agar media containing Bacto Yeast Extract. Some batches of laboratory made autolysates of bakers' yeast gave excellent results but others were unsatisfactory. They were able to rely on a medium prepared with 0.5% Bacto Yeast Extract and 0.5% Bacto Peptone, which was very favorable for isolation and maintenance of the culture. In a survey of ingredients for a medium for the standard plate count for dairy products, Buckbinder and associates[14] compared various yeast extracts and showed that media prepared with Bacto Yeast Extract gave a higher count than similar media prepared with other yeast extracts.

Bacto Yeast Extract is an excellent source of B-complex vitamins and is often used to supply these factors in bacteriological culture media. For example, Snell and Strong[15] used Bacto Yeast Extract for the preparation of the yeast supplement in their medium for riboflavin assay. It has proved to be a valuable ingredient of media used for carrying stock cultures and for preparation of inocula of lactobacilli for microbiological assay of vitamins. This product is also of value in the assay of antibiotics.[16,17]

REFERENCES

1. Tech. Bull. 147, N.Y. Agr. Exp. Sta., 1928.
2. Tech. Bull. 153, N.Y. Agr. Exp. Sta., 1929.
3. J. Dairy Science, 15:383, 1932.
4. J. Dairy Science, 18:647, 1935.
5. Am. J. Pub. Health, 24:396, 1934.
6. J. Bact., 35:429, 1938.
7. J. Bact., 31:603, 1936.
8. J. Bact., 36:201, 1938.
9. Biochem. J., 32:1262, 1938.
10. Biochem. J., 31:349, 1937.
11. J. Bact., 31:595, 1936.
12. Ind. Eng. Chem., Anal. Ed., 12:443, 1940.
13. J. Gen. Microbiol., 1:267, 1947.
14. Pub. Health Reports, 66:327, 1951.
15. Ind. Eng. Chem., Anal. Ed., 11:346, 1939.
16. J. Bact., 47:199, 1944.
17. Science, 98:69, 1943.

PACKAGING

Bacto Yeast Extract

1 lb (454 g)	0127-01-7
1/4 lb (114 g)	0127-02-6
5 lb (2.27 kg	0127-05-3
10 kg	0127-08-0

YEAST EXTRACT, TECHNICAL

Yeast Extract, Technical is used in bacteriological culture media when a standardized yeast extract is not essential.

PACKAGING

Yeast Extract, Technical

10 kg	0886-08-1
500 g	0886-17-0

YEAST-MEDIA FOR CLASSIFICATION
BACTO YEAST MORPHOLOGY AGAR
BACTO YEAST NITROGEN BASE
BACTO YEAST NITROGEN BASE w/o AMINO ACIDS
BACTO YEAST NITROGEN BASE w/o AMINO ACIDS AND AMMONIUM SULFATE
BACTO YEAST CARBON BASE
BACTO VITAMIN FREE YEAST BASE

INTENDED USE
These media have been prepared according to the formulae of Wickerham[1,2,3,4,5] and are recommended for use in classifying yeast according to the following criteria suggested by him.
1. Colonial characteristics and cell morphology using Bacto Yeast Morphology Agar. It is also recommended for the bioassay of antifungal agents when using the agar dilution test method.[6,7]
2. Carbon assimilation using Bacto Yeast Nitrogen Base, and especially for susceptibility testing with antifungal drugs, particularly with the broth dilution method.[6,7]
3. Amino acid and carbohydrate requirements using Bacto Yeast Nitrogen Base w/o Amino Acids.
4. Nitrogen and carbon requirements using Bacto Yeast Nitrogen Base w/o Amino Acids and Ammonium Sulfate.
5. Nitrogen assimilation using Bacto Yeast Carbon Base.
6. Vitamin requirements using Bacto Vitamin Free Yeast Base.

HISTORY/PRINCIPLES
Bacto Yeast Morphology Agar is composed of ingredients of known composition prepared according to the formula suggested by Wickerham. A modified Dalman technique is used for the determination of typical colonial morphology.

Some strains of yeast require the presence of certain vitamins for the assimilation of carbon. Bacto Yeast Nitrogen Base is a suitable medium for such studies.

Bacto Yeast Nitrogen Base w/o Amino Acids, prepared according to Guenther's[8] modification of Wickerham's yeast nitrogen base formulation, is the same formulation as Bacto Yeast Nitrogen Base except that histidine, methionine and tryptophan have been omitted.

Bacto Yeast Nitrogen Base w/o Amino Acids and Ammonium Sulfate uses the same yeast nitrogen base formulation except that the amino acids and ammonium sulfate have been omitted from the medium.

Bacto Yeast Carbon Base tests the ability of yeasts to assimilate nitrogen by the addition of various nitrogen sources. The inclusion of vitamins in this base was found necessary by Wickerham as an aid for the utilization of nitrogen containing compounds by certain yeasts which cannot assimilate these compounds in the absence of vitamins.

Bacto Vitamin Free Yeast Base contains sufficient nitrogen and carbon sources to permit growth of yeast after the addition of the required vitamins.

FORMULAE

	Bacto Yeast Morphology Agar	Bacto Yeast Nitrogen Base	Bacto Yeast Nitrogen Base w/o Amino Acids	Bacto Yeast Nitrogen Base w/o Amino Acids and Ammonium Sulfate	Bacto Yeast Carbon Base	Bacto Vitamin Free Yeast Base
	Ingredients per liter					
Nitrogen Sources						
Ammonium Sulfate	3.5 g	5 g	5 g	—	—	5 g
Bacto Asparagine	1.5 g	—	—	—	—	—
Carbon Source						
Bacto Dextrose	10 g	—	—	—	10 g	10 g
Amino Acids						
L-Histidine Mono-hydrochloride	10 mg	10 mg	—	—	1 mg	10 mg
LD-Methionine	20 mg	20 mg	—	—	2 mg	20 mg
LD-Tryptophan	20 mg	20 mg	—	—	2 mg	20 mg
Vitamins						
Biotin	2 µg	2 µg	2 µg	2 µg	2 µg	—
Calcium Pantothenate	400 µg	400 µg	400 µg	400 µg	400 µg	—
Folic Acid	2 µg	2 µg	2 µg	2 µg	2 µg	—
Inositol	2000 µg	2000 µg	2000 µg	2000 µg	2000 µg	—
Niacin	400 µg	400 µg	400 µg	400 µg	400 µg	—
p-Aminobenzoic Acid, Difco	200 µg	200 µg	200 µg	200 µg	200 µg	—
Pyridoxine Hydrochloride	400 µg	400 µg	400 µg	400 µg	400 µg	—
Riboflavin	200 µg	200 µg	200 µg	200 µg	200 µg	—
Thiamine Hydrochloride	400 µg	400 µg	400 µg	400 µg	400 µg	—
Compounds supplying trace elements						
Boric Acid	500 µg	500 µg	500 µg	500 µg	500 µg	500 µg
Copper Sulfate	40 µg	40 µg	40 µg	40 µg	40 µg	40 µg
Potassium Iodide	100 µg	100 µg	100 µg	100 µg	100 µg	100 µg
Ferric Chloride	200 µg	200 µg	200 µg	200 µg	200 µg	200 µg
Manganese Sulfate	400 µg	400 µg	400 µg	400 µg	400 µg	400 µg
Sodium Molybdate	200 µg	200 µg	200 µg	200 µg	200 µg	200 µg
Zinc Sulfate	400 µg	400 µg	400 µg	400 µg	400 µg	400 µg
Salts						
Potassium Phosphate Monobasic	1 g	1 g	1 g	1 g	1 g	1 g
Magnesium Sulfate	0.5 g	0.5 g	0.5 g	0.5 g	0.5 g	0.5 g
Sodium Chloride	0.1 g	0.1 g	0.1 g	0.1 g	0.1 g	0.1 g
Calcium Chloride	0.1 g	0.1 g	0.1 g	0.1 g	0.1 g	0.1 g
Bacto Agar	18 g	—	—	—	—	—
Amount of final medium from 100 grams dehydrated medium	2.8 liters	14.9 liters	14.9 liters	58.8 liters	8.5 liters	5.9 liters
Final pH ± 0.2 at 25°C	5.6	5.4	5.4	4.5	5.5	5.6

PROCEDURES

Bacto Yeast Morphology Agar

1. To rehydrate, suspend 35 g in 1 liter distilled or deionized water. Heat to boiling to dissolve completely.
2. Sterilize in the autoclave for 15 minutes at 15 lbs pressure (121°C).
3. Pour the sterile medium into plates to a depth of about 1.5 mm. Allow the plates to stand at room temperature for one or two days before using to be assured of a dry surface.

4. Inoculation is made using the Dolman plate technique as described by Wickerham and Rettger.[1] This technique, an excellent method for studying the hyphae of filamentous yeasts, is outlined below.

   a. Make a single streak inoculation near one side of the plate (as from relative positions of 10 o'clock to 2 o'clock). Inoculum should be light and taken from a slant culture.

   b. In addition to the single streak inoculation, two point inoculations are made near the other sides of the plate (as at positions 4 o'clock and 8 o'clock).

   c. A central section of the streak inoculation and one of the point inoculations are covered with cover glasses.

   d. With forceps, remove cover glasses from absolute alcohol, drain momentarily, and burn off excess alcohol by passing over a low flame.

   e. When the cover glass has cooled, place one edge on the agar and allow the cover glass to fall across the central portion of the inoculated streak, and a second cover glass over one point inoculation.

   f. Plate is incubated six or seven days at 25°C.

   g. Observe after incubation, using the high dry objective.

Bacto Yeast Nitrogen Base

1. For best results, the medium should be filter sterilized in 10X strength. Suspend 6.7 g of Bacto Yeast Nitrogen Base and 5 g of Bacto Dextrose, or an equivalent amount of other carbohydrate, in 100 ml distilled or deionized water.

2. It may be necessary to warm the distilled or deionized water slightly to effect complete solution of some of the carbohydrates.

3. Filter sterilize the 10X solution. Keep in refrigerator and use as needed.

4. Final medium is prepared by pipetting, under aseptic conditions, 0.5 ml into 4.5 ml of sterile distilled water in sterile 16 mm cotton-stoppered tubes.

5. Mix solutions thoroughly by shaking.

6. Tubes of media are inoculated *very* lightly, then placed at 25°C.

7. After 6 – 7 days incubation, and again at 20 – 24 days, tubes are shaken to suspend growth, then they are placed against a white card bearing lines approximately 3/4 of a mm wide. The lines are drawn with India ink.

8. If lines cannot be seen through the culture or if the lines appear as diffuse, broad bands, the test is positive.

9. If the lines are distinguishable as such, the test is negative.

10. The two observations indicate which reactions occur rapidly and which occur latently.

Bacto Yeast Nitrogen Base w/o Amino Acids

1. For best results, medium should be filter sterilized in 10X strength. Suspend 6.7 g of Bacto Yeast Nitrogen Base w/o Amino Acids and 5 g of Bacto Dextrose or an equivalent amount of other carbohydrate in 100 ml of distilled or deionized water.

2. Amino acids and other chemicals modifying growth of yeasts are added singly or in combination as required experimentally.

3. Filter sterilize the 10X strength solution. The medium is kept in the refrigerator and used as needed.

4. To prepare final medium, aseptically pipette 0.5 ml of the 10X medium into 4.5 ml sterile distilled water in 16 mm cotton-stoppered tubes.

5. Mix solutions thoroughly by shaking.

6. Tubes of media are inoculated very lightly, then placed at 25°C.

7. Incubate for 6 – 7 days, read for growth and reincubate for a total of 20 – 24 days.

8. Tubes are then shaken to suspend the growth, and are placed against a white card bearing black lines approximately 3/4 mm wide. If the lines cannot be seen through

the culture or if the lines appear as diffuse, broad bands, the test is positive. If the lines are not obscured, the test is negative.

Bacto Yeast Nitrogen Base w/o Amino Acids and Ammonium Sulfate

1. For best results, the medium for use in carbon assimilation studies should be prepared in 10X strength. Suspend 1.7 g of Bacto Yeast Nitrogen Base w/o Amino Acids and Ammonium Sulfate in 100 ml of cold distilled or deionized water and add 5 g of ammonium sulfate, 10 mg L-histidine, 20 mg DL-methionine and 20 mg DL-tryptophane.
2. Carbon compounds for assimilation test are added in 10X concentration singly or in combination as required.
3. Filter sterilize the 10X strength solution. It is then kept in the refrigerator and used as needed.
4. Prepare final medium by aseptically pipetting 0.5 ml of the 10X sterile medium into 4.5 ml sterile distilled water in 16 mm cotton-stoppered tubes.
5. For nitrogen assimilation tests, add to the rehydrated Bacto Yeast Nitrogen Base w/o Amino Acids and Ammonium Sulfate, 1.0 g dextrose, 1 mg L-histidine, 2 mg DL-methionine and 2 mg DL-tryptophane.
6. Add nitrogen compounds for assimilation tests in 10X concentration singly or in combination as required.
7. Filter sterilize the 10X strength solution. It is kept in the refrigerator and used as needed.
8. Prepare final medium by aseptically pipetting 0.5 ml of the 10X sterile medium into 4.5 ml sterile distilled water in 16 mm cotton-stoppered tubes.

Carbon Assimilation Test

1. Culture yeasts to be tested on slants of previously prepared Bacto YM Agar. Incubate for 24 – 48 hours at 25°C.
2. After incubation, inoculate cultures into 10 ml of sterile single strength Bacto Yeast Nitrogen Base w/o Amino Acids and Ammonium Sulfate containing 0.1% dextrose, 0.5% ammonium sulfate, 10 mg L-histidine, 20 mg DL-methionine and 20 mg DL-tryptophane.
3. Transfer cultures by aseptically adding 1 ml of this sterile medium to the slant cultures on Bacto YM Agar and suspending the cells by agitation with a sterile 1 ml graduated pipette.
4. Using same pipette, inoculate 0.2 – 0.4 ml into the complete medium.
5. Incubate cultures for 48 hours at 25°C to adapt the cells to the medium and to reduce their carbohydrate reserves.
6. To prepare the inoculum for assimilation test, dilute cultures with a sufficient amount of previously prepared single strength Bacto Yeast Nitrogen Base w/o Amino Acids and Ammonium Sulfate to give a light transmittance equivalent to 55% when using a Lumetron Photometer at 420 A° with the cell suspension in an 18 mm (outside diameter) tube. One tenth (0.1 ml) of this suspension is used to inoculate each tube.
7. The tubes are read after 7 days incubation at 25°C and again after 20 and 24 days.
8. Shake tubes to suspend the growth uniformly.
9. Read tubes against a white background on which lines of India ink (3/4 mm width) have been drawn. If growth in the tube completely obliterates the line, it is recorded as 3+; if the line appears through the growth as an indistinct band, it is recorded as 2+; if the line can be distinguished as such but the edges of the lines are indistinct; the growth is recorded as 1+; and if the edge can be distinctly seen it is recorded as negative.

Nitrogen Assimilation Test

1. Yeasts are cultured and the inoculum prepared in same manner as for carbon assimilation tests.
2. One tenth (0.1) ml of the inoculum is added to each tube.
3. Incubate cultures for seven days.
4. If growth is observed, a second set of tubes is inoculated with a loop of culture from the first set. This dilutes the residual nitrogen which may have been present in the first inoculum.
5. Incubate second set of tubes for seven days and observe for growth.
6. If growth is obtained in the amount of 2+ or 3+ when observed in the same manner as for carbon assimilation tests, the reaction is positive. If there is no growth or 1+ growth, the reaction is negative.

Bacto Yeast Carbon Base

1. For best results, the medium should be filter sterilized in 10X strength. Dissolve 11.7 g of Bacto Yeast Carbon base and the nitrogen sources as desired in 100 ml distilled or deionized water.
2. It may be necessary to warm the distilled or deionized water slightly to effect complete solution.
3. Filter sterilize the 10X solution. The 10X strength sterile medium is kept in the refrigerator and used as needed.
4. Prepare the final medium by pipetting 0.5 ml into 4.5 ml sterile distilled water in a 16 mm cotton-stoppered tube. The required number of such water blanks are prepared and autoclaved in advance.
5. Mix solutions thoroughly by shaking before inoculation.

One of the most important nitrogen containing compounds being used for nitrogen assimilation is potassium nitrate. This medium may be prepared by dissolving 0.78 g potassium nitrate in 100 ml freshly distilled water containing 11.7 g of Bacto Yeast Carbon Base.

Yeasts which have grown on a rich medium may carry a reserve of nitrogen in the form of protein. Possible errors due to this reserve are eliminated by making two serial transfers in the complete medium. When the first transfer is seven days old, the culture is shaken and one loopful is transferred to a second tube of the complete medium containing the same source of nitrogen. If a positive test is obtained when the second culture is seven days old, the organism under test assimilates this particular nitrogen source.

Bacto Vitamin Free Yeast Base

1. Dissolve 16.7 g in 100 ml distilled or deionized water containing the desired vitamins. It may be necessary to warm the distilled water slightly to effect complete solution. The solution will be 10X strength.
2. Filter sterilize the 10X strength solution. Store in refrigerator and use as needed.
3. Prepare final medium by pipetting, under aseptic conditions, 0.5 ml into 4.5 ml sterile distilled water in 16 mm cotton-stoppered tubes.
4. Shake to mix solutions thoroughly before inoculation.
5. If separate filter sterilization of the basal medium and desired vitamins is preferred, dissolve 16.7 g in 100 ml distilled or deionized water and filter sterilize.
6. Distribute in 0.5 ml amounts in sterile 16 mm cotton-stoppered tubes.
7. Filter sterilize sufficient amounts of the desired vitamins for 100 ml of the final medium in 90 ml distilled water.

8. Final medium is prepared by adding 4.5 ml of sterile vitamin solution to each 0.5 ml of sterile basal medium under aseptic conditions.
9. Shake to mix thoroughly before inoculation.

STORAGE

Dehydrated media	Below 30°C
Prepared media	15 – 30°C

QUALITY CONTROL

Identity Specifications

	Bacto Yeast Morphology Agar	Bacto Yeast Nitrogen Base	Bacto Yeast Nitrogen Base w/o Amino Acids
Dehydrated powder:	free-flowing, homogeneous, beige	free-flowing, homogeneous, off white	free-flowing, homogeneous, off white
% solution:	3.5	0.67	0.67
pH ± 0.2 at 25°C:	5.6	5.4	5.4
Prepared medium:	light amber, slightly opalescent	clear, colorless	clear, colorless

	Bacto Yeast Nitrogen Base w/o Amino Acids and Ammonium Sulfate	Bacto Yeast Carbon Base	Bacto Vitamin Free Yeast Base
Dehydrated powder:	free-flowing, homogeneous, off white	free-flowing, homogeneous, white	free-flowing, homogeneous, white
% solution:	0.17	1.17	1.67
pH ± 0.2 at 25°C:	4.5	5.5	5.6
Prepared medium:	clear, colorless	clear, colorless	clear, colorless

Typical Cultural Response After 6 – 7 Days (Longer if Necessary) at 25 – 30°C

	Bacto Yeast Morphology Agar	Bacto Yeast Nitrogen Base		Bacto Yeast Nitrogen Base w/o Amino Acids	
		plain	w/dextrose	plain	w/amino acids
Kloeckera apiculata ATCC® 9774 (formerly *K. brevis*)	good	none to poor	good	none to poor	good
Saccharomyces uvarum ATCC® 9080 (formerly *S. carlsbergensis*)	good	none to poor	good	none to poor	good

	Bacto Yeast Nitrogen Base w/o Amino Acids and Ammonium Sulfate		Bacto Yeast Carbon Base w/ammonium		Bacto Vitamin Free Yeast Base w/trace elements and vitamins	
	plain	w/additions	plain	sulfate	plain	vitamins
Kloeckera apiculata ATCC® 9774 (formerly *K. brevis*)	none to poor	good	none to poor	good	none to poor	good
Saccharomyces uvarum ATCC® 9080 (formerly *S. carlsbergensis*)	none to poor	good	none to poor	good	none to poor	good

REFERENCES

1. U.S. Dept. Agric. Tech. Bull. No. 1029, 1951.
2. J. Tropical Med. Hyg., 42:176, 1939.
3. J. Bact., 52:293, 1946.
4. *ibid.*, 56:363, 1948.
5. *ibid.*, 46:501, 1943.
6. Shadomy, S., and Espinel Ingroff, A. 1980. Susceptibility Testing with Antifungal Drugs, p. 647–653. In E. H. Lennette, A. Balows, W. J. Hausler, Jr., and J. P. Truant, Manual of Clinical Microbiology, 3rd Ed., American Society for Microbiology, Washington, D.C.
7. Padhye, A. A. 1981. Susceptibility Tests and Blood Level Determinations of Antimycotic Agents, p. 1057–1070. In A. Balows and W. J. Hausler Jr., Diagnostic Procedures for Bacterial, Mycotic and Parasitic Infections, 6th Ed., American Public Health Association, Washington, D.C.
8. Personal Communication.

PACKAGING

Bacto Yeast Morphology Agar	100 g	0393-15-8
Bacto Yeast Nitrogen Base	100 g	0392-15-9
Bacto Yeast Nitrogen Base w/o Amino Acids	100 g	0919-15-3
Bacto Yeast Nitrogen Base w/o Amino Acids and Ammonium Sulfate	100 g	0335-15-9
Bacto Yeast Carbon Base	100 g	0391-15-0
Bacto Vitamin Free Yeast Base	100 g	0394-15-7

YERSINIA MEDIA

BACTO YERSINIA SELECTIVE AGAR BASE
BACTO YERSINIA ANTIMICROBIC SUPPLEMENT CN

INTENDED USE

Bacto Yersinia Selective Agar Base and Bacto Yersinia Antimicrobic Supplement CN are used in the preparation of Yersinia Selective Agar. The complete medium, based on the Cefsulodin-Irgasan-Novobiocin (CIN) Agar formulation of Schiemann, is recommended for use in the selective isolation and cultivation of *Yersinia enterocolitica* from clinical and nonclinical sources.[1-4]

HISTORY

Yersinia enterocolitica is being recognized with increasing frequency as a food and water-borne human pathogen.[5-9] The most common form of yersiniosis in man due to *Y. enterocolitica* is acute gastroenteritis.[10-11] However, *Y. enterocolitica* has also been shown to cause meningitis, septicemia, abscesses of the spleen and colon, peritonitis, lymphadenitis and cholecystitis.[10-12]

Several plating media have been proposed for directly isolating *Y. enterocolitica* from clinical and nonclinical materials.[1-4,13-15] One of these *Yersinia* selective agars, CIN agar, was first described by Schiemann in 1979 and employs cefsulodin, Irgasan®, novobiocin, bile salts and crystal violet as the selective agents.[2] In comparison with MacConkey and Salmonella-Shigella agars, Schiemann found that CIN agar provided better inhibition of normal enteric organisms while also providing improved direct recovery of *Y. enterocolitica* from feces.[2] Schiemann later modified his original formula by substituting 0.5 g of deoxycholate for the bile salts mixture and by reducing the content of novobiocin to 2.5 mg/liter for improved growth of strains of *Y. enterocolitica* serogroup 0:8.[4] The concentration of cefsulodin in the antimicrobic supplement has been reduced from that described by Schiemann to further improve growth and recovery of *Y. enterocolitica*.

PRINCIPLES OF THE PROCEDURE

Bacto Yersinia Selective Agar Base is a selective and differential basal medium which supports good growth of *Y. enterocolitica*. Selectivity of the base is due to the presence of bile salts, crystal violet and Irgasan® which markedly inhibit the growth of gram-positive and a number of gram-negative organisms. Bacto Yersinia Antimicrobic Supplement CN is a lyophilized antimicrobic supplement containing cefsulodin and novobiocin. Supplementation of Bacto Yersinia Selective Agar Base with the cefsulodin and novobiocin additive provides for improved inhibition of normal enteric organisms.

The differential property of the medium is based on fermentation of mannitol. Organisms capable of fermenting mannitol produce a localized pH drop around the colony which, followed by absorption of the neutral red, imparts a red color to the colony. Due to the localized pH drop, a zone of precipitated bile may also be present. Colonies of organisms that do not metabolize mannitol to acid end products remain colorless and translucent.

FORMULAE

BACTO YERSINIA SELECTIVE AGAR BASE
DEHYDRATED
Ingredients per liter

Bacto Yeast Extract	2 g	Sodium Pyruvate	2 g
Bacto Peptone	17 g	Magnesium Sulfate,	
Proteose Peptone, Difco	3 g	Heptahydrate	10 mg
Mannitol	20 g	Bacto Agar	13.5 g
Sodium Deoxycholate	0.5 g	Bacto Neutral Red	30 mg
Sodium Cholate	0.5 g	Bacto Crystal Violet	1 mg
Sodium Chloride	1 g	Irgasan®	4 mg

Final pH 7.4 ± 0.2 at 25°C.

Five hundred grams will make 8.4 liters of final medium.

BACTO YERSINIA ANTIMICROBIC SUPPLEMENT CN
Ingredients per 10 ml vial

Cefsulodin	4 mg
Novobiocin	2.5 mg

PRECAUTIONS

Use aseptic technique in rehydrating Bacto Yersinia Antimicrobic Supplement CN and adding the supplement to the base. Follow proper, established laboratory procedures in handling and disposing of infectious materials.

STORAGE INSTRUCTIONS FOR BACTO YERSINIA SELECTIVE AGAR BASE

Store below 30°C. The powder is very hygroscopic. Keep container tightly closed. The expiration date applies to the product in its intact container when stored as directed.

STORAGE AND REHYDRATION INSTRUCTIONS FOR BACTO YERSINIA ANTIMICROBIC SUPPLEMENT CN

Store at 2 – 8°C. Use immediately after rehydration or within 24 hours. Store rehydrated vials at 2 – 8°C. To rehydrate, aseptically add 10 ml sterile distilled or deionized water and invert gently several times to dissolve the powder. The expiration date applies to the product in its intact container when stored as directed. Do not open or rehydrate vials until ready to use.

PRODUCT DETERIORATION

Do not use the rehydrated supplement if it is contaminated, partially or completely evaporated, or shows other signs of deterioration. Do not use the agar base if it is caked, discolored or shows other signs of deterioration.

SPECIMEN COLLECTION

Fecal specimens should be collected in sterile containers or with a sterile rectal swab and transported immediately to the laboratory. Food and environmental specimens should be collected in sterile containers and transported to the laboratory in accordance with recommended guidelines.[16-17] Fecal specimens intended for *Yersinia* culture may be held at 4°C prior to inoculation onto appropriate media.[11]

PROCEDURE
Materials Provided

Bacto Yersinia Selective Agar Base
Bacto Yersinia Antimicrobic Supplement CN

Materials Not Provided
Specimen containers or sterile swabs
Inoculating loop
Bunsen burner or incinerator
Incubator (22 to 32°C)
User quality control cultures
Sterile Petri dishes

Preparation of Yersinia Selective Agar
1. To rehydrate the base, suspend 59.5 g Bacto Yersinia Selective Agar Base in 1 liter distilled or deionized water and heat to boiling to dissolve completely.
2. Sterilize in the autoclave for 15 minutes at 15 lbs pressure (121°C).
3. Cool to 45 – 50°C.
4. Aseptically add 10 ml rehydrated Bacto Yersinia Antimicrobic Supplement CN.
5. Mix thoroughly, avoiding the formation of air bubbles, and dispense into sterile 90 – 100 mm Petri dishes, approximately 20 ml per dish.

Inoculation and Incubation
Inoculate the specimen directly onto the surface of a Yersinia Selective Agar plate and streak for isolation.

Incubate plates at 22 – 32°C. Examine for growth after 18 – 24 hours and after 48 hours incubation. Enrichment procedures may also be employed with subsequent inoculation onto Yersinia Selective Agar. Appropriate references should be consulted for further information.[3,4,9 – 11,16,17]

RESULTS

Colonies of *Y. enterocolitica*, after 18 – 24 hours incubation, appear translucent or translucent with dark pink centers; edges are entire or irregular. After 48 hours incubation, colonies of *Y. enterocolitica* appear dark pink with a translucent border and may be surrounded by a zone of precipitated bile. Growth of non *Yersinia* organisms is markedly to completely inhibited. Refer to LIMITATIONS OF THE PROCEDURE for further information.

USER QUALITY CONTROL

1. Examine the agar base for color and texture. The powder should be pink, free flowing and homogeneous.
2. Determine the pH of the medium after preparation and cooling to 25°C. The pH should be 7.4 ± 0.2.
3. Examine the lyophilized and rehydrated antimicrobic supplement for evidence of deterioration as described under PRODUCT DETERIORATION.
4. Check the performance of the base and antimicrobic supplement by testing in the complete medium. Plates should be inoculated with approximately 100 colony forming units of the test cultures and incubated aerobically at 22 – 32°C. Examine plates for growth after 18 – 24 hours and after 48 hours incubation. Results should be as stated below:

Expected Results on Yersinia Selective Agar

Organisms	Growth	Colonial Morphology
Yersinia enterocolitica ATCC® 27729	Good	Translucent with dark pink center; bile precipitation may be present.
Escherichia coli ATCC® 25922	Marked to complete inhibition	
Proteus mirabilis ATCC® 29245	Marked to complete inhibition	
Pseudomonas aeruginosa ATCC® 27853	Marked to complete inhibition	
Streptococcus faecalis ATCC® 33186	Marked to complete inhibition	

LIMITATIONS OF THE PROCEDURE

1. Bacto Yersinia Selective Agar Base and Bacto Yersinia Antimicrobic Supplement CN are intended for use in the preparation of Yersinia Selective Agar. Although this medium is selective primarily for *Yersinia*, biochemical testing using pure cultures is necessary for complete identification. Consult appropriate references for further information.[3,4,8,10,11,18,19]
2. Due to the selective properties of this medium, some strains of *Y. enterocolitica* may be encountered that fail to grow or grow poorly on the base or complete medium; similarly, some strains of normal enteric organisms may be encountered that are not inhibited or only partially inhibited on the base or complete medium.

3. Growth of *Yersinia frederiksenii, Yersinia kristensenii* and *Yersinia intermedia* is not inhibited on the base or complete medium. In addition, some strains of *Citrobacter freundii, Serratia liquefaciens* and *Enterobacter agglomerans* have been reported to grow on this medium.[3,4] Colonies of these organisms look similar to those of *Y. enterocolitica* and must be differentiated from *Y. enterocolitica* on the basis of additional characteristics.

REFERENCES

1. Schiemann, D. A. 1979. Synthesis of a selective agar medium for *Yersinia enterocolitica*. Can. J. Microbiol. 25:1298 – 1304.
2. Schiemann, D. A. 1980. *Yersinia enterocolitica*: observations on some growth characteristics and response to selective agents. Can. J. Microbiol. 26:1232 – 1240.
3. Devenish, J. A., and D. A. Schiemann. 1981. An abbreviated scheme for identification of *Yersinia enterocolitica* isolated from food enrichments on CIN (cefsulodin-irgasan-novobiocin) agar. Can. J. Microbiol. 27:937 – 941.
4. Schiemann, D. A. 1982. Development of a two step enrichment procedure for recovery of *Yersinia enterocolitica* from food. Appl. Environ. Microbiol. 43:14 – 27.
5. Black, R. E., R. J. Jackson, T. Tsai, M. Medvesky, M. Shayegani, J. C. Feeley, K. I. E. Macleod, and A. M. Wakelee. 1978. Epidemic *Yersinia enterocolitica* infection due to contaminated chocolate milk. N. Eng. J. Med. 298:76 – 79.
6. Asakawa, Y., S. Akahane, K. Shiozawa, and T. Honma. 1979. Investigations of the source and route of *Yersinia* infections. Contrib. Microbiol. Immunol. 5:115 – 121.
7. Eden, K. V., M. L. Rosenberg, M. Stoopler, B. T. Wood, A. K. Highsmith, P. Skaliy, J. G. Wells, and J. C. Feeley, 1977. Waterborne gastrointestinal illness at a ski resort. Publ. Hlth. Rep., Wash. 92:245 – 250.
8. Ukema, T. E., and W. C. Washington, Jr. 1981. *Yersinia pseudotuberculosis* and *Yersinia enterocolitica*. ASCP Check Sample, Advanced Microbiology No. AMB 81-4, Commission on Continuing Education, American Society of Clinical Pathologists, Chicago.
9. Highsmith, A. K., J. C. Feeley, and G. K. Morris. 1977. *Yersinia enterocolitica*: A review of the bacterium and recommended laboratory methodology Hlth. Lab. Sci. 14:253 – 260.
10. Weissfeld, A. S. 1981. *Yersinia enterocolitica*. Clin. Microbiol. Newsletter 3:91 – 93.
11. Sack, R. B., R. C. Tilton, and A. S. Weissfeld. 1980. Cumitech 12, Laboratory diagnosis of bacterial diarrhea. Coordinating ed., S. J. Rubin. American Society for Microbiology, Washington, D.C.
12. Sonnenwirth, A. C., and R. E. Weaver. 1970. *Yersinia enterocolitica*. N. Engl. J. Med. 283:1468.
13. Bowen, J. H., and S. D. Kominos. 1979. Evaluation of a pectin agar. Am. J. Clin. Pathol. 72:586 – 590.
14. Dudley, M. V., and E. B. Shotts, Jr. 1979. Medium for isolation of *Yersinia enterocolitica*. J. Clin. Microbiol. 10:180 – 183.
15. Soltesz, L. V., C. Schalen, and P. Mardh. 1980. An effective, selective medium for *Yersinia enterocolitica* containing sodium oxalate. Acta. Pathol. Microbiol. Scand. (b) 88:11 – 16.
16. Feeley, J. C., W. H. Lee, and G. K. Morris. 1976. *Yersinia enterocolitica*, p. 351 – 357. *In* M. L. Speck (ed.), Compendium of methods for the microbiological examination of foods. American Public Health Association, Inc., Washington, D.C.
17. Bryan, F. L. 1980. Procedures to use during outbreaks of food-borne disease, p. 40 – 51. *In* E. H. Lennette, A. Balows, W. J. Hausler, Jr., and J. P. Truant (ed.), Manual of clinical microbiology, 3rd Ed. American Society for Microbiology, Washington, D.C.
18. Hawkins, T. M., and D. J. Brenner, 1978. Isolation and identification of *Yersinia enterocolitica*. Center for Disease Control, Atlanta, GA.
19. Martin, W. J., and J. A. Washington. 1980. *Enterobacteriaceae*, p. 195 – 219. *In* E. H. Lennette, A. Balows, W. J. Hausler, Jr., and J. P. Truant (ed.), Manual of clinical microbiology, 3rd Ed. American Society for Microbiology, Washington, D.C.

PACKAGING

Bacto Yersinia Selective Agar Base	500 g	1817-17-2
	10 kg	1817-08-3
Bacto Yersinia Antimicrobic Supplement CN	6 × 10 ml	3196-60-5

INDEX